SI Prefixes

Multiple	*Exponential Form*	*Prefix*	*SI Symbol*
1 000 000 000	10^9	giga	G
1 000 000	10^6	mega	M
1 000	10^3	kilo	k
Submultiple			
0.001	10^{-3}	milli	m
0.000 001	10^{-6}	micro	μ
0.000 000 001	10^{-9}	nano	n

Conversion Factors (FPS) to (SI)

Quantity	*Unit of Measurement (FPS)*	*Equals*	*Unit of Measurement (SI)*
Force	lb		4.4482 N
Mass	slug		14.5938 kg
Length	ft		0.3048 m

Conversion Factors (FPS)

1 ft = 12 in. (inches)
1 mi. (mile) = 5280 ft
1 kip (kilopound) = 1000 lb
1 ton = 2000 lb

STATICS

Engineering Mechanics
STATICS

ELEVENTH EDITION

R. C. Hibbeler

Upper Saddle River, NJ 07458

Library of Congress Cataloging-in-Publication Data on File

Vice President and Editorial Director, ECS: *Marcia Horton*
Associate Editor: *Dee Bernhard*
Editorial Assistant: *Dolores Mars*
Executive Managing Editor: *Vince O'Brien*
Managing Editor: *David A. George*
Production Editor: *Rose Kernan*
Director of Creative Services: *Paul Belfanti*
Manager of Electronic Composition & Digital Content: *Allyson Graesser*
Assistant Manager of Electronic Composition & Digital Content: *Clara Bartunek*
Electronic Composition: *Julita Nazario and Lissette Quinones*
Creative Director: *Juan Lopez*
Art Director: *Jonathan Boylan*
Art Editor: *Xiaohong Zhu*
Manufacturing Manager: *Alexis Heydt-Long*
Manufacturing Buyer: *Lisa McDowell*
Marketing Manager: *Tim Galligan*

About the Cover: Background image—Full-length image of a worker pausing near the top of the George Washington Bridge during its construction, New York, circa 1930. Photo by Hulton Archive/Getty Images. Foreground image—George Washington Bridge and Manhattan Skyline at dusk, New York. Photo by Miles Ertman/Masterfile.

Published by Pearson Prentice Hall
Pearson Education, Inc.
Upper Saddle River, New Jersey 07458

Printed in the United States of America

10 9 8 7 6 5 4 3 2 1

ISBN 0-13-221500-4

Pearson Education Ltd., *London*
Pearson Education Australia Pty. Ltd., *Sydney*
Pearson Education Singapore, Pte. Ltd.
Pearson Education North Asia Ltd., *Hong Kong*
Pearson Education Canada, Inc., *Toronto*
Pearson Educación de Mexico, S.A. de C.V.
Pearson Education—Japan, *Tokyo*
Pearson Education Malaysia, Pte. Ltd.
Pearson Education, Inc., *Upper Saddle River, New Jersey*

To the Student

With the hope that this work will stimulate
an interest in Engineering Mechanics
and provide an acceptable guide to its understanding.

Preface

The main purpose of this book is to provide the student with a clear and thorough presentation of the theory and applications of engineering mechanics. To achieve this objective, the author has by no means worked alone; to a large extent, this book, through its 11 editions, has been shaped by the comments and suggestions of hundreds of reviewers in the teaching profession as well as many of the author's students.

New and Enhanced Features

- **Review Material.** New end-of-chapter review sections have been added in response to student requests. These are designed to help students recall and study key chapter concepts.
- **Illustrations.** Instructor and student feedback to the enhanced illustrations in the tenth edition was overwhelmingly positive. In this edition we have enhanced over 150 additional illustrations for a total of over 350 pieces of photo realistic art. These figures provide a strong connection to the 3-D nature of engineering. Particular attention has also been paid to providing a view of any physical object, its dimensions, and the vectors in a manner that can be easily understood.
- **Problems.** The problems sets have been revised so that instructors can select both design and analysis problems having a wide range of difficulty. Apart from the author, three other professionals have checked all the problems for clarity and accuracy of the solutions. At the end of appropriate chapters, design projects are included.
- **Notes.** Closing notes have been added to appropriate Example Problems. These notes add additional context for the example material.

Hallmark Features

Accuracy. You have told us that accuracy is paramount. As with previous editions, each page of the book has been checked by three accuracy checkers in addition to the author.

Organization and Approach. The contents of each chapter are organized into well-defined sections that contain an explanation of specific topics, illustrative example problems, and a set of homework problems. The topics within each section are placed into subgroups defined by boldface titles. The purpose of this is to present a structured method for introducing each new definition or concept and to make the book convenient for later reference and review.

Chapter Contents. Each chapter begins with an illustration demonstrating a broad-range application of the material within the chapter. A bulleted list of the chapter contents is provided to give a general overview of the material that will be covered.

Free-Body Diagrams. The first step to solving most mechanics problems requires drawing a diagram. By doing so, the student forms the habit of tabulating the necessary data while focusing on the physical aspects of the problem and its associated geometry. If this step is performed correctly, applying the relevant equations of mechanics becomes somewhat methodical since the data can be taken directly from the diagram. This step is particularly important when solving equilibrium problems, and for this reason drawing free-body diagrams is strongly emphasized throughout the book. In particular, special sections and examples are devoted to show how to draw free-body diagrams, and specific homework problems in many sections of the book have been added to develop this practice.

Procedures for Analysis. Found after many of the sections of the book, this unique feature provides the student with a logical and orderly method to follow when applying the theory. The example problems are solved using this outlined method in order to clarify its numerical application. It is to be understood, however, that once the relevant principles have been mastered and enough confidence and judgment have been obtained, the student can then develop his or her own procedures for solving problems.

Photographs. Many photographs are used throughout the book to explain how the principles of mechanics apply to real-world situations. In some sections, photographs have been used to show how engineers must first make an idealized model for analysis and then proceed to draw a free-body diagram of this model in order to apply the theory.

Important Points. This feature provides a review or summary of the most important concepts in a section and highlights the most significant points that should be realized when applying the theory to solve problems.

Conceptual Understanding. Through the use of photographs placed throughout the book, theory is applied in a simplified way in order to illustrate some of its more important conceptual features and instill the physical meaning of many of the terms used in the equations. These simplified applications increase interest in the subject matter and better prepare the student to understand the examples and solve problems.

Example Problems. All the example problems are presented in a concise manner and in a style that is easy to understand.

Homework Problems

- **Free-Body Diagram Problems.** Many sections of the book contain introductory problems that only require drawing the free-body diagram for the specific problems within a problem set. These assignments will impress upon the student the importance of mastering this skill as a requirement for a complete solution of any equilibrium problem.

- **General Analysis and Design Problems.** The majority of problems in the book depict realistic situations encountered in engineering practice. Some of these problems come from actual products used in industry and are stated as such. It is hoped that this realism will both stimulate the student's interest in engineering mechanics and provide a means for developing the skill to reduce any such problem from its physical description to a model or symbolic representation to which the principles of mechanics may be applied.

Throughout the book, there is an approximate balance of problems using either SI or FPS units. Furthermore, in any set, an attempt has been made to arrange the problems in order of increasing difficulty. (Review problems at the end of each chapter are presented in random order.) The answers to all but every fourth problem are listed in the back of the book. To alert the user to a problem without a reported answer, an asterisk (*) is placed before the problem number.

- **Computer Problems.** An effort has been made to include some problems that may be solved using a numerical procedure executed on either a desktop computer or a programmable pocket calculator. Suitable numerical techniques along with associated computer programs are given in Appendix B. The intent here is to broaden the student's capacity for using other forms of mathematical analysis without sacrificing the time needed to focus on the application of the principles of mechanics. Problems of this type, which either can or must be solved using numerical procedures, are identified by a "square" symbol (■) preceding the problem number.

- **Design Projects.** At the end of some of the chapters, design projects have been included. It is felt that this type of assignment should be given only after the student has developed a basic understanding of the subject matter. These projects focus on solving a problem by specifying the geometry of a structure or mechanical object needed for a specific purpose. A force analysis is required and, in many cases, safety and cost issues must be addressed.

Appendices. The appendices provide a source of mathematical formula and numerical analysis needed to solve the problems in the book. Appendix C provides a set of problems typically found on the Fundamentals of Engineering Examination. By providing a partial solution to all the problems, the student is given a chance to further practice his or her skills.

Contents

The book is divided into 11 chapters, in which the principles are applied first to simple, then to more complicated situations. Most often, each principle is applied first to a particle, then to a rigid body subjected to a coplanar system of forces, and finally to a general case of three-dimensional force systems acting on a rigid body.

Chapter 1 begins with an introduction to mechanics and a discussion of units. The notation of a vector and the properties of a concurrent force system are introduced in Chapter 2. This theory is then applied to the equilibrium of a particle in Chapter 3. Chapter 4 contains a general discussion of both concentrated and distributed force systems and the methods used to simplify them. The principles of rigid-body equilibrium are developed in Chapter 5 and then applied to specific problems involving the equilibrium of trusses, frames, and machines in Chapter 6, and to the analysis of internal forces in beams and cables in Chapter 7. Applications to problems involving frictional forces are discussed in Chapter 8, and topics related to the center of gravity and centroid are treated in Chapter 9. If time permits, sections concerning more advanced topics, indicated by stars (★) may be covered. Most of these topics are included in Chapter 10 (area and mass moments of inertia) and Chapter 11 (virtual work and potential energy). Note that this material also provides a suitable reference for basic principles when it is discussed in more advanced courses.

Alternative Coverage. At the discretion of the instructor, some of the material may be presented in a different sequence with no loss of continuity. For example, it is possible to introduce the concept of a force and all the necessary methods of vector analysis by first covering Chapter 2 and Section 4.2. Then after covering the rest of Chapter 4 (force and moment systems), the equilibrium methods of Chapters 3 and 5 can be discussed.

Instructor and Student Resources

Hibbeler's robust supplements package supports students and instructors. The eleventh edition features:

OneKey: Custom Homework, Secure Solutions, and Much More

Hibbeler's OneKey course—available at www.prenhall.com/onekey—offers over 3000 *Statics* and *Dynamics* problems that you can personalize and post for your assignments. Editing the values in a problem guarantees a fresh problem for your students. Then, use solutions powered by Mathcad to generate your own personal solution, and if you choose, post the solutions for your students on-line. OneKey also contains PHGradeAssist—an online assessment tool with approximately 600 algorithmic test bank problems. PHGA generates unique problems for students, grades the answer, and tracks students results automatically.

You'll find Hibbeler's OneKey course contains much more to help you and your students, including

- Student Hints: Each problem contains a student hint that you may choose to provide to your students. This hint is also fully editable should you wish to change it.
- Active Book: A complete, online HTML version of the textbook students can use and refer to while completing homework and assignments.
- An extra bank of student practice problems with solutions.
- Complete bank of .jpg images.
- Complete set of PowerPoint slides.
- Active Learning slides—Perfect for classroom response systems, created by Sudhir Mehta of North Dakota State University and Scott Danielson of Arizona State University.
- Mathcad and MATLAB tutorials.
- Animations and simulations.
- Math review tutorials.
- Mechanics visualization software–Ideal for in-class demonstrations.

Instructors should visit www.prenhall.com/onekey and /or contact your their local sales representative to register and receive more information. You may also send an email requesting information to engineering@prenhall.com.

Instructor's Solutions Manual. (0-13-221502-0) This supplement provides complete solutions supported by problem statements and problem figures. The eleventh edition ISM is fully typeset and has also been triple accuracy checked.

Instructor's Resource CD-ROM. (0-13-221503-9) This supplement offers visual resources in CD-ROM format. These resources are also found on the Hibbeler OneKey course.

***Statics* Study Pack.** (0-13-221501-2) This supplement contains chapter-by-chapter study materials, a Free-Body Diagram Workbook and access to the Problems Website.

***Statics* Practice Problems Workbook.** (0-13-224975-8) This supplement contains more worked problems. Problems are partially solved and designed to help guide students through difficult topics.

Ordering Options. The *Statics* Study Pack and *Statics* Study Pack with OneKey are available as stand-alone items for student purchase and are also available packaged with the texts. The ISBN for each valuepack option is as follows:

- *Engineering Mechanics: Statics* with Study Pack (0-13-229566-0)
- *Engineering Mechanics: Statics* with Study Pack and OneKey (0-13-156147-2)

Acknowledgments

The author has endeavored to write this book so that it will appeal to both the student and instructor. Through the years, many people have helped in its development, and I will always be grateful for their valued suggestions and comments. Specifically, I wish to personally thank the following individuals who have contributed their comments to the *Statics* and *Dynamics* series:

Zhen Chen, *University of Missouri—Columbia*
Ahmad M. Itani, *University of Nevada, Reno*
John B. Ligon, *Michigan Tech University*
G. N. Jazar, *North Dakota State University*
Peter Sandborn, *University of Maryland*
Thomas H. Miller, *Oregon State University*
Richard Bennett, *The University of Tennessee*
Ali Rostami, *Virginia Commonwealth University*
Robert L. Rennaker, *The University of Oklahoma*
Richard R. Neptune, *The University of Texas at Austin*
Zhikun Hou, *Worcester Polytechnic*
David W. Parish, *North Carolina State University*
Michael H. Santare, *University of Delaware*

A particular note of thanks is also given to Professors Will Liddell, Jr. and Henry Kuhlman for their specific help. A special note of thanks is given to the accuracy checkers:

- Scott Hendricks, *Virginia Polytechnic Institute and State University*
- Karim Nohra, *University of South Florida*
- Kurt Norlin, *Laurel Tech Integrated Publishing Services*

who diligently checked all of the text and problems. I should also like to acknowledge the proofreading assistance of my wife, Conny (Cornelie), during the time it has taken to prepare this manuscript for publication.

Lastly, many thanks are extended to all my students and to members of the teaching profession who have freely taken the time to send me their suggestions and comments. Since this list is too long to mention, it is hoped that those who have given help in this manner will accept this anonymous recognition.

I would greatly appreciate hearing from you if at any time you have any comments, suggestions, or problems related to any matters regarding this edition.

RUSSELL CHARLES HIBBELER
hibbeler@bellsouth.net

Contents

1 General Principles 3

2 Force Vectors 17

3 Equilibrium of a Particle 83

4 Force System Resultants 117

5 Equilibrium of a Rigid Body 201

6 Structural Analysis 267

7 Internal Forces 337

8 Friction 393

9 Center of Gravity and Centroid 455

10 Moments of Inertia 519

11 Virtual Work 573

Appendices

STATICS

The design of this rocket and gantry structure requires a basic knowledge of both statics and dynamics, which form the subject matter of engineering mechanics.

1 General Principles

CHAPTER OBJECTIVES

- To provide an introduction to the basic quantities and idealizations of mechanics.
- To give a statement of Newton's Laws of Motion and Gravitation.
- To review the principles for applying the SI system of units.
- To examine the standard procedures for performing numerical calculations.
- To present a general guide for solving problems.

1.1 Mechanics

Mechanics can be defined as that branch of the physical sciences concerned with the state of rest or motion of bodies that are subjected to the action of forces. In general, this subject is subdivided into three branches: *rigid-body mechanics, deformable-body mechanics*, and *fluid mechanics*. This book treats only rigid-body mechanics since it forms a suitable basis for the design and analysis of many types of structural, mechanical, or electrical devices encountered in engineering. Also, rigid-body mechanics provides part of the necessary background for the study of the mechanics of deformable bodies and the mechanics of fluids.

Rigid-body mechanics is divided into two areas: statics and dynamics. *Statics* deals with the equilibrium of bodies, that is, those that are either at rest or move with a constant velocity; whereas *dynamics* is concerned with the accelerated motion of bodies. Although statics can be considered as a special case of dynamics, in which the acceleration is zero, statics deserves separate treatment in engineering education since many objects are designed with the intention that they remain in equilibrium.

Historical Development. The subject of statics developed very early in history because the principles involved could be formulated simply from measurements of geometry and force. For example, the writings of Archimedes (287–212 B.C.) deal with the principle of the lever. Studies of the pulley, inclined plane, and wrench are also recorded in ancient writings—at times when the requirements of engineering were limited primarily to building construction.

Since the principles of dynamics depend on an accurate measurement of time, this subject developed much later. Galileo Galilei (1564–1642) was one of the first major contributors to this field. His work consisted of experiments using pendulums and falling bodies. The most significant contributions in dynamics, however, were made by Issac Newton (1642–1727), who is noted for his formulation of the three fundamental laws of motion and the law of universal gravitational attraction. Shortly after these laws were postulated, important techniques for their application were developed by Euler, D'Alembert, Lagrange, and others.

1.2 Fundamental Concepts

Before we begin our study of engineering mechanics, it is important to understand the meaning of certain fundamental concepts and principles.

Basic Quantities. The following four quantities are used throughout mechanics.

Length. *Length* is needed to locate the position of a point in space and thereby describe the size of a physical system. Once a standard unit of length is defined, one can then quantitatively define distances and geometric properties of a body as multiples of the unit length.

Time. *Time* is conceived as a succession of events. Although the principles of statics are time independent, this quantity does play an important role in the study of dynamics.

Mass. *Mass* is a property of matter by which we can compare the action of one body with that of another. This property manifests itself as a gravitational attraction between two bodies and provides a quantitative measure of the resistance of matter to a change in velocity.

Force. In general, *force* is considered as a "push" or "pull" exerted by one body on another. This interaction can occur when there is direct contact between the bodies, such as a person pushing on a wall, or it can occur through a distance when the bodies are physically separated. Examples of the latter type include gravitational, electrical, and magnetic forces. In any case, a force is completely characterized by its magnitude, direction, and point of application.

Idealizations. Models or idealizations are used in mechanics in order to simplify application of the theory. A few of the more important idealizations will now be defined. Others that are noteworthy will be discussed at points where they are needed.

Particle. A *particle* has a mass, but a size that can be neglected. For example, the size of the earth is insignificant compared to the size of its orbit, and therefore the earth can be modeled as a particle when studying its orbital motion. When a body is idealized as a particle, the principles of mechanics reduce to a rather simplified form since the geometry of the body will not be involved in the analysis of the problem.

Rigid Body. A *rigid body* can be considered as a combination of a large number of particles in which all the particles remain at a fixed distance from one another both before and after applying a load. As a result, the material properties of any body that is assumed to be rigid will not have to be considered when analyzing the forces acting on the body. In most cases the actual deformations occurring in structures, machines, mechanisms, and the like are relatively small, and the rigid-body assumption is suitable for analysis.

Concentrated Force. A *concentrated force* represents the effect of a loading which is assumed to act at a point on a body. We can represent a load by a concentrated force, provided the area over which the load is applied is very small compared to the overall size of the body. An example would be the contact force between a wheel and the ground.

Newton's Three Laws of Motion. The entire subject of rigid-body mechanics is formulated on the basis of Newton's three laws of motion, the validity of which is based on experimental observation. They apply to the motion of a particle as measured from a nonaccelerating reference frame. With reference to Fig. 1–1, they may be briefly stated as follows.

First Law. A particle originally at rest, or moving in a straight line with constant velocity, will remain in this state provided the particle is *not* subjected to an unbalanced force.

Second Law. A particle acted upon by an *unbalanced force* **F** experiences an acceleration **a** that has the same direction as the force and a magnitude that is directly proportional to the force.* If **F** is applied to a particle of mass m, this law may be expressed mathematically as

$$\mathbf{F} = m\mathbf{a} \tag{1–1}$$

Third Law. The mutual forces of action and reaction between two particles are equal, opposite, and collinear.

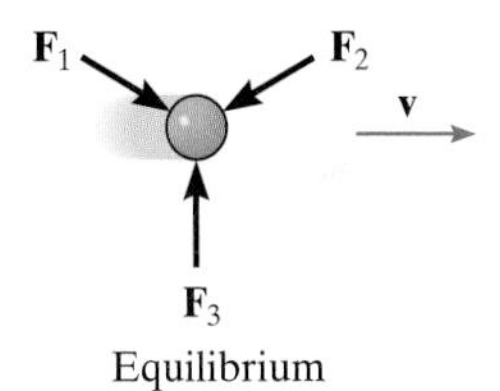

Equilibrium

Accelerated motion

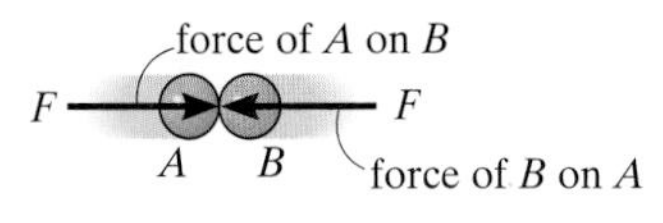

Action – reaction

Fig. 1–1

* Stated another way, the unbalanced force acting on the particle is proportional to the time rate of change of the particle's linear momentum.

Newton's Law of Gravitational Attraction. Shortly after formulating his three laws of motion, Newton postulated a law governing the gravitational attraction between any two particles. Stated mathematically,

$$F = G\frac{m_1 m_2}{r^2} \qquad (1\text{–}2)$$

where

F = force of gravitation between the two particles

G = universal constant of gravitation; according to experimental evidence, $G = 66.73(10^{-12})\ \text{m}^3/(\text{kg}\cdot\text{s}^2)$

m_1, m_2 = mass of each of the two particles

r = distance between the two particles

Weight. According to Eq. 1–2, any two particles or bodies have a mutual attractive (gravitational) force acting between them. In the case of a particle located at or near the surface of the earth, however, the only gravitational force having any sizable magnitude is that between the earth and the particle. Consequently, this force, termed the *weight*, will be the only gravitational force considered in our study of mechanics.

From Eq. 1–2, we can develop an approximate expression for finding the weight W of a particle having a mass $m_1 = m$. If we assume the earth to be a nonrotating sphere of constant density and having a mass $m_2 = M_e$, then if r is the distance between the earth's center and the particle, we have

$$W = G\frac{mM_e}{r^2}$$

Letting $g = GM_e/r^2$ yields

$$\boxed{W = mg} \qquad (1\text{–}3)$$

By comparison with $\mathbf{F} = m\mathbf{a}$, we term g the acceleration due to gravity. Since it depends on r, it can be seen that the weight of a body is *not* an absolute quantity. Instead, its magnitude is determined from where the measurement was made. For most engineering calculations, however, g is determined at sea level and at a latitude of 45°, which is considered the "standard location."

1.3 Units of Measurement

The four basic quantities—force, mass, length and time—are not all independent from one another; in fact, they are *related* by Newton's second law of motion, $\mathbf{F} = m\mathbf{a}$. Because of this, the *units* used to measure these quantities cannot *all* be selected arbitrarily. The equality $\mathbf{F} = m\mathbf{a}$ is maintained only if three of the four units, called *base units*, are *arbitrarily defined* and the fourth unit is then *derived* from the equation.

SI Units. The International System of units, abbreviated SI after the French "Système International d'Unités," is a modern version of the metric system which has received worldwide recognition. As shown in Table 1–1, the SI system specifies length in meters (m), time in seconds (s), and mass in kilograms (kg). The unit of force, called a newton (N), is *derived* from $\mathbf{F} = m\mathbf{a}$. Thus, 1 newton is equal to a force required to give 1 kilogram of mass an acceleration of 1 m/s² ($\text{N} = \text{kg}\cdot\text{m/s}^2$).

If the weight of a body located at the "standard location" is to be determined in newtons, then Eq. 1–3 must be applied. Here $g = 9.806\,65\ \text{m/s}^2$; however, for calculations, the value $g = 9.81\ \text{m/s}^2$ will be used. Thus,

$$W = mg \qquad (g = 9.81\ \text{m/s}^2) \tag{1–4}$$

Therefore, a body of mass 1 kg has a weight of 9.81 N, a 2-kg body weighs 19.62 N, and so on, Fig. 1–2*a*.

U.S. Customary. In the U.S. Customary system of units (FPS) length is measured in feet (ft), force in pounds (lb), and time in seconds (s), Table 1–1. The unit of mass, called a *slug*, is *derived* from $\mathbf{F} = m\mathbf{a}$. Hence, 1 slug is equal to the amount of matter accelerated at 1 ft/s² when acted upon by a force of 1 lb ($\text{slug} = \text{lb}\cdot\text{s}^2/\text{ft}$).

In order to determine the mass of a body having a weight measured in pounds, we must apply Eq. 1–3. If the measurements are made at the "standard location," then $g = 32.2\ \text{ft/s}^2$ will be used for calculations. Therefore,

$$m = \frac{W}{g} \qquad (g = 32.2\ \text{ft/s}^2) \tag{1–5}$$

And so a body weighing 32.2 lb has a mass of 1 slug, a 64.4-lb body has a mass of 2 slugs, and so on, Fig. 1–2*b*.

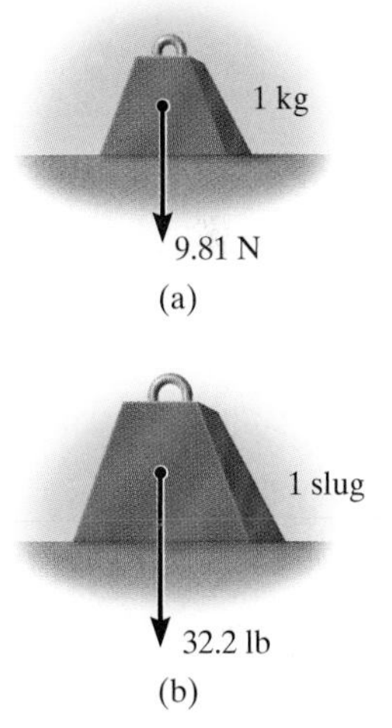

Fig. 1–2

TABLE 1–1 Systems of Units

Name	Length	Time	Mass	Force
International System of Units (SI)	meter (m)	second (s)	kilogram (kg)	newton* (N) $\left(\frac{\text{kg}\cdot\text{m}}{\text{s}^2}\right)$
U.S. Customary (FPS)	foot (ft)	second (s)	slug* $\left(\frac{\text{lb}\cdot\text{s}^2}{\text{ft}}\right)$	pound (lb)

*Derived unit.

Conversion of Units. Table 1–2 provides a set of direct conversion factors between FPS and SI units for the basic quantities. Also, in the FPS system, recall that 1 ft = 12 in. (inches), 5280 ft = 1 mi (mile), 1000 lb = 1 kip (kilo-pound), and 2000 lb = 1 ton.

TABLE 1–2 Conversion Factors

Quantity	Unit of Measurement (FPS)	Equals	Unit of Measurement (SI)
Force	lb		4.448 2 N
Mass	slug		14.593 8 kg
Length	ft		0.304 8 m

1.4 The International System of Units

The SI system of units is used extensively in this book since it is intended to become the worldwide standard for measurement. Consequently, the rules for its use and some of its terminology relevant to mechanics will now be presented.

Prefixes. When a numerical quantity is either very large or very small, the units used to define its size may be modified by using a prefix. Some of the prefixes used in the SI system are shown in Table 1–3. Each represents a multiple or submultiple of a unit which, if applied successively, moves the decimal point of a numerical quantity to every third place.* For example, 4 000 000 N = 4 000 kN (kilo-newton) = 4 MN (mega-newton), or 0.005 m = 5 mm multiple deca (10) or the submultiple centi (0.01), which form part of the metric system. Except for some volume and area measurements, the use of these prefixes is to be avoided in science and engineering.

TABLE 1–3 Prefixes

	Exponential Form	Prefix	SI Symbol
Multiple			
1 000 000 000	10^9	giga	G
1 000 000	10^6	mega	M
1 000	10^3	kilo	k
Submultiple			
0.001	10^{-3}	milli	m
0.000 001	10^{-6}	micro	μ
0.000 000 001	10^{-9}	nano	n

*The kilogram is the only base unit that is defined with a prefix.

Rules for Use. The following rules are given for the proper use of the various SI symbols:

1. A symbol is *never* written with a plural "s," since it may be confused with the unit for second (s).
2. Symbols are always written in lowercase letters, with the following exceptions: symbols for the two largest prefixes shown in Table 1–3, giga and mega, are capitalized as G and M, respectively; and symbols named after an individual are also capitalized, e.g., N.
3. Quantities defined by several units which are multiples of one another are separated by a *dot* to avoid confusion with prefix notation, as indicated by $\text{N} = \text{kg} \cdot \text{m/s}^2 = \text{kg} \cdot \text{m} \cdot \text{s}^{-2}$. Also, $\text{m} \cdot \text{s}$ (meter-second), whereas ms (milli-second).
4. The exponential power represented for a unit having a prefix refers to both the unit *and* its prefix. For example, $\mu\text{N}^2 = (\mu\text{N})^2 = \mu\text{N} \cdot \mu\text{N}$.

   Likewise, mm^2 represents $(\text{mm})^2 = \text{mm} \cdot \text{mm}$.
5. Physical constants or numbers having several digits on either side of the decimal point should be reported with a *space* between every three digits rather than with a comma; e.g., 73 569.213 427. In the case of four digits on either side of the decimal, the spacing is optional; e.g., 8537 or 8 537. Furthermore, always try to use decimals and avoid fractions; that is, write 15.25 *not* $15\frac{1}{4}$.
6. When performing calculations, represent the numbers in terms of their *base or derived units* by converting all prefixes to powers of 10. The final result should then be expressed using a *single prefix*. Also, after calculation, it is best to keep numerical values between 0.1 and 1000; otherwise, a suitable prefix should be chosen. For example,

$$\begin{aligned}(50 \text{ kN})(60 \text{ nm}) &= [50(10^3) \text{ N}][60(10^{-9}) \text{ m}] \\ &= 3000(10^{-6}) \text{ N} \cdot \text{m} = 3(10^{-3}) \text{ N} \cdot \text{m} = 3 \text{ mN} \cdot \text{m}\end{aligned}$$

7. Compound prefixes should not be used; e.g., kμs (kilo-micro-second) should be expressed as ms (milli-second) since $1 \text{ k}\mu\text{s} = 1(10^3)(10^{-6}) \text{ s} = 1(10^{-3}) \text{ s} = 1 \text{ ms}$.
8. With the exception of the base unit the kilogram, in general avoid the use of a prefix in the denominator of composite units. For example, do not write N/mm, but rather kN/m; also, m/mg should be written as Mm/kg.
9. Although not expressed in multiples of 10, the minute, hour, etc., are retained for practical purposes as multiples of the second. Furthermore, plane angular measurement is made using radians (rad). In this book, however, degrees will often be used, where $180° = \pi$ rad.

1.5 Numerical Calculations

Numerical work in engineering practice is most often performed by using handheld calculators and computers. It is important, however, that the answers to any problem be reported with both justifiable accuracy and appropriate significant figures. In this section we will discuss these topics together with some other important aspects involved in all engineering calculations.

Dimensional Homogeneity. The terms of any equation used to describe a physical process must be *dimensionally homogeneous;* that is, each term must be expressed in the same units. Provided this is the case, all the terms of an equation can then be combined if numerical values are substituted for the variables. Consider, for example, the equation $s = vt + \frac{1}{2}at^2$, where, in SI units, s is the position in meters, m, t is time in seconds, s, v is velocity in m/s, and a is acceleration in m/s^2. Regardless of how this equation is evaluated, it maintains its dimensional homogeneity. In the form stated, each of the three terms is expressed in meters [m, $(\text{m}/\cancel{\text{s}})\,\cancel{\text{s}}$, $(\text{m}/\cancel{\text{s}^2})\,\cancel{\text{s}^2}$,] or solving for a, $a = 2s/t^2 - 2v/t$, the terms are each expressed in units of m/s^2 [m/s^2, m/s^2, (m/s)/s].

Since problems in mechanics involve the solution of dimensionally homogeneous equations, the fact that all terms of an equation are represented by a consistent set of units can be used as a partial check for algebraic manipulations of an equation.

Significant Figures. The accuracy of a number is specified by the number of significant figures it contains. A *significant figure* is any digit, including a zero, provided it is not used to specify the location of the decimal point for the number. For example, the numbers 5604 and 34.52 each have four

Computers are often used in engineering for advanced design and analysis.

significant figures. When numbers begin or end with zeros, however, it is difficult to tell how many significant figures are in the number. Consider the number 400. Does it have one (4), or perhaps two (40), or three (400) significant figures? In order to clarify this situation, the number should be reported using powers of 10. Using *engineering notation*, the exponent is displayed in multiples of three in order to facilitate conversion of SI units to those having an appropriate prefix. Thus, 400 expressed to one significant figure would be $0.4(10^3)$. Likewise, 2500 and 0.00546 expressed to three significant figures would be $2.50(10^3)$ and $5.46(10^{-3})$.

Rounding Off Numbers. For numerical calculations, the accuracy obtained from the solution of a problem generally can never be better than the accuracy of the problem data. This is what is to be expected, but often handheld calculators or computers involve more figures in the answer than the number of significant figures used for the data. For this reason, a calculated result should always be "rounded off" to an appropriate number of significant figures.

To convey appropriate accuracy, the following rules for rounding off a number to n significant figures apply:

- If the $n + 1$ digit is *less than* 5, the $n + 1$ digit and others following it are dropped. For example, 2.326 and 0.451 rounded off to $n = 2$ significant figures would be 2.3 and 0.45.

- If the $n + 1$ digit is equal to 5 with zeros following it, then round off the nth digit to an *even number*. For example, $1.245(10^3)$ and 0.8655 rounded off to $n = 3$ significant figures become $1.24(10^3)$ and 0.866.

- If the $n + 1$ digit is *greater than* 5 or equal to 5 with any nonzero digits following it, then increase the nth digit by 1 and drop the $n + 1$ digit and others following it. For example, 0.723 87 and 565.500 3 rounded off to $n = 3$ significant figures become 0.724 and 566.

Calculations. As a general rule, to ensure accuracy of a final result when performing calculations on a pocket calculator, always retain a greater number of digits than the problem data. If possible, try to work out the computations so that numbers which are approximately equal are not subtracted since accuracy is often lost from this calculation.

In engineering we generally round off final answers to *three* significant figures since the data for geometry, loads, and other measurements are often reported with this accuracy.* Consequently, in this book the intermediate calculations for the examples are often worked out to four significant figures and the answers are generally reported to *three* significant figures.

*Of course, some numbers, such as π, e, or numbers used in derived formulas are exact and are therefore accurate to an infinite number of significant figures.

EXAMPLE 1.1

Convert 2 km/h to m/s. How many ft/s is this?

SOLUTION

Since 1 km = 1000 m and 1 h = 3600 s, the factors of conversion are arranged in the following order, so that a cancellation of the units can be applied:

$$2\text{ km/h} = \frac{2\text{ km}}{\text{h}}\left(\frac{1000\text{ m}}{\text{km}}\right)\left(\frac{1\text{ h}}{3600\text{ s}}\right)$$

$$= \frac{2000\text{ m}}{3600\text{ s}} = 0.556\text{ m/s} \qquad \textit{Ans.}$$

From Table 1–2, 1 ft = 0.3048 m. Thus

$$0.556\text{ m/s} = \frac{0.556\text{ m}}{\text{s}}\,\frac{1\text{ ft}}{0.3048\text{ m}}$$

$$= 1.82\text{ ft/s} \qquad \textit{Ans.}$$

NOTE: Remember to round off the final answer to three significant figures.

EXAMPLE 1.2

Convert the quantities $300\text{ lb}\cdot\text{s}$ and 52 slug/ft^3 to appropriate SI units.

SOLUTION

Using Table 1–2, 1 lb = 4.448 2 N.

$$300\text{ lb}\cdot\text{s} = 300\text{ lb}\cdot\text{s}\left(\frac{4.448\text{ 2 N}}{\text{lb}}\right)$$

$$= 1334.5\text{ N}\cdot\text{s} = 1.33\text{ kN}\cdot\text{s} \qquad \textit{Ans.}$$

Also, 1 slug = 14.593 8 kg and 1 ft = 0.304 8 m.

$$52\text{ slug/ft}^3 = \frac{52\text{ slug}}{\text{ft}^3}\left(\frac{14.593\text{ 8 kg}}{1\text{ slug}}\right)\left(\frac{1\text{ ft}}{0.304\text{ 8 m}}\right)^3$$

$$= 26.8(10^3)\text{ kg/m}^3$$

$$= 26.8\text{ Mg/m}^3 \qquad \textit{Ans.}$$

EXAMPLE 1.3

Evaluate each of the following and express with SI units having an appropriate prefix: (a) (50 mN)(6 GN), (b) $(400\text{ mm})(0.6\text{ MN})^2$, (c) $45\text{ MN}^3/900\text{ Gg}$.

SOLUTION

First convert each number to base units, perform the indicated operations, then choose an appropriate prefix (see Rule 6 on p. 9).

Part (a)

$$
\begin{aligned}
(50\text{ mN})(6\text{ GN}) &= [50(10^{-3})\text{ N}][6(10^{9})\text{ N}] \\
&= 300(10^{6})\text{ N}^2 \\
&= 300(10^{6})\,\cancel{\text{N}^2}\left(\frac{1\text{ kN}}{10^3\,\cancel{\text{N}}}\right)\left(\frac{1\text{ kN}}{10^3\,\cancel{\text{N}}}\right) \\
&= 300\text{ kN}^2 \qquad \textit{Ans.}
\end{aligned}
$$

NOTE: Keep in mind the convention $\text{kN}^2 = (\text{kN})^2 = 10^6\text{ N}^2$ (Rule 4 on p. 9).

Part (b)

$$
\begin{aligned}
(400\text{ mm})(0.6\text{ MN})^2 &= [400(10^{-3})\text{ m}][0.6(10^{6})\text{ N}]^2 \\
&= [400(10^{-3})\text{ m}][0.36(10^{12})\text{ N}^2] \\
&= 144(10^{9})\text{ m}\cdot\text{N}^2 \\
&= 144\text{ Gm}\cdot\text{N}^2 \qquad \textit{Ans.}
\end{aligned}
$$

We can also write

$$
\begin{aligned}
144(10^{9})\text{ m}\cdot\text{N}^2 &= 144(10^{9})\text{ m}\cdot\cancel{\text{N}^2}\left(\frac{1\text{ MN}}{10^6\,\cancel{\text{N}}}\right)\left(\frac{1\text{ MN}}{10^6\,\cancel{\text{N}}}\right) \\
&= 0.144\text{ m}\cdot\text{MN}^2
\end{aligned}
$$

Part (c)

$$
\begin{aligned}
45\text{ MN}^3/900\text{ Gg} &= \frac{45(10^{6}\text{ N})^3}{900(10^{6})\text{ kg}} \\
&= 0.05(10^{12})\text{ N}^3/\text{kg} \\
&= 0.05(10^{12})\,\cancel{\text{N}^3}\left(\frac{1\text{ kN}}{10^3\,\cancel{\text{N}}}\right)^3\frac{1}{\text{kg}} \\
&= 0.05(10^{3})\text{ kN}^3/\text{kg} \\
&= 50\text{ kN}^3/\text{kg} \qquad \textit{Ans.}
\end{aligned}
$$

NOTE: Here we have used Rules 4 and 8 on p. 9.

When solving problems, do the work as neatly as possible. Being neat generally stimulates clear and orderly thinking, and vice versa.

1.6 General Procedure for Analysis

The most effective way of learning the principles of engineering mechanics is to *solve problems*. To be successful at this, it is important to always present the work in a *logical* and *orderly manner*, as suggested by the following sequence of steps:

1. Read the problem carefully and try to correlate the actual physical situation with the theory studied.
2. Draw any necessary diagrams and tabulate the problem data.
3. Apply the relevant principles, generally in mathematical form.
4. Solve the necessary equations algebraically as far as practical, then, making sure they are dimensionally homogeneous, use a consistent set of units and complete the solution numerically. Report the answer with no more significant figures than the accuracy of the given data.
5. Study the answer with technical judgment and common sense to determine whether or not it seems reasonable.

Important Points

- Statics is the study of bodies that are at rest or move with constant velocity.
- A particle has a mass but a size that can be neglected.
- A rigid body does not deform under load.
- Concentrated forces are assumed to act at a point on a body.
- Newton's three laws of motion should be memorized.
- Mass is a property of matter that does not change from one location to another.
- Weight refers to the gravitational attraction of the earth on a body or quantity of mass. Its magnitude depends upon the elevation at which the mass is located.
- In the SI system the unit of force, the newton, is a derived unit. The meter, second, and kilogram are base units.
- Prefixes G, M, k, m, μ, n are used to represent large and small numerical quantities. Their exponential size should be known, along with the rules for using the SI units.
- Perform numerical calculations to several significant figures and then report the final answer to three significant figures.
- Algebraic manipulations of an equation can be checked in part by verifying that the equation remains dimensionally homogeneous.
- Know the rules for rounding off numbers.

PROBLEMS

1–1. Represent each of the following combinations of units in the correct SI form using an appropriate prefix: (a) m/ms, (b) μkm, (c) ks/mg, and (d) km $\cdot$ μN.

1–2. Wood has a density of 4.70 slug/ft^3. What is its density expressed in SI units?

1–3. Represent each of the following combinations of units in the correct SI form using an appropriate prefix: (a) Mg/mm, (b) mN/μs, (c) μm $\cdot$ Mg.

***1–4.** Represent each of the following combinations of units in the correct SI form: (a) Mg/ms, (b) N/mm, (c) mN/(kg $\cdot$ μs).

1–5. Represent each of the following with SI units having an appropriate prefix: (a) 8653 ms, (b) 8368 N, (c) 0.893 kg.

1–6. Represent each of the following to three significant figures and express each answer in SI units using an appropriate prefix: (a) 45 320 kN, (b) 568(10^5)mm, and (c) 0.005 63 mg.

1–7. Evaluate (204 mm)(0.004 57 kg)/(34.6 N) to three significant figures and express the answer in SI units using an appropriate prefix.

***1–8.** If a car is traveling at 55 mi/h, determine its speed in kilometers per hour and meters per second.

1–9. Convert: (a) 200 lb $\cdot$ ft to N $\cdot$ m, (b) 350 lb/ft^3 to kN/m^3, (c) 8 ft/h to mm/s. Express the result to three significant figures. Use an appropriate prefix.

1–10. What is the weight in newtons of an object that has a mass of: (a) 10 kg, (b) 0.5 g, (c) 4.50 Mg? Express the result to three significant figures. Use an appropriate prefix.

1–11. If an object has a mass of 40 slugs, determine its mass in kilograms.

***1–12.** The specific weight (wt./vol.) of brass is 520 lb/ft^3. Determine its density (mass/vol.) in SI units. Use an appropriate prefix.

1–13. A concrete column has a diameter of 350 mm and a length of 2 m. If the density (mass/volume) of concrete is 2.45 Mg/m^3, determine the weight of the column in pounds.

1–14. The density (mass/volume) of aluminum is 5.26 slug/ft^3. Determine its density in SI units. Use an appropriate prefix.

1–15. Using Table 1–2, determine your own mass in kilograms, your weight in newtons, and your height in meters.

***1–16.** Two particles have a mass of 8 kg and 12 kg, respectively. If they are 800 mm apart, determine the force of gravity acting between them. Compare this result with the weight of each particle.

1–17. Using the base units of the SI system, show that Eq. 1–2 is a dimensionally homogeneous equation which gives F in newtons. Compute the gravitational force acting between two identical spheres that are touching each other. The mass of each sphere is 150 kg, and the radius is 275 mm.

1–18. Evaluate each of the following to three significant figures and express each answer in SI units using an appropriate prefix: (a) (200 kN)^2, (b) (0.005 mm)^2, (c) (400 m)^3.

1–19. Evaluate each of the following to three significant figures and express each answer in SI units using an appropriate prefix: (a) (684 μm)/(43 ms), (b) (28 ms)(0.0458 Mm)/(348 mg), (c) (2.68 mm)(426 Mg).

***1–20.** Evaluate each of the following to three significant figures and express each answer in SI units using an appropriate prefix: (a) (0.631 Mm)/(8.60 kg)^2, (b) (35 mm)^2(48 kg)^3.

These communication towers are stabilized by cables that exert forces at the points of connection. In this chapter, we will show how to express these forces as Cartesian vectors.

2 Force Vectors

CHAPTER OBJECTIVES

- To show how to add forces and resolve them into components using the Parallelogram Law.
- To express force and position in Cartesian vector form and explain how to determine the vector's magnitude and direction.
- To introduce the dot product in order to determine the angle between two vectors or the projection of one vector onto another.

2.1 Scalars and Vectors

Most of the physical quantities in mechanics can be expressed mathematically by means of scalars and vectors.

Scalar. A quantity characterized by a positive or negative number is called a *scalar*. For example, mass, volume, and length are scalar quantities often used in statics. In this book, scalars are indicated by letters in italic type, such as the scalar A.

Vector. A *vector* is a quantity that has both a magnitude and a direction. In statics the vector quantities frequently encountered are position, force, and moment. For handwritten work, a vector is generally represented by a letter with an arrow written over it, such as $\vec{\mathrm{A}}$. The magnitude is designated $|\vec{\mathrm{A}}|$ or simply A. In this book vectors will be symbolized in boldface type; for example, **A** is used to designate the vector "A." Its magnitude, which is always a positive quantity, is symbolized in italic type, written as $|A|$, or simply A when it is understood that A is a positive scalar.

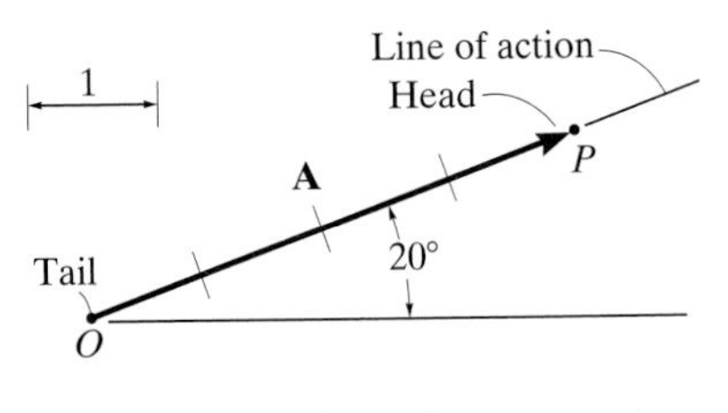

Fig. 2–1

A vector is represented graphically by an arrow, which is used to define its magnitude, direction, and sense. The *magnitude* of the vector is the length of the arrow, the *direction* is defined by the angle between a reference axis and the arrow's line of action, and the *sense* is indicated by the arrowhead. For example, the vector **A** shown in Fig. 2–1 has a magnitude of 4 units, a direction which is 20° measured counterclockwise from the horizontal axis, and a sense which is upward and to the right. The point O is called the *tail* of the vector, the point P the *tip* or *head*.

2.2 Vector Operations

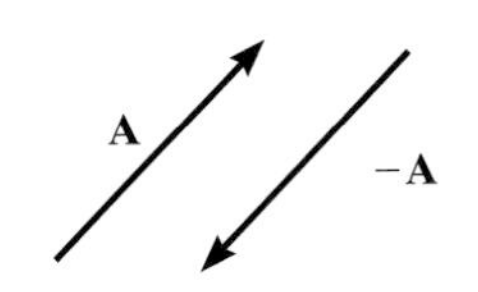

Vector **A** and its negative counterpart

Fig. 2–2

Multiplication and Division of a Vector by a Scalar. The product of vector **A** and scalar a, yielding $a\mathbf{A}$, is defined as a vector having a magnitude $|aA|$. The *sense* of $a\mathbf{A}$ is the *same* as **A** provided a is *positive*; it is *opposite* to **A** if a is *negative*. In particular, the negative of a vector is formed by multiplying the vector by the scalar (-1), Fig. 2–2. Division of a vector by a scalar can be defined using the laws of multiplication, since $\mathbf{A}/a = (1/a)\mathbf{A}$, $a \neq 0$. Graphic examples of these operations are shown in Fig. 2–3.

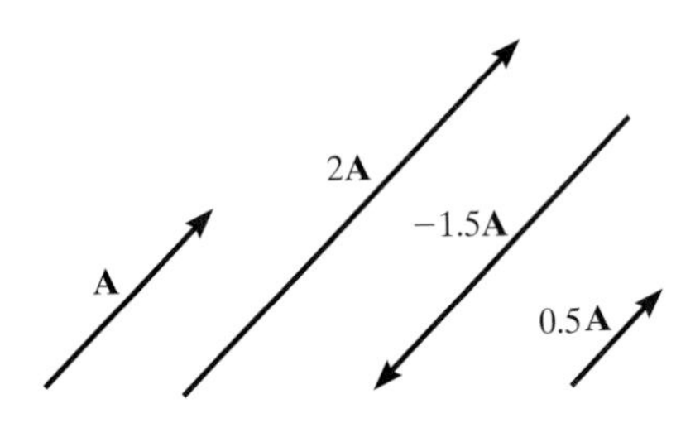

Scalar multiplication and division

Fig. 2–3

Vector Addition. Two vectors **A** and **B** such as force or position, Fig. 2–4*a*, may be added to form a "resultant" vector $\mathbf{R} = \mathbf{A} + \mathbf{B}$ by using the *parallelogram law*. To do this, **A** and **B** are joined at their tails, Fig. 2–4*b*. Parallel lines drawn from the head of each vector intersect at a common point, thereby forming the adjacent sides of a parallelogram. As shown, the resultant **R** is the diagonal of the parallelogram, which extends from the tails of **A** and **B** to the intersection of the lines.

We can also add **B** to **A** using a *triangle construction*, which is a special case of the parallelogram law, whereby vector **B** is added to vector **A** in a "head-to-tail" fashion, i.e., by connecting the head of **A** to the tail of **B**, Fig. 2–4*c*. The resultant **R** extends from the tail of **A** to the head of **B**. In a similar manner, **R** can also be obtained by adding **A** to **B**, Fig. 2–4*d*. By comparison, it is seen that vector addition is commutative; in other words, the vectors can be added in either order, i.e., $\mathbf{R} = \mathbf{A} + \mathbf{B} = \mathbf{B} + \mathbf{A}$.

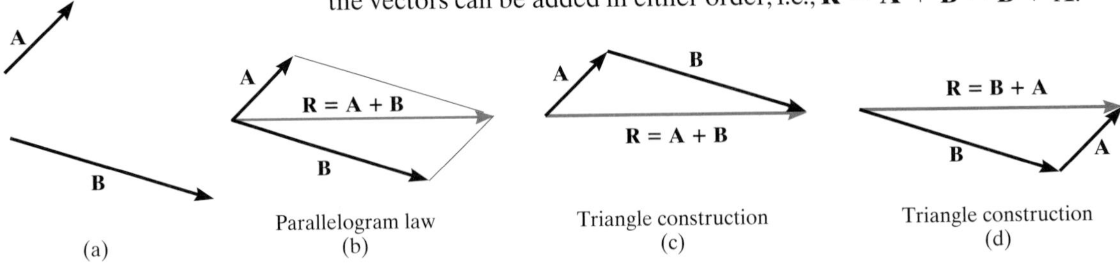

Vector addition

Fig. 2–4

As a special case, if the two vectors **A** and **B** are *collinear*, i.e., both have the same line of action, the parallelogram law reduces to an *algebraic* or *scalar addition* $R = A + B$, as shown in Fig. 2–5.

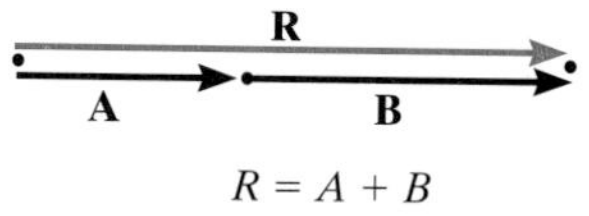

$R = A + B$

Addition of collinear vectors

Fig. 2–5

Vector Subtraction. The resultant *difference* between two vectors **A** and **B** of the same type may be expressed as

$$\mathbf{R}' = \mathbf{A} - \mathbf{B} = \mathbf{A} + (-\mathbf{B})$$

This vector sum is shown graphically in Fig. 2–6. Subtraction is therefore defined as a special case of addition, so the rules of vector addition also apply to vector subtraction.

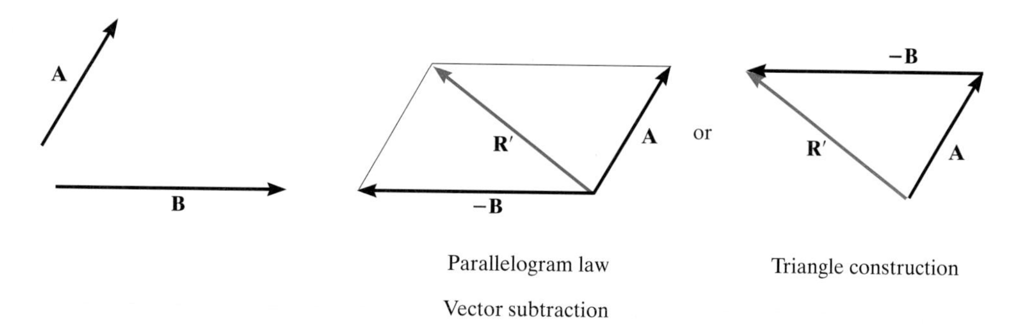

Parallelogram law

Triangle construction

Vector subtraction

Fig. 2–6

Resolution of Vector. A vector may be resolved into two "components" having known lines of action by using the parallelogram law. For example, if **R** in Fig. 2–7*a* is to be resolved into components acting along the lines *a* and *b*, one starts at the *head* of **R** and extends a line *parallel* to *a* until it intersects *b*. Likewise, a line parallel to *b* is drawn from the *head* of **R** to the point of intersection with *a*, Fig. 2–7*a*. The two components **A** and **B** are then drawn such that they extend from the tail of **R** to the points of intersection, as shown in Fig. 2–7*b*.

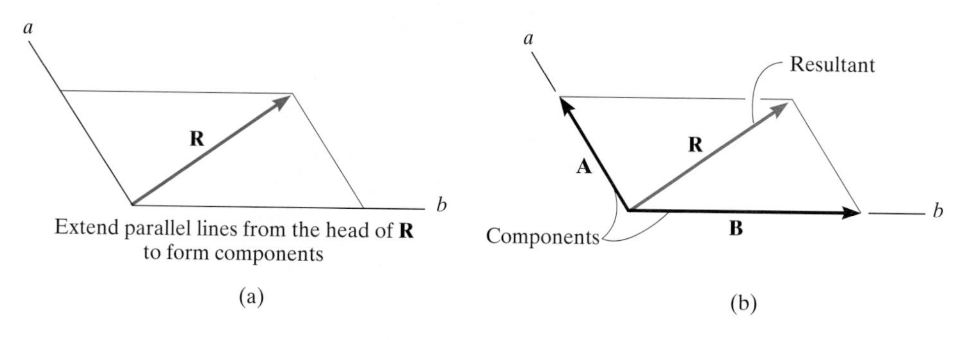

Resolution of a vector

Fig. 2–7

2.3 Vector Addition of Forces

Experimental evidence has shown that a force is a vector quantity since it has a specified magnitude, direction, and sense and it adds according to the parallelogram law. Two common problems in statics involve either finding the resultant force, knowing its components, or resolving a known force into two components. As described in Sec. 2.2, both of these problems require application of the parallelogram law.

If more than two forces are to be added, successive applications of the parallelogram law can be carried out in order to obtain the resultant force. For example, if three forces $\mathbf{F}_1$, $\mathbf{F}_2$, $\mathbf{F}_3$ act at a point O, Fig. 2–8, the resultant of any two of the forces is found—say, $\mathbf{F}_1 + \mathbf{F}_2$—and then this resultant is added to the third force, yielding the resultant of all three forces; i.e., $\mathbf{F}_R = (\mathbf{F}_1 + \mathbf{F}_2) + \mathbf{F}_3$. Using the parallelogram law to add more than two forces, as shown here, often requires extensive geometric and trigonometric calculation to determine the numerical values for the magnitude and direction of the resultant. Instead, problems of this type are easily solved by using the "rectangular-component method," which is explained in Sec. 2.4.

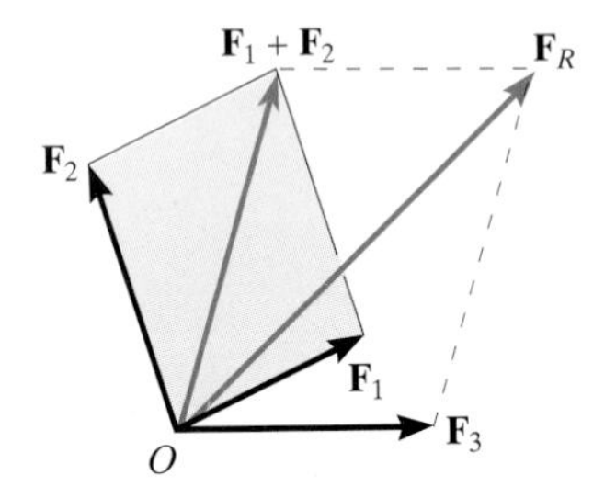

Fig. 2–8

If we know the forces $\mathbf{F}_a$ and $\mathbf{F}_b$ that the two chains a and b exert on the hook, we can find their resultant force $\mathbf{F}_c$ by using the parallelogram law. This requires drawing lines parallel to a and b from the heads of $\mathbf{F}_a$ and $\mathbf{F}_b$ as shown thus forming a parallelogram.

In a similar manner, if the force $\mathbf{F}_c$ along chain c is known, then its two components $\mathbf{F}_a$ and $\mathbf{F}_b$, that act along a and b, can be determined from the parallelogram law. Here we must start at the head of $\mathbf{F}_c$ and construct lines parallel to a and b, thereby forming the parallelogram.

Procedure for Analysis

Problems that involve the addition of two forces can be solved as follows:

Parallelogram Law.

- Make a sketch by placing the vectors together at their tails and then showing the vector addition using the parallelogram law.
- Two "component" forces add according to the parallelogram law, yielding a *resultant* force that forms the diagonal of the parallelogram.
- If a force is to be resolved into *components* along two axes directed from the tail of the force, then start at the head of the force and construct lines parallel to the axes, thereby forming the parallelogram. The sides of the parallelogram represent the components.
- Label all the known and unknown force magnitudes and the angles on the sketch and identify the two unknowns.

Trigonometry.

- Redraw a half portion of the parallelogram to illustrate the triangular head-to-tail addition of the components.
- The magnitude of the resultant force can be determined from the law of cosines, and its direction is determined from the law of sines, Fig. 2–9.
- The magnitudes of two force components are determined from the law of sines, Fig. 2–9.

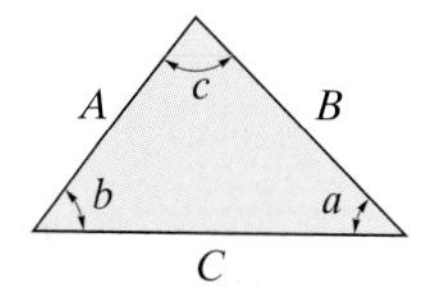

Sine law:

$$\frac{A}{\sin a} = \frac{B}{\sin b} = \frac{C}{\sin c}$$

Cosine law:

$$C = \sqrt{A^2 + B^2 - 2AB \cos c}$$

Fig. 2–9

Important Points

- A scalar is a positive or negative number.
- A vector is a quantity that has a magnitude, direction, and sense.
- Multiplication or division of a vector by a scalar will change the magnitude of the vector. The sense of the vector will change if the scalar is negative.
- As a special case, if the vectors are collinear, the resultant is formed by an algebraic or scalar addition.

EXAMPLE 2.1

The screw eye in Fig. 2–10*a* is subjected to two forces, $\mathbf{F}_1$ and $\mathbf{F}_2$. Determine the magnitude and direction of the resultant force.

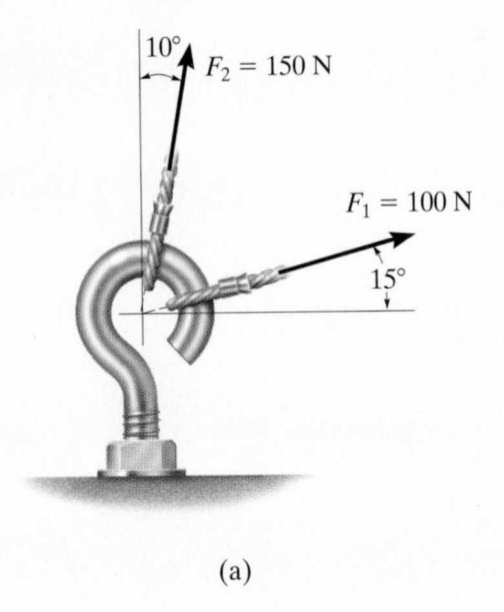

(a)

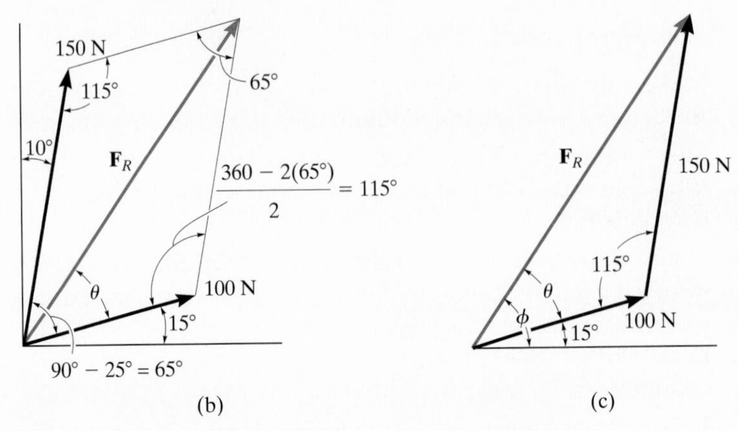

(b) (c)

Fig. 2–10

SOLUTION

Parallelogram Law. The parallelogram law of addition is shown in Fig. 2–10*b*. The two unknowns are the magnitude of $\mathbf{F}_R$ and the angle θ (theta).

Trigonometry. From Fig. 2–10*b*, the vector triangle, Fig. 2–10*c*, is constructed. F_R is determined by using the law of cosines:

$$F_R = \sqrt{(100\text{ N})^2 + (150\text{ N})^2 - 2(100\text{ N})(150\text{ N})\cos 115^\circ}$$
$$= \sqrt{10\,000 + 22\,500 - 30\,000(-0.4226)} = 212.6\text{ N}$$
$$= 213\text{ N} \qquad\qquad \textit{Ans.}$$

The angle θ is determined by applying the law of sines, using the computed value of F_R:

$$\frac{150\text{ N}}{\sin\theta} = \frac{212.6\text{ N}}{\sin 115^\circ}$$

$$\sin\theta = \frac{150\text{ N}}{212.6\text{ N}}(0.9063)$$

$$\theta = 39.8^\circ$$

Thus, the direction ϕ (phi) of $\mathbf{F}_R$, measured from the horizontal, is

$$\phi = 39.8^\circ + 15.0^\circ = 54.8^\circ \measuredangle\phi \qquad\qquad \textit{Ans.}$$

NOTE: The results seem reasonable, since Fig. 2–10*b* shows $\mathbf{F}_R$ to have a magnitude larger than its components and a direction that is between them.

EXAMPLE 2.2

Resolve the 200-lb force acting on the pipe, Fig. 2–11*a*, into components in the (a) *x* and *y* directions, and (b) *x'* and *y* directions.

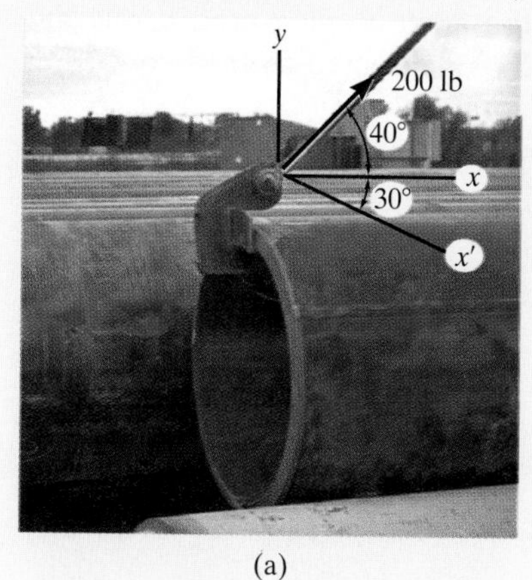

(a)

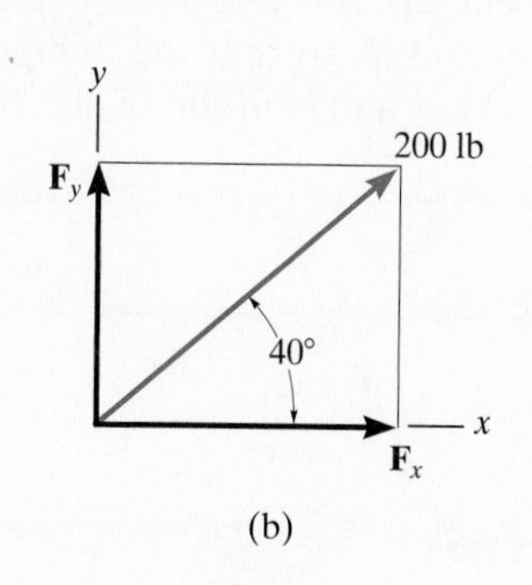

(b)

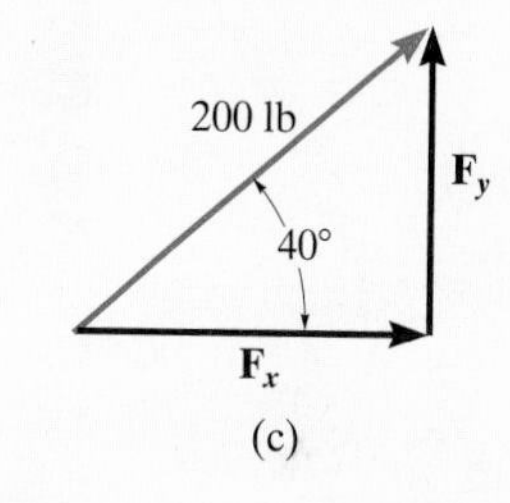

(c)

Fig. 2–11

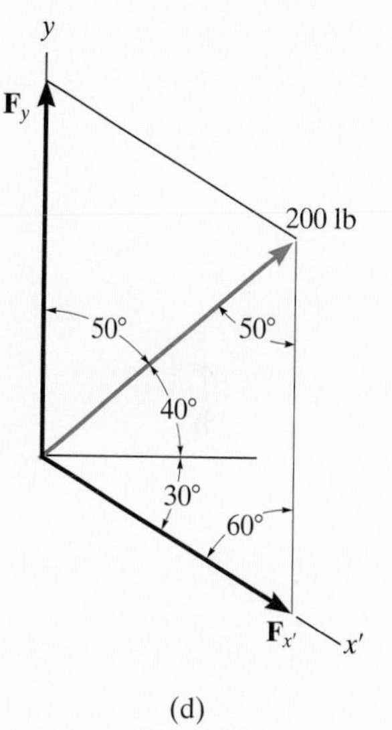

(d)

SOLUTION

In each case the parallelogram law is used to resolve **F** into its two components, and then the vector triangle is constructed to determine the numerical results by trigonometry.

Part (a). The vector addition $\mathbf{F} = \mathbf{F}_x + \mathbf{F}_y$ is shown in Fig. 2–11*b*. In particular, note that the length of the components is scaled along the *x* and *y* axes by first constructing lines from the tip of **F** parallel to the axes in accordance with the parallelogram law. From the vector triangle, Fig. 2–11*c*,

$$F_x = 200 \text{ lb} \cos 40^\circ = 153 \text{ lb} \qquad \textit{Ans.}$$

$$F_y = 200 \text{ lb} \sin 40^\circ = 129 \text{ lb} \qquad \textit{Ans.}$$

Part (b). The vector addition $\mathbf{F} = \mathbf{F}_{x'} + \mathbf{F}_y$ is shown in Fig. 2–11*d*. Note carefully how the parallelogram is constructed. Applying the law of sines and using the data listed on the vector triangle, Fig. 2–11*e*, yields

$$\frac{F_{x'}}{\sin 50^\circ} = \frac{200 \text{ lb}}{\sin 60^\circ}; \qquad F_{x'} = 200 \text{ lb}\left(\frac{\sin 50^\circ}{\sin 60^\circ}\right) = 177 \text{ lb} \qquad \textit{Ans.}$$

$$\frac{F_y}{\sin 70^\circ} = \frac{200 \text{ lb}}{\sin 60^\circ}; \qquad F_y = 200 \text{ lb}\left(\frac{\sin 70^\circ}{\sin 60^\circ}\right) = 217 \text{ lb} \qquad \textit{Ans.}$$

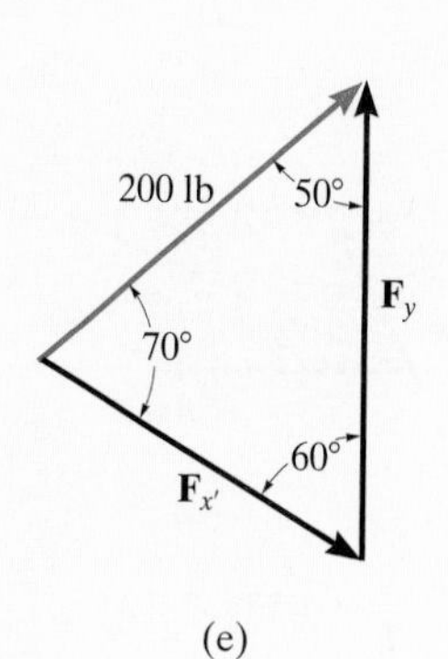

(e)

NOTE: A rough sketch drawn to scale will give some idea of the relative magnitude of the components, as calculated here.

EXAMPLE 2.3

The force **F** acting on the frame shown in Fig. 2–12*a* has a magnitude of 500 N and is to be resolved into two components acting along members *AB* and *AC*. Determine the angle θ, measured *below* the horizontal, so that the component $\mathbf{F}_{AC}$ is directed from *A* toward *C* and has a magnitude of 400 N.

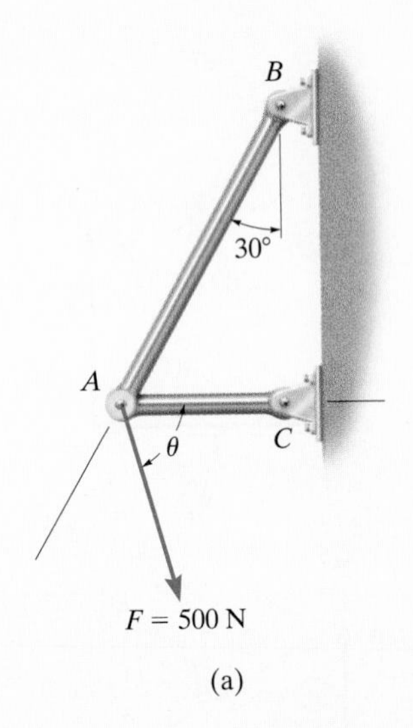

(a)

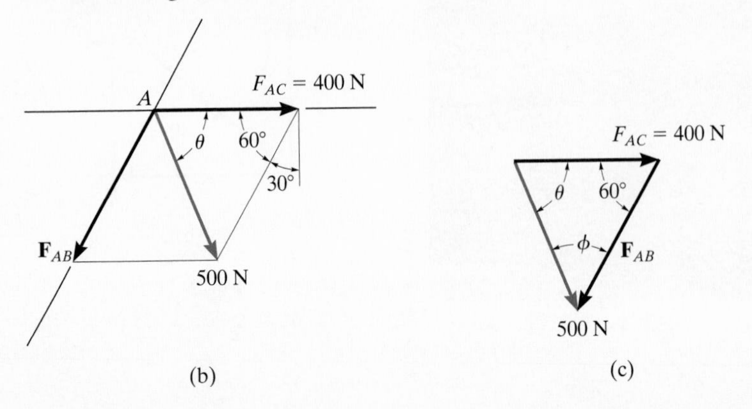

(b) (c)

SOLUTION

By using the parallelogram law, the vector addition of the two components yielding the resultant is shown in Fig. 2–12*b*. Note carefully how the resultant force is resolved into the two components $\mathbf{F}_{AB}$ and $\mathbf{F}_{AC}$, which have specified lines of action. The corresponding vector triangle is shown in Fig. 2–12*c*.

The angle ϕ can be determined by using the law of sines:

$$\frac{400\text{ N}}{\sin\phi} = \frac{500\text{ N}}{\sin 60^\circ}$$

$$\sin\phi = \left(\frac{400\text{ N}}{500\text{ N}}\right)\sin 60^\circ = 0.6928$$

$$\phi = 43.9^\circ$$

Hence,

$$\theta = 180^\circ - 60^\circ - 43.9^\circ = 76.1^\circ \measuredangle\phi \qquad \textit{Ans.}$$

Using this value for θ, apply the law of cosines or the law of sines and show that $\mathbf{F}_{AB}$ has a magnitude of 561 N. This result shows that a component can sometimes have a greater magnitude than the resultant.

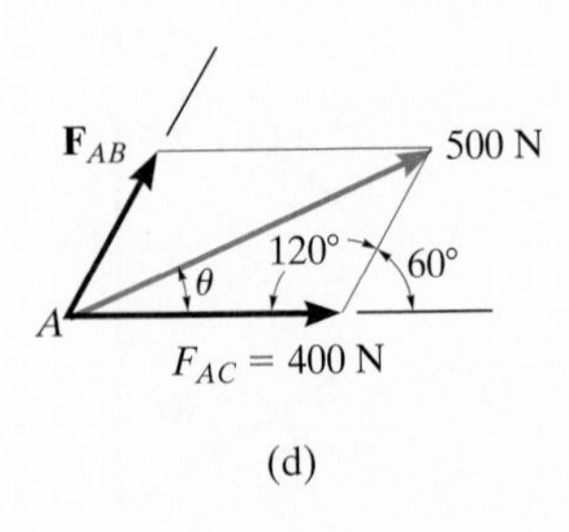

(d)

Fig. 2–12

NOTE: **F** can also be directed at an angle θ *above* the horizontal, as shown in Fig. 2–12*d*, and still produce the required component $\mathbf{F}_{AC}$. Show that in this case $\theta = 16.1^\circ$ and $F_{AB} = 161$ N.

EXAMPLE 2.4

The ring shown in Fig. 2–13*a* is subjected to two forces, $\mathbf{F}_1$ and $\mathbf{F}_2$. If it is required that the resultant force have a magnitude of 1 kN and be directed vertically downward, determine (a) the magnitudes of $\mathbf{F}_1$ and $\mathbf{F}_2$ provided $\theta = 30°$, and (b) the magnitudes of $\mathbf{F}_1$ and $\mathbf{F}_2$ if F_2 is to have a minimum magnitude.

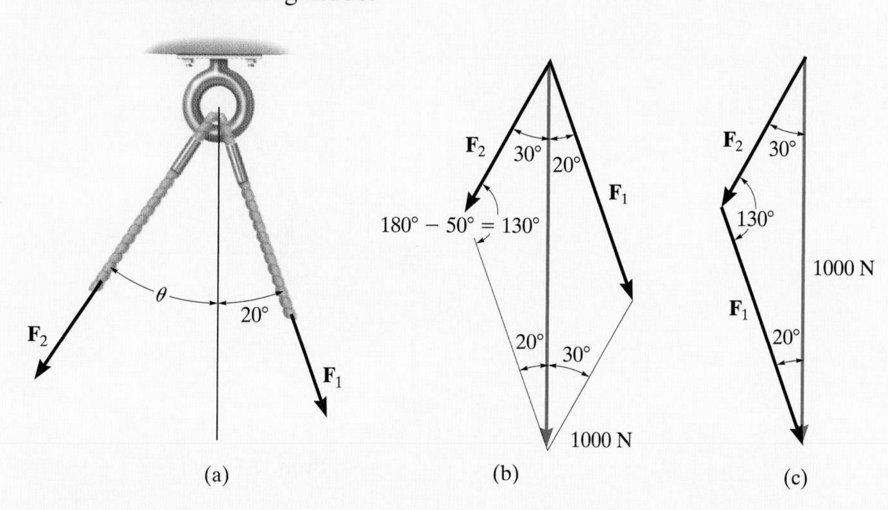

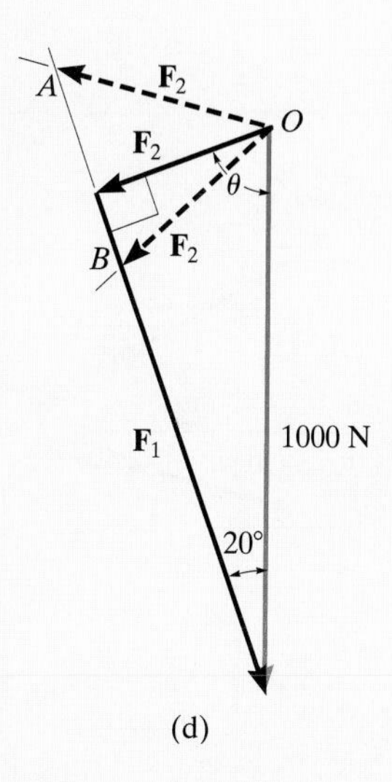

SOLUTION

Part (a). A sketch of the vector addition according to the parallelogram law is shown in Fig. 2–13*b*. From the vector triangle constructed in Fig. 2–13*c*, the unknown magnitudes F_1 and F_2 are determined by using the law of sines:

$$\frac{F_1}{\sin 30°} = \frac{1000 \text{ N}}{\sin 130°}$$

$$F_1 = 653 \text{ N} \qquad \textit{Ans.}$$

$$\frac{F_2}{\sin 20°} = \frac{1000 \text{ N}}{\sin 130°}$$

$$F_2 = 446 \text{ N} \qquad \textit{Ans.}$$

Part (b). If θ is not specified, then by the vector triangle, Fig. 2–13*d*, $\mathbf{F}_2$ may be added to $\mathbf{F}_1$ in various ways to yield the resultant 1000-N force. In particular, the *minimum* length or magnitude of $\mathbf{F}_2$ will occur when its line of action is *perpendicular to* $\mathbf{F}_1$. Any other direction, such as OA or OB, yields a larger value for F_2. Hence, when $\theta = 90° - 20° = 70°$, F_2 is minimum. From the triangle shown in Fig. 2–13*e*, it is seen that

$$F_1 = 1000 \sin 70° \text{N} = 940 \text{ N} \qquad \textit{Ans.}$$

$$F_2 = 1000 \cos 70° \text{N} = 342 \text{ N} \qquad \textit{Ans.}$$

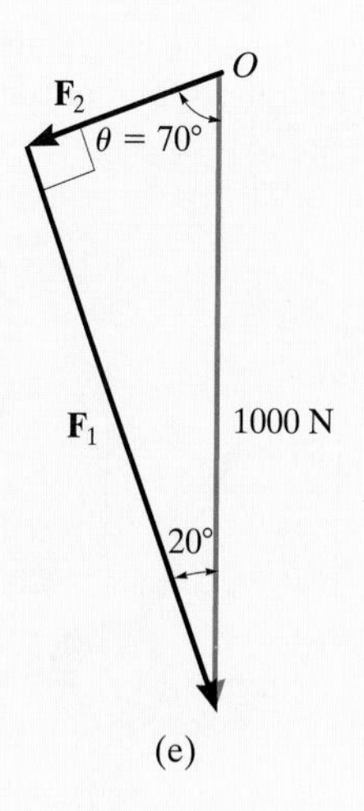

Fig. 2–13

PROBLEMS

2–1. Determine the magnitude of the resultant force $\mathbf{F}_R = \mathbf{F}_1 + \mathbf{F}_2$ and its direction, measured counterclockwise from the positive x axis.

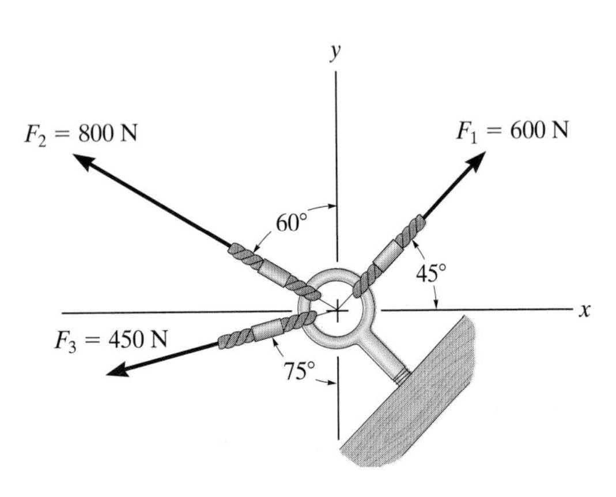

Prob. 2–1

2–2. Determine the magnitude of the resultant force and its direction measured counterclockwise from the positive x axis.

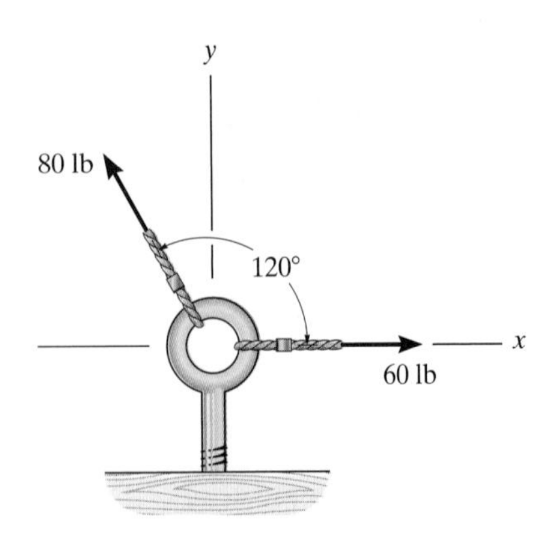

Prob. 2–2

2–3. Determine the magnitude of the resultant force $\mathbf{F}_R = \mathbf{F}_1 + \mathbf{F}_2$ and its direction, measured counterclockwise from the positive x axis.

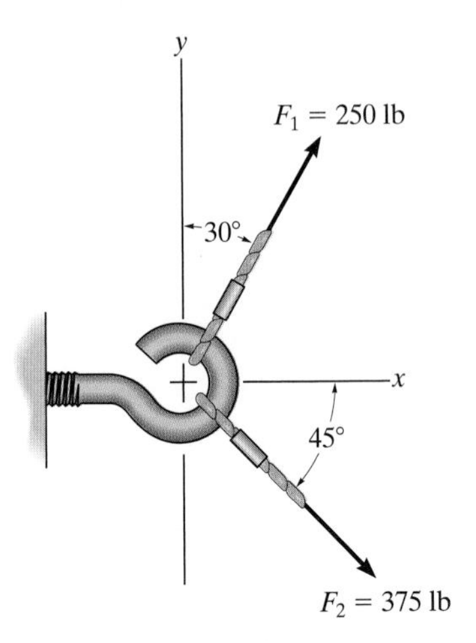

Prob. 2–3

***2–4.** Determine the magnitude of the resultant force $\mathbf{F}_R = \mathbf{F}_1 + \mathbf{F}_2$ and its direction, measured clockwise from the positive u axis.

2–5. Resolve the force $\mathbf{F}_1$ into components acting along the u and v axes and determine the magnitudes of the components.

2–6. Resolve the force $\mathbf{F}_2$ into components acting along the u and v axes and determine the magnitudes of the components.

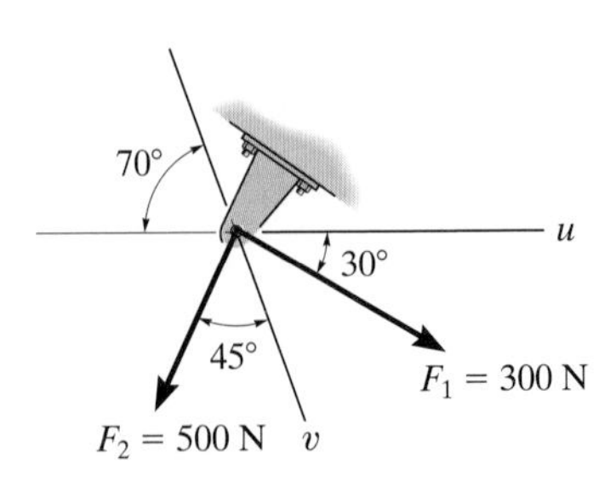

Probs. 2–4/5/6

2–7. Determine the magnitude of the resultant force $\mathbf{F}_R = \mathbf{F}_1 + \mathbf{F}_2$ and its direction measured counterclockwise from the positive u axis.

***2–8.** Resolve the force $\mathbf{F}_1$ into components acting along the u and v axes and determine the components.

2–9. Resolve the force $\mathbf{F}_2$ into components acting along the u and v axes and determine the components.

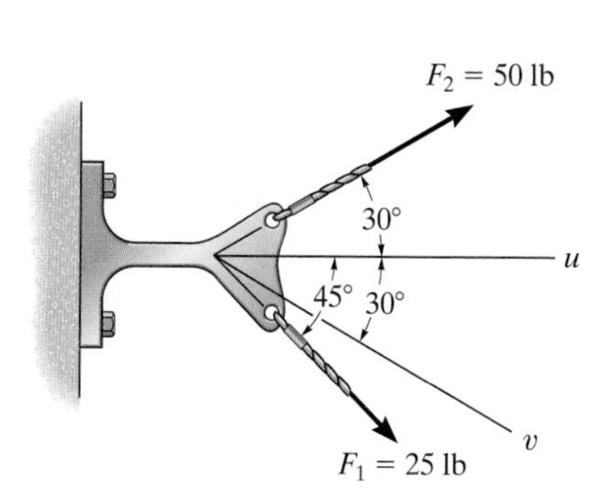

Probs. 2–7/8/9

2–10. Determine the components of the 250-N force acting along the u and v axes.

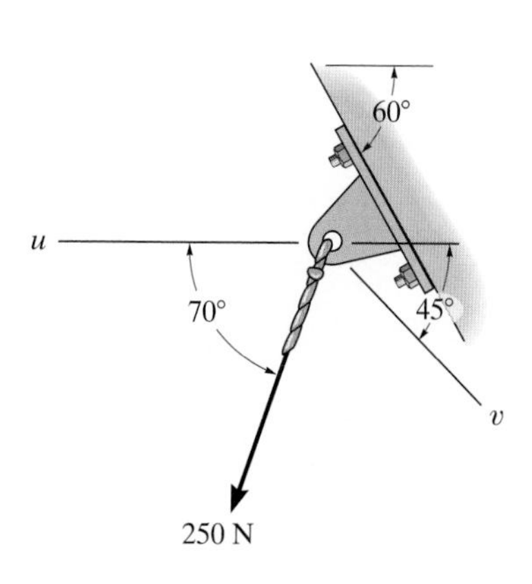

Prob. 2–10

2–11. The force acting on the gear tooth is $F = 20$ lb. Resolve this force into two components acting along the lines aa and bb.

***2–12.** The component of force $\mathbf{F}$ acting along line aa is required to be 30 lb. Determine the magnitude of $\mathbf{F}$ and its component along line bb.

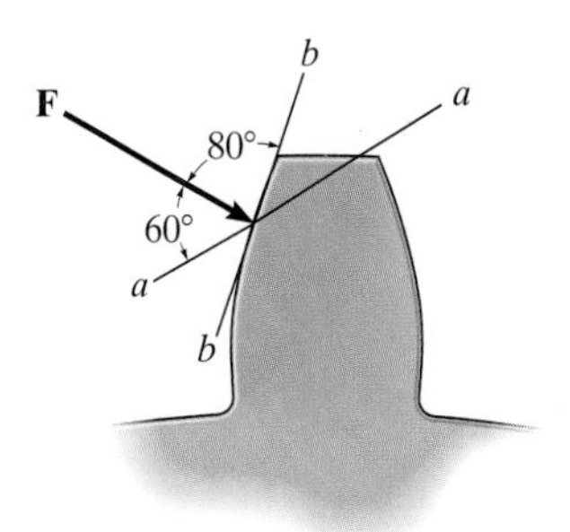

Probs. 2–11/12

2–13. A resultant vertical force of 350 lb is necessary to hold the balloon in place. Resolve this force into components along the tether lines AB and AC, and compute the magnitude of each component.

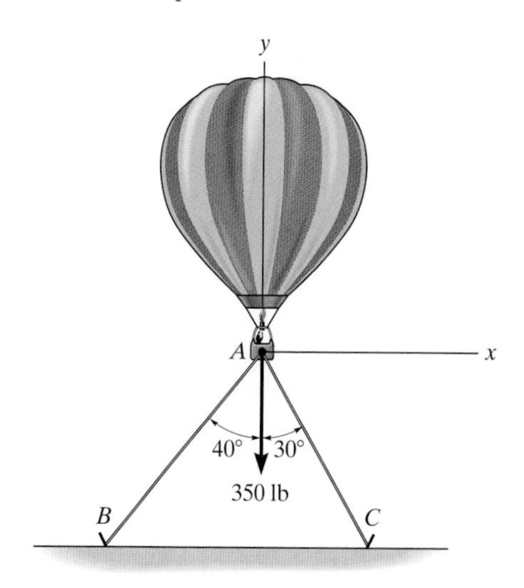

Prob. 2–13

2–14. The post is to be pulled out of the ground using two ropes A and B. Rope A is subjected to a force of 600 lb and is directed at 60° from the horizontal. If the resultant force acting on the post is to be 1200 lb, vertically upward, determine the force T in rope B and the corresponding angle θ.

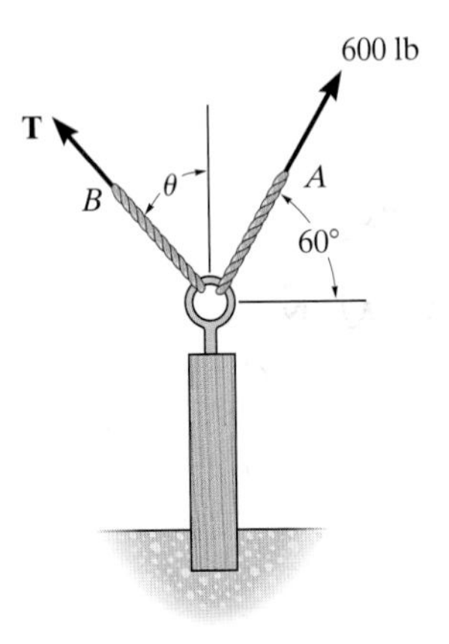

Prob. 2–14

2–15. Resolve the force $\mathbf{F}_1$ into components acting along the u and v axes and determine the magnitudes of the components.

***2–16.** Resolve the force $\mathbf{F}_2$ into components acting along the u and v axes and determine the magnitudes of the components.

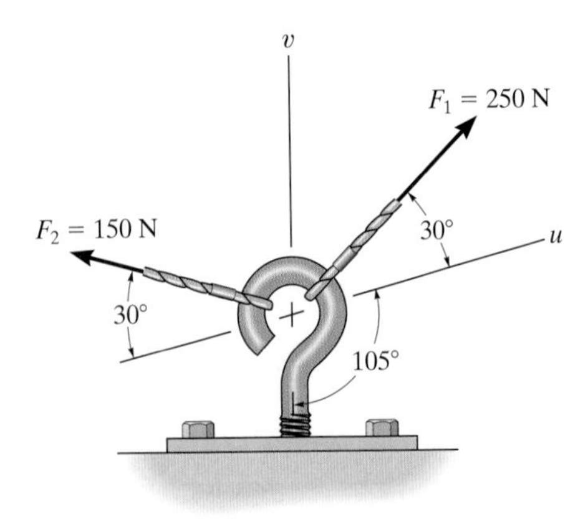

Probs. 2–15/16

2–17. Determine the magnitude and direction of the resultant force $\mathbf{F}_R$. Express the result in terms of the magnitudes of the components $\mathbf{F}_1$ and $\mathbf{F}_2$ and the angle ϕ.

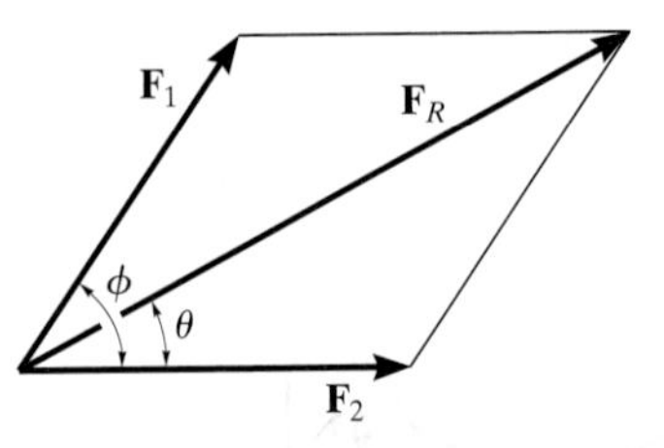

Prob. 2–17

2–18. If the tension in the cable is 400 N, determine the magnitude and direction of the resultant force acting on the pulley. This angle defines the same angle θ of line AB on the tailboard block.

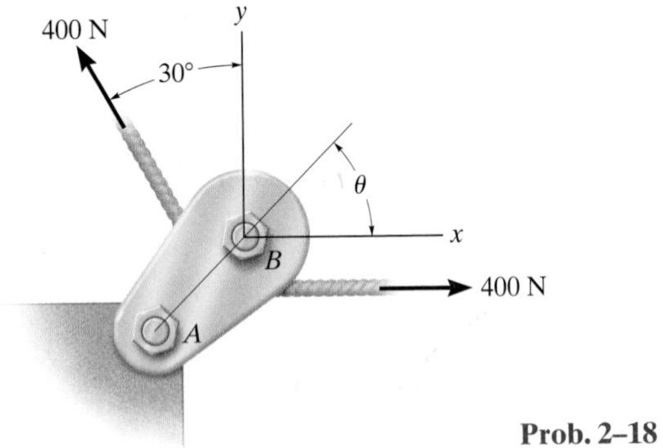

Prob. 2–18

2–19. The riveted bracket supports two forces. Determine the angle θ so that the resultant force is directed along the negative x axis. What is the magnitude of this resultant force?

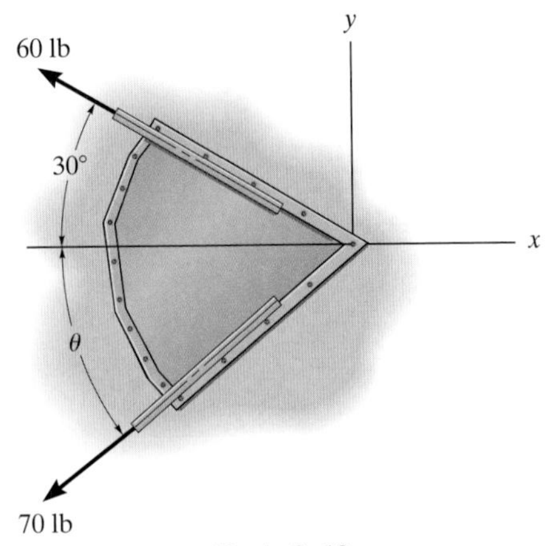

Prob. 2–19

***2–20.** The plate is subjected to the forces acting on members A and B as shown. If $\theta = 60^\circ$, determine the magnitude of the resultant of these forces and its direction measured clockwise from the positive x axis.

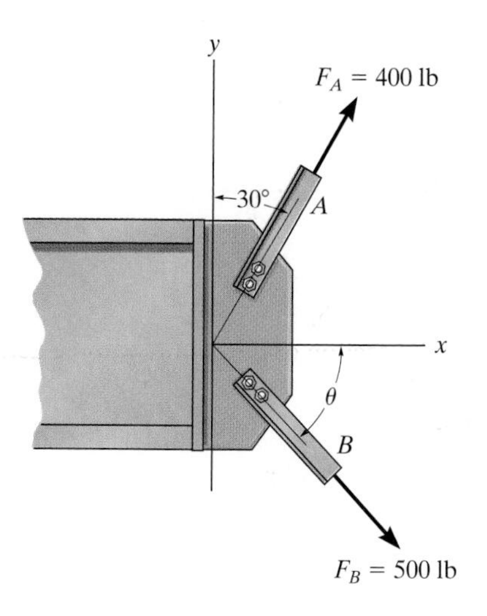

Prob. 2–20

2–21. Determine the angle θ for connecting member B to the plate so that the resultant angle of $\mathbf{F}_A$ and $\mathbf{F}_B$ is directed along the positive x axis. What is the magnitude of the resultant force?

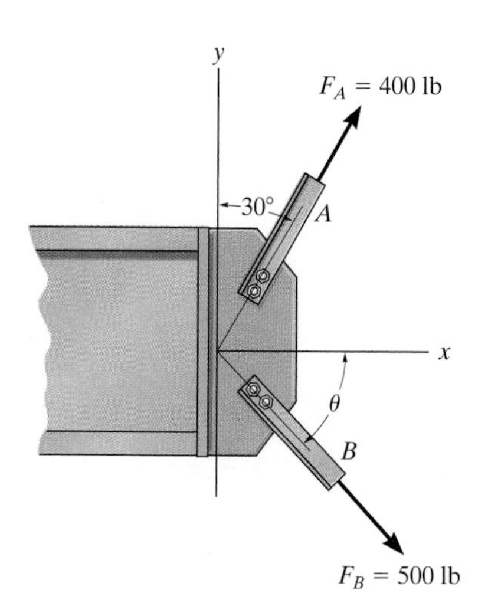

Prob. 2–21

2–22. Determine the magnitude and direction of the resultant $\mathbf{F}_R = \mathbf{F}_1 + \mathbf{F}_2 + \mathbf{F}_3$ of the three forces by first finding the resultant $\mathbf{F}' = \mathbf{F}_1 + \mathbf{F}_2$ and then forming $\mathbf{F}_R = \mathbf{F}' + \mathbf{F}_3$.

2–23. Determine the magnitude and direction of the resultant $\mathbf{F}_R = \mathbf{F}_1 + \mathbf{F}_2 + \mathbf{F}_3$ of the three forces by first finding the resultant $\mathbf{F}' = \mathbf{F}_2 + \mathbf{F}_3$ and then forming $\mathbf{F}_R = \mathbf{F}' + \mathbf{F}_1$.

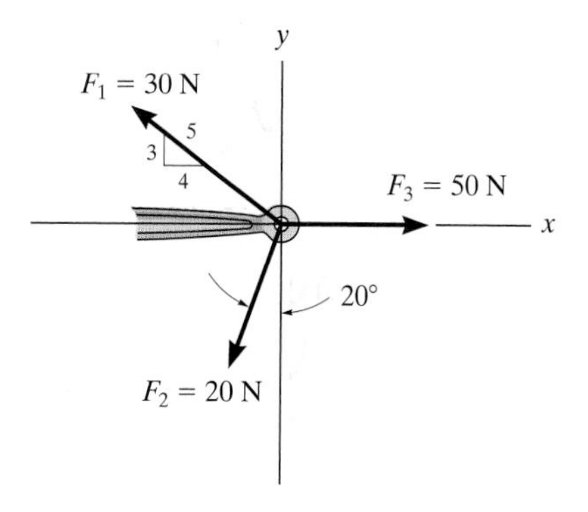

Probs. 2–22/23

***2–24.** Resolve the 50-lb force into components acting along (a) the x and y axes, and (b) the x and y' axes.

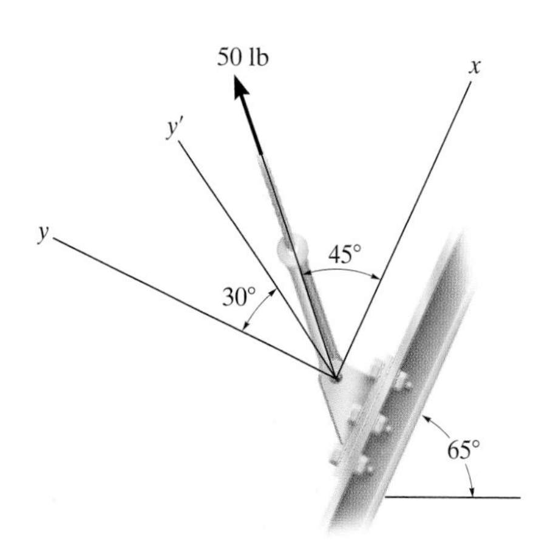

Prob. 2–24

2–25. The boat is to be pulled onto the shore using two ropes. Determine the magnitudes of forces $\mathbf{T}$ and $\mathbf{P}$ acting in each rope in order to develop a resultant force of 80 lb, directed along the keel *aa* as shown. Take $\theta = 40°$.

2–26. The boat is to be pulled onto the shore using two ropes. If the resultant force is to be 80 lb, directed along the keel *aa* as shown, determine the magnitudes of forces $\mathbf{T}$ and $\mathbf{P}$ acting in each rope and the angle θ of $\mathbf{P}$ so that the magnitude of $\mathbf{P}$ is a *minimum*. $\mathbf{T}$ acts at 30° from the keel as shown.

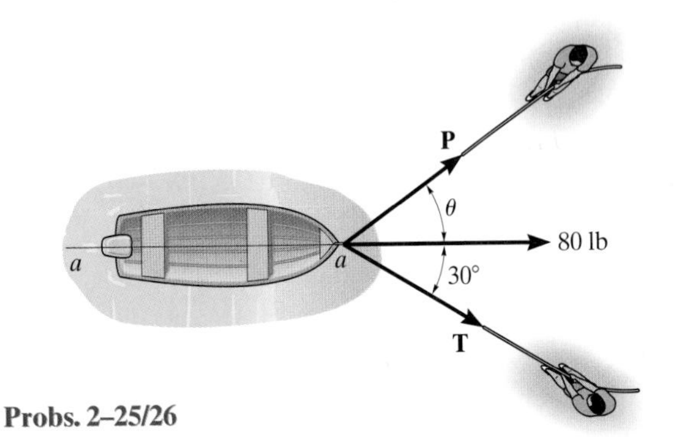

Probs. 2–25/26

2–27. The beam is to be hoisted using two chains. Determine the magnitudes of forces $\mathbf{F}_A$ and $\mathbf{F}_B$ acting on each chain in order to develop a resultant force of 600 N directed along the positive y axis. Set $\theta = 45°$.

***2–28.** The beam is to be hoisted using two chains. If the resultant force is to be 600 N, directed along the positive y axis, determine the magnitudes of forces $\mathbf{F}_A$ and $\mathbf{F}_B$ acting on each chain and the orientation θ of $\mathbf{F}_B$ so that the magnitude of $\mathbf{F}_B$ is a *minimum*. $\mathbf{F}_A$ acts at 30° from the y axis as shown.

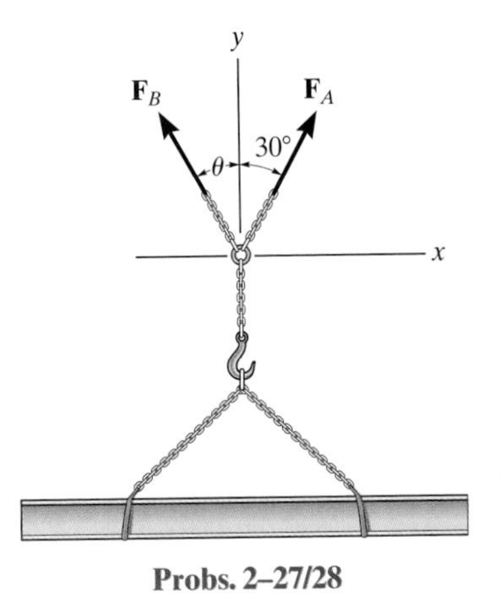

Probs. 2–27/28

2–29. Three chains act on the bracket such that they create a resultant force having a magnitude of 500 lb. If two of the chains are subjected to known forces, as shown, determine the orientation θ of the third chain, measured clockwise from the positive x axis, so that the magnitude of force $\mathbf{F}$ in this chain is a *minimum*. All forces lie in the x–y plane. What is the magnitude of $\mathbf{F}$? *Hint*: First find the resultant of the two known forces. Force $\mathbf{F}$ acts in this direction.

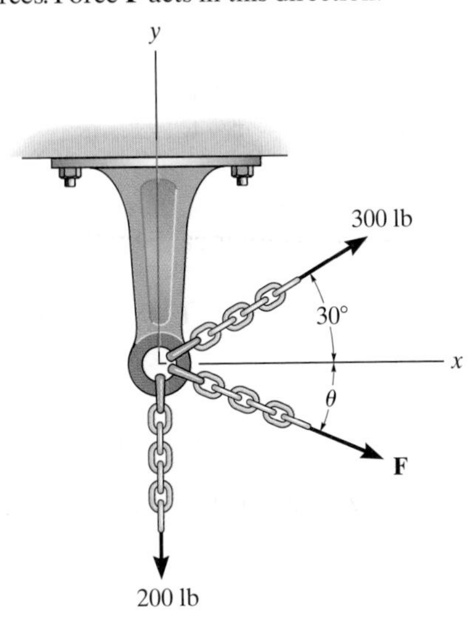

Prob. 2–29

2–30. Three cables pull on the pipe such that they create a resultant force having a magnitude of 900 lb. If two of the cables are subjected to known forces, as shown in the figure, determine the direction θ of the third cable so that the magnitude of force $\mathbf{F}$ in this cable is a *minimum*. All forces lie in the x–y plane. What is the magnitude of $\mathbf{F}$? *Hint*: First find the resultant of the two known forces.

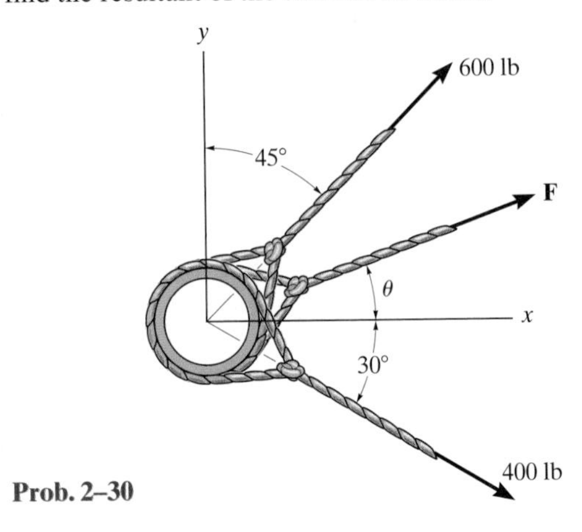

Prob. 2–30

2.4 Addition of a System of Coplanar Forces

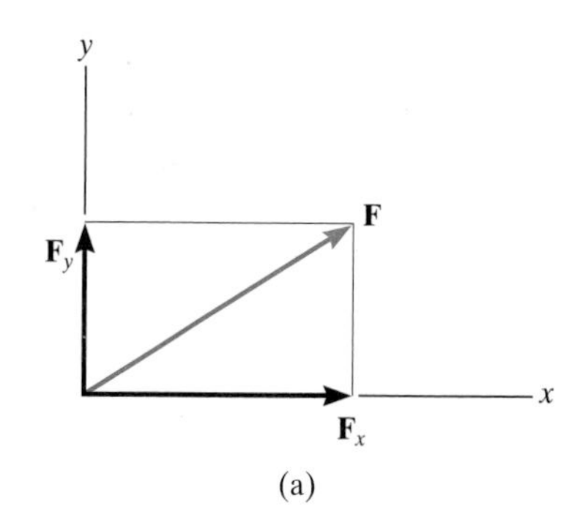

(a)

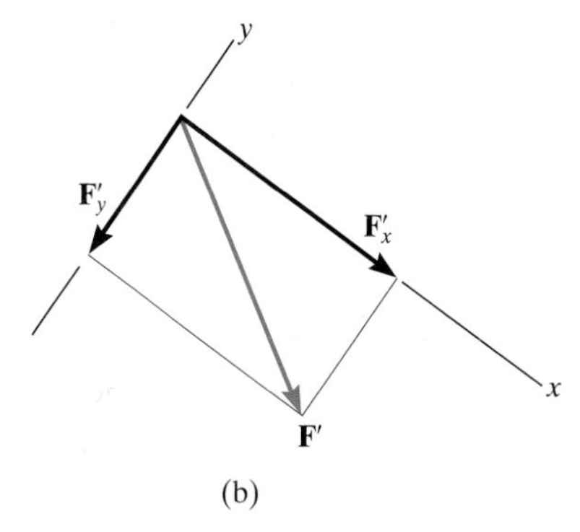

(b)

Fig. 2–14

When the resultant of more than two forces has to be obtained, it is easier to find the components of each force along specified axes, add these components algebraically, and then form the resultant, rather than form the resultant of the forces by successive application of the parallelogram law as discussed in Sec. 2.3.

In this section we will resolve each force into its rectangular components $\mathbf{F}_x$ and $\mathbf{F}_y$, which lie along the x and y axes, respectively, Fig. 2–14*a*. Although the axes are horizontal and vertical, they may in general be directed at any inclination, as long as they remain perpendicular to one another, Fig. 2–14*b*. In either case, by the parallelogram law, we require

$$\mathbf{F} = \mathbf{F}_x + \mathbf{F}_y$$

and

$$\mathbf{F}' = \mathbf{F}'_x + \mathbf{F}'_y$$

As shown in Fig. 2–14, the sense of direction of each force component is represented *graphically* by the *arrowhead*. For *analytical* work, however, we must establish a notation for representing the directional sense of the rectangular components. This can be done in one of two ways.

Scalar Notation. Since the x and y axes have designated positive and negative directions, the magnitude and directional sense of the rectangular components of a force can be expressed in terms of *algebraic scalars*. For example, the components of **F** in Fig. 2–14*a* can be represented by positive scalars F_x and F_y since their sense of direction is along the *positive* x and y axes, respectively. In a similar manner, the components of $\mathbf{F}'$ in Fig. 2–14*b* are F'_x and $-F'_y$. Here the y component is negative, since $\mathbf{F}'_y$ is directed along the negative y axis.

It is important to keep in mind that this positive and negative scalar notation is to be used only for computational purposes, not for graphical representations in figures. Throughout the book, the *head of a vector arrow* in any figure indicates the sense of the vector *graphically*; algebraic signs are not used for this purpose. Thus, the vectors in Figs. 2–14*a* and 2–14*b* are designated by using boldface (vector) notation.* Whenever italic symbols are written near vector arrows in figures, they indicate the *magnitude* of the vector, which is *always* a *positive* quantity.

*Negative signs are used only in figures with boldface notation when showing equal but opposite pairs of vectors as in Fig. 2–2.

Cartesian Vector Notation. It is also possible to represent the components of a force in terms of Cartesian unit vectors. When we do this the methods of vector algebra are easier to apply, and we will see that this becomes particularly advantageous for solving problems in three dimensions.

In two dimensions the *Cartesian unit vectors* **i** and **j** are used to designate the *directions* of the x and y axes, respectively, Fig. 2–15*a*.*

As shown in Fig. 2–15*a*, the *magnitude* of each component of **F** is *always a positive quantity*, which is represented by the (positive) scalars F_x and F_y. Therefore, having established notation to represent the magnitude and the direction of each vector component, we can express **F** in Fig. 2–15*a* as a *Cartesian vector*,

$$\mathbf{F} = F_x\mathbf{i} + F_y\mathbf{j}$$

And in the same way, **F′** in Fig. 2–15*b* can be expressed as

$$\mathbf{F}' = F'_x\mathbf{i} + F'_y(-\mathbf{j})$$

or simply

$$\mathbf{F}' = F'_x\mathbf{i} - F'_y\mathbf{j}$$

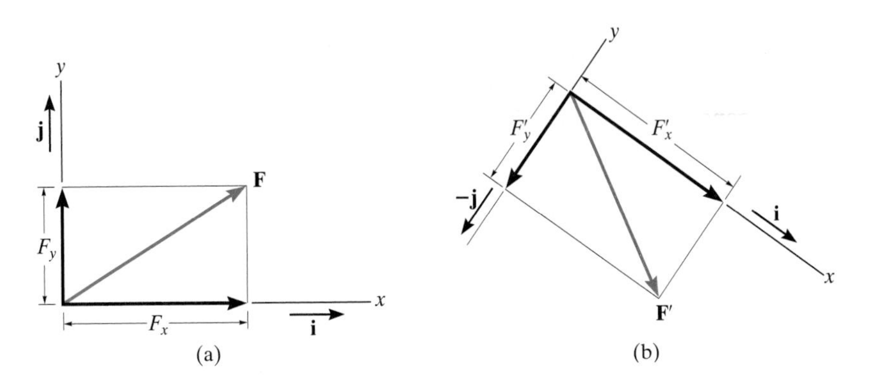

Fig. 2–15

*For handwritten work, unit vectors are usually indicated using a circumflex, e.g., $\hat{i}$ and $\hat{j}$. These vectors have a dimensionless magnitude of unity, and their sense (or arrowhead) will be described analytically by a plus or minus sign, depending on whether they are pointing along the positive or negative x or y axis.

Coplanar Force Resultants. Either of the two methods just described can be used to determine the resultant of several *coplanar forces.* To do this, each force is first resolved into its x and y components, and then the respective components are added using *scalar algebra.* Since they are collinear. The resultant force is then formed by adding the resultants of the x and y components using the parallelogram law. For example, consider the three concurrent forces in Fig. 2–16a, which have x and y components as shown in Fig. 2–16b. To solve this problem using *Cartesian vector notation*, each force is first represented as a Cartesian vector, i.e.,

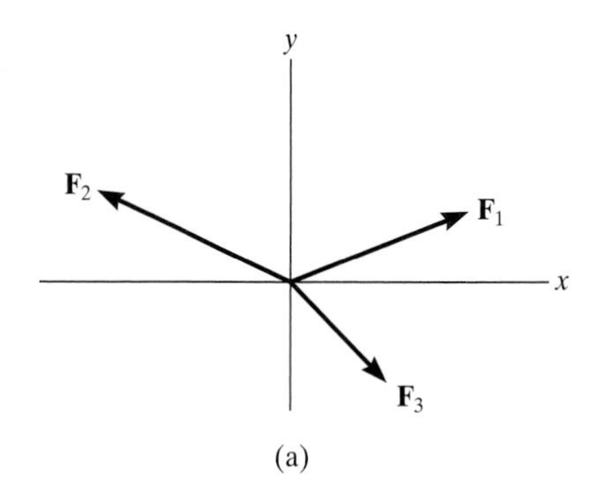

(a)

$$\mathbf{F}_1 = F_{1x}\mathbf{i} + F_{1y}\mathbf{j}$$
$$\mathbf{F}_2 = -F_{2x}\mathbf{i} + F_{2y}\mathbf{j}$$
$$\mathbf{F}_3 = F_{3x}\mathbf{i} - F_{3y}\mathbf{j}$$

The vector resultant is therefore

$$\begin{aligned}\mathbf{F}_R &= \mathbf{F}_1 + \mathbf{F}_2 + \mathbf{F}_3\\ &= F_{1x}\mathbf{i} + F_{1y}\mathbf{j} - F_{2x}\mathbf{i} + F_{2y}\mathbf{j} + F_{3x}\mathbf{i} - F_{3y}\mathbf{j}\\ &= (F_{1x} - F_{2x} + F_{3x})\mathbf{i} + (F_{1y} + F_{2y} - F_{3y})\mathbf{j}\\ &= (F_{Rx})\mathbf{i} + (F_{Ry})\mathbf{j}\end{aligned}$$

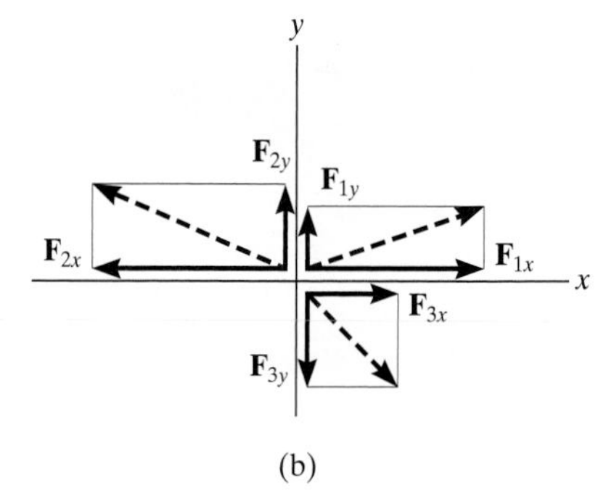

(b)

If *scalar notation* is used, then, from Fig. 2–16b, since x is positive to the right and y is positive upward, we have

$$(\xrightarrow{+}) \qquad F_{Rx} = F_{1x} - F_{2x} + F_{3x}$$
$$(+\uparrow) \qquad F_{Ry} = F_{1y} + F_{2y} - F_{3y}$$

These results are the *same* as the **i** and **j** components of $\mathbf{F}_R$ determined above.

In the general case, the x and y components of the resultant of any number of coplanar forces can be represented symbolically by the algebraic sum of the x and y components of all the forces, i.e.,

$$F_{Rx} = \Sigma F_x \qquad (2\text{–}1)$$
$$F_{Ry} = \Sigma F_y$$

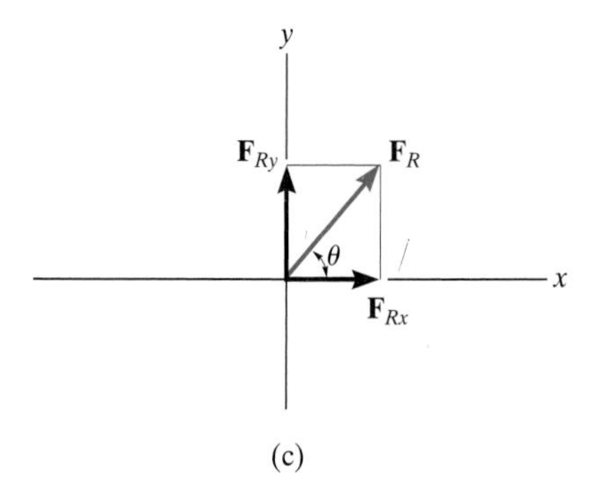

(c)

Fig. 2–16

When applying these equations, it is important to use the *sign convention* established for the components; and that is, components having a directional sense along the positive coordinate axes are considered positive scalars, whereas those having a directional sense along the negative coordinate axes are considered negative scalars. If this convention is followed, then the signs of the resultant components will specify the sense of these components. For example, a positive result indicates that the component has a directional sense which is in the positive coordinate direction.

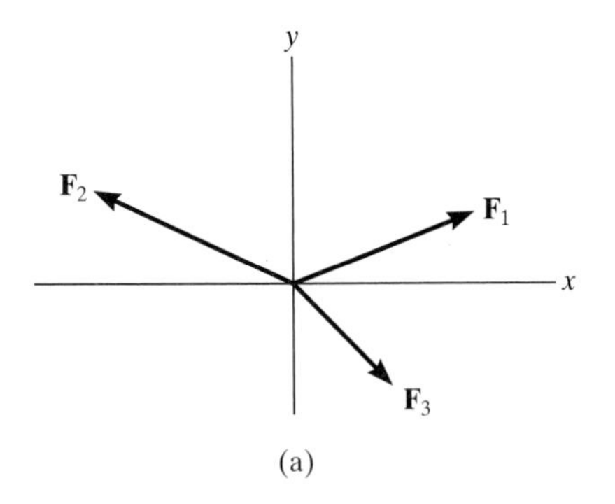

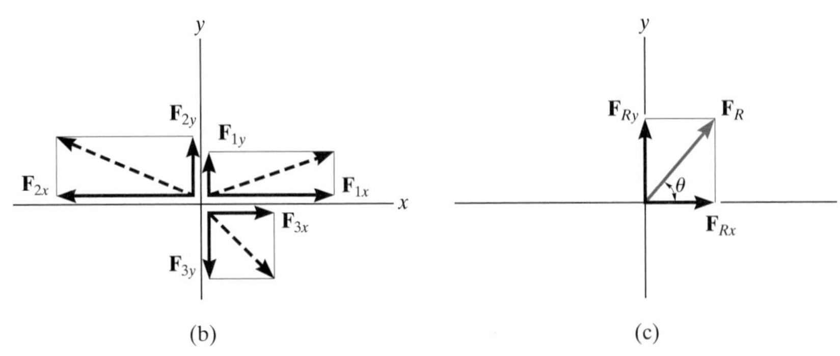

Fig. 2–16

The resultant force of the four cable forces acting on the supporting bracket can be determined by adding algebraically the separate x and y components of each cable force. This resultant $\mathbf{F}_R$ produces the *same pulling effect* on the bracket as all four cables.

Once the resultant components are determined, they may be sketched along the x and y axes in their proper directions, and the resultant force can be determined from vector addition, as shown in Fig. 2–16*c*. From this sketch, the magnitude of $\mathbf{F}_R$ is then found from the Pythagorean theorem; that is,

$$F_R = \sqrt{F_{Rx}^2 + F_{Ry}^2}$$

Also, the direction angle θ, which specifies the orientation of the force, is determined from trigonometry:

$$\theta = \tan^{-1}\left|\frac{F_{Ry}}{F_{Rx}}\right|$$

The above concepts are illustrated numerically in the examples which follow.

Important Points

- The resultant of several coplanar forces can easily be determined if an x, y coordinate system is established and the forces are resolved along the axes.
- The direction of each force is specified by the angle its line of action makes with one of the axes, or by a sloped triangle.
- The orientation of the x and y axes is arbitrary, and their positive direction can be specified by the Cartesian unit vectors **i** and **j**.
- The x and y components of the *resultant force* are simply the algebraic addition of the components of all the coplanar forces.
- The magnitude of the resultant force is determined from the Pythagorean theorem, and when the components are sketched on the x and y axes, the direction can be determined from trigonometry.

EXAMPLE 2.5

Determine the x and y components of $\mathbf{F}_1$ and $\mathbf{F}_2$ acting on the boom shown in Fig. 2–17a. Express each force as a Cartesian vector.

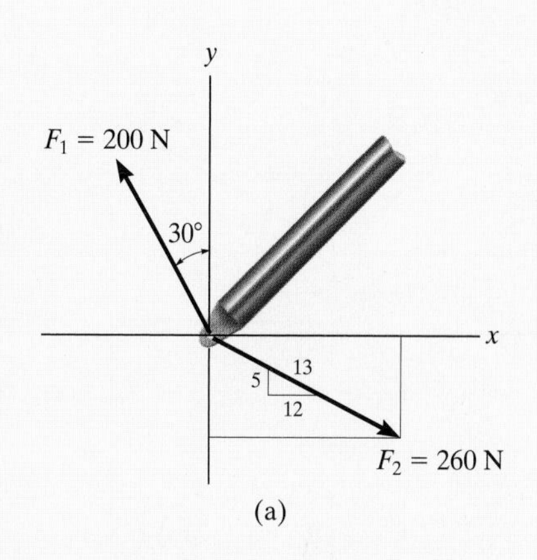

(a)

SOLUTION

Scalar Notation. By the parallelogram law, $\mathbf{F}_1$ is resolved into x and y components, Fig. 2–17b. The magnitude of each component is determined by trigonometry. Since $\mathbf{F}_{1x}$ acts in the $-x$ direction, and $\mathbf{F}_{1y}$ acts in the $+y$ direction, we have

$$F_{1x} = -200 \sin 30° \text{ N} = -100 \text{ N} = 100 \text{ N} \leftarrow \qquad \textit{Ans.}$$

$$F_{1y} = 200 \cos 30° \text{ N} = 173 \text{ N} = 173 \text{ N} \uparrow \qquad \textit{Ans.}$$

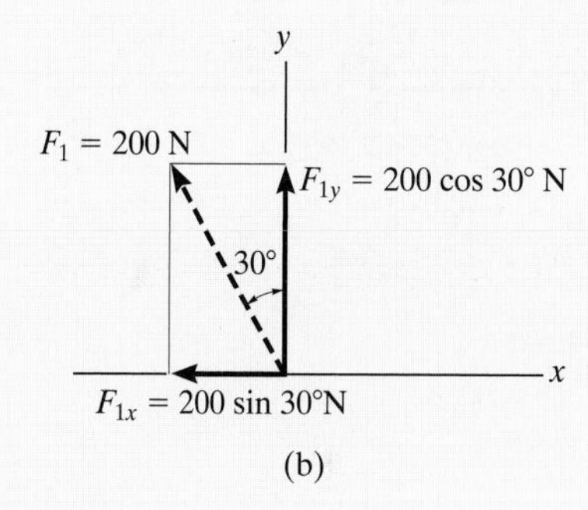

(b)

The force $\mathbf{F}_2$ is resolved into its x and y components as shown in Fig. 2–17c. Here the *slope* of the line of action for the force is indicated. From this "slope triangle" we could obtain the angle θ, e.g., $\theta = \tan^{-1}(\frac{5}{12})$, and then proceed to determine the magnitudes of the components in the same manner as for $\mathbf{F}_1$. An easier method, however, consists of using proportional parts of similar triangles, i.e.,

$$\frac{F_{2x}}{260 \text{ N}} = \frac{12}{13} \qquad F_{2x} = 260 \text{ N}\left(\frac{12}{13}\right) = 240 \text{ N}$$

Similarly,

$$F_{2y} = 260 \text{ N}\left(\frac{5}{13}\right) = 100 \text{ N}$$

Hence, using scalar notation,

$$F_{2x} = 240 \text{ N} = 240 \text{ N} \rightarrow \qquad \textit{Ans.}$$

$$F_{2y} = -100 \text{ N} = 100 \text{ N} \downarrow \qquad \textit{Ans.}$$

Cartesian Vector Notation. Having determined the magnitudes and directions of the components of each force, we can express each force as a Cartesian vector.

$$\mathbf{F}_1 = \{-100\mathbf{i} + 173\mathbf{j}\} \text{ N} \qquad \textit{Ans.}$$

$$\mathbf{F}_2 = \{240\mathbf{i} - 100\mathbf{j}\} \text{ N} \qquad \textit{Ans.}$$

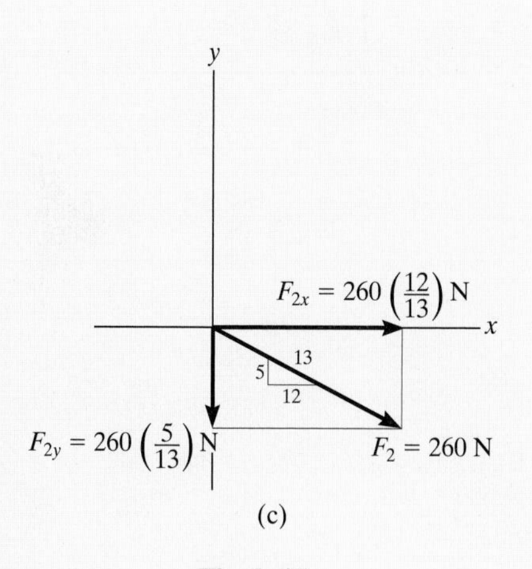

(c)

Fig. 2–17

NOTE: The magnitude of the *horizontal component*, F_{2x}, was obtained by multiplying the force magnitude by the ratio of the *horizontal leg* of the slope triangle divided by the hypotenuse; whereas the magnitude of the *vertical component*, F_{2y}, was obtained by multiplying the force magnitude by the ratio of the *vertical leg* divided by the hypotenuse.

EXAMPLE 2.6

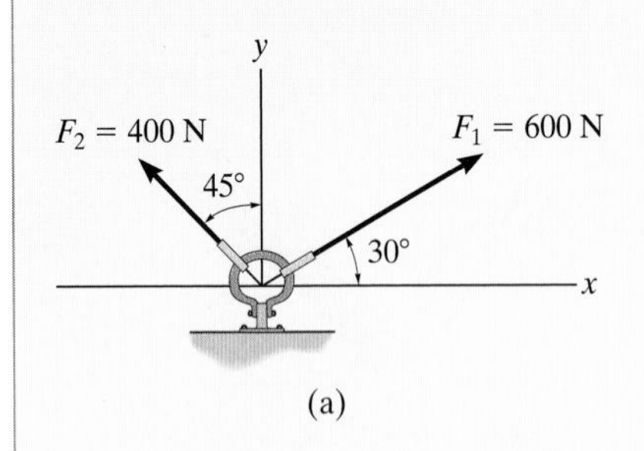

(a)

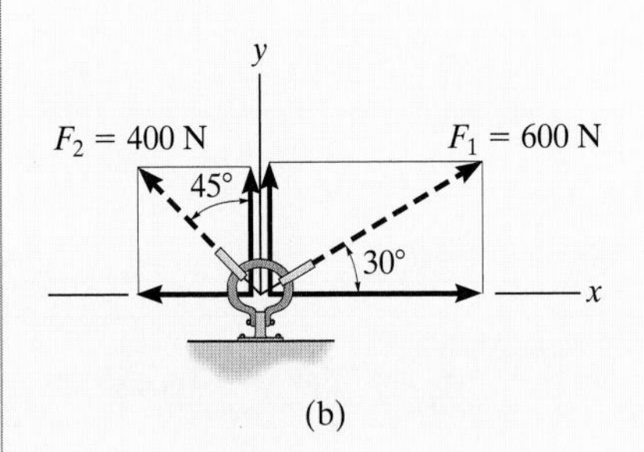

(b)

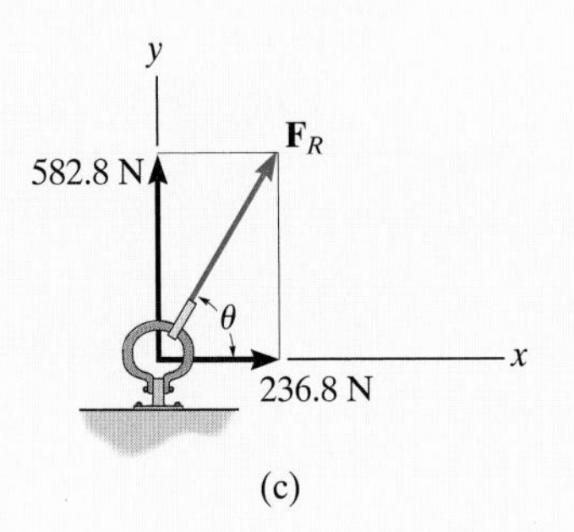

(c)

Fig. 2–18

The link in Fig. 2–18*a* is subjected to two forces $\mathbf{F}_1$ and $\mathbf{F}_2$. Determine the magnitude and orientation of the resultant force.

SOLUTION I

Scalar Notation. This problem can be solved by using the parallelogram law; however, here we will resolve each force into its x and y components, Fig. 2–18*b*, and sum these components algebraically. Indicating the "positive" sense of the x and y force components alongside each equation, we have

$$\overset{+}{\rightarrow} F_{Rx} = \Sigma F_x; \qquad F_{Rx} = 600 \cos 30^\circ \text{ N} - 400 \sin 45^\circ \text{ N}$$
$$= 236.8 \text{ N} \rightarrow$$

$$+\uparrow F_{Ry} = \Sigma F_y; \qquad F_{Ry} = 600 \sin 30^\circ \text{ N} + 400 \cos 45^\circ \text{ N}$$
$$= 582.8 \text{ N}\uparrow$$

The resultant force, shown in Fig. 2–18*c*, has a *magnitude* of

$$F_R = \sqrt{(236.8 \text{ N})^2 + (582.8 \text{ N})^2}$$
$$= 629 \text{ N} \qquad \textit{Ans.}$$

From the vector addition, Fig. 2–18*c*, the direction angle θ is

$$\theta = \tan^{-1}\left(\frac{582.8 \text{ N}}{236.8 \text{ N}}\right) = 67.9^\circ \qquad \textit{Ans.}$$

SOLUTION II

Cartesian Vector Notation. From Fig. 2–18*b*, each force is expressed as a Cartesian vector

$$\mathbf{F}_1 = \{600 \cos 30^\circ \mathbf{i} + 600 \sin 30^\circ \mathbf{j}\} \text{ N}$$
$$\mathbf{F}_2 = \{-400 \sin 45^\circ \mathbf{i} + 400 \cos 45^\circ \mathbf{j}\} \text{ N}$$

Thus,

$$\mathbf{F}_R = \mathbf{F}_1 + \mathbf{F}_2 = (600 \cos 30^\circ \text{ N} - 400 \sin 45^\circ \text{ N})\mathbf{i}$$
$$+ (600 \sin 30^\circ \text{ N} + 400 \cos 45^\circ \text{ N})\mathbf{j}$$
$$= \{236.8\mathbf{i} + 582.8\mathbf{j}\} \text{ N}$$

The magnitude and direction of $\mathbf{F}_R$ are determined in the same manner as shown above.

NOTE: Comparing the two methods of solution, notice that the use of scalar notation is more efficient since the components can be found *directly*, without first having to express each force as a Cartesian vector before adding the components. Later we will show that Cartesian vector analysis is very beneficial for solving three-dimensional problems.

EXAMPLE 2.7

The end of the boom O in Fig. 2–19*a* is subjected to three concurrent and coplanar forces. Determine the magnitude and orientation of the resultant force.

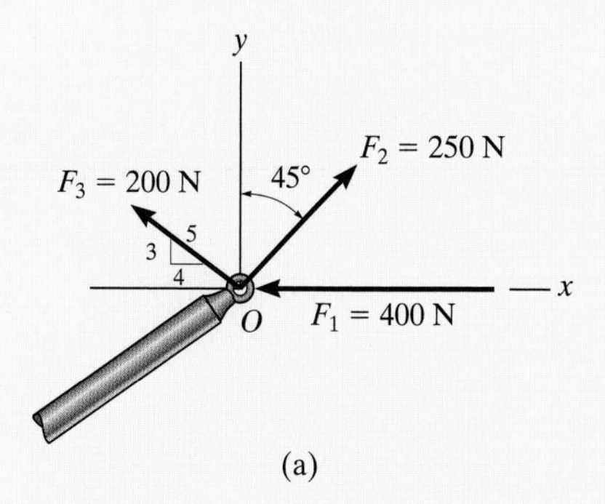

(a)

SOLUTION

Each force is resolved into its x and y components, Fig. 2–19*b*. Summing the x components, we have

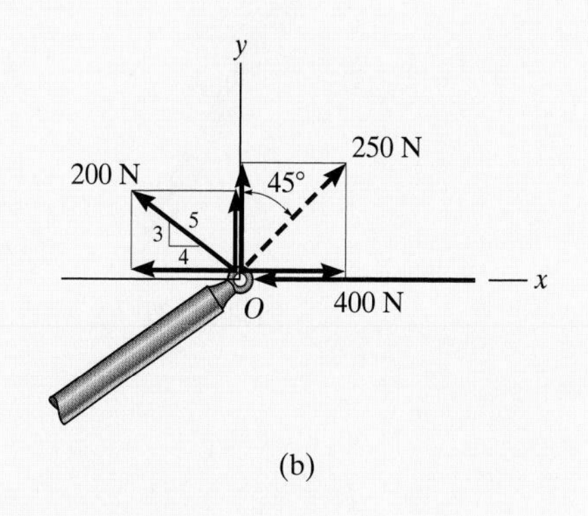

(b)

$$\overset{+}{\rightarrow} F_{Rx} = \Sigma F_x; \qquad F_{Rx} = -400 \text{ N} + 250 \sin 45^\circ \text{ N} - 200\left(\tfrac{4}{5}\right)\text{N}$$
$$= -383.2 \text{ N} = 383.2 \text{ N} \leftarrow$$

The negative sign indicates that F_{Rx} acts to the left, i.e., in the negative x direction as noted by the small arrow. Obviously, this occurs because F_1 and F_3 in Fig. 2–19*b* contribute a greater pull to the left than F_2, which pulls to the right. Summing the y components yields

$$+\uparrow F_{Ry} = \Sigma F_y; \qquad F_{Ry} = 250 \cos 45^\circ \text{ N} + 200\left(\tfrac{3}{5}\right) \text{N}$$
$$= 296.8 \text{ N}\uparrow$$

The resultant force, shown in Fig. 2–19*c*, has a *magnitude* of

$$F_R = \sqrt{(-383.2 \text{ N})^2 + (296.8 \text{ N})^2}$$
$$= 485 \text{ N} \qquad \textit{Ans.}$$

From the vector addition in Fig. 2–19*c*, the direction angle θ is

$$\theta = \tan^{-1}\left(\frac{296.8}{383.2}\right) = 37.8^\circ \qquad \textit{Ans.}$$

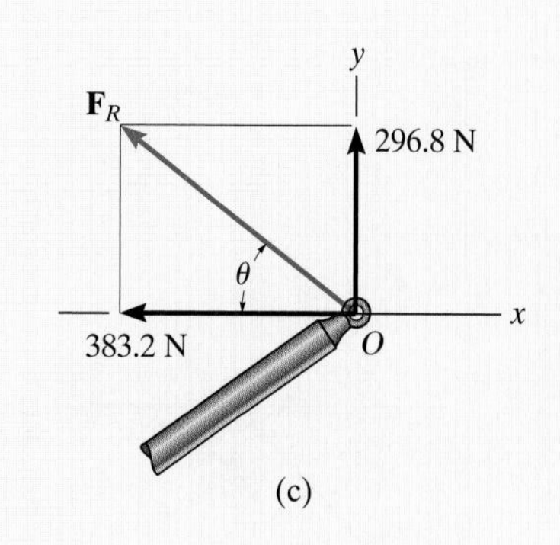

(c)

Fig. 2–19

NOTE: Application of this method is more convenient, compared to two applications of the parallelogram law.

PROBLEMS

2–31. Determine the x and y components of the 800-lb force.

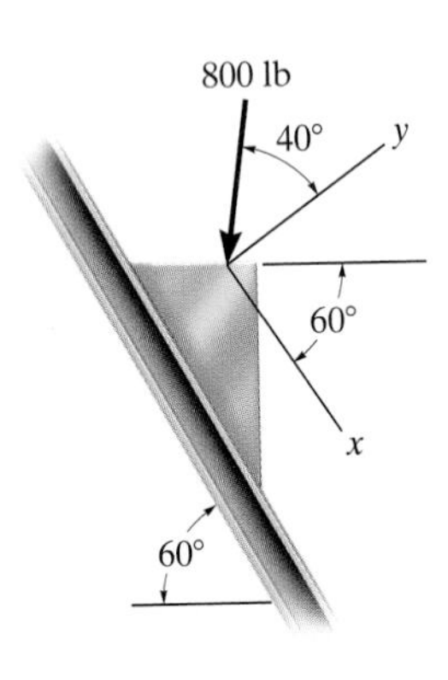

Prob. 2–31

2–33. Determine the magnitude of the resultant force and its direction measured counterclockwise from the positive x axis.

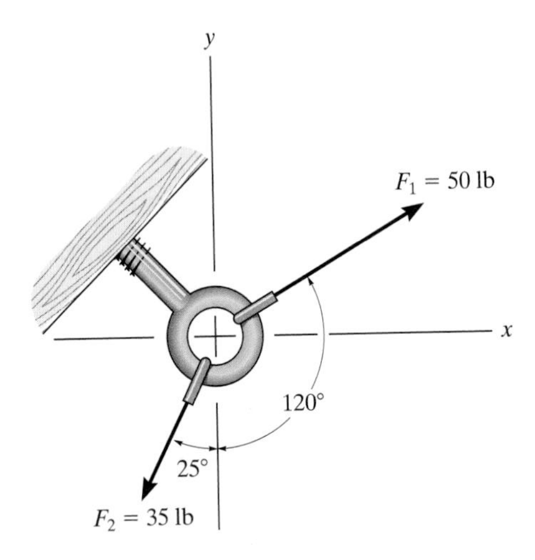

Prob. 2–33

***2–32.** Determine the magnitude of the resultant force and its direction, measured clockwise from the positive x axis.

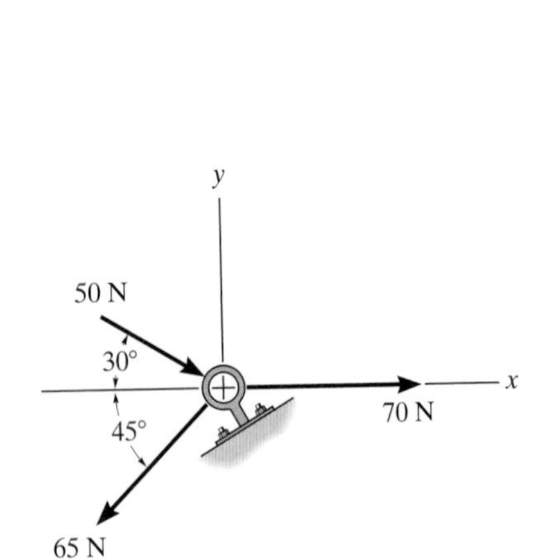

Prob. 2–32

2–34. Determine the magnitude of the resultant force and its direction, measured counterclockwise from the positive x axis.

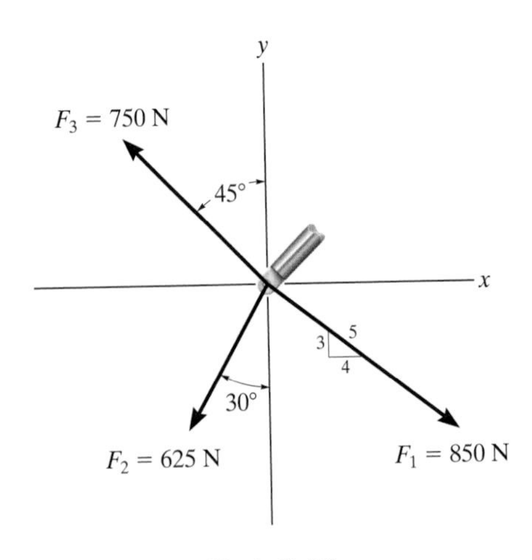

Prob. 2–34

2–35. Three forces act on the bracket. Determine the magnitude and direction θ of $\mathbf{F}_1$ so that the resultant force is directed along the positive x' axis and has a magnitude of 1 kN.

***2–36.** If $F_1 = 300$ N and $\theta = 20°$, determine the magnitude and direction, measured counterclockwise from the x' axis, of the resultant force of the three forces acting on the bracket.

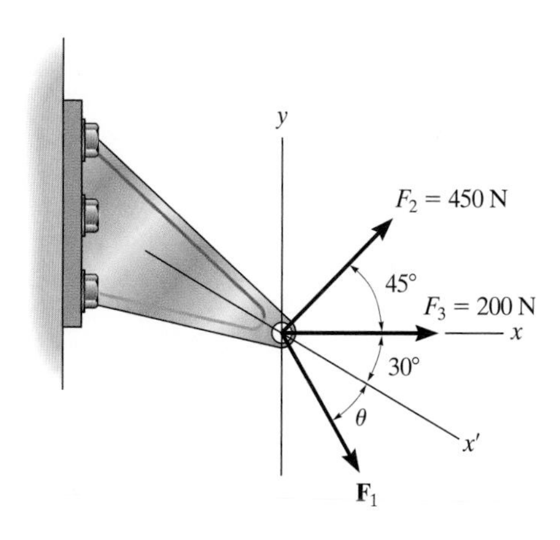

Probs. 2–35/36

2–37. Determine the magnitude of the resultant force and its direction, measured counterclockwise from the positive x axis.

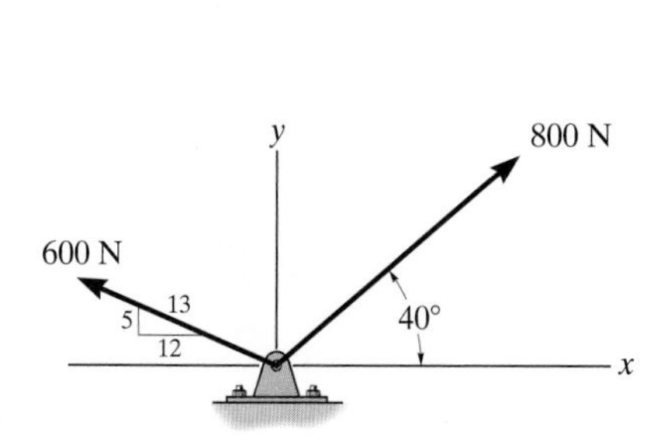

Prob. 2–37

2–38. Determine the magnitude of the resultant force and its direction, measured counterclockwise from the positive x axis.

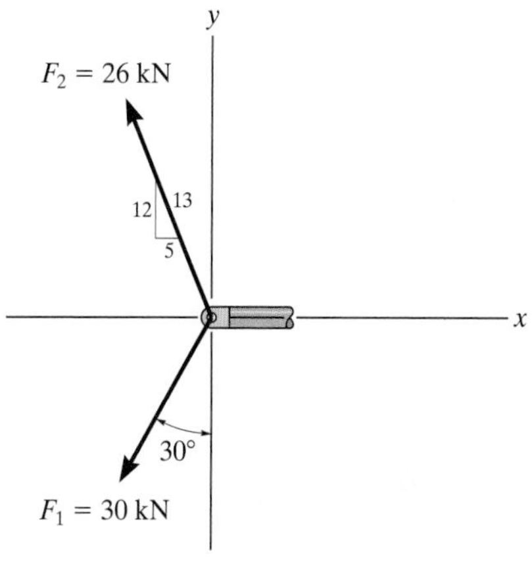

Prob. 2–38

2–39. Determine the magnitude of the resultant force and its direction measured counterclockwise from the positive x axis.

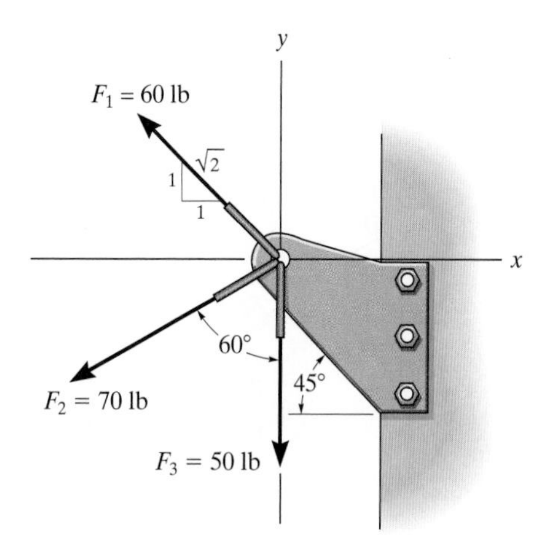

Prob. 2–39

***2–40.** Solve Prob. 2–1 by summing the rectangular or x, y components of the forces to obtain the resultant force.

2–41. Solve Prob. 2–22 by summing the rectangular or x, y components of the forces to obtain the resultant force.

2-42. Determine the magnitude and orientation, measured counterclockwise from the positive y axis, of the resultant force acting on the bracket, if $F_B = 600$ N and $\theta = 20°$.

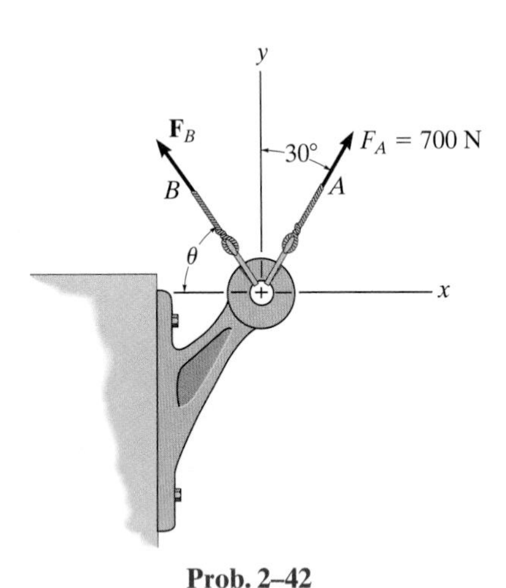

Prob. 2–42

2–43. If $F_1 = 300$ N and $\theta = 10°$, determine the magnitude and direction, measured counterclockwise from the positive x' axis, of the resultant force of the three forces acting on the bracket.

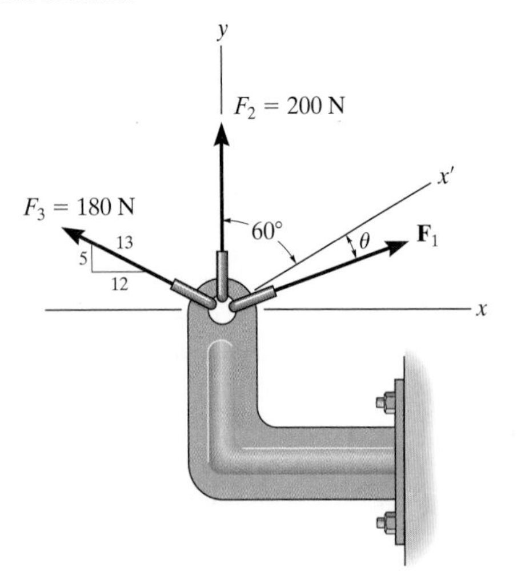

Prob. 2–43

***2–44.** Determine the x and y components of $\mathbf{F}_1$ and $\mathbf{F}_2$.

2–45. Determine the magnitude of the resultant force and its direction, measured counterclockwise from the positive x axis.

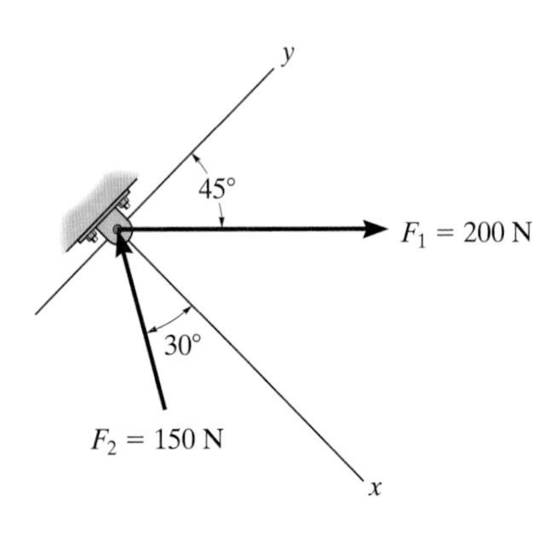

Probs. 2–44/45

2–46. Determine the x and y components of each force acting on the *gusset plate* of the bridge truss. Show that the resultant force is zero.

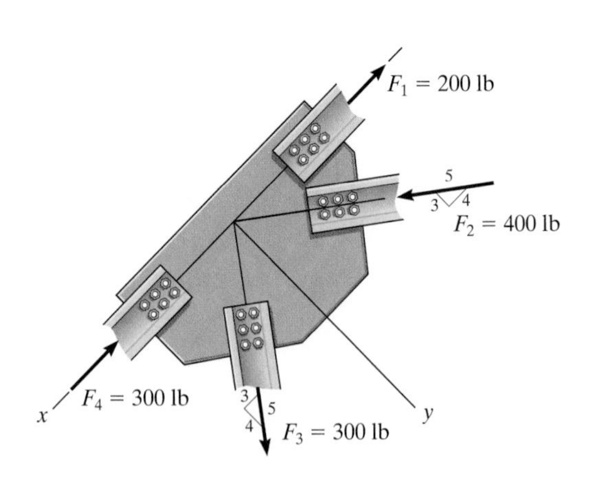

Prob. 2–46

2–47. If $\theta = 60°$ and $F = 20$ kN, determine the magnitude of the resultant force and its direction measured clockwise from the positive x axis.

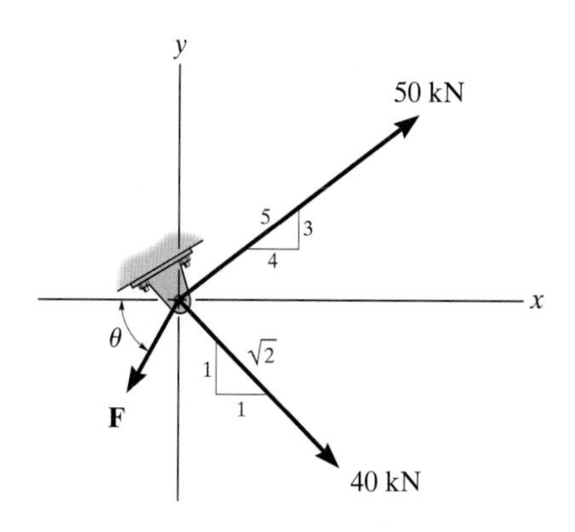

Prob. 2–47

***2–48.** Three forces act on the bracket. Determine the magnitude and direction θ of $\mathbf{F}_1$ so that the resultant force is directed along the positive x' axis and has a magnitude of 800 N.

2–49. If $F_1 = 300$ N and $\theta = 10°$, determine the magnitude and direction, measured counterclockwise from the positive x' axis, of the resultant force acting on the bracket.

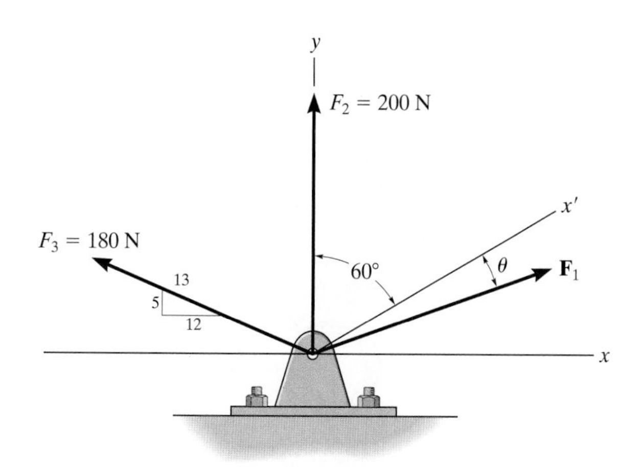

Probs. 2–48/49

2–50. Express each of the three forces acting on the column in Cartesian vector form and compute the magnitude of the resultant force.

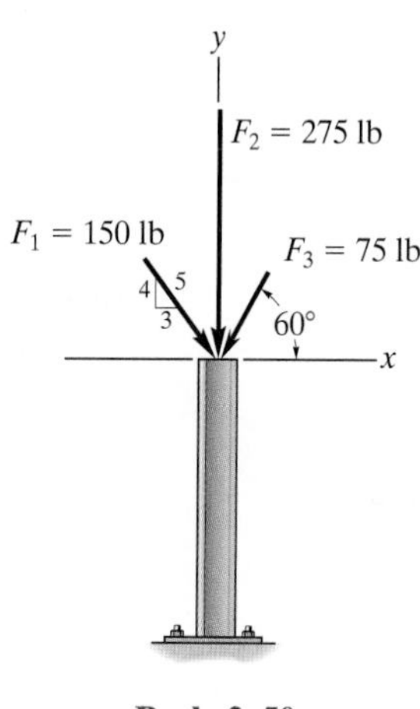

Prob. 2–50

2–51. Determine the magnitude of force $\mathbf{F}$ so that the resultant $\mathbf{F}_R$ of the three forces is as small as possible. What is the minimum magnitude of $\mathbf{F}_R$?

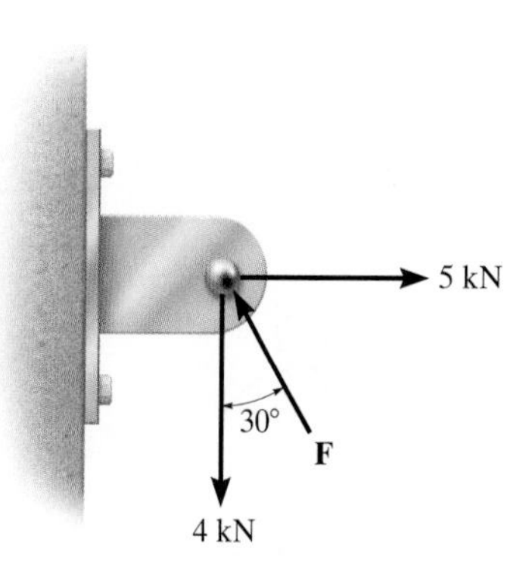

Prob. 2–51

***2–52.** Express each of the three forces acting on the bracket in Cartesian vector form with respect to the x and y axes. Determine the magnitude and direction θ of $\mathbf{F}_1$ so that the resultant force is directed along the positive x' axis and has a magnitude of $F_R = 600$ N.

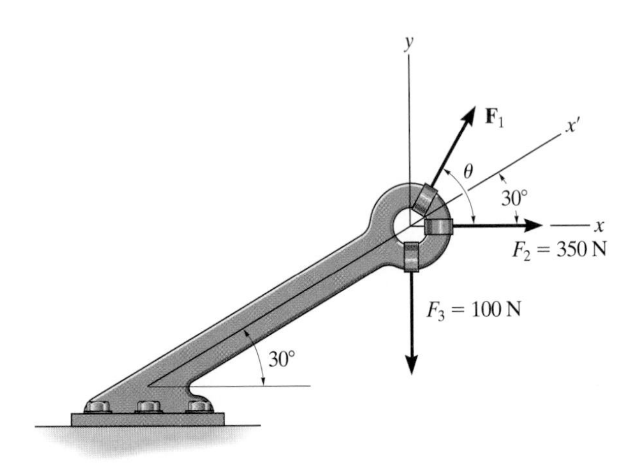

Prob. 2–52

2–53. The three concurrent forces acting on the post produce a resultant force $\mathbf{F}_R = \mathbf{0}$. If $F_2 = \frac{1}{2}F_1$, and $\mathbf{F}_1$ is to be 90° from $\mathbf{F}_2$ as shown, determine the required magnitude F_3 expressed in terms of F_1 and the angle θ.

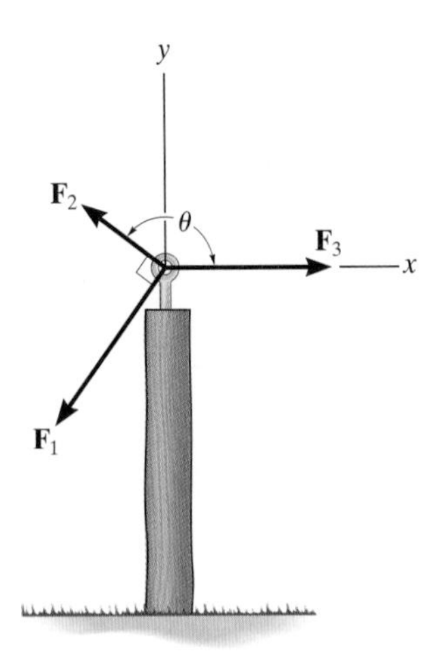

Prob. 2–53

2–54. Three forces act on the bracket. Determine the magnitude and orientation θ of $\mathbf{F}_2$ so that the resultant force is directed along the positive u axis and has a magnitude of 50 lb.

2–55. If $F_2 = 150$ lb and $\theta = 55°$, determine the magnitude and orientation, measured clockwise from the positive x axis, of the resultant force of the three forces acting on the bracket.

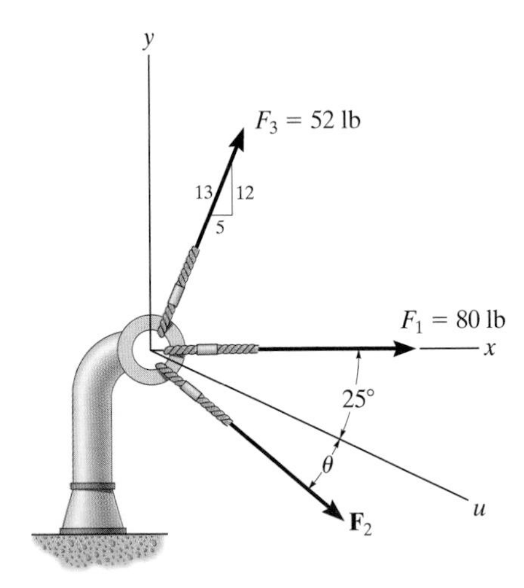

Probs. 2–54/55

***2–56.** Three forces act on the ring. Determine the range of values for the magnitude of $\mathbf{P}$ so that the magnitude of the resultant force does not exceed 2500 N. Force $\mathbf{P}$ is always directed to the right.

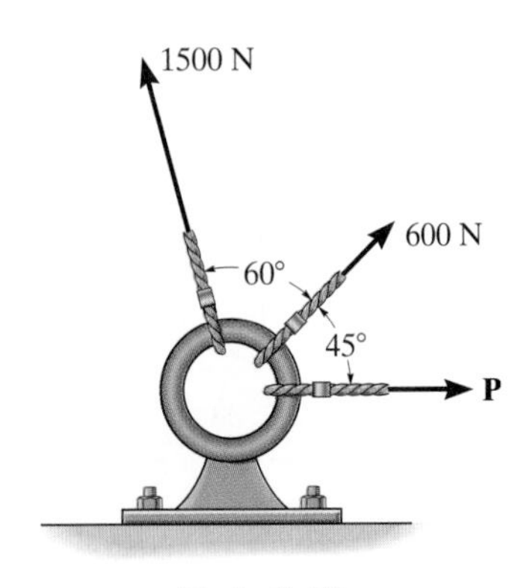

Prob. 2–56

2.5 Cartesian Vectors

The operations of vector algebra, when applied to solving problems in *three dimensions*, are greatly simplified if the vectors are first represented in Cartesian vector form. In this section we will present a general method for doing this; then in the next section we will apply this method to solving problems involving the addition of forces. Similar applications will be illustrated for the position and moment vectors given in later sections of the book.

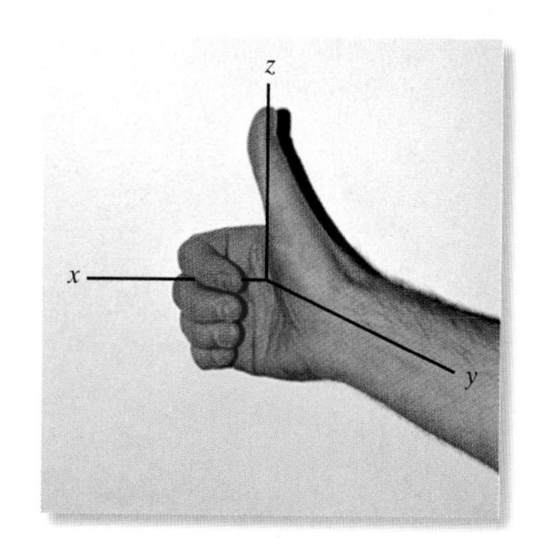

Fig. 2–20

Right-Handed Coordinate System. We will use a right-handed coordinate system to develop the theory of vector algebra that follows. A rectangular or Cartesian coordinate system is said to be *right-handed* provided the thumb of the right hand points in the direction of the positive z axis when the right-hand fingers are curled about this axis and directed from the positive x toward the positive y axis, Fig. 2–20. Furthermore, according to this rule, the z axis for a two-dimensional problem as in Fig. 2–19 would be directed outward, perpendicular to the page.

Rectangular Components of a Vector. A vector $\mathbf{A}$ may have one, two, or three rectangular components along the x, y, z coordinate axes, depending on how the vector is oriented relative to the axes. In general, though, when $\mathbf{A}$ is directed within an octant of the x, y, z frame, Fig. 2–21, then by two successive applications of the parallelogram law, we may resolve the vector into components as $\mathbf{A} = \mathbf{A}' + \mathbf{A}_z$ and then $\mathbf{A}' = \mathbf{A}_x + \mathbf{A}_y$. Combining these equations, $\mathbf{A}$ is represented by the vector sum of its *three* rectangular components,

$$\mathbf{A} = \mathbf{A}_x + \mathbf{A}_y + \mathbf{A}_z \tag{2–2}$$

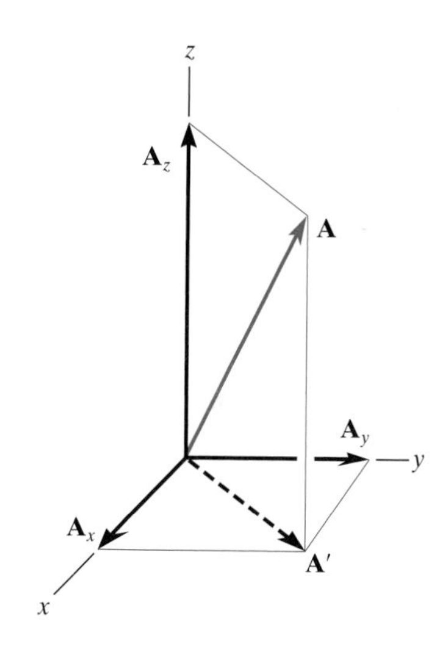

Fig. 2–21

Unit Vector. The direction of $\mathbf{A}$ can be specified using a unit vector, so named since it has a magnitude of 1. If $\mathbf{A}$ is a vector having a magnitude $A \neq 0$, then the unit vector having the *same direction* as $\mathbf{A}$ is represented by

$$\mathbf{u}_A = \frac{\mathbf{A}}{A} \tag{2–3}$$

So that

$$\mathbf{A} = A\mathbf{u}_A \tag{2–4}$$

Since $\mathbf{A}$ is of a certain type, e.g., a force vector, it is customary to use the proper set of units for its description. The magnitude A also has this same set of units; hence, from Eq. 2–3, the *unit vector will be dimensionless* since the units will cancel out. Equation 2–4 shows that vector $\mathbf{A}$ can be expressed in terms of both its magnitude and direction *separately*; i.e., A (a positive scalar) defines the *magnitude* of $\mathbf{A}$, and $\mathbf{u}_A$ (a dimensionless vector) defines the *direction* and sense of $\mathbf{A}$, Fig. 2–22.

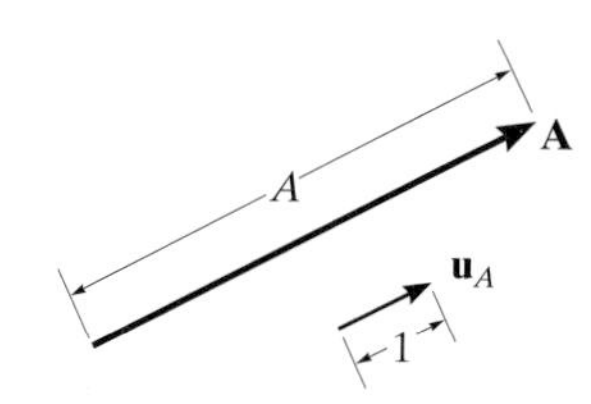

Fig. 2–22

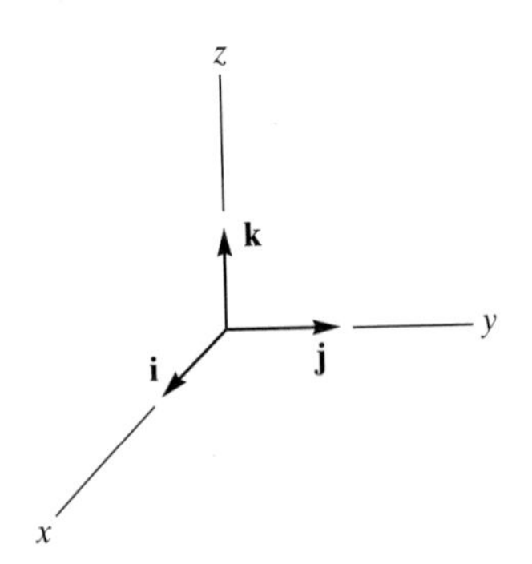

Fig. 2–23

Cartesian Unit Vectors. In three dimensions, the set of Cartesian unit vectors, **i**, **j**, **k**, is used to designate the directions of the x, y, z axes respectively. As stated in Sec. 2.4, the *sense* (or arrowhead) of these vectors will be described analytically by a plus or minus sign, depending on whether they are pointing along the positive or negative x, y, or z axes. The positive Cartesian unit vectors are shown in Fig. 2–23.

Cartesian Vector Representation. Since the three components of **A** in Eq. 2–2 act in the positive **i**, **j**, and **k** directions, Fig. 2–24, we can write **A** in Cartesian vector form as

$$\mathbf{A} = A_x\mathbf{i} + A_y\mathbf{j} + A_z\mathbf{k} \tag{2–5}$$

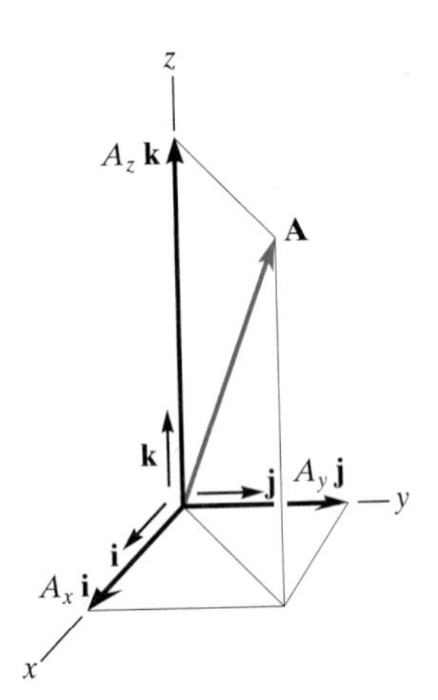

Fig. 2–24

There is a distinct advantage to writing vectors in this manner. Separating the *magnitude* and *direction* of each *component vector* simplifies the operations of vector algebra, particularly in three dimensions.

Magnitude of a Cartesian Vector. It is always possible to obtain the magnitude of **A** provided it is expressed in Cartesian vector form. As shown in Fig. 2–25, from the blue right triangle, $A = \sqrt{A'^2 + A_z^2}$, and from the gray right triangle, $A' = \sqrt{A_x^2 + A_y^2}$. Combining these equations yields

$$A = \sqrt{A_x^2 + A_y^2 + A_z^2} \tag{2–6}$$

Hence, the magnitude of **A** *is equal to the positive square root of the sum of the squares of its components.*

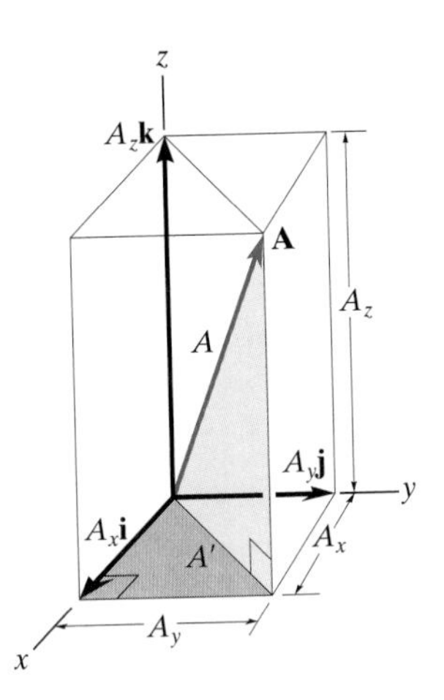

Fig. 2–25

Direction of a Cartesian Vector. The *orientation* of **A** is defined by the *coordinate direction angles* α (alpha), β (beta), and γ (gamma), measured between the *tail* of **A** and the *positive* x, y, z axes located at the tail of **A**, Fig. 2–26. Note that regardless of where **A** is directed, each of these angles will be between 0° and 180°.

To determine α, β, and γ, consider the projection of **A** onto the x, y, z axes, Fig. 2–27. Referring to the blue colored right triangles shown in each figure, we have

$$\cos\alpha = \frac{A_x}{A} \qquad \cos\beta = \frac{A_y}{A} \qquad \cos\gamma = \frac{A_z}{A} \tag{2–7}$$

These numbers are known as the *direction cosines* of **A**. Once they have been obtained, the coordinate direction angles α, β, γ can then be determined from the inverse cosines.

An easy way of obtaining the direction cosines of **A** is to form a unit vector in the direction of **A**, Eq. 2–3. Provided **A** is expressed in Cartesian vector form, $\mathbf{A} = A_x\mathbf{i} + A_y\mathbf{j} + A_z\mathbf{k}$ (Eq. 2–5), then

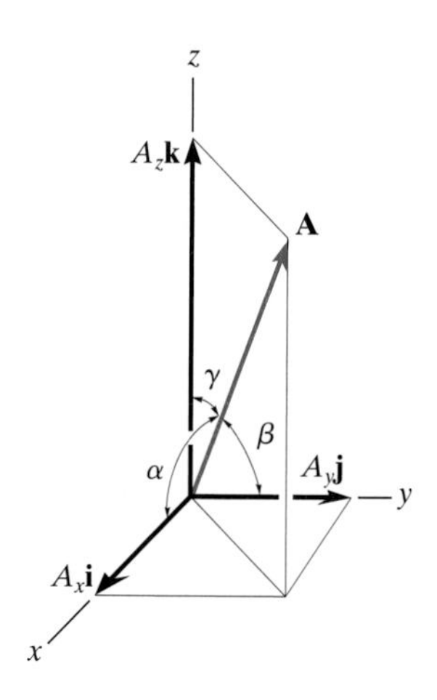

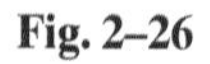
Fig. 2–26

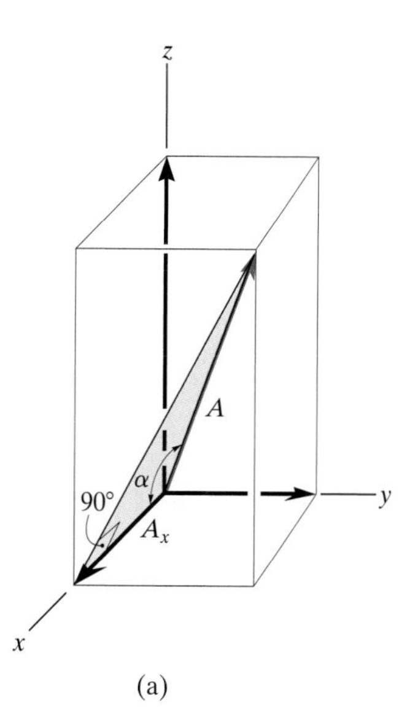

(a)

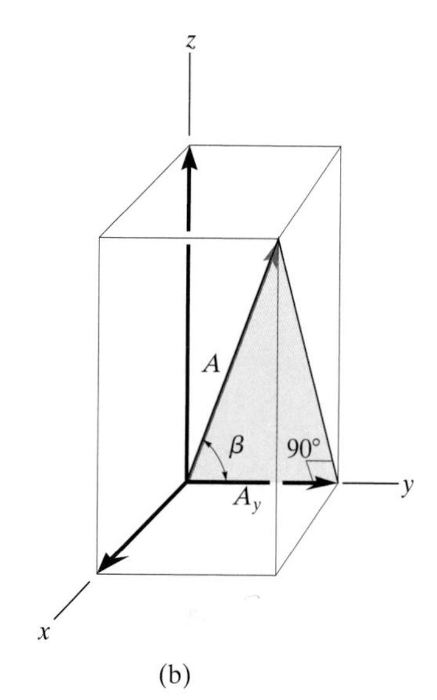

(b)

$$\mathbf{u}_A = \frac{\mathbf{A}}{A} = \frac{A_x}{A}\mathbf{i} + \frac{A_y}{A}\mathbf{j} + \frac{A_z}{A}\mathbf{k} \tag{2–8}$$

where $A = \sqrt{A_x^2 + A_y^2 + A_z^2}$ (Eq. 2–6). By comparison with Eqs. 2–7, it is seen that *the* **i**, **j**, **k** *components of* $\mathbf{u}_A$ *represent the direction cosines of* **A**, i.e.,

$$\mathbf{u}_A = \cos\alpha\mathbf{i} + \cos\beta\mathbf{j} + \cos\gamma\mathbf{k} \tag{2–9}$$

Since the magnitude of a vector is equal to the positive square root of the sum of the squares of the magnitudes of its components, and $\mathbf{u}_A$ has a magnitude of 1, then from Eq. 2–9 an important relation between the direction cosines can be formulated as

$$\cos^2\alpha + \cos^2\beta + \cos^2\gamma = 1 \tag{2–10}$$

Provided vector **A** lies in a known octant, this equation can be used to determine one of the coordinate direction angles if the other two are known.

Finally, if the magnitude and coordinate direction angles of **A** are given, **A** may be expressed in Cartesian vector form as

$$\begin{aligned}\mathbf{A} &= A\mathbf{u}_A \\ &= A\cos\alpha\mathbf{i} + A\cos\beta\mathbf{j} + A\cos\gamma\mathbf{k} \\ &= A_x\mathbf{i} + A_y\mathbf{j} + A_z\mathbf{k}\end{aligned} \tag{2–11}$$

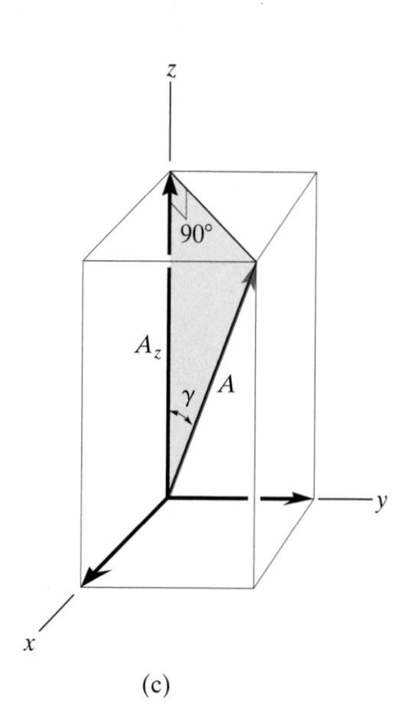

(c)

Fig. 2–27

2.6 Addition and Subtraction of Cartesian Vectors

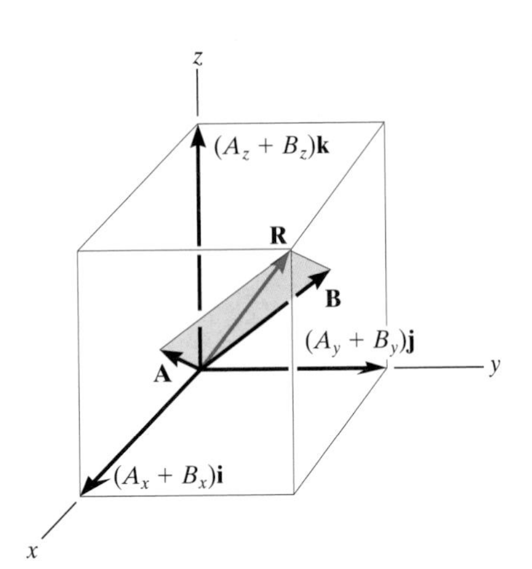

Fig. 2–28

The vector operations of addition and subtraction of two or more vectors are greatly simplified if the vectors are expressed in terms of their Cartesian components. For example, if $\mathbf{A} = A_x\mathbf{i} + A_y\mathbf{j} + A_z\mathbf{k}$ and $\mathbf{B} = B_x\mathbf{i} + B_y\mathbf{j} + B_z\mathbf{k}$, Fig. 2–28, then the resultant vector, $\mathbf{R}$, has components which represent the scalar sums of the $\mathbf{i}, \mathbf{j}, \mathbf{k}$ components of $\mathbf{A}$ and $\mathbf{B}$, i.e.,

$$\mathbf{R} = \mathbf{A} + \mathbf{B} = (A_x + B_x)\mathbf{i} + (A_y + B_y)\mathbf{j} + (A_z + B_z)\mathbf{k}$$

Vector subtraction, being a special case of vector addition, simply requires a scalar subtraction of the respective $\mathbf{i}, \mathbf{j}, \mathbf{k}$ components of either $\mathbf{A}$ or $\mathbf{B}$. For example,

$$\mathbf{R}' = \mathbf{A} - \mathbf{B} = (A_x - B_x)\mathbf{i} + (A_y - B_y)\mathbf{j} + (A_z - B_z)\mathbf{k}$$

Concurrent Force Systems. If the above concept of vector addition is generalized and applied to a system of several concurrent forces, then the force resultant is the vector sum of all the forces in the system and can be written as

$$\mathbf{F}_R = \Sigma\mathbf{F} = \Sigma F_x\mathbf{i} + \Sigma F_y\mathbf{j} + \Sigma F_z\mathbf{k} \tag{2–12}$$

Here ΣF_x, ΣF_y, and ΣF_z represent the algebraic sums of the respective x, y, z or $\mathbf{i}, \mathbf{j}, \mathbf{k}$ components of each force in the system.

The examples which follow illustrate numerically the methods used to apply the above theory to the solution of problems involving force as a vector quantity.

The force $\mathbf{F}$ that the tie-down rope exerts on the ground support at O is directed along the rope. Using the local x, y, z axes, the coordinate direction angles α, β, γ can be measured. The cosines of their values form the components of a unit vector $\mathbf{u}$ which acts in the direction of the rope. If the force has a magnitude F, then the force can be written in Cartesian vector form, as $\mathbf{F} = F\mathbf{u} = F\cos\alpha\mathbf{i} + F\cos\beta\mathbf{j} + F\cos\gamma\mathbf{k}$.

Important Points

- Cartesian vector analysis is often used to solve problems in three dimensions.
- The positive directions of the x, y, z axes are defined by the Cartesian unit vectors $\mathbf{i}, \mathbf{j}, \mathbf{k}$, respectively.
- The *magnitude* of a Cartesian vector is $A = \sqrt{A_x^2 + A_y^2 + A_z^2}$.
- The *direction* of a Cartesian vector is specified using coordinate direction angles which the tail of the vector makes with the positive x, y, z axes, respectively. The components of the unit vector $\mathbf{u} = \mathbf{A}/A$ represent the direction cosines of α, β, γ. Only two of the angles α, β, γ have to be specified. The third angle is determined from the relationship $\cos^2\alpha + \cos^2\beta + \cos^2\gamma = 1$.
- To find the *resultant* of a concurrent force system, express each force as a Cartesian vector and add the $\mathbf{i}, \mathbf{j}, \mathbf{k}$ components of all the forces in the system.

EXAMPLE 2.8

Express the force **F** shown in Fig. 2–29 as a Cartesian vector.

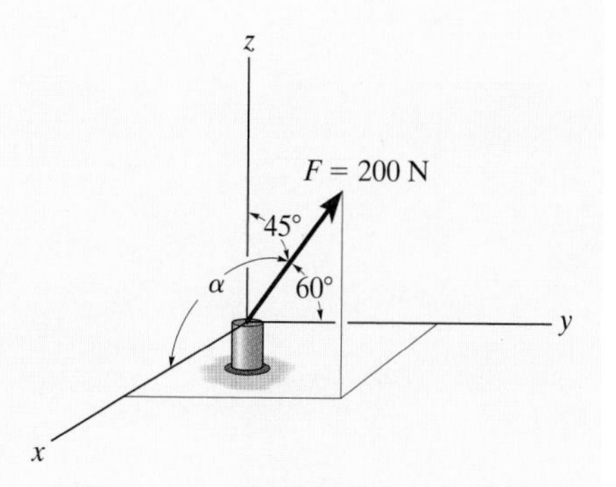

Fig. 2–29

SOLUTION

Since only two coordinate direction angles are specified, the third angle α must be determined from Eq. 2–10; i.e.,

$$\cos^2\alpha + \cos^2\beta + \cos^2\gamma = 1$$
$$\cos^2\alpha + \cos^2 60^\circ + \cos^2 45^\circ = 1$$
$$\cos\alpha = \sqrt{1 - (0.5)^2 - (0.707)^2} = \pm 0.5$$

Hence, two possibilities exist, namely,

$$\alpha = \cos^{-1}(0.5) = 60^\circ \quad \text{or} \quad \alpha = \cos^{-1}(-0.5) = 120^\circ$$

By inspection of Fig. 2–29, it is necessary that $\alpha = 60^\circ$, since $\mathbf{F}_x$ is in the $+x$ direction.

Using Eq. 2–11, with $F = 200$ N, we have

$$\mathbf{F} = F\cos\alpha\mathbf{i} + F\cos\beta\mathbf{j} + F\cos\gamma\mathbf{k}$$
$$= (200\cos 60^\circ \text{ N})\mathbf{i} + (200\cos 60^\circ \text{ N})\mathbf{j} + (200\cos 45^\circ \text{ N})\mathbf{k}$$
$$= \{100.0\mathbf{i} + 100.0\mathbf{j} + 141.4\mathbf{k}\} \text{ N} \qquad \textit{Ans.}$$

NOTE: By applying Eq. 2–6, note that indeed the magnitude of $F = 200$ N.

$$F = \sqrt{F_x^2 + F_y^2 + F_z^2}$$
$$= \sqrt{(100.0)^2 + (100.0)^2 + (141.4)^2} = 200 \text{ N}$$

EXAMPLE 2.9

Determine the magnitude and the coordinate direction angles of the resultant force acting on the ring in Fig. 2–30*a*.

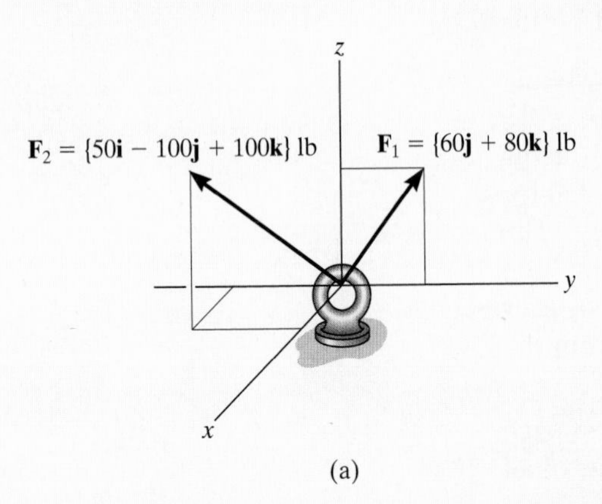

Fig. 2–30

SOLUTION

Since each force is represented in Cartesian vector form, the resultant force, shown in Fig. 2–30*b*, is

$$\mathbf{F}_R = \Sigma\mathbf{F} = \mathbf{F}_1 + \mathbf{F}_2 = \{60\mathbf{j} + 80\mathbf{k}\}\text{ lb} + \{50\mathbf{i} - 100\mathbf{j} + 100\mathbf{k}\}\text{ lb}$$
$$= \{50\mathbf{i} - 40\mathbf{j} + 180\mathbf{k}\}\text{ lb}$$

The magnitude of $\mathbf{F}_R$ is found from Eq. 2–6, i.e.,

$$F_R = \sqrt{(50)^2 + (-40)^2 + (180)^2} = 191.0$$
$$= 191\text{ lb} \qquad \textit{Ans.}$$

The coordinate direction angles α, β, γ are determined from the components of the unit vector acting in the direction of $\mathbf{F}_R$.

$$\mathbf{u}_{F_R} = \frac{\mathbf{F}_R}{F_R} = \frac{50}{191.0}\mathbf{i} - \frac{40}{191.0}\mathbf{j} + \frac{180}{191.0}\mathbf{k}$$
$$= 0.2617\mathbf{i} - 0.2094\mathbf{j} + 0.9422\mathbf{k}$$

so that

$$\cos\alpha = 0.2617 \qquad \alpha = 74.8^\circ \qquad \textit{Ans.}$$
$$\cos\beta = -0.2094 \qquad \beta = 102^\circ \qquad \textit{Ans.}$$
$$\cos\gamma = 0.9422 \qquad \gamma = 19.6^\circ \qquad \textit{Ans.}$$

These angles are shown in Fig. 2–30*b*.

NOTE: In particular, notice that $\beta > 90^\circ$ since the **j** component of $\mathbf{u}_{F_R}$ is negative. This seems reasonable considering how $\mathbf{F}_1$ and $\mathbf{F}_2$ add according to the parallelogram law.

EXAMPLE 2.10

Express the force $\mathbf{F}_1$, shown in Fig. 2–31*a* as a Cartesian vector.

SOLUTION

The angles of 60° and 45° defining the direction of $\mathbf{F}_1$ are *not* coordinate direction angles. The two successive applications of the parallelogram law needed to resolve $\mathbf{F}_1$ into its *x, y, z* components are shown in Fig. 2–31*b*. By trigonometry, the magnitudes of the components are

$$F_{1z} = 100 \sin 60° \text{ lb} = 86.6 \text{ lb}$$
$$F' = 100 \cos 60° \text{ lb} = 50 \text{ lb}$$
$$F_{1x} = 50 \cos 45° \text{ lb} = 35.4 \text{ lb}$$
$$F_{1y} = 50 \sin 45° \text{ lb} = 35.4 \text{ lb}$$

(a)

Realizing that $\mathbf{F}_{1y}$ has a direction defined by $-\mathbf{j}$, we have

$$\mathbf{F}_1 = \{35.4\mathbf{i} - 35.4\mathbf{j} + 86.6\mathbf{k}\} \text{ lb} \qquad \textit{Ans.}$$

To show that the magnitude of this vector is indeed 100 lb, apply Eq. 2–6,

$$F_1 = \sqrt{F_{1x}^2 + F_{1y}^2 + F_{1z}^2}$$
$$= \sqrt{(35.4)^2 + (-35.4)^2 + (86.6)^2} = 100 \text{ lb}$$

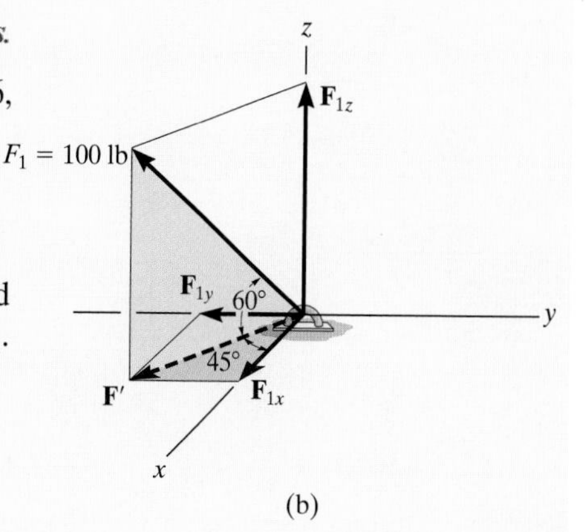

(b)

If needed, the coordinate direction angles of $\mathbf{F}_1$ can be determined from the components of the unit vector acting in the direction of $\mathbf{F}_1$. Hence,

$$\mathbf{u}_1 = \frac{\mathbf{F}_1}{F_1} = \frac{F_{1x}}{F_1}\mathbf{i} + \frac{F_{1y}}{F_1}\mathbf{j} + \frac{F_{1z}}{F_1}\mathbf{k}$$
$$= \frac{35.4}{100}\mathbf{i} - \frac{35.4}{100}\mathbf{j} + \frac{86.6}{100}\mathbf{k}$$
$$= 0.354\mathbf{i} - 0.354\mathbf{j} + 0.866\mathbf{k}$$

so that

$$\alpha_1 = \cos^{-1}(0.354) = 69.3°$$
$$\beta_1 = \cos^{-1}(-0.354) = 111°$$
$$\gamma_1 = \cos^{-1}(0.866) = 30.0°$$

These results are shown in Fig. 2–31*c*.

NOTE: Using this same method, show that $\mathbf{F}_2$ in Fig. 2–31*a* can be written in Cartesian vector form as

$$\mathbf{F}_2 = \{106\mathbf{i} + 184\mathbf{j} - 212\mathbf{k}\} \text{ N} \qquad \textit{Ans.}$$

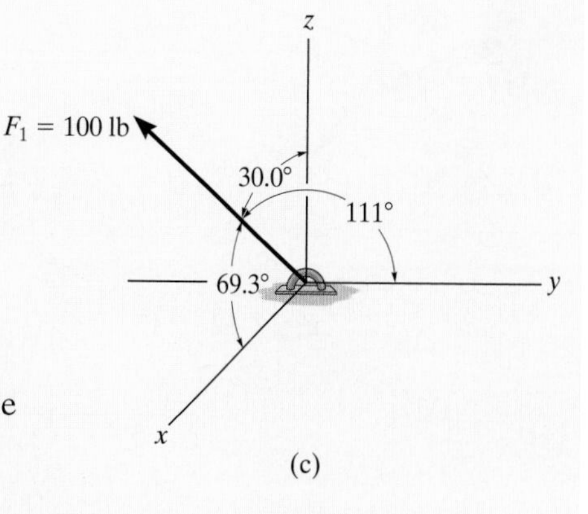

(c)

Fig. 2–31

EXAMPLE 2.11

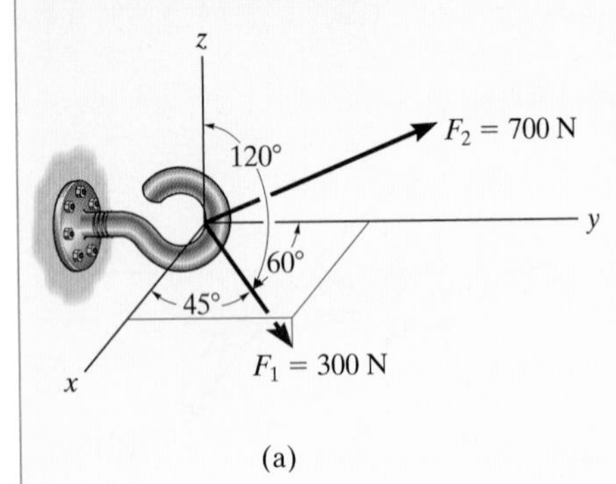

(a)

Two forces act on the hook shown in Fig. 2–32*a*. Specify the coordinate direction angles of $\mathbf{F}_2$ so that the resultant force $\mathbf{F}_R$ acts along the positive *y* axis and has a magnitude of 800 N.

SOLUTION

To solve this problem, the resultant force $\mathbf{F}_R$ and its two components, $\mathbf{F}_1$ and $\mathbf{F}_2$, will each be expressed in Cartesian vector form. Then, as shown in Fig. 2–32*b*, it is necessary that $\mathbf{F}_R = \mathbf{F}_1 + \mathbf{F}_2$.

Applying Eq. 2–11,

$$\mathbf{F}_1 = F_1 \cos \alpha_1 \mathbf{i} + F_1 \cos \beta_1 \mathbf{j} + F_1 \cos \gamma_1 \mathbf{k}$$
$$= 300 \cos 45° \text{ N}\mathbf{i} + 300 \cos 60° \text{ N}\mathbf{j} + 300 \cos 120° \text{ N}\mathbf{k}$$
$$= \{212.1\mathbf{i} + 150\mathbf{j} - 150\mathbf{k}\} \text{ N}$$
$$\mathbf{F}_2 = F_{2x}\mathbf{i} + F_{2y}\mathbf{j} + F_{2z}\mathbf{k}$$

Since the resultant force $\mathbf{F}_R$ has a magnitude of 800 N and acts in the $+\mathbf{j}$ direction.

$$\mathbf{F}_R = (800 \text{ N})(+\mathbf{j}) = \{800\mathbf{j}\} \text{ N}$$

We require

$$\mathbf{F}_R = \mathbf{F}_1 + \mathbf{F}_2$$
$$800\mathbf{j} = 212.1\mathbf{i} + 150\mathbf{j} - 150\mathbf{k} + F_{2x}\mathbf{i} + F_{2y}\mathbf{j} + F_{2z}\mathbf{k}$$
$$800\mathbf{j} = (212.1 + F_{2x})\mathbf{i} + (150 + F_{2y})\mathbf{j} + (-150 + F_{2z})\mathbf{k}$$

To satisfy this equation, the corresponding **i**, **j**, **k** components on the left and right sides must be equal. This is equivalent to stating that the *x, y, z* components of $\mathbf{F}_R$ must be equal to the corresponding *x, y, z* components of $(\mathbf{F}_1 + \mathbf{F}_2)$. Hence,

$$0 = 212.1 + F_{2x} \qquad F_{2x} = -212.1 \text{ N}$$
$$800 = 150 + F_{2y} \qquad F_{2y} = 650 \text{ N}$$
$$0 = -150 + F_{2z} \qquad F_{2z} = 150 \text{ N}$$

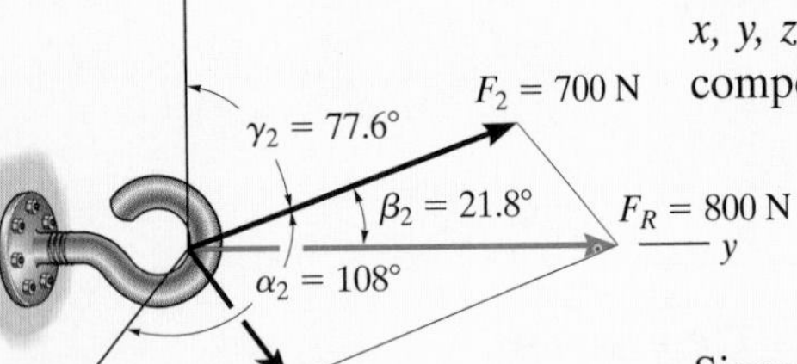

(b)

Fig. 2–32

Since the magnitudes of $\mathbf{F}_2$ and its components are known, we can use Eq. 2–11 to determine α_2, β_2, γ_2.

$$-212.1 = 700 \cos \alpha_2; \qquad \alpha_2 = \cos^{-1}\left(\frac{-212.1}{700}\right) = 108° \qquad \textit{Ans.}$$

$$650 = 700 \cos \beta_2; \qquad \beta_2 = \cos^{-1}\left(\frac{650}{700}\right) = 21.8° \qquad \textit{Ans.}$$

$$150 = 700 \cos \gamma_2; \qquad \gamma_2 = \cos^{-1}\left(\frac{150}{700}\right) = 77.6° \qquad \textit{Ans.}$$

These results are shown in Fig. 2–32*b*.

PROBLEMS

2–57. Determine the magnitude and coordinate direction angles of $\mathbf{F}_1 = \{60\mathbf{i} - 50\mathbf{j} + 40\mathbf{k}\}$ N and $\mathbf{F}_2 = \{-40\mathbf{i} - 85\mathbf{j} + 30\mathbf{k}\}$ N. Sketch each force on an x, y, z reference.

2–58. Express each force in Cartesian vector form.

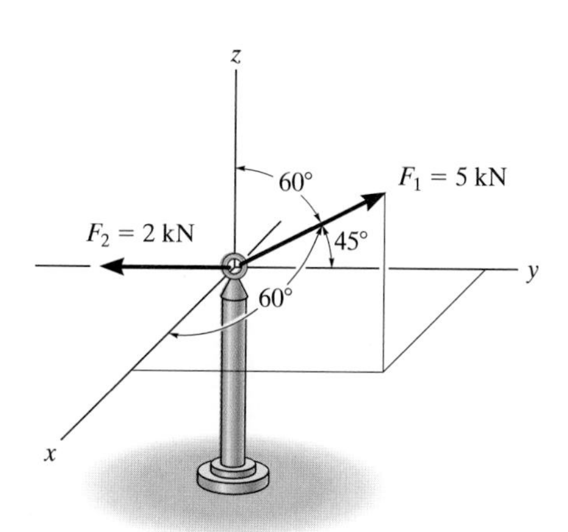

Prob. 2–58

2–59. Determine the magnitude and coordinate direction angles of the force **F** acting on the stake.

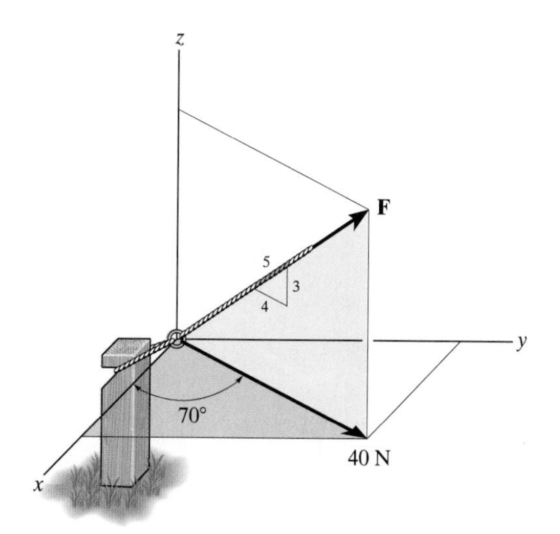

Prob. 2–59

***2–60.** Express each force in Cartesian vector form.

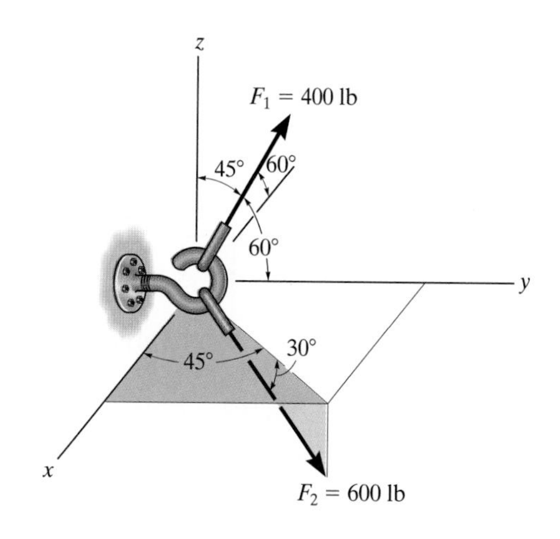

Prob. 2–60

2–61. The stock mounted on the lathe is subjected to a force of 60 N. Determine the coordinate direction angle β and express the force as a Cartesian vector.

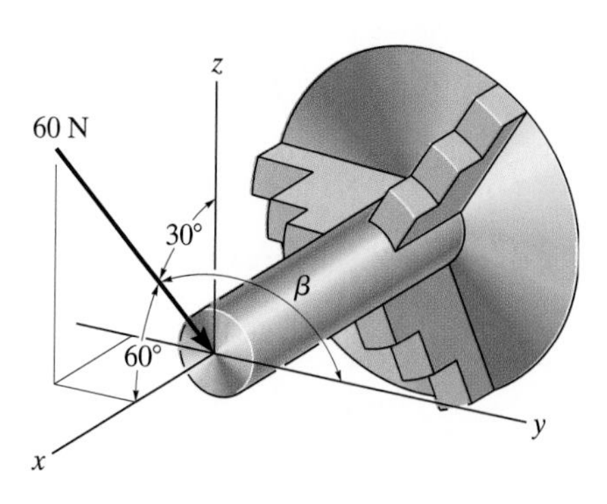

Prob. 2–61

2–62. Determine the magnitude and coordinate direction angles of the resultant force and sketch this vector on the coordinate system.

2–63. Specify the coordinate direction angles of $\mathbf{F}_1$ and $\mathbf{F}_2$ and express each force as a Cartesian vector.

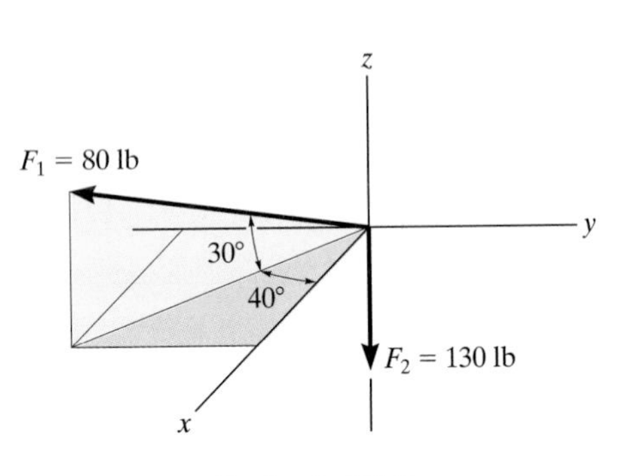

Probs. 2–62/63

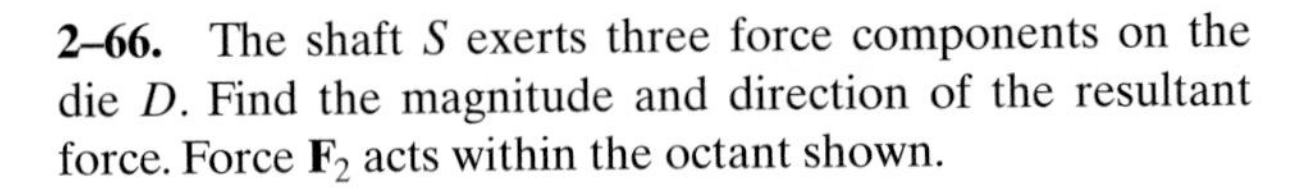

2–66. The shaft S exerts three force components on the die D. Find the magnitude and direction of the resultant force. Force $\mathbf{F}_2$ acts within the octant shown.

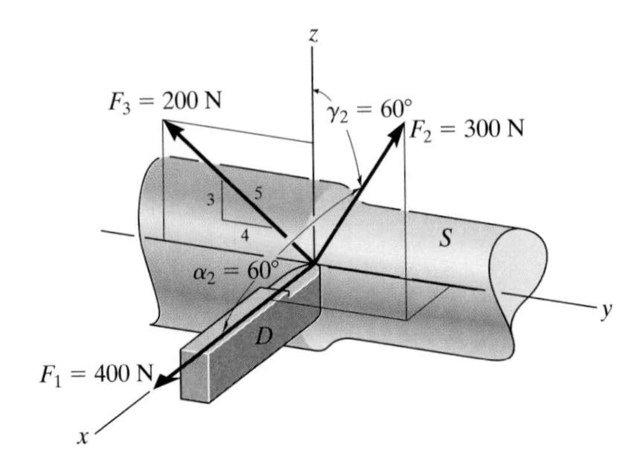

Prob. 2–66

***2–64.** The mast is subjected to the three forces shown. Determine the coordinate direction angles $\alpha_1, \beta_1, \gamma_1$ of $\mathbf{F}_1$ so that the resultant force acting on the mast is $\mathbf{F}_R = \{350\mathbf{i}\}$ N.

2–65. The mast is subjected to the three forces shown. Determine the coordinate direction angles $\alpha_1, \beta_1, \gamma_1$ of $\mathbf{F}_1$ so that the resultant force acting on the mast is zero.

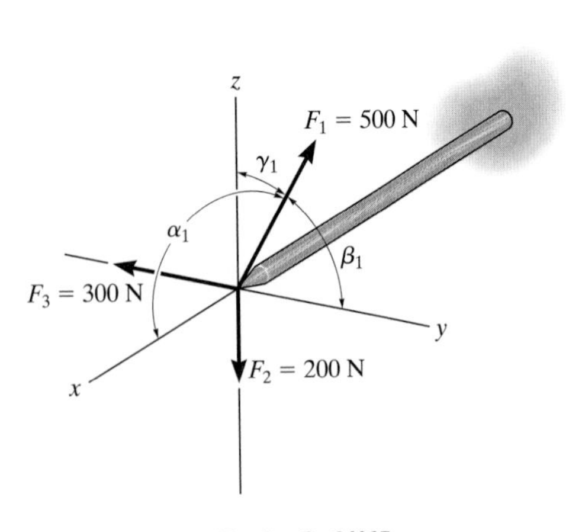

Probs. 2–64/65

2–67. The beam is subjected to the two forces shown. Express each force in Cartesian vector form and determine the magnitude and coordinate direction angles of the resultant force.

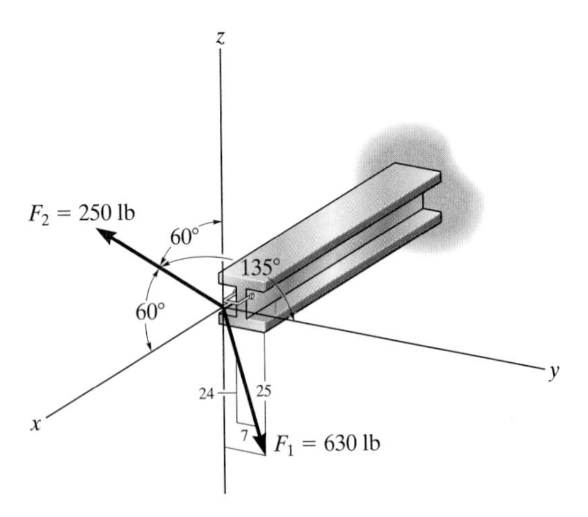

Prob. 2–67

***2–68.** Determine the magnitude and coordinate direction angles of the resultant force and sketch this vector on the coordinate system.

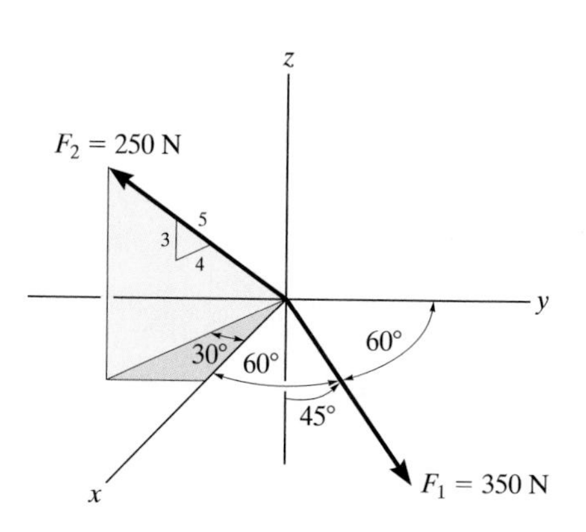

Prob. 2–68

2–69. Determine the magnitude and coordinate direction angles of $\mathbf{F}_3$ so that the resultant of the three forces acts along the positive y axis and has a magnitude of 600 lb.

2–70. Determine the magnitude and coordinate direction angles of $\mathbf{F}_3$ so that the resultant of the three forces is zero.

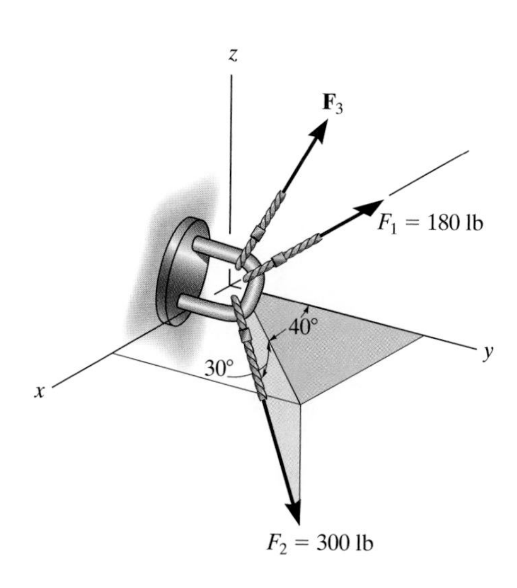

Probs. 2–69/70

2–71. Specify the magnitude F_3 and directions α_3, β_3, and γ_3 of $\mathbf{F}_3$ so that the force of the three forces is $\mathbf{F}_R = \{9\mathbf{j}\}$ kN.

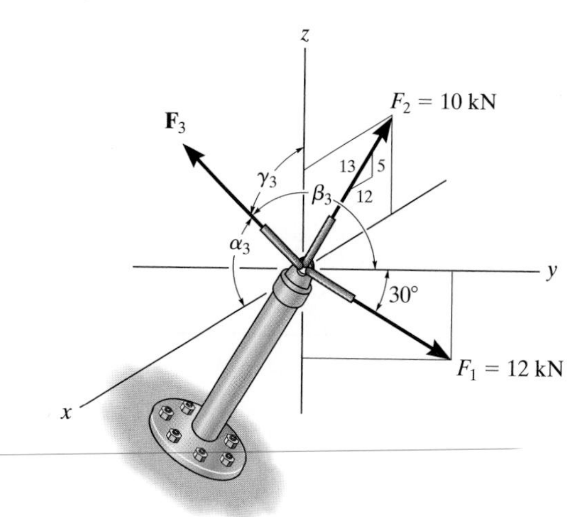

Prob. 2–71

***2–72.** The pole is subjected to the force $\mathbf{F}$, which has components acting along the x, y, z axes as shown. If the magnitude of $\mathbf{F}$ is 3 kN, and $\beta = 30°$ and $\gamma = 75°$, determine the magnitudes of its three components.

2–73. The pole is subjected to the force $\mathbf{F}$ which has components $F_x = 1.5$ kN and $F_z = 1.25$ kN. If $\beta = 75°$, determine the magnitudes of $\mathbf{F}$ and $\mathbf{F}_y$.

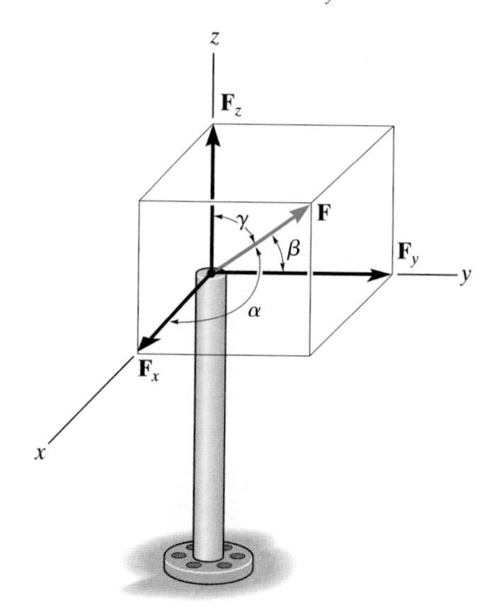

Probs. 2–72/73

2–74. The eye bolt is subjected to the cable force **F** which has a component along the x axis of $F_x = 60$ N, a component along the z axis of $F_z = -80$ N, and a coordinate direction angle of $\beta = 80°$. Determine the magnitude of **F**.

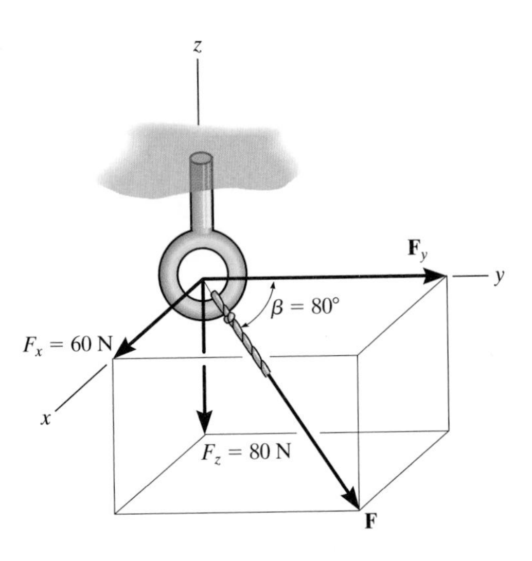

Prob. 2–74

2–75. Three forces act on the hook. If the resultant force $\mathbf{F}_R$ has a magnitude and direction as shown, determine the magnitude and the coordinate direction angles of force $\mathbf{F}_3$.

***2–76.** Determine the coordinate direction angles of $\mathbf{F}_1$ and $\mathbf{F}_R$.

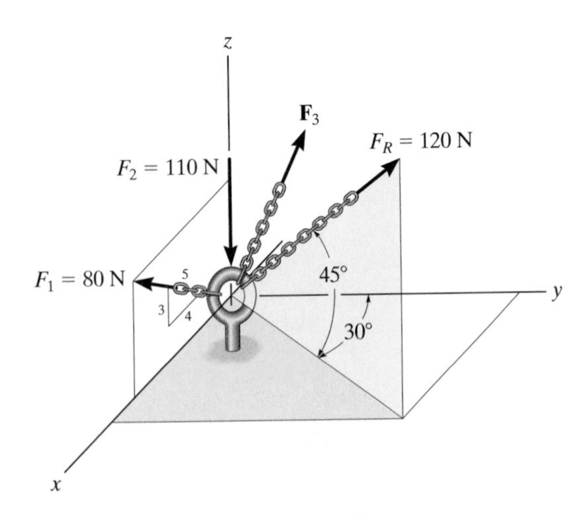

Probs. 2–75/76

2–77. The bolt is subjected to the force **F**, which has components acting along the x, y, z axes as shown. If the magnitude of **F** is 80 N, and $\alpha = 60°$ and $\gamma = 45°$, determine the magnitudes of its components.

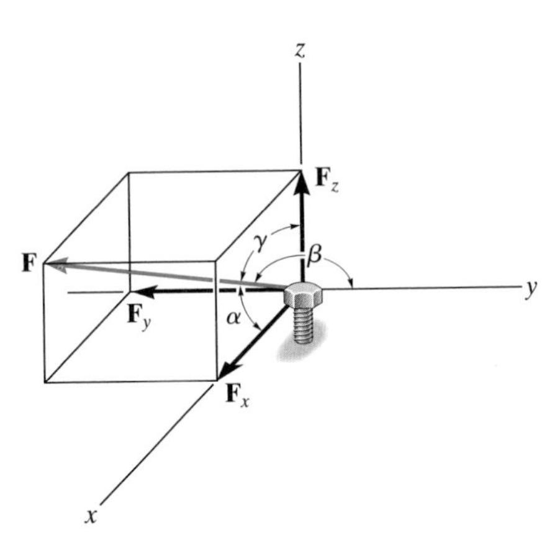

Prob. 2–77

2–78. Two forces $\mathbf{F}_1$ and $\mathbf{F}_2$ act on the bolt. If the resultant force $\mathbf{F}_R$ has a magnitude of 50 lb and coordinate direction angles $\alpha = 110°$ and $\beta = 80°$, as shown, determine the magnitude of $\mathbf{F}_2$ and its coordinate direction angles.

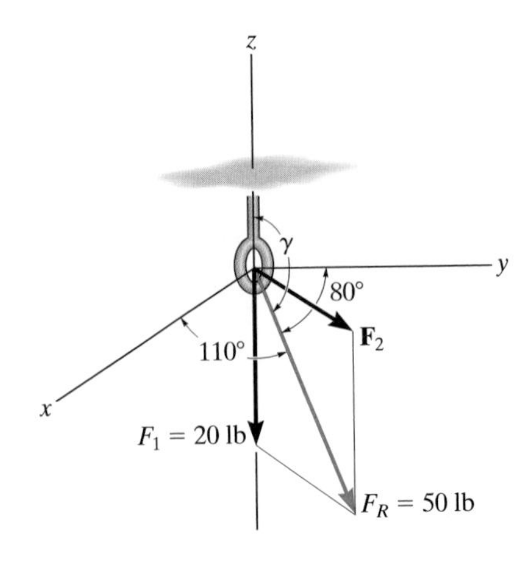

Prob. 2–78

2.7 Position Vectors

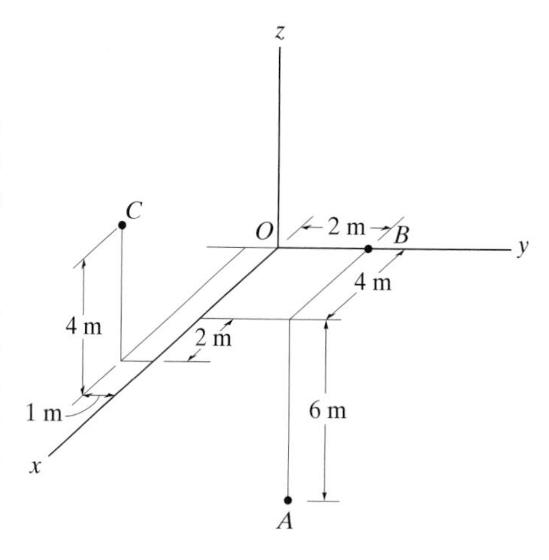

Fig. 2–33

In this section we will introduce the concept of a position vector. It will be shown that this vector is of importance in formulating a Cartesian force vector directed between any two points in space. Later, in Chapter 4, we will use it for finding the moment of a force.

x, y, z Coordinates. Throughout the book we will use a *right-handed* coordinate system to reference the location of points in space. We will also use the convention followed in many technical books, which requires the positive z axis to be directed *upward* (the zenith direction) so that it measures the height of an object or the altitude of a point. The x, y axes then lie in the horizontal plane, Fig. 2–33. Points in space are located relative to the origin of coordinates, O, by successive measurements along the x, y, z axes. For example, in Fig. 2–33 the coordinates of point A are obtained by starting at O and measuring $x_A = +4$ m along the x axis, $y_A = +2$ m along the y axis, and $z_A = -6$ m along the z axis. Thus, $A(4, 2, -6)$. In a similar manner, measurements along the x, y, z axes from O to B yield the coordinates of B, i.e., $B(0, 2, 0)$. Also notice that $C(6, -1, 4)$.

Position Vector. The *position vector* $\mathbf{r}$ is defined as a fixed vector which locates a point in space relative to another point. For example, if $\mathbf{r}$ extends from the origin of coordinates, O, to point $P(x, y, z)$, Fig. 2–34*a*, then $\mathbf{r}$ can be expressed in Cartesian vector form as

$$\mathbf{r} = x\mathbf{i} + y\mathbf{j} + z\mathbf{k}$$

Note how the head-to-tail vector addition of the three components yields vector $\mathbf{r}$, Fig. 2–34*b*. Starting at the origin O, one travels x in the $+\mathbf{i}$ direction, then y in the $+\mathbf{j}$ direction, and finally z in the $+\mathbf{k}$ direction to arrive at point $P(x, y, z)$.

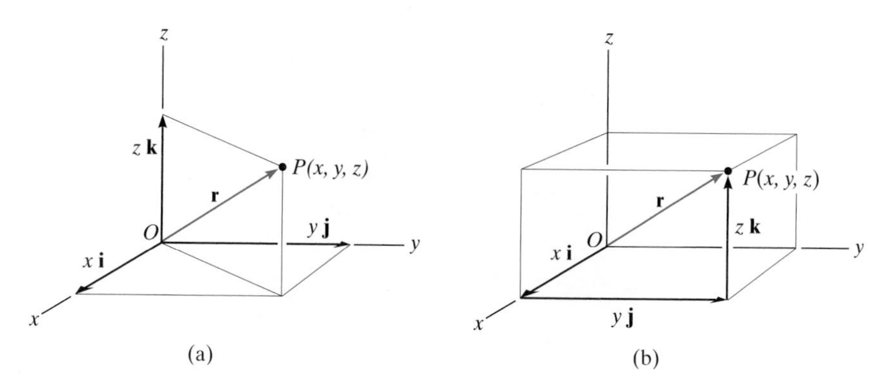

Fig. 2–34

In the more general case, the position vector may be directed from point A to point B in space, Fig. 2–35*a*. This vector is also designated by the symbol $\mathbf{r}$. However, as a matter of convention, we will *sometimes* refer to this vector with *two subscripts* to indicate from and to the point where it is directed. Thus, $\mathbf{r}$ can also be designated as $\mathbf{r}_{AB}$. Also, note that $\mathbf{r}_A$ and $\mathbf{r}_B$ in Fig. 2–35*a* are referenced with only one subscript since they extend from the origin of coordinates.

From Fig. 2–35*a*, by the head-to-tail vector addition, we require

$$\mathbf{r}_A + \mathbf{r} = \mathbf{r}_B$$

Solving for $\mathbf{r}$ and expressing $\mathbf{r}_A$ and $\mathbf{r}_B$ in Cartesian vector form yields

$$\mathbf{r} = \mathbf{r}_B - \mathbf{r}_A = (x_B\mathbf{i} + y_B\mathbf{j} + z_B\mathbf{k}) - (x_A\mathbf{i} + y_A\mathbf{j} + z_A\mathbf{k})$$

or

$$\mathbf{r} = (x_B - x_A)\mathbf{i} + (y_B - y_A)\mathbf{j} + (z_B - z_A)\mathbf{k} \qquad (2\text{–}13)$$

Thus, the $\mathbf{i}$, $\mathbf{j}$, $\mathbf{k}$ *components of the position vector* $\mathbf{r}$ *may be formed by taking the coordinates of the tail of the vector,* $A(x_A, y_A, z_A)$, *and subtracting them from the corresponding coordinates of the head,* $B(x_B, y_B, z_B)$. Again note how the head-to-tail addition of these three components yields $\mathbf{r}$, i.e., going from A to B, Fig. 2–35*b*, one first travels $(x_B - x_A)$ in the $+\mathbf{i}$ direction, then $(y_B - y_A)$ in the $+\mathbf{j}$ direction, and finally $(z_B - z_A)$ in the $+\mathbf{k}$ direction.

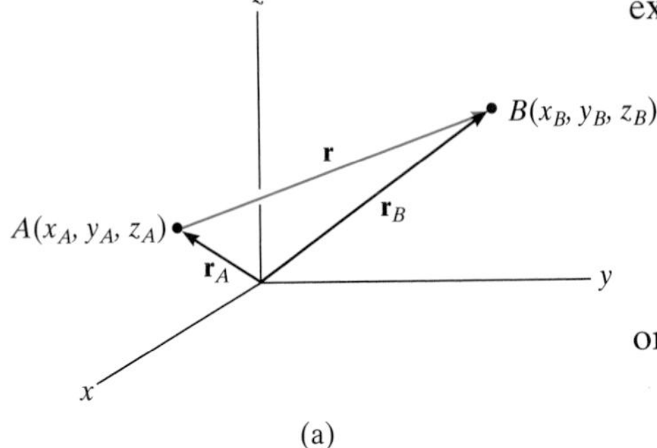

(a)

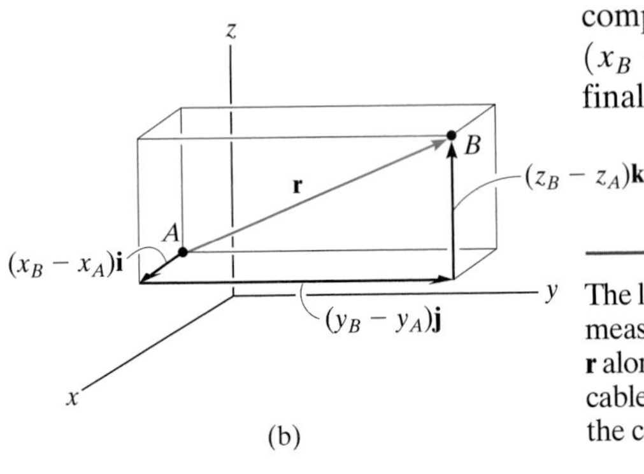

(b)

Fig. 2–35

The length and direction of cable AB used to support the stack can be determined by measuring the coordinates of points A and B using the x, y, z axes. The position vector $\mathbf{r}$ along the cable can then be established. The magnitude r represents the length of the cable, and the direction of the cable is defined by α, β, γ, which are determined from the components of the unit vector found from the position vector, $\mathbf{u} = \mathbf{r}/r$.

EXAMPLE 2.12

An elastic rubber band is attached to points A and B as shown in Fig. 2–36*a*. Determine its length and its direction measured from A toward B.

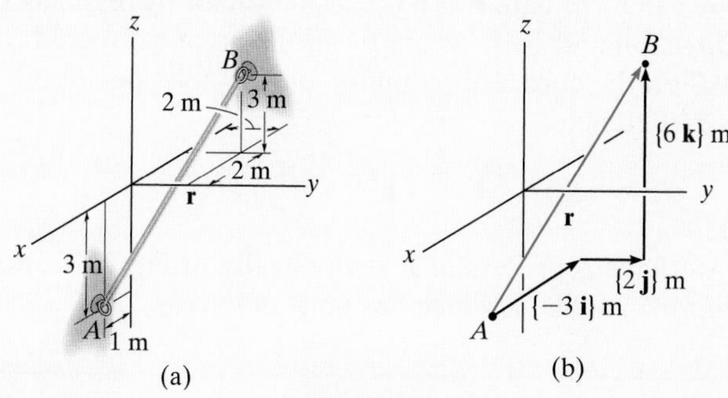

Fig. 2–36

SOLUTION

We first establish a position vector from A to B, Fig. 2–36*b*. In accordance with Eq. 2–13, the coordinates of the tail $A(1\text{ m}, 0, -3\text{ m})$ are subtracted from the coordinates of the head $B(-2\text{ m}, 2\text{ m}, 3\text{ m})$, which yields

$$\mathbf{r} = [-2\text{ m} - 1\text{ m}]\mathbf{i} + [2\text{ m} - 0]\mathbf{j} + [3\text{ m} - (-3\text{ m})]\mathbf{k}$$
$$= \{-3\mathbf{i} + 2\mathbf{j} + 6\mathbf{k}\}\text{ m}$$

These components of $\mathbf{r}$ can also be determined *directly* by realizing from Fig. 2–36*a* that they represent the direction and distance one must go along each axis in order to move from A to B, i.e., along the x axis $\{-3\mathbf{i}\}$ m, along the y axis $\{2\mathbf{j}\}$ m, and finally along the z axis $\{6\mathbf{k}\}$ m.

The magnitude of $\mathbf{r}$ represents the length of the rubber band.

$$r = \sqrt{(-3)^2 + (2)^2 + (6)^2} = 7\text{ m} \qquad \textit{Ans.}$$

Formulating a unit vector in the direction of $\mathbf{r}$, we have

$$\mathbf{u} = \frac{\mathbf{r}}{r} = \frac{-3}{7}\mathbf{i} + \frac{2}{7}\mathbf{j} + \frac{6}{7}\mathbf{k}$$

The components of this unit vector yield the coordinate direction angles

$$\alpha = \cos^{-1}\left(\frac{-3}{7}\right) = 115^\circ \qquad \textit{Ans.}$$

$$\beta = \cos^{-1}\left(\frac{2}{7}\right) = 73.4^\circ \qquad \textit{Ans.}$$

$$\gamma = \cos^{-1}\left(\frac{6}{7}\right) = 31.0^\circ \qquad \textit{Ans.}$$

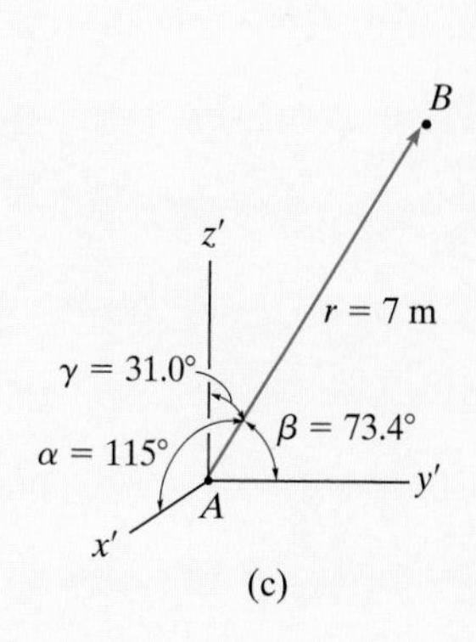

NOTE: These angles are measured from the *positive axes* of a localized coordinate system placed at the tail of $\mathbf{r}$, as shown in Fig. 2–36*c*.

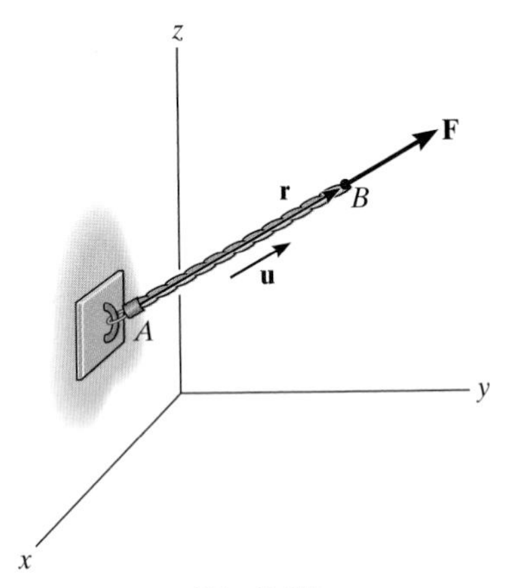

Fig. 2–37

2.8 Force Vector Directed along a Line

Quite often in three-dimensional statics problems, the direction of a force is specified by two points through which its line of action passes. Such a situation is shown in Fig. 2–37, where the force **F** is directed along the cord AB. We can formulate **F** as a Cartesian vector by realizing that it has the *same direction* and *sense* as the position vector **r** directed from point A to point B on the cord. This common direction is specified by the *unit vector* $\mathbf{u} = \mathbf{r}/r$. Hence,

$$\mathbf{F} = F\mathbf{u} = F\left(\frac{\mathbf{r}}{r}\right)$$

Although we have represented **F** symbolically in Fig. 2–37, note that it has *units of force*, unlike **r**, which has units of length.

The force F acting along the chain can be represented as a Cartesian vector by first establishing x, y, z axes and forming a position vector **r** along the length of the chain, then finding the corresponding unit vector $\mathbf{u} = \mathbf{r}/r$ that defines the direction of both the chain and the force. Finally, the magnitude of the force is combined with its direction, $\mathbf{F} = F\mathbf{u}$.

Important Points

- A position vector locates one point in space relative to another point.
- The easiest way to formulate the components of a position vector is to determine the distance and direction that must be traveled along the x, y, z directions—going from the tail to the head of the vector.
- A force **F** acting in the direction of a position vector **r** can be represented in Cartesian form if the unit vector **u** of the position vector is determined and this is multiplied by the magnitude of the force, i.e., $\mathbf{F} = F\mathbf{u} = F(\mathbf{r}/r)$.

EXAMPLE 2.13

The man shown in Fig. 2–38*a* pulls on the cord with a force of 70 lb. Represent this force, acting on the support A, as a Cartesian vector and determine its direction.

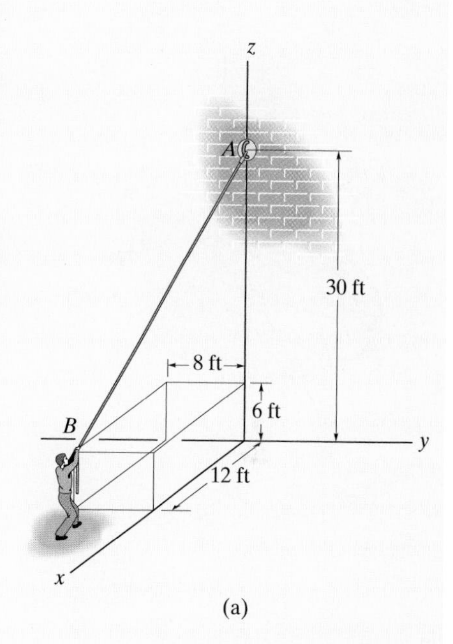

(a)

SOLUTION

Force **F** is shown in Fig. 2–38*b*. The *direction* of this vector, **u**, is determined from the position vector **r**, which extends from A to B, Fig. 2–38*b*. The coordinates of the end points of the cord are $A(0, 0, 30\text{ ft})$ and $B(12\text{ ft}, -8\text{ ft}, 6\text{ ft})$. Forming the position vector by subtracting the corresponding x, y, and z coordinates of A from those of B, we have

$$\mathbf{r} = (12\text{ ft} - 0)\mathbf{i} + (-8\text{ ft} - 0)\mathbf{j} + (6\text{ ft} - 30\text{ ft})\mathbf{k}$$
$$= \{12\mathbf{i} - 8\mathbf{j} - 24\mathbf{k}\}\text{ ft}$$

This result can also be determined *directly* by noting in Fig. 2–38*a*, that one must go from A $\{-24\mathbf{k}\}$ ft, then $\{-8\mathbf{j}\}$ ft, and finally $\{12\mathbf{i}\}$ ft to get to B.

The magnitude of **r**, which represents the *length* of cord AB, is

$$r = \sqrt{(12\text{ ft})^2 + (-8\text{ ft})^2 + (-24\text{ ft})^2} = 28\text{ ft}$$

Forming the unit vector that defines the direction and sense of both **r** and **F** yields

$$\mathbf{u} = \frac{\mathbf{r}}{r} = \frac{12}{28}\mathbf{i} - \frac{8}{28}\mathbf{j} - \frac{24}{28}\mathbf{k}$$

Since **F** has a *magnitude* of 70 lb and a *direction* specified by **u**, then

$$\mathbf{F} = F\mathbf{u} = 70\text{ lb}\left(\frac{12}{28}\mathbf{i} - \frac{8}{28}\mathbf{j} - \frac{24}{28}\mathbf{k}\right)$$
$$= \{30\mathbf{i} - 20\mathbf{j} - 60\mathbf{k}\}\text{lb} \qquad \textit{Ans.}$$

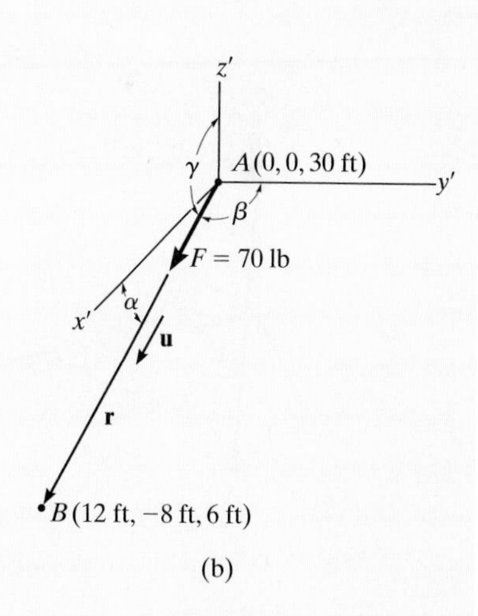

(b)

Fig. 2–38

The coordinate direction angles are measured between **r** (or **F**) and the *positive axes* of a localized coordinate system with origin placed at A, Fig. 2–38*b*. From the components of the unit vector:

$$\alpha = \cos^{-1}\left(\frac{12}{28}\right) = 64.6^\circ \qquad \textit{Ans.}$$

$$\beta = \cos^{-1}\left(\frac{-8}{28}\right) = 107^\circ \qquad \textit{Ans.}$$

$$\gamma = \cos^{-1}\left(\frac{-24}{28}\right) = 149^\circ \qquad \textit{Ans.}$$

NOTE: These results make sense when compared with the angles shown in Fig. 2–38*b*.

EXAMPLE 2.14

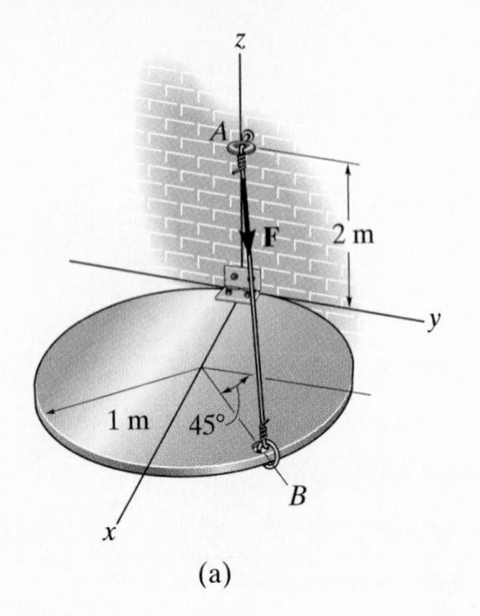

(a)

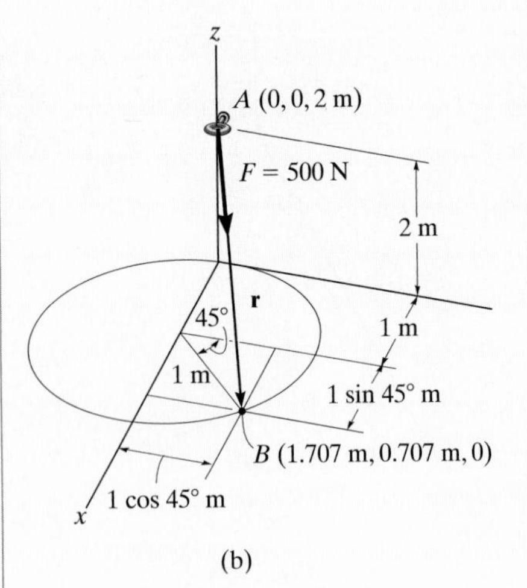

(b)

Fig. 2–39

The circular plate in Fig. 2–39*a* is partially supported by the cable *AB*. If the force of the cable on the hook at *A* is $F = 500$ N, express **F** as a Cartesian vector.

SOLUTION

As shown in Fig. 2–39*b*, **F** has the same direction and sense as the position vector **r**, which extends from *A* to *B*. The coordinates of the end points of the cable are $A(0, 0, 2\text{ m})$ and $B(1.707\text{ m}, 0.707\text{ m}, 0)$, as indicated in the figure. Thus,

$$\mathbf{r} = (1.707\text{ m} - 0)\mathbf{i} + (0.707\text{ m} - 0)\mathbf{j} + (0 - 2\text{ m})\mathbf{k}$$

$$= \{1.707\mathbf{i} + 0.707\mathbf{j} - 2\mathbf{k}\}\text{ m}$$

Note how one can calculate these components *directly* by going from *A*, $\{-2\mathbf{k}\}$ m along the z axis, then $\{1.707\mathbf{i}\}$ m along the x axis, and finally $\{0.707\mathbf{j}\}$ m along the y axis to get to *B*.

The magnitude of **r** is

$$r = \sqrt{(1.707)^2 + (0.707)^2 + (-2)^2} = 2.723\text{ m}$$

Thus,

$$\mathbf{u} = \frac{\mathbf{r}}{r} = \frac{1.707}{2.723}\mathbf{i} + \frac{0.707}{2.723}\mathbf{j} - \frac{2}{2.723}\mathbf{k}$$

$$= 0.6269\mathbf{i} + 0.2597\mathbf{j} - 0.7345\mathbf{k}$$

Since $F = 500$ N and **F** has the direction **u**, we have

$$\mathbf{F} = F\mathbf{u} = 500\text{ N}(0.6269\mathbf{i} + 0.2597\mathbf{j} - 0.7345\mathbf{k})$$

$$= \{313\mathbf{i} + 130\mathbf{j} - 367\mathbf{k}\}\text{ N} \qquad \textit{Ans.}$$

Using these components, notice that indeed the magnitude of **F** is 500 N; i.e.,

$$F = \sqrt{(313)^2 + (130)^2 + (-367)^2} = 500\text{ N}$$

NOTE: From the unit vector, show that the coordinate direction angle $\gamma = 137°$, and indicate this angle on the figure.

EXAMPLE 2.15

The roof is supported by cables as shown in the photo. If the cables exert forces $F_{AB} = 100$ N and $F_{AC} = 120$ N on the wall hook at A as shown in Fig. 2–40a, determine the magnitude of the resultant force acting at A.

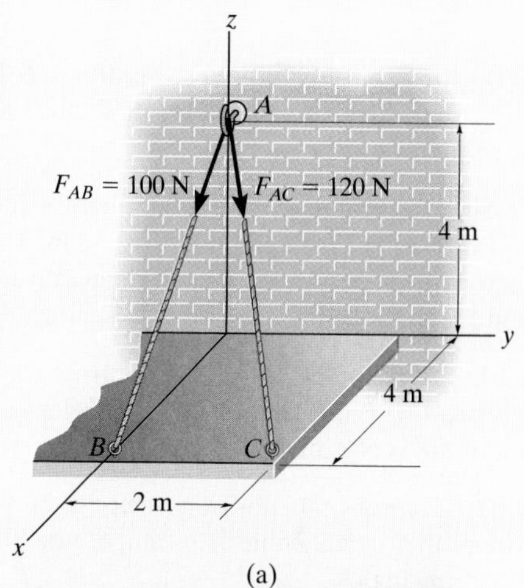

(a)

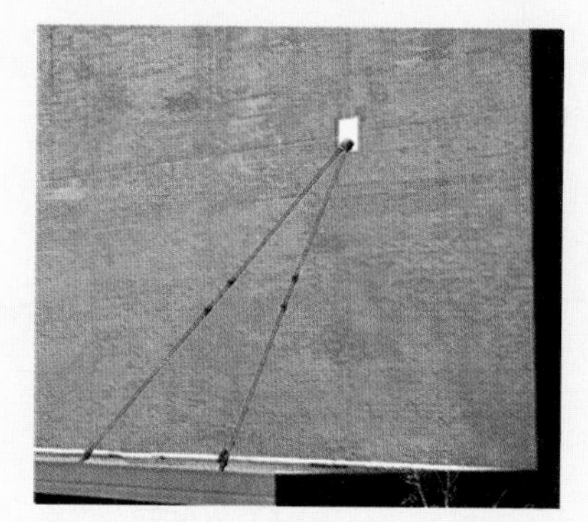

SOLUTION

The resultant force $\mathbf{F}_R$ is shown graphically in Fig. 2–40b. We can express this force as a Cartesian vector by first formulating $\mathbf{F}_{AB}$ and $\mathbf{F}_{AC}$ as Cartesian vectors and then adding their components. The directions of $\mathbf{F}_{AB}$ and $\mathbf{F}_{AC}$ are specified by forming unit vectors $\mathbf{u}_{AB}$ and $\mathbf{u}_{AC}$ along the cables. These unit vectors are obtained from the associated position vectors $\mathbf{r}_{AB}$ and $\mathbf{r}_{AC}$. With reference to Fig. 2–40b, for $\mathbf{F}_{AB}$ we have

$$\mathbf{r}_{AB} = (4\text{ m} - 0)\mathbf{i} + (0 - 0)\mathbf{j} + (0 - 4\text{ m})\mathbf{k}$$
$$= \{4\mathbf{i} - 4\mathbf{k}\}\text{ m}$$
$$r_{AB} = \sqrt{(4)^2 + (-4)^2} = 5.66\text{ m}$$
$$\mathbf{F}_{AB} = 100\text{ N}\left(\frac{\mathbf{r}_{AB}}{r_{AB}}\right) = 100\text{ N}\left(\frac{4}{5.66}\mathbf{i} - \frac{4}{5.66}\mathbf{k}\right)$$
$$\mathbf{F}_{AB} = \{70.7\mathbf{i} - 70.7\mathbf{k}\}\text{ N}$$

For $\mathbf{F}_{AC}$ we have

$$\mathbf{r}_{AC} = (4\text{ m} - 0)\mathbf{i} + (2\text{ m} - 0)\mathbf{j} + (0 - 4\text{ m})\mathbf{k}$$
$$= \{4\mathbf{i} + 2\mathbf{j} - 4\mathbf{k}\}\text{ m}$$
$$r_{AC} = \sqrt{(4)^2 + (2)^2 + (-4)^2} = 6\text{ m}$$
$$\mathbf{F}_{AC} = 120\text{ N}\left(\frac{\mathbf{r}_{AC}}{r_{AC}}\right) = 120\text{ N}\left(\frac{4}{6}\mathbf{i} + \frac{2}{6}\mathbf{j} - \frac{4}{6}\mathbf{k}\right)$$
$$= \{80\mathbf{i} + 40\mathbf{j} - 80\mathbf{k}\}\text{ N}$$

The resultant force is therefore

$$\mathbf{F}_R = \mathbf{F}_{AB} + \mathbf{F}_{AC} = \{70.7\mathbf{i} - 70.7\mathbf{k}\}\text{ N} + \{80\mathbf{i} + 40\mathbf{j} - 80\mathbf{k}\}\text{ N}$$
$$= \{150.7\mathbf{i} + 40\mathbf{j} - 150.7\mathbf{k}\}\text{ N}$$

The magnitude of $\mathbf{F}_R$ is thus

$$F_R = \sqrt{(150.7)^2 + (40)^2 + (-150.7)^2}$$
$$= 217\text{ N} \qquad \textit{Ans.}$$

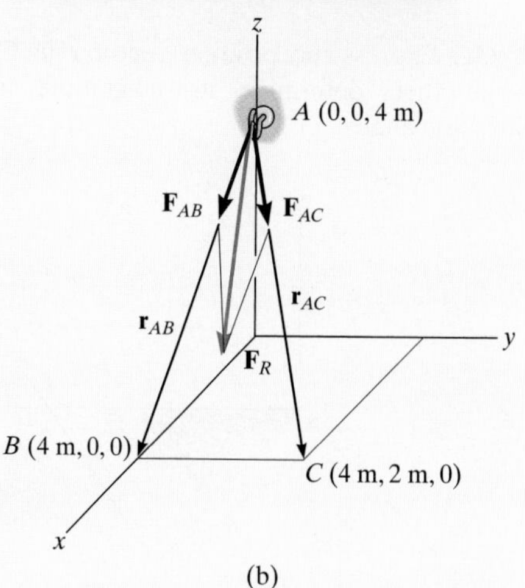

(b)

Fig. 2–40

PROBLEMS

2–79. If $\mathbf{r}_1 = \{3\mathbf{i} - 4\mathbf{j} + 3\mathbf{k}\}$ m, $\mathbf{r}_2 = \{4\mathbf{i} - 5\mathbf{k}\}$ m, $\mathbf{r}_3 = \{3\mathbf{i} - 2\mathbf{j} + 5\mathbf{k}\}$ m, determine the magnitude and direction of $\mathbf{r} = 2\mathbf{r}_1 - \mathbf{r}_2 + 3\mathbf{r}_3$.

***2–80.** Represent the position vector $\mathbf{r}$ acting from point $A(3 \text{ m}, 5 \text{ m}, 6 \text{ m})$ to point $B(5 \text{ m}, -2 \text{ m}, 1 \text{ m})$ in Cartesian vector form. Determine its coordinate direction angles and find the distance between points A and B.

2–81. A position vector extends from the origin to point $A(2 \text{ m}, 3 \text{ m}, 6 \text{ m})$. Determine the angles α, β, γ which the tail of the vector makes with the x, y, z axes, respectively.

2–82. Express the position vector $\mathbf{r}$ in Cartesian vector form; then determine its magnitude and coordinate direction angles.

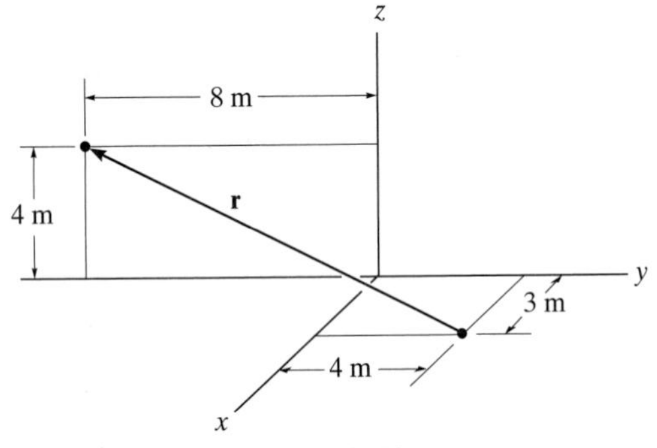

Prob. 2–82

2–83. Express the position vector $\mathbf{r}$ in Cartesian vector form; then determine its magnitude and coordinate direction angles.

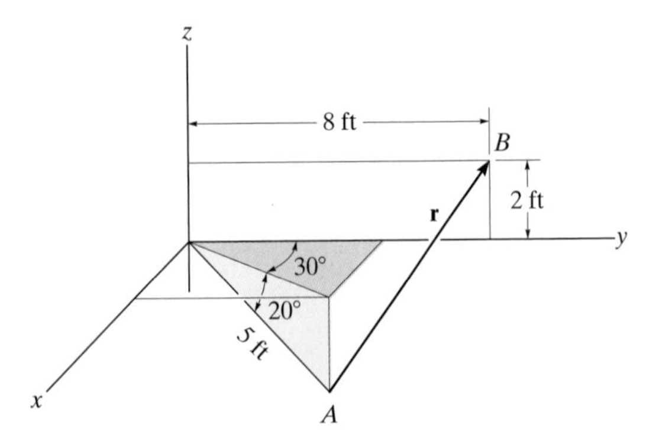

Prob. 2–83

***2–84.** Determine the length of the connecting rod AB by first formulating a Cartesian position vector from A to B and then determining its magnitude.

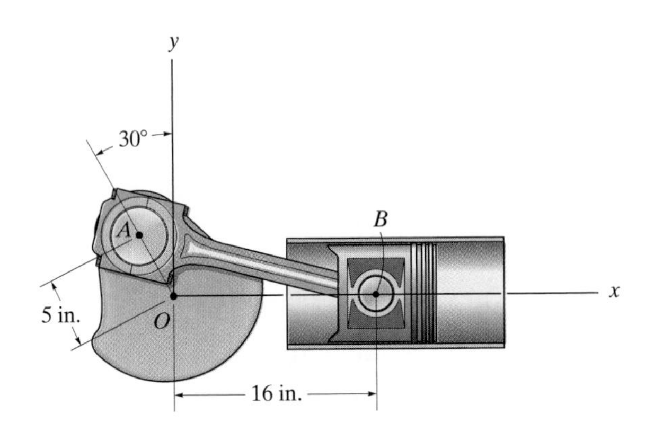

Prob. 2–84

2–85. Determine the length of member AB of the truss by first establishing a Cartesian position vector from A to B and then determining its magnitude.

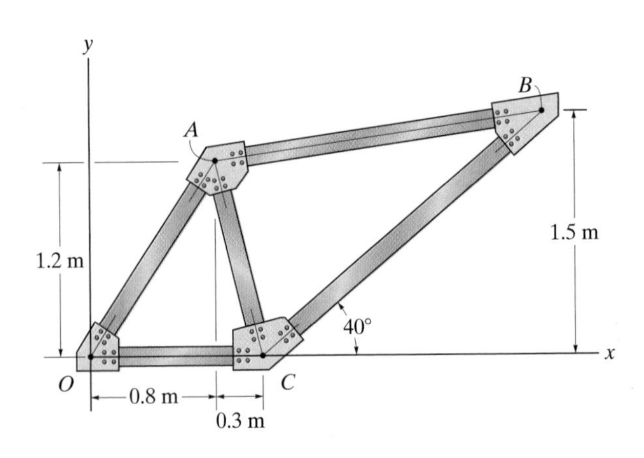

Prob. 2–85

2–86. The positions of point A on the building and point B on the antenna have been measured relative to the electronic distance meter (EDM) at O. Determine the distance between A and B. *Hint:* Formulate a position vector directed from A to B; then determine its magnitude.

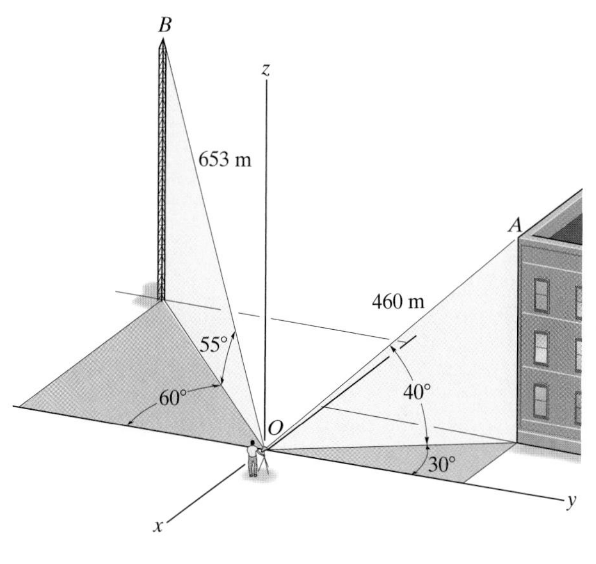

Prob. 2–86

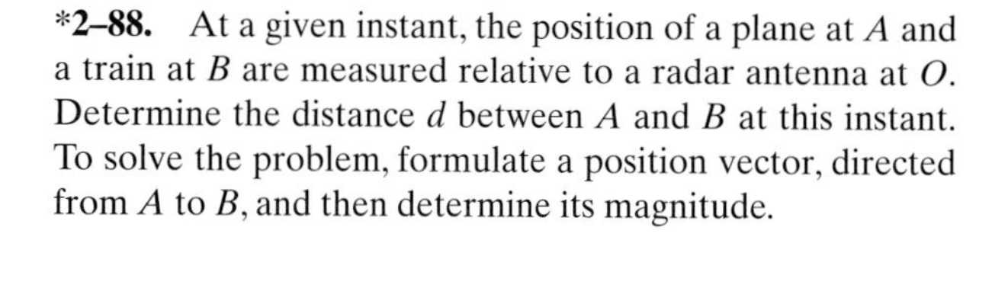

***2–88.** At a given instant, the position of a plane at A and a train at B are measured relative to a radar antenna at O. Determine the distance d between A and B at this instant. To solve the problem, formulate a position vector, directed from A to B, and then determine its magnitude.

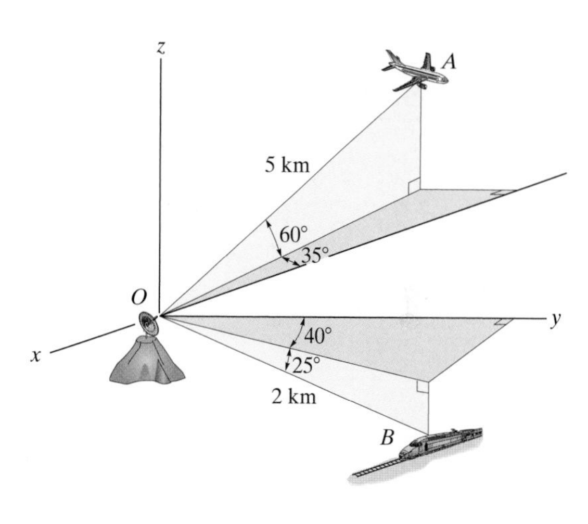

Prob. 2–88

2–87. Determine the lengths of cords ACB and CO. The knot at C is located midway between A and B.

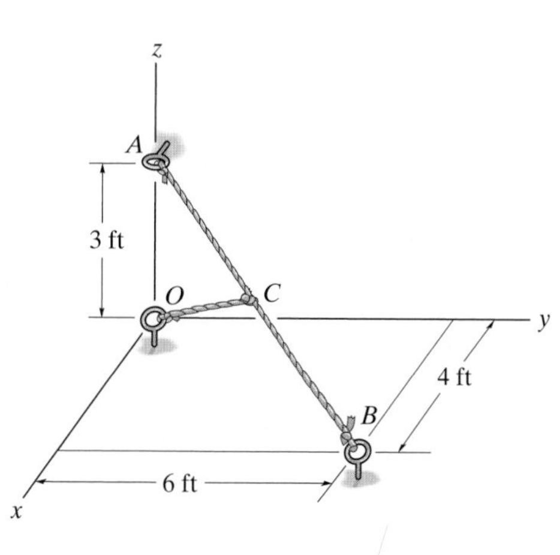

Prob. 2–87

2–89. Determine the lengths of wires AD, BD, and CD. The ring at D is midway between A and B.

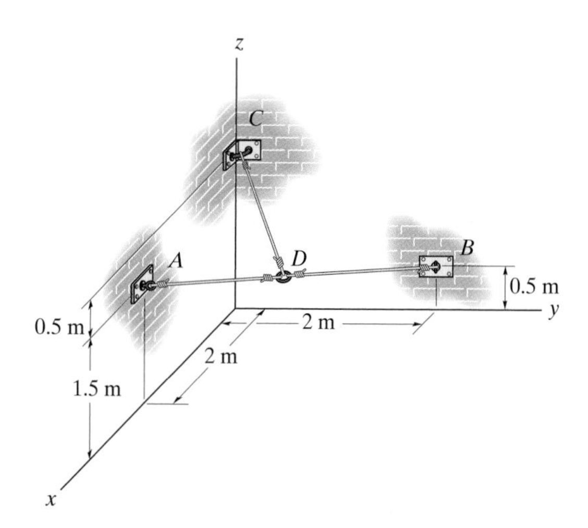

Prob. 2–89

2–90. Determine the magnitude and coordinates on angles of the resultant force.

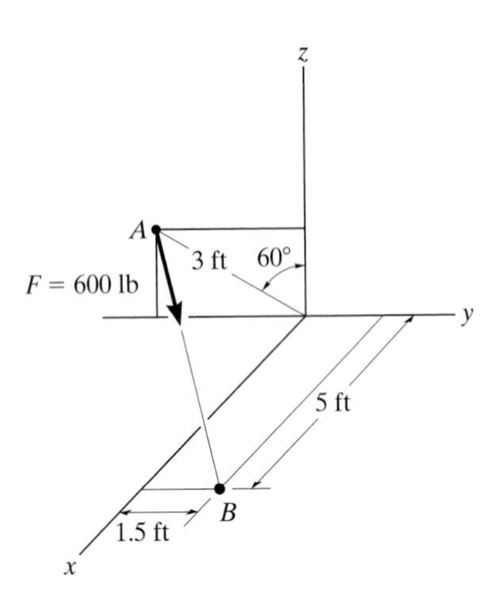

Prob. 2–90

2–91. Express force **F** as a Cartesian vector; then determine its coordinate direction angles.

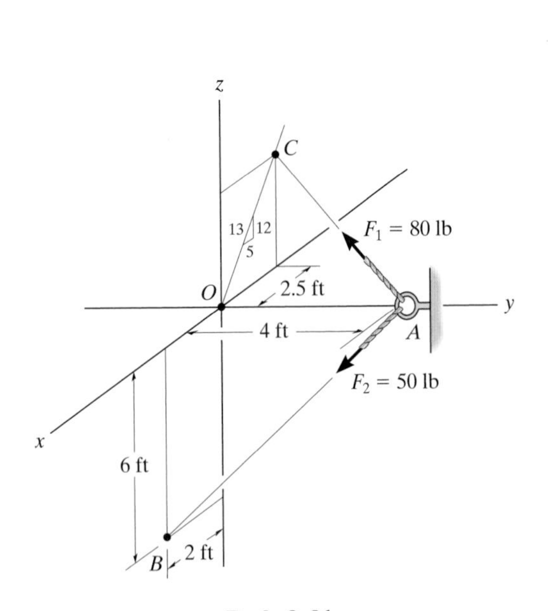

Prob. 2–91

***2–92.** Determine the magnitude and coordinate direction angles of the resultant force acting at point A.

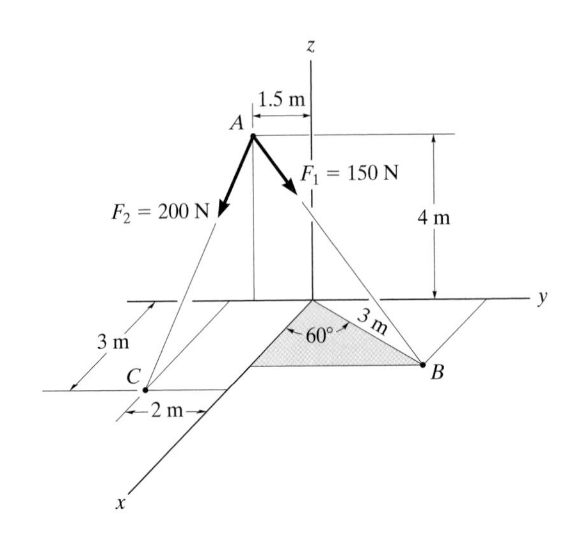

Prob. 2–92

2–93. The plate is suspended using the three cables each exert the forces shown. Express each force as a Cartesian vector.

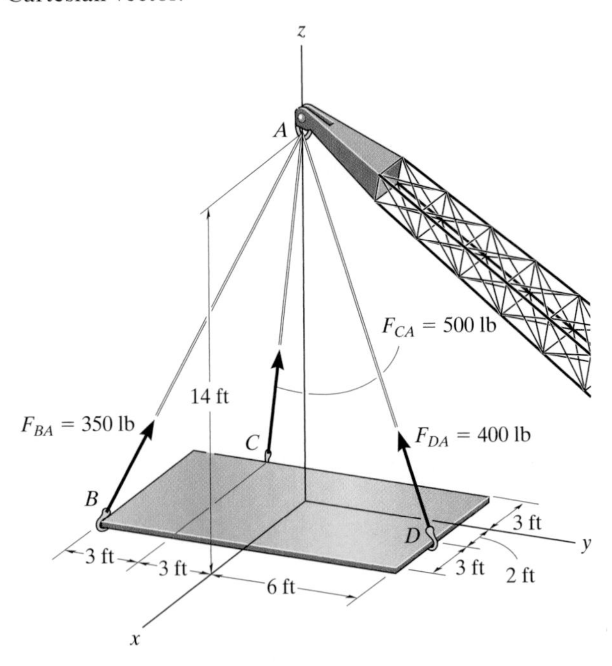

Prob. 2–93

2–94. The engine of the lightweight plane is supported by struts that are connected to the space truss that makes up the structure of the plane. The anticipated loading in two of the struts is shown. Express each of these forces as a Cartesian vector.

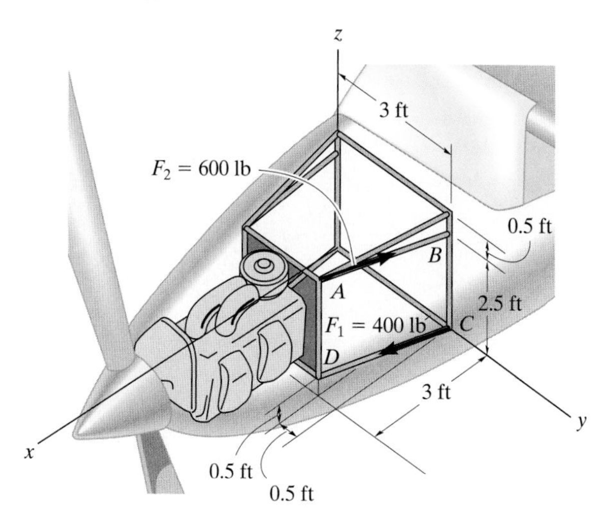

Prob. 2–94

2–95. The window is held open by cable AB. Determine the length of the cable and express the 30-N force acting at A along the cable as a Cartesian vector.

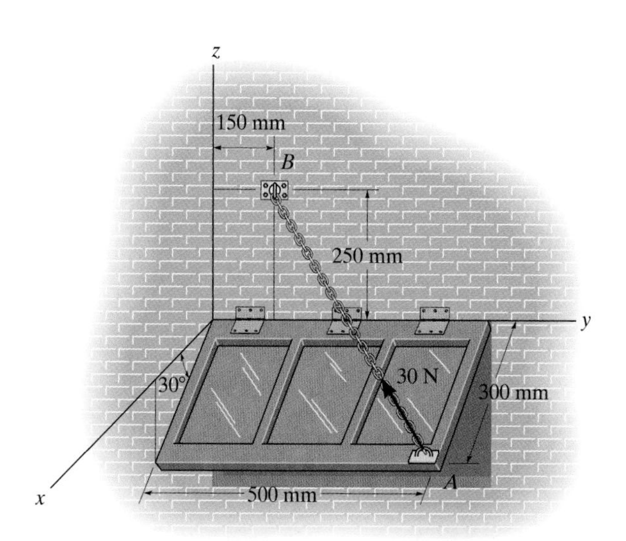

Prob. 2–95

***2–96.** The force acting on the man, caused by his pulling on the anchor cord, is $\mathbf{F} = \{40\mathbf{i} + 20\mathbf{j} - 50\mathbf{k}\}$ N. If the length of the cord is 25 m, determine the coordinates $A(x, y, -z)$ of the anchor.

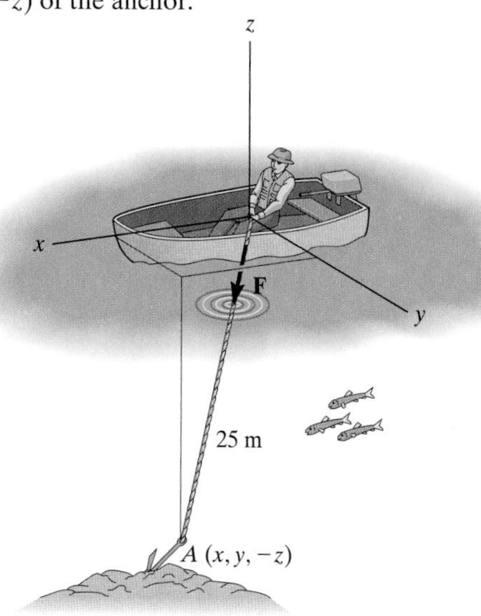

Prob. 2–96

2–97. Express each of the forces in Cartesian vector form and determine the magnitude and coordinate direction angles of the resultant force.

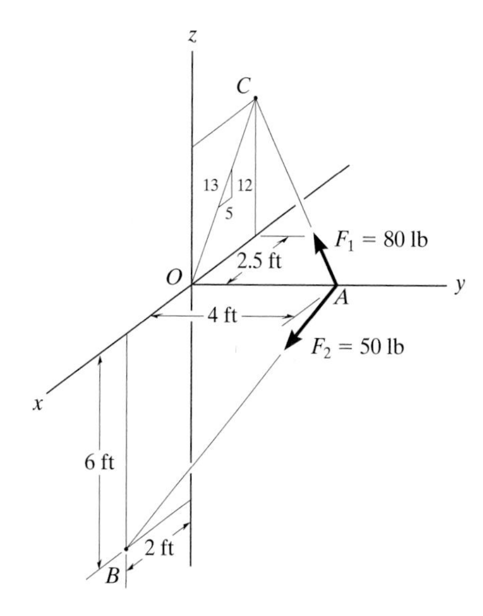

Prob. 2–97

2–98. The cable attached to the tractor at B exerts a force of 350 lb on the framework. Express this force as a Cartesian vector.

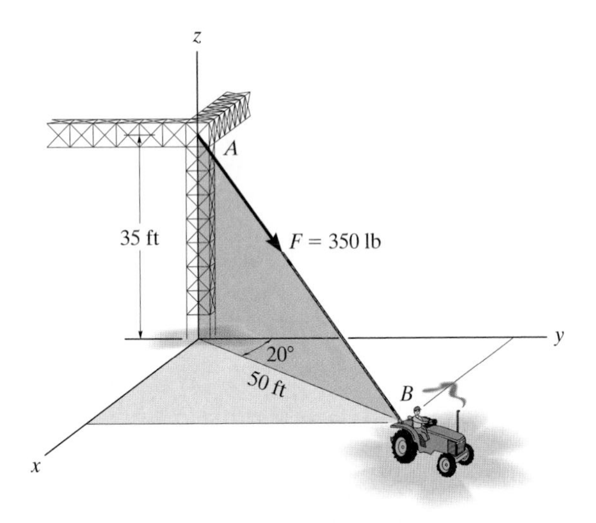

Prob. 2–98

2–99. The cable OA exerts a force on point O of $\mathbf{F} = \{40\mathbf{i} + 60\mathbf{j} + 70\mathbf{k}\}$ N. If the length of the cable is 3 m, what are the coordinates (x, y, z) of point A?

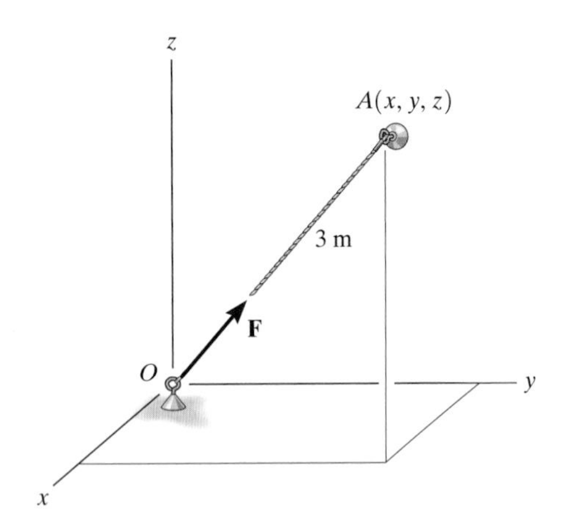

Prob. 2–99

***2–100.** Determine the position $(x, y, 0)$ for fixing cable BA so that the resultant of the forces exerted on the pole is directed along its axis, from B toward O and has a magnitude of 1 kN. Also, what is the magnitude of force $\mathbf{F}_3$?

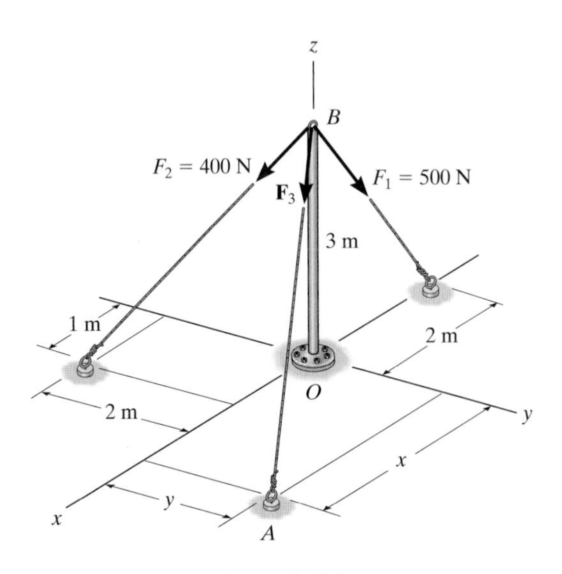

Prob. 2–100

2–101. The cord exerts a force of $\mathbf{F} = \{12\mathbf{i} + 9\mathbf{j} - 8\mathbf{k}\}$ lb on the hook. If the cord is 8 ft long, determine the location x, y of the point of attachment B, and the height z of the hook.

2–102. The cord exerts a force of $F = 30$ lb on the hook. If the cord is 8 ft long, $z = 4$ ft, and the x component of the force is $F_x = 25$ lb, determine the location x, y of the point of attachment B of the cord to the ground.

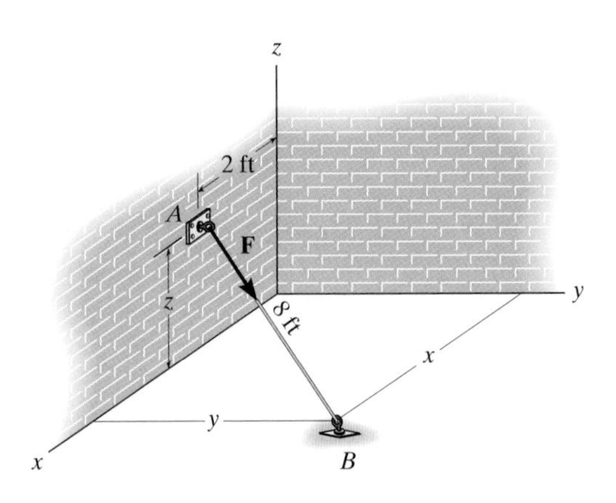

Probs. 2–101/102

2–103. Each of the four forces acting at E has a magnitude of 28 kN. Express each force as a Cartesian vector and determine the resultant force.

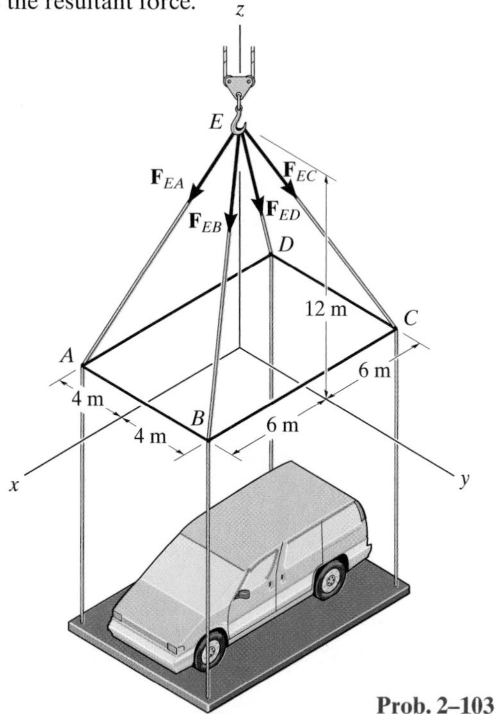

Prob. 2–103

***2–104.** The tower is held in place by three cables. If the force of each cable acting on the tower is shown, determine the magnitude and coordinate direction angles α, β, γ of the resultant force. Take $x = 20$ m, $y = 15$ m.

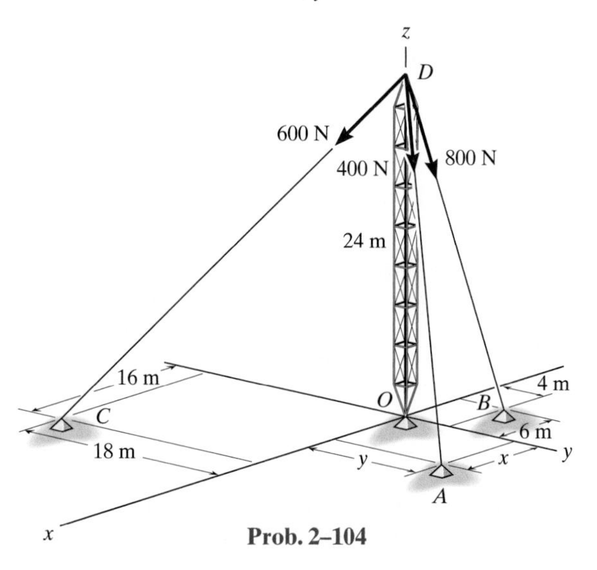

Prob. 2–104

2–105. The chandelier is supported by three chains which are concurrent at point O. If the force in each chain has a magnitude of 60 lb, express each force as a Cartesian vector and determine the magnitude and coordinate direction angles of the resultant force.

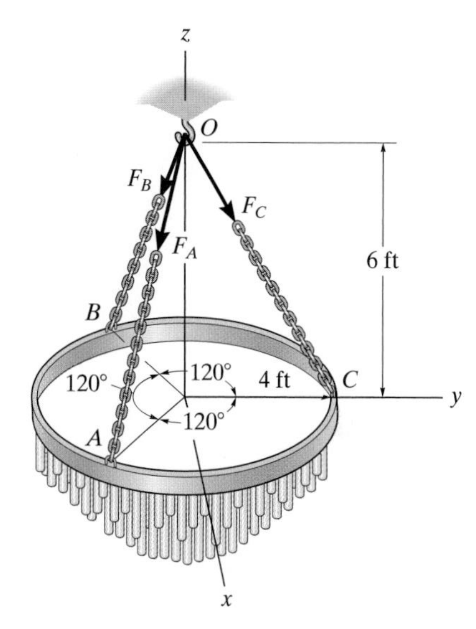

Prob. 2–105

2–106. The chandelier is supported by three chains which are concurrent at point O. If the resultant force at O has a magnitude of 130 lb and is directed along the negative z axis, determine the force in each chain assuming $F_A = F_B = F_C$.

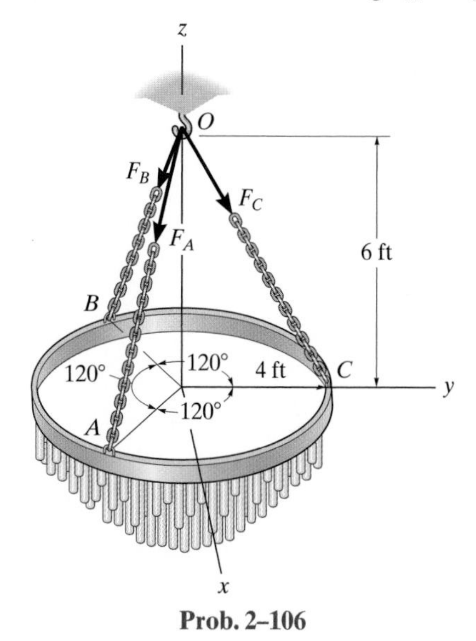

Prob. 2–106

2.9 Dot Product

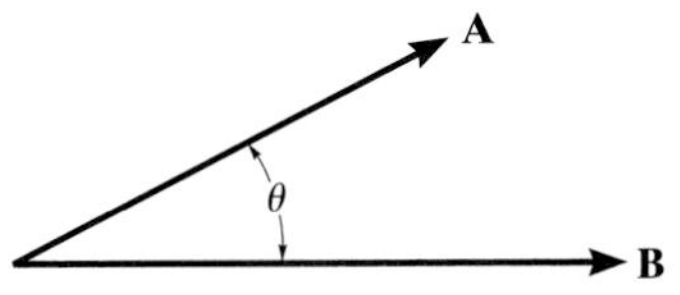

Fig. 2–41

Occasionally in statics one has to find the angle between two lines or the components of a force parallel and perpendicular to a line. In two dimensions, these problems can readily be solved by trigonometry since the geometry is easy to visualize. In three dimensions, however, this is often difficult, and consequently vector methods should be employed for the solution. The dot product defines a particular method for "multiplying" two vectors and is used to solve the above-mentioned problems.

The *dot product* of vectors **A** and **B**, written $\mathbf{A} \cdot \mathbf{B}$, and read "**A** dot **B**," is defined as the product of the magnitudes of **A** and **B** and the cosine of the angle θ between their tails, Fig. 2–41. Expressed in equation form,

$$\mathbf{A} \cdot \mathbf{B} = AB \cos \theta \qquad (2\text{–}14)$$

where $0° \leq \theta \leq 180°$. The dot product is often referred to as the *scalar product* of vectors since the result is a *scalar* and not a vector.

Laws of Operation.

1. Commutative law:

$$\mathbf{A} \cdot \mathbf{B} = \mathbf{B} \cdot \mathbf{A}$$

2. Multiplication by a scalar:

$$a(\mathbf{A} \cdot \mathbf{B}) = (a\mathbf{A}) \cdot \mathbf{B} = \mathbf{A} \cdot (a\mathbf{B}) = (\mathbf{A} \cdot \mathbf{B})a$$

3. Distributive law:

$$\mathbf{A} \cdot (\mathbf{B} + \mathbf{D}) = (\mathbf{A} \cdot \mathbf{B}) + (\mathbf{A} \cdot \mathbf{D})$$

It is easy to prove the first and second laws by using Eq. 2–14. The proof of the distributive law is left as an exercise (see Prob. 2–107).

Cartesian Vector Formulation. Equation 2–14 may be used to find the dot product for each of the Cartesian unit vectors. For example, $\mathbf{i} \cdot \mathbf{i} = (1)(1) \cos 0° = 1$ and $\mathbf{i} \cdot \mathbf{j} = (1)(1) \cos 90° = 0$. In a similar manner,

$$\begin{array}{lll} \mathbf{i} \cdot \mathbf{i} = 1 & \mathbf{j} \cdot \mathbf{j} = 1 & \mathbf{k} \cdot \mathbf{k} = 1 \\ \mathbf{i} \cdot \mathbf{j} = 0 & \mathbf{i} \cdot \mathbf{k} = 0 & \mathbf{k} \cdot \mathbf{j} = 0 \end{array}$$

These results should not be memorized; rather, it should be clearly understood how each is obtained.

Consider now the dot product of two general vectors **A** and **B** which are expressed in Cartesian vector form. We have

$$\begin{aligned} \mathbf{A} \cdot \mathbf{B} &= (A_x\mathbf{i} + A_y\mathbf{j} + A_z\mathbf{k}) \cdot (B_x\mathbf{i} + B_y\mathbf{j} + B_z\mathbf{k}) \\ &= A_xB_x(\mathbf{i} \cdot \mathbf{i}) + A_xB_y(\mathbf{i} \cdot \mathbf{j}) + A_xB_z(\mathbf{i} \cdot \mathbf{k}) \\ &\quad + A_yB_x(\mathbf{j} \cdot \mathbf{i}) + A_yB_y(\mathbf{j} \cdot \mathbf{j}) + A_yB_z(\mathbf{j} \cdot \mathbf{k}) \\ &\quad + A_zB_x(\mathbf{k} \cdot \mathbf{i}) + A_zB_y(\mathbf{k} \cdot \mathbf{j}) + A_zB_z(\mathbf{k} \cdot \mathbf{k}) \end{aligned}$$

Carrying out the dot-product operations, the final result becomes

$$\mathbf{A} \cdot \mathbf{B} = A_x B_x + A_y B_y + A_z B_z \qquad (2\text{–}15)$$

Thus, to determine the dot product of two Cartesian vectors, multiply their corresponding x, y, z components and sum their products algebraically. Since the result is a scalar, be careful *not* to include any unit vectors in the final result.

Applications. The dot product has two important applications in mechanics.

1. *The angle formed between two vectors or intersecting lines.* The angle θ between the tails of vectors **A** and **B** in Fig. 2–41 can be determined from Eq. 2–14 and written as

$$\theta = \cos^{-1}\left(\frac{\mathbf{A} \cdot \mathbf{B}}{AB}\right) \qquad 0^\circ \le \theta \le 180^\circ$$

   Here $\mathbf{A} \cdot \mathbf{B}$ is found from Eq. 2–15. In particular, notice that if $\mathbf{A} \cdot \mathbf{B} = 0$, $\theta = \cos^{-1} 0 = 90^\circ$, so that **A** will be *perpendicular* to **B**.

2. *The components of a vector parallel and perpendicular to a line.* The component of vector **A** parallel to or collinear with the line aa' in Fig. 2–42 is defined by $\mathbf{A}_\parallel$, where $A_\parallel = A\cos\theta$. This component is sometimes referred to as the *projection* of **A** onto the line, since a right angle is formed in the construction. If the *direction* of the line is specified by the unit vector **u**, then, since $u = 1$, we can determine $A_\parallel$ directly from the dot product (Eq. 2–14); i.e.,

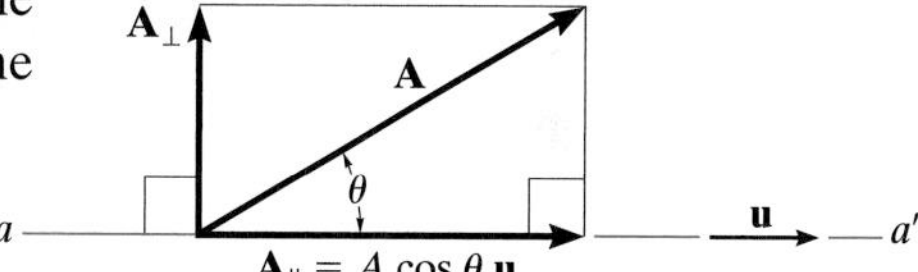

Fig. 2–42

$$A_\parallel = A\cos\theta = \mathbf{A} \cdot \mathbf{u}$$

   Hence, the scalar projection of **A** *along a line is determined from the dot product of* **A** *and the unit vector* **u** *which defines the direction of the line.* Notice that if this result is positive, then $\mathbf{A}_\parallel$ has a directional sense which is the same as **u**, whereas if $A_\parallel$ is a negative scalar, then $\mathbf{A}_\parallel$ has the opposite sense of direction to **u**. The component $\mathbf{A}_\parallel$ represented as a *vector* is therefore

$$\mathbf{A}_\parallel = A\cos\theta\,\mathbf{u} = (\mathbf{A} \cdot \mathbf{u})\mathbf{u}$$

   The component of **A** that is *perpendicular* to line aa' can also be obtained, Fig. 2–42. Since $\mathbf{A} = \mathbf{A}_\parallel + \mathbf{A}_\perp$, then $\mathbf{A}_\perp = \mathbf{A} - \mathbf{A}_\parallel$. There are two possible ways of obtaining $A_\perp$. One way would be to determine θ from the dot product, $\theta = \cos^{-1}(\mathbf{A} \cdot \mathbf{u}/A)$, then $A_\perp = A\sin\theta$. Alternatively, if $A_\parallel$ is known, then by the Pythagorean theorem we can also write $A_\perp = \sqrt{A^2 - A_\parallel^2}$.

Photo A: The angle θ which is made between the rope and the connecting beam A can be determined by using the dot product. Simply formulate position vectors or unit vectors along the beam, $\mathbf{u}_A = \mathbf{r}_A/r_A$, and along the rope, $\mathbf{u}_r = \mathbf{r}_r/r_r$. Since θ is defined between the tails of these vectors we can solve for θ using $\theta = \cos^{-1}(\mathbf{r}_A \cdot \mathbf{r}_r/r_A r_r) = \cos^{-1} \mathbf{u}_A \cdot \mathbf{u}_r$.

Photo B: If the rope exerts a force $\mathbf{F}$ on the joint, the projection of this force along beam A can be determined by first defining the *direction of the beam* using the unit vector $\mathbf{u}_A = \mathbf{r}_A/r_A$ and then formulating the force as a Cartesian vector $\mathbf{F} = F(\mathbf{r}_r/r_r) = F\mathbf{u}_r$. Applying the dot product, the projection is $F_{\parallel} = \mathbf{F} \cdot \mathbf{u}_A$.

A

B

Important Points

- The dot product is used to determine the angle between two vectors or the projection of a vector in a specified direction.
- If the vectors **A** and **B** are expressed in Cartesian form, the dot product is determined by multiplying the respective x, y, z scalar components together and algebraically adding the results, i.e., $\mathbf{A} \cdot \mathbf{B} = A_x B_x + A_y B_y + A_z B_z$.
- From the definition of the dot product, the angle formed between the tails of vectors **A** and **B** is $\theta = \cos^{-1}(\mathbf{A} \cdot \mathbf{B}/AB)$.
- The magnitude of the projection of vector **A** along a line whose direction is specified by **u** is determined from the dot product $A_{\parallel} = \mathbf{A} \cdot \mathbf{u}$.

EXAMPLE 2.16

The frame shown in Fig. 2–43*a* is subjected to a horizontal force $\mathbf{F} = \{300\mathbf{j}\}$ N. Determine the magnitude of the components of this force parallel and perpendicular to member AB.

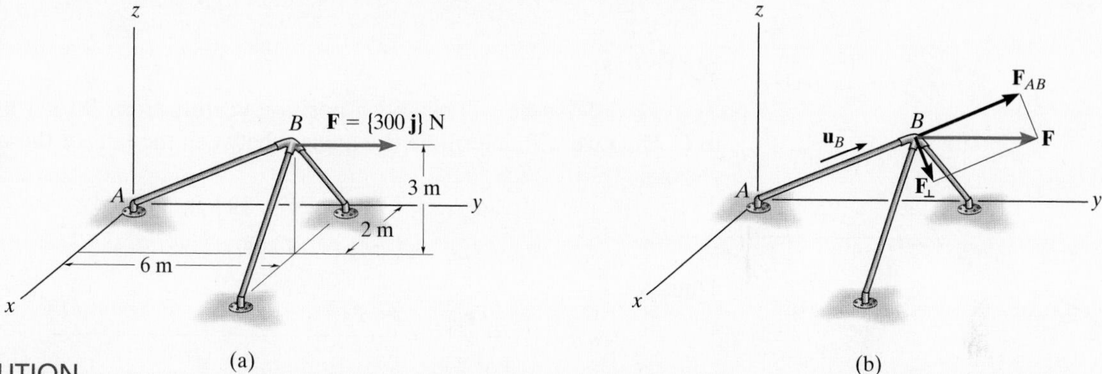

Fig. 2–43

SOLUTION

The magnitude of the component of **F** along AB is equal to the dot product of **F** and the unit vector $\mathbf{u}_B$, which defines the direction of AB, Fig. 2–43*b*. Since

$$\mathbf{u}_B = \frac{\mathbf{r}_B}{r_B} = \frac{2\mathbf{i} + 6\mathbf{j} + 3\mathbf{k}}{\sqrt{(2)^2 + (6)^2 + (3)^2}} = 0.286\mathbf{i} + 0.857\mathbf{j} + 0.429\mathbf{k}$$

then

$$\begin{aligned} F_{AB} &= F\cos\theta = \mathbf{F}\cdot\mathbf{u}_B = (300\mathbf{j})\cdot(0.286\mathbf{i} + 0.857\mathbf{j} + 0.429\mathbf{k}) \\ &= (0)(0.286) + (300)(0.857) + (0)(0.429) \\ &= 257.1 \text{ N} \end{aligned}$$ *Ans.*

Since the result is a positive scalar, $\mathbf{F}_{AB}$ has the same sense of direction as $\mathbf{u}_B$, Fig. 2–43*b*.

Expressing $\mathbf{F}_{AB}$ in Cartesian vector form, we have

$$\begin{aligned} \mathbf{F}_{AB} &= F_{AB}\mathbf{u}_B = (257.1 \text{ N})(0.286\mathbf{i} + 0.857\mathbf{j} + 0.429\mathbf{k}) \\ &= \{73.5\mathbf{i} + 220\mathbf{j} + 110\mathbf{k}\} \text{ N} \end{aligned}$$ *Ans.*

The perpendicular component, Fig. 2–43*b*, is therefore

$$\begin{aligned} \mathbf{F}_\perp &= \mathbf{F} - \mathbf{F}_{AB} = 300\mathbf{j} - (73.5\mathbf{i} + 220\mathbf{j} + 110\mathbf{k}) \\ &= \{-73.5\mathbf{i} + 80\mathbf{j} - 110\mathbf{k}\} \text{ N} \end{aligned}$$

Its magnitude can be determined either from this vector or from the Pythagorean theorem, Fig. 2–43*b*:

$$\begin{aligned} F_\perp &= \sqrt{F^2 - F_{AB}^2} \\ &= \sqrt{(300 \text{ N})^2 - (257.1 \text{ N})^2} \\ &= 155 \text{ N} \end{aligned}$$ *Ans.*

EXAMPLE 2.17

The pipe in Fig. 2–44*a* is subjected to the force of $F = 80$ lb. Determine the angle θ between **F** and the pipe segment *BA*, and the magnitudes of the components of **F**, which are parallel and perpendicular to *BA*.

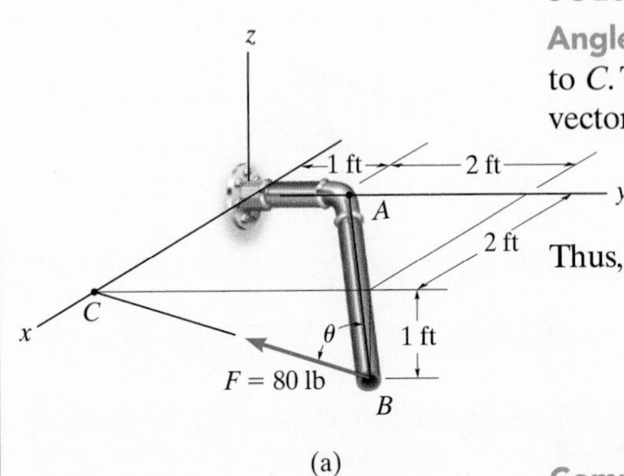

(a)

SOLUTION

Angle θ. First we will establish position vectors from *B* to *A* and *B* to *C*. Then we will determine the angle θ between the tails of these two vectors.

$$\mathbf{r}_{BA} = \{-2\mathbf{i} - 2\mathbf{j} + 1\mathbf{k}\} \text{ ft}$$
$$\mathbf{r}_{BC} = \{-3\mathbf{j} + 1\mathbf{k}\} \text{ ft}$$

Thus,

$$\cos\theta = \frac{\mathbf{r}_{BA}\cdot\mathbf{r}_{BC}}{r_{BA}r_{BC}} = \frac{(-2)(0) + (-2)(-3) + (1)(1)}{3\sqrt{10}}$$
$$= 0.7379$$
$$\theta = 42.5^\circ \qquad \textit{Ans.}$$

Components of F. The force **F** is resolved into components as shown in Fig. 2–44*b*. Since $F_{BA} = \mathbf{F}\cdot\mathbf{u}_{BA}$, we must first formulate the unit vector along *BA* and force **F** as Cartesian vectors.

$$\mathbf{u}_{BA} = \frac{\mathbf{r}_{BA}}{r_{BA}} = \frac{(-2\mathbf{i} - 2\mathbf{j} + 1\mathbf{k})}{3} = -\frac{2}{3}\mathbf{i} - \frac{2}{3}\mathbf{j} + \frac{1}{3}\mathbf{k}$$

$$\mathbf{F} = 80 \text{ lb}\left(\frac{\mathbf{r}_{BC}}{r_{BC}}\right) = 80\left(\frac{-3\mathbf{j} + 1\mathbf{k}}{\sqrt{10}}\right) = -75.89\mathbf{j} + 25.30\mathbf{k}$$

Thus,

$$F_{BA} = \mathbf{F}\cdot\mathbf{u}_{BA} = (-75.89\mathbf{j} + 25.30\mathbf{k})\cdot\left(-\frac{2}{3}\mathbf{i} - \frac{2}{3}\mathbf{j} + \frac{1}{3}\mathbf{k}\right)$$
$$= 0 + 50.60 + 8.43$$
$$= 59.0 \text{ lb} \qquad \textit{Ans.}$$

Since θ was calculated in Fig. 2–44*b*, this same result can also be obtained directly from trigonometry.

$$F_{BA} = 80 \cos 42.5^\circ \text{ lb} = 59.0 \text{ lb} \qquad \textit{Ans.}$$

The perpendicular component can be obtained by trigonometry,

$$F_\perp = F \sin\theta$$
$$= 80 \sin 42.5^\circ \text{ lb}$$
$$= 54.0 \text{ lb} \qquad \textit{Ans.}$$

NOTE: We can also obtain this result using the Pythagorean theorem,

$$F_\perp = \sqrt{F^2 - F_{BA}^2} = \sqrt{(80)^2 - (59.0)^2}$$
$$= 54.0 \text{ lb} \qquad \textit{Ans.}$$

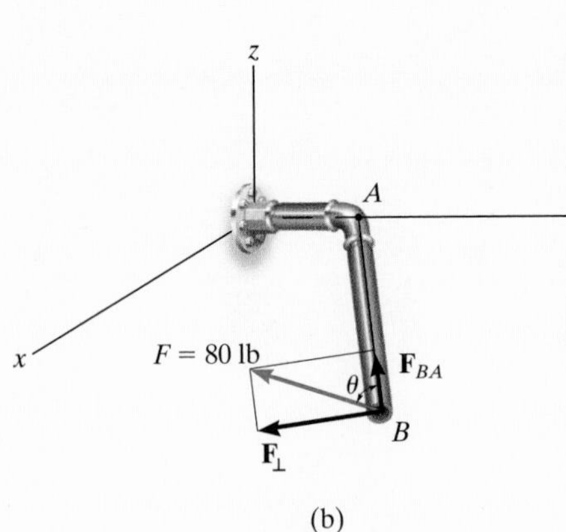

(b)

Fig. 2–44

PROBLEMS

2–107. Given the three vectors **A**, **B**, and **D**, show that $\mathbf{A} \cdot (\mathbf{B} + \mathbf{D}) = (\mathbf{A} \cdot \mathbf{B}) + (\mathbf{A} \cdot \mathbf{D})$.

***2–108.** Cable BC exerts a force of $F = 28$ N on the top of the flagpole. Determine the projection of this force along the z axis of the pole.

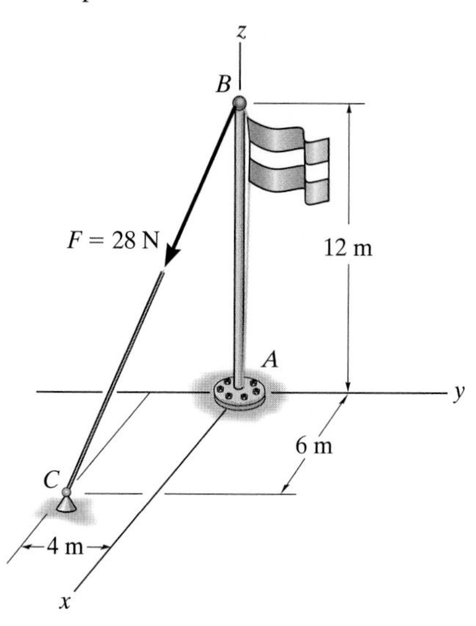

Prob. 2–108

2–109. Determine the angle θ between the tails of the two vectors.

2–110. Determine the magnitude of the projected component of $\mathbf{r}_1$ along $\mathbf{r}_2$, and the projection of $\mathbf{r}_2$ along $\mathbf{r}_1$.

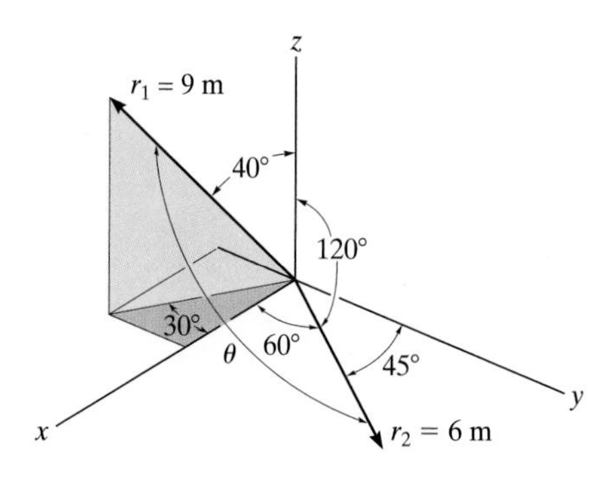

Probs. 2–109/110

2–111. Determine the angles θ and ϕ between the wire segments.

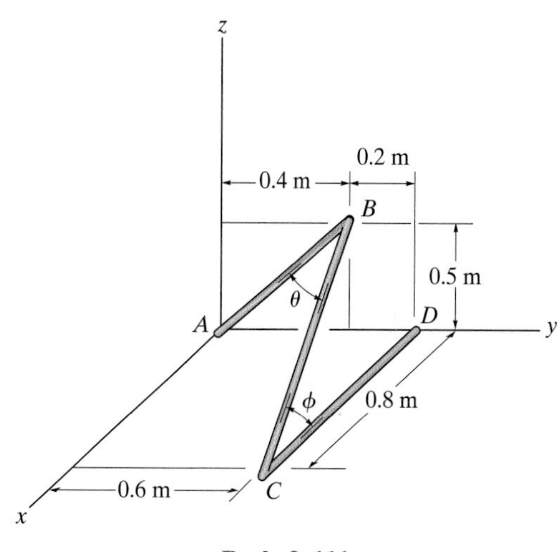

Prob. 2–111

***2–112.** Determine the angle θ between the two cords.

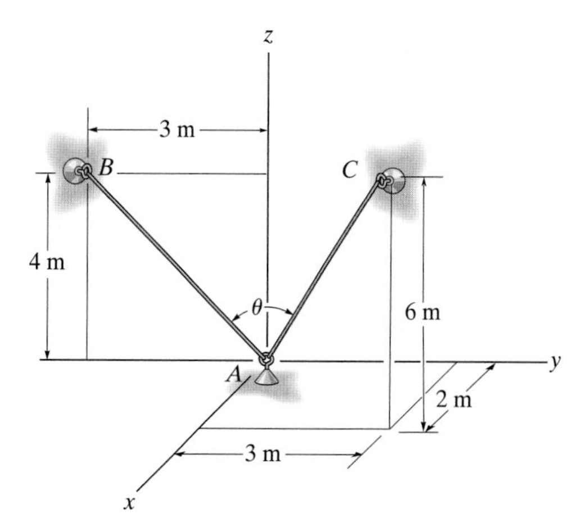

Prob. 2–112

2–113. Determine the angle θ between the two cables.

2–114. Determine the projected component of the force $F = 12$ lb acting in the direction of cable AC. Express the result as a Cartesian vector.

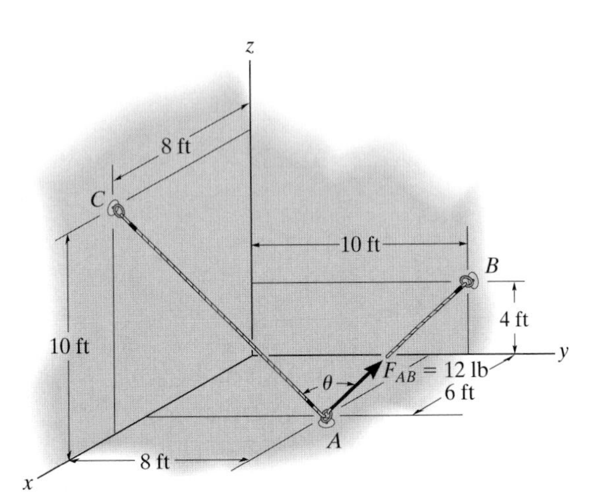

Probs. 2–113/114

2–117. Determine the magnitude of the projected component of the length of cord OA along the Oa axis.

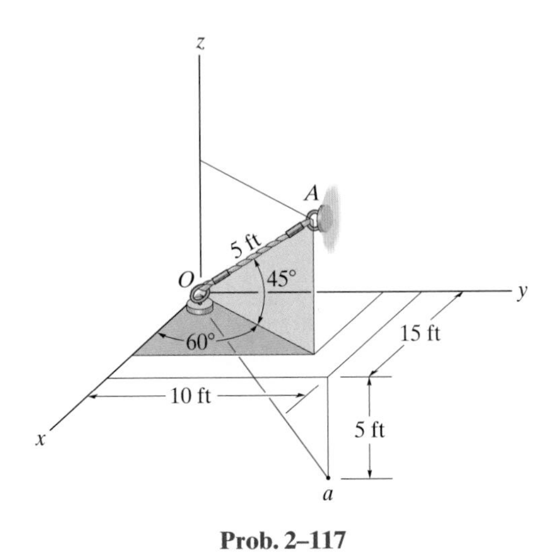

Prob. 2–117

2–115. Determine the components of **F** that act along rod AC and perpendicular to it. Point B is located at the midpoint of the rod.

***2–116.** Determine the components of **F** that act along rod AC and perpendicular to it. Point B is located 3 m along the rod from end C.

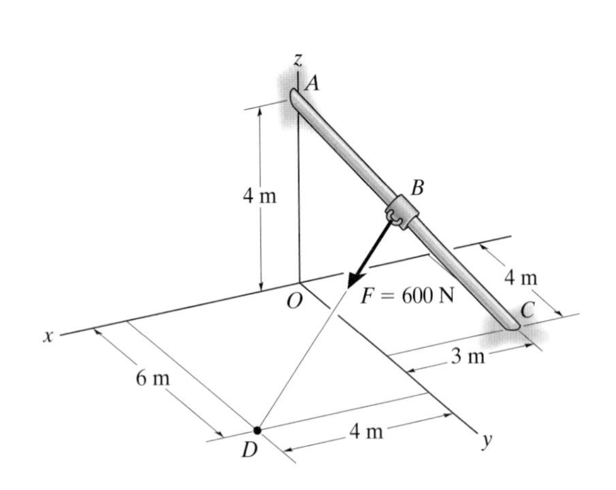

Probs. 2–115/116

2–118. A force of $\mathbf{F} = \{-40\mathbf{k}\}$ lb acts at the end of the pipe. Determine the magnitudes of the components $\mathbf{F}_1$ and $\mathbf{F}_2$ which are directed along the pipe's axis and perpendicular to it.

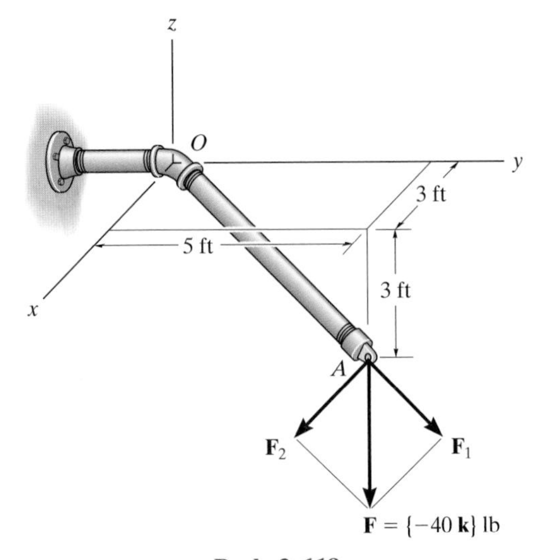

Prob. 2–118

2–119. Determine the projected component of the 80-N force acting along the axis AB of the pipe.

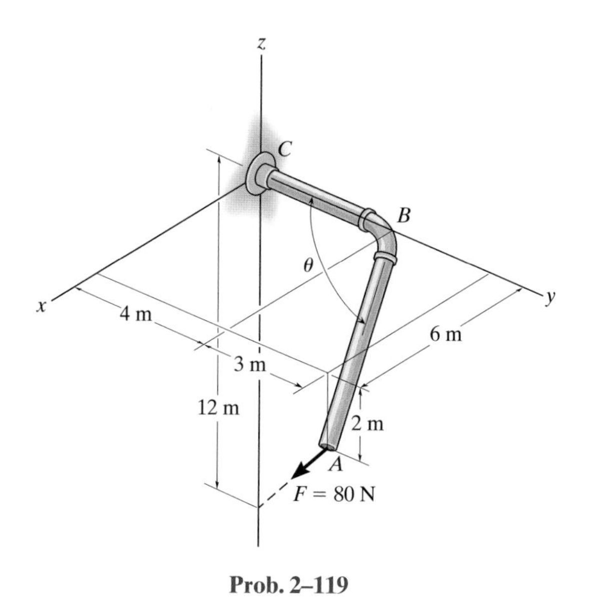

Prob. 2–119

2–122. A force of $F = 80$ N is applied to the handle of the wrench. Determine the angle θ between the tail of the force and the handle AB.

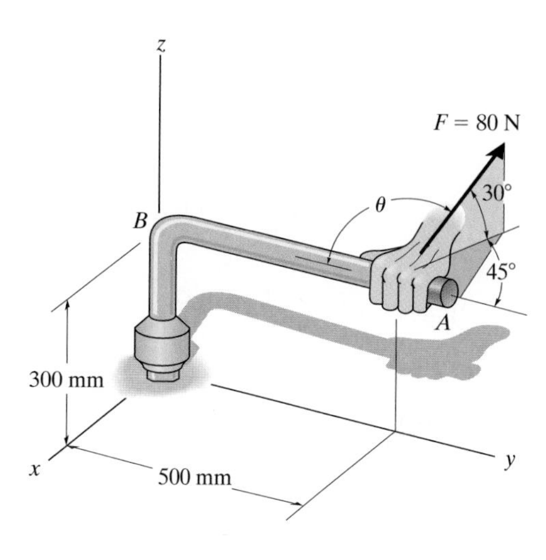

Prob. 2–122

***2–120.** Determine the angles θ and ϕ between the axis OA of the pole and each cable, AB and AC.

2–121. The two cables exert the forces shown on the pole. Determine the magnitude of the projected component of each force acting along the axis OA of the pole.

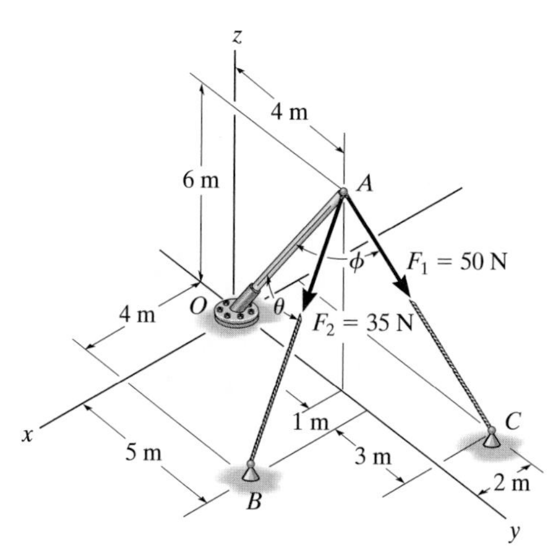

Probs. 2–120/121

2–123. Two cables exert forces on the pipe. Determine the magnitude of the projected component of $\mathbf{F}_1$ along the line of action of $\mathbf{F}_2$.

***2–124.** Determine the angle θ between the two cables attached to the pipe.

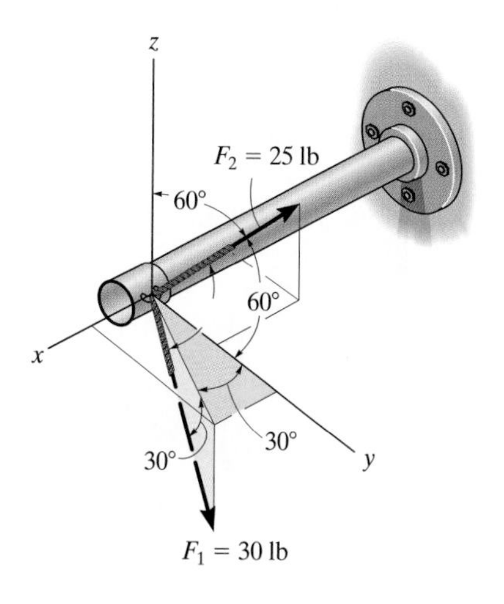

Probs. 2–123/124

2–125. Determine the angle θ between the two cables.

2–126. Determine the projection of the force $\mathbf{F}_1$ along cable AB. Determine the projection of the force $\mathbf{F}_2$ along cable AC.

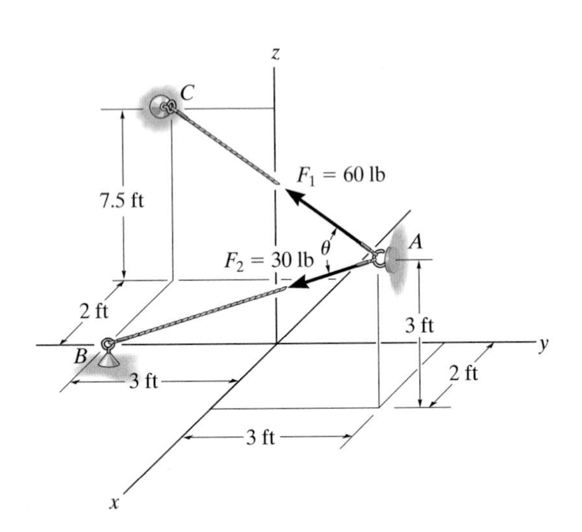

Probs. 2–125/126

***2–128.** Determine the magnitude of the projected component of the 100-lb force acting along the axis BC of the pipe.

2–129. Determine the angle θ between pipe segments BA and BC.

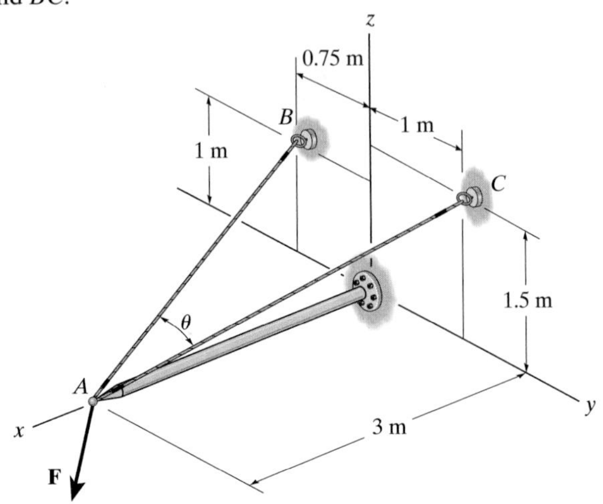

Probs. 2–128/129

2–127. Determine the angle θ between the edges of the sheet-metal bracket.

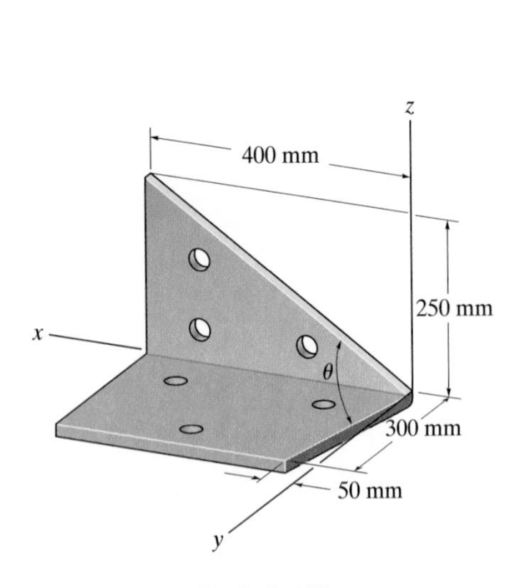

Prob. 2–127

2–130. Determine the angles θ and ϕ made between the axes OA of the flag pole and AB and AC, respectively, of each cable.

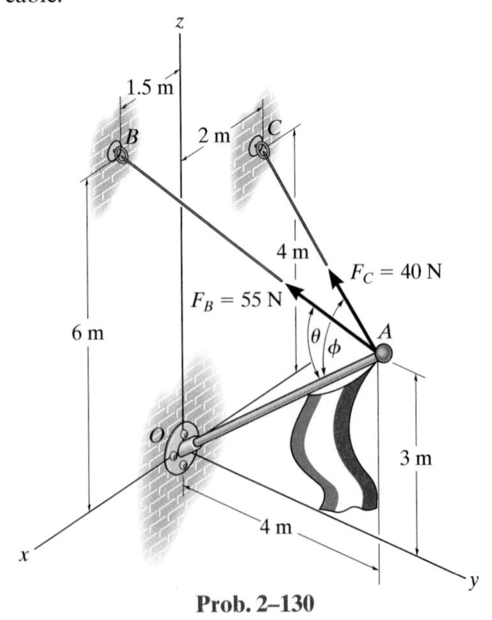

Prob. 2–130

Chapter Review

A scalar is a positive or negative number; e.g., mass and temperature.

A vector has a magnitude and an arrowhead sense of direction; e.g., force and position.
(*See page 17.*)

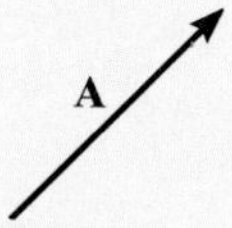

Multiplication or division of a vector by a scalar will change only the magnitude of the vector. If the scalar is negative the sense of the vector will change, so that it acts in the opposite direction.
(*See page 18.*)

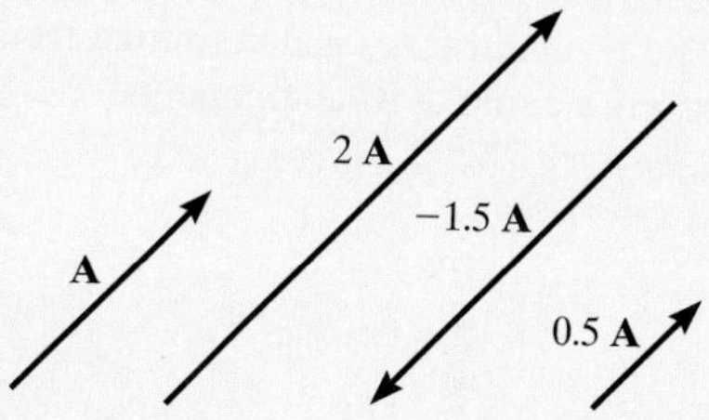

If vectors are collinear, the resultant is formed by an algebraic or scalar addition.
(*See page 19.*)

$$R = A + B$$

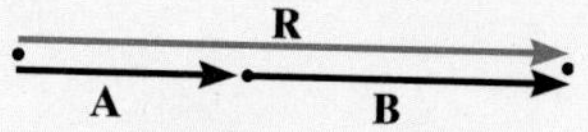

Parallelogram Law

Two forces add according to the parallelogram law. The *components* form the sides of the parallelogram and the *resultant* is the diagonal.

To find the components of a force along any two axes, extend lines from the head of the force, parallel to the axes, to form the components.

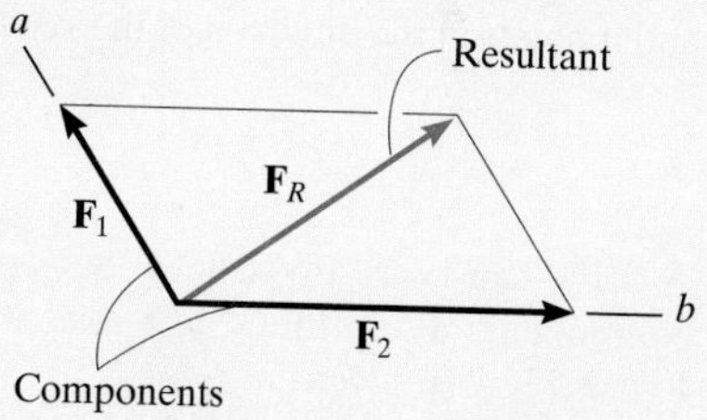

To obtain the components or the resultant, show how the forces add by a tip-to-tail addition using the triangle rule, and then use the law of sines and the law of cosines to calculate their values.
(*See pages 18–21.*)

$$F_R = \sqrt{F_1^{\,2} + F_2^{\,2} - 2\,F_1F_2\cos\theta_R}$$

$$\frac{F_1}{\sin\theta_1} = \frac{F_2}{\sin\theta_2} = \frac{F_R}{\sin\theta_R}$$

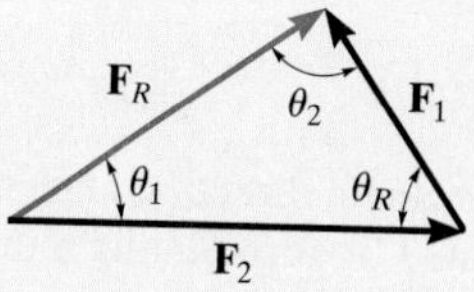

Rectangular Components: Two Dimensions

Vectors $\mathbf{F}_x$ and $\mathbf{F}_y$ are rectangular components of $\mathbf{F}$.

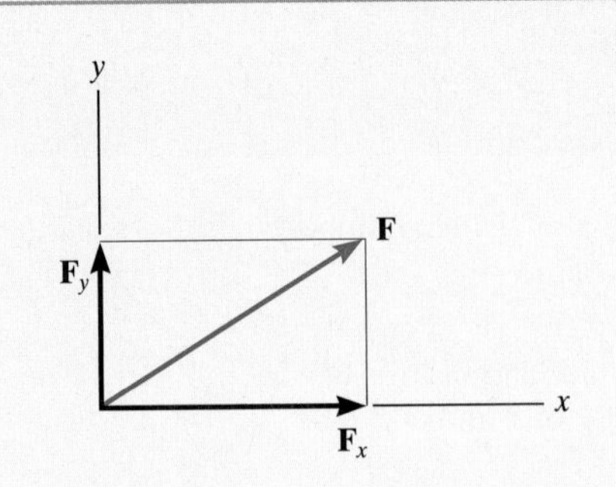

The resultant force is determined from the algebraic sum of its components.
(*See page 31.*)

$$F_{Rx} = \Sigma F_x$$
$$F_{Ry} = \Sigma F_y$$
$$F_R = \sqrt{(F_{Rx})^2 + (F_{Ry})^2}, \quad \theta = \tan^{-1}\left|\frac{F_{Ry}}{F_{Rx}}\right|$$

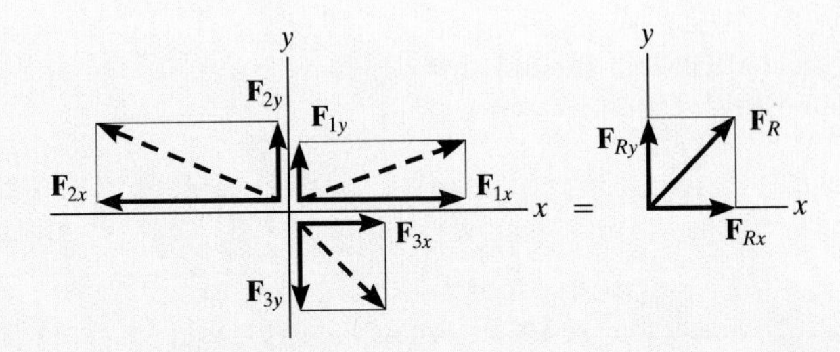

Cartesian Vectors

The unit vector $\mathbf{u}_A$ has a length 1, no units, and it points in the direction of the vector $\mathbf{F}$.

$$\mathbf{u} = \frac{\mathbf{F}}{F}$$

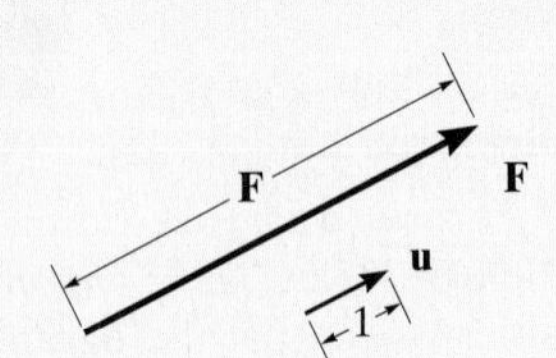

A force can be resolved into its Cartesian components along the x, y, z axes so that $\mathbf{F} = F_x\mathbf{i} + F_y\mathbf{j} + F_z\mathbf{k}$.

The magnitude of $\mathbf{F}$ is determined from the positive square root of the sum of the squares of its components.

$$F = \sqrt{F_x^2 + F_y^2 + F_z^2}$$

The coordinate direction angles α, β, γ are determined by formulating a unit vector in the direction of $\mathbf{F}$. The components of $\mathbf{u}$ represent $\cos\alpha$, $\cos\beta$, $\cos\gamma$.

$$\mathbf{u} = \frac{\mathbf{F}}{F} = \frac{F_x}{F}\mathbf{i} + \frac{F_y}{F}\mathbf{j} + \frac{F_z}{F}\mathbf{k}$$

$$\mathbf{u} = \cos\alpha\mathbf{i} + \cos\beta\mathbf{j} + \cos\gamma\mathbf{k}$$

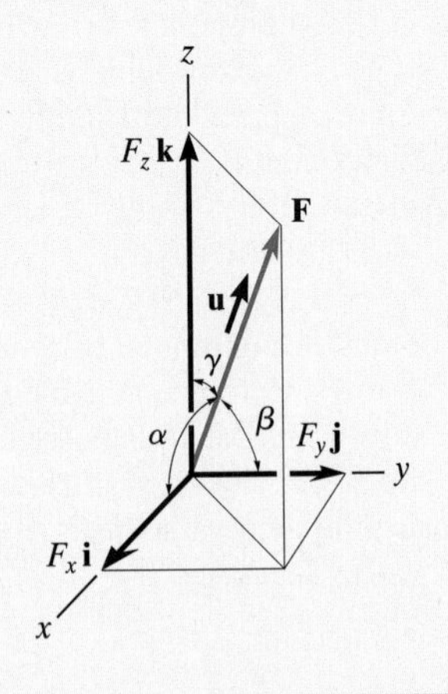

The coordinate direction angles are related so that only two of the three angles are independent of one another.

$$\cos^2 \alpha + \cos^2 \beta + \cos^2 \gamma = 1$$

To find the resultant of a concurrent force system, express each force as a Cartesian vector and add the **i**, **j**, **k** components of all the forces in the system.
(*See pages 43–46.*)

$$\mathbf{F}_R = \Sigma\mathbf{F} = \Sigma F_x\mathbf{i} + \Sigma F_y\mathbf{j} + \Sigma F_z\mathbf{k}$$

Position and Force Vectors

A position vector locates one point in space relative to another. The easiest way to formulate the components of a position vector is to determine the distance and direction that one must travel along the x, y, and z directions—going from the tail to the head of the vector.

$$\mathbf{r} = (x_B - x_A)\mathbf{i} + (y_B - y_A)\mathbf{j} + (z_B - z_A)\mathbf{k}$$

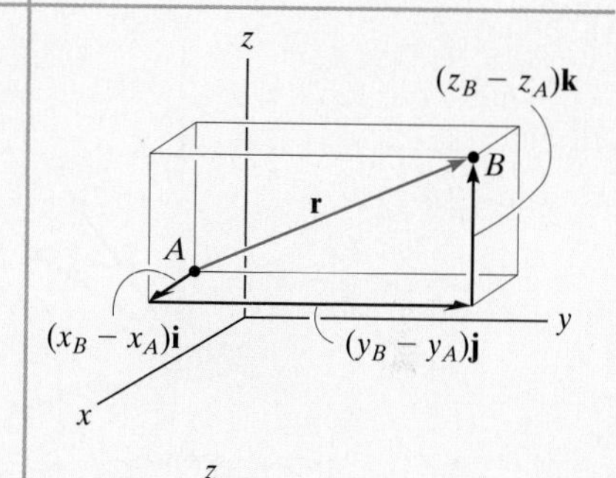

If the line of action of a force passes through points A and B, then the force acts in the same direction **u** as the position vector **r**. The force can then be expressed as a Cartesian vector.
(*See pages 55, 56, and 58.*)

$$\mathbf{F} = F\mathbf{u} = F\left(\frac{\mathbf{r}}{r}\right)$$

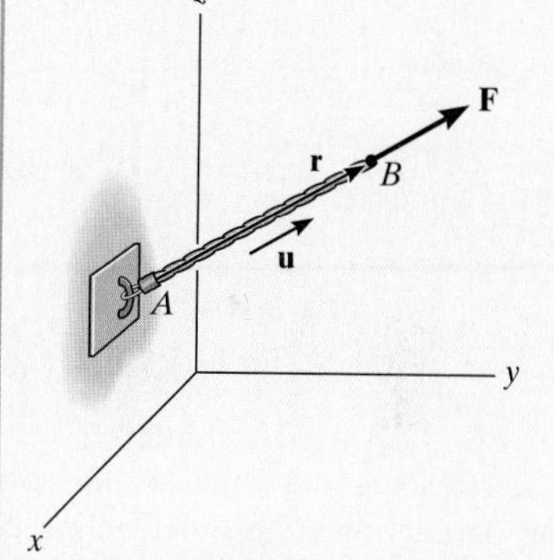

Dot Product

The dot product between two vectors **A** and **B** yields a scalar. If **A** and **B** are expressed in Cartesian vector form, then the dot product is the sum of the products of their x, y, and z components.

$$\mathbf{A} \cdot \mathbf{B} = AB\cos\theta$$
$$= A_xB_x + A_yB_y + A_zB_z$$

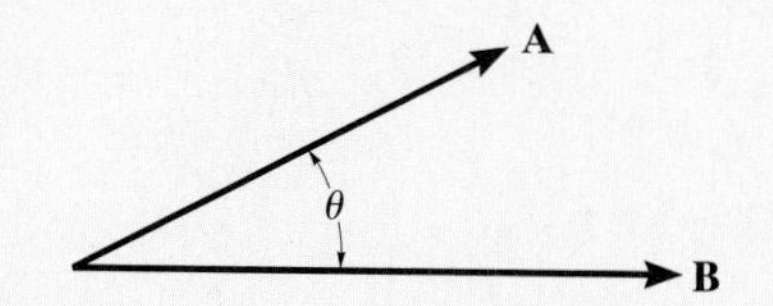

The dot product can be used to determine the angle between **A** and **B**.

$$\theta = \cos^{-1}\left(\frac{\mathbf{A} \cdot \mathbf{B}}{AB}\right)$$

The dot product is also used to determine the projected component of a vector **A** onto an axis aa' defined by its unit vector **u**.
(*See pages 68–69.*)

$$\mathbf{A}_{\parallel} = A\cos\theta\,\mathbf{u} = (\mathbf{A} \cdot \mathbf{u})\mathbf{u}$$

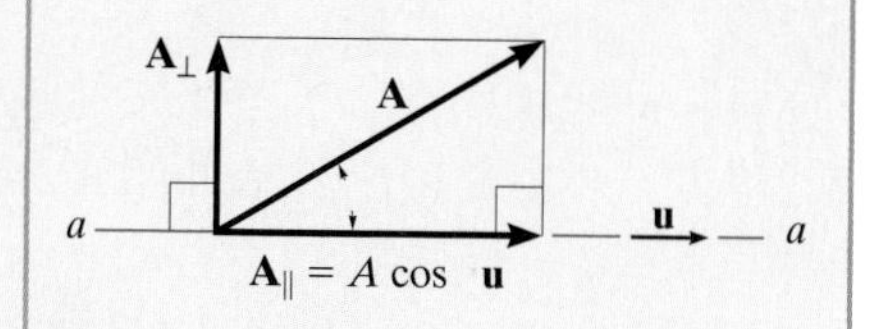

REVIEW PROBLEMS

2–131. Determine the magnitude and coordinate direction angles of $\mathbf{F}_3$ so that the resultant of the three forces acts along the positive y axis and has a magnitude of 600 lb.

***2–132.** Determine the magnitude and coordinate direction angles of $\mathbf{F}_3$ so that the resultant of the three forces is zero.

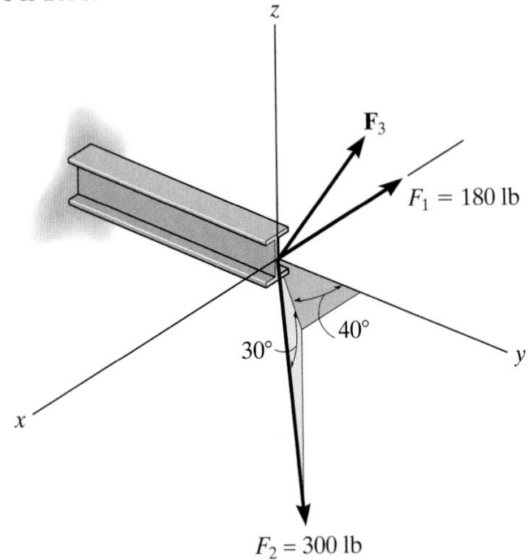

Probs. 2–131/132

2–133. Resolve the force **F** into two components, one acting parallel and the other acting perpendicular to the u axis.

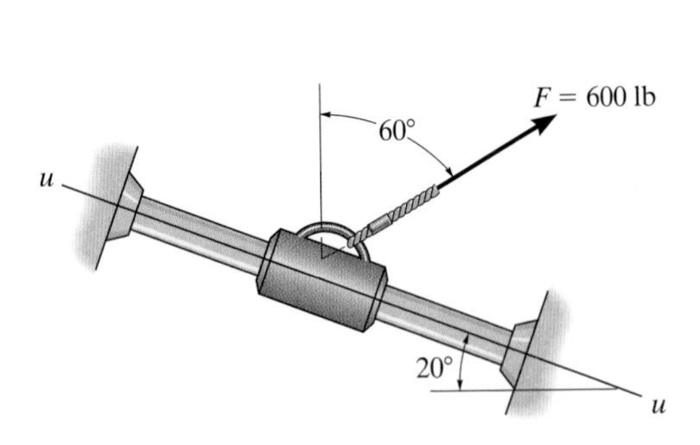

Prob. 2–133

2–134. The force **F** has a magnitude of 80 lb and acts at the midpoint C of the thin rod. Express the force as a Cartesian vector.

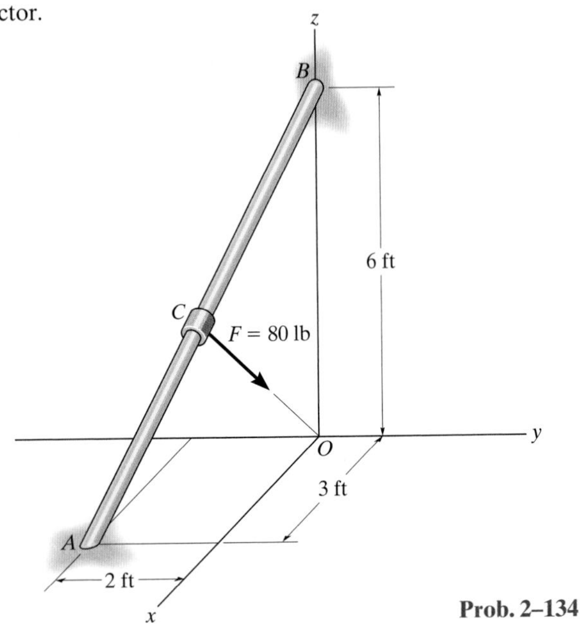

Prob. 2–134

2–135. Determine the magnitude and direction of the resultant $\mathbf{F}_R = \mathbf{F}_1 + \mathbf{F}_2 + \mathbf{F}_3$ of the three forces by first finding the resultant $\mathbf{F}' = \mathbf{F}_1 + \mathbf{F}_3$ and then forming $\mathbf{F}_R = \mathbf{F}' + \mathbf{F}_2$. Specify its direction measured counterclockwise from the positive x axis.

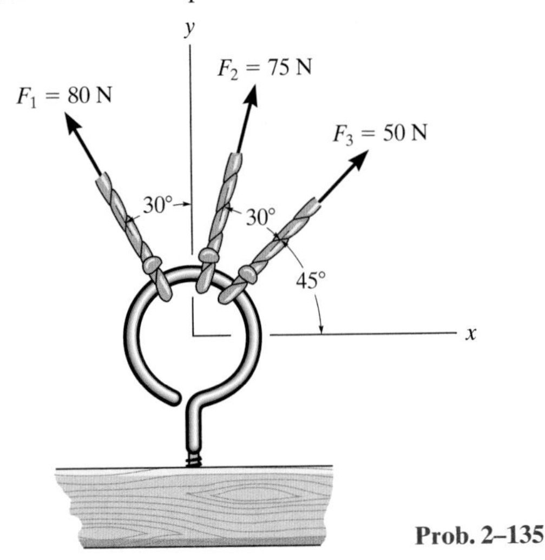

Prob. 2–135

***2–136.** The leg is held in position by the quadriceps AB, which is attached to the pelvis at A. If the force exerted on this muscle by the pelvis is 85 N, in the direction shown, determine the stabilizing force component acting along the positive y axis and the supporting force component acting along the negative x axis.

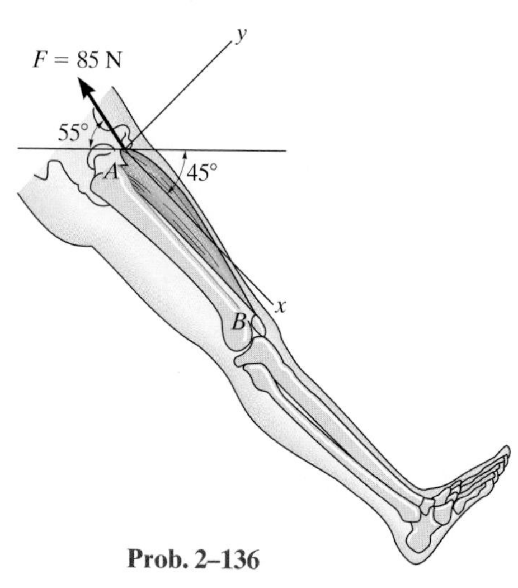

Prob. 2–136

2–137. Determine the magnitudes of the projected components of the force $\mathbf{F} = \{60\mathbf{i} + 12\mathbf{j} - 40\mathbf{k}\}$ N in the direction of the cables AB and AC.

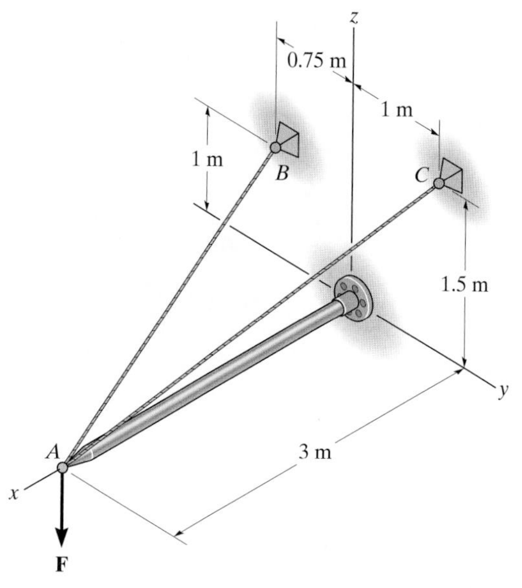

Prob. 2–137

2–138. Determine the magnitude of the projected component of the 100-lb force acting along the axis BC of the pipe.

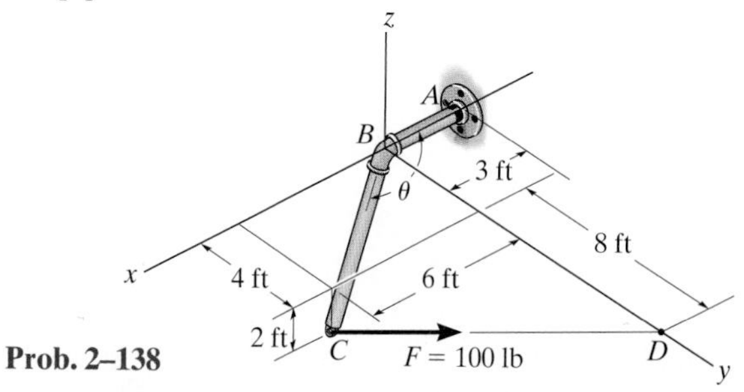

Prob. 2–138

2–139. If $F_1 = F_2 = 30$ lb, determine the angles θ and ϕ so that the resultant force is directed along the positive x axis and has a magnitude of $F_R = 20$ lb.

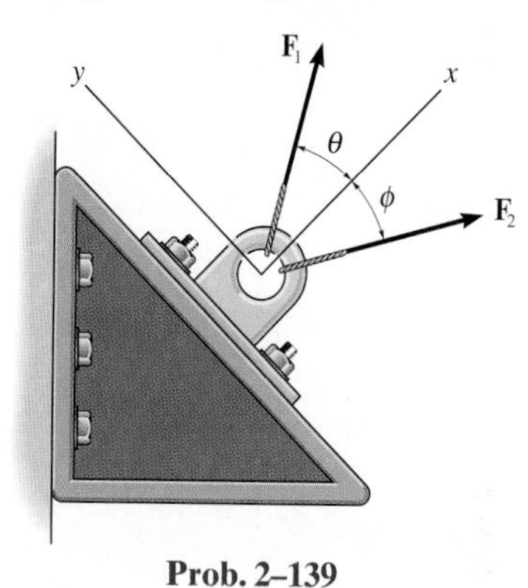

Prob. 2–139

***2–140.** Determine the magnitude of the resultant and its direction measured counterclockwise from axis.

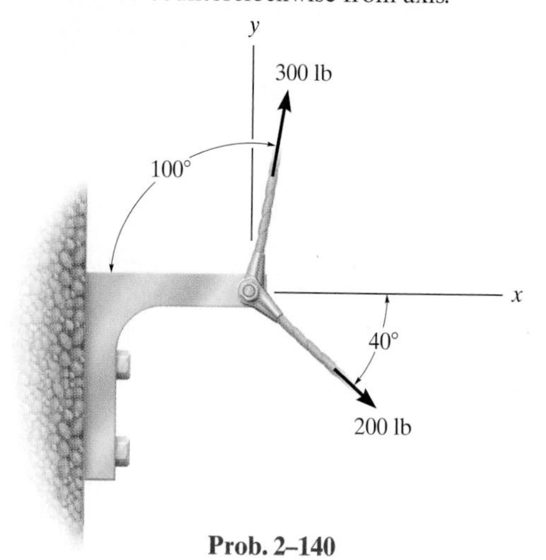

Prob. 2–140

Whenever cables are used for hoisting loads, they must be selected so that they do not fail when they are placed at their points of attachment. In this chapter, we will show how to calculate cable loadings for such cases.

3 Equilibrium of a Particle

CHAPTER OBJECTIVES

- To introduce the concept of the free-body diagram for a particle.
- To show how to solve particle equilibrium problems using the equations of equilibrium.

3.1 Condition for the Equilibrium of a Particle

A particle is in *equilibrium* provided it is at rest if originally at rest or has a constant velocity if originally in motion. Most often, however, the term "equilibrium" or, more specifically, "static equilibrium" is used to describe an object at rest. To maintain equilibrium, it is *necessary* to satisfy Newton's first law of motion, which requires the *resultant force* acting on a particle to be equal to *zero*. This condition may be stated mathematically as

$$\Sigma \mathbf{F} = \mathbf{0} \tag{3–1}$$

where $\Sigma \mathbf{F}$ is the vector *sum of all the forces* acting on the particle.

Not only is Eq. 3–1 a necessary condition for equilibrium, it is also a *sufficient* condition. This follows from Newton's second law of motion, which can be written as $\Sigma \mathbf{F} = m\mathbf{a}$. Since the force system satisfies Eq. 3–1, then $m\mathbf{a} = \mathbf{0}$, and therefore the particle's acceleration $\mathbf{a} = \mathbf{0}$. Consequently the particle indeed moves with constant velocity or remains at rest.

3.2 The Free-Body Diagram

To apply the equation of equilibrium, we must account for *all* the known and unknown forces ($\Sigma \mathbf{F}$) which act *on* the particle. The best way to do this is to draw the particle's *free-body diagram*. This diagram is simply a sketch which shows the particle "free" from its surroundings with *all* the forces that act *on* it.

Before presenting a formal procedure as to how to draw a free-body diagram, we will first consider two types of connections often encountered in particle equilibrium problems.

Springs. If a *linearly elastic spring* is used for support, the length of the spring will change in direct proportion to the force acting on it. A characteristic that defines the "elasticity" of a spring is the *spring constant* or *stiffness* k. The magnitude of force exerted on a linearly elastic spring which has a stiffness k and is deformed (elongated or compressed) a distance s, measured from its *unloaded* position, is

$$F = ks \qquad (3\text{–}2)$$

Here s is determined from the difference in the spring's deformed length l and its undeformed length l_0, i.e., $s = l - l_0$. If s is positive, **F** "pulls" on the spring; whereas if s is negative, **F** must "push" on it. For example, the spring shown in Fig. 3–1 has an undeformed length $l_0 = 0.4$ m and stiffness $k = 500$ N/m. To stretch it so that $l = 0.6$ m, a force $F = ks = (500 \text{ N/m})(0.6 \text{ m} - 0.4 \text{ m}) = 100$ N is needed. Likewise, to compress it to a length $l = 0.2$ m, a force $F = ks = (500 \text{ N/m})(0.2 \text{ m} - 0.4 \text{ m}) = -100$ N is required, Fig. 3–1.

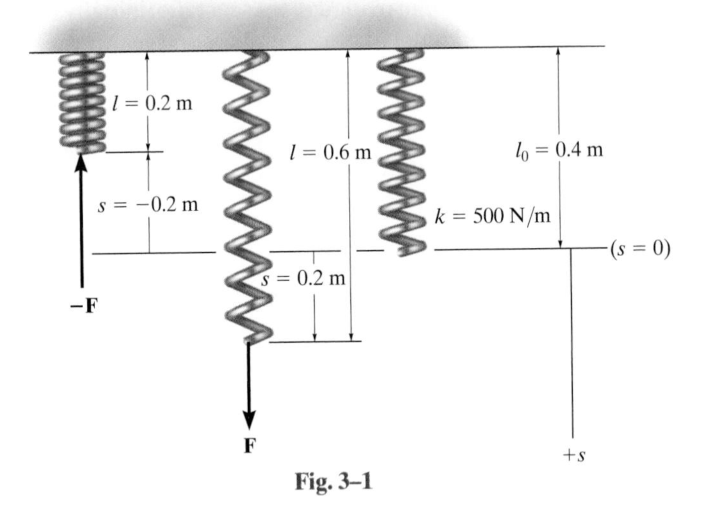

Fig. 3–1

Cables and Pulleys. Throughout this book, except in Sec. 7.4, all cables (or cords) are assumed to have negligible weight and they cannot stretch. Also, a cable can support *only* a tension or "pulling" force, and this force always acts in the direction of the cable. In Chapter 5, it will be shown that the tension force developed in a *continuous cable* which passes over a frictionless pulley must have a *constant* magnitude to keep the cable in equilibrium. Hence, for any angle θ, shown in Fig. 3–2, the cable is subjected to a constant tension T throughout its length.

Procedure for Drawing a Free-Body Diagram

Since we must account for *all the forces acting on the particle* when applying the equations of equilibrium, the importance of first drawing a free-body diagram cannot be overemphasized. To construct a free-body diagram, the following three steps are necessary.

Draw Outlined Shape.

Imagine the particle to be *isolated* or cut "free" from its surroundings by drawing its outlined shape.

Show All Forces.

Indicate on this sketch *all* the forces that act *on the particle*. These forces can be *active forces*, which tend to set the particle in motion, or they can be *reactive forces* which are the result of the constraints or supports that tend to prevent motion. To account for all these forces, it may help to trace around the particle's boundary, carefully noting each force acting on it.

Identify Each Force.

The forces that are *known* should be labeled with their proper magnitudes and directions. Letters are used to represent the magnitudes and directions of forces that are unknown.

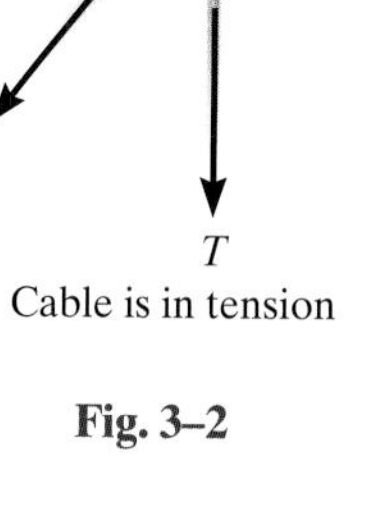

Cable is in tension

Fig. 3–2

Photo A: The bucket is held in equilibrium by the cable, and instinctively we know that the force in the cable must equal the weight of the bucket. By drawing a free-body diagram of the bucket we can understand why this is so. This diagram shows that there are only two forces *acting on the bucket*, namely, its weight **W** and the force **T** of the cable. For equilibrium, the resultant of these forces must be equal to zero, and so $T = W$. The important point is that by *isolating the bucket* the unknown cable force **T** becomes "exposed" and must be considered as a requirement for equilibrium.

Photo B: The spool has a weight W and is suspended from the crane boom. If we wish to obtain the forces in cables AB and AC, then we should consider the free-body diagram of the ring at A. Here the cables AD exert a resultant force of **W** on the ring and the condition of equilibrium is used to obtain $\mathbf{T}_B$ and $\mathbf{T}_C$.

A

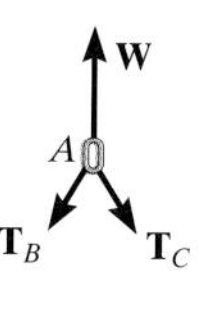

B

EXAMPLE 3.1

The sphere in Fig. 3–3a has a mass of 6 kg and is supported as shown. Draw a free-body diagram of the sphere, the cord CE, and the knot at C.

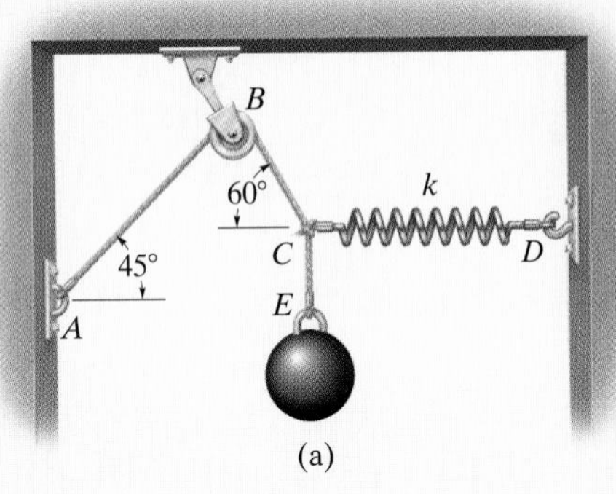

(a)

SOLUTION

Sphere. By inspection, there are only two forces acting on the sphere, namely, its weight and the force of cord CE. The sphere has a weight of 6 kg $(9.81\text{ m/s}^2) = 58.9$ N. The free-body diagram is shown in Fig. 3–3b.

Cord CE. When the cord CE is isolated from its surroundings, its free-body diagram shows only two forces acting on it, namely, the force of the sphere and the force of the knot, Fig. 3–3c. Notice that $\mathbf{F}_{CE}$ shown here is equal but opposite to that shown in Fig. 3–3b, a consequence of Newton's third law. Also, $\mathbf{F}_{CE}$ and $\mathbf{F}_{EC}$ pull on the cord and keep it in tension so that it doesn't collapse. For equilibrium, $F_{CE} = F_{EC}$.

Knot. The knot at C is subjected to three forces, Fig. 3–3d. They are caused by the cords CBA and CE and the spring CD. As required the free-body diagram shows all these forces labeled with their magnitudes and directions. It is important to recognize that the weight of the sphere does not directly act on the knot. Instead, the cord CE subjects the knot to this force.

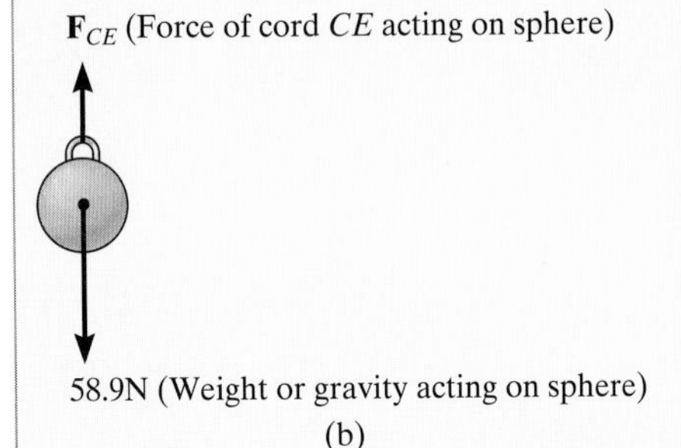

(b)

$\mathbf{F}_{EC}$ (Force of knot acting on cord CE)

$\mathbf{F}_{CE}$ (Force of sphere acting on cord CE)

(c)

$\mathbf{F}_{CBA}$ (Force of cord CBA acting on knot)

60°

C

$\mathbf{F}_{CD}$ (Force of spring acting on knot)

$\mathbf{F}_{CE}$ (Force of cord CE acting on knot)

(d)

Fig. 3–3

3.3 Coplanar Force Systems

If a particle is subjected to a system of coplanar forces that lie in the x–y plane, Fig. 3–4, then each force can be resolved into its $\mathbf{i}$ and $\mathbf{j}$ components. For equilibrium, Eq. 3–1 can be written as

$$\Sigma \mathbf{F} = \mathbf{0}$$
$$\Sigma F_x \mathbf{i} + \Sigma F_y \mathbf{j} = \mathbf{0}$$

For this vector equation to be satisfied, both the x and y components must be equal to zero. Hence,

$$\Sigma F_x = 0 \qquad\qquad (3\text{–}3)$$
$$\Sigma F_y = 0$$

These *scalar equations of equilibrium* require that the *algebraic sum* of the x and y components of all the forces acting on the particle be equal to zero. As a result, Eqs. 3–3 can be solved for at most two unknowns, generally represented as angles and magnitudes of forces shown on the particle's free-body diagram.

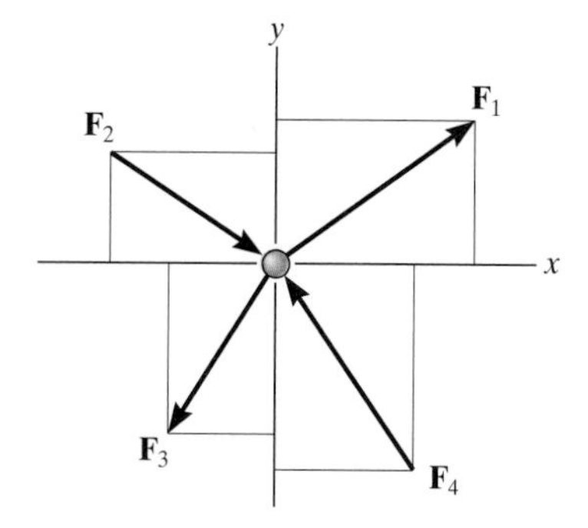

Fig. 3–4

Scalar Notation. Since each of the two equilibrium equations requires the resolution of vector components along a specified x or y axis, we will use scalar notation to represent the components when applying these equations. When doing this, the sense of direction for each component is accounted for by an *algebraic sign* which corresponds to the arrowhead direction of the component along each axis. If a force has an *unknown magnitude*, then the arrowhead sense of the force on the free-body diagram can be *assumed*. Since the magnitude of a force is *always positive*, then if the *solution* yields a *negative scalar*, this indicates that the sense of the force acts in the opposite direction.

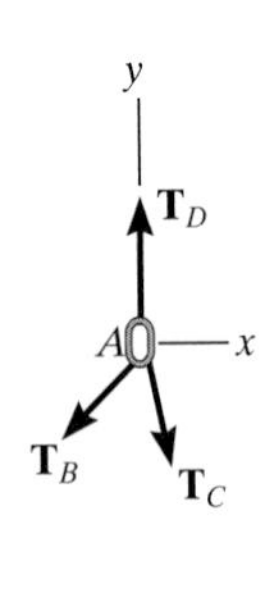

The chains exert three forces on the ring at A. The ring will not move, or will move with constant velocity, provided the summation of these forces along the x and along the y axis on the free-body diagram is zero. If one of the three forces is known, the magnitudes of the other two forces can be obtained from the two equations of equilibrium.

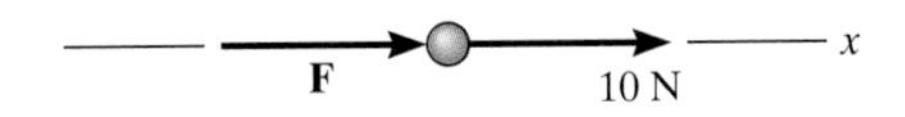

Fig. 3–5

For example, consider the free-body diagram of the particle subjected to the two forces shown in Fig. 3–5. Here it is *assumed* that the *unknown force* **F** acts to the right to maintain equilibrium. Applying the equation of equilibrium along the x axis, we have

$$\overset{+}{\rightarrow} \Sigma F_x = 0; \qquad +F + 10 \text{ N} = 0$$

Both terms are "positive" since both forces act in the positive x direction. When this equation is solved, $F = -10$ N. Here the *negative sign* indicates that **F** must act to the left to hold the particle in equilibrium, Fig. 3–5. Notice that if the $+x$ axis in Fig. 3–5 was directed to the left, both terms in the above equation would be negative, but again, after solving, $F = -10$ N, indicating again **F** would be directed to the left.

Procedure for Analysis

Coplanar force equilibrium problems for a particle can be solved using the following procedure.

Free-Body Diagram.

- Establish the x, y axes in any suitable orientation.
- Label all the known and unknown force magnitudes and directions on the diagram.
- The sense of a force having an unknown magnitude can be assumed.
- All forces on $AFBD$ will be concurrent.

Equations of Equilibrium.

- Apply the equations of equilibrium $\Sigma F_x = 0$ and $\Sigma F_y = 0$.
- Components are positive if they are directed along a positive axis, and negative if they are directed along a negative axis.
- If more than two unknowns exist and the problem involves a spring, apply $F = ks$ to relate the spring force to the deformation s of the spring.
- If the solution yields a negative result, this indicates the sense of the force is the reverse of that shown on the free-body diagram.

EXAMPLE 3.2

Determine the tension in cables AB and AD for equilibrium of the 250-kg engine shown in Fig. 3–6*a*.

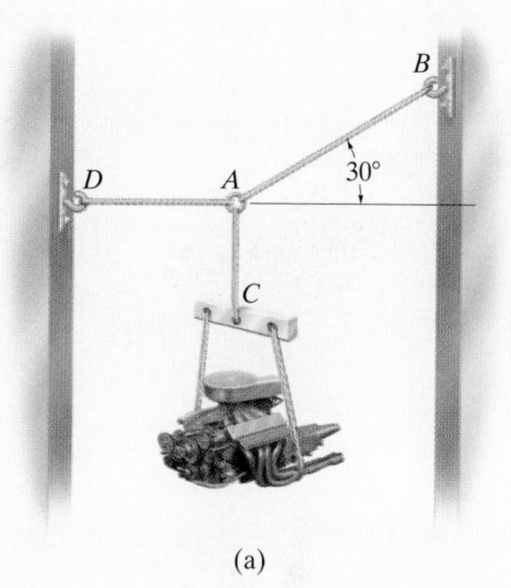

(a)

Fig. 3–6

SOLUTION

Free-Body Diagram. To solve this problem, we will investigate the equilibrium of the ring at A because this "particle" is subjected to the forces of both cables AB and AD. First, however, note that the engine has a weight $(250 \text{ kg})(9.81 \text{ m/s}^2) = 2.452 \text{ kN}$ which is supported by cable CA. Therefore, as shown in Fig. 3–6*b*, there are three concurrent forces *acting on the ring*. The forces $\mathbf{T}_B$ and $\mathbf{T}_D$ have unknown magnitudes but known directions, and cable AC exerts a downward force on A equal to 2.452 kN.

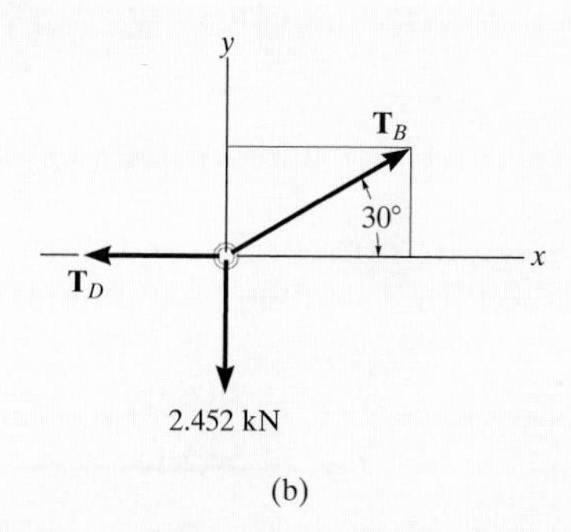

(b)

Equations of Equilibrium. The two unknown magnitudes T_B and T_D can be obtained from the two scalar equations of equilibrium, $\Sigma F_x = 0$ and $\Sigma F_y = 0$. To apply these equations, the x, y axes are established on the free-body diagram and $\mathbf{T}_B$ must be resolved into its x and y components. Thus,

$$\overset{+}{\rightarrow} \Sigma F_x = 0; \qquad T_B \cos 30^\circ - T_D = 0 \qquad (1)$$

$$+\uparrow \Sigma F_y = 0; \qquad T_B \sin 30^\circ - 2.452 \text{ kN} = 0 \qquad (2)$$

Solving Eq. 2 for T_B and substituting into Eq. 1 to obtain T_D yields

$$T_B = 4.90 \text{ kN} \qquad \textit{Ans.}$$

$$T_D = 4.25 \text{ kN} \qquad \textit{Ans.}$$

NOTE: The accuracy of these results, of course, depends on the accuracy of the data, i.e., measurements of geometry and loads. For most engineering work involving a problem such as this, the data as measured to three significant figures would be sufficient. Also, note that here we have neglected the weights of the cables, a reasonable assumption since they would be small in comparison with the weight of the engine.

EXAMPLE 3.3

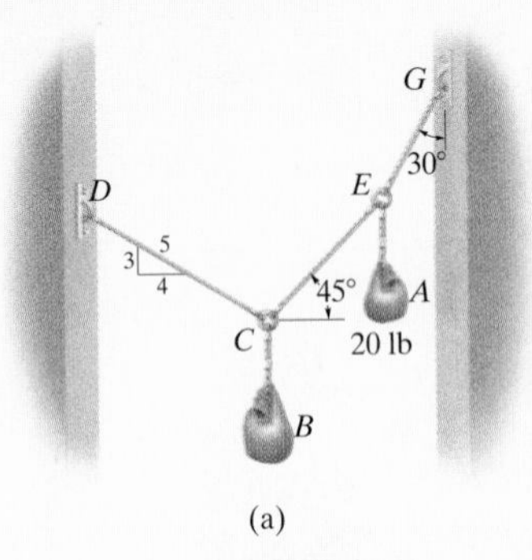

(a)

If the sack at A in Fig. 3–7a has a weight of 20 lb, determine the weight of the sack at B and the force in each cord needed to hold the system in the equilibrium position shown.

SOLUTION

Since the weight of A is known, the unknown tension in the two cords EG and EC can be determined by investigating the equilibrium of the ring at E. Why?

Free-Body Diagram. There are three forces acting on E, as shown in Fig. 3–7b.

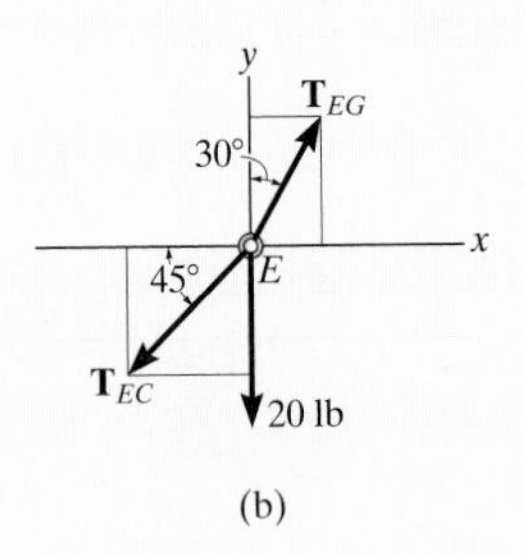

(b)

Equations of Equilibrium. Establishing the x, y axes and resolving each force into its x and y components using trigonometry, we have

$$\overset{+}{\rightarrow} \Sigma F_x = 0; \qquad T_{EG} \sin 30^\circ - T_{EC} \cos 45^\circ = 0 \qquad (1)$$

$$+\uparrow \Sigma F_y = 0; \qquad T_{EG} \cos 30^\circ - T_{EC} \sin 45^\circ - 20 \text{ lb} = 0 \qquad (2)$$

Solving Eq. 1 for T_{EG} in terms of T_{EC} and substituting the result into Eq. 2 allows a solution for T_{EC}. One then obtains T_{EG} from Eq. 1. The results are

$$T_{EC} = 38.6 \text{ lb} \qquad \textit{Ans.}$$

$$T_{EG} = 54.6 \text{ lb} \qquad \textit{Ans.}$$

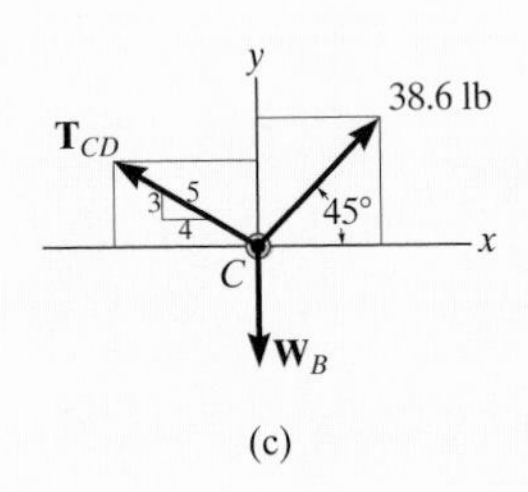

(c)

Using the calculated result for T_{EC}, the equilibrium of the ring at C can now be investigated to determine the tension in CD and the weight of B.

Free-Body Diagram. As shown in Fig. 3–7c, T_{EC} = 38.6 lb "pulls" on C. The reason for this becomes clear when one draws the free-body diagram of cord CE and applies both equilibrium and the principle of action, equal but opposite force reaction (Newton's third law), Fig. 3–7d.

Equations of Equilibrium. Establishing the x, y axes and noting the components of $\mathbf{T}_{CD}$ are proportional to the slope of the cord as defined by the 3–4–5 triangle, we have

$$\overset{+}{\rightarrow} \Sigma F_x = 0; \qquad 38.6 \cos 45^\circ \text{ lb} - \left(\tfrac{4}{5}\right)T_{CD} = 0 \qquad (3)$$

$$\overset{+}{\rightarrow} \Sigma F_y = 0; \qquad \left(\tfrac{3}{5}\right)T_{CD} + 38.6 \sin 45^\circ \text{ lb} - W_B = 0 \qquad (4)$$

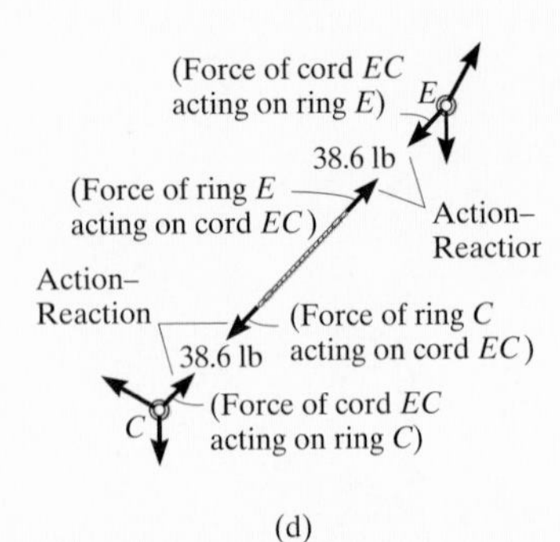

(d)

Fig. 3–7

Solving Eq. 3 and substituting the result into Eq. 4 yields

$$T_{CD} = 34.2 \text{ lb} \qquad \textit{Ans.}$$

$$W_B = 47.8 \text{ lb} \qquad \textit{Ans.}$$

NOTE: Review this example by constructing the free-body diagrams on a separate sheet and identifying each force.

EXAMPLE 3.4

Determine the required length of cord AC in Fig. 3–8*a* so that the 8-kg lamp is suspended in the position shown. The *undeformed* length of spring AB is $l'_{AB} = 0.4$ m, and the spring has a stiffness of $k_{AB} = 300$ N/m.

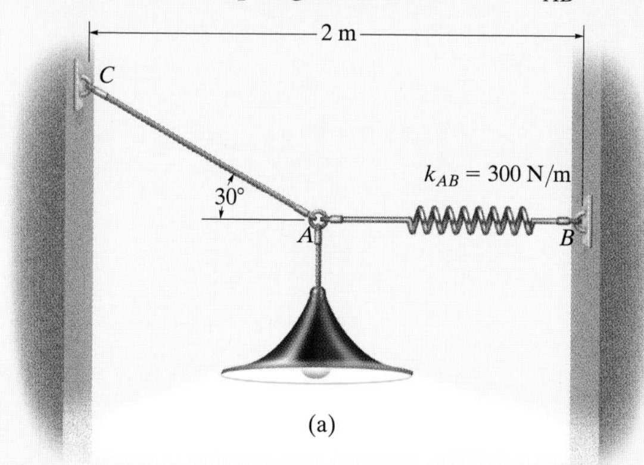

(a)

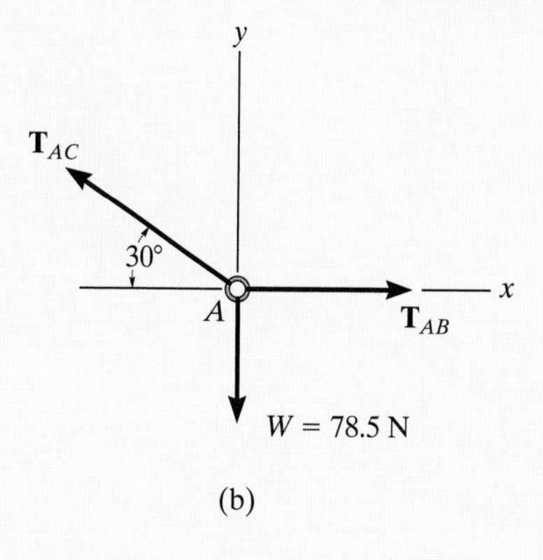

(b)

Fig. 3–8

SOLUTION

If the force in spring AB is known, the stretch of the spring can be found using $F = ks$. From the problem geometry, it is then possible to calculate the required length of AC.

Free-Body Diagram. The lamp has a weight $W = 8(9.81) = 78.5$ N. The free-body diagram of the ring at A is shown in Fig. 3–8*b*.

Equations of Equilibrium. Using the x, y axes,

$$\overset{+}{\rightarrow} \Sigma F_x = 0; \qquad T_{AB} - T_{AC} \cos 30^\circ = 0$$

$$+\uparrow \Sigma F_y = 0; \qquad T_{AC} \sin 30^\circ - 78.5 \text{ N} = 0$$

Solving, we obtain

$$T_{AC} = 157.0 \text{ N}$$
$$T_{AB} = 136.0 \text{ N}$$

The stretch of spring AB is therefore

$$T_{AB} = k_{AB} s_{AB}; \qquad 136.0 \text{ N} = 300 \text{ N/m}(s_{AB})$$
$$s_{AB} = 0.453 \text{ m}$$

so the stretched length is

$$l_{AB} = l'_{AB} + s_{AB}$$
$$l_{AB} = 0.4 \text{ m} + 0.453 \text{ m} = 0.853 \text{ m}$$

The horizontal distance from C to B, Fig. 3–8*a*, requires

$$2 \text{ m} = l_{AC} \cos 30^\circ + 0.853 \text{ m}$$
$$l_{AC} = 1.32 \text{ m} \qquad \textit{Ans.}$$

PROBLEMS

3–1. Determine the magnitudes of $\mathbf{F}_1$ and $\mathbf{F}_2$ so that the particle is in equilibrium.

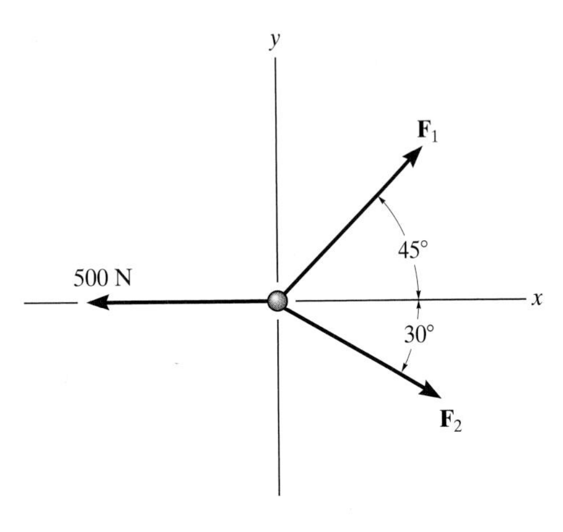

Prob. 3–1

3–2. Determine the magnitude and direction θ of $\mathbf{F}$ so that the particle is in equilibrium.

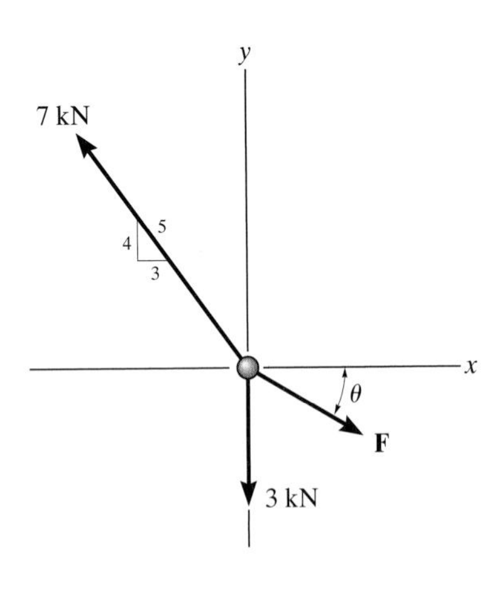

Prob. 3–2

3–3. Determine the magnitude of $\mathbf{F}$ and the orientation θ of the 750-N force so that the particle is in equilibrium.

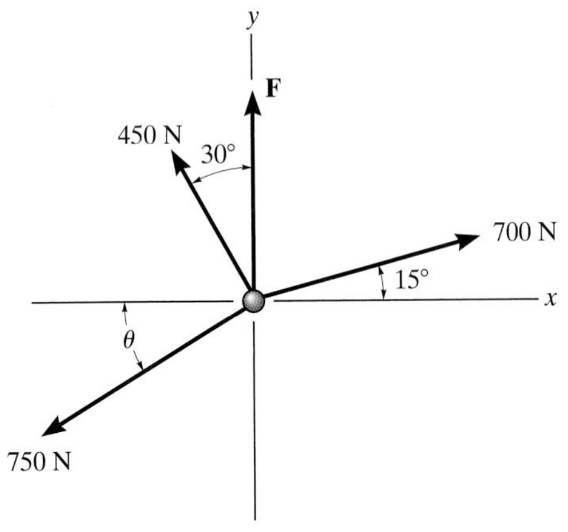

Prob. 3–3

***3–4.** Determine the magnitude and angle θ of $\mathbf{F}$ so that the particle is in equilibrium.

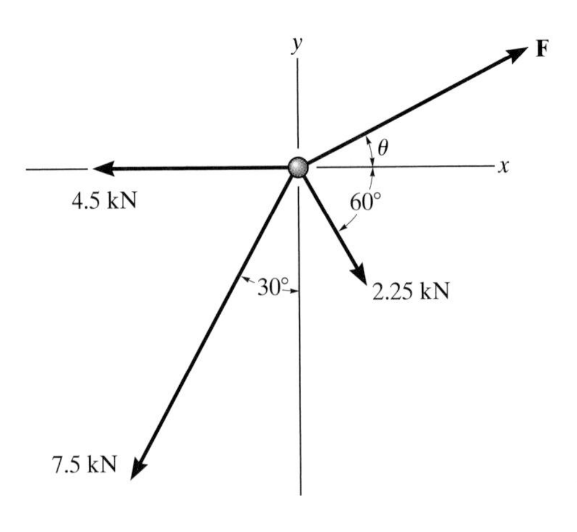

Prob. 3–4

3–5. The members of a truss are connected to the gusset plate. If the forces are concurrent at point O, determine the magnitudes of **F** and **T** for equilibrium. Take $\theta = 30°$.

3–6. The gusset plate is subjected to the forces of four members. Determine the force in member B and its proper orientation θ for equilibrium. The forces are concurrent at point O. Take $F = 12$ kN.

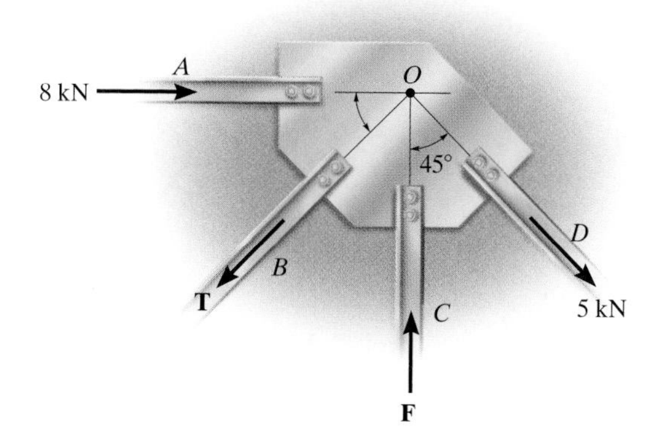

Probs. 3–5/6

3–7. Determine the maximum weight of the engine that can be supported without exceeding a tension of 450 lb in chain AB and 480 lb in chain AC.

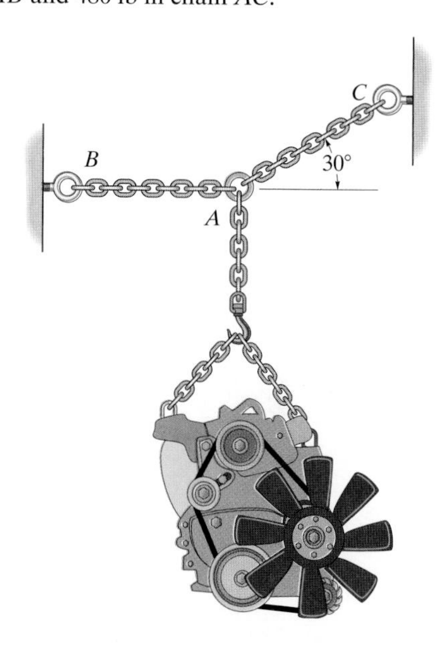

Prob. 3–7

***3–8.** The 200-kg engine is suspended from a vertical chain at A. A second chain is wrapped around the engine and held in position by the spreader bar BC. Determine the compressive force acting along the axis of the bar and the tension forces in segments BA and CA of the chain. *Hint:* Analyze equilibrium first at A, then at B.

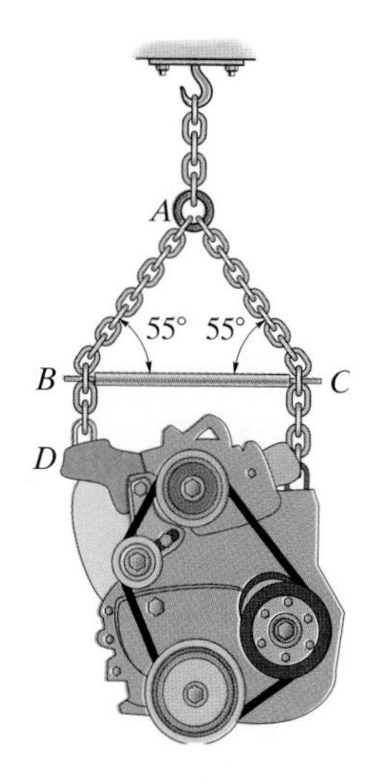

Prob. 3–8

3–9. Cords AB and AC can each sustain a maximum tension of 800 lb. If the drum has a weight of 900 lb, determine the smallest angle θ at which they can be attached to the drum.

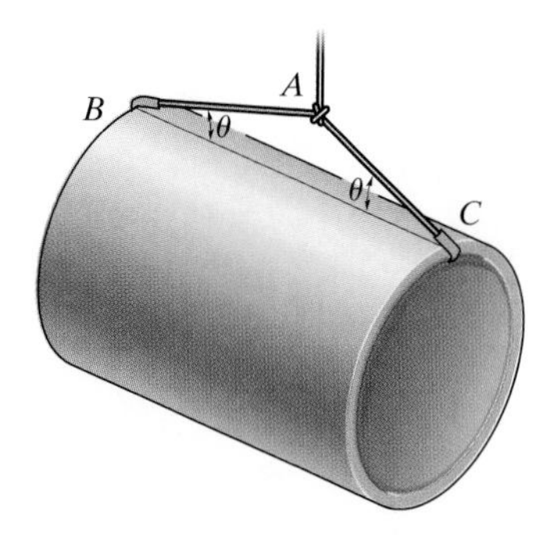

Prob. 3–9

3–10. The 500-lb crate is hoisted using the ropes AB and AC. Each rope can withstand a maximum tension of 2500 lb before it breaks. If AB always remains horizontal, determine the smallest angle θ to which the crate can be hoisted.

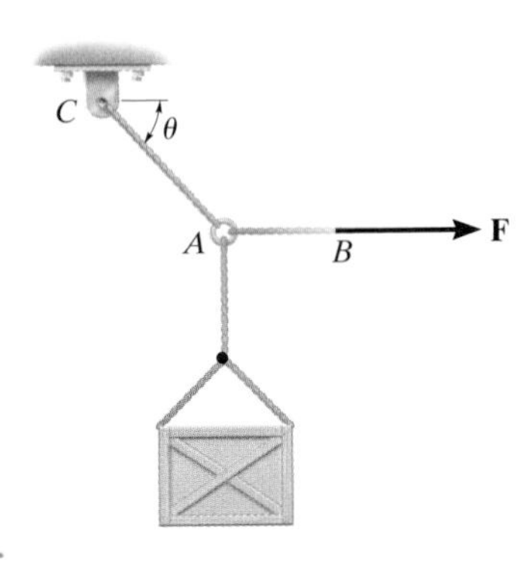

Prob. 3–10

3–11. Two electrically charged pith balls, each having a mass of 0.2 g, are suspended from light threads of equal length. Determine the resultant horizontal force of repulsion, F, acting on each ball if the measured distance between them is $r = 200$ mm.

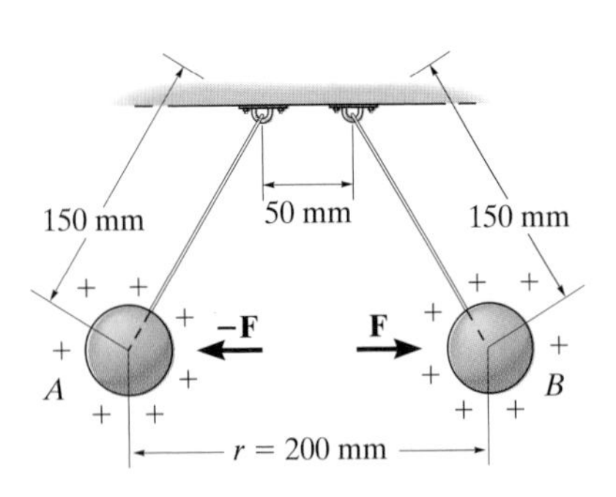

Prob. 3–11

***3–12.** The towing pendant AB is subjected to the force of 50 kN which is developed from a tugboat. Determine the force that is in each of the bridles, BC and BD, if the ship is moving forward with constant velocity.

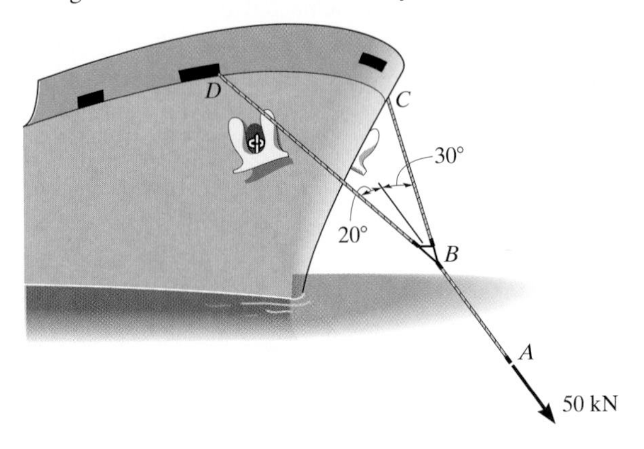

Prob. 3–12

3–13. Determine the stretch in each spring for equilibrium of the 2-kg block. The springs are shown in the equilibrium position.

3–14. The unstretched length of spring AB is 2 m. If the block is held in the equilibrium position shown, determine the mass of the block at D.

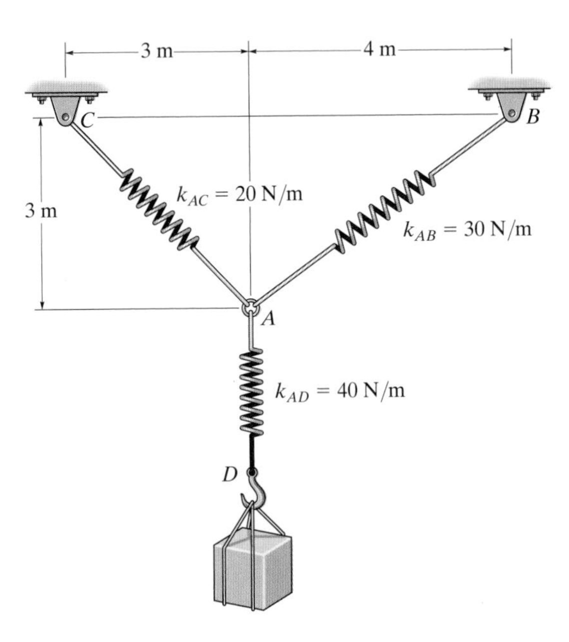

Probs. 3–13/14

3–15. The springs AB and BC have a stiffness of 500 N/m and an unstretched length of 6 m. Determine the horizontal force **F** applied to the cord which is attached to the *small* pulley B so that the displacement of the pulley from the wall is $d = 1.5$ m.

***3–16.** The springs AB and BC have a stiffness of 500 N/m and an unstretched length of 6 m. Determine the displacement d of the cord from the wall when a force $F = 175$ N is applied to the cord.

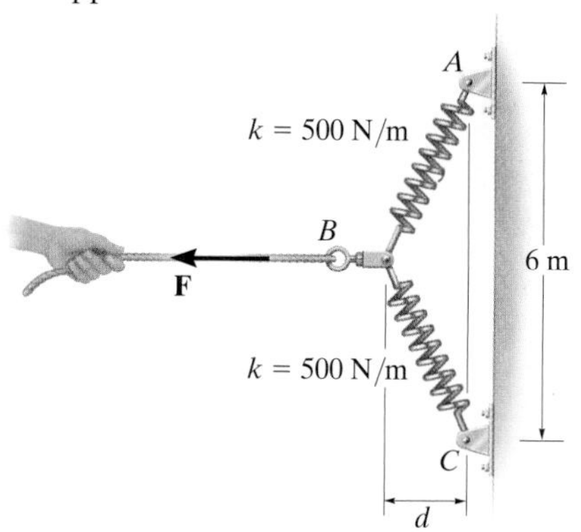

Probs. 3–15/16

3–17. Determine the force in each cable and the force **F** needed to hold the 4-kg lamp in the position shown. *Hint:* First analyze the equilibrium at B; then, using the result for the force in BC, analyze the equilibrium at C.

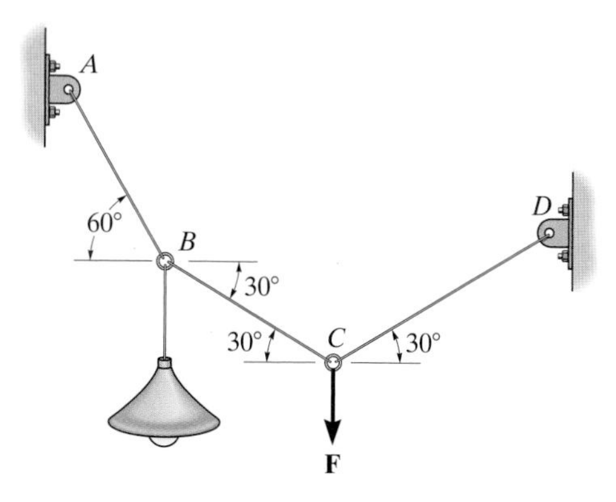

Prob. 3–17

3–18. The motor at B winds up the cord attached to the 65-lb crate with a constant speed. Determine the force in cord CD supporting the pulley and the angle θ for equilibrium. Neglect the size of the pulley at C.

3–19. The cords BCA and CD can each support a maximum load of 100 lb. Determine the maximum weight of the crate that can be hoisted at constant velocity, and the angle θ for equilibrium.

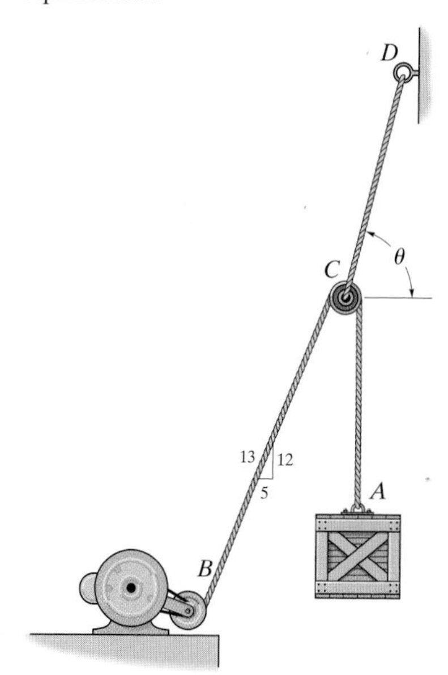

Probs. 3–18/19

***3–20.** The sack has a weight of 15 lb and is supported by the six cords tied together as shown. Determine the tension in each cord and the angle θ for equilibrium. Cord BC is horizontal.

3–21. Each cord can sustain a maximum tension of 200 lb. Determine the largest weight of the sack that can be supported. Also, determine θ of cord DC for equilibrium.

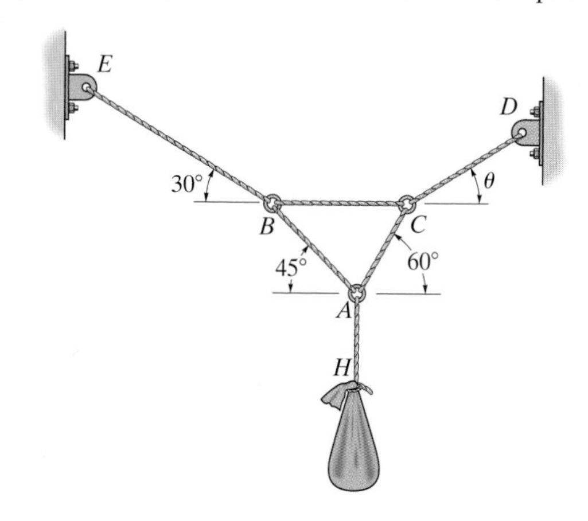

Probs. 3–20/21

3–22. The block has a weight of 20 lb and is being hoisted at uniform velocity. Determine the angle θ for equilibrium and the required force in each cord.

3–23. Determine the maximum weight W of the block that can be suspended in the position shown if each cord can support a maximum tension of 80 lb. Also, what is the angle θ for equilibrium?

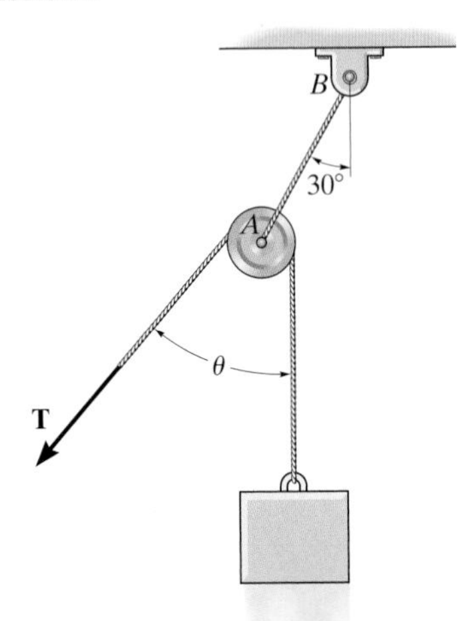

Probs. 3–22/23

3–25. Blocks D and F weigh 5 lb each and block E weighs 8 lb. Determine the sag s for equilibrium. Neglect the size of the pulleys.

3–26. If blocks D and F weigh 5 lb each, determine the weight of block E if the sag $s = 3$ ft. Neglect the size of the pulleys.

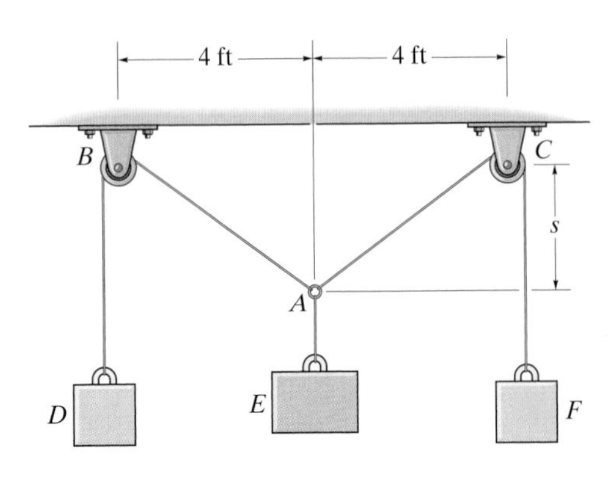

Probs. 3–25/26

***3–24.** Two spheres A and B have an equal mass and are electrostatically charged such that the repulsive force acting between them has a magnitude of 20 mN and is directed along line AB. Determine the angle θ, the tension in cords AC and BC, and the mass m of each sphere.

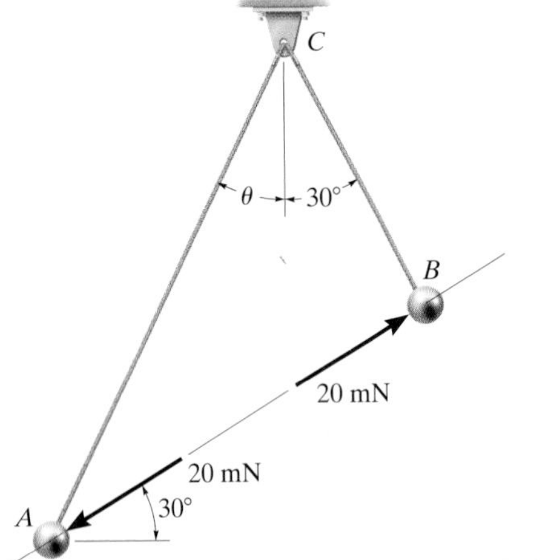

Prob. 3–24

3–27. The 30-kg block is supported by two springs having the stiffness shown. Determine the unstretched length of each spring after the block is removed.

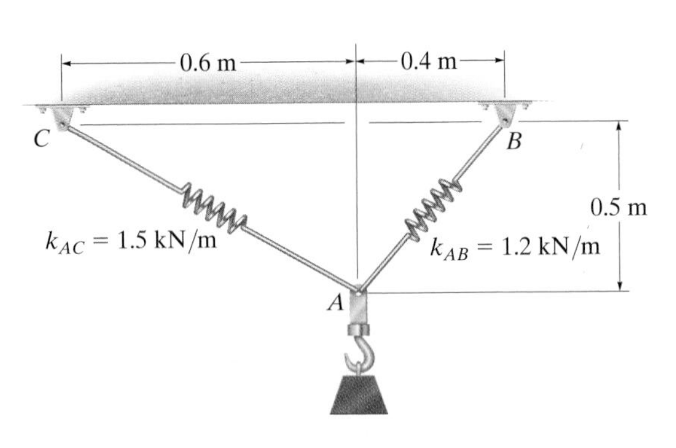

Prob. 3–27

***3–28.** Three blocks are supported using the cords and pulleys. If they have weights of $W_A = W$, $W_B = 0.25W$, and $W_C = W$, determine the angle θ for equilibrium.

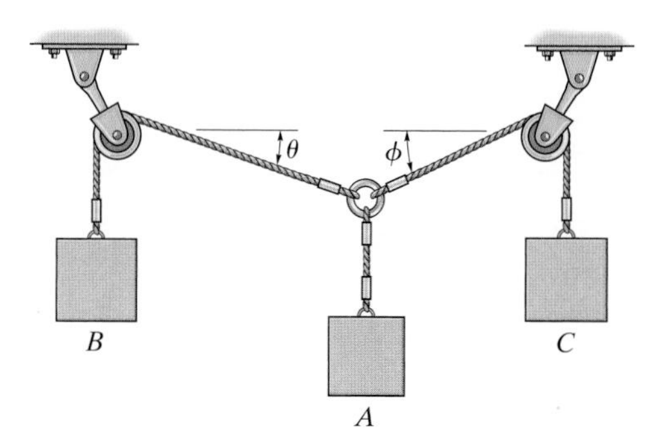

Prob. 3–28

3–29. A continuous cable of total length 4 m is wrapped around the *small* pulleys at A, B, C, and D. If each spring is stretched 300 mm, determine the mass m of each block. Neglect the weight of the pulleys and cords. The springs are unstretched when $d = 2$ m.

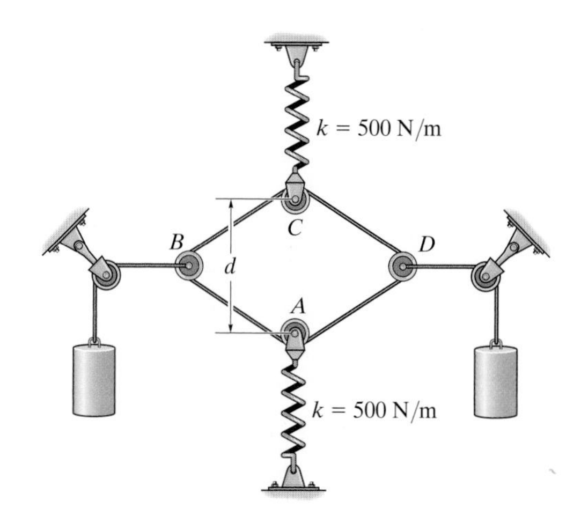

Prob. 3–29

3–30. Prove *Lami's theorem*, which states that if three concurrent forces are in equilibrium, each is proportional to the sine of the angle of the other two; that is, $P/\sin \alpha = Q/\sin \beta = R/\sin \gamma$.

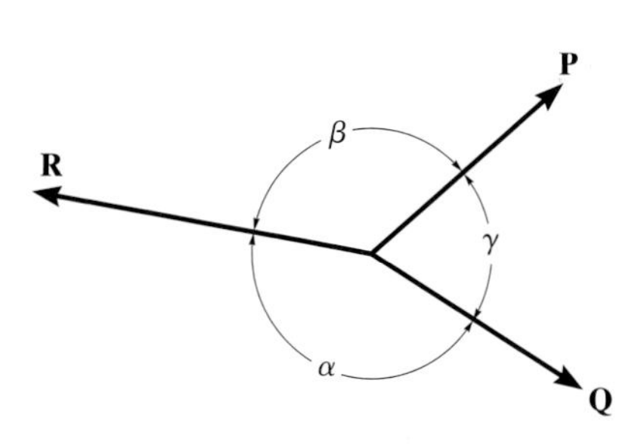

Prob. 3–30

■3–31. A vertical force $P = 10$ lb is applied to the ends of the 2-ft cord AB and spring AC. If the spring has an unstretched length of 2 ft, determine the angle θ for equilibrium. Take $k = 15$ lb/ft.

***3–32.** Determine the unstretched length of spring AC if a force $P = 80$ lb causes the angle $\theta = 60°$ for equilibrium. Cord AB is 2 ft long. Take $k = 50$ lb/ft.

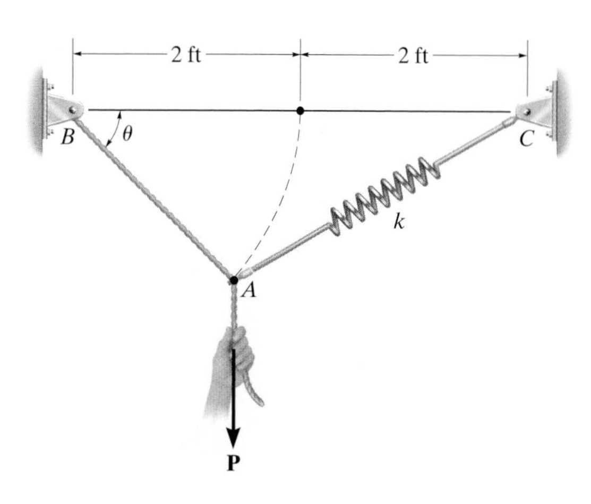

Probs. 3–31/32

■**3–33.** The 20-kg flowerpot is suspended from three wires and supported by the hooks at B and C. Determine the tension in AB and AC for equilibrium.

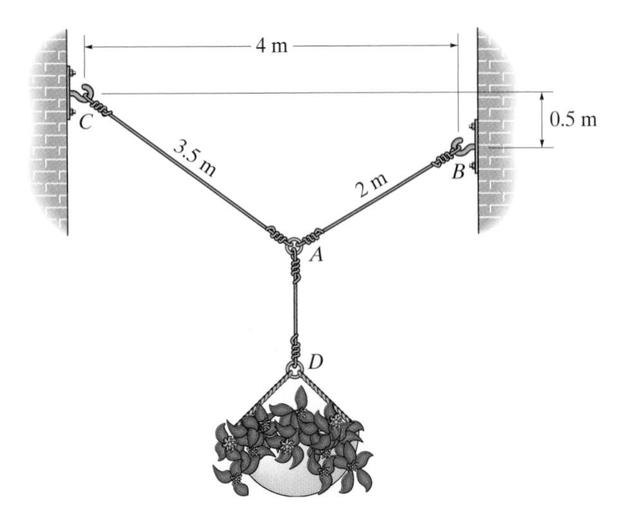

Prob. 3–33

■**3–34.** A car is to be towed using the rope arrangement shown. The towing force required is 600 lb. Determine the minimum length l of rope AB so that the tension in either rope AB or AC does not exceed 750 lb. *Hint:* Use the equilibrium condition at point A to determine the required angle θ for attachment, then determine l using trigonometry applied to triangle ABC.

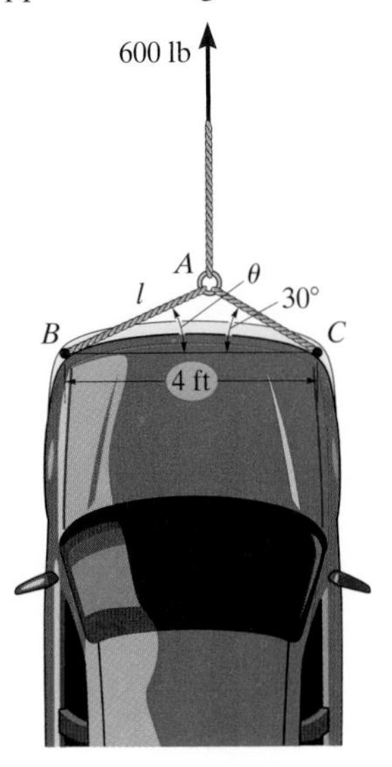

Prob. 3–34

3–35. Determine the mass of each of the two cylinders if they cause a sag of $s = 0.5$ m when suspended from the rings at A and B. Note that $s = 0$ when the cylinders are removed.

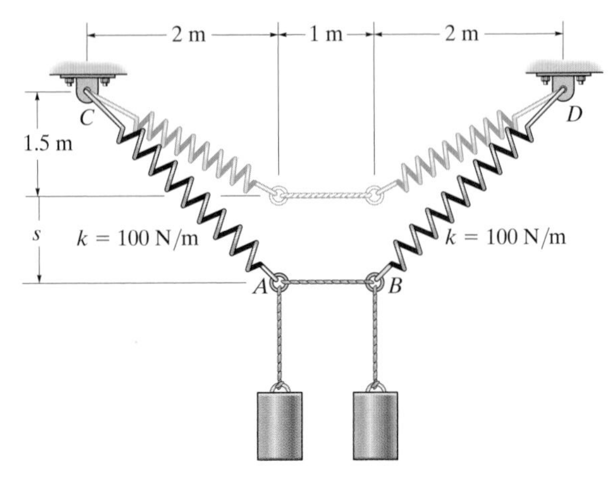

Prob. 3–35

***3–36.** The sling BAC is used to lift the 100-lb load with constant velocity. Determine the force in the sling and plot its value T (ordinate) as a function of its orientation θ, where $0 \leq \theta \leq 90°$.

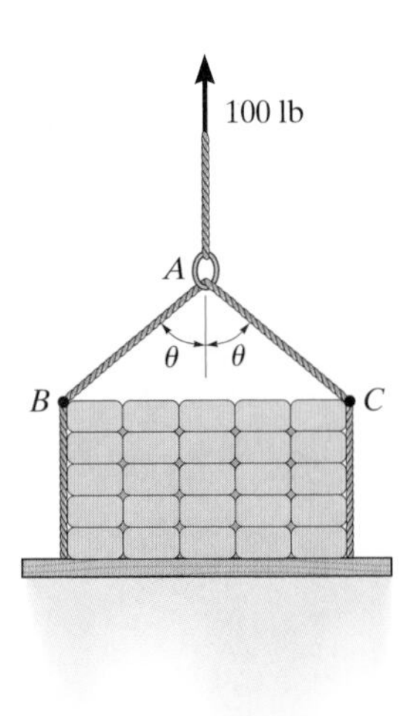

Prob. 3–36

■**3–37.** The 10-lb lamp fixture is suspended from two springs, each having an unstretched length of 4 ft and stiffness of $k = 5$ lb/ft. Determine the angle θ for equilibrium.

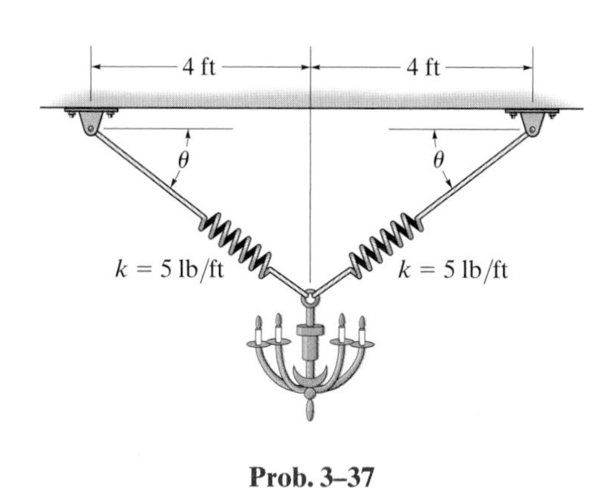

Prob. 3–37

3–38. The 200-lb uniform tank is suspended by means of a 6-ft-long cable, which is attached to the sides of the tank and passes over the small pulley located at O. If the cable can be attached at either points A and B, or C and D, determine which attachment produces the least amount of tension in the cable. What is this tension?

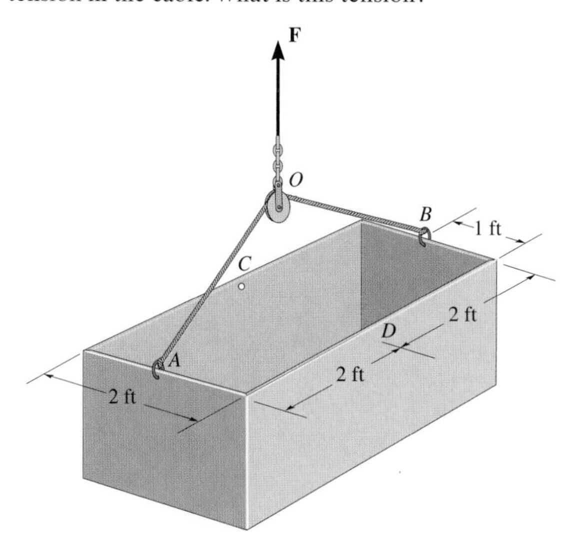

Prob. 3–38

3–39. A 4-kg sphere rests on the smooth parabolic surface. Determine the normal force it exerts on the surface and the mass m_B of block B needed to hold it in the equilibrium position shown.

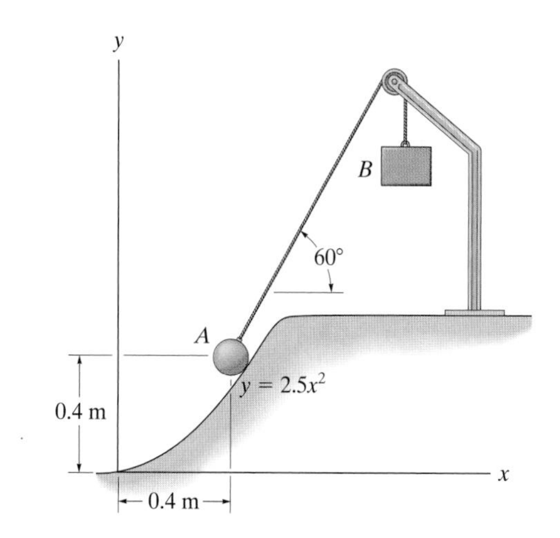

Prob. 3–39

***3–40.** The 30-kg pipe is supported at A by a system of five cords. Determine the force in each cord for equilibrium.

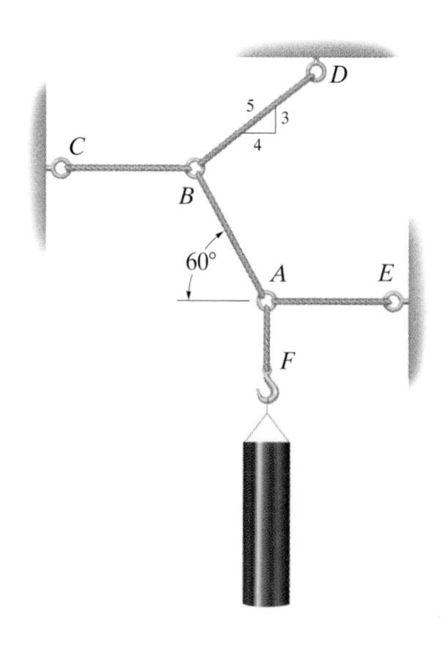

Prob. 3–40

3.4 Three-Dimensional Force Systems

For particle equilibrium we require

$$\Sigma \mathbf{F} = \mathbf{0} \tag{3–4}$$

If the forces are resolved into their respective **i**, **j**, **k** components, Fig. 3–9, then we have

$$\Sigma F_x \mathbf{i} + \Sigma F_y \mathbf{j} + \Sigma F_z \mathbf{k} = \mathbf{0}$$

To ensure equilibrium, we must therefore require that the following three scalar component equations be satisfied:

$$\begin{aligned} \Sigma F_x &= 0 \\ \Sigma F_y &= 0 \\ \Sigma F_z &= 0 \end{aligned} \tag{3–5}$$

These equations represent the *algebraic sums* of the x, y, z force components acting on the particle. Using them we can solve for at most three unknowns, generally represented as angles or magnitudes of forces shown on the particle's free-body diagram.

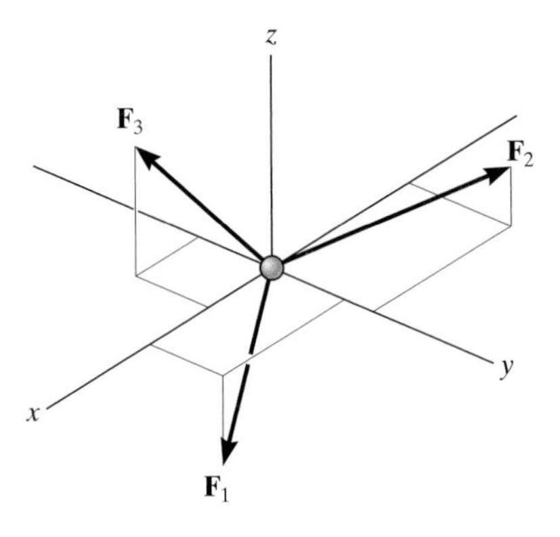

Fig. 3–9

The ring at A is subjected to the force from the hook as well as forces from each of the three chains. If the electromagnet and its load have a weight W, then the force at the hook will be $\mathbf{W}$, and the three scalar equations of equilibrium can be applied to the free-body diagram of the ring in order to determine the chain forces, $\mathbf{F}_B$, $\mathbf{F}_C$ and $\mathbf{F}_D$.

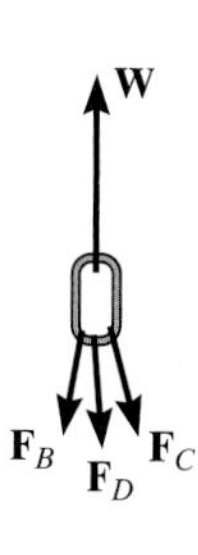

Procedure for Analysis

Three-dimensional force equilibrium problems for a particle can be solved using the following procedure.

Free-Body Diagram.

- Establish the x, y, z axes in any suitable orientation.
- Label all the known and unknown force magnitudes and directions on the diagram.
- The sense of a force having an unknown magnitude can be assumed.

Equations of Equilibrium.

- Use the scalar equations of equilibrium, $\Sigma F_x = 0$, $\Sigma F_y = 0$, $\Sigma F_z = 0$, in cases where it is easy to resolve each force into its x, y, z components.
- If the three-dimensional geometry appears difficult, then first express each force as a Cartesian vector, substitute these vectors into $\Sigma \mathbf{F} = \mathbf{0}$, and then set the $\mathbf{i}, \mathbf{j}, \mathbf{k}$ components equal to zero.
- If the solution yields a negative result, this indicates the sense of the force is the reverse of that shown on the free-body diagram.

EXAMPLE 3.5

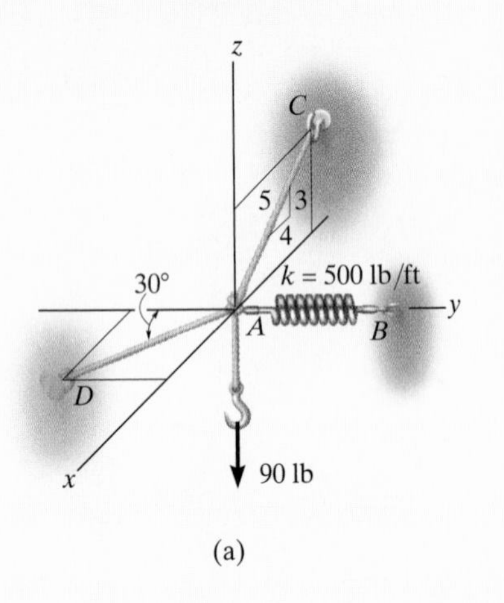

(a)

A 90-lb load is suspended from the hook shown in Fig. 3–10*a*. The load is supported by two cables and a spring having a stiffness $k = 500$ lb/ft. Determine the force in the cables and the stretch of the spring for equilibrium. Cable *AD* lies in the *x–y* plane and cable *AC* lies in the *x–z* plane.

SOLUTION

The stretch of the spring can be determined once the force in the spring is determined.

Free-Body Diagram. The connection at *A* is chosen for the equilibrium analysis since the cable forces are concurrent at this point. The free-body diagram is shown in Fig. 3–10*b*.

Equations of Equilibrium. By inspection, each force can easily be resolved into its *x, y, z* components, and therefore the three scalar equations of equilibrium can be directly applied. Considering components directed along the positive axes as "positive," we have

$$\Sigma F_x = 0; \qquad F_D \sin 30^\circ - \tfrac{4}{5}F_C = 0 \qquad (1)$$

$$\Sigma F_y = 0; \qquad -F_D \cos 30^\circ + F_B = 0 \qquad (2)$$

$$\Sigma F_z = 0; \qquad \tfrac{3}{5}F_C - 90 \text{ lb} = 0 \qquad (3)$$

Solving Eq. 3 for F_C, then Eq. 1 for F_D, and finally Eq. 2 for F_B, yields

$$F_C = 150 \text{ lb} \qquad \textit{Ans.}$$

$$F_D = 240 \text{ lb} \qquad \textit{Ans.}$$

$$F_B = 208 \text{ lb} \qquad \textit{Ans.}$$

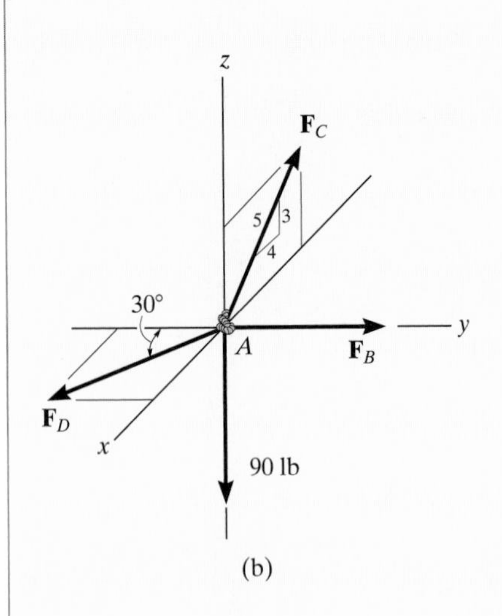

(b)

Fig. 3–10

The stretch of the spring is therefore

$$F_B = ks_{AB}$$

$$208 \text{ lb} = 500 \text{ lb/ft}(s_{AB})$$

$$s_{AB} = 0.416 \text{ ft} \qquad \textit{Ans.}$$

NOTE: Since the results for all the cable forces are positive, each cable is in tension; that is, it pulls on point *A* as expected, Fig. 3–10*b*.

EXAMPLE 3.6

Determine the magnitude and coordinate direction angles of force **F** in Fig. 3–11*a* that are required for equilibrium of particle *O*.

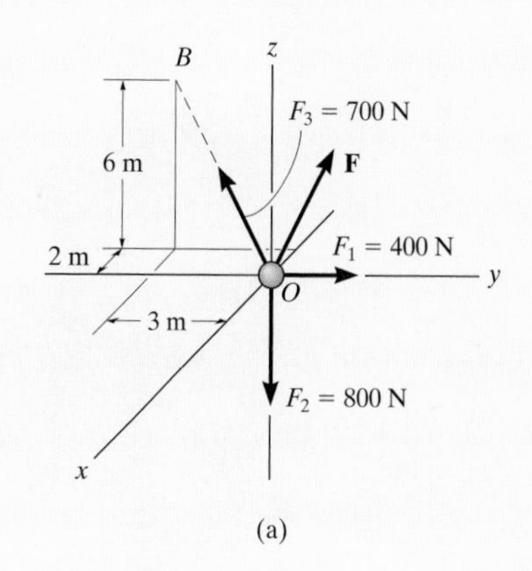

(a)

SOLUTION

Free-Body Diagram. Four forces act on particle *O*, Fig. 3–11*b*.

Equations of Equilibrium. Each of the forces can be expressed in Cartesian vector form, and the equations of equilibrium can be applied to determine the *x, y, z* components of **F**. Noting that the coordinates of *B* are $B(-2\text{ m}, -3\text{ m}, 6\text{ m})$, we have

$$\mathbf{F}_1 = \{400\mathbf{j}\}\text{ N}$$
$$\mathbf{F}_2 = \{-800\mathbf{k}\}\text{ N}$$
$$\mathbf{F}_3 = F_3\left(\frac{\mathbf{r}_B}{r_B}\right) = 700\text{ N}\left[\frac{-2\mathbf{i} - 3\mathbf{j} + 6\mathbf{k}}{\sqrt{(-2)^2 + (-3)^2 + (6)^2}}\right]$$
$$= \{-200\mathbf{i} - 300\mathbf{j} + 600\mathbf{k}\}\text{ N}$$
$$\mathbf{F} = F_x\mathbf{i} + F_y\mathbf{j} + F_z\mathbf{k}$$

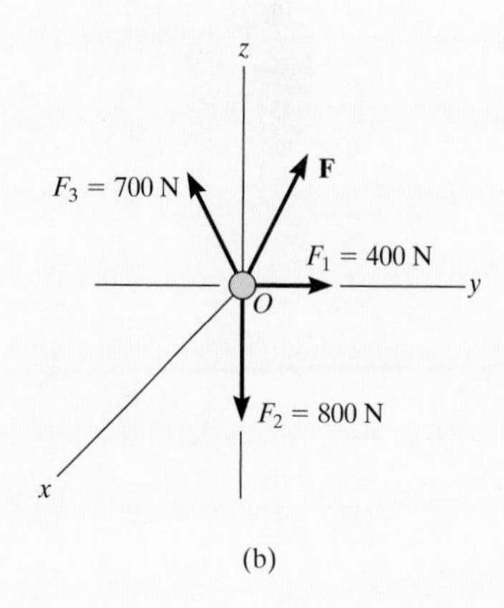

(b)

For equilibrium

$$\Sigma\mathbf{F} = \mathbf{0}; \qquad \mathbf{F}_1 + \mathbf{F}_2 + \mathbf{F}_3 + \mathbf{F} = \mathbf{0}$$
$$400\mathbf{j} - 800\mathbf{k} - 200\mathbf{i} - 300\mathbf{j} + 600\mathbf{k} + F_x\mathbf{i} + F_y\mathbf{j} + F_z\mathbf{k} = \mathbf{0}$$

Equating the respective **i**, **j**, **k** components to zero, we have

$$\Sigma F_x = 0; \qquad -200 + F_x = 0 \qquad F_x = 200\text{ N}$$
$$\Sigma F_y = 0; \qquad 400 - 300 + F_y = 0 \qquad F_y = -100\text{ N}$$
$$\Sigma F_z = 0; \qquad -800 + 600 + F_z = 0 \qquad F_z = 200\text{ N}$$

Thus,

$$\mathbf{F} = \{200\mathbf{i} - 100\mathbf{j} + 200\mathbf{k}\}\text{ N}$$
$$F = \sqrt{(200)^2 + (-100)^2 + (200)^2} = 300\text{ N} \qquad \textit{Ans.}$$
$$\mathbf{u}_F = \frac{\mathbf{F}}{F} = \frac{200}{300}\mathbf{i} - \frac{100}{300}\mathbf{j} + \frac{200}{300}\mathbf{k}$$
$$\alpha = \cos^{-1}\left(\frac{200}{300}\right) = 48.2° \qquad \textit{Ans.}$$
$$\beta = \cos^{-1}\left(\frac{-100}{300}\right) = 109° \qquad \textit{Ans.}$$
$$\gamma = \cos^{-1}\left(\frac{200}{300}\right) = 48.2° \qquad \textit{Ans.}$$

The magnitude and correct direction of **F** are shown in Fig. 3–11*c*.

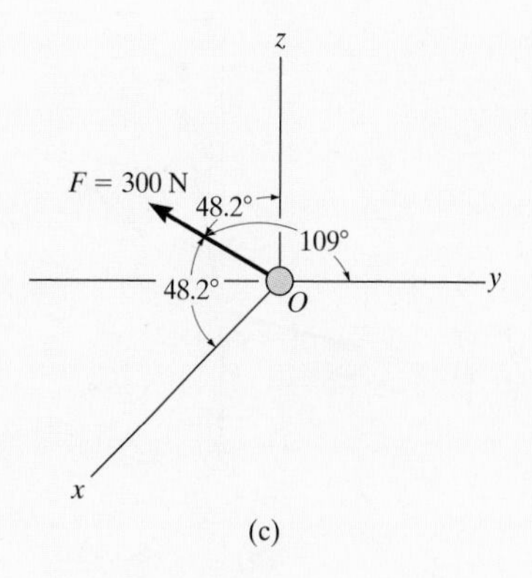

(c)

Fig. 3–11

EXAMPLE 3.7

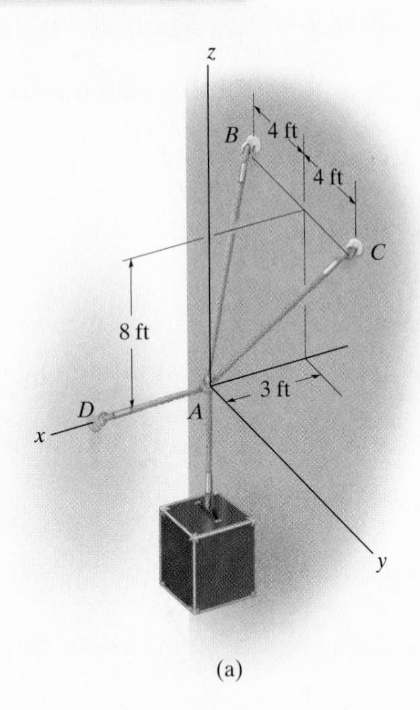

(a)

Determine the force developed in each cable used to support the 40-lb crate shown in Fig. 3–12*a*.

SOLUTION

Free-Body Diagram. As shown in Fig. 3–12*b*, the free-body diagram of point *A* is considered in order to "expose" the three unknown forces in the cables.

Equations of Equilibrium. First we will express each force in Cartesian vector form. Since the coordinates of points *B* and *C* are $B(-3 \text{ ft}, -4 \text{ ft}, 8 \text{ ft})$ and $C(-3 \text{ ft}, 4 \text{ ft}, 8 \text{ ft})$, we have

$$\mathbf{F}_B = F_B\left[\frac{-3\mathbf{i} - 4\mathbf{j} + 8\mathbf{k}}{\sqrt{(-3)^2 + (-4)^2 + (8)^2}}\right]$$

$$= -0.318F_B\mathbf{i} - 0.424F_B\mathbf{j} + 0.848F_B\mathbf{k}$$

$$\mathbf{F}_C = F_C\left[\frac{-3\mathbf{i} + 4\mathbf{j} + 8\mathbf{k}}{\sqrt{(-3)^2 + (4)^2 + (8)^2}}\right]$$

$$= -0.318F_C\mathbf{i} + 0.424F_C\mathbf{j} + 0.484F_C\mathbf{k}$$

$$\mathbf{F}_D = F_D\mathbf{i}$$

$$\mathbf{W} = \{-40\mathbf{k}\} \text{ lb}$$

Equilibrium requires

$$\Sigma\mathbf{F} = \mathbf{0}; \qquad \mathbf{F}_B + \mathbf{F}_C + \mathbf{F}_D + \mathbf{W} = \mathbf{0}$$

$$-0.318F_B\mathbf{i} - 0.424F_B\mathbf{j} + 0.848F_B\mathbf{k} - 0.318F_C\mathbf{i} + 0.424F_C\mathbf{j} + 0.848F_C\mathbf{k} + F_D\mathbf{i} - 40\mathbf{k} = \mathbf{0}$$

Equating the respective **i**, **j**, **k** components to zero yields

$$\Sigma F_x = 0; \qquad -0.318F_B - 0.318F_C + F_D = 0 \qquad (1)$$

$$\Sigma F_y = 0; \qquad -0.424F_B + 0.424F_C = 0 \qquad (2)$$

$$\Sigma F_z = 0; \qquad 0.848F_B + 0.848F_C - 40 = 0 \qquad (3)$$

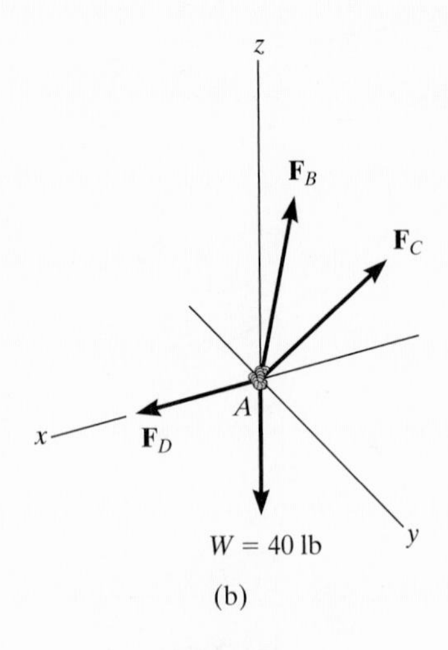

(b)

Fig. 3–12

Equation 2 states that $F_B = F_C$. Thus, solving Eq. 3 for F_B and F_C and substituting the result into Eq. 1 to obtain F_D, we have

$$F_B = F_C = 23.6 \text{ lb} \qquad \textit{Ans.}$$

$$F_D = 15.0 \text{ lb} \qquad \textit{Ans.}$$

NOTE: Do you understand why each cable force is in tension? See Fig. 3–7*d*.

EXAMPLE 3.8

The 100-kg crate shown in Fig. 3–13a is supported by three cords, one of which is connected to a spring. Determine the tension in cords AC and AD and the stretch of the spring.

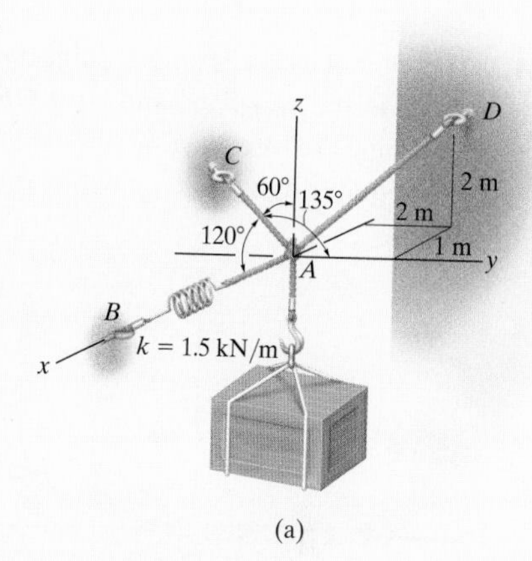

(a)

SOLUTION

Free-Body Diagram. The force in each of the cords can be determined by investigating the equilibrium of point A. The free-body diagram is shown in Fig. 3–13b. The weight of the crate is $W = 100(9.81) = 981$ N.

Equations of Equilibrium. Each vector on the free-body diagram is first expressed in Cartesian vector form. Using Eq. 2–11 for $\mathbf{F}_C$ and noting point $D(-1\text{ m}, 2\text{ m}, 2\text{ m})$ for $\mathbf{F}_D$, we have

$$\mathbf{F}_B = F_B\mathbf{i}$$

$$\mathbf{F}_C = F_C \cos 120°\mathbf{i} + F_C \cos 135°\mathbf{j} + F_C \cos 60°\mathbf{k}$$

$$= -0.5F_C\mathbf{i} - 0.707F_C\mathbf{j} + 0.5F_C\mathbf{k}$$

$$\mathbf{F}_D = F_D\left[\frac{-1\mathbf{i} + 2\mathbf{j} + 2\mathbf{k}}{\sqrt{(-1)^2 + (2)^2 + (2)^2}}\right]$$

$$= -0.333F_D\mathbf{i} + 0.667F_D\mathbf{j} + 0.667F_D\mathbf{k}$$

$$\mathbf{W} = \{-981\mathbf{k}\}\text{ N}$$

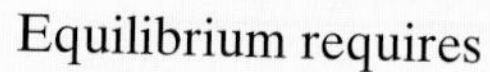

Equilibrium requires

$$\Sigma\mathbf{F} = \mathbf{0}; \qquad \mathbf{F}_B + \mathbf{F}_C + \mathbf{F}_D + \mathbf{W} = \mathbf{0}$$

$$F_B\mathbf{i} - 0.5F_C\mathbf{i} - 0.707F_C\mathbf{j} + 0.5F_C\mathbf{k} - 0.333F_D\mathbf{i} + 0.667F_D\mathbf{j} + 0.667F_D\mathbf{k} - 981\mathbf{k} = \mathbf{0}$$

Equating the respective **i**, **j**, **k** components to zero,

$$\Sigma F_x = 0; \qquad F_B - 0.5F_C - 0.333F_D = 0 \qquad (1)$$

$$\Sigma F_y = 0; \qquad -0.707F_C + 0.667F_D = 0 \qquad (2)$$

$$\Sigma F_z = 0; \qquad 0.5F_C + 0.667F_D - 981 = 0 \qquad (3)$$

Solving Eq. 2 for F_D in terms of F_C and substituting into Eq. 3 yields F_C. F_D is determined from Eq. 2. Finally, substituting the results into Eq. 1 gives F_B. Hence,

$$F_C = 813\text{ N} \qquad \textit{Ans.}$$

$$F_D = 862\text{ N} \qquad \textit{Ans.}$$

$$F_B = 693.7\text{ N}$$

The stretch of the spring is therefore

$$F = ks; \qquad 693.7 = 1500s$$

$$s = 0.462\text{ m} \qquad \textit{Ans.}$$

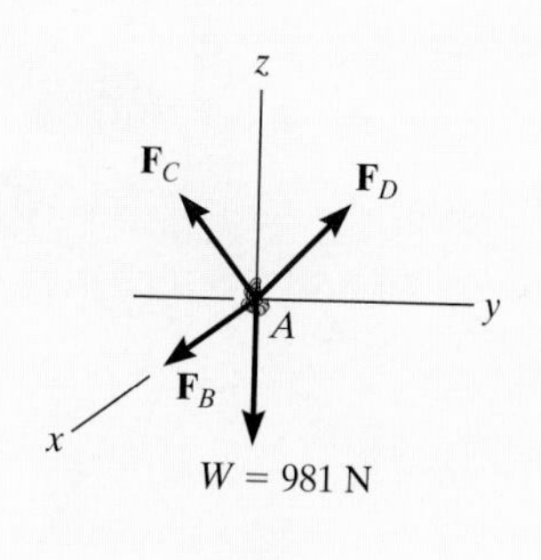

(b)

Fig. 3–13

PROBLEMS

3–41. The joint of a space frame is subjected to four forces. Strut OA lies in the x–y plane and strut OB lies in the y–z plane. Determine the forces acting in each of the three struts required for equilibrium.

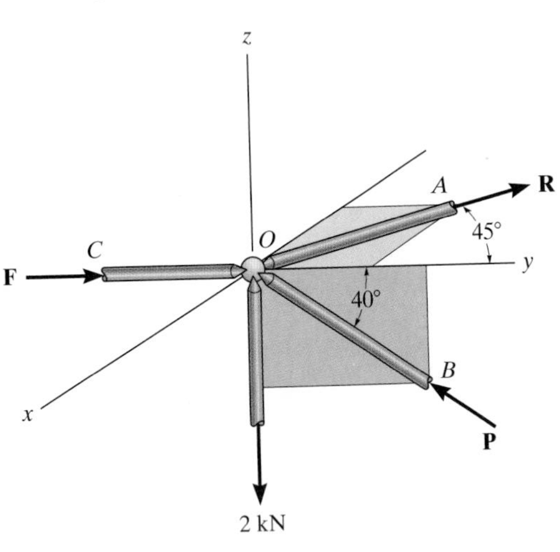

Prob. 3–41

3–42. Determine the magnitudes of $\mathbf{F}_1$, $\mathbf{F}_2$, and $\mathbf{F}_3$ for equilibrium of the particle.

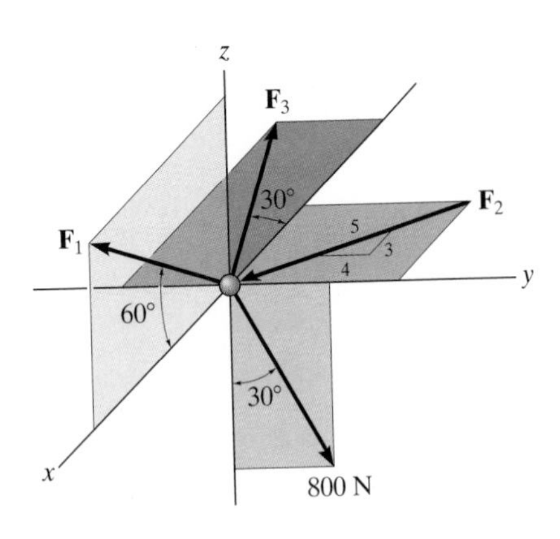

Prob. 3–42

3–43. Determine the magnitudes of $\mathbf{F}_1$, $\mathbf{F}_2$, and $\mathbf{F}_3$ for equilibrium of the particle.

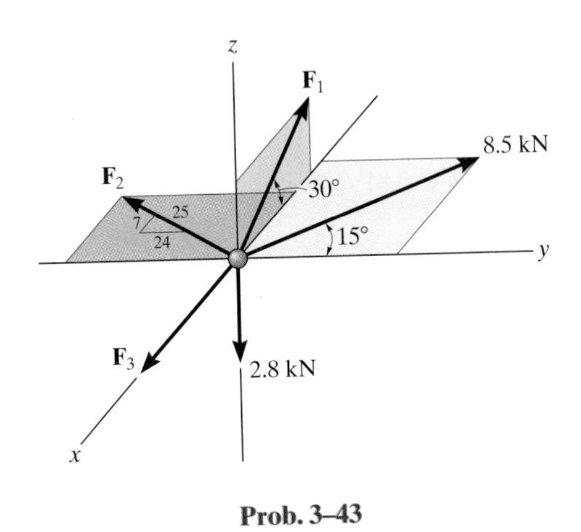

Prob. 3–43

*■**3–44.** Determine the magnitudes of $\mathbf{F}_1$, $\mathbf{F}_2$, and $\mathbf{F}_3$ for equilibrium of the particle. $\mathbf{F} = \{-9\mathbf{i} - 8\mathbf{j} - 5\mathbf{k}\}$ kN.

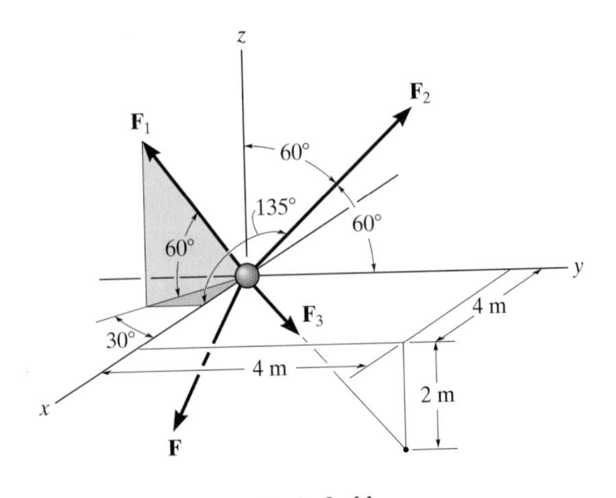

Prob. 3–44

3–45. The three cables are used to support the 800-N lamp. Determine the force developed in each cable for equilibrium.

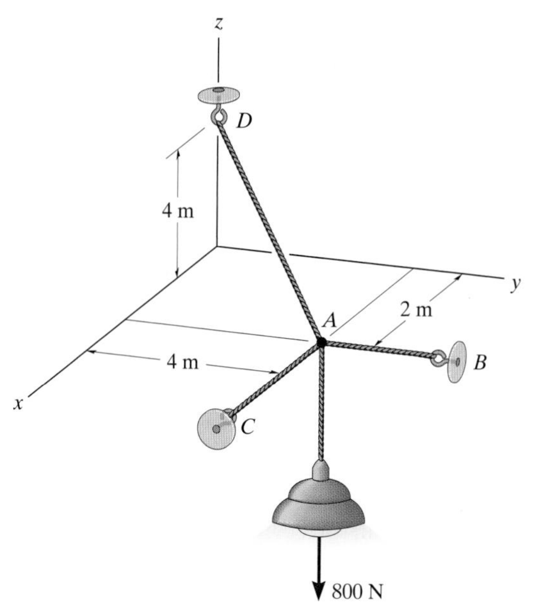

Prob. 3–45

3–46. Determine the force in each cable needed to support the 500-lb load.

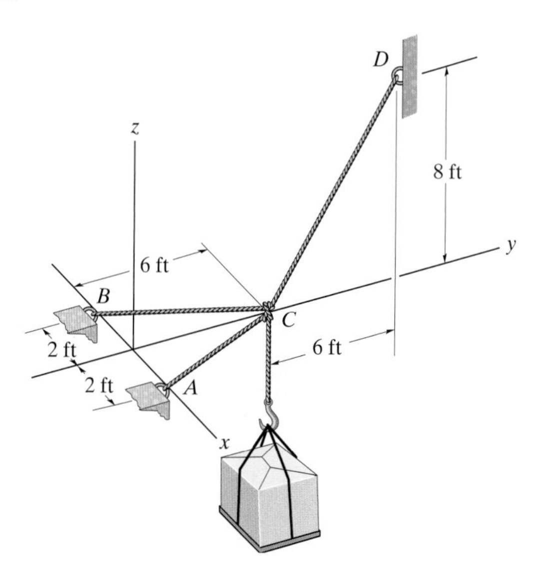

Prob. 3–46

3–47. Determine the stretch in each of the two springs required to hold the 20-kg crate in the equilibrium position shown. Each spring has an unstretched length of 2 m and a stiffness of $k = 300$ N/m.

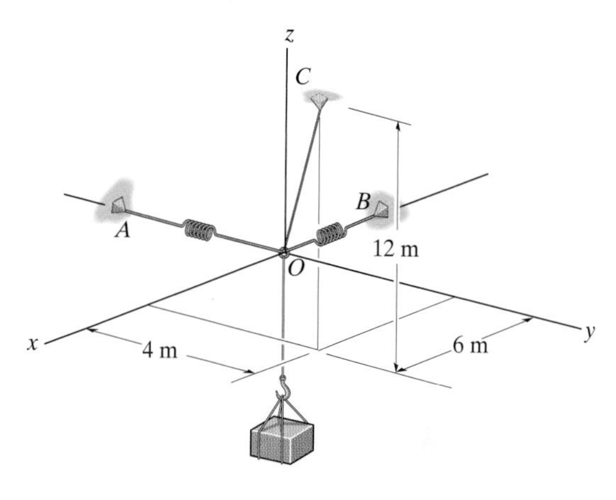

Prob. 3–47

***3–48.** If the bucket and its contents have a total weight of 20 lb, determine the force in the supporting cables DA, DB, and DC.

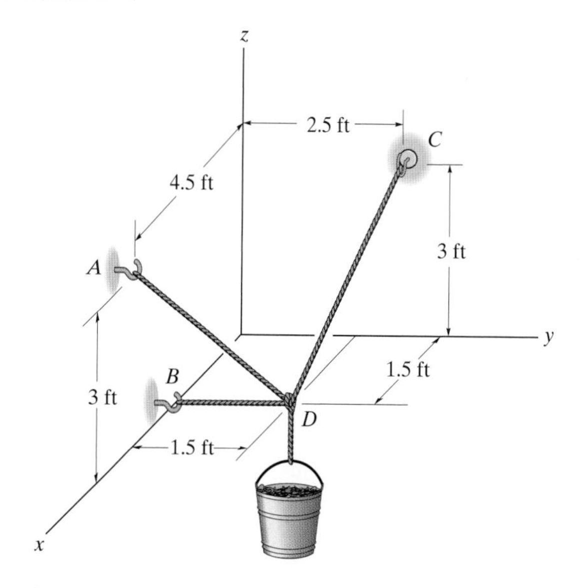

Prob. 3–48

■3–49. The 2500-N crate is to be hoisted with constant velocity from the hold of a ship using the cable arrangement shown. Determine the tension in each of the three cables for equilibrium.

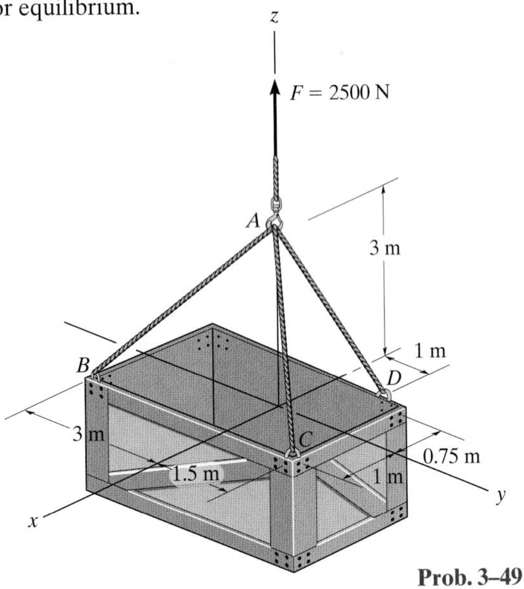

Prob. 3–49

■3–50. The lamp has a mass of 15 kg and is supported by a pole AO and cables AB and AC. If the force in the pole acts along its axis, determine the forces in AO, AB, and AC for equilibrium.

3–51. Cables AB and AC can sustain a maximum tension of 500 N, and the pole can support a maximum compression of 300 N. Determine the maximum weight of the lamp that can be supported in the position shown. The force in the pole acts along the axis of the pole.

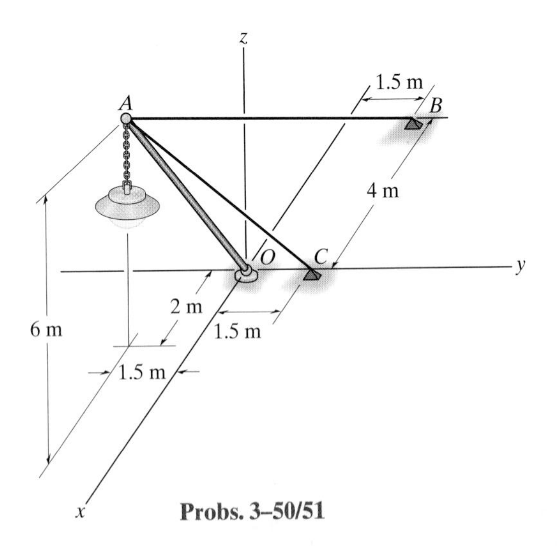

Probs. 3–50/51

*3–52. Determine the tension in cables AB, AC, and AD, required to hold the 60-lb crate in equilibrium.

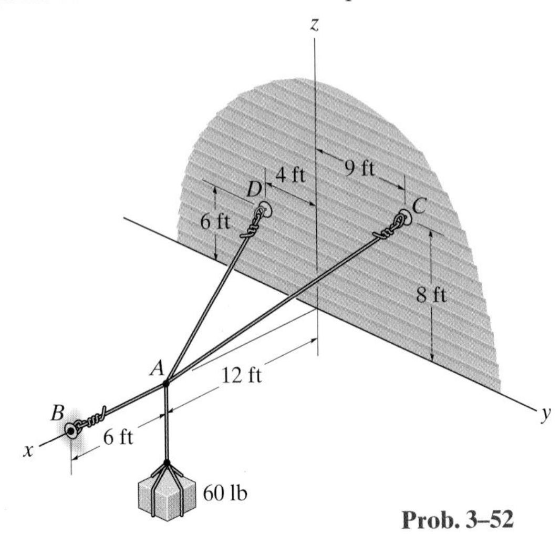

Prob. 3–52

3–53. The bucket has a weight of 20 lb. Determine the tension developed in each cord for equilibrium.

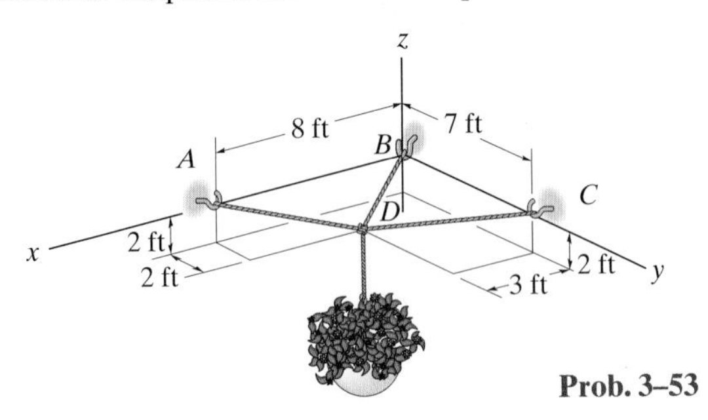

Prob. 3–53

3–54. The mast OA is supported by three cables. If cable AB is subjected to a tension of 500 N, determine the tension in cables AC and AD and the vertical force $\mathbf{F}$ which the mast exerts along its axis on the collar at A.

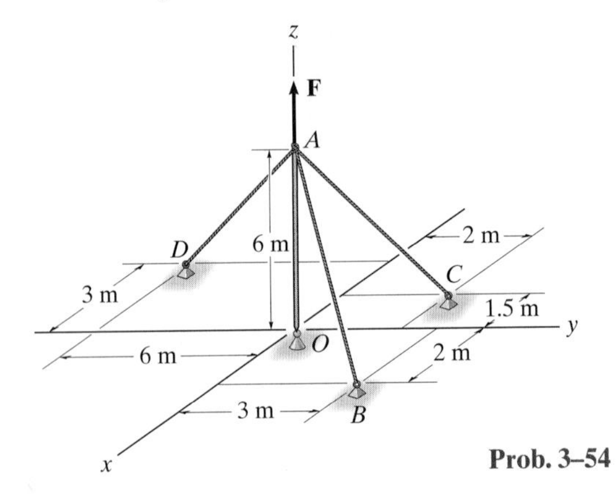

Prob. 3–54

3–55. The ends of the three cables are attached to a ring at A and to the edge of a uniform 150-kg plate. Determine the tension in each of the cables for equilibrium.

***3–56.** The ends of the three cables are attached to a ring at A and to the edge of the uniform plate. Determine the largest mass the plate can have if each cable can support a maximum tension of 15 kN.

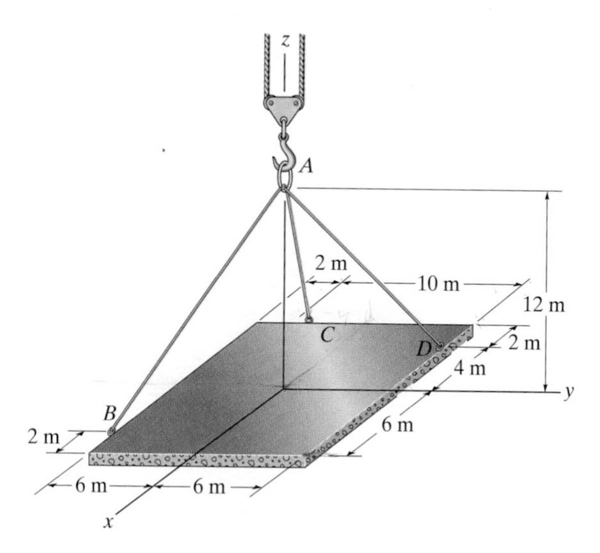

Probs. 3–55/56

3–57. The 500-lb crate is suspended from the cable system shown. Determine the force in each segment of the cable, i.e., AB, AC, CD, CE, and CF. *Hint:* First analyze the equilibrium of point A, then using the result for AC, analyze the equilibrium of point C.

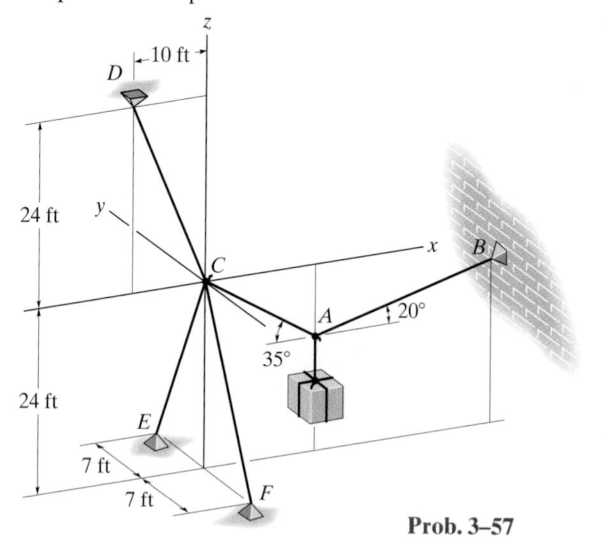

Prob. 3–57

3–58. The 80-lb chandelier is supported by three wires as shown. Determine the force in each wire for equilibrium.

3–59. If each wire can sustain a maximum tension of 120 lb before it fails, determine the greatest weight of the chandelier the wires will support in the position shown.

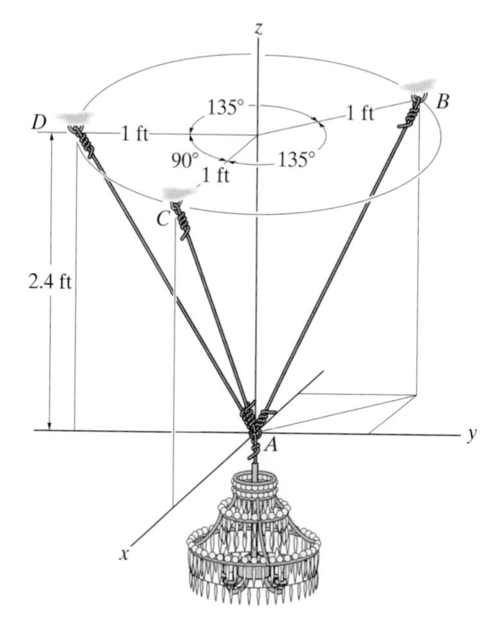

Probs. 3–58/59

***3–60.** Determine the force in each cable used to lift the 9.50-Mg surge arrester at constant velocity.

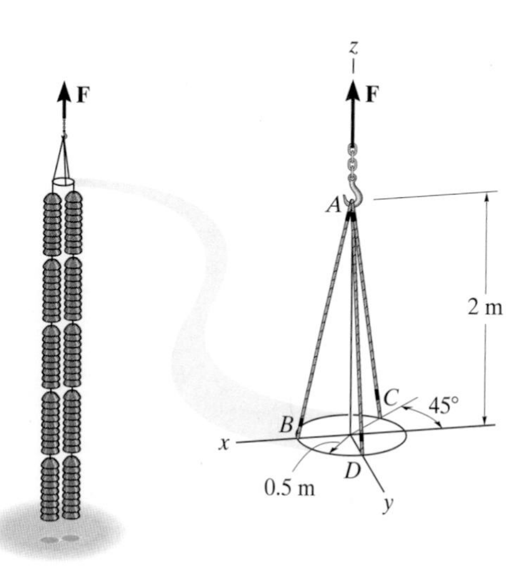

Prob. 3–60

3–61. The 800-lb cylinder is supported by three chains as shown. Determine the force in each chain for equilibrium. Take $d = 1$ ft.

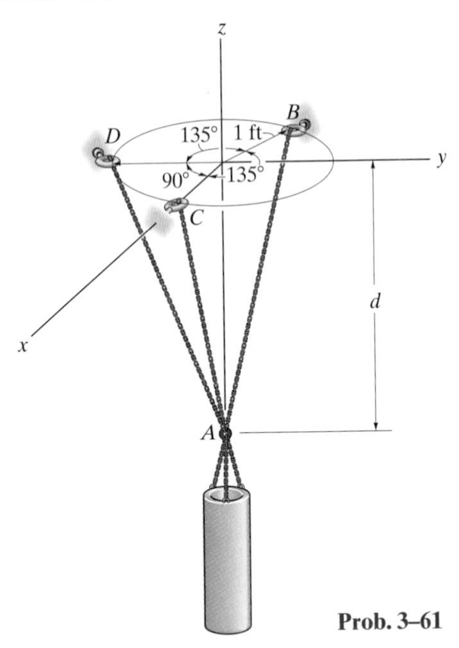

Prob. 3–61

3–62. The triangular frame ABC can be adjusted vertically between the three equal-length cords. If it remains in a horizontal plane, determine the required distance s so that the tension in each of the cords, OA, OB, and OC, equals 20 N. The lamp has a mass of 5 kg.

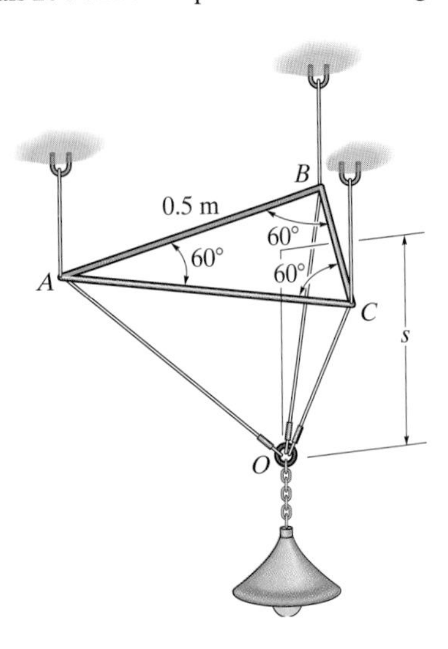

Prob. 3–62

3–63. Determine the force in each cable needed to support the 3500-lb platform. Set $d = 4$ ft.

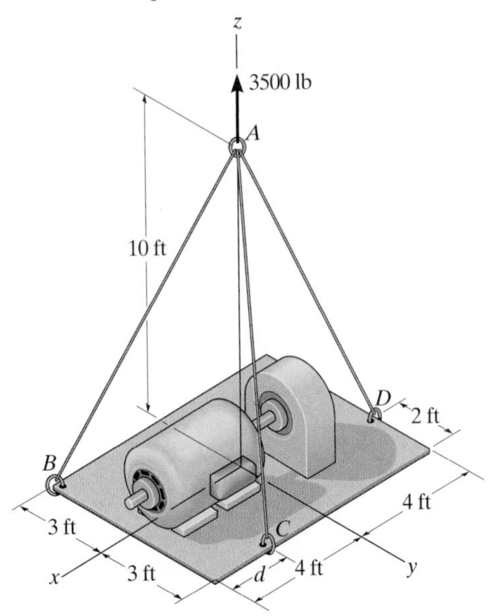

Prob. 3–63

***3–64.** The 25-kg flowerpot is supported at A by the three cords. Determine the force acting in each cord for equilibrium.

3–65. If each cord can sustain a maximum tension of 50 N before it fails, determine the greatest weight of the flowerpot the cords can support.

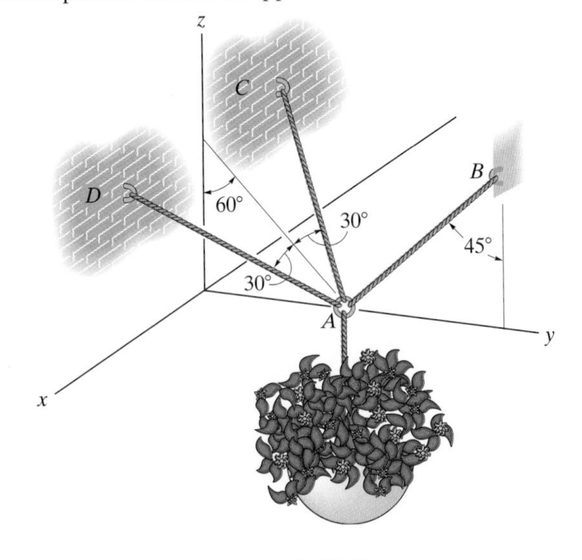

Probs. 3–64/65

Chapter Review

Particle Equilibrium

When a particle is at rest or moves with constant velocity, it is in equilibrium. This requires that all the forces acting on the particle form a zero force resultant.

In order to account for all the forces that act on a particle, it is necessary to draw its free-body diagram. This diagram is an outlined shape of the particle that shows all the forces listed with their known or unknown magnitudes and directions.
(*See page 83.*)

$$\mathbf{F}_R = \Sigma \mathbf{F} = \mathbf{0}$$

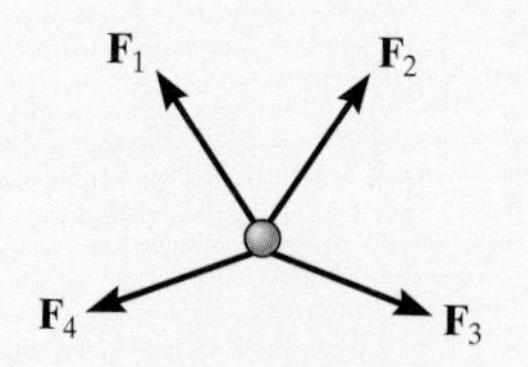

Two Dimensions

The two scalar equations of force equilibrium can be applied when referenced from an established x, y coordinate system.

$$\Sigma F_x = 0$$
$$\Sigma F_y = 0$$

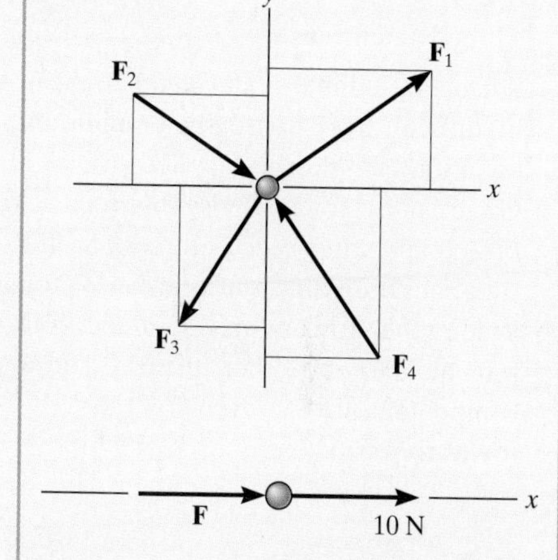

If the solution for a force magnitude yields a negative scalar, then the force acts in the opposite direction to that shown on the free-body diagram.

$$\Sigma F = 0;$$
$$F + 10\text{ N} = 0, F = -10\text{ N}$$

The tension force developed in a *continuous cable* that passes over a frictionless pulley must have a *constant* magnitude to keep the cable in equilibrium.
(*See pages 84–88.*)

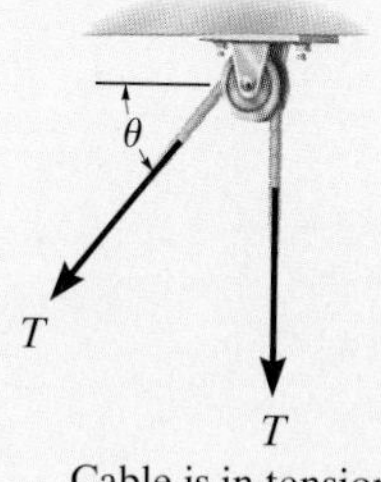

Cable is in tension

If the problem involves a linearly elastic spring, then the stretch or compression s of the spring can be related to the force applied to it. (*See page 84.*)	$F = ks$	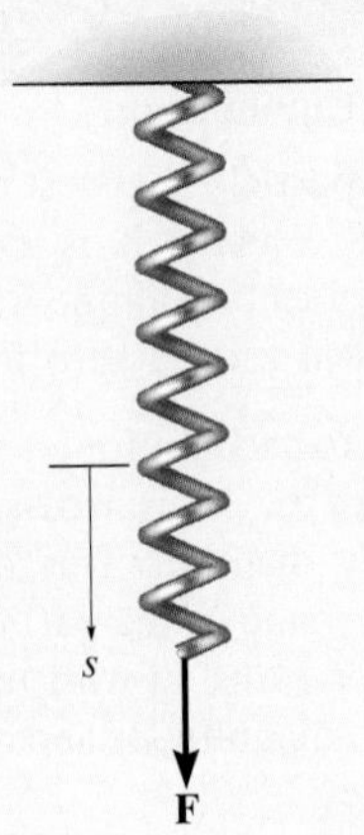 s F
Three Dimensions Since three-dimensional geometry can be difficult to visualize, the equilibrium equation should be applied using a Cartesian vector analysis. This requires first expressing each force in the free-body diagram as a Cartesian vector. When the forces are summed and set equal to zero, then the **i**, **j**, and **k** components are also zero. (*See page 100.*)	$\Sigma \mathbf{F} = \mathbf{0}$ $\Sigma F_x = 0$ $\Sigma F_y = 0$ $\Sigma F_z = 0$	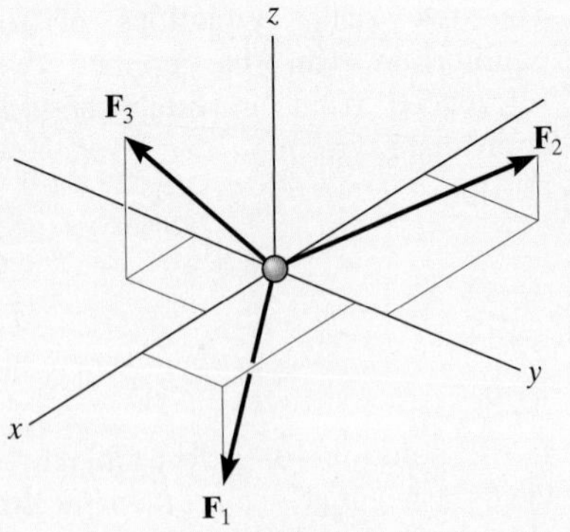 z $\mathbf{F}_3$ $\mathbf{F}_2$ y x $\mathbf{F}_1$

REVIEW PROBLEMS

3–66. The pipe is held in place by the vice. If the bolt exerts a force of 50 lb on the pipe in the direction shown, determine the forces F_A and F_B that the smooth contacts at A and B exert on the pipe.

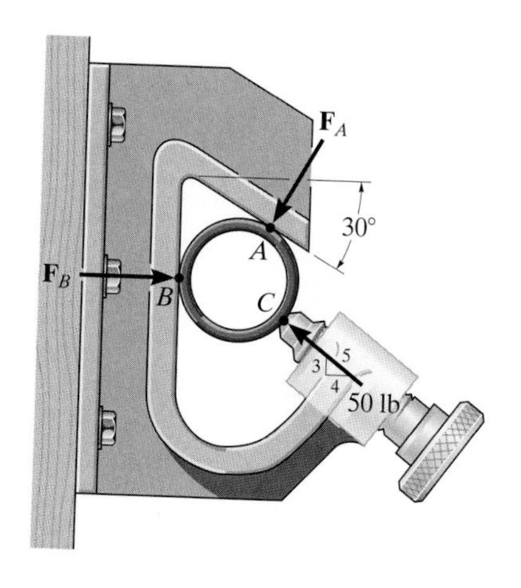

Prob. 3–66

3–67. When y is zero, the springs sustain a force of 60 lb. Determine the magnitude of the applied vertical forces $\mathbf{F}$ and $-\mathbf{F}$ required to pull point A away from point B a distance of $y = 2$ ft. The ends of cords CAD and CBD are attached to rings at C and D.

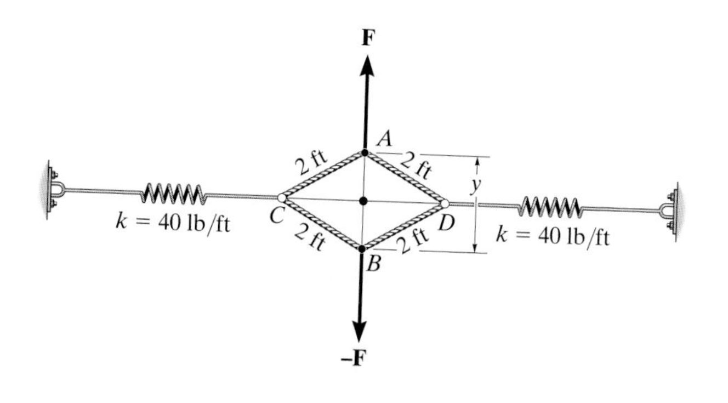

Prob. 3–67

***3–68.** When y is zero, the springs are each stretched 1.5 ft. Determine the distance y if a force of $F = 60$ lb is applied to points A and B as shown. The ends of cords CAD and CBD are attached to rings at C and D.

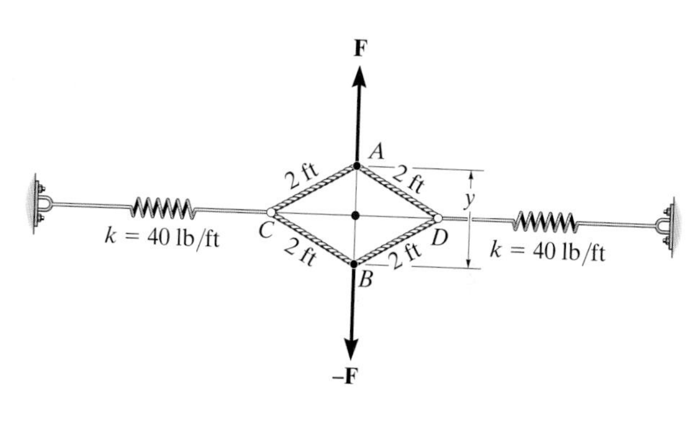

Prob. 3–68

■**3–69.** The 5-ft-long cord AB is attached to the end B of a spring having an unstretched length of 5 ft. The other end of the spring is attached to a roller C so that the spring remains horizontal as it stretches. If a 10-lb weight is suspended from B, determine the angle θ of cord AB for equilibrium.

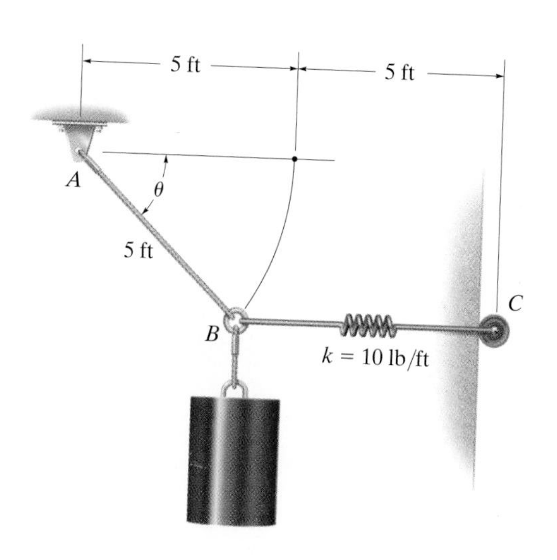

Prob. 3–69

3–70. The uniform 50-kg crate is suspended by using a 2-m-long cord that is attached to the sides of the crate and passes over the small pulley at O. If the cord can be attached at either points A and B, or C and D, determine which attachment produces the least amount of tension in the cord and specify the cord tension in this case.

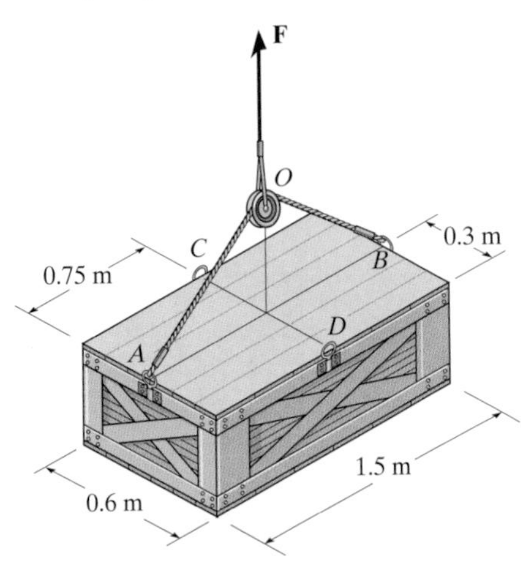

Prob. 3–70

3–71. The man attempts to pull the log at C by using the three ropes. Determine the direction θ in which he should pull on his rope with a force of 80 lb, so that he exerts a maximum force on the log. What is the force on the log for this case? Also, determine the direction in which he should pull in order to maximize the force in the rope attached to B. What is this maximum force?

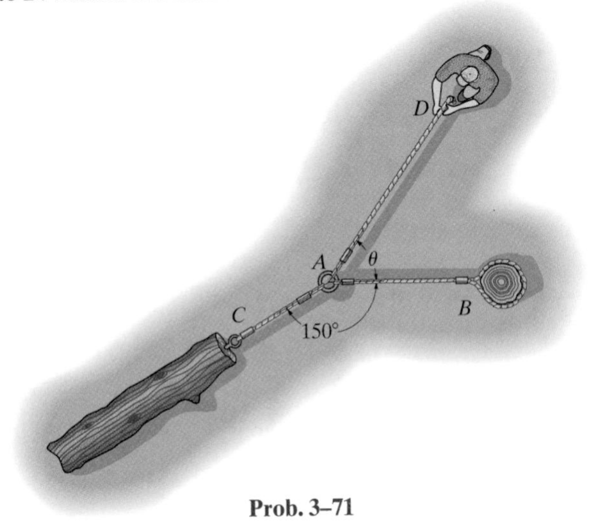

Prob. 3–71

***3–72.** The "scale" consists of a known weight W which is suspended at A from a cord of total length L. Determine the weight w at B if A is at a distance y for equilibrium. Neglect the sizes and weights of the pulleys.

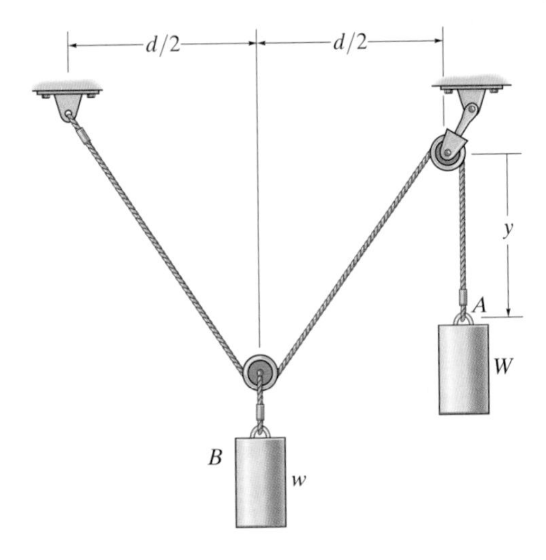

Prob. 3–72

3–73. Determine the maximum weight W that can be supported in the position shown if each cable AC and AB can support a maximum tension of 600 lb before it fails.

Prob. 3–73

3–74. If the spring on rope OB has been stretched 2 in. and fixed in place as shown, determine the tension developed in each of the other three ropes in order to hold the 225-lb weight in equilibrium. Rope OD lies in the x–y plane.

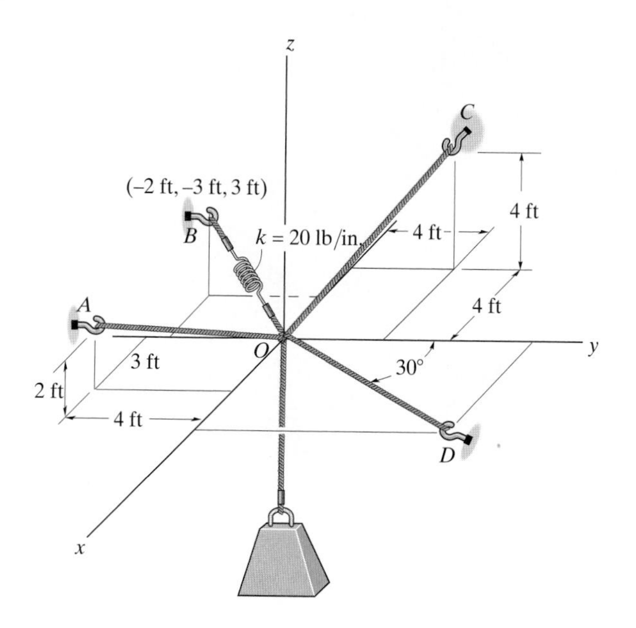

Prob. 3–74

3–75. The joint of a space frame is subjected to four member forces. Member OA lies in the x–y plane and member OB lies in the y–z plane. Determine the forces acting in each of the members required for equilibrium of the joint.

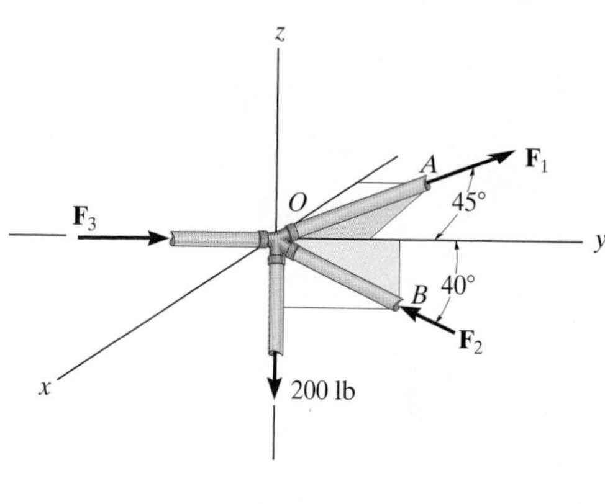

Prob. 3–75

Application of forces to the handle of this wrench will produce a tendency to rotate the wrench about its end. It is important to calculate this effect and, in some cases, to be able to simplify this system to a single resultant force and specify where this resultant acts on the wrench.

4 Force System Resultants

CHAPTER OBJECTIVES

- To discuss the concept of the moment of a force and show how to calculate it in two and three dimensions.
- To provide a method for finding the moment of a force about a specified axis.
- To define the moment of a couple.
- To present methods for determining the resultants of nonconcurrent force systems.
- To indicate how to reduce a simple distributed loading to a resultant force having a specified location.

4.1 Moment of a Force—Scalar Formulation

The *moment* of a force about a point or axis provides a measure of the tendency of the force to cause a body to rotate about the point or axis. For example, consider the horizontal force $\mathbf{F}_x$, which acts perpendicular to the handle of the wrench and is located a distance d_y from point O, Fig. 4–1*a*. It is seen that this force tends to cause the pipe to turn about the z axis. The larger the force or the distance d_y, the greater the turning effect. This tendency for rotation caused by $\mathbf{F}_x$ is sometimes called a *torque*, but most often it is called the *moment of a force* or simply the *moment* $(\mathbf{M}_O)_z$. Note that the *moment axis* (z) is perpendicular to the shaded plane (x–y) which contains both $\mathbf{F}_x$ and d_y and that this axis intersects the plane at point O.

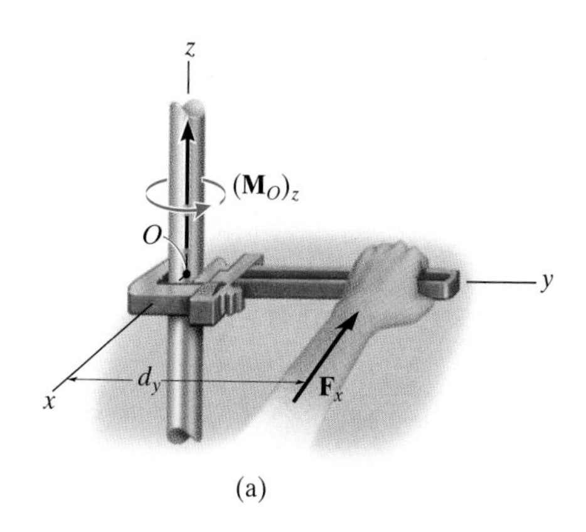

(a)

Fig. 4–1

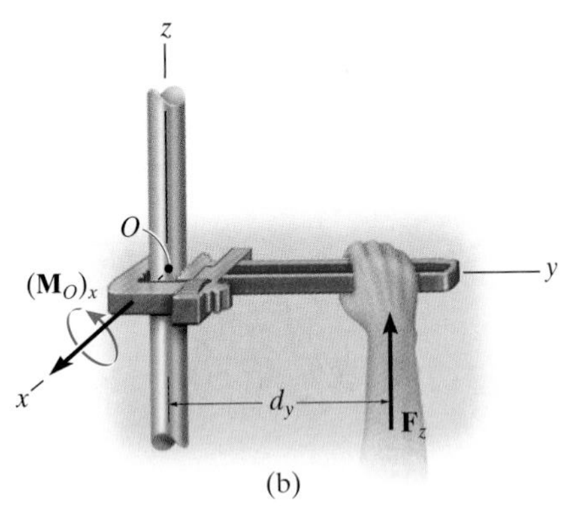

(b)

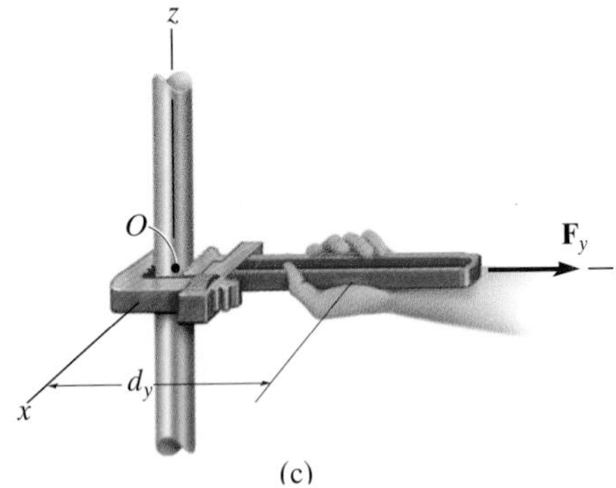

(c)

Fig. 4–1 (cont.)

Now consider applying the force $\mathbf{F}_z$ to the wrench, Fig. 4–1*b*. This force will *not* rotate the pipe about the *z* axis. Instead, it tends to rotate it about the *x* axis. Keep in mind that although it may not be possible to actually "rotate" or turn the pipe in this manner, $\mathbf{F}_z$ still creates the *tendency* for rotation and so the moment $(\mathbf{M}_O)_x$ is produced. As before, the force and distance d_y lie in the shaded plane (*y–z*) which is perpendicular to the moment axis (*x*). Lastly, if a force $\mathbf{F}_y$ is applied to the wrench, Fig. 4–1*c*, no moment is produced about point *O*. This results in a lack of turning since the line of action of the force passes through *O* and therefore no tendency for rotation is possible.

We will now generalize the above discussion and consider the force **F** and point *O* which lie in a shaded plane as shown in Fig. 4–2*a*. The moment $\mathbf{M}_O$ about point *O*, or about an axis passing through *O* and perpendicular to the plane, is a *vector quantity* since it has a specified magnitude and direction.

Magnitude. The magnitude of M_O is

$$M_O = Fd \tag{4–1}$$

where *d* is referred to as the *moment arm* or perpendicular distance from the axis at point *O* to the line of action of the force. Units of moment magnitude consist of force times distance, e.g., N · m or lb · ft.

Direction. The direction of $\mathbf{M}_O$ will be specified by using the "right-hand rule." To do this, the fingers of the right hand are curled such that they follow the sense of rotation, which would occur if the force could rotate about point O, Fig. 4–2a. The *thumb* then *points* along the *moment axis* so that it gives the direction and sense of the moment vector, which is upward and *perpendicular* to the shaded plane containing **F** and *d*.

In three dimensions, $\mathbf{M}_O$ is illustrated by a vector arrow with a curl on it to *distinguish* it from a force vector, Fig. 4–2*a*. Many problems in mechanics, however, involve coplanar force systems that may be conveniently viewed in two dimensions. For example, a two-dimensional view of Fig. 4–2*a* is given in Fig. 4–2*b*. Here $\mathbf{M}_O$ is simply represented by the (counterclockwise) curl, which indicates the action of **F**. The arrowhead on this curl is used to show the *sense of rotation* caused by **F**. Using the right-hand rule, however, realize that the direction and sense of the moment vector in Fig. 4–2*b* are specified by the thumb, which points *out* of the page since the fingers follow the curl. In particular, notice that *this curl or sense of rotation can always be determined by observing in which direction the force would "orbit" about point O* (counterclockwise in Fig. 4–2*b*). In two dimensions we will often refer to finding the moment of a force "about a point" (*O*). Keep in mind, however, that the moment *always acts about an axis* which is perpendicular to the plane containing **F** and *d*, and this axis intersects the plane at the point (*O*), Fig. 4–2*a*.

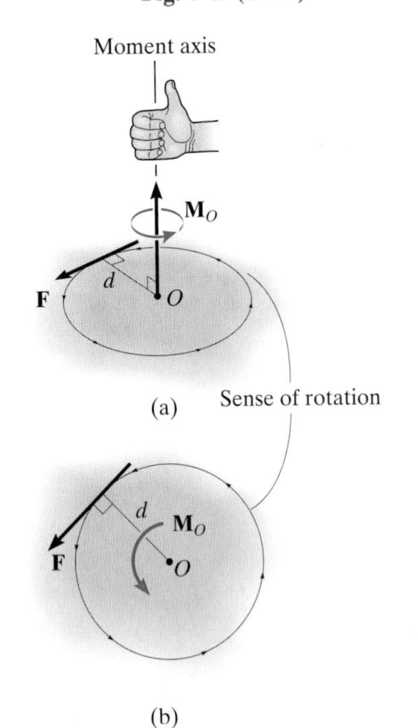

Fig. 4–2

Resultant Moment of a System of Coplanar Forces. If a system of forces lies in an x–y plane, then the moment produced by each force about point O will be directed along the z axis, Fig. 4–3. Consequently, the resultant moment $\mathbf{M}_{R_O}$ of the system can be determined by simply adding the moments of all forces *algebraically* since all the moment vectors are collinear. We can write this vector sum symbolically as

$$\circlearrowleft + M_{R_O} = \Sigma Fd \tag{4–2}$$

Here the counterclockwise curl written alongside the equation indicates that, by the scalar sign convention, the moment of any force will be positive if it is directed along the $+z$ axis, whereas a negative moment is directed along the $-z$ axis.

The following examples illustrate numerical application of Eqs. 4–1 and 4–2.

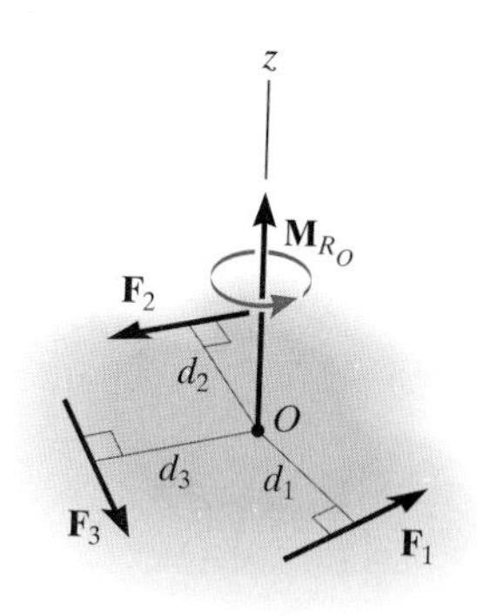

Fig. 4–3

By pushing down on the pry bar the load on the ground at A can be lifted. The turning effect, caused by the applied force, is due to the moment about A. To produce this moment with minimum effort we instinctively know that the force should be applied to the *end* of the bar; however, the *direction* in which this force is applied is also important. This is because a moment is the product of the force and the moment arm. Notice that when the force is at an angle $\theta < 90°$, then the moment arm distance is *shorter* than when the force is applied perpendicular to the bar, $\theta = 90°$, i.e., $d' < d$. Hence the greatest moment is produced when the force is farthest from point A and applied perpendicular to the axis of the bar so as to maximize the moment arm.

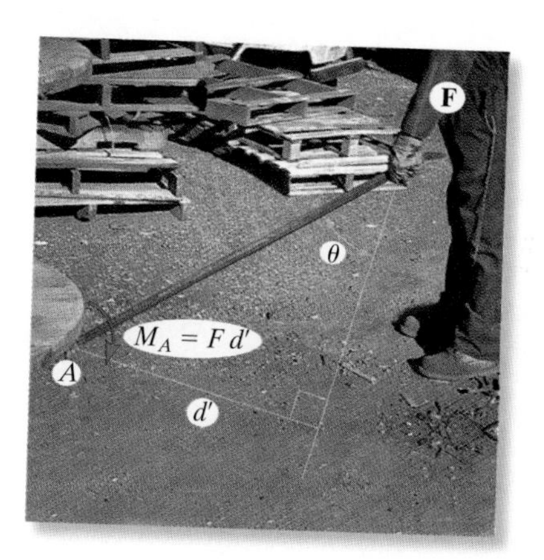

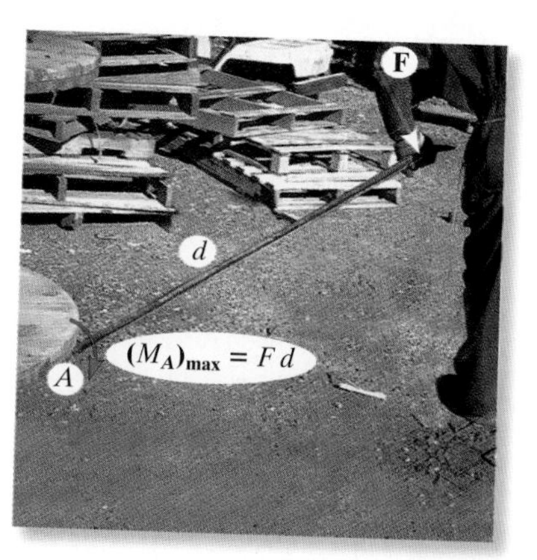

In the photo to the right, the moment of a force does not always cause a rotation. For example, the force $\mathbf{F}$ tends to rotate the beam clockwise about its support at A with a moment $M_A = Fd_A$. The actual rotation would occur if the support at B were removed. In the same manner, $\mathbf{F}$ creates a tendency to rotate the beam counterclockwise about B with a moment $M_B = Fd_B$. Here the support at A prevents the rotation.

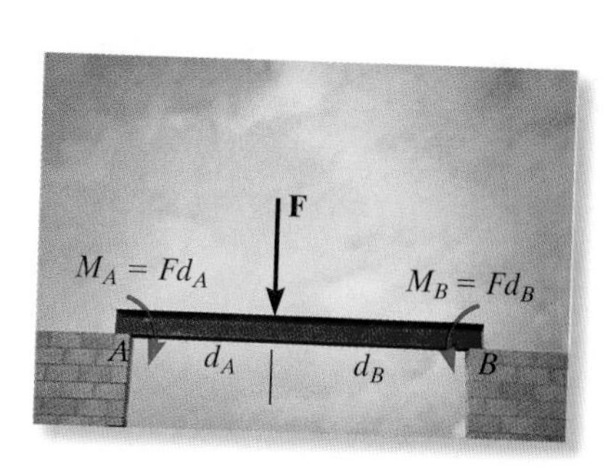

EXAMPLE 4.1

For each case illustrated in Fig. 4–4, determine the moment of the force about point O.

100 N

O

2 m

(a)

SOLUTION (SCALAR ANALYSIS)

The line of action of each force is extended as a dashed line in order to establish the moment arm d. Also illustrated is the tendency of rotation of the member as caused by the force. Furthermore, the orbit of the force is shown as a colored curl. Thus,

Fig. 4–4*a* $\quad M_O = (100\text{ N})(2\text{ m}) = 200\text{ N}\cdot\text{m} \circlearrowright$ *Ans.*

Fig. 4–4*b* $\quad M_O = (50\text{ N})(0.75\text{ m}) = 37.5\text{ N}\cdot\text{m} \circlearrowright$ *Ans.*

Fig. 4–4*c* $\quad M_O = (40\text{ lb})(4\text{ ft} + 2\cos 30^\circ\text{ ft}) = 229\text{ lb}\cdot\text{ft} \circlearrowright$ *Ans.*

Fig. 4–4*d* $\quad M_O = (60\text{ lb})(1\sin 45^\circ\text{ ft}) = 42.4\text{ lb}\cdot\text{ft} \circlearrowleft$ *Ans.*

Fig. 4–4*e* $\quad M_O = (7\text{ kN})(4\text{ m} - 1\text{ m}) = 21.0\text{ kN}\cdot\text{m} \circlearrowleft$ *Ans.*

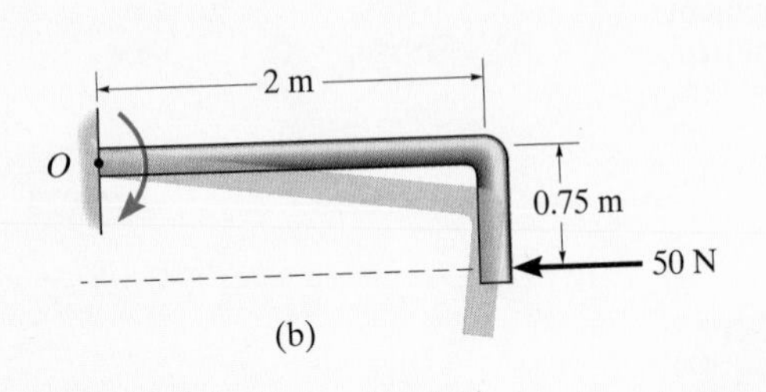

(b)

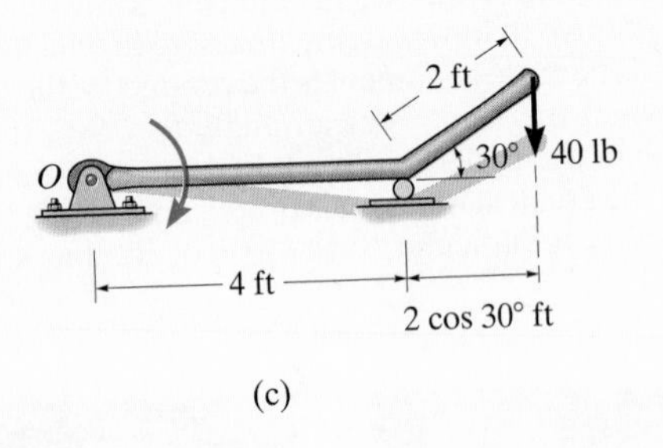

(c)

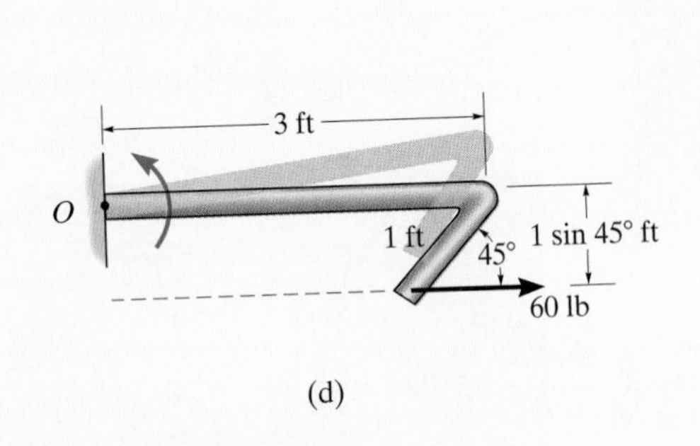

(d)

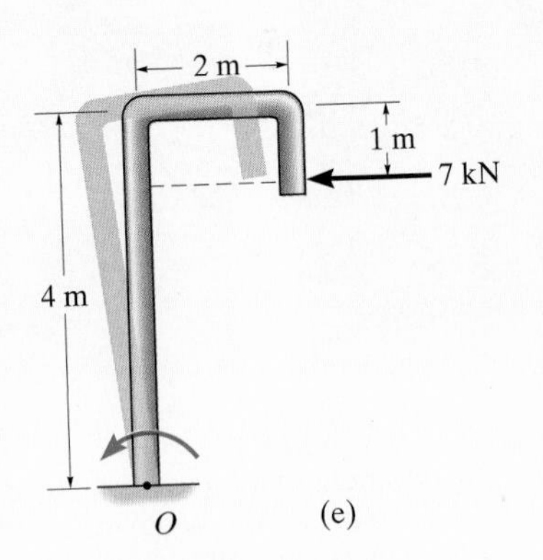

(e)

Fig. 4–4

EXAMPLE 4.2

Determine the moments of the 800-N force acting on the frame in Fig. 4–5 about points *A*, *B*, *C*, and *D*.

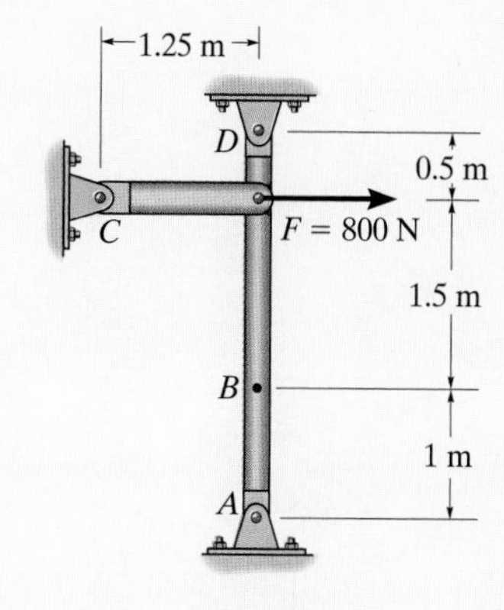

Fig. 4–5

SOLUTION *(SCALAR ANALYSIS)*

In general, $M = Fd$, where d is the moment arm or *perpendicular distance* from the *point* on the moment axis to the *line of action* of the force. Hence,

$$M_A = 800\text{ N}(2.5\text{ m}) = 2000\text{ N}\cdot\text{m} \circlearrowright \qquad \textit{Ans.}$$

$$M_B = 800\text{ N}(1.5\text{ m}) = 1200\text{ N}\cdot\text{m} \circlearrowright \qquad \textit{Ans.}$$

$$M_C = 800\text{ N}(0) = 0 \quad \text{(line of action of } \mathbf{F} \text{ passes through } C) \qquad \textit{Ans.}$$

$$M_D = 800\text{ N}(0.5\text{ m}) = 400\text{ N}\cdot\text{m} \circlearrowleft \qquad \textit{Ans.}$$

The curls indicate the sense of rotation of the moment, which is defined by the direction the force orbits about each point.

EXAMPLE 4.3

Determine the resultant moment of the four forces acting on the rod shown in Fig. 4–6 about point *O*.

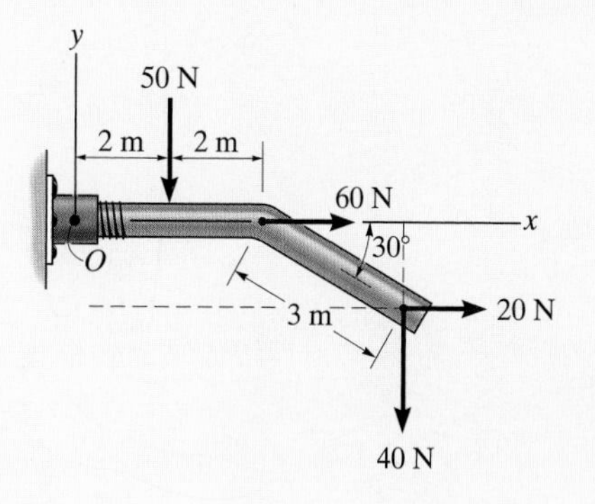

Fig. 4–6

SOLUTION

Assuming that positive moments act in the $+\mathbf{k}$ direction, i.e., counterclockwise, we have

$$\circlearrowleft + M_{R_O} = \Sigma Fd;$$

$$M_{R_O} = -50\text{ N}(2\text{ m}) + 60\text{ N}(0) + 20\text{ N}(3\sin 30^\circ\text{ m})$$
$$-40\text{ N}(4\text{ m} + 3\cos 30^\circ\text{ m})$$

$$M_{R_O} = -334\text{ N}\cdot\text{m} = 334\text{ N}\cdot\text{m} \circlearrowright \qquad \textit{Ans.}$$

For this calculation, note how the moment-arm distances for the 20-N and 40-N forces are established from the extended (dashed) lines of action of each of these forces.

4.2 Cross Product

The moment of a force will be formulated using Cartesian vectors in the next section. Before doing this, however, it is first necessary to expand our knowledge of vector algebra and introduce the cross-product method of vector multiplication.

The *cross product* of two vectors **A** and **B** yields the vector **C**, which is written

$$\mathbf{C} = \mathbf{A} \times \mathbf{B}$$

and is read "**C** equals **A** cross **B**."

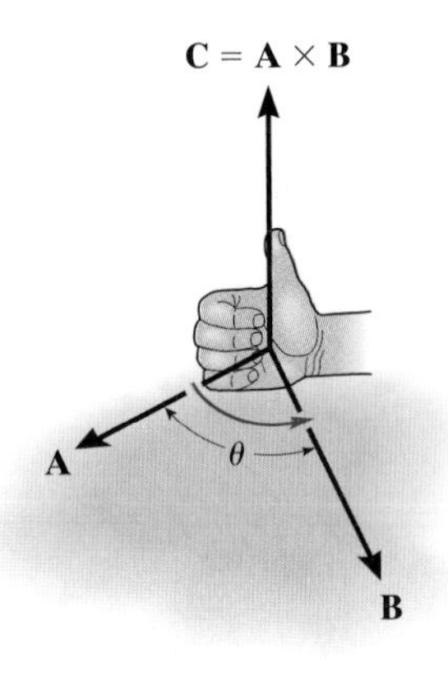

Fig. 4–7

Magnitude. The *magnitude* of **C** is defined as the product of the magnitudes of **A** and **B** and the sine of the angle θ between their tails $(0° \leq \theta \leq 180°)$. Thus, $C = AB \sin \theta$.

Direction. Vector **C** has a *direction* that is perpendicular to the plane containing **A** and **B** such that **C** is specified by the right-hand rule; i.e., curling the fingers of the right hand from vector **A** (cross) to vector **B**, the thumb then points in the direction of **C**, as shown in Fig. 4–7.

Knowing both the magnitude and direction of **C**, we can write

$$\mathbf{C} = \mathbf{A} \times \mathbf{B} = (AB \sin \theta)\mathbf{u}_C \tag{4–3}$$

where the scalar $AB \sin \theta$ defines the *magnitude* of **C** and the unit vector $\mathbf{u}_C$ defines the *direction* of **C**. The terms of Eq. 4–3 are illustrated graphically in Fig. 4–8.

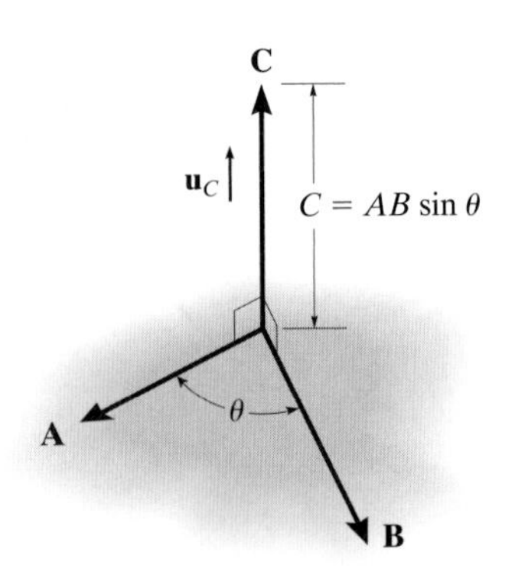

Fig. 4–8

Laws of Operation.

1. The commutative law is *not* valid; i.e.,

$$\mathbf{A} \times \mathbf{B} \neq \mathbf{B} \times \mathbf{A}$$

Rather,

$$\mathbf{A} \times \mathbf{B} = -\mathbf{B} \times \mathbf{A}$$

This is shown in Fig. 4–9 by using the right-hand rule. The cross product $\mathbf{B} \times \mathbf{A}$ yields a vector that acts in the opposite direction to $\mathbf{C}$; i.e., $\mathbf{B} \times \mathbf{A} = -\mathbf{C}$.

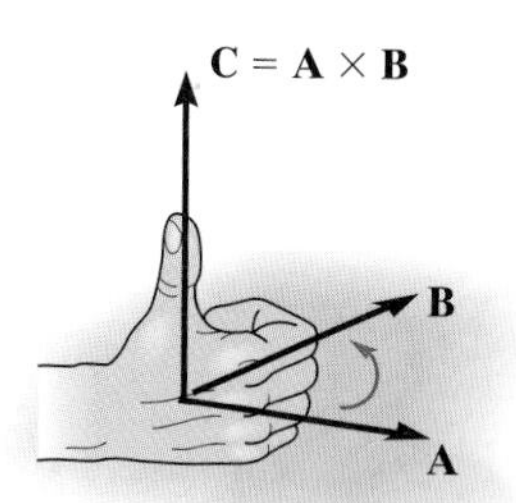

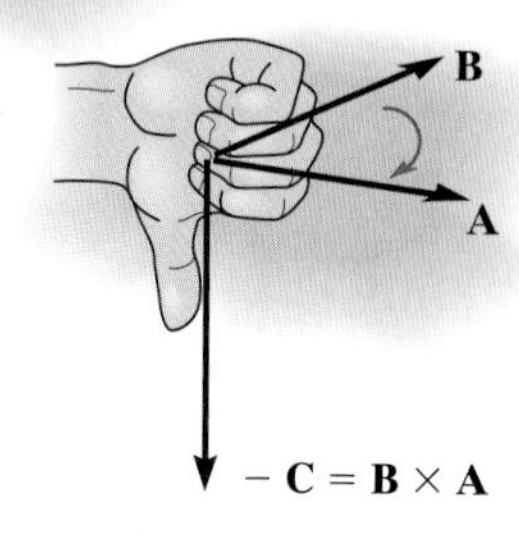

Fig. 4–9

2. Multiplication by a scalar:

$$a(\mathbf{A} \times \mathbf{B}) = (a\mathbf{A}) \times \mathbf{B} = \mathbf{A} \times (a\mathbf{B}) = (\mathbf{A} \times \mathbf{B})a$$

This property is easily shown since the magnitude of the resultant vector ($|a|AB \sin \theta$) and its direction are the same in each case.

3. The distributive law:

$$\mathbf{A} \times (\mathbf{B} + \mathbf{D}) = (\mathbf{A} \times \mathbf{B}) + (\mathbf{A} \times \mathbf{D})$$

The proof of this identity is left as an exercise (see Prob. 4–1). It is important to note that *proper order* of the cross products must be maintained, since they are not commutative.

Cartesian Vector Formulation. Equation 4–3 may be used to find the cross product of a pair of Cartesian unit vectors. For example, to find $\mathbf{i} \times \mathbf{j}$, the magnitude of the resultant vector is $(i)(j)(\sin 90°) = (1)(1)(1) = 1$, and its direction is determined using the right-hand rule. As shown in Fig. 4–10, the resultant vector points in the $+\mathbf{k}$ direction. Thus, $\mathbf{i} \times \mathbf{j} = (1)\mathbf{k}$. In a similar manner,

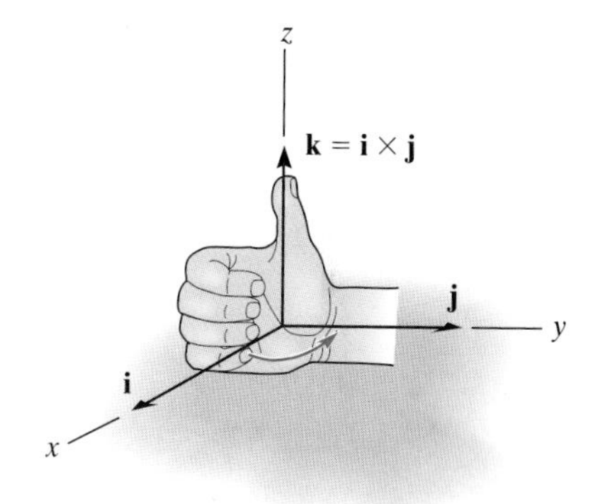

Fig. 4–10

$$\begin{array}{lll} \mathbf{i} \times \mathbf{j} = \mathbf{k} & \mathbf{i} \times \mathbf{k} = -\mathbf{j} & \mathbf{i} \times \mathbf{i} = \mathbf{0} \\ \mathbf{j} \times \mathbf{k} = \mathbf{i} & \mathbf{j} \times \mathbf{i} = -\mathbf{k} & \mathbf{j} \times \mathbf{j} = \mathbf{0} \\ \mathbf{k} \times \mathbf{i} = \mathbf{j} & \mathbf{k} \times \mathbf{j} = -\mathbf{i} & \mathbf{k} \times \mathbf{k} = \mathbf{0} \end{array}$$

These results should *not* be memorized; rather, it should be clearly understood how each is obtained by using the right-hand rule and the definition of the cross product. A simple scheme shown in Fig. 4–11 is helpful for obtaining the same results when the need arises. If the circle is constructed as shown, then "crossing" two unit vectors in a *counterclockwise* fashion around the circle yields the *positive* third unit vector; e.g., $\mathbf{k} \times \mathbf{i} = \mathbf{j}$. Moving *clockwise*, a *negative* unit vector is obtained; e.g., $\mathbf{i} \times \mathbf{k} = -\mathbf{j}$.

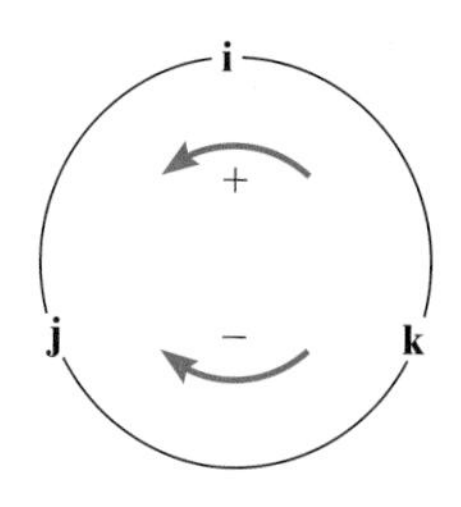

Fig. 4–11

Consider now the cross product of two general vectors **A** and **B** which are expressed in Cartesian vector form. We have

$$\begin{aligned}\mathbf{A} \times \mathbf{B} &= (A_x\mathbf{i} + A_y\mathbf{j} + A_z\mathbf{k}) \times (B_x\mathbf{i} + B_y\mathbf{j} + B_z\mathbf{k}) \\ &= A_xB_x(\mathbf{i} \times \mathbf{i}) + A_xB_y(\mathbf{i} \times \mathbf{j}) + A_xB_z(\mathbf{i} \times \mathbf{k}) \\ &\quad + A_yB_x(\mathbf{j} \times \mathbf{i}) + A_yB_y(\mathbf{j} \times \mathbf{j}) + A_yB_z(\mathbf{j} \times \mathbf{k}) \\ &\quad + A_zB_x(\mathbf{k} \times \mathbf{i}) + A_zB_y(\mathbf{k} \times \mathbf{j}) + A_zB_z(\mathbf{k} \times \mathbf{k})\end{aligned}$$

Carrying out the cross-product operations and combining terms yields

$$\mathbf{A} \times \mathbf{B} = (A_yB_z - A_zB_y)\mathbf{i} - (A_xB_z - A_zB_x)\mathbf{j} + (A_xB_y - A_yB_x)\mathbf{k} \quad (4\text{–}4)$$

This equation may also be written in a more compact determinant form as

$$\mathbf{A} \times \mathbf{B} = \begin{vmatrix} \mathbf{i} & \mathbf{j} & \mathbf{k} \\ A_x & A_y & A_z \\ B_x & B_y & B_z \end{vmatrix} \quad (4\text{–}5)$$

Thus, to find the cross product of any two Cartesian vectors **A** and **B**, it is necessary to expand a determinant whose first row of elements consists of the unit vectors **i**, **j**, and **k** and whose second and third rows represent the *x, y, z* components of the two vectors **A** and **B**, respectively.*

*A determinant having three rows and three columns can be expanded using three minors, each of which is multiplied by one of the three terms in the first row. There are four elements in each minor, e.g.,

$$\begin{vmatrix} A_{11} & A_{12} \\ A_{21} & A_{22} \end{vmatrix}$$

By *definition*, this notation represents the terms $(A_{11}A_{22} - A_{12}A_{21})$, which is simply the product of the two elements of the arrow slanting downward to the right $(A_{11}A_{22})$ *minus* the product of the two elements intersected by the arrow slanting downward to the left $(A_{12}A_{21})$. For a 3×3 determinant, such as Eq. 4–5, the three minors can be generated in accordance with the following scheme:

For element **i**: $\begin{vmatrix} \mathbf{i} & \mathbf{j} & \mathbf{k} \\ A_x & A_y & A_z \\ B_x & B_y & B_z \end{vmatrix} = \mathbf{i}(A_yB_z - A_zB_y)$

For element **j**: $\begin{vmatrix} \mathbf{i} & \mathbf{j} & \mathbf{k} \\ A_x & A_y & A_z \\ B_x & B_y & B_z \end{vmatrix} = -\mathbf{j}(A_xB_z - A_zB_x)$

For element **k**: $\begin{vmatrix} \mathbf{i} & \mathbf{j} & \mathbf{k} \\ A_x & A_y & A_z \\ B_x & B_y & B_z \end{vmatrix} = \mathbf{k}(A_xB_y - A_yB_x)$

Adding the results and noting that the **j** element *must include the minus sign* yields the expanded form of $\mathbf{A} \times \mathbf{B}$ given by Eq. 4–4.

4.3 Moment of a Force—Vector Formulation

The moment of a force **F** about point O, or actually about the moment axis passing through O and perpendicular to the plane containing O and **F**, Fig. 4–12*a*, can be expressed using the vector cross product, namely,

$$\mathbf{M}_O = \mathbf{r} \times \mathbf{F} \tag{4–6}$$

Here **r** represents a position vector drawn *from O* to *any point* lying on the line of action of **F**. We will now show that indeed the moment $\mathbf{M}_O$, when determined by this cross product, has the proper magnitude and direction.

Magnitude. The magnitude of the cross product is defined from Eq. 4–3 as $M_O = rF \sin\theta$, where the angle θ is measured between the *tails* of **r** and **F**. To establish this angle, **r** must be treated as a sliding vector so that θ can be constructed properly, Fig. 4–12*b*. Since the moment arm $d = r\sin\theta$, then

$$M_O = rF\sin\theta = F(r\sin\theta) = Fd$$

which agrees with Eq. 4–1.

Direction. The direction and sense of $\mathbf{M}_O$ in Eq. 4–6 are determined by the right-hand rule as it applies to the cross product. Thus, extending **r** to the dashed position and curling the right-hand fingers from **r** toward **F**, "**r** cross **F**," the thumb is directed upward or perpendicular to the plane containing **r** and **F** and this is in the *same direction* as $\mathbf{M}_O$, the moment of the force about point O, Fig. 4–12*b*. Note that the "curl" of the fingers, like the curl around the moment vector, indicates the sense of rotation caused by the force. Since the cross product is not commutative, it is important that the *proper order* of **r** and **F** be maintained in Eq. 4–6.

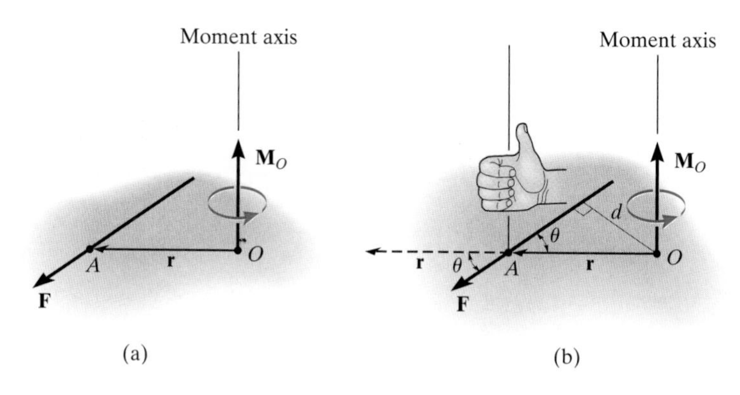

Fig. 4–12

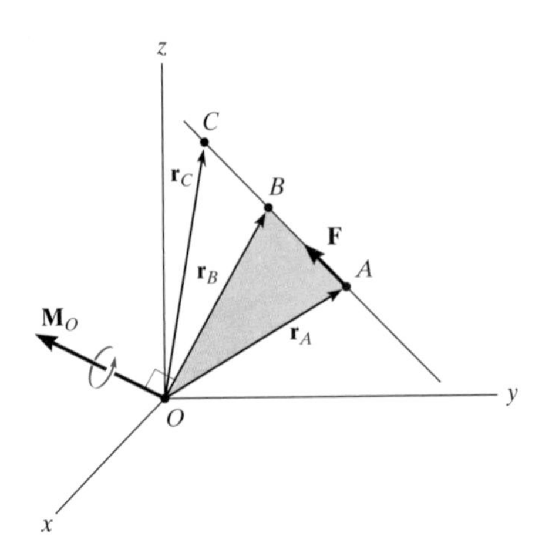

Fig. 4–13

Principle of Transmissibility. Consider the force $\mathbf{F}$ applied at point A in Fig. 4–13. The moment created by $\mathbf{F}$ about O is $\mathbf{M}_O = \mathbf{r}_A \times \mathbf{F}$; however, it was shown that "$\mathbf{r}$" can extend from O to *any point* on the line of action of $\mathbf{F}$. Consequently, $\mathbf{F}$ may be applied at point B or C, and the same moment $\mathbf{M}_O = \mathbf{r}_B \times \mathbf{F} = \mathbf{r}_C \times \mathbf{F}$ will be computed. As a result, $\mathbf{F}$ has the properties of a *sliding vector* and can therefore act at *any point along its line of action and still create the same moment about point O*. We refer to this as the *principle of transmissibility*, and we will discuss this property further in Sec. 4.7.

Cartesian Vector Formulation. If we establish x, y, z coordinate axes, then the position vector $\mathbf{r}$ and force $\mathbf{F}$ can be expressed as Cartesian vectors, Fig. 4–14. Applying Eq. 4–5 we have

$$\mathbf{M}_O = \mathbf{r} \times \mathbf{F} = \begin{vmatrix} \mathbf{i} & \mathbf{j} & \mathbf{k} \\ r_x & r_y & r_z \\ F_x & F_y & F_z \end{vmatrix} \tag{4–7}$$

where

r_x, r_y, r_z represent the x, y, z components of the position vector drawn from point O to *any point* on the line of action of the force

F_x, F_y, F_z represent the x, y, z components of the force vector

If the determinant is expanded, then like Eq. 4–4 we have

$$\mathbf{M}_O = (r_y F_z - r_z F_y)\mathbf{i} - (r_x F_z - r_z F_x)\mathbf{j} + (r_x F_y - r_y F_x)\mathbf{k} \tag{4–8}$$

The physical meaning of these three moment components becomes evident by studying Fig. 4–14*a*. For example, the $\mathbf{i}$ component of $\mathbf{M}_O$ is determined from the moments of $\mathbf{F}_x$, $\mathbf{F}_y$, and $\mathbf{F}_z$ about the x axis. In particular, note that $\mathbf{F}_x$ does *not* create a moment or tendency to cause

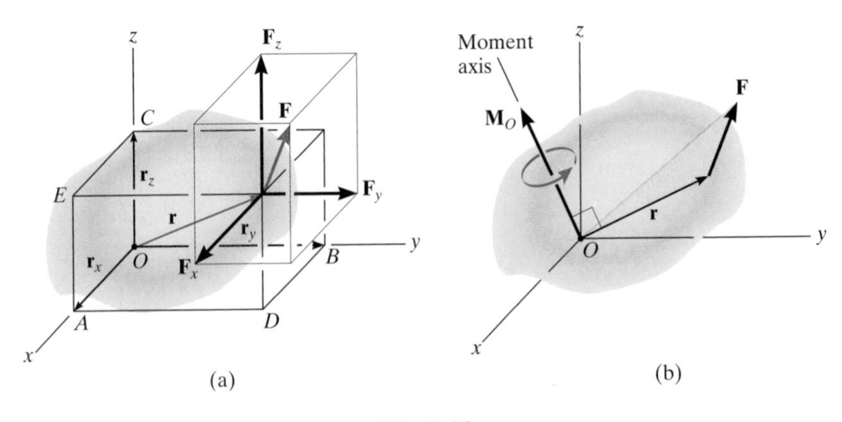

Fig. 4–14

turning about the x axis since this force is *parallel* to the x axis. The line of action of $\mathbf{F}_y$ passes through point E, and so the magnitude of the moment of $\mathbf{F}_y$ about point A on the x axis is r_zF_y. By the right-hand rule this component acts in the negative $\mathbf{i}$ direction. Likewise, $\mathbf{F}_z$ contributes a moment component of $r_yF_z\mathbf{i}$. Thus, $(M_O)_x = (r_yF_z - r_zF_y)$ as shown in Eq. 4–8. As an exercise, establish the $\mathbf{j}$ and $\mathbf{k}$ components of $\mathbf{M}_O$ in this manner and show that indeed the expanded form of the determinant, Eq. 4–8, represents the moment of $\mathbf{F}$ about point O. Once $\mathbf{M}_O$ is determined, realize that it will always be *perpendicular* to the shaded plane containing vectors $\mathbf{r}$ and $\mathbf{F}$, Fig. 4–14*b*.

It will be shown in Example 4.4 that the computation of the moment using the cross product has a distinct advantage over the scalar formulation when solving problems in *three dimensions*. This is because it is generally easier to establish the position vector $\mathbf{r}$ to the force, rather than determining the moment-arm distance d that must be directed *perpendicular* to the line of action of the force.

Resultant Moment of a System of Forces. If a body is acted upon by a system of forces, Fig. 4–15, the resultant moment of the forces about point O can be determined by vector addition resulting from successive applications of Eq. 4–6. This resultant can be written symbolically as

$$\mathbf{M}_{R_O} = \Sigma(\mathbf{r} \times \mathbf{F}) \tag{4–9}$$

and is shown in Fig. 4–15.

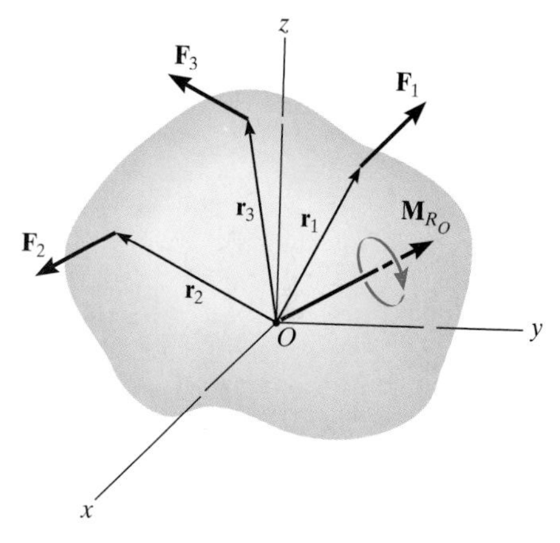

Fig. 4–15

If we pull on cable BC with a force $\mathbf{F}$ at *any point along the cable*, the moment of this force about the base of the utility pole at A will always be the *same*. This is a consequence of the principle of transmissibility. Note that the moment arm, or perpendicular distance from A to the cable, is r_d, and so $M_A = r_dF$. In three dimensions this distance is often difficult to determine, and so we can use the vector cross product to obtain the moment in a more direct manner. For example, $\mathbf{M}_A = \mathbf{r}_{AB} \times \mathbf{F} = \mathbf{r}_{AC} \times \mathbf{F}$. As required, both of these vectors are directed from point A to a point on the line of action of the force.

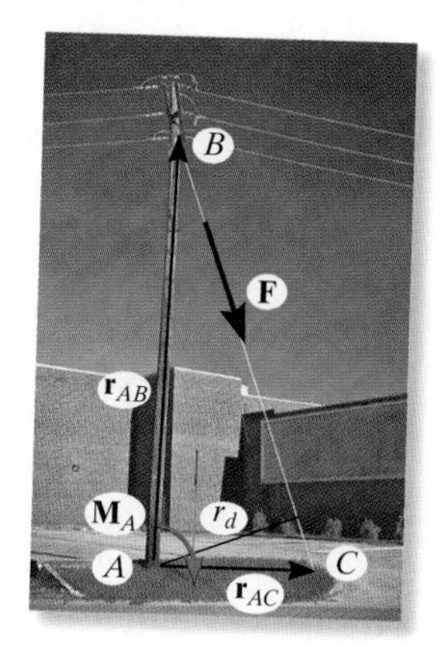

EXAMPLE 4.4

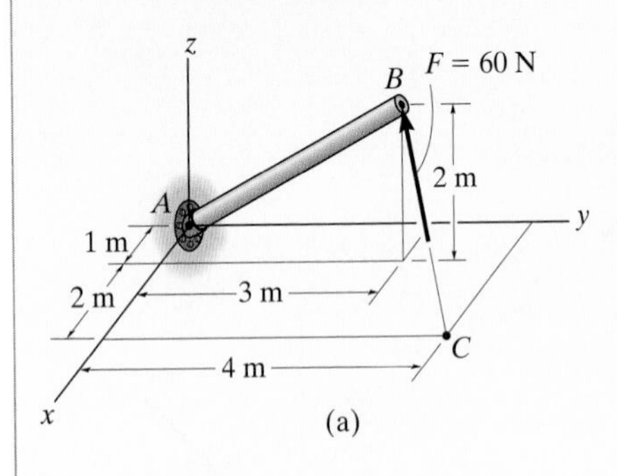

(a)

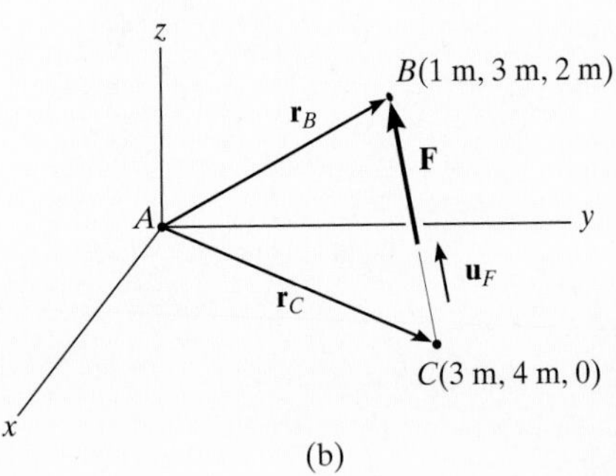

(b)

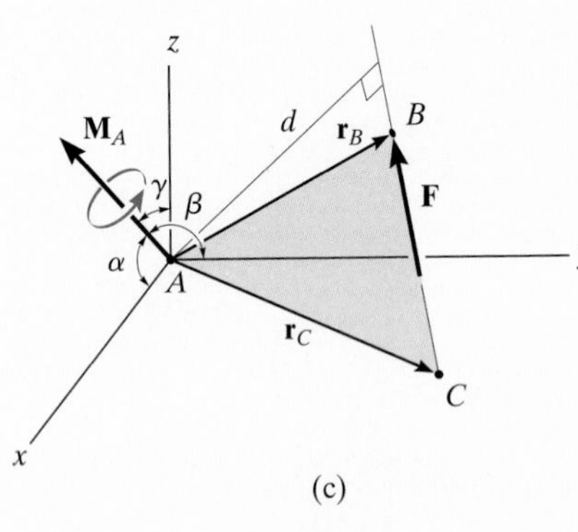

(c)

Fig. 4–16

The pole in Fig. 4–16*a* is subjected to a 60-N force that is directed from *C* to *B*. Determine the magnitude of the moment created by this force about the support at *A*.

SOLUTION *(VECTOR ANALYSIS)*

As shown in Fig. 4–16*b*, either one of two position vectors can be used for the solution, since $\mathbf{M}_A = \mathbf{r}_B \times \mathbf{F}$ or $\mathbf{M}_A = \mathbf{r}_C \times \mathbf{F}$. The position vectors are represented as

$$\mathbf{r}_B = \{1\mathbf{i} + 3\mathbf{j} + 2\mathbf{k}\}\text{ m} \quad \text{and} \quad \mathbf{r}_C = \{3\mathbf{i} + 4\mathbf{j}\}\text{ m}$$

The force has a magnitude of 60 N and a direction specified by the unit vector $\mathbf{u}_F$, directed from *C* to *B*. Thus,

$$\mathbf{F} = (60\text{ N})\mathbf{u}_F = (60\text{ N})\left[\frac{(1-3)\mathbf{i} + (3-4)\mathbf{j} + (2-0)\mathbf{k}}{\sqrt{(-2)^2 + (-1)^2 + (2)^2}}\right]$$
$$= \{-40\mathbf{i} - 20\mathbf{j} + 40\mathbf{k}\}\text{ N}$$

Substituting into the determinant formulation, Eq. 4–7, and following the scheme for determinant expansion as stated in the footnote on page 124, we have

$$\mathbf{M}_A = \mathbf{r}_B \times \mathbf{F} = \begin{vmatrix} \mathbf{i} & \mathbf{j} & \mathbf{k} \\ 1 & 3 & 2 \\ -40 & -20 & 40 \end{vmatrix}$$
$$= [3(40) - 2(-20)]\mathbf{i} - [1(40) - 2(-40)]\mathbf{j} + [1(-20) - 3(-40)]\mathbf{k}$$

or

$$\mathbf{M}_A = \mathbf{r}_C \times \mathbf{F} = \begin{vmatrix} \mathbf{i} & \mathbf{j} & \mathbf{k} \\ 3 & 4 & 0 \\ -40 & -20 & 40 \end{vmatrix}$$
$$= [4(40) - 0(-20)]\mathbf{i} - [3(40) - 0(-40)]\mathbf{j} + [3(-20) - 4(-40)]\mathbf{k}$$

In both cases,

$$\mathbf{M}_A = \{160\mathbf{i} - 120\mathbf{j} + 100\mathbf{k}\}\text{ N}\cdot\text{m}$$

The *magnitude* of $\mathbf{M}_A$ is therefore

$$M_A = \sqrt{(160)^2 + (-120)^2 + (100)^2} = 224\text{ N}\cdot\text{m} \qquad \textit{Ans.}$$

NOTE: As expected, $\mathbf{M}_A$ acts perpendicular to the shaded plane containing vectors $\mathbf{F}$, $\mathbf{r}_B$, and $\mathbf{r}_C$, Fig. 4–16*c*. (How would you find its coordinate direction angles $\alpha = 44.3°$, $\beta = 122°$, $\gamma = 63.4°$?) Had this problem been worked using a scalar approach, where $M_A = Fd$, notice the difficulty that can arise in obtaining the moment arm *d*.

EXAMPLE 4.5

Three forces act on the rod shown in Fig. 4–17*a*. Determine the resultant moment they create about the flange at *O* and determine the coordinate direction angles of the moment axis.

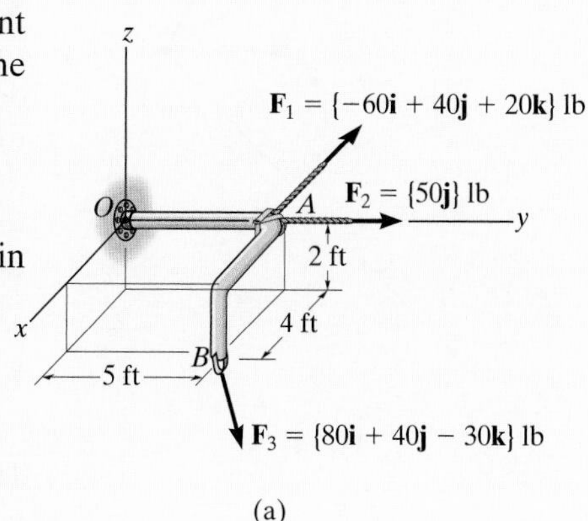

(a)

SOLUTION

Position vectors are directed from point *O* to each force as shown in Fig. 4–17*b*. These vectors are

$$\mathbf{r}_A = \{5\mathbf{j}\}\text{ ft}$$
$$\mathbf{r}_B = \{4\mathbf{i} + 5\mathbf{j} - 2\mathbf{k}\}\text{ ft}$$

The resultant moment about *O* is therefore

$$\begin{aligned}
\mathbf{M}_{R_O} &= \Sigma(\mathbf{r} \times \mathbf{F}) \\
&= \mathbf{r}_A \times \mathbf{F}_1 + \mathbf{r}_A \times \mathbf{F}_2 + \mathbf{r}_B \times \mathbf{F}_3 \\
&= \begin{vmatrix} \mathbf{i} & \mathbf{j} & \mathbf{k} \\ 0 & 5 & 0 \\ -60 & 40 & 20 \end{vmatrix} + \begin{vmatrix} \mathbf{i} & \mathbf{j} & \mathbf{k} \\ 0 & 5 & 0 \\ 0 & 50 & 0 \end{vmatrix} + \begin{vmatrix} \mathbf{i} & \mathbf{j} & \mathbf{k} \\ 4 & 5 & -2 \\ 80 & 40 & -30 \end{vmatrix} \\
&= [5(20) - 40(0)]\mathbf{i} - [0\mathbf{j}] + [0(40) - (-60)(5)]\mathbf{k} + [0\mathbf{i} - 0\mathbf{j} + 0\mathbf{k}] \\
&\quad + [5(-30) - (40)(-2)]\mathbf{i} - [4(-30) - 80(-2)]\mathbf{j} + [4(40) - 80(5)]\mathbf{k} \\
&= \{30\mathbf{i} - 40\mathbf{j} + 60\mathbf{k}\}\text{ lb}\cdot\text{ft} \qquad \textit{Ans.}
\end{aligned}$$

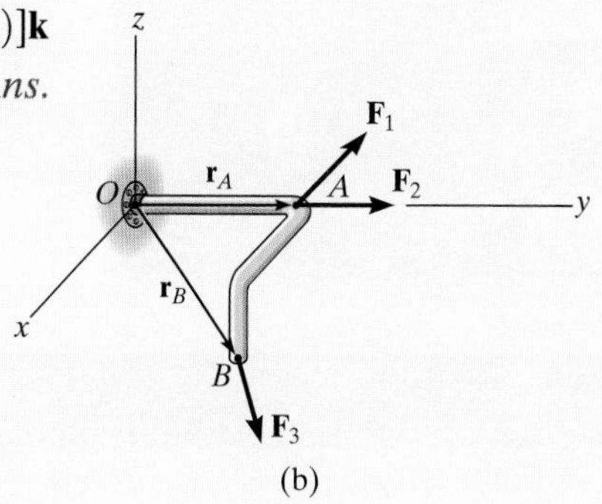

(b)

The moment axis is directed along the line of action of $\mathbf{M}_{R_O}$. Since the magnitude of this moment is

$$M_{R_O} = \sqrt{(30)^2 + (-40)^2 + (60)^2} = 78.10\text{ lb}\cdot\text{ft}$$

the unit vector which defines the direction of the moment axis is

$$\mathbf{u} = \frac{\mathbf{M}_{R_O}}{M_{R_O}} = \frac{30\mathbf{i} - 40\mathbf{j} + 60\mathbf{k}}{78.10} = 0.3841\mathbf{i} - 0.5121\mathbf{j} + 0.7682\mathbf{k}$$

Therefore, the coordinate direction angles of the moment axis are

$$\cos\alpha = 0.3841; \qquad \alpha = 67.4^\circ \qquad \textit{Ans.}$$
$$\cos\beta = -0.5121; \qquad \beta = 121^\circ \qquad \textit{Ans.}$$
$$\cos\gamma = 0.7682; \qquad \gamma = 39.8^\circ \qquad \textit{Ans.}$$

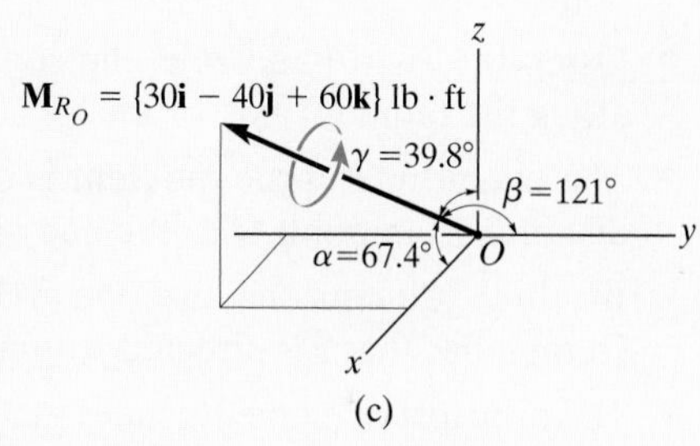

(c)

Fig. 4–17

NOTE: These results are shown in Fig. 4–17*c*. Realize that the three forces tend to cause the rod to rotate about this axis in the manner shown by the curl indicated on the moment vector.

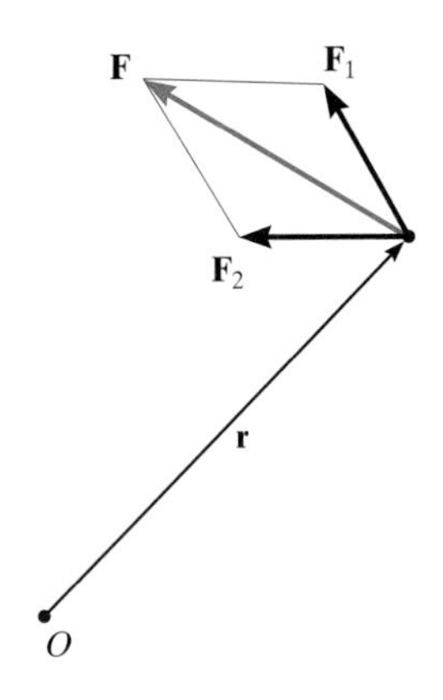

Fig. 4–18

4.4 Principle of Moments

A concept often used in mechanics is the *principle of moments*, which is sometimes referred to as *Varignon's theorem* since it was originally developed by the French mathematician Varignon (1654–1722). It states that *the moment of a force about a point is equal to the sum of the moments of the force's components about the point*. The proof follows directly from the distributive law of the vector cross product. To show this, consider the force **F** and two of its rectangular components, where $\mathbf{F} = \mathbf{F}_1 + \mathbf{F}_2$, Fig. 4–18. We have

$$\mathbf{M}_O = \mathbf{r} \times \mathbf{F}_1 + \mathbf{r} \times \mathbf{F}_2 = \mathbf{r} \times (\mathbf{F}_1 + \mathbf{F}_2) = \mathbf{r} \times \mathbf{F}$$

This concept has important applications to the solution of problems and proofs of theorems that follow, since it is often easier to determine the moments of a force's components rather than the moment of the force itself.

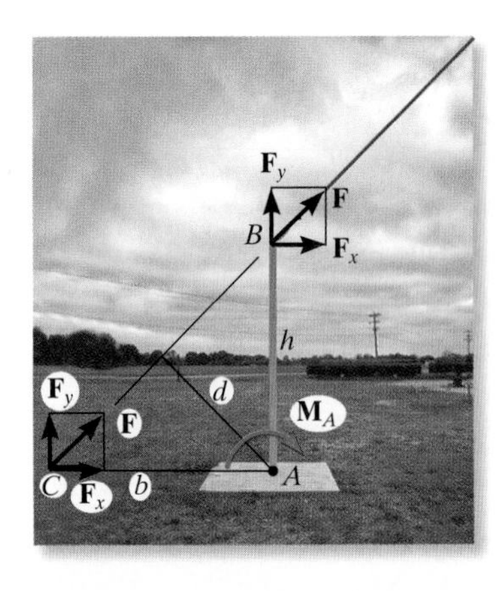

The guy cable exerts a force **F** on the pole and this creates a moment about the base at A of $M_A = Fd$. If the force is replaced by its two components $\mathbf{F}_x$ and $\mathbf{F}_y$ at point B where the cable acts on the pole, then the sum of the moments of these two components about A will yield the *same* resultant moment. For the calculation $\mathbf{F}_y$ will create zero moment about A and so $M_A = F_x h$. This is an application of the *principle of moments*. In addition we can apply the *principle of transmissibility* and slide the force to where its line of action intersects the ground at C. In this case $\mathbf{F}_x$ will create zero moment about A, and so $M_A = F_y b$.

Important Points

- The moment of a force indicates the tendency of a body to turn about an axis passing through a specific point O.
- Using the right-hand rule, the sense of rotation is indicated by the fingers, and the thumb is directed along the moment axis, or line of action of the moment.
- The magnitude of the moment is determined from $M_O = Fd$, where d is the perpendicular or shortest distance from point O to the line of action of the force **F**.
- In three dimensions use the vector cross product to determine the moment, i.e., $\mathbf{M}_O = \mathbf{r} \times \mathbf{F}$. Remember that **r** is directed *from* point O *to any point* on the line of action of **F**.
- The principle of moments states that the moment of a force about a point is equal to the sum of the moments of the force's components about the point. This is a very convenient method to use in two dimensions.

EXAMPLE 4.6

A 200-N force acts on the bracket shown in Fig. 4–19*a*. Determine the moment of the force about point *A*.

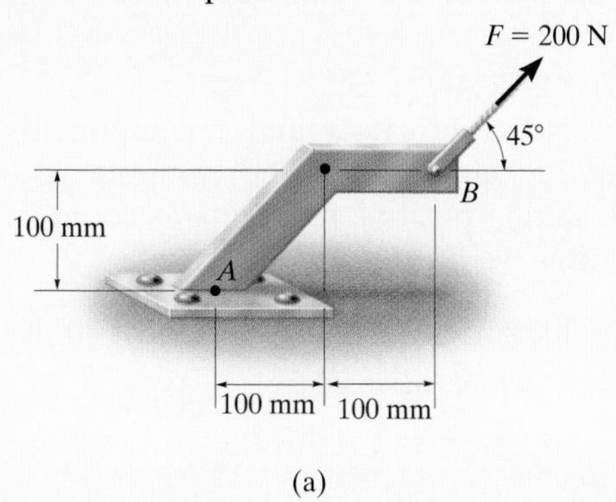

(a)

SOLUTION I

The moment arm *d* can be found by trigonometry, using the construction shown in Fig. 4–19*b*. From the right triangle *BCD*,

$$CB = d = 100 \cos 45° = 70.71 \text{ mm} = 0.070\,71 \text{ m}$$

Thus,

$$M_A = Fd = 200 \text{ N}(0.070\,71 \text{ m}) = 14.1 \text{ N}\cdot\text{m} \circlearrowleft$$

According to the right-hand rule, $\mathbf{M}_A$ is directed in the $+\mathbf{k}$ direction since the force tends to rotate or orbit *counterclockwise* about point *A*. Hence, reporting the moment as a Cartesian vector, we have

$$\mathbf{M}_A = \{14.1\mathbf{k}\} \text{ N}\cdot\text{m} \qquad \textit{Ans.}$$

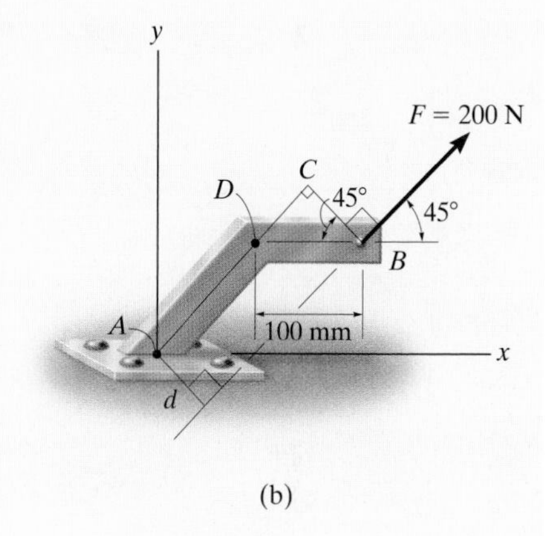

(b)

SOLUTION II

The 200-N force may be resolved into *x* and *y* components, as shown in Fig. 4–19*c*. In accordance with the principle of moments, the moment of **F** computed about point *A* is equivalent to the sum of the moments produced by the two force components. Assuming counterclockwise rotation as positive, i.e., in the $+\mathbf{k}$ direction, we can apply Eq. 4–2 $(M_A = \Sigma Fd)$, in which case

$$\circlearrowleft + M_A = (200 \sin 45° \text{ N})(0.20 \text{ m}) - (200 \cos 45° \text{ N})(0.10 \text{ m})$$
$$= 14.1 \text{ N}\cdot\text{m} \circlearrowleft$$

Thus

$$\mathbf{M}_A = \{14.1\mathbf{k}\} \text{ N}\cdot\text{m} \qquad \textit{Ans.}$$

NOTE: By comparison, it is seen that Solution II provides a more *convenient method* for analysis than Solution I since the moment arm for each component force is easier to establish.

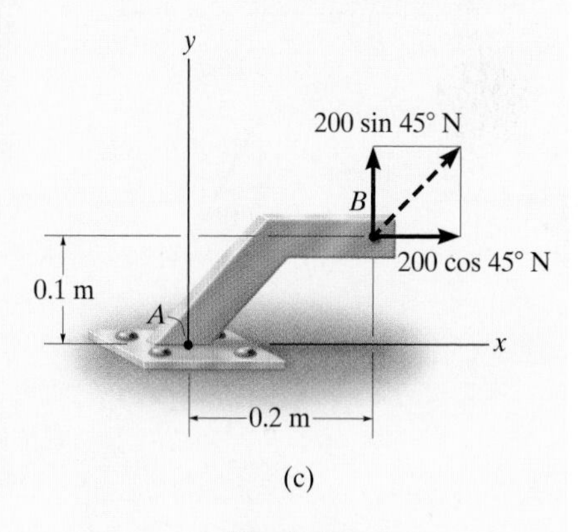

(c)

Fig. 4–19

EXAMPLE 4.7

The force **F** acts at the end of the angle bracket shown in Fig. 4–20*a*. Determine the moment of the force about point *O*.

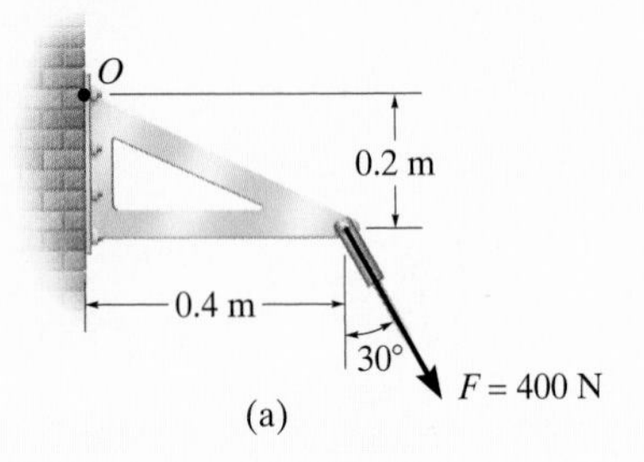

(a)

SOLUTION I (SCALAR ANALYSIS)

The force is resolved into its *x* and *y* components as shown in Fig. 4–20*b*, and the moments of the components are computed about point *O*. Taking positive moments as counterclockwise, i.e., in the $+\mathbf{k}$ direction, we have

$$\circlearrowleft + M_O = 400 \sin 30^\circ \text{ N}(0.2 \text{ m}) - 400 \cos 30^\circ \text{ N}(0.4 \text{ m})$$
$$= -98.6 \text{ N}\cdot\text{m} = 98.6 \text{ N}\cdot\text{m} \circlearrowright$$

or

$$\mathbf{M}_O = \{-98.6\mathbf{k}\} \text{ N}\cdot\text{m} \qquad \textit{Ans.}$$

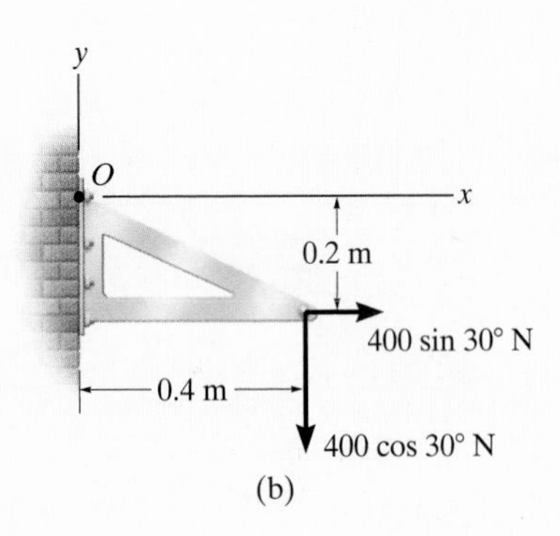

(b)

SOLUTION II (VECTOR ANALYSIS)

Using a Cartesian vector approach, the force and position vectors shown in Fig. 4–20*c* can be represented as

$$\mathbf{r} = \{0.4\mathbf{i} - 0.2\mathbf{j}\} \text{ m}$$
$$\mathbf{F} = \{400 \sin 30^\circ\mathbf{i} - 400 \cos 30^\circ\mathbf{j}\} \text{ N}$$
$$= \{200.0\mathbf{i} - 346.4\mathbf{j}\} \text{ N}$$

The moment is therefore

$$\mathbf{M}_O = \mathbf{r} \times \mathbf{F} = \begin{vmatrix} \mathbf{i} & \mathbf{j} & \mathbf{k} \\ 0.4 & -0.2 & 0 \\ 200.0 & -346.4 & 0 \end{vmatrix}$$
$$= 0\mathbf{i} - 0\mathbf{j} + [0.4(-346.4) - (-0.2)(200.0)]\mathbf{k}$$
$$= \{-98.6\mathbf{k}\} \text{ N}\cdot\text{m} \qquad \textit{Ans.}$$

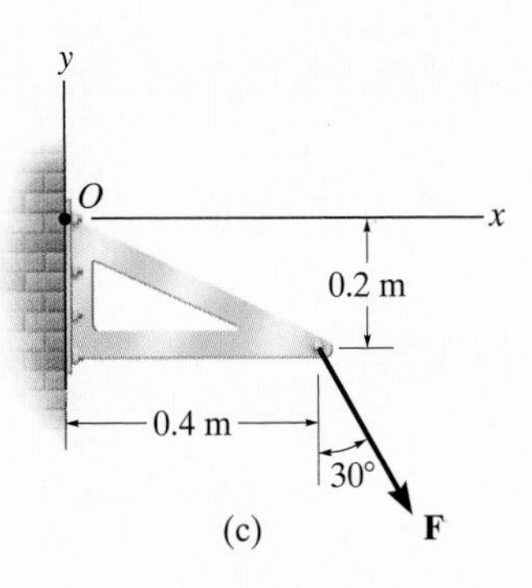

(c)

Fig. 4–20

NOTE: By comparison, it is seen that the scalar analysis (Solution I) provides a more *convenient method* for analysis than Solution II since the direction of the moment and the moment arm for each component force are easy to establish. Hence, this method is generally recommended for solving problems displayed in two dimensions. On the other hand, Cartesian vector analysis is generally recommended only for solving three-dimensional problems, where the moment arms and force components are often more difficult to determine.

PROBLEMS

4–1. If **A**, **B**, and **D** are given vectors, prove the distributive law for the vector cross product, i.e., $\mathbf{A} \times (\mathbf{B} + \mathbf{D}) = (\mathbf{A} \times \mathbf{B}) + (\mathbf{A} \times \mathbf{D})$.

4–2. Prove the triple scalar product identity $\mathbf{A} \cdot (\mathbf{B} \times \mathbf{C}) = (\mathbf{A} \times \mathbf{B}) \cdot \mathbf{C}$.

4–3. Given the three nonzero vectors **A**, **B**, and **C**, show that if $\mathbf{A} \cdot (\mathbf{B} \times \mathbf{C}) = 0$, the three vectors *must* lie in the same plane.

***4–4.** Determine the magnitude and directional sense of the resultant moment of the forces at A and B about point O.

4–5. Determine the magnitude and directional sense of the resultant moment of the forces at A and B about point P.

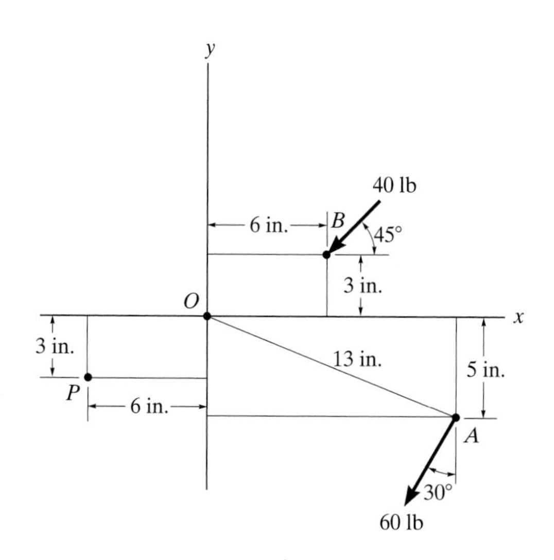

Probs. 4–4/5

4–6. Determine the magnitude of the force **F** that should be applied at the end of the lever such that this force creates a clockwise moment of 15 N · m about point O when $\theta = 30°$.

■**4–7.** If the force $F = 100$ N, determine the angle θ $(0 \leq \theta \leq 90°)$ so that the force develops a clockwise moment about point O of 20 N · m.

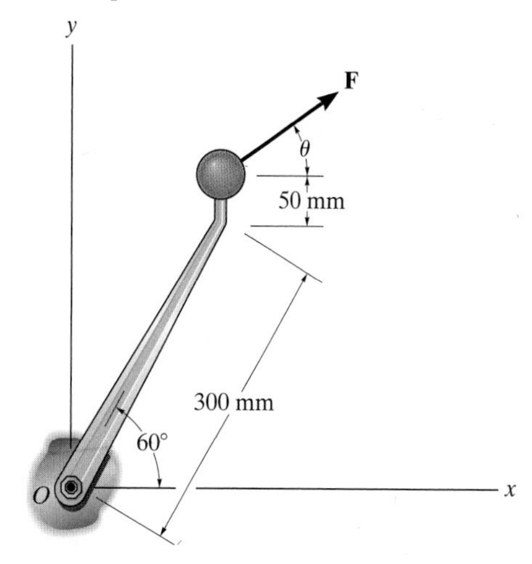

Probs. 4–6/7

***4–8.** Determine the magnitude and directional sense of the resultant moment of the forces about point O.

4–9. Determine the magnitude and directional sense of the resultant moment of the forces about point P.

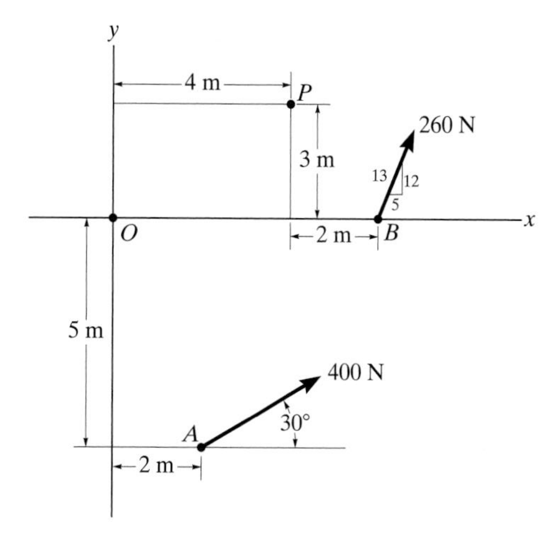

Probs. 4–8/9

4–10. A force of 40 N is applied to the wrench. Determine the moment of this force about point O. Solve the problem using both a scalar analysis and a vector analysis.

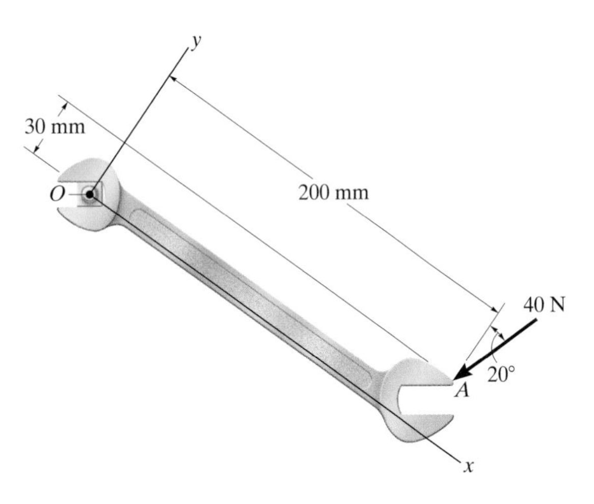

Prob. 4–10

4–11. Determine the magnitude and directional sense of the resultant moment of the forces about point O.

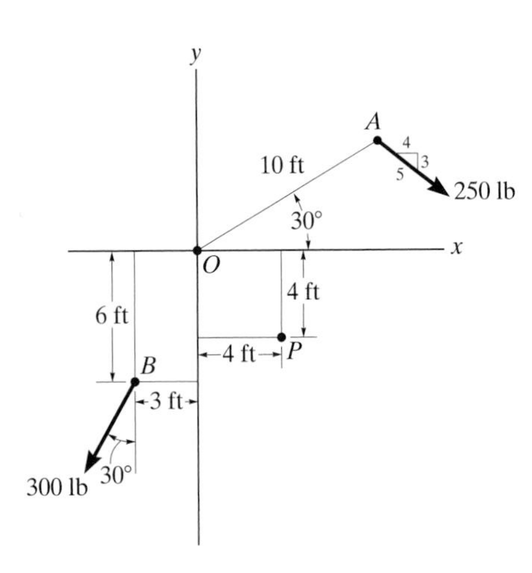

Prob. 4–11

*__4–12.__ To correct a birth defect, the tibia of the leg is straightened using three wires that are attached through holes made in the bone and then to an external brace that is worn by the patient. Determine the moment of each wire force about joint A.

4–13. To correct a birth defect, the tibia of the leg is straightened using three wires that are attached through holes made in the bone and then to an external brace that is worn by the patient. Determine the moment of each wire force about joint B.

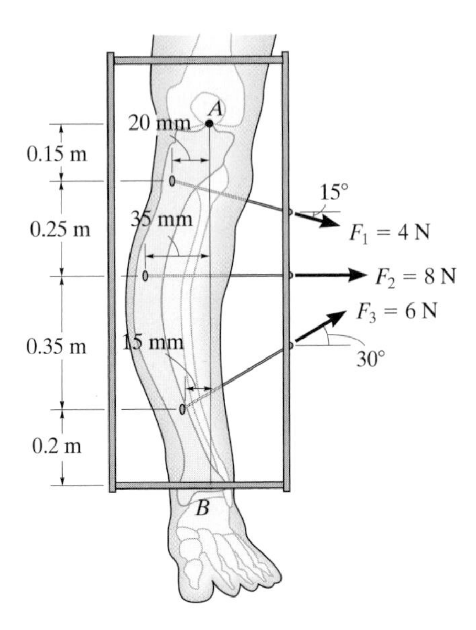

Probs. 4–12/13

4–14. Determine the moment of each force about the bolt located at A. Take $F_B = 40$ lb, $F_C = 50$ lb.

4–15. If $F_B = 30$ lb and $F_C = 45$ lb, determine the resultant moment about the bolt located at A.

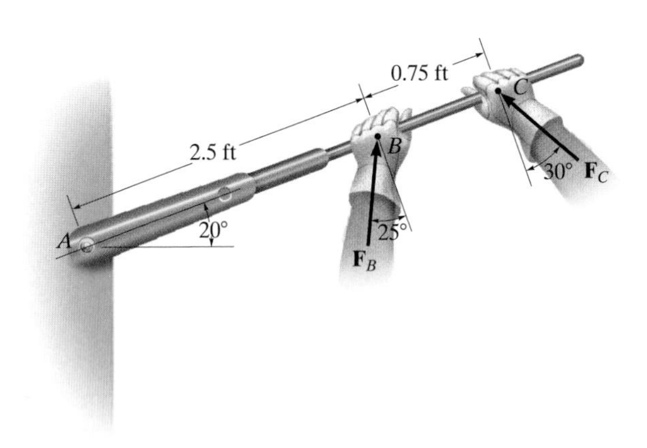

Probs. 4–14/15

*4–16. The elbow joint is flexed using the biceps brachii muscle, which remains essentially vertical as the arm moves in the vertical plane. If this muscle is located a distance of 16 mm from the pivot point A on the humerus, determine the variation of the moment capacity about A if the constant force developed by the muscle is 2.30 kN. Plot these results of M vs. θ for $-60^\circ \leq \theta \leq 80^\circ$.

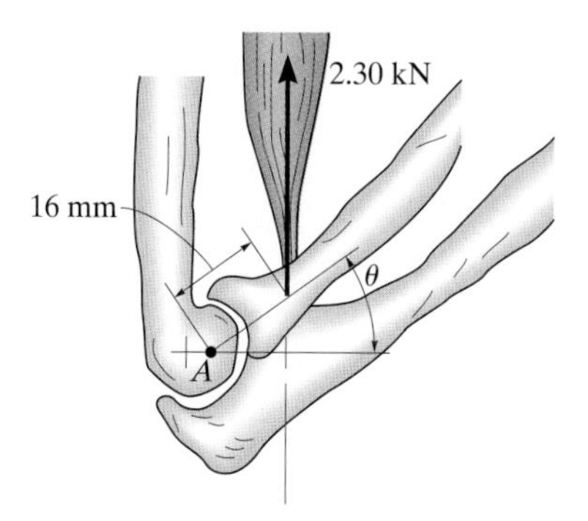

Prob. 4–16

4–17. The Snorkel Co. produces the articulating boom platform that can support a weight of 550 lb. If the boom is in the position shown, determine the moment of this force about points A, B, and C.

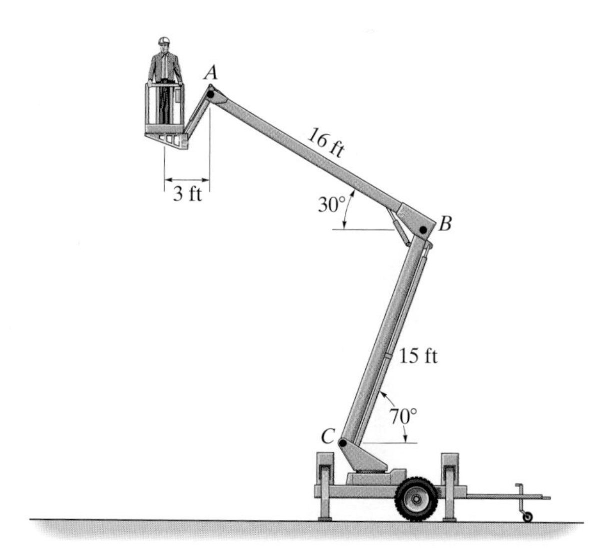

Prob. 4–17

4–18. Determine the direction θ ($0^\circ \leq \theta \leq 180^\circ$) of the force $F = 40$ lb so that it produces (a) the maximum moment about point A and (b) the minimum moment about point A. Compute the moment in each case.

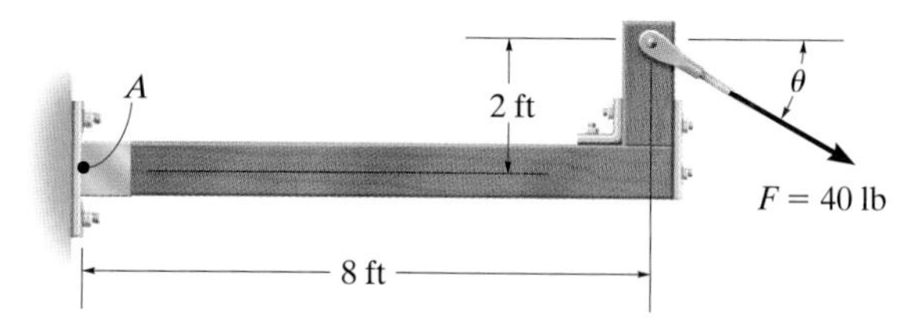

Prob. 4–18

4–19. The rod on the power control mechanism for a business jet is subjected to a force of 80 N. Determine the moment of this force about the bearing at A.

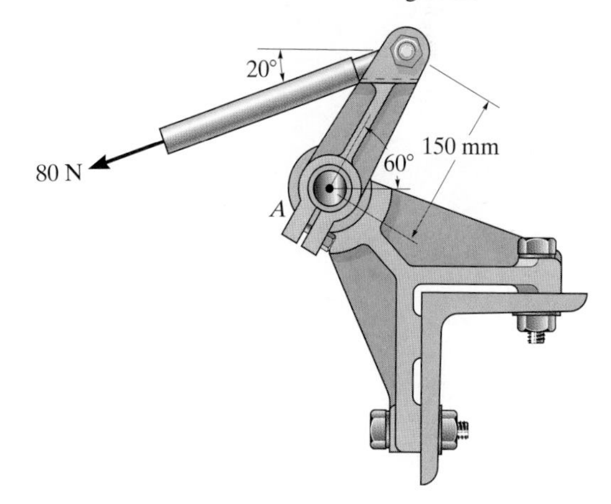

Prob. 4–19

*4–20. The boom has a length of 30 ft, a weight of 800 lb, and mass center at G. If the maximum moment that can be developed by the motor at A is $M = 20(10^3)$ lb · ft, determine the maximum load W, having a mass center at G', that can be lifted. Take $\theta = 30^\circ$.

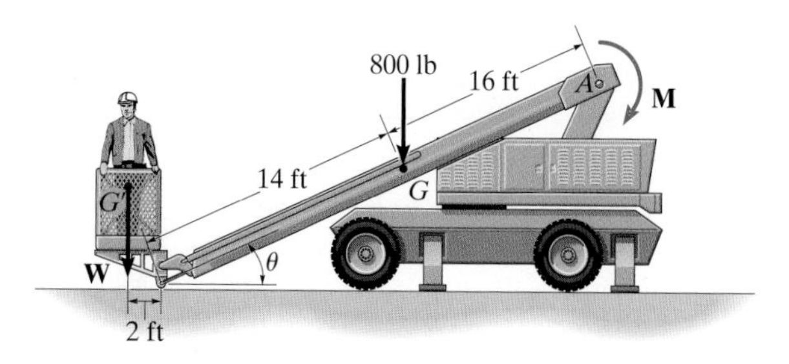

Prob. 4–20

4–21. The tool at A is used to hold a power lawnmower blade stationary while the nut is being loosened with the wrench. If a force of 50 N is applied to the wrench at B in the direction shown, determine the moment it creates about the nut at C. What is the magnitude of force $\mathbf{F}$ at A so that it creates the opposite moment about C?

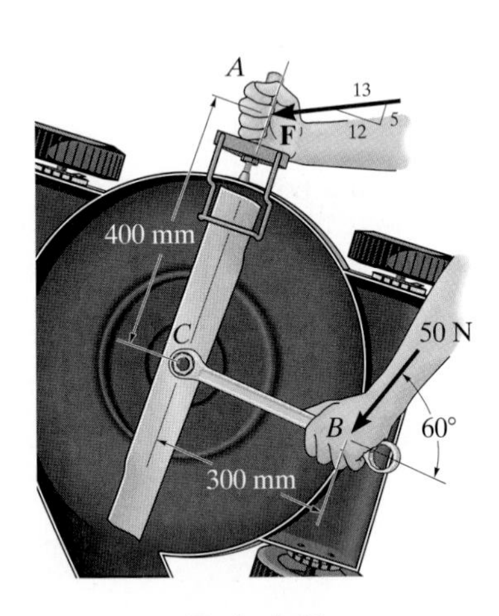

Prob. 4–21

4–22. Determine the clockwise direction θ ($0° \leq \theta \leq 180°$) of the force $F = 80$ lb so that it produces (a) the maximum moment about point A and (b) no moment about point A. Compute the moment in each case.

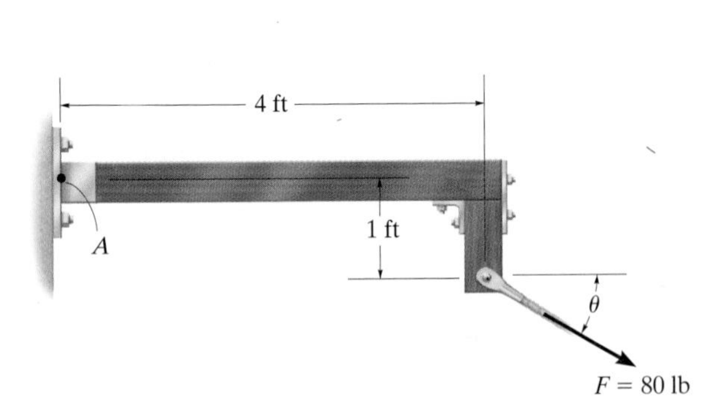

Prob. 4–22

4–23. The Y-type structure is used to support the high voltage transmission cables. If the supporting cables each exert a force of 275 lb on the structure at B, determine the moment of each force about point A. Also, by the principle of transmissibility, locate the forces at points C and D and determine the moments.

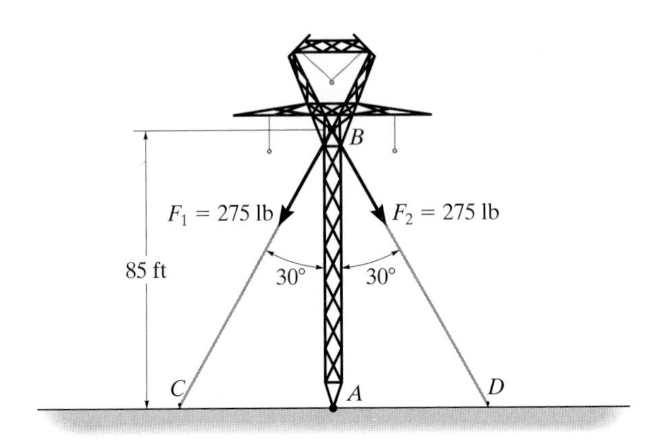

Prob. 4–23

***4–24.** The 70-N force acts on the end of the pipe at B. Determine (a) the moment of this force about point A, and (b) the magnitude and direction of a horizontal force, applied at C, which produces the same moment. Take $\theta = 60°$.

4–25. The 70-N force acts on the end of the pipe at B. Determine the angles θ ($0° \leq \theta \leq 180°$) of the force that will produce maximum and minimum moments about point A. What are the magnitudes of these moments?

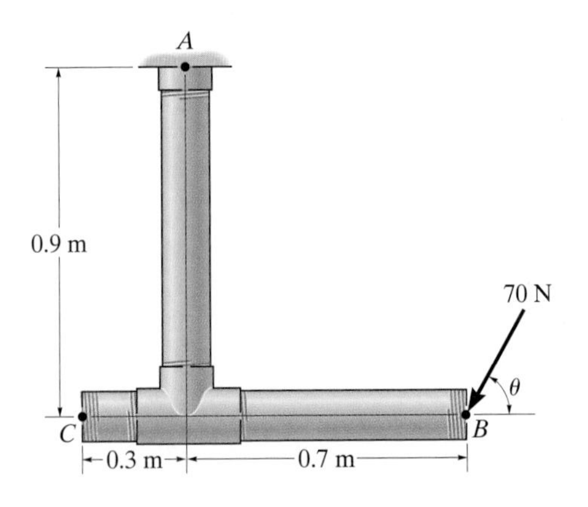

Probs. 4–24/25

4–26. The towline exerts a force of $P = 4$ kN at the end of the 20-m-long crane boom. If $\theta = 30°$, determine the placement x of the hook at A so that this force creates a maximum moment about point O. What is this moment?

4–27. The towline exerts a force of $P = 4$ kN at the end of the 20-m-long crane boom. If $x = 25$ m, determine the position θ of the boom so that this force creates a maximum moment about point O. What is this moment?

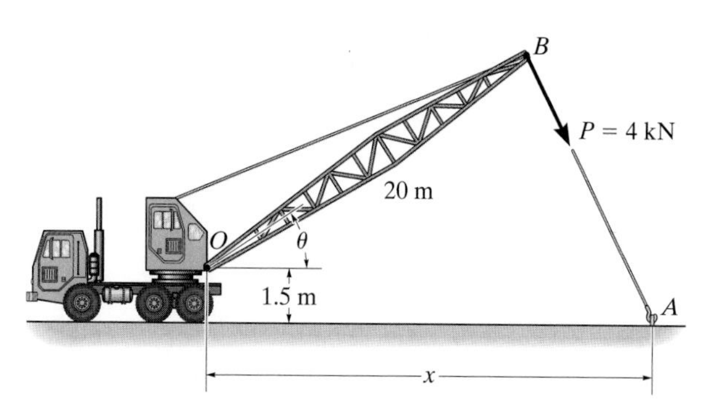

Probs. 4–26/27

***4–28.** Determine the resultant moment of the forces about point A. Solve the problem first by considering each force as a whole, and then by using the principle of moments. Take $F_1 = 250$ N, $F_2 = 300$ N, $F_3 = 500$ N.

4–29. If the resultant moment about point A is 4800 N · m clockwise, determine the magnitude of $\mathbf{F}_3$ if $F_1 = 300$ N and $F_2 = 400$ N.

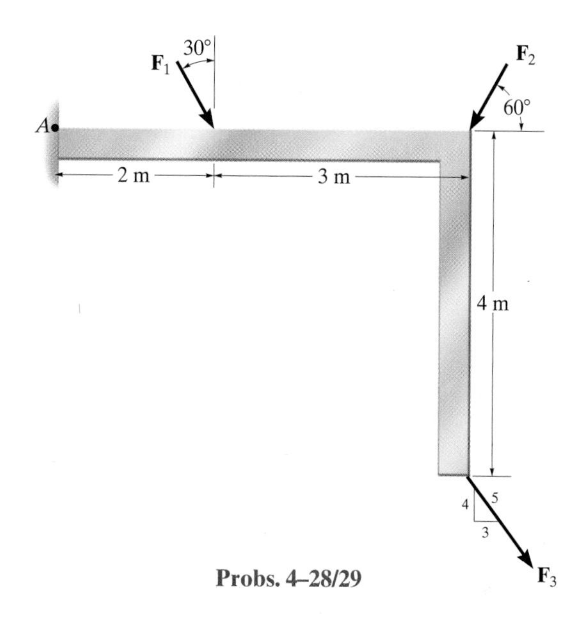

Probs. 4–28/29

4–30. The flat-belt tensioner is manufactured by the Daton Co. and is used with V-belt drives on poultry and livestock fans. If the tension in the belt is 52 lb, when this pulley is not turning, determine the moment of each of these forces about the pin at A.

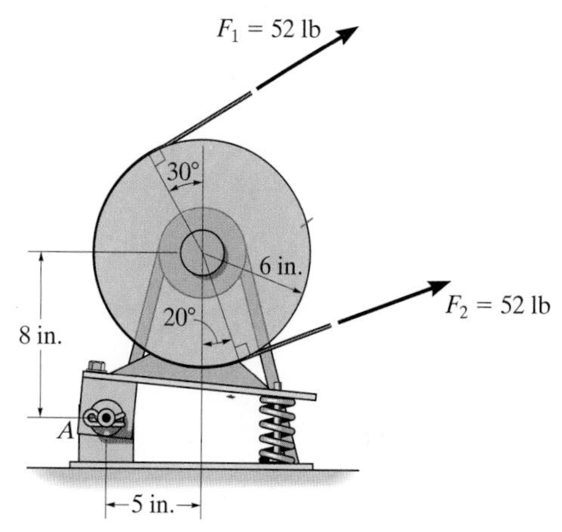

Prob. 4–30

4–31. The worker is using the bar to pull two pipes together in order to complete the connection. If he applies a horizontal force of 80 lb to the handle of the lever, determine the moment of this force about the end A. What would be the tension T in the cable needed to cause the opposite moment about point A?

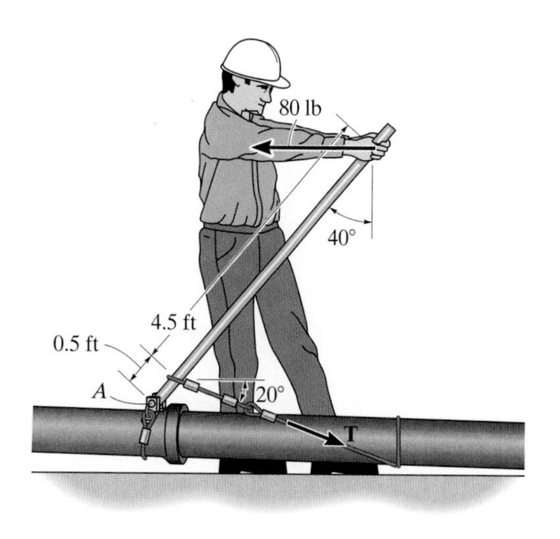

Prob. 4–31

*4–32. If it takes a force of $F = 125$ lb to pull the nail out, determine the smallest vertical force **P** that must be applied to the handle of the crowbar. *Hint:* This requires the moment of **F** about point A to be equal to the moment of **P** about A. Why?

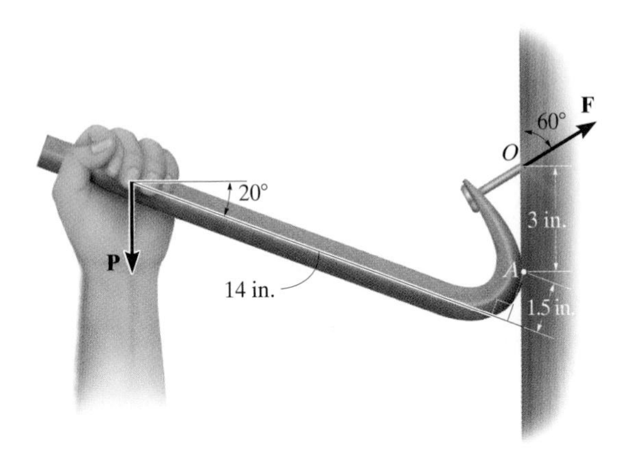

Prob. 4–32

■4–33. The pipe wrench is activated by pulling on the cable segment with a horizontal force of 500 N. Determine the moment M_A produced by the wrench on the pipe when $\theta = 20°$. Neglect the size of the pulley.

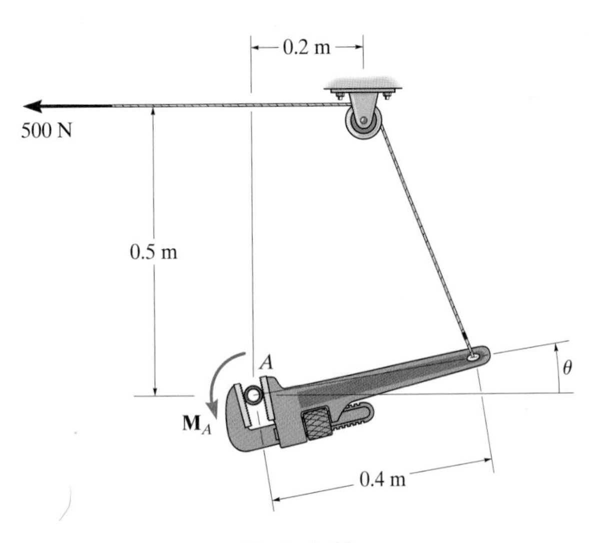

Prob. 4–33

4–34. Determine the moment of the force **F** at A about point O. Express the result as a Cartesian vector.

4–35. Determine the moment of the force **F** at A about point P. Express the result as a Cartesian vector.

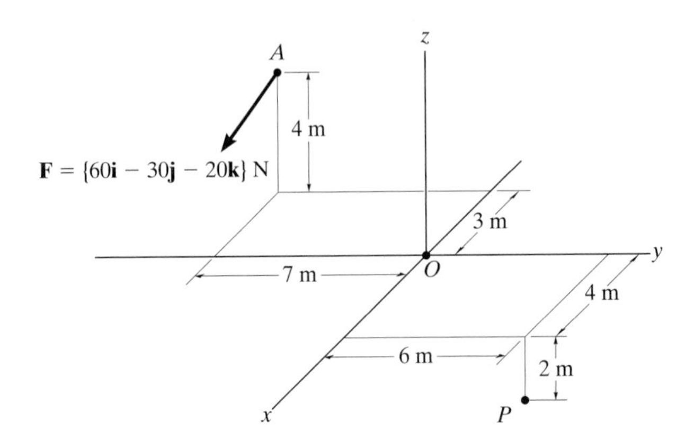

Probs. 4–34/35

*4–36. Determine the moment of the force **F** at A about point O. Express the result as a Cartesian vector.

4–37. Determine the moment of the force **F** at A about point P. Express the result as a Cartesian vector.

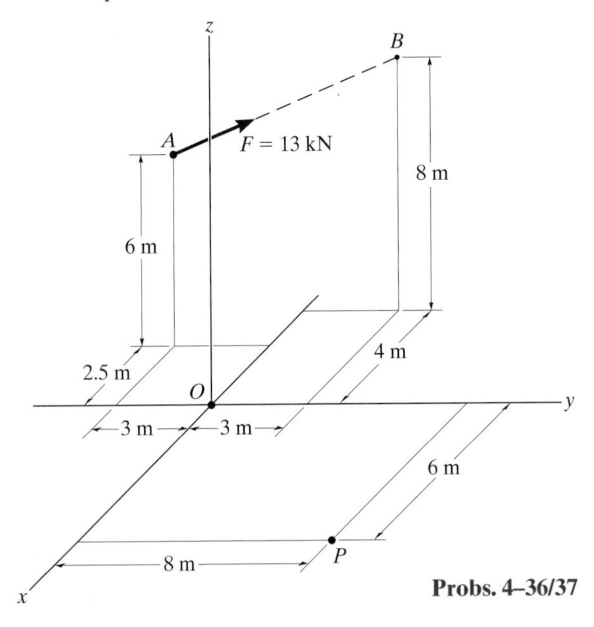

Probs. 4–36/37

4–38. The curved rod lies in the x–y plane and has a radius of 3 m. If a force of $F = 80$ N acts at its end as shown, determine the moment of this force about point O.

4–39. The curved rod lies in the x–y plane and has a radius of 3 m. If a force of $F = 80$ N acts at its end as shown, determine the moment of this force about point B.

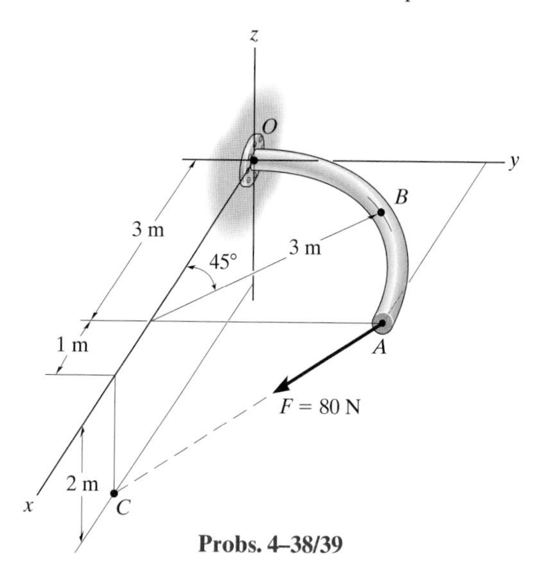

Probs. 4–38/39

4–41. The pole supports a 22-lb traffic light. Using Cartesian vectors, determine the moment of the weight of the traffic light about the base of the pole at A.

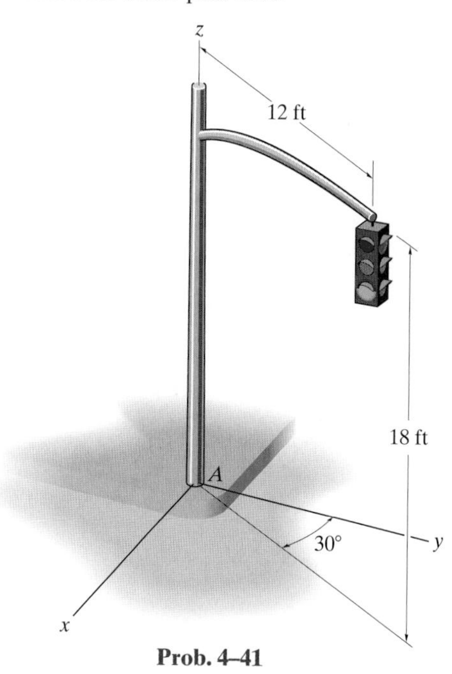

Prob. 4–41

***4–40.** The force $\mathbf{F} = \{600\mathbf{i} + 300\mathbf{j} - 600\mathbf{k}\}$ N acts at the end of the beam. Determine the moment of the force about point A.

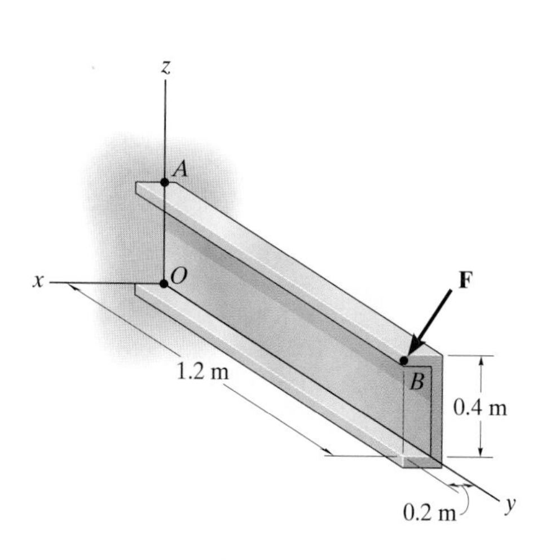

Prob. 4–40

4–42. The man pulls on the rope with a force of $F = 20$ N. Determine the moment that this force exerts about the base of the pole at O. Solve the problem two ways, i.e., by using a position vector from O to A, then O to B.

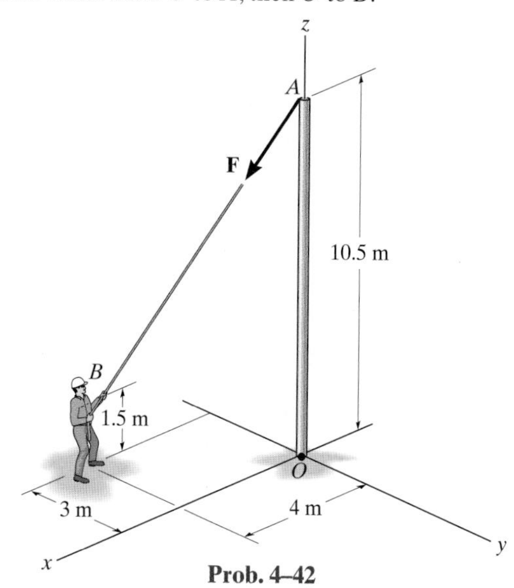

Prob. 4–42

4–43. Determine the smallest force F that must be applied along the rope in order to cause the curved rod, which has a radius of 5 ft, to fail at the support C. This requires a moment of $M = 80$ lb · ft to be developed at C.

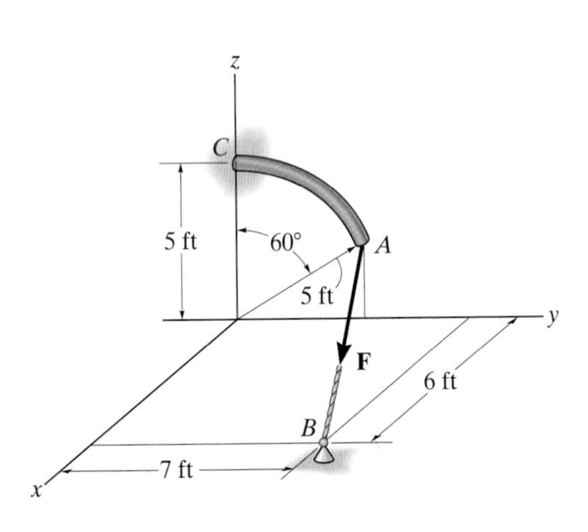

Prob. 4–43

***4–44.** The pipe assembly is subjected to the 80-N force. Determine the moment of this force about point A.

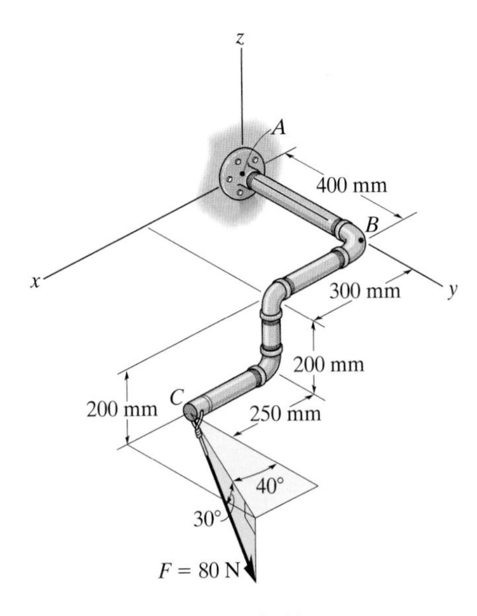

Prob. 4–44

4–45. The pipe assembly is subjected to the 80-N force. Determine the moment of this force about point B.

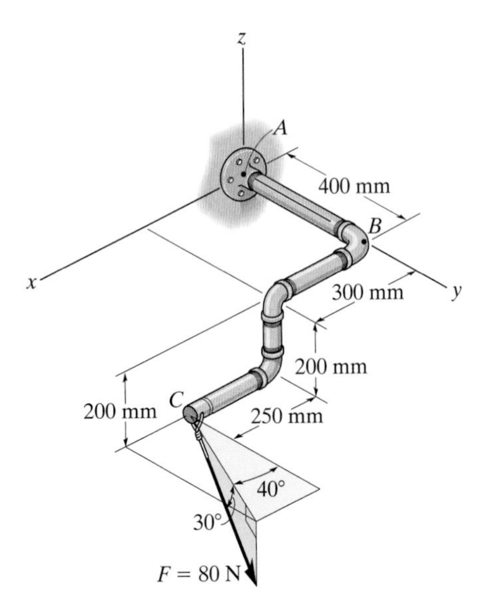

Prob. 4–45

4–46. The x-ray machine is used for medical diagnosis. If the camera and housing at C have a mass of 150 kg and a mass center at G, determine the moment of its weight about point O when it is in the position shown.

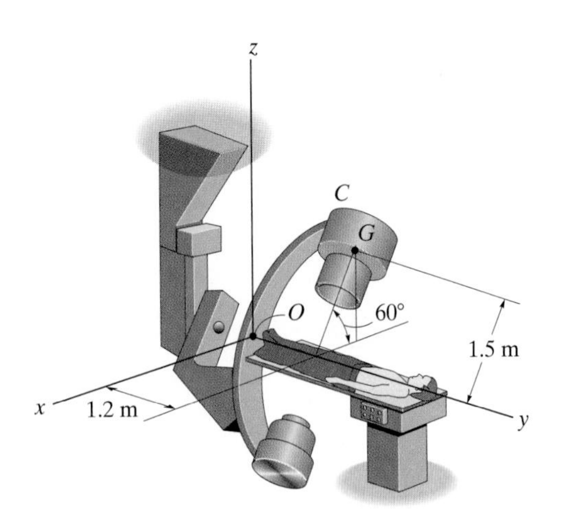

Prob. 4–46

4–47. Using Cartesian vector analysis, determine the resultant moment of the three forces about the base of the column at A. Take $\mathbf{F}_1 = \{400\mathbf{i} + 300\mathbf{j} + 120\mathbf{k}\}$ N.

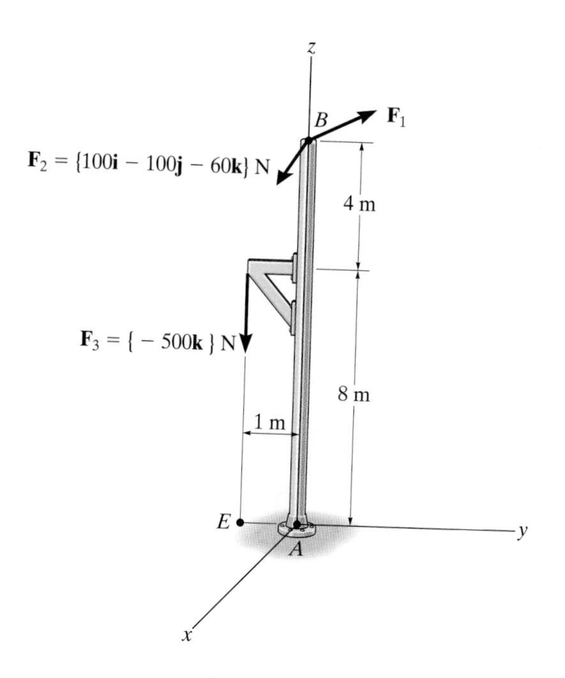

Prob. 4–47

*4–48. A force of $\mathbf{F} = \{6\mathbf{i} - 2\mathbf{j} + 1\mathbf{k}\}$ kN produces a moment of $\mathbf{M}_O = \{4\mathbf{i} + 5\mathbf{j} - 14\mathbf{k}\}$ kN·m about the origin of coordinates, point O. If the force acts at a point having an x coordinate of $x = 1$ m, determine the y and z coordinates.

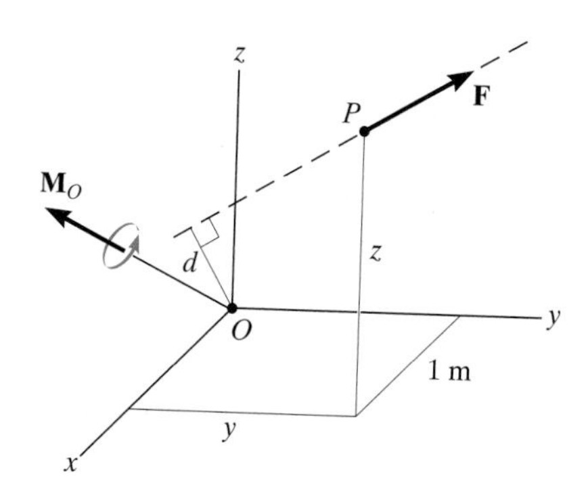

Prob. 4–48

4–49. The force $\mathbf{F} = \{6\mathbf{i} + 8\mathbf{j} + 10\mathbf{k}\}$ N creates a moment about point O of $\mathbf{M}_O = \{-14\mathbf{i} + 8\mathbf{j} + 2\mathbf{k}\}$ N·m. If the force passes through a point having an x coordinate of 1 m, determine the y and z coordinates of the point. Also, realizing that $M_O = Fd$, determine the perpendicular distance d from point O to the line of action of $\mathbf{F}$.

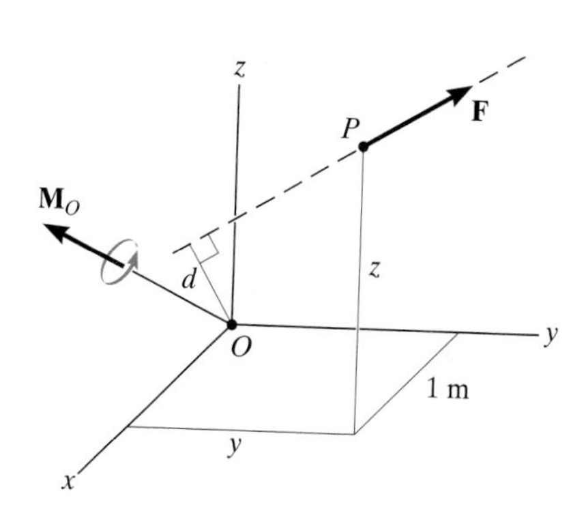

Prob. 4–49

4–50. A force of $\mathbf{F} = \{6\mathbf{i} - 2\mathbf{j} + 1\mathbf{k}\}$ kN produces a moment of $\mathbf{M}_O = \{4\mathbf{i} + 5\mathbf{j} - 14\mathbf{k}\}$ kN·m about the origin of coordinates, point O. If the force acts at a point having an x coordinate of $x = 1$ m, determine the y and z coordinates.

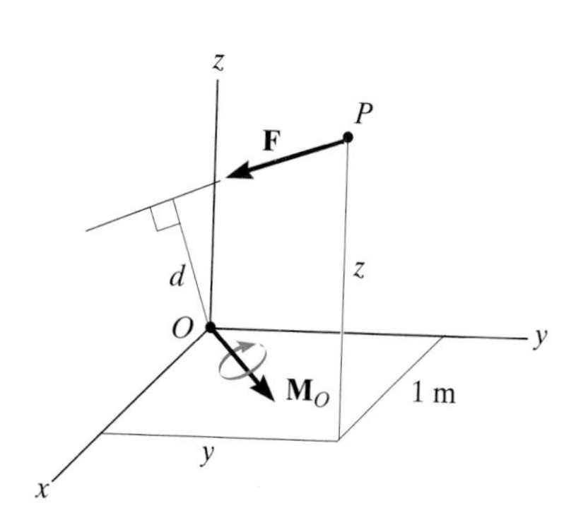

Prob. 4–50

4.5 Moment of a Force about a Specified Axis

Recall that when the moment of a force is computed about a point, the moment and its axis are *always* perpendicular to the plane containing the force and the moment arm. In some problems it is important to find the *component* of this moment along a *specified axis* that passes through the point. To solve this problem either a scalar or vector analysis can be used.

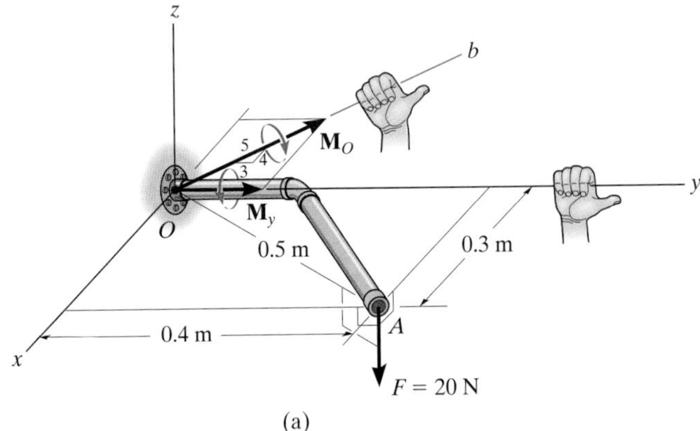

Fig. 4–21 (a)

Scalar Analysis. As a numerical example of this problem, consider the pipe assembly shown in Fig. 4–21*a*, which lies in the horizontal plane and is subjected to the vertical force of $F = 20$ N applied at point A. The moment of this force about point O has a *magnitude* of $M_O = (20 \text{ N})(0.5 \text{ m}) = 10 \text{ N} \cdot \text{m}$, and a *direction* defined by the right-hand rule, as shown in Fig. 4–21*a*. This moment tends to turn the pipe about the Ob axis. For practical reasons, however, it may be necessary to determine the *component* of $\mathbf{M}_O$ about the y axis, $\mathbf{M}_y$, since this component tends to unscrew the pipe from the flange at O. From Fig. 4–21*a*, $\mathbf{M}_y$ has a magnitude of $M_y = \frac{3}{5}(10 \text{ N} \cdot \text{m}) = 6 \text{ N} \cdot \text{m}$ and a sense of direction shown by the vector resolution. Rather than performing this *two-step* process of first finding the moment of the force about point O and then resolving the moment along the y axis, it is also possible to solve this problem *directly*. To do so, it is necessary to determine the perpendicular or moment-arm distance from the line of action of $\mathbf{F}$ to the y axis. From Fig. 4–21*a* this distance is 0.3 m. Thus the *magnitude* of the moment of the force about the y axis is again $M_y = 0.3(20 \text{ N}) = 6 \text{ N} \cdot \text{m}$, and the *direction* is determined by the right-hand rule as shown.

In general, then, *if the line of action of a force* $\mathbf{F}$ *is perpendicular to any specified axis aa*, the magnitude of the moment of $\mathbf{F}$ about the axis can be determined from the equation

$$M_a = Fd_a \qquad (4\text{–}10)$$

Here d_a is the *perpendicular or shortest distance* from the force line of action to the axis. The direction is determined from the thumb of the right hand when the fingers are curled in accordance with the direction of rotation as produced by the force. In particular, realize that a *force will not contribute a moment about a specified axis if the force line of action is parallel to the axis or its line of action passes through the axis.*

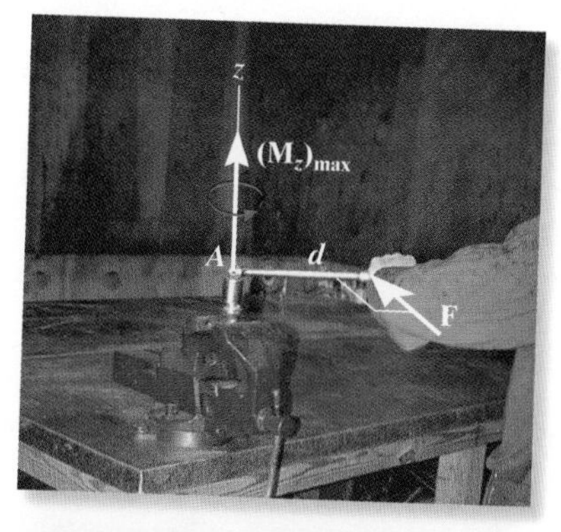

If a horizontal force **F** is applied to the handle of the flex-headed wrench, it tends to turn the socket at A about the z axis. This effect is caused by the moment of **F** about the z axis. The *maximum moment* is determined when the wrench is in the horizontal plane so that full leverage from the handle can be achieved, i.e., $(M_z)_{\max} = Fd$. If the handle is not in the horizontal position, then the moment about the z axis is determined from $M_z = Fd'$, where d' is the perpendicular distance from the force line of action to the axis. We can also determine this moment by first finding the moment of **F** about A, $M_A = Fd$, then finding the projection or component of this moment along z, i.e., $M_z = M_A \cos\theta$.

Vector Analysis. The previous two-step solution of first finding the moment of the force about a point on the axis and then finding the projected component of the moment about the axis can also be performed using a vector analysis, Fig. 4–21*b*. Here the moment about point O is first determined from $\mathbf{M}_O = \mathbf{r}_A \times \mathbf{F} = (0.3\mathbf{i} + 0.4\mathbf{j}) \times (-20\mathbf{k}) = \{-8\mathbf{i} + 6\mathbf{j}\}\ \text{N}\cdot\text{m}$. The component or projection of this moment along the y axis is then determined from the dot product (Sec. 2.9). Since the unit vector for this axis (or line) is $\mathbf{u}_a = \mathbf{j}$, then $M_y = \mathbf{M}_O \cdot \mathbf{u}_a = (-8\mathbf{i} + 6\mathbf{j}) \cdot \mathbf{j} = 6\ \text{N}\cdot\text{m}$. This result, of course, is to be expected, since it represents the **j** component of $\mathbf{M}_O$.

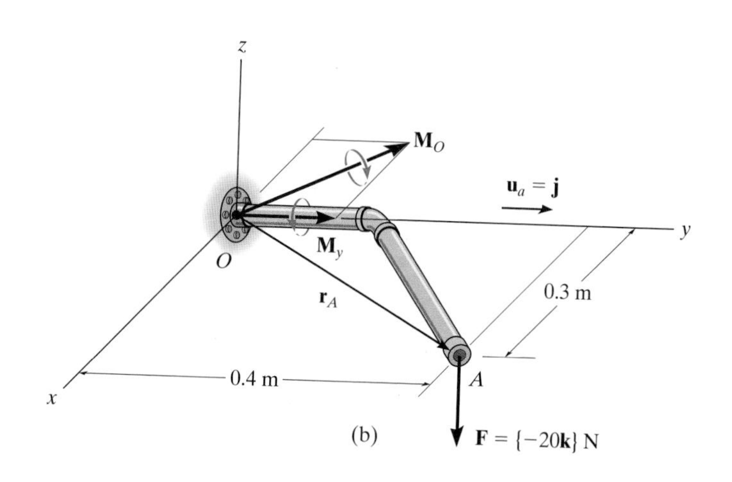

(b)

Fig. 4–21 (cont.)

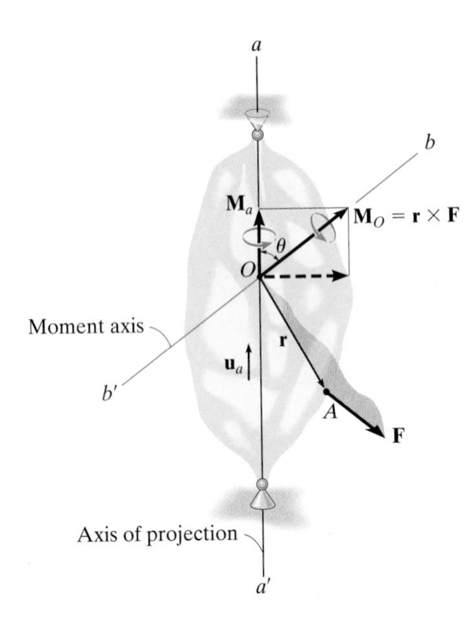

Fig. 4–22

A vector analysis such as this is particularly advantageous for finding the moment of a force about an axis when the force components or the appropriate moment arms are difficult to determine. For this reason, the above two-step process will now be generalized and applied to a body of arbitrary shape. To do so, consider the body in Fig. 4–22, which is subjected to the force **F** acting at point A. Here we wish to determine the effect of **F** in tending to rotate the body about the aa' axis. This tendency for rotation is measured by the moment component $\mathbf{M}_a$. To determine $\mathbf{M}_a$ we first compute the moment of **F** about any *arbitrary point O* that lies on the aa' axis. In this case, $\mathbf{M}_O$ is expressed by the cross product $\mathbf{M}_O = \mathbf{r} \times \mathbf{F}$, where **r** is directed from O to A. Here $\mathbf{M}_O$ acts along the moment axis bb', and so the component or projection of $\mathbf{M}_O$ onto the aa' axis is then $\mathbf{M}_a$. The *magnitude* of $\mathbf{M}_a$ is determined by the dot product, $M_a = M_O \cos\theta = \mathbf{M}_O \cdot \mathbf{u}_a$ where $\mathbf{u}_a$ is a unit vector that defines the direction of the aa' axis. Combining these two steps as a general expression, we have $M_a = (\mathbf{r} \times \mathbf{F}) \cdot \mathbf{u}_a$. Since the dot product is commutative, we can also write

$$M_a = \mathbf{u}_a \cdot (\mathbf{r} \times \mathbf{F})$$

In vector algebra, this combination of dot and cross product yielding the scalar M_a is called the *triple scalar product*. Provided *x, y, z* axes are established and the Cartesian components of each of the vectors can be determined, then the triple scalar product may be written in determinant form as

$$M_a = (u_{a_x}\mathbf{i} + u_{a_y}\mathbf{j} + u_{a_z}\mathbf{k}) \cdot \begin{vmatrix} \mathbf{i} & \mathbf{j} & \mathbf{k} \\ r_x & r_y & r_z \\ F_x & F_y & F_z \end{vmatrix}$$

or simply

$$M_a = \mathbf{u}_a \cdot (\mathbf{r} \times \mathbf{F}) = \begin{vmatrix} u_{a_x} & u_{a_y} & u_{a_z} \\ r_x & r_y & r_z \\ F_x & F_y & F_z \end{vmatrix} \tag{4–11}$$

where

$u_{a_x}, u_{a_y}, u_{a_z}$ represent the *x, y, z* components of the unit vector defining the direction of the aa' axis

r_x, r_y, r_z represent the *x, y, z* components of the position vector drawn from *any point O* on the aa' axis to *any point A* on the line of action of the force

F_x, F_y, F_z represent the *x, y, z* components of the force vector.

When M_a is evaluated from Eq. 4–11, it will yield a positive or negative scalar. The sign of this scalar indicates the sense of direction of $\mathbf{M}_a$ along the aa' axis. If it is positive, then $\mathbf{M}_a$ will have the same sense as $\mathbf{u}_a$, whereas if it is negative, then $\mathbf{M}_a$ will act opposite to $\mathbf{u}_a$.

Once M_a is determined, we can then express $\mathbf{M}_a$ as a Cartesian vector, namely,

$$\mathbf{M}_a = M_a\mathbf{u}_a = [\mathbf{u}_a \cdot (\mathbf{r} \times \mathbf{F})]\mathbf{u}_a \qquad (4\text{–}12)$$

Finally, if the resultant moment of a series of forces is to be computed about the aa' axis, then the moment components of each force are added together *algebraically*, since each component lies along the same axis. Thus the magnitude of $\mathbf{M}_a$ is

$$M_a = \Sigma[\mathbf{u}_a \cdot (\mathbf{r} \times \mathbf{F})] = \mathbf{u}_a \cdot \Sigma(\mathbf{r} \times \mathbf{F})$$

The examples which follow illustrate a numerical application of the above concepts.

Wind blowing on the face of this traffic sign creates a resultant force $\mathbf{F}$ that tends to tip the sign over due to the moment $\mathbf{M}_A$ created about the a–a axis. The moment of $\mathbf{F}$ about a point A that lies on the axis is $\mathbf{M}_A = \mathbf{r} \times \mathbf{F}$. The projection of this moment along the axis, whose direction is defined by the unit vector $\mathbf{u}_a$, is $M_a = \mathbf{u}_a \cdot (\mathbf{r} \times \mathbf{F})$. Had this moment been calculated using scalar methods, then the perpendicular distance from the force line of action to the a–a axis would have to be determined, which in this case would be a more difficult task.

Important Points

- The moment of a force about a specified axis can be determined provided the perpendicular distance d_a from *both* the force line of action and the axis can be determined. $M_a = Fd_a$.
- If vector analysis is used, $M_a = \mathbf{u}_a \cdot (\mathbf{r} \times \mathbf{F})$, where $\mathbf{u}_a$ defines the direction of the axis and $\mathbf{r}$ is directed from *any point* on the axis to *any point* on the line of action of the force.
- If M_a is calculated as a negative scalar, then the sense of direction of $\mathbf{M}_a$ is opposite to $\mathbf{u}_a$.
- The moment $\mathbf{M}_a$ expressed as a Cartesian vector is determined from $\mathbf{M}_a = M_a\mathbf{u}_a$.

EXAMPLE 4.8

The force $\mathbf{F} = \{-40\mathbf{i} + 20\mathbf{j} + 10\mathbf{k}\}$ N acts at point A shown in Fig. 4–23a. Determine the moments of this force about the x and a axes.

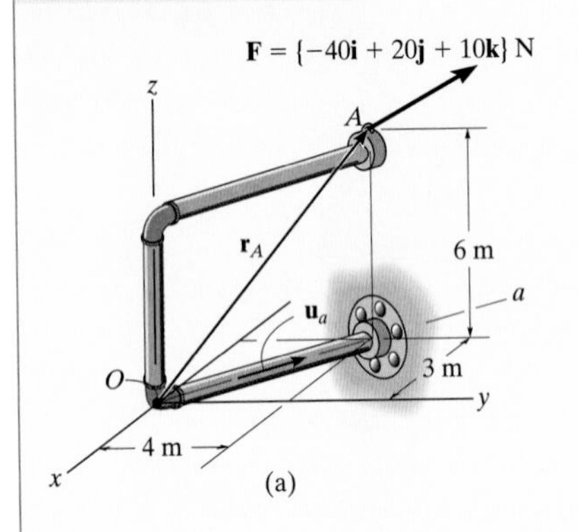

(a)

SOLUTION I (VECTOR ANALYSIS)

We can solve this problem by using the position vector $\mathbf{r}_A$. Why? Since $\mathbf{r}_A = \{-3\mathbf{i} + 4\mathbf{j} + 6\mathbf{k}\}$ m and $\mathbf{u}_x = \mathbf{i}$, then applying Eq. 4–11,

$$M_x = \mathbf{i}\cdot(\mathbf{r}_A \times \mathbf{F}) = \begin{vmatrix} 1 & 0 & 0 \\ -3 & 4 & 6 \\ -40 & 20 & 10 \end{vmatrix}$$

$$= 1[4(10)-6(20)]-0[(-3)(10)-6(-40)]+0[(-3)(20)-4(-40)]$$

$$= -80 \text{ N}\cdot\text{m} \qquad \textit{Ans.}$$

The negative sign indicates that the sense of $\mathbf{M}_x$ is opposite to $\mathbf{i}$.

We can compute M_a also using $\mathbf{r}_A$ because $\mathbf{r}_A$ extends from a point on the a axis to the force. Also, $\mathbf{u}_a = -\frac{3}{5}\mathbf{i} + \frac{4}{5}\mathbf{j}$. Thus,

$$M_a = \mathbf{u}_a\cdot(\mathbf{r}_A \times \mathbf{F}) = \begin{vmatrix} -\frac{3}{5} & \frac{4}{5} & 0 \\ -3 & 4 & 6 \\ -40 & 20 & 10 \end{vmatrix}$$

$$= -\tfrac{3}{5}[4(10) - 6(20)]-\tfrac{4}{5}[(-3)(10)-6(-40)]+0[(-3)(20)-4(-40)]$$

$$= -120 \text{ N}\cdot\text{m} \qquad \textit{Ans.}$$

What does the negative sign indicate?

The moment components are shown in Fig. 4–23b.

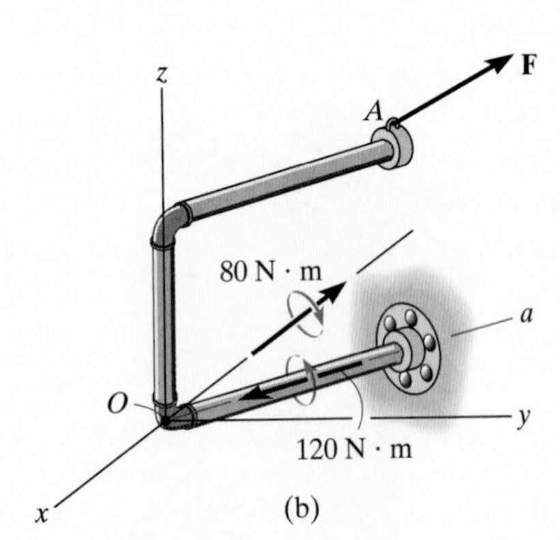

(b)

SOLUTION II (SCALAR ANALYSIS)

Since the force components and moment arms are easy to determine for computing M_x, a scalar analysis can be used to solve this problem. Referring to Fig. 4–23c, only the 10-N and 20-N forces contribute moments about the x axis. (The line of action of the 40-N force is parallel to this axis and hence its moment about the x axis is zero.) Using the right-hand rule, the algebraic sum of the moment components about the x axis is therefore

$$M_x = (10 \text{ N})(4 \text{ m}) - (20 \text{ N})(6 \text{ m}) = -80 \text{ N}\cdot\text{m} \qquad \textit{Ans.}$$

Although not required here, note also that

$$M_y = (10 \text{ N})(3 \text{ m}) - (40 \text{ N})(6 \text{ m}) = -210 \text{ N}\cdot\text{m}$$

$$M_z = (40 \text{ N})(4 \text{ m}) - (20 \text{ N})(3 \text{ m}) = 100 \text{ N}\cdot\text{m}$$

If we were to determine M_a by this scalar method, it would require much more effort since the force components of 40 N and 20 N are *not perpendicular* to the direction of the a axis. The vector analysis yields a more direct solution.

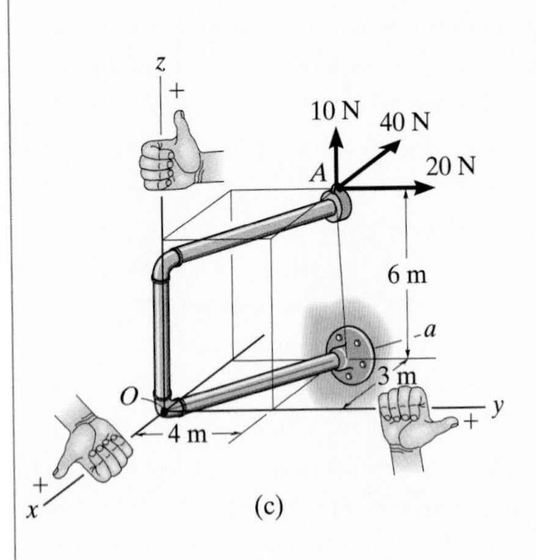

(c)

Fig. 4–23

EXAMPLE 4.9

The rod shown in Fig. 4–24*a* is supported by two brackets at *A* and *B*. Determine the moment $\mathbf{M}_{AB}$ produced by $\mathbf{F} = \{-600\mathbf{i} + 200\mathbf{j} - 300\mathbf{k}\}$ N, which tends to rotate the rod about the *AB* axis.

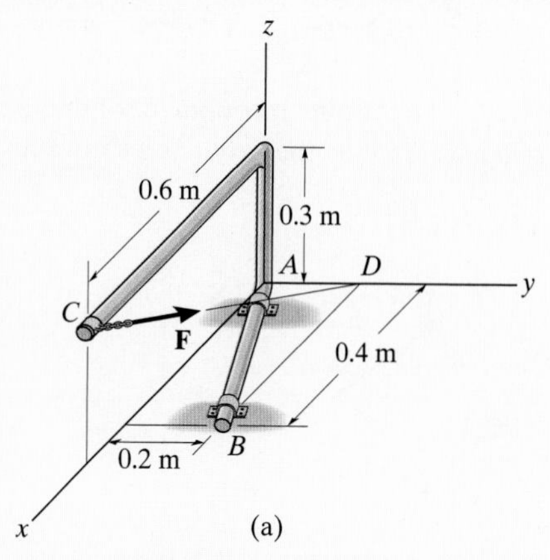

(a)

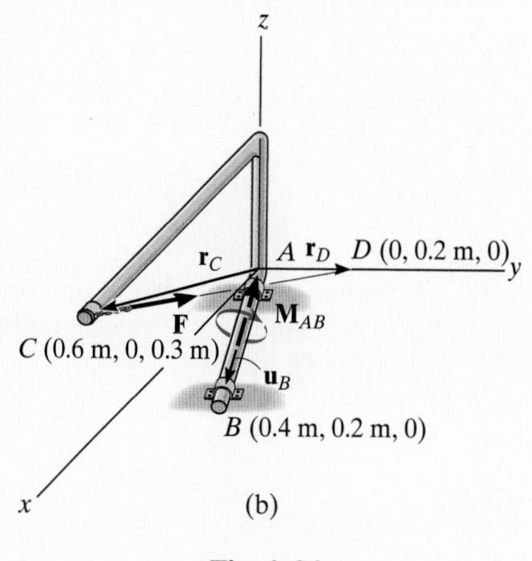

(b)

Fig. 4–24

SOLUTION

A vector analysis using $M_{AB} = \mathbf{u}_B \cdot (\mathbf{r} \times \mathbf{F})$ will be considered for the solution since the moment arm or perpendicular distance from the line of action of **F** to the *AB* axis is difficult to determine. Each of the terms in the equation will now be identified.

Unit vector $\mathbf{u}_B$ defines the direction of the *AB* axis of the rod, Fig. 4–24*b*, where

$$\mathbf{u}_B = \frac{\mathbf{r}_B}{r_B} = \frac{0.4\mathbf{i} + 0.2\mathbf{j}}{\sqrt{(0.4)^2 + (0.2)^2}} = 0.894\mathbf{i} + 0.447\mathbf{j}$$

Vector **r** is directed from *any point* on the *AB* axis to *any point* on the line of action of the force. For example, position vectors $\mathbf{r}_C$ and $\mathbf{r}_D$ are suitable, Fig. 4–24*b*. (Although not shown, $\mathbf{r}_{BC}$ or $\mathbf{r}_{BD}$ can also be used.) For simplicity, we choose $\mathbf{r}_D$, where

$$\mathbf{r}_D = \{0.2\mathbf{j}\} \text{ m}$$

The force is

$$\mathbf{F} = \{-600\mathbf{i} + 200\mathbf{j} - 300\mathbf{k}\} \text{ N}$$

Substituting these vectors into the determinant form and expanding, we have

$$M_{AB} = \mathbf{u}_B \cdot (\mathbf{r}_D \times \mathbf{F}) = \begin{vmatrix} 0.894 & 0.447 & 0 \\ 0 & 0.2 & 0 \\ -600 & 200 & -300 \end{vmatrix}$$

$$= 0.894[0.2(-300) - 0(200)] - 0.447[0(-300) - 0(-600)] + 0[0(200) - 0.2(-600)]$$

$$= -53.67 \text{ N}\cdot\text{m}$$

The negative sign indicates that the sense of $\mathbf{M}_{AB}$ is opposite to that of $\mathbf{u}_B$.

Expressing $\mathbf{M}_{AB}$ as a Cartesian vector yields

$$\mathbf{M}_{AB} = M_{AB}\mathbf{u}_B = (-53.67 \text{ N}\cdot\text{m})(0.894\mathbf{i} + 0.447\mathbf{j})$$

$$= \{-48.0\mathbf{i} - 24.0\mathbf{j}\} \text{ N}\cdot\text{m} \qquad \textit{Ans.}$$

The result is shown in Fig. 4–24*b*.

NOTE: If axis *AB* is defined using a unit vector directed from *B* toward *A*, then in the above formulation $-\mathbf{u}_B$ would have to be used. This would lead to $M_{AB} = +53.67 \text{ N}\cdot\text{m}$. Consequently, $\mathbf{M}_{AB} = M_{AB}(-\mathbf{u}_B)$, and the above result would again be determined.

PROBLEMS

4–51. Determine the moment of the force **F** about the *Oa* axis. Express the result as a Cartesian vector.

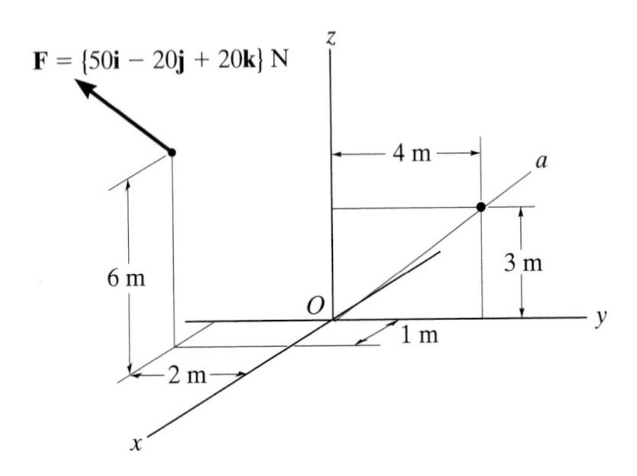

Prob. 4–51

*4–52. Determine the moment of the force **F** about the *aa* axis. Express the result as a Cartesian vector.

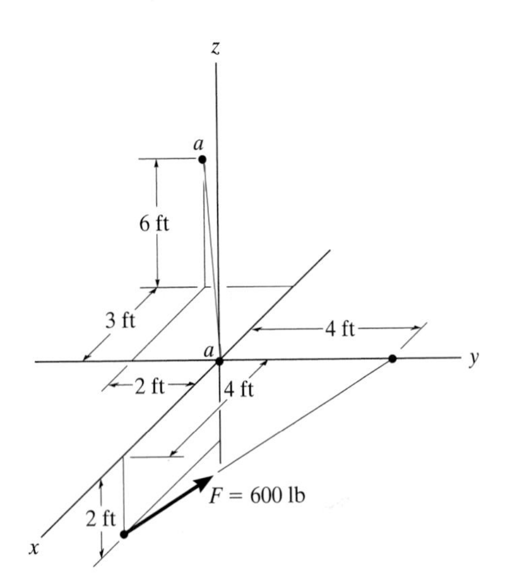

Prob. 4–52

4–53. Determine the resultant moment of the two forces about the *Oa* axis. Express the result as a Cartesian vector.

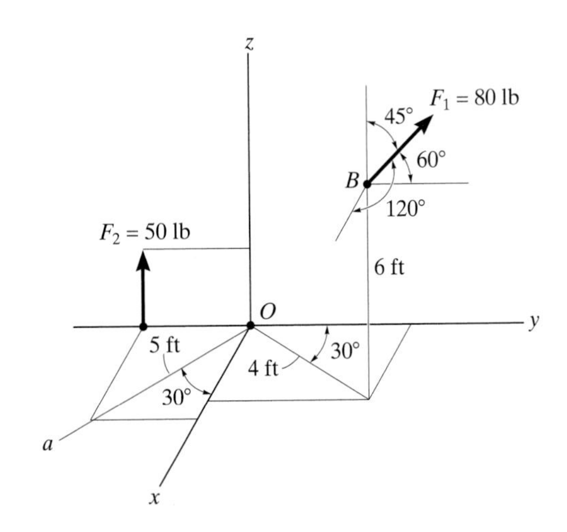

Prob. 4–53

4–54. A force of $\mathbf{F} = \{8\mathbf{i} - 1\mathbf{j} + 1\mathbf{k}\}$ lb is applied to the handle of the box wrench. Determine the component of the moment of this force about the z axis which is effective in loosening the bolt.

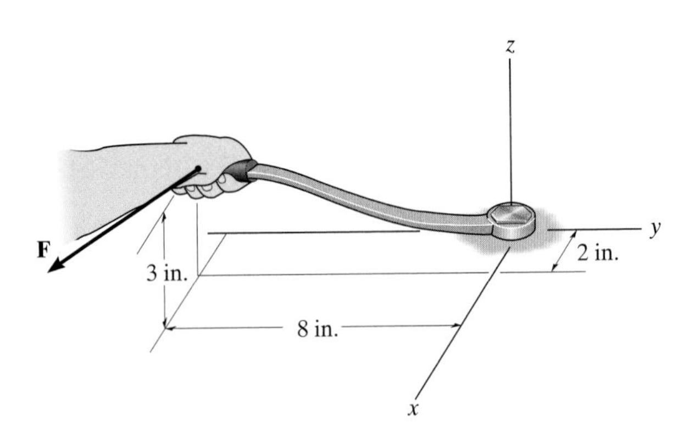

Prob. 4–54

4–55. The 50-lb force acts on the gear in the direction shown. Determine the moment of this force about the y axis.

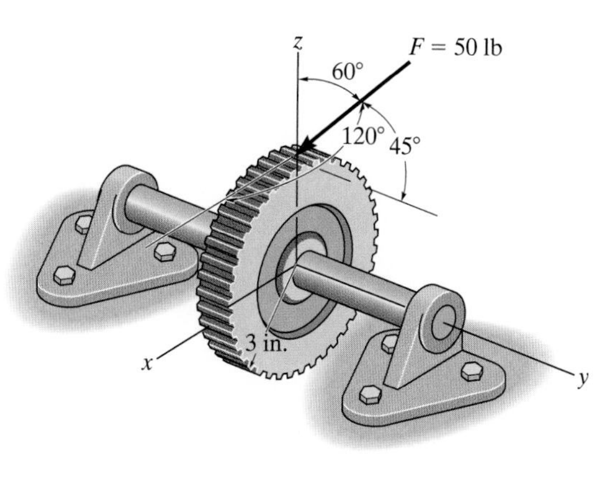

Prob. 4–55

*__4–56.__ The Rollerball skate is an in-line tandem skate that uses two large spherical wheels on each skate, rather than traditional wafer-shape wheels. During skating the two forces acting on the wheel of one skate consist of a 78-lb normal force and a 13-lb friction force. Determine the moment of both of these forces about the axle AB of the wheel.

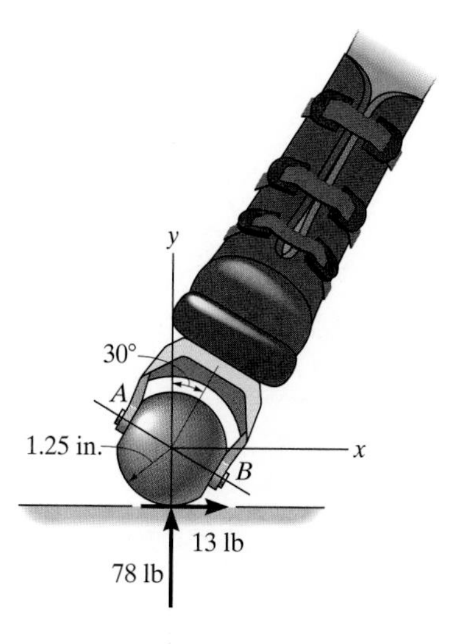

Prob. 4–56

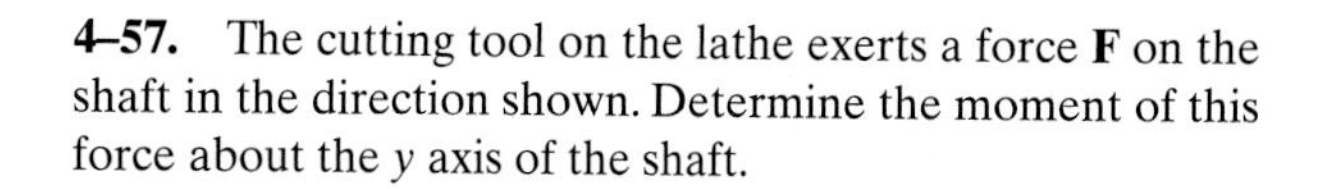

4–57. The cutting tool on the lathe exerts a force **F** on the shaft in the direction shown. Determine the moment of this force about the y axis of the shaft.

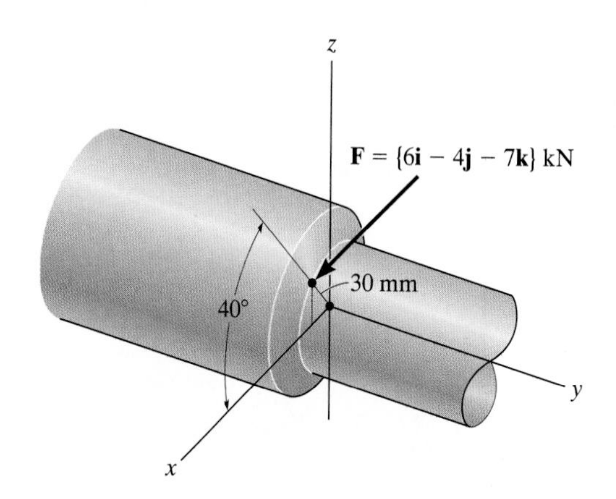

Prob. 4–57

4–58. The hood of the automobile is supported by the strut AB, which exerts a force of $F = 24$ lb on the hood. Determine the moment of this force about the hinged axis y.

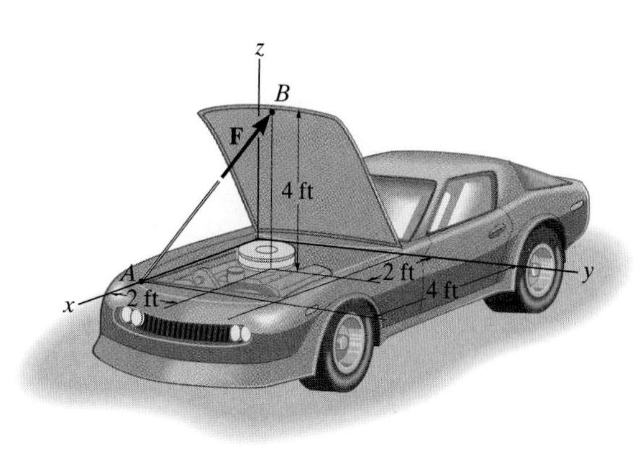

Prob. 4–58

4–59. The lug nut on the wheel of the automobile is to be removed using the wrench and applying the vertical force of $F = 30$ N at A. Determine if this force is adequate, provided 14 N · m of torque about the x axis is initially required to turn the nut. If the 30-N force can be applied at A in any other direction, will it be possible to turn the nut?

***4–60.** Solve Prob. 4–59 if the cheater pipe AB is slipped over the handle of the wrench and the 30-N force can be applied at any point and in any direction on the assembly.

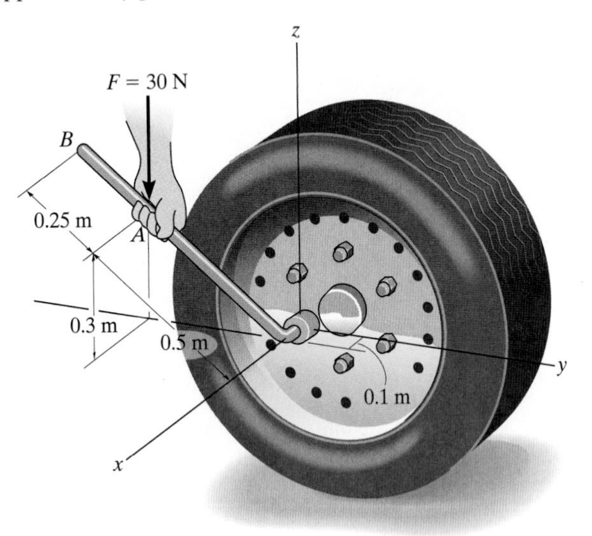

Probs. 4–59/60

4–61. The bevel gear is subjected to the force **F** which is caused from contact with another gear. Determine the moment of this force about the y axis of the gear shaft.

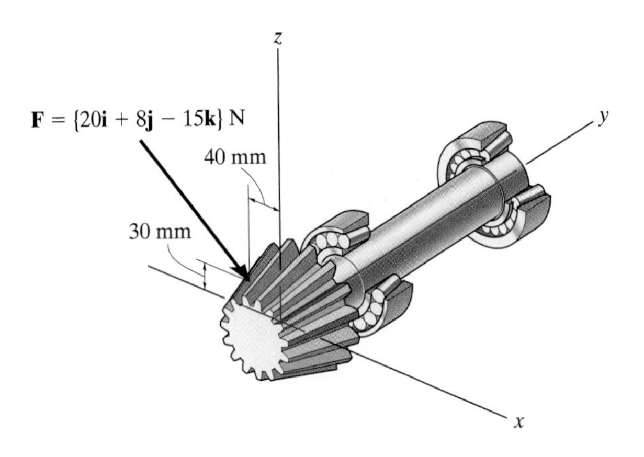

Prob. 4–61

4–62. The wooden shaft is held in a lathe. The cutting tool exerts a force **F** on the shaft in the direction shown. Determine the moment of this force about the x axis of the shaft. Express the result as a Cartesian vector. The distance OA is 25 mm.

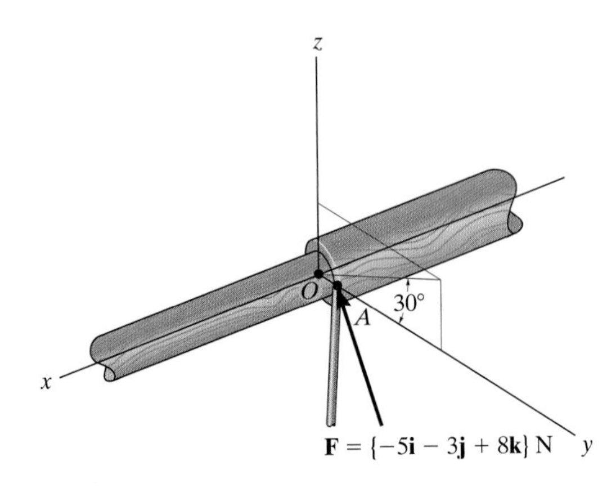

Prob. 4–62

4–63. Determine the magnitude of the moment of the force $\mathbf{F} = \{50\mathbf{i} - 20\mathbf{j} - 80\mathbf{k}\}$ N about the base line CA of the tripod.

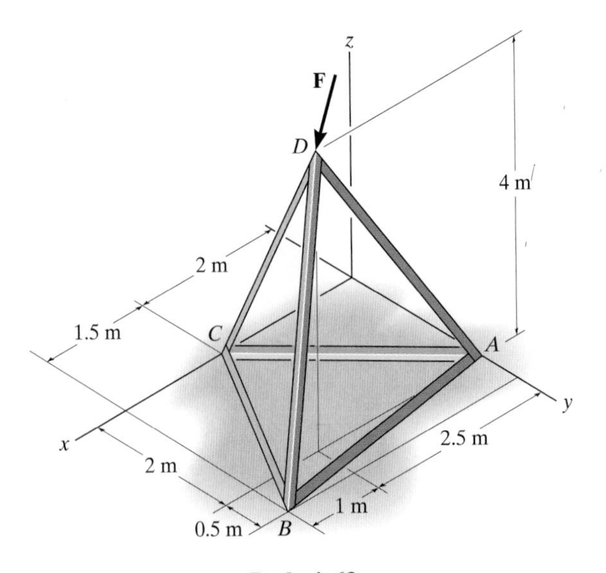

Prob. 4–63

*4–64. The flex-headed ratchet wrench is subjected to a force of $P = 16$ lb, applied perpendicular to the handle as shown. Determine the moment or torque this imparts along the vertical axis of the bolt at A.

4–65. If a torque or moment of 80 lb·in. is required to loosen the bolt at A, determine the force P that must be applied perpendicular to the handle of the flex-headed ratchet wrench.

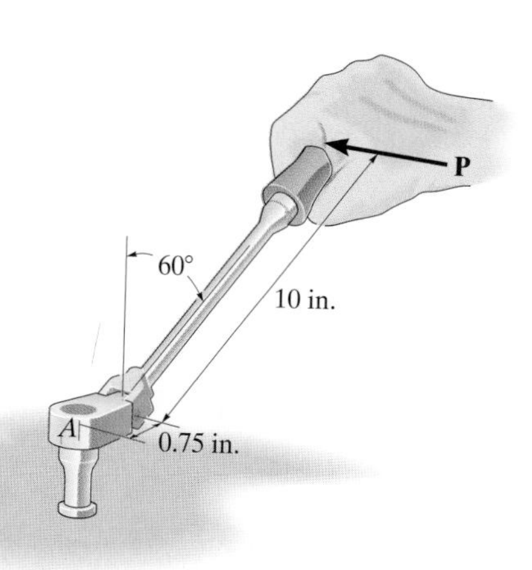

Probs. 4–64/65

4–66. The A-frame is being hoisted into an upright position by the vertical force of $F = 80$ lb. Determine the moment of this force about the y axis when the frame is in the position shown.

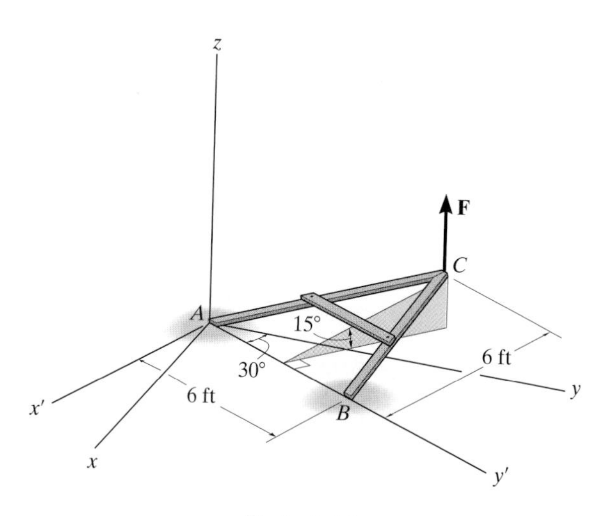

Prob. 4–66

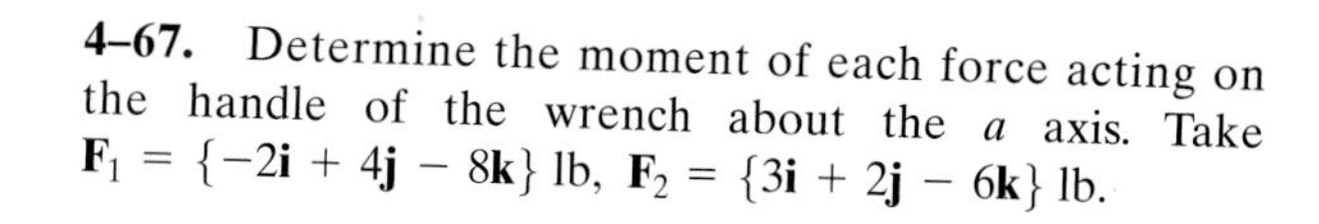
4–67. Determine the moment of each force acting on the handle of the wrench about the a axis. Take $\mathbf{F}_1 = \{-2\mathbf{i} + 4\mathbf{j} - 8\mathbf{k}\}$ lb, $\mathbf{F}_2 = \{3\mathbf{i} + 2\mathbf{j} - 6\mathbf{k}\}$ lb.

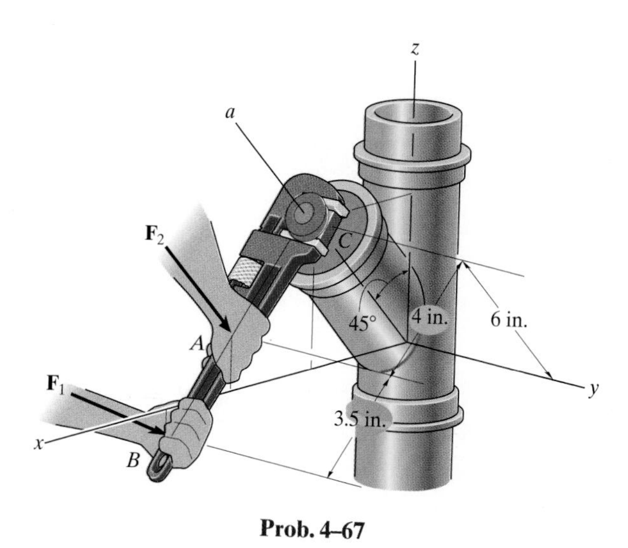

Prob. 4–67

*4–68. Determine the moment of each force acting on the handle of the wrench about the z axis. Take $\mathbf{F}_1 = \{-2\mathbf{i} + 4\mathbf{j} - 8\mathbf{k}\}$ lb, $\mathbf{F}_2 = \{3\mathbf{i} + 2\mathbf{j} - 6\mathbf{k}\}$ lb.

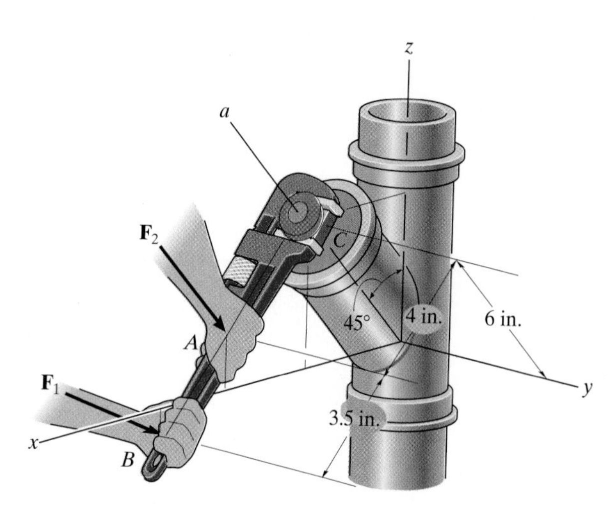

Prob. 4–68

4.6 Moment of a Couple

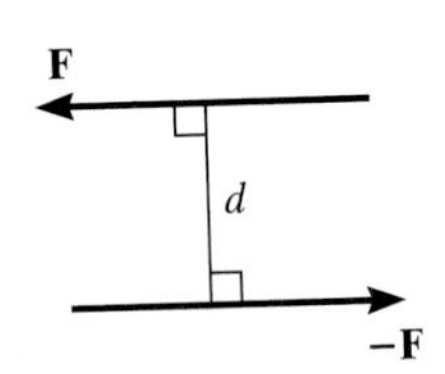

Fig. 4–25

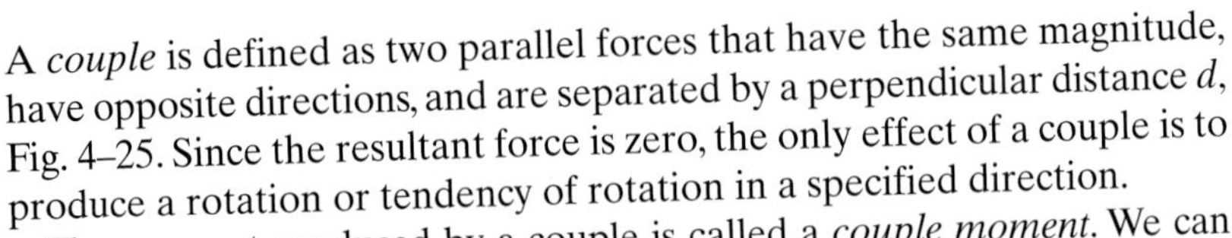

A *couple* is defined as two parallel forces that have the same magnitude, have opposite directions, and are separated by a perpendicular distance d, Fig. 4–25. Since the resultant force is zero, the only effect of a couple is to produce a rotation or tendency of rotation in a specified direction.

The moment produced by a couple is called a *couple moment*. We can determine its value by finding the sum of the moments of both couple forces about *any* arbitrary point. For example, in Fig. 4–26, position vectors $\mathbf{r}_A$ and $\mathbf{r}_B$ are directed from point O to points A and B lying on the line of action of $-\mathbf{F}$ and $\mathbf{F}$. The couple moment computed about O is therefore

$$\mathbf{M} = \mathbf{r}_A \times (-\mathbf{F}) + \mathbf{r}_B \times \mathbf{F}$$

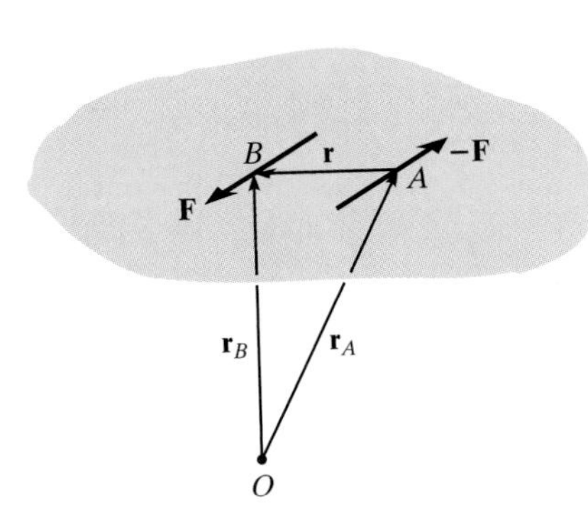

Fig. 4–26

Rather than sum the moments of both forces to determine the couple moment, it is simpler to take moments about a point lying on the line of action of one of the forces. If point A is chosen, then the moment of $-\mathbf{F}$ about A is zero, and we have

$$\mathbf{M} = \mathbf{r} \times \mathbf{F} \tag{4–13}$$

The fact that we obtain the *same result* in both cases can be demonstrated by noting that in the first case we can write $\mathbf{M} = (\mathbf{r}_B - \mathbf{r}_A) \times \mathbf{F}$; and by the triangle rule of vector addition, $\mathbf{r}_A + \mathbf{r} = \mathbf{r}_B$ or $\mathbf{r} = \mathbf{r}_B - \mathbf{r}_A$, so that upon substitution we obtain Eq. 4–13. This result indicates that a couple moment is a *free vector*, i.e., it can act at *any point* since $\mathbf{M}$ depends *only* upon the position vector $\mathbf{r}$ directed *between* the forces and *not* the position vectors $\mathbf{r}_A$ and $\mathbf{r}_B$, directed from the arbitrary point O to the forces. This concept is therefore unlike the moment of a force, which requires a definite point (or axis) about which moments are determined.

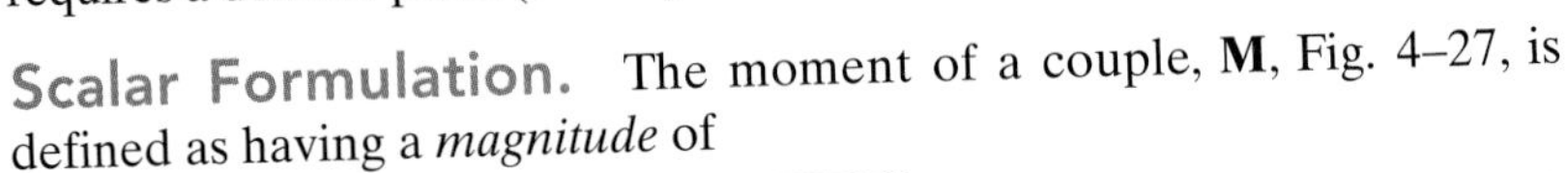

Scalar Formulation. The moment of a couple, $\mathbf{M}$, Fig. 4–27, is defined as having a *magnitude* of

$$M = Fd \tag{4–14}$$

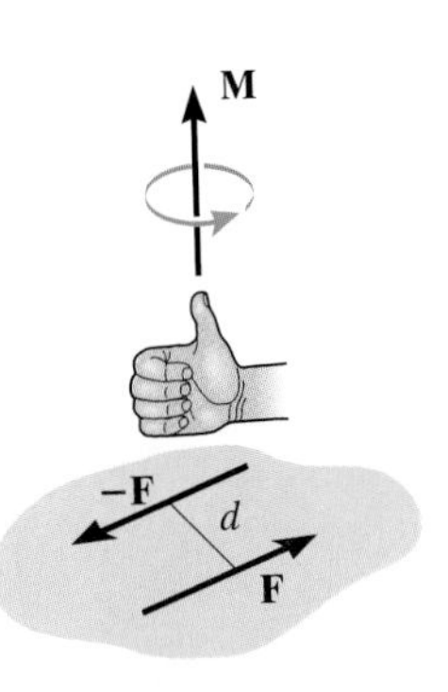

Fig. 4–27

where F is the magnitude of one of the forces and d is the perpendicular distance or moment arm between the forces. The *direction* and sense of the couple moment are determined by the right-hand rule, where the thumb indicates the direction when the fingers are curled with the sense of rotation caused by the two forces. In all cases, $\mathbf{M}$ acts perpendicular to the plane containing these forces.

Vector Formulation. The moment of a couple can also be expressed by the vector cross product using Eq. 4–13, i.e.,

$$\mathbf{M} = \mathbf{r} \times \mathbf{F} \tag{4–15}$$

Application of this equation is easily remembered if one thinks of taking the moments of both forces about a point lying on the line of action of one of the forces. For example, if moments are taken about point A in

Fig. 4–26, the moment of $-\mathbf{F}$ is *zero* about this point, and the moment or $\mathbf{F}$ is defined from Eq. 4–15. Therefore, in the formulation $\mathbf{r}$ is crossed with the force $\mathbf{F}$ to which it is directed.

Equivalent Couples. Two couples are said to be equivalent if they produce the same moment. Since the moment produced by a couple is always perpendicular to the plane containing the couple forces, it is therefore necessary that the forces of equal couples lie either in the same plane or in planes that are *parallel* to one another. In this way, the direction of each couple moment will be the same, that is, perpendicular to the parallel planes.

Resultant Couple Moment. Since couple moments are free vectors, they may be applied at any point P on a body and added vectorially. For example, the two couples acting on different planes of the body in Fig. 4–28*a* may be replaced by their corresponding couple moments $\mathbf{M}_1$ and $\mathbf{M}_2$, Fig. 4–28*b*, and then these free vectors may be moved to the *arbitrary point P* and added to obtain the resultant couple moment $\mathbf{M}_R = \mathbf{M}_1 + \mathbf{M}_2$, shown in Fig. 4–28*c*.

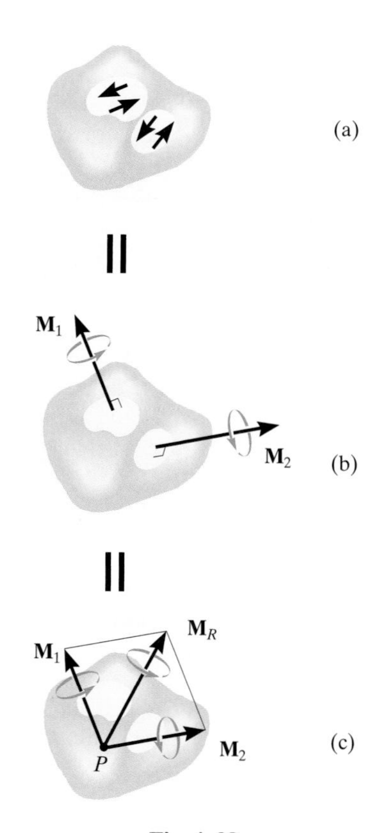

Fig. 4–28

If more than two couple moments act on the body, we may generalize this concept and write the vector resultant as

$$\mathbf{M}_R = \Sigma(\mathbf{r} \times \mathbf{F}) \tag{4–16}$$

These concepts are illustrated numerically in the examples which follow. In general, problems projected in two dimensions should be solved using a scalar analysis since the moment arms and force components are easy to compute.

A moment of 12 N · m is needed to turn the shaft connected to the center of the wheel. To do this it is efficient to apply a couple since this effect produces a pure rotation. The couple forces can be made as small as possible by placing the hands on the *rim* of the wheel, where the spacing is 0.4 m. In this case $12\ \text{N}\cdot\text{m} = F(0.4\ \text{m})$, $F = 30$ N. An equivalent couple moment of 12 N · m can be produced if one grips the wheel within the inner hub, although here much larger forces are needed. If the distance between the hands becomes 0.3 m, then $12\ \text{N}\cdot\text{m} = F'(0.3)$, $F' = 40$ N. Also, realize that if the wheel was connected to the shaft at a point other than at its center, the wheel would still turn when the forces are applied since the 12-N · m couple moment is a *free vector*.

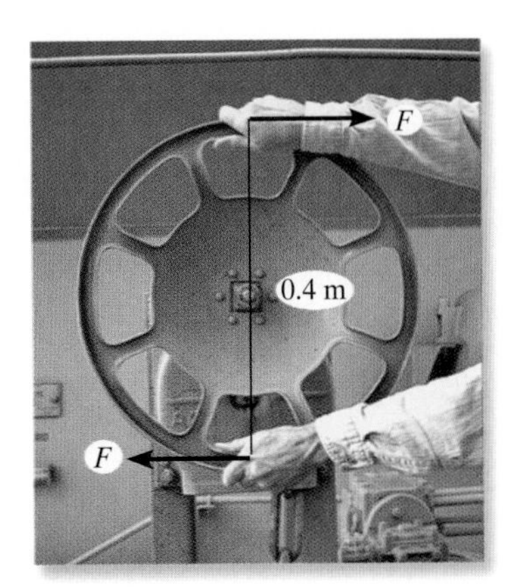

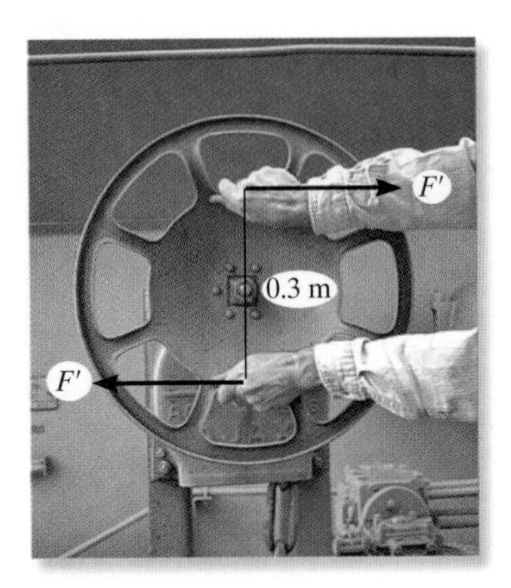

The frictional forces of the floor on the blades of the concrete finishing machine create a couple moment $\mathbf{M}_c$ on the machine that tends to turn it. An equal but opposite couple moment must be applied by the hands of the operator to prevent the turning. Here the couple moment, $M_c = Fd$, is applied on the handle, although it could be applied at any other point on the machine.

Important Points

- A couple moment is produced by two noncollinear forces that are equal but opposite. Its effect is to produce pure rotation, or tendency for rotation in a specified direction.
- A couple moment is a free vector, and as a result it causes the same effect of rotation on a body regardless of where the couple moment is applied to the body.
- The moment of the two couple forces can be computed about *any point*. For convenience, this point is often chosen on the line of action of one of the forces in order to eliminate the moment of this force about the point.
- In three dimensions the couple moment is often determined using the vector formulation, $\mathbf{M} = \mathbf{r} \times \mathbf{F}$, where $\mathbf{r}$ is directed from *any point* on the line of action of one of the forces to *any point* on the line of action of the other force $\mathbf{F}$.
- A resultant couple moment is simply the vector sum of all the couple moments of the system.

EXAMPLE 4.10

A couple acts on the gear teeth as shown in Fig. 4–29*a*. Replace it by an equivalent couple having a pair of forces that act through points *A* and *B*.

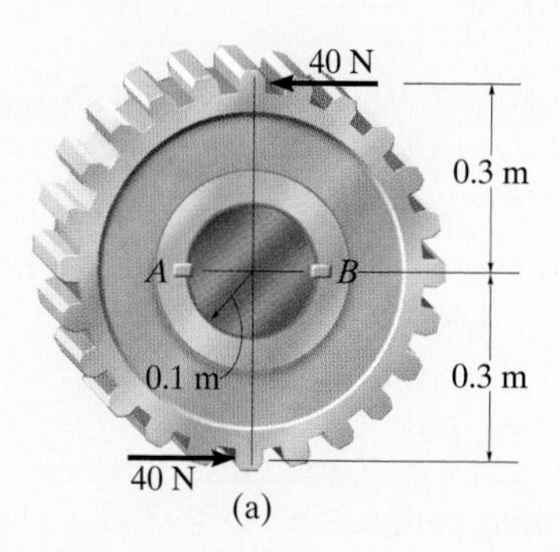

(a)

(b)

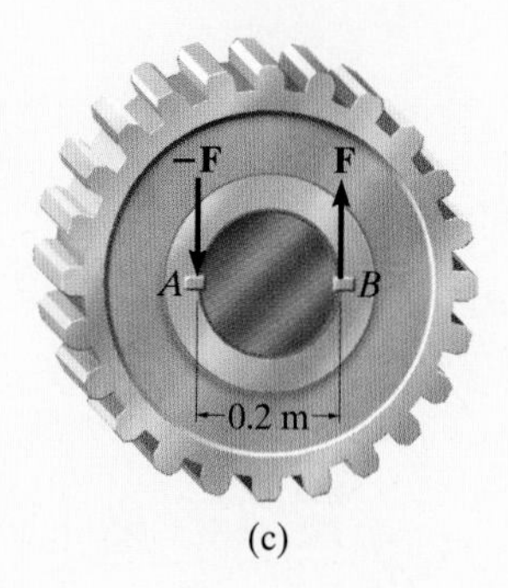

(c)

Fig. 4–29

SOLUTION *(SCALAR ANALYSIS)*

The couple has a magnitude of $M = Fd = 40(0.6) = 24\ \text{N}\cdot\text{m}$ and a direction that is out of the page since the forces tend to rotate counterclockwise. $\mathbf{M}$ is a free vector, and so it can be placed at any point on the gear, Fig. 4–29*b*. To preserve the counterclockwise rotation of $\mathbf{M}$, *vertical* forces acting through points *A* and *B* must be directed as shown in Fig. 4–29*c*. The magnitude of each force is

$$M = Fd \qquad 24\ \text{N}\cdot\text{m} = F(0.2\ \text{m})$$

$$F = 120\ \text{N} \qquad \textit{Ans.}$$

EXAMPLE 4.11

Determine the moment of the couple acting on the member shown in Fig. 4–30*a*.

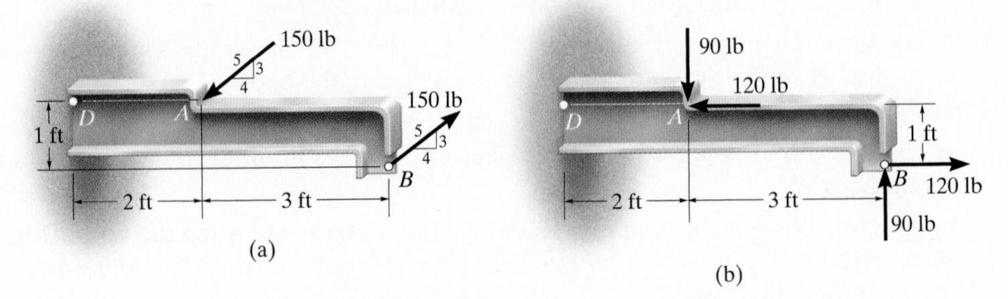

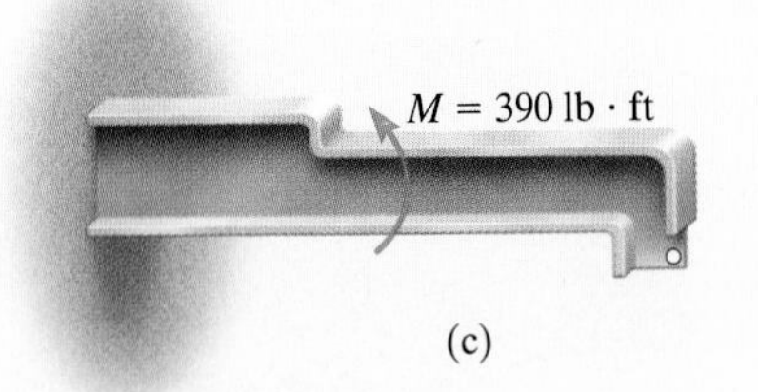

Fig. 4–30

SOLUTION *(SCALAR ANALYSIS)*

Here it is somewhat difficult to determine the perpendicular distance between the forces and compute the couple moment as $M = Fd$. Instead, we can resolve each force into its horizontal and vertical components, $F_x = \frac{4}{5}(150 \text{ lb}) = 120 \text{ lb}$ and $F_y = \frac{3}{5}(150 \text{ lb}) = 90 \text{ lb}$, Fig. 4–30*b*, and then use the principle of moments. The couple moment can be determined about *any point*. For example, if point *D* is chosen, we have for all four forces,

$$\curvearrowleft\!+M = 120 \text{ lb } (0 \text{ ft}) - 90 \text{ lb } (2 \text{ ft}) + 90 \text{ lb } (5 \text{ ft}) + 120 \text{ lb } (1 \text{ ft})$$
$$= 390 \text{ lb} \cdot \text{ft} \curvearrowleft \qquad \textit{Ans.}$$

It is easier, however, to determine the moments about point *A* or *B* in order to *eliminate* the moment of the forces acting at the moment point. For point *A*, Fig. 4–30*b*, we have

$$\curvearrowleft\!+M = 90 \text{ lb } (3 \text{ ft}) + 120 \text{ lb } (1 \text{ ft})$$
$$= 390 \text{ lb} \cdot \text{ft} \curvearrowleft \qquad \textit{Ans.}$$

NOTE: Show that one obtains the same result if moments are summed about point *B*. Notice also that the couple in Fig. 4–30*a* can be replaced by *two* couples in Fig. 4–30*b*. Using $M = Fd$, one couple has a moment of $M_1 = 90 \text{ lb } (3 \text{ ft}) = 270 \text{ lb} \cdot \text{ft}$ and the other has a moment of $M_2 = 120 \text{ lb } (1 \text{ ft}) = 120 \text{ lb} \cdot \text{ft}$. By the right-hand rule, both couple moments are counterclockwise and are therefore directed out of the page. Since these couples are free vectors, they can be moved to any point and added, which yields $M = 270 \text{ lb} \cdot \text{ft} + 120 \text{ lb} \cdot \text{ft} = 390 \text{ lb} \cdot \text{ft} \curvearrowleft$, the same result determined above. **M** is a free vector and can therefore act at any point on the member, Fig. 4–30*c*. Also, realize that the external effects, such as the support reactions on the member, will be the *same* if the member supports the couple, Fig. 4–30*a*, or the couple moment, Fig. 4–30*c*.

EXAMPLE 4.12

Determine the couple moment acting on the pipe shown in Fig. 4–31a. Segment AB is directed 30° below the x–y plane.

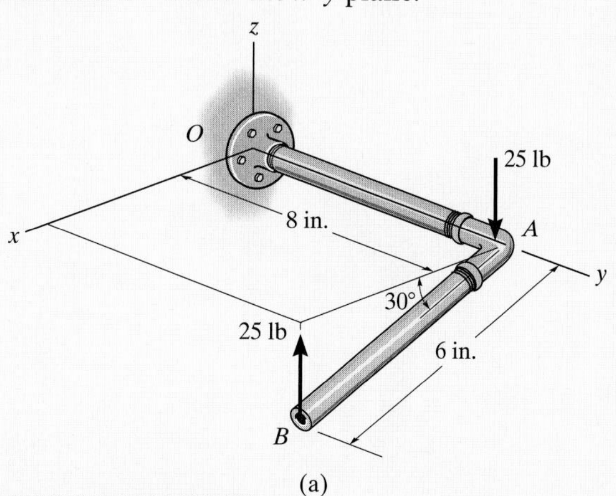

(a)

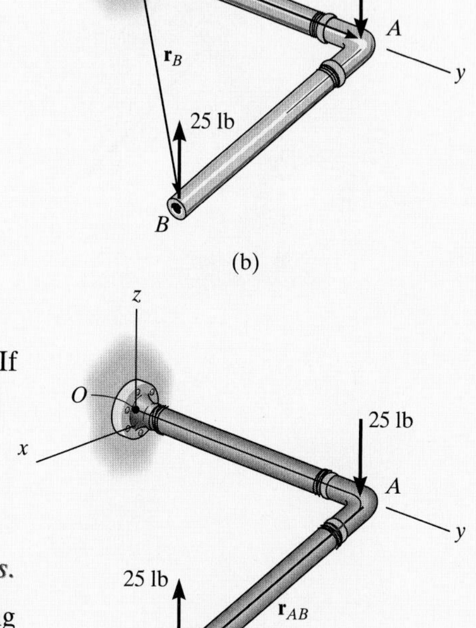

(b)

(c)

SOLUTION I (VECTOR ANALYSIS)

The moment of the two couple forces can be found about *any point*. If point O is considered, Fig. 4–31b, we have

$$
\begin{aligned}
\mathbf{M} &= \mathbf{r}_A \times (-25\mathbf{k}) + \mathbf{r}_B \times (25\mathbf{k}) \\
&= (8\mathbf{j}) \times (-25\mathbf{k}) + (6\cos 30°\mathbf{i} + 8\mathbf{j} - 6\sin 30°\mathbf{k}) \times (25\mathbf{k}) \\
&= -200\mathbf{i} - 129.9\mathbf{j} + 200\mathbf{i} \\
&= \{-130\mathbf{j}\}\ \text{lb}\cdot\text{in.}
\end{aligned}
$$
Ans.

It is *easier* to take moments of the couple forces about a point lying on the line of action of one of the forces, e.g., point A, Fig. 4–31c. In this case the moment of the force A is zero, so that

$$
\begin{aligned}
\mathbf{M} &= \mathbf{r}_{AB} \times (25\mathbf{k}) \\
&= (6\cos 30°\mathbf{i} - 6\sin 30°\mathbf{k}) \times (25k) \\
&= \{-130\mathbf{j}\}\ \text{lb}\cdot\text{in.}
\end{aligned}
$$
Ans.

SOLUTION II (SCALAR ANALYSIS)

Although this problem is shown in three dimensions, the geometry is simple enough to use the scalar equation $M = Fd$. The perpendicular distance between the lines of action of the forces is $d = 6\cos 30° = 5.20$ in., Fig. 4–31d. Hence, taking moments of the forces about either point A or B yields

$$M = Fd = 25\ \text{lb}\ (5.20\ \text{in.}) = 129.9\ \text{lb}\cdot\text{in.}$$

Applying the right-hand rule, $\mathbf{M}$ acts in the $-\mathbf{j}$ direction. Thus,

$$\mathbf{M} = \{-130\mathbf{j}\}\ \text{lb}\cdot\text{in.}$$
Ans.

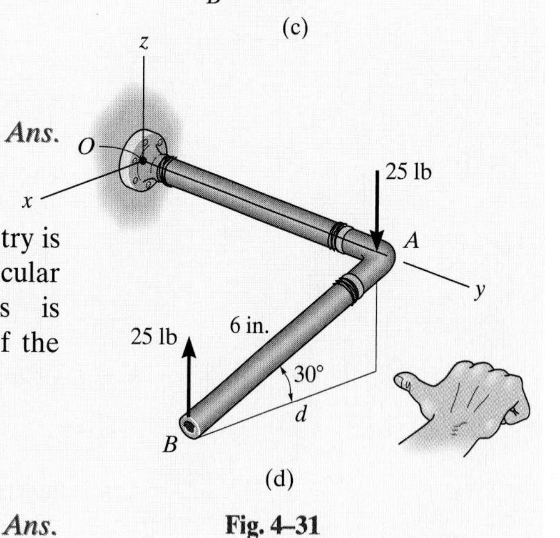

(d)

Fig. 4–31

EXAMPLE 4.13

Replace the two couples acting on the pipe column in Fig. 4–32*a* by a resultant couple moment.

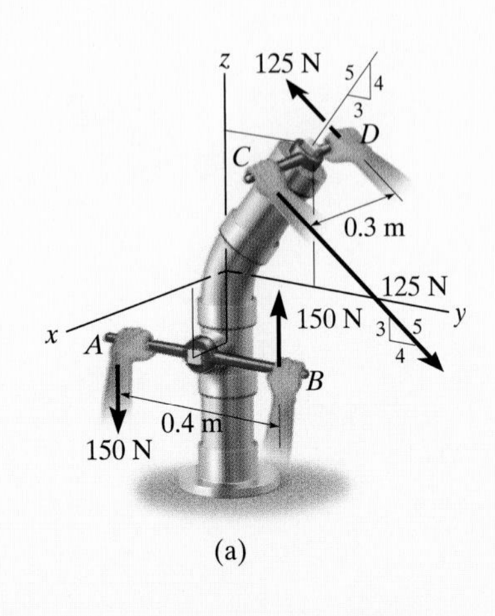

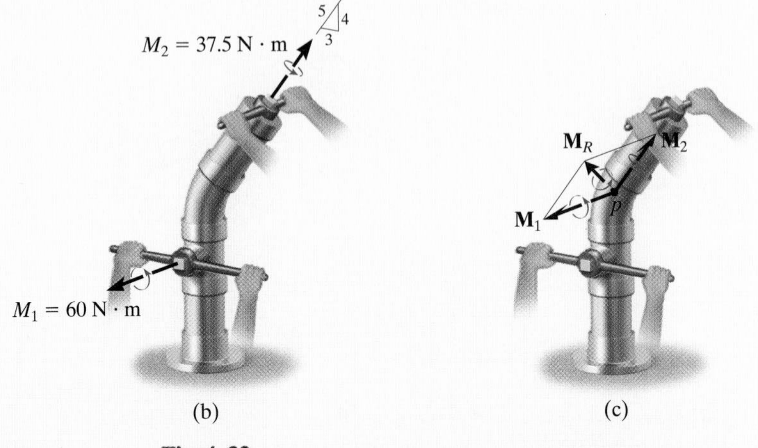

Fig. 4–32

SOLUTION (*VECTOR ANALYSIS*)

The couple moment $\mathbf{M}_1$, developed by the forces at A and B, can easily be determined from a scalar formulation.

$$M_1 = Fd = 150\text{ N}(0.4\text{ m}) = 60\text{ N}\cdot\text{m}$$

By the right-hand rule, $\mathbf{M}_1$ acts in the $+\mathbf{i}$ direction, Fig. 4–32*b*. Hence,

$$\mathbf{M}_1 = \{60\mathbf{i}\}\text{ N}\cdot\text{m}$$

Vector analysis will be used to determine $\mathbf{M}_2$, caused by forces at C and D. If moments are computed about point D, Fig. 4–32*a*, $\mathbf{M}_2 = \mathbf{r}_{DC} \times \mathbf{F}_C$, then

$$\begin{aligned}\mathbf{M}_2 = \mathbf{r}_{DC} \times \mathbf{F}_C &= (0.3\mathbf{i}) \times \left[125\left(\tfrac{4}{5}\right)\mathbf{j} - 125\left(\tfrac{3}{5}\right)\mathbf{k}\right] \\ &= (0.3\mathbf{i}) \times [100\mathbf{j} - 75\mathbf{k}] = 30(\mathbf{i} \times \mathbf{j}) - 22.5(\mathbf{i} \times \mathbf{k}) \\ &= \{22.5\mathbf{j} + 30\mathbf{k}\}\text{ N}\cdot\text{m}\end{aligned}$$

Since $\mathbf{M}_1$ and $\mathbf{M}_2$ are free vectors, they may be moved to some arbitrary point P and added vectorially, Fig. 4–32*c*. The resultant couple moment becomes

$$\mathbf{M}_R = \mathbf{M}_1 + \mathbf{M}_2 = \{60\mathbf{i} + 22.5\mathbf{j} + 30\mathbf{k}\}\text{ N}\cdot\text{m} \qquad \textit{Ans.}$$

NOTE: Try to establish $\mathbf{M}_2$ by using a scalar formulation, Fig. 4–32*b*.

PROBLEMS

4–69. Determine the magnitude and sense of the couple moment.

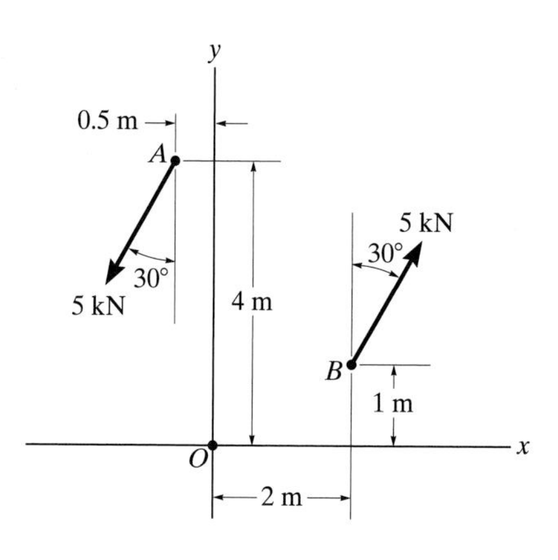

Prob. 4–69

4–70. Determine the magnitude and sense of the couple moment. Each force has a magnitude of $F = 65$ lb..

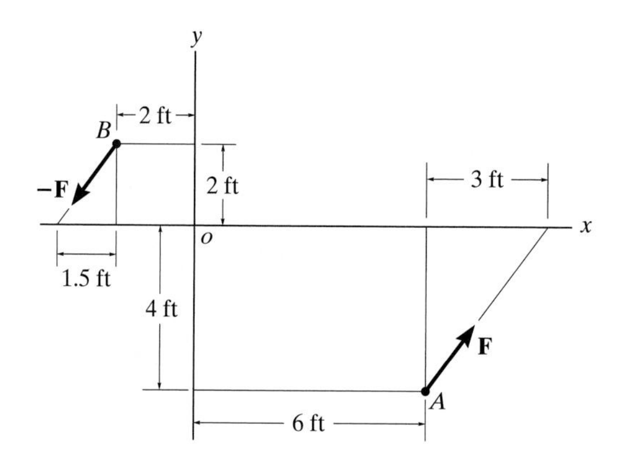

Prob. 4–70

4–71. Determine the magnitude and sense of the couple moment.

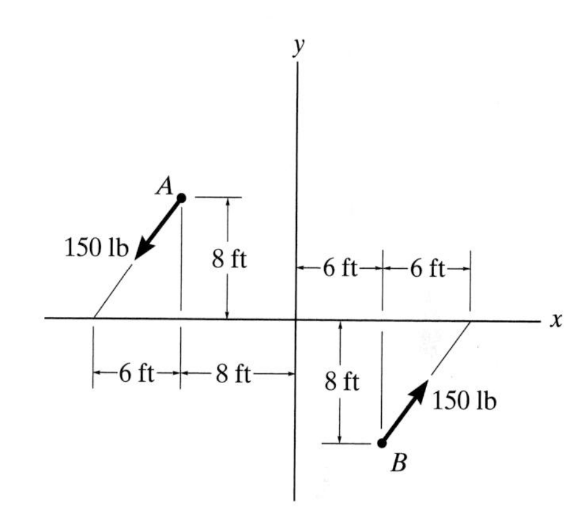

Prob. 4–71

***4–72.** If the couple moment has a magnitude of 300 lb · ft, determine the magnitude F of the couple forces.

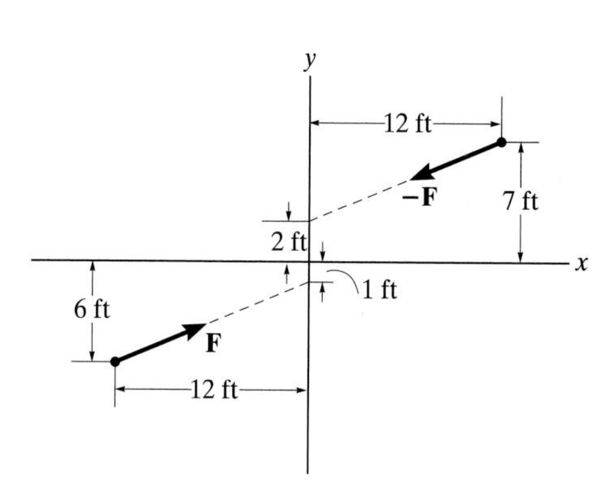

Prob. 4–72

4–73. A clockwise couple $M = 5\text{ N}\cdot\text{m}$ is resisted by the shaft of the electric motor. Determine the magnitude of the reactive forces $-\mathbf{R}$ and $\mathbf{R}$ which act at supports A and B so that the resultant of the two couples is zero.

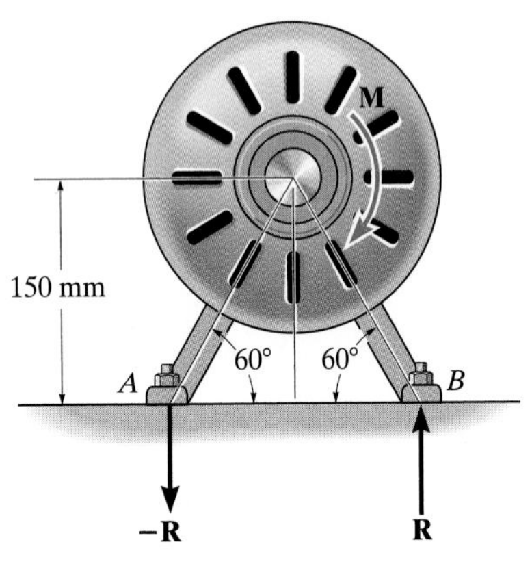

Prob. 4–73

4–74. The resultant couple moment created by the two couples acting on the disk is $\mathbf{M}_R = \{10\mathbf{k}\}$ kip·in. Determine the magnitude of force $\mathbf{T}$.

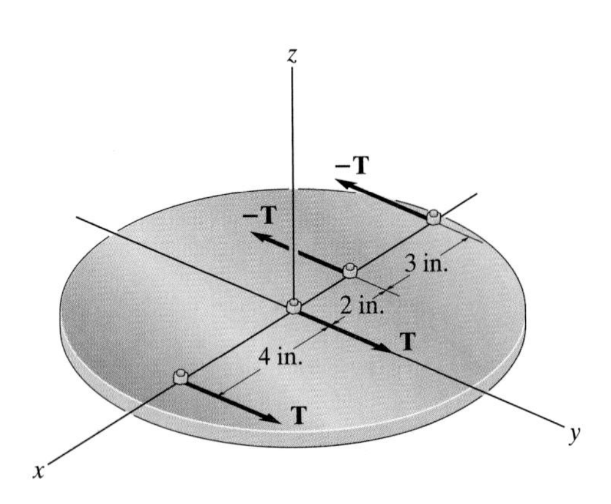

Prob. 4–74

4–75. Three couple moments act on the pipe assembly. Determine the magnitude of $\mathbf{M}_3$ and the bend angle θ so that the resultant couple moment is zero.

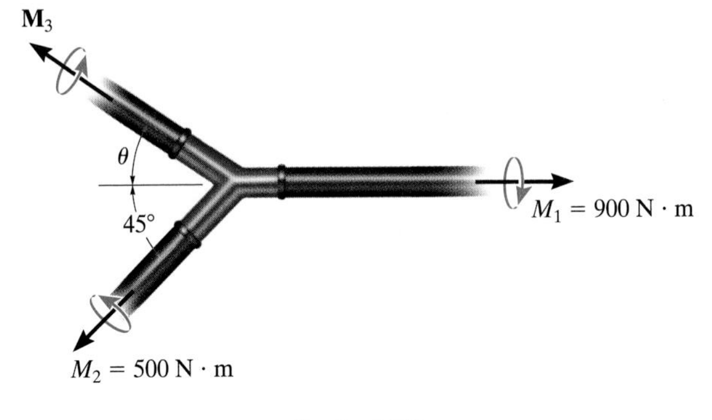

Prob. 4–75

***4–76.** The floor causes a couple moment of $M_A = 40\text{ N}\cdot\text{m}$ and $M_B = 30\text{ N}\cdot\text{m}$ on the brushes of the polishing machine. Determine the magnitude of the couple forces that must be developed by the operator on the handles so that the resultant couple moment on the polisher is zero. What is the magnitude of these forces if the brush at B suddenly stops so that $M_B = 0$?

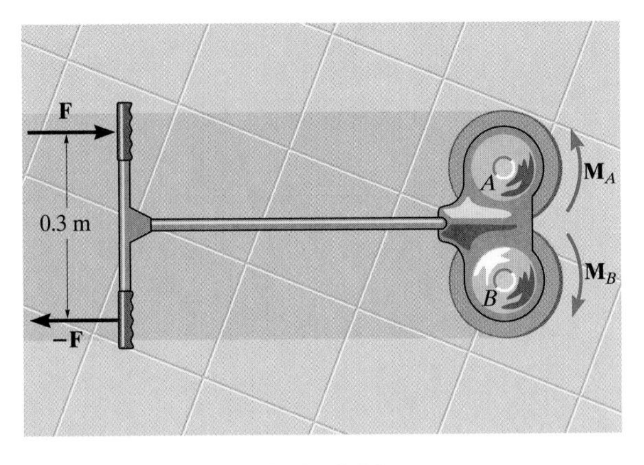

Prob. 4–76

4–77. The ends of the triangular plate are subjected to three couples. Determine the magnitude of the force **F** so that the resultant couple moment is 400 N · m clockwise.

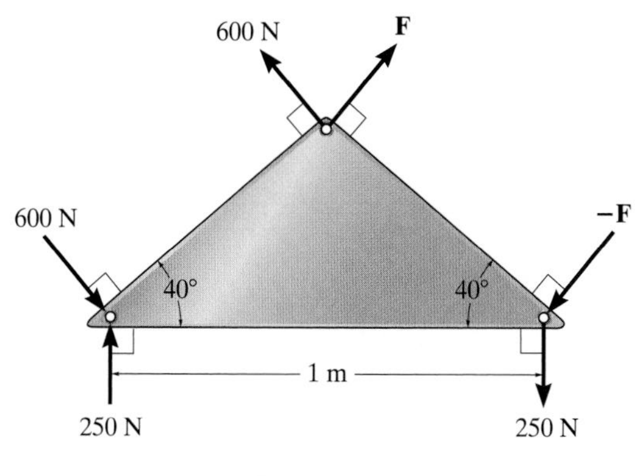

Prob. 4–77

4–78. Two couples act on the beam. Determine the magnitude of **F** so that the resultant couple moment is 450 lb · ft, counterclockwise. Where on the beam does the resultant couple moment act?

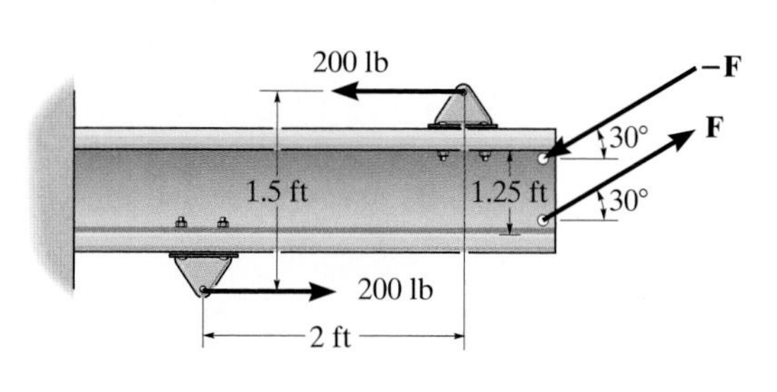

Prob. 4–78

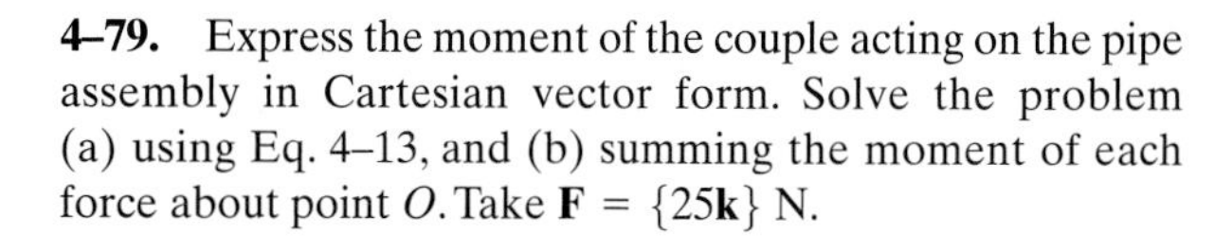

4–79. Express the moment of the couple acting on the pipe assembly in Cartesian vector form. Solve the problem (a) using Eq. 4–13, and (b) summing the moment of each force about point O. Take $\mathbf{F} = \{25\mathbf{k}\}$ N.

***4–80.** If the couple moment acting on the pipe has a magnitude of 400 N · m, determine the magnitude F of the vertical force applied to each wrench.

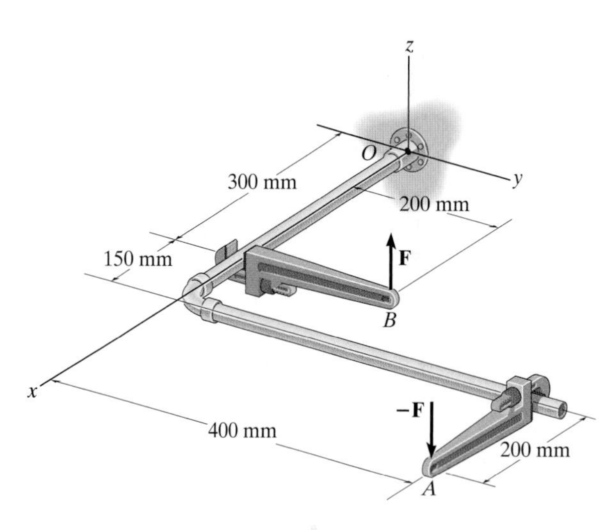

Probs. 4–79/80

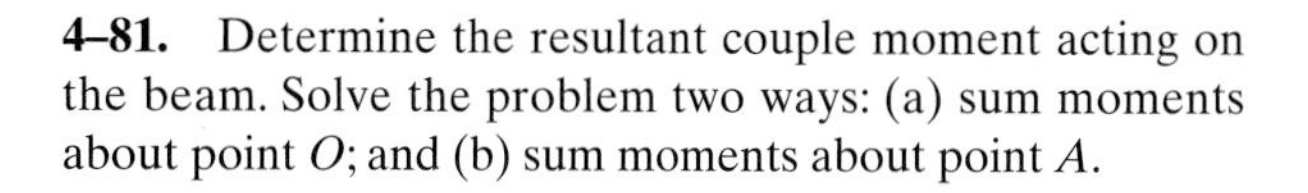

4–81. Determine the resultant couple moment acting on the beam. Solve the problem two ways: (a) sum moments about point O; and (b) sum moments about point A.

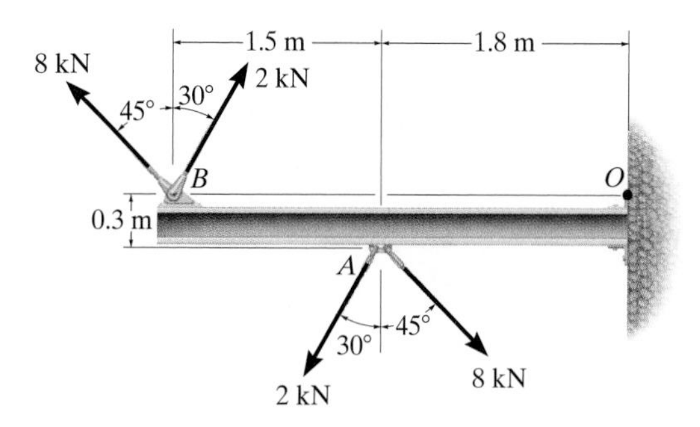

Prob. 4–81

4–82. Two couples act on the beam as shown. Determine the magnitude of **F** so that the resultant couple moment is 300 lb · ft counterclockwise. Where on the beam does the resultant couple act?

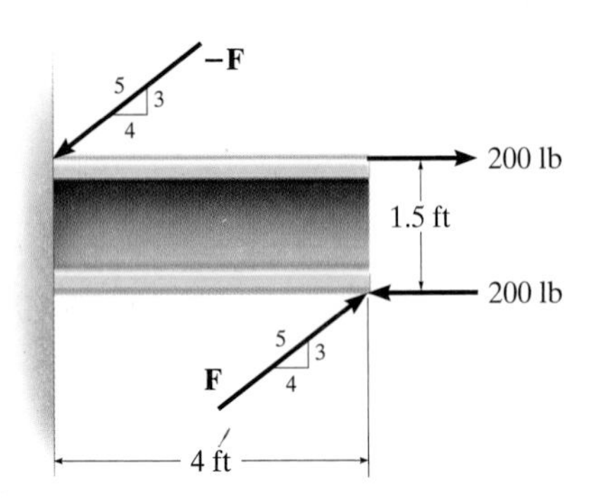

Prob. 4–82

4–83. Two couples act on the frame. If the resultant couple moment is to be zero, determine the distance d between the 80-lb couple forces.

***4–84.** Two couples act on the frame. If $d = 4$ ft, determine the resultant couple moment. Compute the result by resolving each force into x and y components and (a) finding the moment of each couple (Eq. 4–13) and (b) summing the moments of all the force components about point A.

4–85. Two couples act on the frame. If $d = 4$ ft, determine the resultant couple moment. Compute the result by resolving each force into x and y components and (a) finding the moment of each couple (Eq. 4–13) and (b) summing the moments of all the force components about point B.

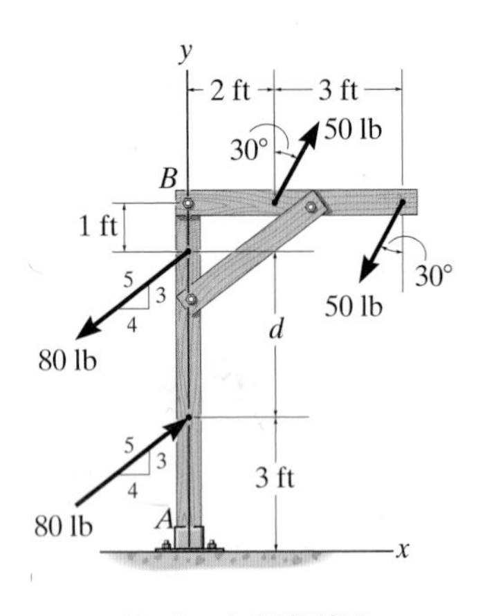

Probs. 4–83/84/85

4–86. Determine the couple moment. Express the result as a Cartesian vector.

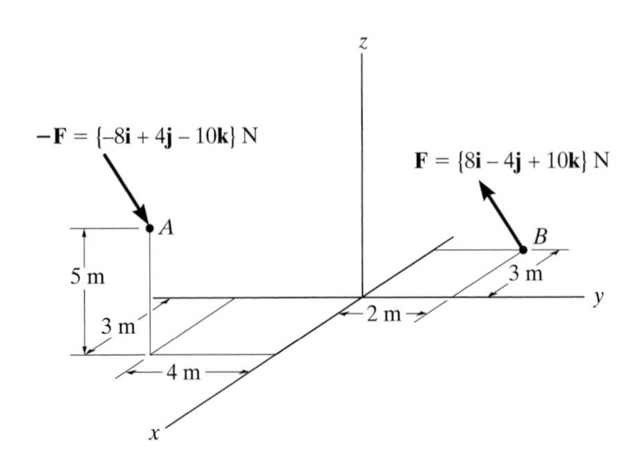

Prob. 4–86

4–87. Determine the couple moment. Express the result as a Cartesian vector.

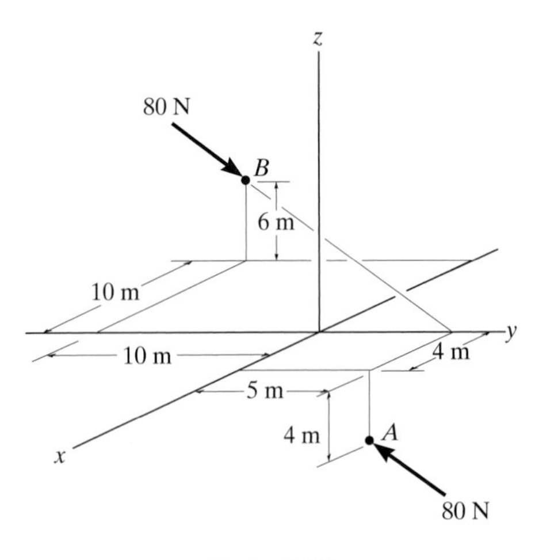

Prob. 4–87

***4–88.** If the resultant couple of the two couples acting on the fire hydrant is $\mathbf{M}_R = \{-15\mathbf{i} + 30\mathbf{j}\}$ N · m, determine the force magnitude P.

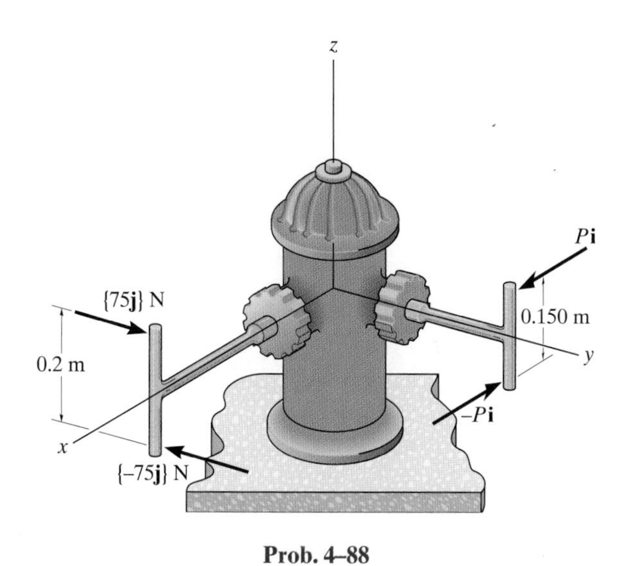

Prob. 4–88

4–89. If the resultant couple of the three couples acting on the triangular block is to be zero, determine the magnitude of forces **F** and **P**.

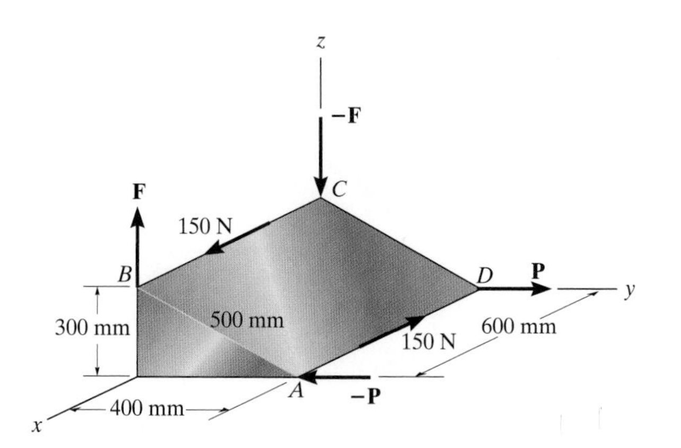

Prob. 4–89

4–90. If $\mathbf{F} = \{100\mathbf{k}\}$ N, determine the couple moment that acts on the assembly. Express the result as a Cartesian vector. Member BA lies in the x–y plane.

4–91. If the magnitude of the resultant couple moment is 15 N · m, determine the magnitude F of the forces applied to the wrenches.

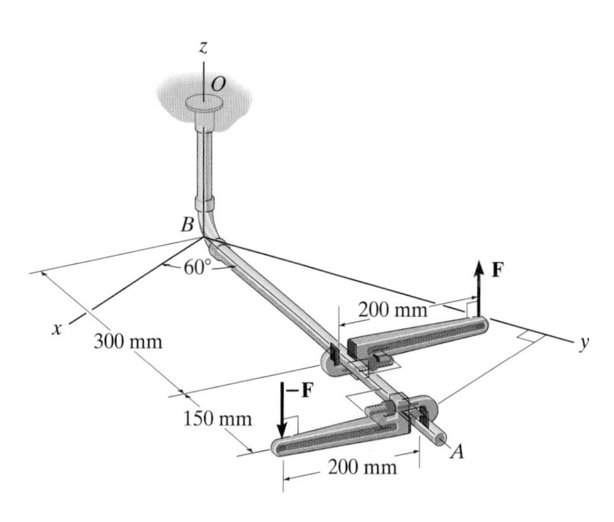

Probs. 4–90/91

***4–92.** The gears are subjected to the couple moments shown. Determine the magnitude and coordinate direction angles of the resultant couple moment.

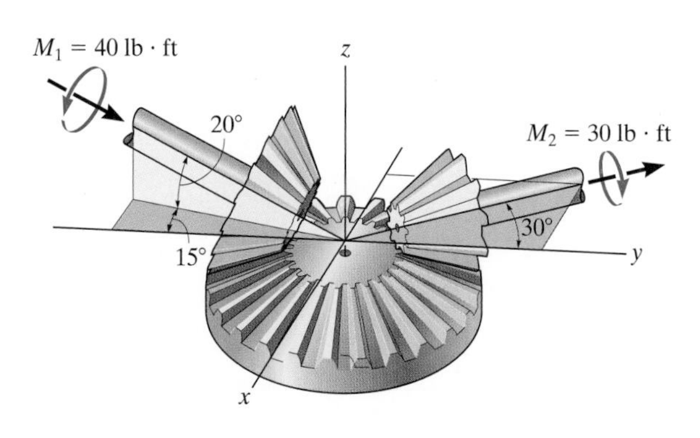

Prob. 4–92

4–93. Express the moment of the couple acting on the rod in Cartesian vector form. What is the magnitude of the couple moment?

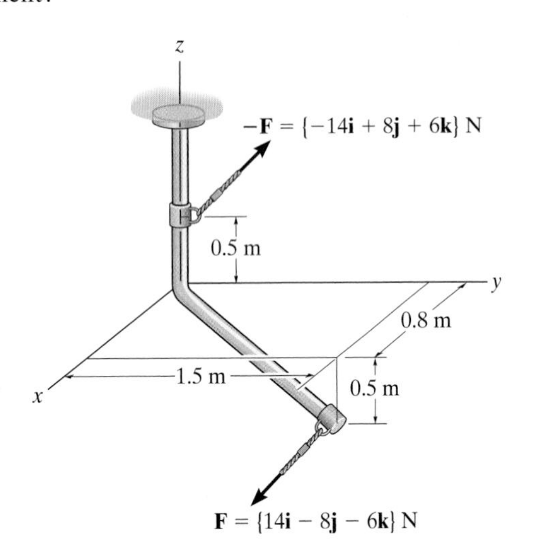

Prob. 4–93

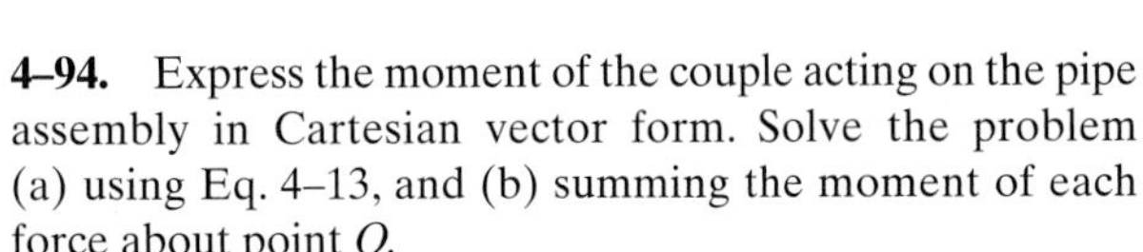

4–94. Express the moment of the couple acting on the pipe assembly in Cartesian vector form. Solve the problem (a) using Eq. 4–13, and (b) summing the moment of each force about point O.

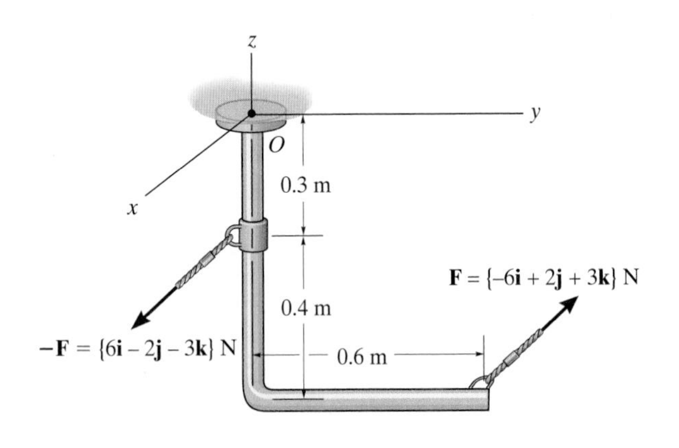

Prob. 4–94

4–95. A couple acts on each of the handles of the dual valve. Determine the magnitude and coordinate direction angles of the resultant couple moment.

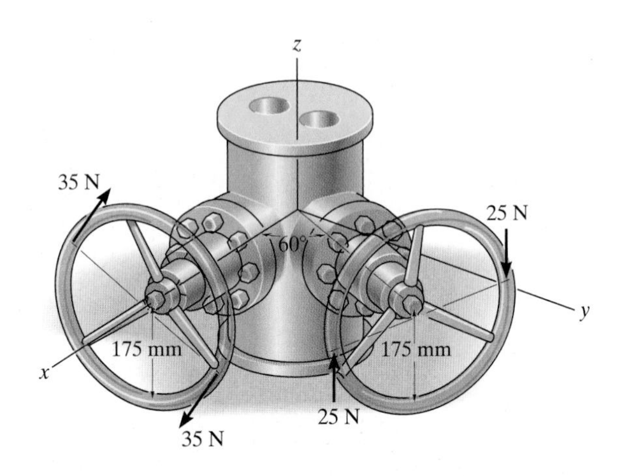

Prob. 4–95

***4–96.** Express the moment of the couple acting on the pipe in Cartesian vector form. What is the magnitude of the couple moment? Take $F = 125$ N.

4–97. If the couple moment acting on the pipe has a magnitude of 300 N · m, determine the magnitude F of the forces applied to the wrenches.

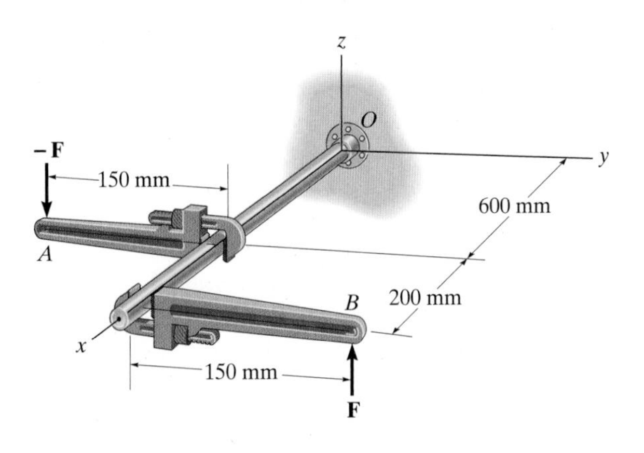

Probs. 4–96/97

4.7 Equivalent System

A force has the effect of both translating and rotating a body, and the amount by which it does so depends upon where and how the force is applied. In the next section we will discuss the method used to *simplify* a system of forces and couple moments acting on a body to a single resultant force and couple moment acting at a specified point O. To do this, however, it is necessary that the force and couple moment system produce the *same* "external" effects of translation and rotation of the body as their resultants. When this occurs these two sets of loadings are said to be *equivalent*.

In this section we wish to show how to maintain this equivalency when a single force is applied to a specific point on a body and when it is located at another point O. Two cases for the location of point O will now be considered.

Point O Is On the Line of Action of the Force. Consider the body shown in Fig. 4–33*a*, which is subjected to the force $\mathbf{F}$ applied to point A. In order to apply the force to point O without altering the external effects on the body, we will first apply equal but opposite forces $\mathbf{F}$ and $-\mathbf{F}$ at O, as shown in Fig. 4–33*b*. The two forces indicated by the slash across them can be canceled, leaving the force at point O as required, Fig. 4–33*c*. By using this construction procedure, an *equivalent system* has been maintained between each of the diagrams, as shown by the equal signs. Note, however, that the force has simply been "transmitted" along its line of action, from point A, Fig. 4–33*a*, to point O, Fig. 4–33*c*. In other words, the force can be considered as a *sliding vector* since it can act at any point O along its line of action. In Sec. 4.3 we referred to this concept as the *principle of transmissibility*. It is important to realize that only the *external effects*, such as the body's motion or the forces needed to support the body if it is stationary, remain *unchanged* after $\mathbf{F}$ is moved. Certainly the *internal effects* depend on where $\mathbf{F}$ is located. For example, when $\mathbf{F}$ acts at A, the internal forces in the body have a high intensity around A; whereas movement of $\mathbf{F}$ away from this point will cause these internal forces to decrease.

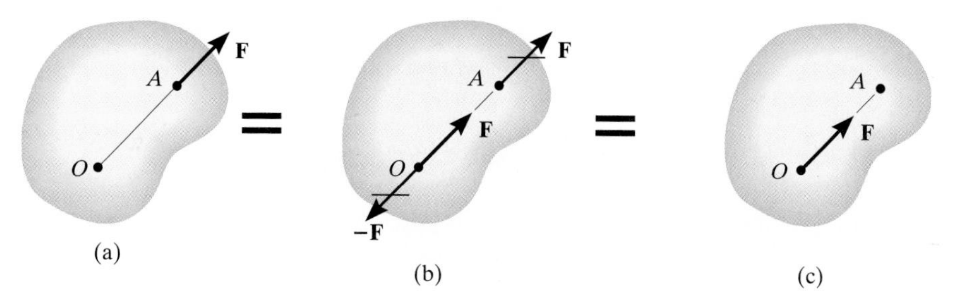

Fig. 4–33

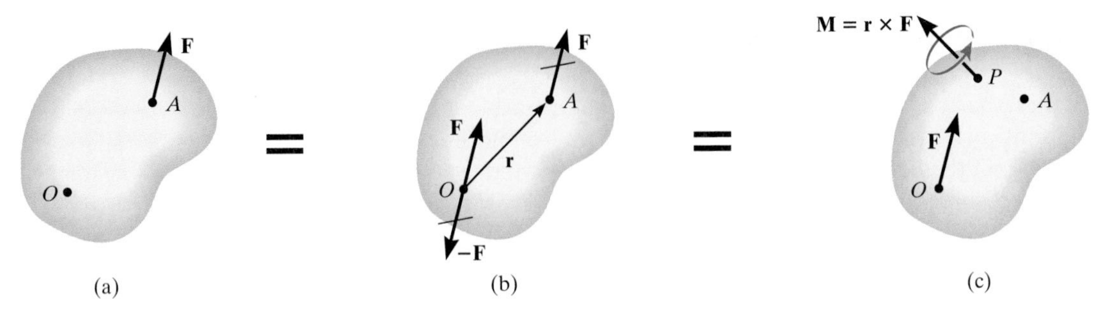

Fig. 4–34

Point O Is Not On the Line of Action of the Force. This case is shown in Fig. 4–34*a*, where **F** is to be moved to point O without altering the external effects on the body. Following the same procedure as before, we first apply equal but opposite forces **F** and $-\mathbf{F}$ at point O, Fig. 4–34*b*. Here the two forces indicated by a slash across them form a couple which has a moment that is perpendicular to **F** and is defined by the cross product $\mathbf{M} = \mathbf{r} \times \mathbf{F}$. Since the couple moment is a *free vector*, it may be applied at *any point* P on the body as shown in Fig. 4–34*c*. In addition to this couple moment, **F** now acts at point O as required.

To summarize these concepts, when the point on the body is *on the line of action of the force*, simply transmit or slide the force along its line of action to the point. When the point is not on the line of action of the force, then move the force to the point and add a couple moment anywhere to the body. This couple moment is found bytaking the moment of the force about the point. When these rules are carried out, equivalent external effects will be produced.

Photo A: Consider the effects on the hand when a stick of negligible weight supports a force **F** at its end. When the force is applied horizontally, the same force is felt at the grip, regardless of where it is applied along its line of action. This is a consequence of the principle of transmissibility.

Photo B: When the force is applied vertically it causes both a downward force **F** to be felt at the grip and a clockwise couple moment or twist of $M = Fd$. These same effects are felt if **F** is applied at the grip and **M** is applied anywhere on the stick. In both cases the systems are equivalent.

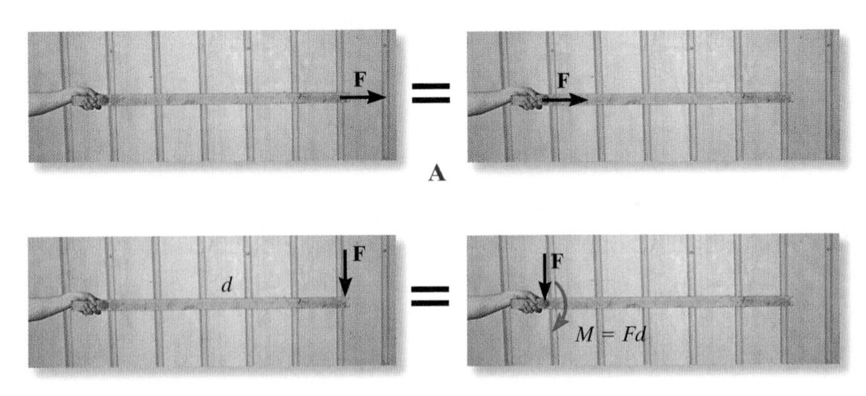

4.8 Resultants of a Force and Couple System

When a rigid body is subjected to a *system* of forces and couple moments, it is often simpler to study the external effects on the body by *replacing* the system by an equivalent single resultant force acting at a specified point O and a resultant couple moment. To show how to determine these resultants we will consider the rigid body in Fig. 4–35*a* and use the concepts discussed in the previous section. Since point O is not on the line of action of the forces, an equivalent effect is produced if the forces are moved to point O *and* the corresponding couple moments $\mathbf{M}_1 = \mathbf{r}_1 \times \mathbf{F}_1$ and $\mathbf{M}_2 = \mathbf{r}_2 \times \mathbf{F}_2$ are applied to the body. Furthermore, the couple moment $\mathbf{M}_c$ is simply moved to point O since it is a free vector. These results are shown in Fig. 4–35*b*. By vector addition, the resultant force is $\mathbf{F}_R = \mathbf{F}_1 + \mathbf{F}_2$, and the resultant couple moment is $\mathbf{M}_{R_O} = \mathbf{M}_c + \mathbf{M}_1 + \mathbf{M}_2$, Fig. 4–35*c*. Since equivalency is maintained between the diagrams in Fig. 4–35, each force and couple system will cause the *same external effects*, i.e., the same translation and rotation of the body. Note that both the magnitude and direction of $\mathbf{F}_R$ are independent of the location of point O; however, $\mathbf{M}_{R_O}$ depends upon this location since the moments $\mathbf{M}_1$ and $\mathbf{M}_2$ are determined using the position vectors $\mathbf{r}_1$ and $\mathbf{r}_2$. Also note that $\mathbf{M}_{R_O}$ is a free vector and can act at *any point* on the body, although point O is generally chosen as its point of application.

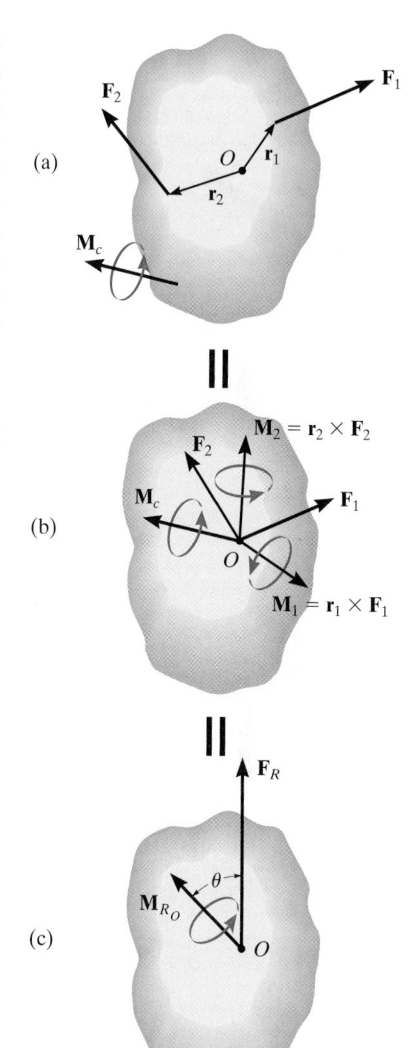

Fig. 4–35

The above method of simplifying any force and couple moment system to a resultant force acting at point O and a resultant couple moment can be generalized and represented by application of the following two equations.

$$\begin{aligned} \mathbf{F}_R &= \Sigma \mathbf{F} \\ \mathbf{M}_{R_O} &= \Sigma \mathbf{M}_c + \Sigma \mathbf{M}_O \end{aligned} \qquad (4\text{–}17)$$

The first equation states that the resultant force of the system is equivalent to the sum of all the forces; and the second equation states that the resultant couple moment of the system is equivalent to the sum of all the couple moments $\Sigma \mathbf{M}_c$, plus the moments about point O of all the forces $\Sigma \mathbf{M}_O$. If the force system lies in the x–y plane and any couple moments are perpendicular to this plane, that is along the z axis, then the above equations reduce to the following three scalar equations.

$$\begin{aligned} F_{R_x} &= \Sigma F_x \\ F_{R_y} &= \Sigma F_y \\ M_{R_o} &= \Sigma M_c + \Sigma M_O \end{aligned} \qquad (4\text{–}18)$$

Note that the resultant force $\mathbf{F}_R$ is equivalent to the vector sum of its two components $\mathbf{F}_{R_x}$ and $\mathbf{F}_{R_y}$.

If the two forces acting on the stick are replaced by an equivalent resultant force and couple momen t at point A, or by the equivalent resultant force and couple moment at point B, then in each case the hand must provide the same resistance to translation and rotation in order to keep the stick in the horizontal position. In other words, the external effects on the stick are the *same* in each case.

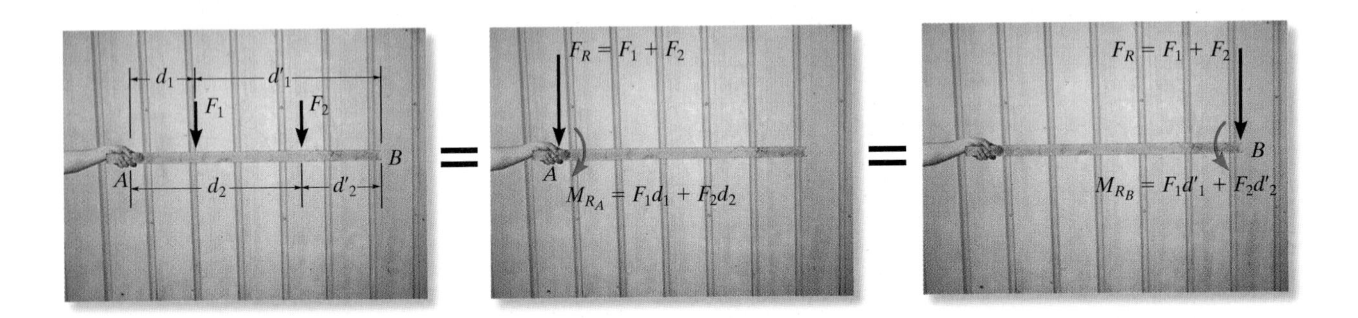

Procedure for Analysis

The following points should be kept in mind when applying Eqs. 4–17 or 4–18.

- Establish the coordinate axes with the origin located at point O and the axes having a selected orientation.

Force Summation.

- If the force system is *coplanar*, resolve each force into its x and y components. If a component is directed along the positive x or y axis, it represents a positive scalar; whereas if it is directed along the negative x or y axis, it is a negative scalar.
- In three dimensions, represent each force as a Cartesian vector before summing the forces.

Moment Summation.

- When determining the moments of a *coplanar* force system about point O, it is generally advantageous to use the principle of moments, i.e., determine the moments of the components of each force rather than the moment of the force itself.
- In three dimensions use the vector cross product to determine the moment of each force about the point. Here the position vectors extend from point O to any point on the line of action of each force.

EXAMPLE 4.14

Replace the forces acting on the brace shown in Fig. 4–36*a* by an equivalent resultant force and couple moment acting at point *A*.

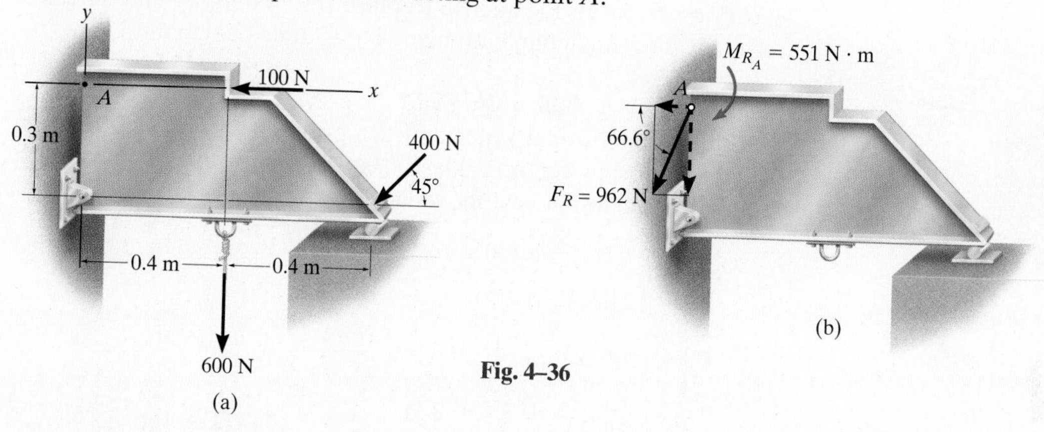

Fig. 4–36

SOLUTION *(SCALAR ANALYSIS)*

The principle of moments will be applied to the 400-N force, whereby the moments of its two rectangular components will be considered.

Force Summation. The resultant force has *x* and *y* components of

$$\overset{+}{\rightarrow} F_{R_x} = \Sigma F_x; \quad F_{R_x} = -100 \text{ N} - 400 \cos 45° \text{ N} = -382.8 \text{ N} = 382.8 \text{ N} \leftarrow$$

$$+\uparrow F_{R_y} = \Sigma F_y; \quad F_{R_y} = -600 \text{ N} - 400 \sin 45° \text{ N} = -882.8 \text{ N} = 882.8 \text{ N} \downarrow$$

As shown in Fig. 4–36*b*, $\mathbf{F}_R$ has a magnitude of

$$F_R = \sqrt{(F_{R_x})^2 + (F_{R_y})^2} = \sqrt{(382.8)^2 + (882.8)^2} = 962 \text{ N} \qquad \textit{Ans.}$$

and a direction of

$$\theta = \tan^{-1}\left(\frac{F_{R_y}}{F_{R_x}}\right) = \tan^{-1}\left(\frac{882.8}{382.8}\right) = 66.6° \quad \theta \nearrow \qquad \textit{Ans.}$$

Moment Summation. The resultant couple moment $\mathbf{M}_{R_A}$ is determined by summing the moments of the forces about point *A*. Assuming that positive moments act counterclockwise, i.e., in the $+\mathbf{k}$ direction, we have

$$\circlearrowleft + M_{R_A} = \Sigma M_A;$$

$$M_{R_A} = 100 \text{ N}(0) - 600 \text{ N}(0.4 \text{ m}) - (400 \sin 45° \text{ N})(0.8 \text{ m}) - (400 \cos 45° \text{ N})(0.3 \text{ m})$$

$$= -551 \text{ N} \cdot \text{m} = 551 \text{ N} \cdot \text{m} \circlearrowright \qquad \textit{Ans.}$$

NOTE: Realize that, when $\mathbf{M}_{R_A}$ and $\mathbf{F}_R$ act on the brace at point *A*, Fig. 4–36*b*, they will produce the *same* external effects or reactions at the supports as those produced by the force system in Fig. 4–36*a*.

EXAMPLE 4.15

A structural member is subjected to a couple moment **M** and forces $\mathbf{F}_1$ and $\mathbf{F}_2$ as shown in Fig. 4–37*a*. Replace this system by an equivalent resultant force and couple moment acting at its base, point *O*.

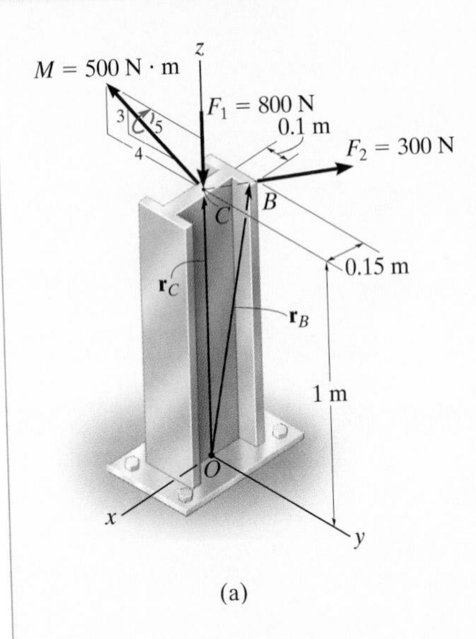

SOLUTION (VECTOR ANALYSIS)

The three-dimensional aspects of the problem can be simplified by using a Cartesian vector analysis. Expressing the forces and couple moment as Cartesian vectors, we have

$$\mathbf{F}_1 = \{-800\mathbf{k}\}\ \text{N}$$

$$\mathbf{F}_2 = (300\ \text{N})\mathbf{u}_{CB} = (300\ \text{N})\left(\frac{\mathbf{r}_{CB}}{r_{CB}}\right)$$

$$= 300\left[\frac{-0.15\mathbf{i} + 0.1\mathbf{j}}{\sqrt{(-0.15)^2 + (0.1)^2}}\right] = \{-249.6\mathbf{i} + 166.4\mathbf{j}\}\ \text{N}$$

$$\mathbf{M} = -500\left(\tfrac{4}{5}\right)\mathbf{j} + 500\left(\tfrac{3}{5}\right)\mathbf{k} = \{-400\mathbf{j} + 300\mathbf{k}\}\ \text{N}\cdot\text{m}$$

Force Summation.

$$\mathbf{F}_R = \Sigma\mathbf{F}; \qquad \mathbf{F}_R = \mathbf{F}_1 + \mathbf{F}_2 = -800\mathbf{k} - 249.6\mathbf{i} + 166.4\mathbf{j}$$

$$= \{-249.6\mathbf{i} + 166.4\mathbf{j} - 800\mathbf{k}\}\ \text{N} \qquad \textit{Ans.}$$

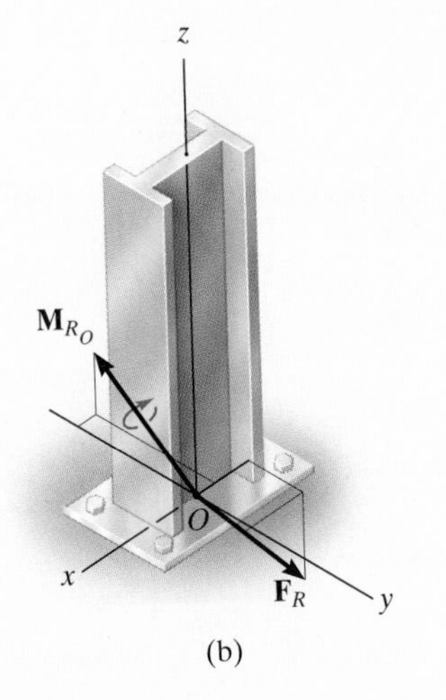

(b)

Fig. 4–37

Moment Summation.

$$\mathbf{M}_{R_O} = \Sigma\mathbf{M}_C + \Sigma\mathbf{M}_O$$

$$\mathbf{M}_{R_O} = \mathbf{M} + \mathbf{r}_C \times \mathbf{F}_1 + \mathbf{r}_B \times \mathbf{F}_2$$

$$\mathbf{M}_{R_O} = (-400\mathbf{j} + 300\mathbf{k}) + (1\mathbf{k}) \times (-800\mathbf{k}) + \begin{vmatrix} \mathbf{i} & \mathbf{j} & \mathbf{k} \\ -0.15 & 0.1 & 1 \\ -249.6 & 166.4 & 0 \end{vmatrix}$$

$$= (-400\mathbf{j} + 300\mathbf{k}) + (\mathbf{0}) + (-166.4\mathbf{i} - 249.6\mathbf{j})$$

$$= \{-166\mathbf{i} - 650\mathbf{j} + 300\mathbf{k}\}\ \text{N}\cdot\text{m} \qquad \textit{Ans.}$$

The results are shown in Fig. 4–37*b*.

4.9 Further Reduction of a Force and Couple System

Simplification to a Single Resultant Force. Consider now a special case for which the system of forces and couple moments acting on a rigid body, Fig. 4–38*a*, reduces at point O to a resultant force $\mathbf{F}_R = \Sigma\mathbf{F}$ and resultant couple moment $\mathbf{M}_{R_O} = \Sigma\mathbf{M}_O$, which are *perpendicular* to one another, Fig. 4–38*b*. Whenever this occurs, we can further simplify the force and couple moment system by moving $\mathbf{F}_R$ to another point P, located either on or off the body so that no resultant couple moment has to be applied to the body, Fig. 4–38*c*. In other words, if the force and couple moment system in Fig. 4–38*a* is reduced to a resultant system at point P, only the force resultant will have to be applied to the body, Fig. 4–38*c*.

The location of point P, measured from point O, can always be determined provided $\mathbf{F}_R$ and $\mathbf{M}_{R_O}$ are known, Fig. 4–38*b*. As shown in Fig. 4–38*c*, P must lie on the *bb* axis, which is perpendicular to both the line of action of $\mathbf{F}_R$ and the *aa* axis. This point is chosen such that the distance d satisfies the scalar equation $M_{R_O} = F_R d$ or $d = M_{R_O}/F_R$. With $\mathbf{F}_R$ so located, it will produce the same external effects on the body as the force and couple moment system in Fig. 4–38*a*, or the force and couple moment resultants in Fig. 4–38*b*.

If a system of forces is either concurrent, coplanar, or parallel, it can always be reduced, as in the above case, to a single resultant force $\mathbf{F}_R$ acting through a specific point. This is because in each of these cases $\mathbf{F}_R$ and $\mathbf{M}_{R_O}$ will always be perpendicular to each other when the force system is simplified at *any* point O.

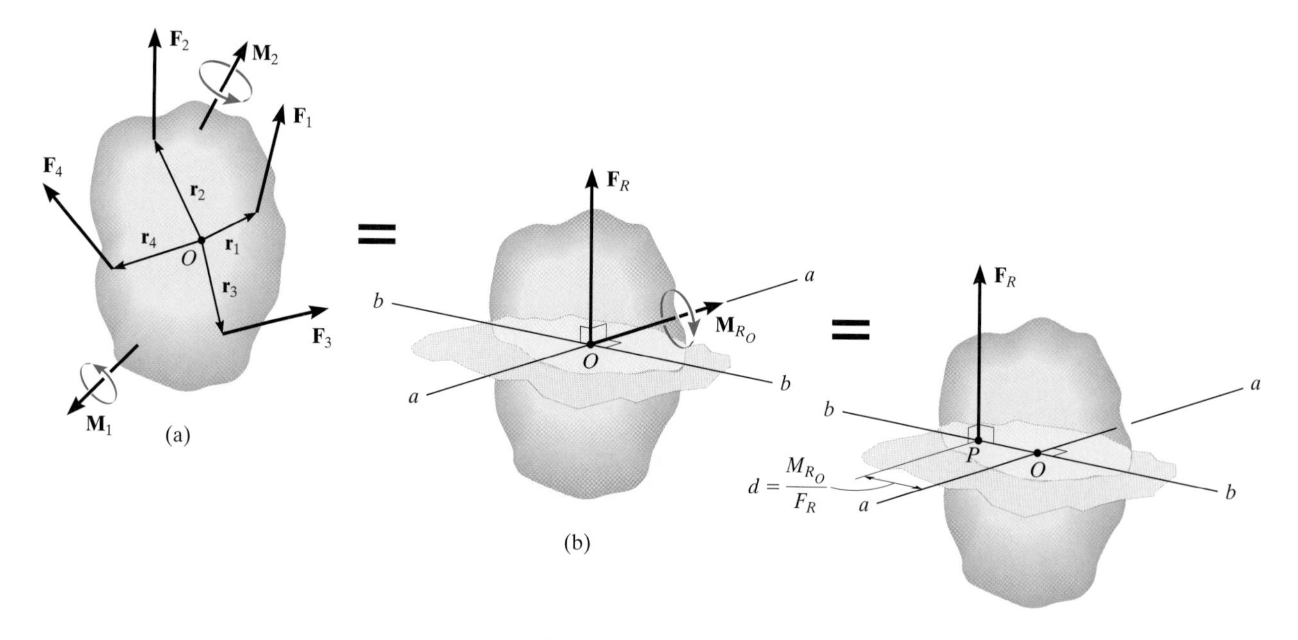

Fig. 4–38

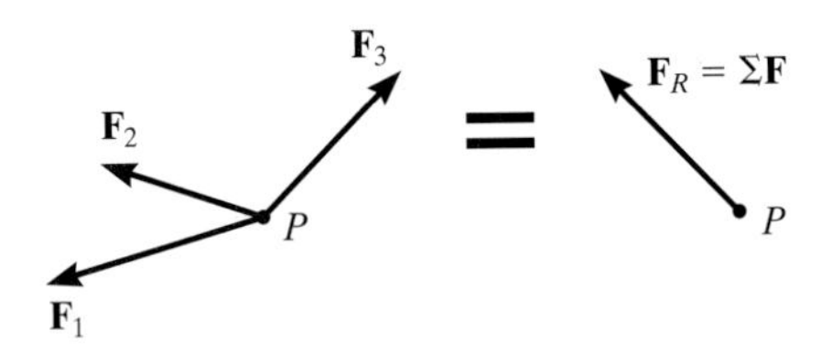

Fig. 4–39

Concurrent Force Systems. A concurrent force system has been treated in detail in Chapter 2. Obviously, all the forces act at a point for which there is no resultant couple moment, so the point P is automatically specified, Fig. 4–39.

Coplanar Force Systems. Coplanar force systems, which may include couple moments directed perpendicular to the plane of the forces as shown in Fig. 4–40*a*, can be reduced to a single resultant force, because when each force in the system is moved to any point O in the x–y plane, it produces a couple moment that is *perpendicular* to the plane, i.e., in the $\pm\mathbf{k}$ direction. The resultant moment $\mathbf{M}_{R_O} = \Sigma\mathbf{M} + \Sigma(\mathbf{r} \times \mathbf{F})$ is thus perpendicular to the resultant force $\mathbf{F}_R$, Fig. 4–40*b*; and so $\mathbf{F}_R$ can be positioned a distance d from O so as to create the same moment $\mathbf{M}_{R_O}$ about O, Fig. 4–40*c*.

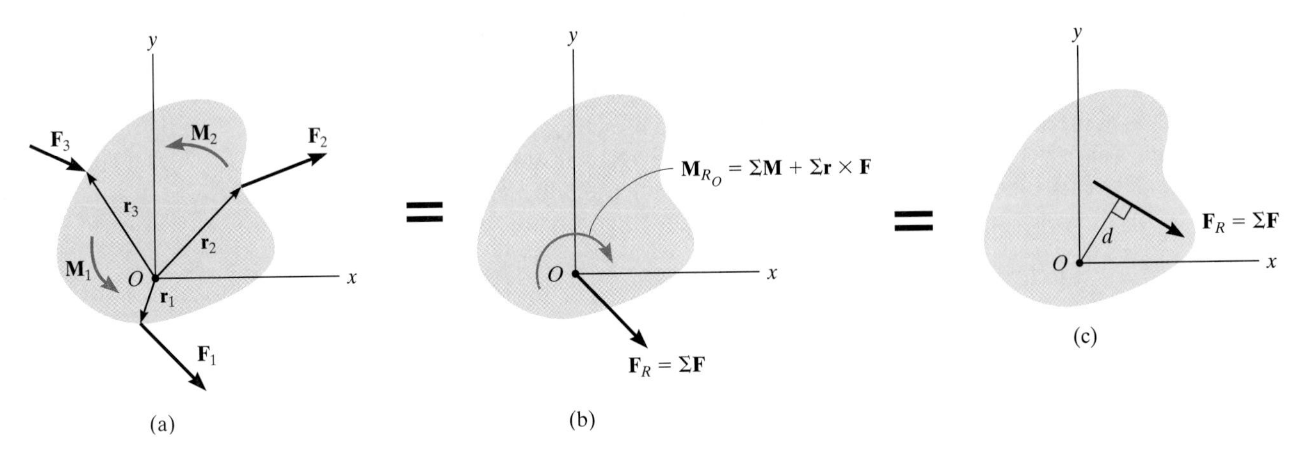

Fig. 4–40

Parallel Force Systems. Parallel force systems, which can include couple moments that are perpendicular to the forces, as shown in Fig. 4–41*a*, can be reduced to a single resultant force because when each force is moved to any point O in the x–y plane, it produces a couple moment that has components only about the x and y axes. The resultant moment $\mathbf{M}_{R_O} = \Sigma\mathbf{M}_O + \Sigma(\mathbf{r} \times \mathbf{F})$ is thus perpendicular to the resultant force $\mathbf{F}_R$, Fig. 4–41*b*; and so $\mathbf{F}_R$ can be moved to a point a distance d away so that it produces the same moment about O.

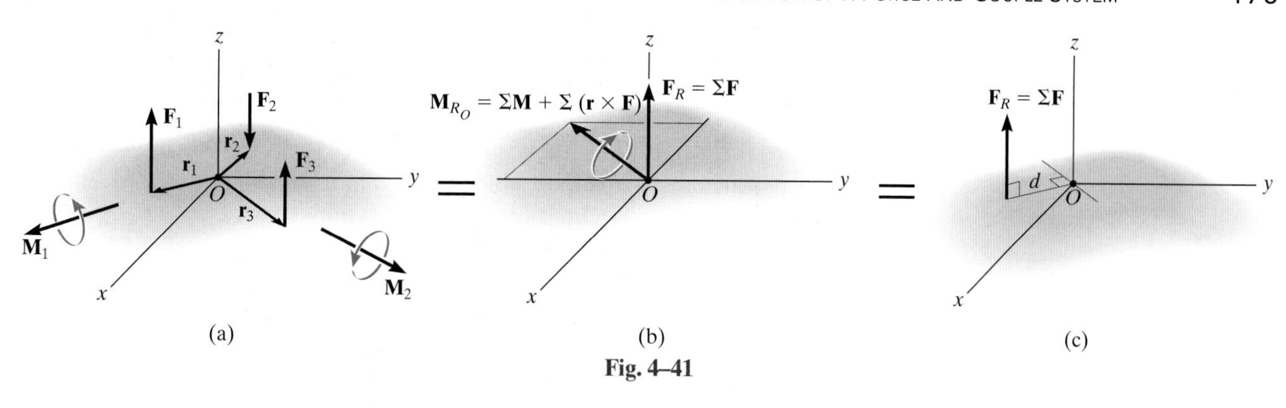

Fig. 4–41

The three parallel forces acting on the stick can be replaced by a single resultant force F_R acting at a distance d from the grip. To be equivalent we require the resultant force to equal the sum of the forces, $F_R = F_1 + F_2 + F_3$, and to find the distance d the moment of the resultant force about the grip must be equal to the moment of all the forces about the grip, $F_R d = F_1 d_1 + F_2 d_2 + F_3 d_3$.

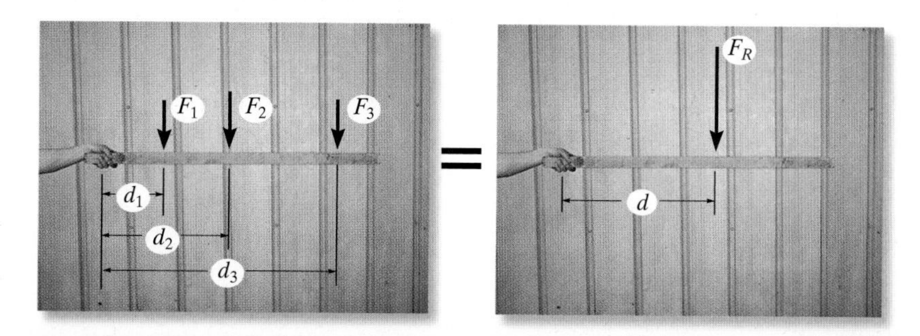

Procedure for Analysis

The technique used to reduce a coplanar or parallel force system to a single resultant force follows a similar procedure outlined in the previous section.

- Establish the x, y, z, axes and locate the resultant force $\mathbf{F}_R$ an arbitrary distance away from the origin of the coordinates.

Force Summation.

- The resultant force is equal to the sum of all the forces in the system.
- For a coplanar force system, resolve each force into its x and y components. Positive components are directed along the positive x and y axes, and negative components are directed along the negative x and y axes.

Moment Summation.

- The moment of the resultant force about point O is equal to the sum of all the couple moments in the system plus the moments about point O of all the forces in the system.
- This moment condition is used to find the location of the resultant force from point O.

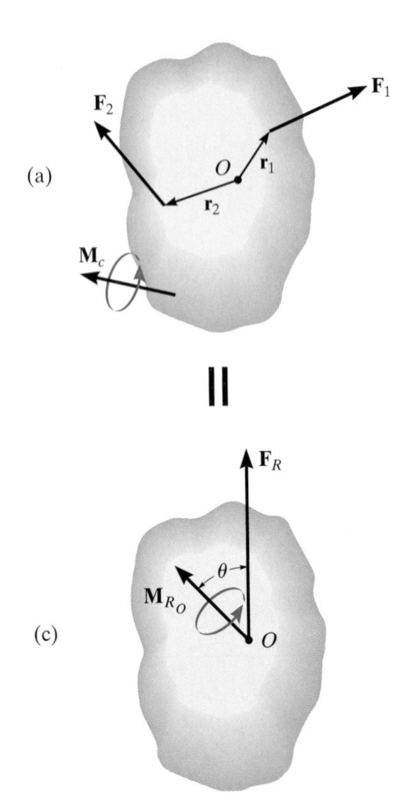

Fig. 4–35 (Repeated)

Reduction to a Wrench. In the general case, the force and couple moment system acting on a body, Fig. 4–35*a*, will reduce to a single resultant force $\mathbf{F}_R$ and couple moment $\mathbf{M}_{R_O}$ at O which are *not* perpendicular. Instead, $\mathbf{F}_R$ will act at an angle θ from $\mathbf{M}_{R_O}$, Fig. 4–35*c*. As shown in Fig. 4–42*a*, however, $\mathbf{M}_{R_O}$ may be resolved into two components: one perpendicular, $\mathbf{M}_\perp$, and the other parallel $\mathbf{M}_\parallel$, to the line of action of $\mathbf{F}_R$. As in the previous discussion, the perpendicular component $\mathbf{M}_\perp$ may be *eliminated* by moving $\mathbf{F}_R$ to point P, as shown in Fig. 4–42*b*. This point lies on axis *bb*, which is perpendicular to both $\mathbf{M}_{R_O}$ and $\mathbf{F}_R$. In order to maintain an equivalency of loading, the distance from O to P is $d = M_\perp / F_R$. Furthermore, when $\mathbf{F}_R$ is applied at P, the moment of $\mathbf{F}_R$ tending to cause rotation of the body *about* O is in the *same direction* as $\mathbf{M}_\perp$, Fig. 4–42*a*. Finally, since $\mathbf{M}_\parallel$ is a free vector, it may be moved to P so that it is collinear with $\mathbf{F}_R$, Fig. 4–42*c*. This combination of a collinear force and couple moment is called a *wrench* or *screw*. The *axis of the wrench* has the same line of action as the force. Hence, the wrench tends to cause both a translation along and a rotation about this axis. Comparing Fig. 4–42*a* to Fig. 4–42*c*, it is seen that a general force and couple moment system acting on a body can be reduced to a wrench. The axis of the wrench and the point through which this axis passes can always be determined.

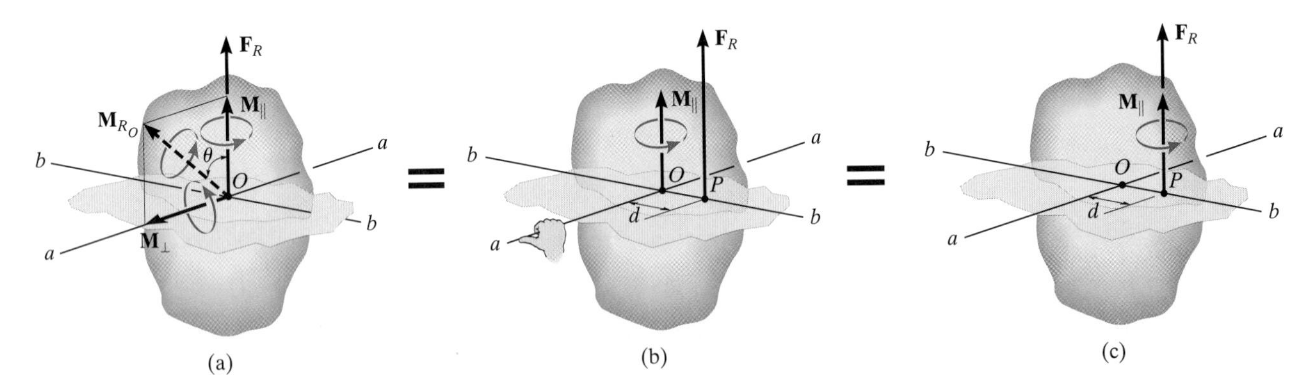

Fig. 4–42

EXAMPLE 4.16

The beam AE in Fig. 4–43a is subjected to a system of coplanar forces. Determine the magnitude, direction, and location on the beam of a resultant force which is equivalent to the given system of forces measured from E.

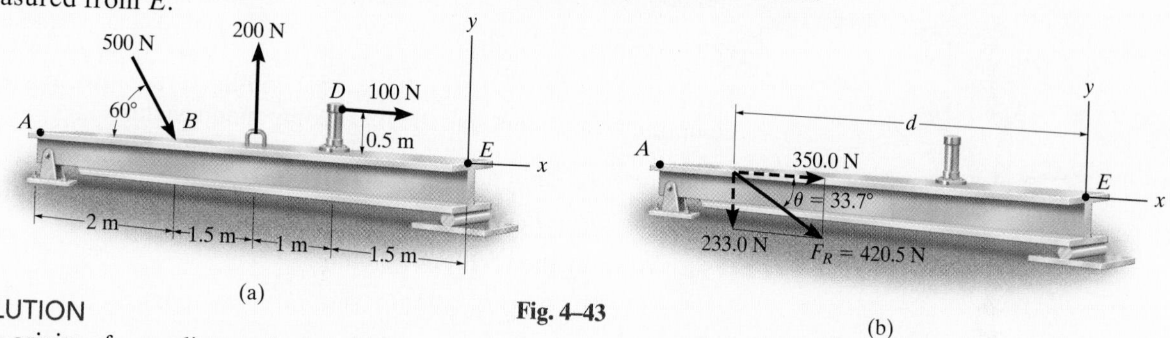

Fig. 4–43

SOLUTION

The origin of coordinates is located at point E as shown in Fig. 4–43a.

Force Summation. Resolving the 500-N force into x and y components and summing the force components yields

$$\xrightarrow{+} F_{R_x} = \Sigma F_x; \qquad F_{R_x} = 500 \cos 60° \text{ N} + 100 \text{ N} = 350.0 \text{ N} \rightarrow$$

$$+\uparrow F_{R_y} = \Sigma F_y; \qquad F_{R_y} = -500 \sin 60° \text{ N} + 200 \text{ N} = -233.0 \text{ N} = 233.0 \text{ N}\downarrow$$

The magnitude and direction of the resultant force are established from the vector addition shown in Fig. 4–43b. We have

$$F_R = \sqrt{(350.0)^2 + (233.0)^2} = 420.5 \text{ N} \qquad \textit{Ans.}$$

$$\theta = \tan^{-1}\left(\frac{233.0}{350.0}\right) = 33.7° \ \theta \qquad \textit{Ans.}$$

Moment Summation. Moments will be summed about point E. Hence, from Figs. 4–43a and 4–43b, we require the moments of the components of $\mathbf{F}_R$ (or the moment of $\mathbf{F}_R$) about point E to equal the moments of the force system about E. Assuming positive moments are counterclockwise, we have

$$\circlearrowleft + M_{R_E} = \Sigma M_E;$$

$$233.0 \text{ N}(d) + 350.0 \text{ N}(0) = (500 \sin 60° \text{ N})(4 \text{ m}) + (500 \cos 60° \text{ N})(0) - (100 \text{ N})(0.5 \text{ m}) - (200 \text{ N})(2.5 \text{ m})$$

$$d = \frac{1182.1}{233.0} = 5.07 \text{ m} \qquad \textit{Ans.}$$

NOTE: Using a clockwise sign convention would yield this same result. Since d is *positive*, $\mathbf{F}_R$ acts to the left of E as shown. Try to solve this problem by summing moments about point A and show $d' = 0.927$ m, measured to the right of A.

EXAMPLE 4.17

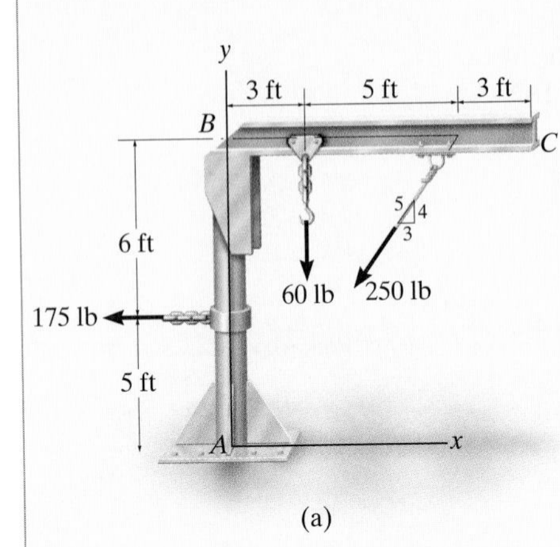

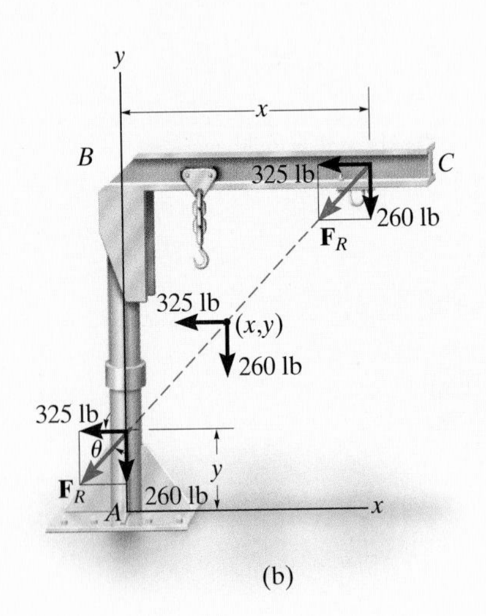

Fig. 4–44

The jib crane shown in Fig. 4–44*a* is subjected to three coplanar forces. Replace this loading by an equivalent resultant force and specify where the resultant's line of action intersects the column *AB* and boom *BC*.

SOLUTION

Force Summation. Resolving the 250-lb force into x and y components and summing the force components yields

$$\xrightarrow{+} F_{R_x} = \Sigma F_x; \quad F_{R_x} = -250 \text{ lb}\left(\tfrac{3}{5}\right) - 175 \text{ lb} = -325 \text{ lb} = 325 \text{ lb} \leftarrow$$

$$+\uparrow F_{R_y} = \Sigma F_y; \quad F_{R_y} = -250 \text{ lb}\left(\tfrac{4}{5}\right) - 60 \text{ lb} = -260 \text{ lb} = 260 \text{ lb} \downarrow$$

As shown by the vector addition in Fig. 4–44*b*,

$$F_R = \sqrt{(325)^2 + (260)^2} = 416 \text{ lb} \qquad \textit{Ans.}$$

$$\theta = \tan^{-1}\left(\frac{260}{325}\right) = 38.7^\circ \ \theta\swarrow \qquad \textit{Ans.}$$

Moment Summation. Moments will be summed about the arbitrary point A. Assuming the line of action of $\mathbf{F}_R$ *intersects* AB, Fig. 4–44*b*, we require the moment of the components of $\mathbf{F}_R$ in Fig. 4–44*b* about A to equal the moments of the force system in Fig. 4–44*a* about A; i.e.,

$$\circlearrowleft + M_{R_A} = \Sigma M_A; \qquad 325 \text{ lb } (y) + 260 \text{ lb } (0)$$

$$= 175 \text{ lb } (5 \text{ ft}) - 60 \text{ lb } (3 \text{ ft}) + 250 \text{ lb}\left(\tfrac{3}{5}\right)(11 \text{ ft}) - 250 \text{ lb}\left(\tfrac{4}{5}\right)(8 \text{ ft})$$

$$y = 2.29 \text{ ft} \qquad \textit{Ans.}$$

By the principle of transmissibility, $\mathbf{F}_R$ can also be treated as intersecting BC, Fig. 4–44*b*, in which case we have

$$\circlearrowleft + M_{R_A} = \Sigma M_A; \qquad 325 \text{ lb } (11 \text{ ft}) - 260 \text{ lb } (x)$$

$$= 175 \text{ lb } (5 \text{ ft}) - 60 \text{ lb } (3 \text{ ft}) + 250 \text{ lb}\left(\tfrac{3}{5}\right)(11 \text{ ft}) - 250 \text{ lb}\left(\tfrac{4}{5}\right)(8 \text{ ft})$$

$$x = 10.9 \text{ ft} \qquad \textit{Ans.}$$

NOTE: We can also solve for these positions by assuming $\mathbf{F}_R$ acts at the arbitrary point (x, y) on its line of action, Fig. 4–44*b*. Summing moments about point A yields

$$\circlearrowleft + M_{R_A} = \Sigma M_A; 325 \text{ lb } (y) - 260 \text{ lb } (x)$$

$$= 175 \text{ lb } (5 \text{ ft}) - 60 \text{ lb } (3 \text{ ft}) + 250 \text{ lb}\left(\tfrac{3}{5}\right)(11 \text{ ft}) - 250 \text{ lb}\left(\tfrac{4}{5}\right)(8 \text{ ft})$$

$$325y - 260x = 745$$

which is the equation of the colored dashed line in Fig. 4–44*b*. To find the points of intersection with the crane along AB, set $x = 0$, then $y = 2.29$ ft, and along BC set $y = 11$ ft, then $x = 10.9$ ft.

EXAMPLE 4.18

The slab in Fig. 4–45*a* is subjected to four parallel forces. Determine the magnitude and direction of a resultant force equivalent to the given force system and locate its point of application on the slab.

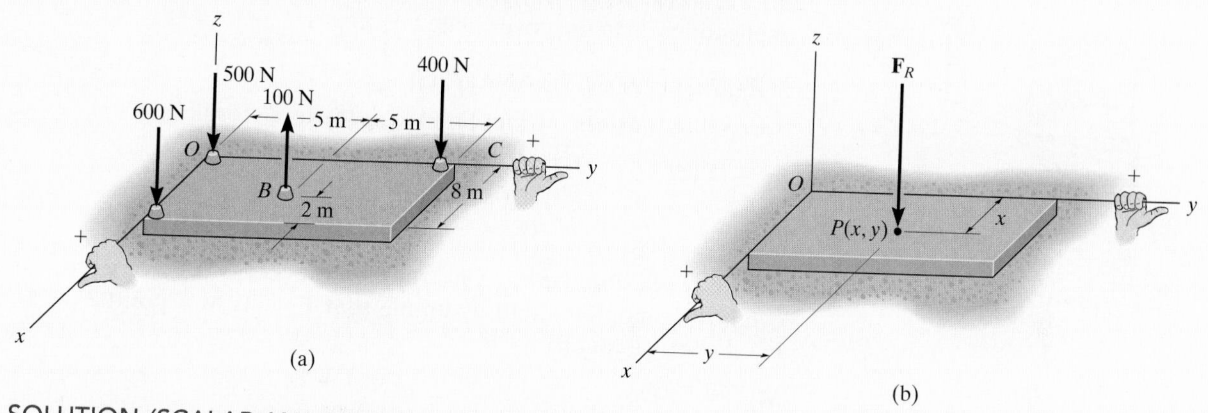

Fig. 4–45

SOLUTION (SCALAR ANALYSIS)

Force Summation. From Fig. 4–45*a*, the resultant force is

$$+\uparrow F_R = \Sigma F; \quad F_R = -600\text{ N} + 100\text{ N} - 400\text{ N} - 500\text{ N}$$
$$= -1400\text{ N} = 1400\text{ N}\downarrow \qquad \textit{Ans.}$$

Moment Summation. We require the moment about the x axis of the resultant force, Fig. 4–45*b*, to be equal to the sum of the moments about the x axis of all the forces in the system, Fig. 4–45*a*. The moment arms are determined from the y coordinates since these coordinates represent the *perpendicular distances* from the x axis to the lines of action of the forces. Using the right-hand rule, where positive moments act in the $+\mathbf{i}$ direction, we have

$$M_{R_x} = \Sigma M_x;$$
$$-(1400\text{ N})y = 600\text{ N}(0) + 100\text{ N}(5\text{ m}) - 400\text{ N}(10\text{ m}) + 500\text{ N}(0)$$
$$-1400y = -3500 \qquad y = 2.50\text{ m} \qquad \textit{Ans.}$$

In a similar manner, assuming that positive moments act in the $+\mathbf{j}$ direction, a moment equation can be written about the y axis using moment arms defined by the x coordinates of each force.

$$M_{R_y} = \Sigma M_y;$$
$$(1400\text{ N})x = 600\text{ N}(8\text{ m}) - 100\text{ N}(6\text{ m}) + 400\text{ N}(0) + 500\text{ N}(0)$$
$$1400x = 4200 \qquad x = 3.00\text{ m} \qquad \textit{Ans.}$$

NOTE: A force of $F_R = 1400$ N placed at point $P(3.00\text{ m}, 2.50\text{ m})$ on the slab, Fig. 4–45*b*, is therefore equivalent to the parallel force system acting on the slab in Fig. 4–45*a*.

EXAMPLE 4.19

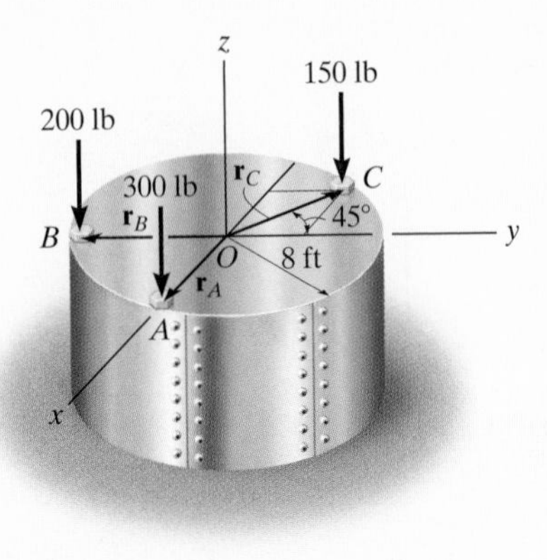

(a)

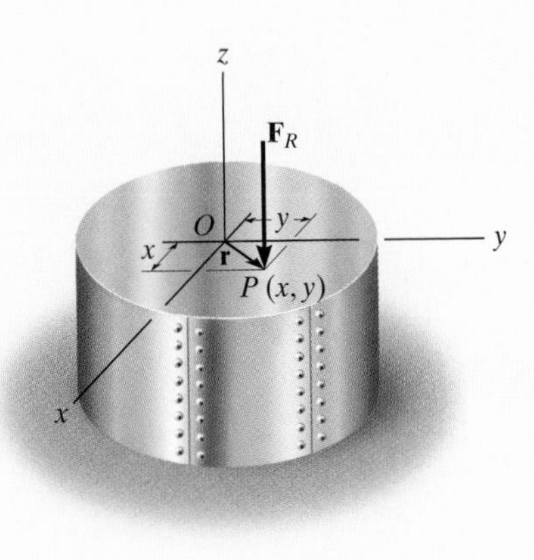

(b)

Fig. 4–46

Three parallel bolting forces act on the rim of the circular cover plate in Fig. 4–46*a*. Determine the magnitude and direction of a resultant force equivalent to the given force system and locate its point of application, *P*, on the cover plate.

SOLUTION (*VECTOR ANALYSIS*)

Force Summation. From Fig. 4–46*a*, the force resultant $\mathbf{F}_R$ is

$$\mathbf{F}_R = \Sigma\mathbf{F}; \qquad \mathbf{F}_R = -300\mathbf{k} - 200\mathbf{k} - 150\mathbf{k}$$
$$= \{-650\mathbf{k}\}\text{ lb} \qquad \textit{Ans.}$$

Moment Summation. Choosing point *O* as a reference for computing moments and assuming that $\mathbf{F}_R$ acts at a point $P(x, y)$, Fig. 4–46*b*, we require

$$\mathbf{M}_{R_O} = \Sigma\mathbf{M}_O;$$
$$\mathbf{r} \times \mathbf{F}_R = \mathbf{r}_A \times (-300\mathbf{k}) + \mathbf{r}_B \times (-200\mathbf{k}) + \mathbf{r}_C \times (-150\mathbf{k})$$
$$(x\mathbf{i} + y\mathbf{j}) \times (-650\mathbf{k}) = (8\mathbf{i}) \times (-300\mathbf{k}) + (-8\mathbf{j}) \times (-200\mathbf{k})$$
$$+ (-8 \sin 45^\circ\mathbf{i} + 8 \cos 45^\circ\mathbf{j}) \times (-150\mathbf{k})$$
$$650x\mathbf{j} - 650y\mathbf{i} = 2400\mathbf{j} + 1600\mathbf{i} - 848.5\mathbf{j} - 848.5\mathbf{i}$$

Equating the corresponding **j** and **i** components yields

$$650x = 2400 - 848.5 \qquad (1)$$
$$-650y = 1600 - 848.5 \qquad (2)$$

Solving these equations, we obtain the coordinates of point *P*,

$$x = 2.39\text{ ft} \qquad y = -1.16\text{ ft} \qquad \textit{Ans.}$$

The negative sign indicates that it was wrong to have assumed a $+y$ position for $\mathbf{F}_R$ as shown in Fig. 4–46*b*.

NOTE: It is also possible to establish Eqs. 1 and 2 directly by summing moments about the *y* and *x* axes. Using the right-hand rule we have

$$M_{R_y} = \Sigma M_y; \qquad 650x = 300\text{ lb }(8\text{ ft}) - 150\text{ lb }(8 \sin 45^\circ\text{ ft})$$
$$M_{R_x} = \Sigma M_x; \qquad -650y = 200\text{ lb }(8\text{ ft}) - 150\text{ lb }(8 \cos 45^\circ\text{ ft})$$

PROBLEMS

4–98. Replace the force at A by an equivalent force and couple moment at point O.

4–99. Replace the force at A by an equivalent force and couple moment at point P.

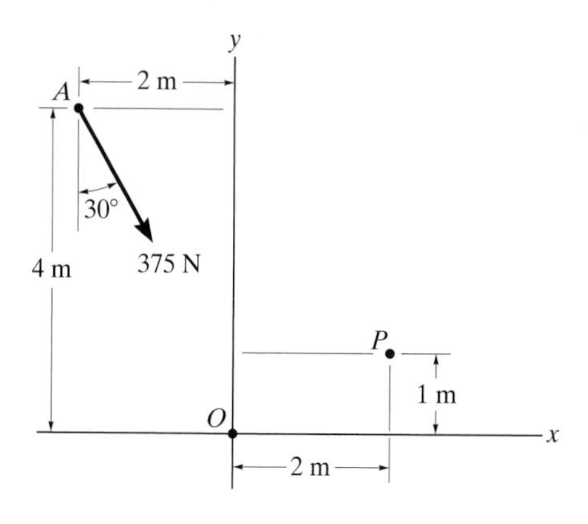

Probs. 4–98/99

4–102. Replace the force system by an equivalent force and couple moment at point O.

4–103. Replace the force system by an equivalent force and couple moment at point P.

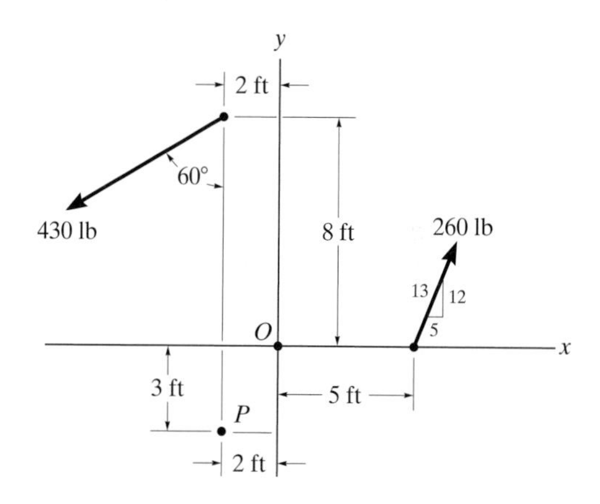

Probs. 4–102/103

***4–100.** Replace the force system by an equivalent resultant force and couple moment at point O.

4–101. Replace the force system by an equivalent resultant force and couple moment at point P.

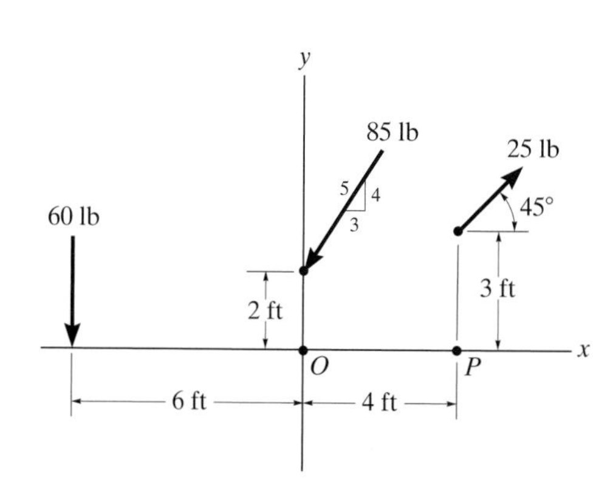

Probs. 4–100/101

***4–104.** Replace the loading system acting on the post by an equivalent resultant force and couple moment at point O.

4–105. Replace the loading system acting on the post by an equivalent resultant force and couple moment at point P.

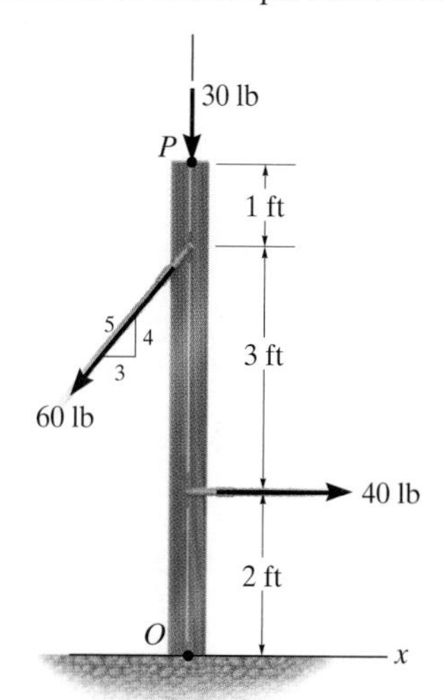

Probs. 4–104/105

4–106. Replace the force and couple system by an equivalent force and couple moment at point O.

4–107. Replace the force and couple system by an equivalent force and couple moment at point P.

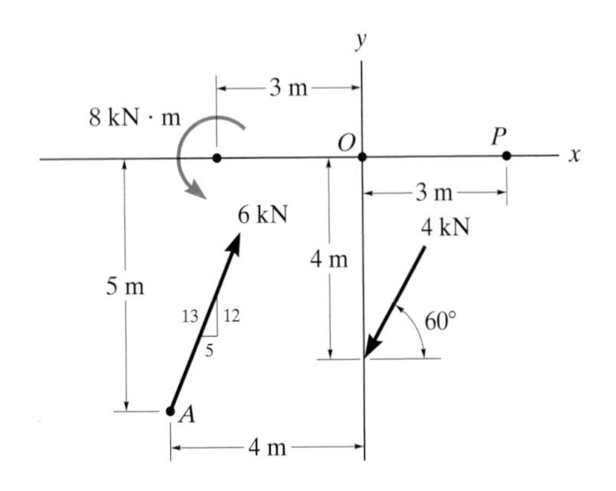

Probs. 4–106/107

***4–108.** Replace the force system by a single force resultant and specify its point of application, measured along the x axis from point O.

4–109. Replace the force system by a single force resultant and specify its point of application, measured along the x axis from point P.

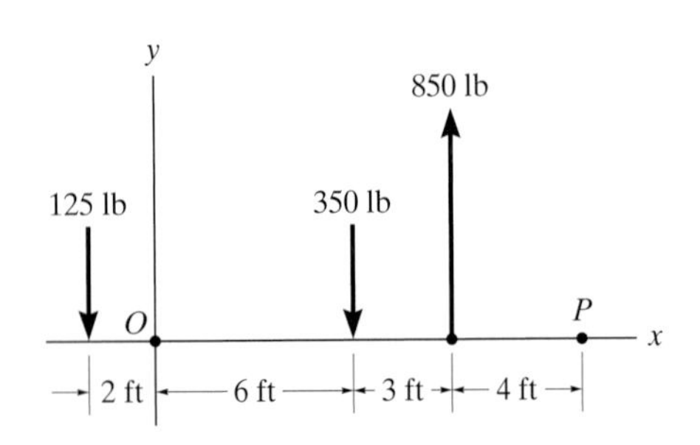

Probs. 4–108/109

4–110. The forces and couple moments which are exerted on the toe and heel plates of a snow ski are $\mathbf{F}_t = \{-50\mathbf{i} + 80\mathbf{j} - 158\mathbf{k}\}$ N, $\mathbf{M}_t = \{-6\mathbf{i} + 4\mathbf{j} + 2\mathbf{k}\}$ N · m, and $\mathbf{F}_h = \{-20\mathbf{i} + 60\mathbf{j} - 250\mathbf{k}\}$ N, $\mathbf{M}_h = \{-20\mathbf{i} + 8\mathbf{j} + 3\mathbf{k}\}$ N · m, respectively. Replace this system by an equivalent force and couple moment acting at point O. Express the results in Cartesian vector form.

4–111. The forces and couple moments that are exerted on the toe and heel plates of a snow ski are $\mathbf{F}_t = \{-50\mathbf{i} + 80\mathbf{j} - 158\mathbf{k}\}$ N, $\mathbf{M}_t = \{-6\mathbf{i} + 4\mathbf{j} + 2\mathbf{k}\}$ N · m, and $\mathbf{F}_h = \{-20\mathbf{i} + 60\mathbf{j} - 250\mathbf{k}\}$ N, $\mathbf{M}_h = \{-20\mathbf{i} + 8\mathbf{j} + 3\mathbf{k}\}$ N · m, respectively. Replace this system by an equivalent force and couple moment acting at point P. Express the results in Cartesian vector form.

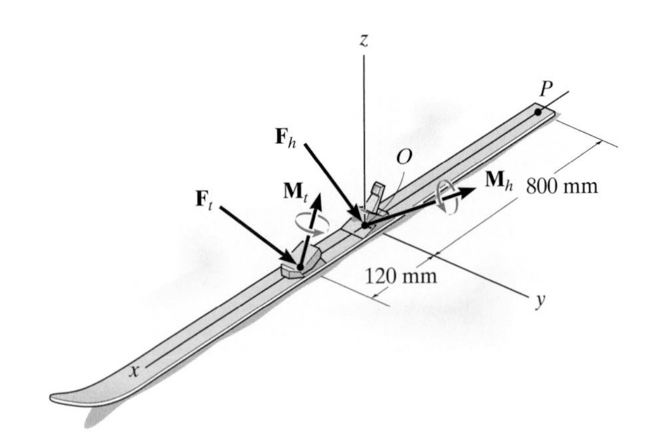

Probs. 4–110/111

***4–112.** Replace the three forces acting on the shaft by a single resultant force. Specify where the force acts, measured from end A.

4–113. Replace the three forces acting on the shaft by a single resultant force. Specify where the force acts, measured from end B.

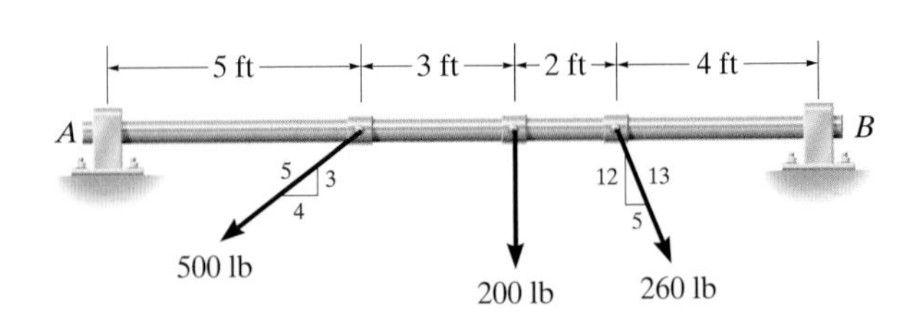

Probs. 4–112/113

4–114. Replace the loading on the frame by a single resultant force. Specify where its line of action intersects member AB, measured from A.

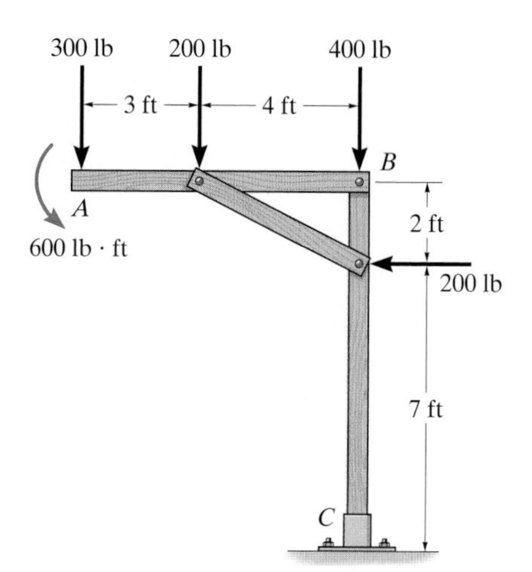

Prob. 4–114

4–115. Replace the loading acting on the beam by a single resultant force. Specify where the force acts, measured from end A.

***4–116.** Replace the loading acting on the beam by a single resultant force. Specify where the force acts, measured from B.

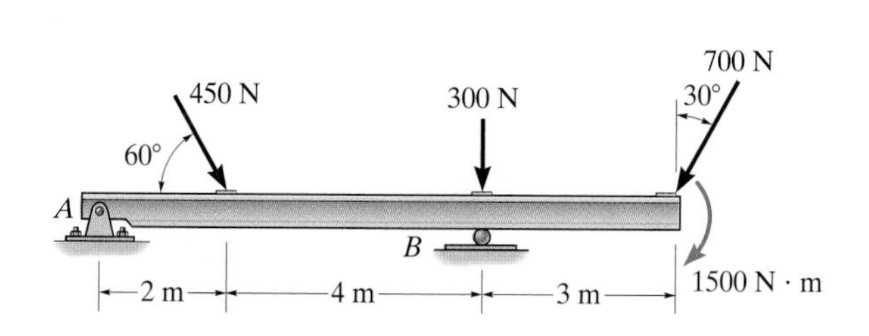

Probs. 4–115/116

4–117. Replace the loading system acting on the beam by an equivalent resultant force and couple moment at point O.

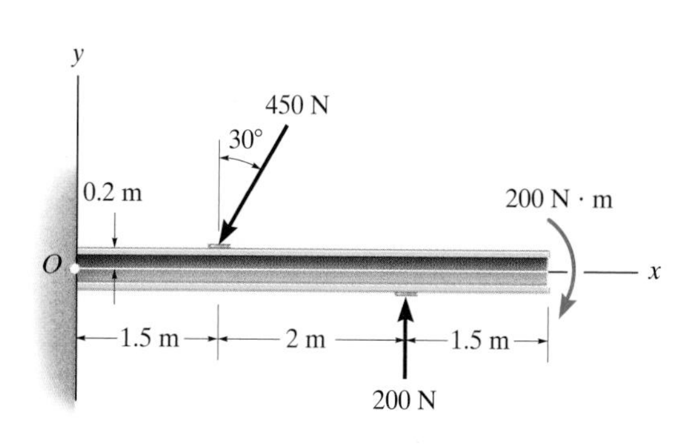

Prob. 4–117

4–118. Determine the magnitude and direction θ of force $\mathbf{F}$ and its placement d on the beam so that the loading system is equivalent to a resultant force of 12 kN acting vertically downward at point A and a clockwise couple moment of 50 kN · m.

4–119. Determine the magnitude and direction θ of force $\mathbf{F}$ and its placement d on the beam so that the loading system is equivalent to a resultant force of 10 kN acting vertically downward at point A and a clockwise couple moment of 45 kN · m.

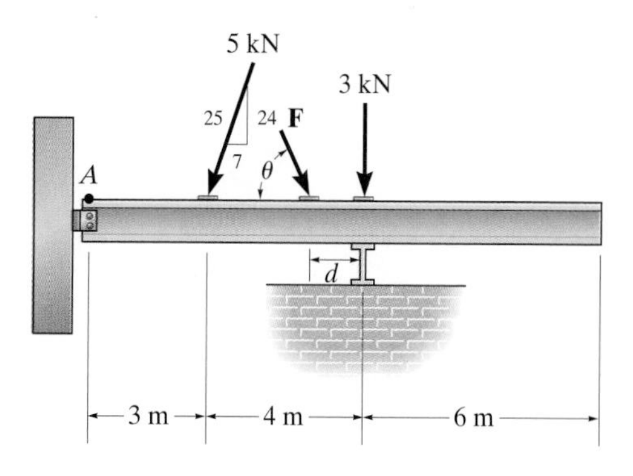

Probs. 4–118/119

***4–120.** Replace the loading on the frame by a single resultant force. Specify where its line of action intersects member AB, measured from A.

4–121. Replace the loading on the frame by a single resultant force. Specify where its line of action intersects member CD, measured from end C.

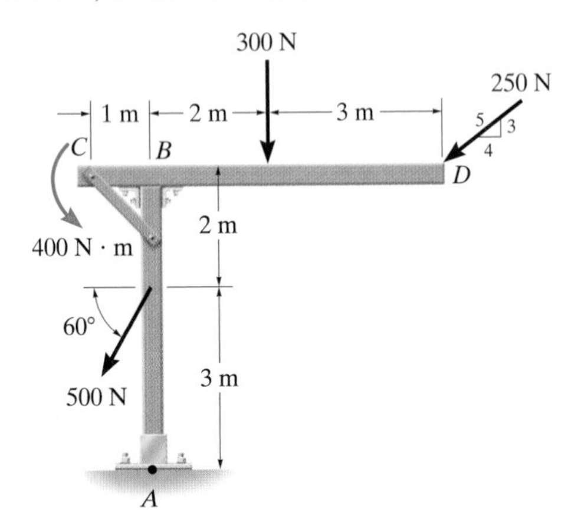

Probs. 4–120/121

4–122. Replace the force system acting on the frame by an equivalent resultant force and specify where the resultant's line of action intersects member AB, measured from point A.

4–123. Replace the force system acting on the frame by an equivalent resultant force and specify where the resultant's line of action intersects member BC, measured from point B.

***4–124.** Replace the force system acting on the frame by an equivalent resultant force and couple moment acting at point A.

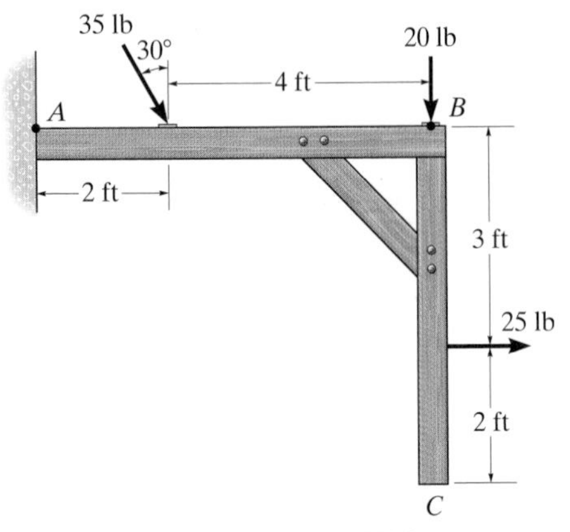

Probs. 4–122/123/124

4–125. Replace the force and couple-moment system by an equivalent resultant force and couple moment at point O. Express the results in Cartesian vector form.

4–126. Replace the force and couple-moment system by an equivalent resultant force and couple moment at point P. Express the results in Cartesian vector form.

4–127. Replace the force and couple-moment system by an equivalent resultant force and couple moment at point Q. Express the results in Cartesian vector form.

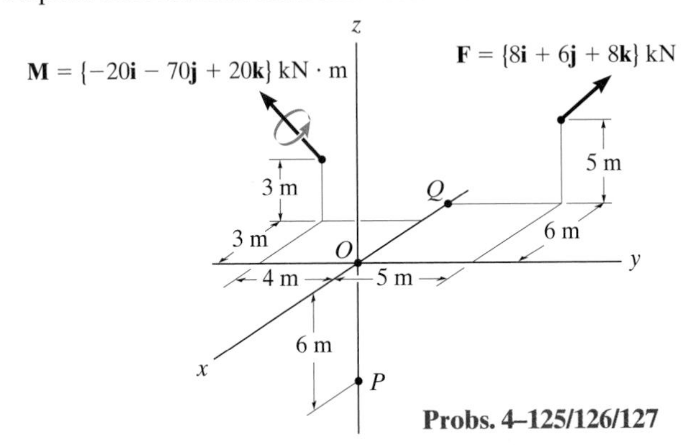

Probs. 4–125/126/127

***4–128.** The belt passing over the pulley is subjected to forces $\mathbf{F}_1$ and $\mathbf{F}_2$, each having a magnitude of 40 N. $\mathbf{F}_1$ acts in the $-\mathbf{k}$ direction. Replace these forces by an equivalent force and couple moment at point A. Express the result in Cartesian vector form. Set $\theta = 0°$ so that $\mathbf{F}_2$ acts in the $-\mathbf{j}$ direction.

4–129. The belt passing over the pulley is subjected to two forces $\mathbf{F}_1$ and $\mathbf{F}_2$, each having a magnitude of 40 N. $\mathbf{F}_1$ acts in the $-\mathbf{k}$ direction. Replace these forces by an equivalent force and couple moment at point A. Express the result in Cartesian vector form. Take $\theta = 45°$.

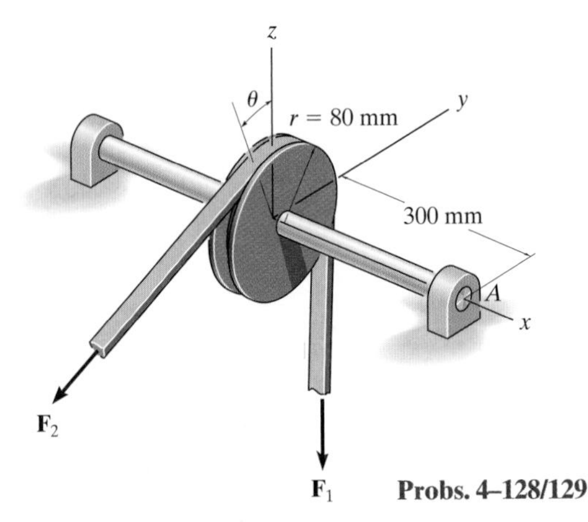

Probs. 4–128/129

4–130. A force and couple act on the pipe assembly. If $F_1 = 50$ N and $F_2 = 80$ N, replace this system by an equivalent resultant force and couple moment acting at O. Express the results in Cartesian vector form.

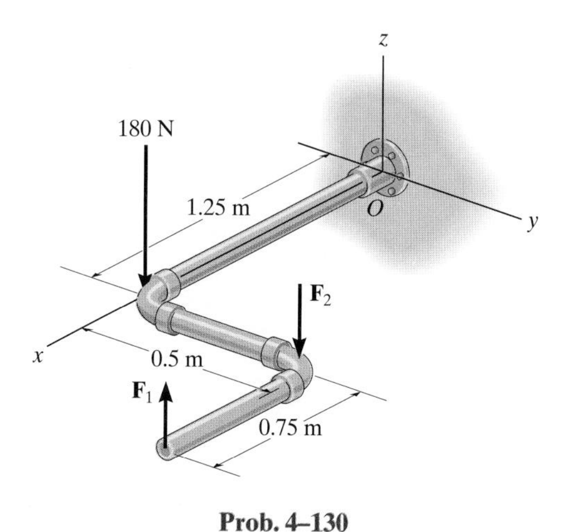

Prob. 4–130

4–131. Handle forces $\mathbf{F}_1$ and $\mathbf{F}_2$ are applied to the electric drill. Replace this system by an equivalent resultant force and couple moment acting at point O. Express the results in Cartesian vector form.

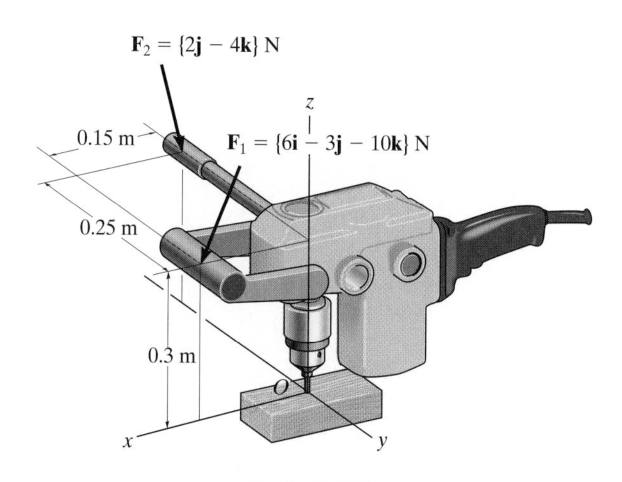

Prob. 4–131

***4–132.** A biomechanical model of the lumbar region of the human trunk is shown. The forces acting in the four muscle groups consist of $F_R = 35$ N for the rectus, $F_O = 45$ N for the oblique, $F_L = 23$ N for the lumbar latissimus dorsi, and $F_E = 32$ N for the erector spinae. These loadings are symmetric with respect to the y–z plane. Replace this system of parallel forces by an equivalent force and couple moment acting at the spine, point O. Express the results in Cartesian vector form.

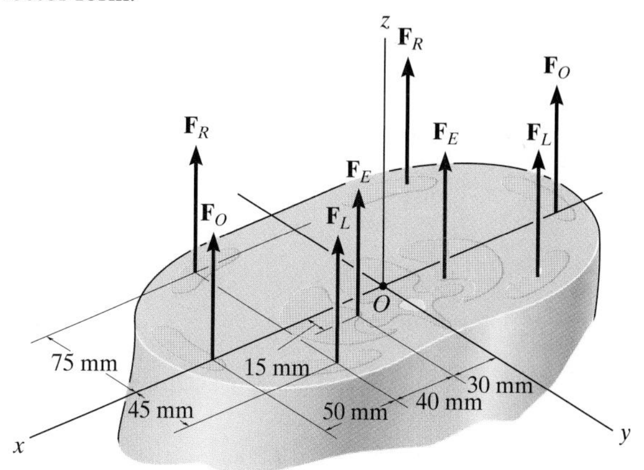

Prob. 4–132

4–133. The building slab is subjected to four parallel column loadings. Determine the equivalent resultant force and specify its location (x, y) on the slab. Take $F_1 = 30$ kN, $F_2 = 40$ kN.

4–134. The building slab is subjected to four parallel column loadings. Determine the equivalent resultant force and specify its location (x, y) on the slab. Take $F_1 = 20$ kN, $F_2 = 50$ kN.

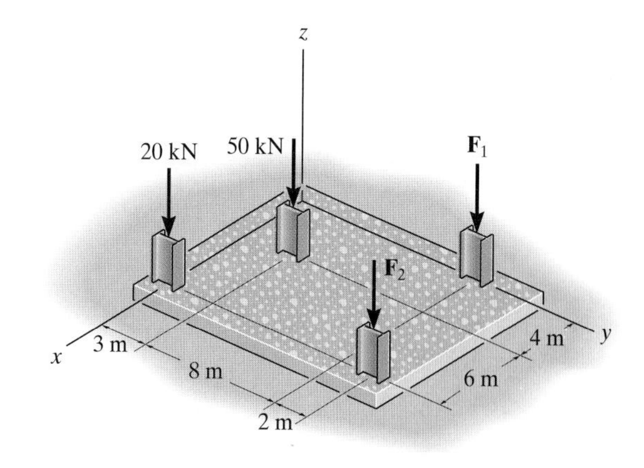

Probs. 4–133/134

4–135. The pipe assembly is subjected to the action of a wrench at B and a couple at A. Determine the magnitude F of the couple forces so that the system can be simplified to a wrench acting at point C.

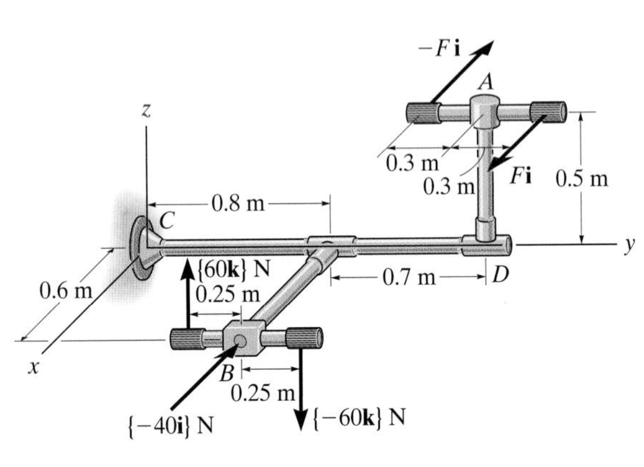

Prob. 4–135

4–137. Replace the three forces acting on the plate by a wrench. Specify the magnitude of the force and couple moment for the wrench and the point $P(x, y)$ where its line of action intersects the plate.

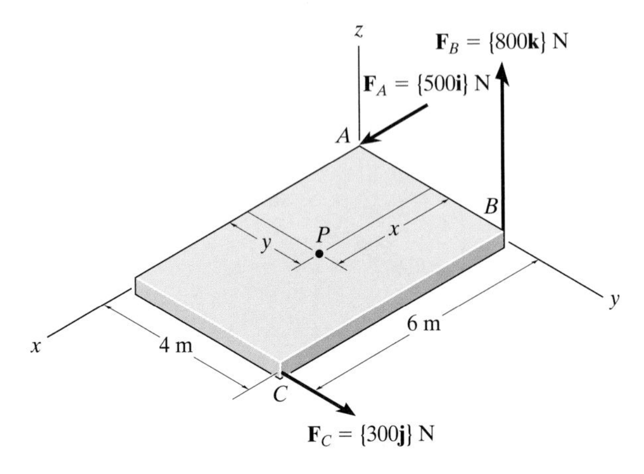

Prob. 4–137

***4–136.** The three forces acting on the block each have a magnitude of 10 lb. Replace this system by a wrench and specify the point where the wrench intersects the z axis, measured from point O.

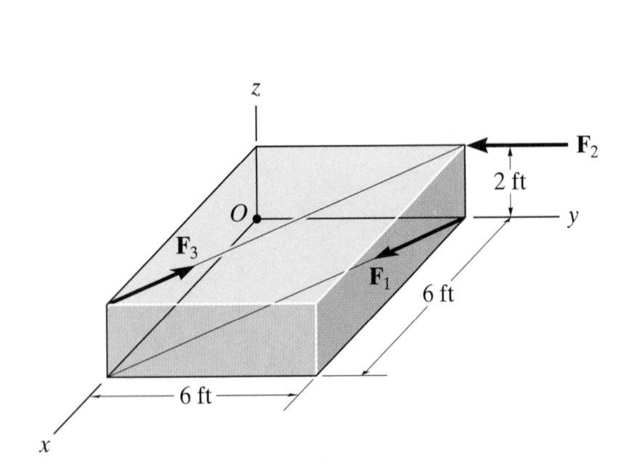

Prob. 4–136

4–138. Replace the three forces acting on the plate by a wrench. Specify the magnitude of the force and couple moment for the wrench and the point $P(y, z)$ where its line of action intersects the plate.

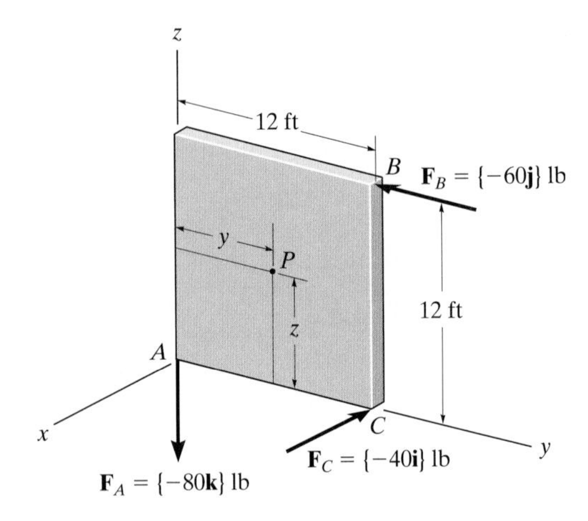

Prob. 4–138

4.10 Reduction of a Simple Distributed Loading

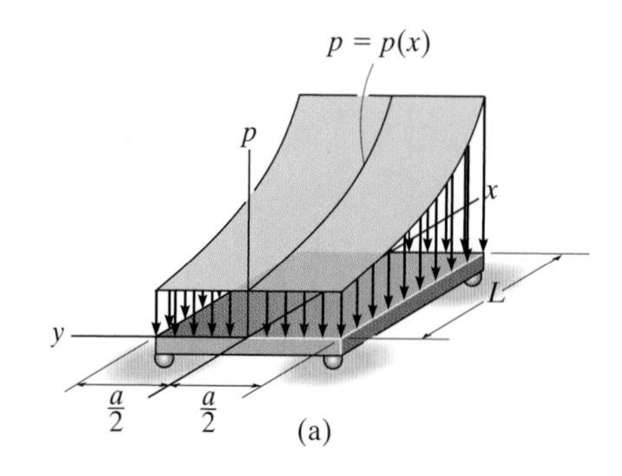

(a)

In many situations a very large surface area of a body may be subjected to *distributed loadings* such as those caused by wind, fluids, or simply the weight of material supported over the body's surface. The *intensity* of these loadings at each point on the surface is defined as the *pressure p* (force per unit area), which can be measured in units of lb/ft^2 or pascals (Pa), where $1 \text{ Pa} = 1 \text{ N/m}^2$.

In this section we will consider the most common case of a distributed pressure loading, which is *uniform* along one axis of a flat rectangular body upon which the loading is applied.* An example of such a loading is shown in Fig. 4–47*a*. The direction of the intensity of the pressure load is indicated by arrows shown on the *load-intensity diagram*. The entire loading on the plate is therefore a system of parallel forces, infinite in number and each acting on a separate differential area of the plate. Here the *loading function*, $p = p(x)$ Pa, is only a function of x since the pressure is uniform along the y axis. If we multiply $p = p(x)$ by the *width* a m of the plate, we obtain $w = [p(x) \text{ N/m}^2]a \text{ m} = w(x) \text{ N/m}$. This loading function, shown in Fig. 4–47*b*, is a measure of load distribution along the line $y = 0$ which is in the plane of symmetry of the loading, Fig. 4–47*a*. As noted, it is measured as a force per unit length, rather than a force per unit area. Consequently, the load-intensity diagram for $w = w(x)$ can be represented by a system of *coplanar* parallel forces, shown in two dimensions in Fig. 4–47*b*. Using the methods of Sec. 4.9, this system of forces can be simplified to a single resultant force $\mathbf{F}_R$ and its location $\bar{x}$ can be specified, Fig. 4–47*c*.

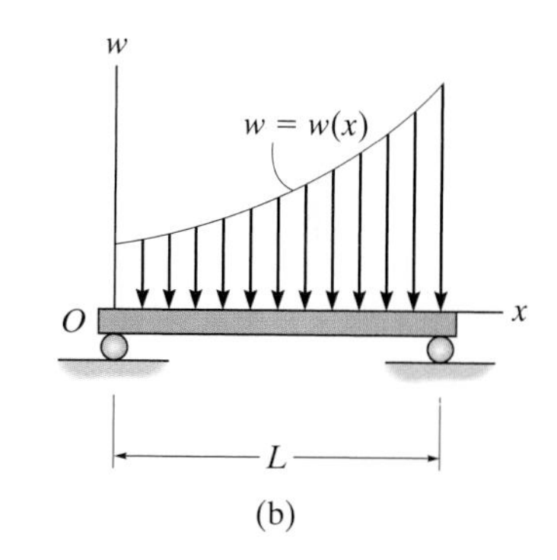

(b)

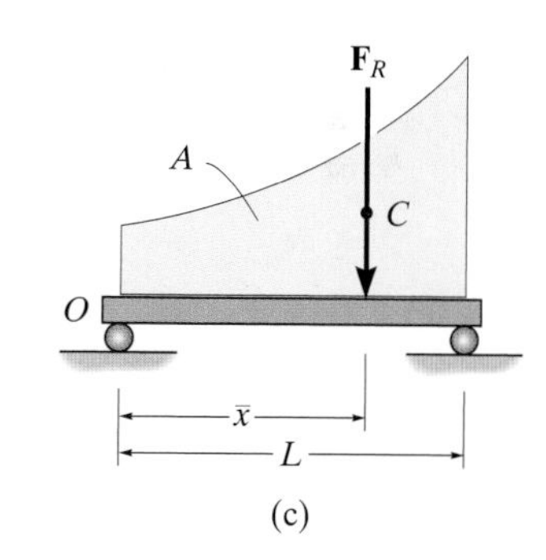

(c)

Magnitude of Resultant Force. From Eq. 4–17 ($F_R = \Sigma F$), the magnitude of $\mathbf{F}_R$ is equivalent to the sum of all the forces in the system. In this case integration must be used since there is an infinite number of parallel forces $d\mathbf{F}$ acting along the plate, Fig. 4–47*b*. Since $d\mathbf{F}$ is acting on an element of length dx and $w(x)$ is a force per unit length, then at the location x, $dF = w(x)\,dx = dA$. In other words, the magnitude of $d\mathbf{F}$ is determined from the colored differential *area dA* under the loading curve. For the entire plate length,

$$+\downarrow F_R = \Sigma F; \qquad F_R = \int_L w(x)\,dx = \int_A dA = A \qquad (4\text{–}19)$$

Hence, the magnitude of the resultant force is equal to the total area A under the loading diagram $w = w(x)$, Fig. 4–47*c*.

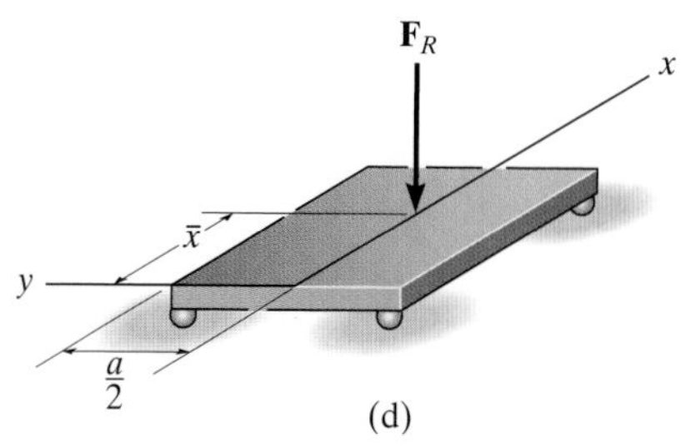

(d)

Fig. 4–47

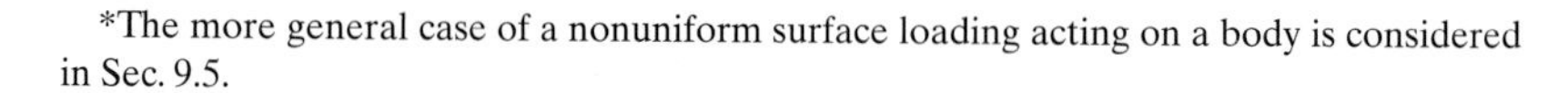
*The more general case of a nonuniform surface loading acting on a body is considered in Sec. 9.5.

Location of Resultant Force. Applying Eq. 4–17 ($M_{R_O} = \Sigma M_O$), the location $\bar{x}$ of the line of action of $\mathbf{F}_R$ can be determined by equating the moments of the force resultant and the force distribution about point O (the y axis). Since $d\mathbf{F}$ produces a moment of $x\,dF = xw(x)\,dx$ about O, Fig. 4–47b, then for the entire plate, Fig. 4–47c,

$$\curvearrowleft + M_{R_O} = \Sigma M_O; \qquad \bar{x}F_R = \int_L xw(x)\,dx$$

Solving for $\bar{x}$, using Eq. 4–19, we can write

$$\bar{x} = \frac{\displaystyle\int_L xw(x)\,dx}{\displaystyle\int_L w(x)\,dx} = \frac{\displaystyle\int_A x\,dA}{\displaystyle\int_A dA} \tag{4–20}$$

This equation represents the x coordinate for the geometric center or *centroid* of the *area* under the distributed-loading diagram $w(x)$. *Therefore, the resultant force has a line of action which passes through the centroid C (geometric center) of the area defined by the distributed-loading diagram $w(x)$*, Fig. 4–47c.

Once $\bar{x}$ is determined, $\mathbf{F}_R$ by symmetry passes through point $(\bar{x}, 0)$ on the surface of the plate, Fig. 4–47d. If we now consider the three-dimensional pressure loading $p(x)$, Fig. 4–47a, we can therefore conclude that *the resultant force has a magnitude equal to the volume under the distributed-loading curve $p = p(x)$ and a line of action which passes through the centroid (geometric center) of this volume.* Detailed treatment of the integration techniques for computing the centroids of volumes or areas is given in Chapter 9. In many cases, however, the distributed-loading diagram is in the shape of a rectangle, triangle, or some other simple geometric form. The centroids for such common shapes do not have to be determined from Eq. 4–20; rather, they can be obtained directly from the tabulation given on the inside back cover.

The beam supporting this stack of lumber is subjected to a *uniform* distributed loading, and so the load-intensity diagram has a rectangular shape. If the load intensity is w_0, then the resultant force is determined from the area of the rectangle, $F_R = w_0 b$. The line of action of this force passes through the centroid or center of this area, $\bar{x} = a + b/2$. This resultant is equivalent to the distributed load, and so both loadings produce the same "external" effects or support reactions on the beam.

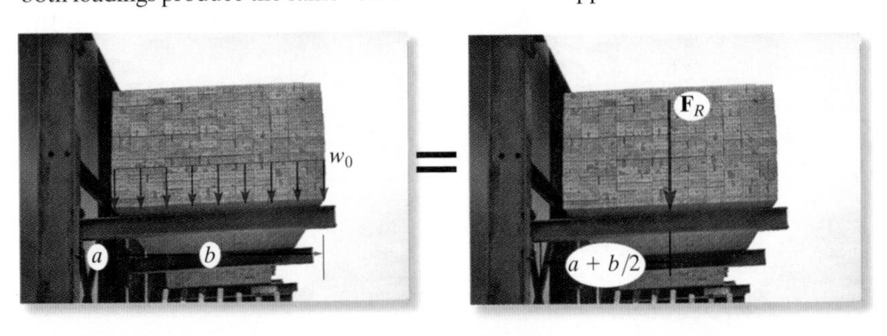

Important Points

- Distributed loadings are defined by using a loading function $w = w(x)$ that indicates the intensity of the loading along the length of the member. This intensity is measured in N/m or lb/ft.
- The external effects caused by a coplanar distributed load acting on a body can be represented by a single resultant force.
- The resultant force is equivalent to the *area* under the distributed loading diagram, and has a line of action that passes through the *centroid* or geometric center of this area.

EXAMPLE 4.20

Determine the magnitude and location of the equivalent resultant force acting on the shaft in Fig. 4–48*a*.

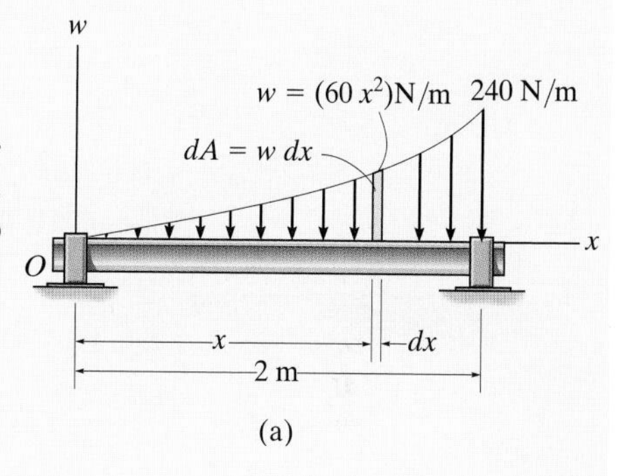

(a)

SOLUTION

Since $w = w(x)$ is given, this problem will be solved by integration. The colored differential area element $dA = w\,dx = 60x^2\,dx$. Applying Eq. 4–19, by summing these elements from $x = 0$ to $x = 2$ m, we obtain the resultant force $\mathbf{F}_R$.

$F_R = \Sigma F;$

$$F_R = \int_A dA = \int_0^2 60x^2\,dx = 60\left[\frac{x^3}{3}\right]_0^2 = 60\left[\frac{2^3}{3} - \frac{0^3}{3}\right]$$

$$= 160 \text{ N} \qquad \textit{Ans.}$$

Since the element of area dA is located an arbitrary distance x from O, the location $\bar{x}$ of $\mathbf{F}_R$ *measured from O*, Fig. 4–48*b*, is determined from Eq. 4–20.

$$\bar{x} = \frac{\int_A x\,dA}{\int_A dA} = \frac{\int_0^2 x(60x^2)\,dx}{160} = \frac{60\left[\frac{x^4}{4}\right]_0^2}{160} = \frac{60\left[\frac{2^4}{4} - \frac{0^4}{4}\right]}{160}$$

$$= 1.5 \text{ m} \qquad \textit{Ans.}$$

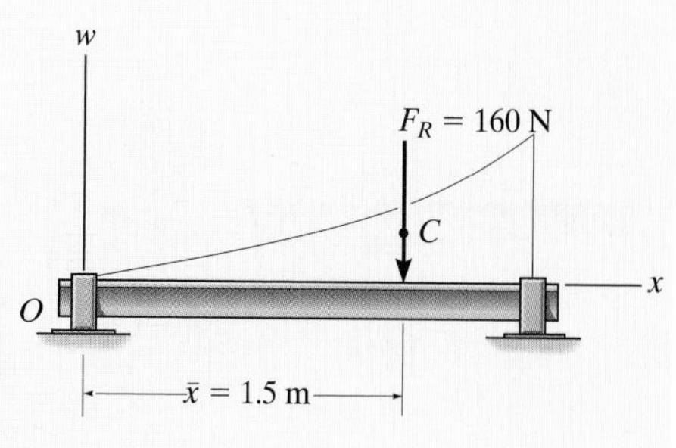

(b)

Fig. 4–48

NOTE: These results may be checked by using the table on the inside back cover, where it is shown that for an exparabolic area of length a, height b, and shape shown in Fig. 4–48*a*,

$$A = \frac{ab}{3} = \frac{2\text{ m}(240\text{ N/m})}{3} = 160\text{ N and } \bar{x} = \frac{3}{4}a = \frac{3}{4}(2\text{ m}) = 1.5\text{ m}$$

EXAMPLE 4.21

A distributed loading of $p = 800x$ Pa acts over the top surface of the beam shown in Fig. 4–49a. Determine the magnitude and location of the equivalent resultant force.

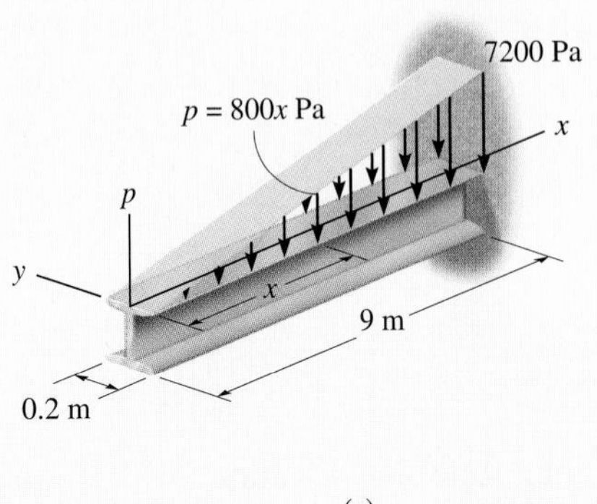

(a)

SOLUTION

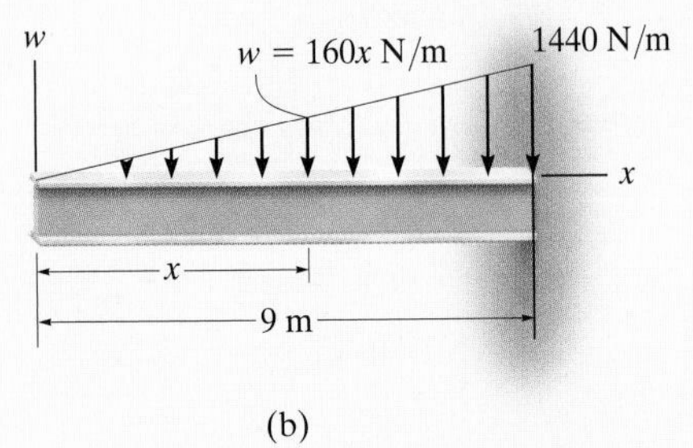

(b)

The loading function $p = 800x$ Pa indicates that the load intensity varies uniformly from $p = 0$ at $x = 0$ to $p = 7200$ Pa at $x = 9$ m. Since the intensity is uniform along the width of the beam (the y axis), the loading may be viewed in two dimensions as shown in Fig. 4–49b. Here

$$w = (800x\ \text{N/m}^2)(0.2\ \text{m})$$
$$= (160x)\ \text{N/m}$$

At $x = 9$ m, note that $w = 1440$ N/m. Although we may again apply Eqs. 4–19 and 4–20 as in Example 4.20, it is simpler to use the table on the inside back cover.

The magnitude of the resultant force is equivalent to the area under the triangle.

$$F_R = \tfrac{1}{2}(9\ \text{m})(1440\ \text{N/m}) = 6480\ \text{N} = 6.48\ \text{kN} \qquad \textit{Ans.}$$

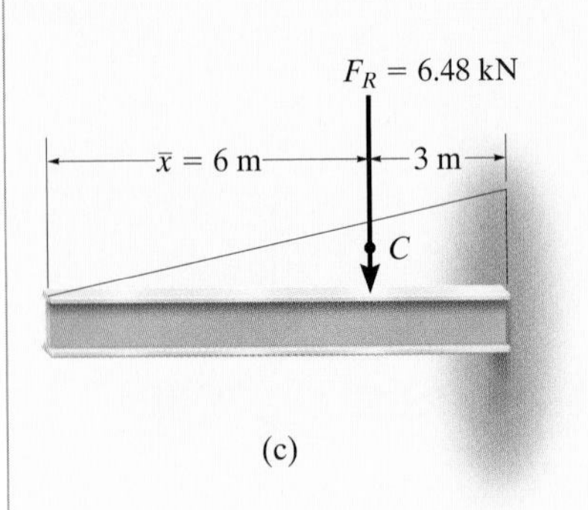

(c)

Fig. 4–49

The line of action of $\mathbf{F}_R$ passes through the *centroid* C of the triangle. Hence,

$$\bar{x} = 9\ \text{m} - \tfrac{1}{3}(9\ \text{m}) = 6\ \text{m} \qquad \textit{Ans.}$$

The results are shown in Fig. 4–49c.

NOTE: We may also view the resultant $\mathbf{F}_R$ as *acting* through the *centroid* of the *volume* of the loading diagram $p = p(x)$ in Fig. 4–49a. Hence $\mathbf{F}_R$ intersects the x–y plane at the point (6 m, 0). Furthermore, the magnitude of $\mathbf{F}_R$ is equal to the volume under the loading diagram; i.e.,

$$F_R = V = \tfrac{1}{2}(7200\ \text{N/m}^2)(9\ \text{m})(0.2\ \text{m}) = 6.48\ \text{kN} \qquad \textit{Ans.}$$

EXAMPLE 4.22

The granular material exerts the distributed loading on the beam as shown in Fig. 4–50*a*. Determine the magnitude and location of the equivalent resultant of this load.

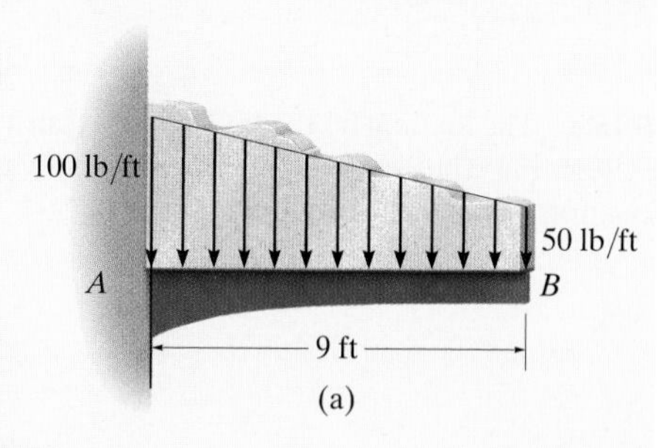

(a)

SOLUTION

The area of the loading diagram is a *trapezoid*, and therefore the solution can be obtained directly from the area and centroid formulas for a trapezoid listed on the inside back cover. Since these formulas are not easily remembered, instead we will solve this problem by using "composite" areas. In this regard, we can divide the trapezoidal loading into a rectangular and triangular loading as shown in Fig. 4–50*b*. The magnitude of the force represented by each of these loadings is equal to its associated *area*,

$$F_1 = \tfrac{1}{2}(9\text{ ft})(50\text{ lb/ft}) = 225\text{ lb}$$
$$F_2 = (9\text{ ft})(50\text{ lb/ft}) = 450\text{ lb}$$

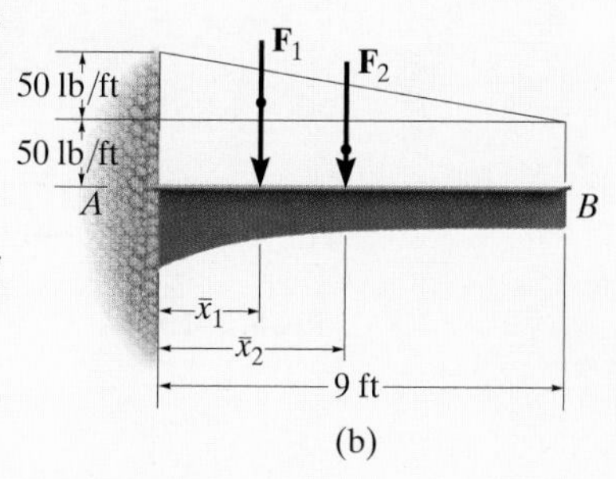

(b)

The lines of action of these parallel forces act through the *centroid* of their associated areas and therefore intersect the beam at

$$\bar{x}_1 = \tfrac{1}{3}(9\text{ ft}) = 3\text{ ft}$$
$$\bar{x}_2 = \tfrac{1}{2}(9\text{ ft}) = 4.5\text{ ft}$$

The two parallel forces $\mathbf{F}_1$ and $\mathbf{F}_2$ can be reduced to a single resultant $\mathbf{F}_R$. The magnitude of $\mathbf{F}_R$ is

$$+\downarrow F_R = \Sigma F; \qquad F_R = 225 + 450 = 675\text{ lb} \qquad \textit{Ans.}$$

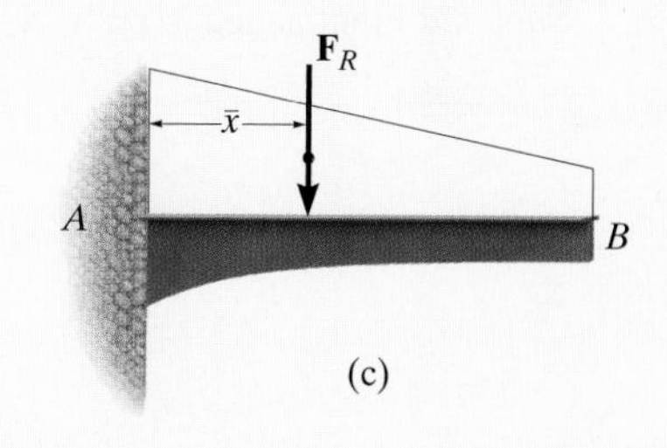

(c)

With reference to point A, Fig. 4–50*b* and 4–50*c*, we can find the location of $\mathbf{F}_R$. We require

$$\curvearrowleft + M_{R_A} = \Sigma M_A; \quad \bar{x}(675) = 3(225) + 4.5(450)$$
$$\bar{x} = 4\text{ ft} \qquad \textit{Ans.}$$

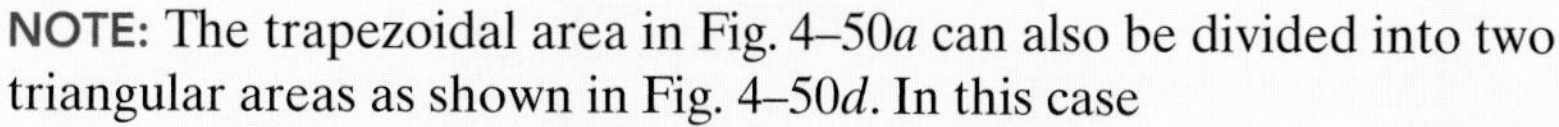

NOTE: The trapezoidal area in Fig. 4–50*a* can also be divided into two triangular areas as shown in Fig. 4–50*d*. In this case

$$F_1 = \tfrac{1}{2}(9\text{ ft})(100\text{ lb/ft}) = 450\text{ lb}$$
$$F_2 = \tfrac{1}{2}(9\text{ ft})(50\text{ lb/ft}) = 225\text{ lb}$$

and

$$\bar{x}_1 = \tfrac{1}{3}(9\text{ ft}) = 3\text{ ft}$$
$$\bar{x}_2 = \tfrac{1}{3}(9\text{ ft}) = 3\text{ ft}$$

NOTE: Using these results, show that again $F_R = 675$ lb and $\bar{x} = 4$ ft.

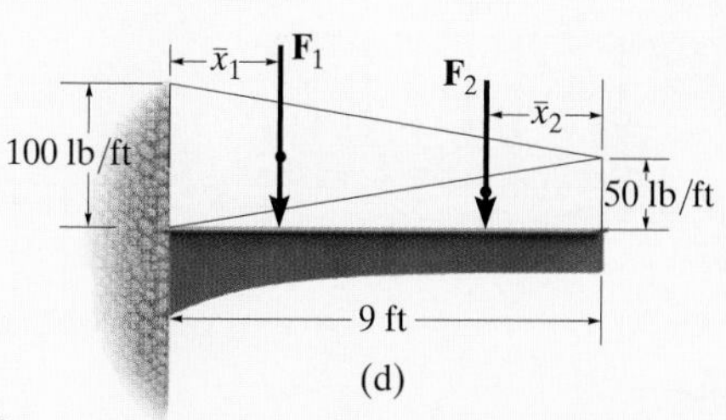

(d)

Fig. 4–50

PROBLEMS

4–139. The loading on the bookshelf is distributed as shown. Determine the magnitude of the equivalent resultant location, measured from point O.

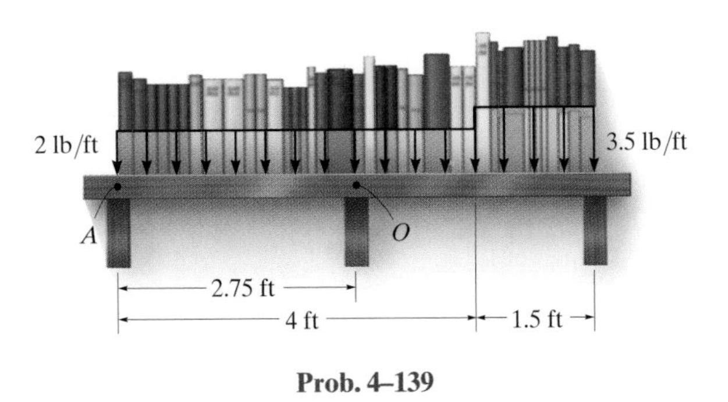

Prob. 4–139

4–141. Replace the loading by an equivalent force and couple moment acting at point O.

4–142. Replace the loading by a single resultant force, and specify the location of the force on the beam measured from point O.

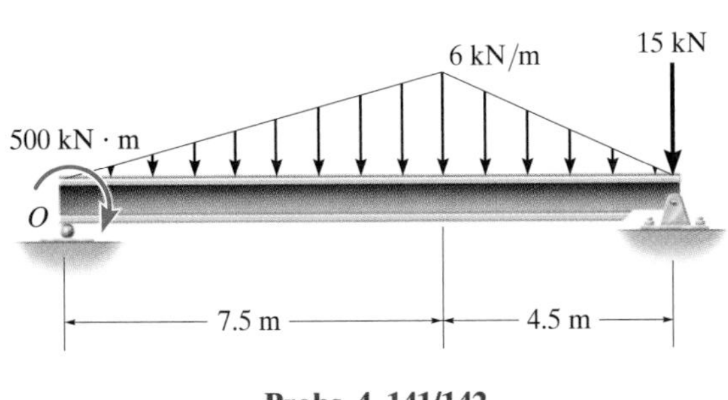

Probs. 4–141/142

***4–140.** Replace the loading by an equivalent resultant force and couple moment acting at point A.

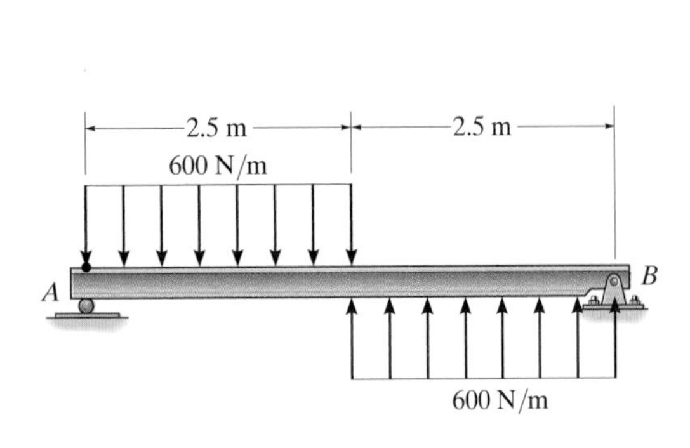

Prob. 4–140

4–143. The column is used to support the floor which exerts a force of 3000 lb on the top of the column. The effect of soil pressure along its side is distributed as shown. Replace this loading by an equivalent resultant force and specify where it acts along the column, measured from its base A.

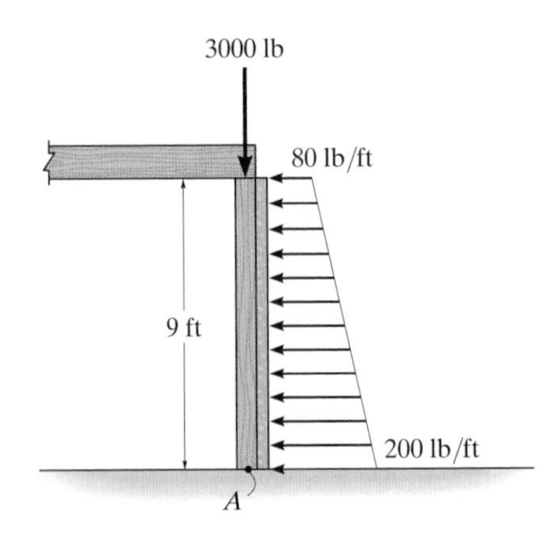

Prob. 4–143

***4–144.** Replace the loading by an equivalent force and couple moment acting at point O.

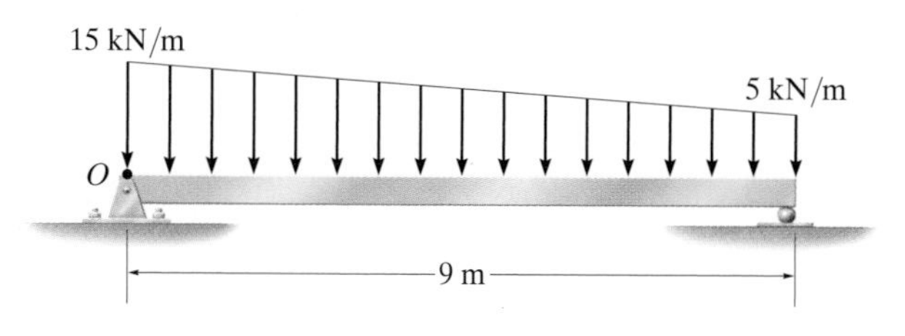

Prob. 4–144

4–145. Replace the distributed loading by an equivalent resultant force, and specify its location on the beam, measured from the pin at C.

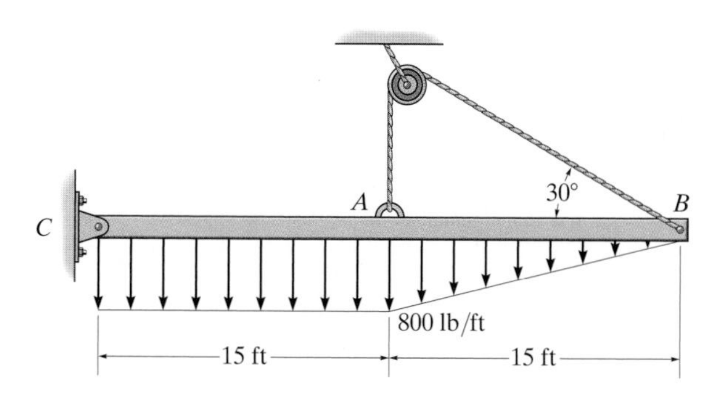

Prob. 4–145

4–146. The beam supports the distributed load caused by the sandbags. Determine the resultant force on the beam and specify its location measured from point A.

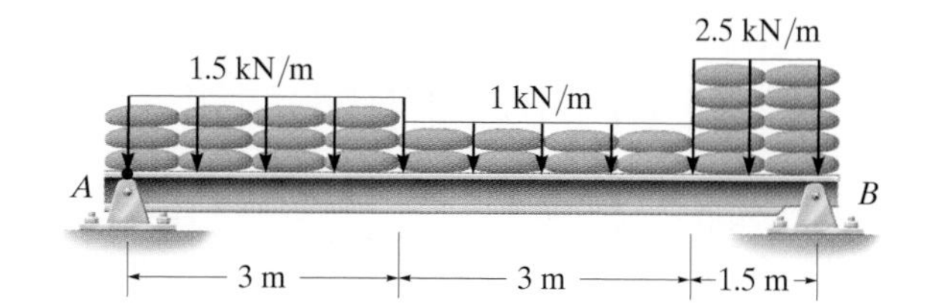

Prob. 4–146

4–147. Determine the length b of the triangular load and its position a on the beam such that the equivalent resultant force is zero and the resultant couple moment is 8 kN · m clockwise.

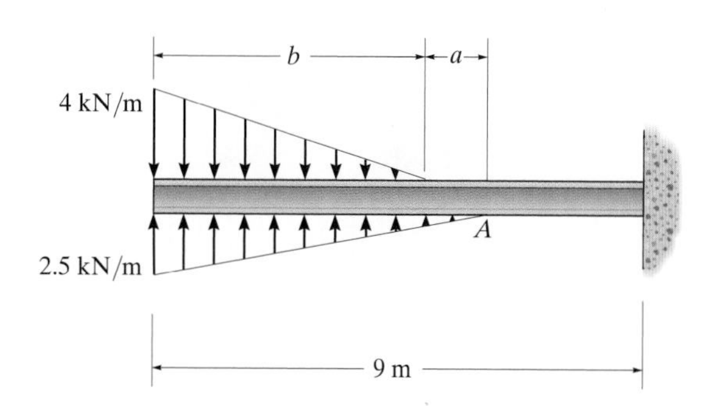

Prob. 4–147

*4–148. Replace the distributed loading by an equivalent resultant force and specify its location, measured from point A.

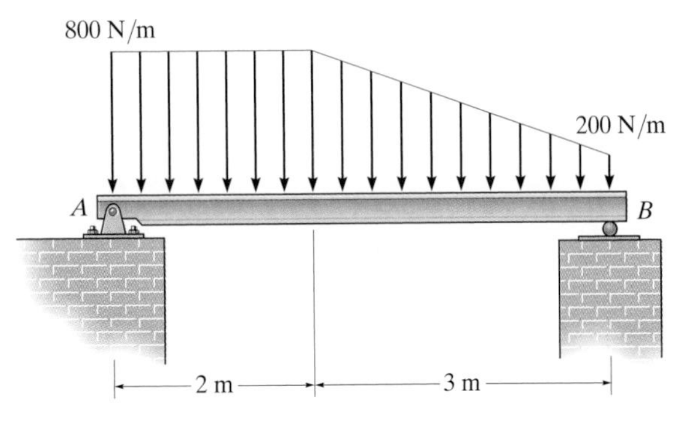

Prob. 4–148

4–149. The distribution of soil loading on the bottom of a building slab is shown. Replace this loading by an equivalent resultant force and specify its location, measured from point O.

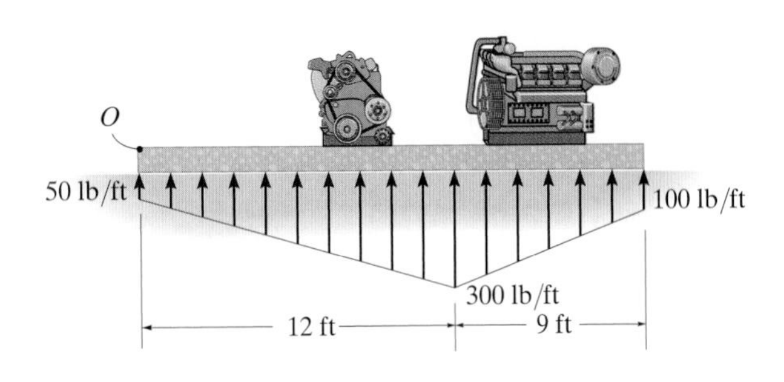

Prob. 4–149

4–150. The beam is subjected to the distributed loading. Determine the length b of the uniform load and its position a on the beam such that the resultant force and couple moment acting on the beam are zero.

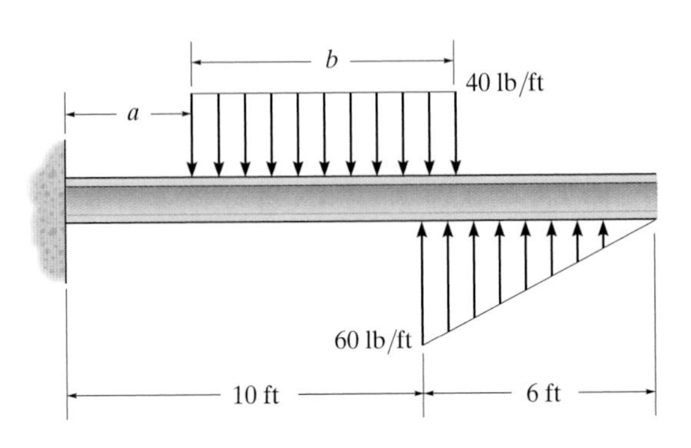

Prob. 4–150

4–151. Replace the loading by an equivalent resultant force and specify its location on the beam, measured from point B.

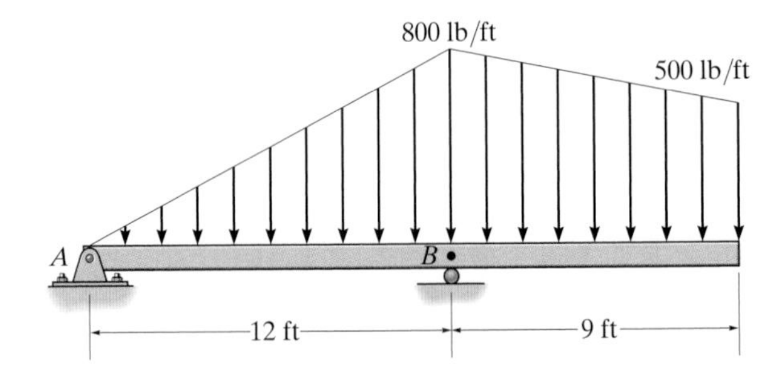

Prob. 4–151

***4–152.** Replace the distributed loading by an equivalent resultant force and specify where its line of action intersects member AB, measured from A.

4–153. Replace the distributed loading by an equivalent resultant force and specify where its line of action intersects member BC, measured from C.

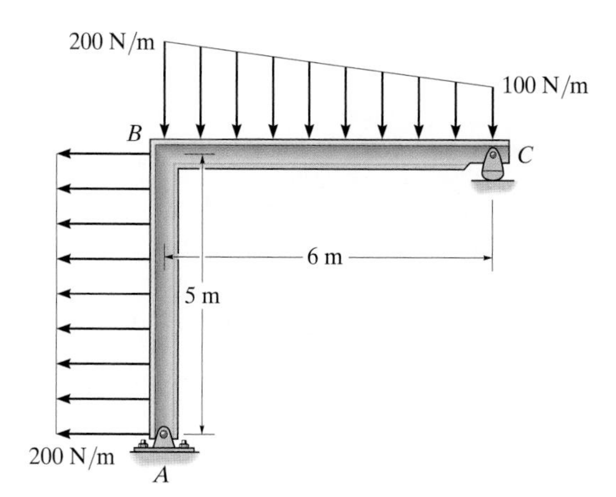

Probs. 4–152/153

4–155. Determine the equivalent resultant force and couple moment at point O.

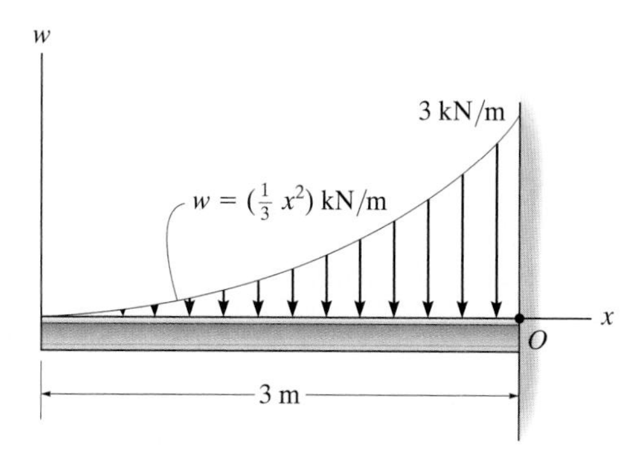

Prob. 4–155

4–154. Replace the loading by an equivalent resultant force and couple moment acting at point O.

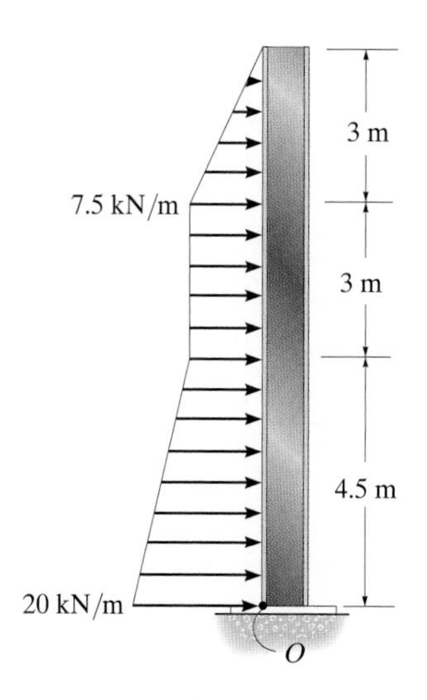

Prob. 4–154

***4–156.** Wind has blown sand over a platform such that the intensity of the load can be approximated by the function $w = (0.5x^3)$ N/m. Simplify this distributed loading to an equivalent resultant force and specify the magnitude and location of the force, measured from A.

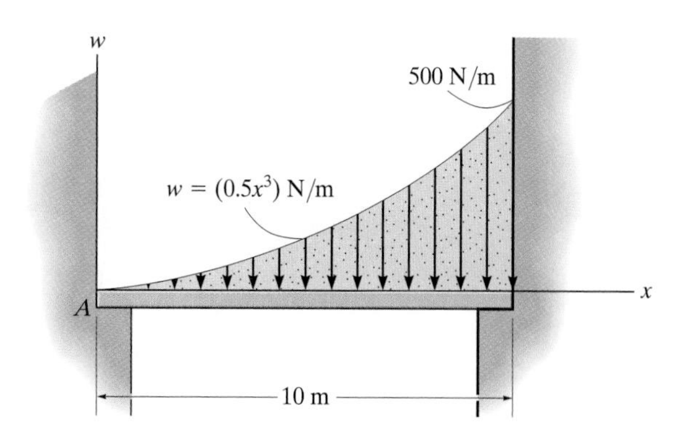

Prob. 4–156

4–157. Determine the equivalent resultant force and its location, measured from point O.

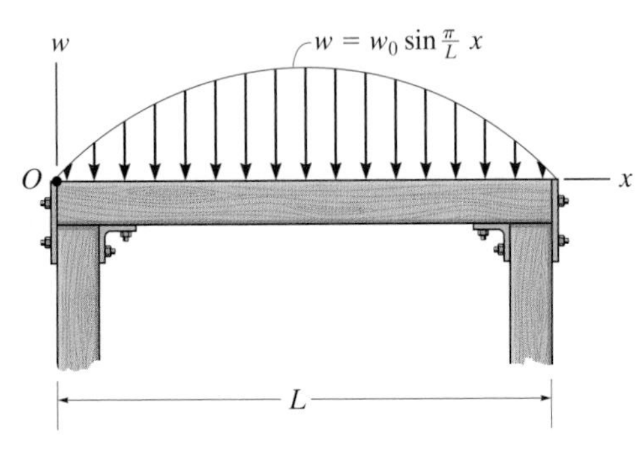

Prob. 4–157

4–158. Determine the equivalent resultant force acting on the bottom of the wing due to air pressure and specify where it acts, measured from point A.

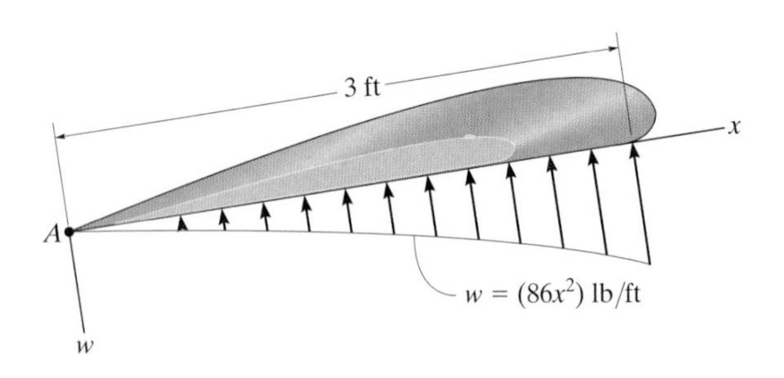

Prob. 4–158

4–159. Currently eighty-five percent of all neck injuries are caused by rear-end car collisions. To alleviate this problem, an automobile seat restraint has been developed that provides additional pressure contact with the cranium. During dynamic tests the distribution of load on the cranium has been plotted and shown to be parabolic. Determine the equivalent resultant force and its location, measured from point A.

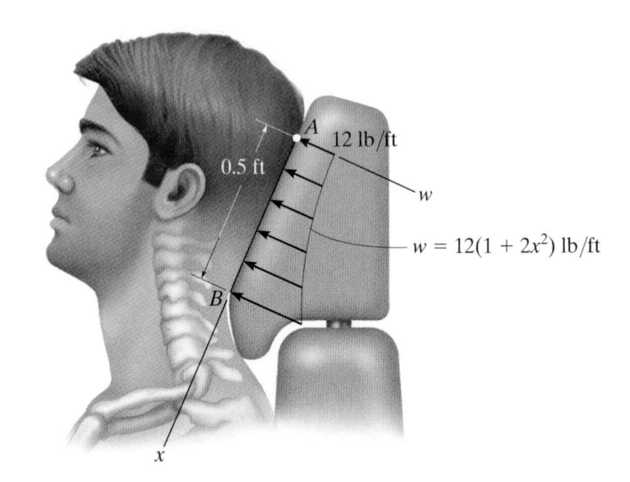

Prob. 4–159

***4–160.** Determine the equivalent resultant force of the distributed loading and its location, measured from point A. Evaluate the integral using Simpson's rule.

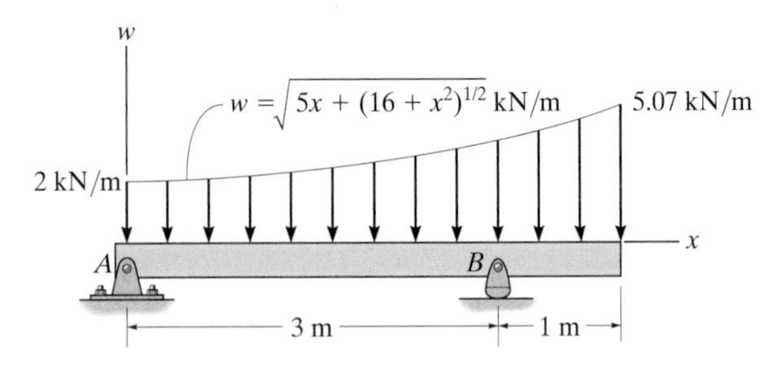

Prob. 4–160

Chapter Review

Moment of Force—Scalar Definition

A force produces a turning effect about a point O that does not lie on its line of action. In scalar form, the moment *magnitude* is the product of the force and the moment arm or perpendicular distance from point O to the line of action of the force.

$$M_O = Fd$$

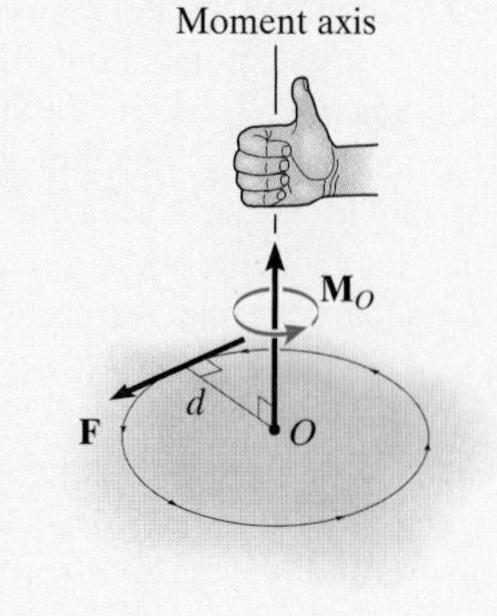

The *direction* of the moment is defined using the right-hand rule. $\mathbf{M}_O$ always acts along an axis perpendicular to the plane containing $\mathbf{F}$ and d, and passes through the point O. If the force acts through the point O, the moment $\mathbf{M}_O$ will be zero.

Rather than finding d, it is normally easier to resolve the force into its x and y components, determine the moment of each component about the point, and then sum the results. This is called the principle of moments.
(*See pages 118–119.*)

$$M_O = Fd = F_x y - F_y x$$

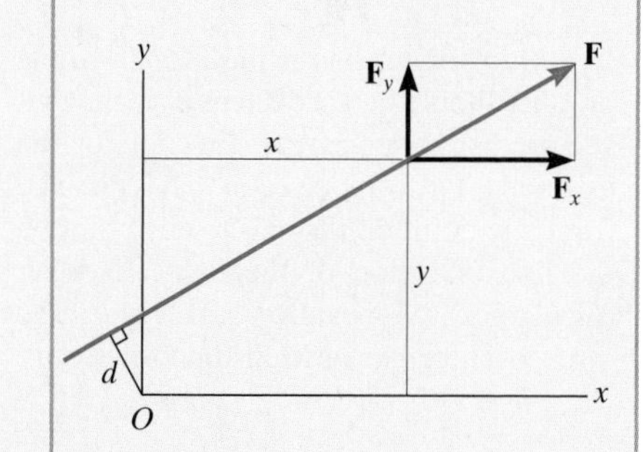

Moment of a Force—Vector Definition

Since three-dimensional geometry is generally more difficult to visualize, the vector cross product can be used to determine the moment.

$$\mathbf{M}_O = \mathbf{r}_A \times \mathbf{F} = \mathbf{r}_B \times \mathbf{F} = \mathbf{r}_C \times \mathbf{F}$$

$\mathbf{r}$ is a position vector that extends from point O to any point A, B, C on the line of action of $\mathbf{F}$.

If the position vector $\mathbf{r}$ and force $\mathbf{F}$ can be expressed as Cartesian vectors, then the cross product can be expressed by the expansion of a determinant.
(*See pages 125–127.*)

$$\mathbf{M}_O = \mathbf{r} \times \mathbf{F} = \begin{vmatrix} \mathbf{i} & \mathbf{j} & \mathbf{k} \\ r_x & r_y & r_z \\ F_x & F_y & F_z \end{vmatrix}$$

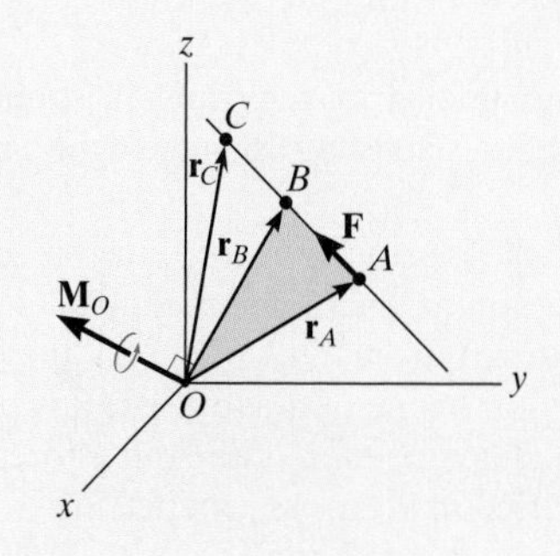

Moment about an Axis

If the moment of a force **F** is to be determined about an arbitrary axis, then the projection of the moment onto the axis must be obtained. Provided the distance d_a that is perpendicular to *both* the line of action of the force and the axis can be found, then the moment of the force about the axis can be determined from a scalar equation.

$$M_a = d_a F$$

It's a scalar equation when the line of action of **F** intersects an axis and the moment of **F** about the axis is zero. Also, when the line of action of **F** is parallel to the axis, the moment of **F** about the axis is zero.

In three dimensions, the scalar triple product should be used. Here $\mathbf{u}_a$ is the unit vector that specifies the direction of the axis, and **r** is a position vector that is directed from any point on the axis to any point on the line of action of the force. If M_a is calculated as a negative scalar, then the sense of direction of $\mathbf{M}_a$ is opposite to $\mathbf{u}_a$. (*See pages 142–145.*)

$$M_a = \mathbf{u}_a \cdot (\mathbf{r} \times \mathbf{F}) = \begin{vmatrix} u_{a_x} & u_{a_y} & u_z \\ r_x & r_y & r_z \\ F_x & F_y & F_z \end{vmatrix}$$

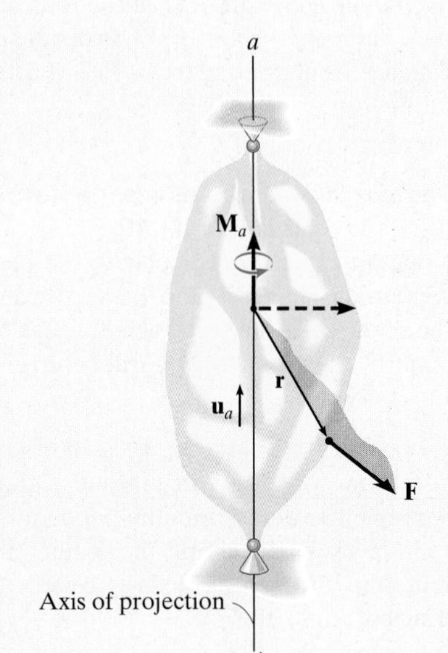

Couple Moment

A couple consists of two equal but opposite forces that act a perpendicular distance d apart. Couples tend to produce a rotation without translation.

$$M_c = Fd$$

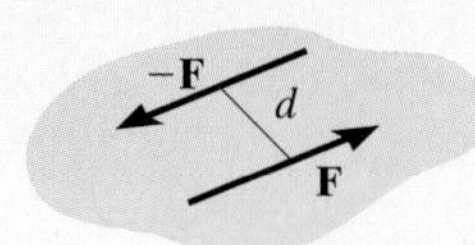

The moment of the couple has a direction that is established using the right-hand rule.

If the vector cross product is used to determine the moment of a couple, then **r** extends from any point on the line of action of one of the forces to any point on the line of action of the other force **F** that is used in the cross product.
(*See pages 152 and 153.*)

$$\mathbf{M}_c = \mathbf{r} \times \mathbf{F}$$

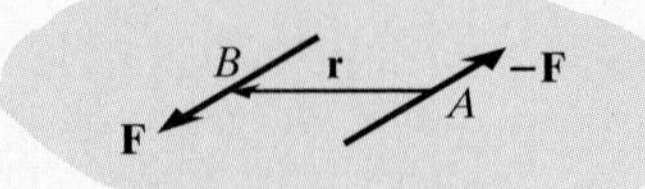

Simplification of a Force and Couple System

Any system of forces and couples can be reduced to a single resultant force and resultant couple moment acting at a point. The resultant force is the sum of all the forces in the system, $F_R = \Sigma F$, and the resultant couple moment is equal to the sum of all the moments and couple moments about the point, $M_{R_O} = \Sigma M_c + \Sigma M_O$

Further simplification to a single resultant force is possible provided the force system is *concurrent*, *coplanar*, or *parallel*. For this case, to find the location of the resultant force from a point, it is necessary to equate the moment of the resultant force about the point to the moment of the forces and couples in the system about the same point.

Equating the moment of a resultant force about a point to the moment of the forces and couples in the system about the same point, for any type of force system that is not concurrent, coplanar, or parallel, would yield a wrench, which consists of the resultant force and a resultant collinear couple moment.

(*See pages 167, 171–172, and 174.*)

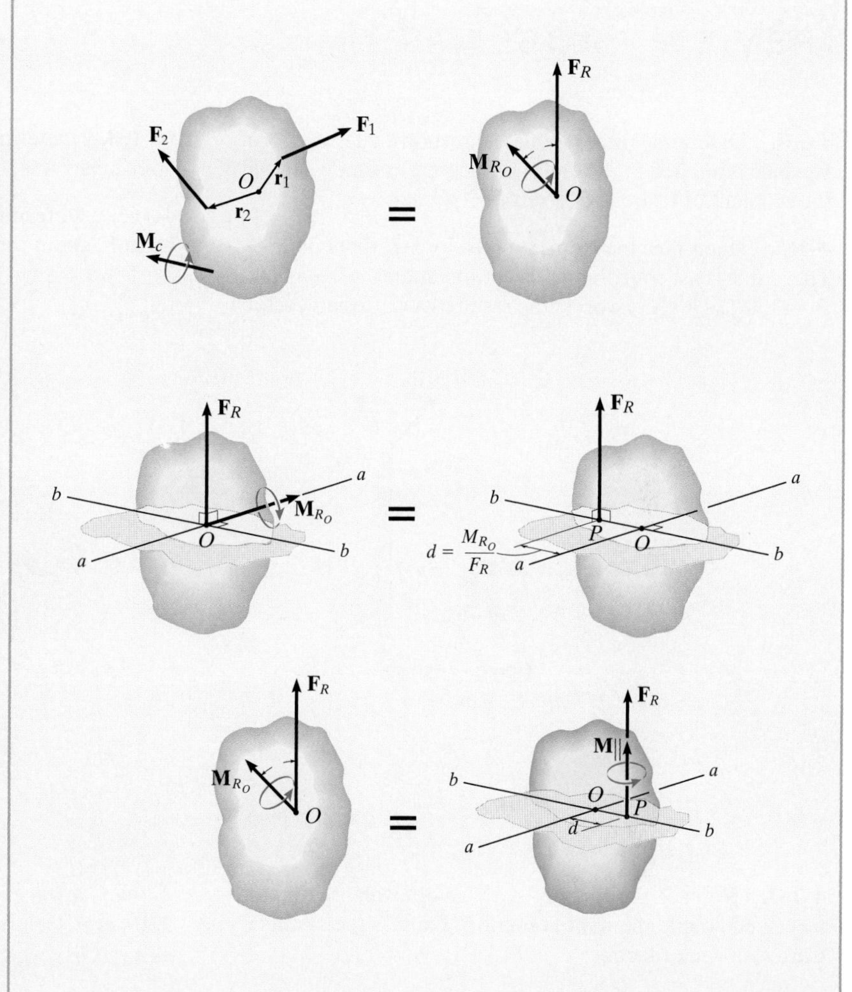

Coplanar Distributed Loading

A resultant force can replace a simple distributed loading, which is equivalent to the *area* under the loading curve. This resultant has a line of action that passes through the *centroid* or geometric center of the area or volume under the loading diagram.

(*See pages 185 and 186.*)

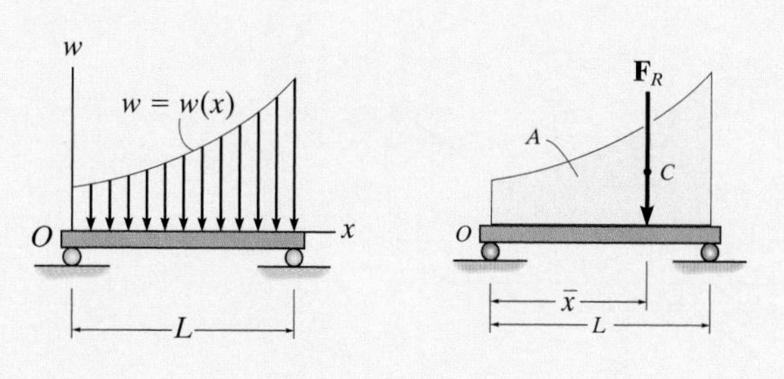

REVIEW PROBLEMS

4–161. Determine the coordinate direction angles α, β, γ of **F**, which is applied to the end A of the pipe assembly, so that the moment of **F** about O is zero.

4–162. Determine the moment of the force **F** about point O. The force has coordinate direction angles of $\alpha = 60°$, $\beta = 120°$, $\gamma = 45°$. Express the result as a Cartesian vector.

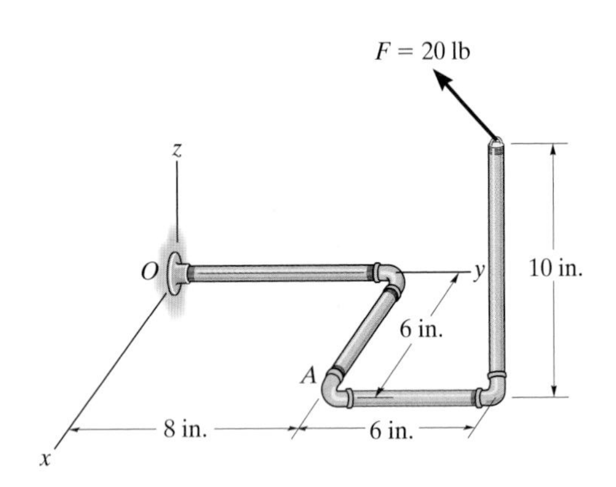

Probs. 4–161/162

***4–164.** Determine the moment of the force $\mathbf{F}_c$ about the door hinge at A. Express the result as a Cartesian vector.

4–165. Determine the magnitude of the moment of the force $\mathbf{F}_c$ about the hinged axis aa of the door.

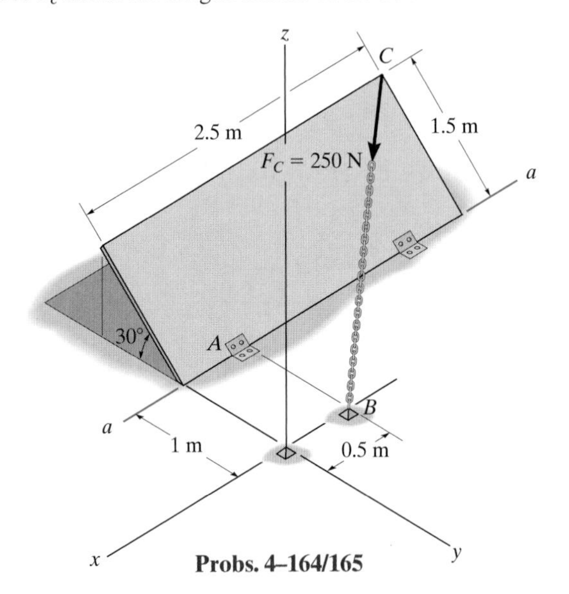

Probs. 4–164/165

4–163. Replace the force at A by an equivalent resultant force and couple moment at point P. Express the results in Cartesian vector form.

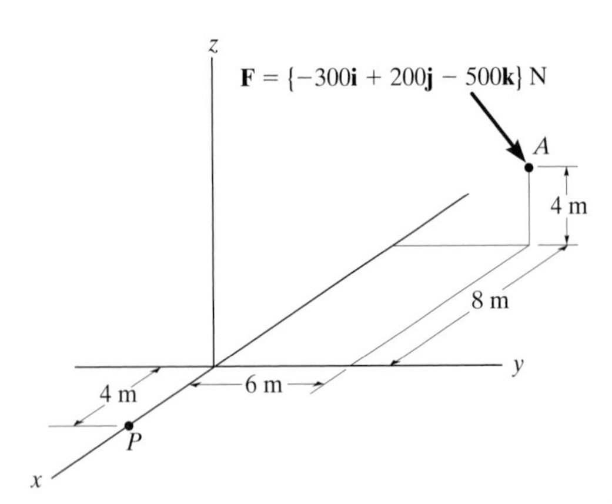

Prob. 4–163

4–166. A force of $F = 80$ N acts vertically downward on the Z-bracket. Determine the moment of this force about the bolt axis (z axis), which is directed at 15° from the vertical.

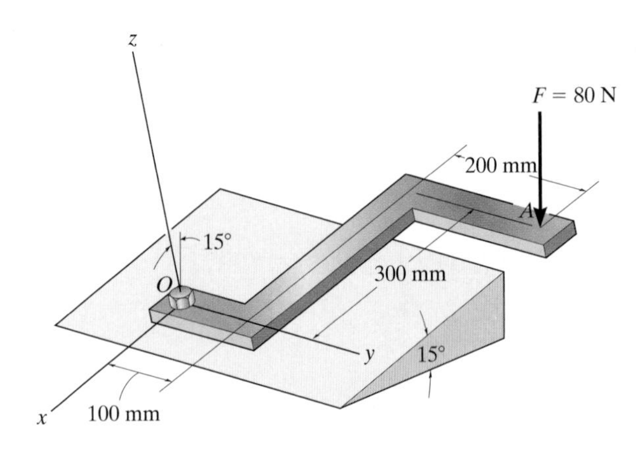

Prob. 4–166

4–167. Replace the force **F** having a magnitude of $F = 50$ lb and acting at point A by an equivalent force and couple moment at point C.

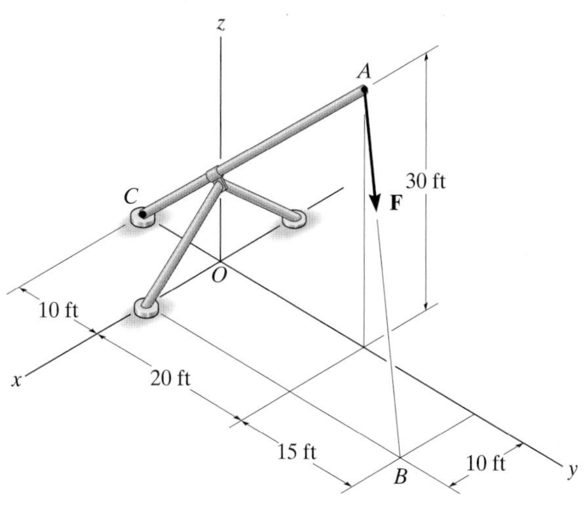

Prob. 4–167

4–169. The horizontal 30-N force acts on the handle of the wrench. Determine the moment of this force about point O. Specify the coordinate direction angles α, β, γ of the moment axis.

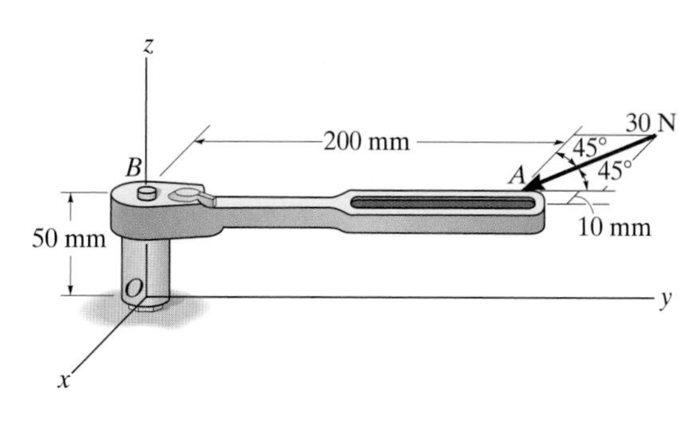

Prob. 4–169

***4–168.** The horizontal 30-N force acts on the handle of the wrench. What is the magnitude of the moment of this force about the z axis?

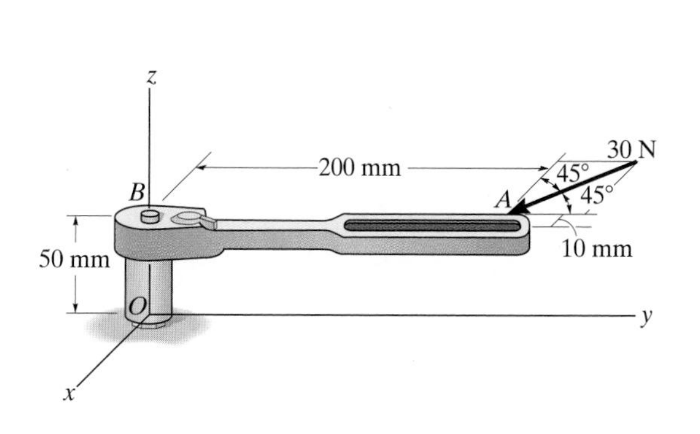

Prob. 4–168

4–170. If the resultant couple moment of the three couples acting on the triangular block is to be zero, determine the magnitudes of forces **F** and **P**.

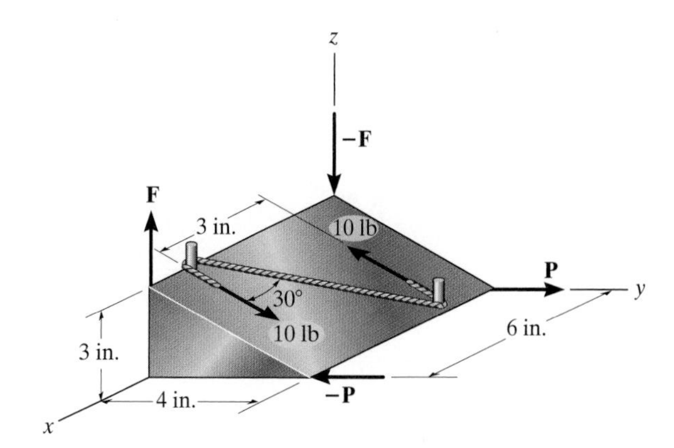

Prob. 4–170

The tower crane is subjected to its weight and the load it supports. In order to calculate the support reactions for the crane, it is necessary to apply the principles of equilibrium.

5 Equilibrium of a Rigid Body

CHAPTER OBJECTIVES

- To develop the equations of equilibrium for a rigid body.
- To introduce the concept of the free-body diagram for a rigid body.
- To show how to solve rigid-body equilibrium problems using the equations of equilibrium.

5.1 Conditions for Rigid-Body Equilibrium

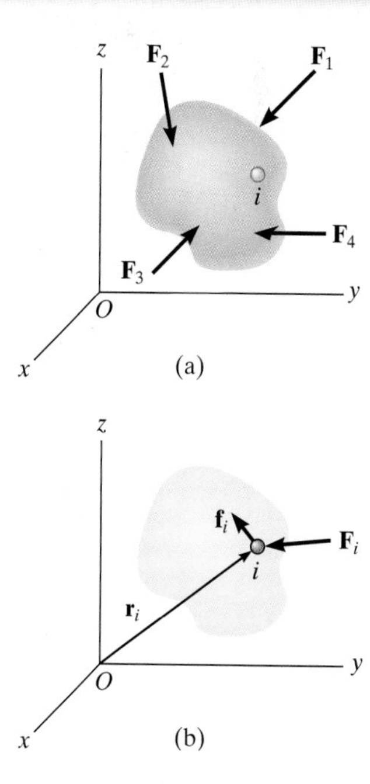

Fig. 5–1

In this section we will develop both the necessary and sufficient conditions required for equilibrium of a rigid body. To do this, consider the rigid body in Fig. 5–1*a*, which is fixed in the *x, y, z* reference and is either at rest or moves with the reference at constant velocity. A free-body diagram of the arbitrary *i*th particle of the body is shown in Fig. 5–1*b*. There are two types of forces which act on it. The resultant *internal force*, $\mathbf{f}_i$, is caused by interactions with adjacent particles. The resultant *external force* $\mathbf{F}_i$ represents, for example, the effects of gravitational, electrical, magnetic, or contact forces between the *i*th particle and adjacent bodies or particles *not* included within the body. If the particle is in equilibrium, then applying Newton's first law we have

$$\mathbf{F}_i + \mathbf{f}_i = \mathbf{0}$$

When the equation of equilibrium is applied to each of the other particles of the body, similar equations will result. If all these equations are added together *vectorially*, we obtain

$$\Sigma\mathbf{F}_i + \Sigma\mathbf{f}_i = \mathbf{0}$$

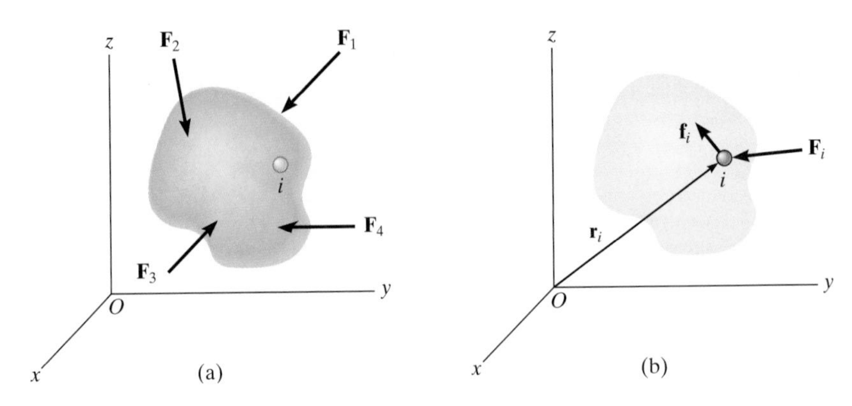

Fig. 5–1

The summation of the internal forces will equal zero since the internal forces between particles within the body will occur in equal but opposite collinear pairs, Newton's third law. Consequently, only the sum of the *external forces* will remain; and therefore, letting $\Sigma \mathbf{F}_i = \Sigma \mathbf{F}$, the above equation can be written as

$$\Sigma \mathbf{F} = \mathbf{0}$$

Let us now consider the moments of the forces acting on the *i*th particle about the arbitrary point *O*, Fig. 5–1*b*. Using the above particle equilibrium equation and the distributive law of the vector cross product we have

$$\mathbf{r}_i \times (\mathbf{F}_i + \mathbf{f}_i) = \mathbf{r}_i \times \mathbf{F}_i + \mathbf{r}_i \times \mathbf{f}_i = \mathbf{0}$$

Similar equations can be written for the other particles of the body, and adding them together vectorially, we obtain

$$\Sigma \mathbf{r}_i \times \mathbf{F}_i + \Sigma \mathbf{r}_i \times \mathbf{f}_i = \mathbf{0}$$

The second term is zero since, as stated above, the internal forces occur in equal but opposite collinear pairs, and therefore the resultant moment of each pair of forces about point *O* is zero. Using the notation $\Sigma \mathbf{M}_O = \Sigma \mathbf{r}_i \times \mathbf{F}_i$, we have

$$\Sigma \mathbf{M}_O = \mathbf{0}$$

Hence the two *equations of equilibrium* for a rigid body can be summarized as follows:

$$\Sigma \mathbf{F} = \mathbf{0}$$
$$\Sigma \mathbf{M}_o = \mathbf{0} \qquad (5\text{–}1)$$

These equations require that a rigid body will remain in equilibrium provided the sum of all the *external forces* acting on the body is equal to zero and the sum of the moments of the external forces about a point is equal to zero. The fact that these conditions are *necessary* for equilibrium has now been proven. They are also *sufficient* for maintaining equilibrium. To show this, let us assume that the body is in equilibrium and the force system acting on the body satisfies Eqs. 5–1. Suppose that an *additional force* $\mathbf{F}'$ is applied to the body. As a result, the equilibrium equations become

$$\Sigma\mathbf{F} + \mathbf{F}' = \mathbf{0}$$

$$\Sigma\mathbf{M}_O + \mathbf{M}'_O = \mathbf{0}$$

where $\mathbf{M}'_O$ is the moment of $\mathbf{F}'$ about O. Since $\Sigma\mathbf{F} = \mathbf{0}$ and $\Sigma\mathbf{M}_O = \mathbf{0}$, then we require $\mathbf{F}' = \mathbf{0}$ (also $\mathbf{M}'_O = \mathbf{0}$). Consequently, the additional force $\mathbf{F}'$ is not required, and indeed Eqs. 5–1 are also sufficient conditions for maintaining equilibrium.

Many types of engineering problems involve symmetric loadings and can be solved by projecting all the forces acting on a body onto a single plane. And so in the next section, the equilibrium of a body subjected to a *coplanar* or *two-dimensional force system* will be considered. Ordinarily the geometry of such problems is not very complex, so a scalar solution is suitable for analysis. The more general discussion of rigid bodies subjected to *three-dimensional force systems* is given in the latter part of this chapter. It will be seen that many of these types of problems are best solved using vector analysis.

EQUILIBRIUM IN TWO DIMENSIONS

5.2 Free-Body Diagrams

Successful application of the equations of equilibrium requires a complete specification of *all* the known and unknown external forces that act *on* the body. The best way to account for these forces is to draw the body's free-body diagram. This diagram is a sketch of the outlined shape of the body, which represents it as being *isolated* or "free" from its surroundings, i.e., a "free body." On this sketch it is necessary to show *all* the forces and couple moments that the surroundings exert *on the body* so that these effects can be accounted for when the equations of equilibrium are applied. For this reason, *a thorough understanding of how to draw a free-body diagram is of primary importance for solving problems in mechanics.*

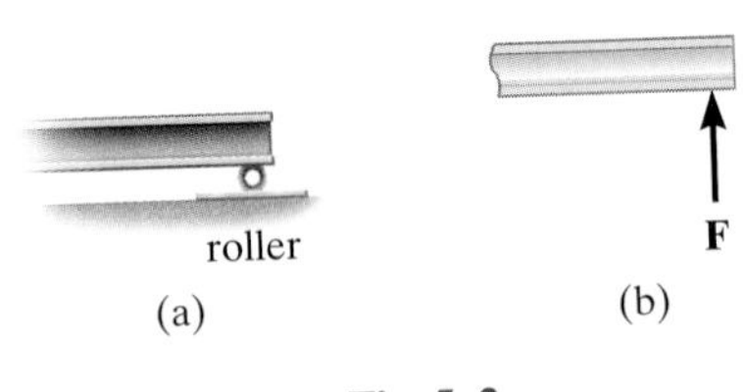

Fig. 5–2

Support Reactions. Before presenting a formal procedure as to how to draw a free-body diagram, we will first consider the various types of reactions that occur at supports and points of support between bodies subjected to coplanar force systems. *As a general rule, if a support prevents the translation of a body in a given direction, then a force is developed on the body in that direction. Likewise, if rotation is prevented, a couple moment is exerted on the body.*

For example, let us consider three ways in which a horizontal member, such as a beam, is supported at its end. One method consists of a *roller* or cylinder, Fig. 5–2*a*. Since this support only prevents the beam from *translating* in the vertical direction, the roller can only exert a *force* on the beam in this direction, Fig. 5–2*b*.

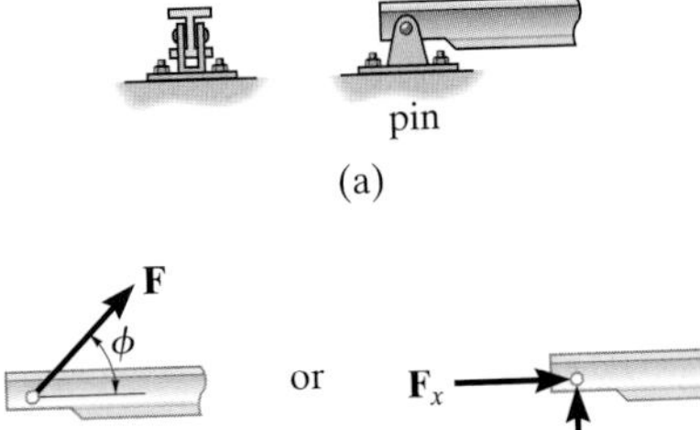

Fig. 5–3

The beam can be supported in a more restrictive manner by using a *pin* as shown in Fig. 5–3*a*. The pin passes through a hole in the beam and two leaves which are fixed to the ground. Here the pin can prevent *translation* of the beam in *any direction* ϕ, Fig. 5–3*b*, and so the pin must exert a *force* **F** on the beam in this direction. For purposes of analysis, it is generally easier to represent this resultant force **F** by its two components $\mathbf{F}_x$ and $\mathbf{F}_y$, Fig. 5–3*c*. If F_x and F_y are known, then F and ϕ can be calculated.

The most restrictive way to support the beam would be to use a *fixed support* as shown in Fig. 5–4*a*. This support will prevent both *translation and rotation* of the beam, and so to do this a *force and couple moment* must be developed on the beam at its point of connection, Fig. 5–4*b*. As in the case of the pin, the force is usually represented by its components $\mathbf{F}_x$ and $\mathbf{F}_y$.

Table 5–1 lists other common types of supports for bodies subjected to coplanar force systems. (In all cases the angle θ is assumed to be known.) Carefully study each of the symbols used to represent these supports and the types of reactions they exert on their contacting members. Although concentrated forces and couple moments are shown in this table, they actually represent the *resultants* of small *distributed surface loads* that exist between each support and its contacting member. It is these *resultants* which will be determined from the equations of equilibrium.

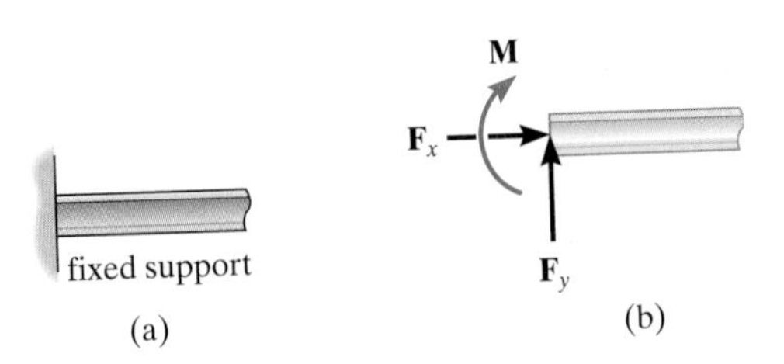

Fig. 5–4

TABLE 5–1 Supports for Rigid Bodies Subjected to Two-Dimensional Force Systems

	Types of Connection	Reaction	Number of Unknowns
(1) 	cable	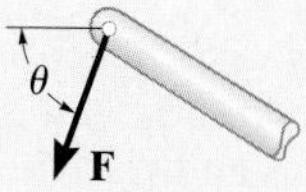 θ, **F**	One unknown. The reaction is a tension force which acts away from the member in the direction of the cable.
(2) 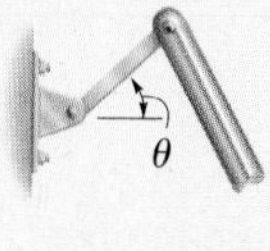	weightless link	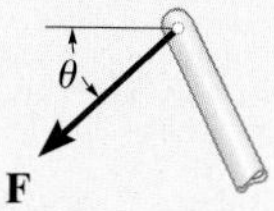 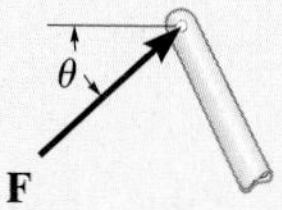 θ, **F** or θ, **F**	One unknown. The reaction is a force which acts along the axis of the link.
(3) 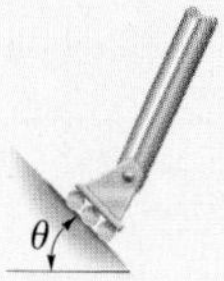	roller	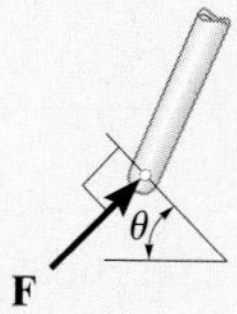 θ, **F**	One unknown. The reaction is a force which acts perpendicular to the surface at the point of contact.
(4) 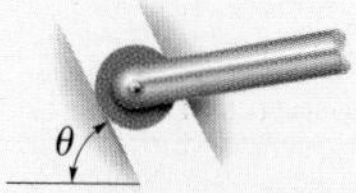	roller or pin in confined smooth slot	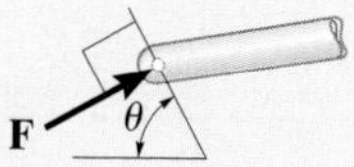 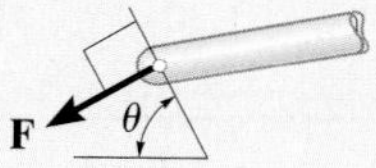 θ, **F** or θ, **F**	One unknown. The reaction is a force which acts perpendicular to the slot.
(5) 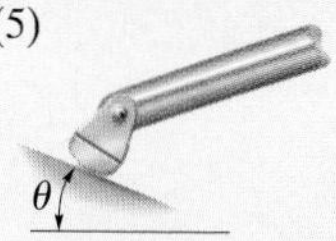	rocker	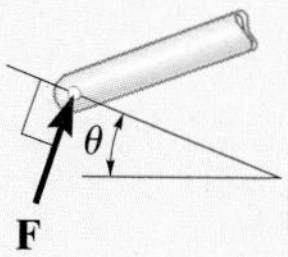 θ, **F**	One unknown. The reaction is a force which acts perpendicular to the surface at the point of contact.
(6) 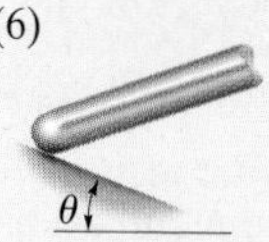	smooth contacting surface	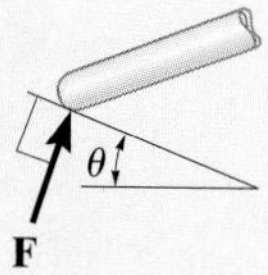 θ, **F**	One unknown. The reaction is a force which acts perpendicular to the surface at the point of contact.
(7) 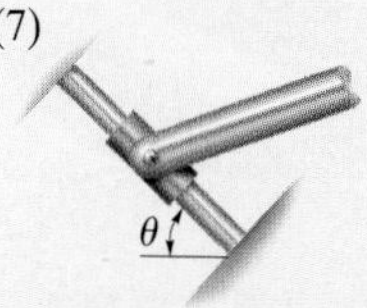	member pin connected to collar on smooth rod	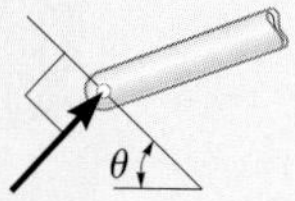 θ or 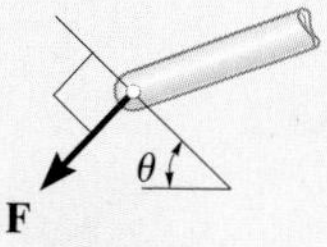 θ, **F**	One unknown. The reaction is a force which acts perpendicular to the rod.

continued

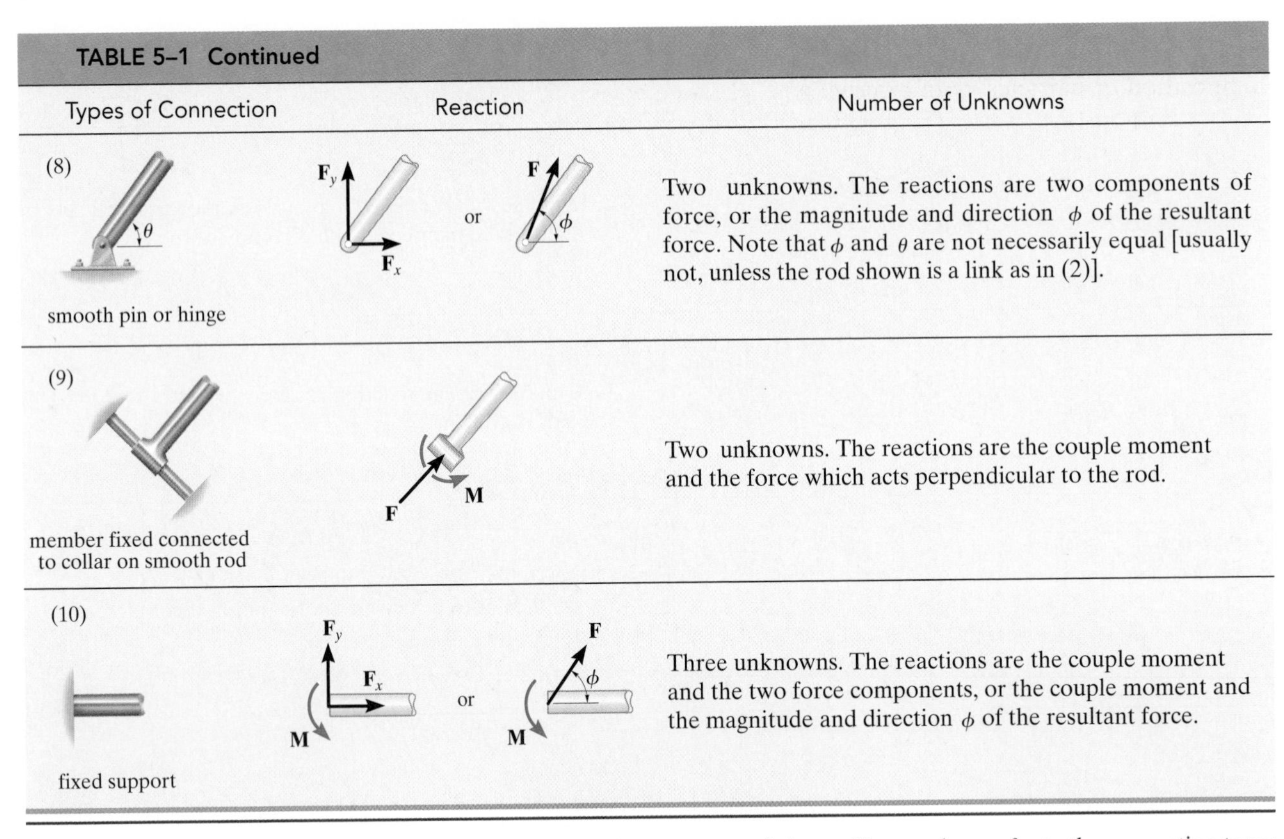

TABLE 5–1 Continued

Types of Connection	Reaction	Number of Unknowns
(8) smooth pin or hinge	$\mathbf{F}_y$, $\mathbf{F}_x$ or $\mathbf{F}$, ϕ	Two unknowns. The reactions are two components of force, or the magnitude and direction ϕ of the resultant force. Note that ϕ and θ are not necessarily equal [usually not, unless the rod shown is a link as in (2)].
(9) member fixed connected to collar on smooth rod	$\mathbf{F}$, $\mathbf{M}$	Two unknowns. The reactions are the couple moment and the force which acts perpendicular to the rod.
(10) fixed support	$\mathbf{F}_y$, $\mathbf{F}_x$, $\mathbf{M}$ or $\mathbf{F}$, ϕ, $\mathbf{M}$	Three unknowns. The reactions are the couple moment and the two force components, or the couple moment and the magnitude and direction ϕ of the resultant force.

Typical examples of actual supports are shown in the following sequence of photos. The numbers refer to the connection types in Table 5–1.

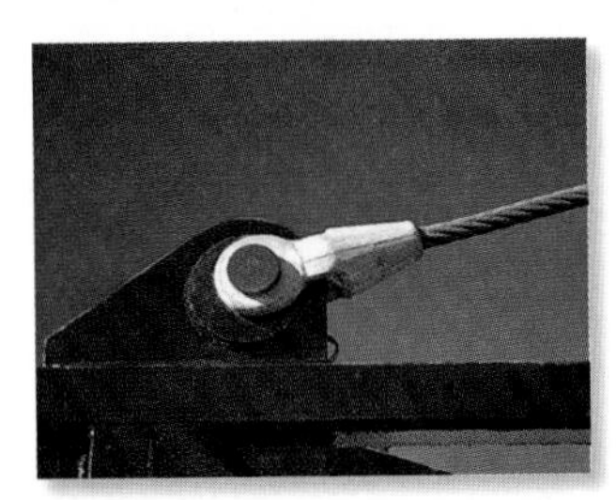

The cable exerts a force on the bracket in the direction of the cable. (1)

The rocker support for this bridge girder allows horizontal movement so the bridge is free to expand and contract due to temperature. (5)

This concrete girder rests on the ledge that is assumed to act as a smooth contacting surface. (6)

This utility building is pin supported at the top of the column. (8)

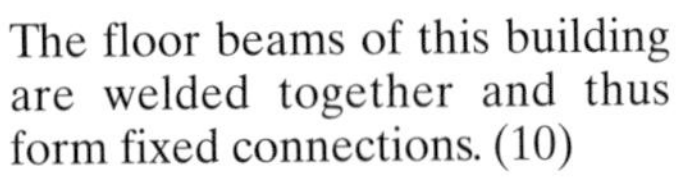

The floor beams of this building are welded together and thus form fixed connections. (10)

External and Internal Forces. Since a rigid body is a composition of particles, both *external* and *internal* loadings may act on it. It is important to realize, however, that if the free-body diagram for the body is drawn, the forces that are *internal* to the body are *not represented* on the free-body diagram. As discussed in Sec. 5.1, these forces always occur in equal but opposite collinear pairs, and therefore their *net effect* on the body is zero.

In some problems, a free-body diagram for a "system" of connected bodies may be used for an analysis. An example would be the free-body diagram of an entire automobile (system) composed of its many parts. Obviously, the connecting forces between its parts would represent *internal forces* which would *not* be included on the free-body diagram of the automobile. To summarize, internal forces act between particles which are contained within the boundary of the free-body diagram. Particles or bodies outside this boundary exert external forces on the system, and these alone must be shown on the free-body diagram.

Weight and the Center of Gravity. When a body is subjected to a gravitational field, then each of its particles has a specified weight. For the entire body it is appropriate to consider these gravitational forces to be represented as a *system of parallel forces* acting on all the particles contained within the boundary of the body. It was shown in Sec. 4.9 that such a system can be reduced to a single resultant force acting through a specified point. We refer to this force resultant as the *weight* **W** of the body and to the location of its point of application as the *center of gravity*. The methods used for its calculation will be developed in Chapter 9.

In the examples and problems that follow, if the weight of the body is important for the analysis, this force will then be reported in the problem statement. Also, when the body is *uniform* or made of homogeneous material, the center of gravity will be located at the body's *geometric center* or *centroid*; however, if the body is nonhomogeneous or has an unusual shape, then the location of its center of gravity will be given.

Idealized Models. When an engineer performs a force analysis of any object, he or she considers a corresponding analytical or idealized model that gives results that approximate as closely as possible the actual situation. To do this, careful choices have to be made so that selection of the type of supports, the material behavior, and the object's dimensions can be justified. This way the engineer can feel confident that any design or analysis will yield results which can be trusted. In complex cases this process may require developing several different models of the object that must be analyzed, but in any case, this selection process requires both skill and experience.

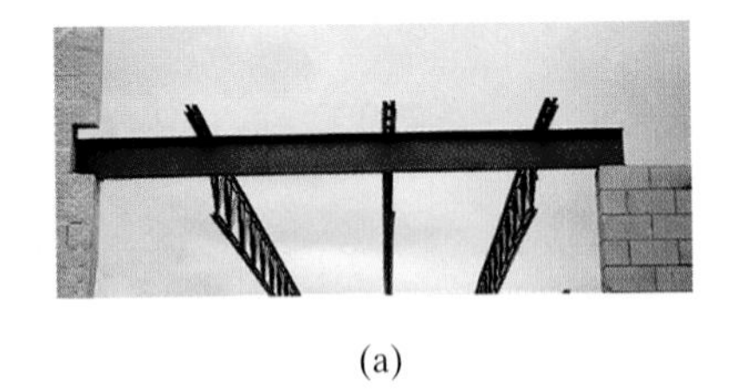

(a)

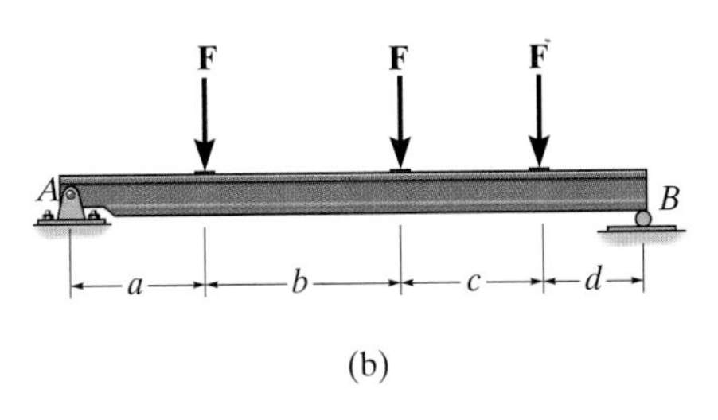

(b)

Fig. 5–5

The cases that follow illustrate what is required to develop a proper model. In Fig. 5–5*a*, the steel beam is to be used to support the roof joists of a building. For a force analysis it is reasonable to assume the material is rigid since only very small deflections will occur when the beam is loaded. A bolted connection at A will allow for any slight rotation that occurs when the load is applied, and so a *pin* can be considered for this support. At B a *roller* can be considered since the support offers no resistance to horizontal movement here. Building code requirements are used to specify the roof loading which results in a calculation of the joist loads **F**. These forces will be larger than any actual loading on the beam since they account for extreme loading cases and for dynamic or vibrational effects. The weight of the beam is generally neglected when it is small compared to the load the beam supports. The idealized model of the beam is shown with average dimensions a, b, c, and d in Fig. 5–5*b*.

As a second case, consider the lift boom in Fig. 5–6*a*. By inspection, it is supported by a pin at A and by the hydraulic cylinder BC, which can be approximated as a weightless link. The material can be assumed rigid, and with its density known, the weight of the boom and the location of its center of gravity G are determined. When a design loading **P** is specified, the idealized model shown in Fig. 5–6*b* can be used for a force analysis. Average dimensions (not shown) are used to specify the location of the loads and the supports.

Idealized models of specific objects will be given in some of the examples throughout the text. It should be realized, however, that each case represents the reduction of a practical situation using simplifying assumptions like the ones illustrated here.

(a)

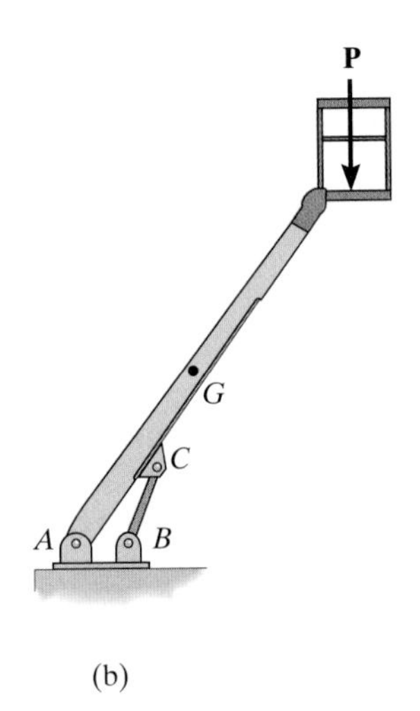

(b)

Fig. 5–6

Procedure for Drawing a Free-Body Diagram

To construct a free-body diagram for a rigid body or group of bodies considered as a single system, the following steps should be performed:

Draw Outlined Shape.

Imagine the body to be *isolated* or cut "free" from its constraints and connections and draw (sketch) its outlined shape.

Show All Forces and Couple Moments.

Identify all the external forces and couple moments that act on the body. Those generally encountered are due to (1) applied loadings, (2) reactions occurring at the supports or at points of contact with other bodies (see Table 5–1), and (3) the weight of the body. To account for all these effects, it may help to trace over the boundary, carefully noting each force or couple moment acting on it.

Identify Each Loading and Give Dimensions.

The forces and couple moments that are known should be labeled with their proper magnitudes and directions. Letters are used to represent the magnitudes and direction angles of forces and couple moments that are *unknown*. Establish an x, y coordinate system so that these unknowns, A_x, B_y, etc., can be identified. Indicate the dimensions of the body necessary for calculating the moments of forces.

Important Points

- No equilibrium problem should be solved without *first* drawing the free-body diagram, so as to account for all the forces and couple moments that act on the body.
- If a support *prevents translation* of a body in a particular direction, then the support exerts a *force* on the body in that direction.
- If *rotation is prevented*, then the support exerts a *couple moment* on the body.
- Study Table 5–1.
- Internal forces are never shown on the free-body diagram since they occur in equal but opposite collinear pairs and therefore cancel out.
- The weight of a body is an external force, and its effect is shown as a single resultant force acting through the body's center of gravity G.
- *Couple moments* can be placed anywhere on the free-body diagram since they are *free vectors. Forces* can act at any point along their lines of action since they are *sliding vectors*.

EXAMPLE 5.1

Draw the free-body diagram of the uniform beam shown in Fig. 5–7*a*. The beam has a mass of 100 kg.

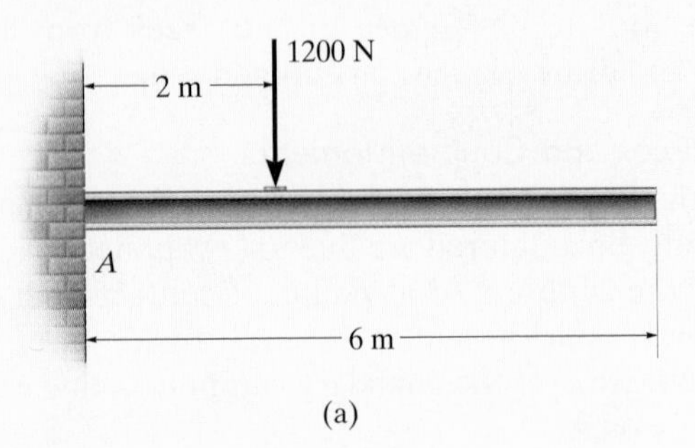

(a)

SOLUTION

The free-body diagram of the beam is shown in Fig. 5–7*b*. Since the support at *A* is a fixed wall, there are three reactions acting *on the beam* at *A*, denoted as $\mathbf{A}_x$, $\mathbf{A}_y$, and $\mathbf{M}_A$ drawn in an arbitrary direction. The magnitudes of these vectors are *unknown*, and their sense has been *assumed*. The weight of the beam, $W = 100(9.81) = 981$ N, acts through the beam's center of gravity *G*, which is 3 m from *A* since the beam is uniform.

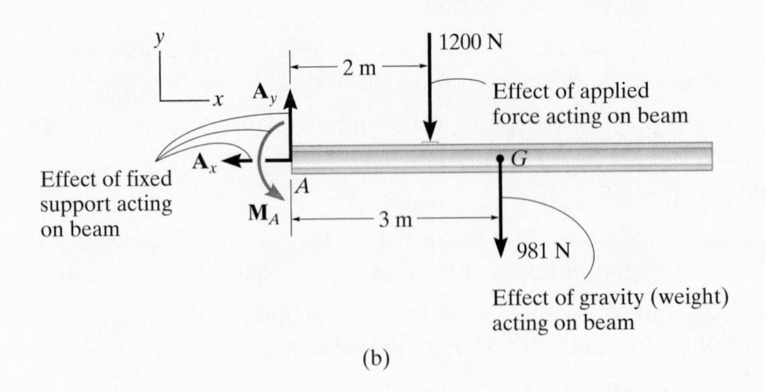

(b)

Fig. 5–7

EXAMPLE 5.2

Draw the free-body diagram of the foot lever shown in Fig. 5–8*a*. The operator applies a vertical force to the pedal so that the spring is stretched 1.5 in. and the force in the short link at B is 20 lb.

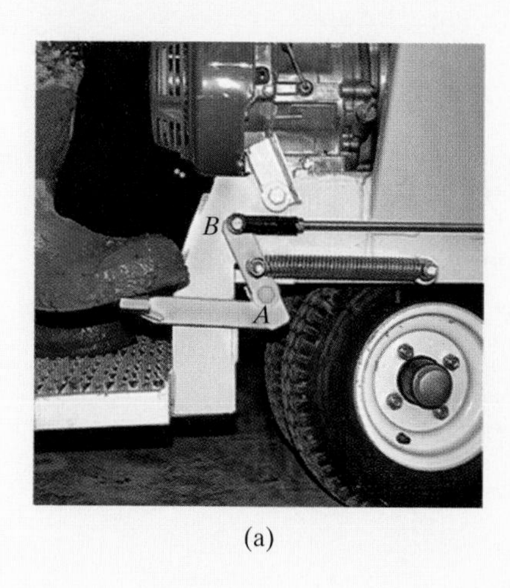

(a)

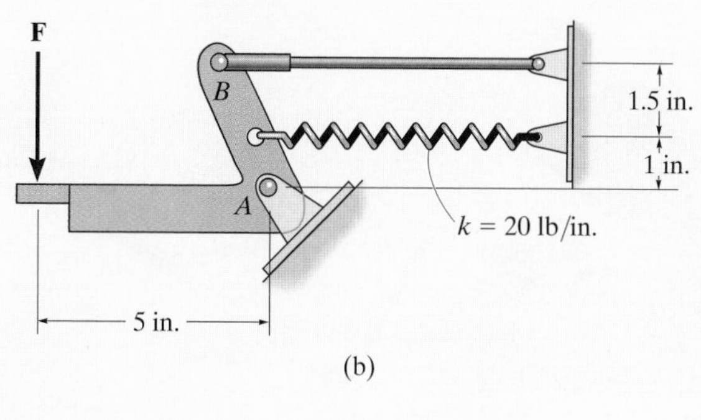

(b)

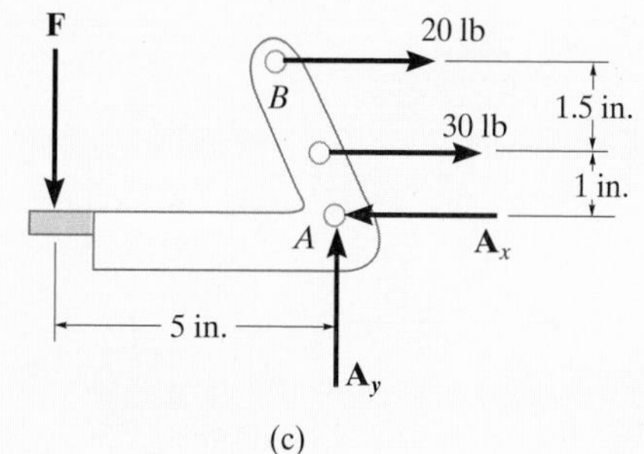

Fig. 5–8 (c)

SOLUTION

By inspection, the lever is loosely bolted to the frame at A. The rod at B is pinned at its ends and acts as a "short link." After making the proper measurements, the idealized model of the lever is shown in Fig. 5–8*b*. From this the free-body diagram must be drawn. As shown in Fig. 5–8*c*, the pin support at A exerts force components $\mathbf{A}_x$ and $\mathbf{A}_y$ on the lever, each force has a known line of action but unknown magnitude. The link at B exerts a force of 20 lb, acting in the direction of the link. In addition the spring also exerts a horizontal force on the lever. If the stiffness is measured and found to be $k = 20$ lb/in., then since the stretch $s = 1.5$ in., using Eq. 3–2, $F_s = ks = 20$ lb/in. (1.5 in.) $= 30$ lb. Finally, the operator's shoe applies a vertical force of $\mathbf{F}$ on the pedal. The dimensions of the lever are also shown on the free-body diagram, since this information will be useful when computing the moments of the forces. As usual, the senses of the unknown forces at A have been assumed. The correct senses will become apparent after solving the equilibrium equations.

EXAMPLE 5.3

Two smooth pipes, each having a mass of 300 kg, are supported by the forks of the tractor in Fig. 5–9*a*. Draw the free-body diagrams for each pipe and both pipes together.

(a)

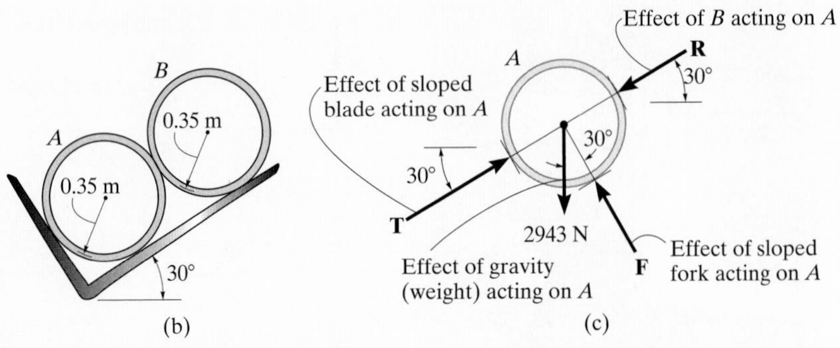

(b) (c)

Fig. 5–9

(d)

SOLUTION

The idealized model from which we must draw the free-body diagrams is shown in Fig. 5–9*b*. Here the pipes are identified, the dimensions have been added, and the physical situation reduced to its simplest form.

The free-body diagram for pipe *A* is shown in Fig. 5–9*c*. Its weight is $W = 300(9.81) = 2943$ N. Assuming all contacting surfaces are *smooth*, the reactive forces **T**, **F**, **R** act in a direction *normal* to the tangent at their surfaces of contact.

The free-body diagram of pipe *B* is shown in Fig. 5–9*d*. Can you identify each of the three forces acting *on this pipe*? In particular, note that **R**, representing the force of *A* on *B*, Fig. 5–9*d*, is equal and opposite to **R** representing the force of *B* on *A*, Fig. 5–9*c*. This is a consequence of Newton's third law of motion.

The free-body diagram of both pipes combined ("system") is shown in Fig. 5–9*e*. Here the contact force **R**, which acts between *A* and *B*, is considered as an *internal* force and hence is not shown on the free-body diagram. That is, it represents a pair of equal but opposite collinear forces which cancel each other.

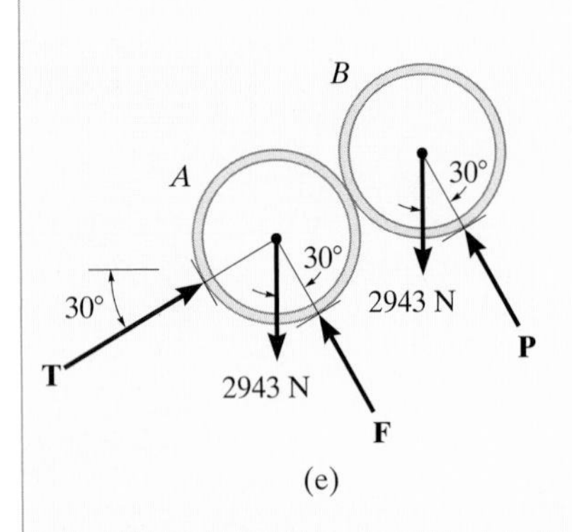

(e)

EXAMPLE 5.4

Draw the free-body diagram of the unloaded platform that is suspended off the edge of the oil rig shown in Fig. 5–10*a*. The platform has a mass of 200 kg.

(a)

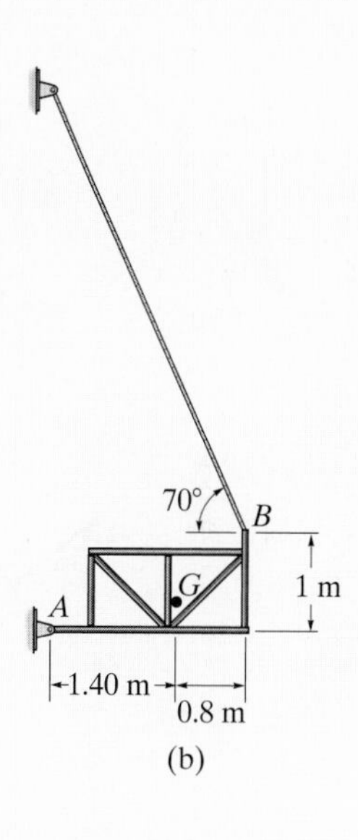

(b)

Fig. 5–10

SOLUTION

The idealized model of the platform will be considered in two dimensions because by observation the loading and the dimensions are all symmetrical about a vertical plane passing through its center, Fig. 5–10*b*. Here the connection at *A* is assumed to be a pin, and the cable supports the platform at *B*. The direction of the cable and average dimensions of the platform are listed, and the center of gravity *G* has been determined. It is from this model that we must proceed to draw the free-body diagram, which is shown in Fig. 5–10*c*. The platform's weight is $200(9.81) = 1962$ N. The force components $\mathbf{A}_x$ and $\mathbf{A}_y$ along with the cable force $\mathbf{T}$ represent the reactions that both pins and both cables exert on the platform, Fig. 5–10*a*. Consequently, after the solution for these reactions, half their magnitude is developed at *A* and half is developed at *B*.

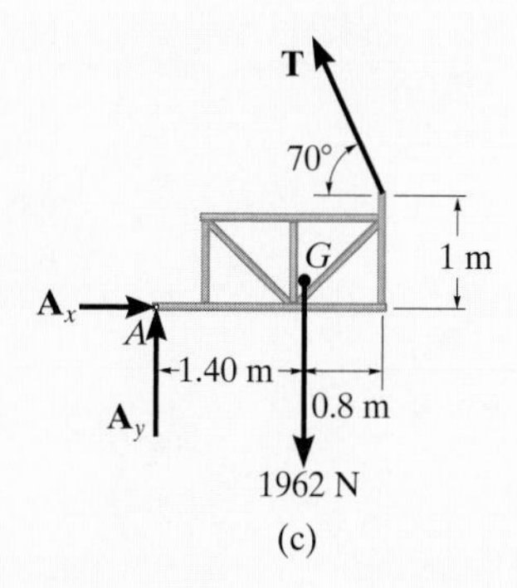

(c)

EXAMPLE 5.5

The free-body diagram of each object in Fig. 5–11 is drawn. Carefully study each solution and identify what each loading represents, as was done in Fig. 5–7*b*.

SOLUTION

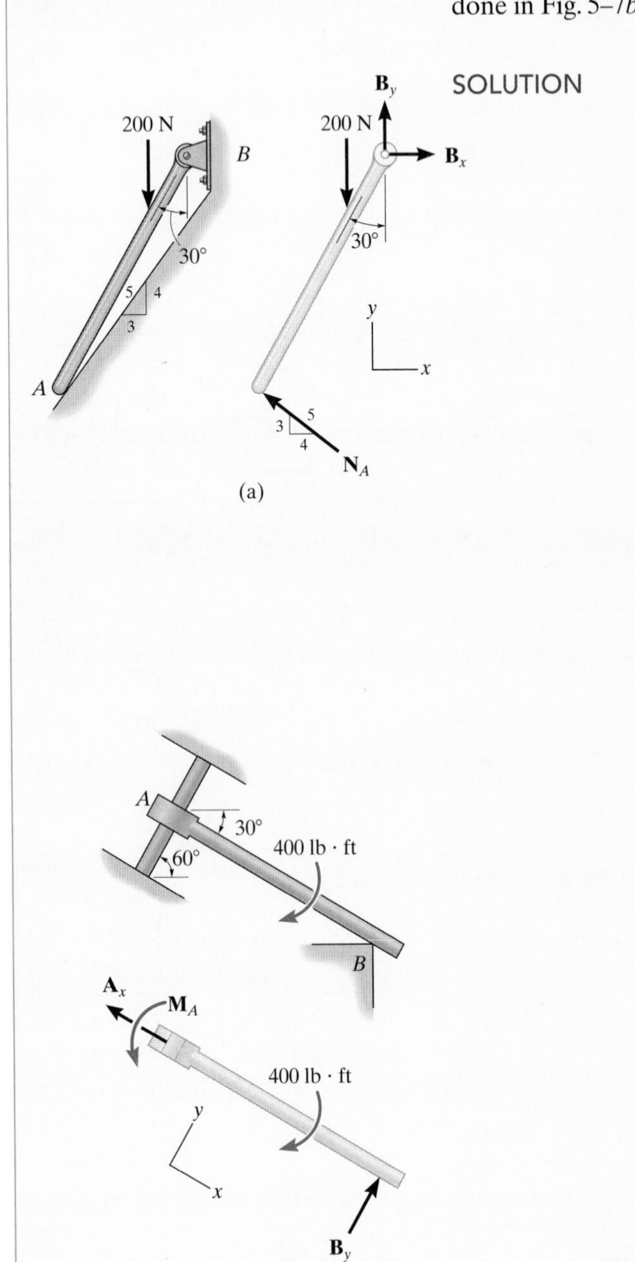

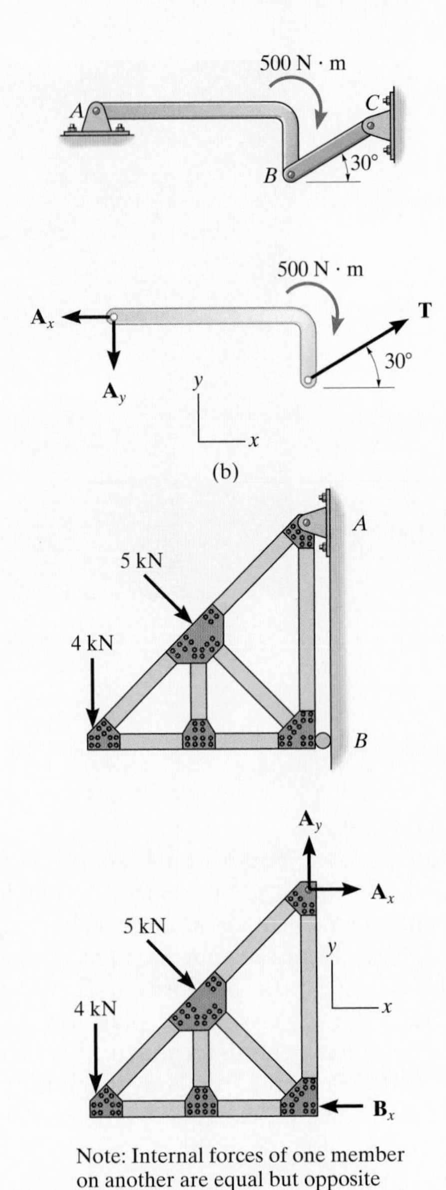

Fig. 5–11

PROBLEMS

5–1. Draw the free-body diagram of the 10-lb sphere resting between the smooth inclined planes. Explain the significance of each force on the diagram. (See Fig. 5–7*b*.)

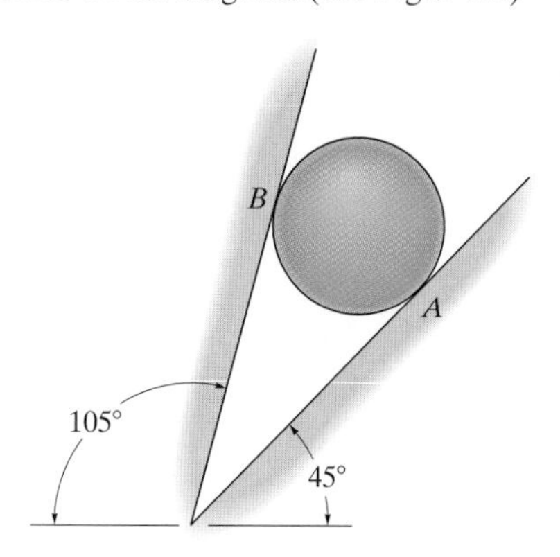

Prob. 5–1

5–2. Draw the free-body diagram of the hand punch, which is pinned at *A* and bears down on the smooth surface at *B*.

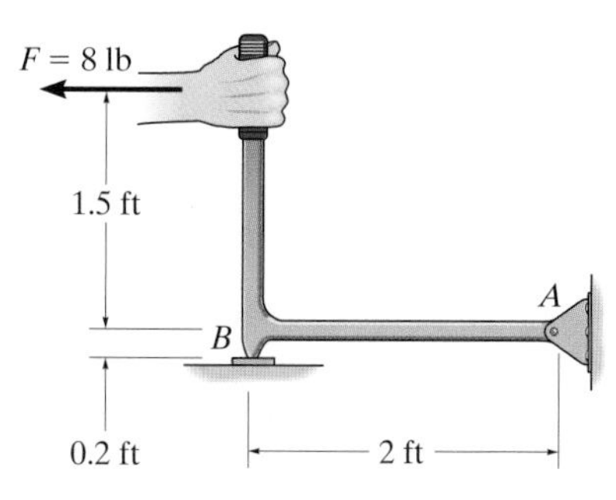

Prob. 5–2

5–3. Draw the free-body diagram of the beam supported at *A* by a fixed support and at *B* by a roller. Explain the significance of each force on the diagram. (See Fig. 5–7*b*.)

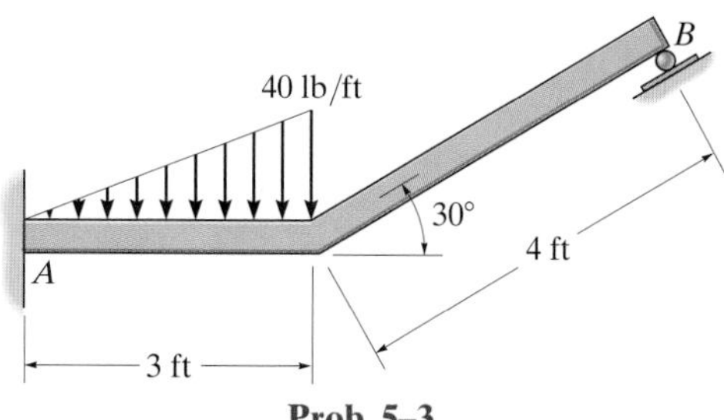

Prob. 5–3

***5–4.** Draw the free-body diagram of the jib crane *AB*, which is pin-connected at *A* and supported by member (link) *BC*.

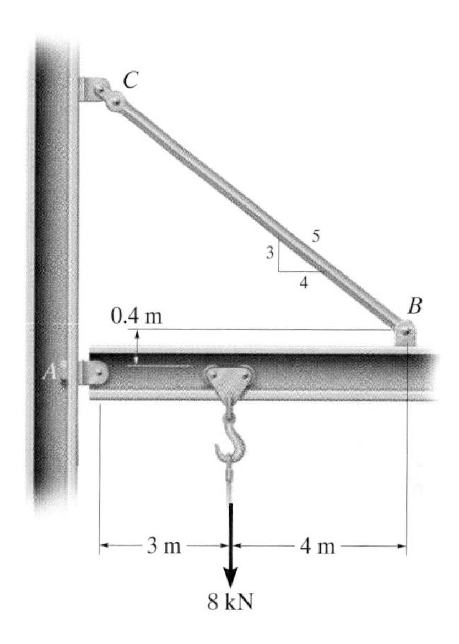

Prob. 5–4

5–5. Draw the free-body diagram of the C-bracket supported at *A*, *B*, and *C* by rollers. Explain the significance of each force on the diagram. (See Fig. 5–7*b*.)

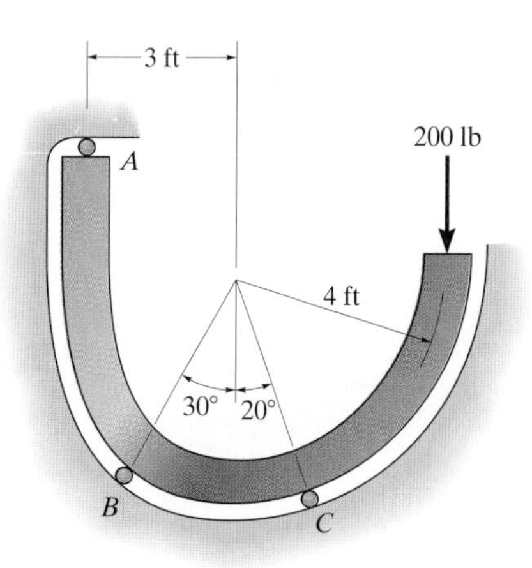

Prob. 5–5

5–6. Draw the free-body diagram of the smooth 20-g rod which rests inside the glass. Explain the significance of each force on the diagram. (See Fig. 5–7*b*.)

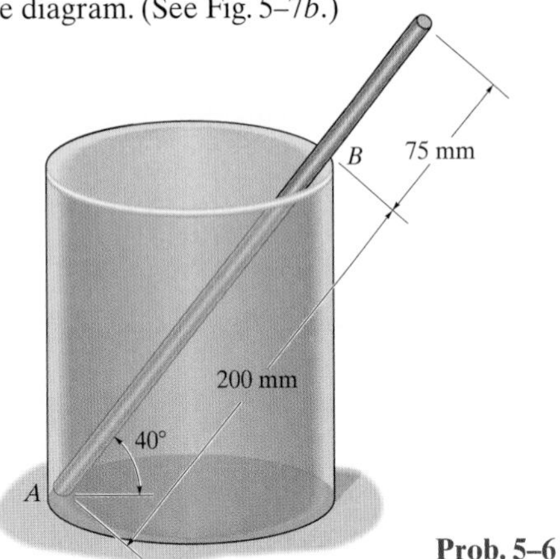

Prob. 5–6

5–7. Draw the free-body diagram of the "spanner wrench" subjected to the 20-lb force. The support at A can be considered a pin, and the surface of contact at B is smooth. Explain the significance of each force on the diagram. (See Fig. 5–7*b*.)

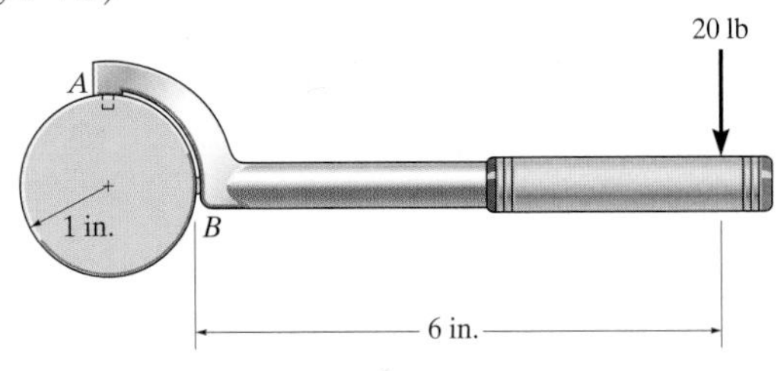

Prob. 5–7

***5–8.** Draw the free-body diagram of the automobile, which is being towed at constant velocity up the incline using the cable at C. The automobile has a mass of 5 Mg and center of mass at G. The tires are free to roll. Explain the significance of each force on the diagram. (See Fig. 5–7*b*.)

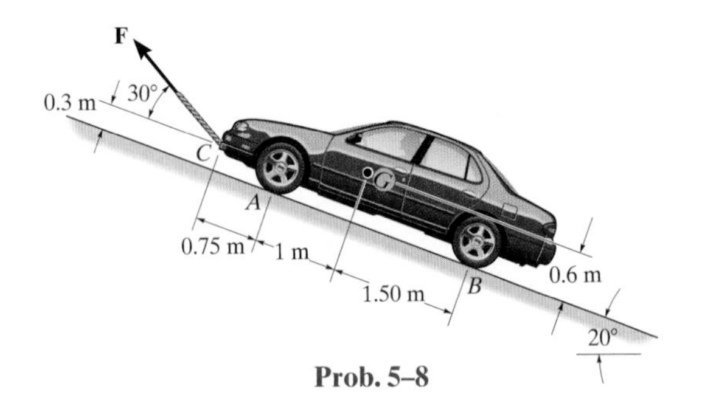

Prob. 5–8

5–9. Draw the free-body diagram of the uniform bar, which has a mass of 100 kg and a center of mass at G. The supports A, B, and C are smooth.

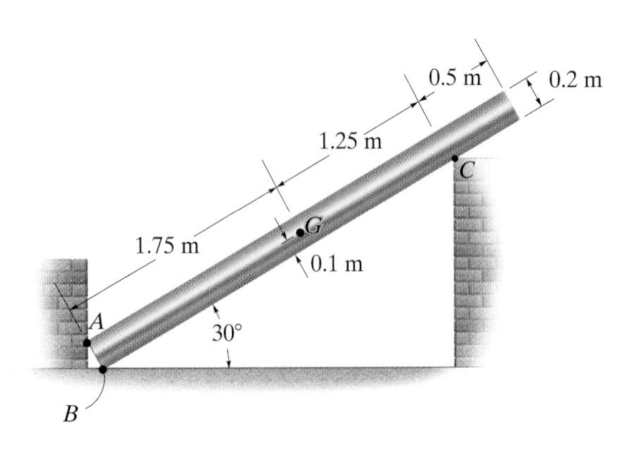

Prob. 5–9

5–10. Draw the free-body diagram of the beam, which is pin-connected at A and rocker-supported at B.

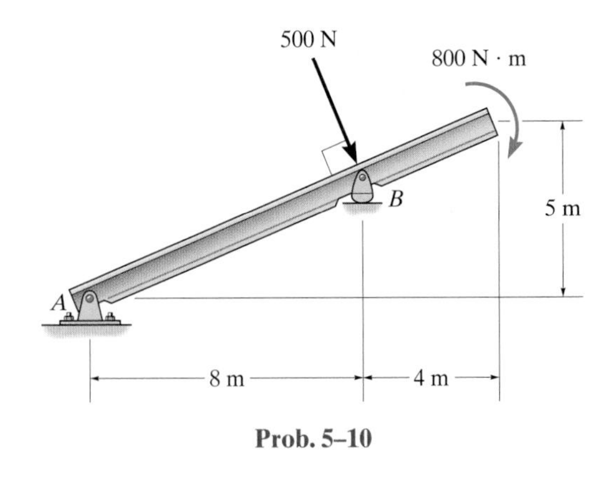

Prob. 5–10

5.3 Equations of Equilibrium

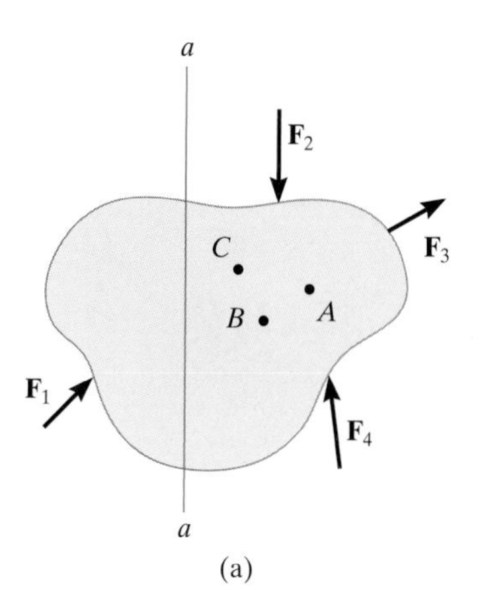

In Sec. 5.1 we developed the two equations which are both necessary and sufficient for the equilibrium of a rigid body, namely, $\Sigma \mathbf{F} = \mathbf{0}$ and $\Sigma \mathbf{M}_O = \mathbf{0}$. When the body is subjected to a system of forces, which all lie in the x–y plane, then the forces can be resolved into their x and y components. Consequently, the conditions for equilibrium in two dimensions are

$$\Sigma F_x = 0 \qquad \Sigma F_y = 0 \qquad \Sigma M_o = 0 \tag{5–2}$$

Here ΣF_x and ΣF_y represent, respectively, the algebraic sums of the x and y components of all the forces acting on the body, and ΣM_O represents the algebraic sum of the couple moments and the moments of all the force components about an axis perpendicular to the x–y plane and passing through the arbitrary point O, which may lie either on or off the body.

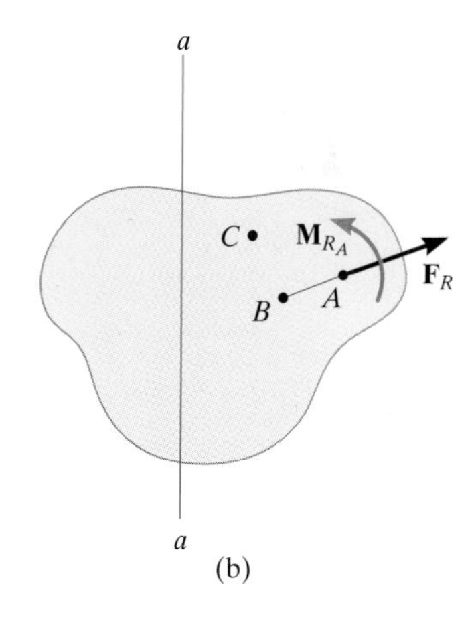

Alternative Sets of Equilibrium Equations. Although Eqs. 5–2 are *most often* used for solving coplanar equilibrium problems, two *alternative* sets of three independent equilibrium equations may also be used. One such set is

$$\Sigma F_a = 0 \qquad \Sigma M_A = 0 \qquad \Sigma M_B = 0 \tag{5–3}$$

When using these equations it is required that a line passing through points A and B is *not perpendicular* to the a axis. To prove that Eqs. 5–3 provide the *conditions* for equilibrium, consider the free-body diagram of an arbitrarily shaped body shown in Fig. 5–12*a*. Using the methods of Sec. 4.8, all the forces on the free-body diagram may be replaced by an equivalent resultant force $\mathbf{F}_R = \Sigma \mathbf{F}$, acting at point A, and a resultant couple moment $\mathbf{M}_{R_A} = \Sigma \mathbf{M}_A$, Fig. 5–12*b*. If $\Sigma M_A = 0$ is satisfied, it is necessary that $\mathbf{M}_{R_A} = \mathbf{0}$. Furthermore, in order that $\mathbf{F}_R$ satisfy $\Sigma F_a = 0$, it must have *no component* along the a axis, and therefore its line of action must be perpendicular to the a axis, Fig. 5–12*c*. Finally, if it is required that $\Sigma M_B = 0$, where B does not lie on the line of action of $\mathbf{F}_R$, then $\mathbf{F}_R = \mathbf{0}$. Since $\Sigma \mathbf{F} = \mathbf{0}$ and $\Sigma \mathbf{M}_A = \mathbf{0}$, indeed the body in Fig. 5–12*a* must be in equilibrium.

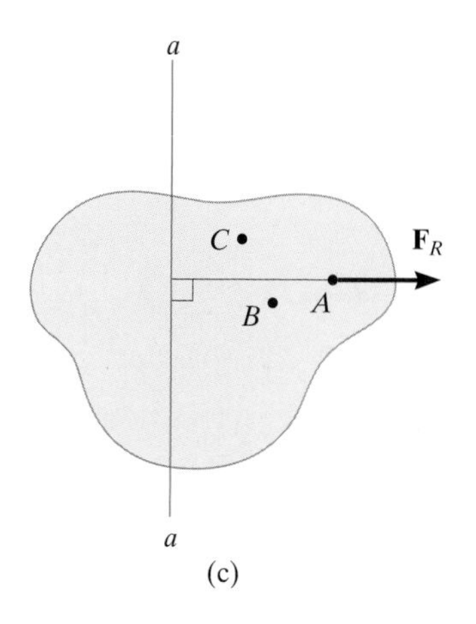

Fig. 5–12

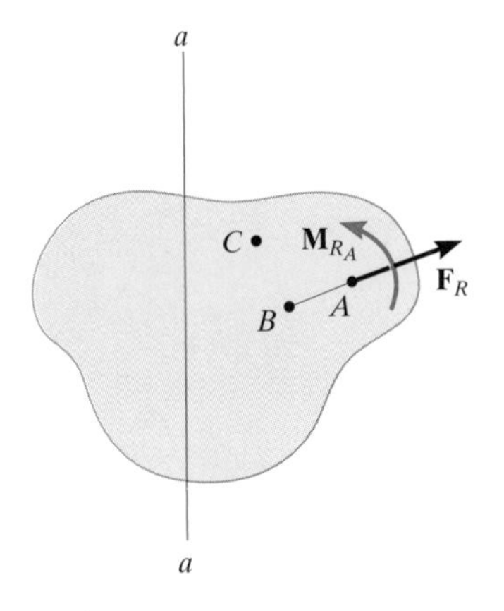

Fig. 5–13

A second alternative set of equilibrium equations is

$$\begin{aligned} \Sigma M_A &= 0 \\ \Sigma M_B &= 0 \\ \Sigma M_C &= 0 \end{aligned} \tag{5–4}$$

Here it is necessary that points A, B, and C do not lie on the same line. To prove that these equations, when satisfied, ensure equilibrium, consider the free-body diagram in Fig. 5–13. If $\Sigma M_A = 0$ is to be satisfied, then $\mathbf{M}_{R_A} = \mathbf{0}$. $\Sigma M_B = 0$ is satisfied if the line of action of $\mathbf{F}_R$ passes through point B as shown. Finally, if we require $\Sigma M_C = 0$, where C does not lie on line AB, it is necessary that $\mathbf{F}_R = \mathbf{0}$, and the body in Fig. 5–12*a* must then be in equilibrium.

Procedure for Analysis

Coplanar force equilibrium problems for a rigid body can be solved using the following procedure.

Free-Body Diagram.

- Establish the x, y coordinate axes in any suitable orientation.
- Draw an outlined shape of the body.
- Show all the forces and couple moments acting on the body.
- Label all the loadings and specify their directions relative to the x, y axes. The sense of a force or couple moment having an *unknown* magnitude but known line of action can be *assumed*.
- Indicate the dimensions of the body necessary for computing the moments of forces.

Equations of Equilibrium.

- Apply the moment equation of equilibrium, $\Sigma M_O = 0$, about a point (O) that lies at the intersection of the lines of action of two unknown forces. In this way, the moments of these unknowns are zero about O, and a *direct solution* for the third unknown can be determined.
- When applying the force equilibrium equations, $\Sigma F_x = 0$ and $\Sigma F_y = 0$, orient the x and y axes along lines that will provide the simplest resolution of the forces into their x and y components.
- If the solution of the equilibrium equations yields a negative scalar for a force or couple moment magnitude, this indicates that the sense is opposite to that which was assumed on the free-body diagram.

EXAMPLE 5.6

Determine the horizontal and vertical components of reaction on the beam caused by the pin at B and the rocker at A as shown in Fig. 5–14a. Neglect the weight of the beam.

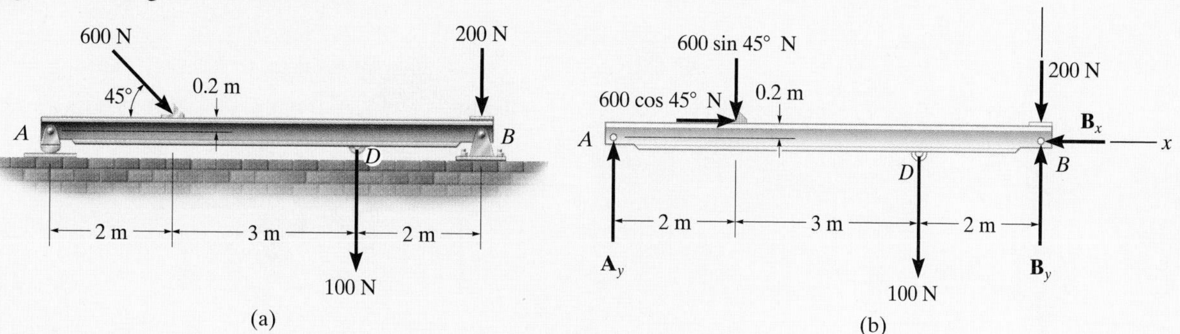

Fig. 5–14

SOLUTION

Free-Body Diagram. Identify each of the forces shown on the free-body diagram of the beam, Fig. 5–14b. (See Example 5.1.) For simplicity, the 600-N force is represented by its x and y components as shown. Also, note that a 200-N force acts on the beam at B and is independent of the force components $\mathbf{B}_x$ and $\mathbf{B}_y$, which represent the effect of the pin on the beam.

Equations of Equilibrium. Summing forces in the x direction yields

$$\overset{+}{\rightarrow} \Sigma F_x = 0; \qquad 600 \cos 45^\circ \text{ N} - B_x = 0$$

$$B_x = 424 \text{ N} \qquad \textit{Ans.}$$

A direct solution for $\mathbf{A}_y$ can be obtained by applying the moment equation $\Sigma M_B = 0$ about point B. For the calculation, it should be apparent that forces 200 N, $\mathbf{B}_x$, and $\mathbf{B}_y$ all create zero moment about B. Assuming counterclockwise rotation about B to be positive (in the $+\mathbf{k}$ direction), Fig. 5–14b, we have

$$\circlearrowleft + \Sigma M_B = 0; \qquad 100 \text{ N}(2 \text{ m}) + (600 \sin 45^\circ \text{ N})(5 \text{ m})$$
$$- (600 \cos 45^\circ \text{ N})(0.2 \text{ m}) - A_y(7 \text{ m}) = 0$$

$$A_y = 319 \text{ N} \qquad \textit{Ans.}$$

Summing forces in the y direction, using this result, gives

$$+\uparrow \Sigma F_y = 0; \quad 319 \text{ N} - 600 \sin 45^\circ \text{ N} - 100 \text{ N} - 200 \text{ N} + B_y = 0$$

$$B_y = 405 \text{ N} \qquad \textit{Ans.}$$

NOTE: We can check this result by summing moments about point A.

$$\circlearrowleft + \Sigma M_A = 0; -(600 \sin 45^\circ \text{ N})(2 \text{ m}) - (600 \cos 45^\circ \text{ N})(0.2 \text{ m})$$
$$-(100 \text{ N})(5 \text{ m}) - (200 \text{ N})(7 \text{ m}) + B_y(7 \text{ m}) = 0$$

$$B_y = 405 \text{ N} \qquad \textit{Ans.}$$

EXAMPLE 5.7

The cord shown in Fig. 5–15*a* supports a force of 100 lb and wraps over the frictionless pulley. Determine the tension in the cord at C and the horizontal and vertical components of reaction at pin A.

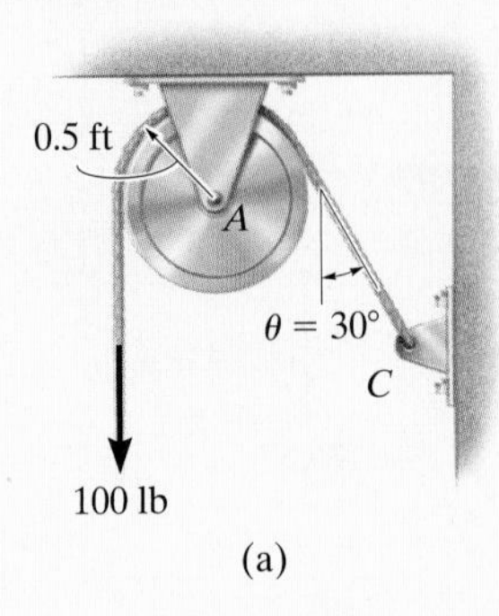

(a)

Fig. 5–15

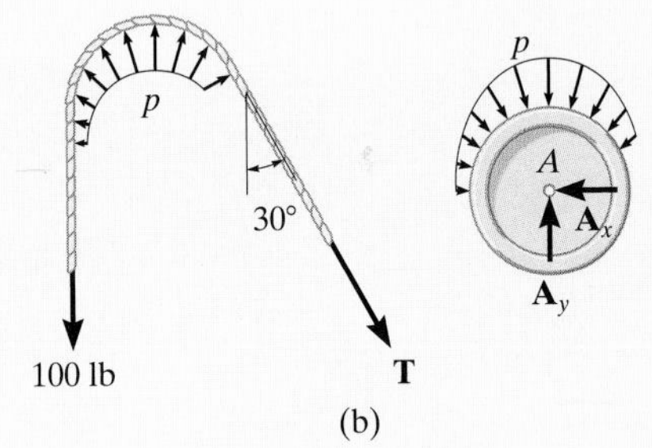

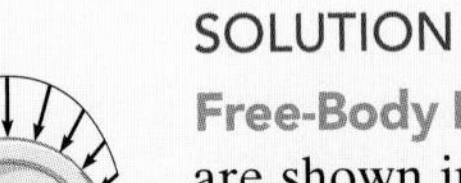

(b)

SOLUTION

Free-Body Diagrams. The free-body diagrams of the cord and pulley are shown in Fig. 5–15*b*. Note that the principle of action, equal but opposite reaction must be carefully observed when drawing each of these diagrams: the cord exerts an unknown load distribution p along part of the pulley's surface, whereas the pulley exerts an equal but opposite effect on the cord. For the solution, however, it is simpler to *combine* the free-body diagrams of the pulley and the contacting portion of the cord, so that the distributed load becomes *internal* to the system and is therefore eliminated from the analysis, Fig. 5–15*c*.

Equations of Equilibrium. Summing moments about point A to eliminate $\mathbf{A}_x$ and $\mathbf{A}_y$, Fig. 5–15*c*, we have

$$\circlearrowright + \Sigma M_A = 0; \qquad 100 \text{ lb } (0.5 \text{ ft}) - T(0.5 \text{ ft}) = 0$$

$$T = 100 \text{ lb} \qquad \textit{Ans.}$$

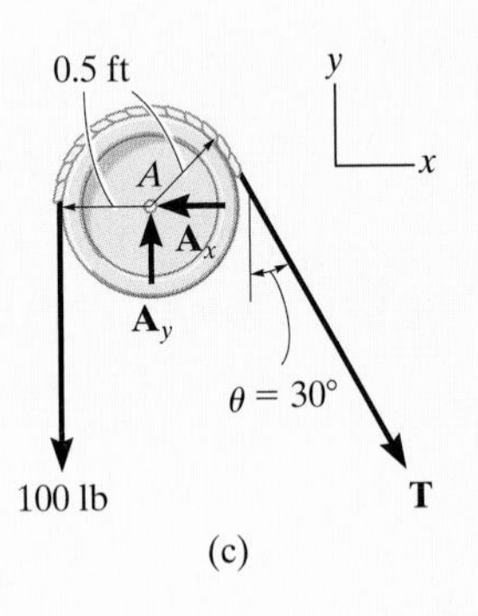

(c)

Using the result for T, a force summation is applied to determine the components of reaction at pin A.

$$\overset{+}{\rightarrow} \Sigma F_x = 0; \qquad -A_x + 100 \sin 30^\circ \text{ lb} = 0$$

$$A_x = 50.0 \text{ lb} \qquad \textit{Ans.}$$

$$+\uparrow \Sigma F_y = 0; \qquad A_y - 100 \text{ lb} - 100 \cos 30^\circ \text{ lb} = 0$$

$$A_y = 187 \text{ lb} \qquad \textit{Ans.}$$

NOTE: It is seen that the tension remains *constant* as the cord passes over the pulley. (This of course is true for *any angle* θ at which the cord is directed and for *any radius r* of the pulley.)

EXAMPLE 5.8

The link shown in Fig. 5–16*a* is pin-connected at *A* and rests against a smooth support at *B*. Compute the horizontal and vertical components of reaction at the pin *A*.

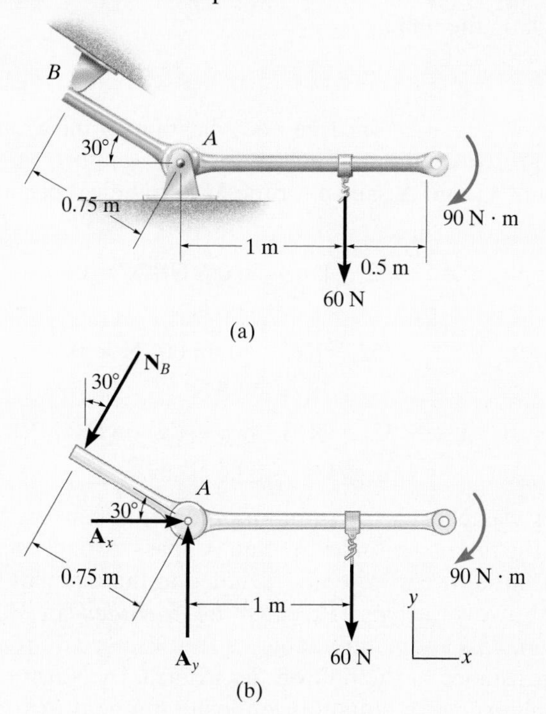

Fig. 5–16

SOLUTION

Free-Body Diagram. As shown in Fig. 5–16*b*, the reaction $\mathbf{N}_B$ is perpendicular to the link at *B*. Also, horizontal and vertical components of reaction are represented at *A*.

Equations of Equilibrium. Summing moments about *A*, we obtain a direct solution for N_B,

$$\circlearrowright + \Sigma M_A = 0; \quad -90\text{ N}\cdot\text{m} - 60\text{ N}(1\text{ m}) + N_B(0.75\text{ m}) = 0$$

$$N_B = 200\text{ N}$$

Using this result,

$$\overset{+}{\rightarrow} \Sigma F_x = 0; \quad A_x - 200 \sin 30^\circ\text{ N} = 0$$

$$A_x = 100\text{ N} \qquad \textit{Ans.}$$

$$+\uparrow \Sigma F_y = 0; \quad A_y - 200 \cos 30^\circ\text{ N} - 60\text{ N} = 0$$

$$A_y = 233\text{ N} \qquad \textit{Ans.}$$

EXAMPLE 5.9

The box wrench in Fig. 5–17*a* is used to tighten the bolt at A. If the wrench does not turn when the load is applied to the handle, determine the torque or moment applied to the bolt and the force of the wrench on the bolt.

(a)

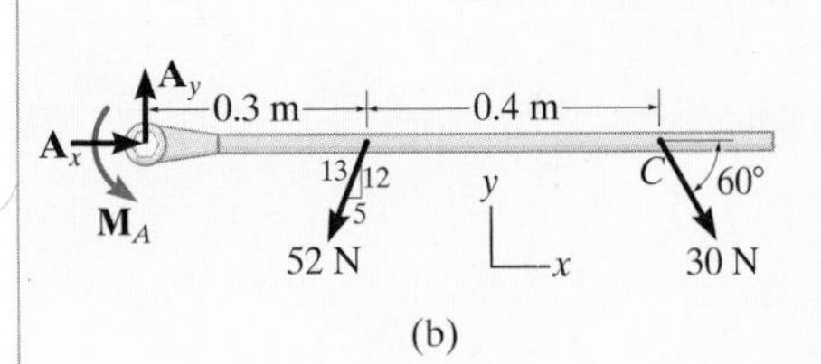

(b)

Fig. 5–17

SOLUTION

Free-Body Diagram. The free-body diagram for the wrench is shown in Fig. 5–17*b*. Since the bolt acts as a "fixed support," it exerts force components $\mathbf{A}_x$ and $\mathbf{A}_y$ and a torque $\mathbf{M}_A$ on the wrench at A.

Equations of Equilibrium.

$$\xrightarrow{+} \Sigma F_x = 0; \quad A_x - 52\left(\tfrac{5}{13}\right)\text{N} + 30 \cos 60^\circ \text{ N} = 0$$

$$A_x = 5.00 \text{ N} \qquad \textit{Ans.}$$

$$+\uparrow \Sigma F_y = 0; \quad A_y - 52\left(\tfrac{12}{13}\right)\text{N} - 30 \sin 60^\circ \text{ N} = 0$$

$$A_y = 74.0 \text{ N} \qquad \textit{Ans.}$$

$$\circlearrowleft + \Sigma M_A = 0; \quad M_A - 52\left(\tfrac{12}{13}\right)\text{N}(0.3 \text{ m}) - (30 \sin 60^\circ \text{ N})(0.7 \text{ m}) = 0$$

$$M_A = 32.6 \text{ N} \cdot \text{m} \qquad \textit{Ans.}$$

Point A was chosen for summing moments because the lines of action of the *unknown* forces $\mathbf{A}_x$ and $\mathbf{A}_y$ pass through this point, and therefore these forces were not included in the moment summation. Realize, however, that $\mathbf{M}_A$ must be *included* in this moment summation. This couple moment is a free vector and represents the twisting resistance of the bolt on the wrench. By Newton's third law, the wrench exerts an equal but opposite moment or torque on the bolt. Furthermore, the resultant force on the wrench is

$$F_A = \sqrt{(5.00)^2 + (74.0)^2} = 74.1 \text{ N} \qquad \textit{Ans.}$$

Because the force components A_x and A_y were calculated as positive quantities, their directional sense is shown correctly on the free-body diagram in Fig. 5–17*b*. Hence

$$\theta = \tan^{-1}\frac{74.0 \text{ N}}{5.00 \text{ N}} = 86.1^\circ \measuredangle \theta$$

Realize that $\mathbf{F}_A$ acts in the opposite direction on the bolt. Why?

NOTE: Although only *three* independent equilibrium equations can be written for a rigid body, it is a good practice to *check* the calculations using a fourth equilibrium equation. For example, the above computations may be verified in part by summing moments about point C:

$$\circlearrowleft + \Sigma M_C = 0; \quad 52\left(\tfrac{12}{13}\right)\text{N}(0.4 \text{ m}) + 32.6 \text{ N} \cdot \text{m} - 74.0 \text{ N}(0.7 \text{ m}) = 0$$

$$19.2 \text{ N} \cdot \text{m} + 32.6 \text{ N} \cdot \text{m} - 51.8 \text{ N} \cdot \text{m} = 0$$

EXAMPLE 5.10

Placement of concrete from the truck is accomplished using the chute shown in the photos, Fig. 5–18*a*. Determine the force that the hydraulic cylinder and the truck frame exert on the chute to hold it in the position shown. The chute and wet concrete contained along its length have a uniform weight of 35 lb/ft.

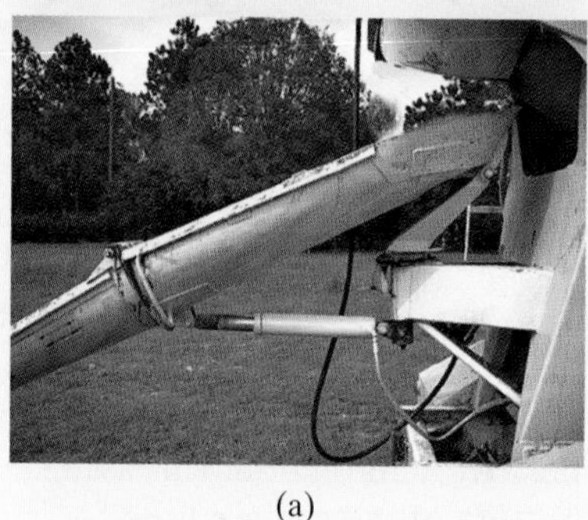

(a)

SOLUTION

The idealized model of the chute is shown in Fig. 5–18*b*. Here the dimensions are given, and it is assumed the chute is pin connected to the frame at *A* and the hydraulic cylinder *BC* acts as a short link.

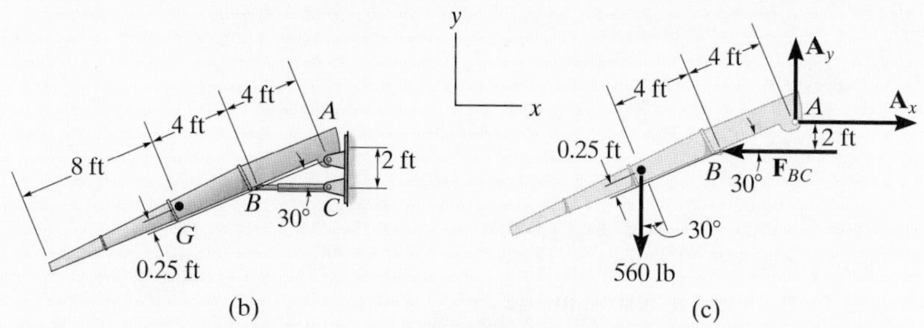

Fig. 5–18

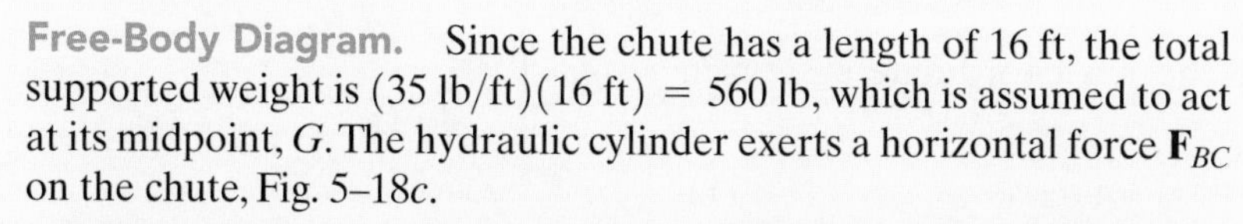

Free-Body Diagram. Since the chute has a length of 16 ft, the total supported weight is $(35\text{ lb/ft})(16\text{ ft}) = 560\text{ lb}$, which is assumed to act at its midpoint, *G*. The hydraulic cylinder exerts a horizontal force $\mathbf{F}_{BC}$ on the chute, Fig. 5–18*c*.

Equations of Equilibrium. A direct solution for $\mathbf{F}_{BC}$ is possible by summing moments about the pin at *A*. To do this we will use the principle of moments and resolve the weight into components parallel and perpendicular to the chute. We have,

$$\circlearrowright +\Sigma M_A = 0;$$
$$-F_{BC}(2\text{ ft}) + 560 \cos 30^\circ\text{ lb}\,(8\text{ ft}) + 560 \sin 30^\circ\text{ lb}\,(0.25\text{ ft}) = 0$$
$$F_{BC} = 1975\text{ lb} \qquad \textit{Ans.}$$

Summing forces to obtain A_x and A_y, we obtain

$$\overset{+}{\rightarrow} \Sigma F_x = 0; \qquad -A_x + 1975\text{ lb} = 0$$
$$A_x = 1975\text{ lb} \qquad \textit{Ans.}$$

$$+\uparrow \Sigma F_y = 0; \qquad A_y - 560\text{ lb} = 0$$
$$A_y = 560\text{ lb} \qquad \textit{Ans.}$$

NOTE: To verify this solution we can sum moments about point *B*.

$$\circlearrowright +\Sigma M_B = 0; \qquad -1975\text{ lb}\,(2\text{ ft}) + 560\text{ lb}\,(4 \cos 30^\circ\text{ ft}) +$$
$$560 \cos 30^\circ\text{ lb}\,(4\text{ ft}) + 560 \sin 30^\circ\text{ lb}\,(0.25\text{ ft}) = 0$$

EXAMPLE 5.11

The uniform smooth rod shown in Fig. 5–19*a* is subjected to a force and couple moment. If the rod is supported at *A* by a smooth wall and at *B* and *C* either at the top or bottom by rollers, determine the reactions at these supports. Neglect the weight of the rod.

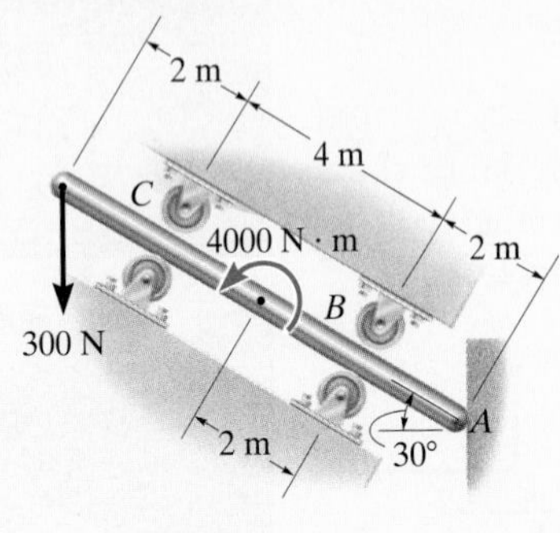

(a)

SOLUTION

Free-Body Diagram. As shown in Fig. 5–19*b*, all the support reactions act normal to the surface of contact since the contacting surfaces are smooth. The reactions at *B* and *C* are shown acting in the positive *y*′ direction. This assumes that only the rollers located on the bottom of the rod are used for support.

Equations of Equilibrium. Using the *x*, *y* coordinate system in Fig. 5–19*b*, we have

$$\xrightarrow{+} \Sigma F_x = 0; \quad C_{y'} \sin 30° + B_{y'} \sin 30° - A_x = 0 \quad (1)$$

$$+\uparrow \Sigma F_y = 0; \quad -300 \text{ N} + C_{y'} \cos 30° + B_{y'} \cos 30° = 0 \quad (2)$$

$$\circlearrowright + \Sigma M_A = 0; \quad -B_{y'}(2 \text{ m}) + 4000 \text{ N}\cdot\text{m} - C_{y'}(6 \text{ m}) + (300 \cos 30° \text{ N})(8 \text{ m}) = 0 \quad (3)$$

When writing the moment equation, it should be noticed that the line of action of the force component 300 sin 30° N passes through point *A*, and therefore this force is not included in the moment equation.

Solving Eqs. 2 and 3 simultaneously, we obtain

$$B_{y'} = -1000.0 \text{ N} = -1 \text{ kN} \qquad \textit{Ans.}$$

$$C_{y'} = 1346.4 \text{ N} = 1.35 \text{ kN} \qquad \textit{Ans.}$$

Since $B_{y'}$ is a negative scalar, the sense of $\mathbf{B}_{y'}$ is opposite to that shown on the free-body diagram in Fig. 5–19*b*. Therefore, the top roller at *B* serves as the support rather than the bottom one. Retaining the negative sign for $B_{y'}$ (Why?) and substituting the results into Eq. 1, we obtain

$$1346.4 \sin 30° \text{ N} - 1000.0 \sin 30° \text{ N} - A_x = 0$$

$$A_x = 173 \text{ N} \qquad \textit{Ans.}$$

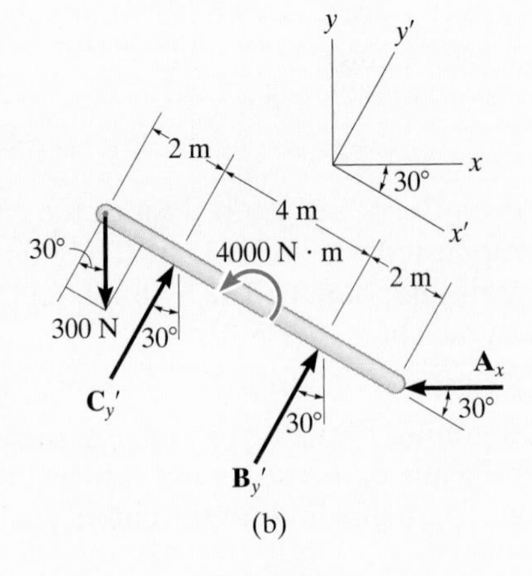

(b)

Fig. 5–19

EXAMPLE 5.12

(a)

The uniform truck ramp shown in Fig. 5–20*a* has a weight of 400 lb and is pinned to the body of the truck at each end and held in the position shown by the two side cables. Determine the tension in the cables.

SOLUTION

The idealized model of the ramp, which indicates all necessary dimensions and supports, is shown in Fig. 5–20*b*. Here the center of gravity is located at the midpoint since the ramp is approximately uniform.

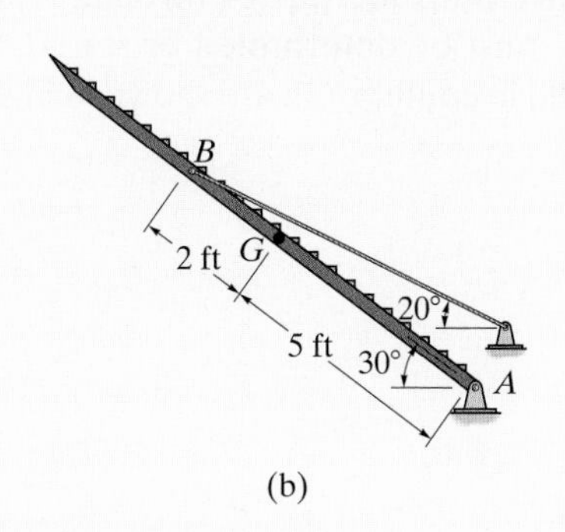

(b)

Free-Body Diagram. Working from the idealized model, the ramp's free-body diagram is shown in Fig. 5–20*c*.

Equations of Equilibrium. Summing moments about point A will yield a direct solution for the cable tension. Using the principle of moments, there are several ways of determining the moment of $\mathbf{T}$ about A. If we use x and y components, with $\mathbf{T}$ applied at B, we have

$$\circlearrowleft + \Sigma M_A = 0; \quad -T\cos 20^\circ(7\sin 30^\circ\text{ ft}) + T\sin 20^\circ(7\cos 30^\circ\text{ ft}) + 400\text{ lb}\,(5\cos 30^\circ\text{ ft}) = 0$$

$$T = 1425\text{ lb}$$

By the principle of transmissibility, we can locate $\mathbf{T}$ at C, even though this point is not on the ramp, Fig. 5–20*c*. In this case the horizontal component of $\mathbf{T}$ does not create a moment about A. First we must determine d using the sine law.

$$\frac{d}{\sin 10^\circ} = \frac{7\text{ ft}}{\sin 20^\circ}; \qquad d = 3.554\text{ ft}$$

$$\circlearrowleft + \Sigma M_A = 0; \quad -T\sin 20^\circ(3.554\text{ ft}) + 400\text{ lb}\,(5\cos 30^\circ\text{ ft}) = 0$$

$$T = 1425\text{ lb}$$

The simplest way to compute the moment of $\mathbf{T}$ about A is to resolve it into components parallel and perpendicular to the ramp at B. Then the moment of the parallel component is zero about A, so that

$$\circlearrowleft + \Sigma M_A = 0; \quad -T\sin 10^\circ(7\text{ ft}) + 400\text{ lb}\,(5\cos 30^\circ\text{ ft}) = 0$$

$$T = 1425\text{ lb}$$

Since there are two cables supporting the ramp,

$$T' = \frac{T}{2} = 712\text{ lb} \qquad \textit{Ans.}$$

Fig. 5–20 (c)

NOTE: As an exercise, show that $A_x = 1339$ lb and $A_y = 887.4$ lb.

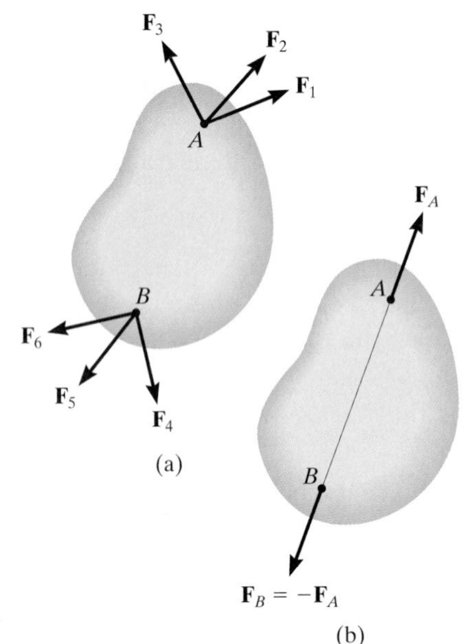

Two-force member

Fig. 5–21

5.4 Two- and Three-Force Members

The solution to some equilibrium problems can be simplified by recognizing members that are subjected to only two or three forces.

Two-Force Members. When a member is subjected to *no couple moments* and forces are applied at only two points on the member, the member is called a *two-force member*. An example is shown in Fig. 5–21*a*. The forces at A and B are summed to obtain their respective *resultants* $\mathbf{F}_A$ and $\mathbf{F}_B$, Fig. 5–21*b*. These two forces will maintain *translational or force equilibrium* $(\Sigma\mathbf{F} = \mathbf{0})$ provided $\mathbf{F}_A$ is of equal magnitude and opposite direction to $\mathbf{F}_B$. Furthermore, *rotational or moment equilibrium* $(\Sigma\mathbf{M}_O = \mathbf{0})$ is satisfied if $\mathbf{F}_A$ is *collinear* with $\mathbf{F}_B$. As a result, the line of action of both forces is known since it always passes through A and B. Hence, only the force magnitude must be determined or stated. Other examples of two-force members held in equilibrium are shown in Fig. 5–22.

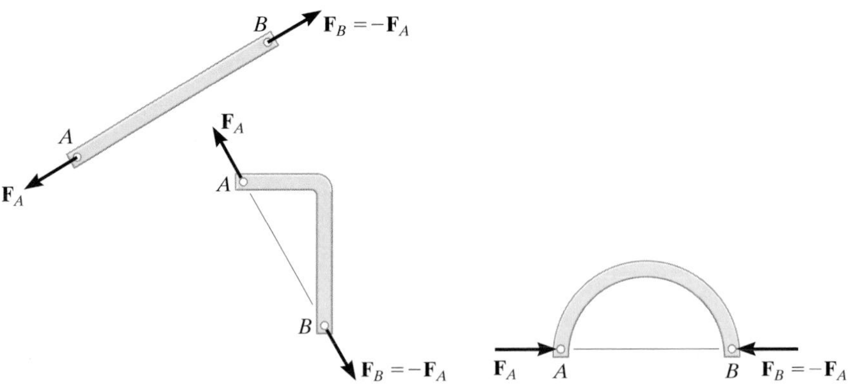

Two-force members

Fig. 5–22

Three-Force Members. If a member is subjected to only three forces, then it is necessary that the forces be either *concurrent* or *parallel* for the member to be in equilibrium. To show the concurrency requirement, consider the body in Fig. 5–23*a* and suppose that any two of the three forces acting on the body have lines of action that intersect at point O. To satisfy moment equilibrium about O, i.e., $\Sigma M_O = 0$, the third force must also pass through O, which then makes the force system *concurrent*. If two of the three forces are parallel, Fig. 5–23*b*, the point of concurrency, O, is considered to be at "infinity," and the third force must be parallel to the other two forces to intersect at this "point."

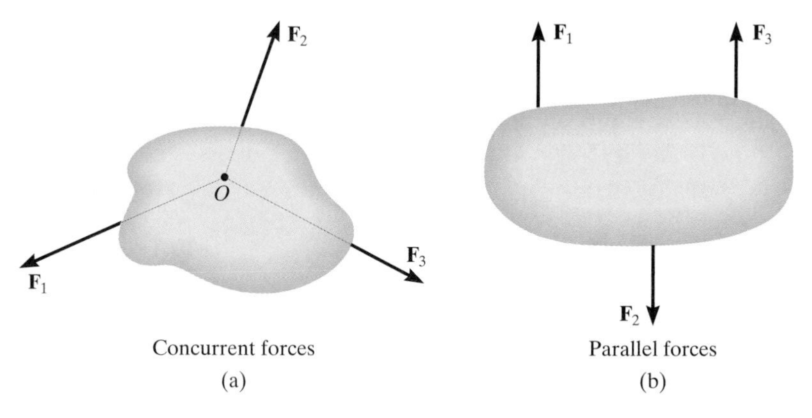

Concurrent forces
(a)

Parallel forces
(b)

Three-force members

Fig. 5–23

Many mechanical elements act as two- or three-force members, and the ability to recognize them in a problem will considerably simplify an equilibrium analysis.

- The bucket link AB on the back-hoe is a typical example of a two-force member since it is pin connected at its ends and, provided its weight is neglected, no other force acts on this member.
- The hydraulic cylinder BC is pin connected at its ends. It is a two-force member. The boom ABD is subjected to the weight of the suspended motor at D, the force of the hydraulic cylinder at B, and the force of the pin at A. If the boom's weight is neglected, it is a three-force member.
- The dump bed of the truck operates by extending the telescopic hydraulic cylinder AB. If the weight of AB is neglected, we can classify it as a two-force member since it is pin connected at its end points.

EXAMPLE 5.13

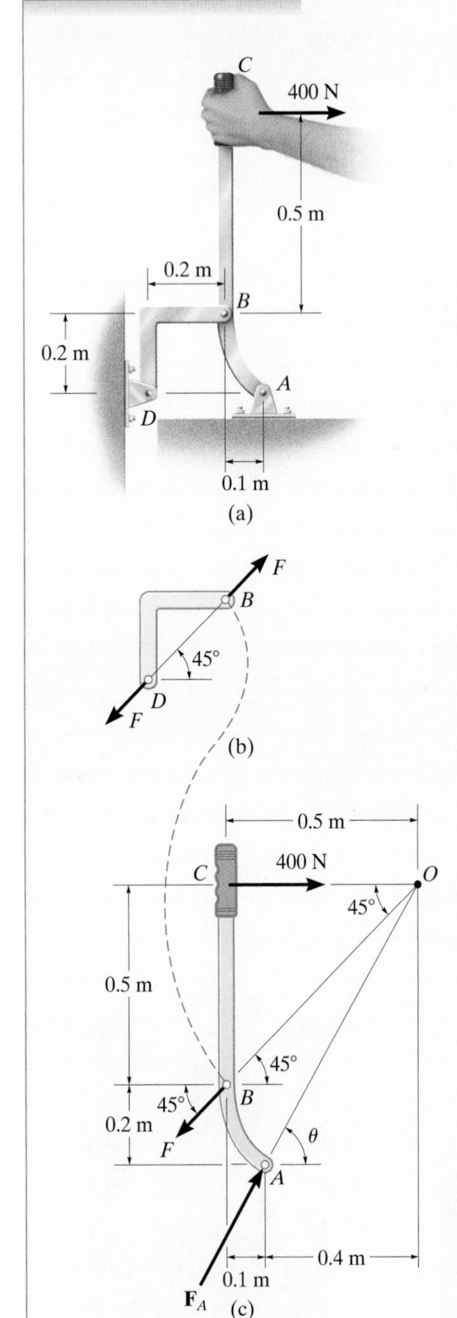

Fig. 5–24

The lever ABC is pin-supported at A and connected to a short link BD as shown in Fig. 5–24*a*. If the weight of the members is negligible, determine the force of the pin on the lever at A.

SOLUTION

Free-Body Diagrams. As shown by the free-body diagram, Fig. 5–24*b*, the short link BD is a *two-force member*, so the *resultant forces* at pins D and B must be equal, opposite, and collinear. Although the magnitude of the force is unknown, the line of action is known since it passes through B and D.

Lever ABC is a *three-force member*, and therefore, in order to satisfy moment equilibrium, the three nonparallel forces acting on it must be concurrent at O, Fig. 5–24*c*. In particular, note that the force F on the lever at B is equal but opposite to the force F acting at B on the link. Why? The distance CO must be 0.5 m since the lines of action of $\mathbf{F}$ and the 400-N force are known.

Equations of Equilibrium. By requiring the force system to be concurrent at O, since $\Sigma M_O = 0$, the angle θ which defines the line of action of $\mathbf{F}_A$ can be determined from trigonometry,

$$\theta = \tan^{-1}\left(\frac{0.7}{0.4}\right) = 60.3^\circ \quad \measuredangle\theta \qquad \textit{Ans.}$$

Using the x, y axes and applying the force equilibrium equations, we can obtain F_A and F.

$$\stackrel{+}{\rightarrow} \Sigma F_x = 0; \qquad F_A \cos 60.3^\circ - F \cos 45^\circ + 400 \text{ N} = 0$$

$$+\uparrow \Sigma F_y = 0; \qquad F_A \sin 60.3^\circ - F \sin 45^\circ = 0$$

Solving, we get

$$F_A = 1.07 \text{ kN} \qquad \textit{Ans.}$$

$$F = 1.32 \text{ kN}$$

NOTE: We can also solve this problem by representing the force at A by its two components $\mathbf{A}_x$ and $\mathbf{A}_y$ and applying $\Sigma M_A = 0$, $\Sigma F_x = 0$, $\Sigma F_y = 0$ to the lever. Once A_x and A_y are determined, how would you find F_A and θ?

PROBLEMS

5–11. Determine the reactions at the supports in Prob. 5–1.

***5–12.** Determine the magnitude of the resultant force acting at pin A of the handpunch in Prob. 5–2.

5–13. Determine the reactions at the supports at C in Prob. 5–5.

5–14. Determine the reactions on the rod in Prob. 5–6.

5–15. Determine the support reactions on the spanner wrench in Prob. 5–7.

***5–16.** Determine the reactions on the car in Prob. 5–8.

5–17. Determine the reactions at the points of contact at A, B, and C of the bar in Prob. 5–9.

5–18. Determine the reactions at the pin A and at the roller at B of the beam in Prob. 5–10.

5–19. Determine the magnitude of the reactions on the beam at A and B. Neglect the thickness of the beam.

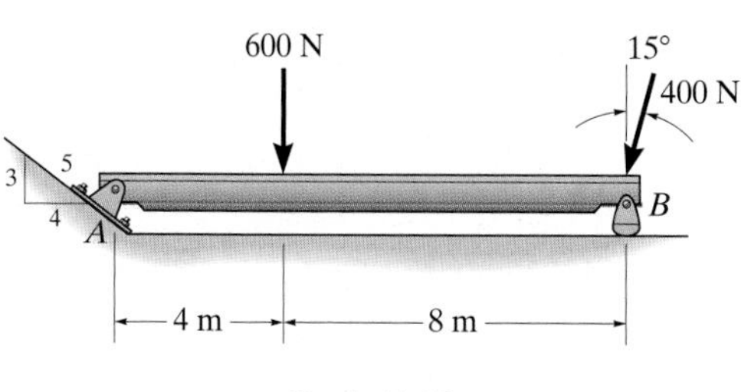

Prob. 5–19

***5–20.** Determine the reactions at the supports.

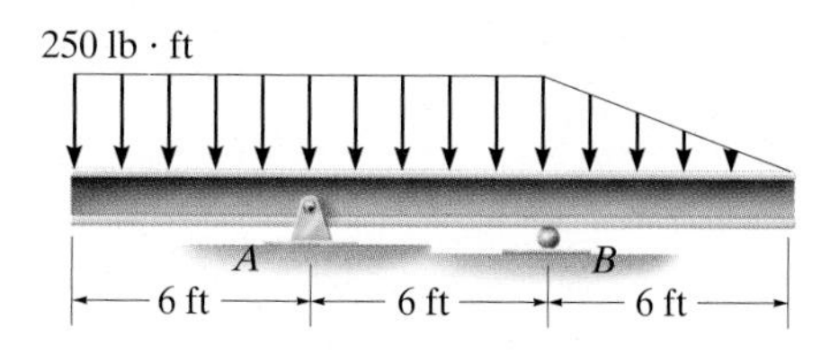

Prob. 5–20

5–21. When holding the 5-lb stone in equilibrium, the humerus H, assumed to be smooth, exerts normal forces $\mathbf{F}_C$ and $\mathbf{F}_A$ on the radius C and ulna A as shown. Determine these forces and the force $\mathbf{F}_B$ that the biceps B exerts on the radius for equilibrium. The stone has a center of mass at G. Neglect the weight of the arm.

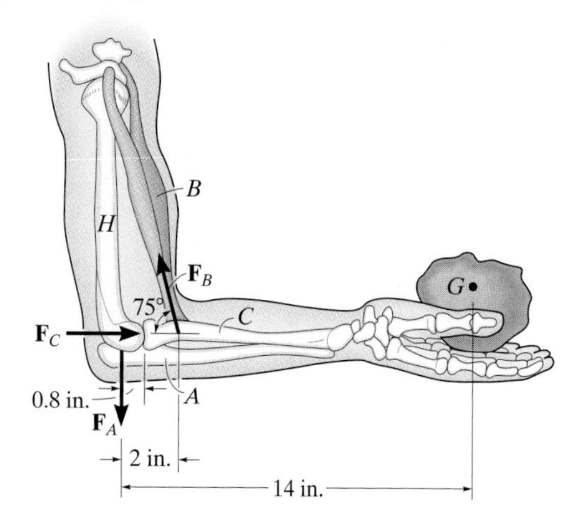

Prob. 5–21

5–22. The uniform door has a weight of 100 lb and a center of gravity at G. Determine the reactions at the hinges if the hinge at A supports only a horizontal reaction on the door, whereas the hinge at B exerts both horizontal and vertical reactions.

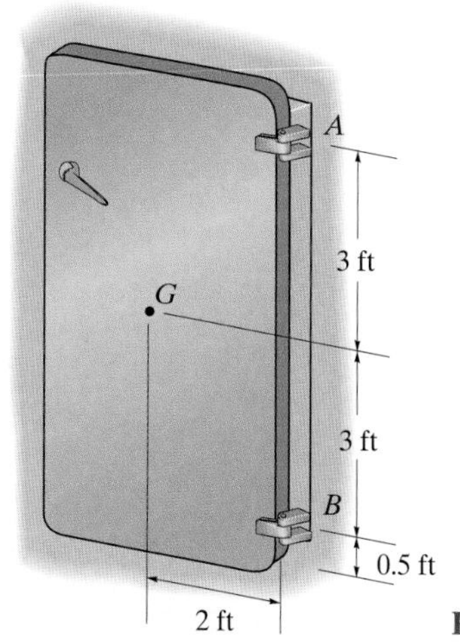

Prob. 5–22

5–23. The ramp of a ship has a weight of 200 lb and a center of gravity at G. Determine the cable force in CD needed to just start lifting the ramp, (i.e., so the reaction at B becomes zero). Also, determine the horizontal and vertical components of force at the hinge (pin) at A.

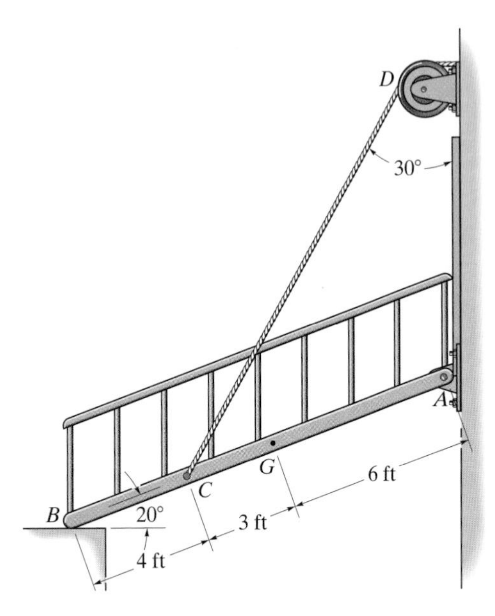

Prob. 5–23

*■**5–24.** The 1.4-Mg drainpipe is held in the tines of the fork lift. Determine the normal forces at A and B as functions of the blade angle θ and plot the results of force (ordinate) versus θ (abscissa) for $0 \leq \theta \leq 90°$.

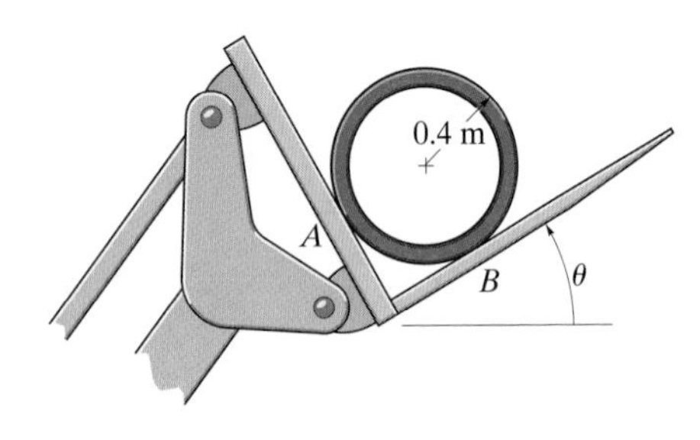

Prob. 5–24

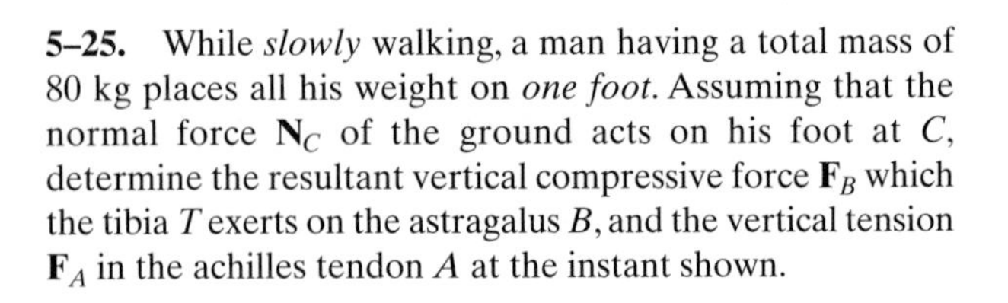

5–25. While *slowly* walking, a man having a total mass of 80 kg places all his weight on *one foot*. Assuming that the normal force $\mathbf{N}_C$ of the ground acts on his foot at C, determine the resultant vertical compressive force $\mathbf{F}_B$ which the tibia T exerts on the astragalus B, and the vertical tension $\mathbf{F}_A$ in the achilles tendon A at the instant shown.

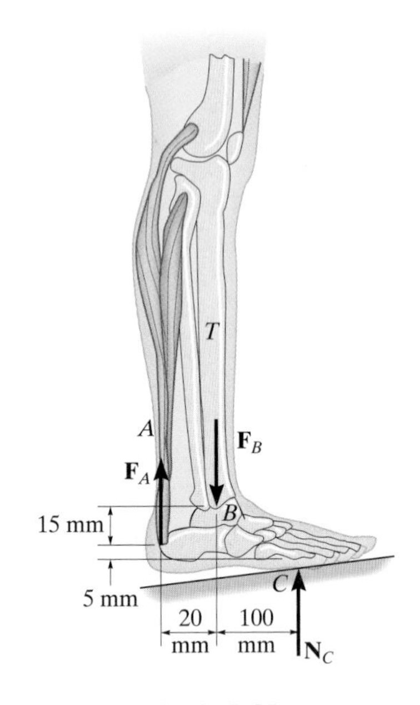

Prob. 5–25

5–26. Determine the reactions at the roller A and pin B.

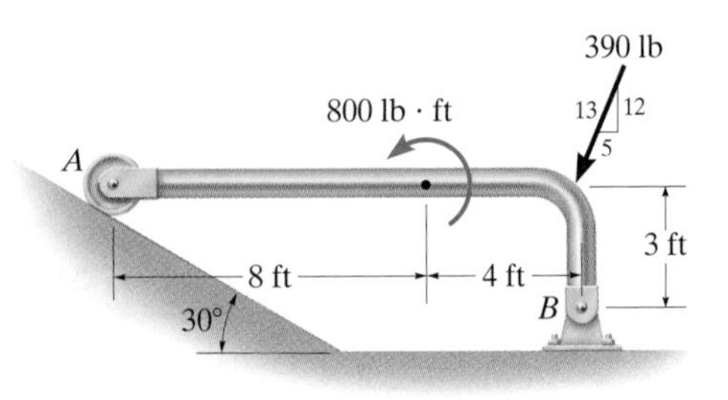

Prob. 5–26

5–27. The platform assembly has a weight of 250 lb and center of gravity at G_1. If it is intended to support a maximum load of 400 lb placed at point G_2, determine the smallest counterweight W that should be placed at B in order to prevent the platform from tipping over.

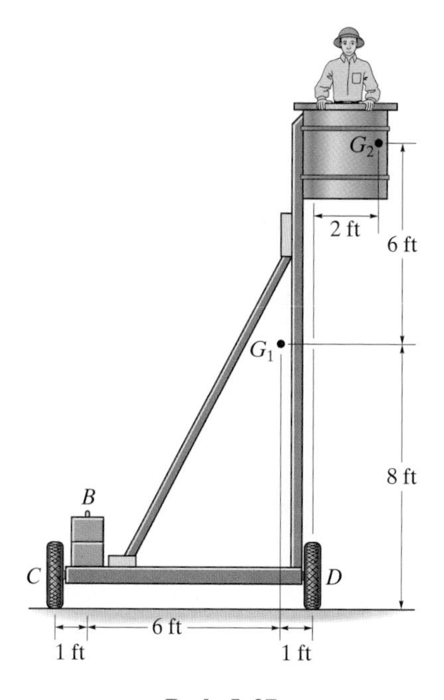

Prob. 5–27

5–29. The device is used to hold an elevator door open. If the spring has a stiffness of $k = 40$ N/m and it is compressed 0.2 m, determine the horizontal and vertical components of reaction at the pin A and the resultant force at the wheel bearing B.

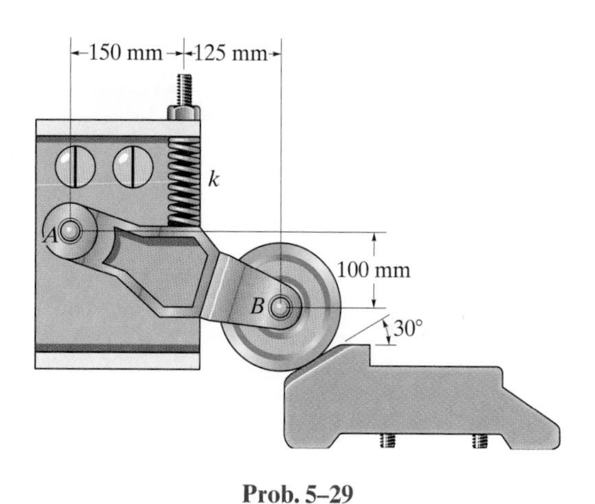

Prob. 5–29

***5–28.** The articulated crane boom has a weight of 125 lb and mass center at G. If it supports a load of 600 lb, determine the force acting at the pin A and the compression in the hydraulic cylinder BC when the boom is in the position shown.

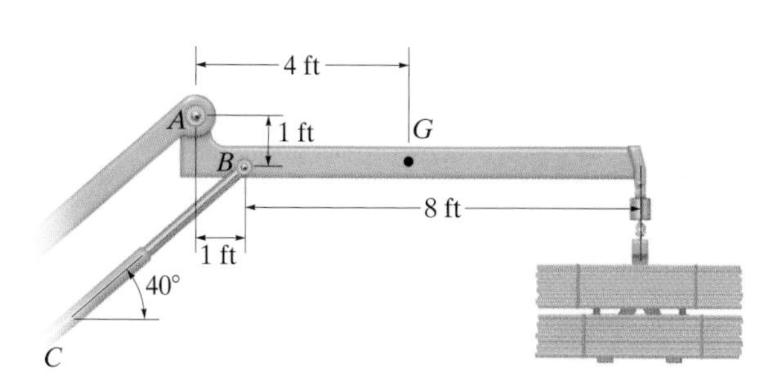

Prob. 5–28

5–30. Determine the reactions on the bent rod which is supported by a smooth surface at B and by a collar at A, which is fixed to the rod and is free to slide over the fixed inclined rod.

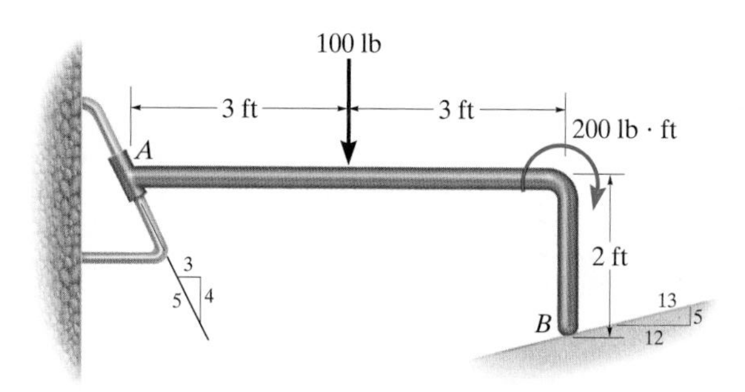

Prob. 5–30

5–31. The cantilevered jib crane is used to support the load of 780 lb. If the trolley T can be placed anywhere between $1.5 \text{ ft} \leq x \leq 7.5 \text{ ft}$, determine the maximum magnitude of reaction at the supports A and B. Note that the supports are collars that allow the crane to rotate freely about the vertical axis. The collar at B supports a force in the vertical direction, whereas the one at A does not.

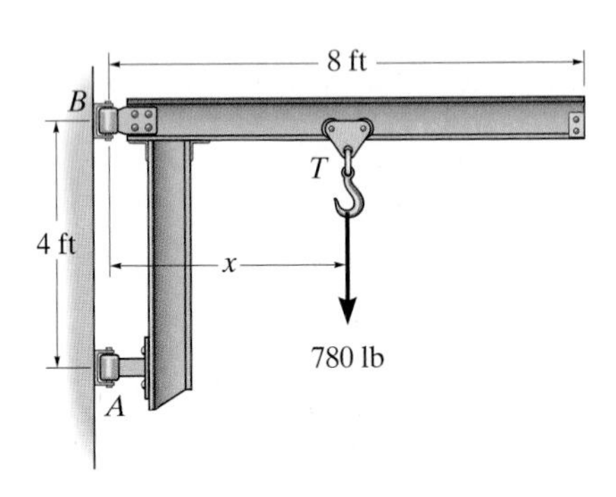

Prob. 5–31

5–33. The power pole supports the three lines, each line exerting a vertical force on the pole due to its weight as shown. Determine the reactions at the fixed support D. If it is possible for wind or ice to snap the lines, determine which line(s) when removed create(s) a condition for the greatest moment reaction at D.

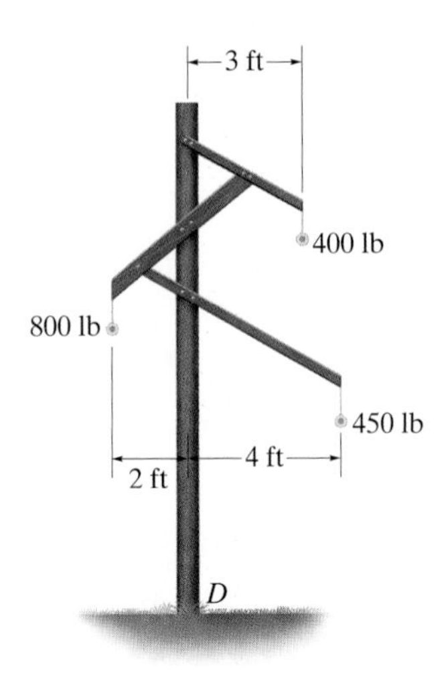

Prob. 5–33

***5–32.** The uniform rod AB has a weight of 15 lb. Determine the force in the cable when the rod is in the position shown.

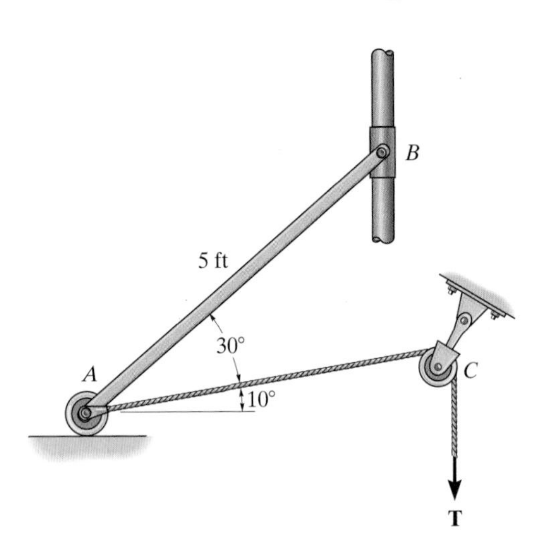

Prob. 5–32

5–34. The picnic table has a weight of 50 lb and a center of gravity at G_T. If a man weighing 225 lb has a center of gravity at G_M and sits down in the centered position shown, determine the vertical reaction at each of the two legs at B. Neglect the thickness of the legs. What can you conclude from the results?

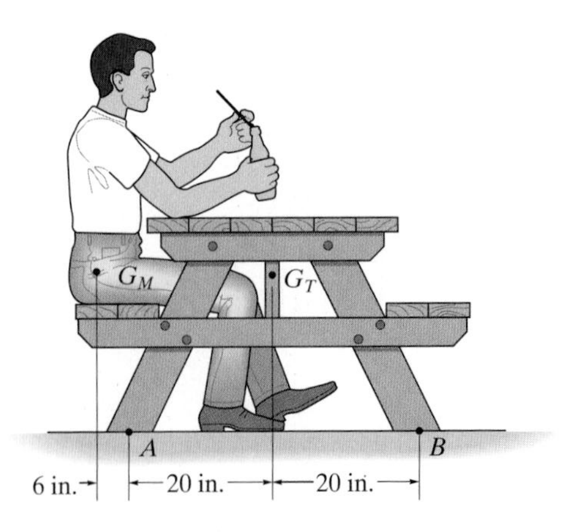

Prob. 5–34

5–35. If the wheelbarrow and its contents have a mass of 60 kg and center of mass at G, determine the magnitude of the resultant force which the man must exert on *each* of the two handles in order to hold the wheelbarrow in equilibrium.

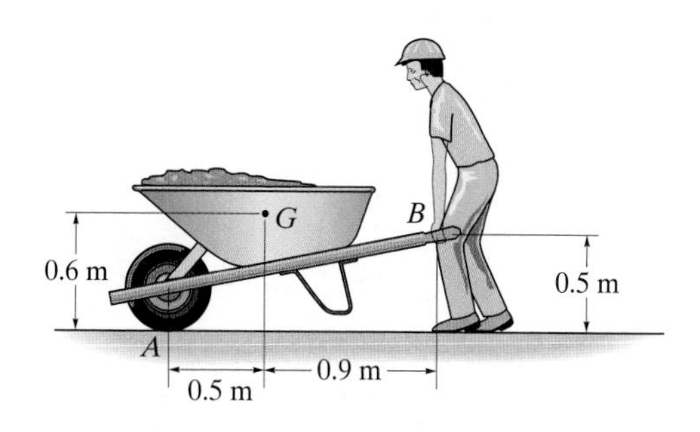

Prob. 5–35

5–37. When no force is applied to the brake pedal of the lightweight truck, the retainer spring AB keeps the pedal in contact with the smooth brake light switch at C. If the force on the switch is 3 N, determine the unstretched length of the spring if the stiffness of the spring is $k = 80$ N/m.

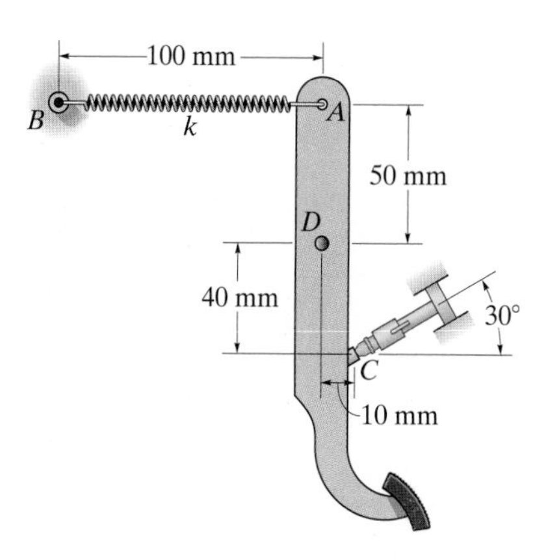

Prob. 5–37

***5–36.** The man has a weight W and stands at the center of the plank. If the planes at A and B are smooth, determine the tension in the cord in terms of W and θ.

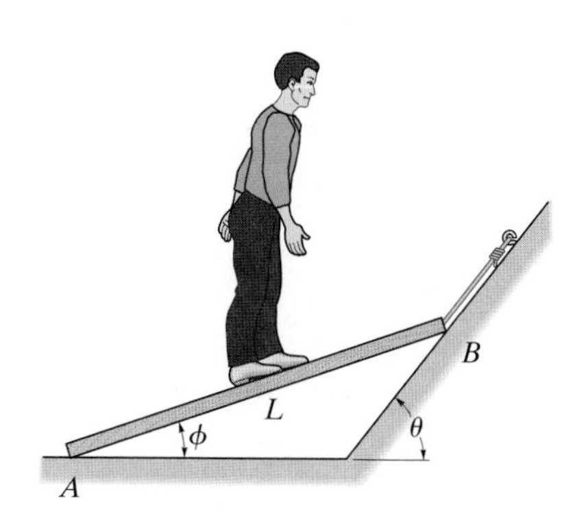

Prob. 5–36

5–38. The telephone pole of negligible thickness is subjected to the force of 80 lb directed as shown. It is supported by the cable BCD and can be assumed pinned at its base A. In order to provide clearance for a sidewalk right of way, where D is located, the strut CE is attached at C, as shown by the dashed lines (cable segment CD is removed). If the tension in CD' is to be twice the tension in BCD, determine the height h for placement of the strut CE.

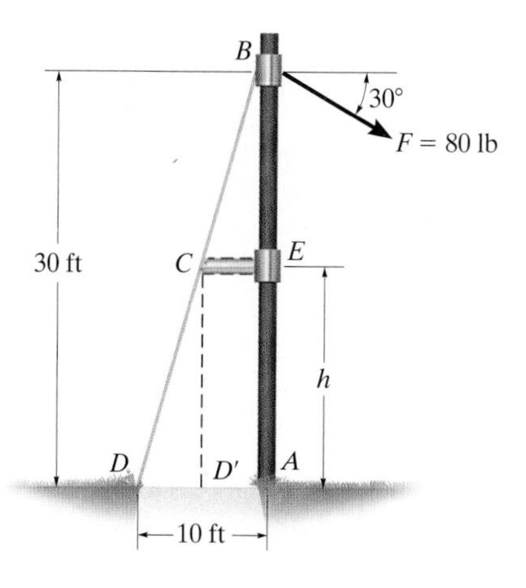

Prob. 5–38

5–39. The worker uses the hand truck to move material down the ramp. If the truck and its contents are held in the position shown and have a weight of 100 lb with center of gravity at G, determine the resultant normal force of both wheels on the ground A and the magnitude of the force required at the grip B.

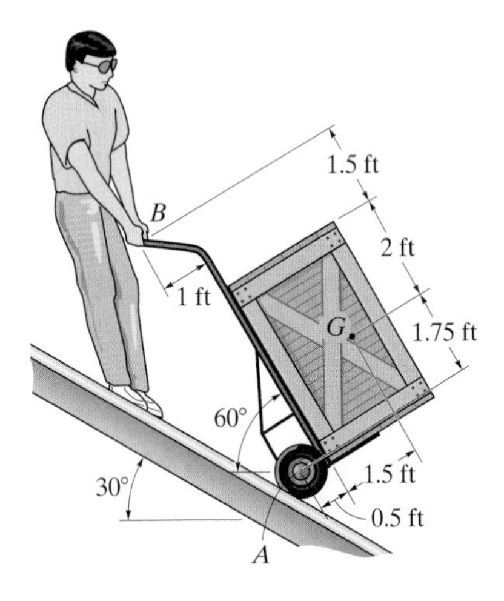

Prob. 5–39

***5–40.** Soil pressure acting on the concrete retaining wall is represented as a loading per foot length of wall. If concrete has a specific weight of 150 lb/ft^3, determine the magnitudes of the soil distribution, w_1 and w_2, and the frictional force **F** for equilibrium.

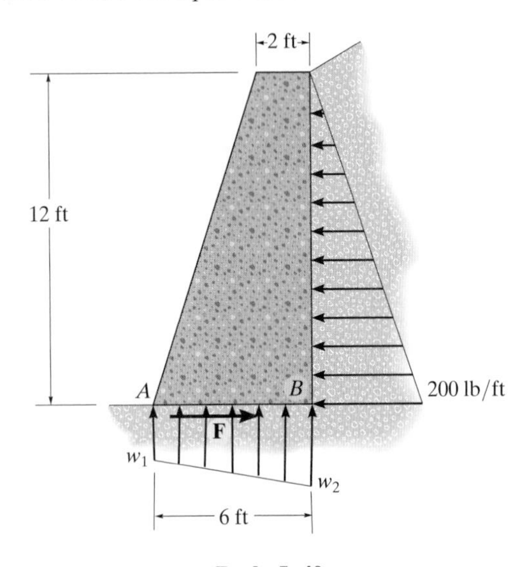

Prob. 5–40

5–41. The shelf supports the electric motor which has a mass of 15 kg and mass center at G_m. The platform upon which it rests has a mass of 4 kg and mass center at G_p. Assuming that a single bolt B holds the shelf up and the bracket bears against the smooth wall at A, determine this normal force at A and the horizontal and vertical components of reaction of the bolt on the bracket.

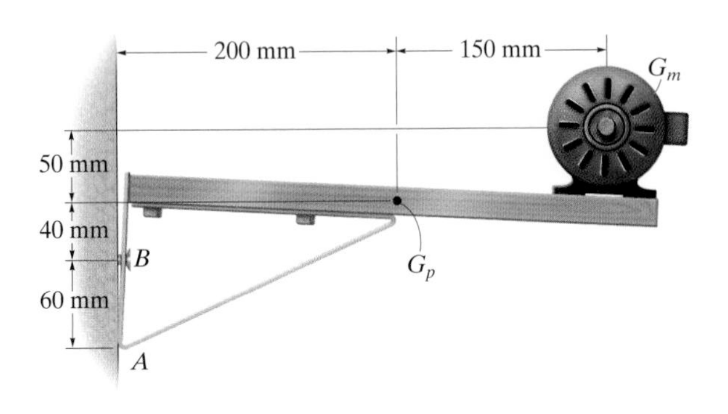

Prob. 5–41

5–42. A cantilever beam, having an extended length of 3 m, is subjected to a vertical force of 500 N. Assuming that the wall resists this load with linearly varying distributed loads over the 0.15-m length of the beam portion inside the wall, determine the intensities w_1 and w_2 for equilibrium.

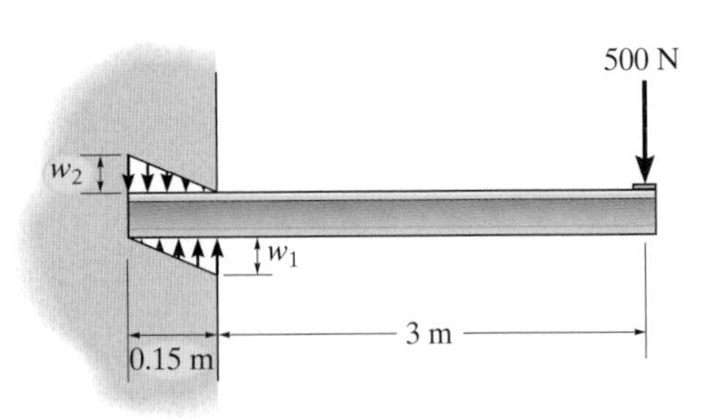

Prob. 5–42

5–43. The upper portion of the crane boom consists of the jib AB, which is supported by the pin at A, the guy line BC, and the backstay CD, each cable being separately attached to the mast at C. If the 5-kN load is supported by the hoist line, which passes over the pulley at B, determine the magnitude of the resultant force the pin exerts on the jib at A for equilibrium, the tension in the guy line BC, and the tension T in the hoist line. Neglect the weight of the jib. The pulley at B has a radius of 0.1 m.

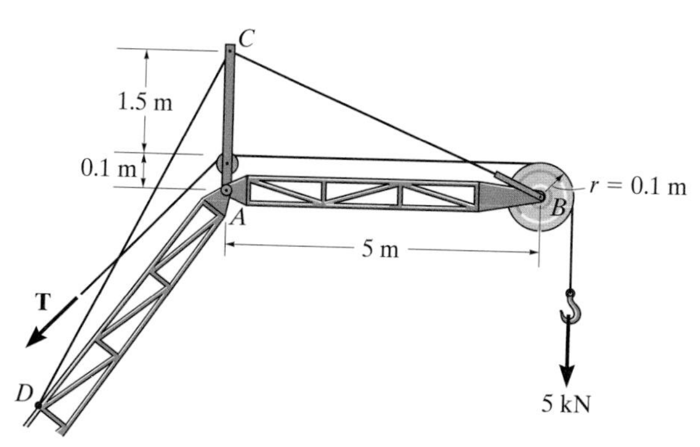

Prob. 5–43

***5–44.** The mobile crane has a weight of 120,000 lb and center of gravity at G_1; the boom has a weight of 30,000 lb and center of gravity at G_2. Determine the smallest angle of tilt θ of the boom, without causing the crane to overturn if the suspended load is $W = 40{,}000$ lb. Neglect the thickness of the tracks at A and B.

5–45. The mobile crane has a weight of 120,000 lb and center of gravity at G_1; the boom has a weight of 30,000 lb and center of gravity at G_2. If the suspended load has a weight of $W = 16{,}000$ lb, determine the normal reactions at the tracks A and B. For the calculation, neglect the thickness of the tracks and take $\theta = 30°$.

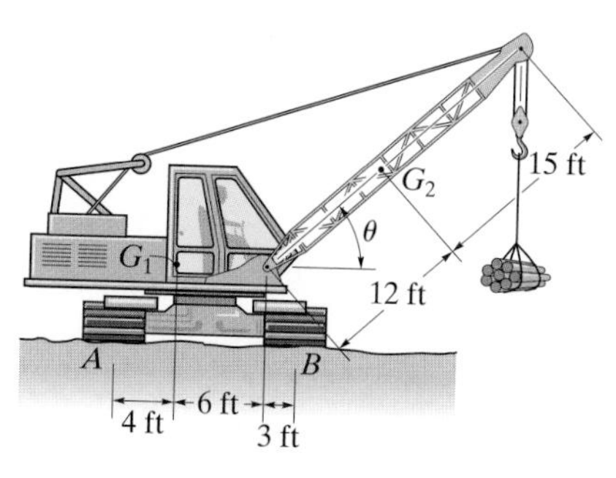

Probs. 5–44/45

5–46. The man attempts to support the load of boards having a weight W and a center of gravity at G. If he is standing on a smooth floor, determine the smallest angle θ at which he can hold them up in the position shown. Neglect his weight.

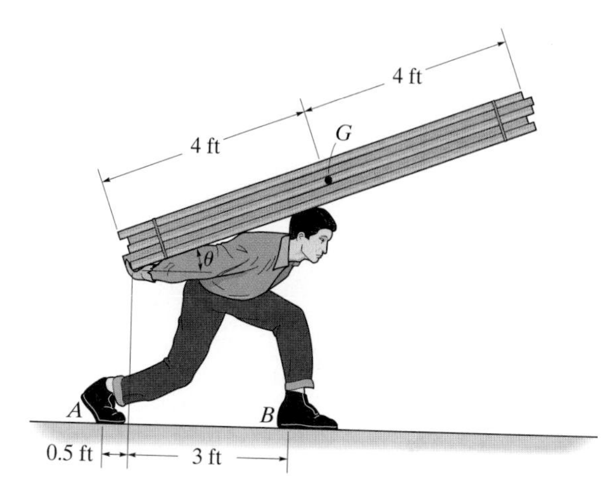

Prob. 5–46

5–47. The motor has a weight of 850 lb. Determine the force that each of the chains exerts on the supporting hooks at A, B, and C. Neglect the size of the hooks and the thickness of the beam.

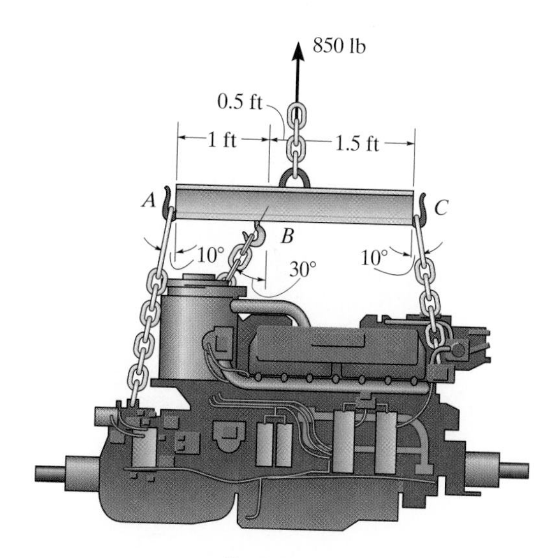

Prob. 5–47

***5–48.** The boom supports the two vertical loads. Neglect the size of the collars at D and B and the thickness of the boom, and compute the horizontal and vertical components of force at the pin A and the force in cable CB. Set $F_1 = 800$ N and $F_2 = 350$ N.

5–49. The boom is intended to support two vertical loads, $\mathbf{F}_1$ and $\mathbf{F}_2$. If the cable CB can sustain a maximum load of 1500 N before it fails, determine the critical loads if $F_1 = 2F_2$. Also, what is the magnitude of the maximum reaction at pin A?

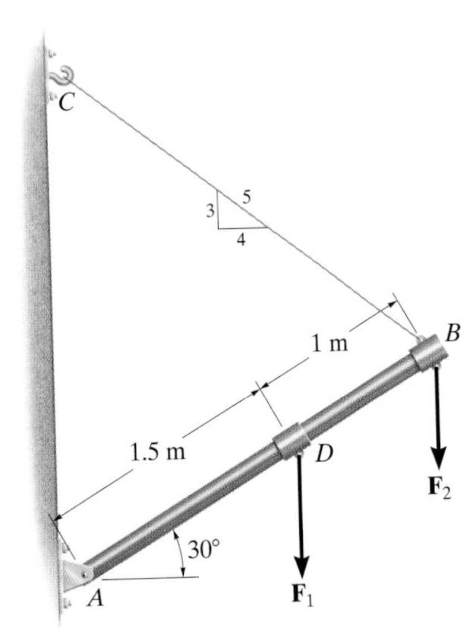

Probs. 5–48/49

5–50. The uniform rod of length L and weight W is supported on the smooth planes. Determine its position θ for equilibrium. Neglect the thickness of the rod.

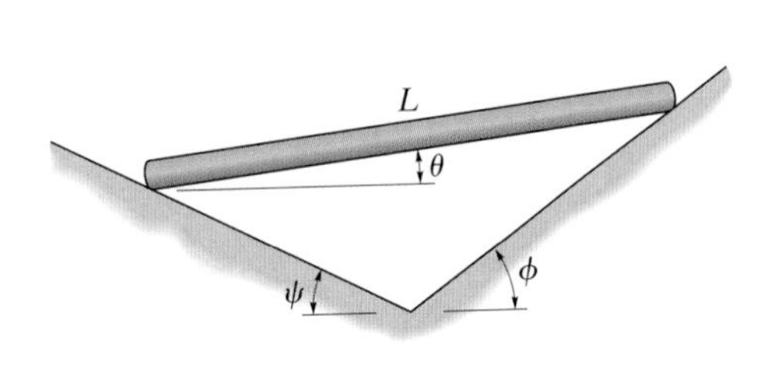

Prob. 5–50

5–51. The toggle switch consists of a cocking lever that is pinned to a fixed frame at A and held in place by the spring which has an unstretched length of 200 mm. Determine the magnitude of the resultant force at A and the normal force on the peg at B when the lever is in the position shown.

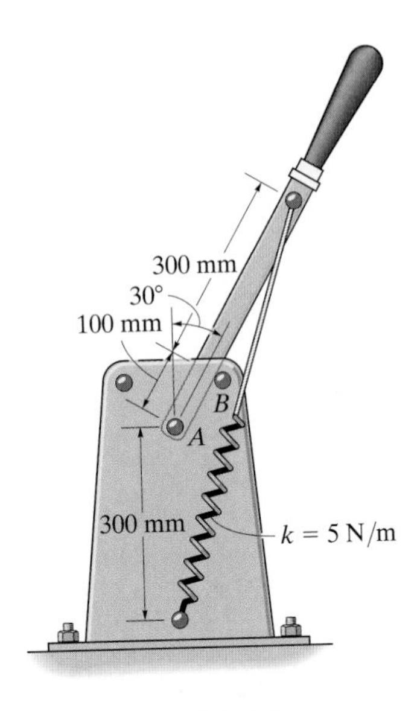

Prob. 5–51

***5–52.** The rigid beam of negligible weight is supported horizontally by two springs and a pin. If the springs are uncompressed when the load is removed, determine the force in each spring when the load $\mathbf{P}$ is applied. Also, compute the vertical deflection of end C. Assume the spring stiffness k is large enough so that only small deflections occur. *Hint:* The beam rotates about A so the deflections in the springs can be related.

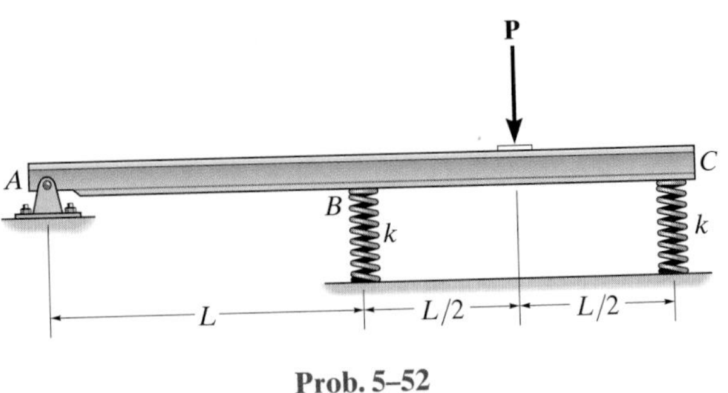

Prob. 5–52

5–53. The rod supports a weight of 200 lb and is pinned at its end A. If it is also subjected to a couple moment of 100 lb · ft, determine the angle θ for equilibrium. The spring has an unstretched length of 2 ft and a stiffness of $k = 50$ lb/ft.

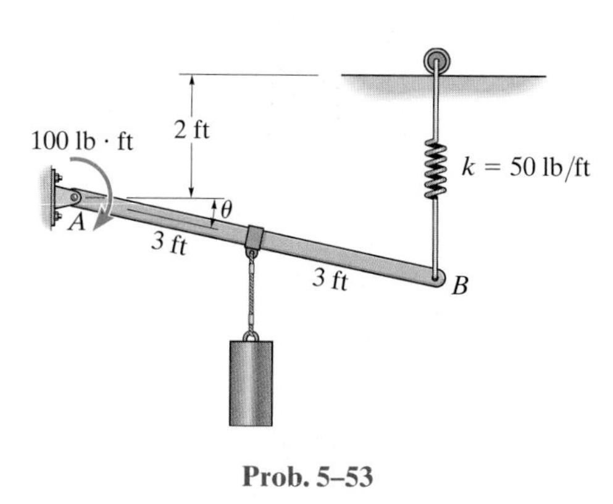

Prob. 5–53

5–54. The smooth pipe rests against the wall at the points of contact A, B, and C. Determine the reactions at these points needed to support the vertical force of 45 lb. Neglect the pipe's thickness in the calculation.

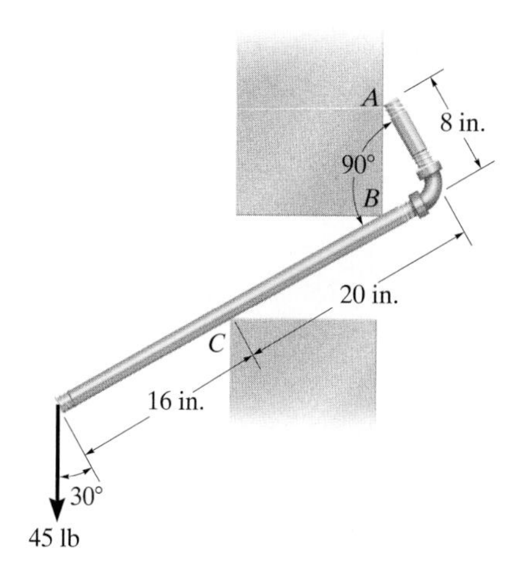

Prob. 5–54

5–55. The rigid metal strip of negligible weight is used as part of an electromagnetic switch. If the stiffness of the springs at A and B is $k = 5$ N/m, and the strip is originally horizontal when the springs are unstretched, determine the smallest force needed to close the contact gap at C.

***5–56.** The rigid metal strip of negligible weight is used as part of an electromagnetic switch. Determine the maximum stiffness k of the springs at A and B so that the contact at C closes when the vertical force developed there is 0.5 N. Originally the strip is horizontal as shown.

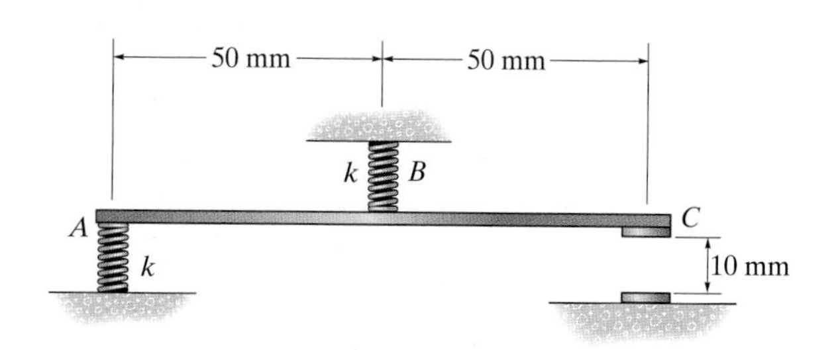

Probs. 5–55/56

5–57. Determine the distance d for placement of the load **P** for equilibrium of the smooth bar in the position θ as shown. Neglect the weight of the bar.

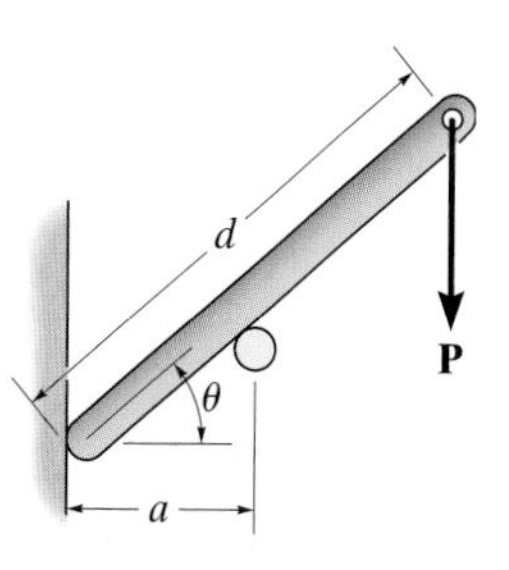

Prob. 5–57

5–58. The wheelbarrow and its contents have a mass m and center of mass at G. Determine the greatest angle of tilt θ without causing the wheelbarrow to tip over.

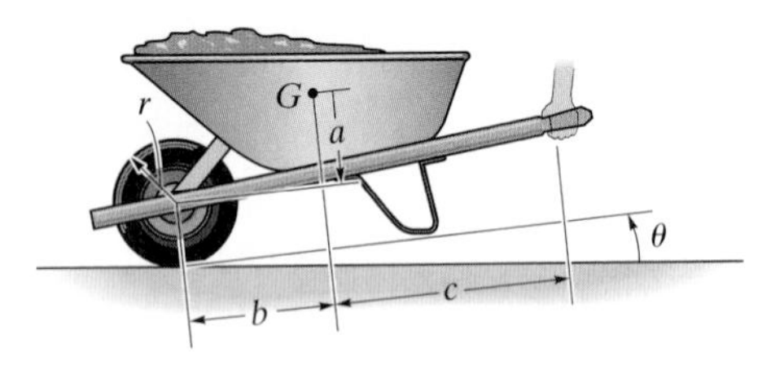

Prob. 5–58

5–59. Determine the force P needed to pull the 50-kg roller over the smooth step. Take $\theta = 60°$.

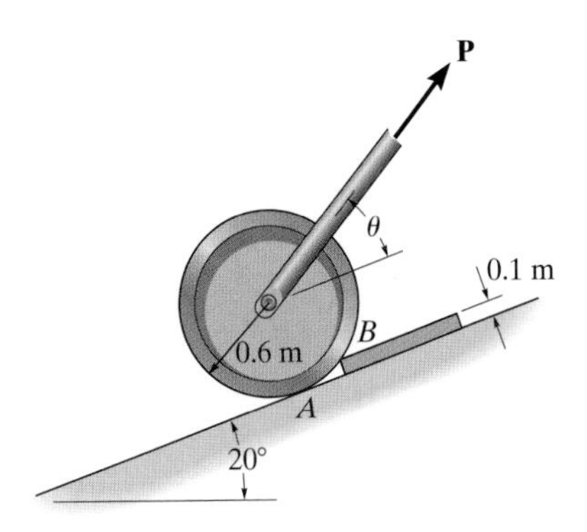

Prob. 5–59

***5–60.** Determine the magnitude and direction θ of the minimum force P needed to pull the 50-kg roller over the smooth step.

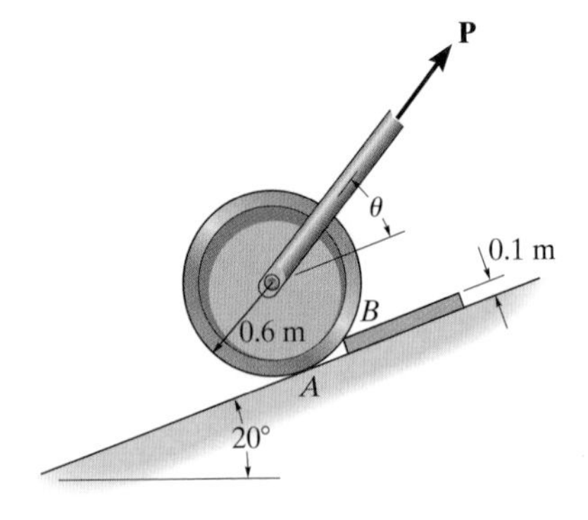

Prob. 5–60

5–61. A uniform glass rod having a length L is placed in the smooth hemispherical bowl having a radius r. Determine the angle of inclination θ for equilibrium.

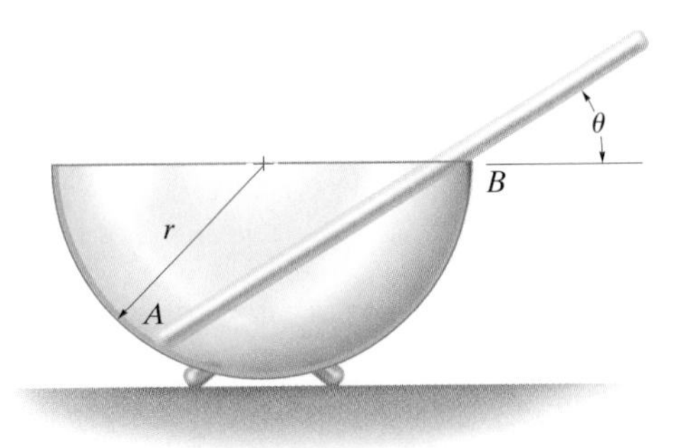

Prob. 5–61

5–62. The disk B has a mass of 20 kg and is supported on the smooth cylindrical surface by a spring having a stiffness of $k = 400$ N/m and unstretched length of $l_0 = 1$ m. The spring remains in the horizontal position since its end A is attached to the small roller guide which has negligible weight. Determine the angle θ to the nearest degree for equilibrium of the roller.

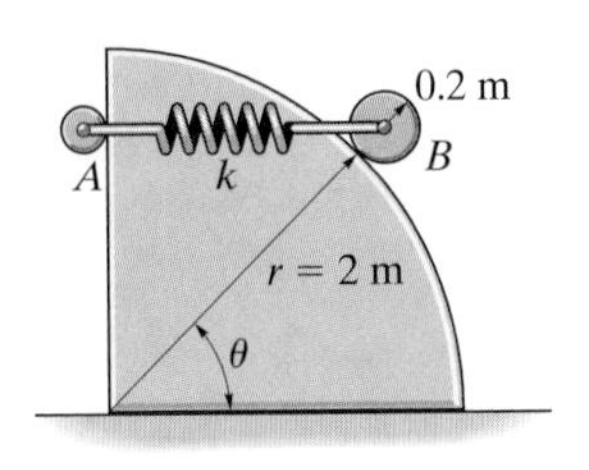

Prob. 5–62

EQUILIBRIUM IN THREE DIMENSIONS

5.5 Free-Body Diagrams

The first step in solving three-dimensional equilibrium problems, as in the case of two dimensions, is to draw a free-body diagram of the body (or group of bodies considered as a system). Before we show this, however, it is necessary to discuss the types of reactions that can occur at the supports.

Support Reactions. The reactive forces and couple moments acting at various types of supports and connections, when the members are viewed in three dimensions, are listed in Table 5–2. It is important to recognize the symbols used to represent each of these supports and to understand clearly how the forces and couple moments are developed by each support. As in the two-dimensional case, *a force is developed by a support that restricts the translation of the attached member, whereas a couple moment is developed when rotation of the attached member is prevented.* For example, in Table 5–2, item (4), the ball-and-socket joint prevents any translation of the connecting member; therefore, a force must act on the member at the point of connection. This force has three components having unknown magnitudes, F_x, F_y, F_z. Provided these components are known, one can obtain the magnitude of force. $F = \sqrt{F_x^2 + F_y^2 + F_z^2}$, and the force's orientation defined by the coordinate direction angles α, β, γ, Eqs. 2–7.* Since the connecting member is allowed to rotate freely about *any* axis, no couple moment is resisted by a ball-and-socket joint.

It should be noted that the *single* bearing supports in items (5) and (7), the *single* pin (8), and the *single* hinge (9) are shown to support both force and couple-moment components. If, however, these supports are used in conjunction with *other* bearings, pins, or hinges to hold a rigid body in equilibrium and the supports are *properly aligned* when connected to the body, then the *force reactions* at these supports *alone* may be adequate for supporting the body. In other words, the couple moments become redundant and are not shown on the free-body diagram. The reason for this should become clear after studying the examples which follow.

*The three unknowns may also be represented as an unknown force magnitude F and two unknown coordinate direction angles. The third direction angle is obtained using the identity $\cos^2\alpha + \cos^2\beta + \cos^2\gamma = 1$, Eq. 2–10.

TABLE 5–2 Supports for Rigid Bodies Subjected to Three-Dimensional Force Systems

Types of Connection	Reaction	Number of Unknowns
(1) cable	$\mathbf{F}$	One unknown. The reaction is a force which acts away from the member in the known direction of the cable.
(2) smooth surface support	$\mathbf{F}$	One unknown. The reaction is a force which acts perpendicular to the surface at the point of contact.
(3) roller	$\mathbf{F}$	One unknown. The reaction is a force which acts perpendicular to the surface at the point of contact.
(4) ball and socket	$\mathbf{F}_z$, $\mathbf{F}_x$, $\mathbf{F}_y$	Three unknowns. The reactions are three rectangular force components.
(5) single journal bearing	$\mathbf{M}_z$, $\mathbf{F}_z$, $\mathbf{M}_x$, $\mathbf{F}_x$	Four unknowns. The reactions are two force and two couple-moment components which act perpendicular to the shaft.

continued

TABLE 5–2 Continued

Types of Connection	Reaction	Number of Unknowns
(6)  single journal bearing with square shaft	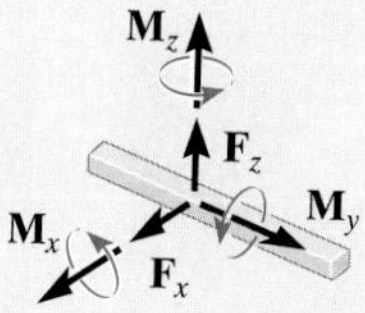 $\mathbf{M}_z$, $\mathbf{F}_z$, $\mathbf{M}_y$, $\mathbf{M}_x$, $\mathbf{F}_x$	Five unknowns. The reactions are two force and three couple-moment components.
(7)  single thrust bearing	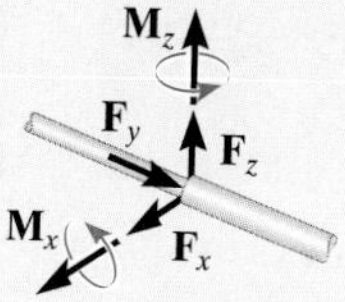 $\mathbf{M}_z$, $\mathbf{F}_y$, $\mathbf{F}_z$, $\mathbf{M}_x$, $\mathbf{F}_x$	Five unknowns. The reactions are three force and two couple-moment components.
(8) 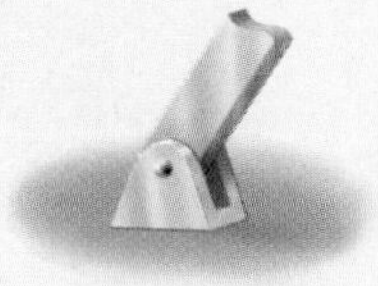 single smooth pin	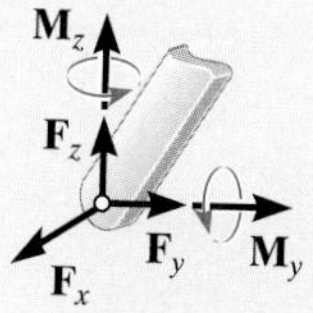 $\mathbf{M}_z$, $\mathbf{F}_z$, $\mathbf{F}_y$, $\mathbf{M}_y$, $\mathbf{F}_x$	Five unknowns. The reactions are three force and two couple-moment components.
(9)  single hinge	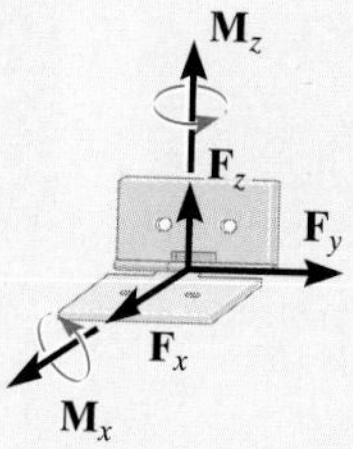 $\mathbf{M}_z$, $\mathbf{F}_z$, $\mathbf{F}_y$, $\mathbf{F}_x$, $\mathbf{M}_x$	Five unknowns. The reactions are three force and two couple-moment components.
(10)  fixed support	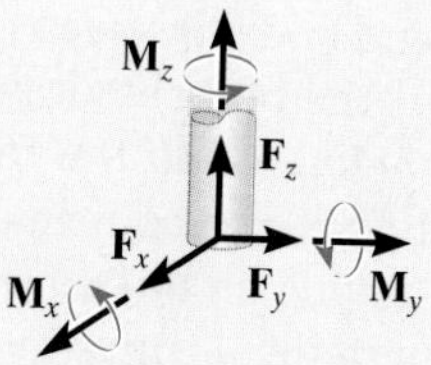 $\mathbf{M}_z$, $\mathbf{F}_z$, $\mathbf{F}_x$, $\mathbf{M}_x$, $\mathbf{F}_y$, $\mathbf{M}_y$	Six unknowns. The reactions are three force and three couple-moment components.

Typical examples of actual supports that are referenced to Table 5–2 are shown in the following sequence of photos.

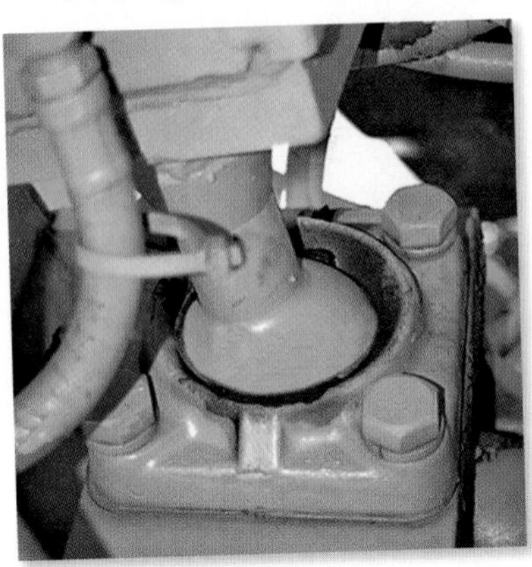

This ball-and-socket joint provides a connection for the housing of an earth grader to its frame. (4)

This journal bearing supports the end of the shaft. (5)

This thrust bearing is used to support the drive shaft on a machine. (7)

This pin is used to support the end of the strut used on a tractor. (8)

Free-Body Diagrams. The general procedure for establishing the free-body diagram of a rigid body has been outlined in Sec. 5.2. Essentially it requires first "isolating" the body by drawing its outlined shape. This is followed by a careful *labeling* of *all* the forces and couple moments in reference to an established x, y, z coordinate system. As a general rule, *components of reaction* having an *unknown magnitude* are shown acting on the free-body diagram in the *positive sense*. In this way, if any negative values are obtained, they will indicate that the components act in the negative coordinate directions.

(a)

(b)

It is a mistake to support a door using a single hinge (a) since the hinge must develop a force $\mathbf{C}_y$ to support the weight $\mathbf{W}$ of the door and a couple moment $\mathbf{M}$ to support the moment of $\mathbf{W}$, i.e., $M = Wd$. If instead two properly aligned hinges are used, (b) then the weight is carried by both hinges, $A_y + B_y = W$, and the moment of the door is resisted by the two hinge forces $\mathbf{F}_x$ and $-\mathbf{F}_x$. These forces form a couple, such that $F_x d' = Wd$. In other words, no couple moments are produced by the hinges on the door provided they are in *proper alignment*. Instead, the forces $\mathbf{F}_x$ and $-\mathbf{F}_x$ resist the rotation caused by $\mathbf{W}$.

EXAMPLE 5.14

Several examples of objects along with their associated free-body diagrams are shown in Fig. 5–25. In all cases, the *x, y, z* axes are established and the unknown reaction components are indicated in the positive sense. The weight of the objects is neglected.

SOLUTION

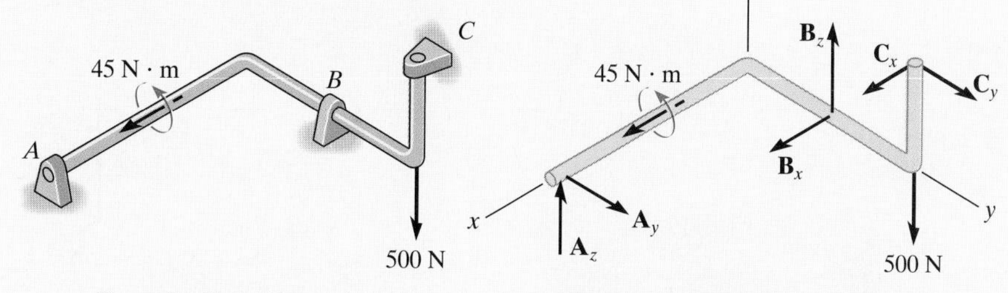

Properly aligned journal bearings at *A*, *B*, *C*.

The force reactions developed by the bearings are sufficient for equilibrium since they prevent the shaft from rotating about each of the coordinate axes.

(a)

Fig. 5–25

continued

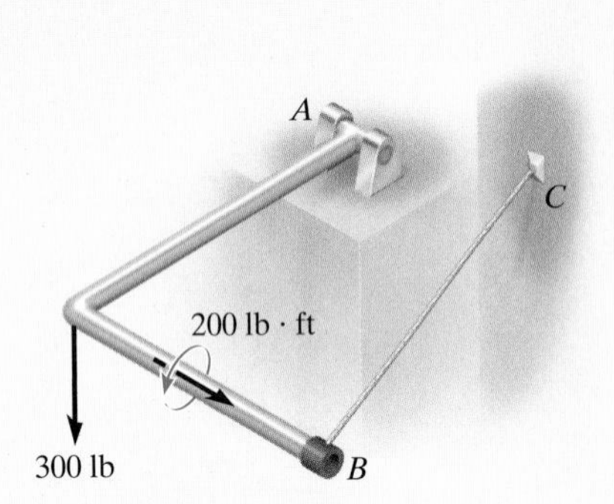

Pin at A and cable BC.

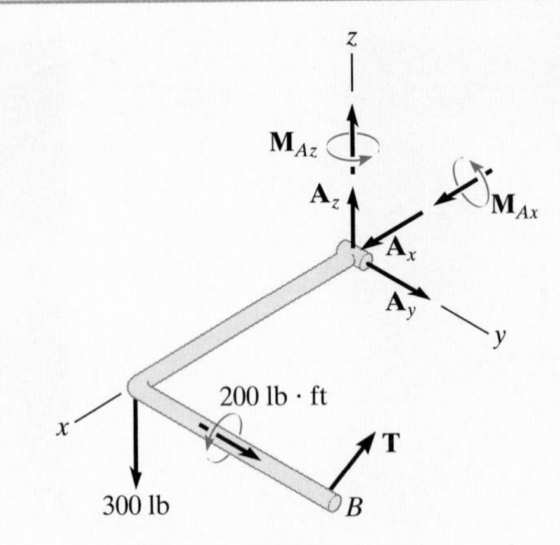

Moment components are developed by the pin on the rod to prevent rotation about the x and z axes.

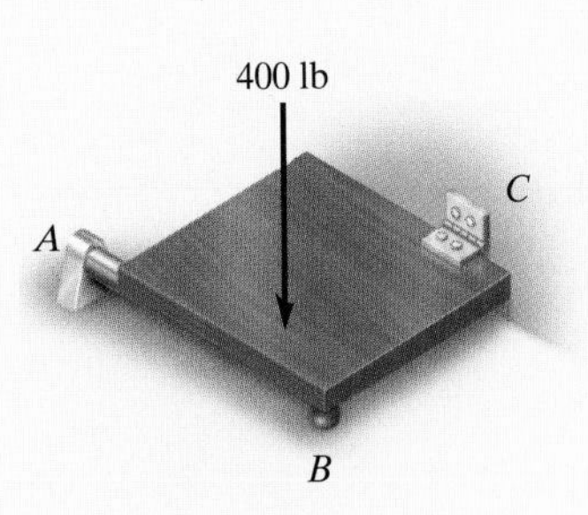

Properly aligned journal bearing at A and hinge at C. Roller at B.

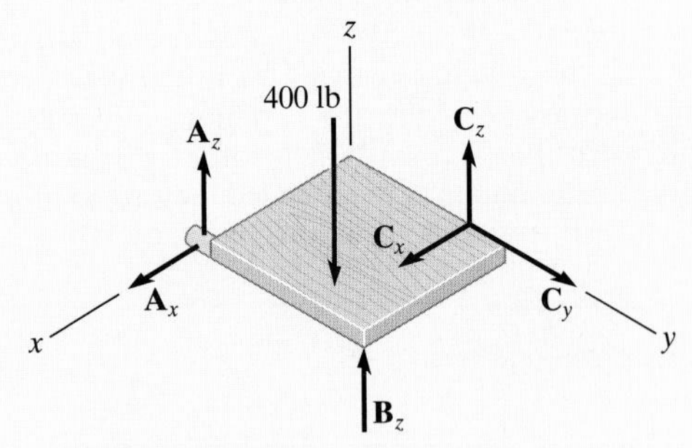

Only force reactions are developed by the bearing and hinge on the plate to prevent rotation about each coordinate axis. No moments at the hinge are developed.

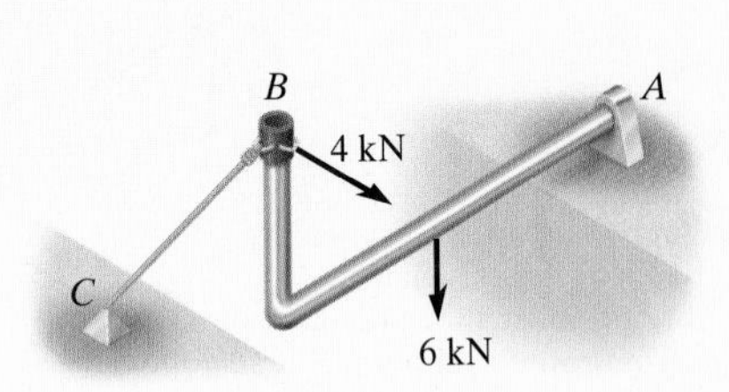

Thrust bearing at A and cable BC

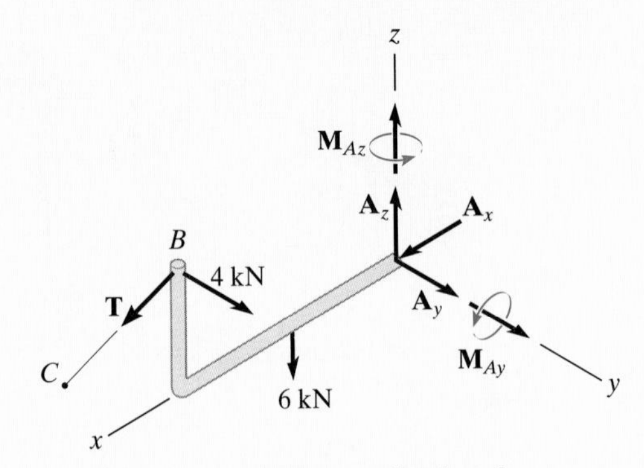

Moment components are developed by the bearing on the rod in order to prevent rotation about the y and z axes.

Fig. 5–25 (cont.)

5.6 Equations of Equilibrium

As stated in Sec. 5.1, the conditions for equilibrium of a rigid body subjected to a three-dimensional force system require that both the *resultant* force and *resultant* couple moment acting on the body be equal to *zero*.

Vector Equations of Equilibrium. The two conditions for equilibrium of a rigid body may be expressed mathematically in vector form as

$$\Sigma \mathbf{F} = \mathbf{0}$$
$$\Sigma \mathbf{M}_o = \mathbf{0} \quad (5\text{–}5)$$

where $\Sigma \mathbf{F}$ is the vector sum of all the external forces acting on the body and $\Sigma \mathbf{M}_O$ is the sum of the couple moments and the moments of all the forces about any point O located either on or off the body.

Scalar Equations of Equilibrium. If all the applied external forces and couple moments are expressed in Cartesian vector form and substituted into Eqs. 5–5, we have

$$\Sigma \mathbf{F} = \Sigma F_x \mathbf{i} + \Sigma F_y \mathbf{j} + \Sigma F_z \mathbf{k} = \mathbf{0}$$
$$\Sigma \mathbf{M}_O = \Sigma M_x \mathbf{i} + \Sigma M_y \mathbf{j} + \Sigma M_z \mathbf{k} = \mathbf{0}$$

Since the **i**, **j**, and **k** components are independent from one another, the above equations are satisfied provided

$$\Sigma F_x = 0$$
$$\Sigma F_y = 0 \quad (5\text{–}6a)$$
$$\Sigma F_z = 0$$

and

$$\Sigma M_x = 0$$
$$\Sigma M_y = 0 \quad (5\text{–}6b)$$
$$\Sigma M_z = 0$$

These *six scalar equilibrium equations* may be used to solve for at most six unknowns shown on the free-body diagram. Equations 5–6*a* express the fact that the sum of the external force components acting in the x, y, and z directions must be zero, and Eqs. 5–6*b* require the sum of the moment components about the x, y, and z axes to be zero.

5.7 Constraints for a Rigid Body

To ensure the equilibrium of a rigid body, it is not only necessary to satisfy the equations of equilibrium, but the body must also be properly held or constrained by its supports. Some bodies may have more supports than are necessary for equilibrium, whereas others may not have enough or the supports may be arranged in a particular manner that could cause the body to collapse. Each of these cases will now be discussed.

Redundant Constraints. When a body has redundant supports, that is, more supports than are necessary to hold it in equilibrium, it becomes statically indeterminate. *Statically indeterminate* means that there will be more unknown loadings on the body than equations of equilibrium available for their solution. For example, the two-dimensional problem, Fig. 5–26*a*, and the three-dimensional problem, Fig. 5–26*b*, shown together with their free-body diagrams, are both statically indeterminate because of additional support reactions. In the two-dimensional case, there are five unknowns, that is, M_A, A_x, A_y, B_y, and C_y, for which only three equilibrium equations can be written ($\Sigma F_x = 0$, $\Sigma F_y = 0$, and $\Sigma M_O = 0$, Eqs. 5–2). The three-dimensional problem has eight unknowns, for which only six equilibrium equations can be written, Eqs. 5–6. The additional equations needed to solve indeterminate problems of the type shown in Fig. 5–26 are generally obtained from the deformation conditions at the points of support. These equations involve the physical properties of the body which are studied in subjects dealing with the mechanics of deformation, such as "mechanics of materials."*

500 N
2 kN · m
A
B
C
y
x
$\mathbf{A}_y$
500 N
2 kN · m
$\mathbf{A}_x$
$\mathbf{M}_A$
$\mathbf{B}_y$
$\mathbf{C}_y$

(a)

Fig. 5–26

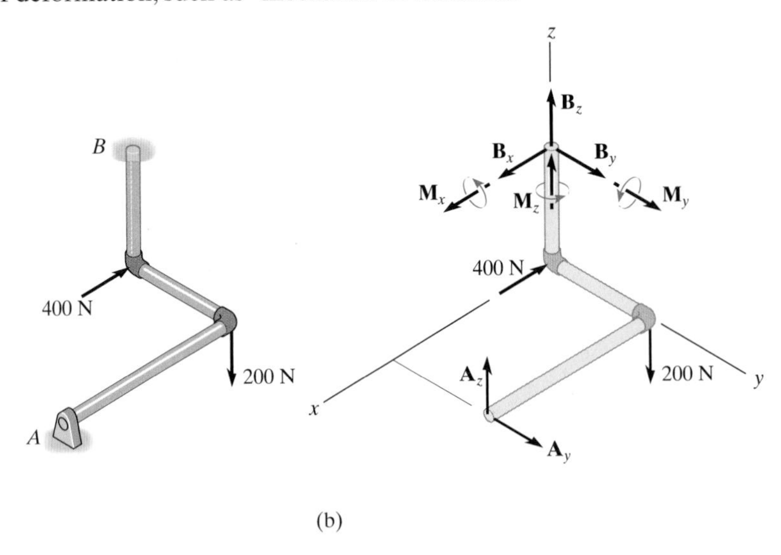

(b)

*See R. C. Hibbeler, *Mechanics of Materials*, 6th edition (Pearson Education/Prentice Hall, Inc., 2005).

Improper Constraints. In some cases, there may be as many unknown forces on the body as there are equations of equilibrium; however, *instability* of the body can develop because of *improper constraining* by the supports. In the case of three-dimensional problems, the body is improperly constrained if the support reactions *all intersect a common axis*. For two-dimensional problems, this axis is *perpendicular* to the plane of the forces and therefore appears as a point. Hence, when all the reactive forces are *concurrent* at this point, the body is improperly constrained. Examples of both cases are given in Fig. 5–27. From the free-body diagrams it is seen that the summation of moments about the x axis, Fig. 5–27a, or point O, Fig. 5–27b, will *not* be equal to zero; thus rotation about the x axis or point O will take place.* Furthermore, in both cases, it becomes *impossible* to solve *completely* for all the unknowns since one can write a moment equation that *does not* involve any of the unknown support reactions, and as a result, this reduces the number of available equilibrium equations by one.

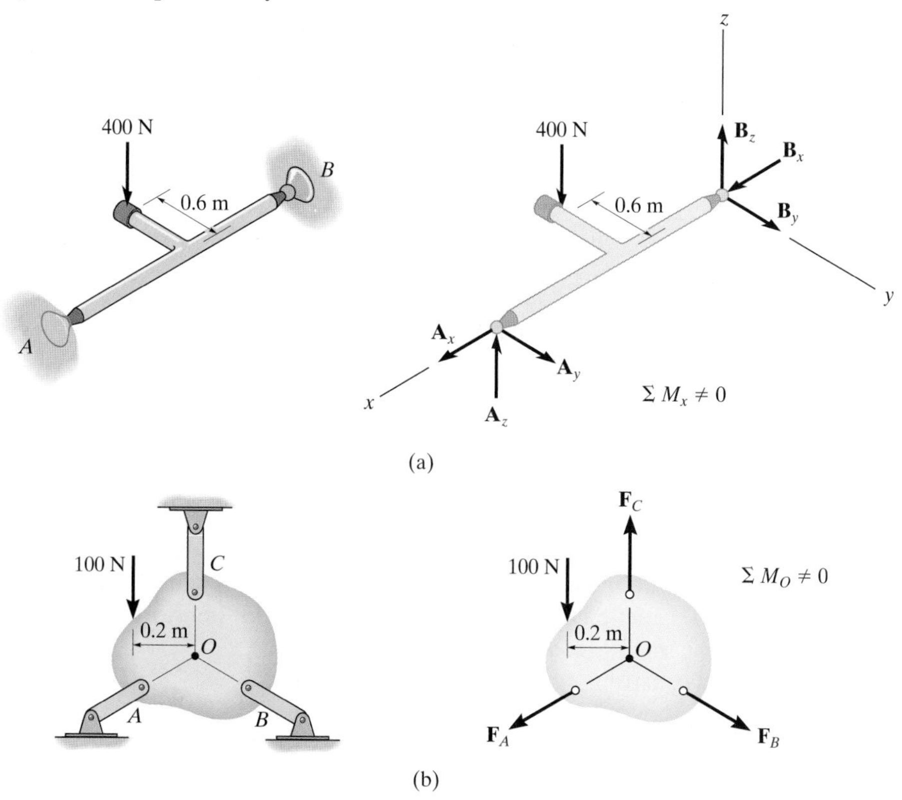

Fig. 5–27

*For the three-dimensional problem, $\Sigma M_x = (400\text{ N})(0.6\text{ m}) \neq 0$, and for the two-dimensional problem, $\Sigma M_O = (100\text{ N})(0.2\text{ m}) \neq 0$.

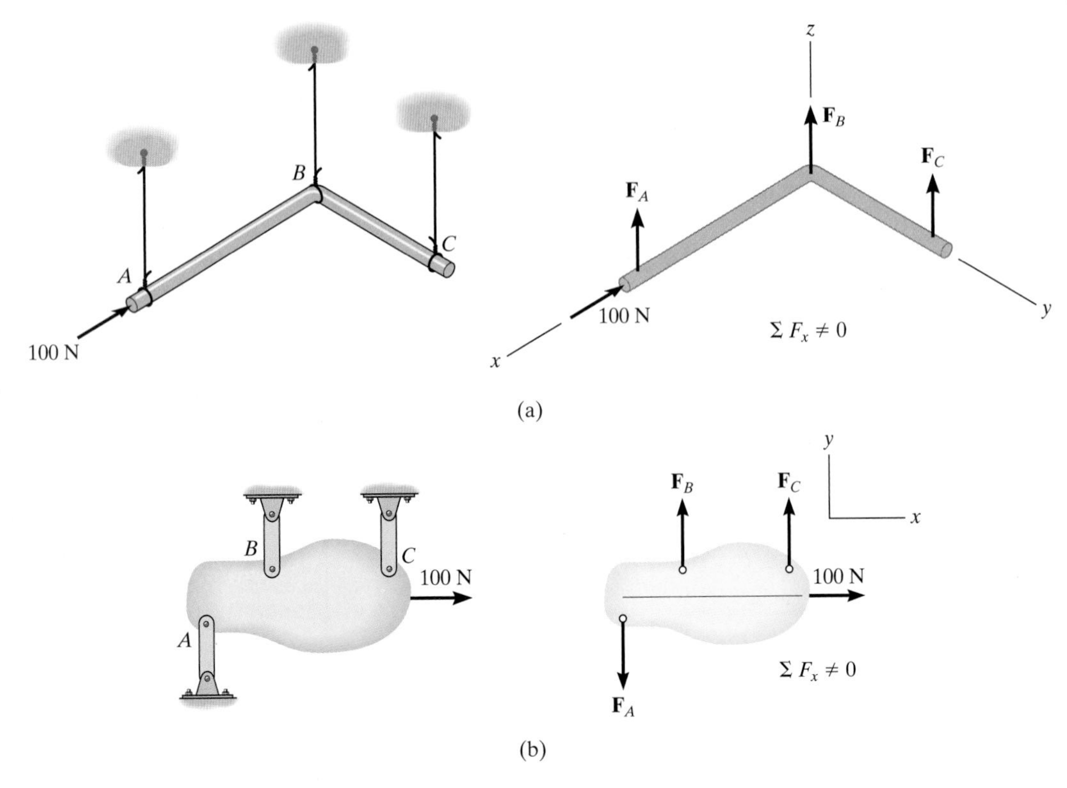

Fig. 5–28

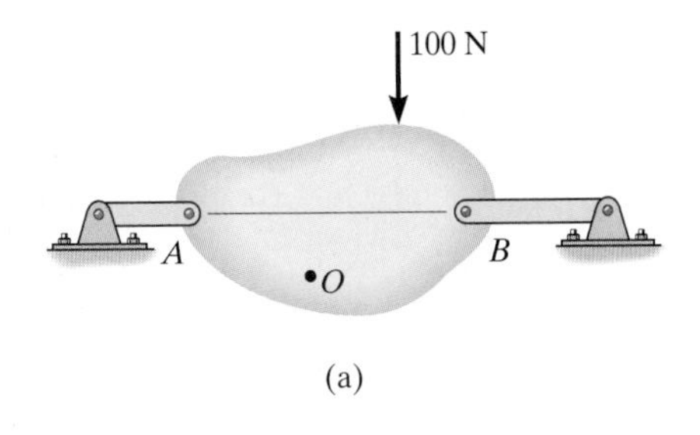

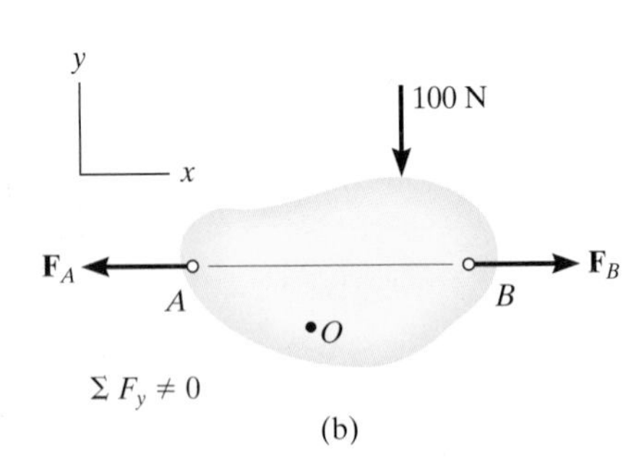

Fig. 5–29

Another way in which improper constraining leads to instability occurs when the *reactive forces* are all *parallel*. Three- and two-dimensional examples of this are shown in Fig. 5–28. In both cases, the summation of forces along the x axis will not equal zero.

In some cases, a body may have *fewer* reactive forces than equations of equilibrium that must be satisfied. The body then becomes only *partially constrained*. For example, consider the body shown in Fig. 5–29*a* with its corresponding free-body diagram in Fig. 5–29*b*. If O is a point not located on the line AB, the equation $\Sigma F_x = 0$ gives $F_A = F_B$, however, $\Sigma F_y = 0$ will not be satisfied for the loading conditions and therefore equilibrium will not be maintained.

Proper constraining therefore requires that:

1. The lines of action of the reactive forces do not intersect points on a common axis.
2. The reactive forces must not all be parallel to one another.

When the minimum number of reactive forces is needed to properly constrain the body in question, the problem will be statically determinate, and therefore the equations of equilibrium can be used to determine *all* the reactive forces.

Important Points

- Always draw the free-body diagram first.
- If a support *prevents translation* of a body in a specific direction, then it exerts a *force* on the body in that direction.
- If *rotation about an axis is prevented*, then the support exerts a *couple moment* on the body about the axis.
- If a body is subjected to more unknown reactions than available equations of equilibrium, then the problem is *statically indeterminate*.
- To avoid instability of a body require that the lines of action of the reactive forces do not intersect a common axis and are not parallel to one another.

Procedure for Analysis

Three-dimensional equilibrium problems for a rigid body can be solved using the following procedure.

Free-Body Diagram.

- Draw an outlined shape of the body.
- Show all the forces and couple moments acting on the body.
- Establish the origin of the x, y, z axes at a convenient point and orient the axes so that they are parallel to as many of the external forces and moments as possible.
- Label all the loadings and specify their directions relative to the x, y, z axes. In general, show all the unknown components having a positive sense along the x, y, z axes if the sense cannot be determined.
- Indicate the dimensions of the body necessary for computing the moments of forces.

Equations of Equilibrium.

- If the x, y, z force and moment components seem easy to determine, then apply the six scalar equations of equilibrium; otherwise use the vector equations.
- It is not necessary that the set of axes chosen for force summation coincide with the set of axes chosen for moment summation. Also, any set of nonorthogonal axes may be chosen for this purpose.
- Choose the direction of an axis for moment summation such that it intersects the lines of action of as many unknown forces as possible. The moments of forces passing through points on this axis and forces which are parallel to the axis will then be zero.
- If the solution of the equilibrium equations yields a negative scalar for a force or couple moment magnitude, it indicates that the sense is opposite to that assumed on the free-body diagram.

EXAMPLE 5.15

The homogeneous plate shown in Fig. 5–30*a* has a mass of 100 kg and is subjected to a force and couple moment along its edges. If it is supported in the horizontal plane by means of a roller at *A*, a ball-and-socket joint at *B*, and a cord at *C*, determine the components of reaction at the supports.

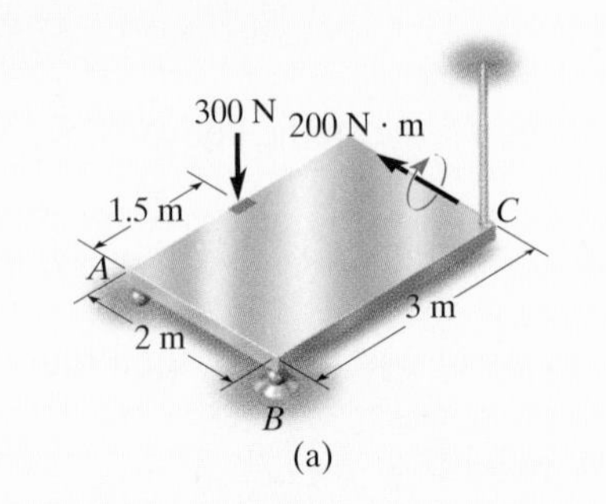

(a)

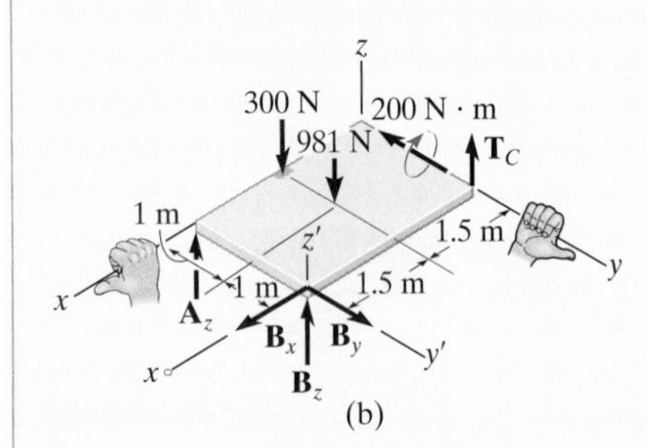

(b)

Fig. 5–30

SOLUTION (*SCALAR ANALYSIS*)

Free-Body Diagram. There are five unknown reactions acting on the plate, as shown in Fig. 5–30*b*. Each of these reactions is assumed to act in a positive coordinate direction.

Equations of Equilibrium. Since the three-dimensional geometry is rather simple, a *scalar analysis* provides a *direct solution* to this problem. A force summation along each axis yields

$$\Sigma F_x = 0; \qquad B_x = 0 \qquad \textit{Ans.}$$

$$\Sigma F_y = 0; \qquad B_y = 0 \qquad \textit{Ans.}$$

$$\Sigma F_z = 0; \qquad A_z + B_z + T_C - 300\text{ N} - 981\text{ N} = 0 \qquad (1)$$

Recall that the moment of a force about an axis is equal to the product of the force magnitude and the perpendicular distance (moment arm) from the line of action of the force to the axis. The sense of the moment is determined by the right-hand rule. Also, forces that are parallel to an axis or pass through it create no moment about the axis. Hence, summing moments of the forces on the free-body diagram, with positive moments acting along the positive *x* or *y* axis, we have

$$\Sigma M_x = 0; \qquad T_C(2\text{ m}) - 981\text{ N}(1\text{ m}) + B_z(2\text{ m}) = 0 \qquad (2)$$

$$\Sigma M_y = 0;$$

$$300\text{ N}(1.5\text{ m}) + 981\text{ N}(1.5\text{ m}) - B_z(3\text{ m}) - A_z(3\text{ m}) - 200\text{ N}\cdot\text{m} = 0 \qquad (3)$$

The components of force at *B* can be eliminated if the x', y', z' axes are used. We obtain

$$\Sigma M_{x'} = 0; \qquad 981\text{ N}(1\text{ m}) + 300\text{ N}(2\text{ m}) - A_z(2\text{ m}) = 0 \qquad (4)$$

$$\Sigma M_{y'} = 0;$$

$$-300\text{ N}(1.5\text{ m}) - 981\text{ N}(1.5\text{ m}) - 200\text{ N}\cdot\text{m} + T_C(3\text{ m}) = 0 \qquad (5)$$

Solving Eqs. 1 through 3 or the more convenient Eqs. 1, 4, and 5 yields

$$A_z = 790\text{ N} \qquad B_z = -217\text{ N} \qquad T_C = 707\text{ N} \qquad \textit{Ans.}$$

The negative sign indicates that $\mathbf{B}_z$ acts downward.

NOTE: The solution of this problem does not require the use of a summation of moments about the *z* axis. The plate is partially constrained since the supports cannot prevent it from turning about the *z* axis if a force is applied to it in the *x*–*y* plane.

EXAMPLE 5.16

The windlass shown in Fig. 5–31*a* is supported by a thrust bearing at A and a smooth journal bearing at B, which are properly aligned on the shaft. Determine the magnitude of the vertical force **P** that must be applied to the handle to maintain equilibrium of the 100-kg bucket. Also calculate the reactions at the bearings.

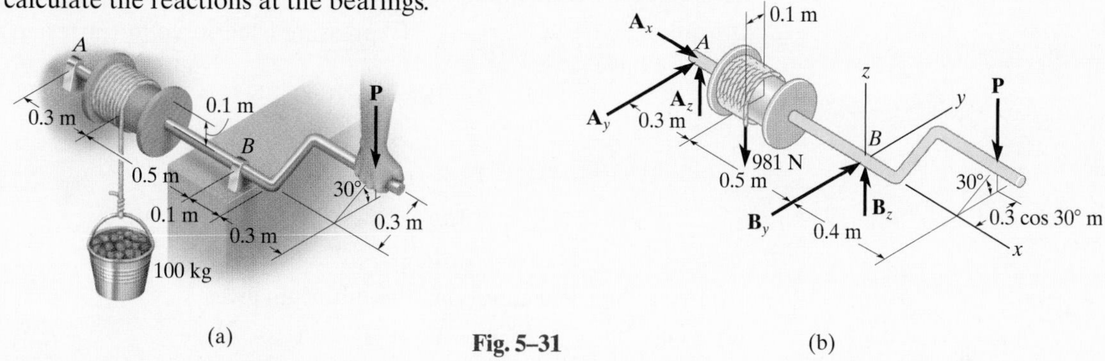

Fig. 5–31

SOLUTION *(SCALAR ANALYSIS)*

Free-Body Diagram. Since the bearings at A and B are aligned correctly, *only* force reactions occur at these supports, Fig. 5–31*b*. Why are there no moment reactions?

Equations of Equilibrium. Summing moments about the x axis yields a direct solution for **P**. Why? For a scalar moment summation, it is necessary to determine the moment of each force as the product of the force magnitude and the *perpendicular distance* from the x axis to the line of action of the force. Using the right-hand rule and assuming positive moments act in the $+\mathbf{i}$ direction, we have

$$\Sigma M_x = 0; \quad 981\text{ N}(0.1\text{ m}) - P(0.3\cos 30^\circ\text{ m}) = 0$$

$$P = 377.6\text{ N} \qquad \textit{Ans.}$$

Using this result for P and summing moments about the y and z axes yields

$$\Sigma M_y = 0; \quad -981\text{ N}(0.5\text{ m}) + A_z(0.8\text{ m}) + (377.6\text{ N})(0.4\text{ m}) = 0$$

$$A_z = 424.3\text{ N} \qquad \textit{Ans.}$$

$$\Sigma M_z = 0; \quad -A_y(0.8\text{ m}) = 0 \qquad A_y = 0$$

The reactions at B are determined by a force summation using these results.

$$\Sigma F_x = 0; \quad A_x = 0$$

$$\Sigma F_y = 0; \quad 0 + B_y = 0 \qquad B_y = 0$$

$$\Sigma F_z = 0; \quad 424.3 - 981 + B_z - 377.6 = 0 \qquad B_z = 934\text{ N} \qquad \textit{Ans.}$$

NOTE: As expected, the force on the handle is much smaller than the weight of the bucket.

EXAMPLE 5.17

Determine the tension in cables BC and BD and the reactions at the ball-and-socket joint A for the mast shown in Fig. 5–32a.

(a)

SOLUTION (VECTOR ANALYSIS)

Free-Body Diagram. There are five unknown force magnitudes shown on the free-body diagram, Fig. 5–32b.

Equations of Equilibrium. Expressing each force in Cartesian vector form, we have

$$\mathbf{F} = \{-1000\mathbf{j}\}\text{ N}$$

$$\mathbf{F}_A = A_x\mathbf{i} + A_y\mathbf{j} + A_z\mathbf{k}$$

$$\mathbf{T}_C = 0.707T_C\mathbf{i} - 0.707T_C\mathbf{k}$$

$$\mathbf{T}_D = T_D\left(\frac{\mathbf{r}_{BD}}{r_{BD}}\right) = -\frac{3}{9}T_D\mathbf{i} + \frac{6}{9}T_D\mathbf{j} - \frac{6}{9}T_D\mathbf{k}$$

Applying the force equation of equilibrium gives

$$\Sigma\mathbf{F} = \mathbf{0}; \qquad \mathbf{F} + \mathbf{F}_A + \mathbf{T}_C + \mathbf{T}_D = \mathbf{0}$$

$$(A_x + 0.707T_C - \frac{3}{9}T_D)\mathbf{i} + (-1000 + A_y + \frac{6}{9}T_D)\mathbf{j} + (A_z - 0.707T_C - \frac{6}{9}T_D)\mathbf{k} = \mathbf{0}$$

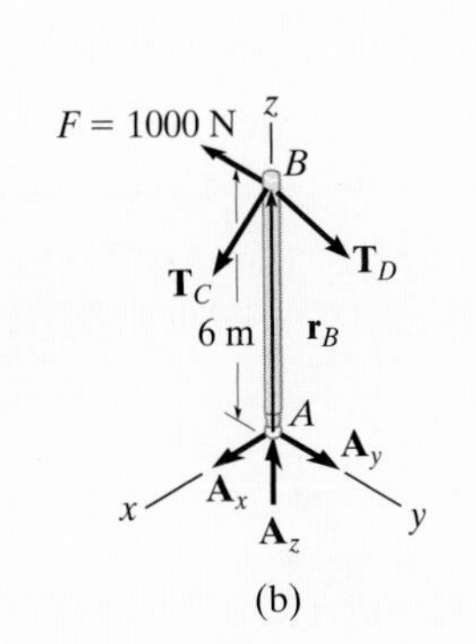

(b)

$$\Sigma F_x = 0; \qquad A_x + 0.707T_C - \frac{3}{9}T_D = 0 \qquad (1)$$

$$\Sigma F_y = 0; \qquad A_y + \frac{6}{9}T_D - 1000 = 0 \qquad (2)$$

$$\Sigma F_z = 0; \qquad A_z - 0.707T_C - \frac{6}{9}T_D = 0 \qquad (3)$$

Summing moments about point A, we have

$$\Sigma\mathbf{M}_A = \mathbf{0}; \qquad \mathbf{r}_B \times (\mathbf{F} + \mathbf{T}_C + \mathbf{T}_D) = \mathbf{0}$$

$$6\mathbf{k} \times (-1000\mathbf{j} + 0.707T_C\mathbf{i} - 0.707T_C\mathbf{k} - \frac{3}{9}T_D\mathbf{i} + \frac{6}{9}T_D\mathbf{j} - \frac{6}{9}T_D\mathbf{k}) = \mathbf{0}$$

Evaluating the cross product and combining terms yields

$$(-4T_D + 6000)\mathbf{i} + (4.24T_C - 2T_D)\mathbf{j} = \mathbf{0}$$

$$\Sigma M_x = 0; \qquad -4T_D + 6000 = 0 \qquad (4)$$

$$\Sigma M_y = 0; \qquad 4.24T_C - 2T_D = 0 \qquad (5)$$

The moment equation about the z axis, $\Sigma M_z = 0$, is automatically satisfied. Why? Solving Eqs. 1 through 5 we have

$$T_C = 707\text{ N} \qquad T_D = 1500\text{ N} \qquad \textit{Ans.}$$

$$A_x = 0\text{ N} \qquad A_y = 0\text{ N} \qquad A_z = 1500\text{ N} \qquad \textit{Ans.}$$

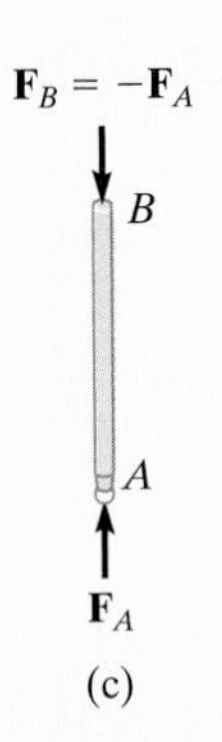

(c)

Fig. 5–32

NOTE: Since the mast is a two-force member, Fig. 5–32c, the value $A_x = A_y = 0$ could have been determined *by inspection*.

EXAMPLE 5.18

Rod AB shown in Fig. 5–33a is subjected to the 200-N force. Determine the reactions at the ball-and-socket joint A and the tension in cables BD and BE.

SOLUTION (VECTOR ANALYSIS)

Free-Body Diagram. Fig. 5–33b.

Equations of Equilibrium. Representing each force on the free-body diagram in Cartesian vector form, we have

$$\mathbf{F}_A = A_x\mathbf{i} + A_y\mathbf{j} + A_z\mathbf{k}$$
$$\mathbf{T}_E = T_E\mathbf{i}$$
$$\mathbf{T}_D = T_D\mathbf{j}$$
$$\mathbf{F} = \{-200\mathbf{k}\}\text{ N}$$

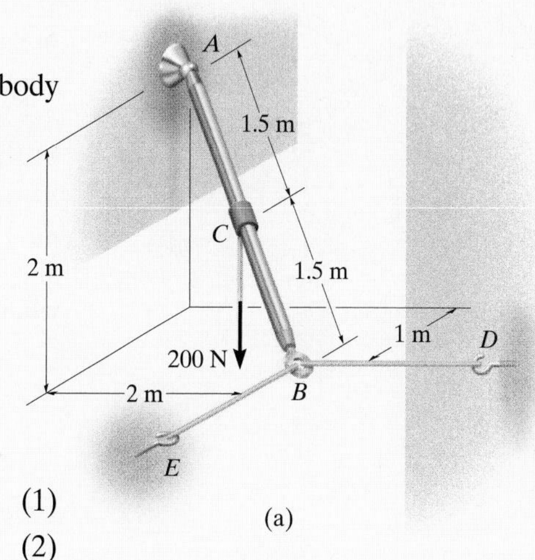

(a)

Applying the force equation of equilibrium.

$\Sigma\mathbf{F} = \mathbf{0};\qquad \mathbf{F}_A + \mathbf{T}_E + \mathbf{T}_D + \mathbf{F} = \mathbf{0}$

$$(A_x + T_E)\mathbf{i} + (A_y + T_D)\mathbf{j} + (A_z - 200)\mathbf{k} = \mathbf{0}$$

$\Sigma F_x = 0;\qquad A_x + T_E = 0 \qquad (1)$

$\Sigma F_y = 0;\qquad A_y + T_D = 0 \qquad (2)$

$\Sigma F_z = 0;\qquad A_z - 200 = 0 \qquad (3)$

Summing moments about point A yields

$\Sigma\mathbf{M}_A = \mathbf{0};\qquad \mathbf{r}_C \times \mathbf{F} + \mathbf{r}_B \times (\mathbf{T}_E + \mathbf{T}_D) = \mathbf{0}$

Since $\mathbf{r}_C = \frac{1}{2}\mathbf{r}_B$, then

$$(0.5\mathbf{i} + 1\mathbf{j} - 1\mathbf{k}) \times (-200\mathbf{k}) + (1\mathbf{i} + 2\mathbf{j} - 2\mathbf{k}) \times (T_E\mathbf{i} + T_D\mathbf{j}) = \mathbf{0}$$

Expanding and rearranging terms gives

$$(2T_D - 200)\mathbf{i} + (-2T_E + 100)\mathbf{j} + (T_D - 2T_E)\mathbf{k} = \mathbf{0}$$

$\Sigma M_x = 0;\qquad 2T_D - 200 = 0 \qquad (4)$

$\Sigma M_y = 0;\qquad -2T_E + 100 = 0 \qquad (5)$

$\Sigma M_z = 0;\qquad T_D - 2T_E = 0 \qquad (6)$

Solving Eqs. 1 through 6, we get

$T_D = 100\text{ N}$ *Ans.*

$T_E = 50\text{ N}$ *Ans.*

$A_x = -50\text{ N}$ *Ans.*

$A_y = -100\text{ N}$ *Ans.*

$A_z = 200\text{ N}$ *Ans.*

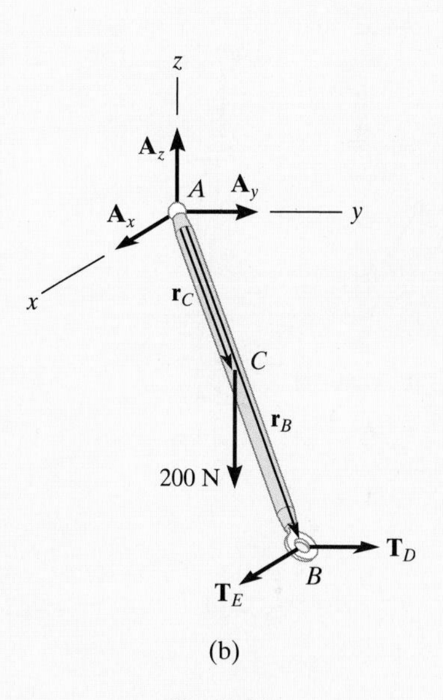

(b)

Fig. 5–33

NOTE: The negative sign indicates that $\mathbf{A}_x$ and $\mathbf{A}_y$ have a sense which is opposite to that shown on the free-body diagram, Fig. 5–33b.

EXAMPLE 5.19

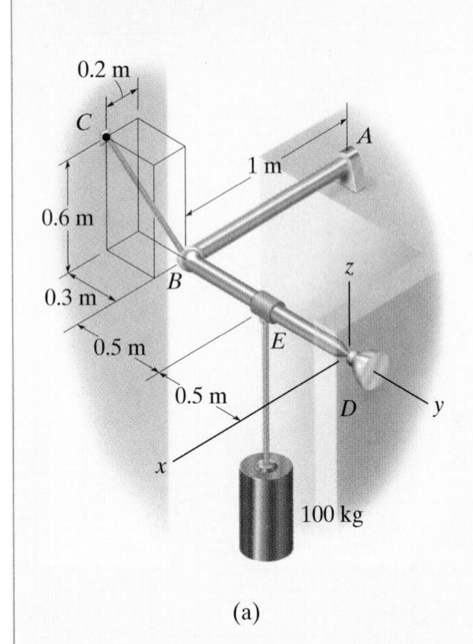

(a)

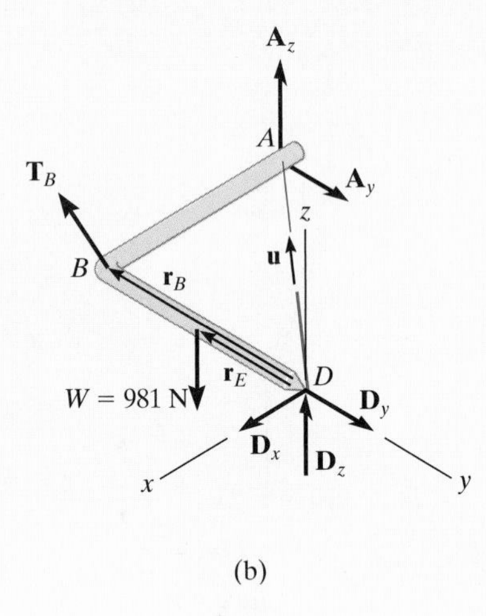

(b)

Fig. 5–34

The bent rod in Fig. 5–34*a* is supported at A by a journal bearing, at D by a ball-and-socket joint, and at B by means of cable BC. Using only *one equilibrium equation*, obtain a direct solution for the tension in cable BC. The bearing at A is capable of exerting force components only in the z and y directions since it is properly aligned on the shaft.

SOLUTION *(VECTOR ANALYSIS)*

Free-Body Diagram. As shown in Fig. 5–34*b*, there are six unknowns: three force components caused by the ball-and-socket joint, two caused by the bearing, and one caused by the cable.

Equations of Equilibrium. The cable tension $\mathbf{T}_B$ may be obtained *directly* by summing moments about an axis passing through points D and A. Why? The direction of the axis is defined by the unit vector $\mathbf{u}$, where

$$\mathbf{u} = \frac{\mathbf{r}_{DA}}{r_{DA}} = -\frac{1}{\sqrt{2}}\mathbf{i} - \frac{1}{\sqrt{2}}\mathbf{j}$$
$$= -0.707\mathbf{i} - 0.707\mathbf{j}$$

Hence, the sum of the moments about this axis is zero provided

$$\Sigma M_{DA} = \mathbf{u} \cdot \Sigma(\mathbf{r} \times \mathbf{F}) = 0$$

Here $\mathbf{r}$ represents a position vector drawn from *any point* on the axis DA to any point on the line of action of force $\mathbf{F}$ (see Eq. 4–11). With reference to Fig. 5–34*b*, we can therefore write

$$\mathbf{u} \cdot (\mathbf{r}_B \times \mathbf{T}_B + \mathbf{r}_E \times \mathbf{W}) = \mathbf{0}$$
$$(-0.707\mathbf{i} - 0.707\mathbf{j}) \cdot \left[(-1\mathbf{j}) \times \left(\tfrac{0.2}{0.7}T_B\mathbf{i} - \tfrac{0.3}{0.7}T_B\mathbf{j} + \tfrac{0.6}{0.7}T_B\mathbf{k}\right) + (-0.5\mathbf{j}) \times (-981\mathbf{k})\right] = \mathbf{0}$$
$$(-0.707\mathbf{i} - 0.707\mathbf{j}) \cdot [(-0.857T_B + 490.5)\mathbf{i} + 0.286T_B\mathbf{k}] = \mathbf{0}$$
$$-0.707(-0.857T_B + 490.5) + 0 + 0 = 0$$

$$T_B = \frac{490.5}{0.857} = 572 \text{ N} \qquad \textit{Ans.}$$

The advantage of using Cartesian vectors for this solution should be noted. It would be especially tedious to determine the perpendicular distance from the DA axis to the line of action of $\mathbf{T}_B$ using scalar methods.

NOTE: In a similar manner, we can obtain D_z $(= 490.5 \text{ N})$ by summing moments about an axis passing through AB. Also, A_z $(= 0)$ is obtained by summing moments about the y axis.

PROBLEMS

5–63. Determine the x, y, z components of reaction at the fixed wall A. The 150-N force is parallel to the z axis and the 200-N force is parallel to the y axis.

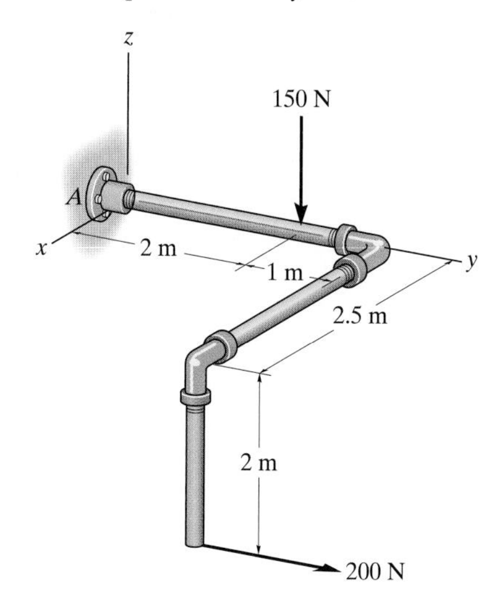

Prob. 5–63

***5–64.** The wing of the jet aircraft is subjected to a thrust of $T = 8$ kN from its engine and the resultant lift force $L = 45$ kN. If the mass of the wing is 2.1 Mg and the mass center is at G, determine the x, y, z components of reaction where the wing is fixed to the fuselage at A.

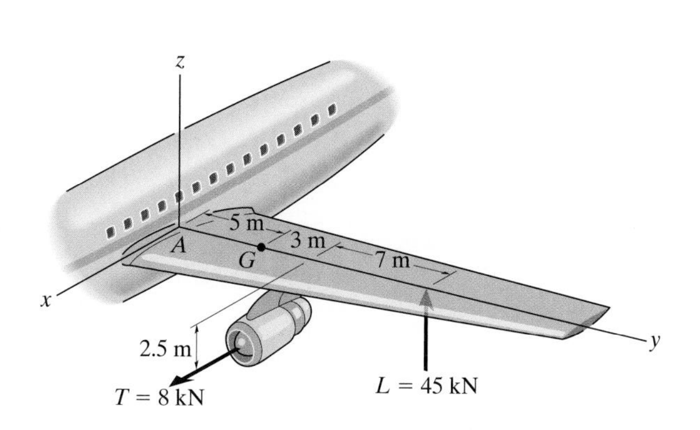

Prob. 5–64

5–65. The uniform concrete slab has a weight of 5500 lb. Determine the tension in each of the three parallel supporting cables when the slab is held in the horizontal plane as shown.

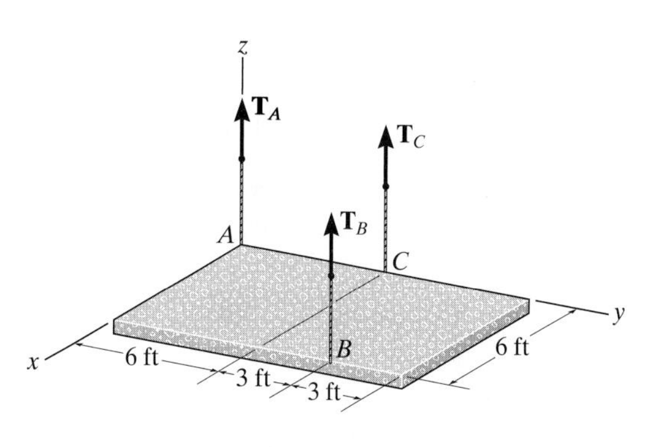

Prob. 5–65

5–66. The air-conditioning unit is hoisted to the roof of a building using the three cables. If the tensions in the cables are $T_A = 250$ lb, $T_B = 300$ lb, and $T_C = 200$ lb, determine the weight of the unit and the location (x, y) of its center of gravity G.

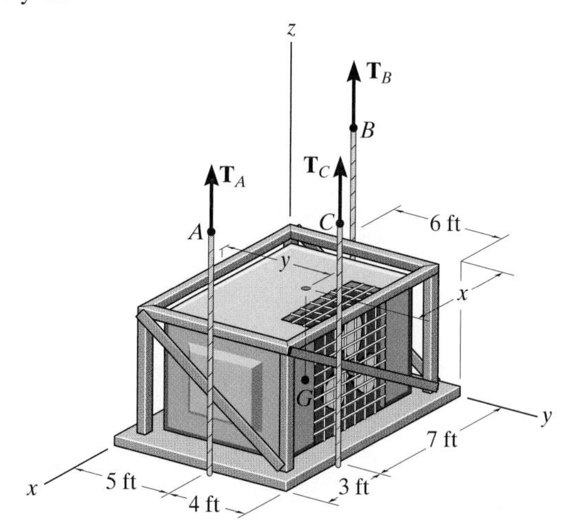

Prob. 5–66

5–67. The platform truck supports the three loadings shown. Determine the normal reactions on each of its three wheels.

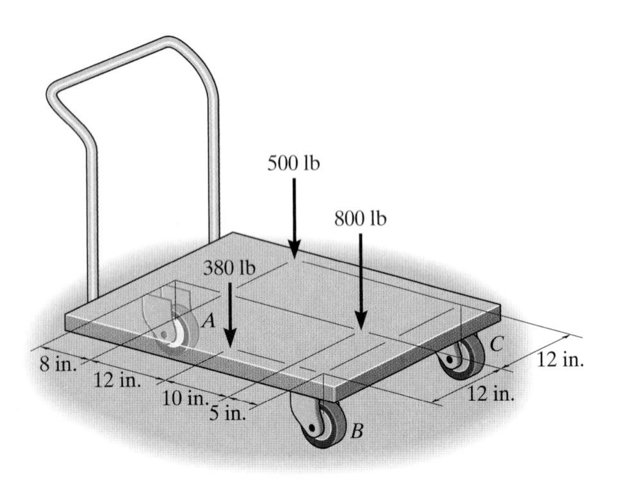

Prob. 5–67

5–69. If the cable can be subjected to a maximum tension of 300 lb, determine the maximum force F which may be applied to the plate. Compute the x, y, z components of reaction at the hinge A for this loading.

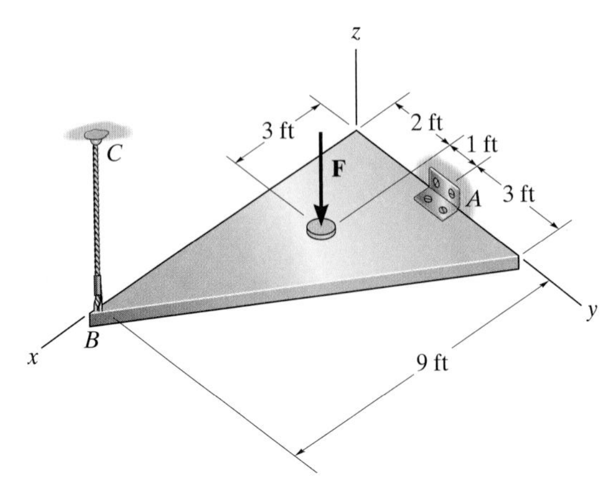

Prob. 5–69

***5–68.** Due to an unequal distribution of fuel in the wing tanks, the centers of gravity for the airplane fuselage A and wings B and C are located as shown. If these components have weights $W_A = 45\,000$ lb, $W_B = 8000$ lb, and $W_C = 6000$ lb, determine the normal reactions of the wheels D, E, and F on the ground.

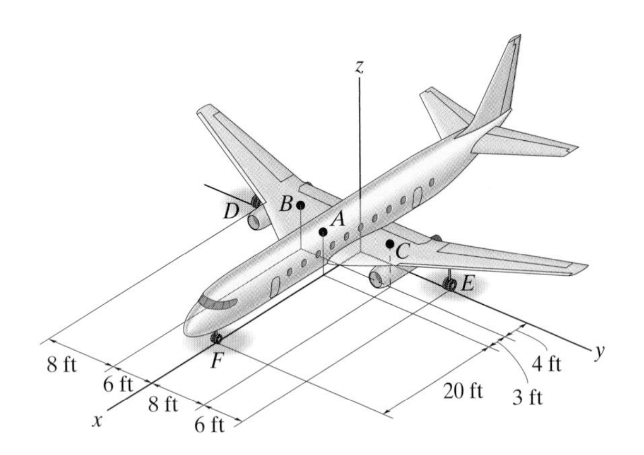

Prob. 5–68

5–70. The boom AB is held in equilibrium by a ball-and-socket joint A and a pulley and cord system as shown. Determine the x, y, z components of reaction at A and the tension in cable DEC if $\mathbf{F} = \{-1500\mathbf{k}\}$ lb.

5–71. The cable CED can sustain a maximum tension of 800 lb before it fails. Determine the greatest vertical force F that can be applied to the boom. Also, what are the x, y, z components of reaction at the ball-and-socket joint A?

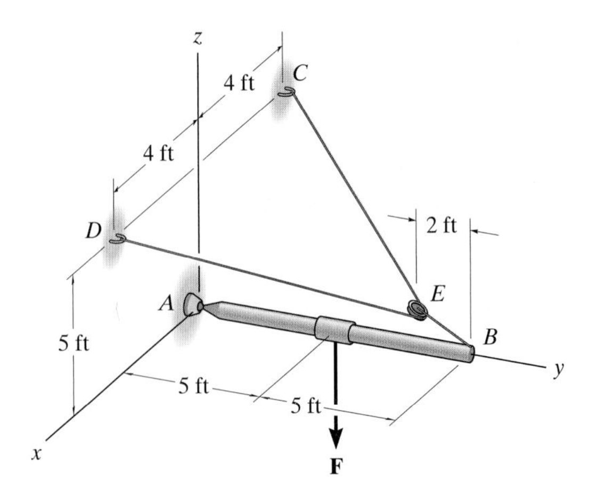

Probs. 5–70/71

*5–72. The uniform table has a weight of 20 lb and is supported by the framework shown. Determine the smallest vertical force **P** that can be applied to its surface that will cause it to tip over. Where should this force be applied?

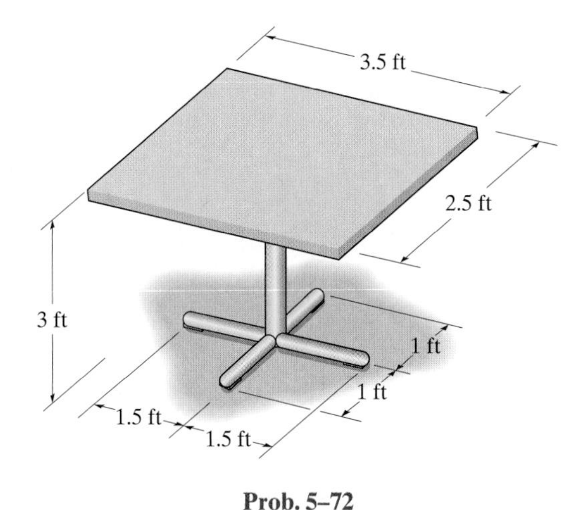

Prob. 5–72

5–74. The 2-kg ball rests between the 45° grooves *A* and *B* of the 10° incline and against a vertical wall at *C*. If all three surfaces of contact are smooth, determine the reactions of the surfaces on the ball. *Hint:* Use the *x, y, z* axes, with origin at the center of the ball, and the *z* axis inclined as shown.

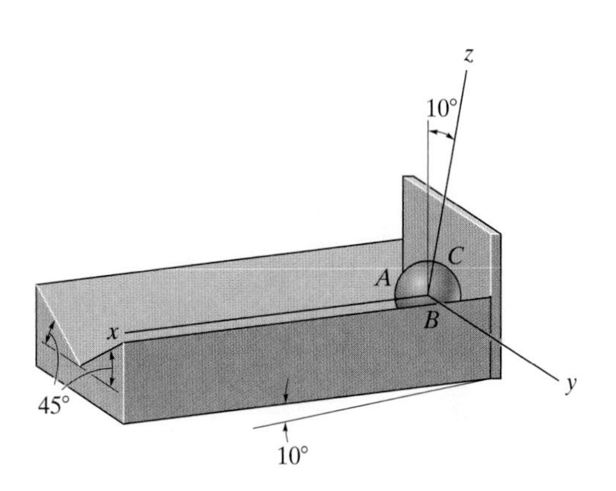

Prob. 5–74

5–73. The windlass is subjected to a load of 150 lb. Determine the horizontal force **P** needed to hold the handle in the position shown, and the components of reaction at the ball-and-socket joint *A* and the smooth journal bearing *B*. The bearing at *B* is in proper alignment and exerts only force reactions on the windlass.

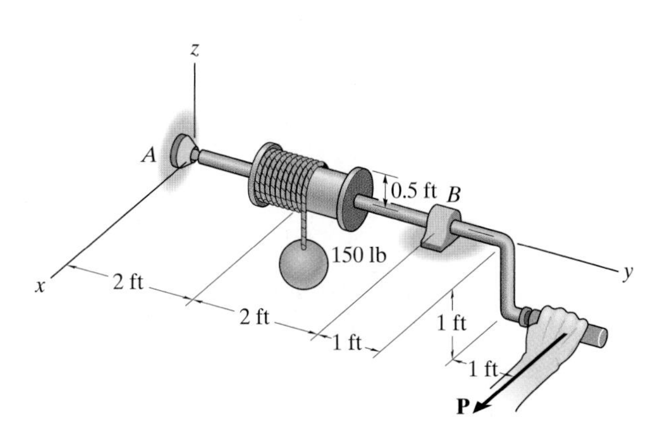

Prob. 5–73

5–75. Member *AB* is supported by a cable *BC* and at *A* by a *square* rod which fits loosely through the square hole at the end joint of the member as shown. Determine the components of reaction at *A* and the tension in the cable needed to hold the 800-lb cylinder in equilibrium.

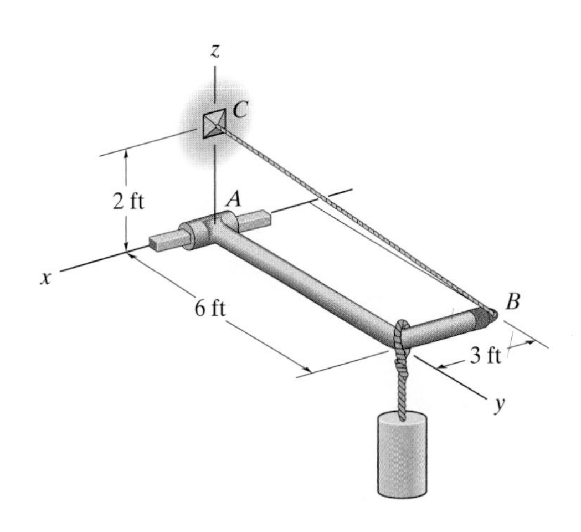

Prob. 5–75

*5–76. The pipe assembly supports the vertical loads shown. Determine the components of reaction at the ball-and-socket joint A and the tension in the supporting cables BC and BD.

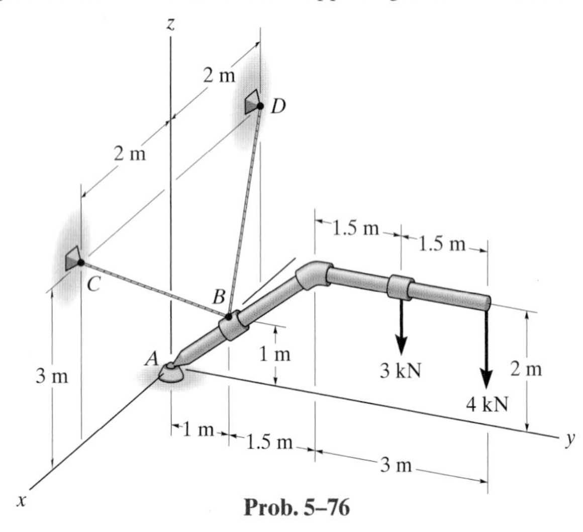

Prob. 5–76

5–77. The hatch door has a weight of 80 lb and center of gravity at G. If the force $\mathbf{F}$ applied to the handle at C has coordinate direction angles of $\alpha = 60°$, $\beta = 45°$, and $\gamma = 60°$, determine the magnitude of $\mathbf{F}$ needed to hold the door slightly open as shown. The hinges are in proper alignment and exert only force reactions on the door. Determine the components of these reactions if A exerts only x and z components of force and B exerts x, y, z force components.

5–78. The hatch door has a weight of 80 lb and center of gravity at G. If the force $\mathbf{F}$ applied to the handle at C has coordinate direction angles of $\alpha = 60°$, $\beta = 45°$, and $\gamma = 60°$, determine the magnitude of $\mathbf{F}$ needed to hold the door slightly open as shown. If the hinge at A becomes loose from its attachment and is ineffective, what are the x, y, z components of reaction at hinge B?

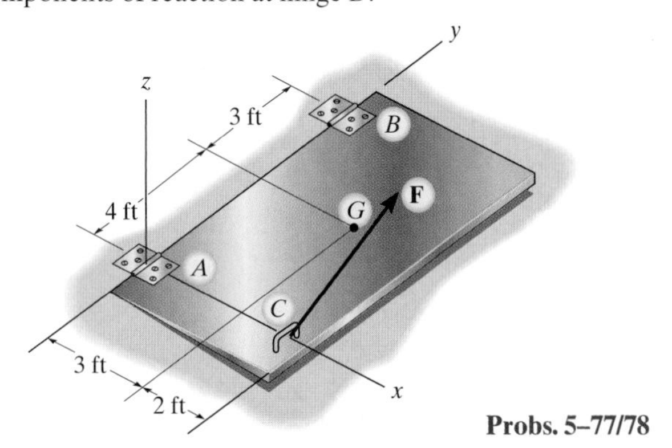

Probs. 5–77/78

5–79. The bent rod is supported at A, B, and C by smooth journal bearings. Compute the x, y, z components of reaction at the bearings if the rod is subjected to forces $F_1 = 300$ lb and $F_2 = 250$ lb. F_1 lies in the y–z plane. The bearings are in proper alignment and exert only force reactions on the rod.

*5–80. The bent rod is supported at A, B, and C by smooth journal bearings. Determine the magnitude of $\mathbf{F}_2$ which will cause the reaction $\mathbf{C}_y$ at the bearing C to be equal to zero. The bearings are in proper alignment and exert only force reactions on the rod. Set $F_1 = 300$ lb.

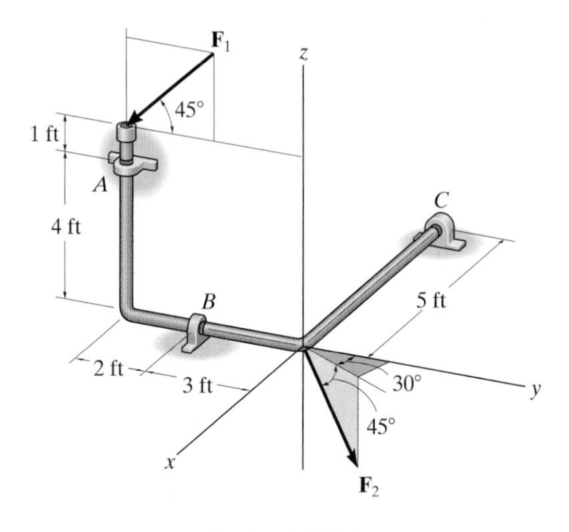

Probs. 5–79/80

5–81. Determine the tension in cables BD and CD and the x, y, z components of reaction at the ball-and-socket joint at A.

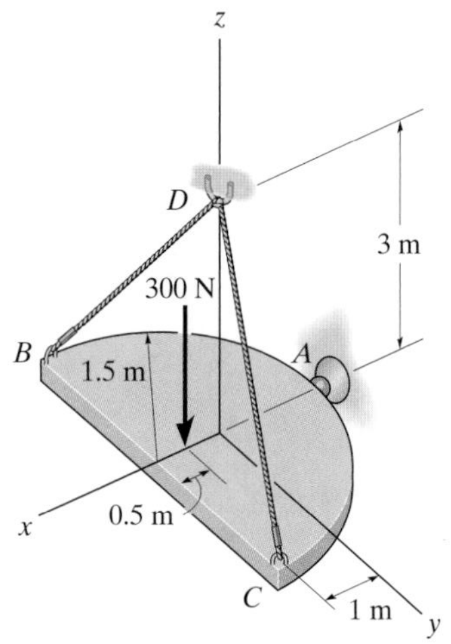

Prob. 5–81

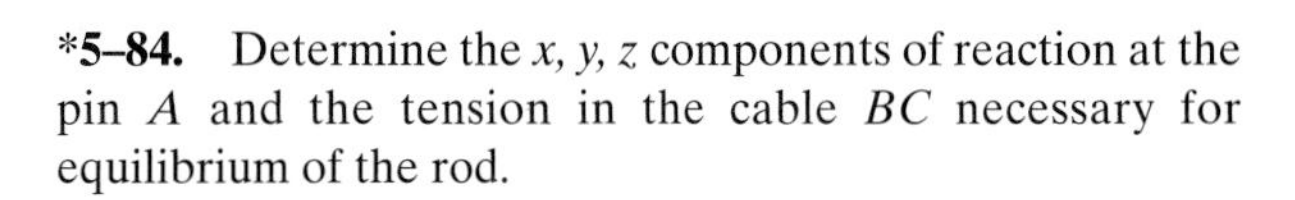

5–82. Determine the tensions in the cables and the components of reaction acting on the smooth collar at A necessary to hold the 50-lb sign in equilibrium. The center of gravity for the sign is at G.

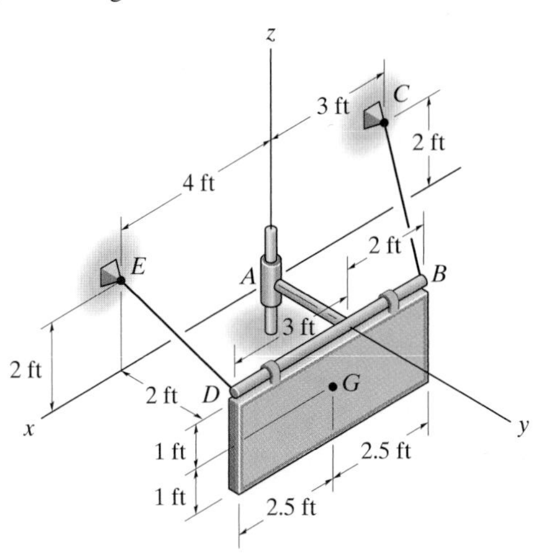

Prob. 5–82

5–83. The member is supported by a pin at A and a cable BC. If the load at D is 300 lb, determine the x, y, z components of reaction at these supports.

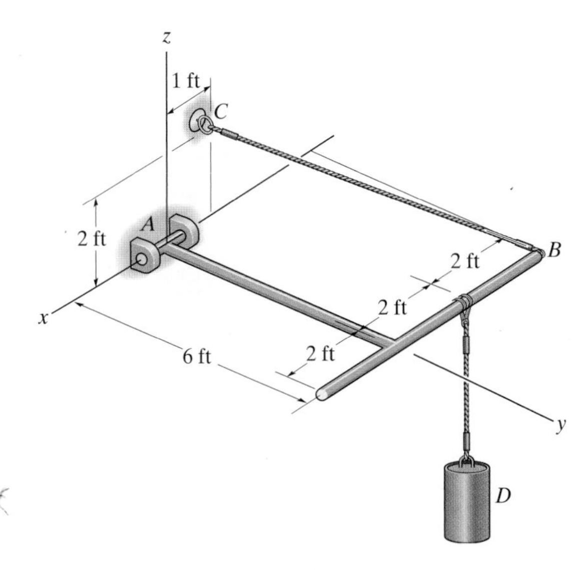

Prob. 5–83

***5–84.** Determine the x, y, z components of reaction at the pin A and the tension in the cable BC necessary for equilibrium of the rod.

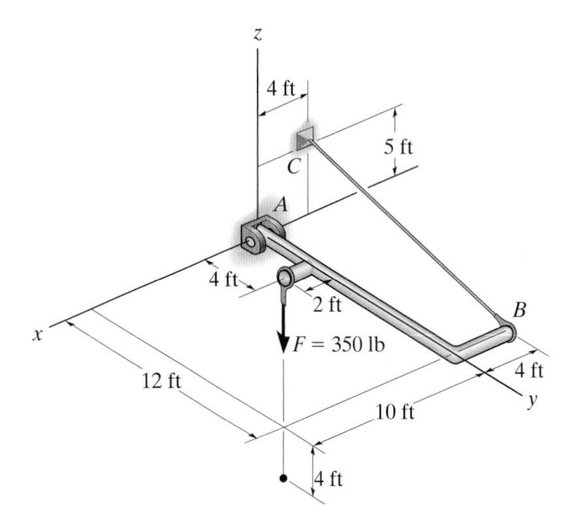

Prob. 5–84

5–85. Rod AB is supported by a ball-and-socket joint at A and a cable at the smooth wall at B. Determine the x, y, z components of reaction at these supports if the rod is subjected to a 50-lb vertical force as shown.

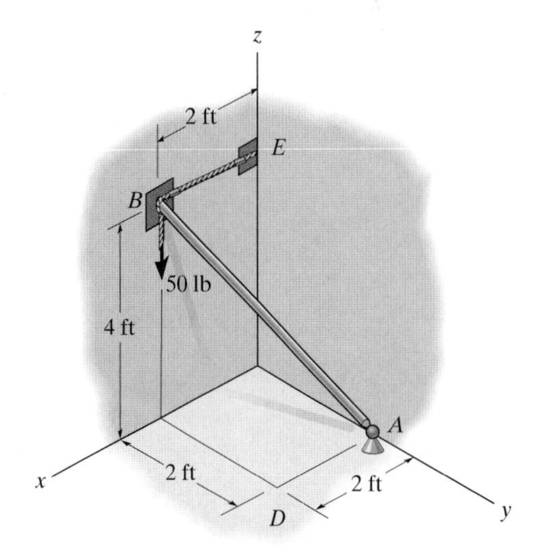

Prob. 5–85

5–86. The member is supported by a square rod which fits loosely through a smooth square hole of the attached collar at A and by a roller at B. Determine the x, y, z components of reaction at these supports when the member is subjected to the loading shown.

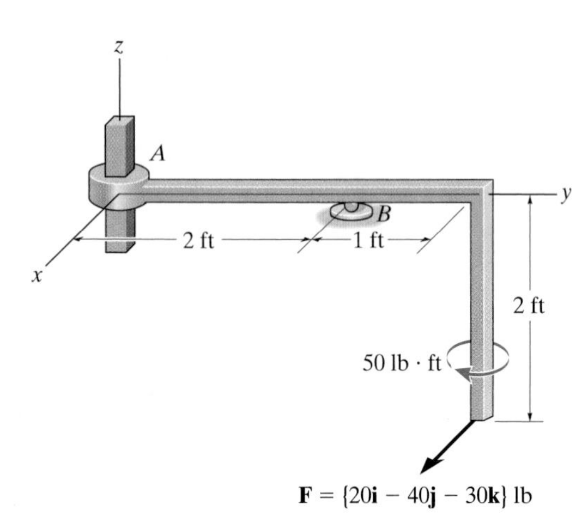

Prob. 5–86

5–87. The platform has a mass of 3 Mg and center of mass located at G. If it is lifted with constant velocity using the three cables, determine the force in each of the cables.

***5–88.** The platform has a mass of 2 Mg and center of mass located at G. If it is lifted using the three cables, determine the force in each of the cables. Solve for each force by using a single moment equation of equilibrium.

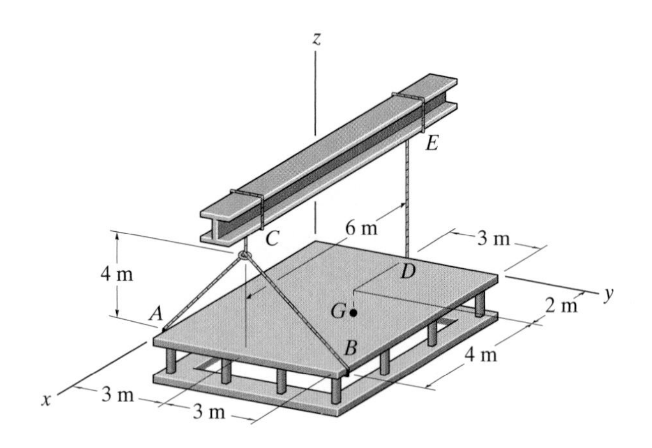

Probs. 5–87/88

5–89. The cables exert the forces shown on the pole. Assuming the pole is supported by a ball-and-socket joint at its base, determine the components of reaction at A. The forces of 140 lb and 75 lb lie in a horizontal plane.

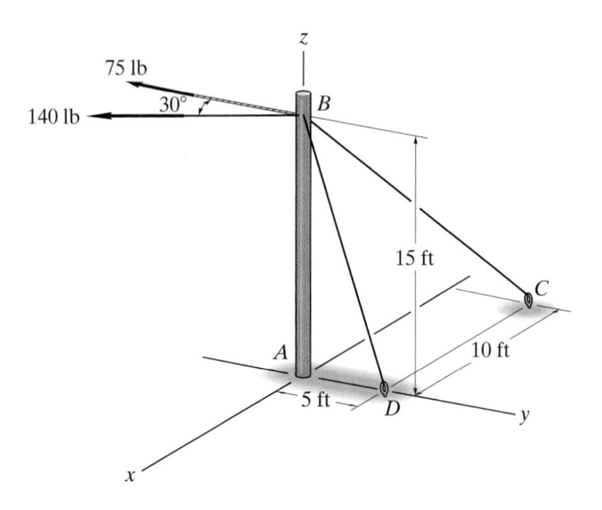

Prob. 5–89

5–90. The silo has a weight of 3500 lb and a center of gravity at G. Determine the vertical component of force that each of the three struts at A, B, and C exerts on the silo if it is subjected to a resultant wind loading of 250 lb which acts in the direction shown.

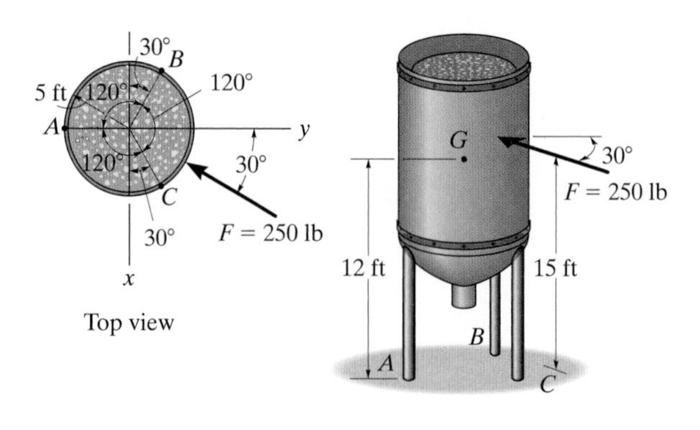

Prob. 5–90

Chapter Review

Equilibrium

A body in equilibrium does not accelerate, so it either moves at constant velocity, or does not move at all or is at rest.
(*See page 201.*)

$$\Sigma \mathbf{F} = \mathbf{0}$$
$$\Sigma \mathbf{M} = \mathbf{0}$$

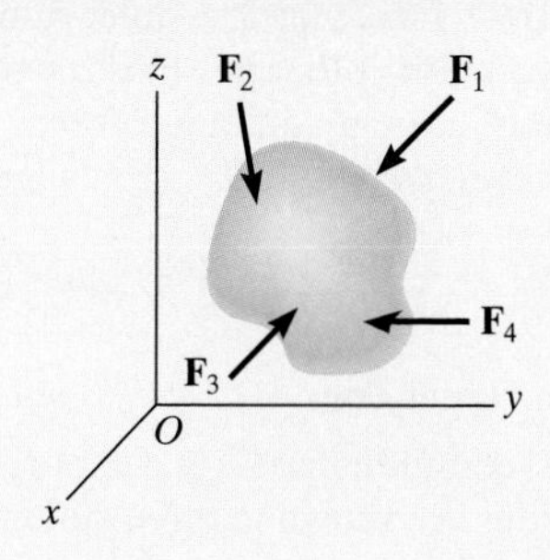

Two Dimensions

Before analyzing the equilibrium of a body, it is first necessary to draw its free-body diagram. This is an outlined shape of the body, which shows all the forces and couple moments that act on it.

Couple moments can be placed anywhere on a free-body diagram since they are free vectors. Forces can act at any point along their line of action since they are sliding vectors.

Angles used to resolve forces, and dimensions used to take moments of the forces, should also be shown on the free-body diagram.

Some common types of supports and their reactions are shown below in two dimensions.

Remember that a support will exert a force on the body in a particular direction if it prevents translation of the body in that direction, and it will exert a couple moments on the body if it prevents rotation.
(*See pages 203–204.*)

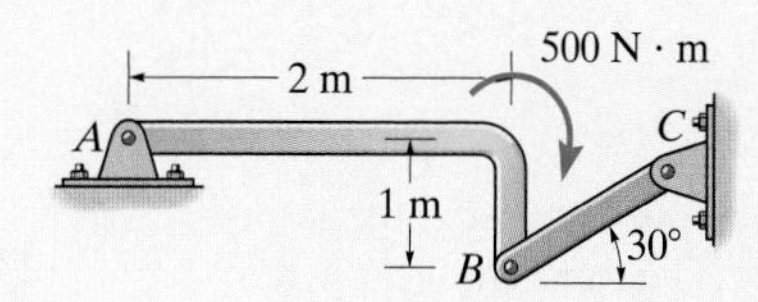

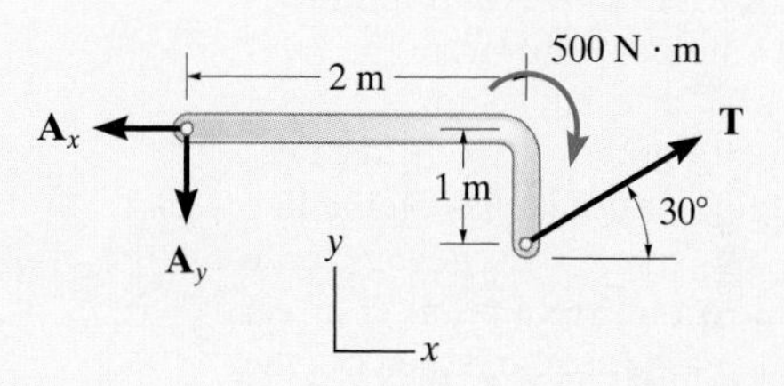

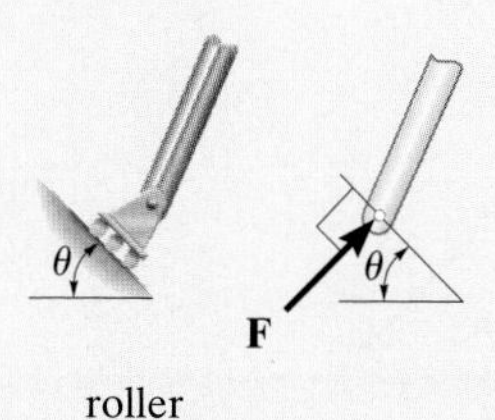

roller

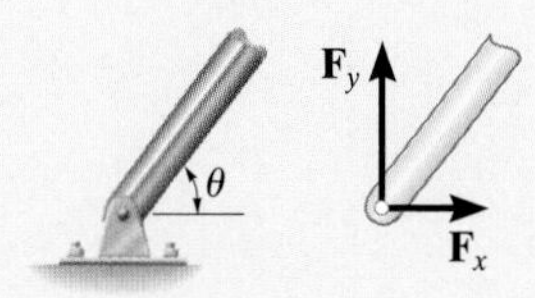

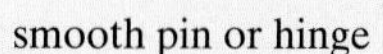

smooth pin or hinge

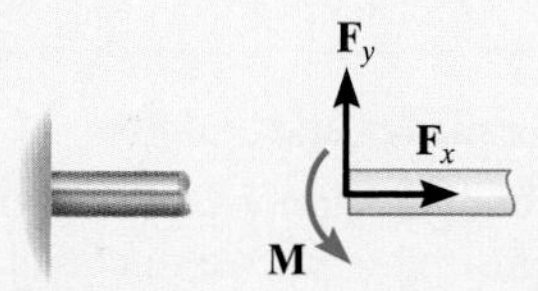

fixed support

The three scalar equations of equilibrium can be applied when solving problems in two dimensions, since the geometry is easy to visualize.

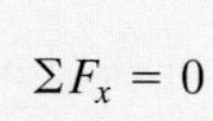

$$\Sigma F_x = 0$$
$$\Sigma F_y = 0$$
$$\Sigma M_O = 0$$

For the most direct solution, try to sum forces along an axis that will eliminate as many unknown forces as possible. Sum moments about a point A that passes through the line of action of as many unknown forces as possible. (*See pages 217 and 218.*)

$\Sigma F_x = 0;$

$$A_x - P_2 = 0 \qquad A_x = P_2$$

$\Sigma M_A = 0;$

$$P_2 d_2 + B_y d_B - P_1 d_1 = 0$$

$$B_y = \frac{P_1 d_1 - P_2 d_2}{d_B}$$

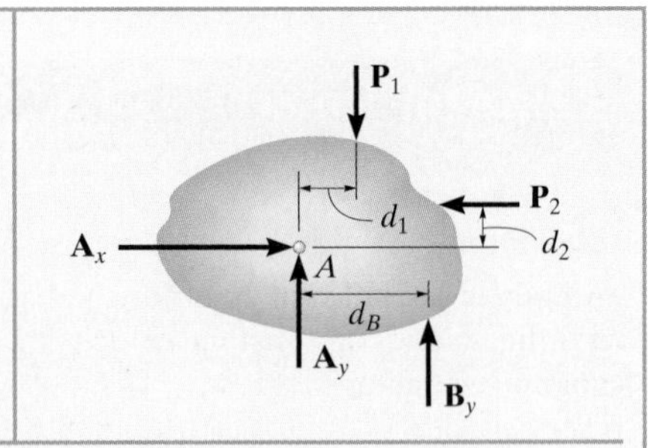

Three Dimensions

Some common types of supports and their reactions are shown in three dimensions.

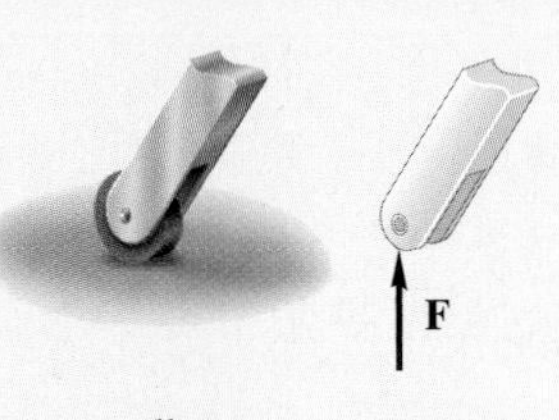

roller

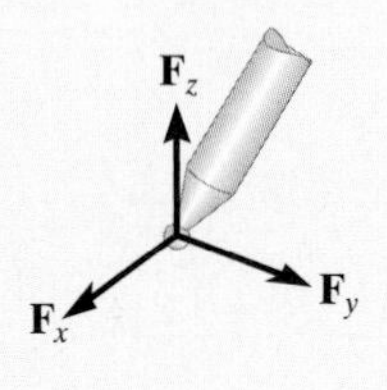

ball and socket

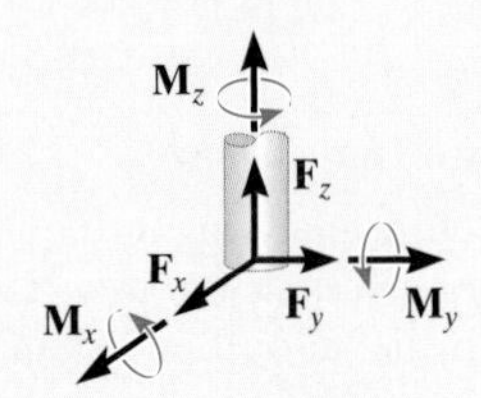

fixed support

In three dimensions, it is often advantageous to use a Cartesian vector analysis when applying the equations of equilibrium. To do this, first express each known and unknown force and couple moment shown on the free-body diagram as a Cartesian vector. Then set the force summation equal to zero. Take moments about a point O that lies on the line of action of as many unknown force components as possible. From point O direct position vectors to each force, and then use the cross product to determine the moment of each force. The six scalar equations of equilibrium are established by setting the respective **i**, **j**, and **k** components of these force and moment summations equal to zero. (*See page 245.*)

$$\Sigma \mathbf{F} = \mathbf{0}$$
$$\Sigma \mathbf{M}_O = \mathbf{0}$$

$$\Sigma F_x = 0$$
$$\Sigma F_y = 0$$
$$\Sigma F_z = 0$$

$$\Sigma M_x = 0$$
$$\Sigma M_y = 0$$
$$\Sigma M_z = 0$$

Determinacy and Stability

If a body is supported by a minimum number of constraints to ensure equilibrium, then it is statically determinate. If it has more constraints than required, then it is statically indeterminate.

To properly constrain the body, the reactions must not all be parallel to one another or concurrent. (*See pages 246–248.*)

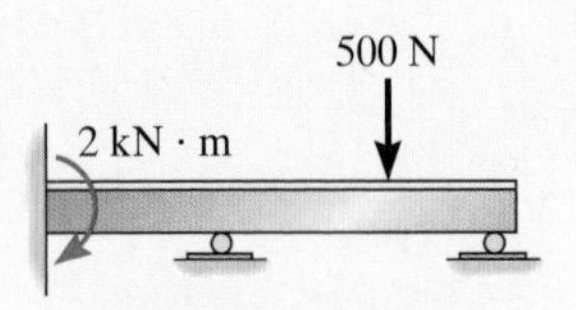

Statically indeterminate, five reactions, three equilibrium equations

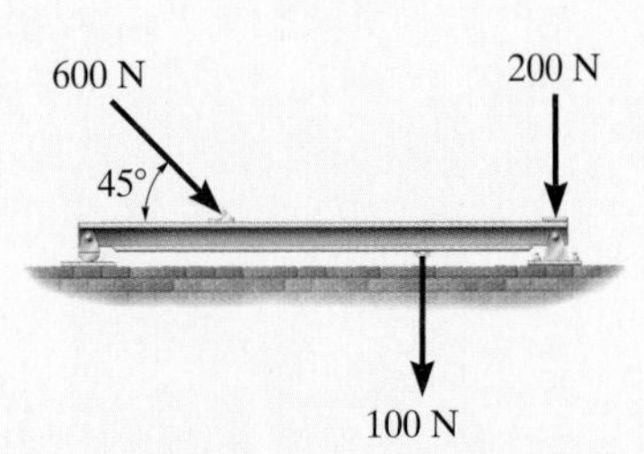

Proper constraint, statically determinant

REVIEW PROBLEMS

5–91. The shaft assembly is supported by two smooth journal bearings A and B and a short link DC. If a couple moment is applied to the shaft as shown, determine the components of force reaction at the bearings and the force in the link. The link lies in a plane parallel to the y–z plane and the bearings are properly aligned on the shaft.

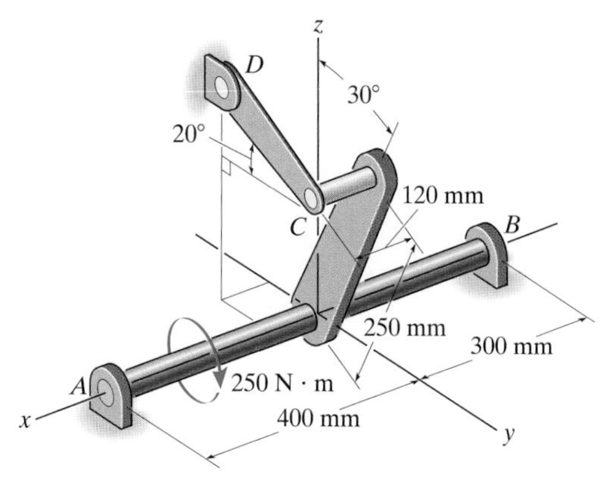

Prob. 5–91

***5–92.** If neither the pin at A nor the roller at B can support a load no greater than 6 kN, determine the maximum intensity of the distributed load w, measured in kN/m, so that failure of a support does not occur.

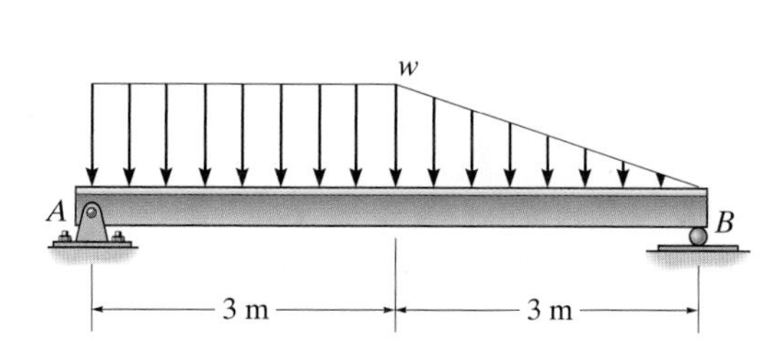

Prob. 5–92

5–93. If the maximum intensity of the distributed load acting on the beam is $w = 4$ kN/m, determine the reactions at the pin A and roller B.

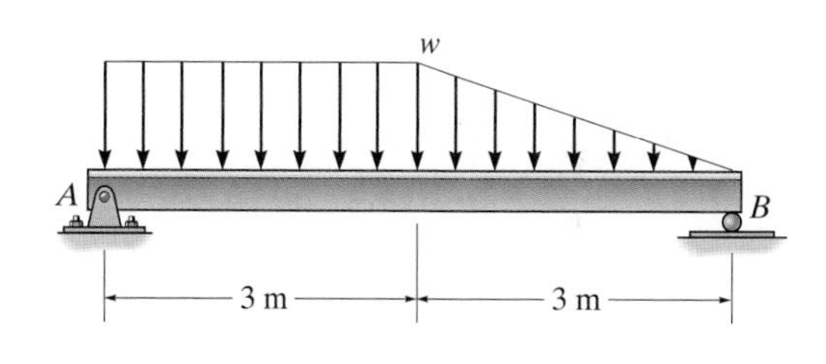

Prob. 5–93

5–94. Determine the normal reaction at the roller A and horizontal and vertical components at pin B for equilibrium of the member.

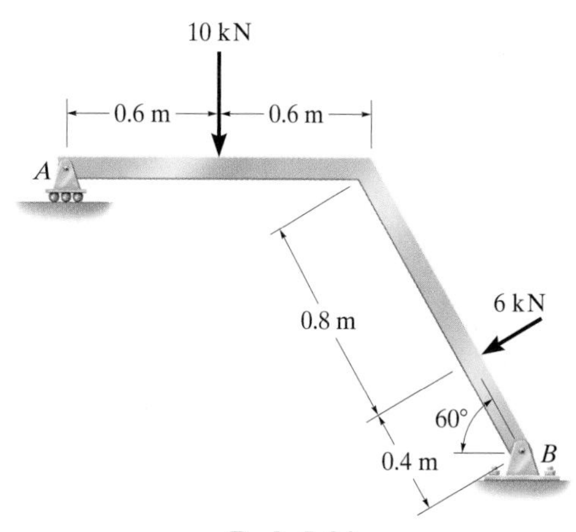

Prob. 5–94

5–95. The symmetrical shelf is subjected to a uniform load of 4 kPa. Support is provided by a bolt (or pin) located at each end A and A' and by the symmetrical brace arms, which bear against the smooth wall on both sides at B and B'. Determine the force resisted by each bolt at the wall and the normal force at B for equilibrium.

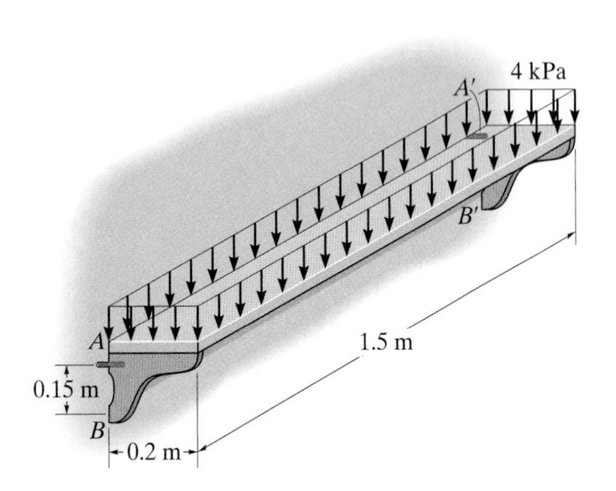

Prob. 5–95

5–97. The uniform ladder rests along the wall of a building at A and on the roof at B. If the ladder has a weight of 25 lb and the surfaces at A and B are assumed smooth, determine the angle θ for equilibrium.

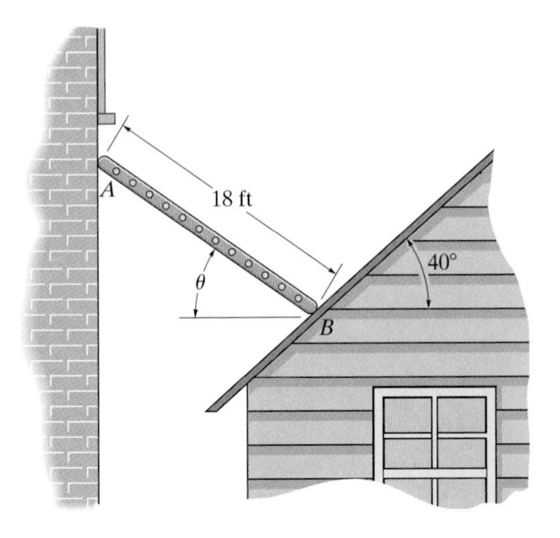

Prob. 5–97

***5–96.** A uniform beam having a weight of 200 lb supports a vertical load of 800 lb. If the ground pressure varies linearly as shown, determine the load intensities w_1 and w_2, measured in lb/ft, necessary for equilibrium.

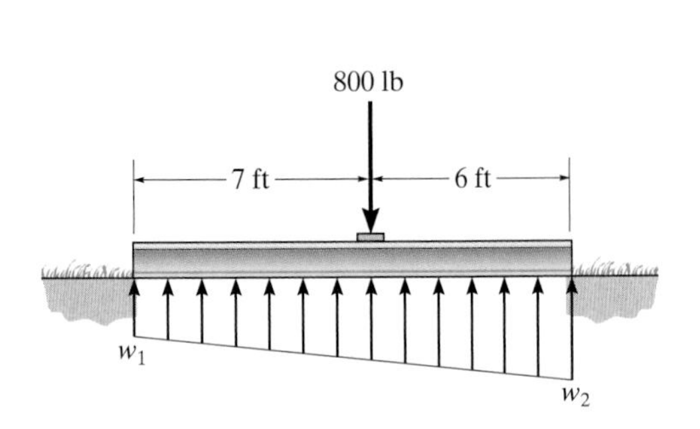

Prob. 5–96

5–98. Determine the x, y, z components of reaction at the ball supports B and C and the ball-and-socket A (not shown) for the uniformly loaded plate.

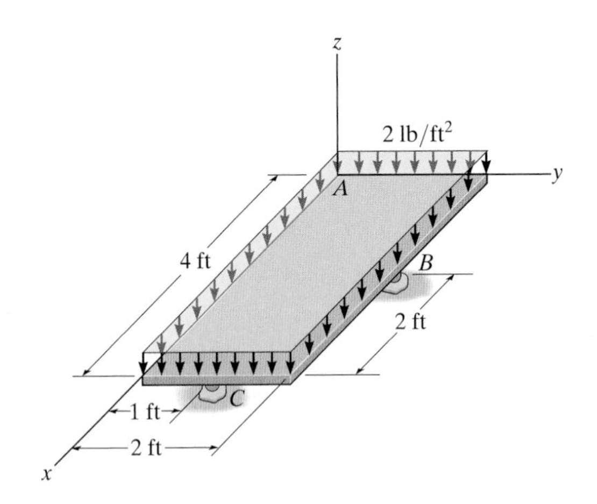

Prob. 5–98

5–99. A vertical force of 80 lb acts on the crankshaft. Determine the horizontal equilibrium force P that must be applied to the handle and the x, y, z components of force at the smooth journal bearing A and the thrust bearing B. The bearings are properly aligned and exert only force reactions on the shaft.

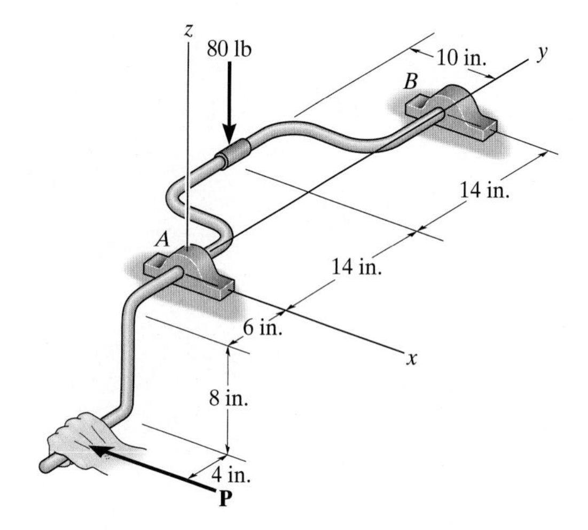

Prob. 5–99

***5–100.** The horizontal beam is supported by springs at its ends. If the stiffness of the spring at A is $k_A = 5\text{ kN/m}$, determine the required stiffness of the spring at B so that if the beam is loaded with the 800-N force, it remains in the horizontal position both before and after loading.

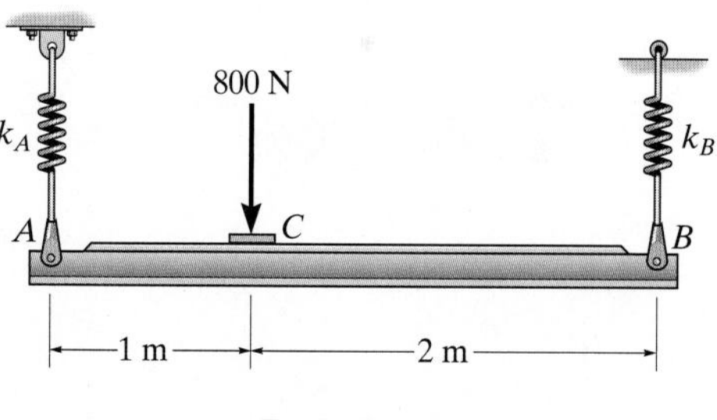

Prob. 5–100

The forces within the members of these truss bridges must be determined if they are to be properly designed.

6 Structural Analysis

CHAPTER OBJECTIVES

- To show how to determine the forces in the members of a truss using the method of joints and the method of sections.
- To analyze the forces acting on the members of frames and machines composed of pin-connected members.

6.1 Simple Trusses

A *truss* is a structure composed of slender members joined together at their end points. The members commonly used in construction consist of wooden struts or metal bars. The joint connections are usually formed by bolting or welding the ends of the members to a common plate, called a *gusset plate,* as shown in Fig. 6–1*a*, or by simply passing a large bolt or pin through each of the members, Fig. 6–1*b*. Gusset plates are used for the member connections of the bridge trusses shown on the chapter opening photo.

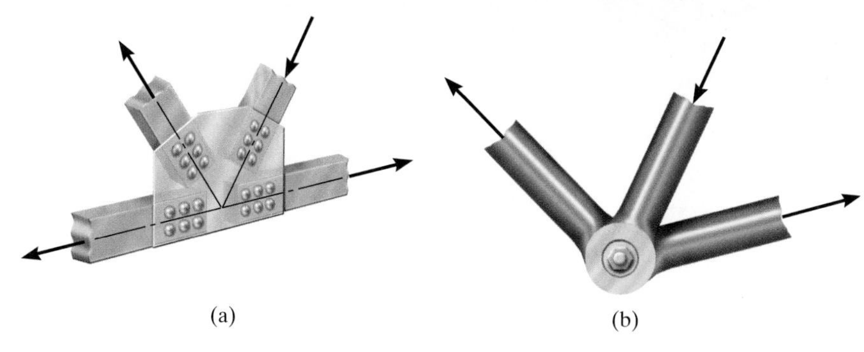

Fig. 6–1

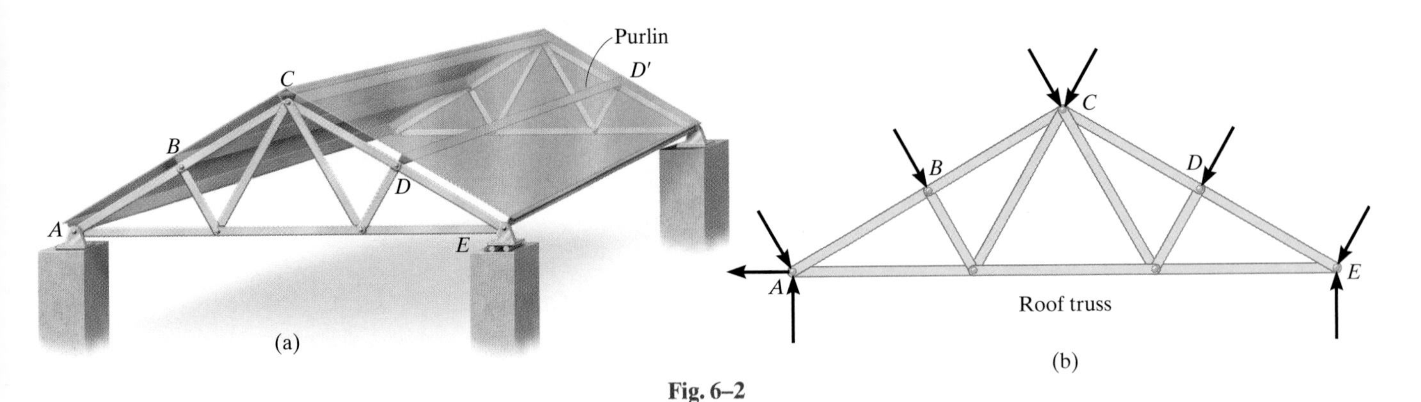

Fig. 6–2

Planar Trusses. *Planar* trusses lie in a single plane and are often used to support roofs and bridges. The truss $ABCDE$, shown in Fig. 6–2*a*, is an example of a typical roof-supporting truss. In this figure, the roof load is transmitted to the truss *at the joints* by means of a series of *purlins,* such as DD'. Since the imposed loading acts in the same plane as the truss, Fig. 6–2*b*, the analysis of the forces developed in the truss members is two-dimensional.

In the case of a bridge, such as shown in Fig. 6–3*a*, the load on the deck is first transmitted to *stringers,* then to *floor beams,* and finally to the *joints B, C,* and *D* of the two supporting side trusses. Like the roof truss, the bridge truss loading is also coplanar, Fig. 6–3*b*.

When bridge or roof trusses extend over large distances, a rocker or roller is commonly used for supporting one end, e.g., joint E in Figs. 6–2*a* and 6–3*a*. This type of support allows freedom for expansion or contraction of the members due to temperature or application of loads.

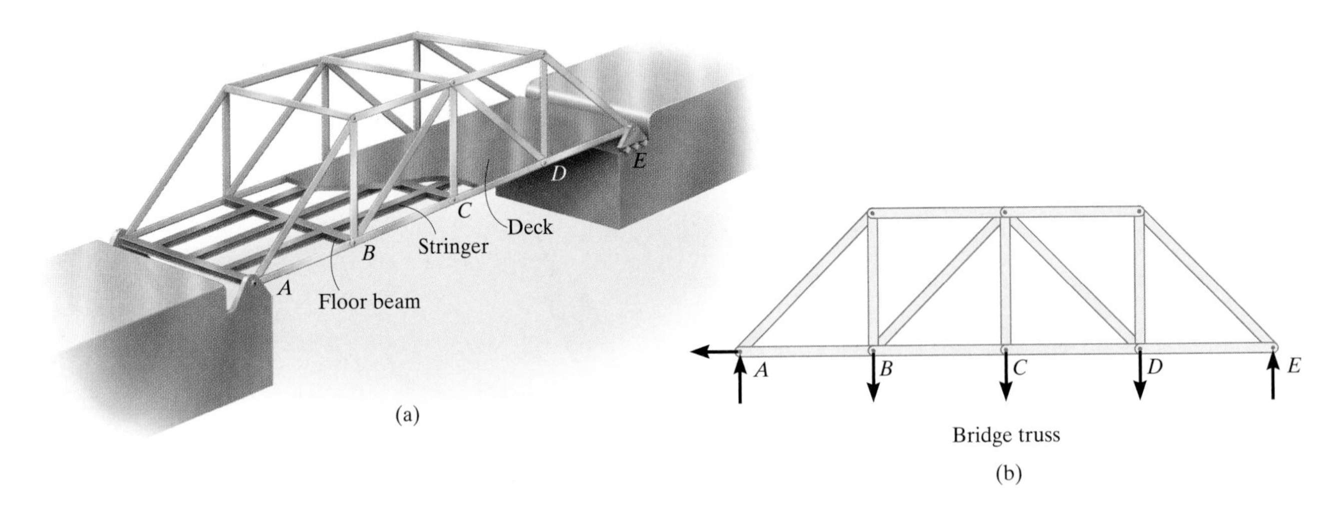

Fig. 6–3

Assumptions for Design. To design both the members and the connections of a truss, it is first necessary to determine the *force* developed in each member when the truss is subjected to a given loading. In this regard, two important assumptions will be made:

1. *All loadings are applied at the joints.* In most situations, such as for bridge and roof trusses, this assumption is true. Frequently in the force analysis the weight of the members is neglected since the forces supported by the members are usually large in comparison with their weight. If the member's weight is to be included in the analysis, it is generally satisfactory to apply it as a vertical force, with half of its magnitude applied at each end of the member.
2. *The members are joined together by smooth pins.* In cases where bolted or welded joint connections are used, this assumption is satisfactory provided the center lines of the joining members are *concurrent,* as in Fig. 6–1*a*.

Because of these two assumptions, *each truss member acts as a two-force member,* and therefore the forces at the ends of the member must be directed along the axis of the member. If the force tends to *elongate* the member, it is a *tensile force* (T), Fig. 6–4*a*; whereas if it tends to *shorten* the member, it is a *compressive force* (C), Fig. 6–4*b*. In the actual design of a truss it is important to state whether the nature of the force is tensile or compressive. Often, compression members must be made *thicker* than tension members because of the buckling or column effect that occurs when a member is in compression.

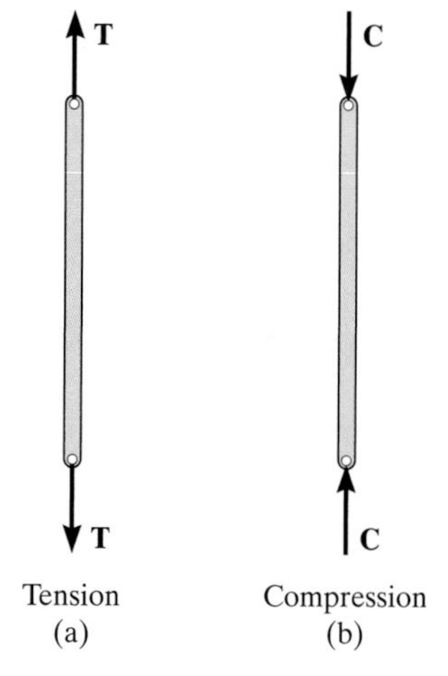

Fig. 6–4

Simple Truss. To prevent collapse, the form of a truss must be rigid. Obviously, the four-bar shape $ABCD$ in Fig. 6–5 will collapse unless a diagonal member, such as AC, is added for support. The simplest form that is rigid or stable is a *triangle.* Consequently, a *simple truss* is constructed by *starting* with a basic triangular element, such as ABC in Fig. 6–6, and connecting two members (AD and BD) to form an additional element. As each additional element consisting of two members and a joint is placed on the truss, it is possible to construct a simple truss.

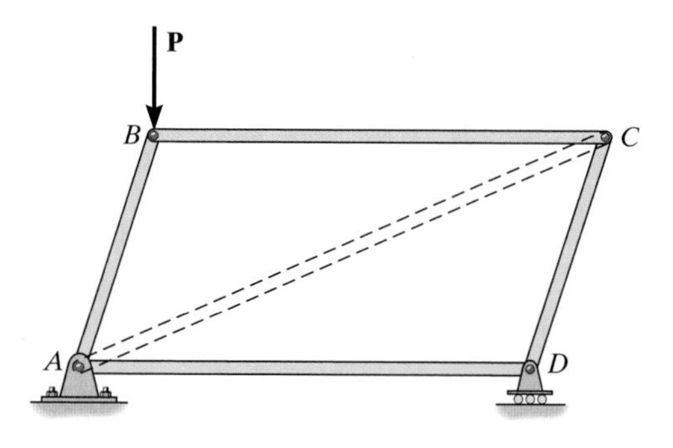

Fig. 6–5

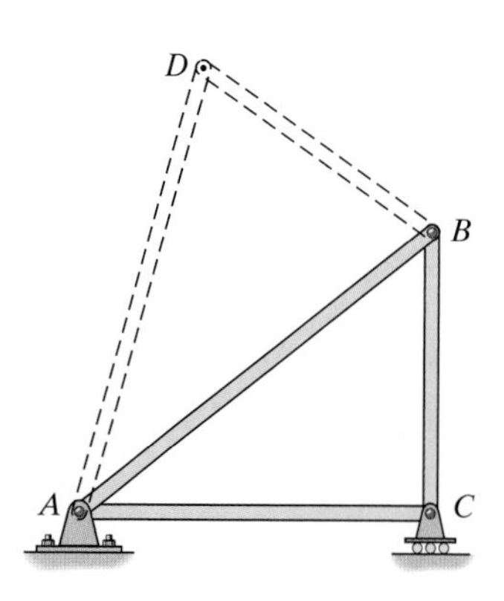

Fig. 6–6

These Howe trusses are used to support the roof of the metal building. This form was named after William Howe, who patented the design in 1840. Note how the members come together at a common point on the gusset plate and how the roof purlins transmit the load to the joints.

6.2 The Method of Joints

In order to analyze or design a truss, we must obtain the force in each of its members. If we were to consider a free-body diagram of the entire truss, then the forces in the members would be *internal forces,* and they could not be obtained from an equilibrium analysis. Instead, if we consider the equilibrium of a joint of the truss then a member force becomes an *external force* on the joint's free-body diagram, and the equations of equilibrium can be applied to obtain its magnitude. This forms the basis for the *method of joints.*

Because the truss members are all straight two-force members lying in the same plane, the force system acting at each joint is *coplanar and concurrent.* Consequently, rotational or moment equilibrium is automatically satisfied at the joint (or pin), and it is only necessary to satisfy $\Sigma F_x = 0$ and $\Sigma F_y = 0$ to ensure equilibrium.

When using the method of joints, it is *first* necessary to draw a free-body diagram of a joint before applying the equilibrium equations. To do this, recall that the *line of action* of each member force acting on the joint is *specified* from the geometry of the truss since the force in a member passes along the axis of the member. As an example, consider the pin at joint B of the truss in Fig. 6–7a. Three forces act on the pin, namely, the 500-N force and the forces exerted by members BA and BC. The free-body diagram is shown in Fig. 6–7b. As shown, $\mathbf{F}_{BA}$ is "pulling" on the pin, which means that member BA is in *tension;* whereas $\mathbf{F}_{BC}$ is "pushing" on the pin, and consequently member BC is in *compression.* These effects are clearly demonstrated by isolating the joint with small segments of the member connected to the pin, Fig. 6–7c. The pushing or pulling on these small segments indicates the effect of the member being either in compression or tension.

The analysis always should start at a joint having at least one known force and at most two unknown forces, as in Fig. 6–7b. In this way, application of $\Sigma F_x = 0$ and $\Sigma F_y = 0$ yields two algebraic equations which can be solved for the two unknowns. When applying these equations, the correct sense of an unknown member force can be determined using one of two possible methods:

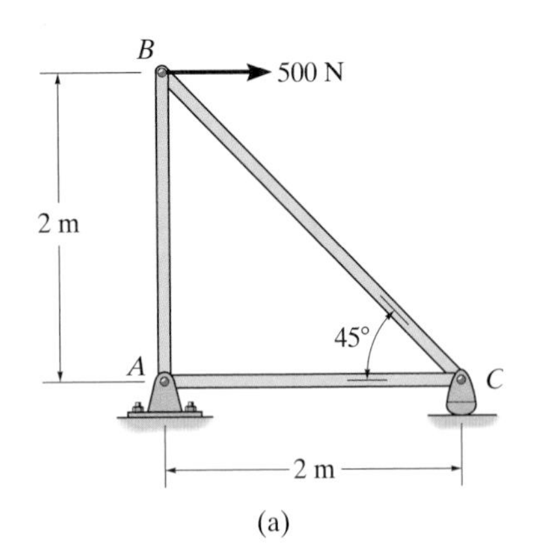

(a)

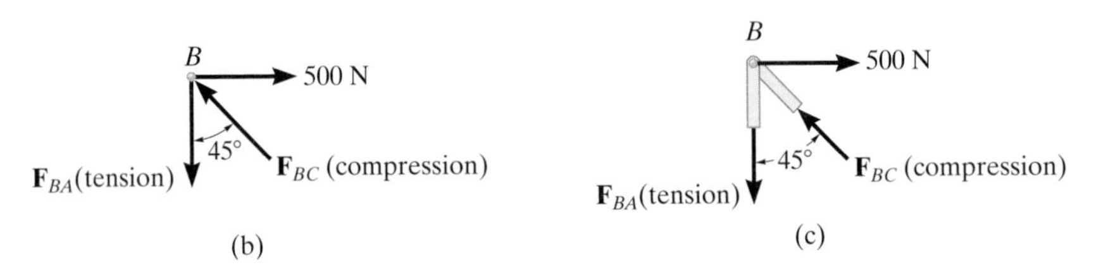

(b) (c)

Fig. 6–7

- *Always assume* the *unknown member forces* acting on the joint's free-body diagram to be in *tension,* i.e., "pulling" on the pin. If this is done, then numerical solution of the equilibrium equations will yield *positive scalars for members in tension and negative scalars for members in compression.* Once an unknown member force is found, use its *correct* magnitude and sense (T or C) on subsequent joint free-body diagrams.
- The *correct* sense of direction of an unknown member force can, in many cases, be determined "by inspection." For example, $\mathbf{F}_{BC}$ in Fig. 6–7*b* must push on the pin (compression) since its horizontal component, $F_{BC} \sin 45°$, must balance the 500-N force ($\Sigma F_x = 0$). Likewise, $\mathbf{F}_{BA}$ is a tensile force since it balances the vertical component, $F_{BC} \cos 45°$ ($\Sigma F_y = 0$). In more complicated cases, the sense of an unknown member force can be *assumed;* then, after applying the equilibrium equations, the assumed sense can be verified from the numerical results. A *positive* answer indicates that the sense is *correct,* whereas a *negative* answer indicates that the sense shown on the free-body diagram must be *reversed.* This is the method we will use in the example problems which follow.

Procedure For Analysis

The following procedure provides a typical means for analyzing a truss using the method of joints.

- Draw the free-body diagram of a joint having at least one known force and at most two unknown forces. (If this joint is at one of the supports, then it may be necessary to know the external reactions at the truss support.)
- Use one of the two methods described above for establishing the sense of an unknown force.
- Orient the x and y axes such that the forces on the free-body diagram can be easily resolved into their x and y components and then apply the two force equilibrium equations $\Sigma F_x = 0$ and $\Sigma F_y = 0$. Solve for the two unknown member forces and verify their correct sense.
- Continue to analyze each of the other joints. Again, it is necessary to choose a joint having at most two unknowns and at least one known force.
- Once the force in a member is found from the analysis of a joint at one of its ends, the result can be used to analyze the forces acting on the joint at its other end. Remember that a member in *compression* "pushes" on the joint and a member in *tension* "pulls" on the joint.

EXAMPLE 6.1

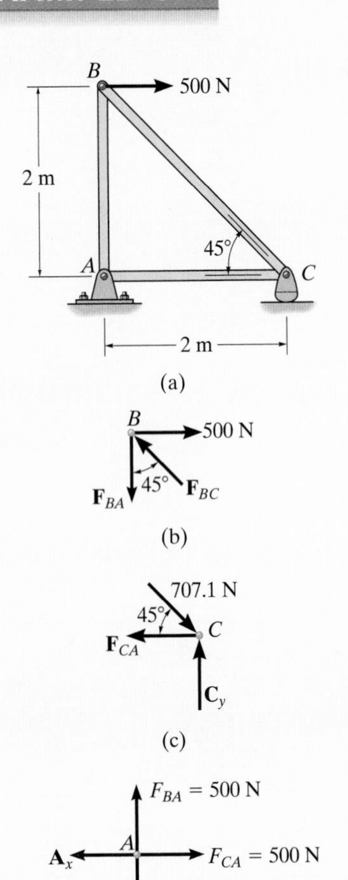

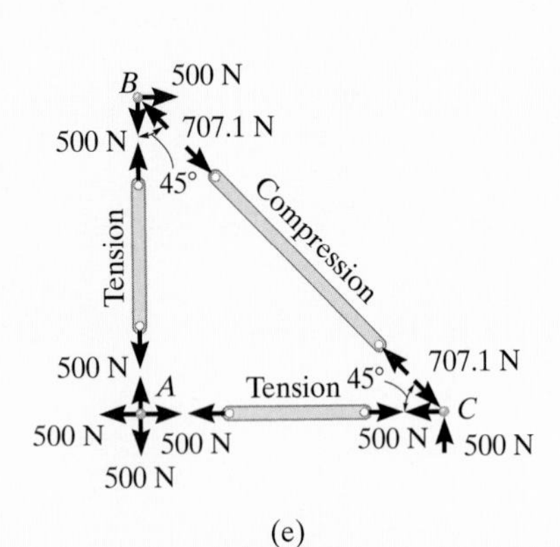

Fig. 6–8

Determine the force in each member of the truss shown in Fig. 6–8*a* and indicate whether the members are in tension or compression.

SOLUTION

By inspection of Fig. 6–8*a*, there are two unknown member forces at joint *B*, two unknown member forces and an unknown reaction force at joint *C*, and two unknown member forces and two unknown reaction forces at joint *A*. Since we should have no more than two unknowns at the joint and at least one known force acting there, we will begin the analysis at joint *B*.

Joint B. The free-body diagram of the pin at *B* is shown in Fig. 6–8*b*. Applying the equations of joint equilibrium, we have

$$\xrightarrow{+} \Sigma F_x = 0; \quad 500\text{ N} - F_{BC} \sin 45^\circ = 0 \qquad F_{BC} = 707.1\text{ N (C)} \quad \textit{Ans.}$$

$$+\uparrow \Sigma F_y = 0; \quad F_{BC} \cos 45^\circ - F_{BA} = 0 \qquad F_{BA} = 500\text{ N (T)} \quad \textit{Ans.}$$

Since the force in member *BC* has been calculated, we can proceed to analyze joint *C* in order to determine the force in member *CA* and the support reaction at the rocker.

Joint C. From the free-body diagram of joint *C*, Fig. 6–8*c*, we have

$$\xrightarrow{+} \Sigma F_x = 0; \quad -F_{CA} + 707.1 \cos 45^\circ \text{ N} = 0 \qquad F_{CA} = 500\text{ N (T)} \quad \textit{Ans.}$$

$$+\uparrow \Sigma F_y = 0; \quad C_y - 707.1 \sin 45^\circ \text{ N} = 0 \qquad C_y = 500\text{ N} \quad \textit{Ans.}$$

Joint A. Although it is not necessary, we can determine the support reactions at joint *A* using the results of $F_{CA} = 500$ N and $F_{BA} = 500$ N. From the free-body diagram, Fig. 6–8*d*, we have

$$\xrightarrow{+} \Sigma F_x = 0; \quad 500\text{ N} - A_x = 0 \quad A_x = 500\text{ N}$$

$$+\uparrow \Sigma F_y = 0; \quad 500\text{ N} - A_y = 0 \quad A_y = 500\text{ N}$$

NOTE: The results of the analysis are summarized in Fig. 6–8*e*. Note that the free-body diagram of each pin shows the effects of all the connected members and external forces applied to the pin, whereas the free-body diagram of each member shows only the effects of the end pins on the member.

EXAMPLE 6.2

Determine the forces acting in all the members of the truss shown in Fig. 6–9*a*.

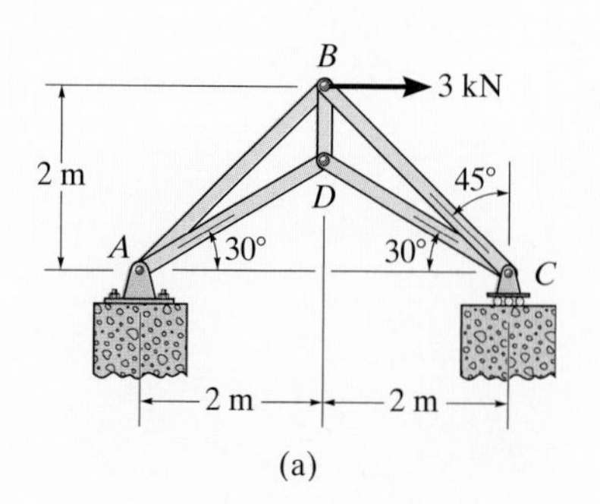

(a)

SOLUTION

By inspection, there are more than two unknowns at each joint. Consequently, the support reactions on the truss must first be determined. Show that they have been correctly calculated on the free-body diagram in Fig. 6–9*b*. We can now begin the analysis at joint *C*. Why?

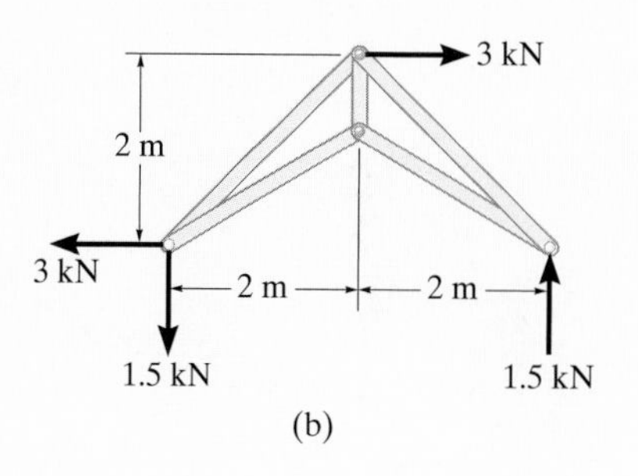

(b)

Joint C. From the free-body diagram, Fig. 6–9*c*,

$$\xrightarrow{+} \Sigma F_x = 0; \qquad -F_{CD} \cos 30^\circ + F_{CB} \sin 45^\circ = 0$$

$$+\uparrow \Sigma F_y = 0; \qquad 1.5 \text{ kN} + F_{CD} \sin 30^\circ - F_{CB} \cos 45^\circ = 0$$

These two equations must be solved *simultaneously* for each of the two unknowns. Note, however, that a *direct solution* for one of the unknown forces may be obtained by applying a force summation along an axis that is *perpendicular* to the direction of the other unknown force. For example, summing forces along the y' axis, which is perpendicular to the direction of $\mathbf{F}_{CD}$, Fig. 6–9*d*, yields a direct solution for F_{CB}.

$$+\nearrow \Sigma F_{y'} = 0;$$

$$1.5 \cos 30^\circ \text{ kN} - F_{CB} \sin 15^\circ = 0 \quad F_{CB} = 5.02 \text{ kN} \quad \text{(C)} \qquad \textit{Ans.}$$

In a similar fashion, summing forces along the y'' axis, Fig. 6–9*e*, yields a direct solution for F_{CD}.

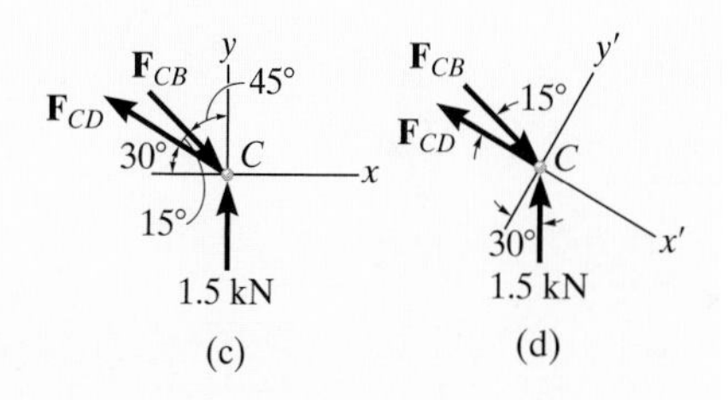

(c) (d)

$$+\nearrow \Sigma F_{y''} = 0;$$

$$1.5 \cos 45^\circ \text{ kN} - F_{CD} \sin 15^\circ = 0 \qquad F_{CD} = 4.10 \text{ kN} \quad \text{(T)} \quad \textit{Ans.}$$

Joint D. We can now proceed to analyze joint *D*. The free-body diagram is shown in Fig. 6–9*f*.

$$\xrightarrow{+} \Sigma F_x = 0; \qquad -F_{DA} \cos 30^\circ + 4.10 \cos 30^\circ \text{ kN} = 0$$

$$F_{DA} = 4.10 \text{ kN} \quad \text{(T)} \qquad \textit{Ans.}$$

$$+\uparrow \Sigma F_y = 0; \qquad F_{DB} - 2(4.10 \sin 30^\circ \text{ kN}) = 0$$

$$F_{DB} = 4.10 \text{ kN} \quad \text{(T)} \qquad \textit{Ans.}$$

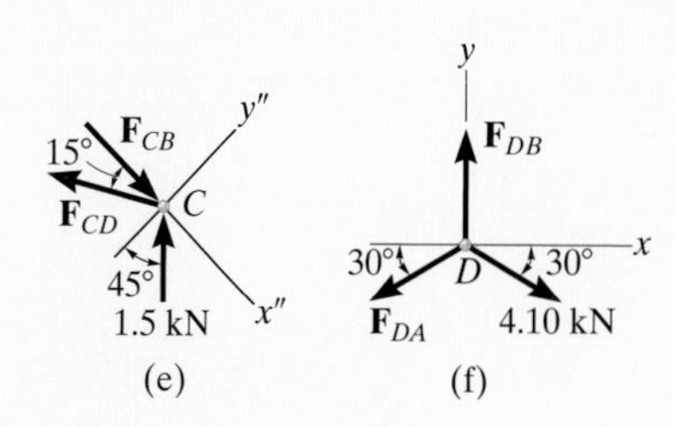

(e) (f)

Fig. 6–9

NOTE: The force in the last member, *BA*, can be obtained from joint *B* or joint *A*. As an exercise, draw the free-body diagram of joint *B*, sum the forces in the horizontal direction, and show that $F_{BA} = 0.776$ kN (C).

EXAMPLE 6.3

Determine the force in each member of the truss shown in Fig. 6–10*a*. Indicate whether the members are in tension or compression.

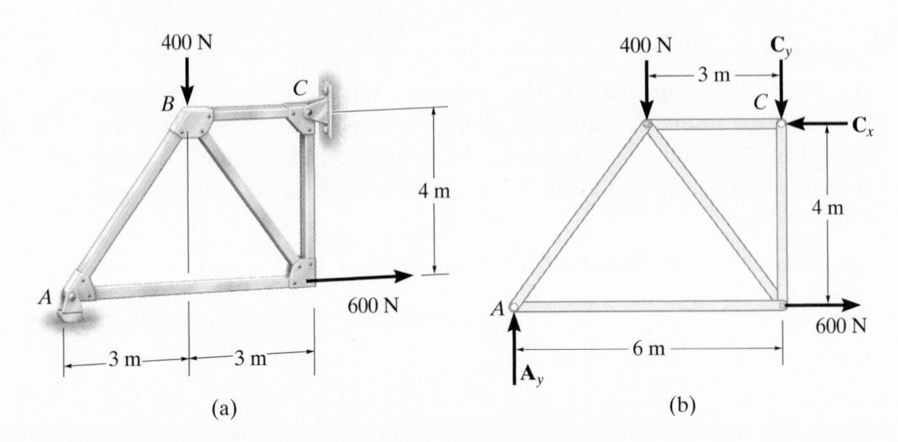

Fig. 6–10

SOLUTION

Support Reactions. No joint can be analyzed until the support reactions are determined. Why? A free-body diagram of the entire truss is given in Fig. 6–10*b*. Applying the equations of equilibrium, we have

$$\xrightarrow{+} \Sigma F_x = 0; \qquad 600 \text{ N} - C_x = 0 \qquad C_x = 600 \text{ N}$$

$$\circlearrowleft + \Sigma M_C = 0; \qquad -A_y(6 \text{ m}) + 400 \text{ N}(3 \text{ m}) + 600 \text{ N}(4 \text{ m}) = 0$$

$$A_y = 600 \text{ N}$$

$$+\uparrow \Sigma F_y = 0; \qquad 600 \text{ N} - 400 \text{ N} - C_y = 0 \qquad C_y = 200 \text{ N}$$

The analysis can now start at either joint *A* or *C*. The choice is arbitrary since there are one known and two unknown member forces acting on the pin at each of these joints.

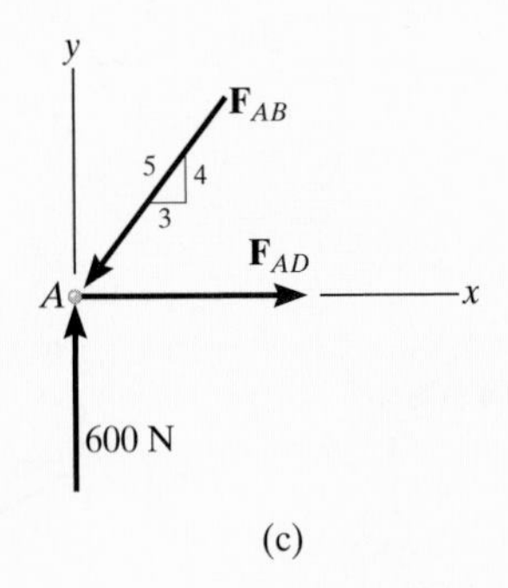

(c)

Joint A (Fig. 6–10*c*). As shown on the free-body diagram, there are three forces that act on the pin at joint *A*. The inclination of $\mathbf{F}_{AB}$ is determined from the geometry of the truss. By inspection, can you see why this force is assumed to be compressive and $\mathbf{F}_{AD}$ tensile? Applying the equations of equilibrium, we have

$$+\uparrow \Sigma F_y = 0; \qquad 600 \text{ N} - \tfrac{4}{5} F_{AB} = 0 \qquad F_{AB} = 750 \text{ N (C)} \qquad \textit{Ans.}$$

$$\xrightarrow{+} \Sigma F_x = 0; \qquad F_{AD} - \tfrac{3}{5}(750 \text{ N}) = 0 \qquad F_{AD} = 450 \text{ N (T)} \qquad \textit{Ans.}$$

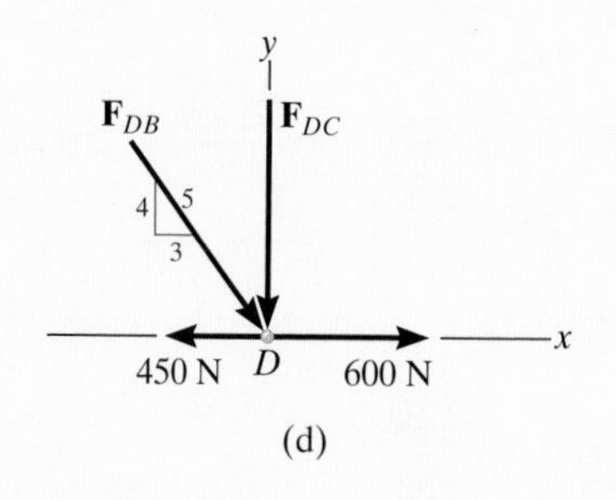

(d)

Joint D (Fig. 6–10*d*). The pin at this joint is chosen next since, by inspection of Fig. 6–10*a*, the force in *AD* is known and the unknown forces in *DB* and *DC* can be determined. Summing forces in the horizontal direction, Fig. 6–10*d*, we have

$$\xrightarrow{+} \Sigma F_x = 0; \qquad -450\text{ N} + \tfrac{3}{5}F_{DB} + 600\text{ N} = 0 \quad F_{DB} = -250\text{ N}$$

The negative sign indicates that $\mathbf{F}_{DB}$ acts in the *opposite sense* to that shown in Fig. 6–10*d*.* Hence,

$$F_{DB} = 250\text{ N} \qquad \text{(T)} \qquad\qquad \textit{Ans.}$$

To determine $\mathbf{F}_{DC}$, we can either correct the sense of $\mathbf{F}_{DB}$ and then apply $\Sigma F_y = 0$, or apply this equation and retain the negative sign for F_{DB}, i.e.,

$$+\uparrow \Sigma F_y = 0; \quad -F_{DC} - \tfrac{4}{5}(-250\text{ N}) = 0 \qquad F_{DC} = 200\text{ N} \quad \text{(C)} \quad \textit{Ans.}$$

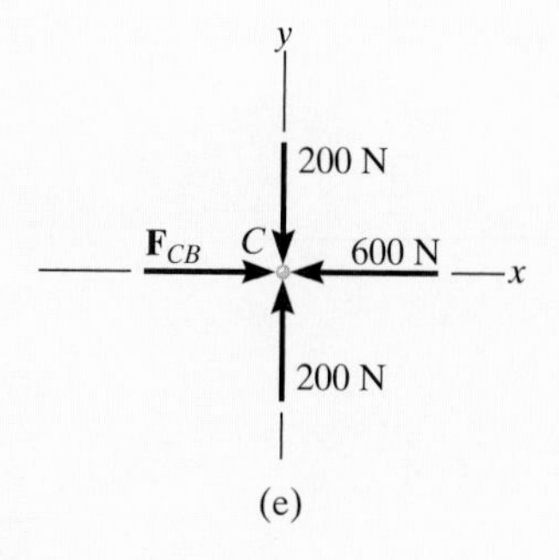

(e)

Joint C (Fig. 6–10*e*).

$$\xrightarrow{+} \Sigma F_x = 0; \qquad F_{CB} - 600\text{ N} = 0 \qquad F_{CB} = 600\text{ N} \quad \text{(C)} \qquad \textit{Ans.}$$

$$+\uparrow \Sigma F_y = 0; \qquad 200\text{ N} - 200\text{ N} \equiv 0 \quad \text{(check)}$$

NOTE: The analysis is summarized in Fig. 6–10*f*, which shows the correct free-body diagram for each pin and member.

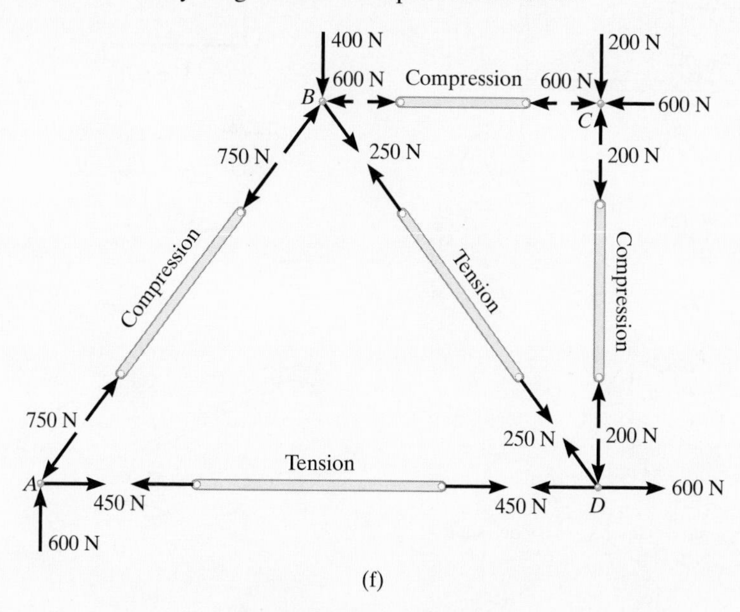

(f)

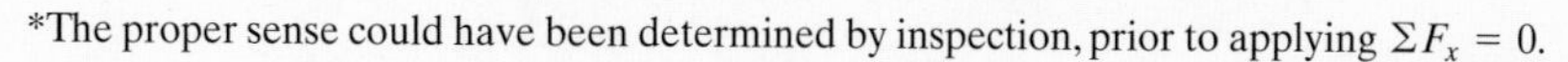
*The proper sense could have been determined by inspection, prior to applying $\Sigma F_x = 0$.

6.3 Zero-Force Members

Truss analysis using the method of joints is greatly simplified if we can first determine those members which support *no loading*. These *zero-force members* are used to increase the stability of the truss during construction and to provide support if the applied loading is changed.

The zero-force members of a truss can generally be determined *by inspection* of each of its joints. For example, consider the truss shown in Fig. 6–11*a*. If a free-body diagram of the pin at joint A is drawn, Fig. 6–11*b*, it is seen that members AB and AF are zero-force members. On the other hand, notice that we could not have come to this conclusion if we had considered the free-body diagrams of joints F or B simply because there are five unknowns at each of these joints. In a similar manner, consider the free-body diagram of joint D, Fig. 6–11*c*. Here again it is seen that DC and DE are zero-force members. As a general rule, *if only two members form a truss joint and no external load or support reaction is applied to the joint, the members must be zero-force members.* The load on the truss in Fig. 6–11*a* is therefore supported by only five members as shown in Fig. 6–11*d*.

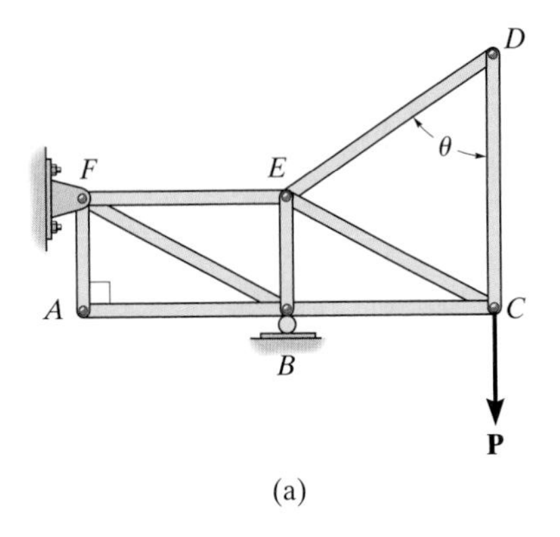

(a)

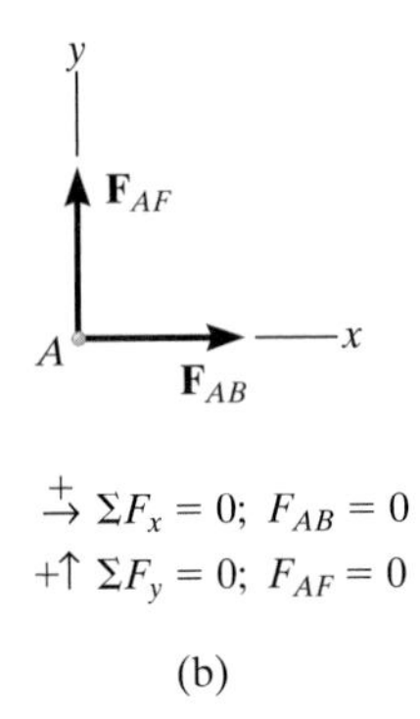

(b)

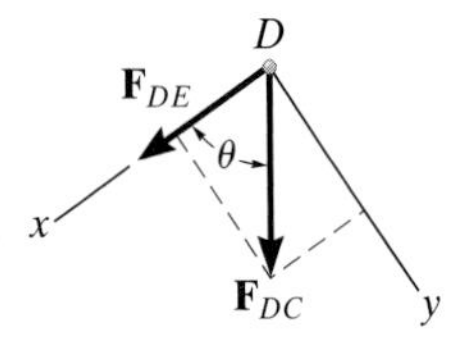

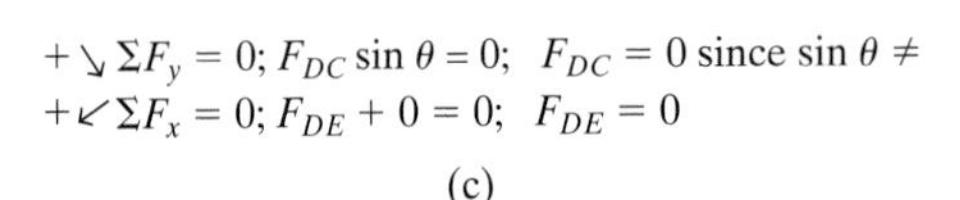

(c)

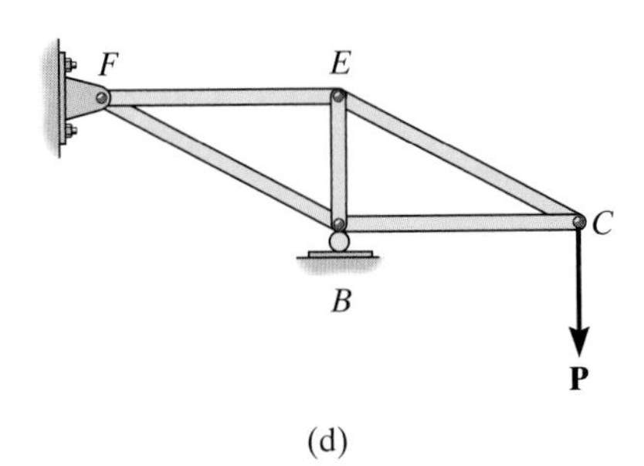

(d)

Fig. 6–11

Now consider the truss shown in Fig. 6–12*a*. The free-body diagram of the pin at joint D is shown in Fig. 6–12*b*. By orienting the y axis along members DC and DE and the x axis along member DA, it is seen that DA is a zero-force member. Note that this is also the case for member CA, Fig. 6–12*c*. In general, *if three members form a truss joint for which two of the members are collinear, the third member is a zero-force member provided no external force or support reaction is applied to the joint.* The truss shown in Fig. 6–12*d* is therefore suitable for supporting the load **P**.

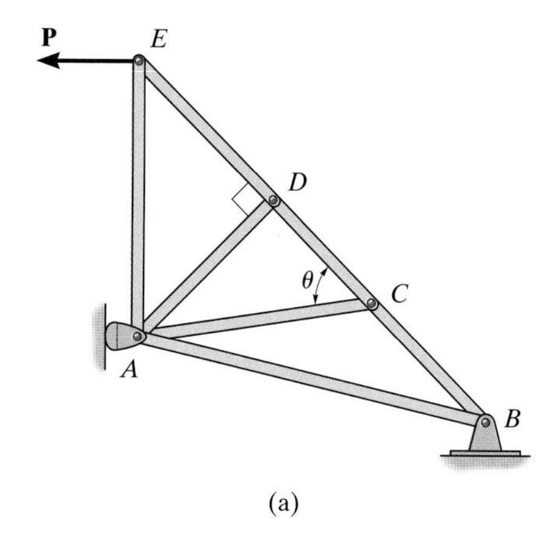

(a)

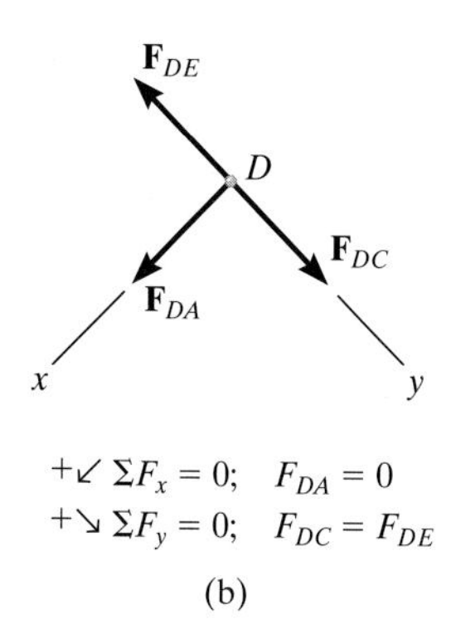

$+\swarrow \Sigma F_x = 0; \quad F_{DA} = 0$

$+\searrow \Sigma F_y = 0; \quad F_{DC} = F_{DE}$

(b)

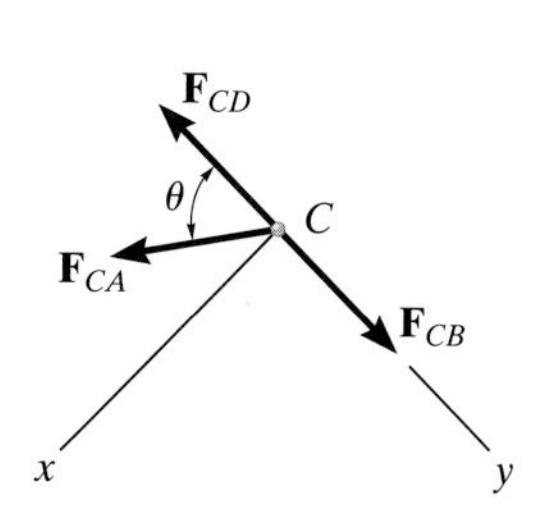

$+\swarrow \Sigma F_x = 0; \quad F_{CA} \sin\theta = 0; \quad F_{CA} = 0 \text{ since } \sin\theta \neq 0;$

$+\searrow \Sigma F_y = 0; \quad F_{CB} = F_{CD}$

(c)

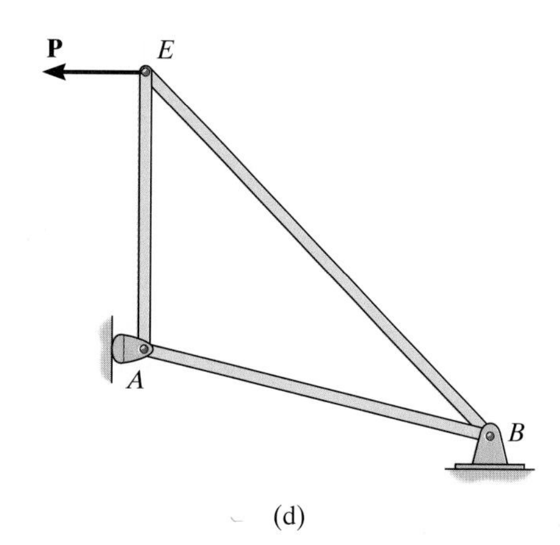

(d)

Fig. 6–12

EXAMPLE 6.4

Using the method of joints, determine all the zero-force members of the *Fink roof truss* shown in Fig. 6–13*a*. Assume all joints are pin connected.

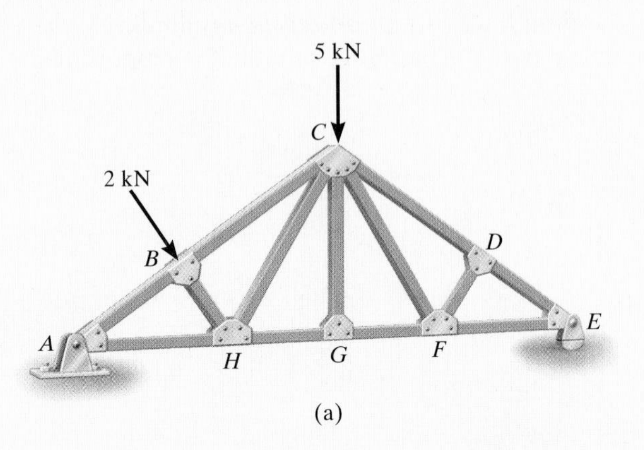

(a)

Fig. 6–13

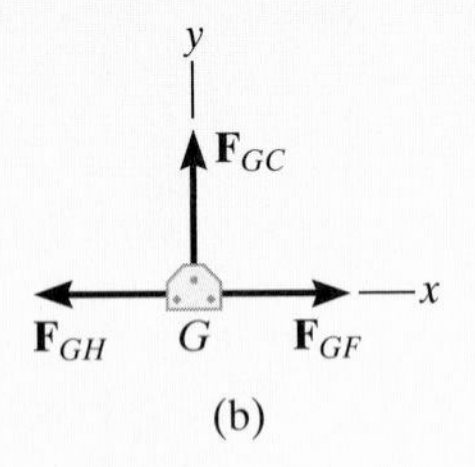

(b)

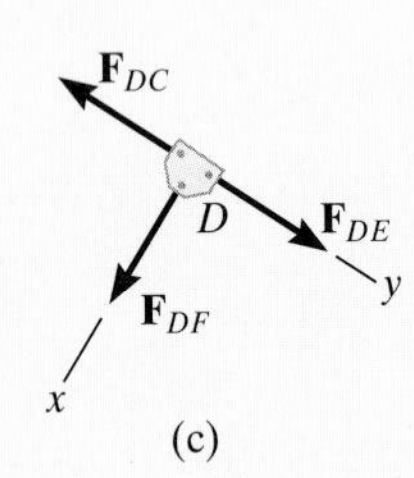

(c)

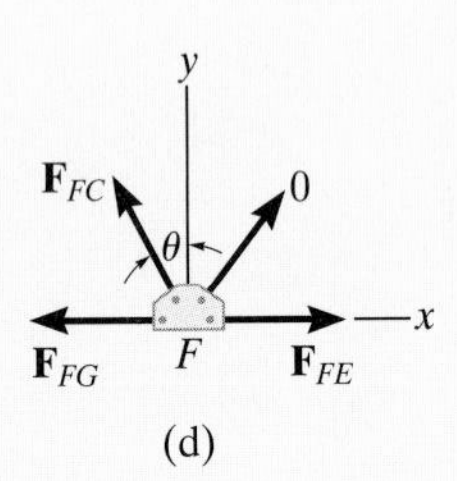

(d)

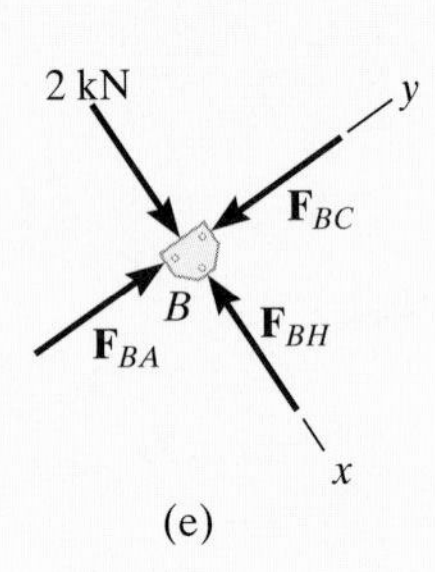

(e)

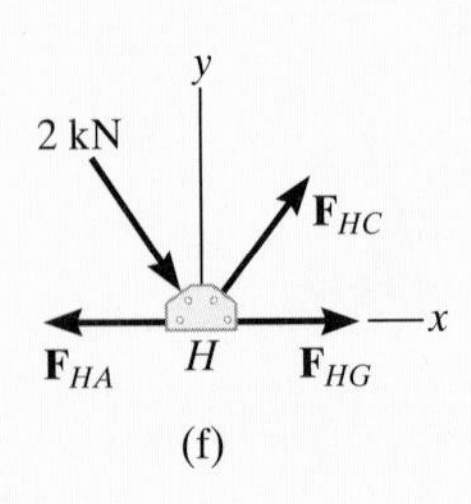

(f)

SOLUTION

Look for joint geometries that have three members for which two are collinear. We have

Joint G (Fig. 6–13*b*).

$$+\uparrow \Sigma F_y = 0; \qquad F_{GC} = 0 \qquad \textit{Ans.}$$

Realize that we could not conclude that *GC* is a zero-force member by considering joint *C*, where there are five unknowns. The fact that *GC* is a zero-force member means that the 5-kN load at *C* must be supported by members *CB*, *CH*, *CF*, and *CD*.

Joint D (Fig. 6–13*c*).

$$+\swarrow \Sigma F_x = 0; \qquad F_{DF} = 0 \qquad \textit{Ans.}$$

Joint F (Fig. 6–13*d*).

$$+\uparrow \Sigma F_y = 0; \qquad F_{FC} \cos\theta = 0 \qquad \text{Since } \theta \neq 90^\circ, \qquad F_{FC} = 0 \qquad \textit{Ans.}$$

NOTE: If joint *B* is analyzed, Fig. 6–13*e*,

$$+\searrow \Sigma F_x = 0; \qquad 2\text{ kN} - F_{BH} = 0 \quad F_{BH} = 2\text{ kN} \qquad \text{(C)}$$

Also, F_{HC} must satisfy $\Sigma F_y = 0$, Fig. 6–13*f*, and therefore *HC* is *not* a zero-force member.

PROBLEMS

6–1. Determine the force in each member of the truss and state if the members are in tension or compression. Set $P_1 = 7$ kN, $P_2 = 7$ kN.

6–2. Determine the force in each member of the truss and state if the members are in tension or compression. Set $P_1 = 8$ kN, $P_2 = 10$ kN.

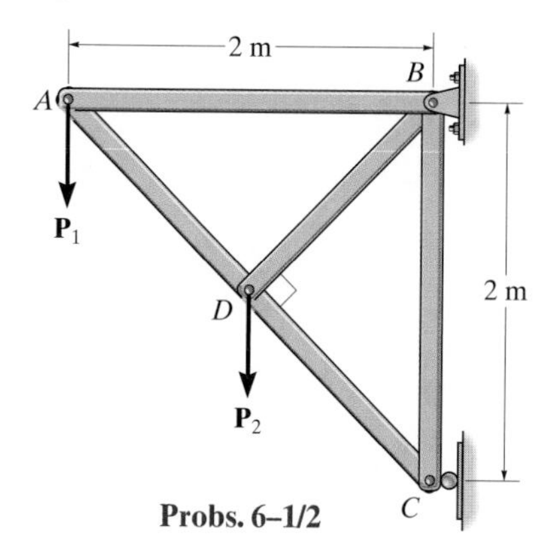

Probs. 6–1/2

6–3. The truss, used to support a balcony, is subjected to the loading shown. Approximate each joint as a pin and determine the force in each member. State whether the members are in tension or compression. Set $P_1 = 600$ lb, $P_2 = 400$ lb.

***6–4.** The truss, used to support a balcony, is subjected to the loading shown. Approximate each joint as a pin and determine the force in each member. State whether the members are in tension or compression. Set $P_1 = 800$ lb, $P_2 = 0$.

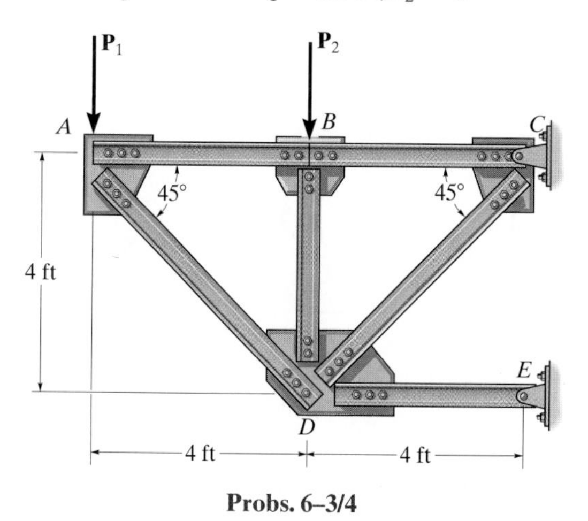

Probs. 6–3/4

6–5. Determine the force in each member of the truss and state if the members are in tension or compression. Set $P_1 = 20$ kN, $P_2 = 10$ kN.

6–6. Determine the force in each member of the truss and state if the members are in tension or compression. Set $P_1 = 40$ kN, $P_2 = 20$ kN.

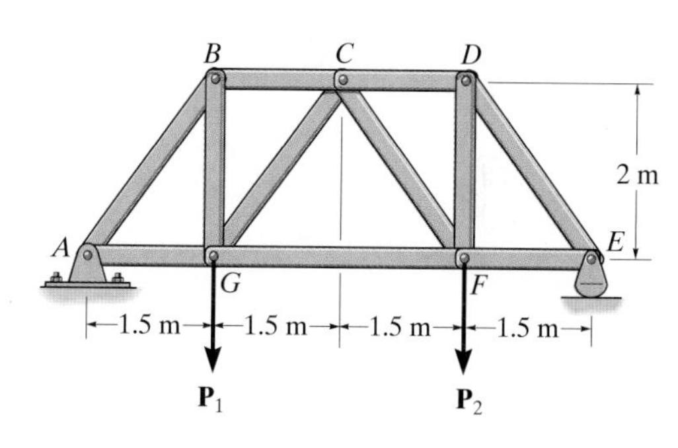

Probs. 6–5/6

6–7. Determine the force in each member of the truss and state if the members are in tension or compression.

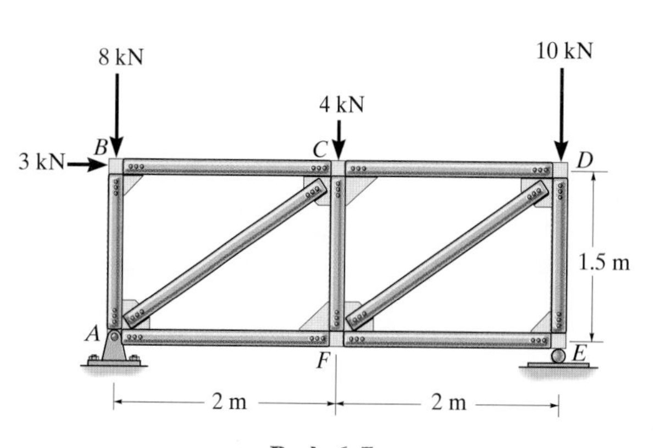

Prob. 6–7

*6–8. Determine the force in each member of the truss in terms of the external loading and state if the members are in tension or compression.

6–9. The maximum allowable tensile force in the members of the truss is $(F_t)_{max} = 1500$ lb, and the maximum allowable compressive force is $(F_c)_{max} = 800$ lb. Determine the maximum magnitude P of the two loads that can be applied to the truss. Take $a = 8$ ft.

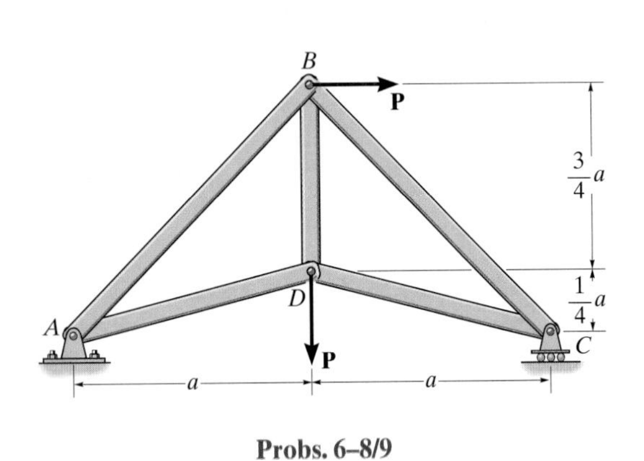

Probs. 6–8/9

*6–12. Determine the force in each member of the truss and state if the members are in tension or compression. Set $P_1 = 10$ kN, $P_2 = 15$ kN.

6–13. Determine the force in each member of the truss and state if the members are in tension or compression. Set $P_1 = 0$, $P_2 = 20$ kN.

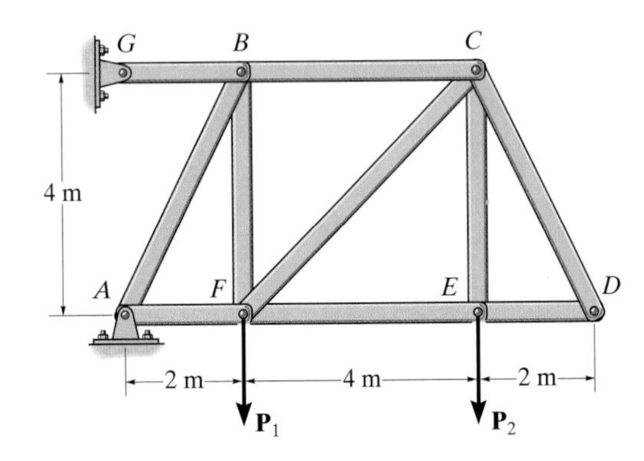

Probs. 6–12/13

6–10. Determine the force in each member of the truss and state if the members are in tension or compression. Set $P_1 = 0$, $P_2 = 1000$ lb.

6–11. Determine the force in each member of the truss and state if the members are in tension or compression. Set $P_1 = 500$ lb, $P_2 = 1500$ lb.

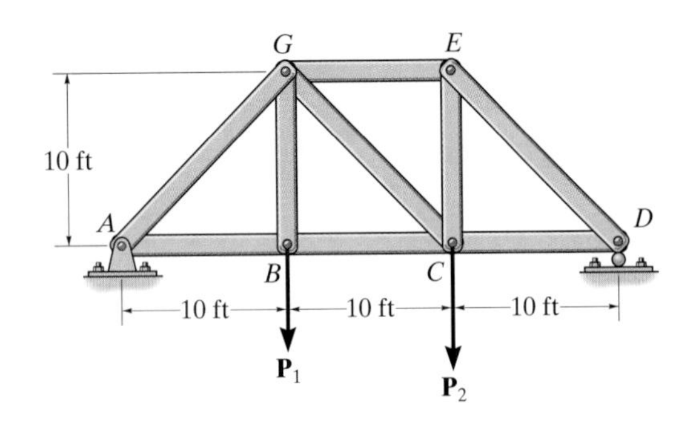

Probs. 6–10/11

6–14. Determine the force in each member of the truss and state if the members are in tension or compression. Set $P_1 = 100$ lb, $P_2 = 200$ lb, $P_3 = 300$ lb.

6–15. Determine the force in each member of the truss and state if the members are in tension or compression. Set $P_1 = 400$ lb, $P_2 = 400$ lb, $P_3 = 0$.

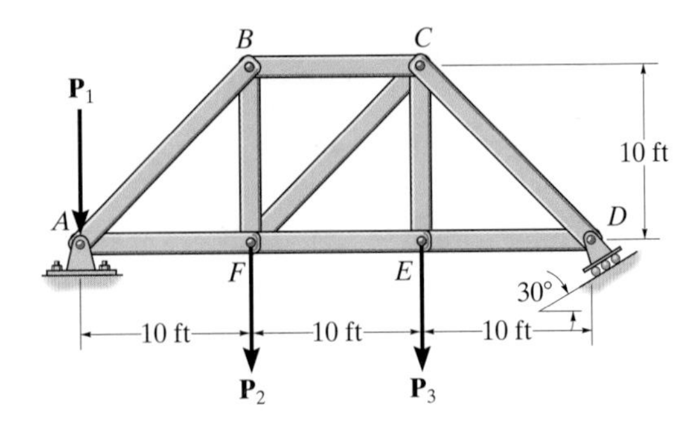

Probs. 6–14/15

*6–16. Determine the force in each member of the truss in terms of the load P and state if the members are in tension or compression.

6–17. The maximum allowable tensile force in the members of the truss is $(F_t)_{max} = 5$ kN, and the maximum allowable compressive force is $(F_c)_{max} = 3$ kN. Determine the maximum magnitude of the load **P** that can be applied to the truss. Take $d = 2$ m.

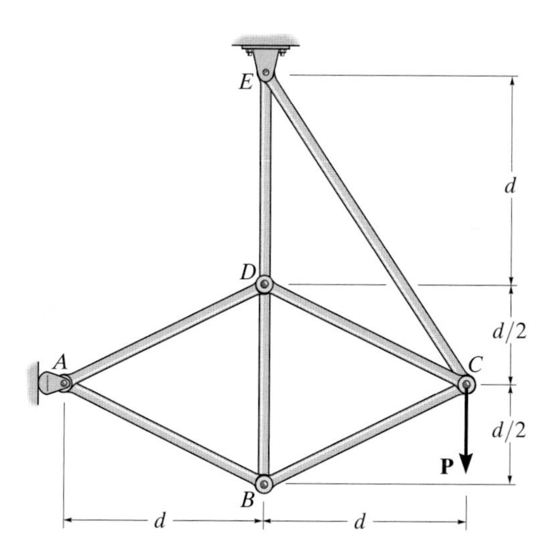

Probs. 6–16/17

6–18. Determine the force in each member of the truss and state if the members are in tension or compression. *Hint:* The horizontal force component at A must be zero. Why?

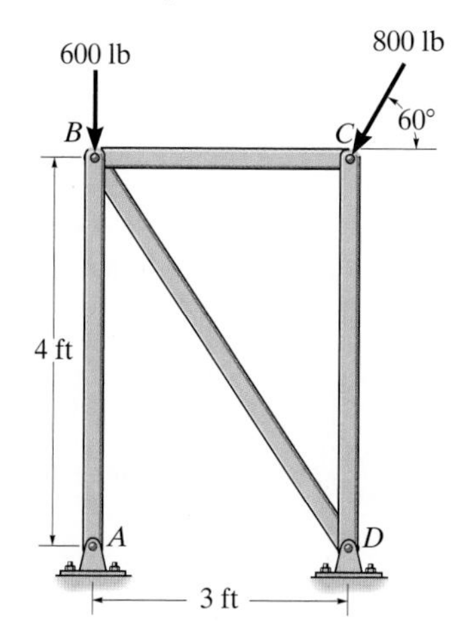

Prob. 6–18

6–19. Determine the force in each member of the truss and state if the members are in tension or compression. *Hint:* The resultant force at the pin E acts along member ED. Why?

*6–20. Each member of the truss is uniform and has a mass of 8 kg/m. Remove the external loads of 3 kN and 2 kN and determine the approximate force in each member due to the weight of the truss. State if the members are in tension or compression. Solve the problem by *assuming* the weight of each member can be represented as a vertical force, half of which is applied at each end of the member.

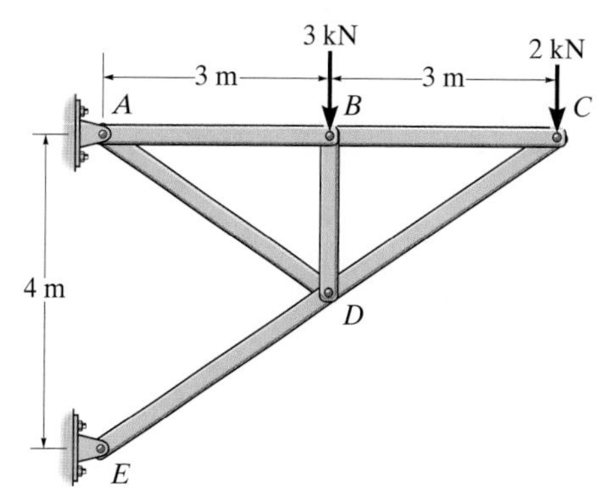

Probs. 6–19/20

6–21. Determine the force in each member of the truss in terms of the external loading and state if the members are in tension or compression.

6–22. The maximum allowable tensile force in the members of the truss is $(F_t)_{max} = 2$ kN, and the maximum allowable compressive force is $(F_c)_{max} = 1.2$ kN. Determine the maximum magnitude P of the two loads that can be applied to the truss. Take $L = 2$ m and $\theta = 30°$.

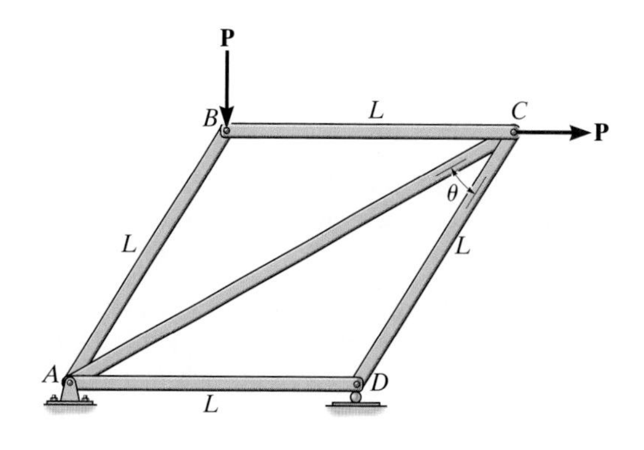

Probs. 6–21/22

6–23. The *Fink truss* supports the loads shown. Determine the force in each member and state if the members are in tension or compression. Approximate each joint as a pin.

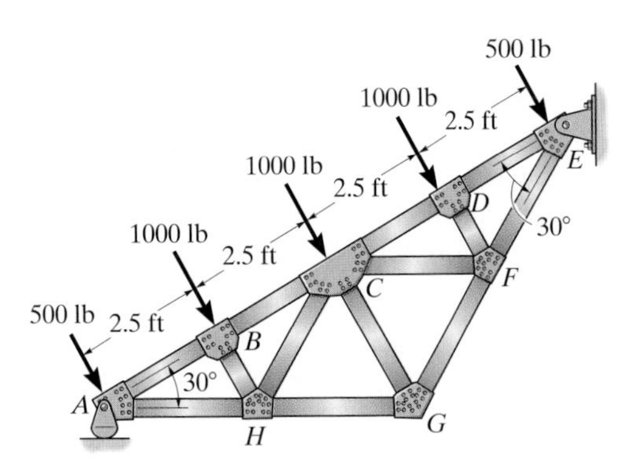

Prob. 6–23

***6–24.** Determine the force in each member of the double scissors truss in terms of the load P and state if the members are in tension or compression.

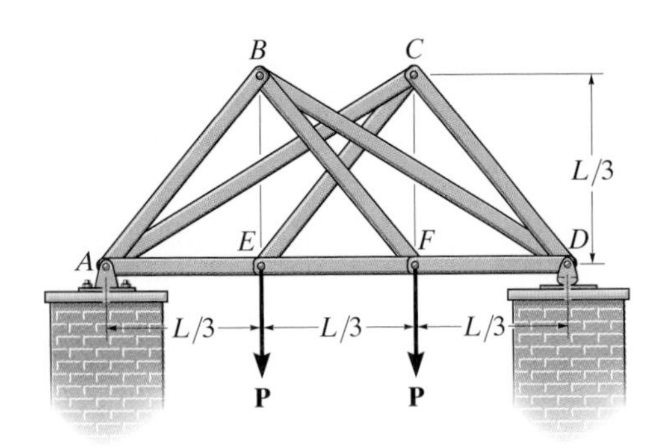

Prob. 6–24

6–25. Determine the force in each member of the truss and state if the members are in tension or compression. *Hint:* The vertical component of force at C must equal zero. Why?

6–26. Each member of the truss is uniform and has a mass of 8 kg/m. Remove the external loads of 6 kN and 8 kN and determine the approximate force in each member due to the weight of the truss. State if the members are in tension or compression. Solve the problem by *assuming* the weight of each member can be represented as a vertical force, half of which is applied at each end of the member.

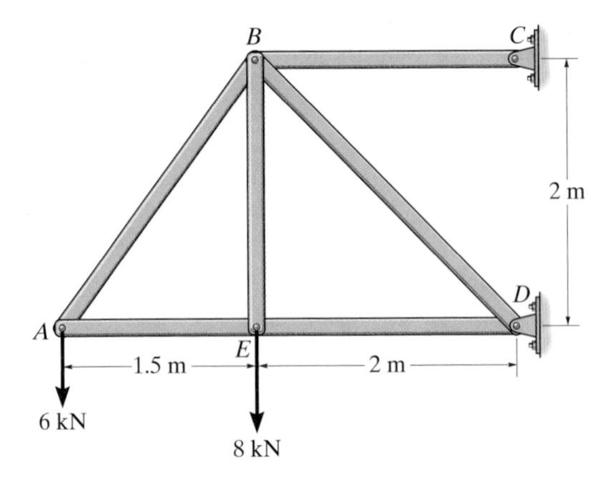

Probs. 6–25/26

6–27. Determine the force in each member of the truss and state if the members are in tension or compression. Set $P_1 = 4$ kN, $P_2 = 0$.

***6–28.** Determine the force in each member of the truss and state if the members are in tension or compression. Set $P_1 = 2$ kN, $P_2 = 4$ kN.

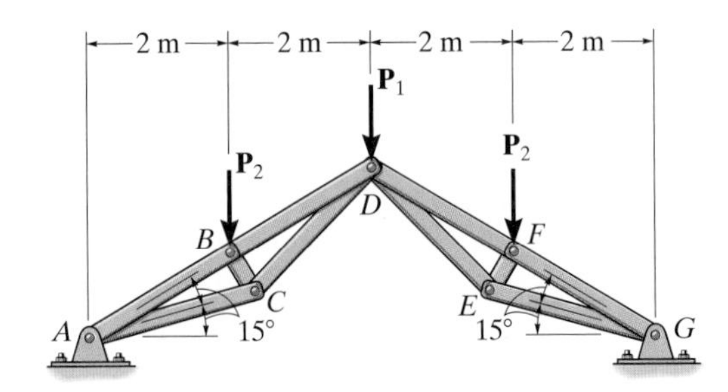

Probs. 6–27/28

6–29. Determine the force in each member of the truss and state if the members are in tension or compression.

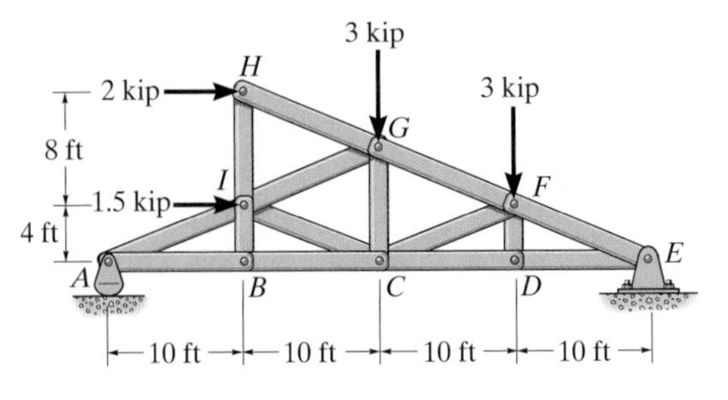

Prob. 6–29

6.4 The Method of Sections

When we need to find the force in only a few members of a truss, we can analyze the truss using the *method of sections*. It is based on the principle that if a body is in equilibrium then any part of the body is also in equilibrium. For example, consider the two truss members shown on the left in Fig. 6–14. If the forces within the members are to be determined, then an imaginary section indicated by the blue line, can be used to cut each member into two parts and thereby "expose" each internal force as "external" to the free-body diagrams shown on the right. Clearly, it can be seen that equilibrium requires that the member in tension (T) be subjected to a "pull," whereas the member in compression (C) is subjected to a "push."

The method of sections can also be used to "cut" or section the members of an entire truss. If the section passes through the truss and the free-body diagram of either of its two parts is drawn, we can then apply the equations of equilibrium to that part to determine the member forces at the "cut section." Since only *three* independent equilibrium equations ($\Sigma F_x = 0$, $\Sigma F_y = 0$, $\Sigma M_O = 0$) can be applied to the isolated part of the truss, try to select a section that, in general, passes through not more than *three* members in which the forces are unknown. For example, consider the truss in Fig. 6–15*a*. If the force in member GC is to be determined, section *aa* would be appropriate. The free-body diagrams of the two parts are shown in Figs. 6–15*b* and 6–15*c*. In particular, note that the line of action of each member force is specified from the *geometry* of the truss, since the force in a member passes along its axis. Also, the member forces acting on one part of the truss are equal but opposite to those acting on the other part—Newton's third law. As noted above, the members assumed to be in *tension* (BC and GC) are subjected to a "pull," whereas the member in *compression* (GF) is subjected to a "push."

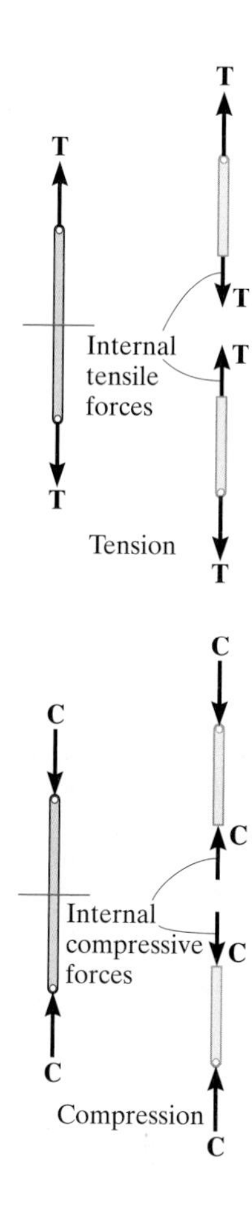

Fig. 6–14

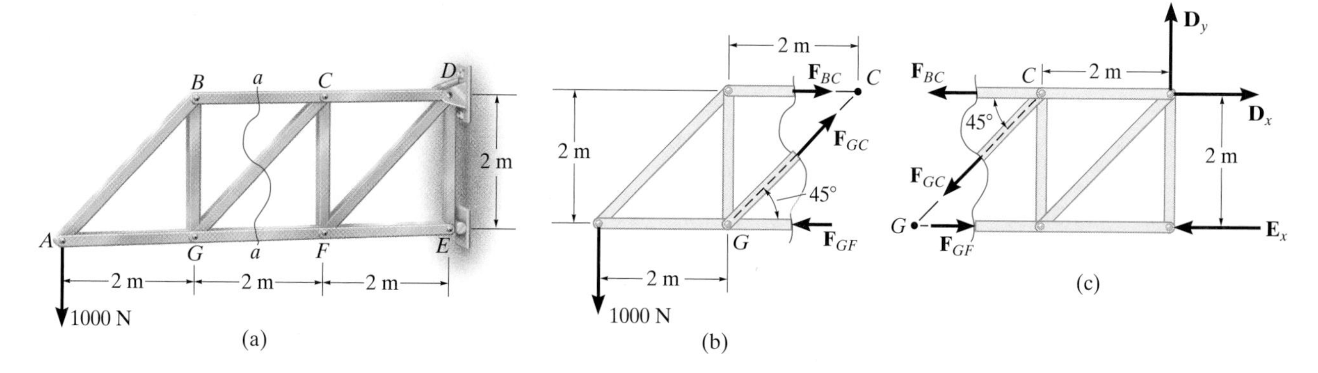

Fig. 6–15

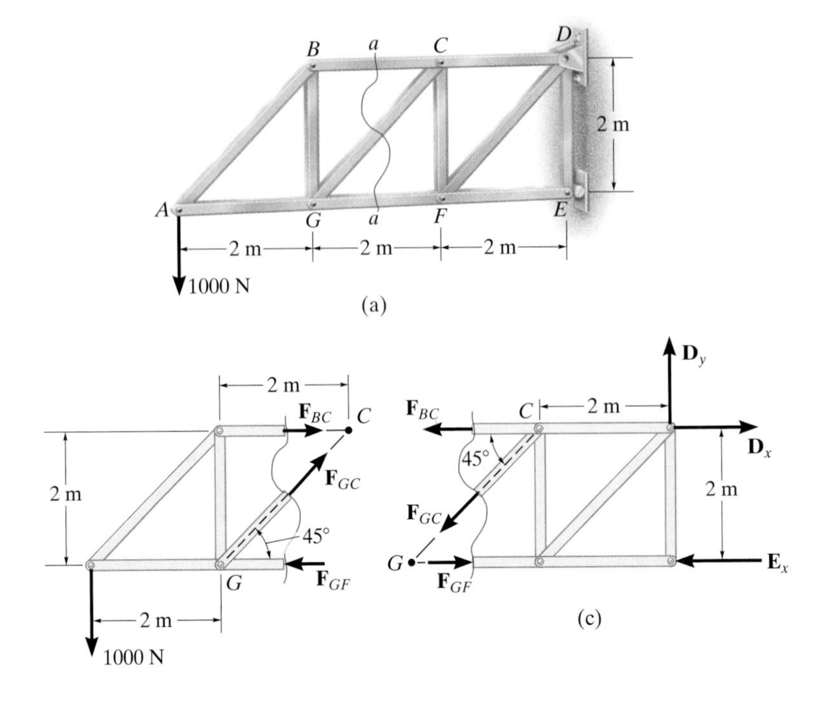

Fig. 6–15

Two Pratt trusses are used to construct this pedestrian bridge.

The three unknown member forces $\mathbf{F}_{BC}$, $\mathbf{F}_{GC}$, and $\mathbf{F}_{GF}$ can be obtained by applying the three equilibrium equations to the free-body diagram in Fig. 6–15*b*. If, however, the free-body diagram in Fig. 6–15*c* is considered, the three support reactions $\mathbf{D}_x$, $\mathbf{D}_y$ and $\mathbf{E}_x$ will have to be known, because only three equations of equilibrium are available. (This, of course, is done in the usual manner by considering a free-body diagram of the *entire truss.*)

When applying the equilibrium equations, one should consider ways of writing the equations so as to yield a *direct solution* for each of the unknowns, rather than having to solve simultaneous equations. For example, summing moments about C in Fig. 6–15*b* would yield a direct solution for $\mathbf{F}_{GF}$ since $\mathbf{F}_{BC}$ and $\mathbf{F}_{GC}$ create zero moment about C. Likewise, $\mathbf{F}_{BC}$ can be directly obtained by summing moments about G. Finally, $\mathbf{F}_{GC}$ can be found directly from a force summation in the vertical direction since $\mathbf{F}_{GF}$ and $\mathbf{F}_{BC}$ have no vertical components. This ability to *determine directly* the force in a particular truss member is one of the main advantages of using the method of sections.*

*By comparison, if the method of joints were used to determine, say, the force in member GC, it would be necessary to analyze joints A, B, and G in sequence.

As in the method of joints, there are two ways in which one can determine the correct sense of an unknown member force:

- *Always assume* that the unknown member forces at the cut section are in *tension,* i.e., "pulling" on the member. By doing this, the numerical solution of the equilibrium equations will yield *positive scalars for members in tension and negative scalars for members in compression.*
- The correct sense of an unknown member force can in many cases be determined "by inspection." For example, $\mathbf{F}_{BC}$ is a tensile force as represented in Fig. 6–15*b* since moment equilibrium about G requires that $\mathbf{F}_{BC}$ create a moment opposite to that of the 1000-N force. Also, $\mathbf{F}_{GC}$ is tensile since its vertical component must balance the 1000-N force which acts downward. In more complicated cases, the sense of an unknown member force may be *assumed.* If the solution yields a *negative* scalar, it indicates that the force's sense is *opposite* to that shown on the free-body diagram. This is the method we will use in the example problems which follow.

Procedure for Analysis

The forces in the members of a truss may be determined by the method of sections using the following procedure.

Free-Body Diagram.

- Make a decision as to how to "cut" or section the truss through the members where forces are to be determined.
- Before isolating the appropriate section, it may first be necessary to determine the truss's *external* reactions. Then three equilibrium equations are available to solve for member forces at the cut section.
- Draw the free-body diagram of that part of the sectioned truss which has the least number of forces acting on it.
- Use one of the two methods described above for establishing the sense of an unknown member force.

Equations of Equilibrium.

- Moments should be summed about a point that lies at the intersection of the lines of action of two unknown forces, so that the third unknown force is determined directly from the moment equation.
- If two of the unknown forces are *parallel,* forces may be summed *perpendicular* to the direction of these unknowns to determine *directly* the third unknown force.

EXAMPLE 6.5

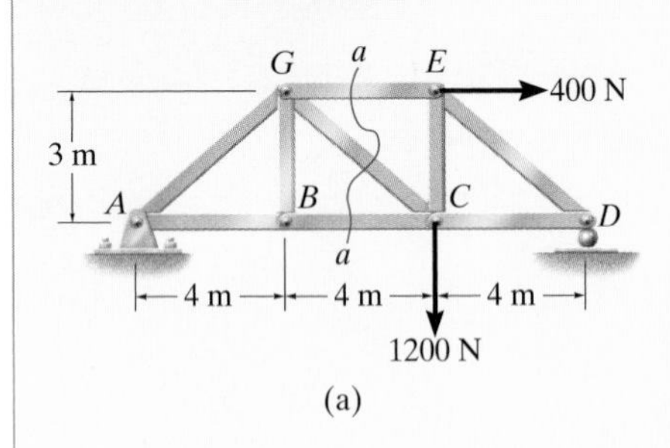

(a)

Determine the force in members GE, GC, and BC of the truss shown in Fig. 6–16a. Indicate whether the members are in tension or compression.

SOLUTION

Section aa in Fig. 6–16a has been chosen since it cuts through the *three* members whose forces are to be determined. In order to use the method of sections, however, it is *first* necessary to determine the external reactions at A or D. Why? A free-body diagram of the entire truss is shown in Fig. 6–16b. Applying the equations of equilibrium, we have

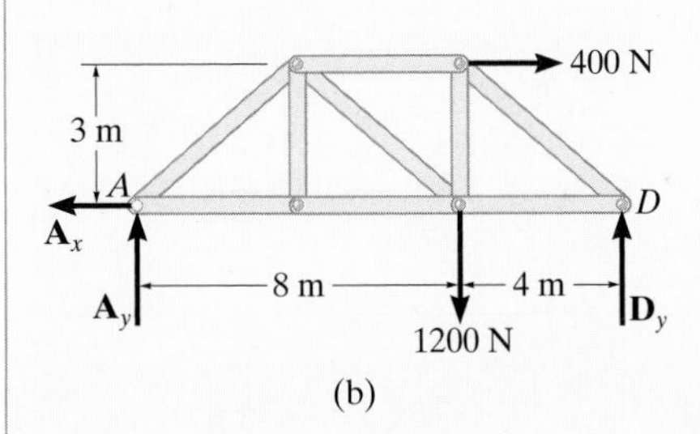

(b)

$$\xrightarrow{+} \Sigma F_x = 0; \qquad 400\text{ N} - A_x = 0 \qquad A_x = 400\text{ N}$$

$$\circlearrowleft + \Sigma M_A = 0; \qquad -1200\text{ N}(8\text{ m}) - 400\text{ N}(3\text{ m}) + D_y(12\text{ m}) = 0$$

$$D_y = 900\text{ N}$$

$$+\uparrow \Sigma F_y = 0; \qquad A_y - 1200\text{ N} + 900\text{ N} = 0 \qquad A_y = 300\text{ N}$$

Free-Body Diagram. The free-body diagram of the left portion of the sectioned truss is shown in Fig. 6–16c. For the analysis this diagram will be used since it involves the least number of forces.

Equations of Equilibrium. Summing moments about point G eliminates $\mathbf{F}_{GE}$and $\mathbf{F}_{GC}$ and yields a direct soluion for F_{BC}.

$$\circlearrowleft + \Sigma M_G = 0; \qquad -300\text{ N}(4\text{ m}) - 400\text{ N}(3\text{ m}) + F_{BC}(3\text{ m}) = 0$$

$$F_{BC} = 800\text{ N} \quad \text{(T)} \qquad \textit{Ans.}$$

In the same manner, by summing moments about point C we obtain a direct solution for F_{GE}.

$$\circlearrowleft + \Sigma M_C = 0; \qquad -300\text{ N}(8\text{ m}) + F_{GE}(3\text{ m}) = 0$$

$$F_{GE} = 800\text{ N} \quad \text{(C)} \qquad \textit{Ans.}$$

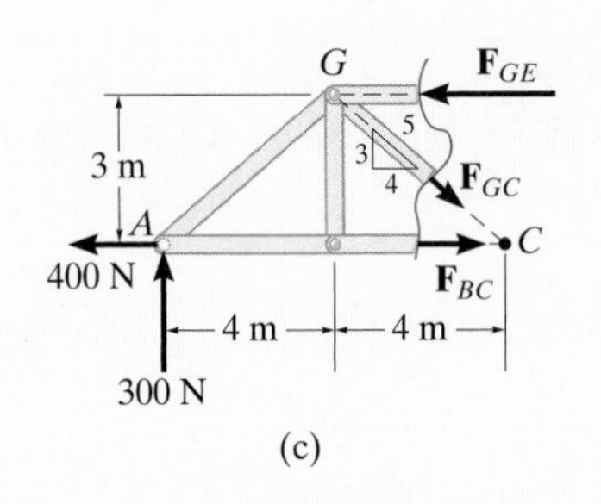

(c)

Fig. 6–16

Since $\mathbf{F}_{BC}$ and $\mathbf{F}_{GE}$ have no vertical components, summing forces in the y direction directly yields F_{GC}, i.e.,

$$+\uparrow \Sigma F_y = 0; \qquad 300\text{ N} - \tfrac{3}{5}F_{GC} = 0$$

$$F_{GC} = 500\text{ N} \quad \text{(T)} \qquad \textit{Ans.}$$

NOTE: As an exercise, obtain these results by applying the equations of equilibrium to the free-body diagram of the right portion of the sectioned truss. You should be able to tell, by inspection, the proper direction for each member force. For example, $\Sigma M_C = 0$ requires $\mathbf{F}_{GE}$ to be compressive to balance the moment of the 300-N force about C.

EXAMPLE 6.6

Determine the force in member *CF* of the bridge truss shown in Fig. 6–17*a*. Indicate whether the member is in tension or compression. Assume each member is pin connected.

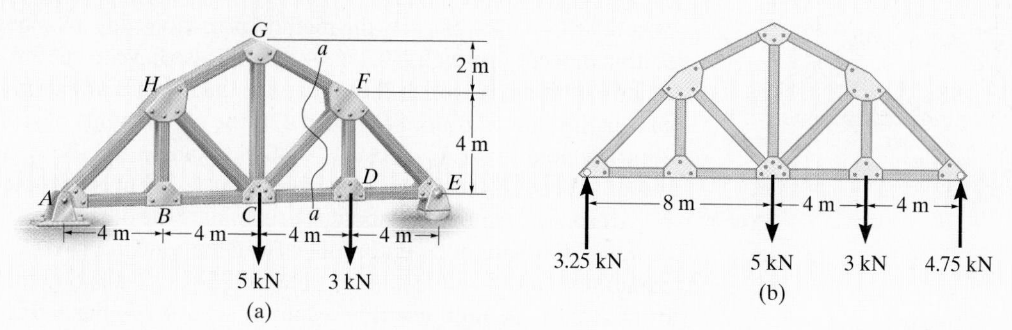

Fig. 6–17

SOLUTION

Free-Body Diagram. Section *aa* in Fig. 6–17*a* will be used since this section will "expose" the internal force in member *CF* as "external" on the free-body diagram of either the right or left portion of the truss. It is first necessary, however, to determine the external reactions on either the left or right side. Verify the results shown on the free-body diagram in Fig. 6–17*b*.

The free-body diagram of the right portion of the truss, which is the easiest to analyze, is shown in Fig. 6–17*c*. There are three unknowns, F_{FG}, F_{CF}, and F_{CD}.

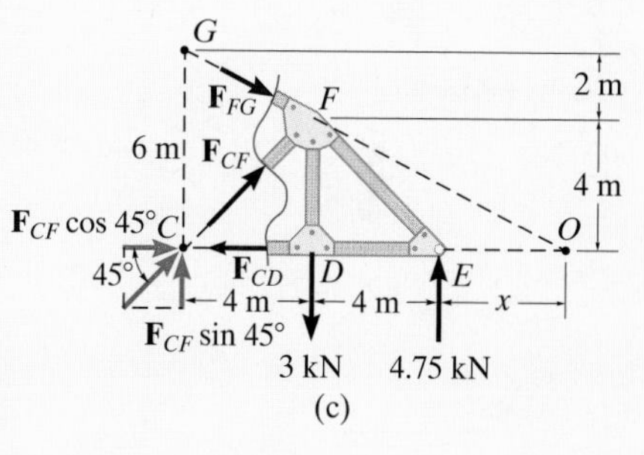

Equations of Equilibrium. The most direct method for solving this problem requires application of the moment equation about a point that eliminates two of the unknown forces. Hence, to obtain $\mathbf{F}_{CF}$, we will eliminate $\mathbf{F}_{FG}$ and $\mathbf{F}_{CD}$ by summing moments about point *O*, Fig. 6–17*c*. Note that the location of point *O* measured from *E* is determined from proportional triangles, i.e., $4/(4 + x) = 6/(8 + x)$, $x = 4$ m. Or, stated in another manner, the slope of member *GF* has a drop of 2 m to a horizontal distance of 4 m. Since *FD* is 4 m, Fig. 6–17*c*, then from *D* to *O* the distance must be 8 m.

An easy way to determine the moment of $\mathbf{F}_{CF}$ about point *O* is to use the principle of transmissibility and move $\mathbf{F}_{CF}$ to point *C*, and then resolve $\mathbf{F}_{CF}$ into its two rectangular components. We have

$\circlearrowleft + \Sigma M_O = 0;$

$$-F_{CF} \sin 45°(12 \text{ m}) + (3 \text{ kN})(8 \text{ m}) - (4.75 \text{ kN})(4 \text{ m}) = 0$$

$$F_{CF} = 0.589 \text{ kN} \quad \text{(C)} \qquad \textit{Ans.}$$

EXAMPLE 6.7

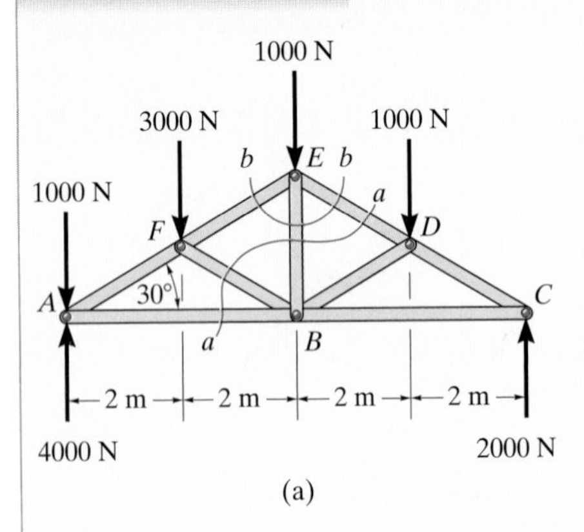

Determine the force in member EB of the roof truss shown in Fig. 6–18a. Indicate whether the member is in tension or compression.

SOLUTION

Free-Body Diagrams. By the method of sections, any imaginary vertical section that cuts through EB, Fig. 6–18a, will also have to cut through three other members for which the forces are unknown. For example, section aa cuts through ED, EB, FB, and AB. If the components of reaction at A are calculated first ($A_x = 0$, $A_y = 4000$ N) and a free-body diagram of the left side of this section is considered, Fig. 6–18b, it is possible to obtain $\mathbf{F}_{ED}$ by summing moments about B to eliminate the other three unknowns; however, $\mathbf{F}_{EB}$ cannot be determined from the remaining two equilibrium equations. One possible way of obtaining $\mathbf{F}_{EB}$ is first to determine $\mathbf{F}_{ED}$ from section aa, then use this result on section bb, Fig. 6–18a, which is shown in Fig. 6–18c. Here the force system is concurrent and our sectioned free-body diagram is the same as the free-body diagram for the pin at E (method of joints).

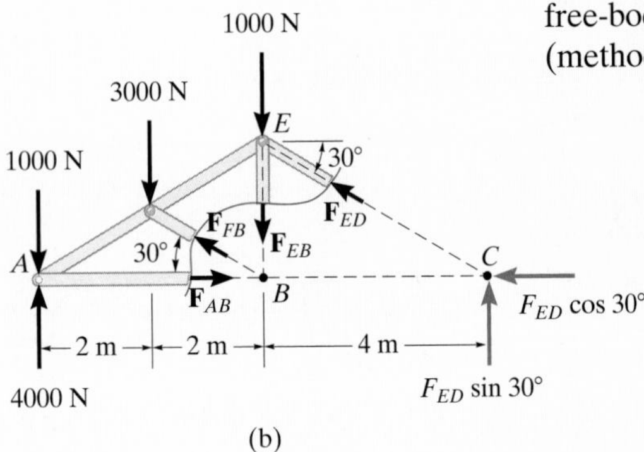

Fig. 6–18

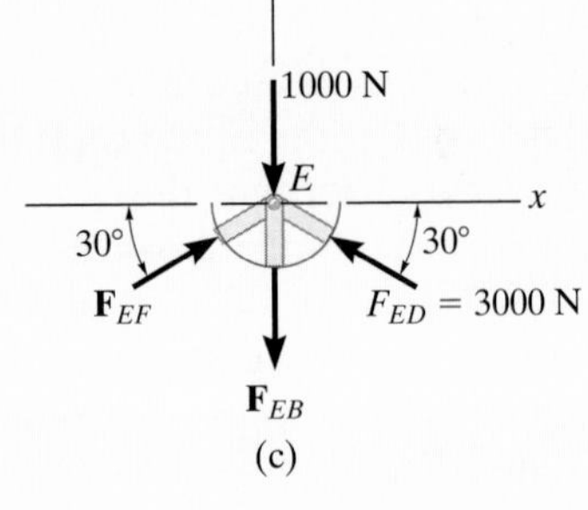

Equations of Equilibrium. In order to determine the moment of $\mathbf{F}_{ED}$ about point B, Fig. 6–18b, we will resolve the force into its rectangular components and, by the principle of transmissibility, extend it to point C as shown. The moments of 1000 N, F_{AB}, F_{FB}, F_{EB}, and $F_{ED}\cos 30°$ are all zero about B. Therefore,

$$\circlearrowleft +\Sigma M_B = 0; \quad 1000\text{ N}(4\text{ m}) + 3000\text{ N}(2\text{ m}) - 4000\text{ N}(4\text{ m}) + F_{ED}\sin 30°(4) = 0$$

$$F_{ED} = 3000\text{ N} \quad \text{(C)}$$

Considering now the free-body diagram of section bb, Fig. 6–18c, we have

$$\xrightarrow{+} \Sigma F_x = 0; \quad F_{EF}\cos 30° - 3000\cos 30°\text{ N} = 0$$

$$F_{EF} = 3000\text{ N} \quad \text{(C)}$$

$$+\uparrow \Sigma F_y = 0; \quad 2(3000\sin 30°\text{ N}) - 1000\text{ N} - F_{EB} = 0$$

$$F_{EB} = 2000\text{ N} \quad \text{(T)} \qquad \textit{Ans.}$$

PROBLEMS

6–30. The *Howe bridge truss* is subjected to the loading shown. Determine the force in members *DE*, *EH*, and *HG*, and state if the members are in tension or compression.

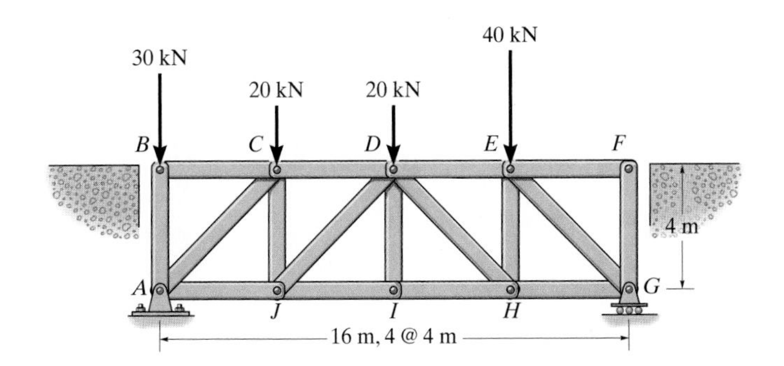

Prob. 6–30

6–31. The *Pratt bridge truss* is subjected to the loading shown. Determine the force in members *LD*, *LK*, *CD*, and *KD*, and state if the members are in tension or compression.

***6–32.** The *Pratt bridge truss* is subjected to the loading shown. Determine the force in members *JI*, *JE*, and *DE*, and state if the members are in tension or compression.

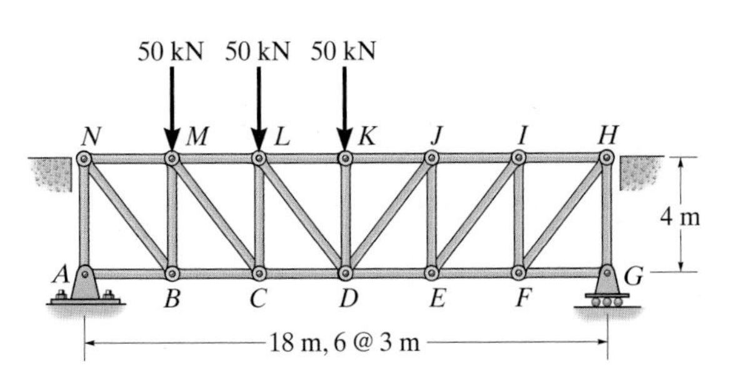

Probs. 6–31/32

6–33. The roof truss supports the vertical loading shown. Determine the force in members *BC*, *CK*, and *KJ* and state if these members are in tension or compression.

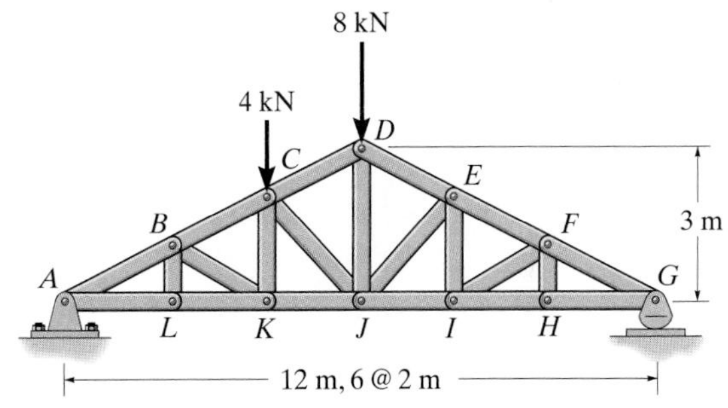

Prob. 6–33

6–34. Determine the force in members *CD*, *CJ*, *KJ*, and *DJ* of the truss which serves to support the deck of a bridge. State if these members are in tension or compression.

6–35. Determine the force in members *EI* and *JI* of the truss which serves to support the deck of a bridge. State if these members are in tension or compression.

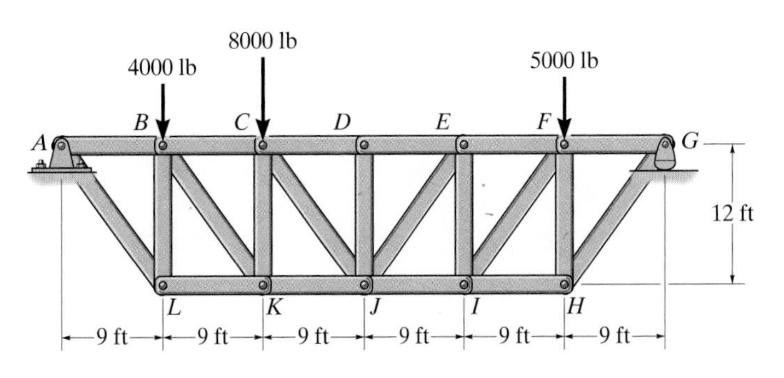

Probs. 6–34/35

***6–36.** Determine the force in members *BE*, *EF*, and *CB*, and state if the members are in tension or compression.

6–37. Determine the force in members *BF*, *BG*, and *AB*, and state if the members are in tension or compression.

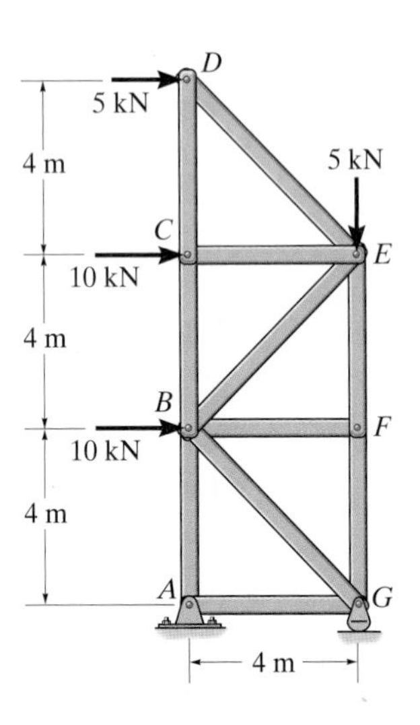

Probs. 6–36/37

6–38. Determine the force developed in members *GB* and *GF* of the bridge truss and state if these members are in tension or compression.

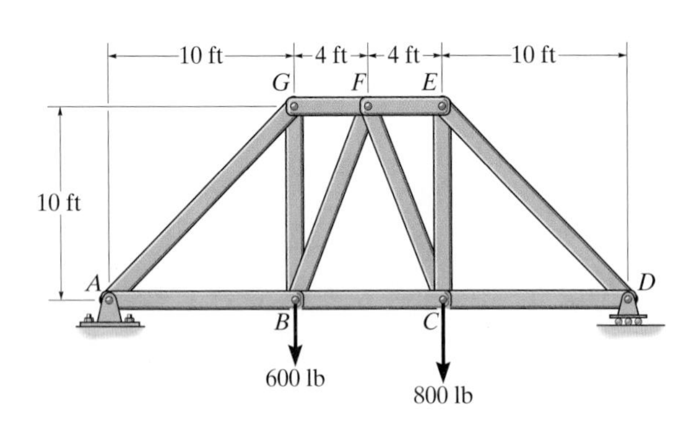

Prob. 6–38

6–39. Determine the force members *BC*, *FC*, and *FE*, and state if the members are in tension or compression.

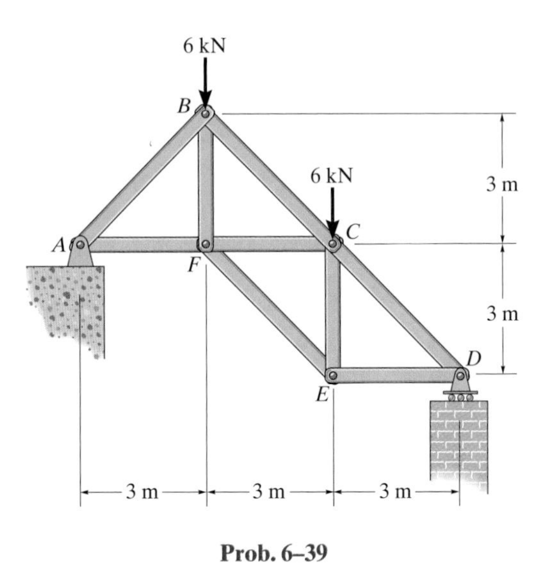

Prob. 6–39

***6–40.** Determine the force in members *IC* and *CG* of the truss and state if these members are in tension or compression. Also, indicate all zero-force members.

6–41. Determine the force in members *JE* and *GF* of the truss and state if these members are in tension or compression. Also, indicate all zero-force members.

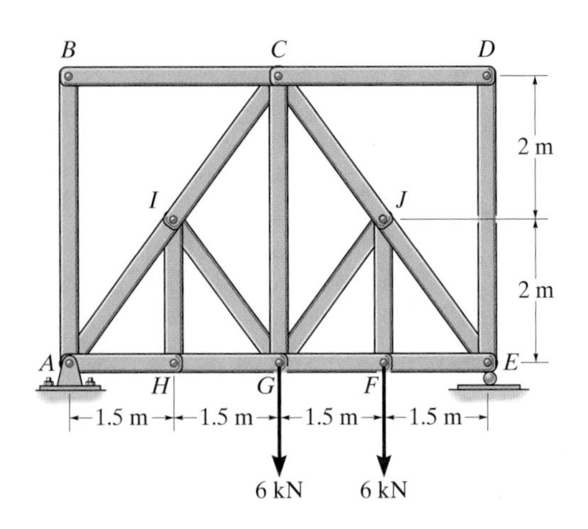

Probs. 6–40/41

6–42. Determine the force in members BC, HC, and HG. After the truss is sectioned use a single equation of equilibrium for the calculation of each force. State if these members are in tension or compression.

6–43. Determine the force in members CD, CF, and CG and state if these members are in tension or compression.

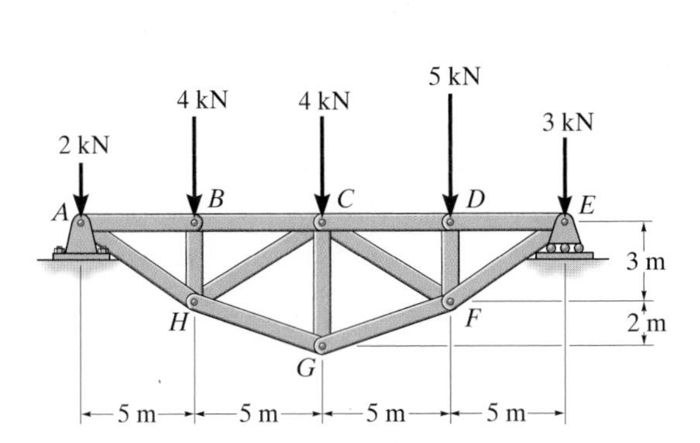

Probs. 6–42/43

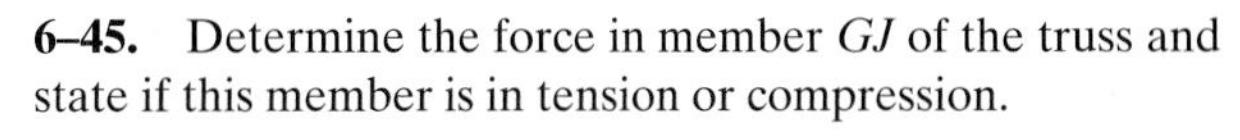

6–45. Determine the force in member GJ of the truss and state if this member is in tension or compression.

6–46. Determine the force in member GC of the truss and state if this member is in tension or compression.

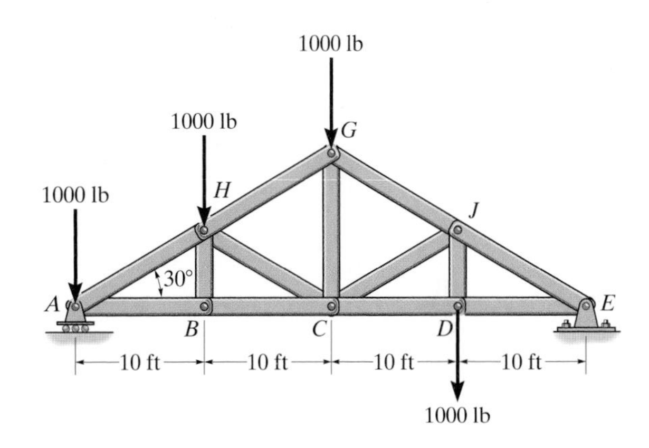

Probs. 6–45/46

***6–44.** Determine the force in members OE, LE, and LK of the *Baltimore truss* and state if the members are in tension or compression.

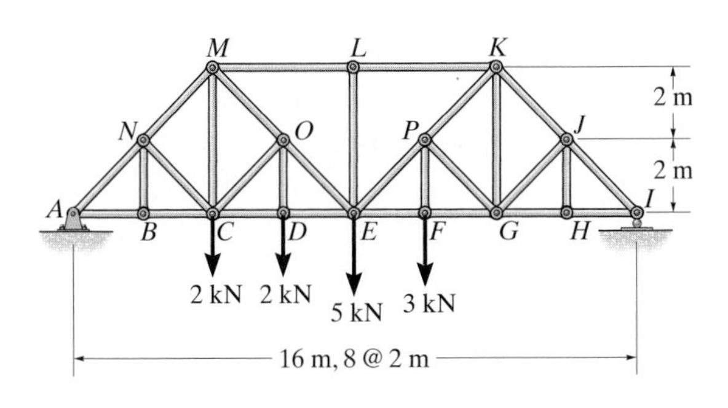

Prob. 6–44

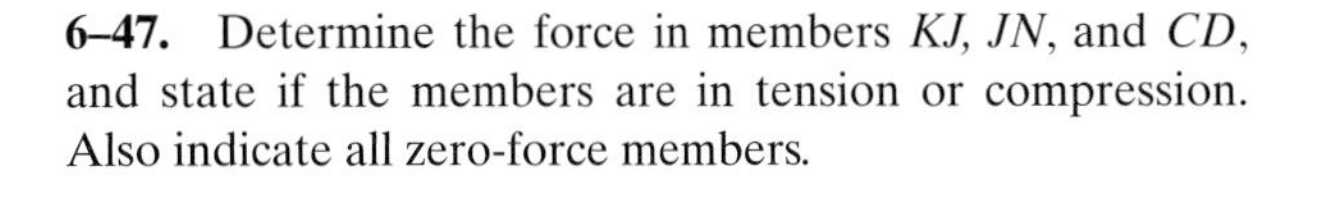

6–47. Determine the force in members KJ, JN, and CD, and state if the members are in tension or compression. Also indicate all zero-force members.

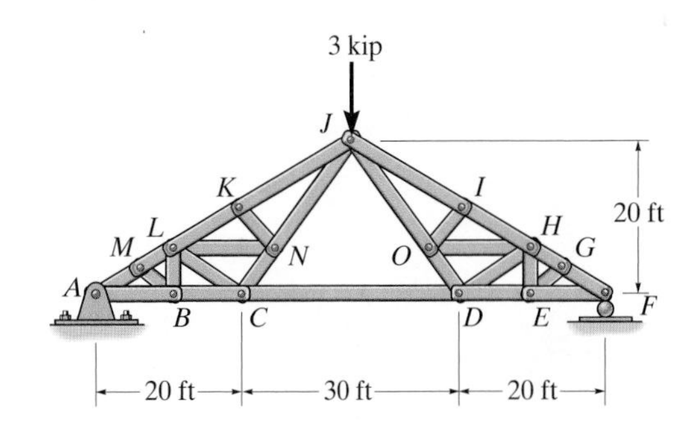

Prob. 6–47

*6–48. Determine the force in members BG, HG, and BC of the truss and state if the members are in tension or compression.

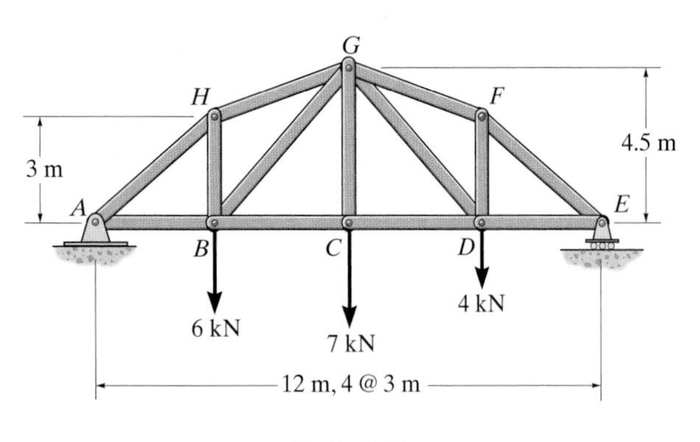

Prob. 6–48

6–49. The skewed truss carries the load shown. Determine the force in members CB, BE, and EF and state if these members are in tension or compression. Assume that all joints are pinned.

6–50. The skewed truss carries the load shown. Determine the force in members AB, BF, and EF and state if these members are in tension or compression. Assume that all joints are pinned.

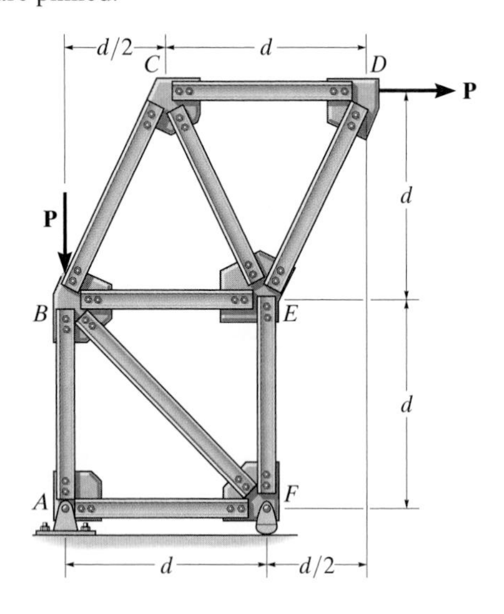

Probs. 6–49/50

6–51. Determine the force developed in members BC and CH of the roof truss and state if the members are in tension or compression.

*6–52. Determine the force in members CD and GF of the truss and state if the members are in tension or compression. Also indicate all zero-force members.

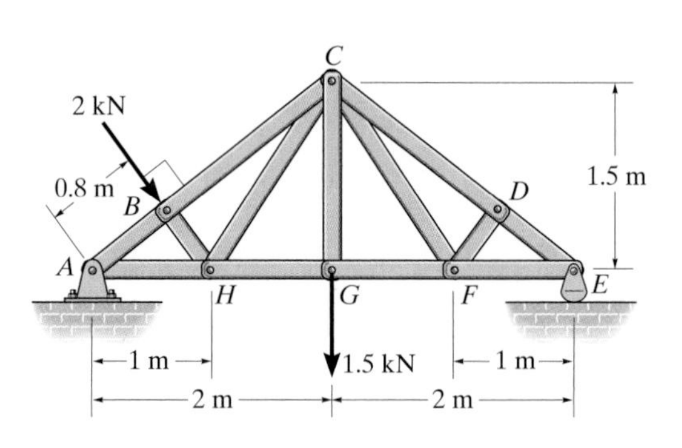

Probs. 6–51/52

6–53. Determine the force in members DE, DL, and ML of the roof truss and state if the members are in tension or compression.

6–54. Determine the force in members EF and EL of the roof truss and state if the members are in tension or compression.

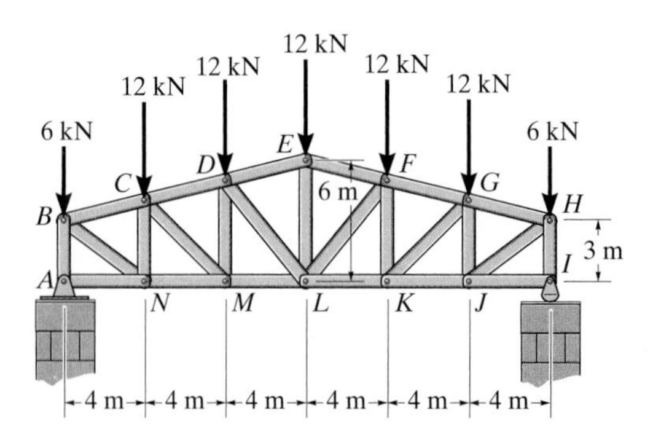

Probs. 6–53/54

*6.5 Space Trusses

A *space truss* consists of members joined together at their ends to form a stable three-dimensional structure. The simplest element of a space truss is a *tetrahedron*, formed by connecting six members together, as shown in Fig. 6–19. Any additional members added to this basic element would be redundant in supporting the force **P**. A *simple space truss* can be built from this basic tetrahedral element by adding three additional members and a joint, forming a system of multiconnected tetrahedrons.

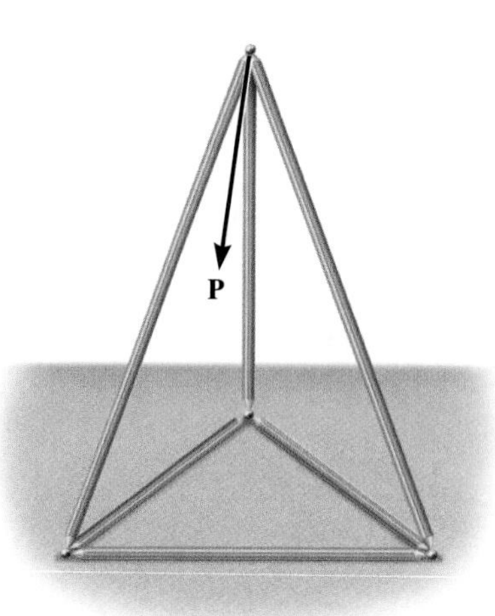

Fig. 6–19

Assumptions for Design. The members of a space truss may be treated as two-force members provided the external loading is applied at the joints and the joints consist of ball-and-socket connections. These assumptions are justified if the welded or bolted connections of the joined members intersect at a common point and the weight of the members can be neglected. In cases where the weight of a member is to be included in the analysis, it is generally satisfactory to apply it as a vertical force, half of its magnitude applied at each end of the member.

Typical roof-supporting space truss. Notice the use of ball-and-socket joints for the connections.

Procedure for Analysis

Either the method of joints or the method of sections can be used to determine the forces developed in the members of a simple space truss.

Method of Joints.

Generally, if the forces in *all* the members of the truss must be determined, the method of joints is most suitable for the analysis. When using the method of joints, it is necessary to solve the three scalar equilibrium equations $\Sigma F_x = 0$, $\Sigma F_y = 0$, $\Sigma F_z = 0$ at each joint. The solution of many simultaneous equations can be avoided if the force analysis begins at a joint having at least one known force and at most three unknown forces. If the three-dimensional geometry of the force system at the joint is hard to visualize, it is recommended that a Cartesian vector analysis be used for the solution.

Method of Sections.

If only a *few* member forces are to be determined, the method of sections may be used. When an imaginary section is passed through a truss and the truss is separated into two parts, the force system acting on one of the parts must satisfy the *six* scalar equilibrium equations: $\Sigma F_x = 0$, $\Sigma F_y = 0$, $\Sigma F_z = 0$, $\Sigma M_x = 0$, $\Sigma M_y = 0$, $\Sigma M_z = 0$ (Eqs. 5–6). By proper choice of the section and axes for summing forces and moments, many of the unknown member forces in a space truss can be computed *directly*, using a single equilibrium equation.

EXAMPLE 6.8

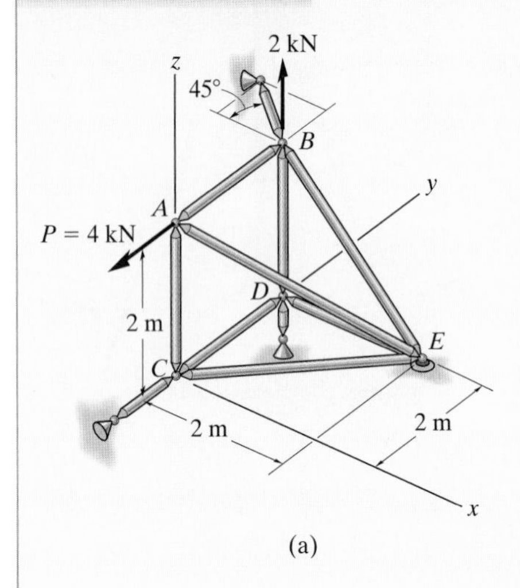

(a)

Determine the forces acting in the members of the space truss shown in Fig. 6–20*a*. Indicate whether the members are in tension or compression.

SOLUTION

Since there are one known force and three unknown forces acting at joint *A*, the force analysis of the truss will begin at this joint.

Joint A (Fig. 6–20*b*). Expressing each force that acts on the free-body diagram of joint *A* in vector notation, we have

$$\mathbf{P} = \{-4\mathbf{j}\}\text{ kN},\quad \mathbf{F}_{AB} = F_{AB}\mathbf{j},\quad \mathbf{F}_{AC} = -F_{AC}\mathbf{k},$$

$$\mathbf{F}_{AE} = F_{AE}\left(\frac{\mathbf{r}_{AE}}{r_{AE}}\right) = F_{AE}(0.577\mathbf{i} + 0.577\mathbf{j} - 0.577\mathbf{k})$$

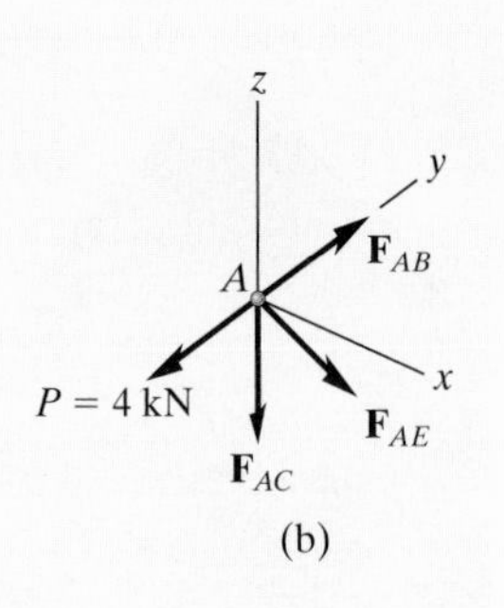

(b)

For equilibrium,

$$\Sigma \mathbf{F} = \mathbf{0}; \qquad \mathbf{P} + \mathbf{F}_{AB} + \mathbf{F}_{AC} + \mathbf{F}_{AE} = \mathbf{0}$$

$$-4\mathbf{j} + F_{AB}\mathbf{j} - F_{AC}\mathbf{k} + 0.577F_{AE}\mathbf{i} + 0.577F_{AE}\mathbf{j} - 0.577F_{AE}\mathbf{k} = \mathbf{0}$$

$$\Sigma F_x = 0; \qquad 0.577F_{AE} = 0$$

$$\Sigma F_y = 0; \qquad -4 + F_{AB} + 0.577F_{AE} = 0$$

$$\Sigma F_z = 0; \qquad -F_{AC} - 0.577F_{AE} = 0$$

$$F_{AC} = F_{AE} = 0 \qquad \textit{Ans.}$$

$$F_{AB} = 4\text{ kN}\quad(\text{T}) \qquad \textit{Ans.}$$

Since F_{AB} is known, joint *B* may be analyzed next.

Joint B (Fig. 6–20*c*).

$$\Sigma F_x = 0; \qquad -R_B\cos 45^\circ + 0.707F_{BE} = 0$$

$$\Sigma F_y = 0; \qquad -4 + R_B\sin 45^\circ = 0$$

$$\Sigma F_z = 0; \qquad 2 + F_{BD} - 0.707F_{BE} = 0$$

$$R_B = F_{BE} = 5.66\text{ kN}\quad(\text{T}),\qquad F_{BD} = 2\text{ kN}\quad(\text{C}) \qquad \textit{Ans.}$$

(c)

Fig. 6–20

The *scalar* equations of equilibrium may also be applied directly to the force systems on the free-body diagrams of joints *D* and *C* since the force components are easily determined. Show that

$$F_{DE} = F_{DC} = F_{CE} = 0 \qquad \textit{Ans.}$$

PROBLEMS

6–55. Two space trusses are used to equally support the uniform 50-kg sign. Determine the force developed in members AB, AC, and BC of truss $ABCD$ and state if the members are in tension or compression. Horizontal short links support the truss at joints B and D and there is a ball-and-socket joint at C.

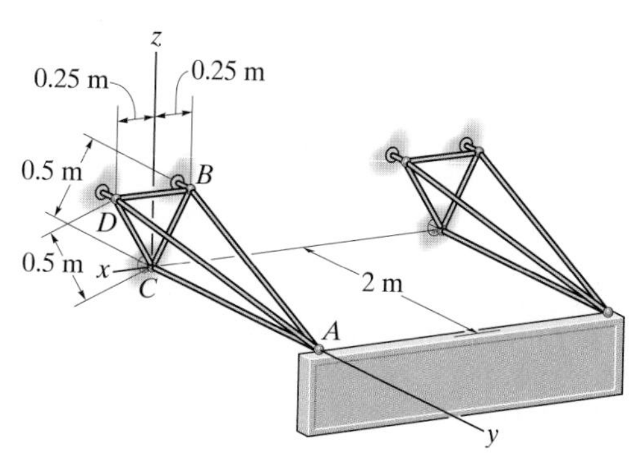

Prob. 6–55

***6–56.** Determine the force in each member of the space truss and state if the members are in tension or compression. The truss is supported by short links at B, C, and D.

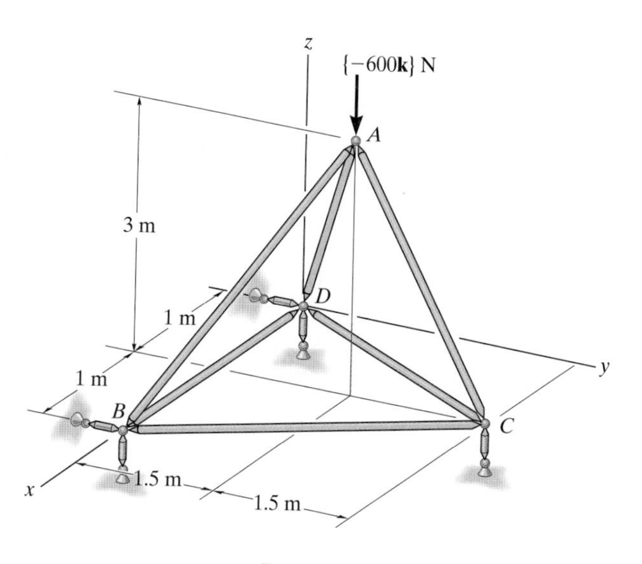

Prob. 6–56

6–57. Determine the force in each member of the space truss and state if the members are in tension or compression. The truss is supported by short links at A, B, and C.

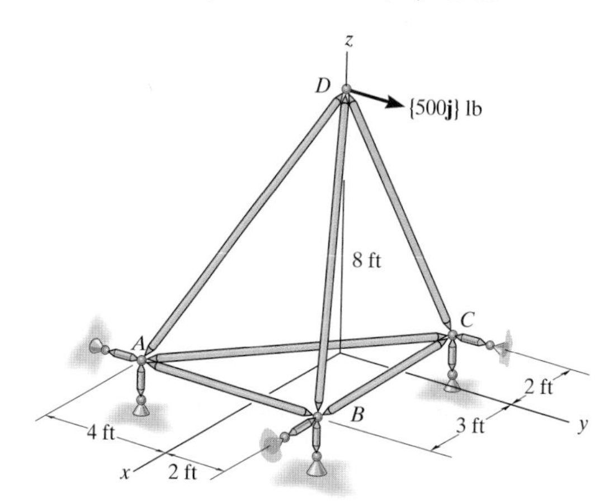

Prob. 6–57

6–58. The space truss is supported by a ball-and-socket joint at D and short links at C and E. Determine the force in each member and state if the members are in tension or compression. Take $\mathbf{F}_1 = \{-500\mathbf{k}\}$ lb and $\mathbf{F}_2 = \{400\mathbf{j}\}$ lb.

6–59. The space truss is supported by a ball-and-socket joint at D and short links at C and E. Determine the force in each member and state if the members are in tension or compression. Take $\mathbf{F}_1 = \{200\mathbf{i} + 300\mathbf{j} - 500\mathbf{k}\}$ lb and $\mathbf{F}_2 = \{400\mathbf{j}\}$ lb.

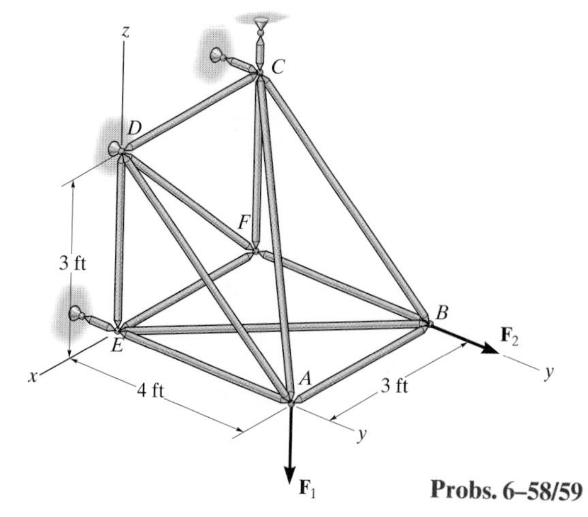

Probs. 6–58/59

*6–60. Determine the force in each member of the space truss and state if the members are in tension or compression. The truss is supported by a ball-and-socket joints at A, B, and E. Set $\mathbf{F} = \{-200\mathbf{i} + 400\mathbf{j}\}$ N. *Hint:* The support reaction at E acts along member EC. Why?

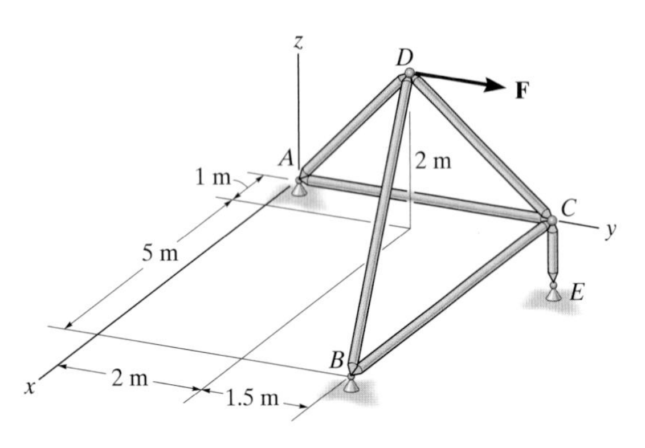

Prob. 6–60

6–61. Determine the force in each member of the space truss and state if the members are in tension or compression. The truss is supported by ball-and-socket joints at C, D, E, and G.

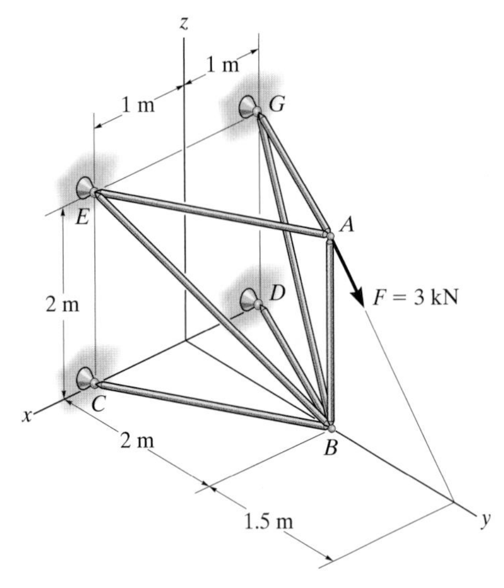

Prob. 6–61

6–62. Determine the force in members BD, AD, and AF of the space truss and state if the members are in tension or compression. The truss is supported by short links at A, B, D, and F.

6–63. Determine the force in members CF, EF, and DF of the space truss and state if the members are in tension or compression. The truss is supported by short links at A, B, D, and F.

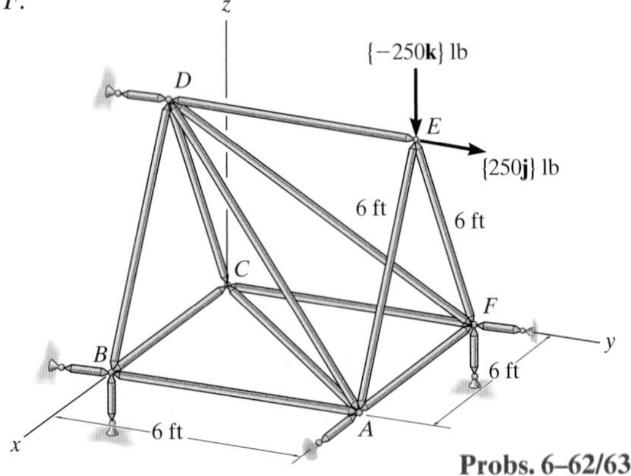

Probs. 6–62/63

*6–64. Determine the force developed in each member of the space truss and state if the members are in tension or compression. The crate has a weight of 150 lb.

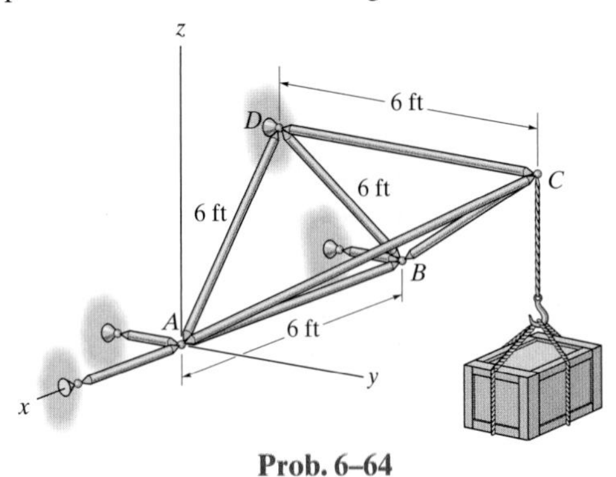

Prob. 6–64

6–65. The space truss is used to support vertical forces at joints B, C, and D. Determine the force in each member and state if the members are in tension or compression.

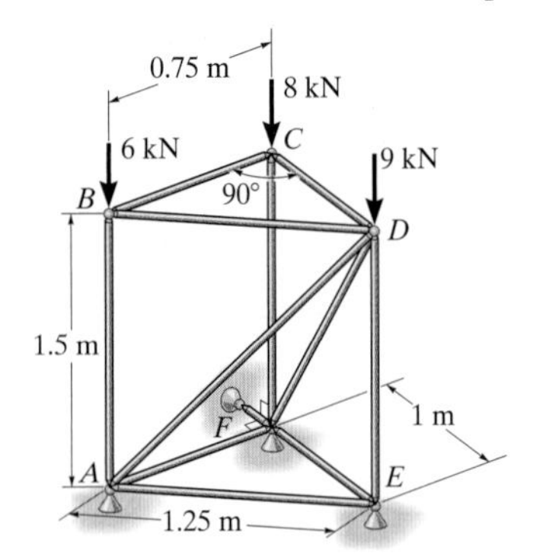

Prob. 6–65

6.6 Frames and Machines

Frames and machines are two common types of structures which are often composed of pin-connected *multiforce members*, i.e., members that are subjected to more than two forces. *Frames* are generally stationary and are used to support loads, whereas *machines* contain moving parts and are designed to transmit and alter the effect of forces. Provided a frame or machine is properly constrained and contains no more supports or members than are necessary to prevent collapse, the forces acting at the joints and supports can be determined by applying the equations of equilibrium to each member. Once the forces at the joints are obtained, it is then possible to *design* the size of the members, connections, and supports using the theory of mechanics of materials and an appropriate engineering design code.

Free-Body Diagrams. In order to determine the forces acting at the joints and supports of a frame or machine, the structure must be disassembled and the free-body diagrams of its parts must be drawn. The following important points *must* be observed:

- Isolate each part by drawing its *outlined shape*. Then show all the forces and/or couple moments that act on the part. Make sure to *label* or *identify* each known and unknown force and couple moment with reference to an established x, y coordinate system. Also, indicate any dimensions used for taking moments. Most often the equations of equilibrium are easier to apply if the forces are represented by their rectangular components. As usual, the sense of an unknown force or couple moment can be assumed.

- Identify all the two-force members in the structure and represent their free-body diagrams as having two equal but opposite collinear forces acting at their points of application. (See Sec. 5.4.) By recognizing the two-force members, we can avoid solving an unnecessary number of equilibrium equations.

- Forces common to any two *contacting* members act with equal magnitudes but opposite sense on the respective members. If the two members are treated as a *"system" of connected members*, then these forces are *"internal"* and are *not shown* on the *free-body diagram of the system*; however, if the free-body diagram of *each member* is drawn, the forces are *"external"* and *must* be shown on each of the free-body diagrams.

The following examples graphically illustrate application of these points in drawing the free-body diagrams of a dismembered frame or machine. In all cases, the weight of the members is neglected.

EXAMPLE 6.9

For the frame shown in Fig. 6–21*a*, draw the free-body diagram of (a) each member, (b) the pin at *B*, and (c) the two members connected together.

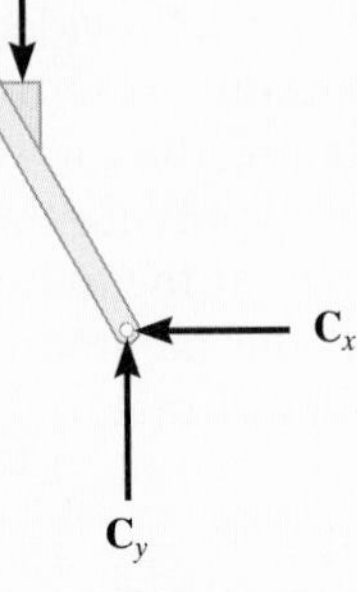

(b)

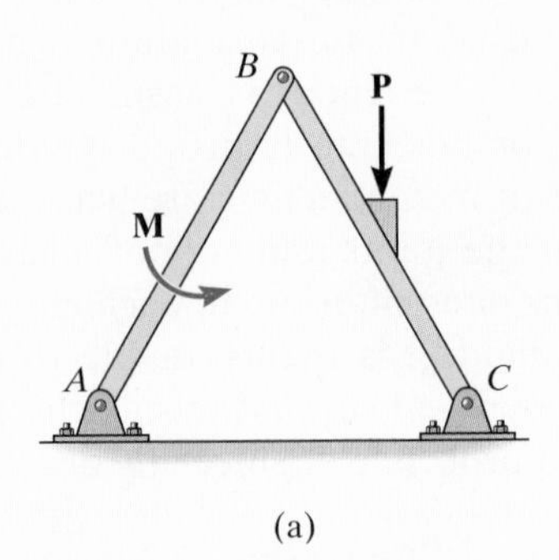

(a)

SOLUTION

Part (a). By inspection, members *BA* and *BC* are *not* two-force members. Instead, as shown on the free-body diagrams, Fig. 6–21*b*, *BC* is subjected to *not* five but *three forces*, namely, the resultant forces from the pins at *B* and *C* and the external force **P**. Likewise, *AB* is subjected to the *resultant* forces from the pins at *A* and *B* and the external couple moment **M**.

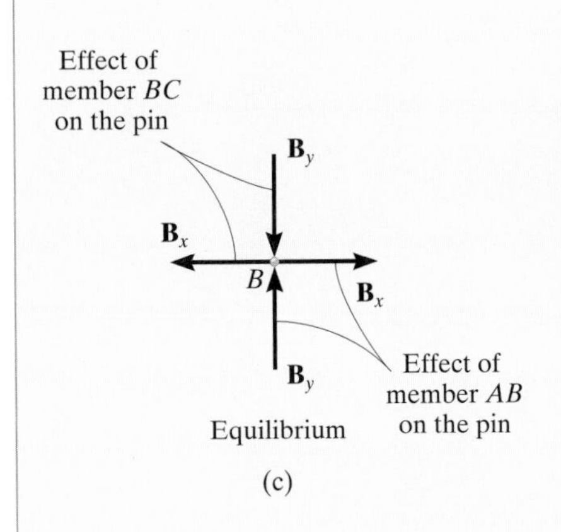

(c)

Part (b). It can be seen in Fig. 6–21*a* that the pin at *B* is subjected to only *two forces*, i.e., the force of member *BC* on the pin and the force of member *AB* on the pin. For *equilibrium* these forces and therefore their respective components must be equal but opposite, Fig. 6–21*c*. Notice carefully how Newton's third law is applied between the pin and its contacting members, i.e., the effect of the pin on the two members, Fig. 6–21*b*, and the equal but opposite effect of the two members on the pin, Fig. 6–21*c*. Also note that $\mathbf{B}_x$ and $\mathbf{B}_y$ shown equal but opposite in Fig. 6–21*b* on members *AB* and *BC*, is *not* the effect of Newton's third law; instead, this results from the *equilibrium* analysis of the pin, Fig. 6–21*c*.

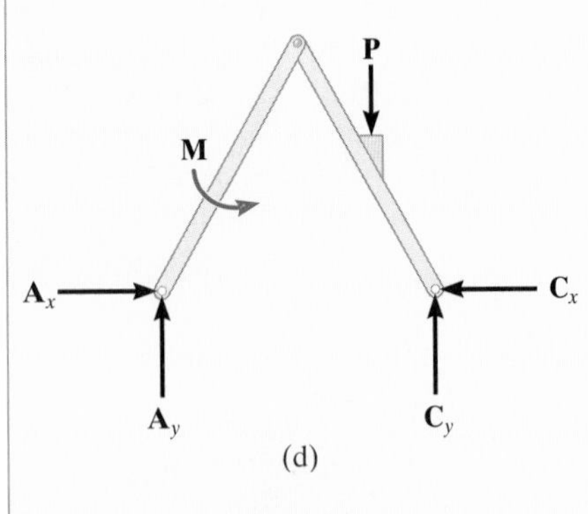

(d)

Fig. 6–21

Part (c). The free-body diagram of both members connected together, yet removed from the supporting pins at *A* and *C*, is shown in Fig. 6–21*d*. The force components $\mathbf{B}_x$ and $\mathbf{B}_y$ are *not shown* on this diagram since they form equal but opposite collinear pairs of *internal* forces (Fig. 6–21*b*) and therefore cancel out. Also, to be consistent when later applying the equilibrium equations, the unknown force components at *A* and *C* must act in the *same sense* as those shown in Fig. 6–21*b*. Here the couple moment **M** can be applied at any point on the frame in order to determine the reactions at *A* and *C*. Note, however, that it must act on member *AB* in Fig. 6–21*b* and *not* on member *BC*.

EXAMPLE 6.10

A constant tension in the conveyor belt is maintained by using the device shown in Fig. 6–22*a*. Draw the free-body diagrams of the frame and the cylinder that the belt surrounds. The suspended block has a weight of W.

(a)

Fig. 6–22

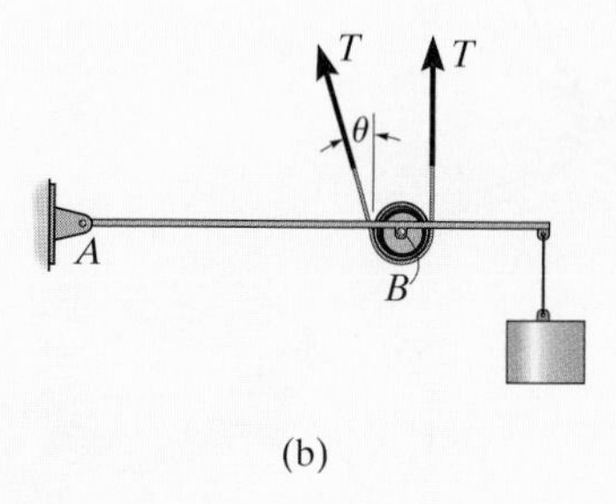

(b)

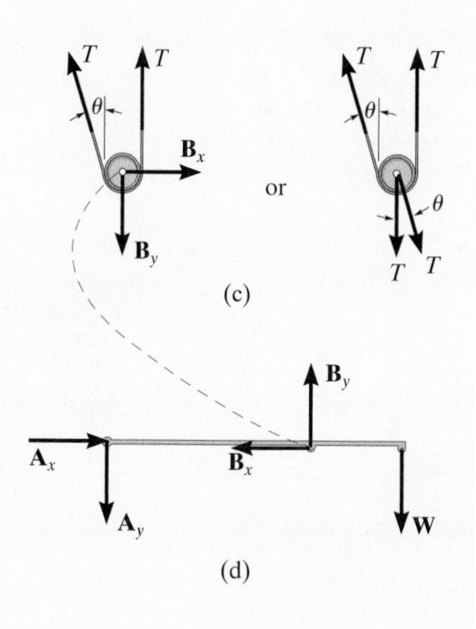

(c)

(d)

SOLUTION

The idealized model of the device is shown in Fig. 6–22*b*. Here the angle θ is assumed to be known. Notice that the tension in the belt is the same on each side of the cylinder, since the cylinder is free to turn. From this model, the free-body diagrams of the cylinder and frame are shown in Figs. 6–22*c* and 6–22*d*, respectively. Note that the force that the pin at B exerts on the cylinder can be represented by either its horizontal and vertical components $\mathbf{B}_x$ and $\mathbf{B}_y$, which can be determined by using the force equations of equilibrium applied to the cylinder, or by the two components T, which provide equal but opposite couple moments on the cylinder and thus keep it from turning. Also, realize that once the pin reactions at A have been determined, half of their values act on each side of the frame since pin connections occur on each side, Fig. 6–22*a*.

EXAMPLE 6.11

Draw the free-body diagram of each part of the smooth piston and link mechanism used to crush recycled cans, which is shown in Fig. 6–23*a*.

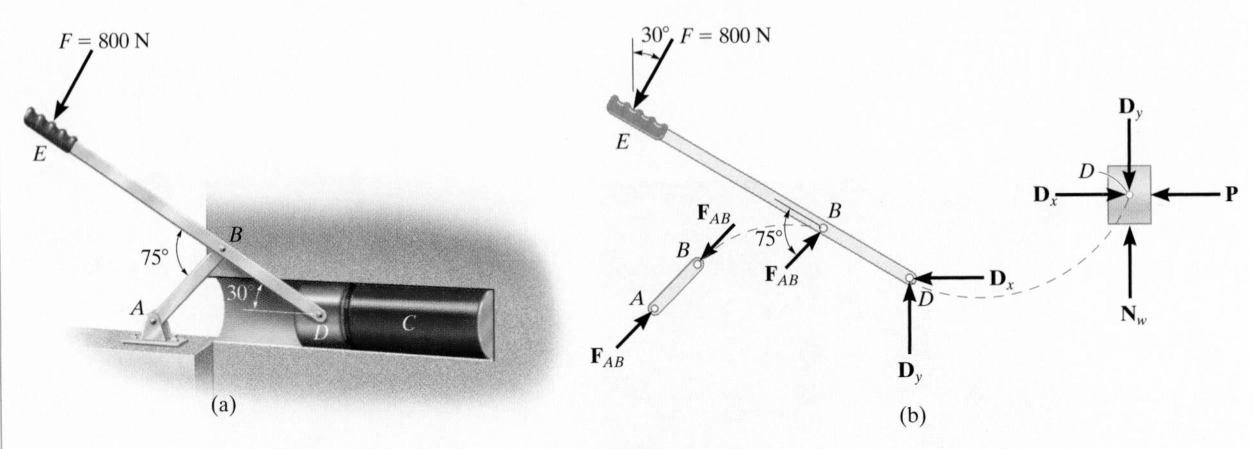

Fig. 6–23

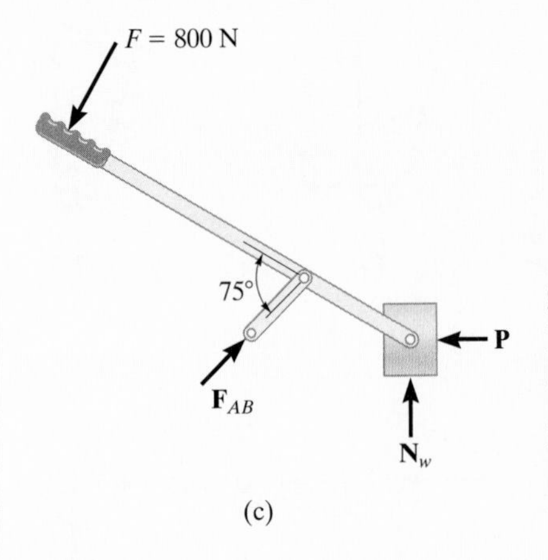

SOLUTION

By inspection, member *AB* is a two-force member. The free-body diagrams of the parts are shown in Fig. 6–23*b*. Since the pins at *B* and *D connect only two parts together,* the forces there are shown as equal but opposite on the separate free-body diagrams of their connected members. In particular, four components of force act on the piston: $\mathbf{D}_x$ and $\mathbf{D}_y$ represent the effect of the pin (or lever *EBD*), $\mathbf{N}_w$ is the *resultant force* of the floor, and $\mathbf{P}$ is the resultant compressive force caused by the can *C*.

NOTE: A free-body diagram of the entire assembly is shown in Fig. 6–23*c*. Here the forces between the components are internal and are not shown on the free-body diagram.

EXAMPLE 6.12

For the frame shown in Fig. 6–24*a*, draw the free-body diagrams of (a) the entire frame including the pulleys and cords, (b) the frame without the pulleys and cords, and (c) each of the pulleys.

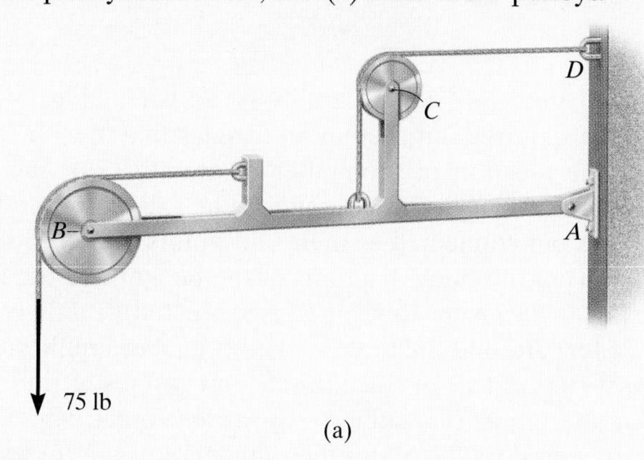

(a)

SOLUTION

Part (a). When the entire frame including the pulleys and cords is considered, the interactions at the points where the pulleys and cords are connected to the frame become pairs of *internal forces* which cancel each other and therefore are not shown on the free-body diagram, Fig. 6–24*b*.

Part (b). When the cords and pulleys are removed, their effect *on the frame* must be shown, Fig. 6–24*c*.

Part (c). The force components $\mathbf{B}_x$, $\mathbf{B}_y$, $\mathbf{C}_x$, $\mathbf{C}_y$ of the pins on the pulleys, Fig. 6–24*d*, are equal but opposite to the force components exerted by the pins on the frame, Fig. 6–24*c*. Why?

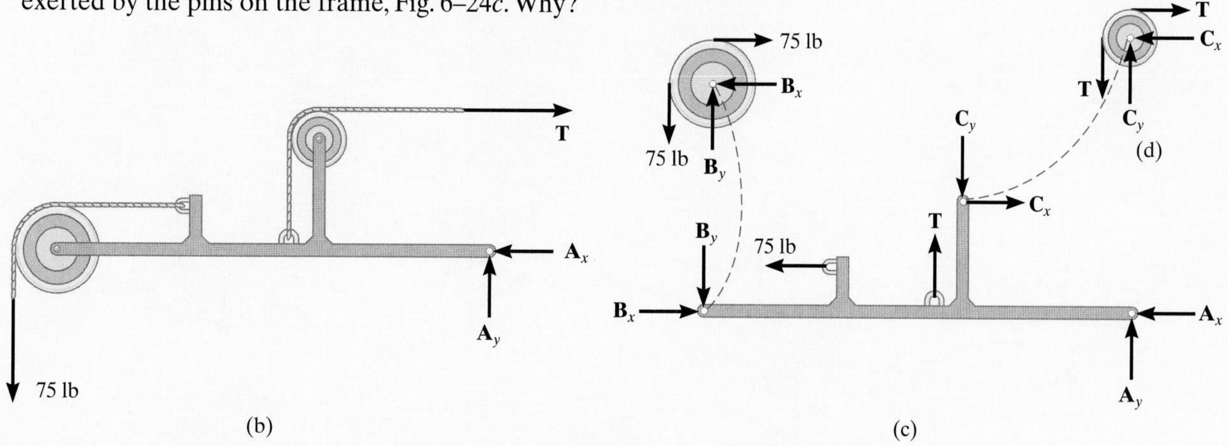

(b) (c)

Fig. 6–24

EXAMPLE 6.13

(a)

Fig. 6–25

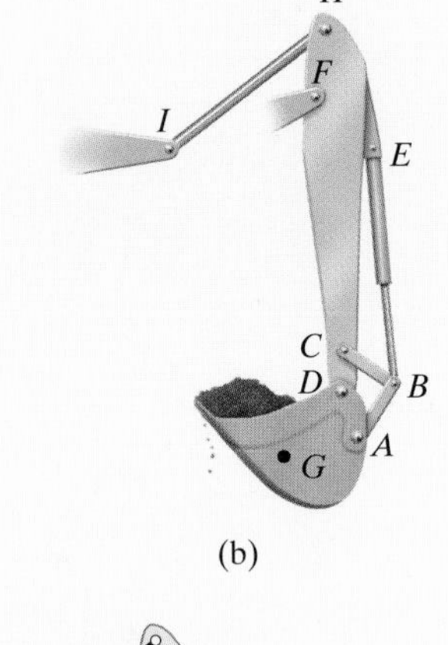

(b)

Draw the free-body diagrams of the bucket and the vertical boom of the back hoe shown in the photo, Fig. 6–25*a*. The bucket and its contents have a weight *W*. Neglect the weight of the members.

SOLUTION

The idealized model of the assembly is shown in Fig. 6–25*b*. Not shown are the required dimensions and angles that must be obtained, along with the location of the center of gravity *G* of the load. By inspection, members *AB*, *BC*, *BE*, and *HI* are all two-force members since they are pin connected at their end points and no other forces act on them. The free-body diagrams of the bucket and the boom are shown in Fig. 6–25*c*. Note that pin *C* is subjected to only two forces, the force of link *BC* and the force of the boom. For equilibrium, these forces must be equal in magnitude but opposite in direction, Fig. 6–25*d*. The pin at *B* is subjected to three forces, Fig. 6–25*e*. The force $\mathbf{F}_{BE}$ is caused by the hydraulic cylinder, and the forces $\mathbf{F}_{BA}$ and $\mathbf{F}_{BC}$ are caused by the links. These three forces are related by the two equations of force equilibrium applied to the pin.

NOTE: The free-body diagram of the entire assembly is shown in Fig. 6–25*f*.

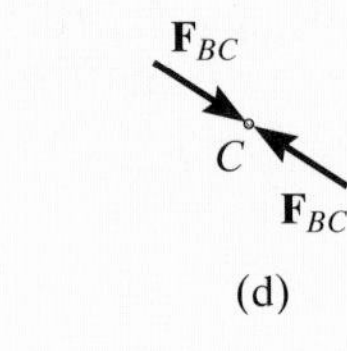

(d)

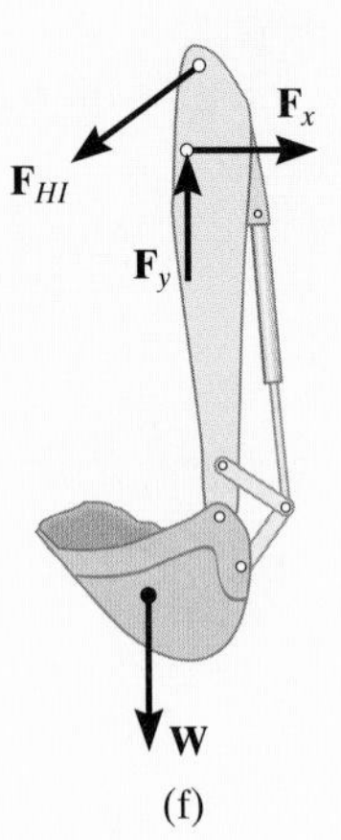

(f)

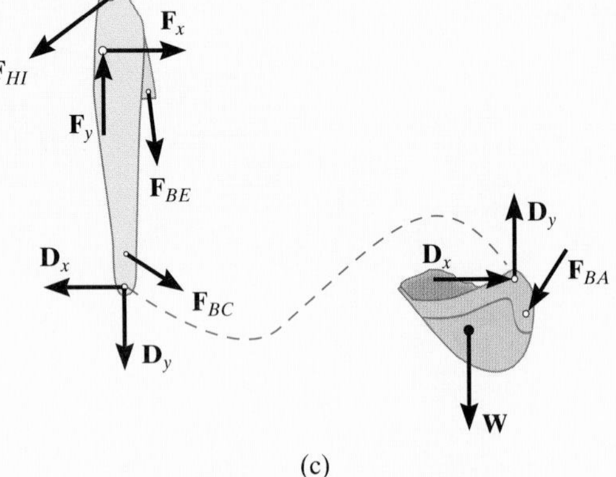

(c)

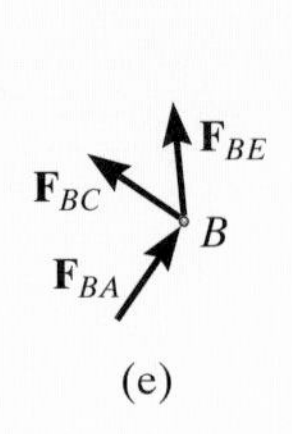

(e)

Before proceeding, it is recommended to cover the solutions to the previous examples and attempt to draw the requested free-body diagrams. When doing so, make sure the work is neat and that all the forces and couple moments are properly labeled.

Equations of Equilibrium. Provided the structure (frame or machine) is properly supported and contains no more supports or members than are necessary to prevent its collapse, then the unknown forces at the supports and connections can be determined from the equations of equilibrium. If the structure lies in the x–y plane, then for *each* free-body diagram drawn the loading must satisfy $\Sigma F_x = 0$, $\Sigma F_y = 0$, and $\Sigma M_O = 0$. The selection of the free-body diagrams used for the analysis is *completely arbitrary*. They may represent each of the members of the structure, a portion of the structure, or its entirety. For example, consider finding the six components of the pin reactions at A, B, and C for the frame shown in Fig. 6–26*a*. If the frame is dismembered, as it is in Fig. 6–26*b*, these unknowns can be determined by applying the three equations of equilibrium to each of the two members (total of six equations). The free-body diagram of the *entire frame* can also be used for part of the analysis, Fig. 6–26*c*. Hence, if so desired, all six unknowns can be determined by applying the three equilibrium equations to the entire frame, Fig. 6–26*c*, and also to either one of its members. Furthermore, the answers can be checked in part by applying the three equations of equilibrium to the remaining "second" member. In general, then, this problem can be solved by writing *at most* six equilibrium equations using free-body diagrams of the members and/or the combination of connected members. Any more than six equations written would *not* be unique from the original six and would only serve to check the results.

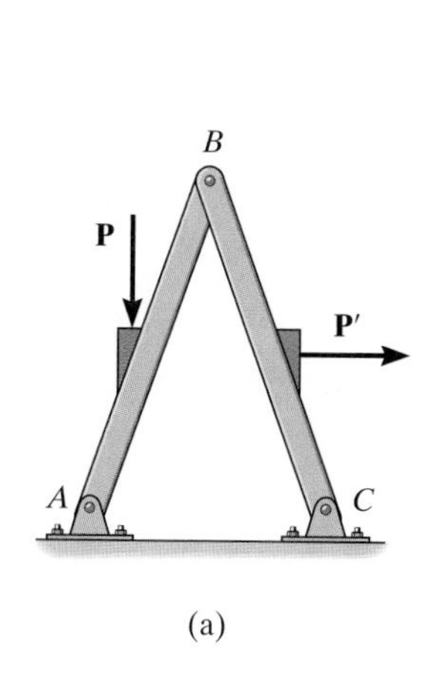

(a)

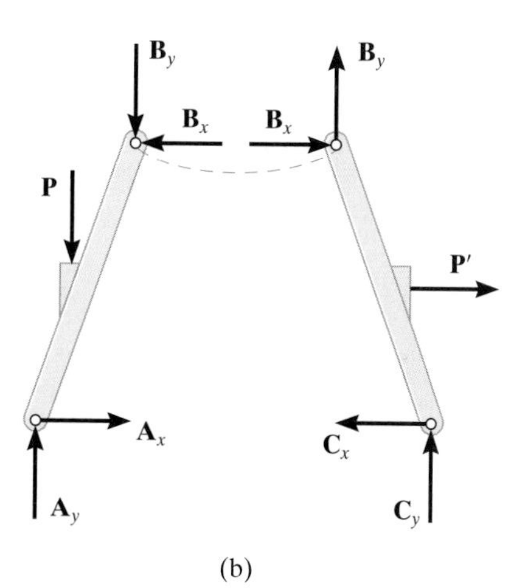

(b)

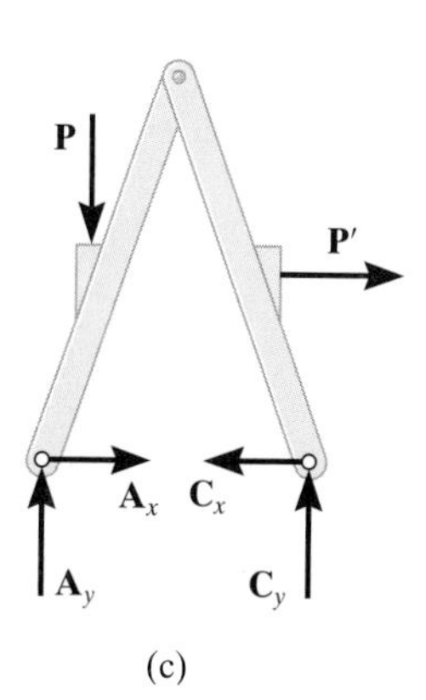

(c)

Fig. 6–26

Procedure For Analysis

The joint reactions on frames or machines (structures) composed of multiforce members can be determined using the following procedure.

Free-Body Diagram.

- Draw the free-body diagram of the entire structure, a portion of the structure, or each of its members. The choice should be made so that it leads to the most direct solution of the problem.
- When the free-body diagram of a group of members of a structure is drawn, the forces at the connected parts of this group are internal forces and are not shown on the free-body diagram of the group.
- Forces common to two members which are in contact act with equal magnitude but opposite sense on the respective free-body diagrams of the members.
- Two-force members, regardless of their shape, have equal but opposite collinear forces acting at the ends of the member.
- In many cases it is possible to tell by inspection the proper sense of the unknown forces acting on a member; however, if this seems difficult, the sense can be assumed.
- A couple moment is a free vector and can act at any point on the free-body diagram. Also, a force is a sliding vector and can act at any point along its line of action.

Equations of Equilibrium.

- Count the number of unknowns and compare it to the total number of equilibrium equations that are available. In two dimensions, there are three equilibrium equations that can be written for each member.
- Sum moments about a point that lies at the intersection of the lines of action of as many unknown forces as possible.
- If the solution of a force or couple moment magnitude is found to be negative, it means the sense of the force is the reverse of that shown on the free-body diagram.

EXAMPLE 6.14

Determine the horizontal and vertical components of force which the pin at C exerts on member BC of the frame in Fig. 6–27*a*.

SOLUTION I

Free-Body Diagrams. By inspection it can be seen that AB is a two-force member. The free-body diagrams are shown in Fig. 6–27*b*.

Equations of Equilibrium. The *three unknowns*, C_x, C_y, and F_{AB}, can be determined by applying the three equations of equilibrium to member CB.

$$\circlearrowleft + \Sigma M_C = 0;\ 2000\text{ N}(2\text{ m}) - (F_{AB}\sin 60^\circ)(4\text{ m}) = 0\quad F_{AB} = 1154.7\text{ N}$$

$$\xrightarrow{+} \Sigma F_x = 0;\ 1154.7\cos 60^\circ\text{ N} - C_x = 0 \qquad C_x = 577\text{ N} \qquad \textit{Ans.}$$

$$+\uparrow \Sigma F_y = 0;\ 1154.7\sin 60^\circ\text{ N} - 2000\text{ N} + C_y = 0 \quad C_y = 1000\text{ N} \qquad \textit{Ans.}$$

SOLUTION II

Free-Body Diagrams. If one does not recognize that AB is a two-force member, then more work is involved in solving this problem. The free-body diagrams are shown in Fig. 6–27*c*.

Equations of Equilibrium. The *six unknowns*, A_x, A_y, B_x, B_y, C_x, and C_y, are determined by applying the three equations of equilibrium to each member.

Member AB

$$\circlearrowleft + \Sigma M_A = 0; \quad B_x(3\sin 60^\circ\text{ m}) - B_y(3\cos 60^\circ\text{ m}) = 0 \qquad (1)$$

$$\xrightarrow{+} \Sigma F_x = 0; \quad A_x - B_x = 0 \qquad (2)$$

$$+\uparrow \Sigma F_y = 0; \quad A_y - B_y = 0 \qquad (3)$$

Member BC

$$\circlearrowleft + \Sigma M_C = 0; \quad 2000\text{ N}(2\text{ m}) - B_y(4\text{ m}) = 0 \qquad (4)$$

$$\xrightarrow{+} \Sigma F_x = 0; \quad B_x - C_x = 0 \qquad (5)$$

$$+\uparrow \Sigma F_y = 0; \quad B_y - 2000\text{ N} + C_y = 0 \qquad (6)$$

The results for C_x and C_y can be determined by solving these equations in the following sequence: 4, 1, 5, then 6. The results are

$$B_y = 1000\text{ N}$$
$$B_x = 577\text{ N}$$
$$C_x = 577\text{ N} \qquad \textit{Ans.}$$
$$C_y = 1000\text{ N} \qquad \textit{Ans.}$$

By comparison, Solution I is simpler since the requirement that F_{AB} in Fig. 6–27*b* be equal, opposite, and collinear at the ends of member AB automatically satisfies Eqs. 1, 2, and 3 above and therefore eliminates the need to write these equations. *As a result, always identify the two-force members before starting the analysis!*

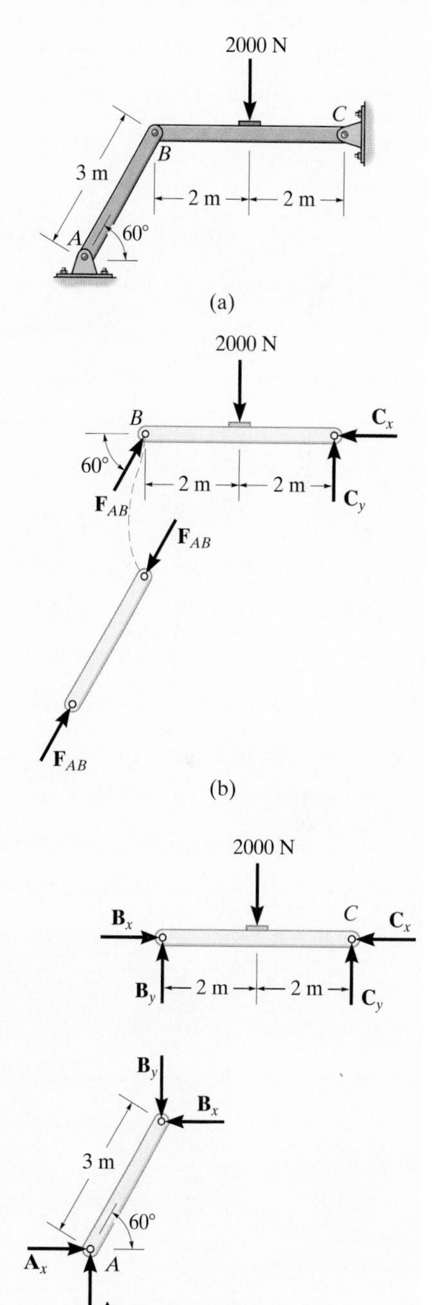

Fig. 6–27

EXAMPLE 6.15

The compound beam shown in Fig. 6–28*a* is pin connected at *B*. Determine the reactions at its supports. Neglect its weight and thickness.

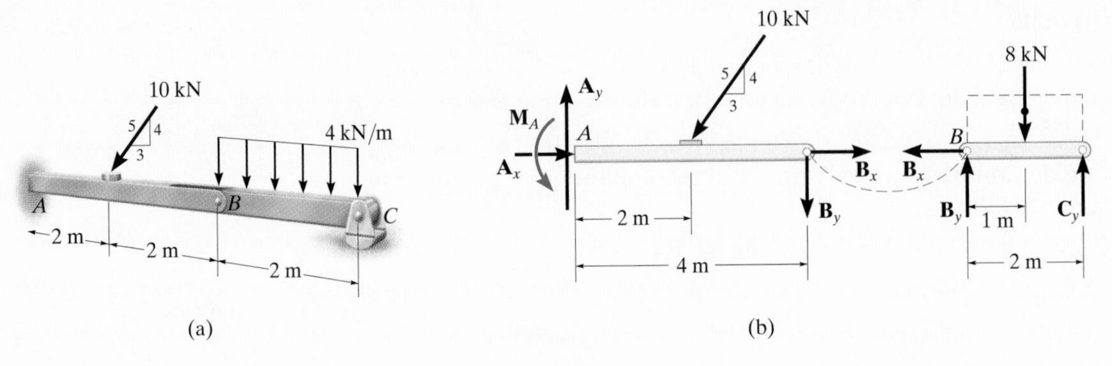

Fig. 6–28

SOLUTION

Free-Body Diagrams. By inspection, if we consider a free-body diagram of the entire beam *ABC*, there will be three unknown reactions at *A* and one at *C*. These four unknowns cannot all be obtained from the three equations of equilibrium, and so it will become necessary to dismember the beam into its two segments as shown in Fig. 6–28*b*.

Equations of Equilibrium. The six unknowns are determined as follows:

Segment BC

$$\xrightarrow{+} \Sigma F_x = 0; \qquad B_x = 0$$

$$\circlearrowleft + \Sigma M_B = 0; \qquad -8\text{ kN}(1\text{ m}) + C_y(2\text{ m}) = 0$$

$$+\uparrow \Sigma F_y = 0; \qquad B_y - 8\text{ kN} + C_y = 0$$

Segment AB

$$\xrightarrow{+} \Sigma F_x = 0; \qquad A_x - (10\text{ kN})\left(\tfrac{3}{5}\right) + B_x = 0$$

$$\circlearrowleft + \Sigma M_A = 0; \qquad M_A - (10\text{ kN})\left(\tfrac{4}{5}\right)(2\text{ m}) - B_y(4\text{ m}) = 0$$

$$+\uparrow \Sigma F_y = 0; \qquad A_y - (10\text{ kN})\left(\tfrac{4}{5}\right) - B_y = 0$$

Solving each of these equations successively, using previously calculated results, we obtain

$A_x = 6\text{ kN}$ $\quad A_y = 12\text{ kN}$ $\quad M_A = 32\text{ kN}\cdot\text{m}$ *Ans.*

$B_x = 0$ $\quad B_y = 4\text{ kN}$

$C_y = 4\text{ kN}$ *Ans.*

EXAMPLE 6.16

Determine the horizontal and vertical components of force which the pin at C exerts on member $ABCD$ of the frame shown in Fig. 6–29*a*.

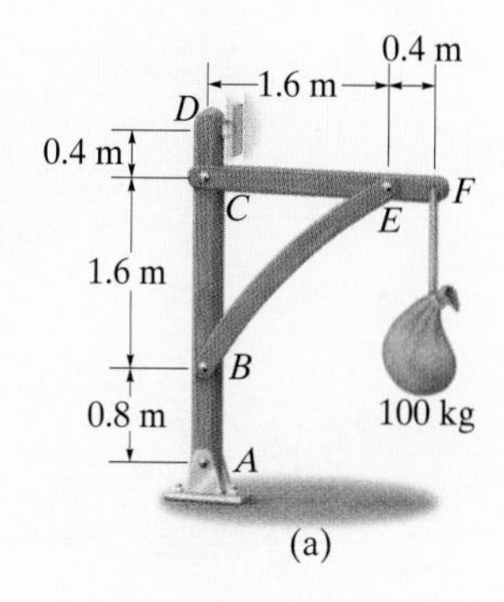

(a)

SOLUTION

Free-Body Diagrams. By inspection, the three components of reaction that the supports exert on $ABCD$ can be determined from a free-body diagram of the entire frame, Fig. 6–29*b*. Also, the free-body diagram of each frame member is shown in Fig. 6–29*c*. Notice that member BE is a two-force member. As shown by the colored dashed lines, the forces at B, C, and E have equal magnitudes but opposite directions on the separate free-body diagrams.

Equations of Equilibrium. The six unknowns A_x, A_y, F_B, C_x, C_y, and D_x will be determined from the equations of equilibrium applied to the entire frame and then to member CEF. We have

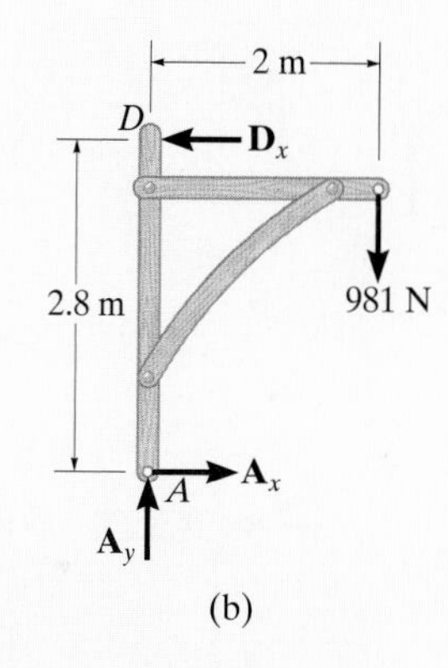

(b)

Entire Frame

$$\circlearrowright+\Sigma M_A = 0; \quad -981\text{ N}(2\text{ m}) + D_x(2.8\text{ m}) = 0 \qquad D_x = 700.7\text{ N}$$

$$\xrightarrow{+}\Sigma F_x = 0; \quad A_x - 700.7\text{ N} = 0 \qquad A_x = 700.7\text{ N}$$

$$+\uparrow\Sigma F_y = 0; \quad A_y - 981\text{ N} = 0 \qquad A_y = 981\text{ N}$$

Member CEF

$$\circlearrowright+\Sigma M_C = 0; \quad -981\text{ N}(2\text{ m}) - (F_B \sin 45^\circ)(1.6\text{ m}) = 0$$

$$F_B = -1734.2\text{ N}$$

$$\xrightarrow{+}\Sigma F_x = 0; \quad -Cx - (-1734.2\cos 45^\circ\text{ N}) = 0$$

$$C_x = 1226\text{ N} \qquad \textit{Ans.}$$

$$+\uparrow\Sigma F_y = 0; \quad C_y - (-1734.2\sin 45^\circ\text{ N}) - 981\text{ N} = 0$$

$$C_y = -245\text{ N} \qquad \textit{Ans.}$$

Since the magnitudes of $\mathbf{F}_B$ and $\mathbf{C}_y$ were calculated as negative quantities, they were assumed to be acting in the wrong sense on the free-body diagrams, Fig. 6–29*c*. The correct sense of these forces might have been determined "by inspection" *before* applying the equations of equilibrium to member CEF. As shown in Fig. 6–29*c*, moment equilibrium about point E on member CEF indicates that $\mathbf{C}_y$ must actually act *downward* to counteract the moment created by the 981-N force about E. Similarly, summing moments about C, it is seen that the vertical component of $\mathbf{F}_B$ must actually act *upward,* and so $\mathbf{F}_B$ must act upward and to the right.

NOTE: The above calculations can be checked by applying the three equilibrium equations to member $ABCD$, Fig. 6–29*c*.

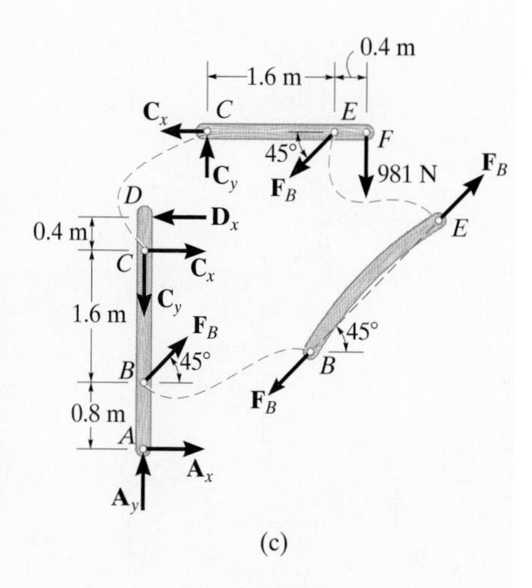

(c)

Fig. 6–29

EXAMPLE 6.17

The smooth disk shown in Fig. 6–30*a* is pinned at *D* and has a weight of 20 lb. Neglecting the weights of the other members, determine the horizontal and vertical components of reaction at pins *B* and *D*.

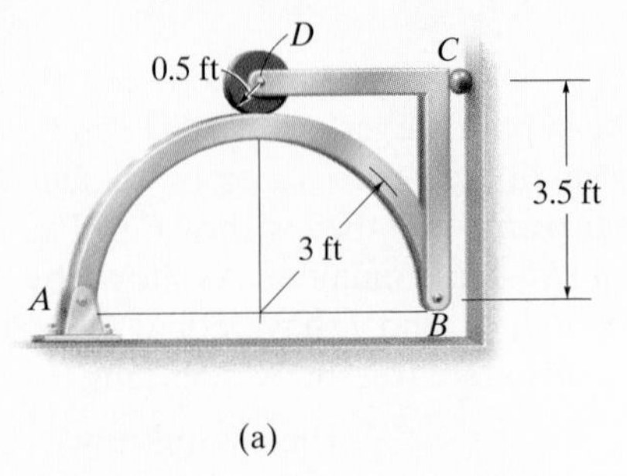

(a)

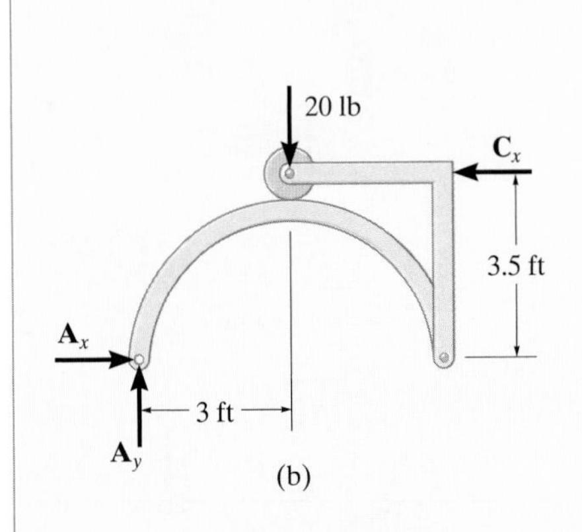

(b)

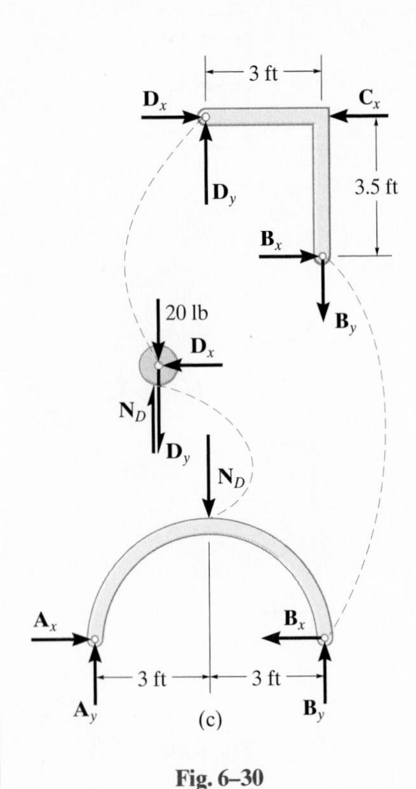

(c)

Fig. 6–30

SOLUTION

Free-Body Diagrams. By inspection, the three components of reaction at the supports can be determined from a free-body diagram of the entire frame, Fig. 6–30*b*. Also, free-body diagrams of the members are shown in Fig. 6–30*c*.

Equations of Equilibrium. The eight unknowns can of course be obtained by applying the eight equilibrium equations to each member—three to member *AB*, three to member *BCD*, and two to the disk. (Moment equilibrium is automatically satisfied for the disk.) If this is done, however, all the results can be obtained only from a simultaneous solution of some of the equations. (Try it and find out.) To avoid this situation, it is best to first determine the three support reactions on the *entire* frame; then, using these results, the remaining five equilibrium equations can be applied to two other parts in order to solve successively for the other unknowns.

Entire Frame

$$\zeta+\Sigma M_A = 0; \quad -20\text{ lb}\,(3\text{ ft}) + C_x(3.5\text{ ft}) = 0 \quad C_x = 17.1\text{ lb}$$

$$\overset{+}{\rightarrow}\Sigma F_x = 0; \quad A_x - 17.1\text{ lb} = 0 \quad A_x = 17.1\text{ lb}$$

$$+\uparrow\Sigma F_y = 0; \quad A_y - 20\text{ lb} = 0 \quad A_y = 20\text{ lb}$$

Member AB

$$\overset{+}{\rightarrow}\Sigma F_x = 0; \quad 17.1\text{ lb} - B_x = 0 \quad B_x = 17.1\text{ lb} \quad \textit{Ans.}$$

$$\zeta+\Sigma M_B = 0; \quad -20\text{ lb}\,(6\text{ ft}) + N_D(3\text{ ft}) = 0 \quad N_D = 40\text{ lb}$$

$$+\uparrow\Sigma F_y = 0; \quad 20\text{ lb} - 40\text{ lb} + B_y = 0 \quad B_y = 20\text{ lb} \quad \textit{Ans.}$$

Disk

$$\overset{+}{\rightarrow}\Sigma F_x = 0; \quad D_x = 0 \quad \textit{Ans.}$$

$$+\uparrow\Sigma F_y = 0; \quad 40\text{ lb} - 20\text{ lb} - D_y = 0 \quad D_y = 20\text{ lb} \quad \textit{Ans.}$$

EXAMPLE 6.18

Determine the tension in the cables and also the force **P** required to support the 600-N force using the frictionless pulley system shown in Fig. 6–31*a*.

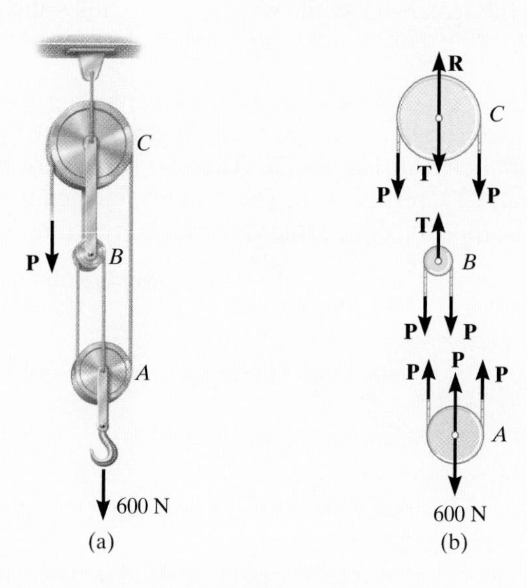

Fig. 6–31

SOLUTION

Free-Body Diagram. A free-body diagram of each pulley *including* its pin and a portion of the contacting cable is shown in Fig. 6–31*b*. Since the cable is *continuous* and the pulleys are frictionless, the cable has a *constant tension* P acting throughout its length (see Example 5.7). The link connection between pulleys B and C is a two-force member, and therefore it has an unknown tension T acting on it. Notice that the *principle of action, equal but opposite reaction* must be carefully observed for forces **P** and **T** when the *separate* free-body diagrams are drawn.

Equations of Equilibrium. The three unknowns are obtained as follows:

Pulley A

$+\uparrow \Sigma F_y = 0; \qquad 3P - 600\text{ N} = 0 \qquad P = 200\text{ N}$ *Ans.*

Pulley B

$+\uparrow \Sigma F_y = 0; \qquad T - 2P = 0 \qquad T = 400\text{ N}$ *Ans.*

Pulley C

$+\uparrow \Sigma F_y = 0; \qquad R - 2P - T = 0 \qquad R = 800\text{ N}$ *Ans.*

EXAMPLE 6.19

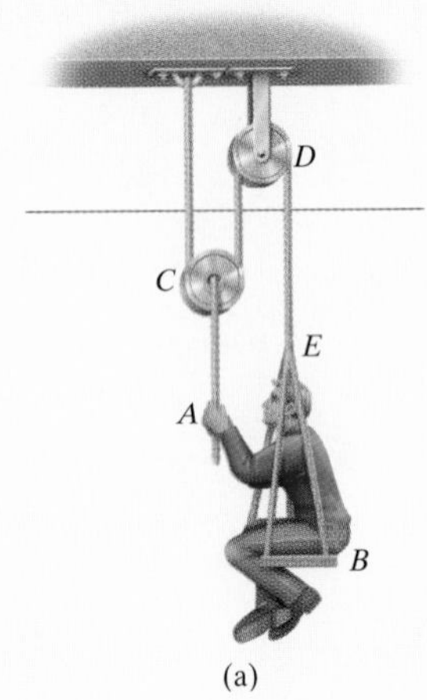

(a)

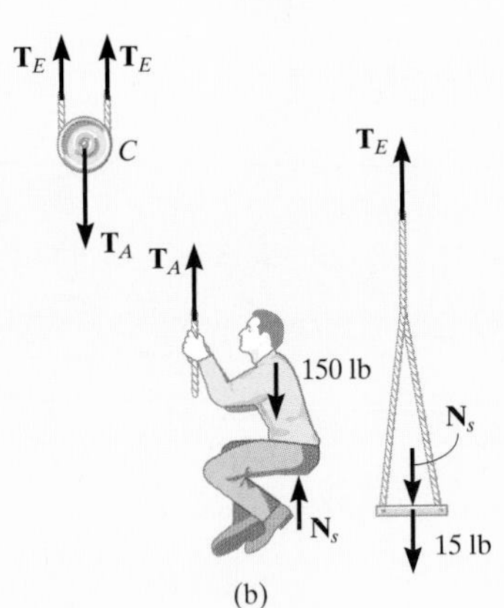

(b)

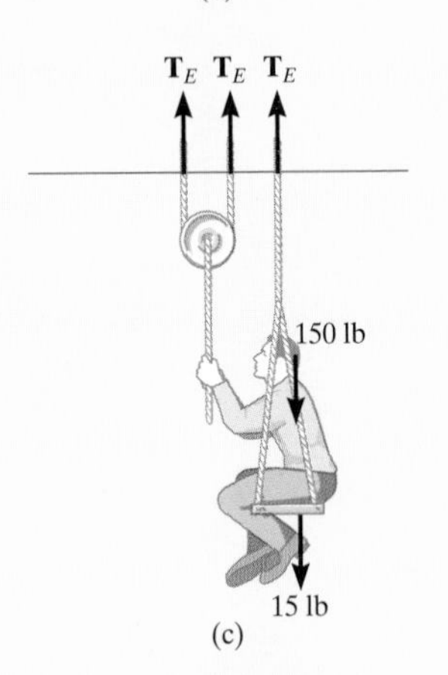

(c)

Fig. 6–32

A man having a weight of 150 lb supports himself by means of the cable and pulley system shown in Fig. 6–32*a*. If the seat has a weight of 15 lb, determine the force that he must exert on the cable at A and the force he exerts on the seat. Neglect the weight of the cables and pulleys.

SOLUTION I

Free-Body Diagrams. The free-body diagrams of the man, seat, and pulley C are shown in Fig. 6–32*b*. The *two* cables are subjected to tensions $\mathbf{T}_A$ and $\mathbf{T}_E$, respectively. The man is subjected to three forces: his weight, the tension $\mathbf{T}_A$ of cable AC, and the reaction $\mathbf{N}_s$ of the seat.

Equations of Equilibrium. The three unknowns are obtained as follows:

Man

$$+\uparrow \Sigma F_y = 0; \qquad T_A + N_s - 150 \text{ lb} = 0 \qquad (1)$$

Seat

$$+\uparrow \Sigma F_y = 0; \qquad T_E - N_s - 15 \text{ lb} = 0 \qquad (2)$$

Pulley C

$$+\uparrow \Sigma F_y = 0; \qquad 2T_E - T_A = 0 \qquad (3)$$

Here T_E can be determined by adding Eqs. 1 and 2 to eliminate N_s and then using Eq. 3. The other unknowns are then obtained by resubstitution of T_E.

$$T_A = 110 \text{ lb} \qquad \textit{Ans.}$$
$$T_E = 55 \text{ lb}$$
$$N_s = 40 \text{ lb} \qquad \textit{Ans.}$$

SOLUTION II

Free-Body Diagrams. By using the blue section shown in Fig. 6–32*a*, the man, pulley, and seat can be considered as a *single system*, Fig. 6–32*c*. Here $\mathbf{N}_s$ and $\mathbf{T}_A$ are *internal* forces and hence are not included on this "combined" free-body diagram.

Equations of Equilibrium. Applying $\Sigma F_y = 0$ yields a *direct* solution for T_E.

$$+\uparrow \Sigma F_y = 0; \; 3T_E - 15 \text{ lb} - 150 \text{ lb} = 0 \qquad T_E = 55 \text{ lb}$$

The other unknowns can be obtained from Eqs. 2 and 3.

EXAMPLE 6.20

The hand exerts a force of 8 lb on the grip of the spring compressor shown in Fig. 6–33*a*. Determine the force in the spring needed to maintain equilibrium of the mechanism.

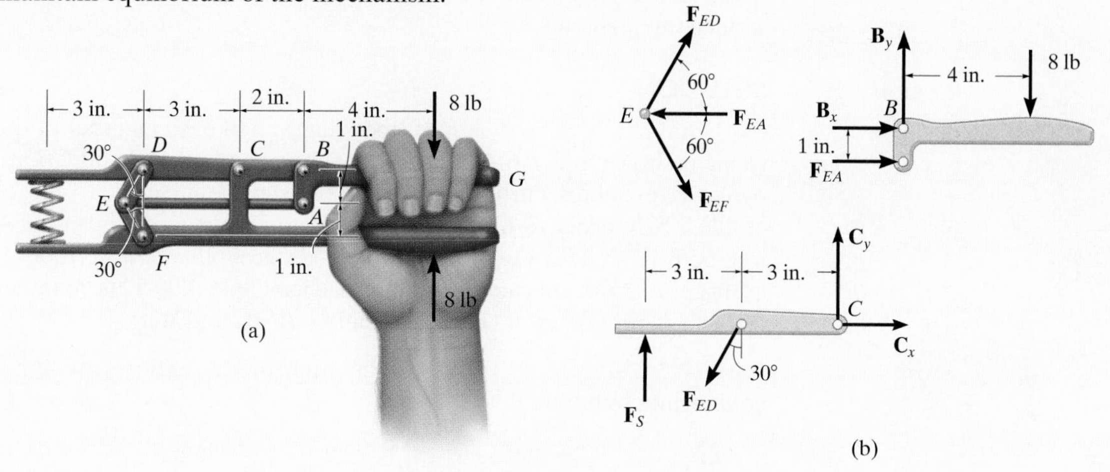

Fig. 6–33

SOLUTION

Free-Body Diagrams. By inspection, members *EA, ED*, and *EF* are all two-force members. The free-body diagrams for parts *DC* and *ABG* are shown in Fig. 6–33*b*. The pin at *E* has also been included here since *three* force interactions occur on this pin. They represent the effects of members *ED, EA*, and *EF*. Note carefully how equal and opposite force reactions occur between each of the parts.

Equations of Equilibrium. By studying the free-body diagrams, the most direct way to obtain the spring force is to apply the equations of equilibrium in the following sequence:

Lever ABG

$$\circlearrowleft+\Sigma M_B = 0; \quad F_{EA}(1\text{ in.}) - 8\text{ lb }(4\text{ in.}) = 0 \quad F_{EA} = 32\text{ lb}$$

Pin E

$$+\uparrow\Sigma F_y = 0; \quad F_{ED}\sin 60^\circ - F_{EF}\sin 60^\circ = 0 \quad F_{ED} = F_{EF} = F$$

$$\overset{+}{\rightarrow}\Sigma F_x = 0; \quad 2F\cos 60^\circ - 32\text{ lb} = 0 \quad F = 32\text{ lb}$$

Arm DC

$$\circlearrowleft+\Sigma M_C = 0; \quad -F_s(6\text{ in.}) + 32\cos 30^\circ\text{ lb }(3\text{ in.}) = 0$$

$$F_s = 13.9\text{ lb} \qquad \textit{Ans.}$$

EXAMPLE 6.21

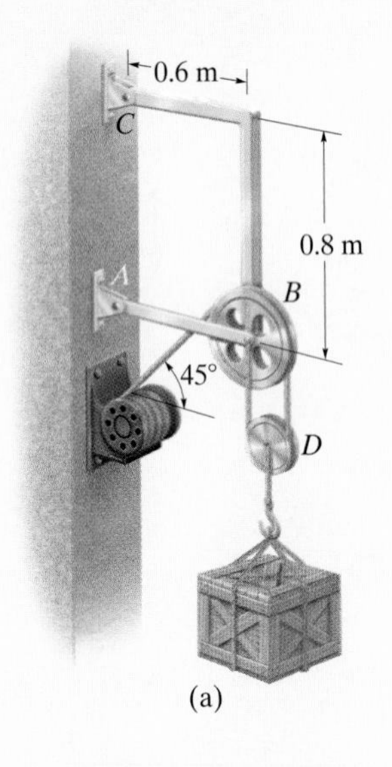

(a)

The 100-kg block is held in equilibrium by means of the pulley and continuous cable system shown in Fig. 6–34*a*. If the cable is attached to the pin at *B*, compute the forces which this pin exerts on each of its connecting members.

SOLUTION

Free-Body Diagrams. A free-body diagram of each member of the frame is shown in Fig. 6–34*b*. By inspection, members *AB* and *CB* are two-force members. Furthermore, the cable must be subjected to a force of 490.5 N in order to hold pulley *D* and the block in equilibrium. A free-body diagram of the pin at *B* is needed since *four interactions* occur at this pin. These are caused by the attached cable (490.5 N), member *AB* ($\mathbf{F}_{AB}$), member *CB* ($\mathbf{F}_{CB}$), and pulley *B* ($\mathbf{B}_x$ and $\mathbf{B}_y$).

Equations of Equilibrium. Applying the equations of force equilibrium to pulley *B*, we have

$$\xrightarrow{+} \Sigma F_x = 0; \quad B_x - 490.5 \cos 45° \text{ N} = 0 \quad B_x = 346.8 \text{ N} \qquad \textit{Ans.}$$

$$+\uparrow \Sigma F_y = 0; \quad B_y - 490.5 \sin 45° \text{ N} - 490.5 \text{ N} = 0$$

$$B_y = 837.3 \text{ N} \qquad \textit{Ans.}$$

Using these results, equilibrium of the pin requires that

$$+\uparrow \Sigma F_y = 0; \quad \tfrac{4}{5}F_{CB} - 837.3 \text{ N} - 490.5 \text{ N} \qquad F_{CB} = 1660 \text{ N} \qquad \textit{Ans.}$$

$$\xrightarrow{+} \Sigma F_x = 0; \quad F_{AB} - \tfrac{3}{5}(1660 \text{ N}) - 346.8 \text{ N} = 0 \; F_{AB} = 1343 \text{ N} \qquad \textit{Ans.}$$

NOTE: It may be noted that the two-force member *CB* is subjected to bending as caused by the force $\mathbf{F}_{CB}$. From the standpoint of design, it would be better to make this member *straight* (from *C* to *B*) so that the force $\mathbf{F}_{CB}$ would create only tension in the member.

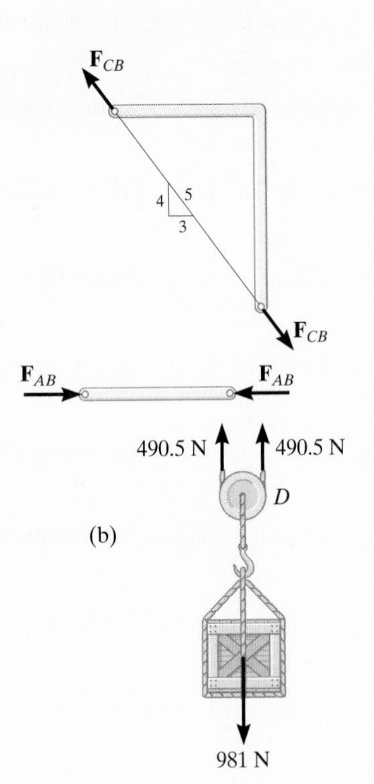

(b)

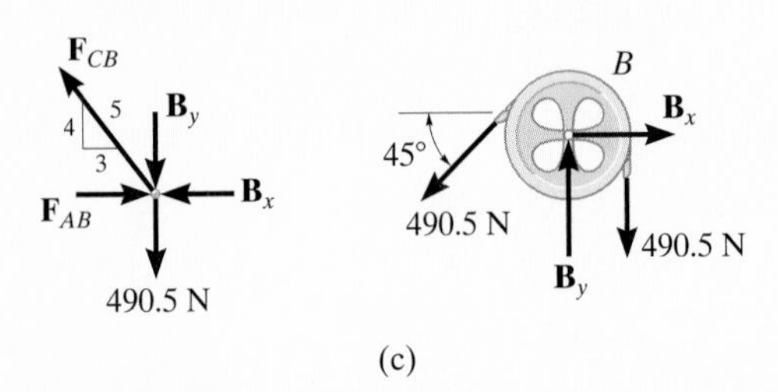

(c)

Fig. 6–34

PROBLEMS

Before solving the following problems, it is suggested that a brief review be made of all the previous examples. This may be done by covering each solution, trying to locate the two-force members, drawing the free-body diagrams, and conceiving ways of applying the equations of equilibrium to obtain the solution.

6–66. A force of $P = 8$ lb is applied to the handles of the pliers. Determine the force developed on the smooth bolt B and the reaction that pin A exerts on its attached members.

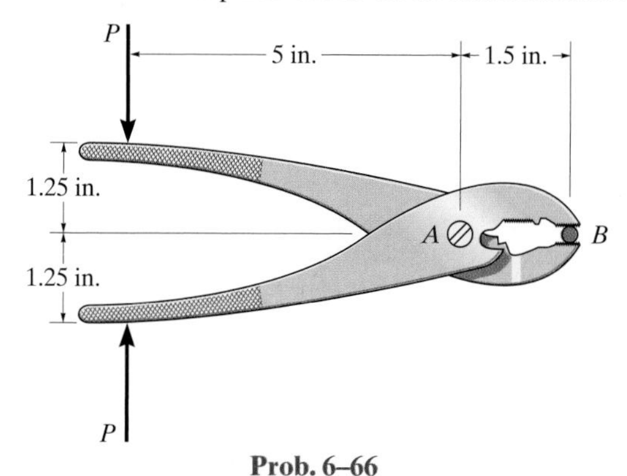

Prob. 6–66

6–67. The eye hook has a positive locking latch when it supports the load because its two parts are pin-connected at A and they bear against one another along the smooth surface at B. Determine the resultant force at the pin and the normal force at B when the eye hook supports a load of 800 lb.

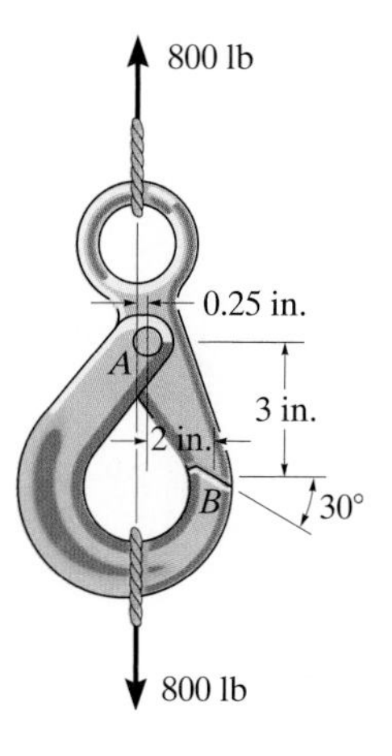

Prob. 6–67

***6–68.** Determine the force **P** needed to hold the 20-lb block in equilibrium.

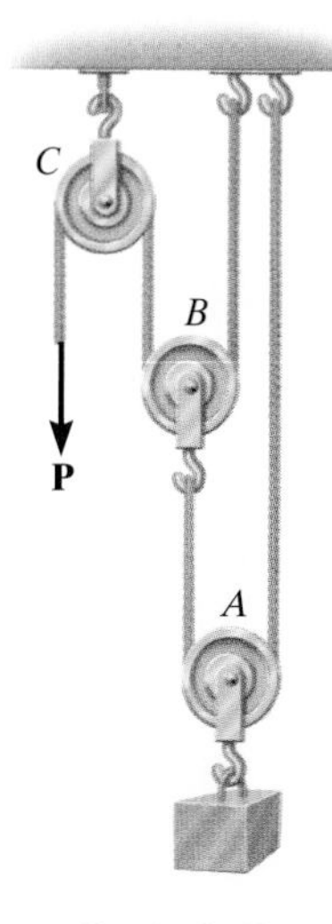

Prob. 6–68

6–69. The link is used to hold the rod in place. Determine the required axial force on the screw at E if the largest force to be exerted on the rod at B, C, or D is to be 100 N. Also, find the magnitude of the force reaction at pin A. Assume all surfaces of contact are smooth.

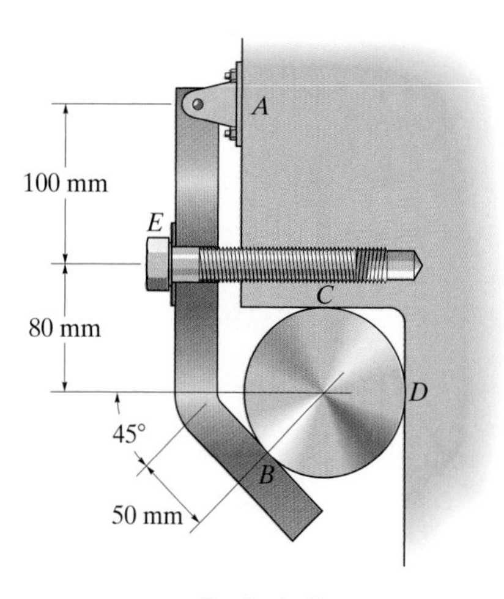

Prob. 6–69

6–70. The 150-lb man attempts to lift himself and the 10-lb seat using the rope and pulley system shown. Determine the force at A needed to do so, and also find his reaction on the seat.

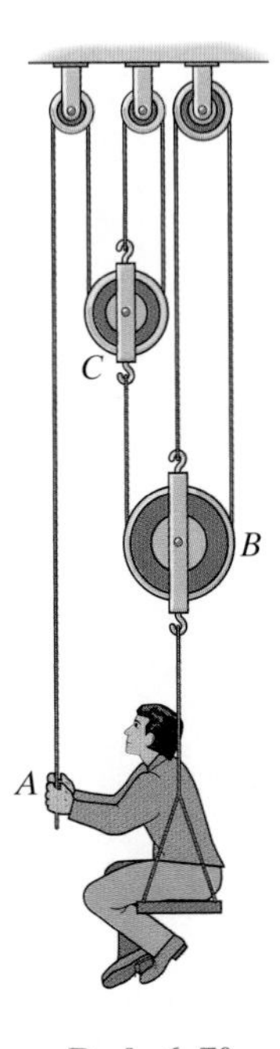

Prob. 6–70

6–71. Determine the horizontal and vertical components of force that pins A and C exert on the frame.

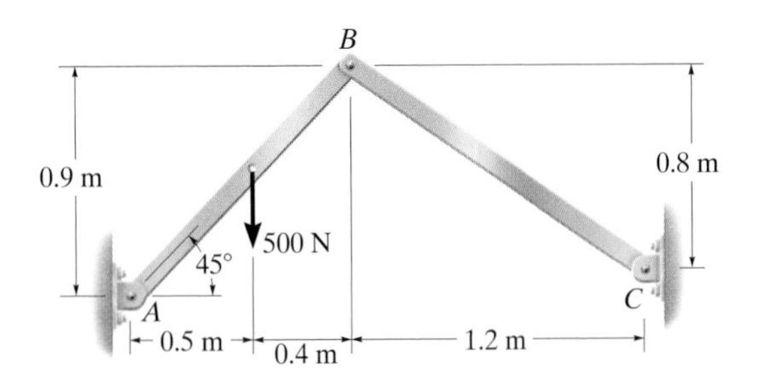

Prob. 6–71

***6–72.** Determine the horizontal and vertical components of force that pins A and C exert on the frame.

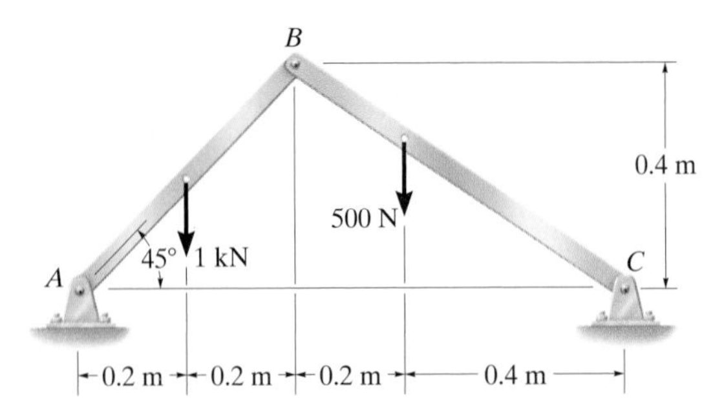

Prob. 6–72

6–73. The truck exerts the three forces shown on the girders of the bridge. Determine the reactions at the supports when the truck is in the position shown. The girders are connected together by a short vertical link DC.

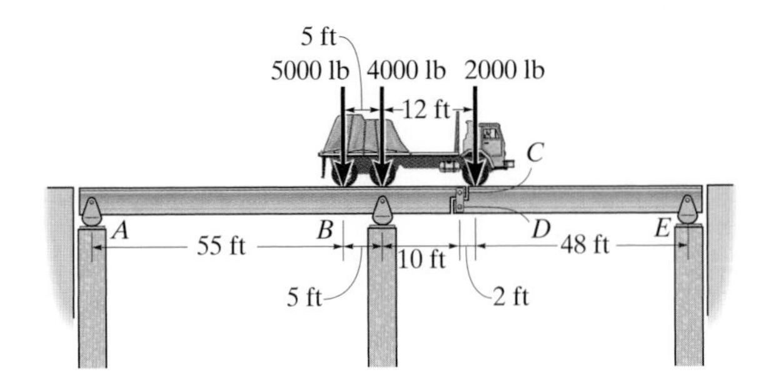

Prob. 6–73

6–74. Determine the greatest force P that can be applied to the frame if the largest force resultant acting at A can have a magnitude of 2 kN.

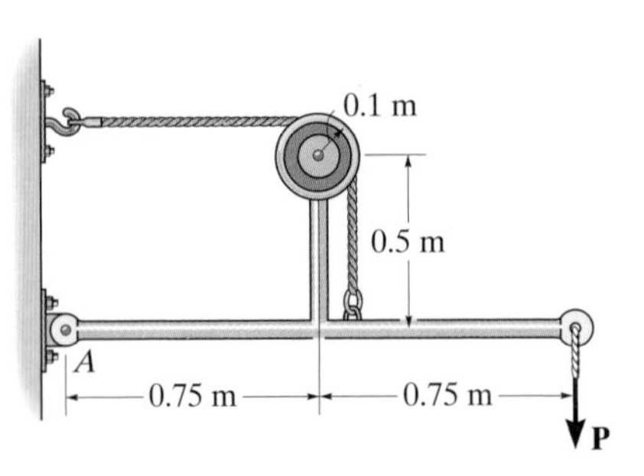

Prob. 6–74

6–75. The compound beam is pin supported at B and supported by rockers at A and C. There is a hinge (pin) at D. Determine the reactions at the supports.

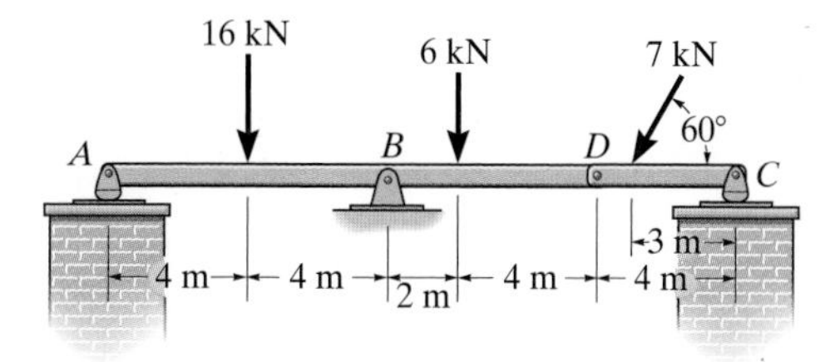

Prob. 6–75

***6–76.** The compound beam is fixed supported at A and supported by rockers at B and C. If there are hinges (pins) at D and E, determine the reactions at the supports A, B, and C.

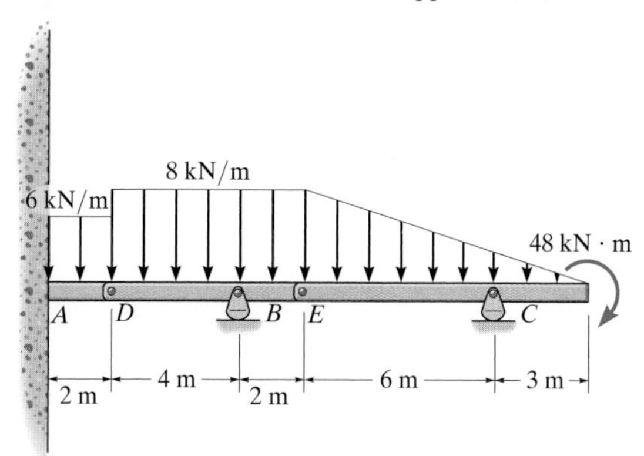

Prob. 6–76

6–77. Determine the reactions at supports A and B.

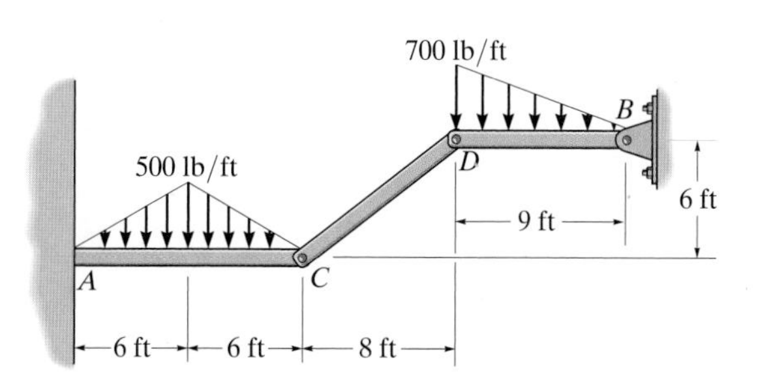

Prob. 6–77

6–78. Determine the horizontal and vertical components of force at C which member ABC exerts on member CEF.

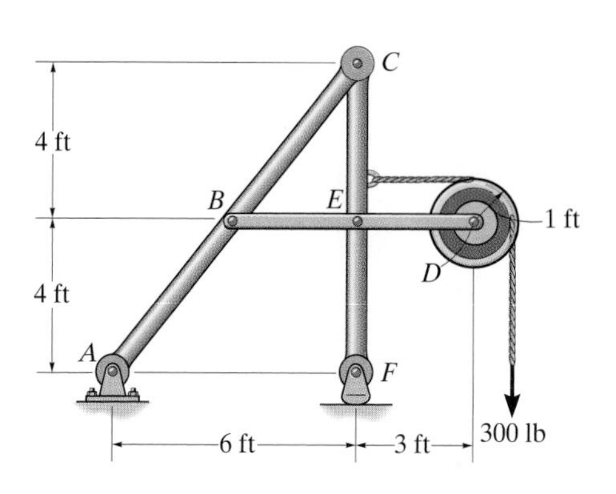

Prob. 6–78

6–79. Determine the horizontal and vertical components of force that the pins at A, B, and C exert on their connecting members.

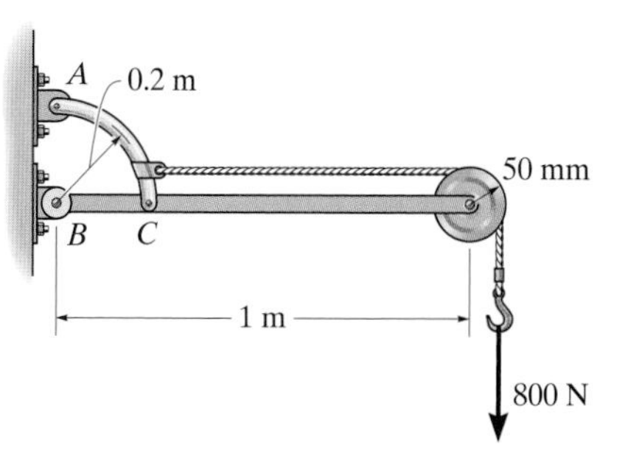

Prob. 6–79

*6–80. Operation of exhaust and intake valves in an automobile engine consists of the cam C, push rod DE, rocker arm EFG which is pinned at F, and a spring and valve, V. If the compression in the spring is 20 mm when the valve is open as shown, determine the normal force acting on the cam lobe at C. Assume the cam and bearings at H, I, and J are smooth. The spring has a stiffness of 300 N/m.

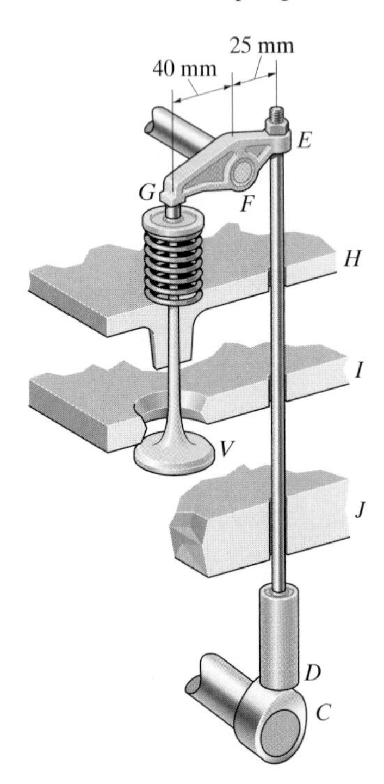

Prob. 6–80

6–81. Determine the force P on the cord, and the angle θ that the pulley-supporting link AB makes with the vertical. Neglect the mass of the pulleys and the link. The block has a weight of 200 lb and the cord is attached to the pin at B. The pulleys have radii of $r_1 = 2$ in. and $r_2 = 1$ in.

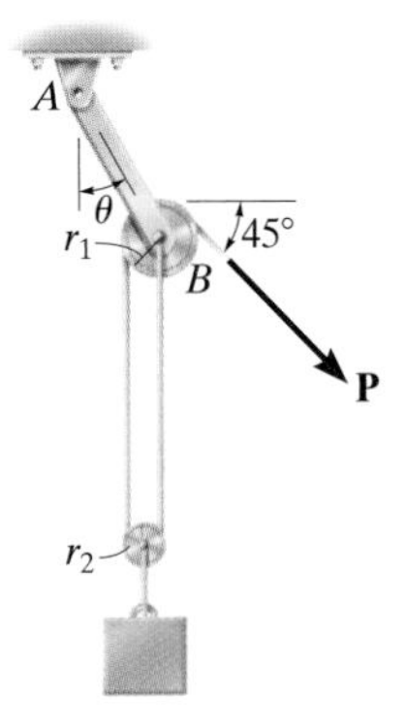

Prob. 6–81

6–82. The nail cutter consists of the handle and the two cutting blades. Assuming the blades are pin connected at B and the surface at D is smooth, determine the normal force on the fingernail when a force of 1 lb is applied to the handles as shown. The pin AC slides through a smooth hole at A and is attached to the bottom member at C.

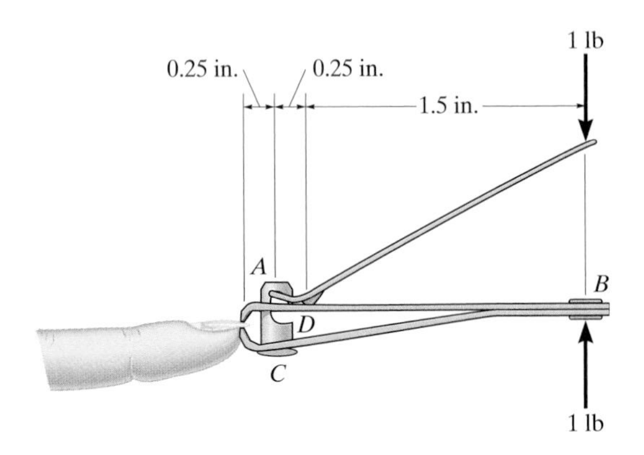

Prob. 6–82

6–83. The wall crane supports a load of 700 lb. Determine the horizontal and vertical components of reaction at the pins A and D. Also, what is the force in the cable at the winch W?

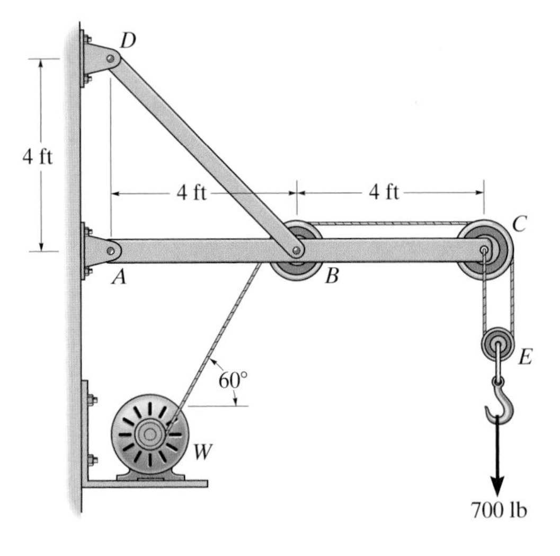

Prob. 6–83

***6–84.** Determine the force that the smooth roller C exerts on beam AB. Also, what are the horizontal and vertical components of reaction at pin A? Neglect the weight of the frame and roller.

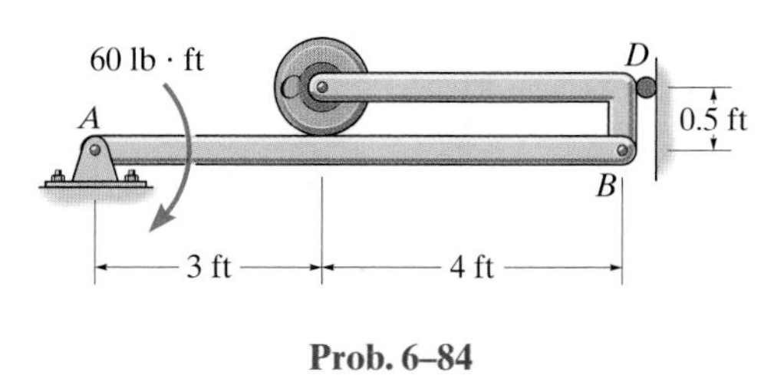

Prob. 6–84

6–85. Determine the horizontal and vertical components of force which the pins exert on member ABC.

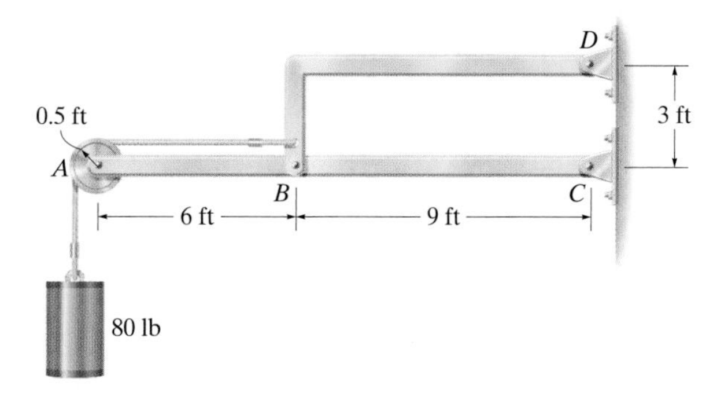

Prob. 6–85

6–86. The floor beams AB and BC are stiffened using the two tie rods CD and AD. Determine the force along each rod when the floor beams are subjected to a uniform load of 80 lb/ft. Assume the three contacting members at B are smooth and the joints at A, C, and D are pins. *Hint:* Members AD, CD, and BD are two-force members.

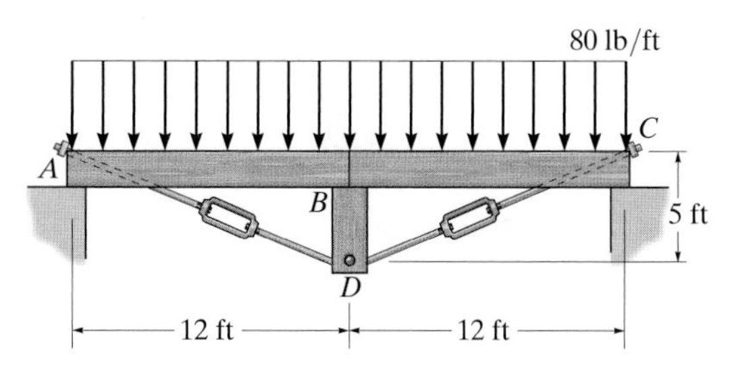

Prob. 6–86

6–87. Determine the horizontal and vertical components of force at pins B and C.

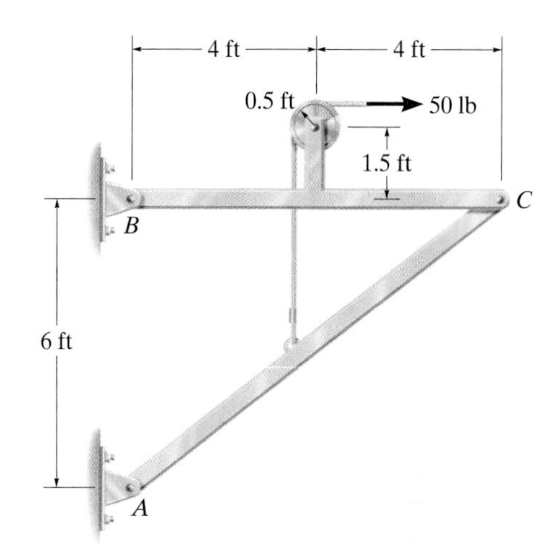

Prob. 6–87

***6–88.** The skid steer loader has a mass of 1.18 Mg, and in the position shown the center of mass is at G_1. If there is a 300-kg stone in the bucket, with center of mass at G_2, determine the reactions of each pair of wheels A and B on the ground and the force in the hydraulic cylinder CD and at the pin E. There is a similar linkage on each side of the loader.

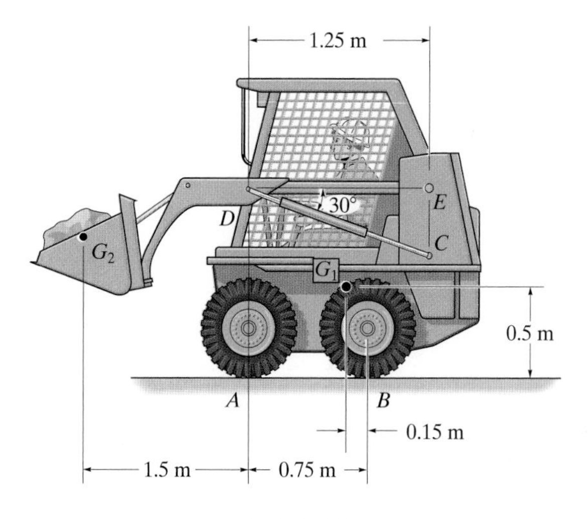

Prob. 6–88

6–89. Determine the horizontal and vertical components of force at each pin. The suspended cylinder has a weight of 80 lb.

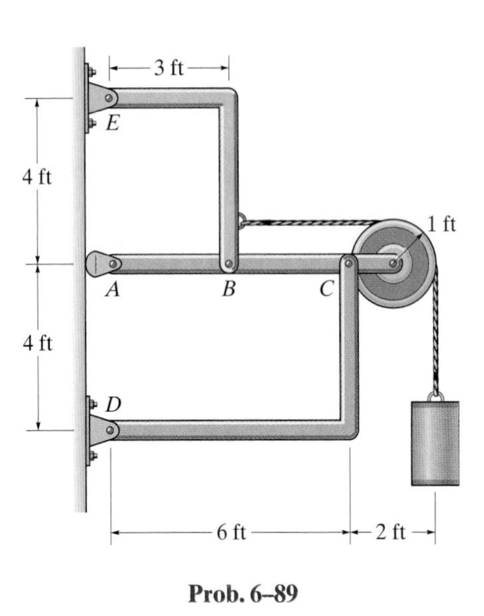

Prob. 6–89

6–90. The two-member frame is pin connected at C, D, and E. The cable is attached to A, passes over the smooth peg at B, and is attached to a 100-lb load. Determine the horizontal and vertical reactions at each pin.

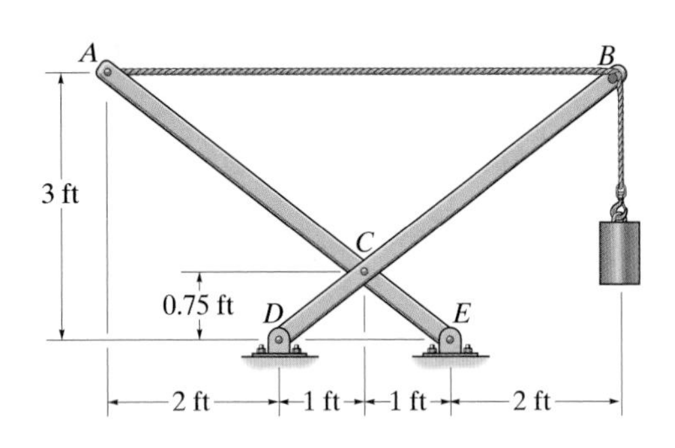

Prob. 6–90

6–91. Determine the horizontal and vertical components of force which the pins at A, B, and C exert on member ABC of the frame.

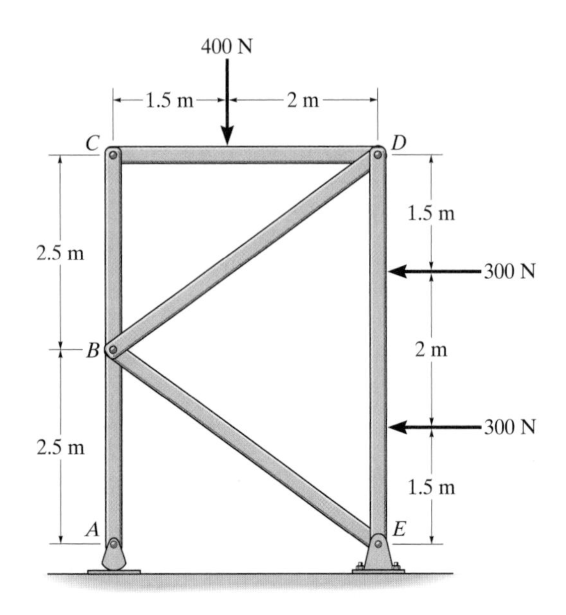

Prob. 6–91

***6–92.** The derrick is pin-connected to the pivot at A. Determine the largest mass that can be supported by the derrick if the maximum force that can be sustained by the pin at A is 18 kN.

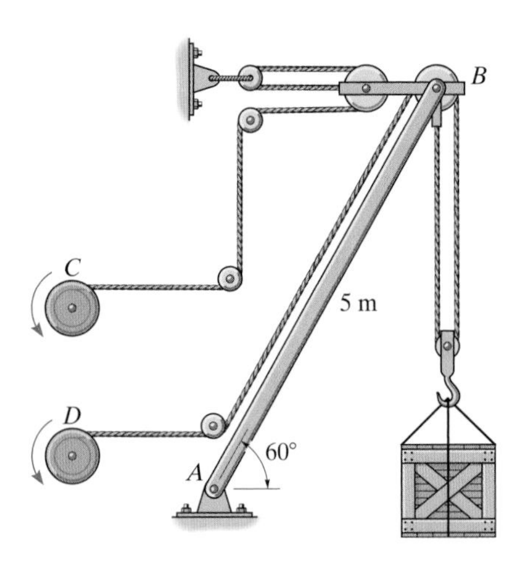

Prob. 6–92

6–93. Determine the required mass of the suspended cylinder if the tension in the chain wrapped around the freely turning gear is to be 2 kN. Also, what is the magnitude of the resultant force on pin A?

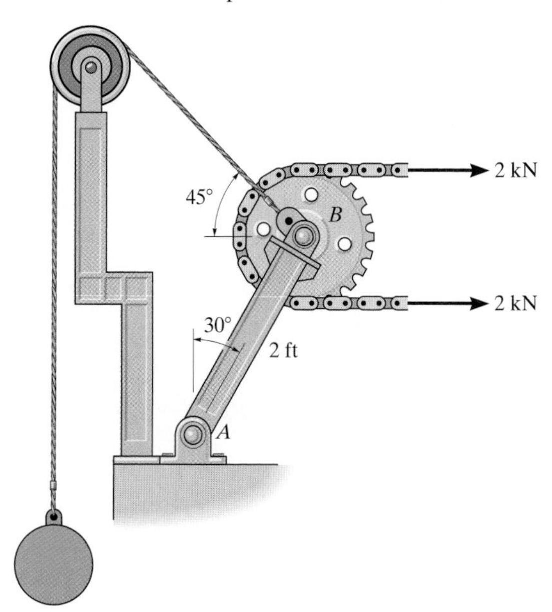

Prob. 6–93

6–94. The tongs consist of two jaws pinned to links at A, B, C, and D. Determine the horizontal and vertical components of force exerted on the 500-lb stone at F and G in order to lift it.

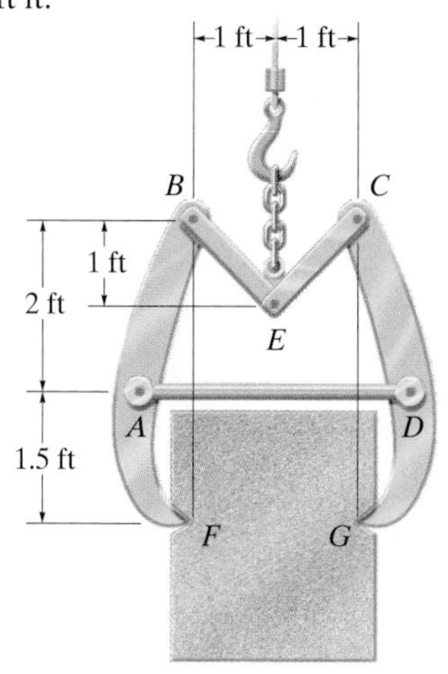

Prob. 6–94

6–95. Determine the force P on the cable if the spring is compressed 0.5 in. when the mechanism is in the position shown. The spring has a stiffness of $k = 800$ lb/ft.

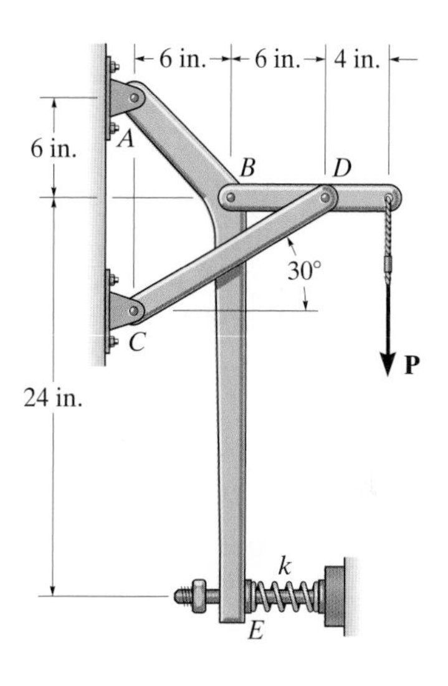

Prob. 6–95

***6–96.** The scale consists of five pin-connected members. Determine the load W on the pan EG if a weight of 3 lb is suspended from the hook at A.

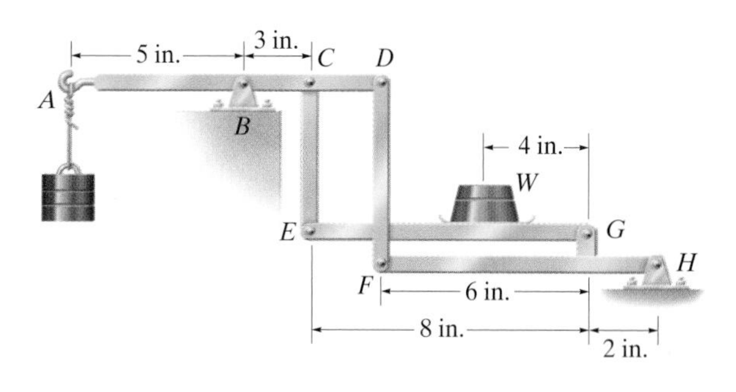

Prob. 6–96

6–97. The machine shown is used for forming metal plates. It consists of two toggles ABC and DEF, which are operated by the hydraulic cylinder H. The toggles push the movable bar G forward, pressing the plate p into the cavity. If the force which the plate exerts on the head is $P = 12$ kN, determine the force F in the hydraulic cylinder when $\theta = 30°$.

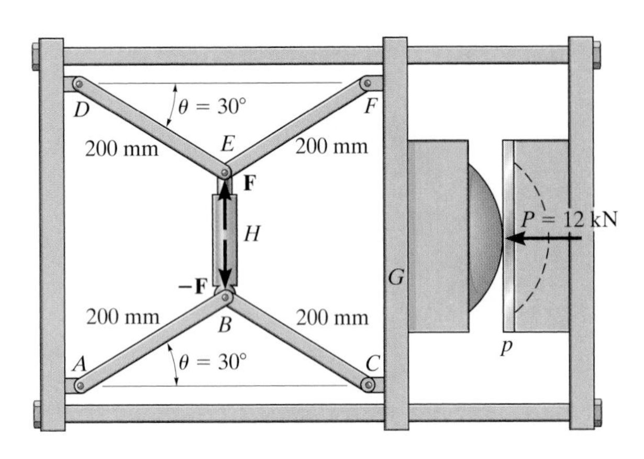

Prob. 6–97

6–98. Determine the horizontal and vertical components of force at pins A and C of the two-member frame.

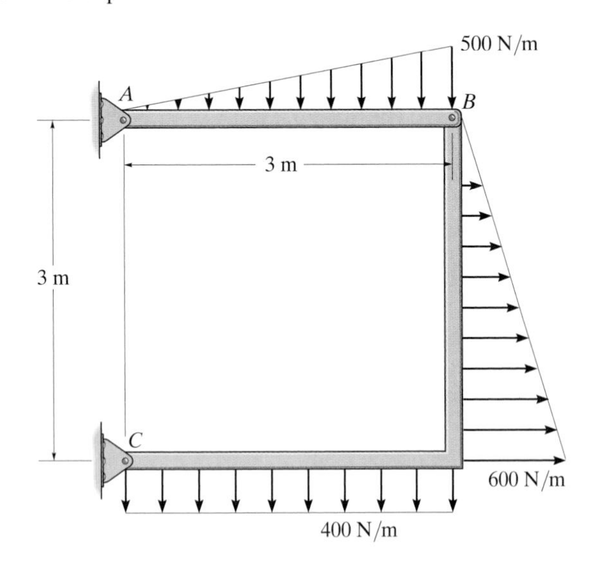

Prob. 6–98

6–99. The truck rests on the scale, which consists of a series of compound levers. If a mass of 15 kg is placed on the pan P and it is required that $x = 0.480$ m to balance the "beam" ABC, determine the mass of the truck. There are pins at all lettered points. Take $EF = 0.2$ m, $FD = 3$ m, $KG = GH = 2.5$ m, and $HI = KJ = 0.1$ m. Is it necessary for the truck to be symmetrically placed on the scale? Explain.

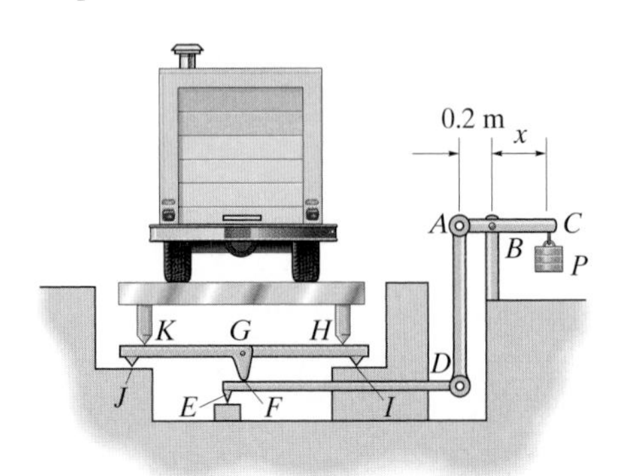

Prob. 6–99

***6–100.** By squeezing on the hand brake of the bicycle, the rider subjects the brake cable to a tension of 50 lb. If the caliper mechanism is pin-connected to the bicycle frame at B, determine the normal force each brake pad exerts on the rim of the wheel. Is this the force that stops the wheel from turning? Explain.

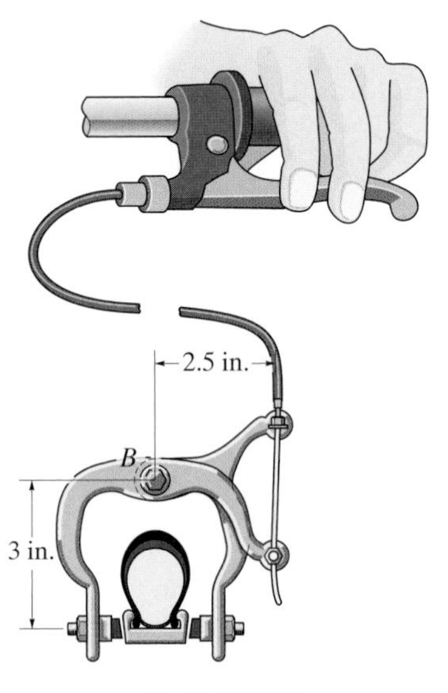

Prob. 6–100

6–101. If a force of $P = 6$ lb is applied perpendicular to the handle of the mechanism, determine the magnitude of force $\mathbf{F}$ for equilibrium. The members are pin-connected at A, B, C, and D.

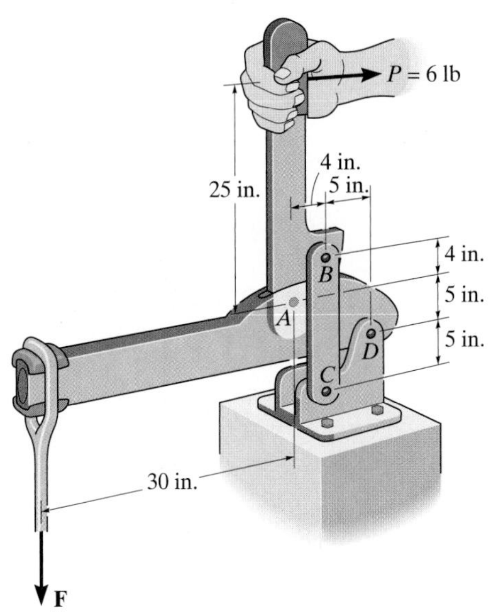

Prob. 6–101

6–102. The pillar crane is subjected to the load having a mass of 500 kg. Determine the force developed in the tie rod AB and the horizontal and vertical reactions at the pin support C when the boom is tied in the position shown.

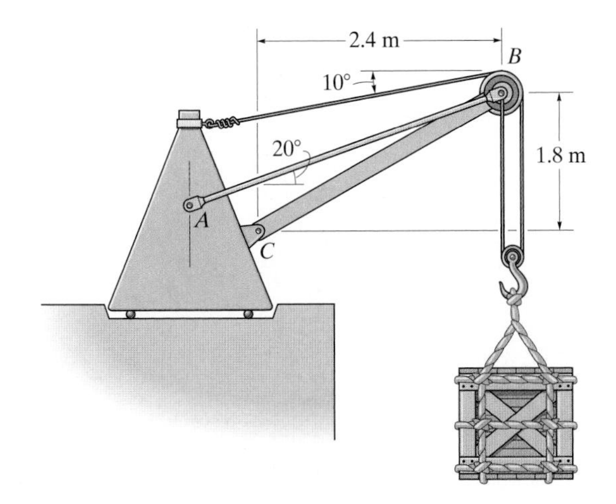

Prob. 6–102

6–103. The tower truss has a weight of 575 lb and a center of gravity at G. The rope system is used to hoist it into the vertical position. If rope CB is attached to the top of the shear leg AC and a second rope CD is attached to the truss, determine the required tension in BC to hold the truss in the position shown. The base of the truss and the shear leg bears against the stake at A, which can be considered as a pin. Also, compute the compressive force acting along the shear leg.

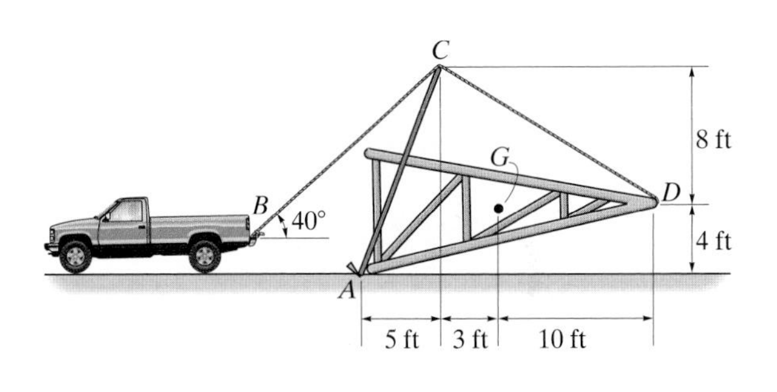

Prob. 6–103

*■**6–104.** The constant moment of 50 N · m is applied to the crank shaft. Determine the compressive force P that is exerted on the piston for equilibrium as a function of θ. Plot the results of P (ordinate) versus θ (abscissa) for $0° \le \theta \le 90°$.

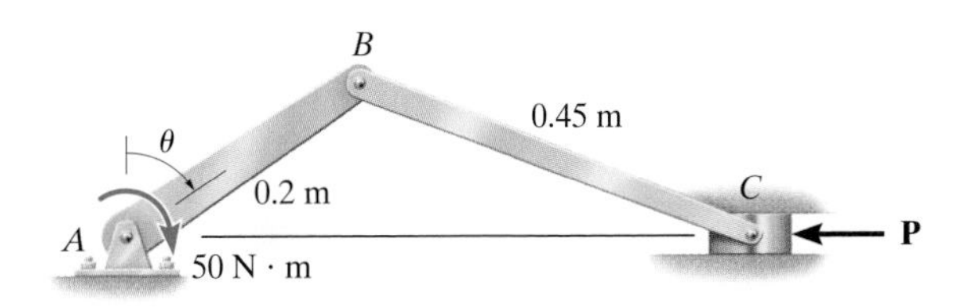

Prob. 6–104

6–105. Five coins are stacked in the smooth plastic container shown. If each coin weighs 0.0235 lb, determine the normal reactions of the bottom coin on the container at points A and B.

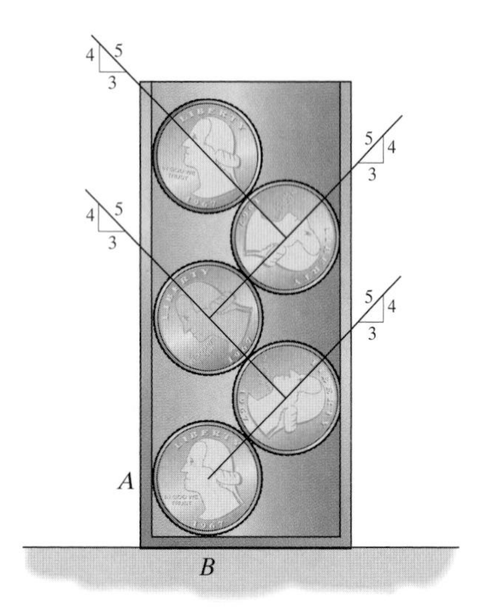

Prob. 6–105

6–106. Determine the horizontal and vertical components of force at pin B and the normal force the pin at C exerts on the smooth slot. Also, determine the moment and horizontal and vertical reactions of force at A. There is a pulley at E.

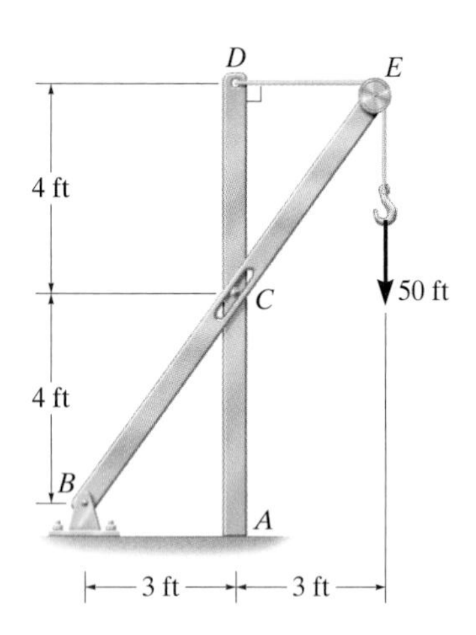

Prob. 6–106

6–107. A 5-lb force is applied to the handles of the vise grip. Determine the compressive force developed on the smooth bolt shank A at the jaws.

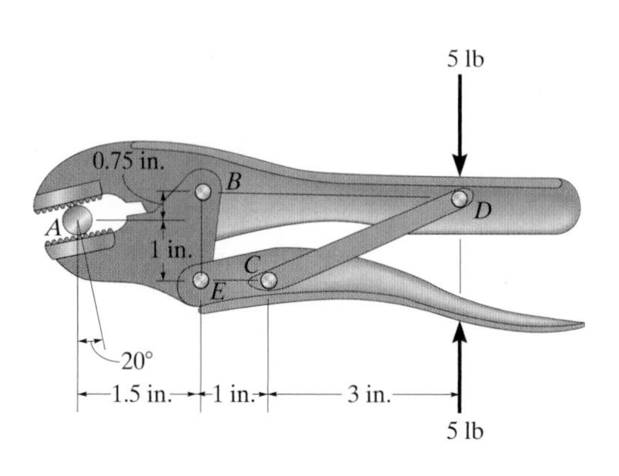

Prob. 6–107

***6–108.** If a force of 10 lb is applied to the grip of the clamp, determine the compressive force F that the wood block exerts on the clamp.

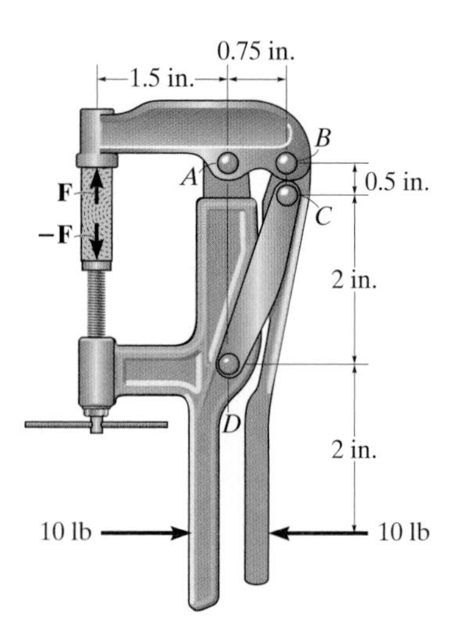

Prob. 6–108

6–109. The hoist supports the 125-kg engine. Determine the force in member DB and in the hydraulic cylinder H of member FB.

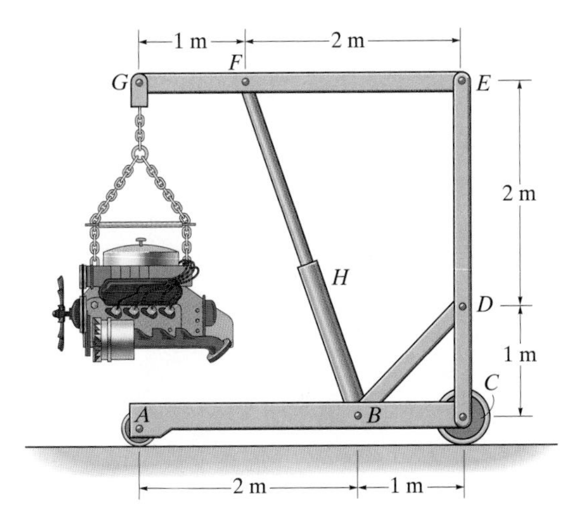

Prob. 6–109

6–110. The flat-bed trailer has a weight of 7000 lb and center of gravity at G_T. It is pin-connected to the cab at D. The cab has a weight of 6000 lb and center of gravity at G_C. Determine the range of values x for the position of the 2000-lb load L so that when it is placed over the rear axle, no axle is subjected to more than 5500 lb. The load has a center of gravity at G_L.

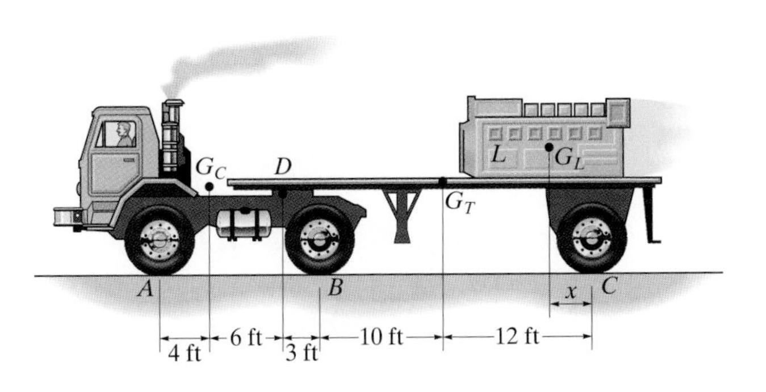

Prob. 6–110

6–111. Determine the force created in the hydraulic cylinders EF and AD in order to hold the shovel in equilibrium. The shovel load has a mass of 1.25 Mg and a center of gravity at G. All joints are pin connected.

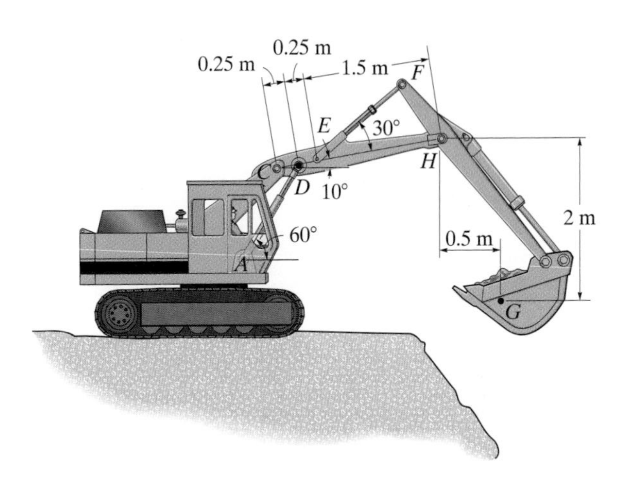

Prob. 6–111

***6–112.** The aircraft-hangar door opens and closes slowly by means of a motor which draws in the cable AB. If the door is made in two sections (bifold) and each section has a uniform weight W and length L, determine the force in the cable as a function of the door's position θ. The sections are pin-connected at C and D and the bottom is attached to a roller that travels along the vertical track.

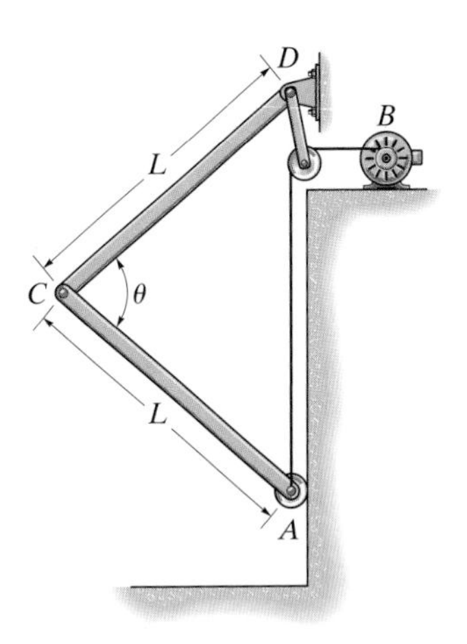

Prob. 6–112

6–113. A man having a weight of 175 lb attempts to lift himself using one of the two methods shown. Determine the total force he must exert on bar AB in each case and the normal reaction he exerts on the platform at C. Neglect the weight of the platform.

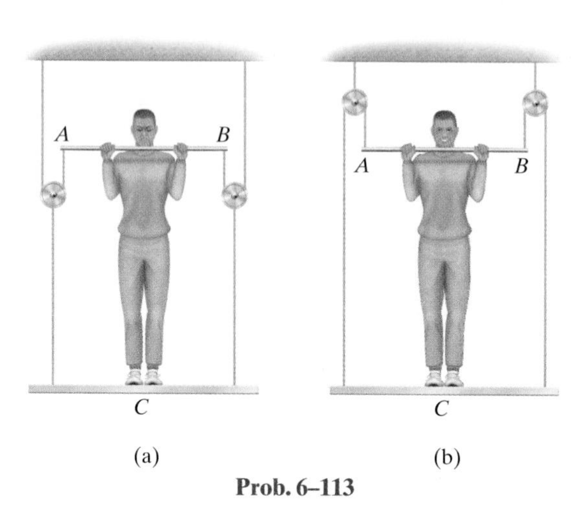

Prob. 6–113

6–114. A man having a weight of 175 lb attempts to lift himself using one of the two methods shown. Determine the total force he must exert on bar AB in each case and the normal reaction he exerts on the platform at C. The platform has a weight of 30 lb.

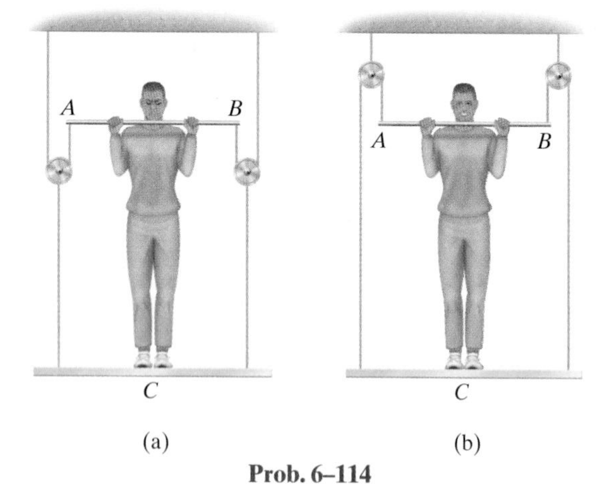

Prob. 6–114

6–115. The piston C moves vertically between the two smooth walls. If the spring has a stiffness of $k = 15$ lb/in., and is unstretched when $\theta = 0°$, determine the couple $\mathbf{M}$ that must be applied to AB to hold the mechanism in equilibrium when $\theta = 30°$.

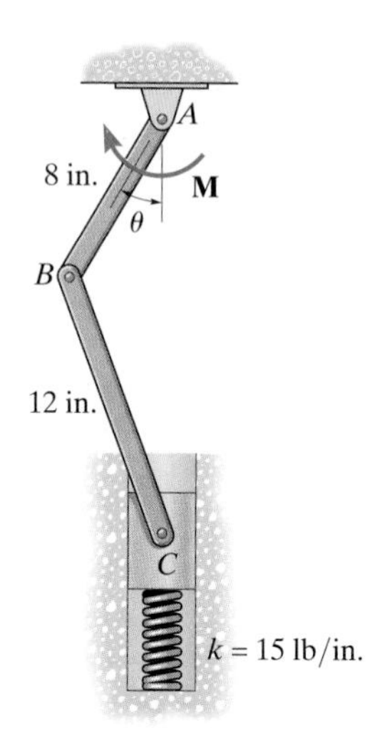

Prob. 6–115

***6–116.** The compound shears are used to cut metal parts. Determine the vertical cutting force exerted on the rod R if a force of $F = 20$ lb is applied at the grip G. The lobe CDE is in smooth contact with the head of the shear blade at E.

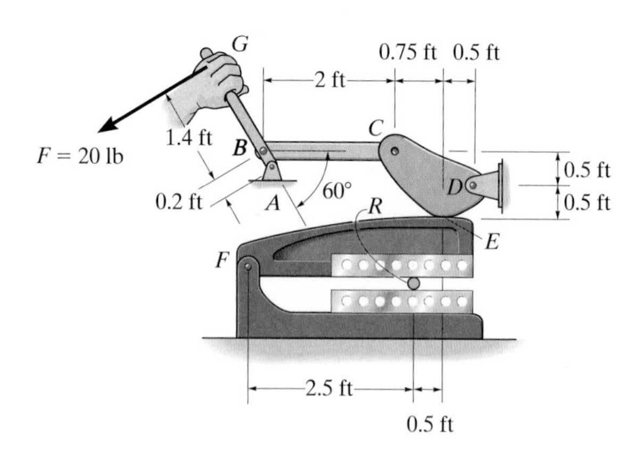

Prob. 6–116

6–117. The handle of the sector press is fixed to gear G, which in turn is in mesh with the sector gear C. Note that AB is pinned at its ends to gear C and the underside of the table EF, which is allowed to move vertically due to the smooth guides at E and F. If the gears exert tangential forces between them, determine the compressive force developed on the cylinder S when a vertical force of 40 N is applied to the handle of the press.

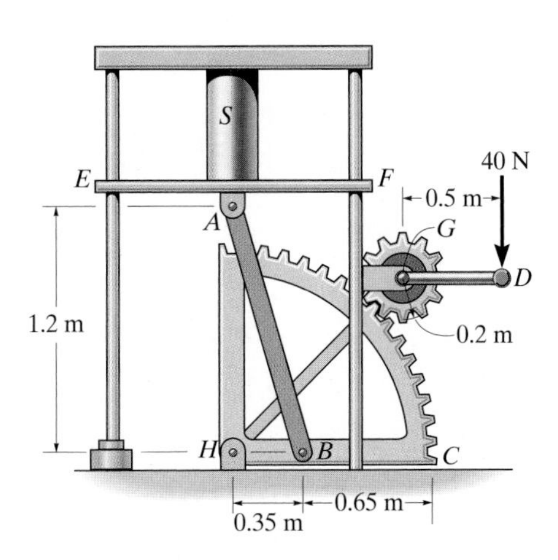

Prob. 6–117

6–118. The mechanism is used to hide kitchen appliances under a cabinet by allowing the shelf to rotate downward. If the mixer weighs 10 lb, is centered on the shelf, and has a mass center at G, determine the stretch in the spring necessary to hold the shelf in the equilibrium position shown. There is a similar mechanism on each side of the shelf, so that each mechanism supports 5 lb of the load. The springs each have a stiffness of $k = 4$ lb/in. spring.

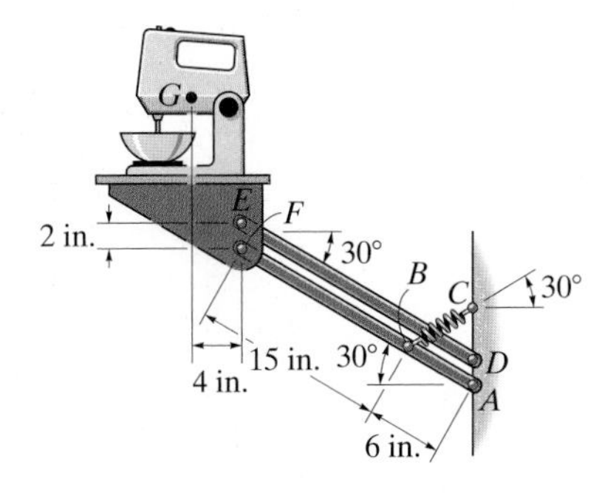

Prob. 6–118

6–119. If each of the three links of the mechanism has a weight of 25 lb, determine the angle θ for equilibrium. The spring, which always remains horizontal, is unstretched when $\theta = 0°$.

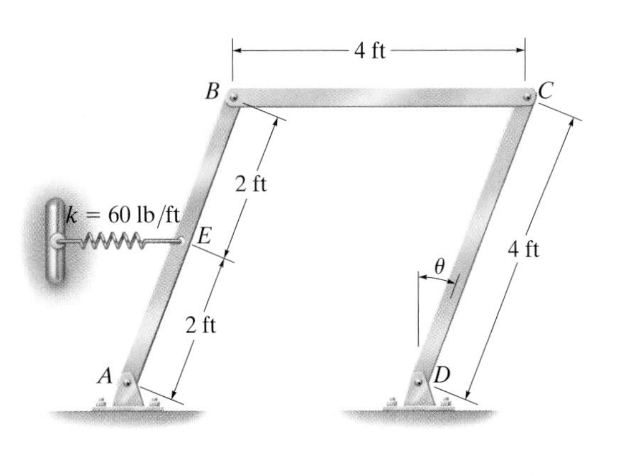

Prob. 6–119

***6–120.** Determine the required force P that must be applied at the blade of the pruning shears so that the blade exerts a normal force of 20 lb on the twig at E.

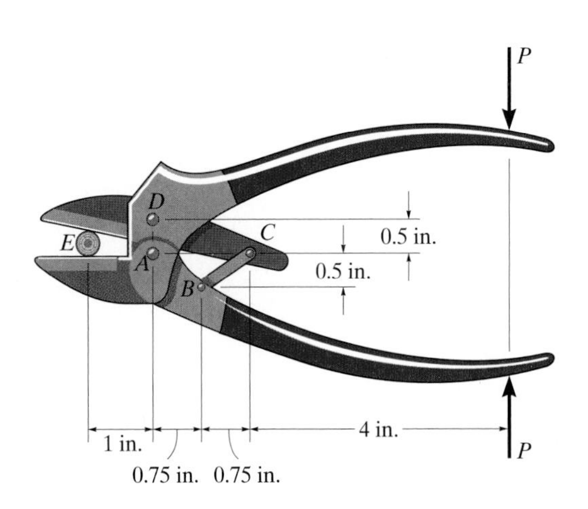

Prob. 6–120

6–121. The three power lines exert the forces shown on the truss joints, which in turn are pin-connected to the poles AH and EG. Determine the force in the guy cable AI and the pin reaction at the support H.

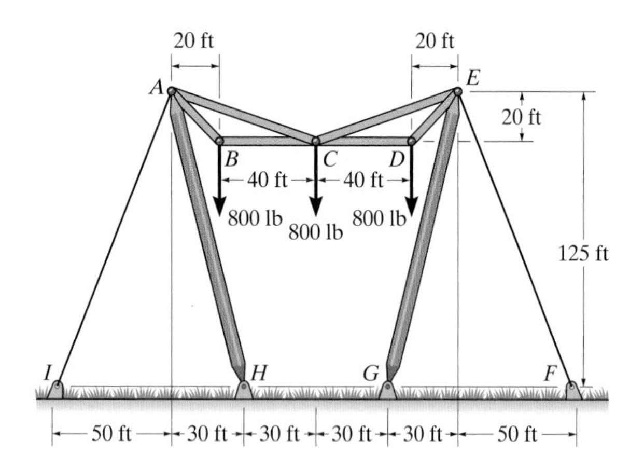

Prob. 6–121

6–123. The kinetic sculpture requires that each of the three pinned beams be in perfect balance at all times during its slow motion. If each member has a uniform weight of 2 lb/ft and length of 3 ft, determine the necessary counterweights W_1, W_2, and W_3 which must be added to the ends of each member to keep the system in balance for any position. Neglect the size of the counterweights.

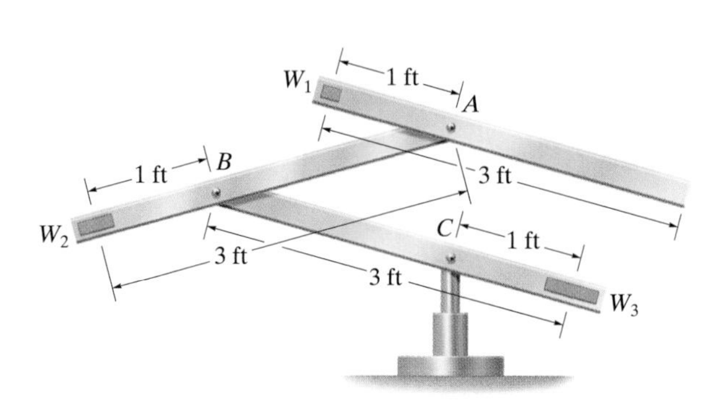

Prob. 6–123

6–122. The hydraulic crane is used to lift the 1400-lb load. Determine the force in the hydraulic cylinder AB and the force in links AC and AD when the load is held in the position shown.

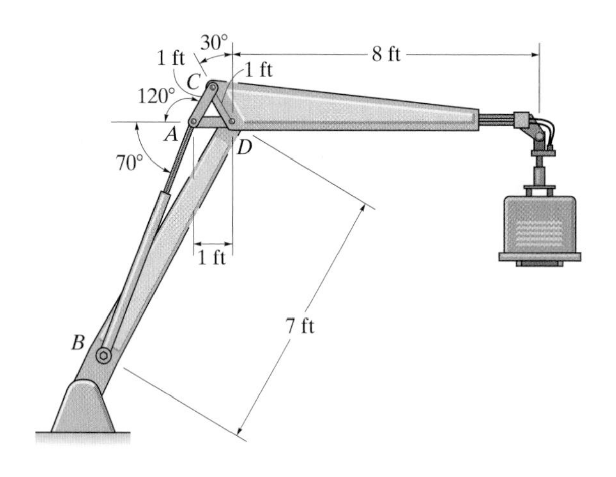

Prob. 6–122

***6–124.** The three-member frame is connected at its ends using ball-and-socket joints. Determine the x, y, z components of reaction at B and the tension in member ED. The force acting at D is $\mathbf{F} = \{135\mathbf{i} + 200\mathbf{j} - 180\mathbf{k}\}$ lb.

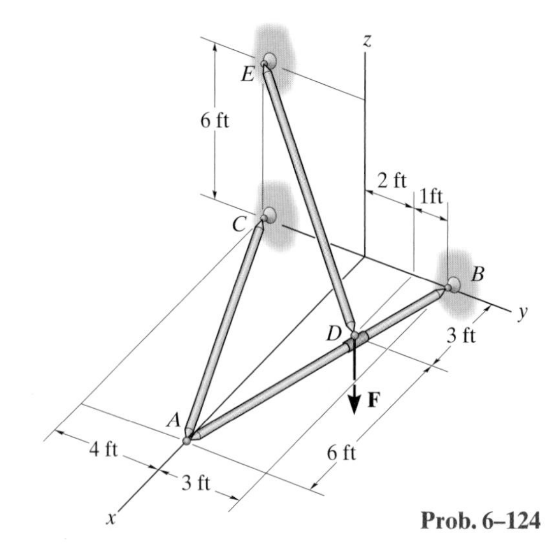

Prob. 6–124

6–125. The four-member "A" frame is supported at A and E by smooth collars and at G by a pin. All the other joints are ball-and-sockets. If the pin at G will fail when the resultant force there is 800 N, determine the largest vertical force P that can be supported by the frame. Also, what are the x, y, z force components which member BD exerts on members EDC and ABC? The collars at A and E and the pin at G only exert force components on the frame.

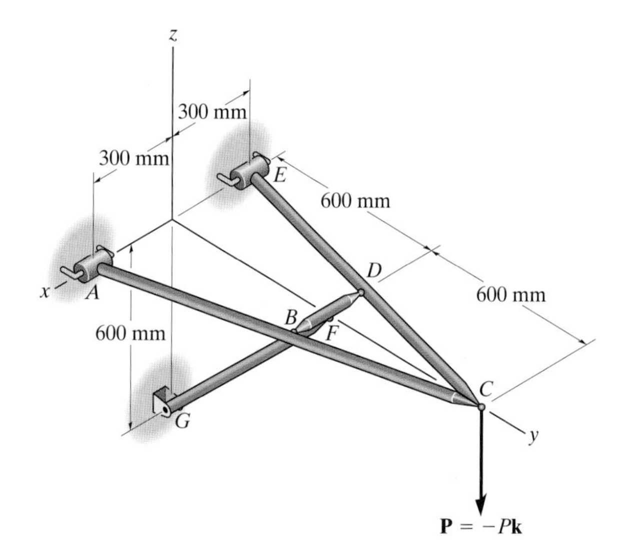

Prob. 6–125

6–126. The structure is subjected to the loading shown. Member AD is supported by a cable AB and roller at C and fits through a smooth circular hole at D. Member ED is supported by a roller at D and a pole that fits in a smooth snug circular hole at E. Determine the x, y, z components of reaction at E and the tension in cable AB.

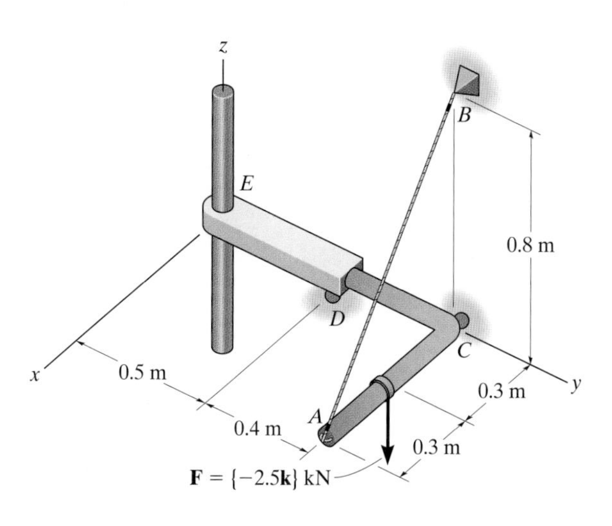

Prob. 6–126

6–127. The structure is subjected to the loadings shown. Member AB is supported by a ball-and-socket at A and smooth collar at B. Member CD is supported by a pin at C. Determine the x, y, z components of reaction at A and C.

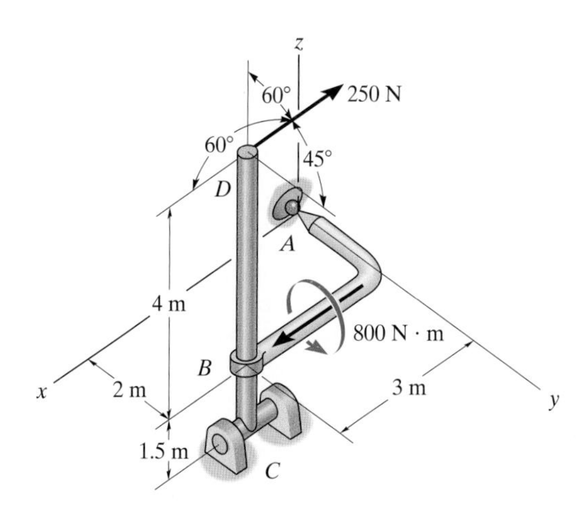

Prob. 6–127

Chapter Review

Simple Truss

A simple truss consists of triangular elements connected together by pin joints. The forces within its members can be determined by assuming the members are all two-force members, connected concurrently at each joint. They are either in tension or compression or carry no force.

(*See pages 267–269.*)

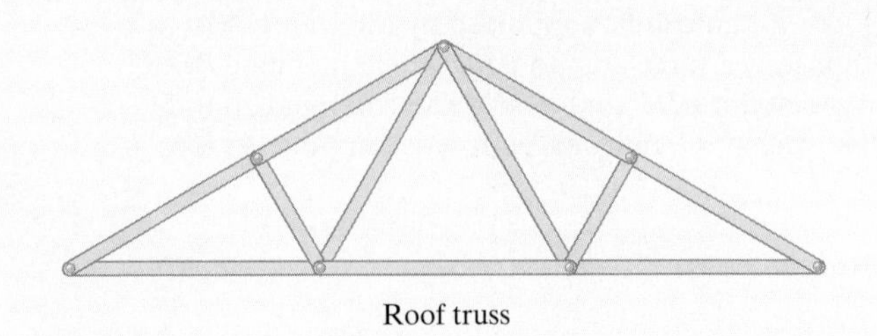

Roof truss

Method of Joints

The method of joints states that if a truss is in equilibrium, then each of its joints is also in equilibrium. For a coplanar truss, the concurrent force system at each joint must satisfy force equilibrium.

$$\Sigma F_x = 0$$

$$\Sigma F_y = 0$$

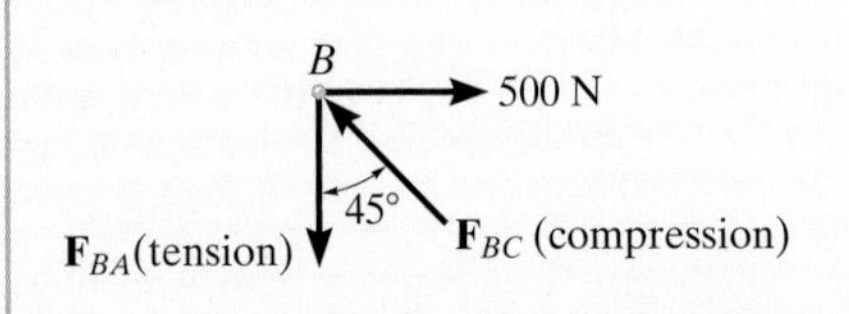

To obtain a numerical solution for the forces in the members, select a joint that has a free-body diagram with at most two unknown forces and one known force. (This may require first finding the reactions at the supports.)

Once a member force is determined, use its value and apply it to an adjacent joint.

Remember that forces that are found to *pull* on the joint are *tension forces*, and those that *push* on the joint are *compression forces*.

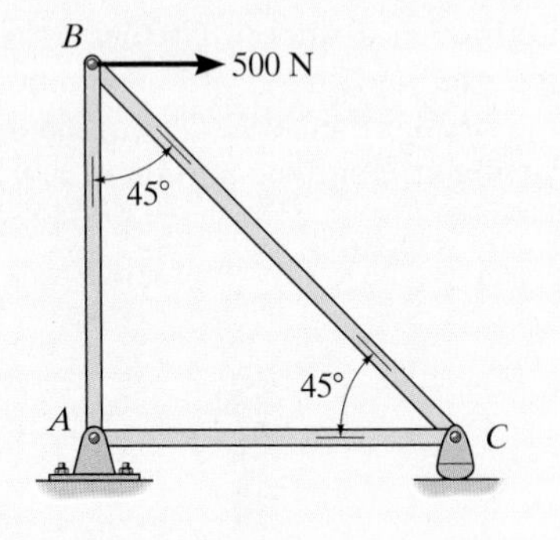

To avoid a simultaneous solution of two equations, try to sum forces in a direction that is perpendicular to one of the unknowns. This will allow a direct solution for the other unknown.

$$\Sigma F_{y'} = 0$$

$$1.5 \text{ kN} \cos 30° - F_{CB} \sin 15° = 0$$

$$F_{CB} = 5.02 \text{ kN (C)}$$

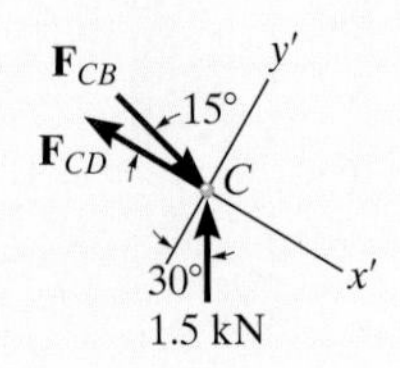

To further simplify the analysis, first identify all the zero-force members.

(*See pages 270–271.*)

Method of Sections

The method of sections states that if a truss is in equilibrium, then each section of the truss is also in equilibrium. Pass a section through the member whose force is to be determined. Then draw the free-body diagram of the sectioned part having the least number of forces on it.

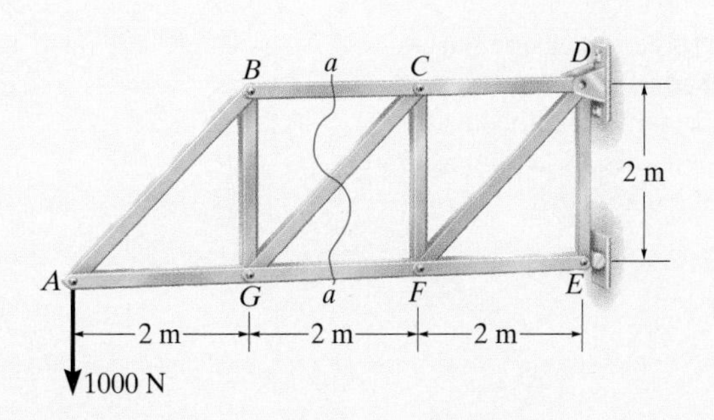

Sectioned members subjected to *pulling* are in *tension*, and those that are subjected to *pushing* are in *compression*.

Three equations of equilibrium are available to determine the unknowns.

$$\Sigma F_x = 0$$
$$\Sigma F_y = 0$$
$$\Sigma M_O = 0$$

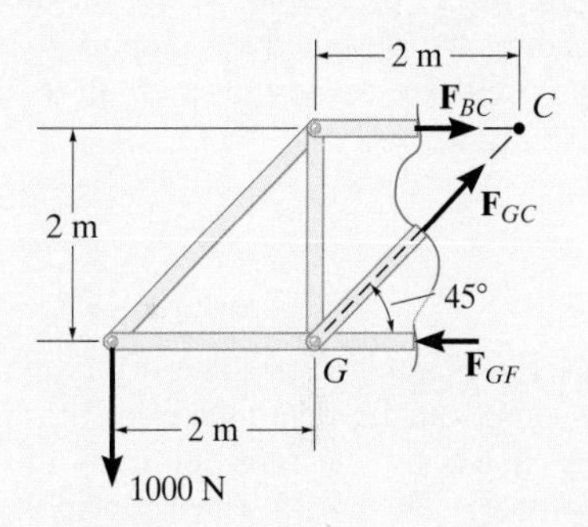

If possible, sum forces in a direction that is perpendicular to two of the three unknown forces. This will yield a direct solution for the third force.

$$+\uparrow \Sigma F_y = 0$$
$$-1000 \text{ N} + F_{GC} \sin 45^\circ = 0$$
$$F_{GC} = 1.41 \text{ kN (T)}$$

Sum moments about a point that passes through the line of action of two of the three unknown forces, so that the third unknown force can be determined directly.
(See pages 283–285.)

$$\circlearrowright + \Sigma M_C = 0$$
$$1000 \text{ N}(4 \text{ m}) - F_{GF}(2 \text{ m}) = 0$$
$$F_{GF} = 2 \text{ kN (C)}$$

Space Truss

A space truss is a three-dimensional truss built from tetrahedral elements, and is analyzed using the same methods as for planar trusses. The joints are assumed to be ball and socket connections.
(*See page 293.*)

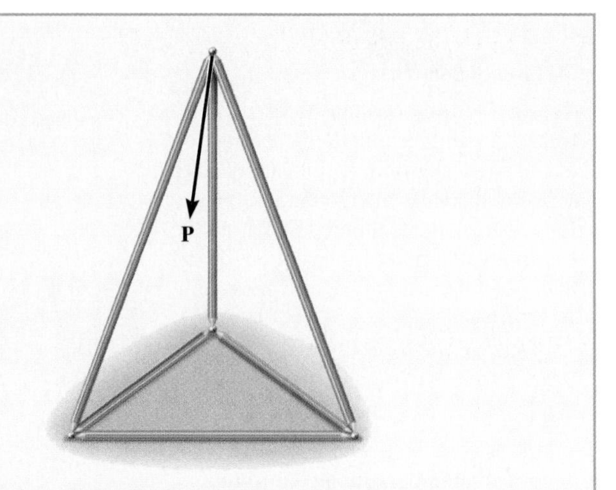

Frames and Machines

Frames and machines are structures that contain one or more multiforce members, that is, members with three or more forces or couples acting on them. Frames are designed to support loads, and machines transmit and alter the effect of forces.

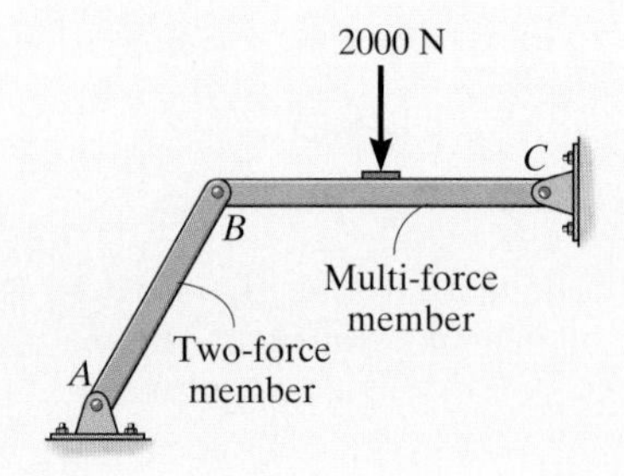

The forces acting at the joints of a frame or machine can be determined by drawing the free-body diagrams of each of its members or parts. The principle of action–reaction should be carefully observed when drawing these forces on each adjacent member or pin. For a coplanar force system, there are three equilibrium equations available for each member.

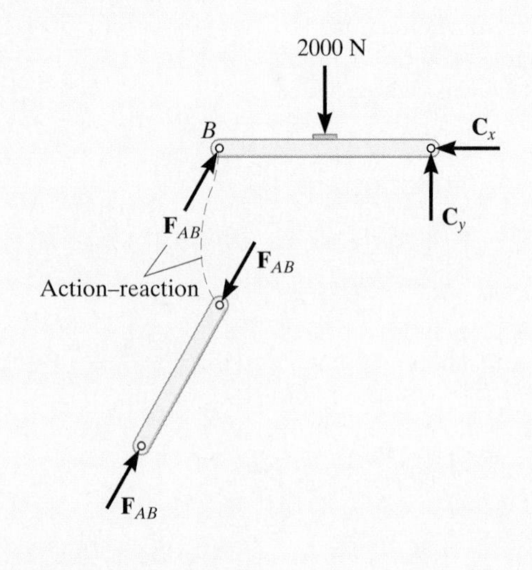

To simplify the analysis, be sure to recognize all two-force members. They have equal but opposite collinear forces at their ends.
(*See pages 297, 303–304.*)

REVIEW PROBLEMS

***6–128.** Determine the resultant forces at pins B and C on member ABC of the four-member frame.

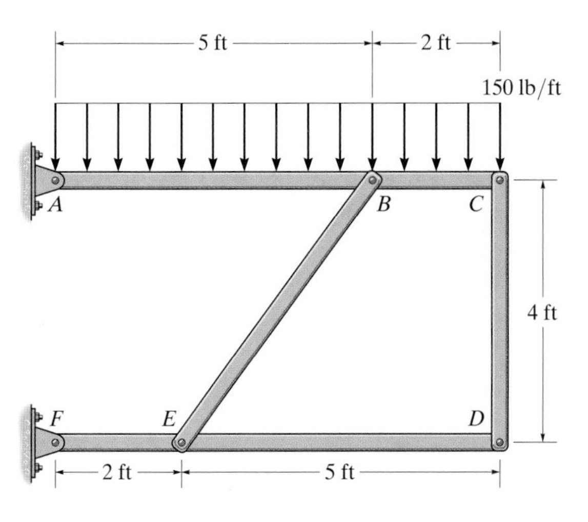

Prob. 6–128

6–129. The mechanism consists of identical meshed gears A and B and arms which are fixed to the gears. The spring attached to the ends of the arms has an unstretched length of 100 mm and a stiffness of $k = 250$ N/m. If a torque of $M = 6$ N·m is applied to gear A, determine the angle θ through which each arm rotates. The gears are each pinned to fixed supports at their centers.

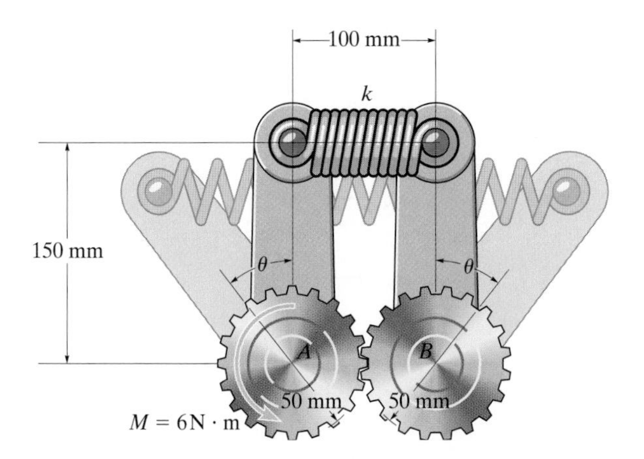

Prob. 6–129

6–130. Determine the force in each member of the truss and state if the members are in tension or compression.

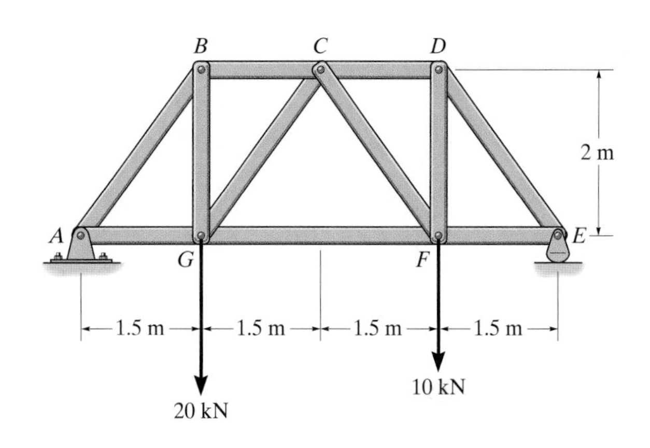

Prob. 6–130

6–131. The spring has an unstretched length of 0.3 m. Determine the angle θ for equilibrium if the uniform links each have a mass of 5 kg.

***6–132.** The spring has an unstretched length of 0.3 m. Determine the mass m of each uniform link if the angle $\theta = 20°$ for equilibrium.

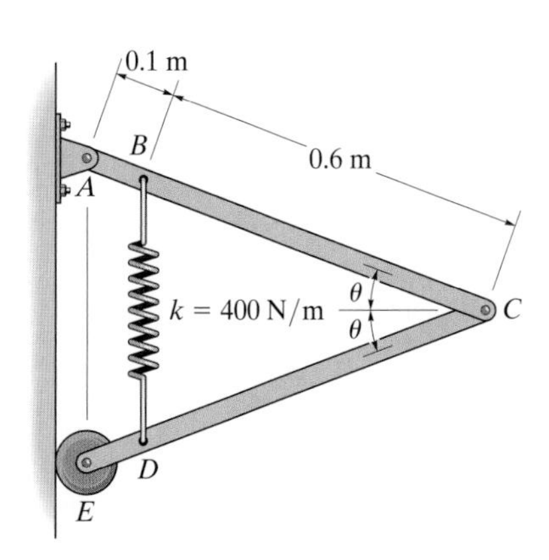

Probs. 6–131/132

6–133. Determine the horizontal and vertical components of force that the pins A and B exert on the two-member frame. Set $F = 0$.

6–134. Determine the horizontal and vertical components of force that pins A and B exert on the two-member frame. Set $F = 500$ N.

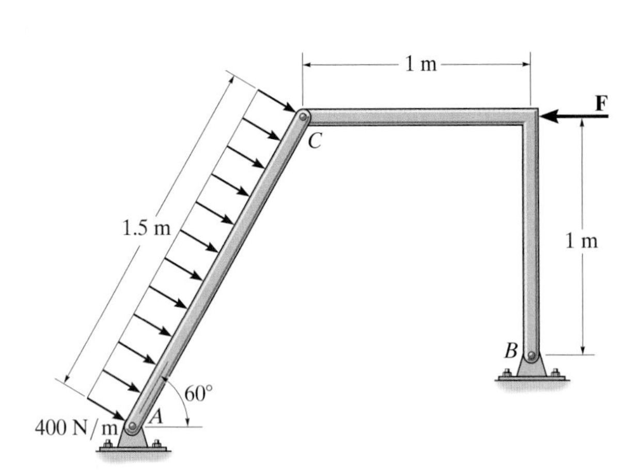

Probs. 6–133/134

***6–136.** Determine the force in each member of the truss and state if the members are in tension or compression.

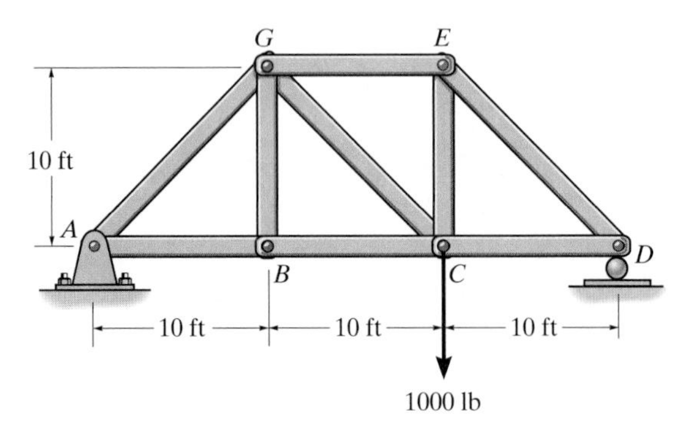

Prob. 6–136

6–135. Determine the force in each member of the truss and indicate whether the members are in tension or compression.

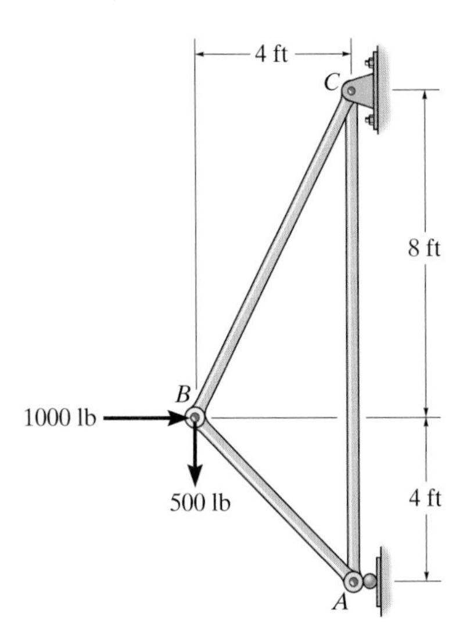

Prob. 6–135

6–137. Determine the force in members AB, AD, and AC of the space truss and state if the members are in tension or compression.

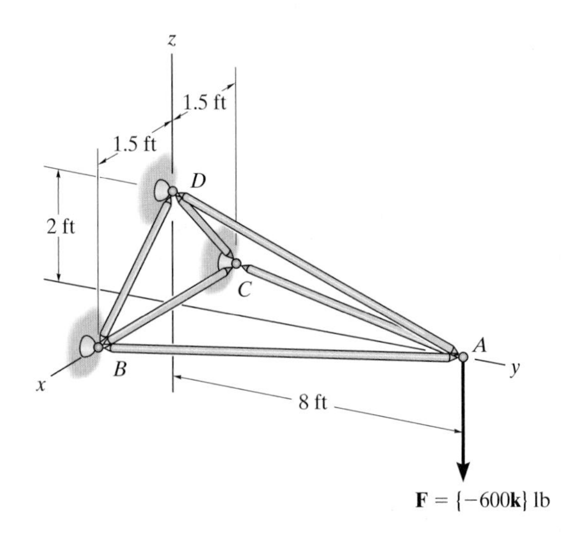

Prob. 6–137

Design Projects

6–1D Design of a Bridge Truss

A bridge having a horizontal top cord is to span between the two piers A and B having an arbitrary height. It is required that a pin-connected truss be used, consisting of steel members bolted together to steel gusset plates, such as the one shown in the figure. The end supports are assumed to be a pin at A and a roller at B. A vertical loading of 5 kN is to be supported within the middle 3 m of the span. This load can be applied in part to several joints on the top cord within this region, or to a single joint at the middle of the top cord. The force of the wind and the weight of the members are to be neglected.

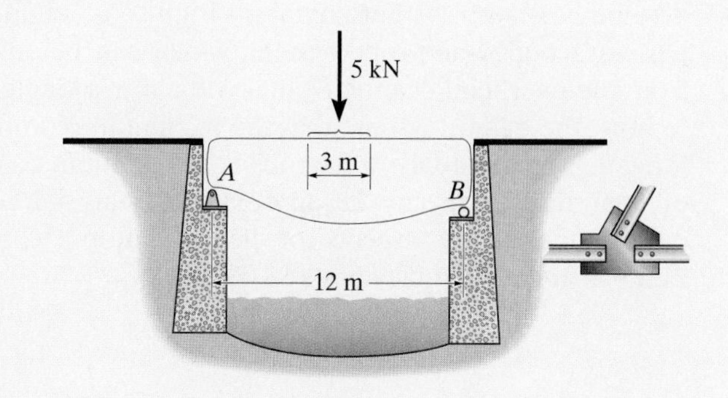

Prob. 6–1D

Assume the maximum tensile force in each member cannot exceed 4.25 kN; and regardless of the length of the member, the maximum compressive force cannot exceed 3.5 kN. Design the most economical truss that will support the loading. The members cost \$7.50/m, and the gusset plates cost \$8.00 each. Submit your cost analysis for the materials, along with a scaled drawing of the truss, identifying on this drawing the tensile and compressive force in each member. Also, include your calculations of the complete force analysis.

6–2D Design of a Cart Lift

A hand cart is used to move a load from one loading dock to another. Any dock will have a different elevation relative to the bed of a truck that backs up to it. It is necessary that the loading platform on the hand cart will bring the load resting on it up to the elevation of each truck bed as shown. The maximum elevation difference between the frame of the hand cart and a truck bed is 1 ft. Design a hand-operated mechanical system that will allow the load to be lifted this distance from the frame of the hand cart. Assume the operator can exert a (comfortable) force of 20 lb to make the lift, and that the maximum load, centered on the loading platform, is 400 lb. Submit a scaled drawing of your design, and explain how it works based on a force analysis.

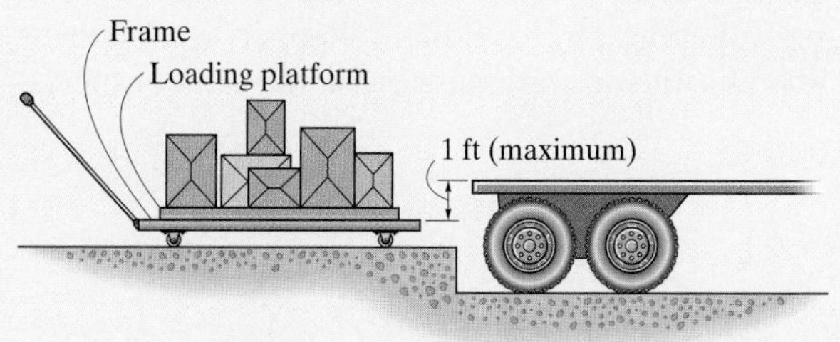

Prob. 6–2D

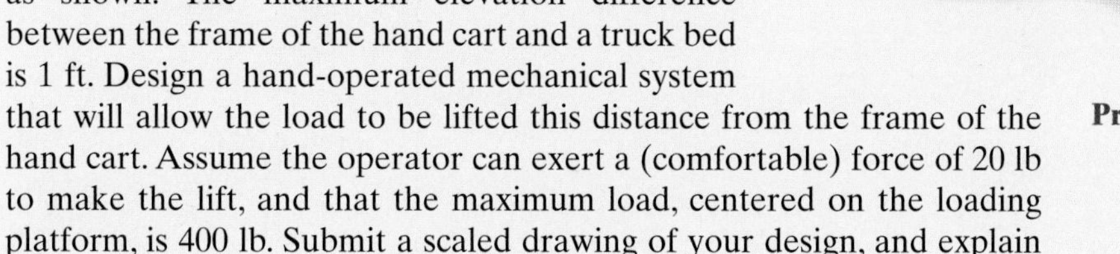

6–3D Design of a Pulley System

The steel beam AB, having a length of 5 m and a mass of 700 kg, is to be hoisted in its horizontal position to a height of 4 m. Design a pulley-and-rope system, which can be suspended from the overhead beam CD, that will allow a single worker to hoist the beam. Assume that the maximum (comfortable) force that he can apply to the rope is 180 N. Submit a drawing of your design, specify its approximate material cost, and discuss the safety aspects of its operation. Rope costs \$1.25/m and each pulley costs \$8.00.

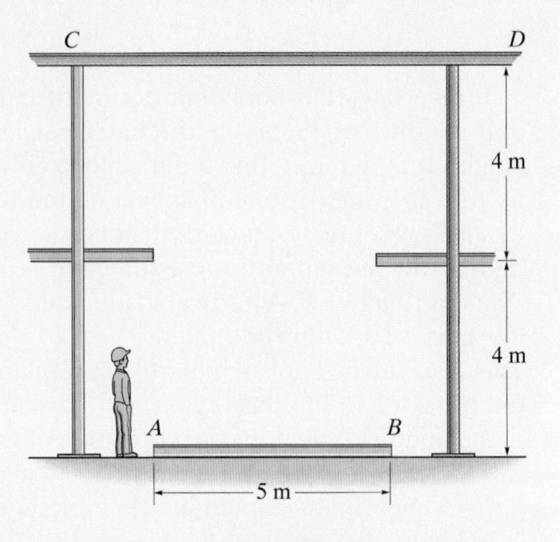

Prob. 6–3D

6–4D Design of a Tool used to Position a Suspended Load

Heavy loads are suspended from an overhead pulley and each load must be positioned over a depository. Design a tool that can be used to shorten or lengthen the pulley cord AB a small amount in order to make the location adjustment. Assume the worker can apply a maximum (comfortable) force of 25 lb to the tool, and the maximum force allowed in cord AB is 500 lb. Submit a scale drawing of the tool, and a brief paragraph to explain how it works using a force analysis. Include a discussion on the safety aspects of its use.

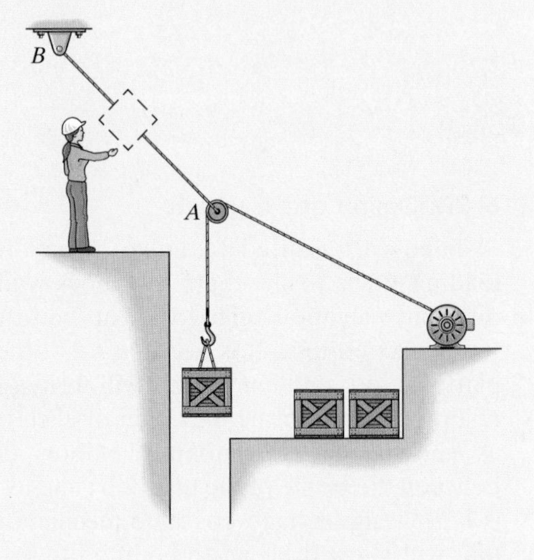

Prob. 6–4D

6–5D Design of a Fence-Post Remover

A farmer wishes to remove several fence posts. Each post is buried 18 in. in the ground and will require a maximum vertical pulling force of 175 lb to remove it. He can use his truck to develop the force, but he needs to devise a method for their removal without breaking the posts. Design a method that can be used, considering that the only materials available are a strong rope and several pieces of wood having various sizes and lengths. Submit a sketch of your design and discuss the safety and reliability of its use. Also, provide a force analysis to show how it works and why it will cause minimal damage to a post when it is removed.

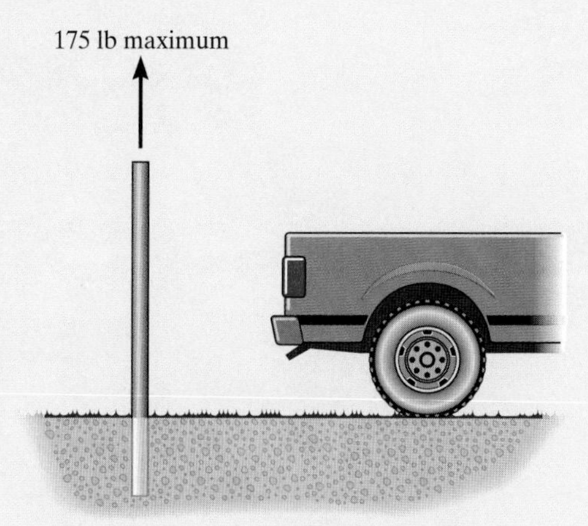

Prob. 6–5D

The design and analysis of any structural member requires knowledge of the internal loadings acting within it. In this chapter, we will discuss how engineers determine these loadings.

7 Internal Forces

CHAPTER OBJECTIVES

- To show how to use the method of sections for determining the internal loadings in a member.
- To generalize this procedure by formulating equations that can be plotted so that they describe the internal shear and moment throughout a member.
- To analyze the forces and study the geometry of cables supporting a load.

7.1 Internal Forces Developed in Structural Members

The design of any structural or mechanical member requires an investigation of the loading acting within the member in order to be sure the material can resist this loading. These internal loadings can be determined by using the *method of sections*. To illustrate this method, consider the "simply supported" beam shown in Fig. 7–1*a*, which is subjected to the forces $\mathbf{F}_1$ and $\mathbf{F}_2$ and the *support reactions* $\mathbf{A}_x$, $\mathbf{A}_y$, and $\mathbf{B}_y$, Fig. 7–1*b*. If the *internal loadings* acting on the cross section at C are to be determined, then an

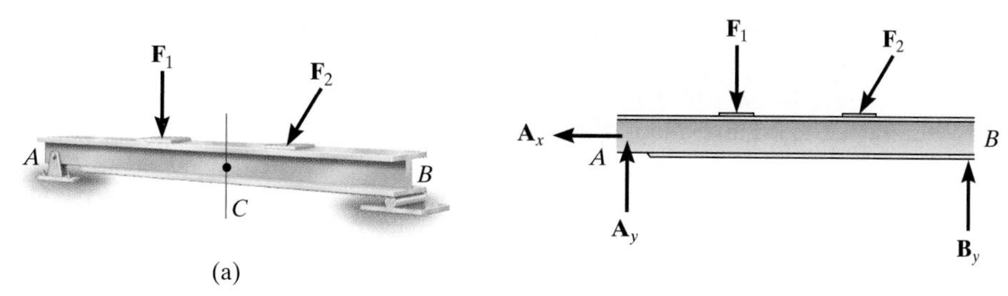

Fig. 7–1

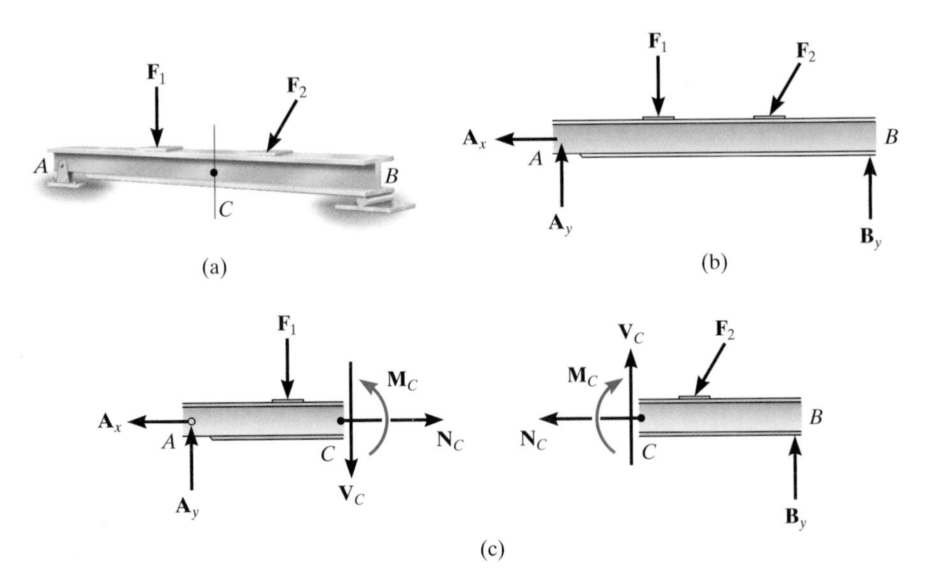

Fig. 7–1 (cont.)

imaginary section is passed through the beam, cutting it into two segments. By doing this the internal loadings at the section become *external* on the free-body diagram of each segment, Fig. 7–1*c*. Since both segments (*AC* and *CB*) were in equilibrium *before* the beam was sectioned, equilibrium of each segment is maintained provided rectangular force components $\mathbf{N}_C$ and $\mathbf{V}_C$ and a resultant couple moment $\mathbf{M}_C$ are developed at the section. Note that these loadings must be equal in magnitude and opposite in direction on each of the segments (Newton's third law). The magnitude of each of these loadings can now be determined by applying the three equations of equilibrium to either segment *AC* or *CB*. A *direct solution* for $\mathbf{N}_C$ is obtained by applying $\Sigma F_x = 0$; $\mathbf{V}_C$ is obtained directly from $\Sigma F_y = 0$; and $\mathbf{M}_C$ is determined by summing moments about point *C*, $\Sigma M_C = 0$, in order to eliminate the moments of the unknowns $\mathbf{N}_C$ and $\mathbf{V}_C$.

The designer of this shop crane realized the need for additional reinforcement around the joint in order to prevent severe internal bending of the joint when a large load is suspended from the chain hoist.

In mechanics, the force components **N**, acting normal to the beam at the cut section, and **V**, acting tangent to the section, are termed the *normal or axial force* and the *shear force*, respectively. The couple moment **M** is referred to as the *bending moment*, Fig. 7–2*a*. In three dimensions, a general internal force and couple moment resultant will act at the section. The *x, y, z* components of these loadings are shown in Fig. 7–2*b*. Here $\mathbf{N}_y$ is the *normal force*, and $\mathbf{V}_x$ and $\mathbf{V}_z$ are *shear force components*. $\mathbf{M}_y$ is a *torsional or twisting moment*, and $\mathbf{M}_x$ and $\mathbf{M}_z$ are *bending moment components*. For most applications, these *resultant loadings* will act at the geometric center or centroid (*C*) of the section's cross-sectional area. Although the magnitude for each loading generally will be different at various points along the axis of the member, the method of sections can always be used to determine their values.

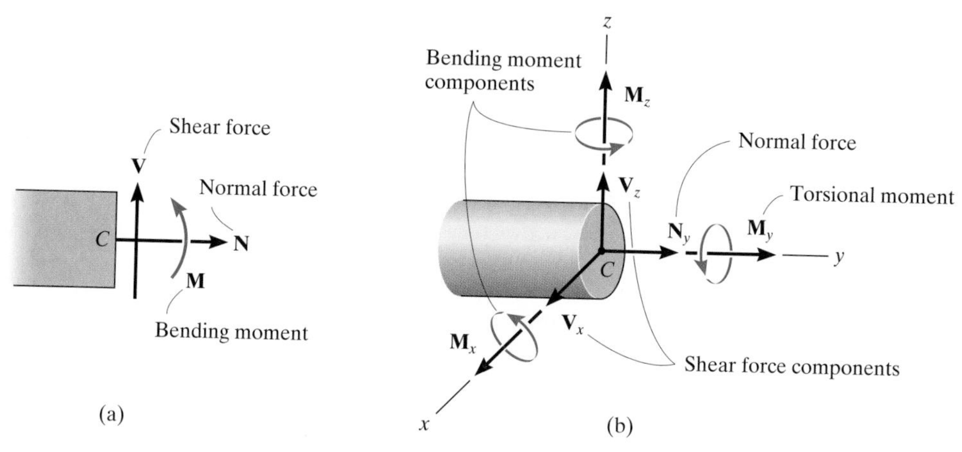

Fig. 7–2

Free-Body Diagrams. Trusses are composed of two-force members that only support normal loads. On the other hand, frames and machines are composed of *multiforce members*, and so each of these members will generally be subjected to internal normal, shear, and bending loadings. For example, consider the frame shown in Fig. 7–3*a*. If the blue section is passed through the frame to determine the internal loadings at points *H*, *G*, and *F*, the resulting free-body diagram of the top portion of this section is shown in Fig. 7–3*b*. At each point where a member is sectioned there is an unknown normal force, shear force, and bending moment. As a result, we cannot apply the *three* equations of equilibrium to this section in order to obtain these *nine unknowns*. Instead, to solve this problem we must *first dismember* the frame and determine the reactions at the connections of the members using the techniques of Sec. 6.6. Once this is done, *each member* may then be sectioned at its appropriate point, and the three equations of equilibrium can be applied to determine **N**, **V**, and **M**. For example, the free-body diagram of segment *DG*, Fig. 7–3*c*, can be used to determine the internal loadings at *G* provided the reactions of the pin, $\mathbf{D}_x$ and $\mathbf{D}_y$, are known.

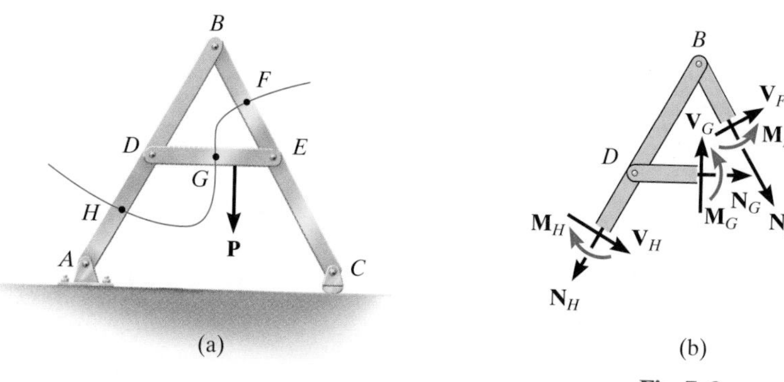

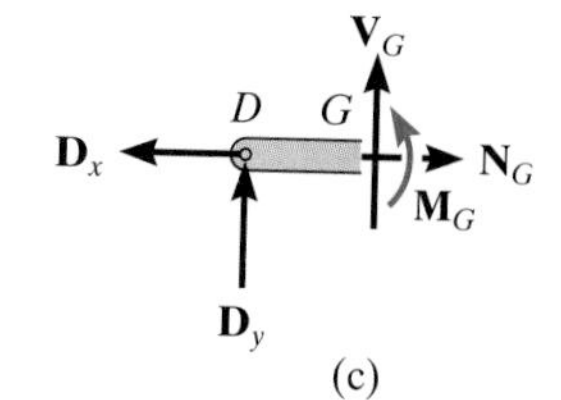

Fig. 7–3

In each case, the link on the backhoe is a two-force member. In the top photo it is subjected to both bending and axial load at its center. By making the member straight, as in the bottom photo, then only an axial force acts within the member.

Procedure for Analysis

The method of sections can be used to determine the internal loadings at a specific location in a member using the following procedure.

Support Reactions.

- Before the member is "cut" or sectioned, it may first be necessary to determine its support reactions, so that the equilibrium equations are used only to solve for the internal loadings when the member is sectioned.
- If the member is part of a frame or machine, the reactions at its connections are determined using the methods of Sec. 6.6.

Free-Body Diagram.

- Keep all distributed loadings, couple moments, and forces acting on the member in their *exact locations*, then pass an imaginary section through the member, perpendicular to its axis at the point where the internal loading is to be determined.
- After the section is made, draw a free-body diagram of the segment that has the least number of loads on it, and indicate the x, y, z components of the force and couple moment resultants at the section.
- If the member is subjected to a *coplanar* system of forces, only **N**, **V**, and **M** act at the section.
- In many cases it may be possible to tell by inspection the proper sense of the unknown loadings; however, if this seems difficult, the sense can be assumed.

Equations of Equilibrium.

- Moments should be summed at the section about axes passing through the *centroid* or geometric center of the member's cross-sectional area in order to eliminate the unknown normal and shear forces and thereby obtain direct solutions for the moment components.
- If the solution of the equilibrium equations yields a negative scalar, the assumed sense of the quantity is opposite to that shown on the free-body diagram.

EXAMPLE 7.1

The bar is fixed at its end and is loaded as shown in Fig. 7–4*a*. Determine the internal normal force at points *B* and *C*.

SOLUTION

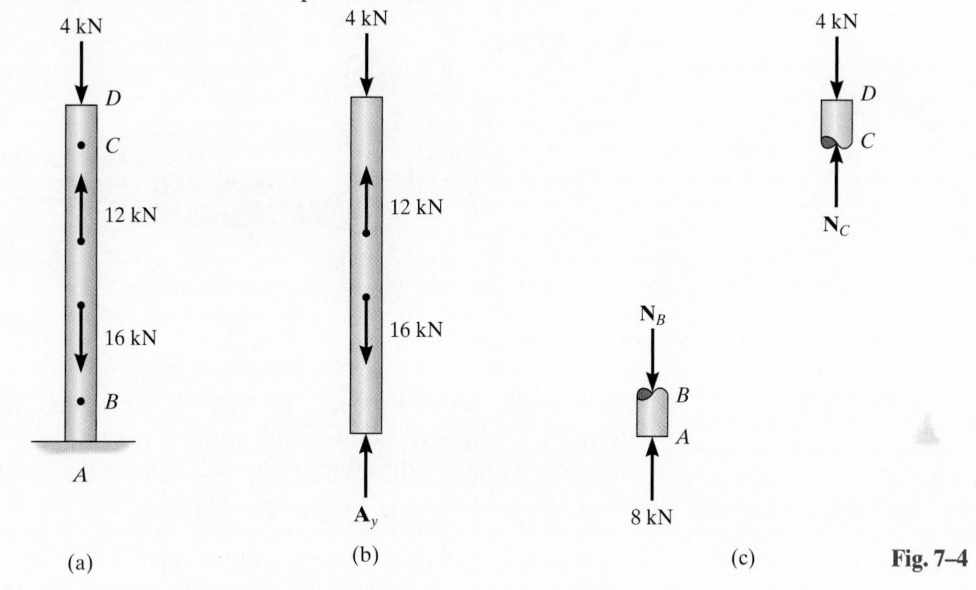

Fig. 7–4

Support Reactions. A free-body diagram of the entire bar is shown in Fig. 7–4*b*. By inspection, only a normal force $\mathbf{A}_y$ acts at the fixed support since the loads are applied symmetrically along the bar's axis. ($A_x = 0$, $M_A = 0$.)

$$+\uparrow \Sigma F_y = 0; \quad A_y - 16\text{ kN} + 12\text{ kN} - 4\text{ kN} = 0 \qquad A_y = 8\text{ kN}$$

Free-Body Diagrams. The internal forces at *B* and *C* will be found using the free-body diagrams of the sectioned bar shown in Fig. 7–4*c*. No shear or moment act on the sections since they are not required for equilibrium. In particular, segments *AB* and *DC* will be chosen here, since they contain the *least* number of forces.

Equations of Equilibrium.

Segment AB

$$+\uparrow \Sigma F_y = 0; \quad 8\text{ kN} - N_B = 0 \qquad N_B = 8\text{ kN} \qquad \textit{Ans.}$$

Segment DC

$$+\uparrow \Sigma F_y = 0; \quad N_C - 4\text{ kN} = 0 \qquad N_C = 4\text{ kN} \qquad \textit{Ans.}$$

NOTE: Try working this problem in the following manner: Determine N_B from segment *BD*. (Note that this approach *does not require* solution for the support reaction at *A*.) Using the result for N_B, isolate segment *BC* to determine N_C.

EXAMPLE 7.2

The circular shaft is subjected to three concentrated torques as shown in Fig. 7–5*a*. Determine the internal torques at points *B* and *C*.

SOLUTION

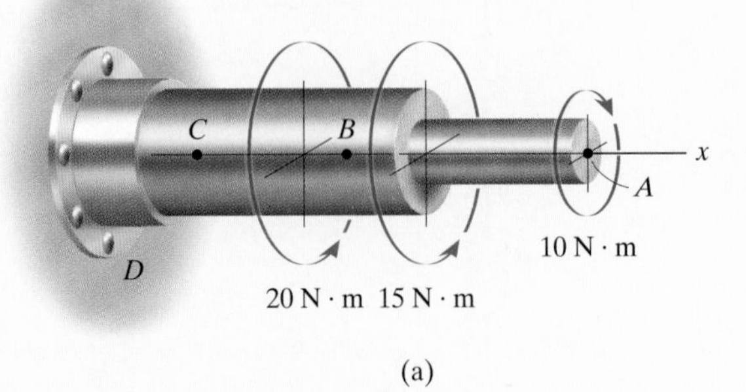

(a)

Support Reactions. Since the shaft is subjected only to collinear torques, a torque reaction occurs at the support, Fig. 7–5*b*. Using the right-hand rule to define the positive directions of the torques, we require

$$\Sigma M_x = 0; \qquad -10\text{ N}\cdot\text{m} + 15\text{ N}\cdot\text{m} + 20\text{ N}\cdot\text{m} - T_D = 0$$
$$T_D = 25\text{ N}\cdot\text{m}$$

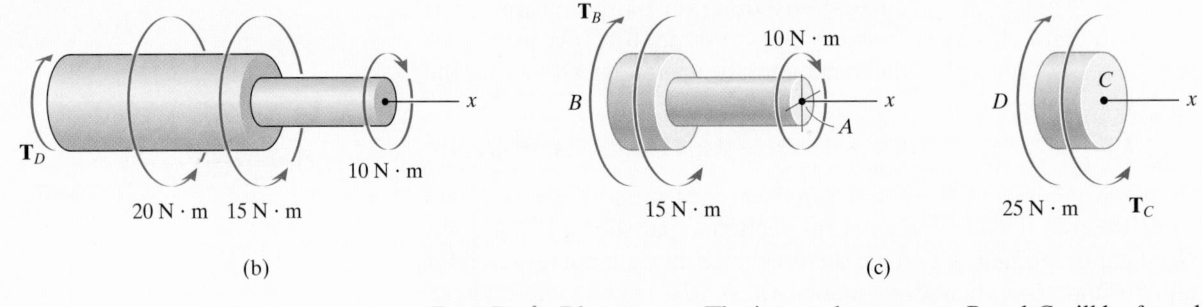

(b) (c)

Fig. 7–5

Free-Body Diagrams. The internal torques at *B* and *C* will be found using the free-body diagrams of the shaft segments *AB* and *CD* shown in Fig. 7–5*c*.

Equations of Equilibrium. Applying the equation of moment equilibrium along the shaft's axis, we have

Segment AB

$$\Sigma M_x = 0; \qquad -10\text{ N}\cdot\text{m} + 15\text{ N}\cdot\text{m} - T_B = 0 \qquad T_B = 5\text{ N}\cdot\text{m} \quad \textit{Ans.}$$

Segment CD

$$\Sigma M_x = 0; \qquad T_C - 25\text{ N}\cdot\text{m} = 0 \qquad T_C = 25\text{ N}\cdot\text{m} \quad \textit{Ans.}$$

NOTE: Try to solve for T_C by using segment *CA*. Note that this approach *does not require* a solution for the support reaction at *D*.

EXAMPLE 7.3

The beam supports the loading shown in Fig. 7–6*a*. Determine the internal normal force, shear force, and bending moment acting just to the left, point B, and just to the right, point C, of the 6-kN force.

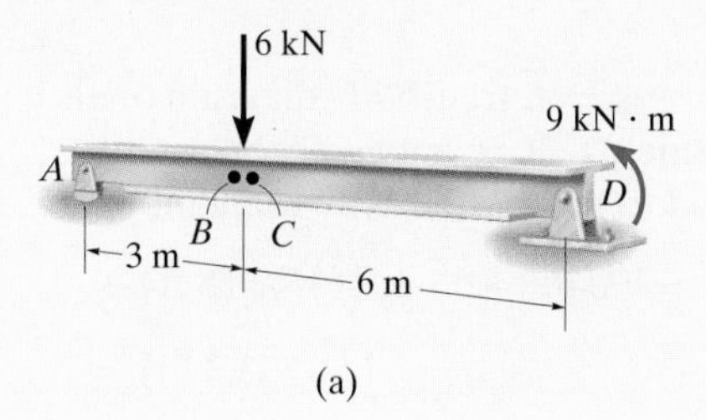

(a)

SOLUTION

Support Reactions. The free-body diagram of the beam is shown in Fig. 7–6*b*. When determining the *external reactions*, realize that the 9-kN · m couple moment is a free vector and therefore it can be placed *anywhere* on the free-body diagram of the entire beam. Here we will only determine $\mathbf{A}_y$, since segments AB and AC will be used for the analysis.

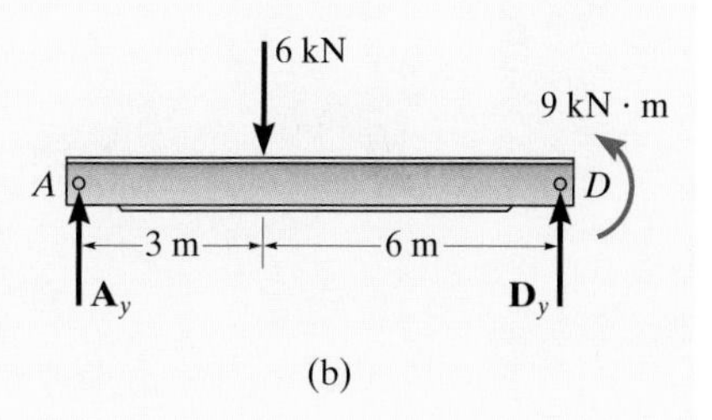

(b)

$$\circlearrowleft+\Sigma M_D = 0; \quad 9\text{ kN}\cdot\text{m} + (6\text{ kN})(6\text{ m}) - A_y(9\text{ m}) = 0$$
$$A_y = 5\text{ kN}$$

Free-Body Diagrams. The free-body diagrams of the left segments AB and AC of the beam are shown in Figs. 7–6*c* and 7–6*d*. In this case the 9-kN · m couple moment is *not included* on these diagrams since it must be kept in its *original position* until *after* the section is made and the appropriate body is isolated. In other words, the free-body diagrams of the left segments of the beam do not show the couple moment since this moment does not actually act on these segments.

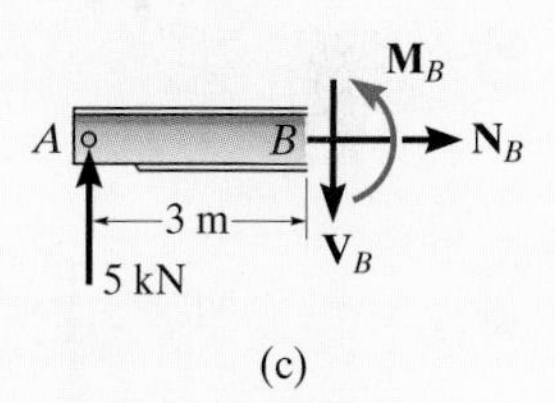

(c)

Equations of Equilibrium.

Segment AB

$$\xrightarrow{+}\Sigma F_x = 0; \qquad N_B = 0 \qquad \textit{Ans.}$$
$$+\uparrow\Sigma F_y = 0; \qquad 5\text{ kN} - V_B = 0 \qquad V_B = 5\text{ kN} \qquad \textit{Ans.}$$
$$\circlearrowleft+\Sigma M_B = 0; \qquad -(5\text{ kN})(3\text{ m}) + M_B = 0 \qquad M_B = 15\text{ kN}\cdot\text{m} \qquad \textit{Ans.}$$

Segment AC

$$\xrightarrow{+}\Sigma F_x = 0; \qquad N_C = 0 \qquad \textit{Ans.}$$
$$+\uparrow\Sigma F_y = 0; \qquad 5\text{ kN} - 6\text{ kN} + V_C = 0 \qquad V_C = 1\text{ kN} \qquad \textit{Ans.}$$
$$\circlearrowleft+\Sigma M_C = 0; \qquad -(5\text{ kN})(3\text{ m}) + M_C = 0 \qquad M_C = 15\text{ kN}\cdot\text{m} \qquad \textit{Ans.}$$

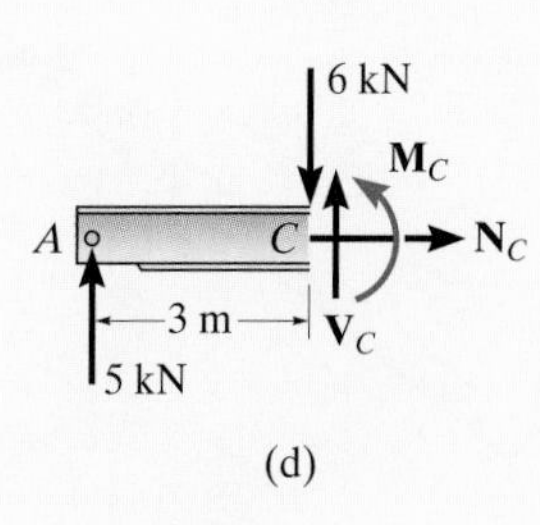

(d)

Fig. 7–6

Here the moment arm for the 5-kN force in both cases is approximately 3 m since B and C are "almost" coincident.

EXAMPLE 7.4

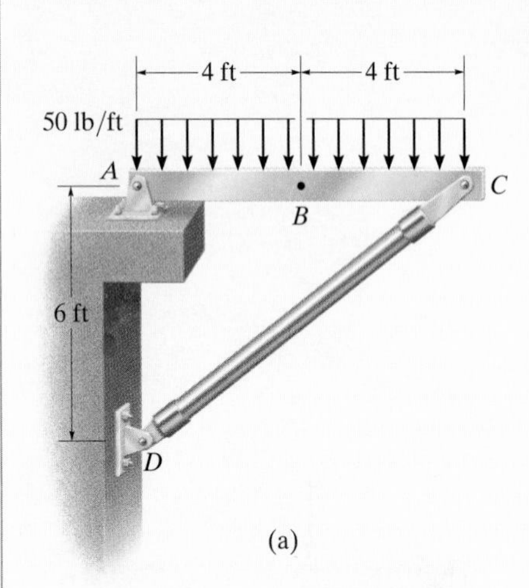

(a)

Determine the internal normal force, shear force, and bending moment acting at point B of the two-member frame shown in Fig. 7–7*a*.

SOLUTION

Support Reactions. A free-body diagram of each member is shown in Fig. 7–7*b*. Since CD is a two-force member, the equations of equilibrium need to be applied only to member AC.

$$\circlearrowleft + \Sigma M_A = 0; \quad -400 \text{ lb } (4 \text{ ft}) + \left(\tfrac{3}{5}\right)F_{DC}(8 \text{ ft}) = 0 \qquad F_{DC} = 333.3 \text{ lb}$$

$$\xrightarrow{+} \Sigma F_x = 0; \quad -A_x + \left(\tfrac{4}{5}\right)(333.3 \text{ lb}) = 0 \qquad A_x = 266.7 \text{ lb}$$

$$+\uparrow \Sigma F_y = 0; \quad A_y - 400 \text{ lb} + \tfrac{3}{5}(333.3 \text{ lb}) = 0 \qquad A_y = 200 \text{ lb}$$

Free-Body Diagrams. Passing an imaginary section perpendicular to the axis of member AC through point B yields the free-body diagrams of segments AB and BC shown in Fig. 7–7*c*. When constructing these diagrams it is important to keep the distributed loading exactly as it is until *after* the section is made. Only then can it be replaced by a single resultant force. Why? Also, notice that $\mathbf{N}_B$, $\mathbf{V}_B$, and $\mathbf{M}_B$ act with equal magnitude but opposite direction on each segment—Newton's third law.

Equations of Equilibrium. Applying the equations of equilibrium to segment AB, we have

$$\xrightarrow{+} \Sigma F_x = 0; \quad N_B - 266.7 \text{ lb} = 0 \qquad N_B = 267 \text{ lb} \qquad \textit{Ans.}$$

$$+\uparrow \Sigma F_y = 0; \quad 200 \text{ lb} - 200 \text{ lb} - V_B = 0 \qquad V_B = 0 \qquad \textit{Ans.}$$

$$\circlearrowleft + \Sigma M_B = 0; \quad M_B - 200 \text{ lb } (4 \text{ ft}) + 200 \text{ lb } (2 \text{ ft}) = 0$$

$$M_B = 400 \text{ lb} \cdot \text{ft} \qquad \textit{Ans.}$$

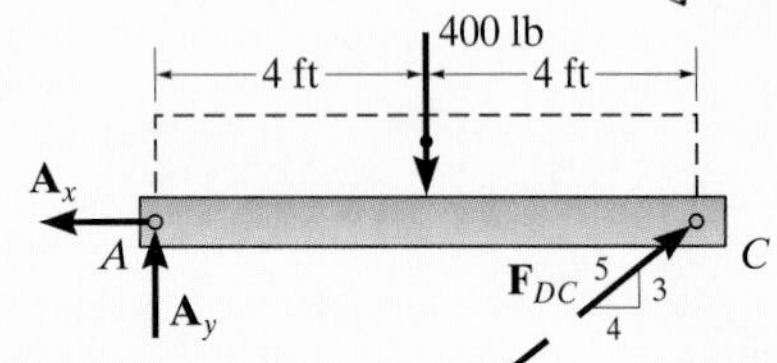

(b)

NOTE: As an exercise, try to obtain these same results using segment BC.

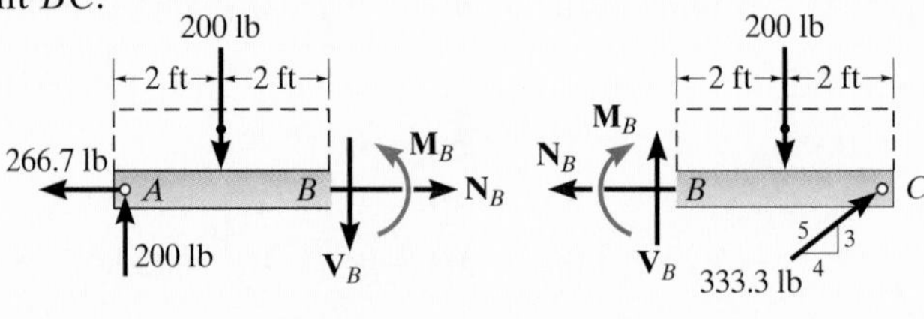

(c)

Fig. 7–7

EXAMPLE 7.5

Determine the normal force, shear force, and bending moment acting at point E of the frame loaded as shown in Fig. 7–8*a*.

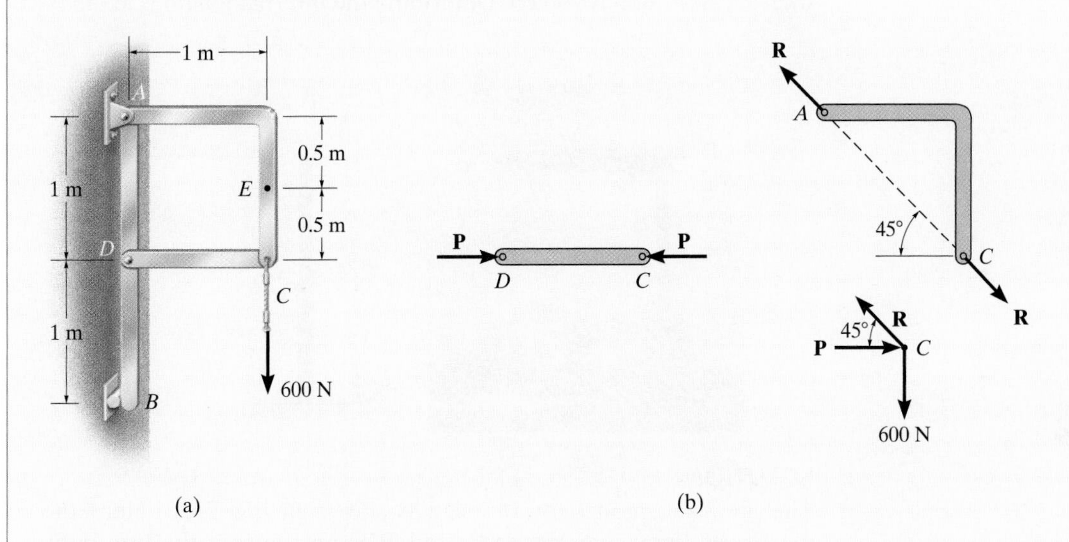

(a) (b)

SOLUTION

Support Reactions. By inspection, members AC and CD are two-force members, Fig. 7–8*b*. In order to determine the internal loadings at E, we must first determine the force $\mathbf{R}$ at the end of member AC. To do this we must analyze the equilibrium of the pin at C. Why?

Summing forces in the vertical direction on the pin, Fig. 7–8*b*, we have

$$+\uparrow \Sigma F_y = 0; \quad R \sin 45^\circ - 600 \text{ N} = 0 \qquad R = 848.5 \text{ N}$$

Free-Body Diagram. The free-body diagram of segment CE is shown in Fig. 7–8*c*.

Equations of Equilibrium.

$$\overset{+}{\rightarrow} \Sigma F_x = 0; \quad 848.5 \cos 45^\circ \text{ N} - V_E = 0 \qquad V_E = 600 \text{ N} \qquad \textit{Ans.}$$

$$+\uparrow \Sigma F_y = 0; \quad -848.5 \sin 45^\circ \text{ N} + N_E = 0 \qquad N_E = 600 \text{ N} \qquad \textit{Ans.}$$

$$\circlearrowright + \Sigma M_E = 0; \quad 848.5 \cos 45^\circ \text{ N}(0.5 \text{ m}) - M_E = 0 \quad M_E = 300 \text{ N}\cdot\text{m} \quad \textit{Ans.}$$

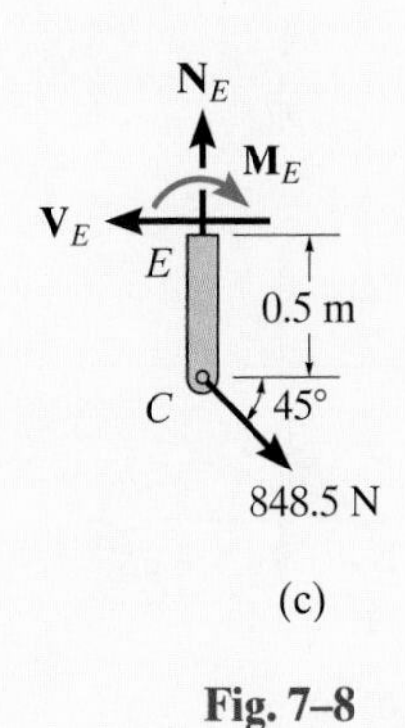

(c)

Fig. 7–8

NOTE: These results indicate a poor design. Member AC should be *straight* (from A to C) so that bending within the member is *eliminated*. If AC is straight then the internal force would only create tension in the member. See Example 6.21.

EXAMPLE 7.6

The uniform sign shown in Fig. 7–9*a* has a mass of 650 kg and is supported on the fixed column. Design codes indicate that the expected maximum uniform wind loading that will occur in the area where it is located is 900 Pa. Determine the internal loadings at *A*.

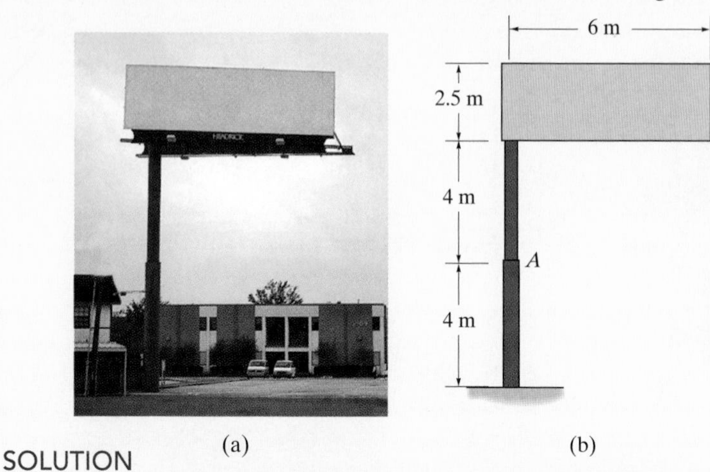

(a) (b)

SOLUTION

The idealized model for the sign is shown in Fig. 7–9*b*. Here the necessary dimensions are indicated. We can consider the free-body diagram of a section above point *A* since it does not involve the support reactions.

Free-Body Diagram. The sign has a weight of $W = 650(9.81) = 6.376$ kN, and the wind creates a resultant force of $F_w = 900\text{ N/m}^2(6\text{ m})(2.5\text{ m}) = 13.5$ kN perpendicular to the face of the sign. These loadings are shown on the free-body diagram, Fig. 7–9*c*.

Equations of Equilibrium. Since the problem is three dimensional, a vector analysis will be used.

$$\Sigma \mathbf{F} = \mathbf{0}; \qquad \mathbf{F}_A - 13.5\mathbf{i} - 6.376\mathbf{k} = \mathbf{0}$$

$$\mathbf{F}_A = \{13.5\mathbf{i} + 6.38\mathbf{k}\}\text{ kN} \qquad \textit{Ans.}$$

$$\Sigma \mathbf{M}_A = \mathbf{0}; \qquad \mathbf{M}_A + \mathbf{r} \times (\mathbf{F}_w + \mathbf{W}) = \mathbf{0}$$

$$\mathbf{M}_A + \begin{vmatrix} \mathbf{i} & \mathbf{j} & \mathbf{k} \\ 0 & 3 & 5.25 \\ -13.5 & 0 & 6.376 \end{vmatrix} = \mathbf{0}$$

$$\mathbf{M}_A = \{-19.1\mathbf{i} + 70.9\mathbf{j} + 40.5\mathbf{k}\}\text{ kN}\cdot\text{m} \qquad \textit{Ans.}$$

NOTE: Here $\mathbf{F}_{A_z} = \{6.38\mathbf{k}\}$ kN represents the normal force *N*, whereas $\mathbf{F}_{A_x} = \{13.5\mathbf{i}\}$ kN is the shear force. Also, the torsional moment is $\mathbf{M}_{A_z} = \{40.5\mathbf{k}\}$ kN · m, and the bending moment is determined from its components $\mathbf{M}_{A_x} = \{-19.1\mathbf{i}\}$ kN · m and $\mathbf{M}_{A_y} = \{-70.9\mathbf{j}\}$ kN · m; i.e., $M_b = \sqrt{M_x^2 + M_y^2}$.

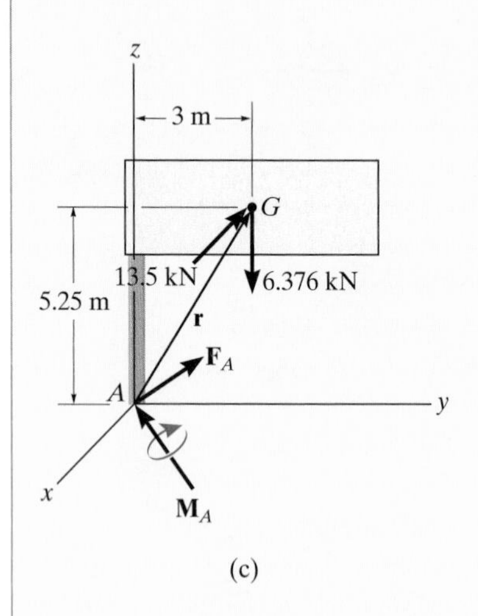

(c)

Fig. 7–9

PROBLEMS

7–1. The column is fixed to the floor and is subjected to the loads shown. Determine the internal normal force, shear force, and moment at points A and B.

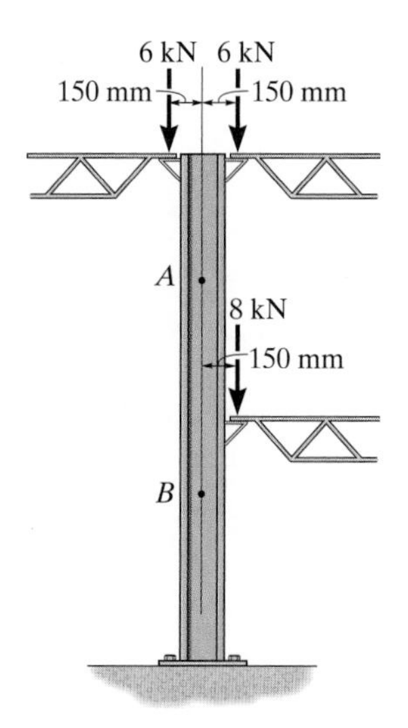

Prob. 7–1

7–2. The axial forces act on the shaft as shown. Determine the internal normal force at points A and B.

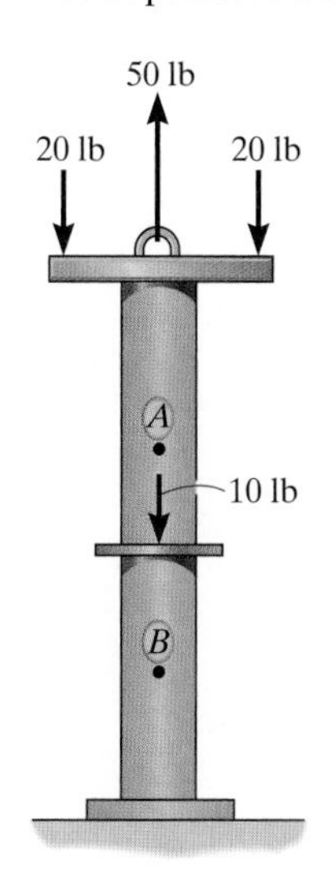

Prob. 7–2

7–3. The shaft is supported by smooth bearings at A and B and subjected to the torques shown. Determine the internal torque at points C, D, and E.

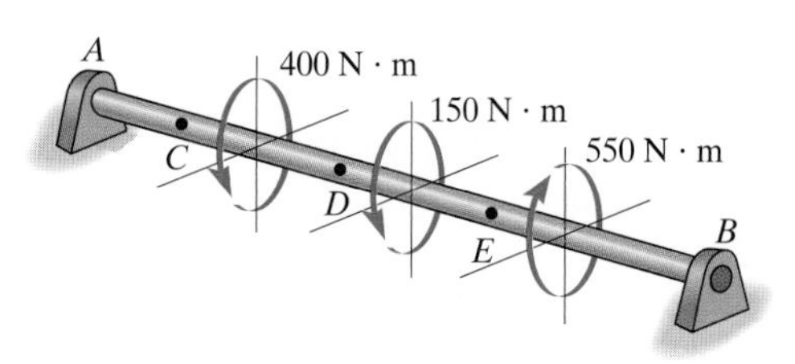

Prob. 7–3

***7–4.** Three torques act on the shaft. Determine the internal torque at points A, B, C, and D.

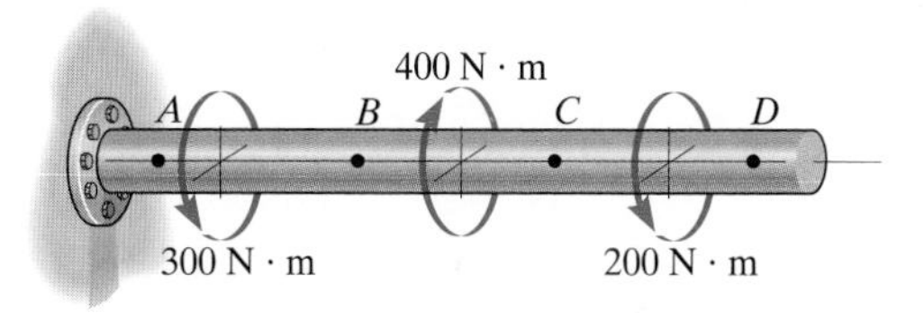

Prob. 7–4

7–5. The shaft is supported by a journal bearing at A and a thrust bearing at B. Determine the normal force, shear force, and moment at a section passing through (a) point C, which is just to the right of the bearing at A, and (b) point D, which is just to the left of the 3000-lb force.

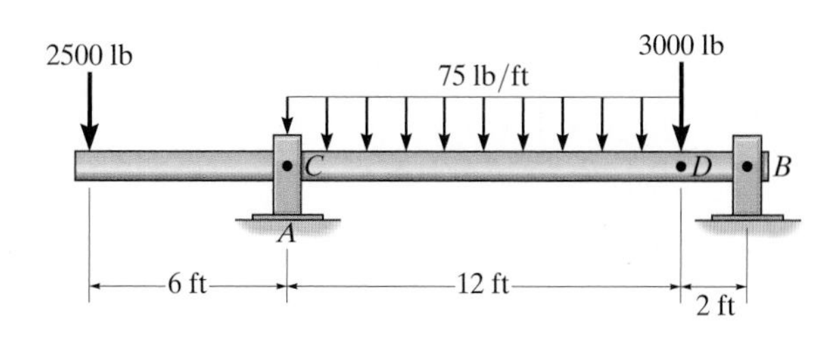

Prob. 7–5

7–6. Determine the internal normal force, shear force, and moment at point C.

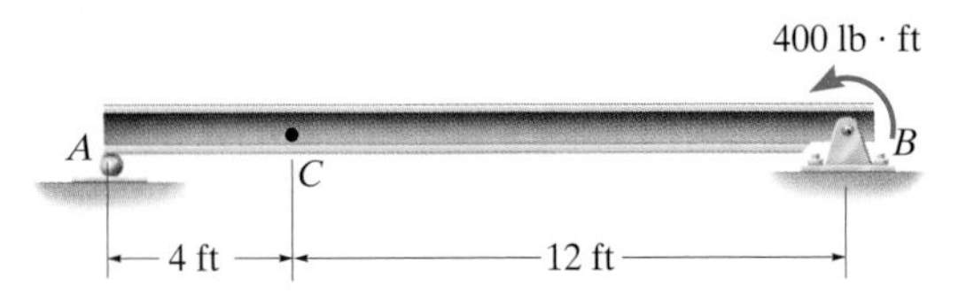

Prob. 7–6

7–7. Determine the internal normal force, shear force, and moment at point C.

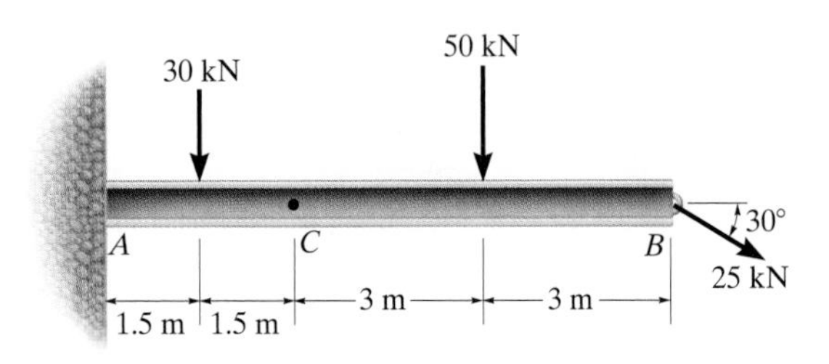

Prob. 7–7

***7–8.** Determine the normal force, shear force, and moment at a section passing through point C. Assume the support at A can be approximated by a pin and B as a roller.

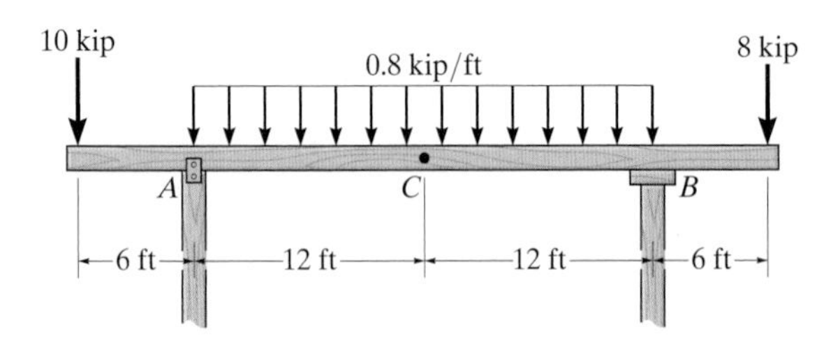

Prob. 7–8

7–9. The beam AB will fail if the maximum internal moment at D reaches 800 N · m or the normal force in member BC becomes 1500 N. Determine the largest load w it can support.

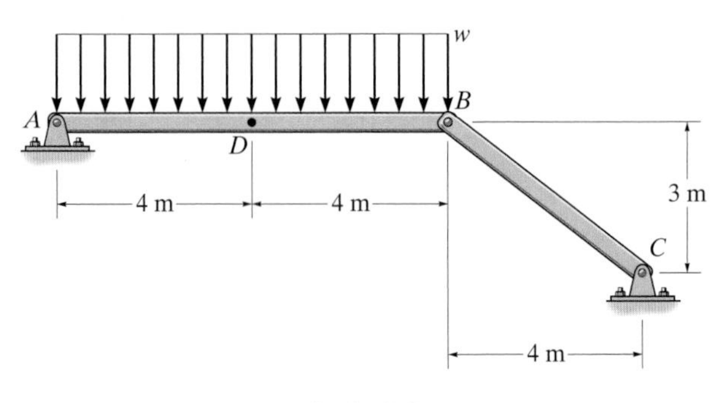

Prob. 7–9

7–10. Determine the shear force and moment acting at a section passing through point C in the beam.

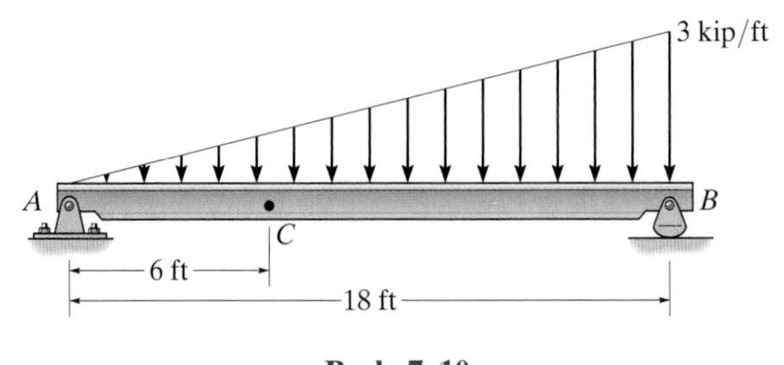

Prob. 7–10

7–11. Determine the internal normal force, shear force, and moment at points E and D of the compound beam.

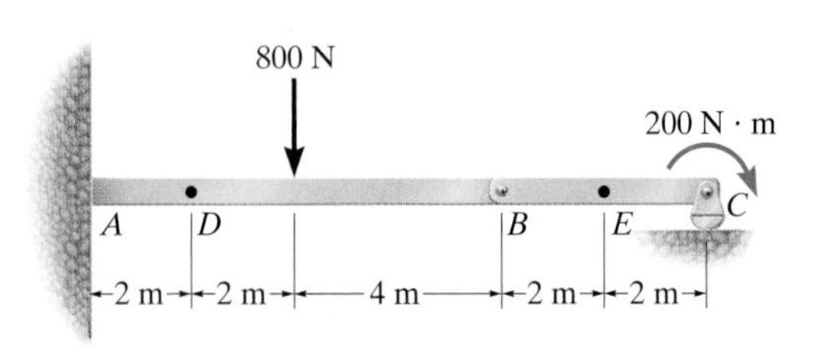

Prob. 7–11

*7–12. The boom DF of the jib crane and the column DE have a uniform weight of 50 lb/ft. If the hoist and load weigh 300 lb, determine the normal force, shear force, and moment in the crane at sections passing through points A, B, and C. *Hint*: Treat the boom tip, beyond the hoist, as weightless.

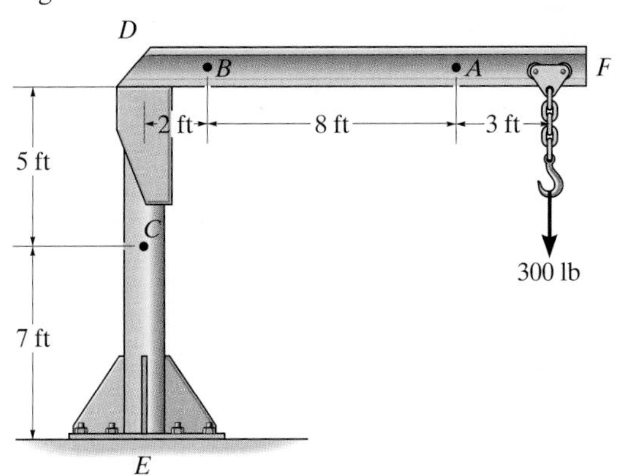

Prob. 7–12

7–13. Determine the internal normal force, shear force, and moment at point C.

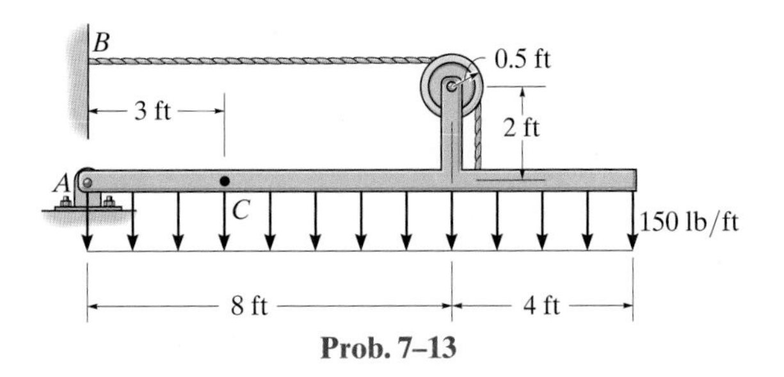

Prob. 7–13

7–14. Determine the normal force, shear force, and moment at a section passing through point D of the two-member frame.

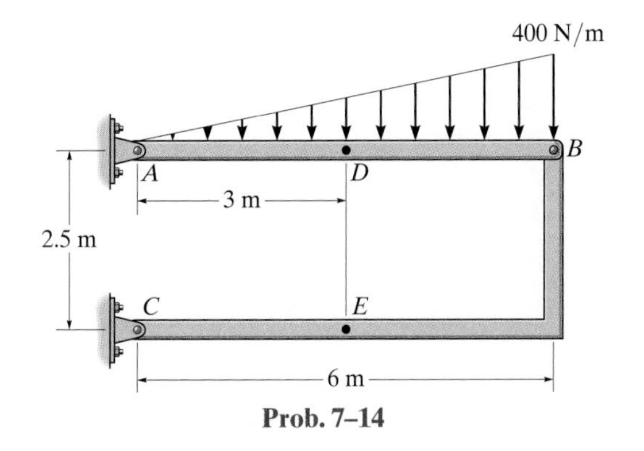

Prob. 7–14

7–15. The beam weighs 280 lb/ft. Determine the internal normal force, shear force, and moment at point C.

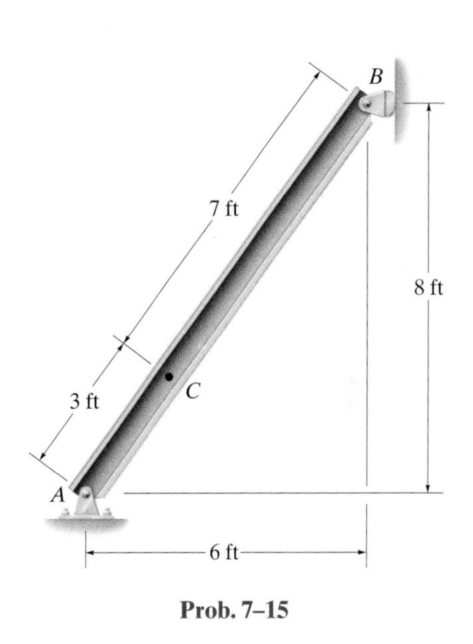

Prob. 7–15

*7–16. Determine the internal normal force, shear force, and moment at points C and D of the beam.

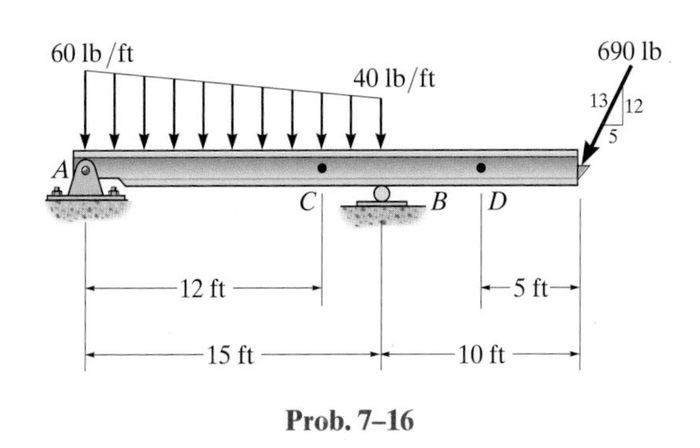

Prob. 7–16

7–17. Determine the normal force, shear force, and moment acting at a section passing through point C.

7–18. Determine the normal force, shear force, and moment acting at a section passing through point D.

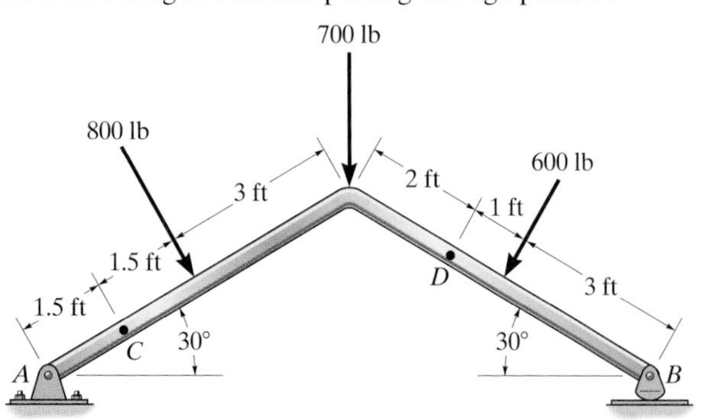

Probs. 7–17/18

7–19. Determine the normal force, shear force, and moment at a section passing through point C. Take $P = 8$ kN.

***7–20.** The cable will fail when subjected to a tension of 2 kN. Determine the largest vertical load P the frame will support and calculate the internal normal force, shear force, and moment at a section passing through point C for this loading.

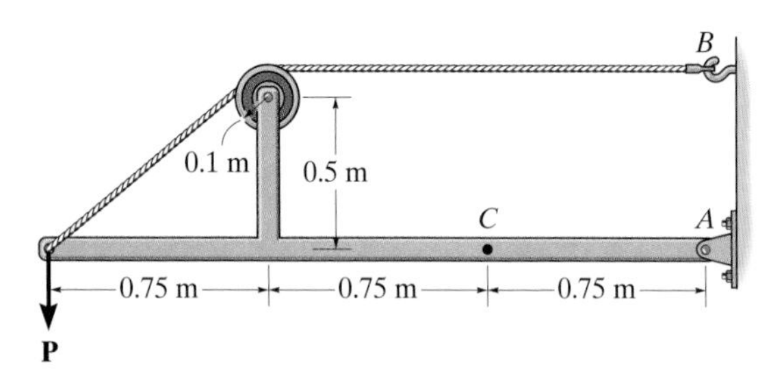

Probs. 7–19/20

7–21. Determine the internal shear force and moment acting at point C of the beam.

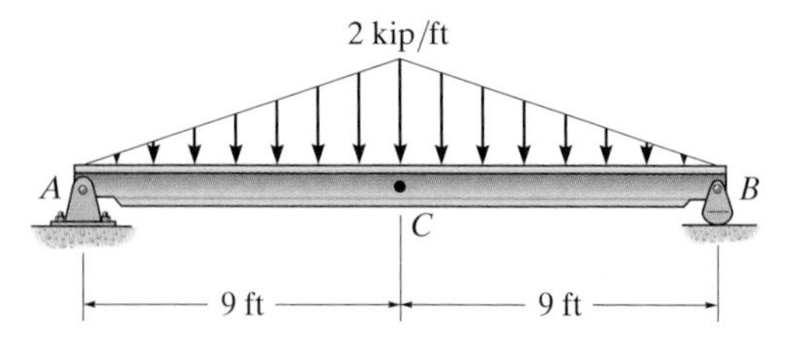

Prob. 7–21

7–22. Determine the internal shear force and moment acting at point D of the beam.

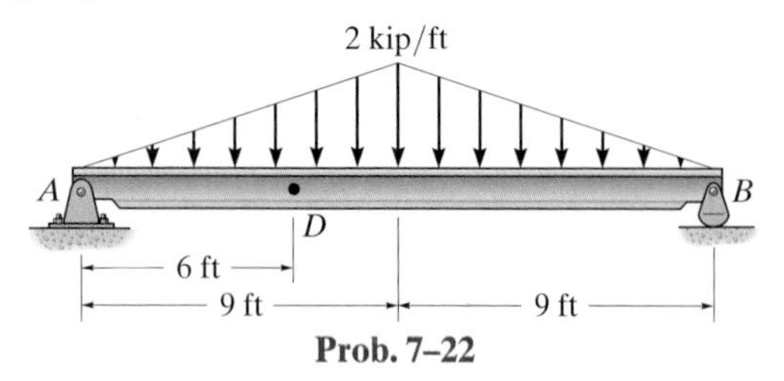

Prob. 7–22

7–23. The shaft is supported by a journal bearing at A and a thrust bearing at B. Determine the internal normal force, shear force, and moment at (a) point C, which is just to the right of the bearing at A, and (b) point D, which is just to the left of the 3000-lb force.

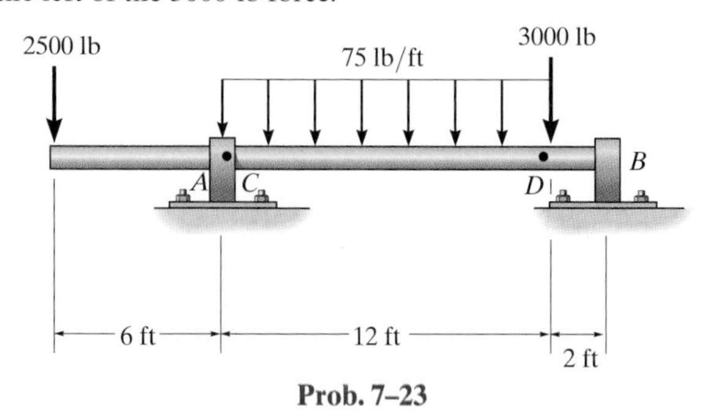

Prob. 7–23

***7–24.** The jack AB is used to straighten the bent beam DE using the arrangement shown. If the axial compressive force in the jack is 5000 lb, determine the internal moment developed at point C of the top beam. Neglect the weight of the beams.

7–25. Solve Prob. 7–24 assuming that each beam has a uniform weight of 150 lb/ft.

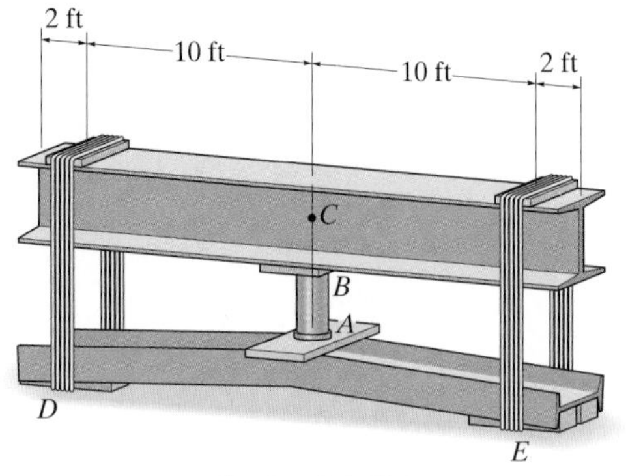

Probs. 7–24/25

7–26. Determine the normal force, shear force, and moment in the beam at sections passing through points D and E. Point E is just to the right of the 3-kip load.

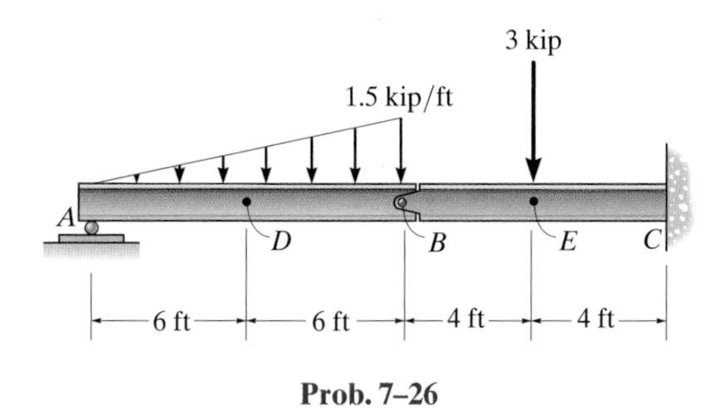

Prob. 7–26

7–27. Determine the normal force, shear force, and moment at a section passing through point D of the two-member frame.

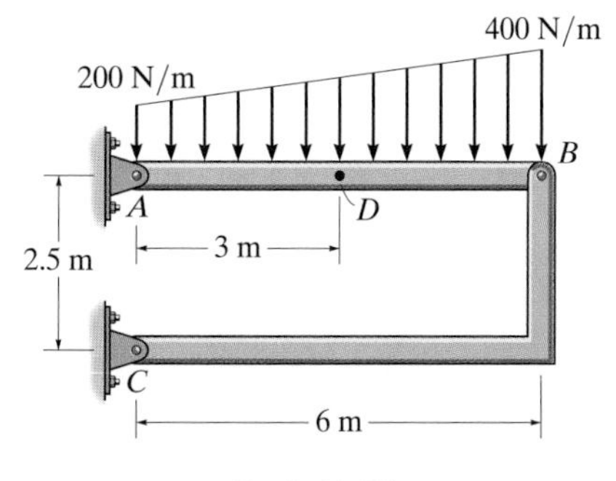

Prob. 7–27

***7–28.** Determine the normal force, shear force, and moment at sections passing through points E and F. Member BC is pinned at B and there is a smooth slot in it at C. The pin at C is fixed to member CD.

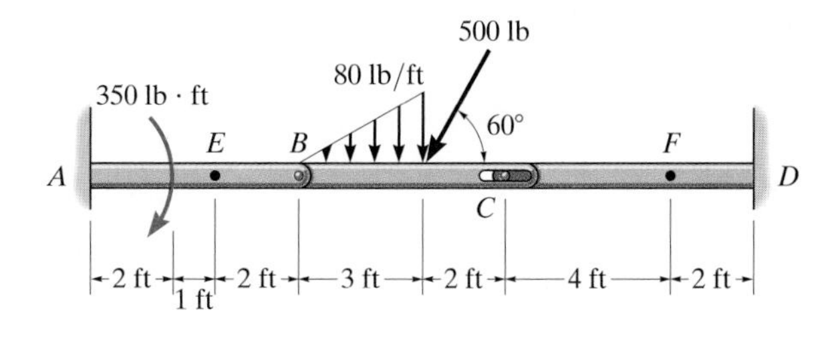

Prob. 7–28

7–29. The bolt shank is subjected to a tension of 80 lb. Determine the internal normal force, shear force, and moment at point C.

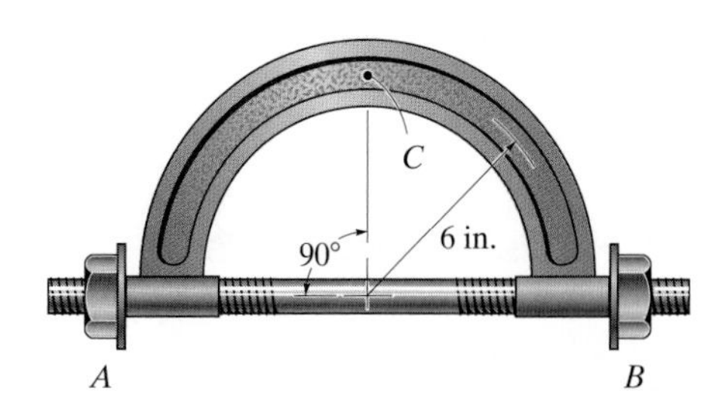

Prob. 7–29

7–30. Determine the normal force, shear force, and moment acting at sections passing through points B and C on the curved rod.

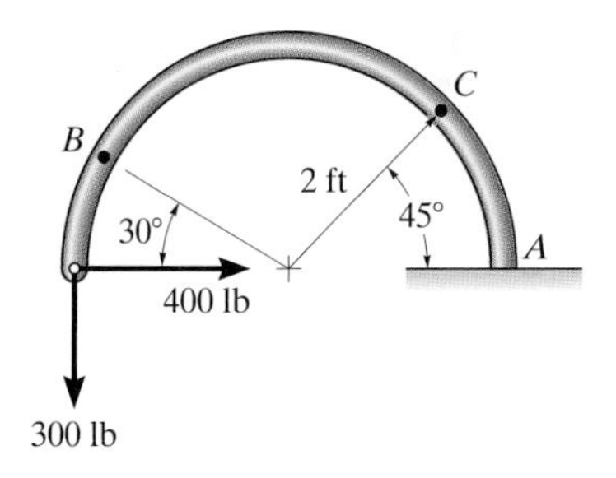

Prob. 7–30

7–31. The cantilevered rack is used to support each end of a smooth pipe that has a total weight of 300 lb. Determine the normal force, shear force, and moment that act in the arm at its fixed support A along a vertical section.

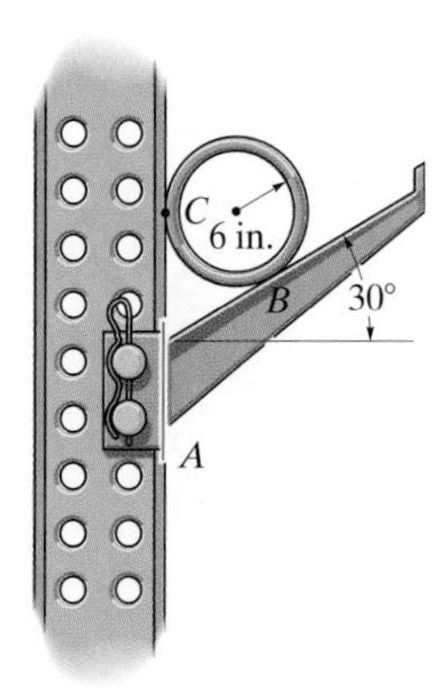

Prob. 7–31

*7–32. Determine the normal force, shear force, and moment at a section passing through point D of the two-member frame.

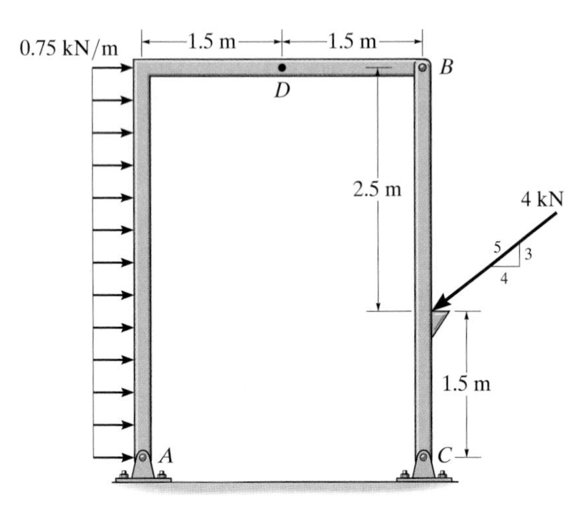

Prob. 7–32

7–33. Determine the internal normal force, shear force, and moment acting at point A of the smooth hook.

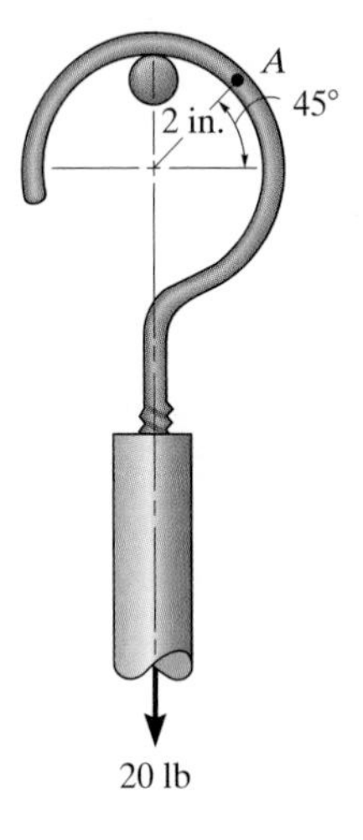

Prob. 7–33

7–34. Determine the internal normal force, shear force, and moment acting at points B and C on the curved rod.

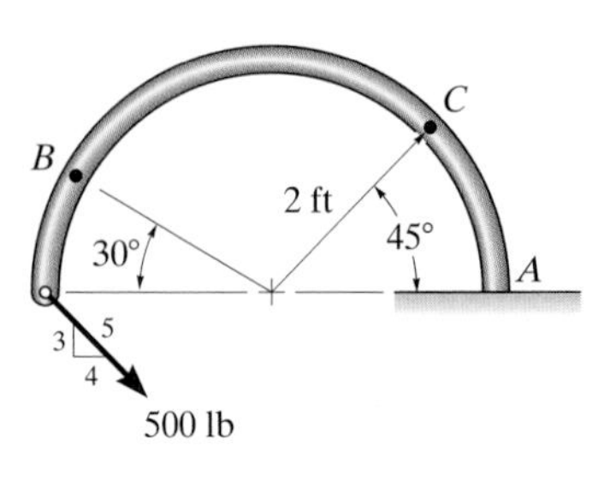

Prob. 7–34

7–35. Determine the ratio of a/b for which the shear force will be zero at the midpoint C of the beam.

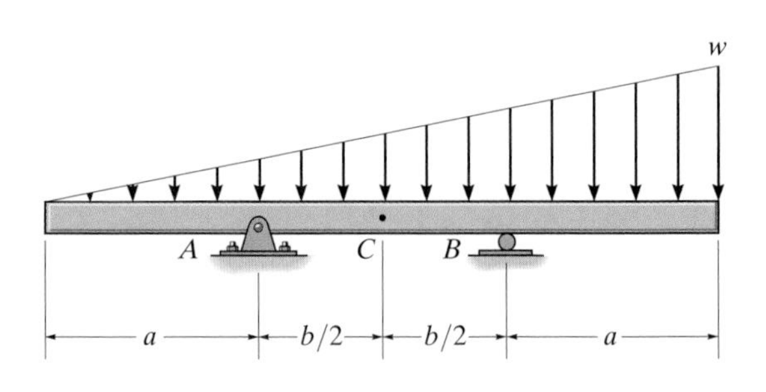

Prob. 7–35

*7–36. The semicircular arch is subjected to a uniform distributed load along its axis of w_0 per unit length. Determine the internal normal force, shear force, and moment in the arch at $\theta = 45°$.

7–37. Solve Prob. 7–36 for $\theta = 120°$.

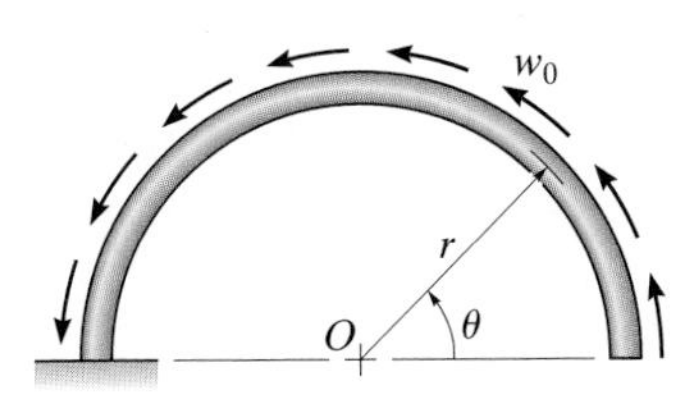

Probs. 7–36/37

7–38. Determine the x, y, z components of internal loading at a section passing through point C in the pipe assembly. Neglect the weight of the pipe. Take $\mathbf{F}_1 = \{350\mathbf{j} - 400\mathbf{k}\}$ lb and $\mathbf{F}_2 = \{150\mathbf{i} - 300\mathbf{k}\}$ lb.

7–39. Determine the x, y, z components of internal loading at a section passing through point C in the pipe assembly. Neglect the weight of the pipe. Take $\mathbf{F}_1 = \{-80\mathbf{i} + 200\mathbf{j} - 300\mathbf{k}\}$ lb and $\mathbf{F}_2 = \{250\mathbf{i} - 150\mathbf{j} - 200\mathbf{k}\}$ lb.

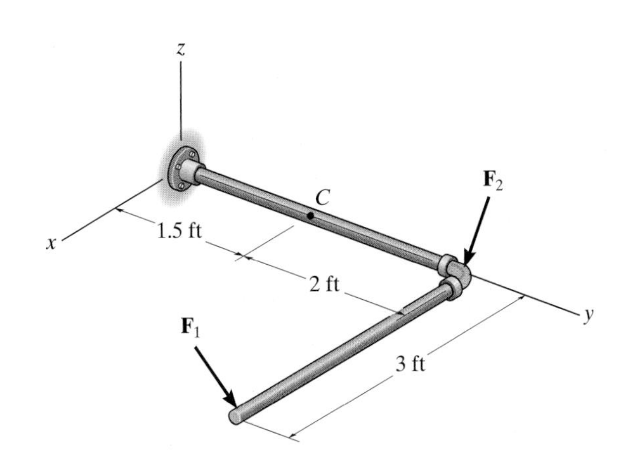

Probs. 7–38/39

*7–40. Determine the x, y, z components of internal loading in the rod at point D. $\mathbf{F} = \{7\mathbf{i} - 12\mathbf{j} - 5\mathbf{k}\}$ kN.

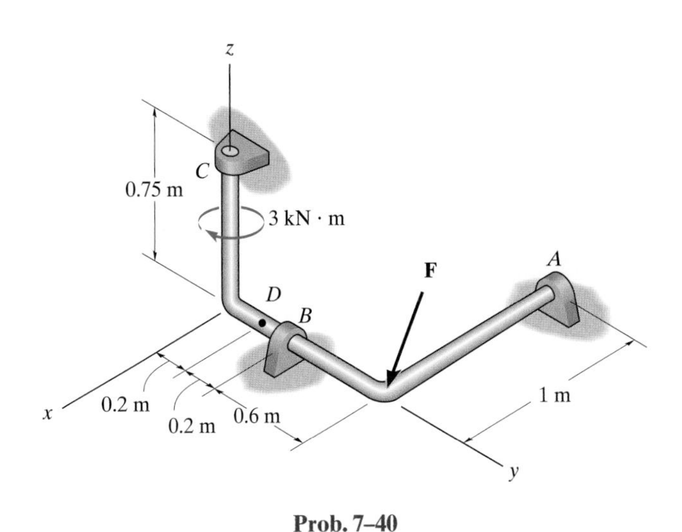

Prob. 7–40

7–41. Determine the x, y, z components of internal loading in the rod at point E. $\mathbf{F} = \{7\mathbf{i} - 12\mathbf{j} - 5\mathbf{k}\}$ kN.

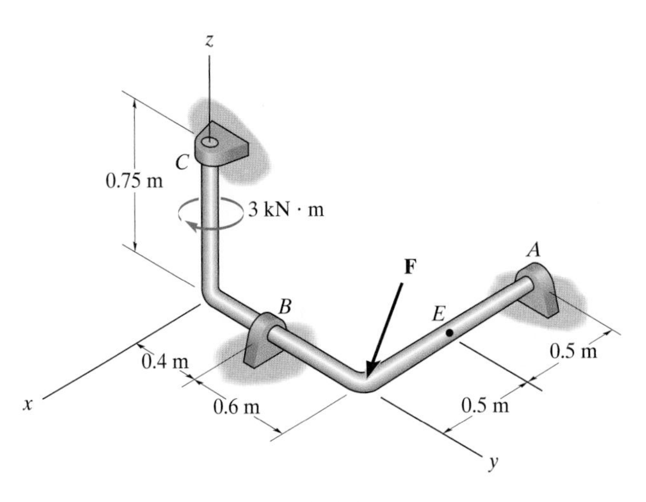

Prob. 7–41

*7.2 Shear and Moment Equations and Diagrams

To save on material the beams used to support the roof of this shelter were tapered since the roof loading will produce a larger internal moment at the beams' centers than at their ends.

Beams are structural members designed to support loadings applied perpendicular to their axes. In general, beams are long, straight bars having a constant cross-sectional area. Often they are classified as to how they are supported. For example, a *simply supported beam* is pinned at one end and roller-supported at the other, Fig. 7–10*a*, whereas a *cantilevered beam* is fixed at one end and free at the other. The actual design of a beam requires a detailed knowledge of the *variation* of the internal shear force V and bending moment M acting at *each point* along the axis of the beam. After this force and bending-moment analysis is complete, the theory of mechanics of materials and an appropriate engineering design code then can be used to determine the beam's required cross-sectional area.

The *variations* of V and M as functions of the position x along the beam's axis can be obtained by using the method of sections discussed in Sec. 7.1. Here, however, it is necessary to section the beam at an arbitrary distance x from one end rather than at a specified point. If the results are plotted, the graphical variations of V and M as functions of x are termed the *shear diagram* and *bending-moment diagram*, respectively, Figs. 7–10*b* and 7–10*c*.

In general, the internal shear and bending-moment functions generally will be discontinuous, or their slopes will be discontinuous at points where a distributed load changes or where concentrated forces or couple moments are applied. Because of this, these functions must be determined for *each segment* of the beam located between any two discontinuities of loading. For example, sections located at x_1, x_2, and x_3 will have to be used to describe the variation of V and M along the length of the beam in Fig. 7–10. These functions will be valid *only* within regions from O to a for x_1, from a to b for x_2, and from b to L for x_3.

The internal normal force will not be considered in the following discussion for two reasons. In most cases, the loads applied to a beam act perpendicular to the beam's axis and hence produce only an internal shear force and bending moment. For design purposes, the beam's resistance to shear, and particularly to bending, is more important than its ability to resist a normal force.

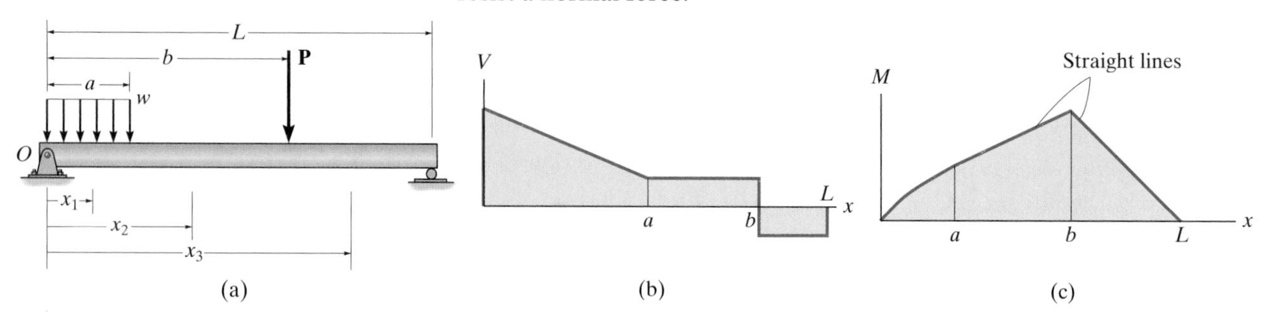

Fig. 7–10

Sign Convention. Before presenting a method for determining the shear and bending moment as functions of x and later plotting these functions (shear and bending-moment diagrams), it is first necessary to establish a *sign convention* so as to define a "positive" and "negative" shear force and bending moment acting in the beam. [This is analogous to assigning coordinate directions x positive to the right and y positive upward when plotting a function $y = f(x)$.] Although the choice of a sign convention is arbitrary, here we will choose the one used for the majority of engineering applications. It is illustrated in Fig. 7–11. Here the positive directions are denoted by an internal *shear force* that causes *clockwise rotation* of the member on which it acts, and by an internal *moment* that causes *compression or pushing on the upper part* of the member. Also, positive moment would tend to bend the member if it were elastic, concave upward. Loadings that are opposite to these are considered negative.

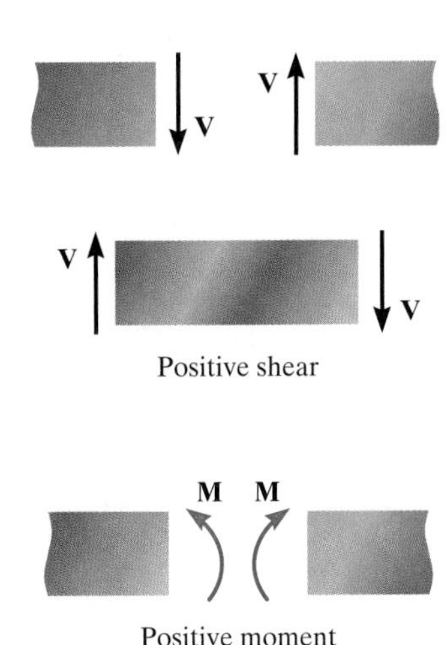

Positive shear

Positive moment

Beam sign convention

Fig. 7–11

Procedure for Analysis

The shear and bending-moment diagrams for a beam can be constructed using the following procedure.

Support Reactions.

- Determine all the reactive forces and couple moments acting on the beam and resolve all the forces into components acting perpendicular and parallel to the beam's axis.

Shear and Moment Functions.

- Specify separate coordinates x having an origin at the beam's *left end* and extending to regions of the beam *between* concentrated forces and/or couple moments, or where there is no discontinuity of distributed loading.
- Section the beam perpendicular to its axis at each distance x and draw the free-body diagram of one of the segments. Be sure **V** and **M** are shown acting in their *positive sense*, in accordance with the sign convention given in Fig. 7–11.
- The shear V is obtained by summing forces perpendicular to the beam's axis.
- The moment M is obtained by summing moments about the sectioned end of the segment.

Shear and Moment Diagrams.

- Plot the shear diagram (V versus x) and the moment diagram (M versus x). If computed values of the functions describing V and M are *positive*, the values are plotted above the x axis, whereas *negative* values are plotted below the x axis.
- Generally, it is convenient to plot the shear and bending-moment diagrams directly below the free-body diagram of the beam.

EXAMPLE 7.7

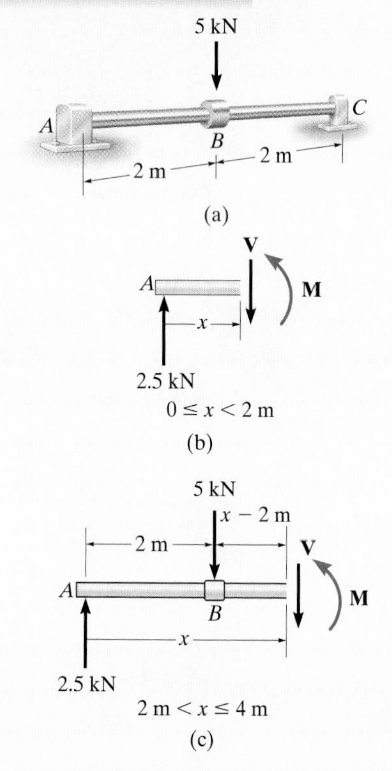

Draw the shear and bending-moment diagrams for the shaft shown in Fig. 7–12*a*. The support at A is a thrust bearing and the support at C is a journal bearing.

SOLUTION

Support Reactions. The support reactions have been computed, as shown on the shaft's free-body diagram, Fig. 7–12*d*.

Shear and Moment Functions. The shaft is sectioned at an arbitrary distance x from point A, extending within the region AB, and the free-body diagram of the left segment is shown in Fig. 7–12*b*. The unknowns **V** and **M** are assumed to act in the *positive sense* on the right-hand face of the segment according to the established sign convention. Why? Applying the equilibrium equations yields

$$+\uparrow \Sigma F_y = 0; \qquad V = 2.5 \text{ kN} \qquad (1)$$

$$\circlearrowleft + \Sigma M = 0; \qquad M = 2.5x \text{ kN}\cdot\text{m} \qquad (2)$$

A free-body diagram for a left segment of the shaft extending a distance x within the region BC is shown in Fig. 7–12*c*. As always, **V** and **M** are shown acting in the positive sense. Hence,

$$+\uparrow \Sigma F_y = 0; \qquad 2.5 \text{ kN} - 5 \text{ kN} - V = 0$$

$$V = -2.5 \text{ kN} \qquad (3)$$

$$\circlearrowleft + \Sigma M = 0; \qquad M + 5 \text{ kN}(x - 2 \text{ m}) - 2.5 \text{ kN}(x) = 0$$

$$M = (10 - 2.5x) \text{ kN}\cdot\text{m} \qquad (4)$$

Shear and Moment Diagrams. When Eqs. 1 through 4 are plotted within the regions in which they are valid, the shear and bending-moment diagrams shown in Fig. 7–12*d* are obtained. The shear diagram indicates that the internal shear force is always 2.5 kN (positive) within shaft segment AB. Just to the right of point B, the shear force changes sign and remains at a constant value of -2.5 kN for segment BC. The moment diagram starts at zero, increases linearly to point B at $x = 2$ m, where $M_{max} = 2.5 \text{ kN}(2 \text{ m}) = 5 \text{ kN}\cdot\text{m}$, and thereafter decreases back to zero.

NOTE: It is seen in Fig. 7–12*d* that the graph of the shear and moment diagrams is discontinuous at points of concentrated force, i.e., points A, B, and C. For this reason, as stated earlier, it is necessary to express both the shear and bending-moment functions separately for regions between concentrated loads. It should be realized, however, that all loading discontinuities are mathematical, arising from the *idealization of a concentrated force and couple moment.* Physically, loads are always applied over a finite area, and if the load variation could actually be accounted for, the shear and bending-moment diagrams would then be continuous over the shaft's entire length.

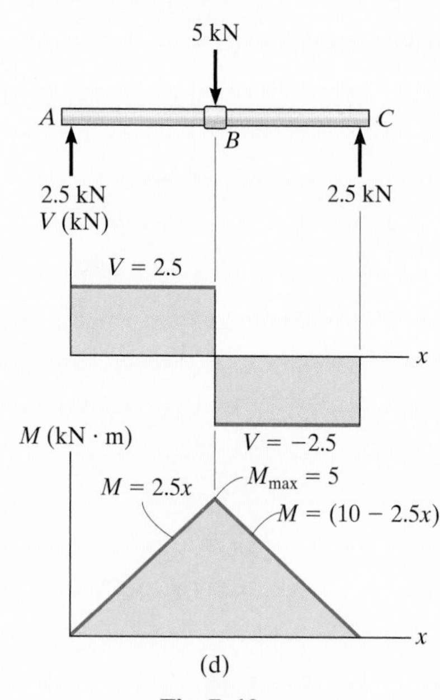

Fig. 7–12

EXAMPLE 7.8

Draw the shear and bending-moment diagrams for the beam shown in Fig. 7–13*a*.

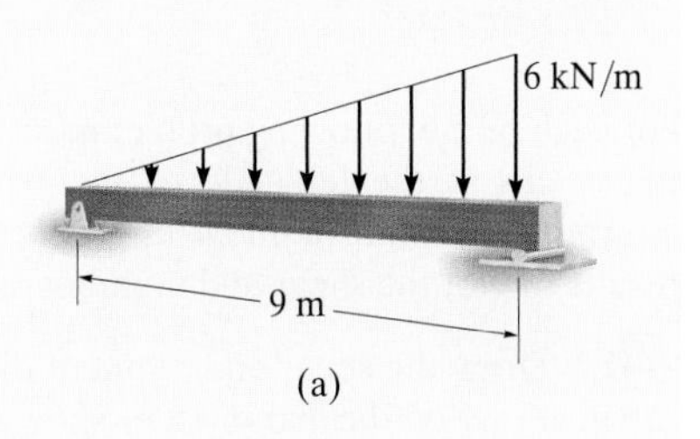

(a)

SOLUTION

Support Reactions. The support reactions have been computed as shown on the beam's free-body diagram, Fig. 7–13*c*.

Shear and Moment Functions. A free-body diagram for a left segment of the beam having a length x is shown in Fig. 7–13*b*. The distributed loading acting on this segment has an intensity of $\frac{2}{3}x$ at its end and is replaced by a resultant force *after* the segment is isolated as a free-body diagram. The *magnitude* of the resultant force is equal to $\frac{1}{2}(x)\left(\frac{2}{3}x\right) = \frac{1}{3}x^2$. This force *act through the centroid* of the distributed loading area, a distance $\frac{1}{3}x$ from the right end. Applying the two equations of equilibrium yields

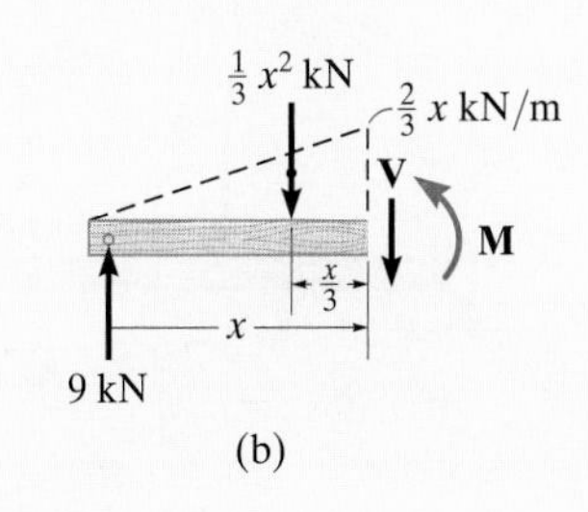

(b)

$$+\uparrow \Sigma F_y = 0; \qquad 9 - \frac{1}{3}x^2 - V = 0$$

$$V = \left(9 - \frac{x^2}{3}\right) \text{kN} \qquad (1)$$

$$\circlearrowleft + \Sigma M = 0; \qquad M + \frac{1}{3}x^2\left(\frac{x}{3}\right) - 9x = 0$$

$$M = \left(9x - \frac{x^3}{9}\right) \text{kN} \cdot \text{m} \qquad (2)$$

Shear and Moment Diagrams. The shear and bending-moment diagrams shown in Fig. 7–13*c* are obtained by plotting Eqs. 1 and 2.

The point of *zero shear* can be found using Eq. 1:

$$V = 9 - \frac{x^2}{3} = 0$$

$$x = 5.20 \text{ m}$$

NOTE: It will be shown in Sec. 7.3 that this value of x happens to represent the point on the beam where the *maximum moment* occurs. Using Eq. 2, we have

$$M_{max} = \left(9(5.20) - \frac{(5.20)^3}{9}\right) \text{kN} \cdot \text{m}$$

$$= 31.2 \text{ kN} \cdot \text{m}$$

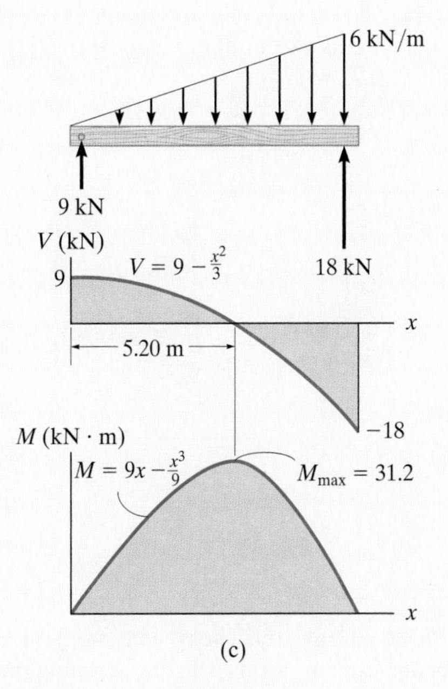

(c)

Fig. 7–13

PROBLEMS

For each of the following problems, establish the x axis with the origin at the left side of the beam, and obtain the internal shear and moment as a function of x. Use these results to plot the shear and moment diagrams.

7–42. Draw the shear and moment diagrams for the shaft (a) in terms of the parameters shown; (b) set $P = 9$ kN, $a = 2$ m, $L = 6$ m. There is a thrust bearing at A and a journal bearing at B.

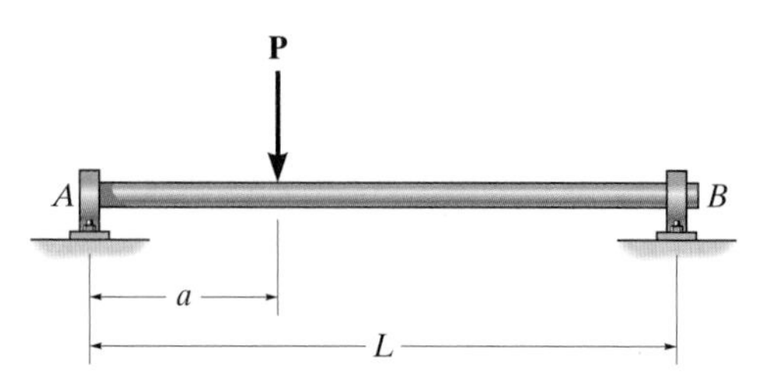

Prob. 7–42

7–43. Draw the shear and moment diagrams for the beam (a) in terms of the parameters shown; (b) set $P = 800$ lb, $a = 5$ ft, $L = 12$ ft.

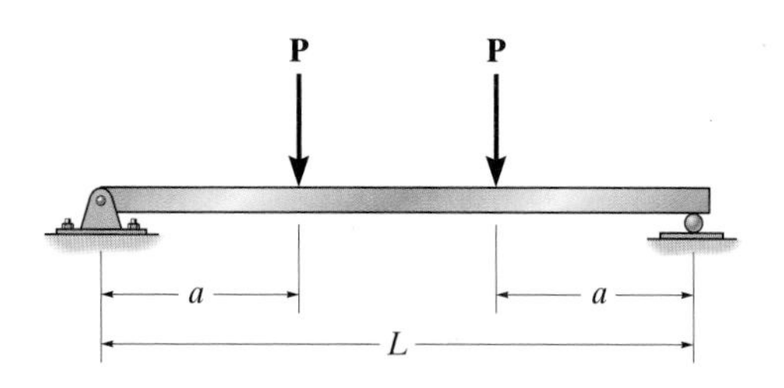

Prob. 7–43

***7–44.** Draw the shear and moment diagrams for the beam (a) in terms of the parameters shown; (b) set $M_0 = 500$ N·m, $L = 8$ m.

7–45. If $L = 9$ m, the beam will fail when the maximum shear force is $V_{max} = 5$ kN or the maximum bending moment is $M_{max} = 2$ kN·m. Determine the magnitude M_0 of the largest couple moments it will support.

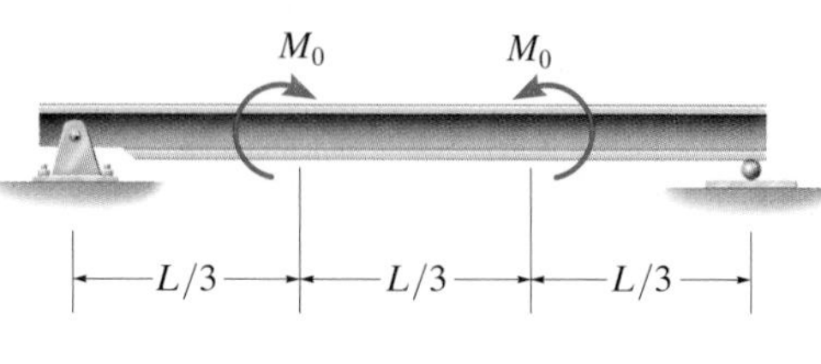

Probs. 7–44/45

7–46. The shaft is supported by a thrust bearing at A and a journal bearing at B. Draw the shear and moment diagrams for the shaft (a) in terms of the parameters shown; (b) set $w = 500$ lb/ft, $L = 10$ ft.

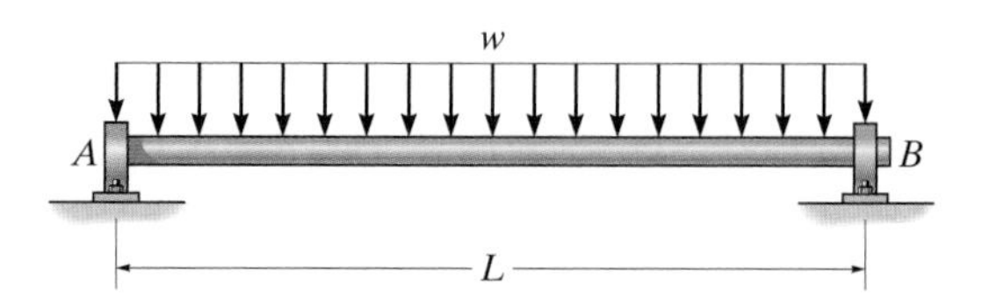

Prob. 7–46

7–47. The shaft is supported by a thrust bearing at A and a journal bearing at B. If $L = 10$ ft, the shaft will fail when the maximum moment is $M_{max} = 5$ kip·ft. Determine the largest uniform distributed load w the shaft will support.

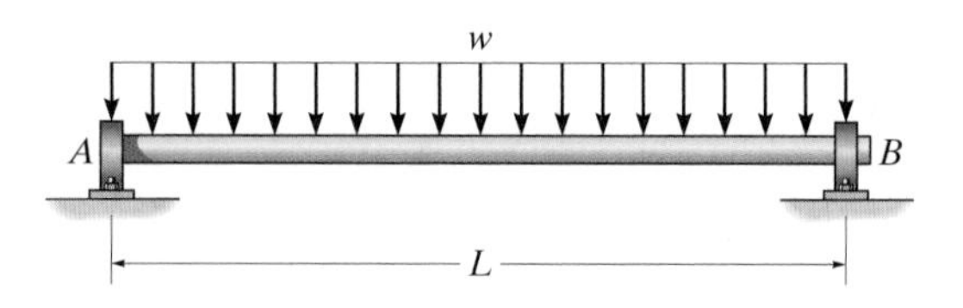

Prob. 7–47

*7–48. Draw the shear and moment diagrams for the beam.

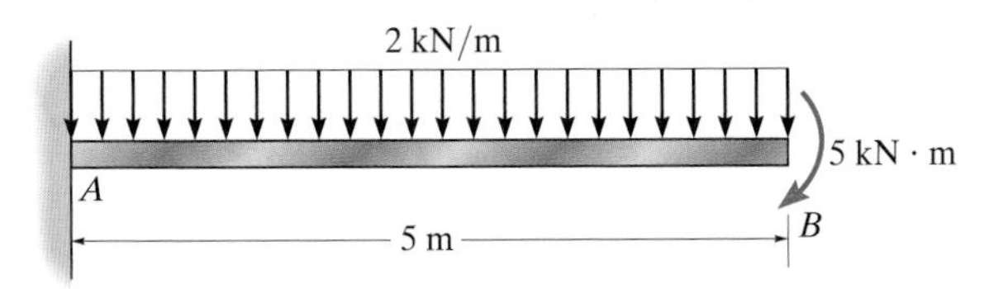

Prob. 7–48

7–49. Draw the shear and moment diagrams for the beam.

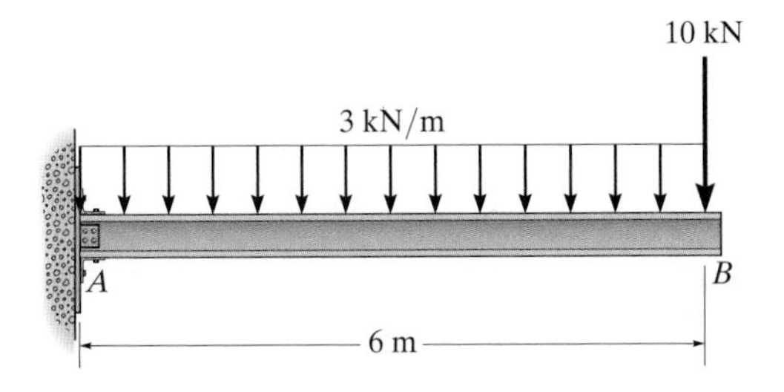

Prob. 7–49

7–50. Draw the shear and moment diagrams for the beam.

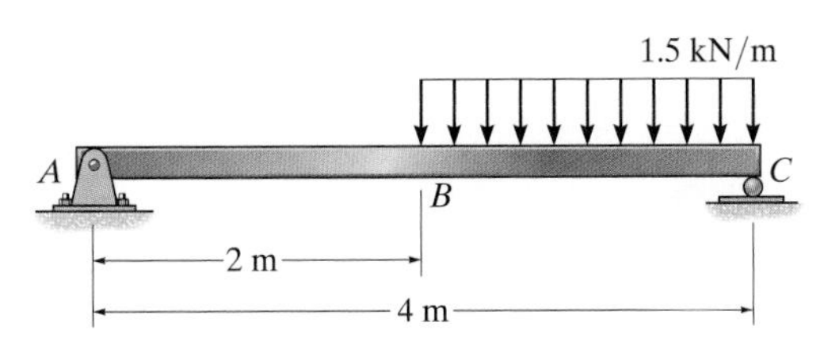

Prob. 7–50

7–51. Draw the shear and moment diagrams for the beam.

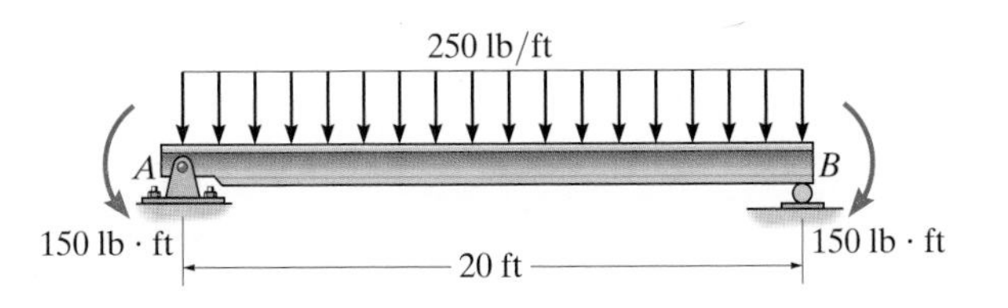

Prob. 7–51

*7–52. Draw the shear and moment diagrams for the beam.

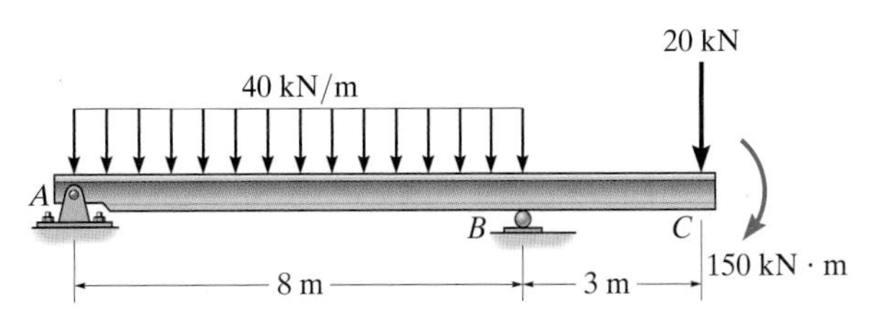

Prob. 7–52

7–53. Draw the shear and moment diagrams for the beam.

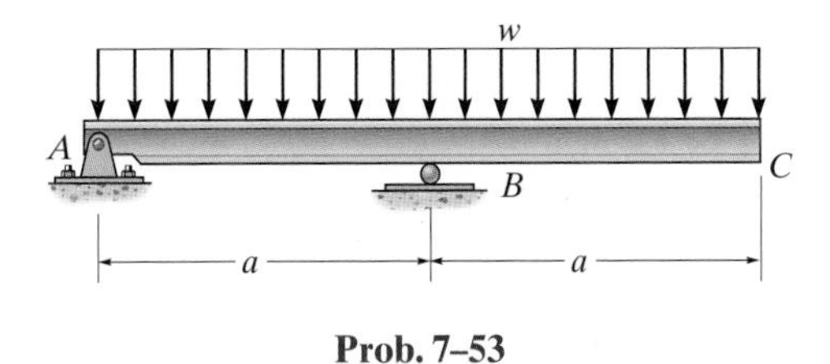

Prob. 7–53

7–54. Draw the shear and bending-moment diagrams for beam ABC. Note that there is a pin at B.

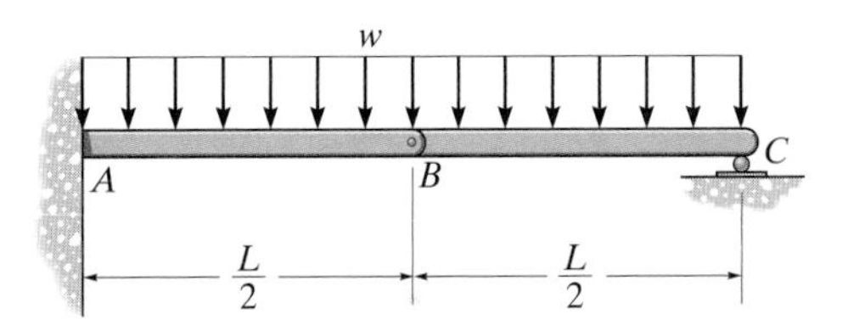

Prob. 7–54

7–55. The beam has a depth of 2 ft and is subjected to a uniform distributed loading of 50 lb/ft which acts at an angle of 30° from the vertical as shown. Determine the internal normal force, shear force, and moment in the beam as a function of x. *Hint*: The moment loading is to be determined from a point along the centerline of the beam (x axis).

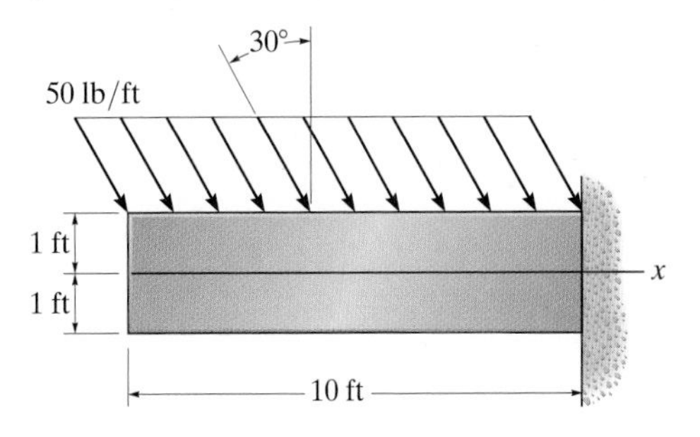

Prob. 7–55

*7–56. Draw the shear and moment diagrams for the beam (a) in terms of the parameters shown; (b) set $w = 250$ lb/ft, $L = 12$ ft.

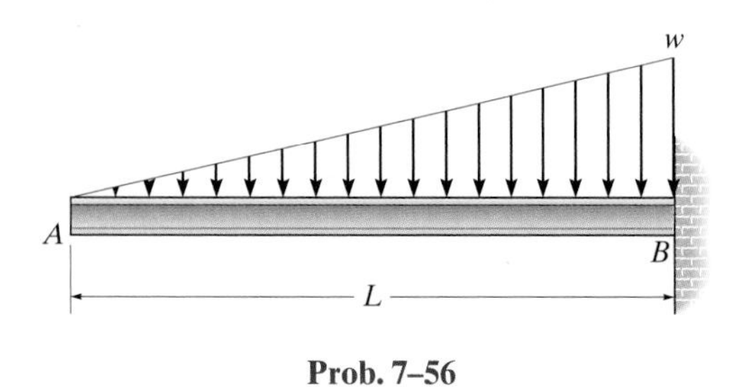

Prob. 7–56

7–57. If $L = 18$ ft, the beam will fail when the maximum shear force is $V_{max} = 800$ lb, or the maximum moment is $M_{max} = 1200$ lb · ft. Determine the largest intensity w of the distributed loading it will support.

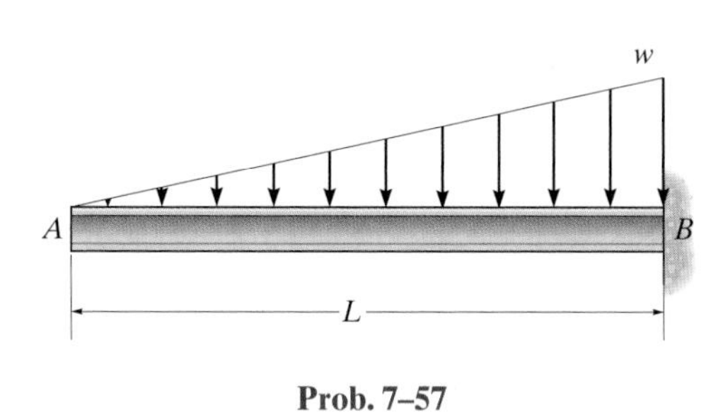

Prob. 7–57

7–58. The beam will fail when the maximum internal moment is M_{max}. Determine the position x of the concentrated force **P** and its smallest magnitude that will cause failure.

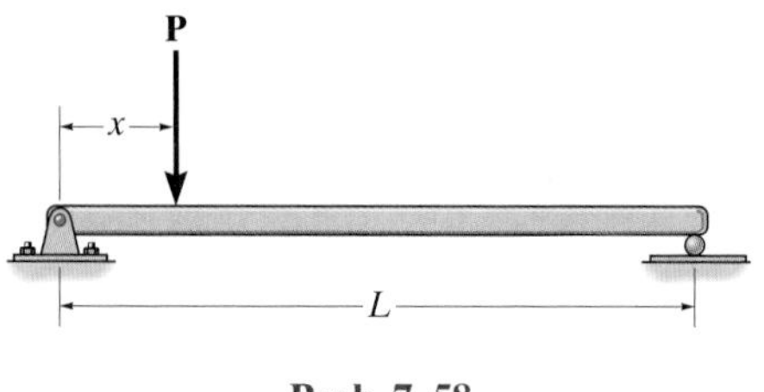

Prob. 7–58

7–59. Draw the shear and moment diagrams for the beam.

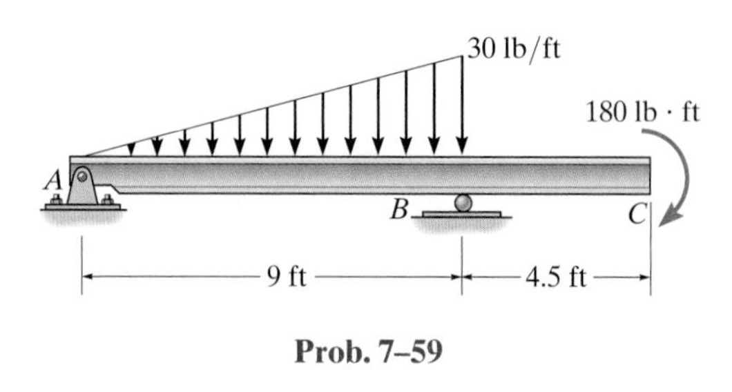

Prob. 7–59

*7–60. The cantilevered beam is made of material having a specific weight γ. Determine the shear and moment in the beam as a function of x.

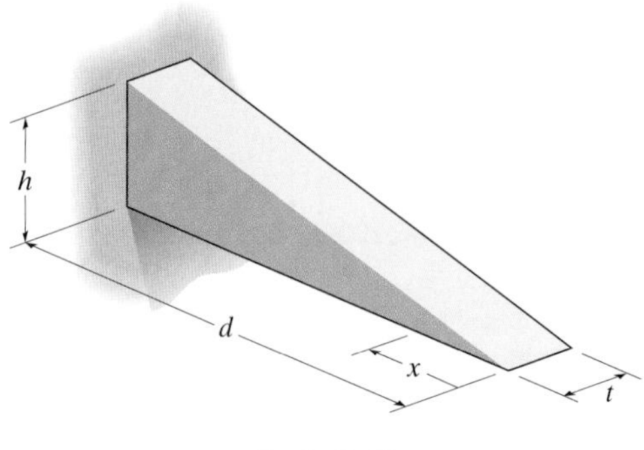

Prob. 7–60

7–61. Draw the shear and moment diagrams for the beam.

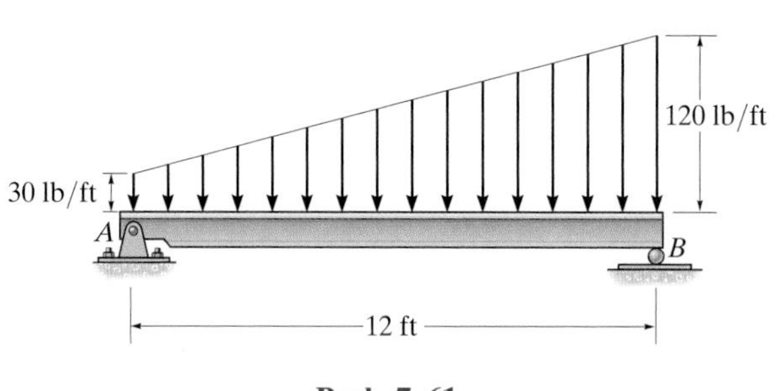

Prob. 7–61

7–62. Draw the shear and moment diagrams for the beam.

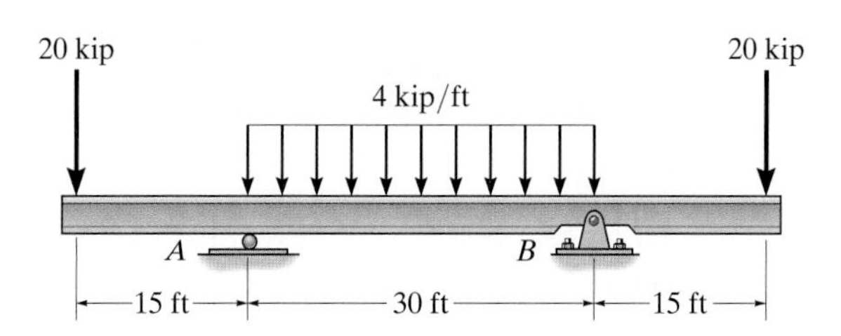

Prob. 7–62

***7–64.** Determine the normal force, shear force, and moment in the curved rod as a function of θ.

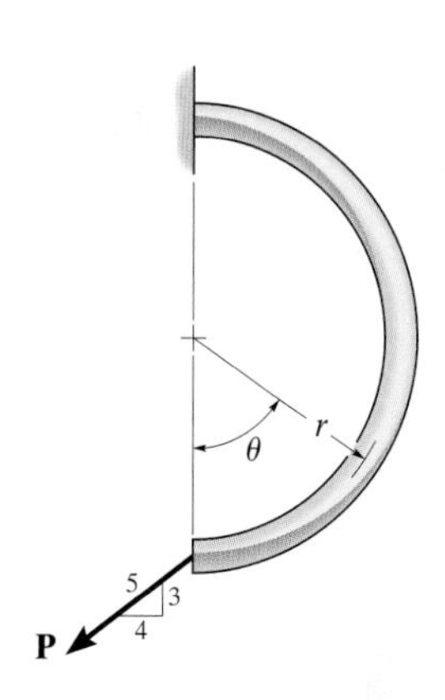

Prob. 7–64

7–63. Express the x, y, z components of internal loading in the rod as a function of y, where $0 \leq y \leq 4$ ft.

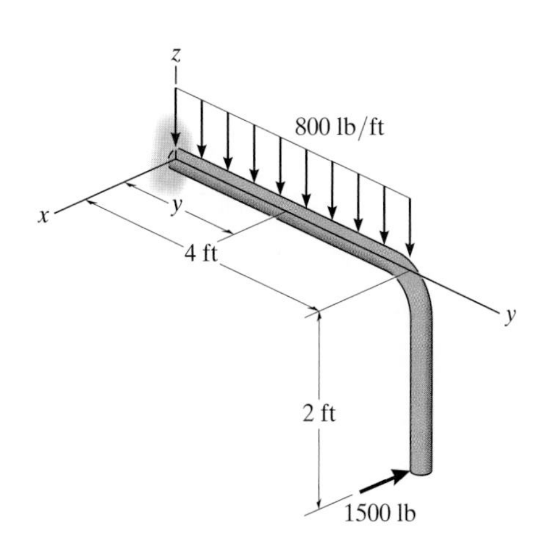

Prob. 7–63

7–65. The quarter circular rod lies in the horizontal plane and supports a vertical force **P** at its end. Determine the magnitudes of the components of the internal shear force, moment, and torque acting in the rod as a function of the angle θ.

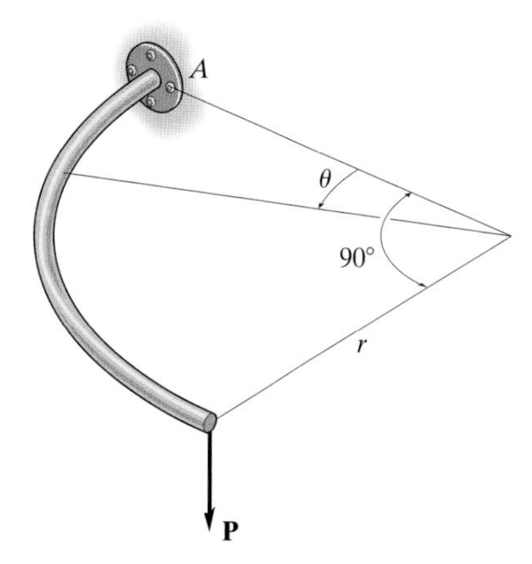

Prob. 7–65

*7.3 Relations between Distributed Load, Shear, and Moment

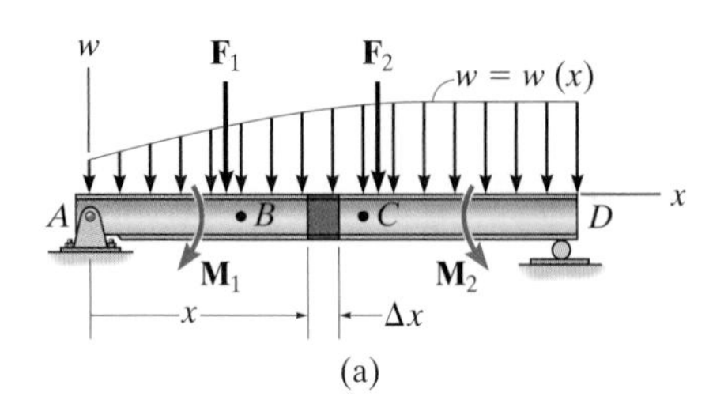

(a)

In cases where a beam is subjected to several concentrated forces, couple moments, and distributed loads, the method of constructing the shear and bending-moment diagrams discussed in Sec. 7.2 may become quite tedious. In this section a simpler method for constructing these diagrams is discussed—a method based on differential relations that exist between the load, shear, and bending moment.

Distributed Load. Consider the beam AD shown in Fig. 7–14*a*, which is subjected to an arbitrary load $w = w(x)$ and a series of concentrated forces and couple moments. In the following discussion, the *distributed load* will be considered *positive* when the *loading acts downward* as shown. A free-body diagram for a small segment of the beam having a length Δx is chosen at a point x along the beam which is *not* subjected to a concentrated force or couple moment, Fig. 7–14*b*. Hence any results obtained will not apply at points of concentrated loading. The internal shear force and bending moment shown on the free-body diagram are assumed to act in the *positive sense* according to the established sign convention. Note that both the shear force and moment acting on the right-hand face must be increased by a small, finite amount in order to keep the segment in equilibrium. The distributed loading has been replaced by a resultant force $\Delta F = w(x)\,\Delta x$ that acts at a fractional distance $k(\Delta x)$ from the right end, where $0 < k < 1$ [for example, if $w(x)$ is *uniform*, $k = \frac{1}{2}$]. Applying the equations of equilibrium, we have

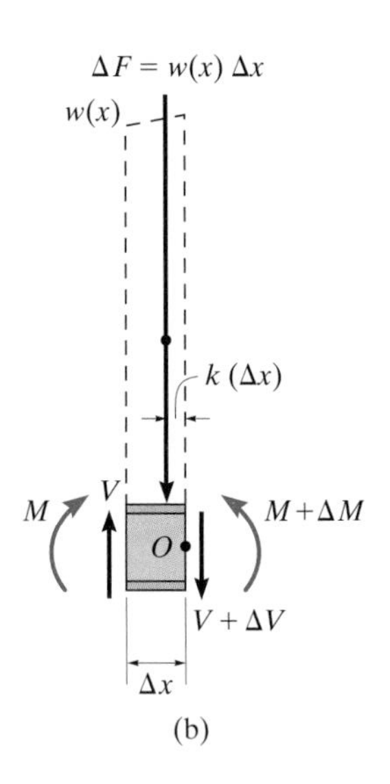

(b)

Fig. 7–14

$$+\uparrow \Sigma F_y = 0; \qquad V - w(x)\,\Delta x - (V + \Delta V) = 0$$
$$\Delta V = -w(x)\,\Delta x$$
$$\circlearrowleft + \Sigma M_O = 0; \qquad -V\,\Delta x - M + w(x)\,\Delta x[k(\Delta x)] + (M + \Delta M) = 0$$
$$\Delta M = V\,\Delta x - w(x)k(\Delta x)^2$$

Dividing by Δx and taking the limit as $\Delta x \rightarrow 0$, these two equations become

$$\frac{dV}{dx} = -w(x)$$

Slope of shear diagram = Negative of distributed load intensity (7–1)

$$\frac{dM}{dx} = V$$

Slope of moment diagram = Shear (7–2)

These two equations provide a convenient means for plotting the shear and moment diagrams for a beam. At a specific point in a beam, Eq. 7–1 states that the *slope of the shear diagram is equal to the negative of the intensity of the distributed load*, while Eq. 7–2 states that the *slope of the moment diagram is equal to the shear*. In particular, if the shear is equal to zero, $dM/dx = 0$, and therefore *a point of zero shear corresponds to a point of maximum (or possibly minimum) moment.*

Equations 7–1 and 7–2 may also be rewritten in the form $dV = -w(x)\,dx$ and $dM = V\,dx$. Noting that $w(x)\,dx$ and $V\,dx$ represent differential areas under the distributed-loading and shear diagrams, respectively, we can integrate these areas between two points B and C along the beam, Fig. 7–14*a*, and write

This concrete beam is used to support the roof. Its size and the placement of steel reinforcement within it can be determined once the shear and moment diagrams have been established.

$$\Delta V_{BC} = -\int w(x)\,dx \tag{7-3}$$

Change in shear = Negative of area under loading curve

and

$$\Delta M_{BC} = \int V\,dx \tag{7-4}$$

Change in moment = Area under shear diagram

Equation 7–3 states that the *change in shear between points B and C is equal to the negative of the area under the distributed-loading curve between these points*. Similarly, from Eq. 7–4, the *change in moment between B and C is equal to the area under the shear diagram within region BC*. Because two integrations are involved, first to determine the change in shear, Eq. 7–3, then to determine the change in moment, Eq. 7–4, we can state that if the loading curve $w = w(x)$ is a polynomial of degree n, then $V = V(x)$ will be a curve of degree $n + 1$, and $M = M(x)$ will be a curve of degree $n + 2$.

As stated previously, the above equations do not apply at points where a *concentrated* force or couple moment acts. These two special cases create *discontinuities* in the shear and moment diagrams, and as a result, each deserves separate treatment.

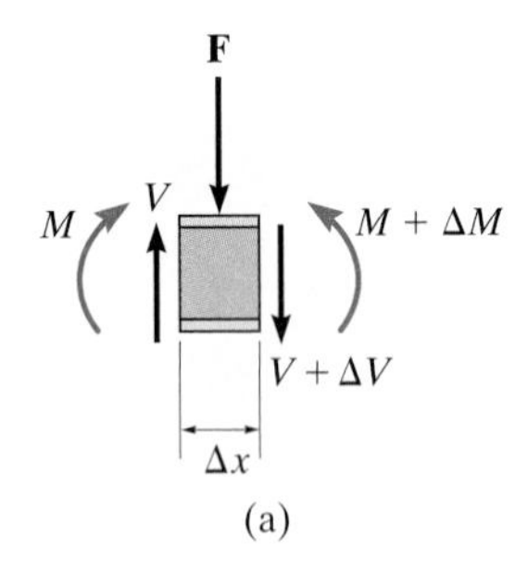

(a)

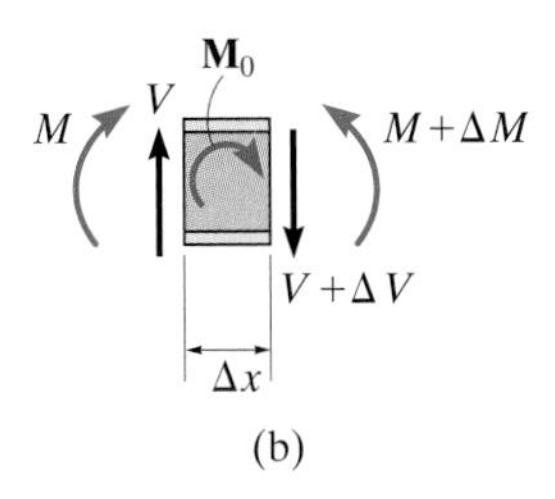

(b)

Fig. 7–15

Force. A free-body diagram of a small segment of the beam in Fig. 7–14*a*, taken from under one of the forces, is shown in Fig. 7–15*a*. Here it can be seen that force equilibrium requires

$$+\uparrow \Sigma F_y = 0; \qquad \Delta V = -F \qquad (7\text{–}5)$$

Thus, the *change in shear is negative*, so that on the shear diagram the shear will "jump" *downward when* **F** *acts downward* on the beam. Likewise, the jump in shear (ΔV) is upward when **F** acts upward.

Couple Moment. If we remove a segment of the beam in Fig. 7–14*a* that is located at the couple moment, the free-body diagram shown in Fig. 7–15*b* results. In this case letting $\Delta x \rightarrow 0$, moment equilibrium requires

$$\circlearrowright + \Sigma M = 0; \qquad \Delta M = M_0 \qquad (7\text{–}6)$$

Thus, the *change in moment is positive*, or the moment diagram will "jump" *upward if* $\mathbf{M}_0$ *is clockwise*. Likewise, the jump ΔM is downward when $\mathbf{M}_0$ is counterclockwise.

The examples which follow illustrate application of the above equations for the construction of the shear and moment diagrams. After working through these examples, it is recommended that Examples 7.7 and 7.8 be solved using this method.

Each outrigger such as AB supporting this crane acts as a beam which is fixed to the frame of the crane at one end and subjected to a force **F** on the footing at its other end. A proper design requires that the outrigger is able to resist its maximum internal shear and moment. The shear and moment diagrams indicate that the shear will be constant throughout its length and the maximum moment occurs at the support A.

Important Points

- The slope of the shear diagram is equal to the negative of the intensity of the distributed loading, where positive distributed loading is downward, i.e., $dV/dx = -w(x)$.
- If a concentrated force acts downward on the beam, the shear will jump downward by the amount of the force.
- The change in the shear ΔV between two points is equal to *the negative of the area* under the distributed-loading curve between the points.
- The slope of the moment diagram is equal to the shear (i.e., $dM/dx = V$).
- The change in the moment ΔM between two points is equal to the *area* under the shear diagram between the two points.
- If a *clockwise* couple moment acts on the beam, the shear will not be affected, however, the moment diagram will jump *upward* by the amount of the moment.
- Points of *zero shear* represent points of *maximum or minimum moment* since $dM/dx = 0$.

EXAMPLE 7.9

Draw the shear and moment diagrams for the beam shown in Fig. 7–16*a*.

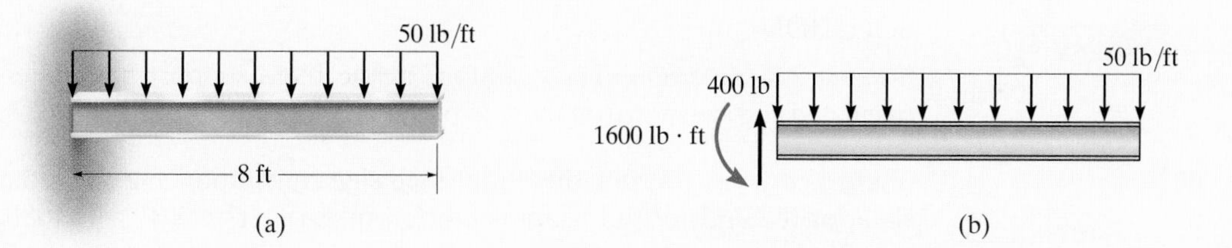

SOLUTION

Support Reactions. The reactions at the fixed support have been calculated and are shown on the free-body diagram of the beam, Fig. 7–16*b*.

Shear Diagram. The shear at the end points is plotted first, Fig. 7–16*c*. From the sign convention, Fig. 7–11, $V = +400$ at $x = 0$ and $V = 0$ at $x = 8$. Since $dV/dx = -w = -50$, a straight, *negative* sloping line connects the end points.

Moment Diagram. From our sign convention, Fig. 7–11, the moments at the beam's end points, $M = -1600$ at $x = 0$ and $M = 0$ at $x = 8$, are plotted first, Fig. 7–16*d*. Successive values of shear taken from the shear diagram, Fig. 7–16*c*, indicate that the *slope* $dM/dx = V$ of the moment diagram, Fig. 7–16*d*, is always positive yet *linearly decreasing* from $dM/dx = 400$ at $x = 0$ to $dM/dx = 0$ at $x = 8$.

NOTE: Due to the integrations, w a constant yields V a sloping line (first-degree curve) and M a parabola (second-degree curve).

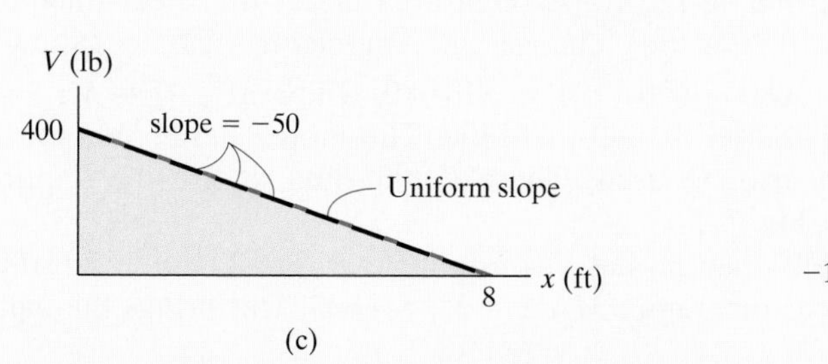

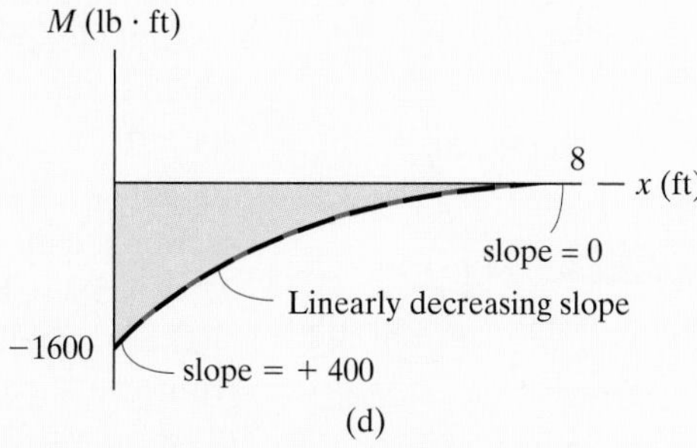

Fig. 7–16

EXAMPLE 7.10

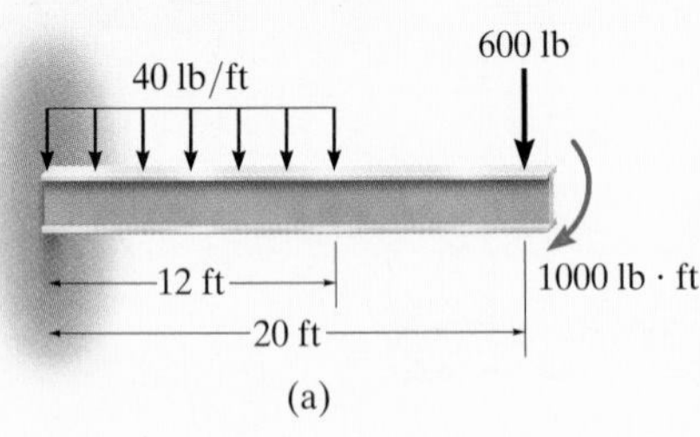

(a)

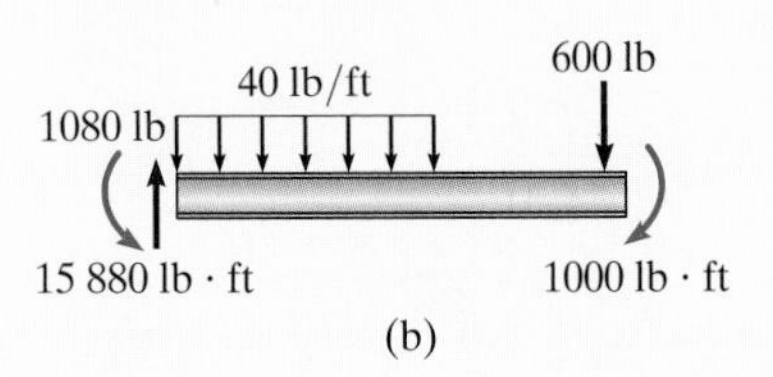

(b)

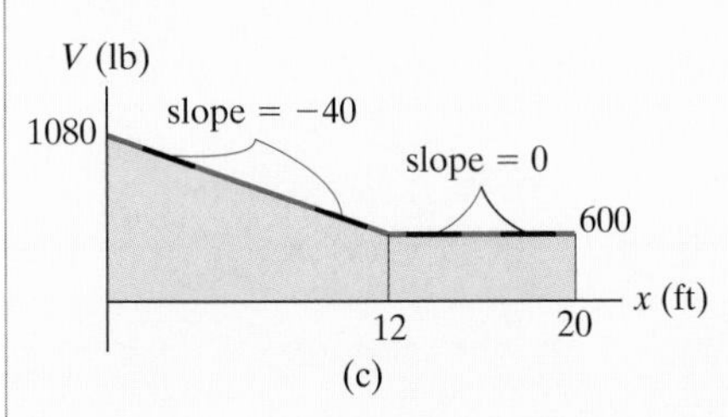

(c)

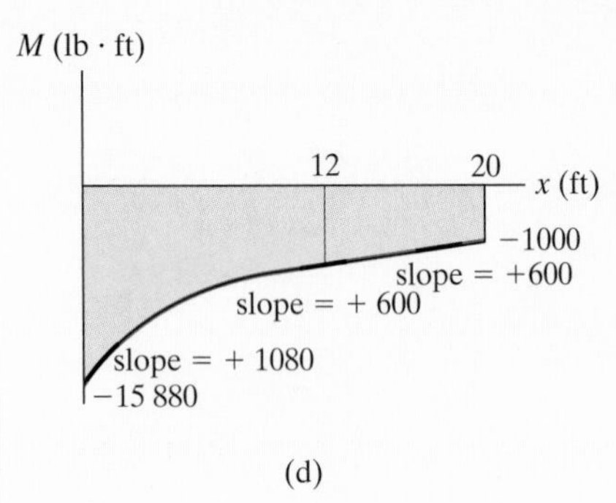

(d)

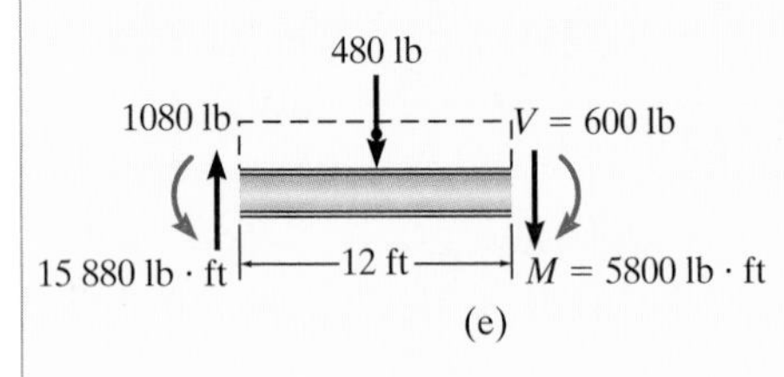

(e)

Fig. 7–17

Draw the shear and moment diagrams for the cantilevered beam shown in Fig. 7–17*a*.

SOLUTION

Support Reactions. The reactions at the fixed support have been calculated and are shown on the free-body diagram of the beam, Fig. 7–17*b*.

Shear Diagram. Using the established sign convention, Fig. 7–11, the shear at the ends of the beam is plotted first; i.e., $x = 0$, $V = +1080$; $x = 20$, $V = +600$, Fig. 7–17*c*.

Since the uniform distributed load is downward and *constant*, the slope of the shear diagram is $dV/dx = -w = -40$ for $0 \le x < 12$ as indicated.

The magnitude of shear at $x = 12$ is $V = +600$. This can be determined by first finding the area under the load diagram between $x = 0$ and $x = 12$. This represents the change in shear. That is, $\Delta V = -\int w(x)\,dx = -40(12) = -480$. Thus $V|_{x=12} = V|_{x=0} + (-480)$ $= 1080 - 480 = 600$. Also, we can obtain this value by using the method of sections, Fig. 7–17*e*, where for equilibrium $V = +600$.

Since the load between $12 < x \le 20$ is $w = 0$, the slope $dV/dx = 0$ as indicated. This brings the shear to the required value of $V = +600$ at $x = 20$.

Moment Diagram. Again, using the established sign convention, the moments at the ends of the beam are plotted first; i.e., $x = 0$, $M = -15\,880$; $x = 20$, $M = -1000$, Fig. 7–17*d*.

Each value of shear gives the slope of the moment diagram since $dM/dx = V$. As indicated, at $x = 0$, $dM/dx = +1080$; and at $x = 12$, $dM/dx = +600$. For $0 \le x < 12$, specific values of the shear diagram are positive but linearly decreasing. Hence, the moment diagram is parabolic with a linearly decreasing positive slope.

The magnitude of moment at $x = 12$ is -5800. This can be found by first determining the trapezoidal area under the shear diagram, which represents the change in moment, $\Delta M = \int V\,dx =$ $600(12) + \frac{1}{2}(1080 - 600)(12) = +10\,080$. Thus, $M|_{x=12} = M|_{x=0} +$ $10\,080 = -15\,880 + 10\,080 = -5800$. The more "basic" method of sections can also be used, where equilibrium at $x = 12$ requires $M = -5800$, Fig. 7–17*e*.

The moment diagram has a constant slope for $12 < x \le 20$ since, from the shear diagram, $dM/dx = V = +600$. This brings the value of $M = -1000$ at $x = 20$, as required.

EXAMPLE 7.11

Draw the shear and moment diagrams for the shaft in Fig. 7–18*a*. The support at *A* is a thrust bearing and the support at *B* is a journal bearing.

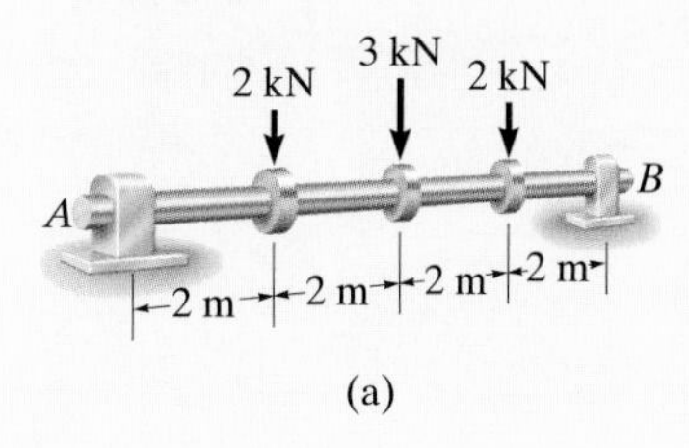

(a)

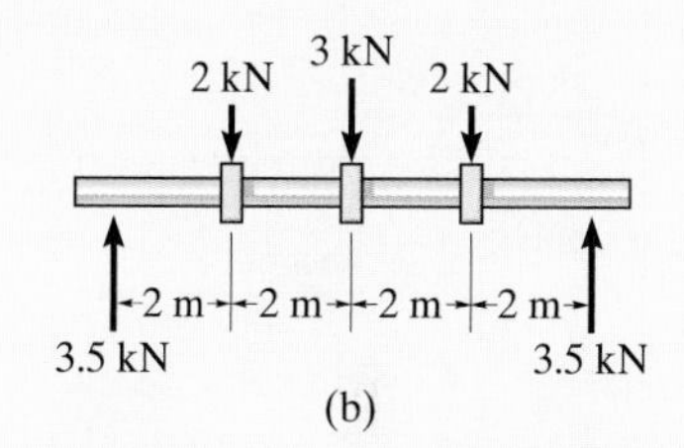

(b)

SOLUTION

Support Reactions. The reactions at the supports are shown on the free-body diagram in Fig. 7–18*b*.

Shear Diagram. The end points $x = 0$, $V = +3.5$ and $x = 8$, $V = -3.5$ are plotted first, as shown in Fig. 7–18*c*.

Since there is no distributed load on the shaft, the slope of the shear diagram throughout the shaft's length is zero; i.e., $dV/dx = -w = 0$. There is a discontinuity or "jump" of the shear diagram, however, at each concentrated force. From Eq. 7–5, $\Delta V = -F$, the change in shear is negative when the force acts downward and positive when the force acts upward. Stated another way, the "jump" follows the force, i.e., a downward force causes a downward jump, and vice versa. Thus, the 2-kN force at $x = 2$ m changes the shear from 3.5 kN to 1.5 kN; the 3-kN force at $x = 4$ m changes the shear from 1.5 kN to -1.5 kN, etc. We can *also* obtain numerical values for the shear at a specified point in the shaft by using the method of sections, as for example, $x = 2^+$ m, $V = 1.5$ kN in Fig. 7–18*e*.

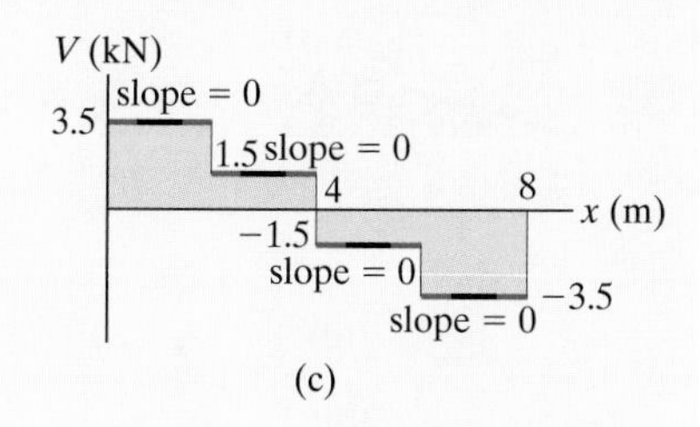

(c)

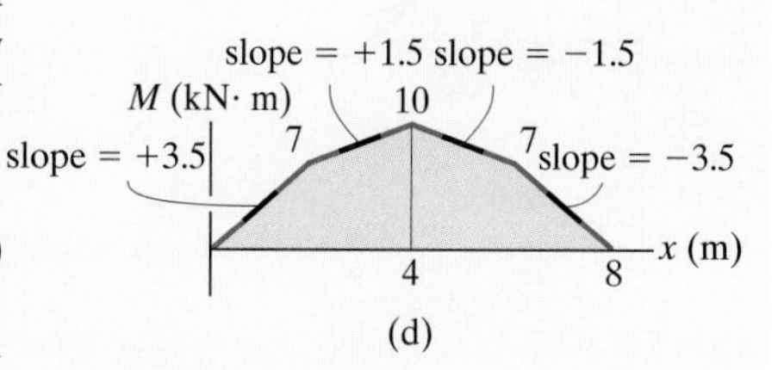

(d)

Moment Diagram. The end points $x = 0$, $M = 0$ and $x = 8$, $M = 0$ are plotted first, as shown in Fig. 7–18*d*.

Since the shear is constant in each region of the shaft, the moment diagram has a corresponding constant positive or negative slope as indicated on the diagram. Numerical values for the change in moment at any point can be computed from the *area* under the shear diagram. For example, at $x = 2$ m, $\Delta M = \int V\,dx = 3.5(2) = 7$. Thus, $M|_{x=2} = M|_{x=0} + 7 = 0 + 7 = 7$. Also, by the method of sections, we can determine the moment at a specified point, as for example, $x = 2^+$ m, $M = 7$ kN·m, Fig. 7–18*e*.

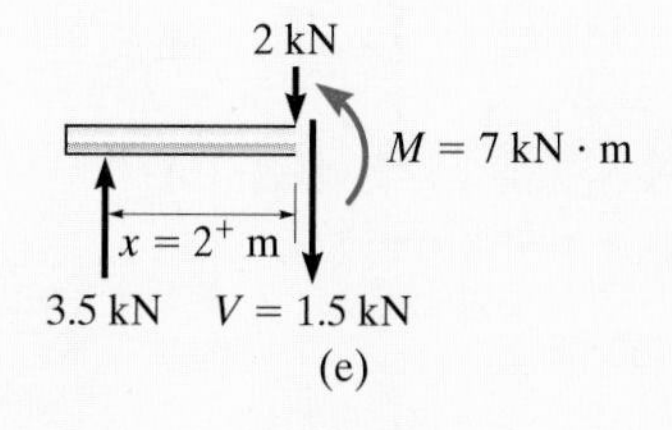

(e)

Fig. 7–18

EXAMPLE 7.12

Sketch the shear and moment diagrams for the beam shown in Fig. 7–19*a*.

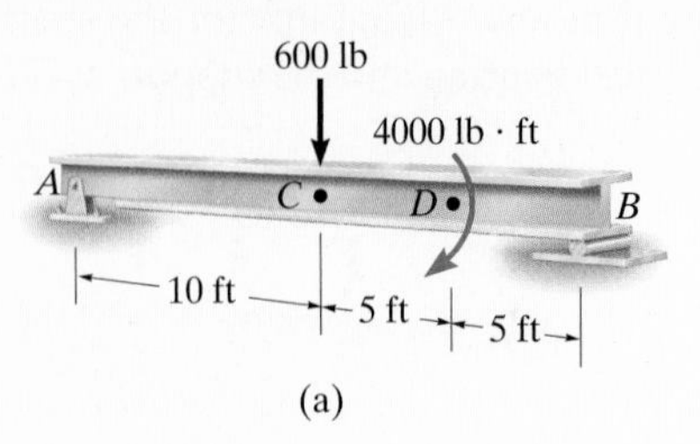

(a)

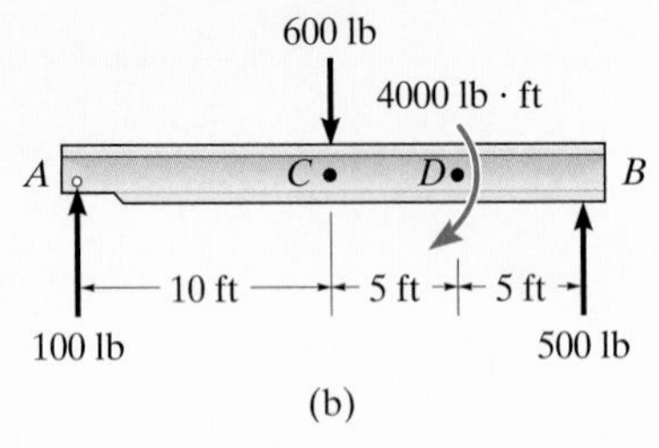

(b)

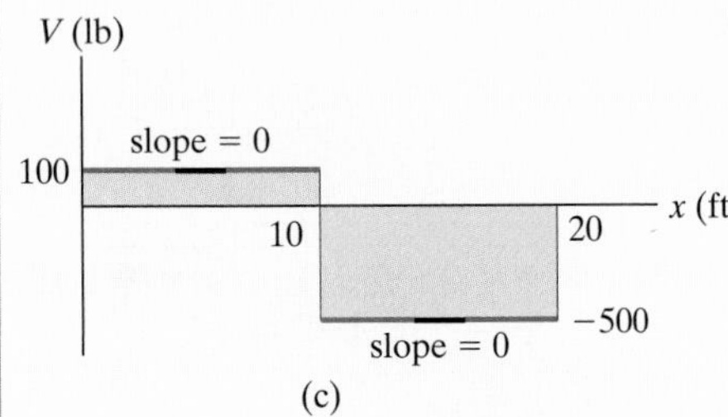

(c)

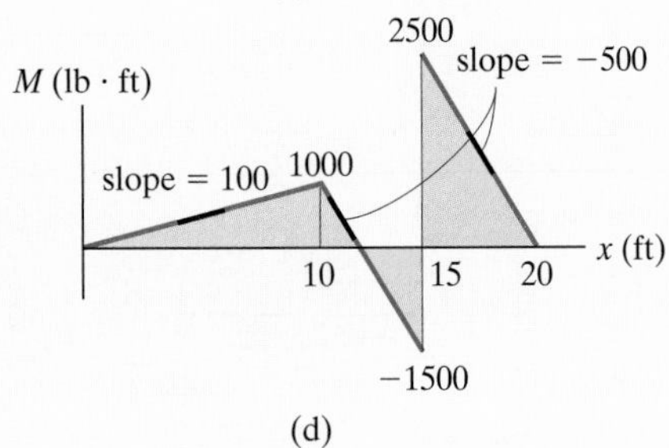

(d)

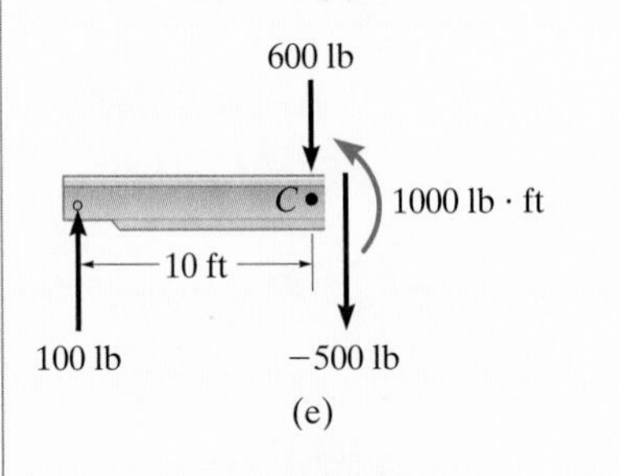

(e)

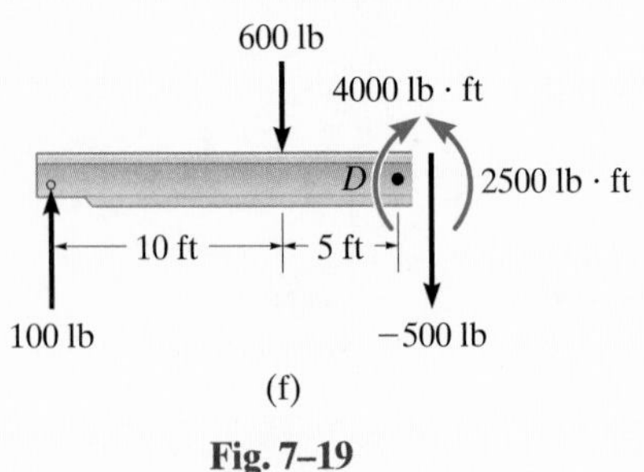

(f)

Fig. 7–19

SOLUTION

Support Reactions. The reactions are calculated and indicated on the free-body diagram, Fig. 7–19*b*.

Shear Diagram. As in Example 7.11, the shear diagram can be constructed by "following the load" on the free-body diagram. In this regard, beginning at A, the reaction is up so $V_A = +100$ lb, Fig. 7–19*c*. No load acts between A and C, so the shear remains constant; i.e., $dV/dx = -w(x) = 0$. At C the 600-lb force acts downward, so the shear jumps down 600 lb, from 100 lb to −500 lb. Again the shear is constant (no load) and ends at −500 lb, point B. Notice that no jump or discontinuity in shear occurs at D, the point where the 4000-lb · ft couple moment is applied, Fig. 7–19*b*. This is because, for force equilibrium, $\Delta V = 0$ in Fig. 7–15*b*.

Moment Diagram. The moment at each end of the beam is zero. These two points are plotted first, Fig. 7–19*d*. The slope of the moment diagram from A to C is constant since $dM/dx = V = +100$. The value of the moment at C can be determined by the method of sections, Fig. 7–19*e*, where $M_C = +1000$ lb · ft; or by first computing the rectangular area under the shear diagram between A and C to obtain the change in moment $\Delta M_{AC} = (100\text{ lb})(10\text{ ft}) = 1000$ lb · ft. Since $M_A = 0$, then $M_C = 0 + 1000$ lb · ft $= 1000$ lb · ft. From C to D the slope of the moment diagram is $dM/dx = V = -500$, Fig. 7–19*c*. The area under the shear diagram between points C and D is $\Delta M_{CD} = (-500\text{ lb})(5\text{ ft}) = -2500$ lb · ft, so that $M_D = M_C + \Delta M_{CD}$ $\Delta M_{CD} = 1000 - 2500 = -1500$ lb · ft. A jump in the moment diagram occurs at point D, which is caused by the concentrated couple moment of 4000 lb · ft. From Eq. 7–6, the jump is *positive* since the couple moment is *clockwise*. Thus, at $x = 15^+$ ft, the moment is $M_D = -1500 + 4000 = 2500$ lb · ft. This value can *also* be determined by the method of sections, Fig. 7–19*f*. From point D the slope of $dM/dx = -500$ is maintained until the diagram closes to zero at B, Fig. 7–19*d*.

PROBLEMS

7–66. Draw the shear and moment diagrams for the beam.

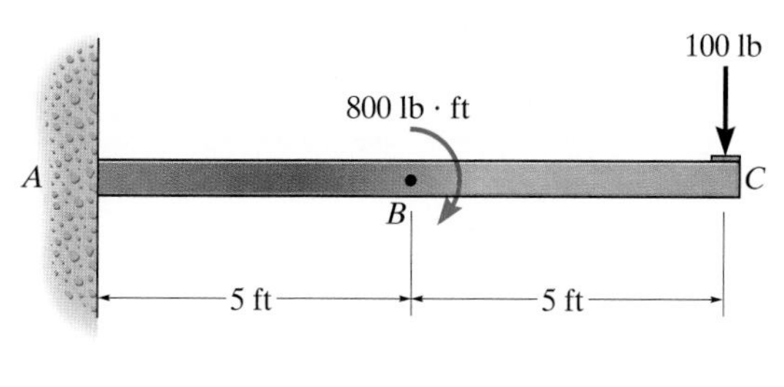

Prob. 7–66

7–67. Draw the shear and moment diagrams for the beam.

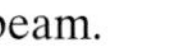

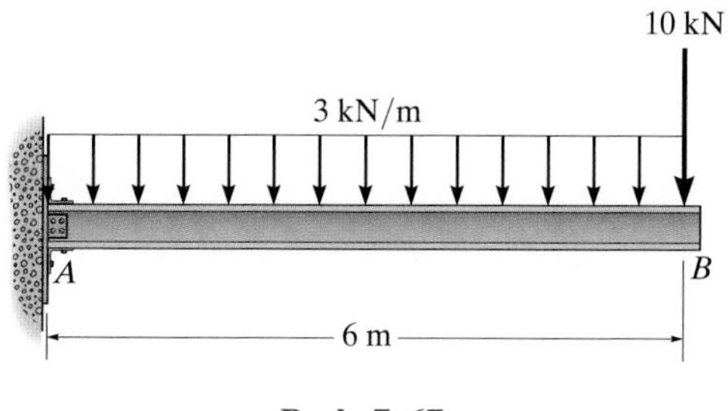

Prob. 7–67

***7–68.** Draw the shear and moment diagrams for the beam.

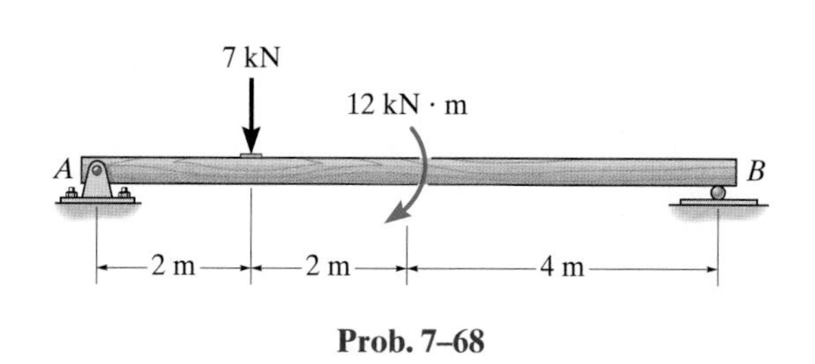

Prob. 7–68

7–69. Draw the shear and moment diagrams for the beam.

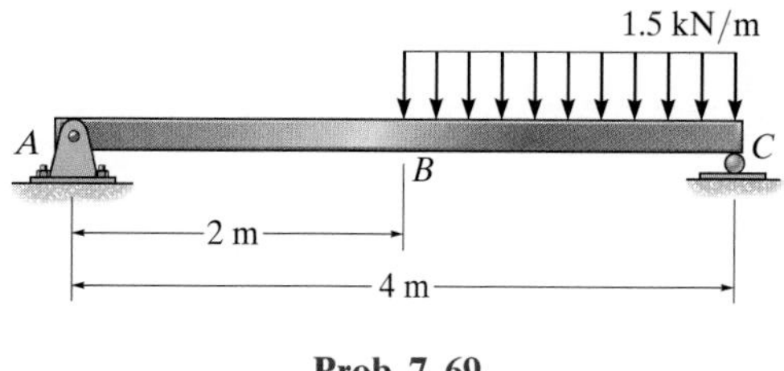

Prob. 7–69

7–70. Draw the shear and moment diagrams for the beam.

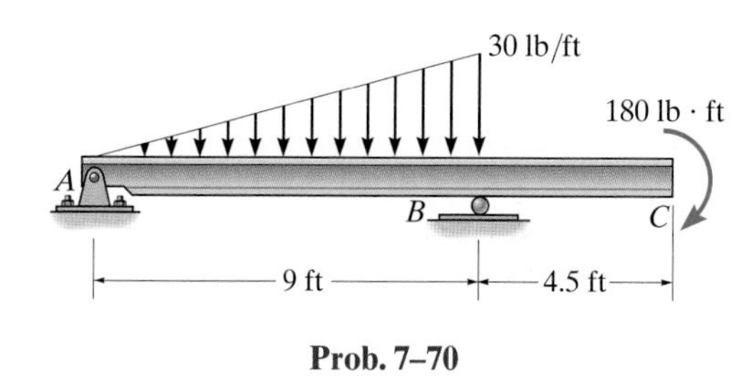

Prob. 7–70

7–71. Draw the shear and moment diagrams for the beam.

30 lb/ft

120 lb/ft

A

B

12 ft

Prob. 7–71

***7–72.** Draw the shear and moment diagrams for the shaft. The support at A is a journal bearing and at B it is a thrust bearing.

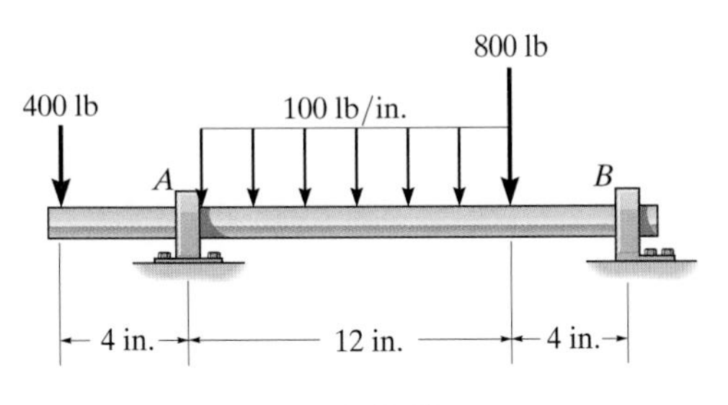

Prob. 7–72

7–73. Draw the shear and moment diagrams for the beam.

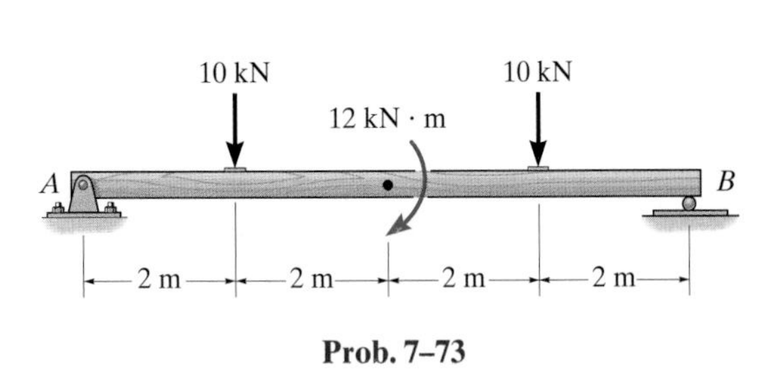

Prob. 7–73

7–74. Draw the shear and moment diagrams for the shaft. The support at A is a journal bearing and at B it is a thrust bearing.

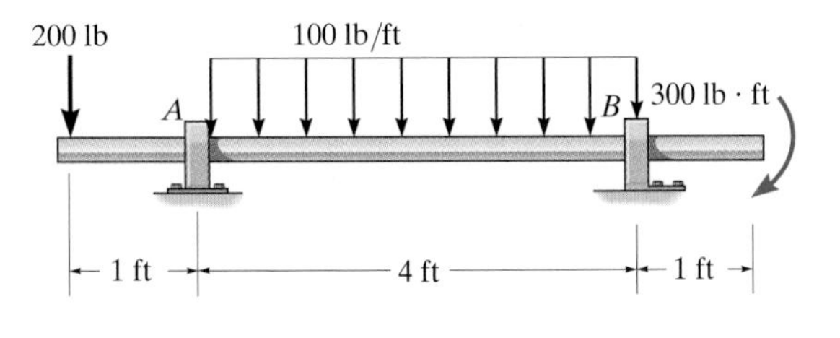

Prob. 7–74

7–75. Draw the shear and moment diagrams for the beam.

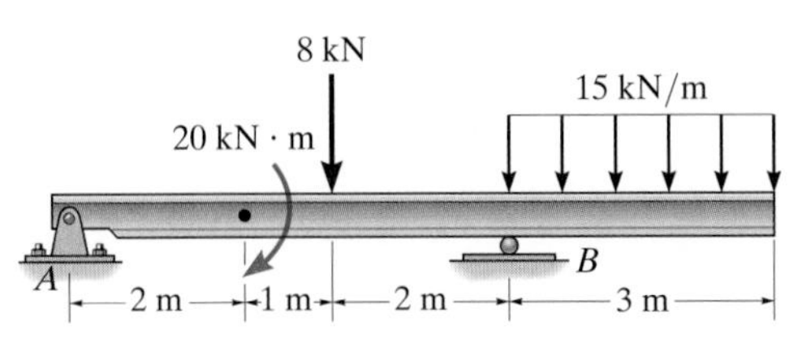

Prob. 7–75

***7–76.** Draw the shear and moment diagrams for the shaft. The support at A is a thrust bearing and at B it is a journal bearing.

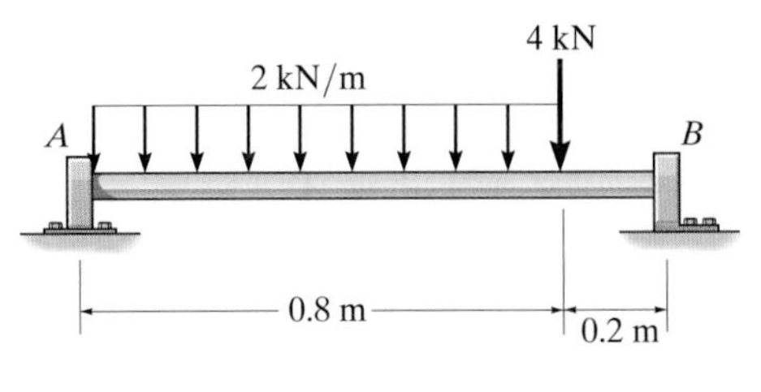

Prob. 7–76

7–77. Draw the shear and moment diagrams for the beam.

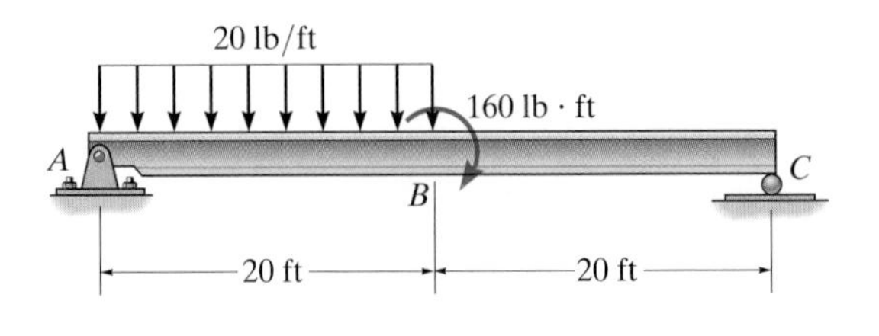

Prob. 7–77

7–78. The beam will fail when the maximum moment is $M_{max} = 30 \text{ kip} \cdot \text{ft}$ or the maximum shear is $V_{max} = 8 \text{ kip}$. Determine the largest distributed load w the beam will support.

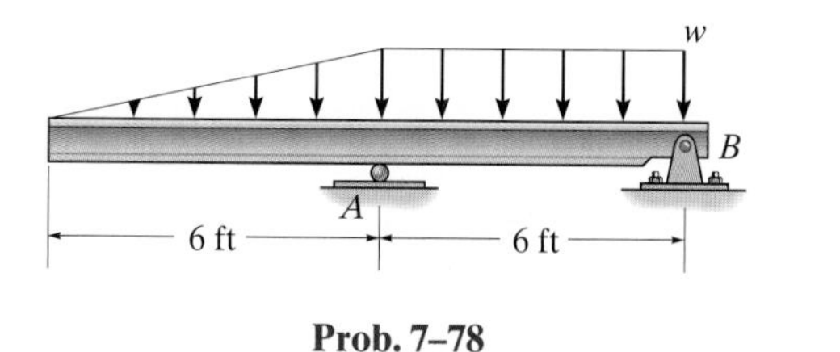

Prob. 7–78

7–79. The beam consists of two segments pin connected at B. Draw the shear and moment diagrams for the beam.

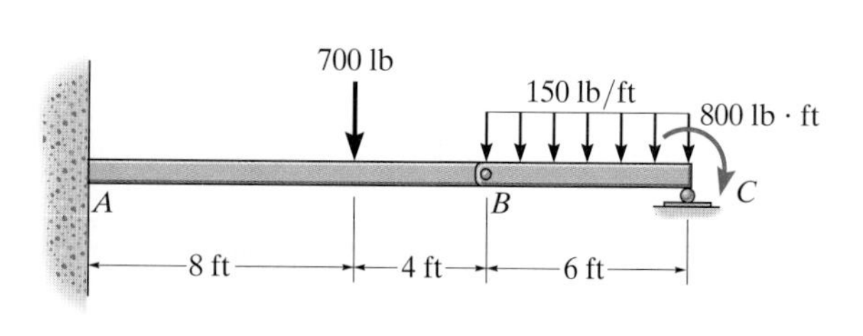

Prob. 7–79

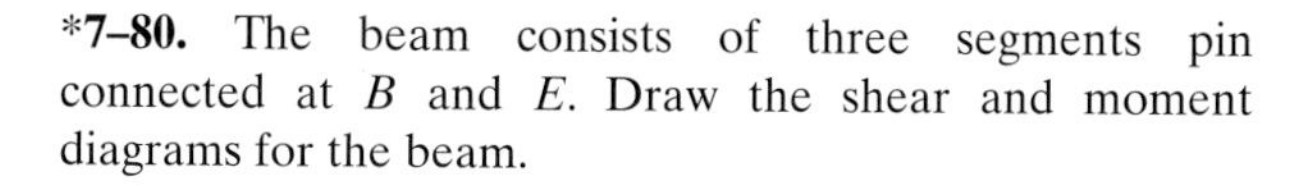

***7–80.** The beam consists of three segments pin connected at B and E. Draw the shear and moment diagrams for the beam.

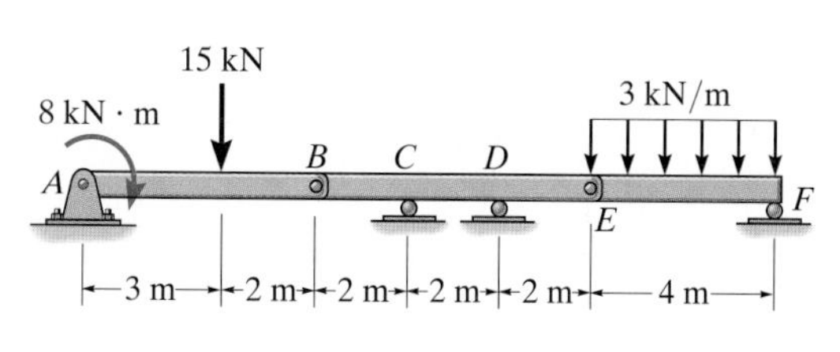

Prob. 7–80

7–81. Draw the shear and moment diagrams for the beam.

w_0

A

B

L

L

Prob. 7–81

7–82. Draw the shear and moment diagrams for the beam.

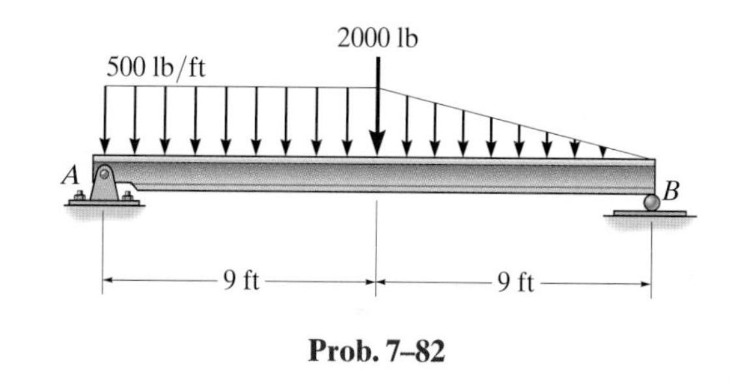

Prob. 7–82

7–83. Draw the shear and moment diagrams for the beam.

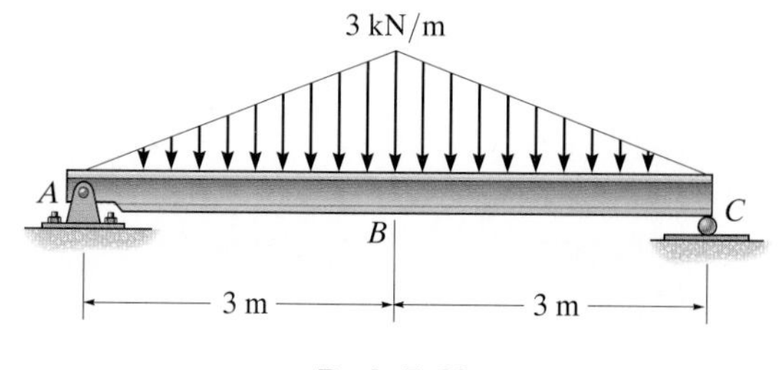

Prob. 7–83

*7–84. Draw the shear and moment diagrams for the beam.

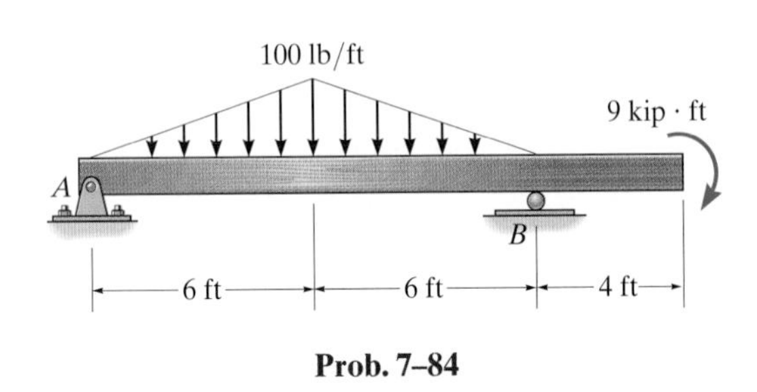

Prob. 7–84

7–85. Draw the shear and moment diagrams for the beam.

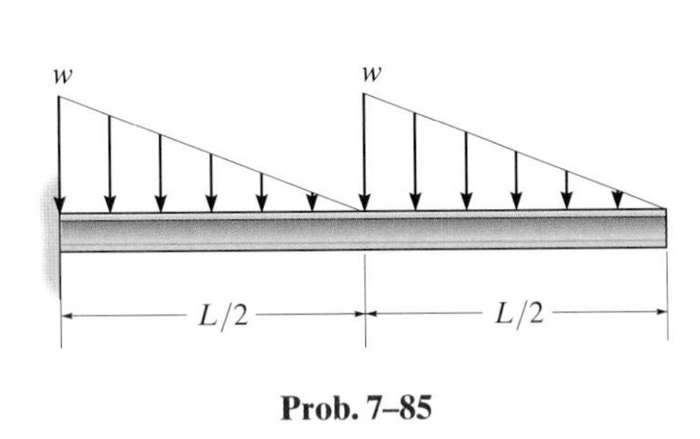

Prob. 7–85

7–86. Draw the shear and moment diagrams for the beam.

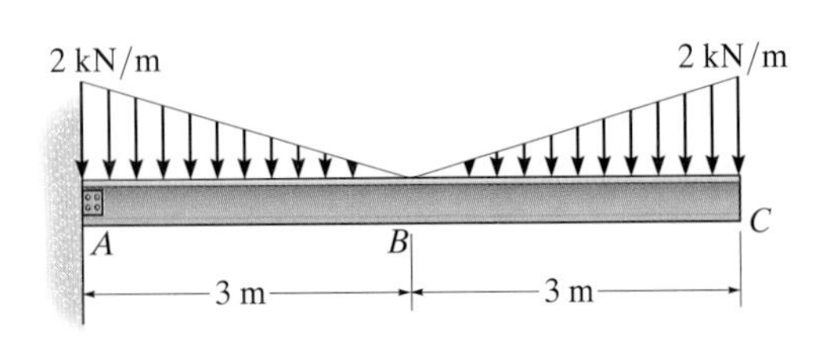

Prob. 7–86

7–87. Draw the shear and moment diagrams for the beam.

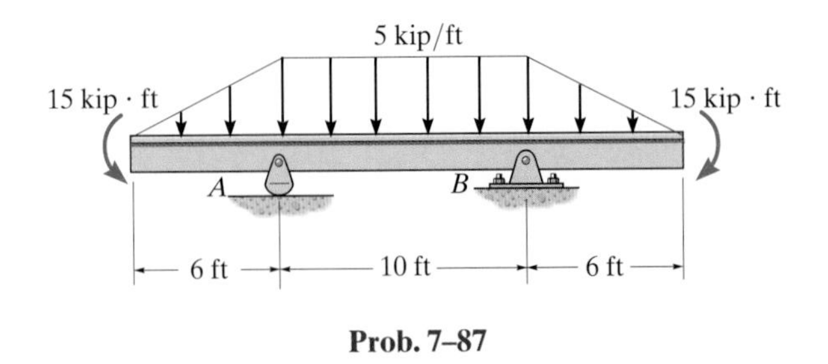

Prob. 7–87

*7–88. Draw the shear and moment diagrams for the beam.

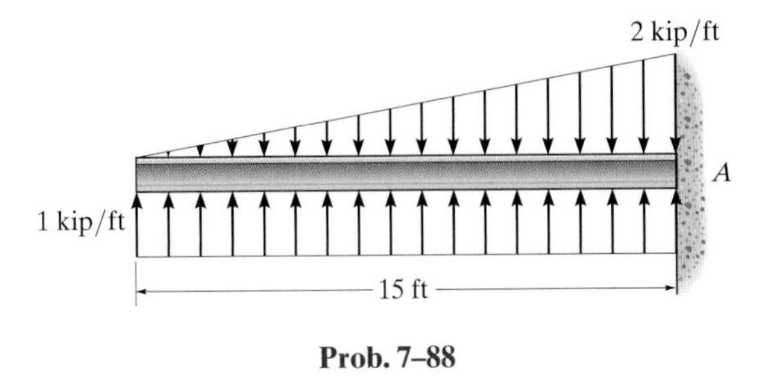

Prob. 7–88

*7.4 Cables

Flexible cables and chains combine strength with lightness and often are used in engineering structures for support and to transmit loads from one member to another. When used to support suspension bridges and trolley wheels, cables form the main load-carrying element of the structure. In the force analysis of such systems, the weight of the cable itself may be neglected because it is often small compared to the load it carries. On the other hand, when cables are used as transmission lines and guys for radio antennas and derricks, the cable weight may become important and must be included in the structural analysis. Three cases are considered in the analysis that follows: (1) a cable subjected to concentrated loads; (2) a cable subjected to a distributed load; and (3) a cable subjected to its own weight. In each case, provided the loading is coplanar with the cable, the requirements for equilibrium are formulated in an identical manner.

Each of the cable segments remains approximately straight as they support the weight of these traffic lights.

When deriving the necessary relations between the force in the cable and its slope, we will make the assumption that the cable is *perfectly flexible* and *inextensible*. Due to its flexibility, the cable offers no resistance to bending, and therefore, the tensile force acting in the cable is always tangent to the cable at points along its length. Being inextensible, the cable has a constant length both before and after the load is applied. As a result, once the load is applied, the geometry of the cable remains fixed, and the cable or a segment of it can be treated as a rigid body.

Cable Subjected to Concentrated Loads. When a cable of negligible weight supports several concentrated loads, the cable takes the form of several straight-line segments, each of which is subjected to a constant tensile force. Consider, for example, the cable shown in Fig. 7–20, where the distances h, L_1, L_2, and L_3 and the loads $\mathbf{P}_1$ and $\mathbf{P}_2$ are known. The problem here is to determine the *nine unknowns* consisting of the tension in each of the *three* segments, the *four* components of reaction at A and B, and the sags y_C and y_D at the *two* points C and D. For the solution we can write *two* equations of force equilibrium at each of points A, B, C, and D. This results in a total of *eight equations*.* To complete the solution, we need to know something about the geometry of the cable in order to obtain the necessary ninth equation. For example, if the cable's total *length* L is specified, then the Pythagorean theorem can be used to relate each of the three segmental lengths, written in terms of h, y_C, y_D, L_1, L_2, and L_3, to the total length L. Unfortunately, this type of problem cannot be solved easily by hand. Another possibility, however, is to specify one of the sags, either y_C or y_D, instead of the cable length. By doing this, the equilibrium equations are then sufficient for obtaining the unknown forces and the remaining sag. Once the sag at each point of loading is obtained, the length of the cable can be determined by trigonometry. The following example illustrates a procedure for performing the equilibrium analysis for a problem of this type.

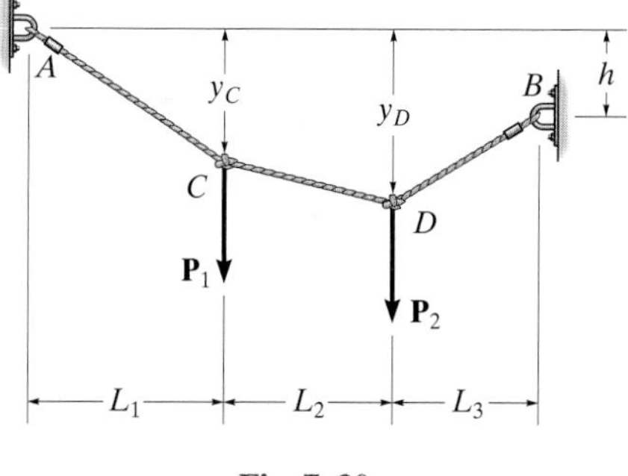

Fig. 7–20

*As will be shown in the following example, the eight equilibrium equations *also* can be written for the entire cable, or any part thereof. But *no more* than *eight* equations are available.

EXAMPLE 7.13

Determine the tension in each segment of the cable shown in Fig. 7–21*a*.

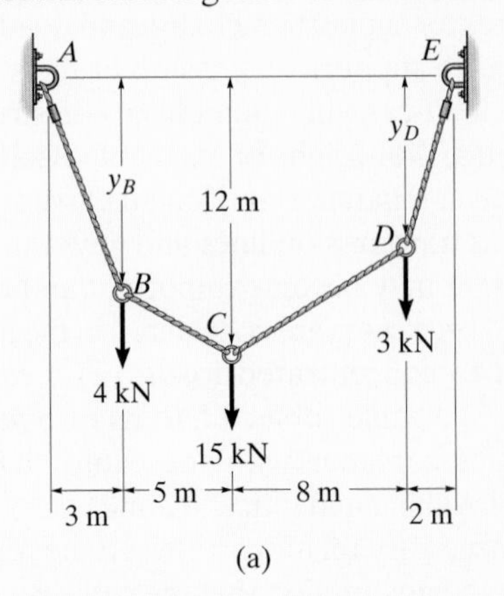

(a)

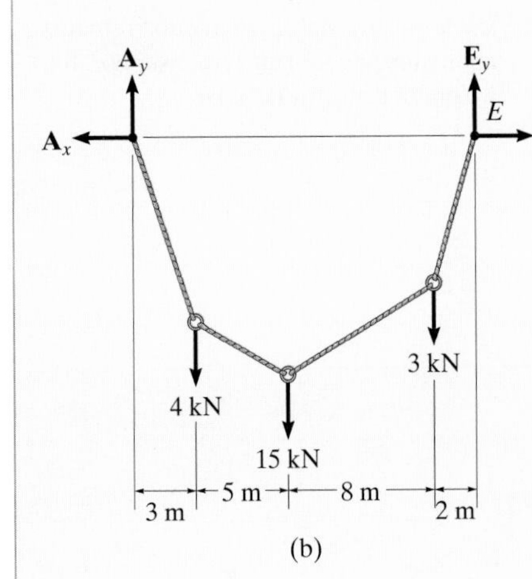

(b)

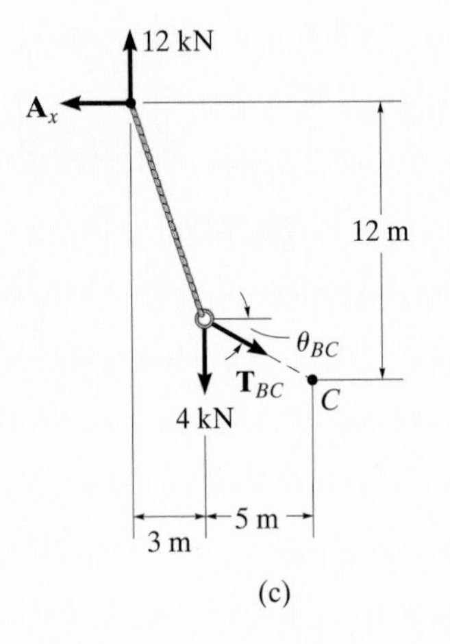

(c)

Fig. 7–21

SOLUTION

By inspection, there are four unknown external reactions (A_x, A_y, E_x, and E_y) and four unknown cable tensions, one in each cable segment. These eight unknowns along with the two unknown sags y_B and y_D can be determined from *ten* available equilibrium equations. One method is to apply these equations as force equilibrium ($\Sigma F_x = 0$, $\Sigma F_y = 0$) to each of the five points A through E. Here, however, we will take a more direct approach.

Consider the free-body diagram for the entire cable, Fig. 7–21*b*. Thus,

$$\overset{+}{\rightarrow}\Sigma F_x = 0; \qquad -A_x + E_x = 0$$

$$\circlearrowleft + \Sigma M_E = 0;$$

$$-A_y(18\text{ m}) + 4\text{ kN}\,(15\text{ m}) + 15\text{ kN}\,(10\text{ m}) + 3\text{ kN}\,(2\text{ m}) = 0$$

$$A_y = 12\text{ kN}$$

$$+\uparrow\Sigma F_y = 0; \qquad 12\text{ kN} - 4\text{ kN} - 15\text{ kN} - 3\text{ kN} + E_y = 0$$

$$E_y = 10\text{ kN}$$

Since the sag $y_C = 12$ m is known, we will now consider the leftmost section, which cuts cable BC, Fig. 7–21*c*.

$$\circlearrowleft + \Sigma M_C = 0; \qquad A_x(12\text{ m}) - 12\text{ kN}\,(8\text{ m}) + 4\text{ kN}\,(5\text{ m}) = 0$$

$$A_x = E_x = 6.33\text{ kN}$$

$$\overset{+}{\rightarrow}\Sigma F_x = 0; \qquad T_{BC}\cos\theta_{BC} - 6.33\text{ kN} = 0$$

$$+\uparrow\Sigma F_y = 0; \qquad 12\text{ kN} - 4\text{ kN} - T_{BC}\sin\theta_{BC} = 0$$

Thus,

$$\theta_{BC} = 51.6^\circ$$

$$T_{BC} = 10.2\text{ kN} \qquad \textit{Ans.}$$

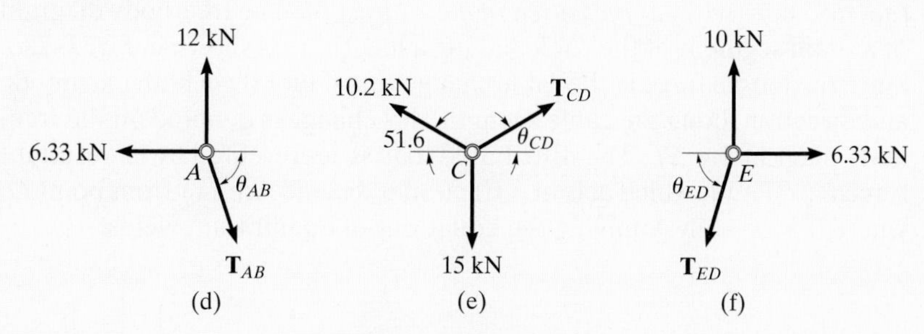

Fig. 7–21 (cont.)

Proceeding now to analyze the equilibrium of points A, C, and E in sequence, we have

Point A (Fig. 7–21*d*).

$$\xrightarrow{+} \Sigma F_x = 0; \qquad T_{AB} \cos \theta_{AB} - 6.33 \text{ kN} = 0$$

$$+\uparrow \Sigma F_y = 0; \qquad -T_{AB} \sin \theta_{AB} + 12 \text{ kN} = 0$$

$$\theta_{AB} = 62.2^\circ$$

$$T_{AB} = 13.6 \text{ kN} \qquad \textit{Ans.}$$

Point C (Fig. 7–21*e*).

$$\xrightarrow{+} \Sigma F_x = 0; \qquad T_{CD} \cos \theta_{CD} - 10.2 \cos 51.6^\circ \text{ kN} = 0$$

$$+\uparrow \Sigma F_y = 0; \qquad T_{CD} \sin \theta_{CD} + 10.2 \sin 51.6^\circ \text{ kN} - 15 \text{ kN} = 0$$

$$\theta_{CD} = 47.9^\circ$$

$$T_{CD} = 9.44 \text{ kN} \qquad \textit{Ans.}$$

Point E (Fig. 7–21*f*).

$$\xrightarrow{+} \Sigma F_x = 0; \qquad 6.33 \text{ kN} - T_{ED} \cos \theta_{ED} = 0$$

$$+\uparrow \Sigma F_y = 0; \qquad 10 \text{ kN} - T_{ED} \sin \theta_{ED} = 0$$

$$\theta_{ED} = 57.7^\circ$$

$$T_{ED} = 11.8 \text{ kN} \qquad \textit{Ans.}$$

NOTE: By comparison, the maximum cable tension is in segment AB since this segment has the greatest slope (θ) and it is required that for any cable segment the horizontal component $T \cos \theta = A_x = E_x$ (a constant). Also, since the slope angles that the cable segments make with the horizontal have now been determined, it is possible to determine the sags y_B and y_D, Fig. 7–21*a*, using trigonometry.

Cable Subjected to a Distributed Load. Consider the weightless cable shown in Fig. 7–22*a*, which is subjected to a loading function $w = w(x)$ *as measured in the x direction*. The free-body diagram of a small segment of the cable having a length Δs is shown in Fig. 7–22*b*. Since the tensile force in the cable changes continuously in both magnitude and direction along the cable's length, this change is denoted on the free-body diagram by ΔT. The distributed load is represented by its resultant force $w(x)(\Delta x)$, which acts at a fractional distance $k(\Delta x)$ from point O, where $0 < k < 1$. Applying the equations of equilibrium yields

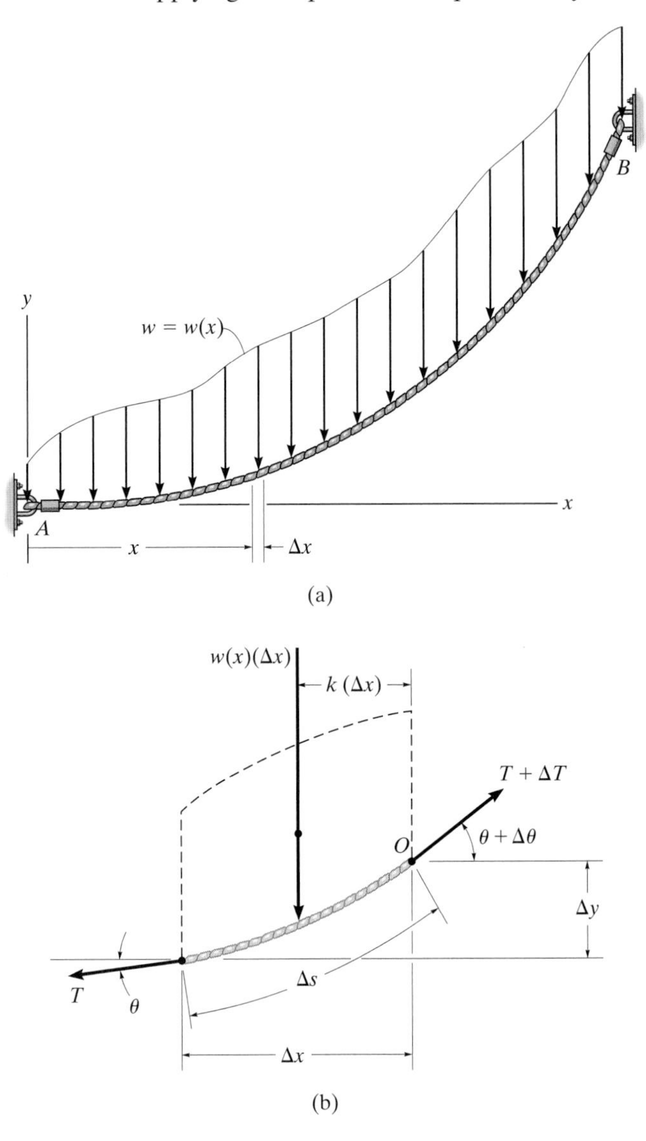

Fig. 7–22

$\xrightarrow{+} \Sigma F_x = 0;\qquad -T\cos\theta + (T + \Delta T)\cos(\theta + \Delta\theta) = 0$

$+\uparrow \Sigma F_y = 0;\qquad -T\sin\theta - w(x)(\Delta x) + (T + \Delta T)\sin(\theta + \Delta\theta) = 0$

$\circlearrowleft + \Sigma M_O = 0;\qquad w(x)(\Delta x)k(\Delta x) - T\cos\theta\,\Delta y + T\sin\theta\,\Delta x = 0$

Dividing each of these equations by Δx and taking the limit as $\Delta x \to 0$, and hence $\Delta y \to 0$, $\Delta\theta \to 0$, and $\Delta T \to 0$, we obtain

$$\frac{d(T\cos\theta)}{dx} = 0 \qquad (7\text{–}7)$$

$$\frac{d(T\sin\theta)}{dx} - w(x) = 0 \qquad (7\text{–}8)$$

$$\frac{dy}{dx} = \tan\theta \qquad (7\text{–}9)$$

The cable and suspenders are used to support the uniform load of a gas pipe which crosses the river.

Integrating Eq. 7–7, we have

$$T\cos\theta = \text{constant} = F_H \qquad (7\text{–}10)$$

Here F_H represents the horizontal component of tensile force at *any point* along the cable.

Integrating Eq. 7–8 gives

$$T\sin\theta = \int w(x)\,dx \qquad (7\text{–}11)$$

Dividing Eq. 7–11 by Eq. 7–10 eliminates T. Then, using Eq. 7–9, we can obtain the slope

$$\tan\theta = \frac{dy}{dx} = \frac{1}{F_H}\int w(x)\,dx$$

Performing a second integration yields

$$\boxed{y = \frac{1}{F_H}\int\left(\int w(x)\,dx\right)dx} \qquad (7\text{–}12)$$

This equation is used to determine the curve for the cable, $y = f(x)$. The horizontal force component F_H and the two constants, say C_1 and C_2, resulting from the integration are determined by applying the boundary conditions for the cable.

EXAMPLE 7.14

The cable of a suspension bridge supports half of the uniform road surface between the two columns at A and B, as shown in Fig. 7–23a. If this distributed loading is w_0, determine the maximum force developed in the cable and the cable's required length. The span length L and sag h are known.

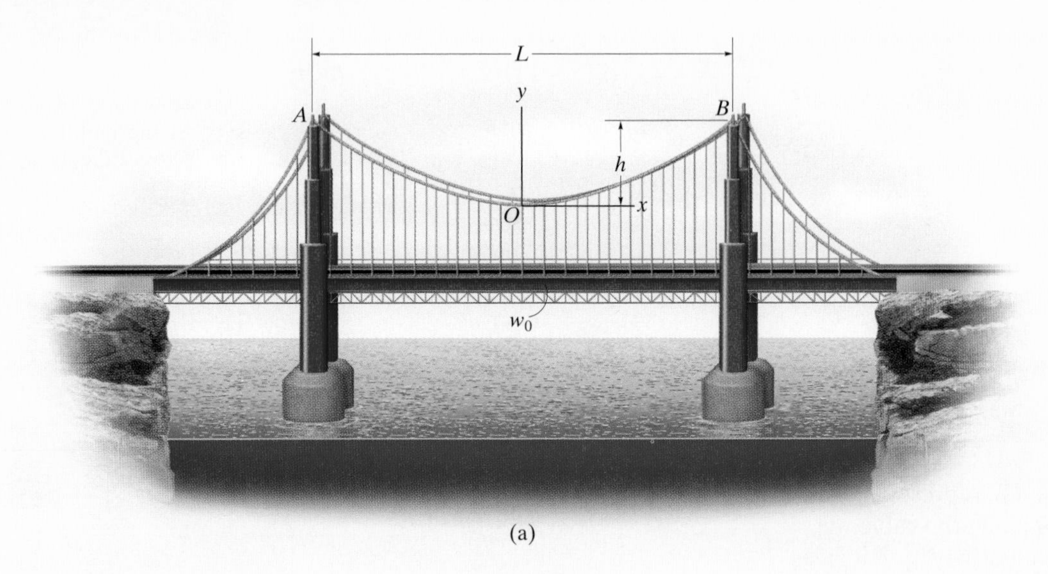

(a)

Fig. 7–23

SOLUTION

We can determine the unknowns in the problem by first finding the curve that defines the shape of the cable by using Eq. 7–12. For reasons of symmetry, the origin of coordinates has been placed at the cable's center. Noting that $w(x) = w_0$, we have

$$y = \frac{1}{F_H}\int\left(\int w_0\,dx\right)dx$$

Performing the two integrations gives

$$y = \frac{1}{F_H}\left(\frac{w_0 x^2}{2} + C_1 x + C_2\right) \qquad (1)$$

The constants of integration may be determined by using the boundary conditions $y = 0$ at $x = 0$ and $dy/dx = 0$ at $x = 0$. Substituting into Eq. 1 yields $C_1 = C_2 = 0$. The curve then becomes

$$y = \frac{w_0}{2F_H}x^2 \qquad (2)$$

This is the equation of a *parabola*. The constant F_H may be obtained by using the boundary condition $y = h$ at $x = L/2$. Thus,

$$F_H = \frac{w_0 L^2}{8h} \tag{3}$$

Therefore, Eq. 2 becomes

$$y = \frac{4h}{L^2}x^2 \tag{4}$$

Since F_H is known, the tension in the cable may be determined using Eq. 7–10, written as $T = F_H/\cos\theta$. For $0 \le \theta < \pi/2$, the maximum tension will occur when θ is *maximum*, i.e., at point B, Fig. 7–23*a*. From Eq. 2, the slope at this point is

$$\left.\frac{dy}{dx}\right|_{x=L/2} = \tan\theta_{\max} = \left.\frac{w_0}{F_H}x\right|_{x=L/2}$$

or

$$\theta_{\max} = \tan^{-1}\left(\frac{w_0 L}{2F_H}\right) \tag{5}$$

Therefore,

$$T_{\max} = \frac{F_H}{\cos(\theta_{\max})} \tag{6}$$

Using the triangular relationship shown in Fig. 7–23*b*, which is based on Eq. 5, Eq. 6 may be written as

$$T_{\max} = \frac{\sqrt{4F_H^2 + w_0^2 L^2}}{2}$$

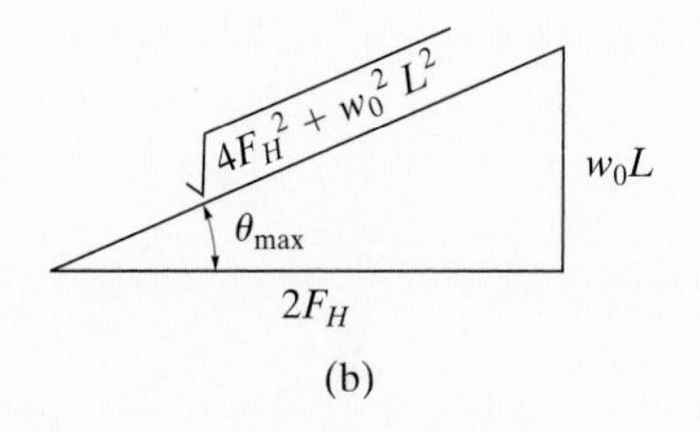

Fig. 7–23 (cont.)

Substituting Eq. 3 into the above equation yields

$$T_{\max} = \frac{w_0 L}{2}\sqrt{1 + \left(\frac{L}{4h}\right)^2} \qquad \textit{Ans.}$$

For a differential segment of cable length ds, we can write

$$ds = \sqrt{(dx)^2 + (dy)^2} = \sqrt{1 + \left(\frac{dy}{dx}\right)^2}\,dx$$

Hence, the total length of the cable, $\mathscr{L}$, can be determined by integration. Using Eq. 4, we have

$$\mathscr{L} = \int ds = 2\int_0^{L/2}\sqrt{1 + \left(\frac{8h}{L^2}x\right)^2}\,dx \tag{7}$$

Integrating yields

$$\mathscr{L} = \frac{L}{2}\left[\sqrt{1 + \left(\frac{4h}{L}\right)^2} + \frac{L}{4h}\sinh^{-1}\left(\frac{4h}{L}\right)\right] \qquad \textit{Ans.}$$

Over time the forces the cables exert on this telephone pole have caused it to tilt. Proper bracing of the pole should be required.

Cable Subjected to Its Own Weight. When the weight of the cable becomes important in the force analysis, the loading function along the cable becomes a function of the arc length s rather than the projected length x. A generalized loading function $w = w(s)$ acting along the cable is shown in Fig. 7–24*a*. The free-body diagram for a segment of the cable is shown in Fig. 7–24*b*. Applying the equilibrium equations to the force system on this diagram, one obtains relationships identical to those given by Eqs. 7–7 through 7–9, but with ds replacing dx. Therefore, it may be shown that

$$T \cos \theta = F_H$$

$$T \sin \theta = \int w(s)\, ds \qquad (7\text{–}13)$$

$$\frac{dy}{dx} = \frac{1}{F_H} \int w(s)\, ds \qquad (7\text{–}14)$$

To perform a direct integration of Eq. 7–14, it is necessary to replace dy/dx by ds/dx. Since

$$ds = \sqrt{dx^2 + dy^2}$$

then

$$\frac{dy}{dx} = \sqrt{\left(\frac{ds}{dx}\right)^2 - 1}$$

Therefore,

$$\frac{ds}{dx} = \left\{1 + \frac{1}{F_H^2}\left(\int w(s)\, ds\right)^2\right\}^{1/2}$$

Separating the variables and integrating yields

$$x = \int \frac{ds}{\left\{1 + \frac{1}{F_H^2}\left(\int w(s)\, ds\right)^2\right\}^{1/2}} \qquad (7\text{–}15)$$

The two constants of integration, say C_1 and C_2, are found using the boundary conditions for the cable.

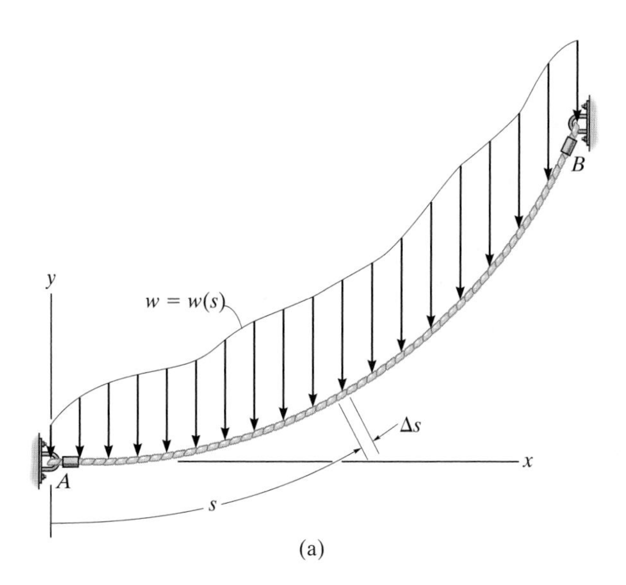

(a)

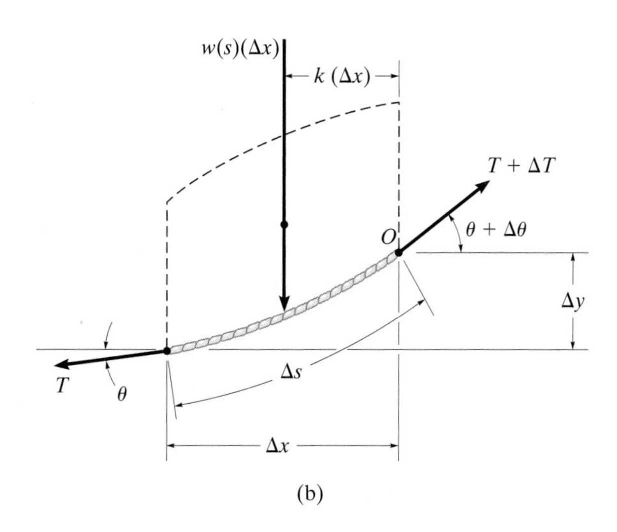

(b)

Fig. 7–24

EXAMPLE 7.15

Determine the deflection curve, the length, and the maximum tension in the uniform cable shown in Fig. 7–25. The cable weighs $w_0 = 5\ \text{N/m}$.

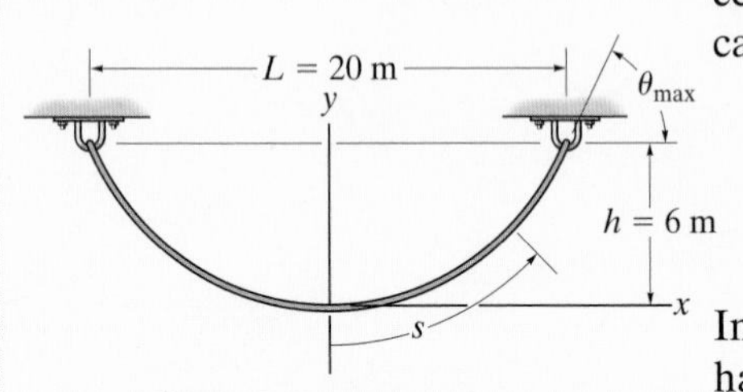

Fig. 7–25

SOLUTION

For reasons of symmetry, the origin of coordinates is located at the center of the cable. The deflection curve is expressed as $y = f(x)$. We can determine it by first applying Eq. 7–15, where $w(s) = w_0$.

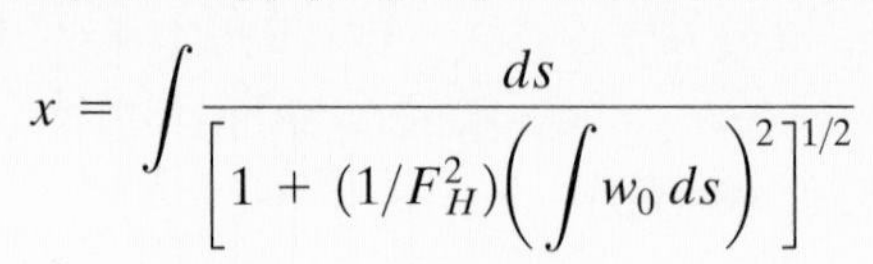

$$x = \int \frac{ds}{\left[1 + (1/F_H^2)\left(\int w_0\, ds\right)^2\right]^{1/2}}$$

Integrating the term under the integral sign in the denominator, we have

$$x = \int \frac{ds}{[1 + (1/F_H^2)(w_0 s + C_1)^2]^{1/2}}$$

Substituting $u = (1/F_H)(w_0 s + C_1)$ so that $du = (w_0/F_H)\, ds$, a second integration yields

$$x = \frac{F_H}{w_0}(\sinh^{-1} u + C_2)$$

or

$$x = \frac{F_H}{w_0}\left\{\sinh^{-1}\left[\frac{1}{F_H}(w_0 s + C_1)\right] + C_2\right\} \tag{1}$$

To evaluate the constants note that, from Eq. 7–14,

$$\frac{dy}{dx} = \frac{1}{F_H}\int w_0\, ds \quad \text{or} \quad \frac{dy}{dx} = \frac{1}{F_H}(w_0 s + C_1)$$

Since $dy/dx = 0$ at $s = 0$, then $C_1 = 0$. Thus,

$$\frac{dy}{dx} = \frac{w_0 s}{F_H} \tag{2}$$

The constant C_2 may be evaluated by using the condition $s = 0$ at $x = 0$ in Eq. 1, in which case $C_2 = 0$. To obtain the deflection curve, solve for s in Eq. 1, which yields

$$s = \frac{F_H}{w_0}\sinh\left(\frac{w_0}{F_H}x\right) \tag{3}$$

Now substitute into Eq. 2, in which case

$$\frac{dy}{dx} = \sinh\left(\frac{w_0}{F_H}x\right)$$

Hence,

$$y = \frac{F_H}{w_0}\cosh\left(\frac{w_0}{F_H}x\right) + C_3 \quad (4)$$

If the boundary condition $y = 0$ at $x = 0$ is applied, the constant $C_3 = -F_H/w_0$, and therefore the deflection curve becomes

$$y = \frac{F_H}{w_0}\left[\cosh\left(\frac{w_0}{F_H}x\right) - 1\right]$$

This equation defines the shape of a *catenary curve*. The constant F_H is obtained by using the boundary condition that $y = h$ at $x = L/2$, in which case

$$h = \frac{F_H}{w_0}\left[\cosh\left(\frac{w_0 L}{2F_H}\right) - 1\right] \quad (5)$$

Since $w_0 = 5$ N/m, $h = 6$ m, and $L = 20$ m, Eqs. 4 and 5 become

$$y = \frac{F_H}{5\text{ N/m}}\left[\cosh\left(\frac{5\text{ N/m}}{F_H}x\right) - 1\right] \quad (6)$$

$$6\text{ m} = \frac{F_H}{5\text{ N/m}}\left[\cosh\left(\frac{50\text{ N}}{F_H}\right) - 1\right] \quad (7)$$

Equation 7 can be solved for F_H by using a trial-and-error procedure. The result is

$$F_H = 45.9\text{ N}$$

and therefore the deflection curve, Eq. 6, becomes

$$y = 9.19[\cosh(0.109x) - 1]\text{ m} \qquad \textit{Ans.}$$

Using Eq. 3, with $x = 10$ m, the half-length of the cable is

$$\frac{\mathcal{L}}{2} = \frac{45.9\text{ N}}{5\text{ N/m}}\sinh\left[\frac{5\text{ N/m}}{45.9\text{ N}}(10\text{ m})\right] = 12.1\text{ m}$$

Hence,

$$\mathcal{L} = 24.2\text{ m} \qquad \textit{Ans.}$$

Since $T = F_H/\cos\theta$, Eq. 7–13, the maximum tension occurs when θ is maximum, i.e., at $s = \mathcal{L}/2 = 12.1$ m. Using Eq. 2 yields

$$\left.\frac{dy}{dx}\right|_{s=12.1\text{ m}} = \tan\theta_{max} = \frac{5\text{ N/m}(12.1\text{ m})}{45.9\text{ N}} = 1.32$$

$$\theta_{max} = 52.8^\circ$$

Thus,

$$T_{max} = \frac{F_H}{\cos\theta_{max}} = \frac{45.9\text{ N}}{\cos 52.8^\circ} = 75.9\text{ N} \qquad \textit{Ans.}$$

PROBLEMS

Neglect the weight of the cable in the following problems, unless specified.

7–89. Determine the force P needed to hold the cable in the position shown, i.e., so segment BC remains horizontal. Also, compute the sag y_B and the maximum tension in the cable.

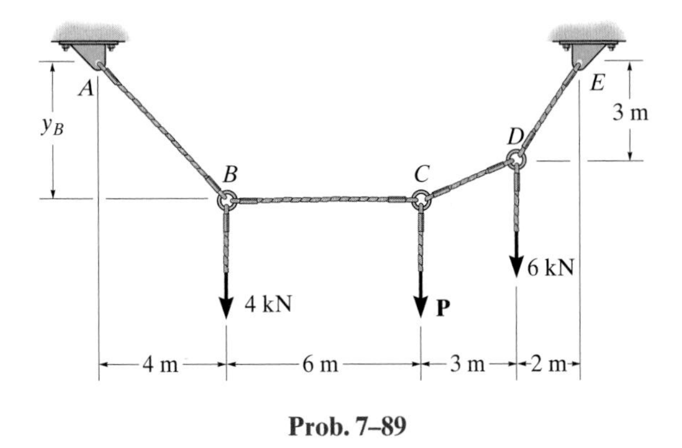

Prob. 7–89

7–90. Cable $ABCD$ supports the 10-kg lamp E and the 15-kg lamp F. Determine the maximum tension in the cable and the sag of point B.

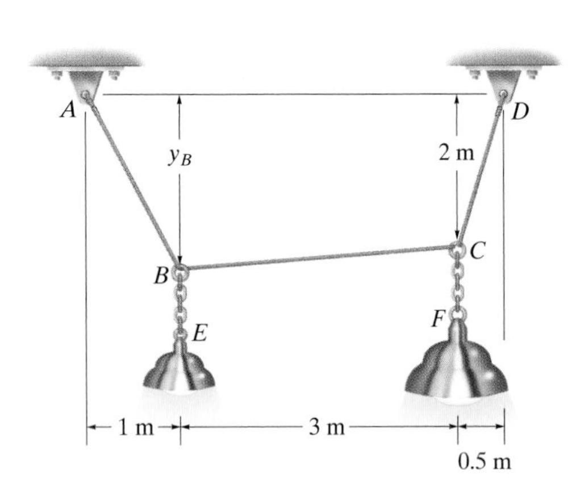

Prob. 7–90

7–91. The cable supports the three loads shown. Determine the sags y_B and y_D of points B and D. Take $P_1 = 400$ lb, $P_2 = 250$ lb.

***7–92.** The cable supports the three loads shown. Determine the magnitude of $\mathbf{P}_1$ if $P_2 = 300$ lb and $y_B = 8$ ft. Also find the sag y_D.

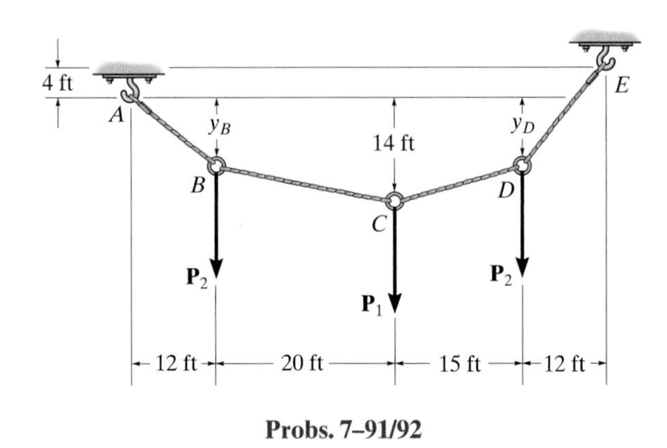

Probs. 7–91/92

7–93. The cable supports the loading shown. Determine the distance x_B the force at point B acts from A. Set $P = 40$ lb.

7–94. The cable supports the loading shown. Determine the magnitude of the horizontal force $\mathbf{P}$ so that $x_B = 6$ ft.

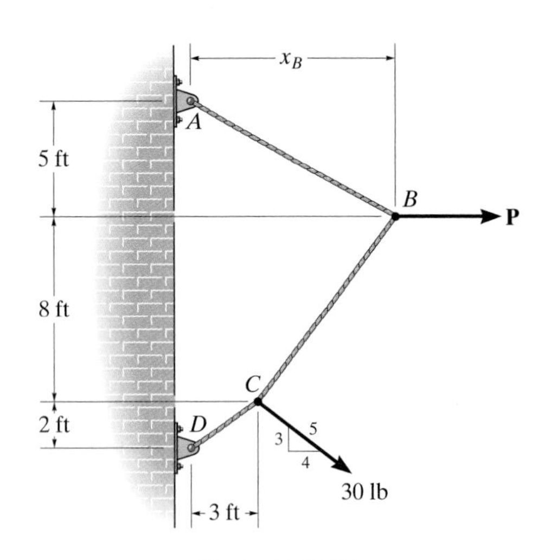

Probs. 7–93/94

7–95. Determine the forces P_1 and P_2 needed to hold the cable in the position shown, i.e., so segment CD remains horizontal. Also, compute the sag y_D and the maximum tension in the cable.

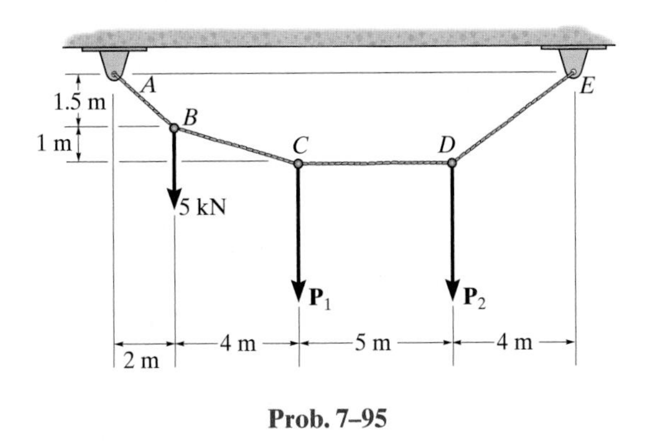

Prob. 7–95

***7–96.** The cable supports the loading shown. Determine the distance x_B from the wall to point B.

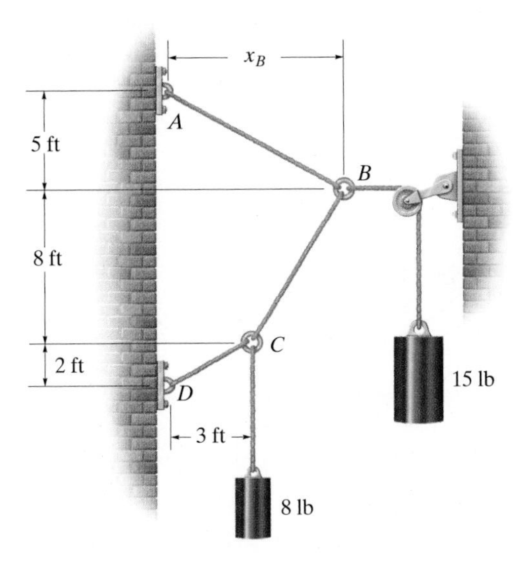

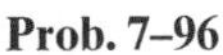

Prob. 7–96

7–97. Determine the maximum uniform loading w, measured in lb/ft, that the cable can support if it is capable of sustaining a maximum tension of 3000 lb before it will break.

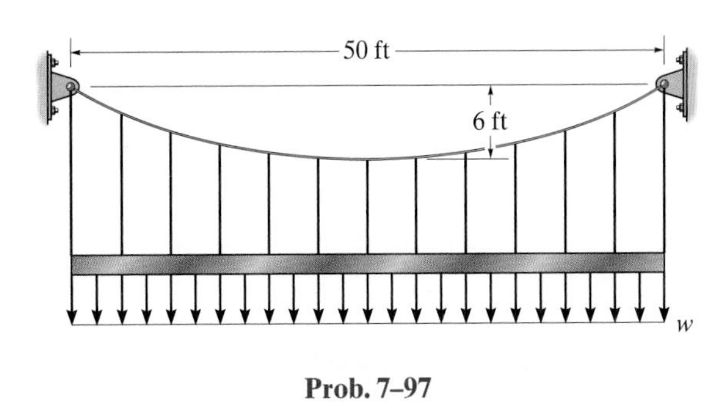

Prob. 7–97

7–98. The cable is subjected to a uniform loading of $w = 250$ lb/ft. Determine the maximum and minimum tension in the cable.

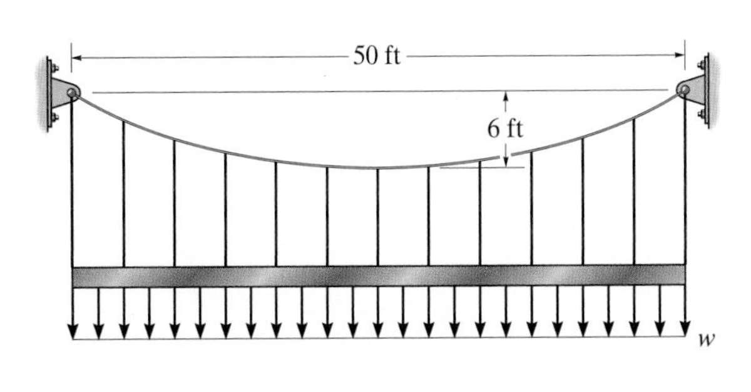

Prob. 7–98

7–99. The cable is subjected to the triangular loading. If the slope of the cable at A is zero, determine the equation of the curve $y = f(x)$ which defines the cable shape AB, and the maximum tension developed in the cable.

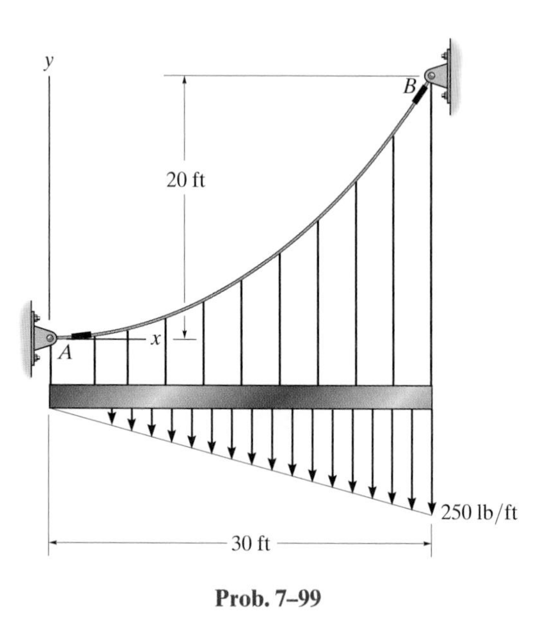

Prob. 7–99

7–101. The cable is subjected to the triangular loading. If the slope of the cable at point O is zero, determine the equation of the curve $y = f(x)$ which defines the cable shape OB, and the maximum tension developed in the cable.

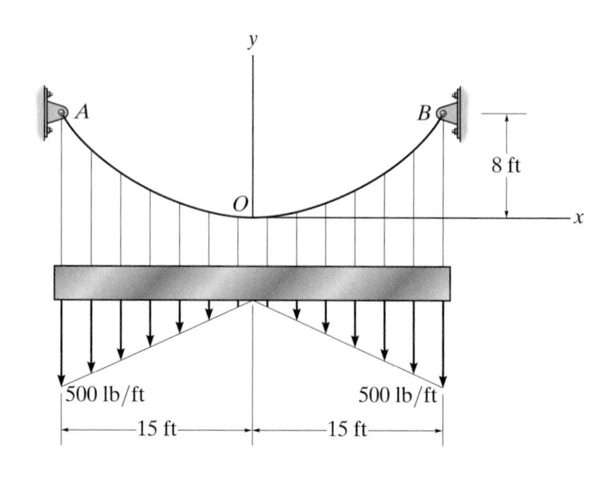

Prob. 7–101

***7–100.** The cable supports a girder which weighs 850 lb/ft. Determine the tension in the cable at points A, B, and C.

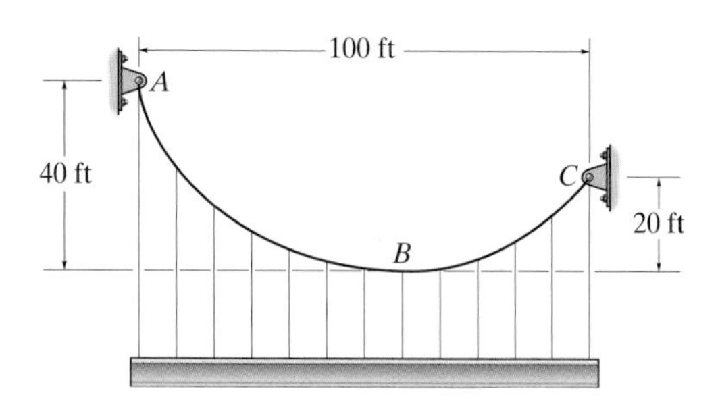

Prob. 7–100

7–102. The cable is subjected to the parabolic loading $w = 150(1 - (x/50)^2)$ lb/ft, where x is in ft. Determine the equation $y = f(x)$ which defines the cable shape AB and the maximum tension in the cable.

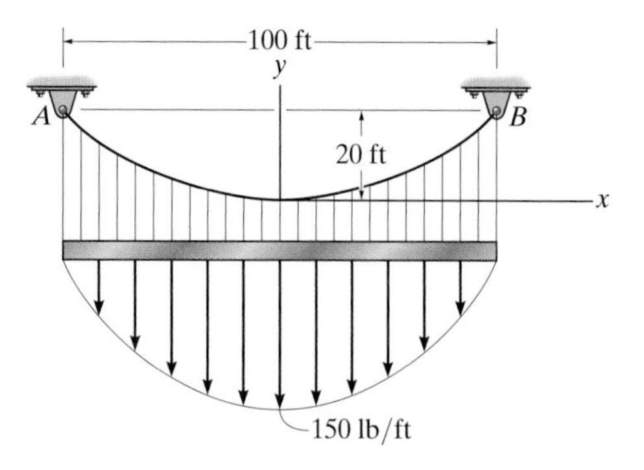

Prob. 7–102

7–103. The cable will break when the maximum tension reaches $T_{max} = 10$ kN. Determine the minimum sag h if it supports the uniform distributed load of $w = 600$ N/m.

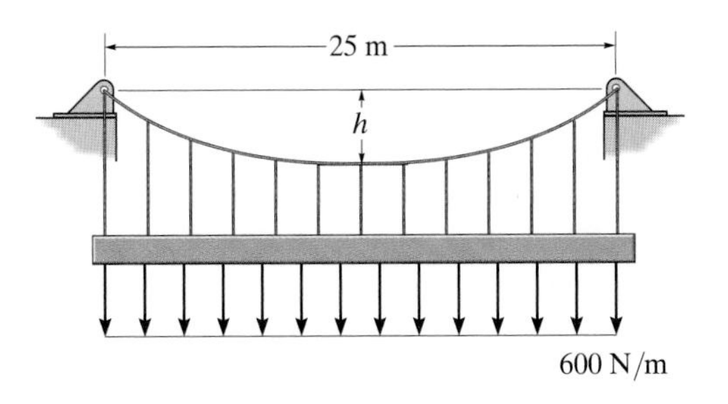

Prob. 7–103

***7–104.** A fiber optic cable is suspended over the poles so that the angle at the supports is $\theta = 22°$. Determine the minimum tension in the cable and the sag. The cable has a mass of 0.9 kg/m and the supports are at the same elevation.

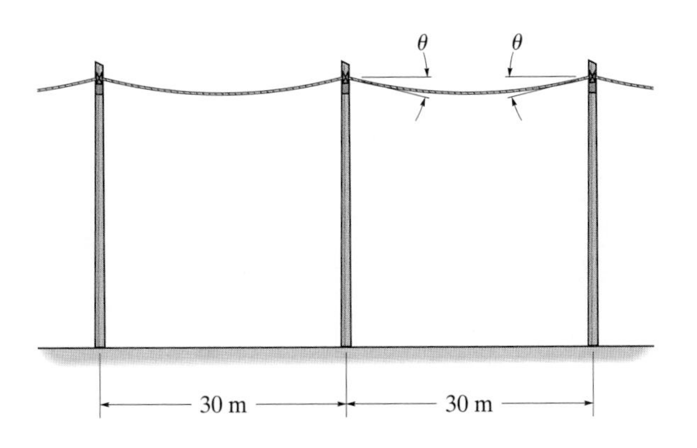

Prob. 7–104

■7–105. A cable has a weight of 3 lb/ft and is supported at points that are 500 ft apart and at the same elevation. If it has a length of 600 ft, determine the sag.

7–106. Show that the deflection curve of the cable discussed in Example 7.15 reduces to Eq. (4) in Example 7.14 when the *hyperbolic cosine function* is expanded in terms of a series and only the first two terms are retained. (The answer indicates that the *catenary* may be replaced by a *parabola* in the analysis of problems in which the sag is small. In this case, the cable weight is assumed to be uniformly distributed along the horizontal.)

7–107. A uniform cord is suspended between two points having the same elevation. Determine the sag-to-span ratio so that the maximum tension in the cord equals the cord's total weight.

***■7–108.** A cable has a weight of 2 lb/ft. If it can span 100 ft and has a sag of 12 ft, determine the length of the cable. The ends of the cable are supported from the same elevation.

7–109. The transmission cable having a weight of 20 lb/ft is strung across the river as shown. Determine the required force that must be applied to the cable at its points of attachment to the towers at B and C.

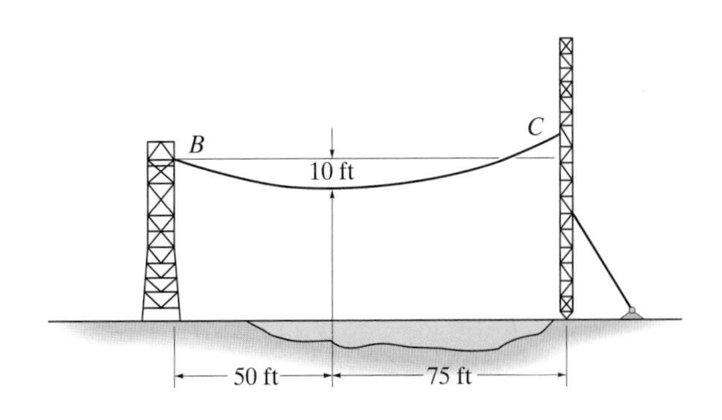

Prob. 7–109

7–110. Determine the maximum tension developed in the cable if it is subjected to a uniform load of 600 N/m.

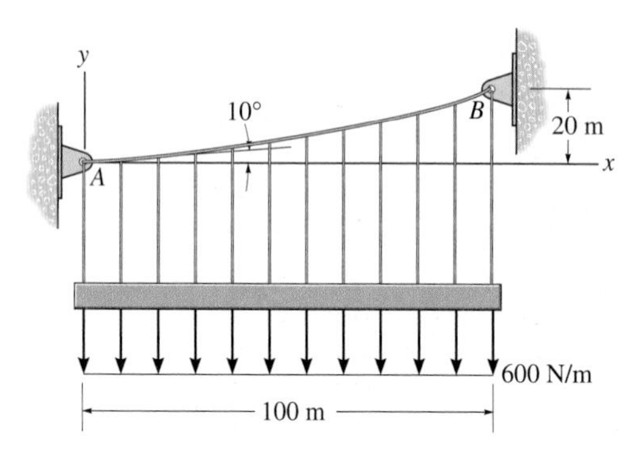

Prob. 7–110

■**7–111.** A 40-m-long chain has a total mass of 100 kg and is suspended between two points 10 m apart at the same elevation. Determine the maximum tension and the sag in the chain.

*■**7–112.** The cable has a mass of 0.5 kg/m, and is 25 m long. Determine the vertical and horizontal components of force it exerts on the top of the tower.

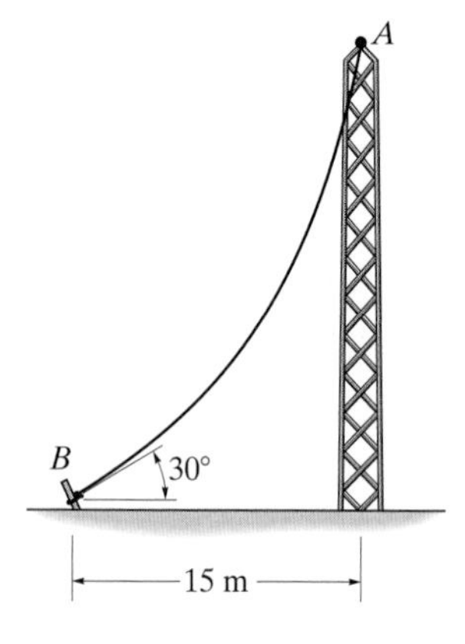

Prob. 7–112

■**7–113.** A 50-ft cable is suspended between two points a distance of 15 ft apart and at the same elevation. If the minimum tension in the cable is 200 lb, determine the total weight of the cable and the maximum tension developed in the cable.

7–114. The 80-ft-long chain is fixed at its ends and hoisted at its midpoint B using a crane. If the chain has a weight of 0.5 lb/ft, determine the minimum height h of the hook in order to lift the chain *completely* off the ground. What is the horizontal force at pin A or C when the chain is in this position? *Hint:* When h is a minimum, the slope at A and C is zero.

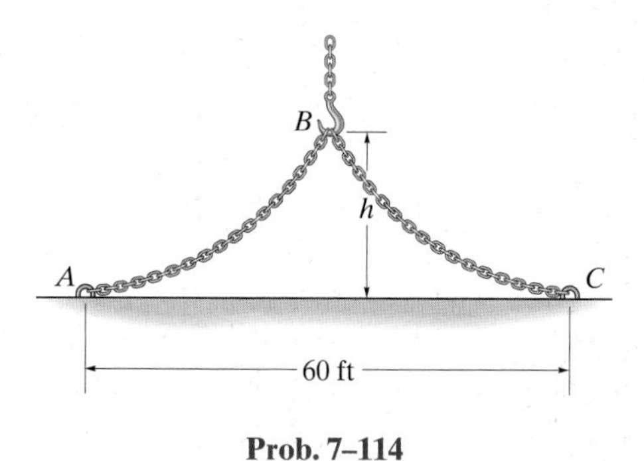

Prob. 7–114

■**7–115.** A steel tape used for measurement in surveying has a length of 100.00 ft and a total weight of 2 lb. How much horizontal tension must be applied to the tape so that the distance marked on the ground is 99.90 ft? In practice the calculation should also include the effects of elastic stretching and temperature changes on the tape's length.

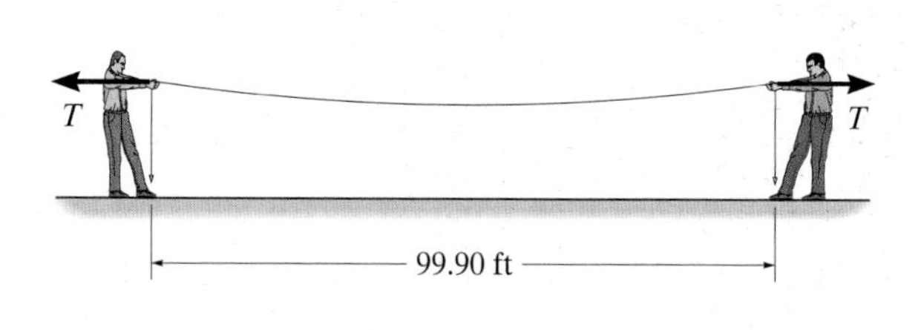

Prob. 7–115

Chapter Review

Internal Loadings

If a coplanar force system acts on a member, then in general a resultant internal *normal force N, shear force V*, and *bending moment M* will act at any cross section along the member.

The resultant internal normal force, shear force, and bending moment are determined using the method of sections. To find them, the member is sectioned at the point where the internal loadings are to be determined. A free-body diagram of one of the sectioned parts is then drawn.

The resultant normal force is determined by summing forces normal to the cross section. The resultant shear force is found by summing forces tangent to the cross section, and the resultant bending moment is found by summing moments about the centroid of the cross-sectional area.

$$\Sigma F_x = 0$$
$$\Sigma F_y = 0$$
$$\Sigma M_C = 0$$

If the member is subjected to a three-dimensional loading, then, in general, a *torsional loading* will also act on the cross section. It can be determined by summing moments about an axis that is perpendicular to the cross section and passes through its centroid.

(*See pages 337–339.*)

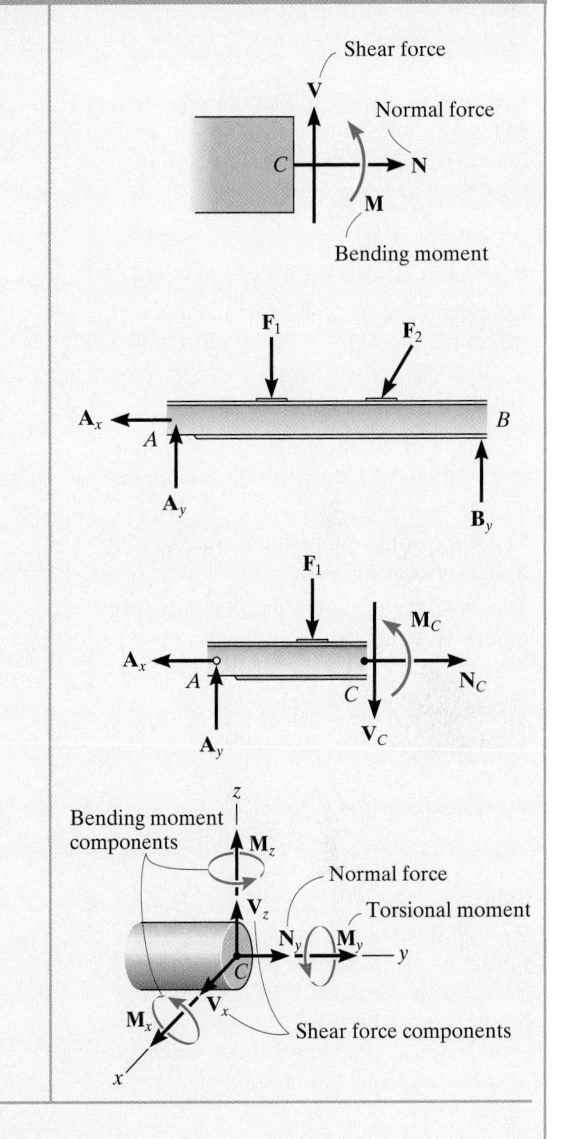

Shear and Moment Diagrams

To construct the shear and moment diagrams for a member, it is necessary to section the member at an arbitrary point, located a distance x from one end.

If the external loading consists of changes in the distributed load, or a series of concentrated forces and couple moments act on the member, then different expressions for V and M must be determined within regions between these different loadings.

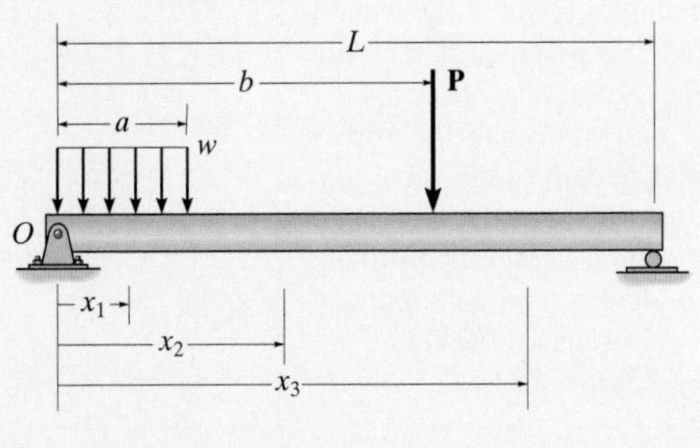

The unknown shear and moment are indicated on the cross section in the positive direction according to the established sign convention.

Application of the equilibrium equations will give the shear and moment as a function of x, which can then be plotted. (*See pages 354 and 355.*)

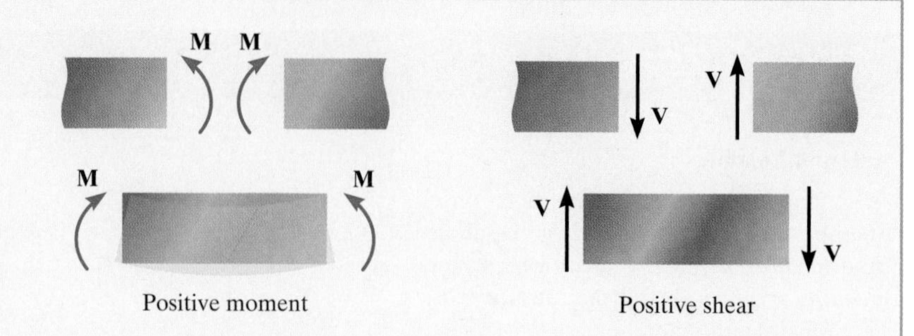

Positive moment

Positive shear

Relations between Shear and Moment

It is possible to plot the shear and moment diagrams quickly by using differential relationships that exist between the distributed loading w and V and M.

The slope of the shear diagram is equal to the negative of the distributed loading at any point.

$$\frac{dV}{dx} = -w$$

The slope of the moment diagram is equal to the shear at any point.

$$\frac{dM}{dx} = V$$

The change in shear between any two points is equal to the area under the distributed loading between the points.

$$\Delta V = -\int w\,dx$$

The change in the moment is equal to the area under the shear diagram between the points.
(*See pages 362–364.*)

$$\Delta M = \int V\,dx$$

Cables

When a flexible and inextensible cable is subjected to a series of concentrated forces, then the analysis of the cable can be performed by using the equations of equilibrium applied to free-body diagrams of either segments or points of application of the loading.

If external distributed loads or the weight of the cable are to be considered, then the forces and shape of the cable must be determined by first analyzing the forces on a differential segment of the cable and then integrating this result. The two constants, say C_1 and C_2, resulting from the integration are determined by applying the boundary conditions for the cable.
(*See pages 376–377 and 380.*)

$$y = \frac{1}{F_H}\int\left(\int w(x)\,dx\right)dx$$

Distributed load

$$x = \int \frac{ds}{\left\{1 + \frac{1}{F_H^2}\left(\int w(s)\,ds\right)^2\right\}^{1/2}}$$

Cable weight

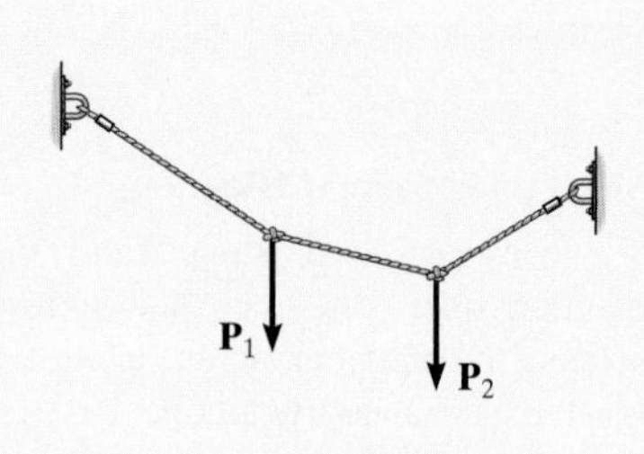

REVIEW PROBLEMS

*■7–116. A 100-lb cable is attached between two points at a distance 50 ft apart having equal elevations. If the maximum tension developed in the cable is 75 lb, determine the length of the cable and the sag.

7–117. Determine the distance a between the supports in terms of the beam's length L so that the moment in the *symmetric* beam is zero at the beam's center.

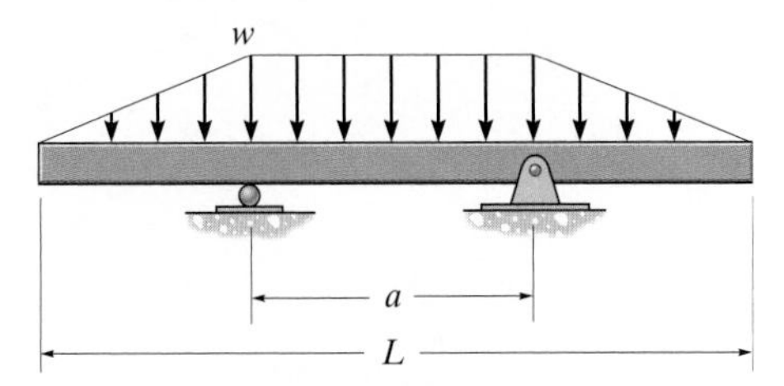

Prob. 7–117

7–118. Determine the internal normal force, shear force, and moment at point D.

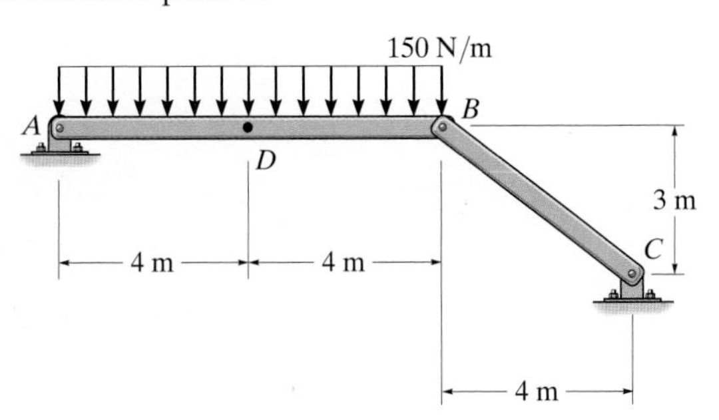

Prob. 7–118

7–119. The beam is supported by a pin at C and a rod AB. Determine the internal normal force, shear force, and moment at point D.

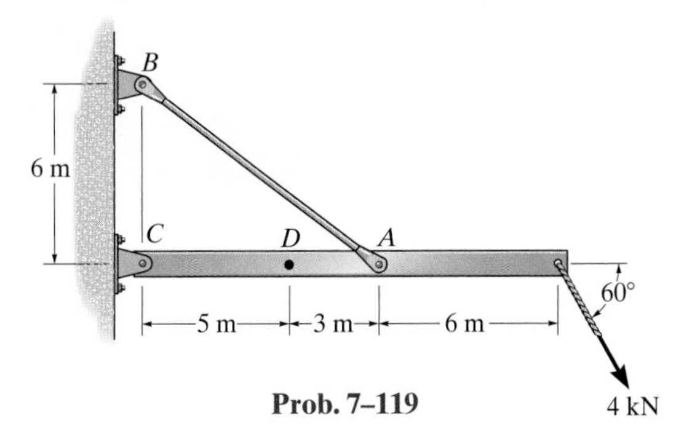

Prob. 7–119

***7–120.** Express the shear and moment acting in the pipe as a function of y, where $0 \le y \le 4$ ft.

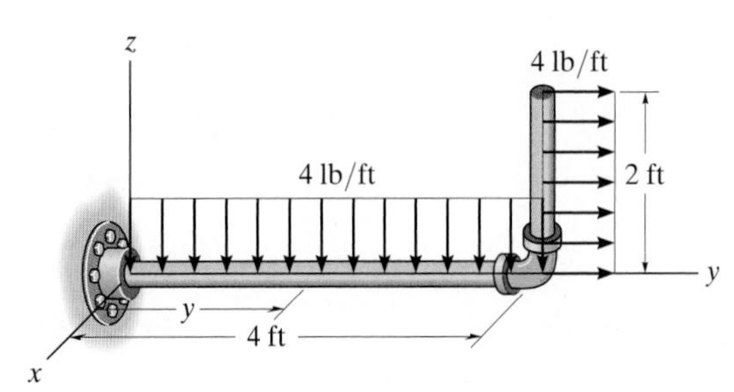

Prob. 7–120

7–121. Determine the normal force, shear force, and moment at points B and C of the beam.

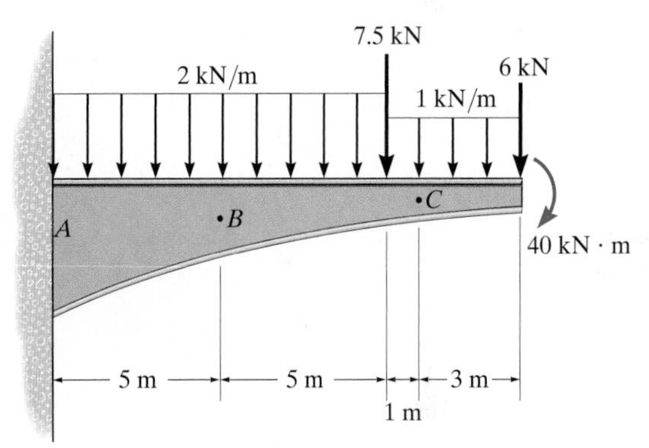

Prob. 7–121

7–122. A chain is suspended between points at the same elevation and spaced a distance of 60 ft apart. If it has a weight of 0.5 lb/ft and the sag is 3 ft, determine the maximum tension in the chain.

7–123. Draw the shear and moment diagrams for the beam.

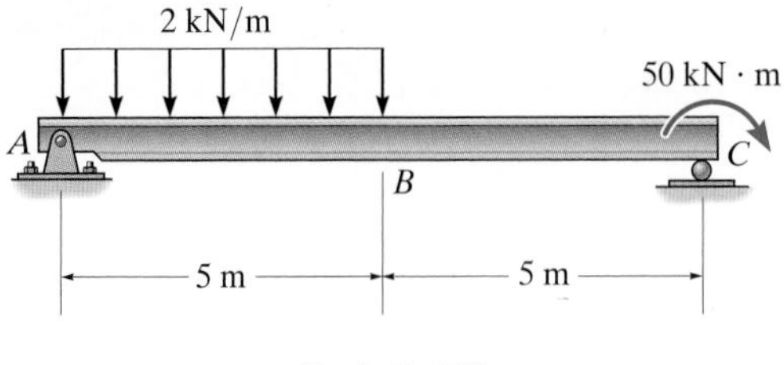

Prob. 7–123

The effective design of a brake system, such as the one for this bicycle, requires efficient capacity for the mechanism to resist frictional forces. In this chapter, we will study the nature of friction and show how these forces are considered in engineering analysis.

8 Friction

CHAPTER OBJECTIVES

- To introduce the concept of dry friction and show how to analyze the equilibrium of rigid bodies subjected to this force.
- To present specific applications of frictional force analysis on wedges, screws, belts, and bearings.
- To investigate the concept of rolling resistance.

8.1 Characteristics of Dry Friction

Friction can be defined as a force of resistance acting on a body that prevents or retards slipping of the body relative to a second body or surface with which it is in contact. This force always acts *tangent* to the surface at points of contact with other bodies and is directed so as to oppose the possible or existing motion of the body relative to these points.

In general, two types of friction can occur between surfaces. *Fluid friction* exists when the contacting surfaces are separated by a film of fluid (gas or liquid). The nature of fluid friction is studied in fluid mechanics since it depends upon knowledge of the velocity of the fluid and the fluid's ability to resist shear force. In this book, we will focus only on the effects of *dry friction*. This type of friction is often called *Coulomb friction* since its characteristics were studied extensively by C. A. Coulomb in 1781. Dry friction occurs between the contacting surfaces of bodies when there is no lubricating fluid.

The heat generated by the abrasive action of friction can be noticed when using this grinder to sharpen a metal blade.

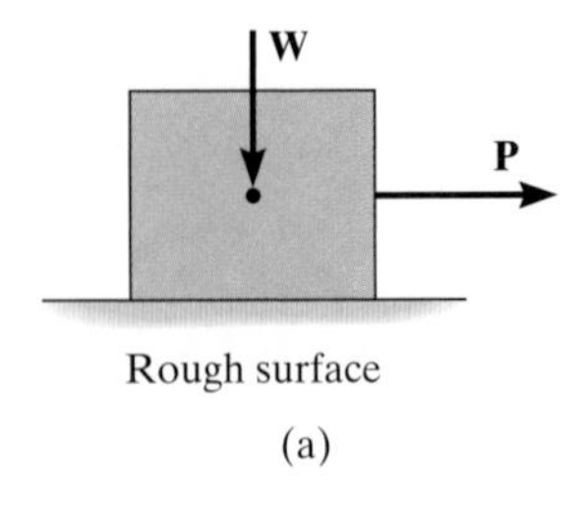

Rough surface

(a)

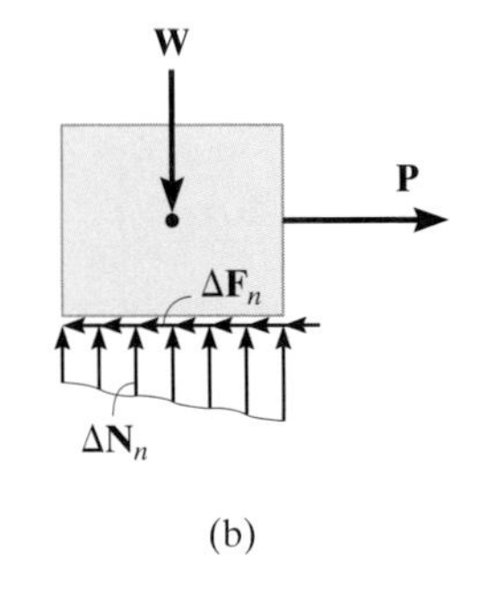

(b)

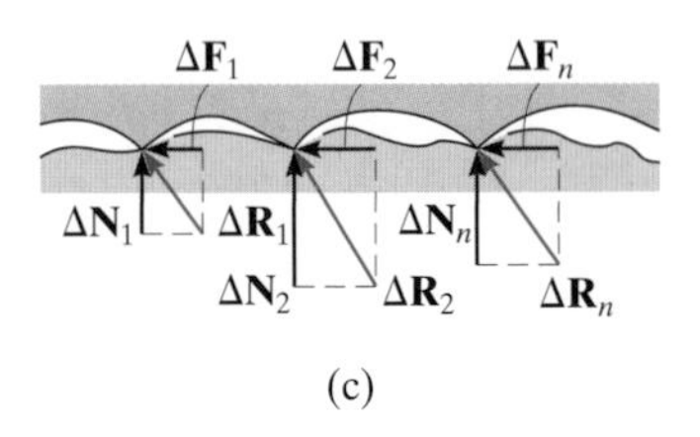

(c)

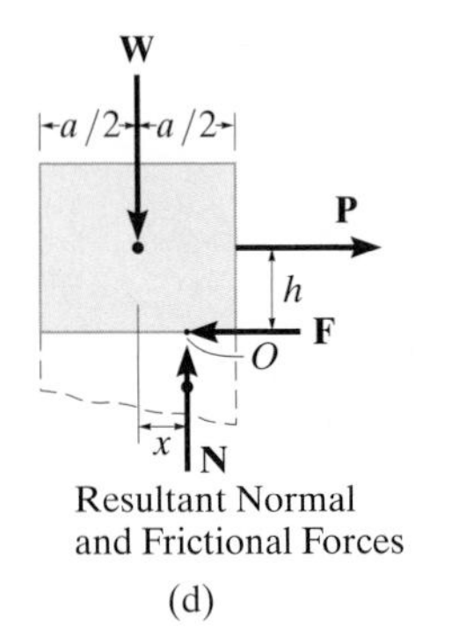

Resultant Normal and Frictional Forces

(d)

Fig. 8–1

Theory of Dry Friction. The theory of dry friction is best explained by considering the effects caused by pulling horizontally on a block of uniform weight **W** which is resting on a rough horizontal surface, Fig. 8–1*a*. To properly develop a full understanding of the nature of friction, it is necessary to consider the surfaces of contact to be *nonrigid or deformable*. The upper portion of the block, however, can be considered rigid. As shown on the free-body diagram of the block, Fig. 8–1*b*, the floor exerts a *distribution* of both *normal force* $\Delta\mathbf{N}_n$ and *frictional force* $\Delta\mathbf{F}_n$ along the contacting surface. For equilibrium, the normal forces must act *upward* to balance the block's weight **W**, and the frictional forces act to the left to prevent the applied force **P** from moving the block to the right. Close examination of the contacting surfaces between the floor and block reveals how these frictional and normal forces develop, Fig. 8–1*c*. It can be seen that many microscopic irregularities exist between the two surfaces and, as a result, reactive forces $\Delta\mathbf{R}_n$ are developed at each of the protuberances.* These forces act at all points of contact, and, as shown, each reactive force contributes both a frictional component $\Delta\mathbf{F}_n$ and a normal component $\Delta\mathbf{N}_n$.

Equilibrium. For simplicity in the following analysis, the effect of the *distributed* normal and frictional loadings will be indicated by their *resultants* **N** and **F**, which are represented on the free-body diagram as shown in Fig. 8-1*d*. Clearly, the distribution of $\Delta\mathbf{F}_n$ in Fig. 8–1*b* indicates that **F** always acts *tangent to the contacting surface, opposite* to the direction of **P**. On the other hand, the normal force **N** is determined from the distribution of $\Delta\mathbf{N}_n$ in Fig. 8–1*b* and is directed upward to balance the block's weight **W**. Notice that **N** acts a distance x to the right of the line of action of **W**, Fig. 8–1*d*. This location, which coincides with the centroid or geometric center of the loading diagram in Fig. 8–1*b*, is necessary in order to balance the "tipping effect" caused by **P**. For example, if **P** is applied at a height h from the surface, Fig. 8–1*d*, then moment equilibrium about point O is satisfied if $Wx = Ph$ or $x = Ph/W$. In particular, the block will be on the verge of *tipping* if N acts at the right corner of the block, $x = a/2$.

*Besides mechanical interactions as explained here, which is referred to as a classical approach, a detailed treatment of the nature of frictional forces must also include the effects of temperature, density, cleanliness, and atomic or molecular attraction between the contacting surfaces. See J. Krim, *Scientific American*, October, 1996.

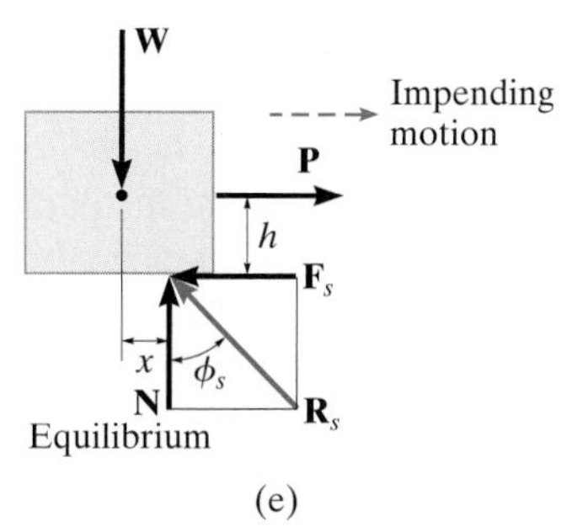

(e)

Impending Motion. In cases where h is small or the surfaces of contact are rather "slippery," the frictional force **F** may *not* be great enough to balance **P**, and consequently the block will tend to slip *before* it can tip. In other words, as P is slowly increased, F correspondingly increases until it attains a certain *maximum value* F_s, called the *limiting static frictional force*, Fig. 8–1*e*. When this value is reached, the block is in *unstable equilibrium* since any further increase in P will cause deformations and fractures at the points of surface contact, and consequently the block will begin to move. Experimentally, it has been determined that the limiting static frictional force F_s is *directly proportional* to the resultant normal force N. This may be expressed mathematically as

$$F_s = \mu_s N \tag{8–1}$$

where the constant of proportionality, μ_s (mu "sub" s), is called the *coefficient of static friction*.

Thus, when the block is on the *verge of sliding*, the normal force **N** and frictional force $\mathbf{F}_s$ combine to create a resultant $\mathbf{R}_s$, Fig. 8–1*e*. The angle ϕ_s that $\mathbf{R}_s$ makes with **N** is called the *angle of static friction*. From the figure,

$$\phi_s = \tan^{-1}\left(\frac{F_s}{N}\right) = \tan^{-1}\left(\frac{\mu_s N}{N}\right) = \tan^{-1}\mu_s$$

Tabular Values of μ_s. Typical values for μ_s, found in many engineering handbooks, are given in Table 8–1. Although this coefficient is generally less than 1, be aware that in some cases it is possible, as in the case of aluminum on aluminum, for μ_s to be greater than 1. Physically this means, of course, that in this case the frictional force is greater than the corresponding normal force. Furthermore, it should be noted that μ_s is dimensionless and depends only on the characteristics of the two surfaces in contact. A wide range of values is given for each value of μ_s since experimental testing was done under variable conditions of roughness and cleanliness of the contacting surfaces. For applications, therefore, it is important that both caution and judgment be exercised when selecting a coefficient of friction for a given set of conditions. When a more accurate calculation of F_s is required, the coefficient of friction should be determined directly by an experiment that involves the two materials to be used.

TABLE 8–1
Typical Values for μ_s

Contact Materials	Coefficient of Static Friction (μ_s)
Metal on ice	0.03–0.05
Wood on wood	0.30–0.70
Leather on wood	0.20–0.50
Leather on metal	0.30–0.60
Aluminum on aluminum	1.10–1.70

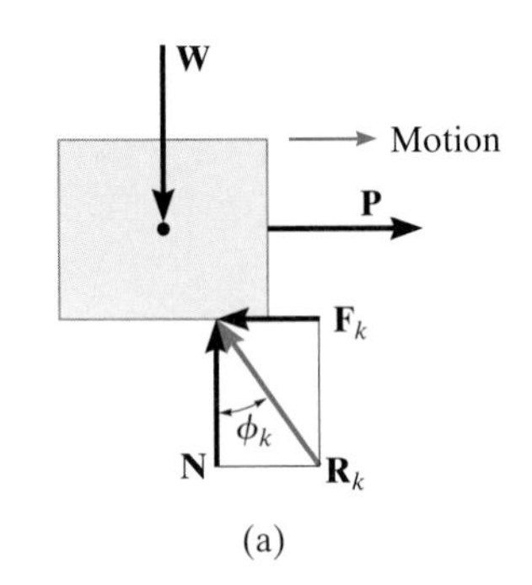

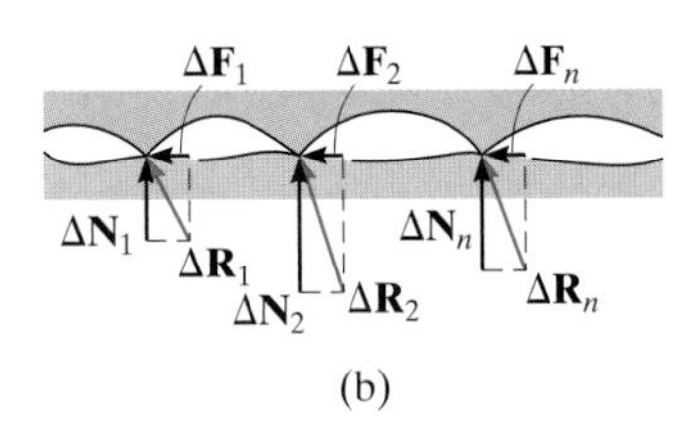

Fig. 8–2

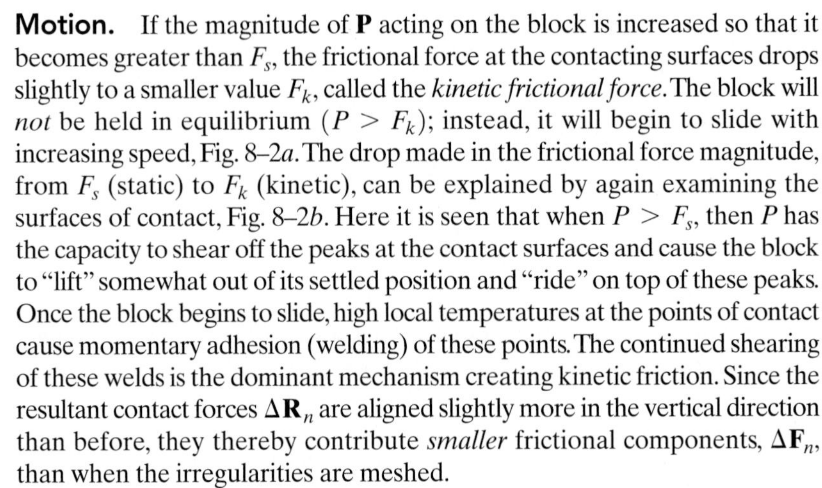

Motion. If the magnitude of **P** acting on the block is increased so that it becomes greater than F_s, the frictional force at the contacting surfaces drops slightly to a smaller value F_k, called the *kinetic frictional force*. The block will *not* be held in equilibrium ($P > F_k$); instead, it will begin to slide with increasing speed, Fig. 8–2*a*. The drop made in the frictional force magnitude, from F_s (static) to F_k (kinetic), can be explained by again examining the surfaces of contact, Fig. 8–2*b*. Here it is seen that when $P > F_s$, then P has the capacity to shear off the peaks at the contact surfaces and cause the block to "lift" somewhat out of its settled position and "ride" on top of these peaks. Once the block begins to slide, high local temperatures at the points of contact cause momentary adhesion (welding) of these points. The continued shearing of these welds is the dominant mechanism creating kinetic friction. Since the resultant contact forces $\Delta \mathbf{R}_n$ are aligned slightly more in the vertical direction than before, they thereby contribute *smaller* frictional components, $\Delta \mathbf{F}_n$, than when the irregularities are meshed.

Experiments with sliding blocks indicate that the magnitude of the resultant frictional force $\mathbf{F}_k$ is directly proportional to the magnitude of the resultant normal force $\mathbf{N}$. This may be expressed mathematically as

$$F_k = \mu_k N \tag{8–2}$$

Here the constant of proportionality, μ_k, is called the *coefficient of kinetic friction*. Typical values for μ_k are approximately 25 percent *smaller* than those listed in Table 8–1 for μ_s.

As shown in Fig. 8–2*a*, in this case, the resultant $\mathbf{R}_k$ has a line of action defined by ϕ_k. This angle is referred to as the *angle of kinetic friction*, where

$$\phi_k = \tan^{-1}\left(\frac{F_k}{N}\right) = \tan^{-1}\left(\frac{\mu_k N}{N}\right) = \tan^{-1}\mu_k$$

By comparison, $\phi_s \geq \phi_k$.

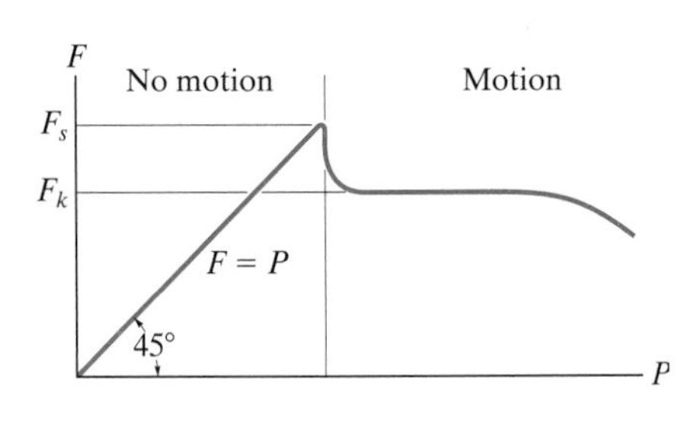

Fig. 8–3

The above effects regarding friction can be summarized by reference to the graph in Fig. 8–3, which shows the variation of the frictional force F versus the applied load P. Here the frictional force is categorized in three different ways: namely, F is a *static frictional force* if equilibrium is maintained; F is a *limiting static frictional force* F_s when it reaches a maximum value needed to maintain equilibrium; and finally, F is termed a *kinetic-frictional force* F_k when sliding occurs at the contacting surface. Notice also from the graph that for very large values of P or for high speeds, because of aerodynamic effects, F_k and likewise μ_k begin to decrease.

Characteristics of Dry Friction. As a result of *experiments* that pertain to the foregoing discussion, we can state the following rules which apply to bodies subjected to dry friction.

- The frictional force acts *tangent* to the contacting surfaces in a direction *opposed* to the *relative motion* or tendency for motion of one surface against another.
- The maximum static frictional force F_s that can be developed is independent of the area of contact, provided the normal pressure is not very low nor great enough to severely deform or crush the contacting surfaces of the bodies.
- The maximum static frictional force is generally greater than the kinetic frictional force for any two surfaces of contact. However, if one of the bodies is moving with a *very low velocity* over the surface of another, F_k becomes approximately equal to F_s, i.e., $\mu_s \approx \mu_k$.
- When *slipping* at the surface of contact is *about to occur*, the maximum static frictional force is proportional to the normal force, such that $F_s = \mu_s N$.
- When *slipping* at the surface of contact is *occurring*, the kinetic frictional force is proportional to the normal force, such that $F_k = \mu_k N$.

8.2 Problems Involving Dry Friction

If a rigid body is in equilibrium when it is subjected to a system of forces that includes the effect of friction, the force system must satisfy not only the equations of equilibrium but *also* the laws that govern the frictional forces.

Types of Friction Problems. In general, there are three types of mechanics problems involving dry friction. They can easily be classified once free-body diagrams are drawn and the total number of unknowns are identified and compared with the total number of available equilibrium equations. Each type of problem will now be explained and illustrated graphically by examples. In all these cases the geometry and dimensions for the problem are assumed to be known.

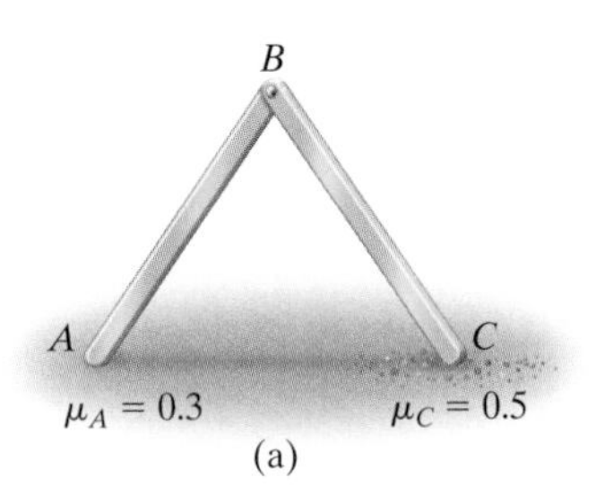

(a)

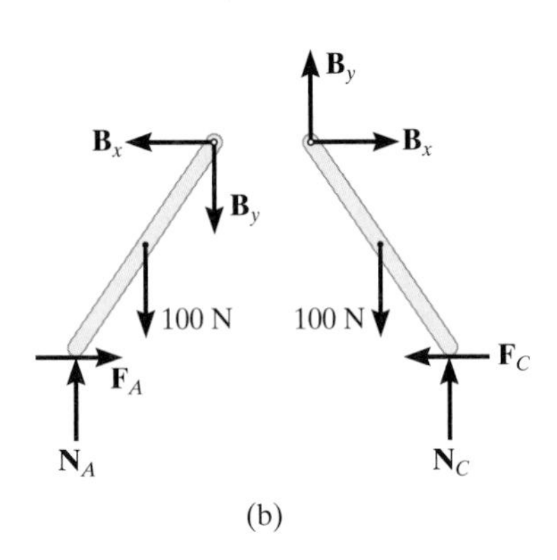

(b)

Fig. 8–4

Equilibrium. Problems in this category are strictly equilibrium problems which require *the total number of unknowns to be equal to the total number of available equilibrium equations*. Once the frictional forces are determined from the solution, however, their numerical values must be checked to be sure they satisfy the inequality $F \leq \mu_s N$; otherwise, slipping will occur and the body will not remain in equilibrium. A problem of this type is shown in Fig. 8–4*a*. Here we must determine the frictional forces at A and C to check if the equilibrium position of the two-member frame can be maintained. If the bars are uniform and have known weights of 100 N each, then the free-body diagrams are as shown in Fig. 8–4*b*. There are six unknown force components which can be determined *strictly* from the six equilibrium equations (three for each member). Once F_A, N_A, F_C, and N_C are determined, then the bars will remain in equilibrium provided $F_A \leq 0.3N_A$ and $F_C \leq 0.5N_C$ are satisfied.

Impending Motion at All Points. In this case *the total number of unknowns will equal the total number of available equilibrium equations plus the total number of available frictional equations*, $F = \mu N$. In particular, if *motion is impending* at the points of contact, then $F_s = \mu_s N$; whereas if the body is *slipping*, then $F_k = \mu_k N$. For example, consider the problem of finding the smallest angle θ at which the 100-N bar in Fig. 8–5*a* can be placed against the wall without slipping. The free-body diagram is shown in Fig. 8–5*b*. Here there are *five* unknowns: F_A, N_A, F_B, N_B, θ. For the solution there are *three* equilibrium equations and *two* static frictional equations which apply at *both* points of contact, so that $F_A = 0.3N_A$ and $F_B = 0.4N_B$.

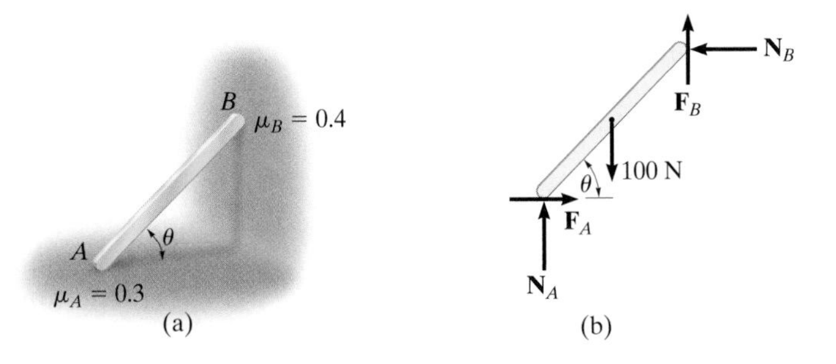

Fig. 8–5

Impending Motion at Some Points. Here *the total number of unknowns will be less than the number of available equilibrium equations plus the total number of frictional equations or conditional equations for tipping*. As a result, several possibilities for motion or impending motion will exist and the problem will involve a determination of the kind of motion which actually occurs. For example, consider the two-member frame shown in Fig. 8–6*a*. In this problem we wish to determine the horizontal force P needed to cause movement. If each member has a weight of 100 N, then the free-body diagrams are as shown in Fig. 8–6*b*. There are *seven* unknowns: N_A, F_A, N_C, F_C, B_x, B_y, P. For a unique solution we must satisfy the *six* equilibrium equations (three for each member) and only *one* of two possible static frictional equations. This means that as P increases it will either cause slipping at A and no slipping at C, so that $F_A = 0.3N_A$ and $F_C \leq 0.5N_C$; or slipping occurs at C and no slipping at A, in which case $F_C = 0.5N_C$ and $F_A \leq 0.3N_A$. The actual situation can be determined by calculating P for each case and then choosing the case for which P is *smaller*. If in both cases the *same value* for P is calculated, which in practice would be highly improbable, then slipping at both points occurs simultaneously; i.e., the *seven unknowns* will satisfy *eight equations*.

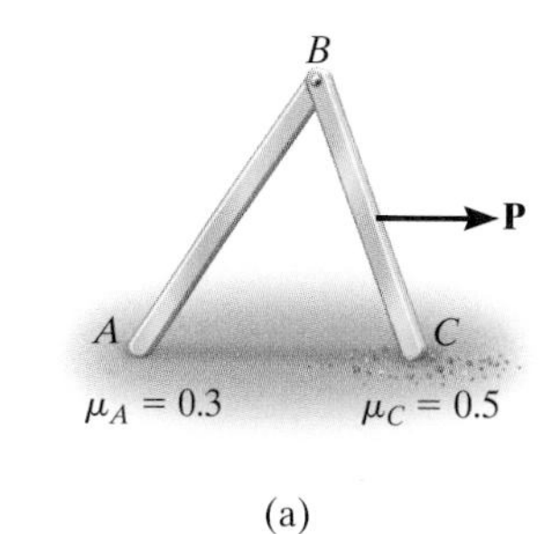

(a)

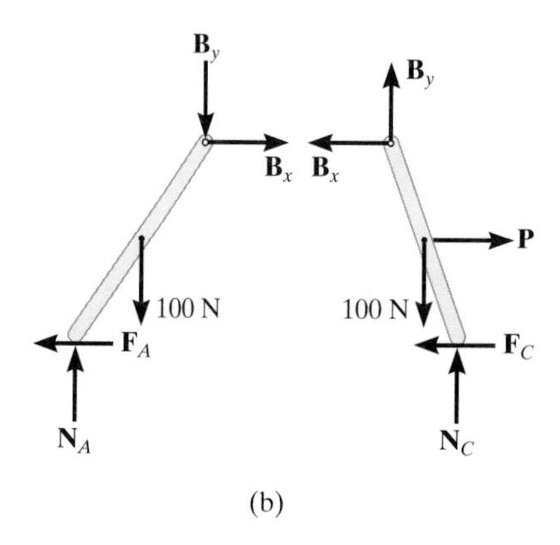

(b)

Fig. 8–6

Consider pushing on the uniform crate that has a weight W and sits on the rough surface. As shown on the first free-body diagram, if the magnitude of $\mathbf{P}$ is small, the crate will remain in equilibrium. As P increases the crate will either be on the verge of slipping on the surface ($F = \mu_s W$), or if the surface is very rough (large μ_s) then the resultant normal force will shift to the corner, $x = b/2$, as shown on the second free-body diagram, and the crate will tip over. The crate has a greater chance of tipping if $\mathbf{P}$ is applied at a greater height h above the surface, or if the crate's width b is smaller.

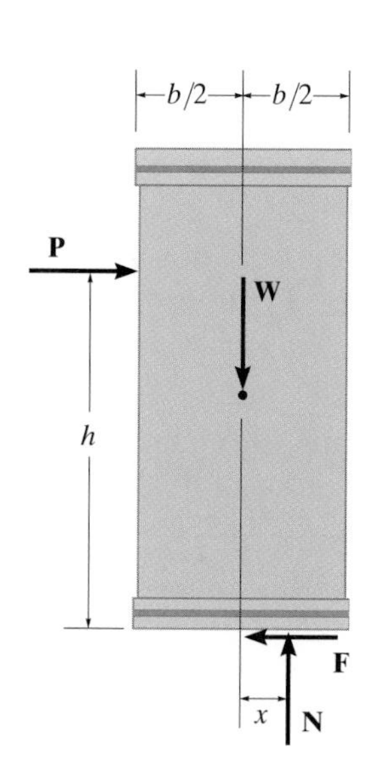

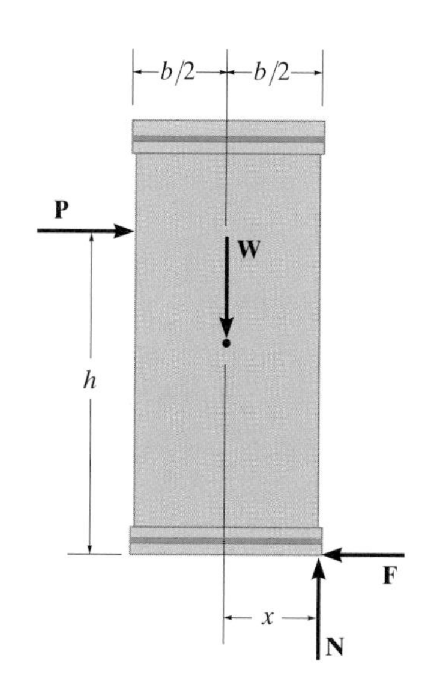

Equilibrium Versus Frictional Equations. It was stated earlier that the frictional force *always* acts so as to either oppose the relative motion or impede the motion of a body over its contacting surface. Realize, however, that we can *assume* the sense of the frictional force in problems which require F to be an "equilibrium force" and satisfy the inequality $F < \mu_s N$. The correct sense is made known *after* solving the equations of equilibrium for F. For example, if F is a negative scalar the sense of $\mathbf{F}$ is the reverse of that which was assumed. This convenience of *assuming* the sense of $\mathbf{F}$ is possible because the equilibrium equations equate to zero the *components of vectors* acting in the *same direction*. In cases where the frictional equation $F = \mu N$ is used in the solution of a problem, however, the convenience of *assuming* the sense of $\mathbf{F}$ is *lost*, since the frictional equation relates only the *magnitudes* of two perpendicular vectors. Consequently, $\mathbf{F}$ *must always* be shown acting with its *correct sense* on the free-body diagram whenever the frictional equation is used for the solution of a problem.

Procedure for Analysis

Equilibrium problems involving dry friction can be solved using the following procedure.

Free-Body Diagrams.

- Draw the necessary free-body diagrams, and unless it is stated in the problem that impending motion or slipping occurs, *always* show the frictional forces as unknowns (i.e., *do not assume* $F = \mu N$).
- Determine the number of unknowns and compare this with the number of available equilibrium equations.
- If there are more unknowns than equations of equilibrium, it will be necessary to apply the frictional equation at some, if not all, points of contact to obtain the extra equations needed for a complete solution.
- If the equation $F = \mu N$ is to be used, it will be necessary to show $\mathbf{F}$ acting in the proper direction on the free-body diagram.

Equations of Equilibrium and Friction.

- Apply the equations of equilibrium and the necessary frictional equations (or conditional equations if tipping is possible) and solve for the unknowns.
- If the problem involves a three-dimensional force system such that it becomes difficult to obtain the force components or the necessary moment arms, apply the equations of equilibrium using Cartesian vectors.

EXAMPLE 8.1

The uniform crate shown in Fig. 8–7*a* has a mass of 20 kg. If a force $P = 80$ N is applied to the crate, determine if it remains in equilibrium. The coefficient of static friction is $\mu = 0.3$.

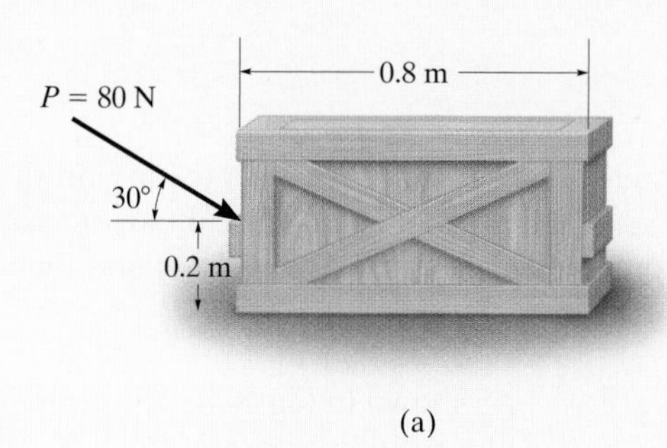

(a)

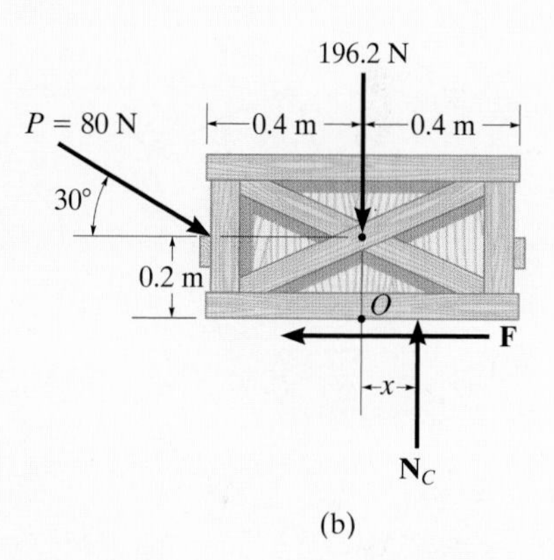

(b)

Fig. 8–7

SOLUTION

Free-Body Diagram. As shown in Fig. 8–7*b*, the *resultant* normal force $\mathbf{N}_C$ must act a distance x from the crate's center line in order to counteract the tipping effect caused by **P**. There are *three unknowns*, F, N_C, and x, which can be determined strictly from the *three* equations of equilibrium.

Equations of Equilibrium.

$$\xrightarrow{+} \Sigma F_x = 0; \qquad 80 \cos 30^\circ \text{ N} - F = 0$$

$$+\uparrow \Sigma F_y = 0; \qquad -80 \sin 30^\circ \text{ N} + N_C - 196.2 \text{ N} = 0$$

$$\circlearrowleft + \Sigma M_O = 0; \; 80 \sin 30^\circ \text{ N}(0.4 \text{ m}) - 80 \cos 30^\circ \text{ N}(0.2 \text{ m}) + N_C(x) = 0$$

Solving,

$$F = 69.3 \text{ N}$$

$$N_C = 236 \text{ N}$$

$$x = -0.00908 \text{ m} = -9.08 \text{ mm}$$

Since x is negative it indicates the *resultant* normal force acts (slightly) to the *left* of the crate's center line. No tipping will occur since $x \leq 0.4$ m. Also, the *maximum* frictional force which can be developed at the surface of contact is $F_{max} = \mu_s N_C = 0.3(236 \text{ N}) = 70.8$ N. Since $F = 69.3 \text{ N} < 70.8$ N, the crate will *not slip*, although it is very close to doing so.

EXAMPLE 8.2

(a)

It is observed that when the bed of the dump truck is raised to an angle of $\theta = 25°$ the vending machines begin to slide off the bed, Fig. 8–8*a*. Determine the static coefficient of friction between a vending machine and the surface of the truckbed.

SOLUTION

An idealized model of a vending machine resting on the truckbed is shown in Fig. 8–8*b*. The dimensions have been measured and the center of gravity has been located. We will assume that the machine weighs W.

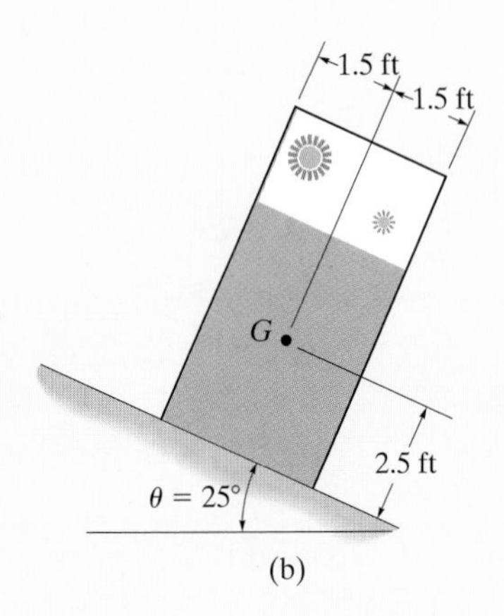

(b)

Free-Body Diagram. As shown in Fig. 8–8*c*, the dimension x is used to locate the position of the resultant normal force **N**. There are four unknowns, N, F, μ_s, and x.

Equations of Equilibrium.

$$+\searrow\Sigma F_x = 0; \qquad W \sin 25° - F = 0 \qquad (1)$$

$$+\nearrow\Sigma F_y = 0; \qquad N - W \cos 25° = 0 \qquad (2)$$

$$\circlearrowleft+\Sigma M_O = 0; \quad -W \sin\theta(2.5\text{ ft}) + W\cos\theta(x) = 0 \qquad (3)$$

Since slipping impends at $\theta = 25°$, using the first two equations, we have

$$F_s = \mu_s N; \qquad W \sin 25° = \mu_s(W \cos 25°)$$

$$\mu_s = \tan 25° = 0.466 \qquad \textit{Ans.}$$

The angle of $\theta = 25°$ is referred to as the *angle of repose*, and by comparison, it is equal to the angle of static friction, $\theta = \phi_s$. Notice from the calculation that θ is independent of the weight of the vending machine, and so knowing θ provides a convenient method for determining the coefficient of static friction.

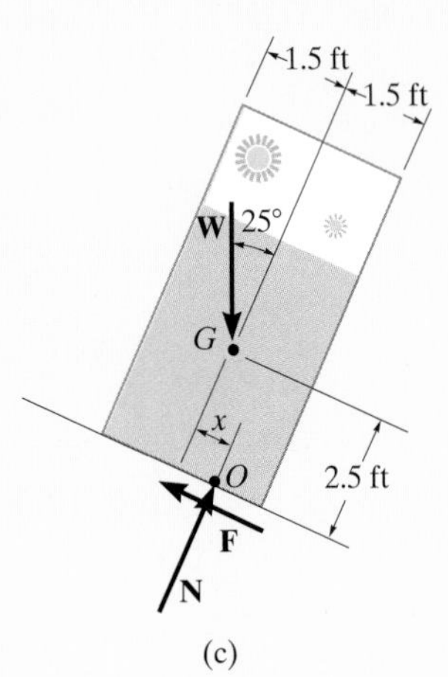

(c)

Fig. 8–8

NOTE: From Eq. 3, with $\theta = 25°$, we find $x = 1.17$ ft. Since $1.17\text{ ft} < 1.5\text{ ft}$, indeed the vending machine will slip before it can tip as observed in Fig. 8–8*a*.

EXAMPLE 8.3

The uniform rod having a weight W and length l is supported at its ends against the surface at A and B in Fig. 8–9a. If the rod is on the verge of slipping when $\theta = 30°$, determine the coefficient of static friction μ_s at A and B. Neglect the thickness of the rod for the calculation.

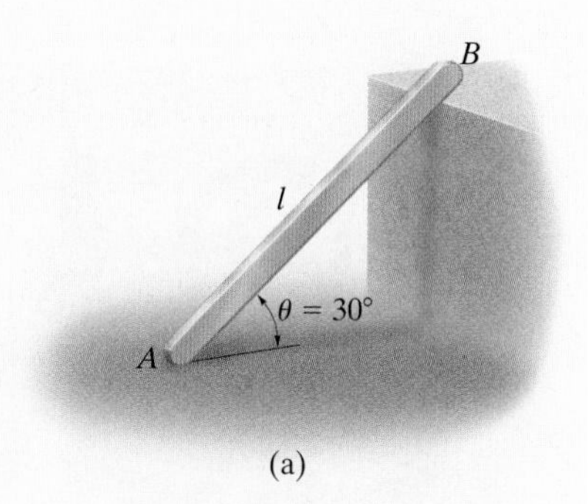

(a)

SOLUTION

Free-Body Diagram. As shown in Fig. 8–9b, there are *five* unknowns: F_A, N_A, F_B, N_B, and μ_s. These can be determined from the *three* equilibrium equations and *two* frictional equations applied at points A and B. The frictional forces must be drawn with their correct sense so that they oppose the tendency for motion of the rod. Why? (Refer to p. 400.)

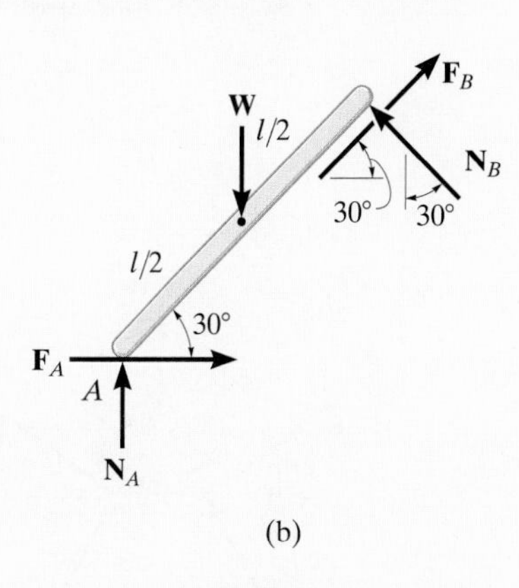

(b)

Fig. 8–9

Equations of Friction and Equilibrium. Writing the frictional equations,

$$F = \mu_s N; \qquad F_A = \mu_s N_A$$
$$F_B = \mu_s N_B$$

Using these results and applying the equations of equilibrium yields

$$\xrightarrow{+} \Sigma F_x = 0; \qquad \mu_s N_A + \mu_s N_B \cos 30° - N_B \sin 30° = 0 \qquad (1)$$

$$+\uparrow \Sigma F_y = 0; \qquad N_A - W + N_B \cos 30° + \mu_s N_B \sin 30° = 0 \qquad (2)$$

$$\circlearrowleft + \Sigma M_A = 0; \qquad N_B l - W\left(\frac{l}{2}\right)\cos 30° = 0 \qquad (3)$$

$$N_B = 0.4330W$$

From Eqs. 1 and 2,

$$\mu_s N_A = 0.2165W - (0.3750W)\mu_s$$
$$N_A = 0.6250W - (0.2165W)\mu_s$$

By division,

$$0.6250\mu_s - 0.2165\mu_s^2 = 0.2165 - 0.375\mu_s$$

or,

$$\mu_s^2 - 4.619\mu_s + 1 = 0$$

Solving for the smallest root,

$$\mu_s = 0.228$$ 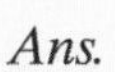 *Ans.*

EXAMPLE 8.4

The concrete pipes are stacked in the yard as shown in Fig. 8–10*a*. Determine the minimum coefficient of static friction at each point of contact so that the pile does not collapse.

(a)

SOLUTION

Free-body Diagrams. Recognize that the coefficient of static friction between two pipes, at A and B, and between a pipe and the ground, at C, will be different since the contacting surfaces are different. We will assume each pipe has an outer radius r and weight W. The free-body diagrams for two of the pipes are shown in Fig. 8–10*b*. There are six unknowns, N_A, F_A, N_B, F_B, N_C, F_C. (Note that when collapse is about to occur the normal force at D is zero.) Since only the six equations of equilibrium are necessary to obtain the unknowns, the sense of direction of the frictional forces can be verified from the solution.

Equations of Equilibrium. For the top pipe, we have

$$\circlearrowleft + \Sigma M_O = 0; \quad -F_A(r) + F_B(r) = 0; \quad F_A = F_B = F$$

$$\xrightarrow{+} \Sigma F_x = 0; \quad N_A \sin 30^\circ - F \cos 30^\circ - N_B \sin 30^\circ + F \cos 30^\circ = 0$$

$$N_A = N_B = N$$

$$+\uparrow \Sigma F_y = 0; \quad 2N \cos 30^\circ + 2F \sin 30^\circ - W = 0 \qquad (1)$$

For the bottom pipe, using $F_A = F$ and $N_A = N$, we have

$$\circlearrowleft + \Sigma M_{O'} = 0; \quad F_C(r) - F(r) = 0; \qquad F_C = F$$

$$\xrightarrow{+} \Sigma F_x = 0; \quad -N \sin 30^\circ + F \cos 30^\circ + F = 0 \qquad (2)$$

$$+\uparrow \Sigma F_y = 0; \quad N_C - W - N \cos 30^\circ - F \sin 30^\circ = 0 \qquad (3)$$

From Eq. 2, $F = 0.2679N$, so that between the pipes

$$(\mu_s)_{\min} = \frac{F}{N} = 0.268 \qquad \textit{Ans.}$$

Using this result in Eq. 1,

$$N = 0.5W$$

From Eq. 3,

$$N_C - W - (0.5W)\cos 30^\circ - 0.2679(0.5W)\sin 30^\circ = 0$$

$$N_C = 1.5W$$

At the ground, the smallest required coefficient of static friction would be

$$(\mu_s')_{\min} = \frac{F}{N_C} = \frac{0.2679(0.5W)}{1.5W} = 0.0893 \qquad \textit{Ans.}$$

NOTE: A greater coefficient of static friction is required between the pipes than that required at the ground; and so it is likely that if slipping would occur between the pipes the bottom two pipes would roll away from one another without slipping as the top pipe falls downward.

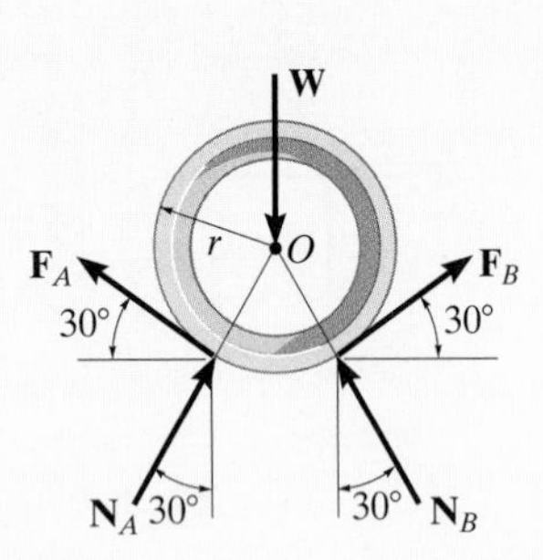

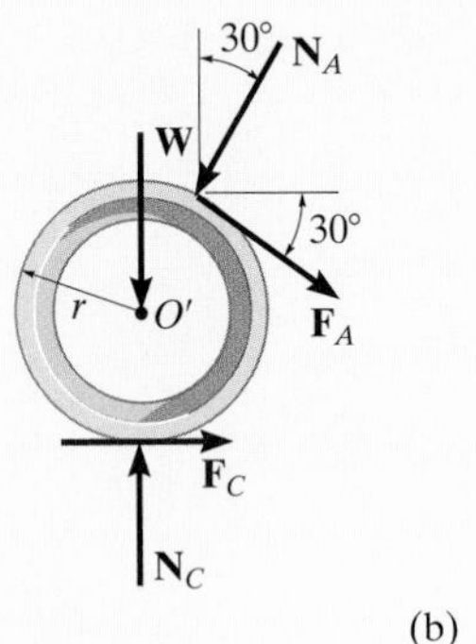

(b)

Fig. 8–10

EXAMPLE 8.5

Beam AB is subjected to a uniform load of 200 N/m and is supported at B by post BC, Fig. 8–11*a*. If the coefficients of static friction at B and C are $\mu_B = 0.2$ and $\mu_C = 0.5$, determine the force **P** needed to pull the post out from under the beam. Neglect the weight of the members and the thickness of the post.

(a)

SOLUTION

Free-Body Diagrams. The free-body diagram of beam AB is shown in Fig. 8–11*b*. Applying $\Sigma M_A = 0$, we obtain $N_B = 400$ N. This result is shown on the free-body diagram of the post, Fig. 8–11*c*. Referring to this member, the *four* unknowns F_B, P, F_C, and N_C are determined from the *three* equations of equilibrium and *one* frictional equation applied either at B or C.

Equations of Equilibrium and Friction.

$$\overset{+}{\rightarrow}\Sigma F_x = 0; \qquad P - F_B - F_C = 0 \qquad (1)$$

$$+\uparrow \Sigma F_y = 0; \qquad N_C - 400\text{ N} = 0 \qquad (2)$$

$$\circlearrowleft + \Sigma M_C = 0; \qquad -P(0.25\text{ m}) + F_B(1\text{ m}) = 0 \qquad (3)$$

(Post Slips Only at B.) This requires $F_C \leq \mu_C N_C$ and

$$F_B = \mu_B N_B; \qquad F_B = 0.2(400\text{ N}) = 80\text{ N}$$

Using this result and solving Eqs. 1 through 3, we obtain

$$P = 320\text{ N}$$
$$F_C = 240\text{ N}$$
$$N_C = 400\text{ N}$$

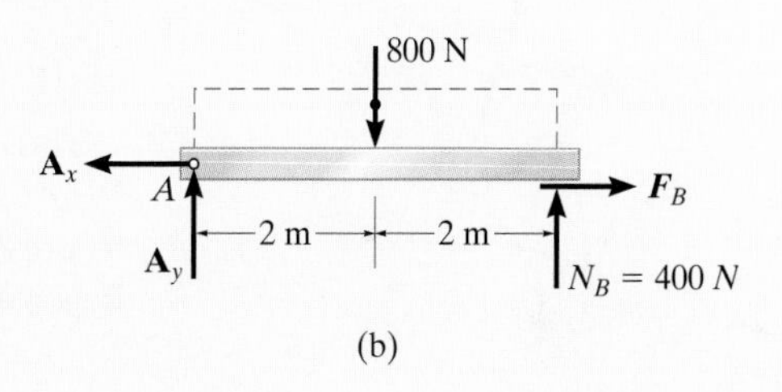

(b)

Since $F_C = 240\text{ N} > \mu_C N_C = 0.5(400\text{ N}) = 200$ N, the other case of movement must be investigated.

(Post Slips Only at C.) Here $F_B \leq \mu_B N_B$ and

$$F_C = \mu_C N_C; \qquad F_C = 0.5 N_C \qquad (4)$$

Solving Eqs. 1 through 4 yields

$$P = 267\text{ N}$$
$$N_C = 400\text{ N} \qquad \textit{Ans.}$$
$$F_C = 200\text{ N}$$
$$F_B = 66.7\text{ N}$$

(c)

Obviously, this case occurs first since it requires a *smaller* value for P.

Fig. 8–11

EXAMPLE 8.6

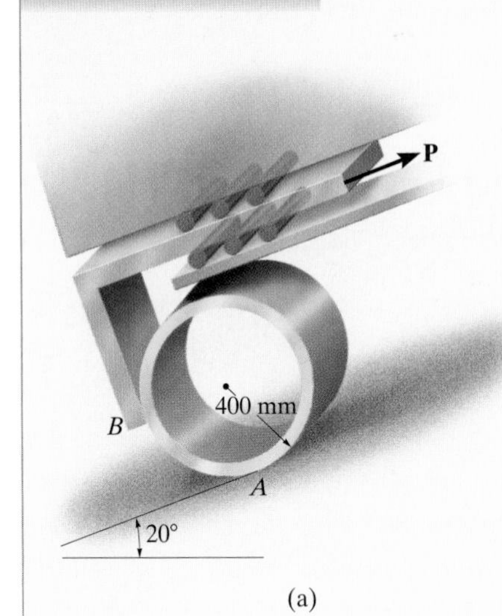

(a)

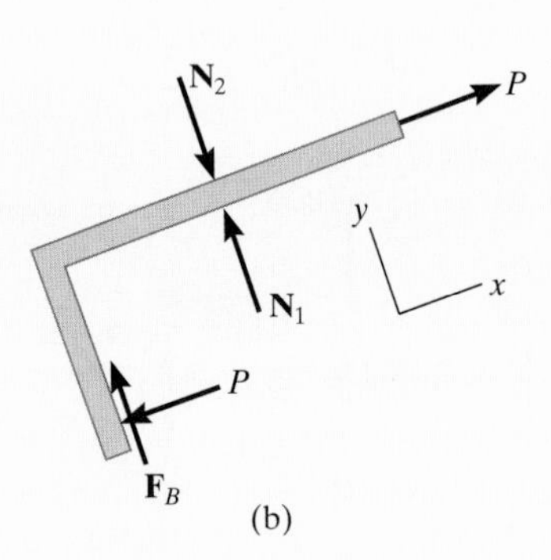

(b)

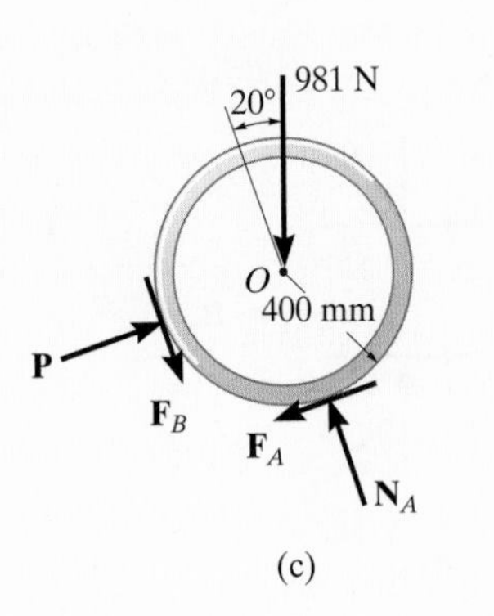

(c)

Fig. 8–12

Determine the normal force P that must be exerted on the rack to begin pushing the 100-kg pipe shown in Fig. 8–12*a* up the 20° incline. The coefficients of static friction at the points of contact are $(\mu_s)_A = 0.15$ and $(\mu_s)_B = 0.4$.

SOLUTION

Free-Body Diagram. As shown in Fig. 8–12*b*, the rack must exert a force P on the pipe due to force equilibrium in the x direction. There are four unknowns P, F_A, N_A, and F_B acting on the pipe Fig. 8–12*c*. These can be determined from the *three* equations of equilibrium and *one* frictional equation, which apply either at A or B. If slipping begins to occur only at B, the pipe will begin to roll up the incline; whereas if slipping occurs only at A, the pipe will begin to *slide* up the incline. Here we must find N_B.

Equations of Equilibrium and Friction. (Fig. 8–12*c*)

$$+\nearrow\Sigma F_x = 0; \qquad -F_A + P - 981 \sin 20^\circ \text{ N} = 0 \qquad (1)$$

$$+\nwarrow\Sigma F_y = 0; \qquad N_A - F_B - 981 \cos 20^\circ \text{ N} = 0 \qquad (2)$$

$$\circlearrowright+\Sigma M_O = 0; \qquad F_B(400 \text{ mm}) - F_A(400 \text{ mm}) = 0 \qquad (3)$$

(Pipe Rolls up Incline.) In this case, $F_A \le 0.15N_A$ and

$$(F_s)_B = (\mu_s)_B N_B; \qquad F_B = 0.4P \qquad (4)$$

The direction of the frictional force at B must be specified correctly. Why? Since the spool is being forced up the incline, $\mathbf{F}_B$ acts downward to prevent any clockwise rolling motion of the pipe, Fig. 8–12*c*. Solving Eqs. 1 through 4, we have

$$N_A = 1146 \text{ N} \quad F_A = 224 \text{ N} \quad F_B = 224 \text{ N} \quad P = 559 \text{ N}$$

The assumption regarding no slipping at A should be checked.

$$F_A \le (\mu_s)_A N_A; \qquad 224 \text{ N} \overset{?}{\le} 0.15(1146 \text{ N}) = 172 \text{ N}$$

The inequality does *not apply*, and therefore slipping occurs at A and not at B. Hence, the other case of motion will occur.

(Pipe Slides up Incline.) In this case, $P \le 0.4N_B$ and

$$(F_s)_A = (\mu_s)_A N_A; \qquad F_A = 0.15N_A \qquad (5)$$

Solving Eqs. 1 through 3 and 5 yields

$$N_A = 1085 \text{ N} \quad F_A = 163 \text{ N} \quad F_B = 163 \text{ N} \quad P = 498 \text{ N} \qquad \textit{Ans.}$$

NOTE: The validity of the solution ($P = 498$ N) can be checked by testing the assumption that indeed no slipping occurs at B.

$$F_B \le (\mu_s)_B P; \qquad 163 \text{ N} < 0.4(498 \text{ N}) = 199 \text{ N} \qquad \text{(check)}$$

PROBLEMS

8–1. If the horizontal force $P = 80$ lb, determine the normal and frictional forces acting on the 300-lb crate. Take $\mu_s = 0.3$, $\mu_k = 0.2$.

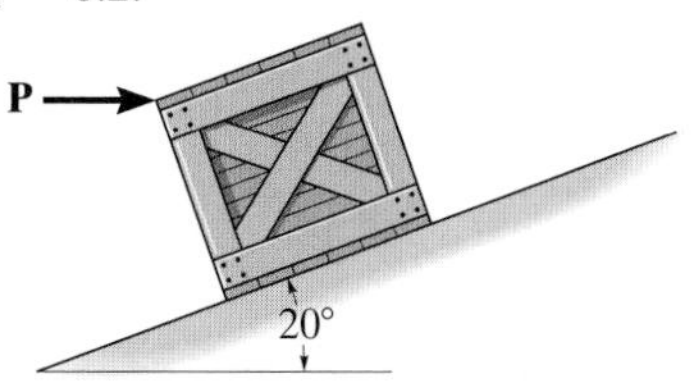

Prob. 8–1

8–2. Determine the magnitude of force **P** needed to start towing the 40-kg crate. Also determine the location of the resultant normal force acting on the crate, measured from point A. Take $\mu_s = 0.3$.

8–3. Determine the friction force on the 40-kg crate, and the resultant normal force and its position x, measured from point A, if the force $P = 300$ N. Take $\mu_s = 0.5$ and $\mu_k = 0.2$.

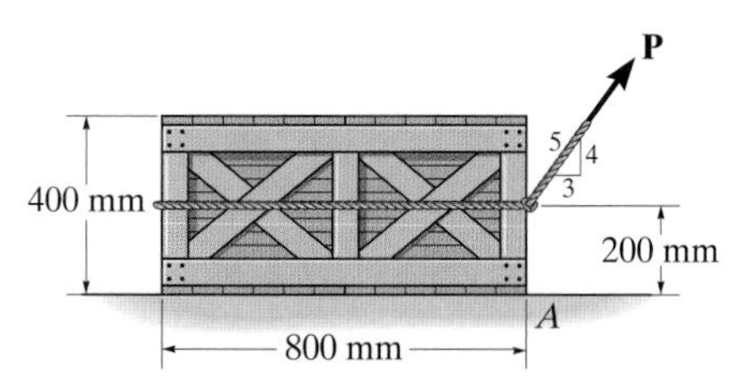

Probs. 8–2/3

***8–4.** The loose-fitting collar is supported by the pipe for which the coefficient of static friction at the points of contact A and B is $\mu_s = 0.2$. Determine the largest dimension d so the rod will not slip when the load **P** is applied.

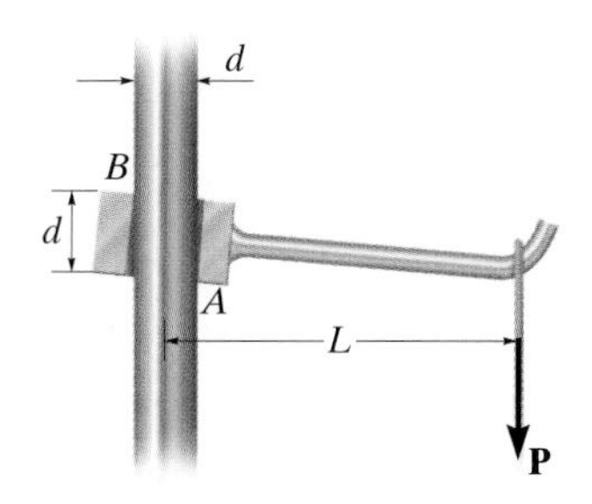

Prob. 8–4

8–5. The spool of wire having a mass of 150 kg rests on the ground at A and against the wall at B. Determine the force P required to begin pulling the wire horizontally off the spool. The coefficient of static friction between the spool and its points of contact is $\mu_s = 0.25$.

8–6. The spool of wire having a mass of 150 kg rests on the ground at A and against the wall at B. Determine the forces acting on the spool at A and B if $P = 800$ N. The coefficient of static friction between the spool and the ground at point A is $\mu_s = 0.35$. The wall at B is smooth.

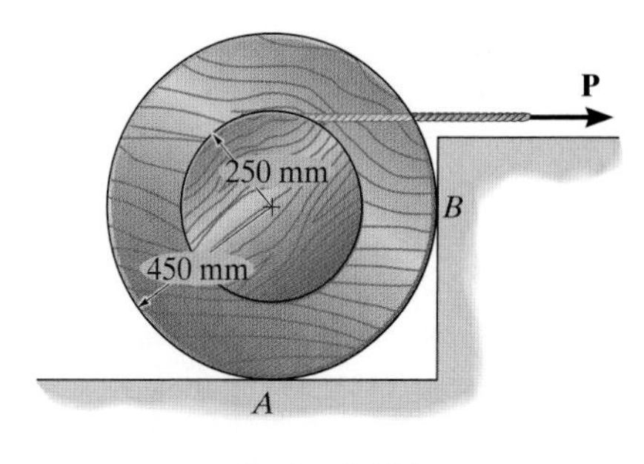

Probs. 8–5/6

8–7. The crate has a mass of 350 kg and is subjected to a towing force **P** acting at a 20° angle with the horizontal. If the coefficient of static friction is $\mu_s = 0.5$, determine the magnitude of **P** to just start the crate moving down the plane.

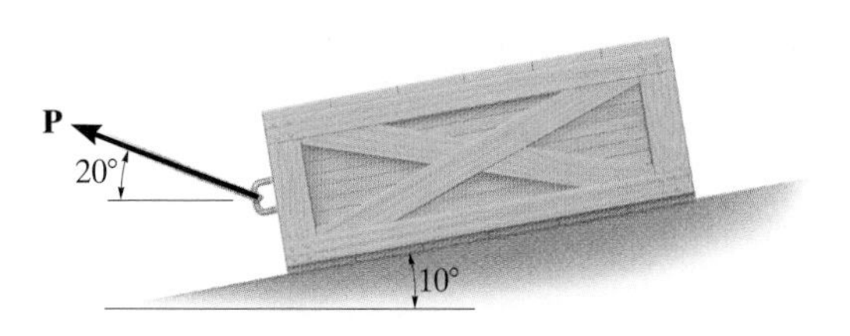

Prob. 8–7

***8–8.** The winch on the truck is used to hoist the garbage bin onto the bed of the truck. If the loaded bin has a weight of 8500 lb and center of gravity at G, determine the force in the cable needed to begin the lift. The coefficients of static friction at A and B are $\mu_A = 0.3$ and $\mu_B = 0.2$, respectively. Neglect the height of the support at A.

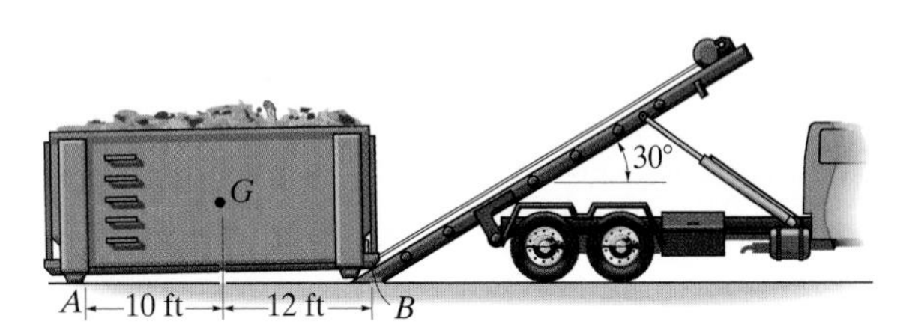

Prob. 8–8

8–9. The motorcyclist travels with constant velocity along a straight, horizontal, banked road. If he aligns his bike so that the tires are perpendicular to the road at A, determine the frictional force at A. The man has a mass of 60 kg and a mass center at G_C, and the motorcycle has a mass of 120 kg and a mass center at G_m. If the coefficient of static friction at A is $\mu_A = 0.4$, will the bike slip?

Prob. 8–9

8–10. The block brake is used to stop the wheel from rotating when the wheel is subjected to a couple moment $\mathbf{M}_0$. If the coefficient of static friction between the wheel and the block is μ_s, determine the smallest force P that should be applied.

8–11. Show that the brake in Prob. 8–10 is self locking, i.e., $P \leq 0$, provided $b/c \leq \mu_s$.

***8–12.** Solve Prob. 8–10 if the couple moment $\mathbf{M}_0$ is applied counterclockwise.

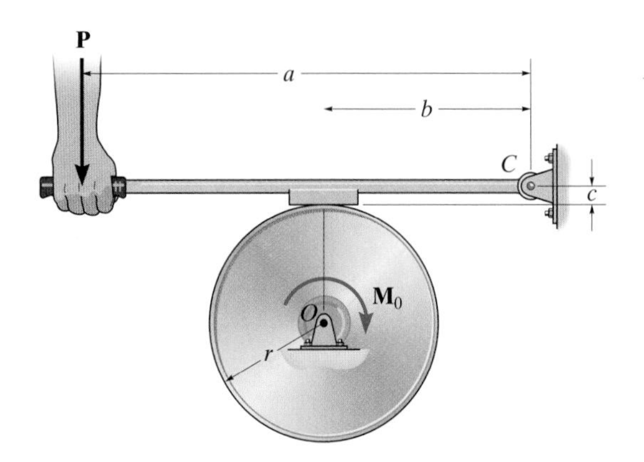

Probs. 8–10/11/12

8–13. The block brake consists of a pin-connected lever and friction block at B. The coefficient of static friction between the wheel and the lever is $\mu_s = 0.3$, and a torque of $5\ \text{N}\cdot\text{m}$ is applied to the wheel. Determine if the brake can hold the wheel stationary when the force applied to the lever is (a) $P = 30$ N, (b) $P = 70$ N.

8–14. Solve Prob. 8–13 if the $5\text{-N}\cdot\text{m}$ torque is applied counter-clockwise.

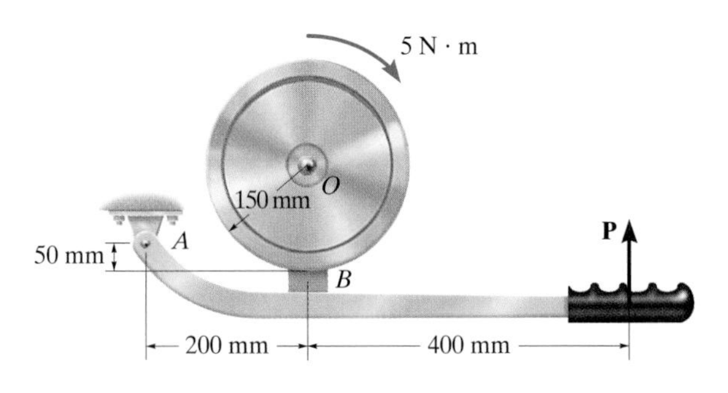

Probs. 8–13/14

8–15. The doorstop of negligible weight is pin connected at A and the coefficient of static friction at B is $\mu_s = 0.3$. Determine the required distance s from A to the floor so that the stop will resist opening of the door for any force P applied to the handle.

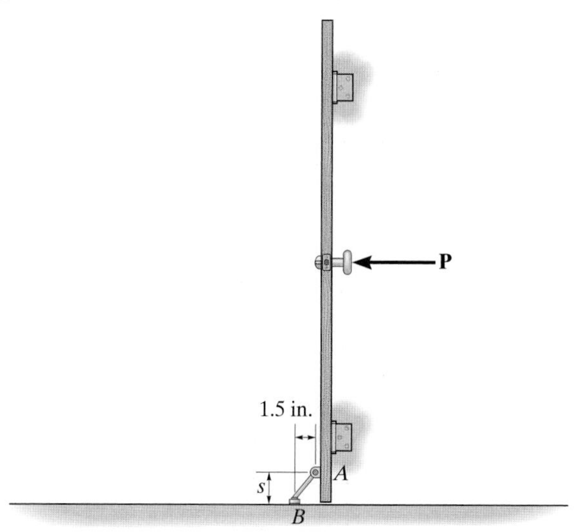

Prob. 8–15

8–17. The uniform hoop of weight W is suspended from the peg at A and a horizontal force $\mathbf{P}$ is slowly applied at B. If the hoop begins to slip at A when $\theta = 30°$, determine the coefficient of static friction between the hoop and the peg.

8–18. The uniform hoop of weight W is suspended from the peg at A and a horizontal force $\mathbf{P}$ is slowly applied at B. If the coefficient of static friction between the hoop and peg is $\mu_s = 0.2$, determine if it is possible for the angle $\theta = 30°$ before the hoop begins to slip.

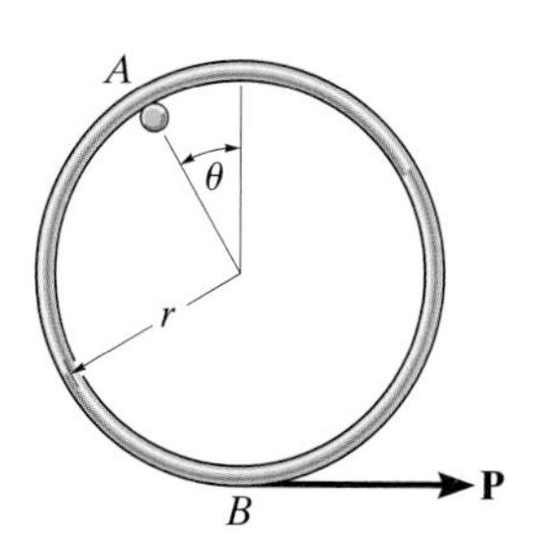

Prob. 8–17/18

***8–16.** The chair has a weight of 10 lb and center of gravity at G. It is propped against the door as shown. If the coefficient of static friction at A is $\mu_A = 0.3$, determine the smallest force P that must be applied to the handle to open the door.

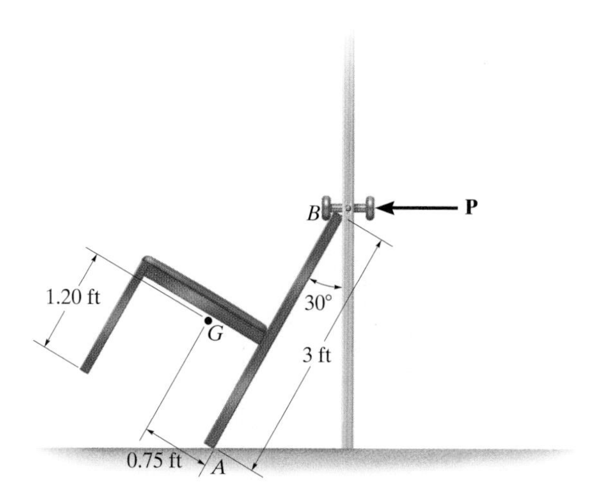

Prob. 8–16

8–19. The coefficient of static friction between the shoes at A and B of the tongs and the pallet is $\mu_s' = 0.5$, and between the pallet and the floor $\mu_s = 0.4$. If a horizontal towing force of $P = 300$ N is applied to the tongs, determine the largest mass that can be towed.

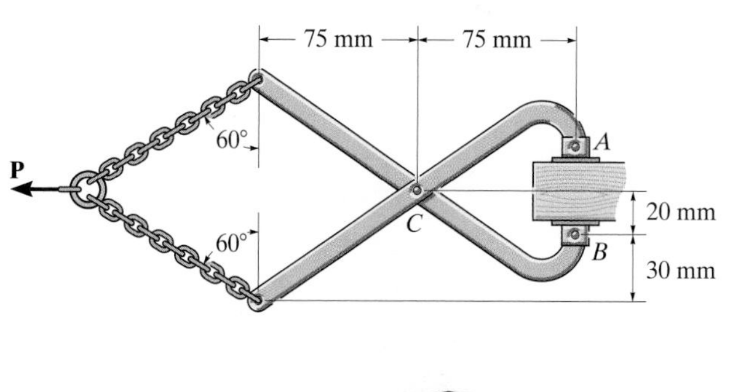

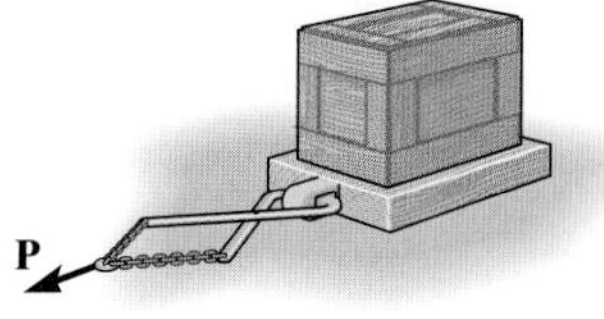

Prob. 8–19

***8–20.** The pipe is hoisted using the tongs. If the coefficient of static friction at A and B is μ_s, determine the smallest dimension b so that any pipe of inner diameter d can be lifted.

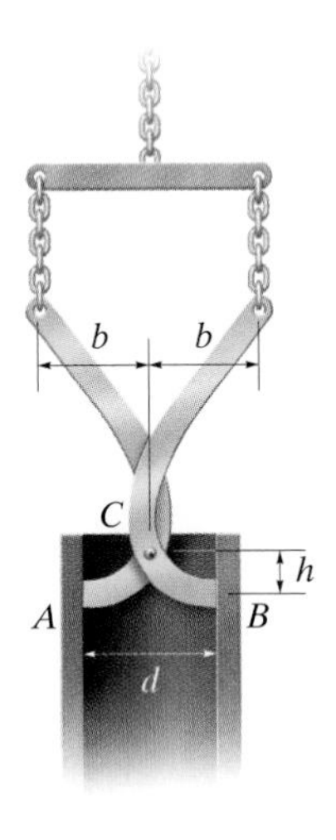

Prob. 8–20

8–21. A very thin bookmark having a width of 1 in. is in the middle of a 10-lb dictionary. If the pages are 8 in. by 10 in., determine the force P needed to start to pull the bookmark out. The coefficient of static friction between the bookmark and the paper is $\mu_s = 0.7$. Assume the pressure on each page and the bookmark is uniform.

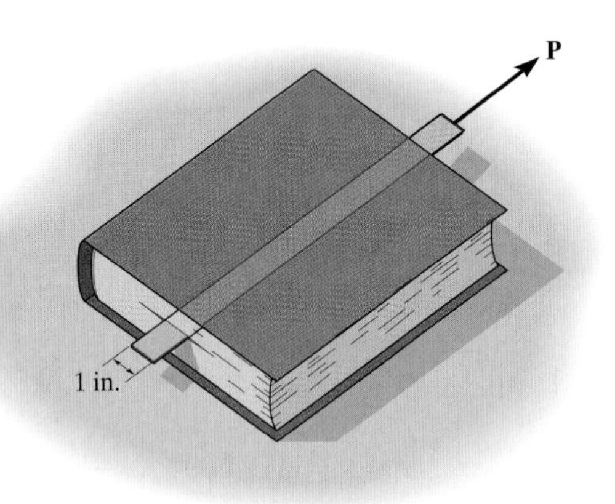

Prob. 8–21

8–22. The uniform dresser has a weight of 90 lb and rests on a tile floor for which $\mu_s = 0.25$. If the man pushes on it in the horizontal direction $\theta = 0°$, determine the smallest magnitude of force **F** needed to move the dresser. Also, if the man has a weight of 150 lb, determine the smallest coefficient of static friction between his shoes and the floor so that he does not slip.

8–23. The uniform dresser has a weight of 90 lb and rests on a tile floor for which $\mu_s = 0.25$. If the man pushes on it in the direction $\theta = 30°$, determine the smallest magnitude of force **F** needed to move the dresser. Also, if the man has a weight of 150 lb, determine the smallest coefficient of static friction between his shoes and the floor so that he does not slip.

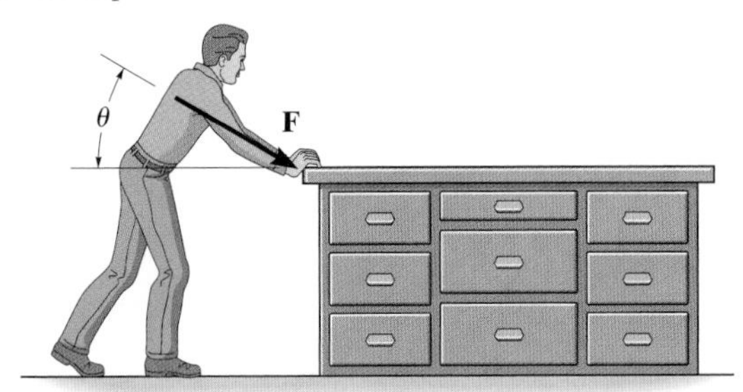

Probs. 8–22/23

***8–24.** The cam is subjected to a couple moment of 5 N · m. Determine the minimum force P that should be applied to the follower in order to hold the cam in the position shown. The coefficient of static friction between the cam and the follower is $\mu_s = 0.4$. The guide at A is smooth.

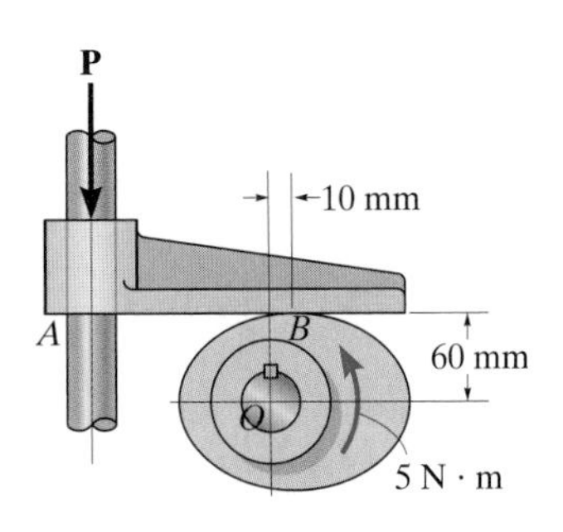

Prob. 8–24

8–25. The board can be adjusted vertically by tilting it up and sliding the smooth pin A along the vertical guide G. When placed horizontally, the bottom C then bears along the edge of the guide, where $\mu_s = 0.4$. Determine the largest dimension d which will support any applied force $\mathbf{F}$ without causing the board to slip downward.

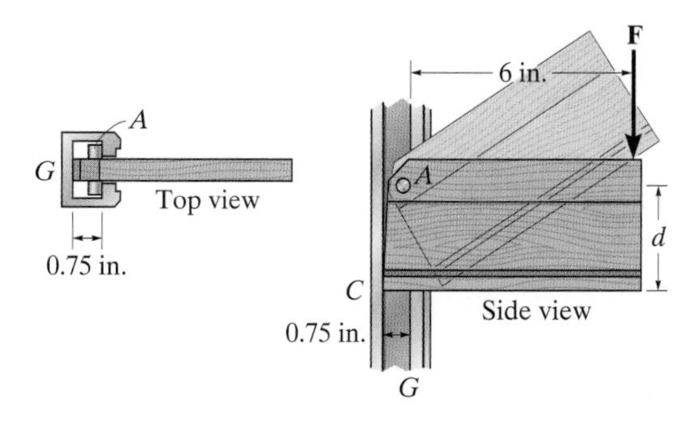

Prob. 8–25

8–26. The homogeneous semicylinder has a mass m and mass center at G. Determine the largest angle θ of the inclined plane upon which it rests so that it does not slip down the plane. The coefficient of static friction between the plane and the cylinder is $\mu_s = 0.3$. Also, what is the angle ϕ for this case?

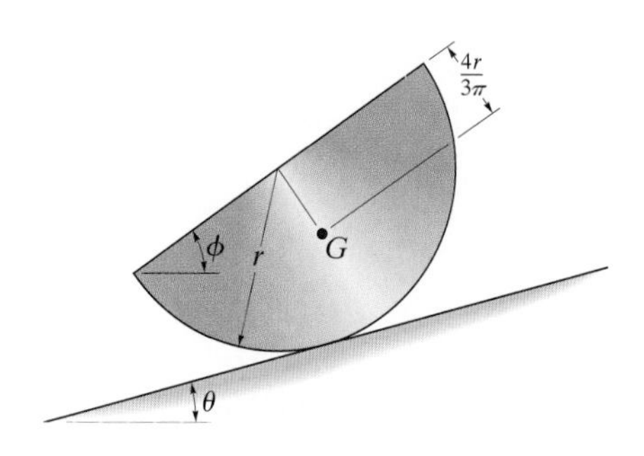

Prob. 8–26

8–27. A chain having a length of 20 ft and weight of 8 lb/ft rests on a street for which the coefficient of static friction is $\mu_s = 0.2$. If a crane is used to hoist the chain, determine the force P it applies to the chain if the length of chain remaining on the ground begins to slip when the horizontal component of P becomes 10 lb. What length of chain remains on the ground?

Prob. 8–27

***8–28.** The fork lift has a weight of 2400 lb and a center of gravity at G. If the rear wheels are powered, whereas the front wheels are free to roll, determine the maximum number of 300-lb crates the fork lift can push forward. The coefficient of static friction between the wheels and the ground is $\mu_s = 0.4$, and between each crate and the ground $\mu_s' = 0.35$.

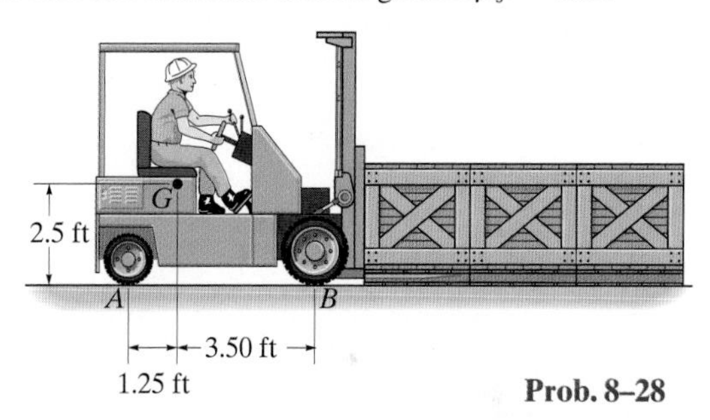

Prob. 8–28

8–29. The brake is to be designed to be self locking, that is, it will not rotate when no load $\mathbf{P}$ is applied to it when the disk is subjected to a clockwise couple moment $\mathbf{M}_0$. Determine the distance d of the lever that will allow this to happen. The coefficient of static friction at B is $\mu_s = 0.5$.

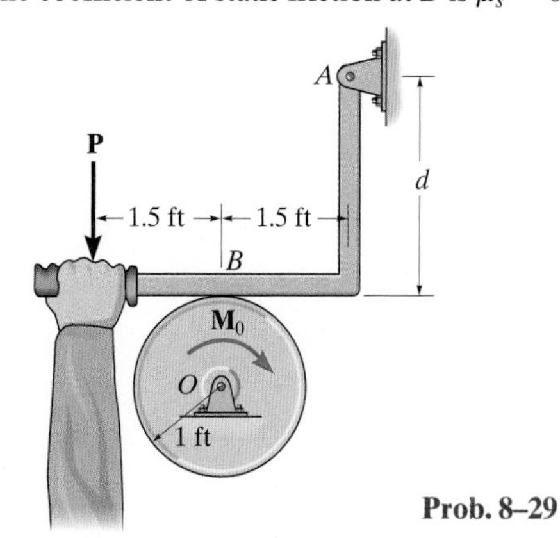

Prob. 8–29

8–30. The 800-lb concrete pipe is being lowered from the truck bed when it is in the position shown. If the coefficient of static friction at the points of support A and B is $\mu_s = 0.4$, determine where it begins to slip first: at A or B, or both at A and B.

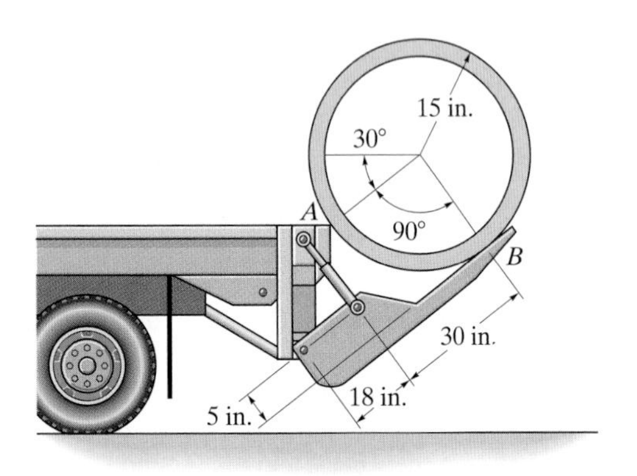

Prob. 8–30

8–31. A 5-kg wedge is placed in the grooved slot of an inclined plane. Determine the maximum angle θ for the incline without causing the wedge to slip. The coefficient of static friction between the wedge and the surfaces of contact is $\mu_s = 0.2$.

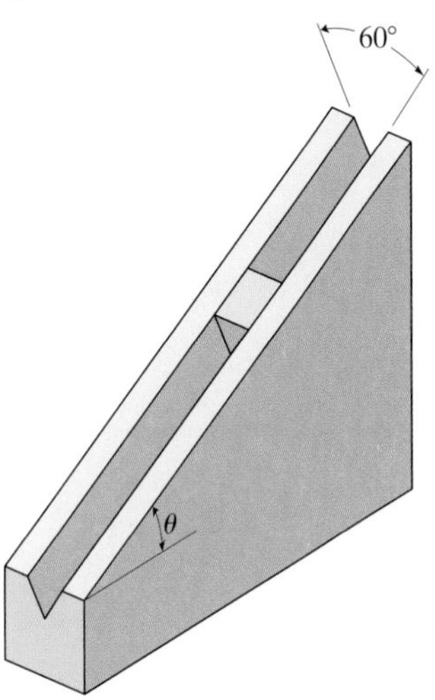

Prob. 8–31

***8–32.** A roll of paper has a uniform weight of 0.75 lb and is suspended from the wire hanger so that it rests against the wall. If the hanger has a negligible weight and the bearing at O can be considered frictionless, determine the force P needed to start turning the roll if $\theta = 30°$. The coefficient of static friction between the wall and the paper is $\mu_s = 0.25$.

8–33. A roll of paper has a uniform weight of 0.75 lb and is suspended from the wire hanger so that it rests against the wall. If the hanger has a negligible weight and the bearing at O can be considered frictionless, determine the minimum force P and the associated angle θ needed to start turning the roll. The coefficient of static friction between the wall and the paper is $\mu_s = 0.25$.

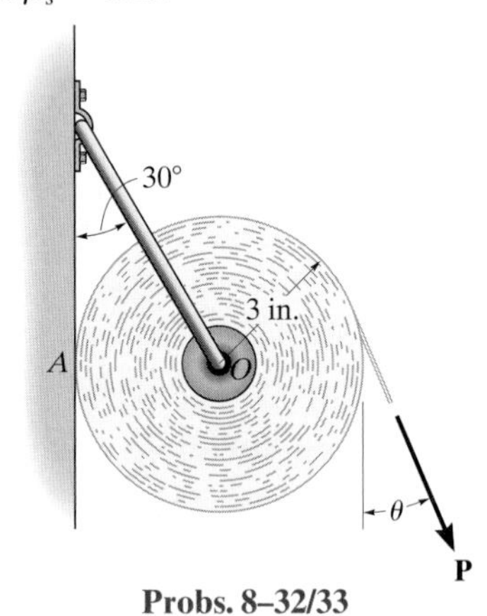

Probs. 8–32/33

8–34. The door brace AB is to be designed to prevent opening the door. If the brace forms a pin connection under the doorknob and the coefficient of static friction with the floor is $\mu_s = 0.5$, determine the largest length L the brace can have to prevent the door from being opened. Neglect the weight of the brace.

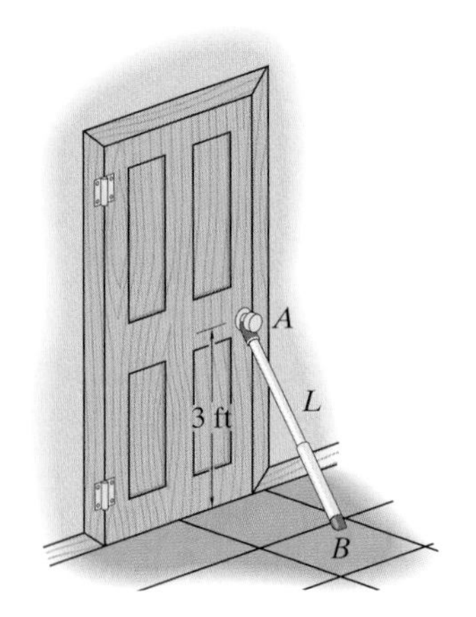

Prob. 8–34

8–35. The man has a weight of 200 lb, and the coefficient of static friction between his shoes and the floor is $\mu_s = 0.5$. Determine where he should position his center of gravity G at d in order to exert the maximum horizontal force on the door. What is this force?

Prob. 8–35

***8–36.** In an effort to move the two 100-lb crates, which are stacked on top of one another, the man pushes horizontally on them at the bottom of crate A as shown. Determine the smallest force P that must be applied in order to cause impending motion. Explain what happens. The coefficient of static friction between the crates is $\mu_s = 0.8$, and between the bottom crate and the floor $\mu_s' = 0.3$.

8–37. The man having a weight of 150 lb pushes horizontally on the bottom of crate A, which is stacked on top of crate B. Each crate has a weight of 100 lb. If the coefficient of static friction between each crate is $\mu_s = 0.8$ and between the bottom crate, his shoes, and the floor is $\mu_s' = 0.3$, determine if he can cause impending motion.

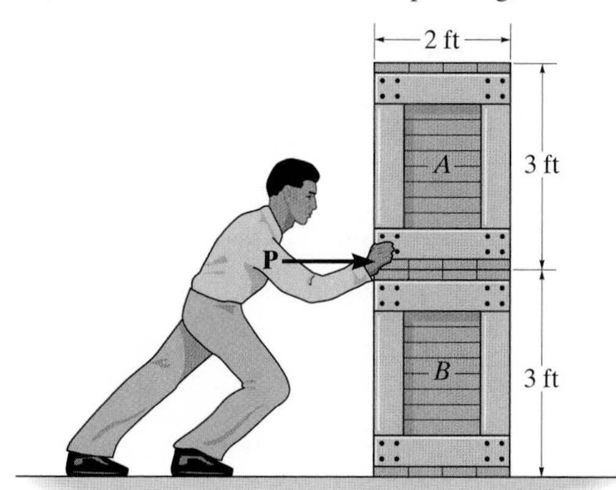

Probs. 8–36/37

8–38. The crate has a weight of 200 lb and a center of gravity at G. Determine the horizontal force P required to tow it. Also, determine the location of the resultant normal force measured from A. Take $h = 4$ ft and $\mu_s = 0.4$.

8–39. The crate has a weight of 200 lb and a center of gravity at G. Determine the height h of the tow rope so that the crate slips and tips at the same time. What horizontal force P is required to do this? Take $\mu_s = 0.4$.

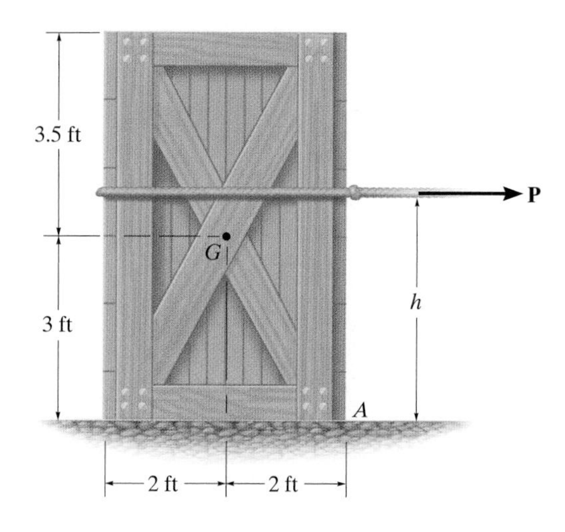

Probs. 8–38/39

***8–40.** Determine the smallest force the man must exert on the rope in order to move the 80-kg crate. Also, what is the angle θ at this moment? The coefficient of static friction between the crate and the floor is $\mu_s = 0.3$.

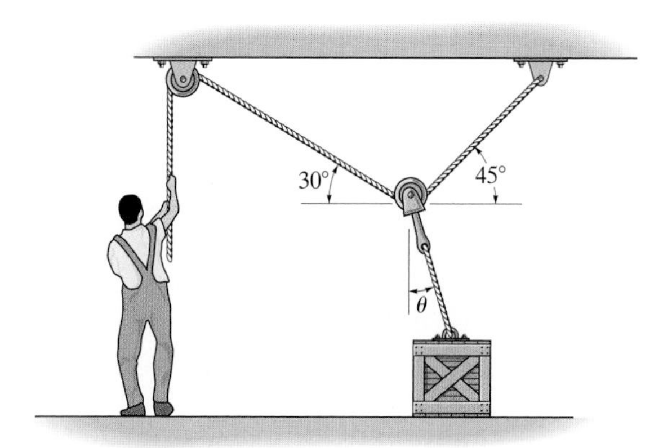

Prob. 8–40

8–41. The symmetrical crab hook is used to lift packages by means of friction developed between the shoes A and B and a package. Determine the smallest coefficient of static friction at the shoes so that the package of weight W can be lifted.

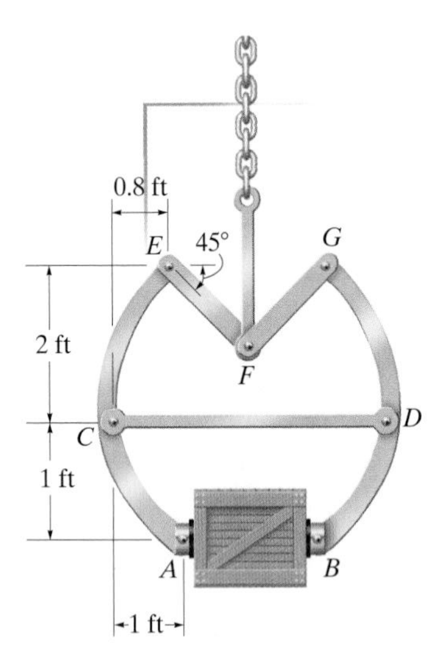

Prob. 8–41

8–42. The friction hook is made from a fixed frame which is shown colored and a cylinder of negligible weight. A piece of paper is placed between the smooth wall and the cylinder. If $\theta = 20°$, determine the smallest coefficient of static friction μ at all points of contact so that any weight W of paper p can be held.

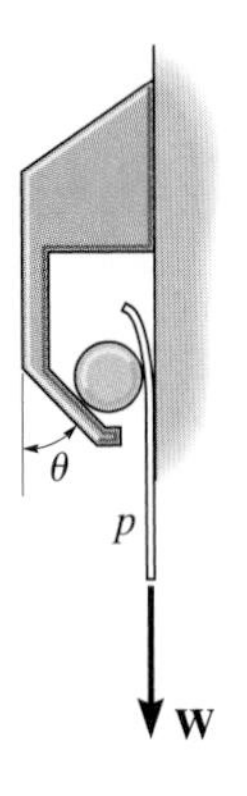

Prob. 8–42

8–43. The crate has a weight of 300 lb and a center of gravity at G. If the coefficient of static friction between the crate and the floor is $\mu_s = 0.2$, determine if the 200-lb man can *push* the crate to the left. The coefficient of static friction between his shoes and the floor is $\mu_s' = 0.4$. Assume the man exerts only a horizontal force on the crate.

***8–44.** The crate has a weight of 300 lb and a center of gravity at G. If the coefficient of static friction between the crate and the floor is $\mu_s = 0.2$, determine the smallest weight of the man so he can push the crate to the left. The coefficient of static friction between his shoes and the floor is $\mu_s' = 0.35$. Assume the man exerts only a horizontal force on the crate.

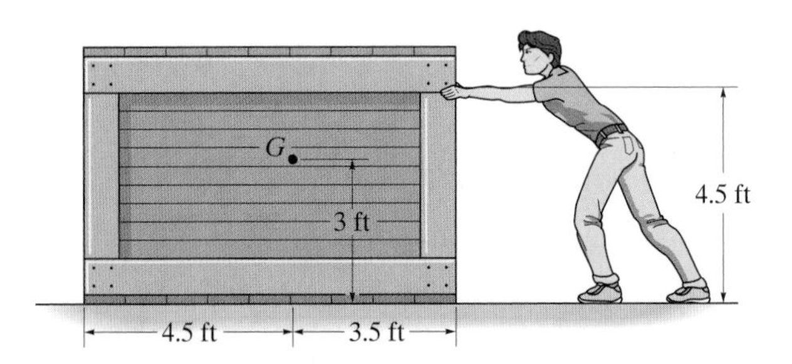

Probs. 8–43/44

8–45. The wheel weighs 20 lb and rests on a surface for which $\mu_B = 0.2$. A cord wrapped around it is attached to the top of the 30-lb homogeneous block. If the coefficient of static friction at D is $\mu_D = 0.3$, determine the smallest vertical force that can be applied tangentially to the wheel which will cause motion to impend.

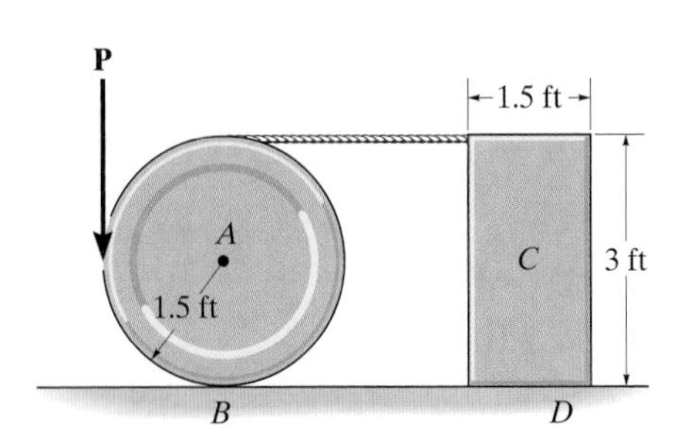

Prob. 8–45

8–46. Determine the smallest couple moment which can be applied to the 20-lb wheel that will cause impending motion. The cord is attached to the 30-lb block, and the coefficients of static friction are $\mu_B = 0.2$ and $\mu_D = 0.3$.

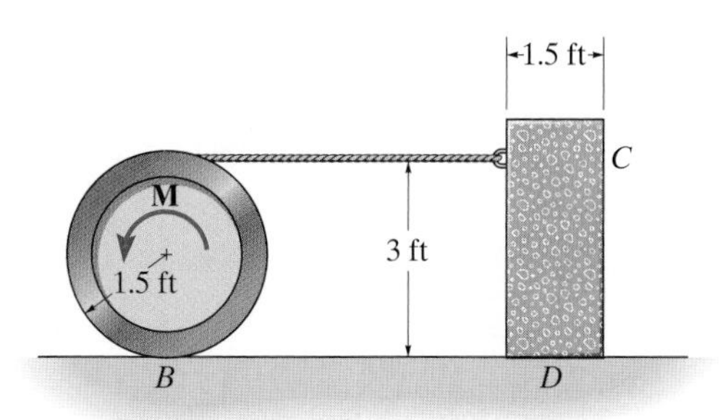

Prob. 8–46

8–47. The beam AB has a negligible mass and thickness and is subjected to a triangular distributed loading. It is supported at one end by a pin and at the other end by a post having a mass of 50 kg and negligible thickness. Determine the minimum force P needed to move the post. The coefficients of static friction at B and C are $\mu_B = 0.4$ and $\mu_C = 0.2$, respectively.

***8–48.** The beam AB has a negligible mass and thickness and is subjected to a triangular distributed loading. It is supported at one end by a pin and at the other end by a post having a mass of 50 kg and negligible thickness. Determine the two coefficients of static friction at B and at C so that when the magnitude of the applied force is increased to $P = 150$ N, the post slips at both B and C simultaneously.

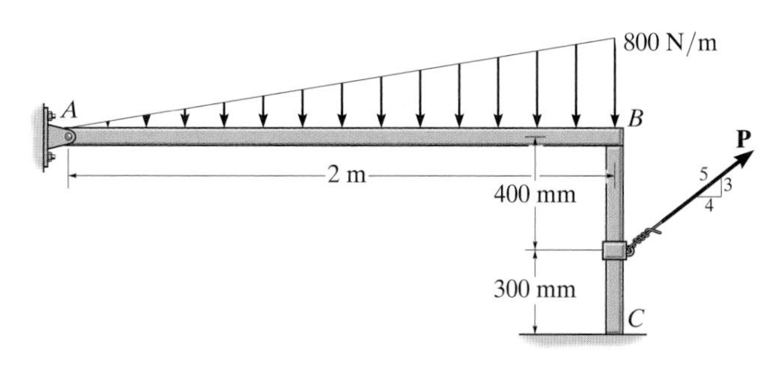

Probs. 8–47/48

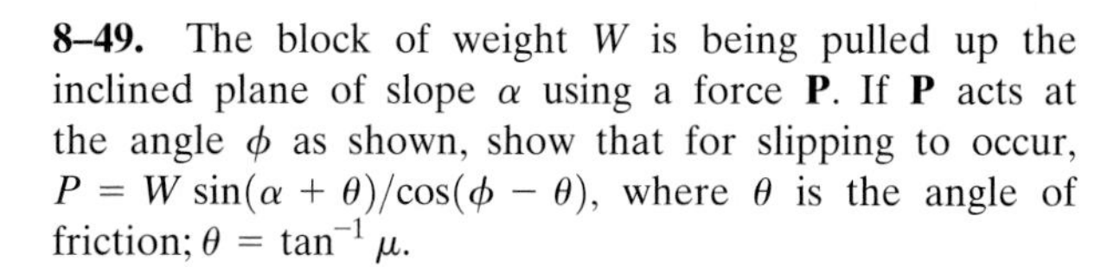

8–49. The block of weight W is being pulled up the inclined plane of slope α using a force **P**. If **P** acts at the angle ϕ as shown, show that for slipping to occur, $P = W\sin(\alpha + \theta)/\cos(\phi - \theta)$, where θ is the angle of friction; $\theta = \tan^{-1}\mu$.

8–50. Determine the angle ϕ at which **P** should act on the block so that the magnitude of **P** is as small as possible to begin pushing the block up the incline. What is the corresponding value of P? The block weighs W and the slope α is known.

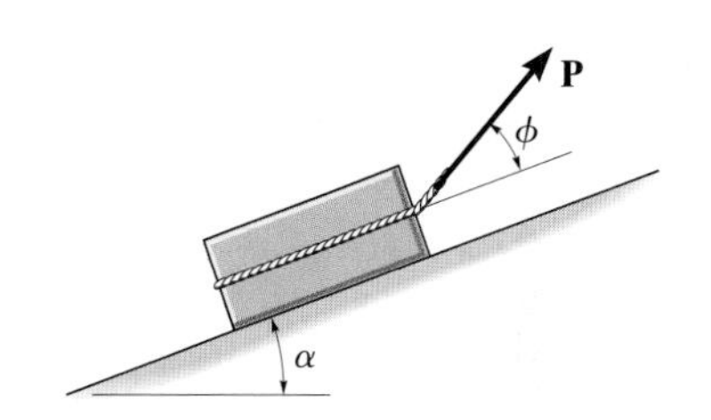

Probs. 8–49/50

8–51. Two blocks A and B, each having a mass of 6 kg, are connected by the linkage shown. If the coefficient of static friction at the contacting surfaces is $\mu_s = 0.5$, determine the largest vertical force P that may be applied to pin C of the linkage without causing the blocks to move. Neglect the weight of the links.

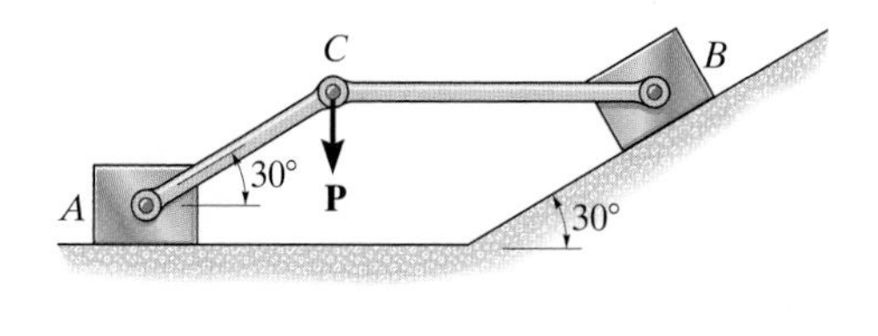

Prob. 8–51

***8–52.** Block C has a mass of 50 kg and is confined between two walls by smooth rollers. If the block rests on top of the 40-kg spool, determine the minimum cable force P needed to move the spool. The cable is wrapped around the spool's inner core. The coefficients of static friction at A and B are $\mu_A = 0.3$ and $\mu_B = 0.6$.

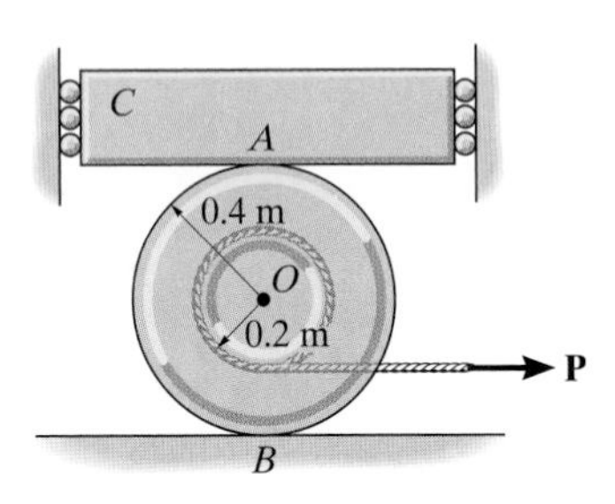

Prob. 8–52

8–53. The 50-lb board is placed across the channel and a 100-lb boy attempts to walk across. If the coefficient of static friction at A and B is $\mu_s = 0.4$, determine if he can make the crossing; and if not, how far will he get from A before the board slips?

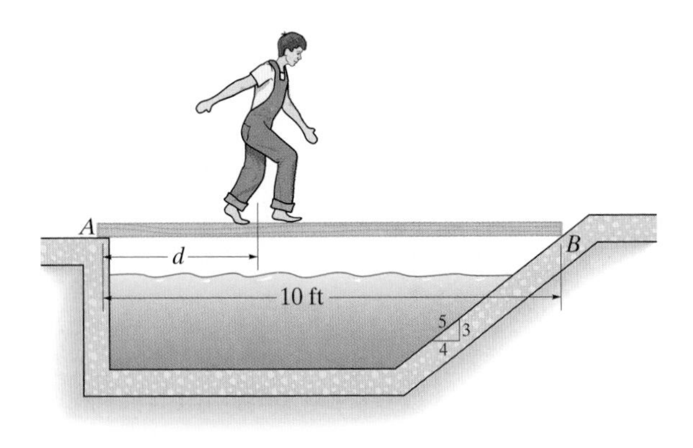

Prob. 8–53

8–54. Determine the minimum force P needed to push the tube E up the incline. The tube has a mass of 75 kg and the roller D has a mass of 100 kg. The force acts parallel to the plane, and the coefficients of static friction at the contacting surfaces are $\mu_A = 0.3$, $\mu_B = 0.25$, and $\mu_C = 0.4$. Each cylinder has a radius of 150 mm.

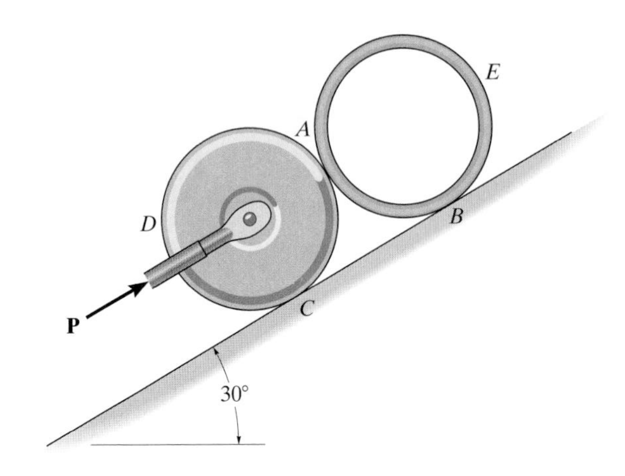

Prob. 8–54

8–55. The concrete pipe at A rests on top of B and C. If the coefficient of static friction between the pipes is μ_s and at the ground μ_s', determine their smallest values so that the pipes will not slip. Each pipe has a radius r and weight W, and the angle between the centers as indicated is θ.

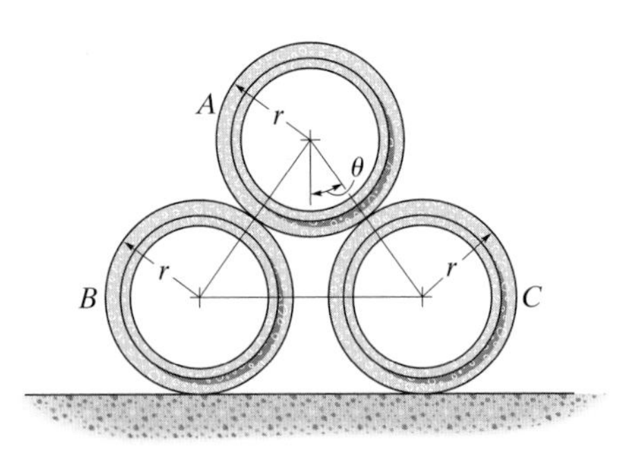

Prob. 8–55

***8–56.** The uniform pole has a weight W and length L. Its end B is tied to a supporting cord, and end A is placed against the wall, for which the coefficient of static friction is μ_s. Determine the largest angle θ at which the pole can be placed without slipping.

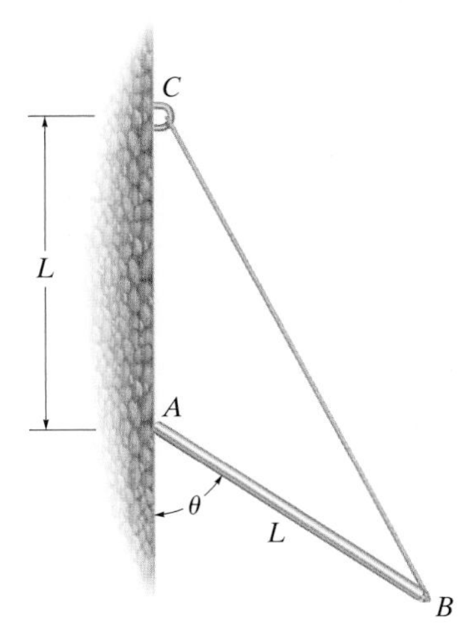

Prob. 8–56

8–57. The carpenter slowly pushes the uniform board horizontally over the top of the saw horse. The board has a uniform weight of 3 lb/ft, and the saw horse has a weight of 15 lb and a center of gravity at G. Determine if the saw horse will stay in position, slip, or tip if the board is pushed forward when $d = 10$ ft. The coefficients of static friction are shown in the figure.

8–58. The carpenter slowly pushes the uniform board horizontally over the top of the saw horse. The board has a uniform weight of 3 lb/ft, and the saw horse has a weight of 15 lb and a center of gravity at G. Determine if the saw horse will stay in position, slip, or tip if the board is pushed forward when $d = 14$ ft. The coefficients of static friction are shown in the figure.

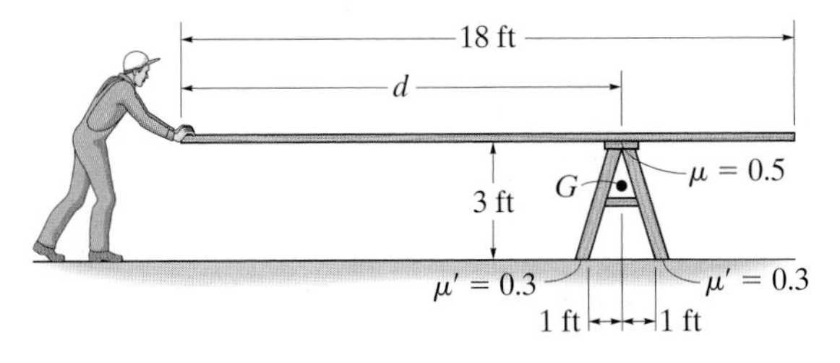

Probs. 8–57/58

8–59. The 45-kg disk rests on the surface for which the coefficient of static friction is $\mu_A = 0.2$. Determine the largest couple moment M that can be applied to the bar without causing motion.

***8–60.** The 45-kg disk rests on the surface for which the coefficient of static friction is $\mu_A = 0.15$. If $M = 50\ \text{N}\cdot\text{m}$, determine the friction force at A.

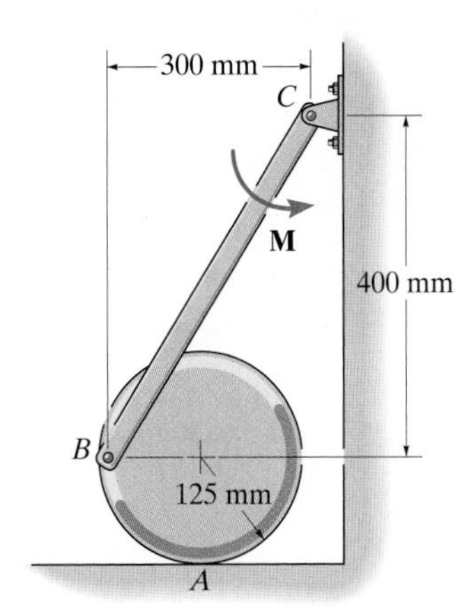

Probs. 8–59/60

■8–61. The 6-lb block is attached to a light rod AD that pivots at pin A. If the coefficient of static friction between the plane and the block is $\mu_s = 0.4$, determine the minimum angle θ at which the block may be placed on the plane without slipping. Neglect the *size* of the block in the calculation.

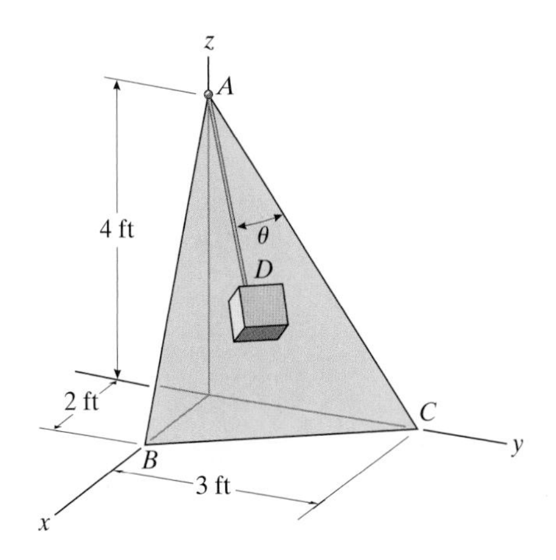

Prob. 8–61

8.3 Wedges

Wedges are often used to adjust the elevation of structural or mechanical parts. Also, they provide stability for objects such as this tank.

A *wedge* is a simple machine that is often used to transform an applied force into much larger forces, directed at approximately right angles to the applied force. Wedges also can be used to slightly move or adjust heavy loads.

Consider, for example, the wedge shown in Fig. 8–13*a*, which is used to *lift* a block of weight **W** by applying a force **P** to the wedge. Free-body diagrams of the block and wedge are shown in Fig. 8–13*b*. Here we have excluded the weight of the wedge since it is usually *small* compared to the weight of the block. Also, note that the frictional forces $\mathbf{F}_1$ and $\mathbf{F}_2$ must oppose the motion of the wedge. Likewise, the frictional force $\mathbf{F}_3$ of the wall on the block must act downward so as to oppose the block's upward motion. The locations of the resultant normal forces are not important in the force analysis since neither the block nor wedge will "tip." Hence the moment equilibrium equations will not be considered. There are seven unknowns, consisting of the applied force **P**, needed to cause motion of the wedge, and six normal and frictional forces. The seven available equations consist of two force equilibrium equations ($\Sigma F_x = 0$, $\Sigma F_y = 0$) applied to the wedge and block (four equations total) and the frictional equation $F = \mu N$ applied at each surface of contact (three equations total).

If the block is to be *lowered*, the frictional forces will all act in a sense opposite to that shown in Fig. 8–13*b*. The applied force **P** must act to the right to hold the block as shown, provided the coefficient of friction is very *small* or the wedge angle θ is *large*. Otherwise, **P** may have a reverse sense of direction in order to *pull* on the wedge to remove it. If **P** is *not applied*, or $\mathbf{P} = \mathbf{0}$, and friction forces hold the block in place, then the wedge is referred to as *self-locking*.

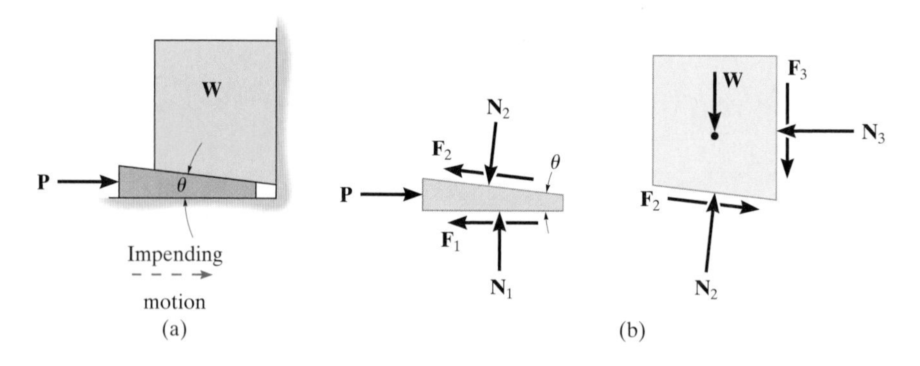

Fig. 8–13

EXAMPLE 8.7

The uniform stone in Fig. 8–14*a* has a mass of 500 kg and is held in the horizontal position using a wedge at *B*. If the coefficient of static friction is $\mu_s = 0.3$ at, the surfaces of contact, determine the minimum force **P** needed to remove the wedge. Is the wedge self-locking? Assume that the stone does not slip at *A*.

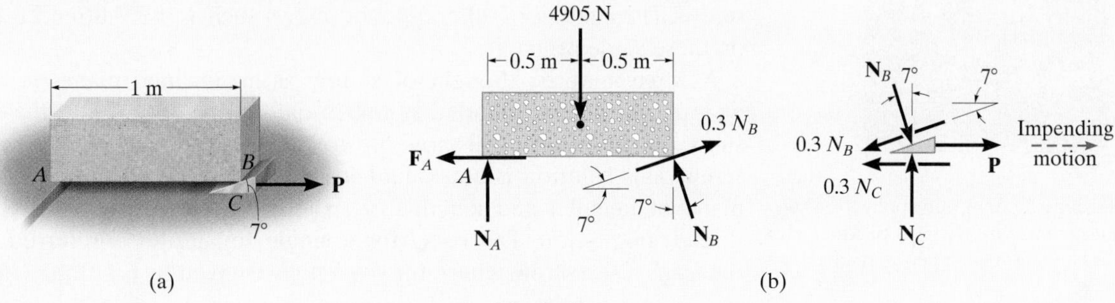

Fig. 8–14

SOLUTION

The minimum force *P* requires $F = \mu_s N$ at the surfaces of contact with the wedge. The free-body diagrams of the stone and wedge are shown in Fig. 8–14*b*. On the wedge the friction force opposes the motion, and on the stone at *A*, $F_A \leq \mu_s N_A$, since slipping does not occur there. There are five unknowns F_A, N_A, N_B, N_C, and *P*. Three equilibrium equations for the stone and two for the wedge are available for solution. From the free-body diagram of the stone,

$$\circlearrowright + \Sigma M_A = 0; \qquad -4905\text{ N}(0.5\text{ m}) + (N_B \cos 7^\circ\text{ N})(1\text{ m}) + (0.3N_B \sin 7^\circ\text{ N})(1\text{ m}) = 0$$

$$N_B = 2383.1\text{ N}$$

Using this result for the wedge, we have

$$\xrightarrow{+} \Sigma F_x = 0;$$

$$2383.1 \sin 7^\circ\text{ N} - 0.3(2383.1 \cos 7^\circ\text{ N}) + P - 0.3N_C = 0$$

$$+\uparrow \Sigma F_y = 0;$$

$$N_C - 2383.1 \cos 7^\circ\text{ N} - 0.3(2383.1 \sin 7^\circ\text{ N}) = 0$$

$$N_C = 2452.5\text{ N}$$

$$P = 1154.9\text{ N} = 1.15\text{ kN} \qquad \textit{Ans.}$$

NOTE: Since *P* is positive, indeed the wedge must be pulled out. If *P* was zero, the wedge would remain in place (self-locking) and the frictional forces developed at *B* and *C* would satisfy $F_B < \mu_s N_B$ and $F_C < \mu_s N_C$.

Square-threaded screws find applications on valves, jacks, and vises, where particularly large forces must be developed along the axis of the screw.

8.4 Frictional Forces on Screws

In most cases screws are used as fasteners; however, in many types of machines they are incorporated to transmit power or motion from one part of the machine to another. A *square-threaded screw* is commonly used for the latter purpose, especially when large forces are applied along its axis. In this section we will analyze the forces acting on square-threaded screws. The analysis of other types of screws, such as the V-thread, is based on these same principles.

A *screw* may be thought of simply as an inclined plane or wedge wrapped around a cylinder. A nut initially at position A on the screw shown in Fig. 8–15*a* will move up to B when rotated 360° around the screw. This rotation is equivalent to translating the nut up an inclined plane of height l and length $2\pi r$, where r is the mean radius of the thread, Fig. 8–15*b*. The rise l for a single revolution is referred to as the *lead* of the screw, where the *lead angle* is given by $\theta = \tan^{-1}(l/2\pi r)$.

Frictional Analysis. When a screw is subjected to large axial loads, the frictional forces developed on the thread become important if we are to determine the moment $\mathbf{M}$* needed to turn the screw. Consider, for example, the square-threaded jack screw shown in Fig. 8–16, which supports the vertical load $\mathbf{W}$. The reactive forces of the jack to this load are actually distributed over the circumference of the screw thread in contact with the screw hole in the jack, that is, within region h shown in Fig. 8–16. For simplicity, this portion of thread can be imagined as being unwound from the screw and represented as a simple block resting on an inclined plane having the screw's lead angle θ, Fig. 8–17*a*. Here the inclined plane represents the inside *supporting thread* of the jack base. Three forces act on the block or screw. The force $\mathbf{W}$ is the total axial load applied to the screw. The horizontal force $\mathbf{S}$ is caused by the applied moment $\mathbf{M}$, such that by summing moments about the axis of the screw, $M = Sr$, where r is the screw's mean radius. As a result of $\mathbf{W}$ and $\mathbf{S}$, the inclined plane exerts a resultant force $\mathbf{R}$ on the block, which is shown to have components acting normal, $\mathbf{N}$, and tangent, $\mathbf{F}$, to the contacting surfaces.

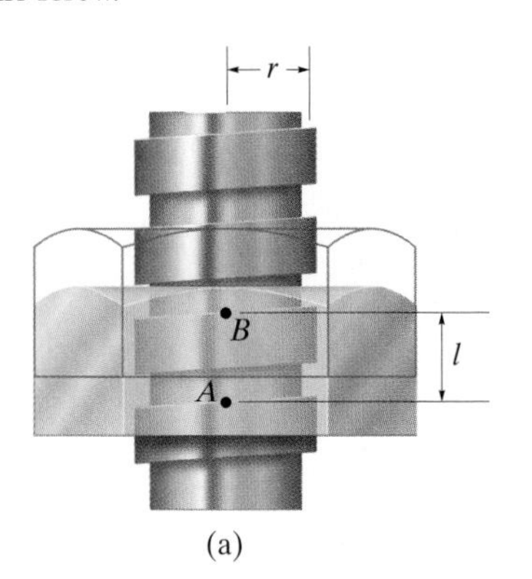

(a)

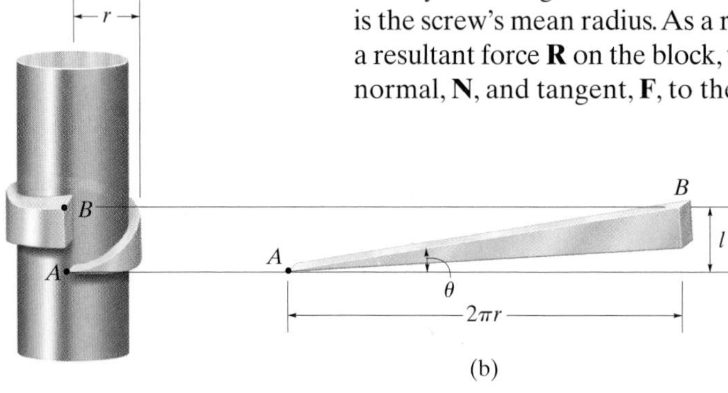

(b)

Fig. 8–15

*For applications, $\mathbf{M}$ is developed by applying a horizontal force $\mathbf{P}$ at a right angle to the end of a lever that would be fixed to the screw.

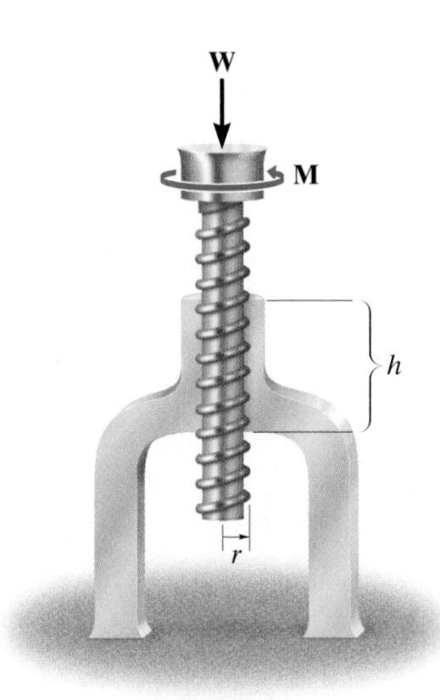

Fig. 8–16

Upward Screw Motion. Provided M is great enough, the screw (and hence the block) can either be brought to the verge of upward impending motion or motion can be occurring. Under these conditions, **R** acts at an angle $(\theta + \phi)$ from the vertical as shown in Fig. 8–17*a*, where $\phi = \tan^{-1}(F/N) = \tan^{-1}(\mu N/N) = \tan^{-1}\mu$. Applying the two force equations of equilibrium to the block, we obtain

$$\xrightarrow{+} \Sigma F_x = 0; \qquad S - R\sin(\theta + \phi) = 0$$

$$+\uparrow \Sigma F_y = 0; \qquad R\cos(\theta + \phi) - W = 0$$

Eliminating R and solving for S, then substituting this value into the equation $M = Sr$, yields

$$M = Wr\tan(\theta + \phi) \tag{8–3}$$

As indicated, M is the moment necessary to cause upward impending motion of the screw, provided $\phi = \phi_s = \tan^{-1}\mu_s$ (the angle of static friction). If ϕ is replaced by $\phi_k = \tan^{-1}\mu_k$ (the angle of kinetic friction), Eq. 8–3 will give a smaller value M necessary to maintain uniform upward motion of the screw.

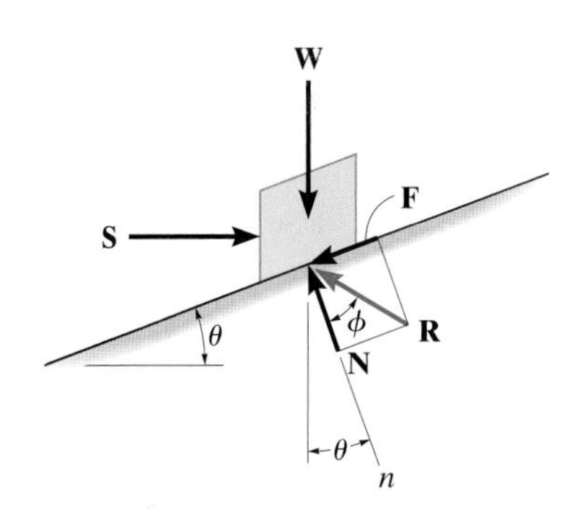

Upward screw motion
(a)

Downward Screw Motion ($\theta > \phi$). If the surface of the screw is very *slippery*, it may be possible for the screw to rotate downward if the magnitude of the moment is reduced to, say, $M' < M$. As shown in Fig. 8–17*b*, this causes the effect of $\mathbf{M}'$ to become $\mathbf{S}'$, and it requires the angle ϕ (ϕ_s or ϕ_k) to lie on the opposite side of the normal n to the plane supporting the block, such that $\theta > \phi$. For this case, Eq. 8–3 becomes

$$M' = Wr\tan(\theta - \phi) \tag{8–4}$$

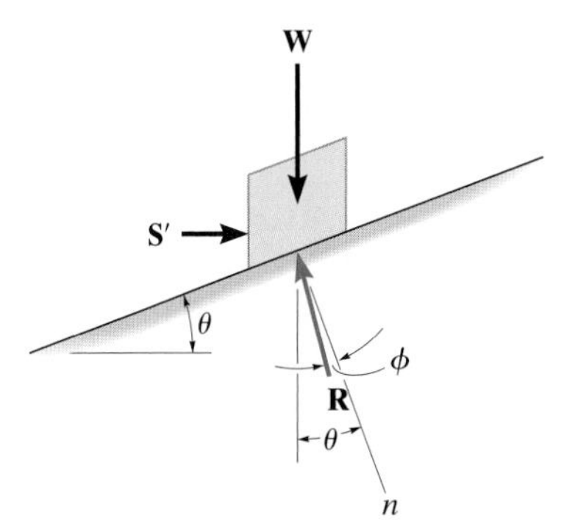

Downward screw motion ($\theta > \phi$)
(b)

Fig. 8–17

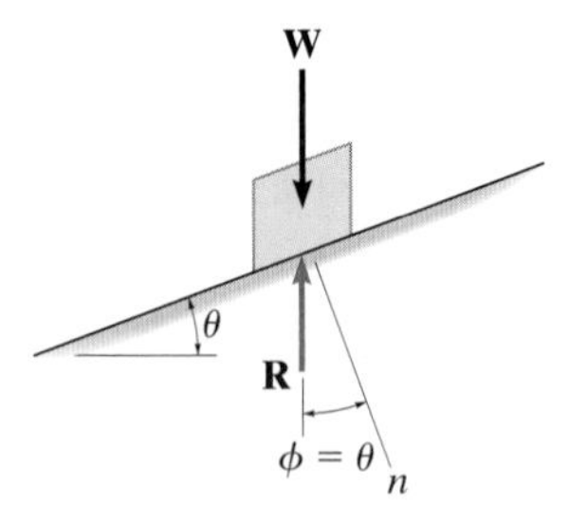

Self-locking screw ($\theta = \phi$)
(on the verge of rotating downward)

(c)

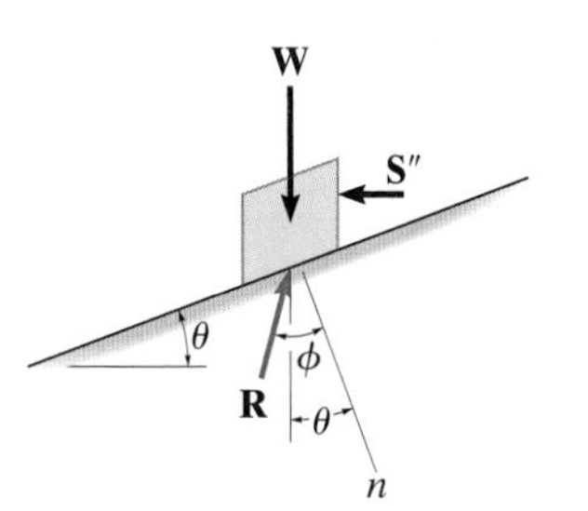

Downward screw motion ($\theta < \phi$)

(d)

Self-Locking Screw. If the moment **M** (or its effect **S**) is *removed*, the screw will remain *self-locking;* i.e., it will support the load **W** by *friction forces alone* provided $\phi \geq \theta$. To show this, consider the necessary limiting case when $\phi = \theta$, Fig. 8–17*c*. Here vertical equilibrium is maintained since **R** is vertical and thus balances **W**.

Downward Screw Motion ($\theta < \phi$). When the surface of the screw is very rough, the screw will not rotate downward as stated above. Instead, the direction of the applied moment must be reversed in order to cause the motion. The free-body diagram shown in Fig. 8–17d is representative of this case. Here **S**″ is caused by the applied (reverse) moment **M**″. Hence Eq. 8–3 becomes

$$M'' = Wr\tan(\phi - \theta) \tag{8–5}$$

Each of the above cases should be thoroughly understood before proceeding to solve problems.

Fig. 8–17 (cont.)

EXAMPLE 8.8

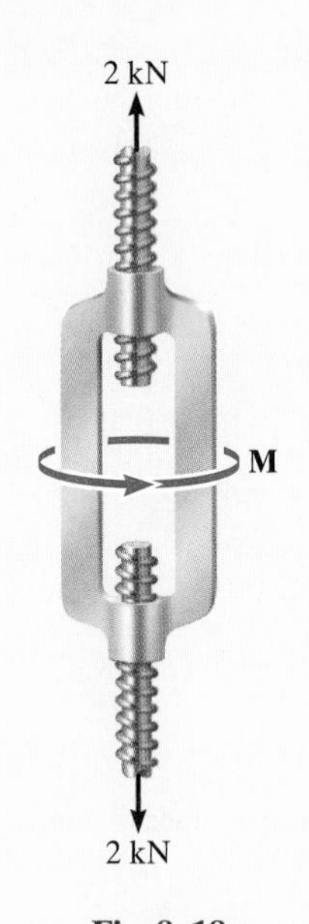

Fig. 8–18

The turnbuckle shown in Fig. 8–18 has a square thread with a mean radius of 5 mm and a lead of 2 mm. If the coefficient of static friction between the screw and the turnbuckle is $\mu_s = 0.25$, determine the moment **M** that must be applied to draw the end screws closer together.

SOLUTION

The moment may be obtained by using Eq. 8–3. Why? Since friction at *two screws* must be overcome, this requires

$$M = 2[Wr\tan(\theta + \phi)] \tag{1}$$

Here $W = 2000$ N, $r = 5$ mm, $\phi_s = \tan^{-1}\mu_s = \tan^{-1}(0.25) = 14.04°$, and $\theta = \tan^{-1}(l/2\pi r) = \tan^{-1}(2\text{ mm}/[2\pi(5\text{ mm})]) = 3.64°$. Substituting these values into Eq. 1 and solving gives

$$M = 2[(2000\text{ N})(5\text{ mm})\tan(14.04° + 3.64°)]$$
$$= 6374.7\text{ N}\cdot\text{mm} = 6.37\text{ N}\cdot\text{m} \qquad \textit{Ans.}$$

NOTE: When the moment is *removed*, the turnbuckle will be self-locking; i.e., it will not unscrew since $\phi_s > \theta$.

PROBLEMS

8–62. Determine the force P needed to lift the 100-lb load. Smooth rollers are placed between the wedges. The coefficient of static friction between A and C and between B and D is $\mu_s = 0.3$. Neglect the weight of each wedge.

Prob. 8–62

8–63. The wedge is used to level the floor of a building. For the floor loading shown, determine the horizontal force P that must be applied to move the wedge forward. The coefficient of static friction between the wedge and the two surfaces of contact is $\mu_s = 0.25$. Neglect the size and weight of the wedge and the thickness of the beam.

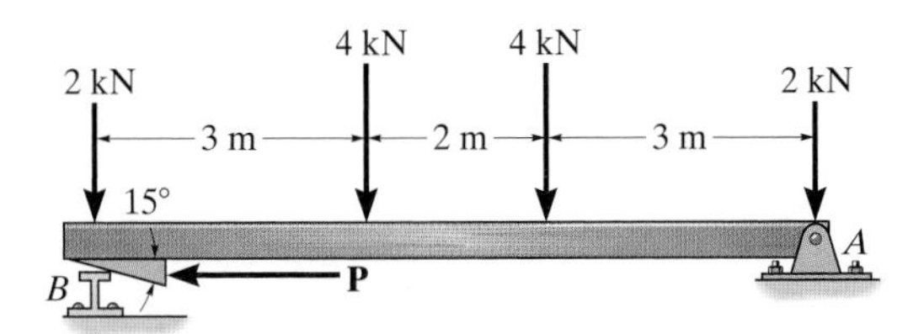

Prob. 8–63

***8–64.** The three stone blocks have weights of $W_A = 600$ lb, $W_B = 150$ lb, and $W_C = 500$ lb. Determine the smallest horizontal force P that must be applied to block C in order to move this block. The coefficient of static friction between the blocks is $\mu_s = 0.3$, and between the floor and each block $\mu_s' = 0.5$.

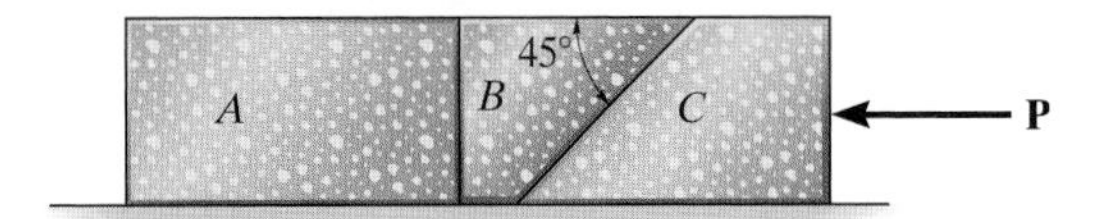

Prob. 8–64

8–65. If the spring is compressed 60 mm and the coefficient of static friction between the tapered stub S and the slider A is $\mu_{SA} = 0.5$, determine the horizontal force **P** needed to move the slider forward. The stub is free to move without friction within the fixed collar C. The coefficient of static friction between A and surface B is $\mu_{AB} = 0.4$. Neglect the weights of the slider and stub.

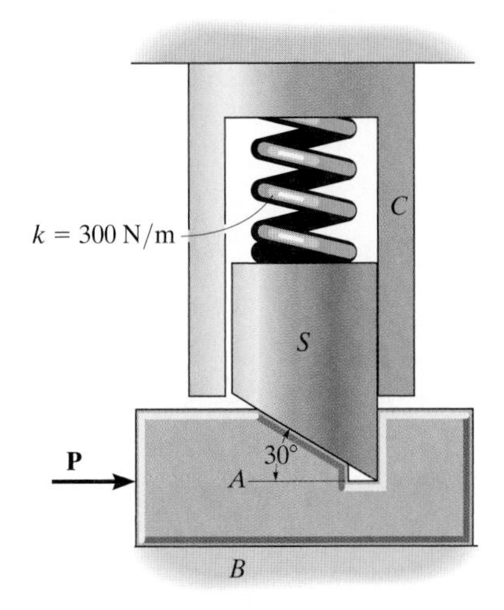

Prob. 8–65

8–66. The coefficient of static friction between wedges B and C is $\mu_s = 0.6$ and between the surfaces of contact B and A and C and D, $\mu_s' = 0.4$. If the spring is compressed 200 mm when in the position shown, determine the smallest force P needed to move wedge C to the left. Neglect the weight of the wedges.

8–67. The coefficient of static friction between the wedges B and C is $\mu_s = 0.6$ and between the surfaces of contact B and A and C and D, $\mu_s' = 0.4$. If $P = 50$ N, determine the largest allowable compression of the spring without causing wedge C to move to the left. Neglect the weight of the wedges.

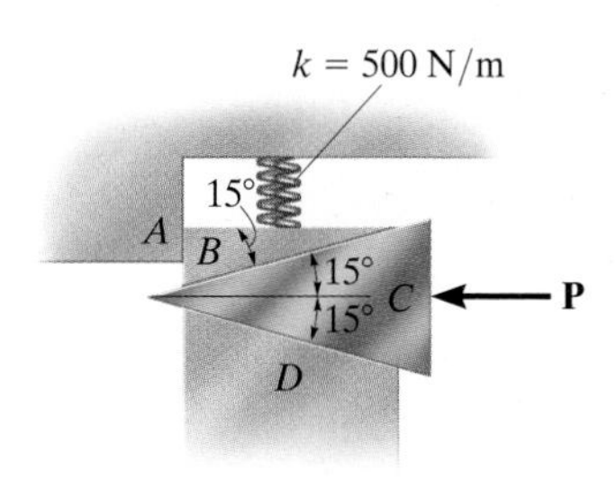

Probs. 8–66/67

***8–68.** The wedge blocks are used to hold the specimen in a tension testing machine. Determine the design angle θ of the wedges so that the specimen will not slip regardless of the applied load. The coefficients of static friction are $\mu_A = 0.1$ at A and $\mu_B = 0.6$ at B. Neglect the weight of the blocks.

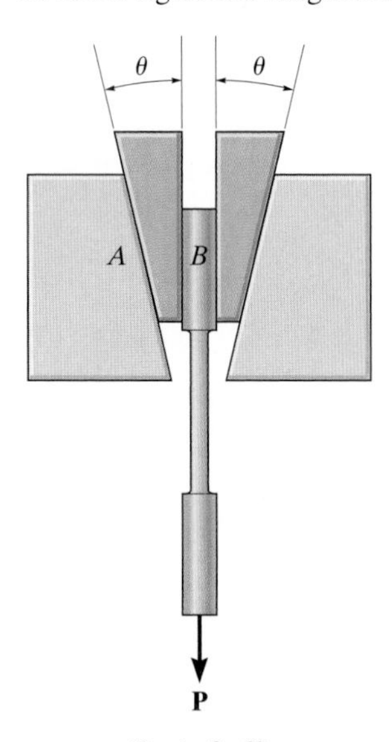

Prob. 8–68

8–69. The wedge is used to level the member. Determine the reversed horizontal force $-\mathbf{P}$ that must be applied to pull the wedge out to the left. The coefficient of static friction between the wedge and the two surfaces of contact is $\mu_s = 0.15$. Neglect the weight of the wedge.

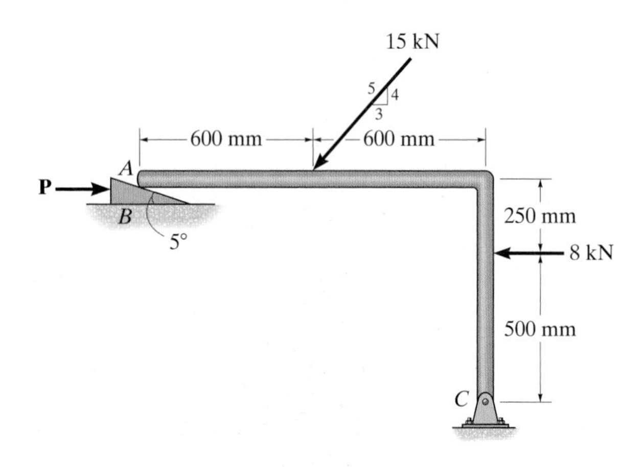

Prob. 8–69

8–70. If the coefficient of static friction between all the surfaces of contact is μ_s, determine the force $\mathbf{P}$ that must be applied to the wedge in order to lift the brace that supports the load $\mathbf{F}$.

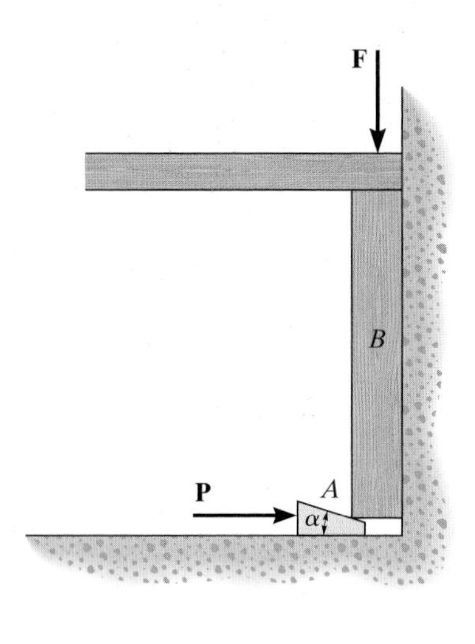

Prob. 8–70

8–71. The column is used to support the upper floor. If a force $F = 80$ N is applied perpendicular to the handle to tighten the screw, determine the compressive force in the column. The square-threaded screw on the jack has a coefficient of static friction of $\mu_s = 0.4$, mean diameter of 25 mm, and a lead of 3 mm.

***8–72.** If the force **F** is removed from the handle of the jack in Prob. 8–71, determine if the screw is self-locking.

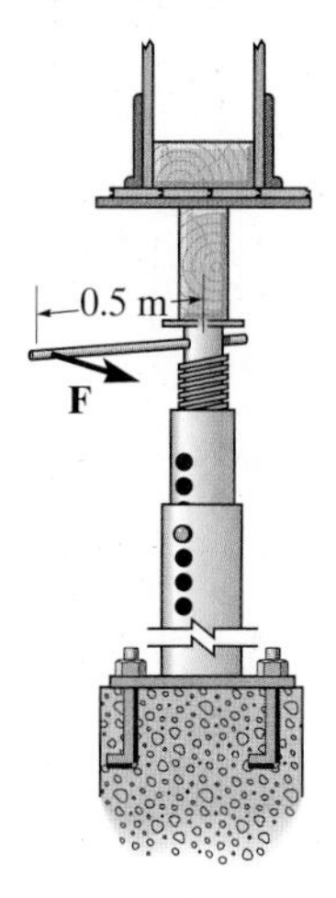

Probs. 8–71/72

8–73. The vise is used to grip the pipe. If a horizontal force of 25 lb is applied perpendicular to the end of the 10-in. handle, determine the compressive force F developed in the pipe. The square threads have a mean diameter of 1.5 in. and a lead of 0.2 in. How much force must be applied perpendicular to the handle to loosen the vise? Take $\mu_s = 0.3$.

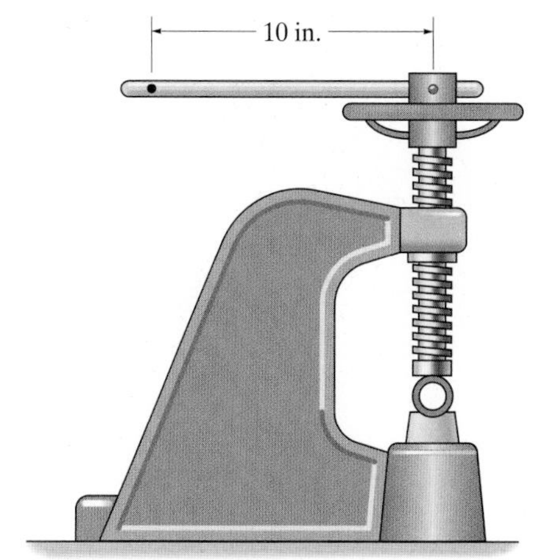

Prob. 8–73

8–74. Determine the couple forces F that must be applied to the handle of the machinist's vise in order to create a compressive force of 400 N in the block. Neglect friction at the bearing A. The guide at B is smooth so that the axial force on the screw is 400 N. The single square-threaded screw has a mean radius of 6 mm and a lead of 8 mm, and the coefficient of static friction is $\mu_s = 0.27$.

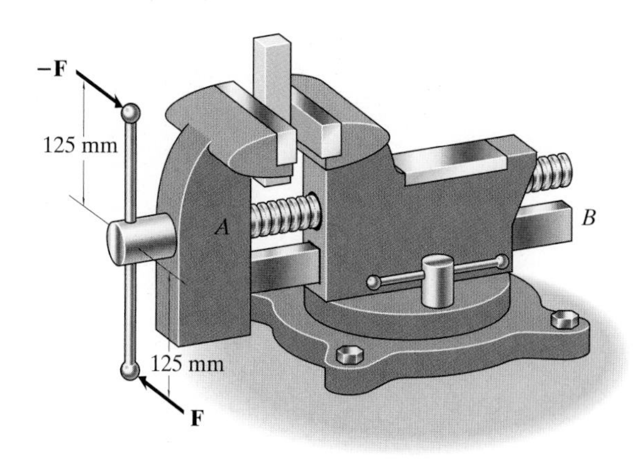

Prob. 8–74

8–75. If couple forces of $F = 35$ N are applied to the handle of the machinist's vise, determine the compressive force developed in the block. Neglect friction at the bearing A. The guide at B is smooth. The single square-threaded screw has a mean radius of 6 mm and a lead of 8 mm, and the coefficient of static friction is $\mu_s = 0.27$.

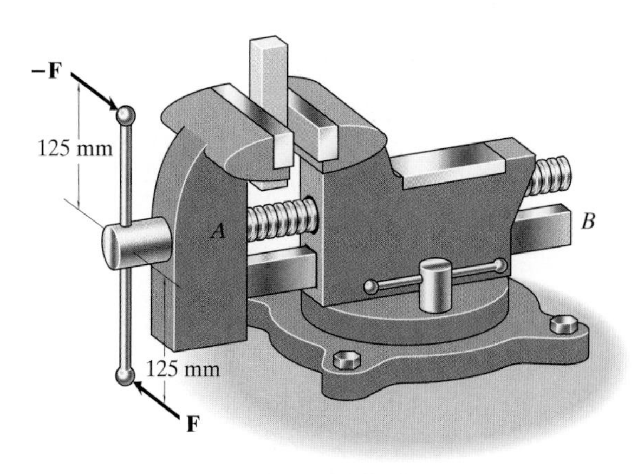

Prob. 8–75

***8–76.** The machine part is held in place using the double-end clamp. The bolt at B has square threads with a mean radius of 4 mm and a lead of 2 mm, and the coefficient of static friction with the nut is $\mu_s = 0.5$. If a torque of $M = 0.4\ \text{N}\cdot\text{m}$ is applied to the nut to tighten it, determine the normal force of the clamp at the smooth contacts A and C.

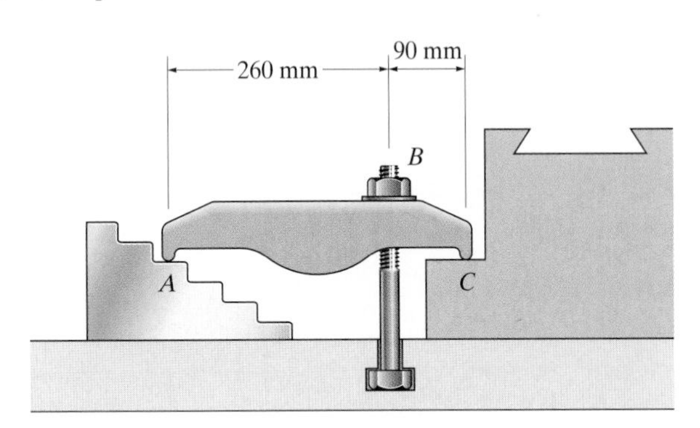

Prob. 8–76

8–77. Determine the clamping force on the board A if the screw of the "C" clamp is tightened with a twist of $M = 8\ \text{N}\cdot\text{m}$. The single square-threaded screw has a mean radius of 10 mm, a lead of 3 mm, and the coefficient of static friction is $\mu_s = 0.35$.

8–78. If the required clamping force at the board A is to be 50 N, determine the torque M that must be applied to the handle of the "C" clamp to tighten it down. The single square-threaded screw has a mean radius of 10 mm, a lead of 3 mm, and the coefficient of static friction is $\mu_s = 0.35$.

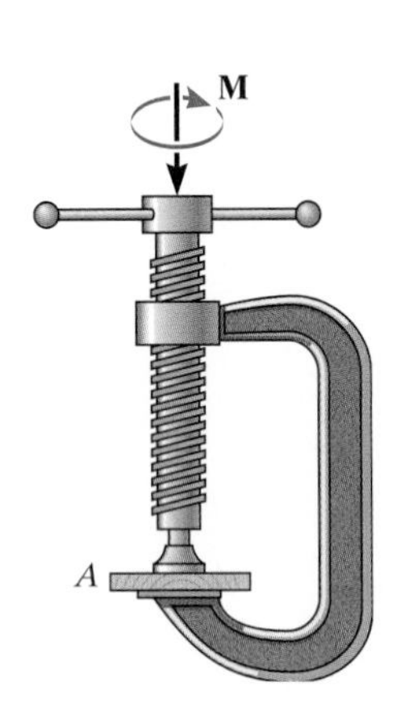

Probs. 8–77/78

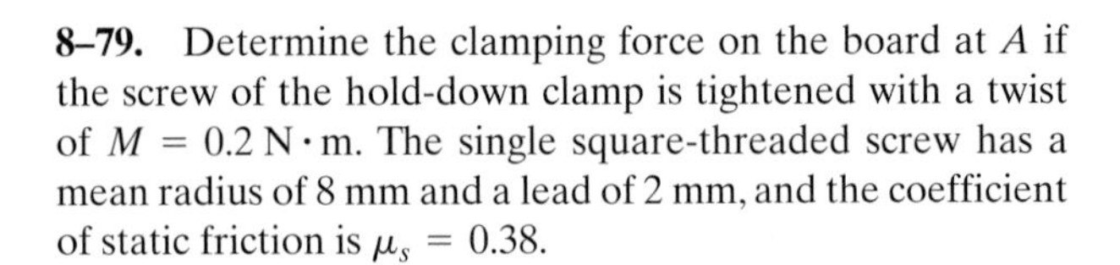

8–79. Determine the clamping force on the board at A if the screw of the hold-down clamp is tightened with a twist of $M = 0.2\ \text{N}\cdot\text{m}$. The single square-threaded screw has a mean radius of 8 mm and a lead of 2 mm, and the coefficient of static friction is $\mu_s = 0.38$.

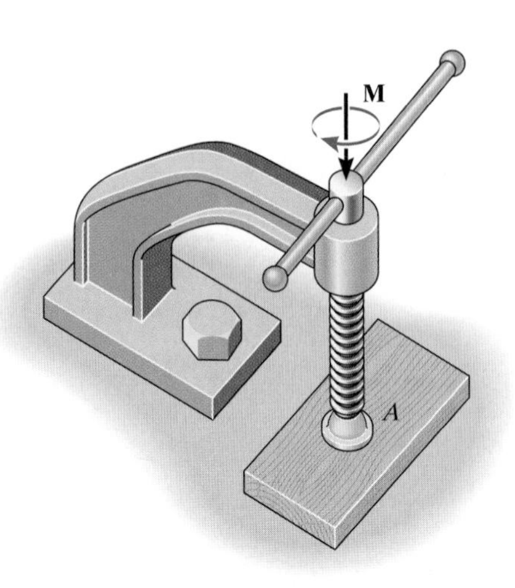

Prob. 8–79

***8–80.** If the required clamping force at the board A is to be 70 N, determine the torque M that must be applied to the handle of the hold-down clamp to tighten it down. The single square-threaded screw has a mean radius of 8 mm and a lead of 2 mm, and the coefficient of static friction is $\mu_s = 0.38$.

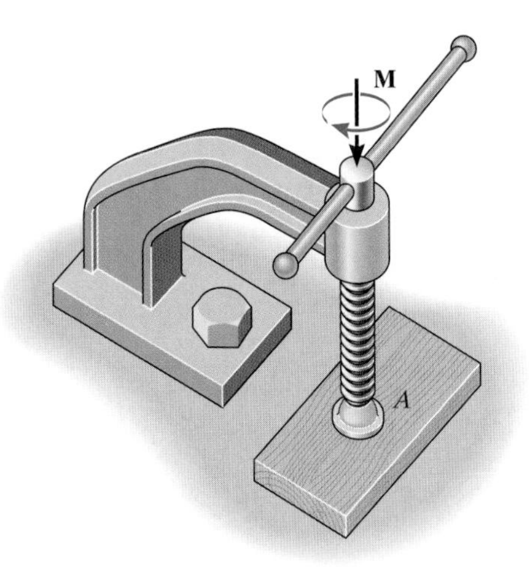

Prob. 8–80

8–81. The fixture clamp consist of a square-threaded screw having a coefficient of static friction of $\mu_s = 0.3$, mean diameter of 3 mm, and a lead of 1 mm. The five points indicated are pin connections. Determine the clamping force at the smooth blocks D and E when a torque of $M = 0.08 \text{ N} \cdot \text{m}$ is applied to the handle of the screw.

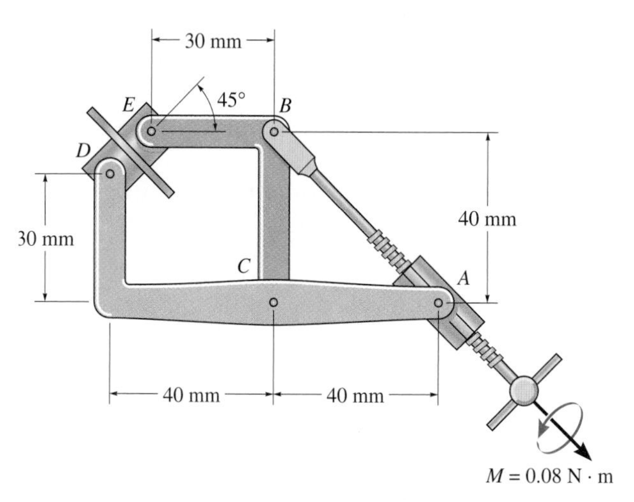

Prob. 8–81

8–82. The clamp provides pressure from several directions on the edges of the board. If the square-threaded screw has a lead of 3 mm, radius of 10 mm, and the coefficient of static friction is $\mu_s = 0.4$, determine the horizontal force developed on the board at A and the vertical forces developed at B and C if a torque of $M = 1.5 \text{ N} \cdot \text{m}$ is applied to the handle to tighten it further. The blocks at B and C are pin-connected to the board.

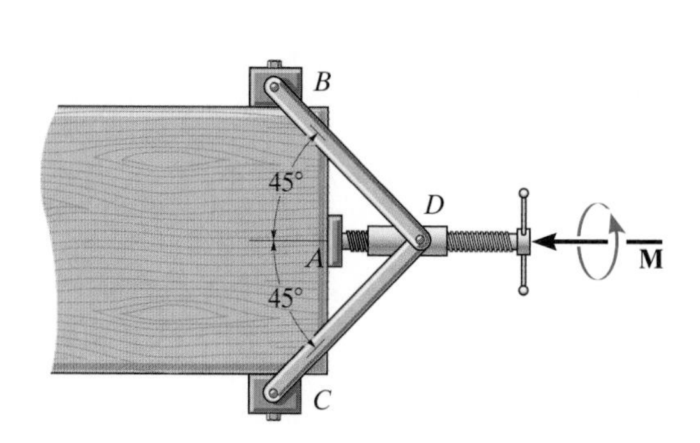

Prob. 8–82

8–83. The two blocks under the double wedge are brought together using a left and right square-threaded screw. If the mean diameter is 20 mm, the lead is 5 mm, and the coefficient of static friction is $\mu_s = 0.4$, determine the torque needed to draw the blocks together. The coefficient of static friction between each block and its surfaces of contact is $\mu_s' = 0.4$.

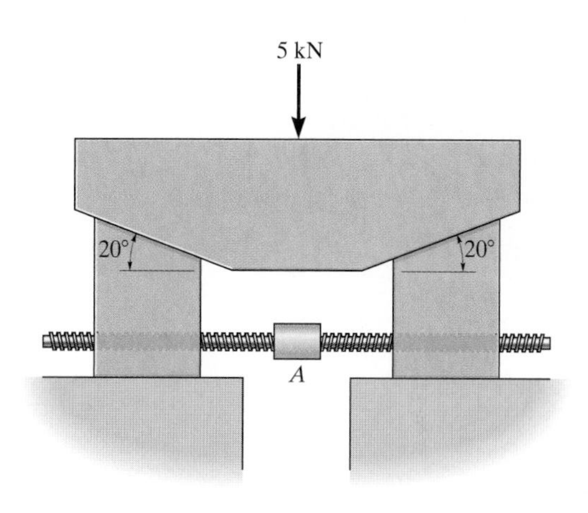

Prob. 8–83

***8–84.** The two blocks under the double wedge are brought together using a left and right square-threaded screw. If the mean diameter is 20 mm, the lead is 5 mm, and the coefficient of static friction is $\mu_s = 0.4$, determine the torque needed to spread the blocks apart. The coefficient of static friction between each block and its surfaces of contact is $\mu_s' = 0.4$.

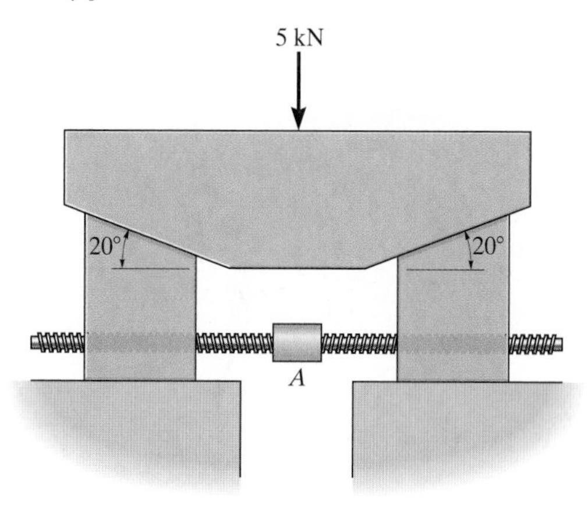

Prob. 8–84

8.5 Frictional Forces on Flat Belts

Flat or V-belts are often used to transmit the torque developed by a motor to a fan or blower.

Whenever belt drives or band brakes are designed, it is necessary to determine the frictional forces developed between the belt and its contacting surface. In this section we will analyze the frictional forces acting on a flat belt, although the analysis of other types of belts, such as the V-belt, is based on similar principles.

Here we will consider the flat belt shown in Fig. 8–19*a*, which passes over a fixed curved surface, such that the total angle of belt to surface contact in radians is β and the coefficient of friction between the two surfaces is μ. We will determine the tension T_2 in the belt which is needed to pull the belt counterclockwise over the surface and thereby overcome both the frictional forces at the surface of contact and the known tension T_1. Obviously, $T_2 > T_1$.

Frictional Analysis. A free-body diagram of the belt segment in contact with the surface is shown in Fig. 8–19*b*. Here the normal force $\mathbf{N}$ and the frictional force $\mathbf{F}$, acting at different points along the belt, will vary both in magnitude and direction. Due to this *unknown* force distribution, the analysis of the problem will proceed on the basis of initially studying the forces acting on a differential element of the belt.

A free-body diagram of an element having a length ds is shown in Fig. 8–19*c*. Assuming either impending motion or motion of the belt, the magnitude of the frictional force $dF = \mu\, dN$. This force opposes the sliding motion of the belt and thereby increases the magnitude of the tensile force acting in the belt by dT. Applying the two force equations of equilibrium, we have

$$\searrow+\Sigma F_x = 0; \qquad T\cos\left(\frac{d\theta}{2}\right) + \mu\, dN - (T + dT)\cos\left(\frac{d\theta}{2}\right) = 0$$

$$+\nearrow\Sigma F_y = 0; \qquad dN - (T + dT)\sin\left(\frac{d\theta}{2}\right) - T\sin\left(\frac{d\theta}{2}\right) = 0$$

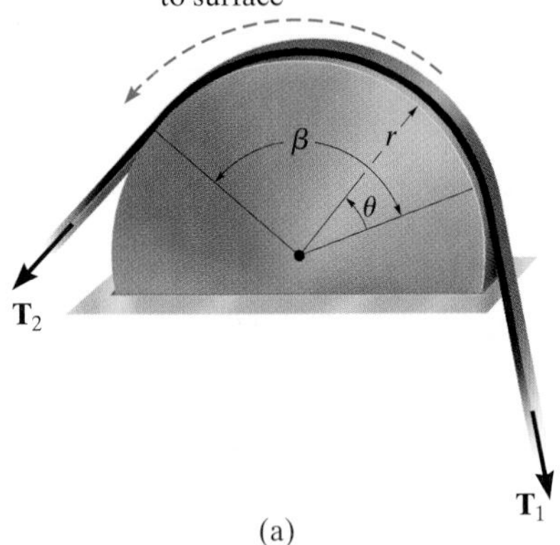

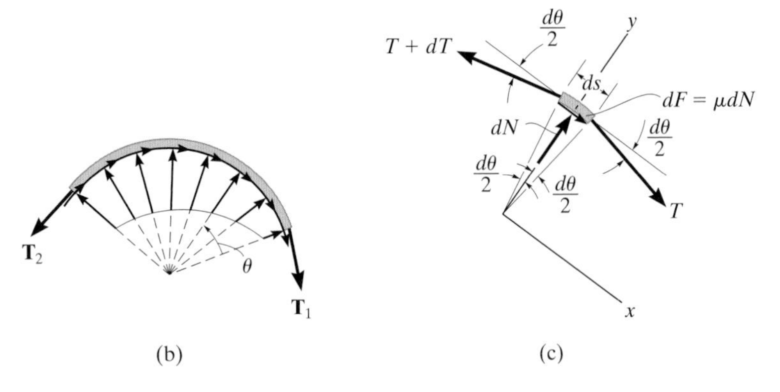

Fig. 8–19

Since $d\theta$ is of *infinitesimal size*, $\sin(d\theta/2)$ and $\cos(d\theta/2)$ can be replaced by $d\theta/2$ and 1, respectively. Also, the *product* of the two infinitesimals dT and $d\theta/2$ may be neglected when compared to infinitesimals of the first order. The above two equations therefore reduce to

$$\mu\, dN = dT$$

and

$$dN = T\, d\theta$$

Eliminating dN yields

$$\frac{dT}{T} = \mu\, d\theta$$

Integrating this equation between all the points of contact that the belt makes with the drum, and noting that $T = T_1$ at $\theta = 0$ and $T = T_2$ at $\theta = \beta$, yields

$$\int_{T_1}^{T_2} \frac{dT}{T} = \mu \int_0^{\beta} d\theta$$

$$\ln\frac{T_2}{T_1} = \mu\beta$$

Solving for T_2, we obtain

$$T_2 = T_1 e^{\mu\beta} \qquad (8\text{–}6)$$

where

T_2, T_1 = belt tensions; T_1 opposes the direction of motion (or impending motion) of the belt measured relative to the surface, while T_2 acts in the direction of the relative belt motion (or impending motion); because of friction, $T_2 > T_1$

μ = coefficient of static or kinetic friction between the belt and the surface of contact

β = angle of belt to surface contact, measured in radians

e = 2.718..., base of the natural logarithm

Note that T_2 is *independent* of the *radius* of the drum, and instead it is a function of the angle of belt to surface contact, β. Furthermore, as indicated by the integration, this equation is valid for flat belts placed on *any shape* of contacting surface. For application, Eq. 8–6 is valid only when *impending motion* or *motion* occurs.

EXAMPLE 8.9

The maximum tension that can be developed in the cord shown in Fig. 8–20*a* is 500 N. If the pulley at *A* is free to rotate and the coefficient of static friction at the fixed drums *B* and *C* is $\mu_s = 0.25$, determine the largest mass of the cylinder that can be lifted by the cord. Assume that the force **F** applied at the end of the cord is directed vertically downward, as shown.

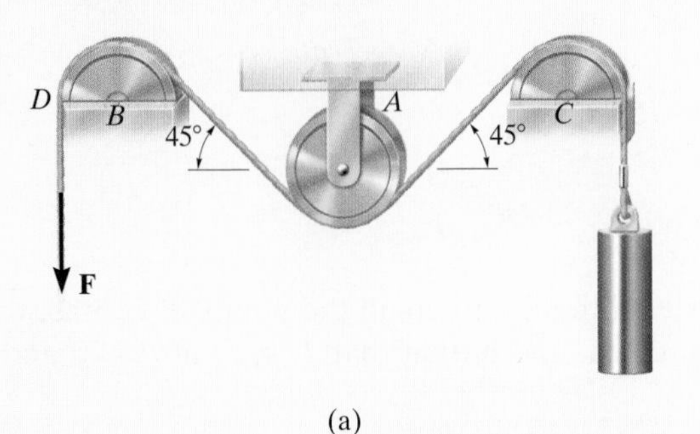

(a)

SOLUTION

Lifting the cylinder, which has a weight $W = mg$, causes the cord to move counterclockwise over the drums at *B* and *C*; hence, the maximum tension T_2 in the cord occurs at *D*. Thus, $F = T_2 = 500$ N. A section of the cord passing over the drum at *B* is shown in Fig. 8–20*b*. Since $180° = \pi$ rad, the angle of contact between the drum and the cord is $\beta = (135°/180°)\pi = 3\pi/4$ rad. Using Eq. 8–6, we have

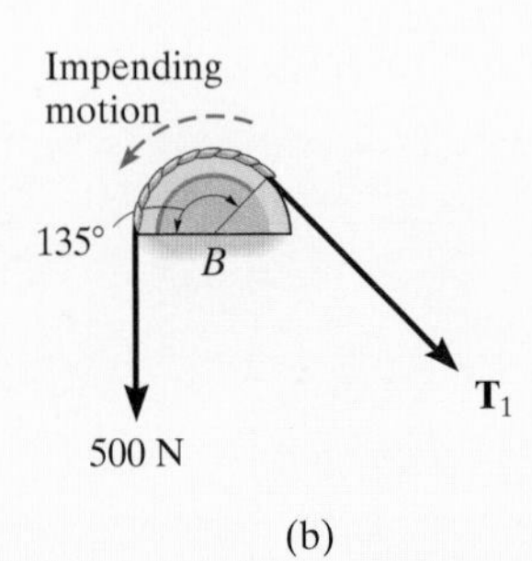

(b)

$$T_2 = T_1 e^{\mu_s \beta}; \qquad 500 \text{ N} = T_1 e^{0.25[(3/4)\pi]}$$

Hence,

$$T_1 = \frac{500 \text{ N}}{e^{0.25[(3/4)\pi]}} = \frac{500 \text{ N}}{1.80} = 277.4 \text{ N}$$

Since the pulley at *A* is free to rotate, equilibrium requires that the tension in the cord remains the *same* on both sides of the pulley.

The section of the cord passing over the drum at *C* is shown in Fig. 8–20*c*. The weight $W < 277.4$ N. Why? Applying Eq. 8–6, we obtain

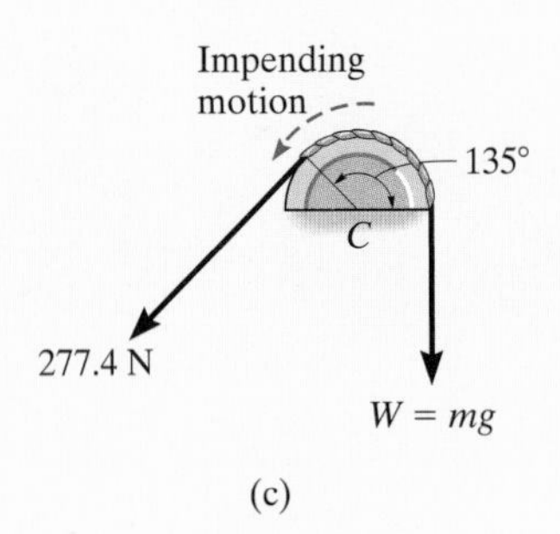

(c)

$$T_2 = T_1 e^{\mu_s \beta}; \qquad 277.4 \text{ N} = W e^{0.25[(3/4)\pi]}$$

$$W = 153.9 \text{ N}$$

so that

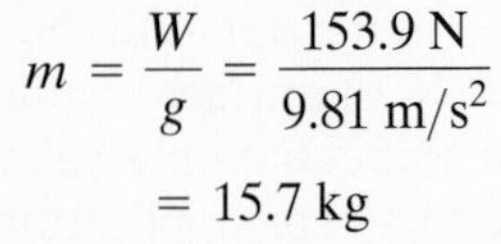

$$m = \frac{W}{g} = \frac{153.9 \text{ N}}{9.81 \text{ m/s}^2}$$

$$= 15.7 \text{ kg} \qquad \textit{Ans.}$$

Fig. 8–20

PROBLEMS

8–85. The cord supporting the 6-kg cylinder passes around three pegs, A, B, C, where $\mu_s = 0.2$. Determine the range of values for the magnitude of the horizontal force $\mathbf{P}$ for which the cylinder will not move up or down.

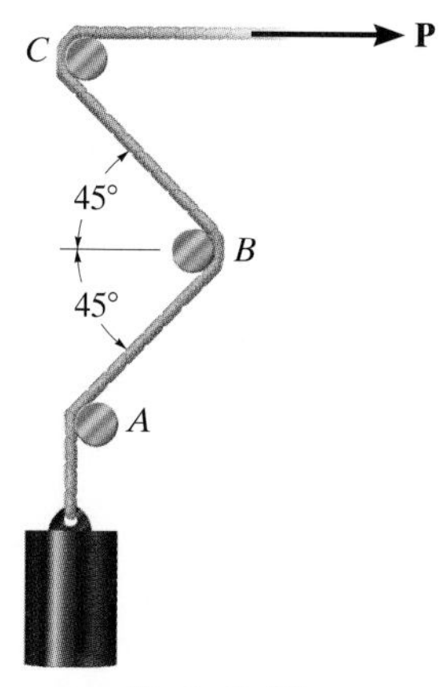

Prob. 8–85

8–86. The truck, which has a mass of 3.4 Mg, is to be lowered down the slope by a rope that is wrapped around a tree. If the wheels are free to roll and the man at A can resist a pull of 300 N, determine the minimum number of turns the rope should be wrapped around the tree to lower the truck at a constant speed. The coefficient of kinetic friction between the tree and rope is $\mu_k = 0.3$.

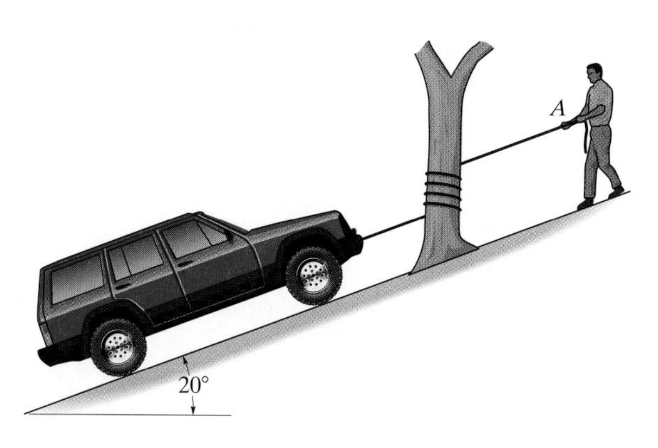

Prob. 8–86

8–87. The wheel is subjected to a torque of $M = 50\ \mathrm{N\cdot m}$. If the coefficient of kinetic friction between the band brake and the rim of the wheel is $\mu_k = 0.3$, determine the smallest horizontal force P that must be applied to the lever to stop the wheel.

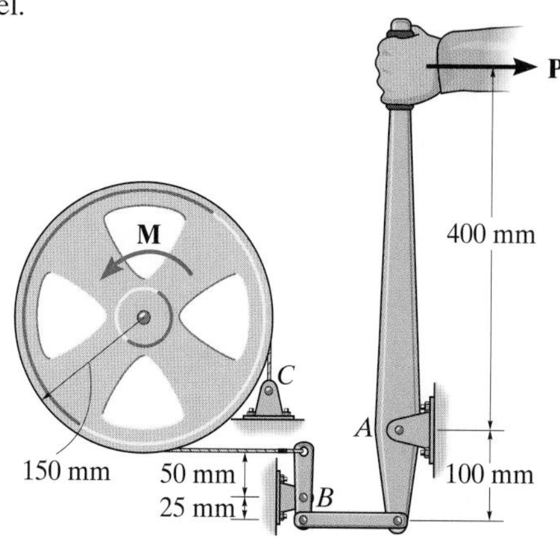

Prob. 8–87

***8–88.** The cylinder A has a mass of 75 kg. Determine the smallest force P applied to the handle of the lever required for equilibrium. The coefficient of static friction between the belt and the wheel is $\mu_s = 0.3$. The drum is pin connected at its center, B.

8–89. Determine the largest mass of cylinder A that can be supported from the drum if a force $P = 20$ N is applied to the handle of the lever. The coefficient of static friction between the belt and the wheel is $\mu_s = 0.3$. The drum is pin supported at its center, B.

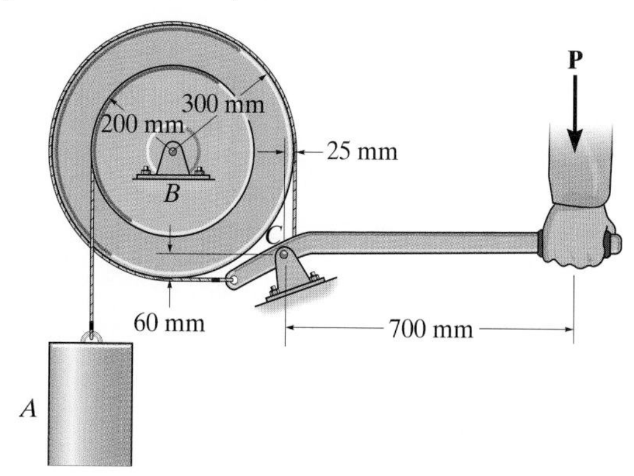

Probs. 8–88/89

8–90. The uniform bar AB is supported by a rope that passes over a frictionless pulley at C and a fixed peg at D. If the coefficient of static friction between the rope and the peg is $\mu_D = 0.3$, determine the smallest distance x from the end of the bar at which a 20-N force may be placed and not cause the bar to move.

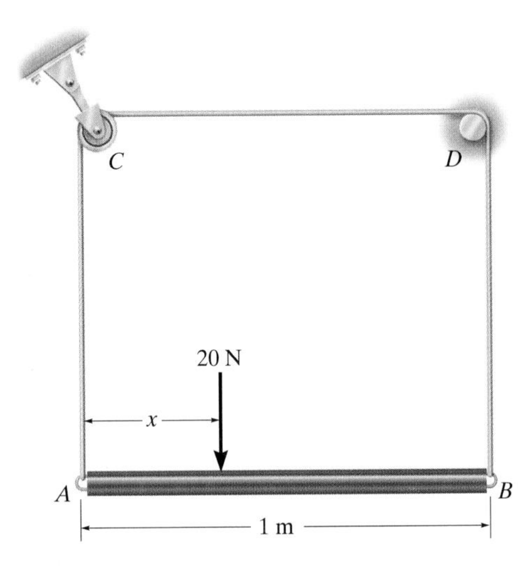

Prob. 8–90

8–91. Determine the smallest lever force P needed to prevent the wheel from rotating if it is subjected to a torque of $M = 250\ \text{N}\cdot\text{m}$. The coefficient of static friction between the belt and the wheel is $\mu_s = 0.3$. The wheel is pin connected at its center, B.

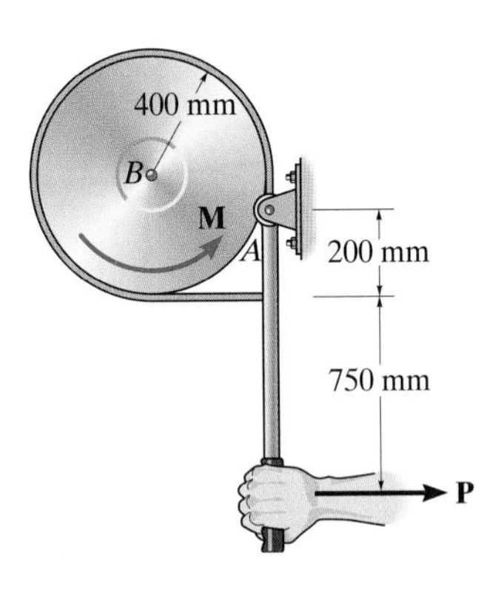

Prob. 8–91

***8–92.** Determine the torque M that can be resisted by the band brake if a force of $P = 30$ N is applied to the handle of the lever. The coefficient of static friction between the belt and the wheel is $\mu_s = 0.3$. The wheel is pin connected at its center, B.

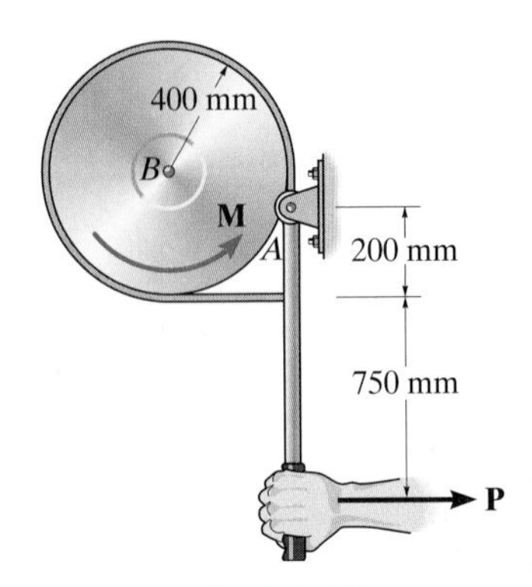

Prob. 8–92

8–93. Blocks A and B weigh 50 lb and 30 lb, respectively. Using the coefficients of static friction indicated, determine the greatest weight of block D without causing motion.

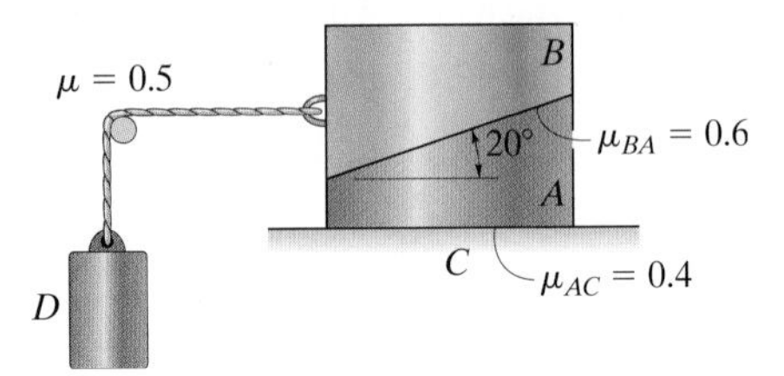

Prob. 8–93

8–94. Blocks A and B weigh 75 lb each, and D weighs 30 lb. Using the coefficients of static friction indicated, determine the frictional force between blocks A and B and between block A and the floor C.

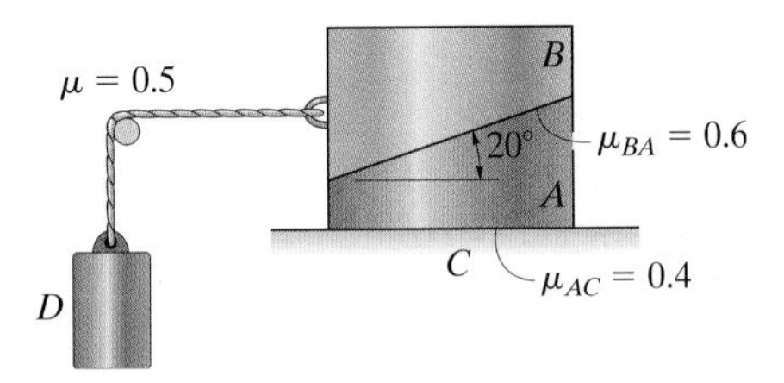

Prob. 8–94

8–95. Show that the frictional relationship between the belt tensions, the coefficient of friction μ, and the angular contacts α and β for the V-belt is $T_2 = T_1 e^{\mu\beta/\sin(\alpha/2)}$.

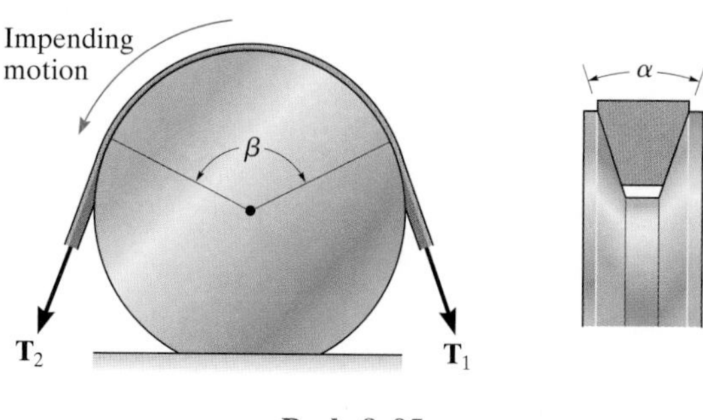

Prob. 8–95

***8–96.** The 60° V-fan-belt of an automobile engine passes around the hub H of a generator G and over the housing F to a fan. If the generator locks, and the maximum tension the belt can sustain is $T_{max} = 175$ lb, determine the maximum possible torque M resisted by the axle as the belt slips over the hub. Assume that slipping of the belt occurs only at H and that the coefficient of static friction for the hub is $\mu_s = 0.25$. Use the result of Prob. 8–95.

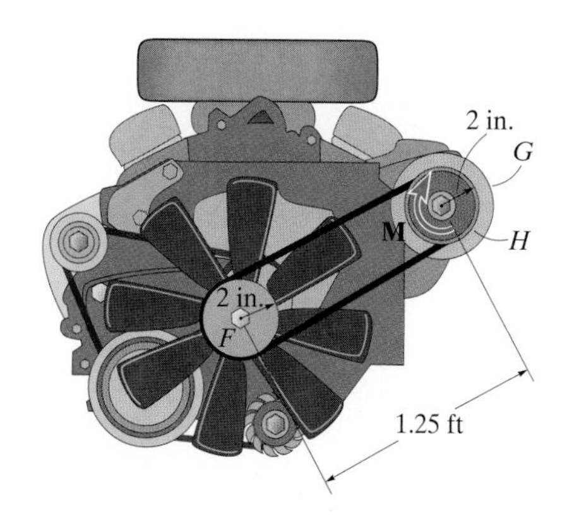

Prob. 8–96

8–97. A cable is attached to the 20-kg plate B, passes over a fixed peg at C, and is attached to the block at A. Using the coefficients of static friction shown, determine the smallest mass of block A so that it will prevent sliding motion of B down the plane.

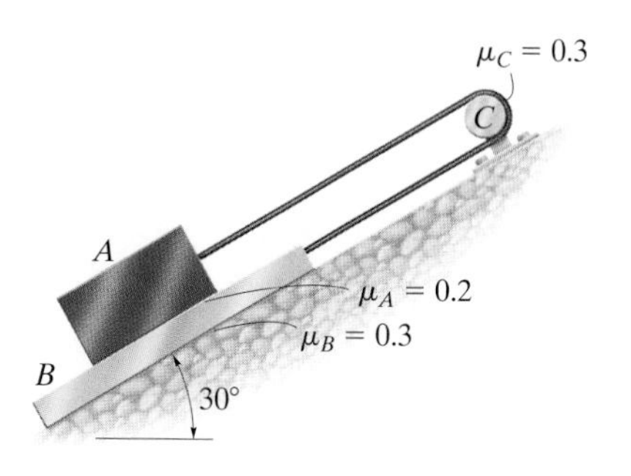

Prob. 8–97

8–98. The simple band brake is constructed so that the ends of the friction strap are connected to the pin at A and the lever arm at B. If the wheel is subjected to a torque of $M = 80$ lb·ft, determine the smallest force P applied to the lever that is required to hold the wheel stationary. The coefficient of static friction between the strap and wheel is $\mu_s = 0.5$.

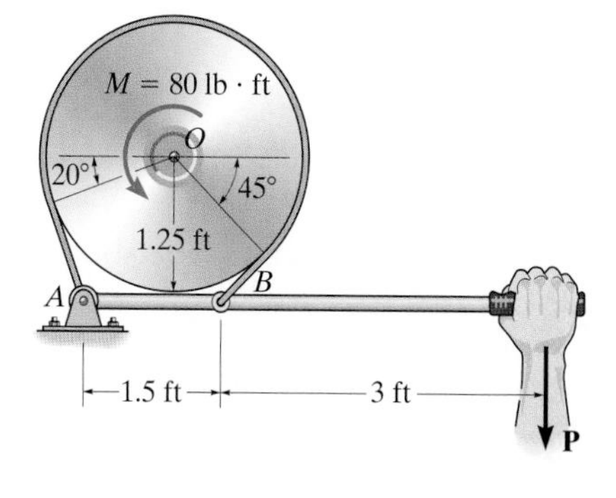

Prob. 8–98

8–99. The uniform 50-lb beam is supported by the rope which is attached to the end of the beam, wraps over the rough peg, and is then connected to the 100-lb block. If the coefficient of static friction between the beam and the block, and between the rope and the peg, is $\mu_s = 0.4$, determine the maximum distance that the block can be placed from A and still remain in equilibrium. Assume the block will not tip.

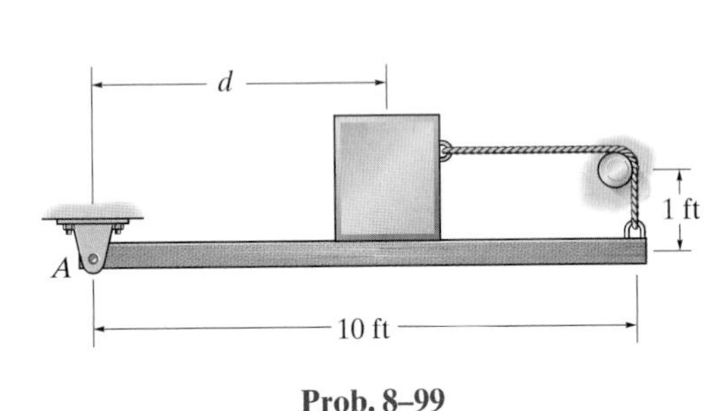

Prob. 8–99

8–101. A cord having a weight of 0.5 lb/ft and a total length of 10 ft is suspended over a peg P as shown. If the coefficient of static friction between the peg and cord is $\mu_s = 0.5$, determine the longest length h which one side of the suspended cord can have without causing motion. Neglect the size of the peg and the length of cord draped over it.

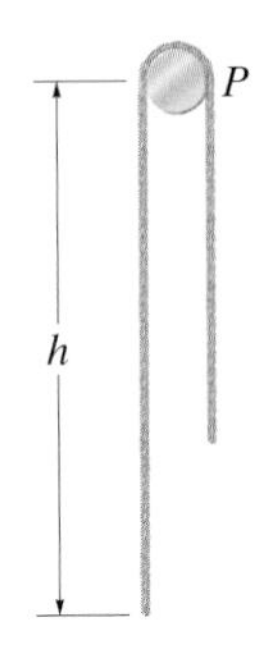

Prob. 8–101

***8–100.** The uniform concrete pipe has a weight of 800 lb and is unloaded slowly from the truck bed using the rope and skids shown. If the coefficient of kinetic friction between the rope and pipe is $\mu_k = 0.3$, determine the force the worker must exert on the rope to lower the pipe at constant speed. There is a pulley at B, and the pipe does not slip on the skids. The lower portion of the rope is parallel to the skids.

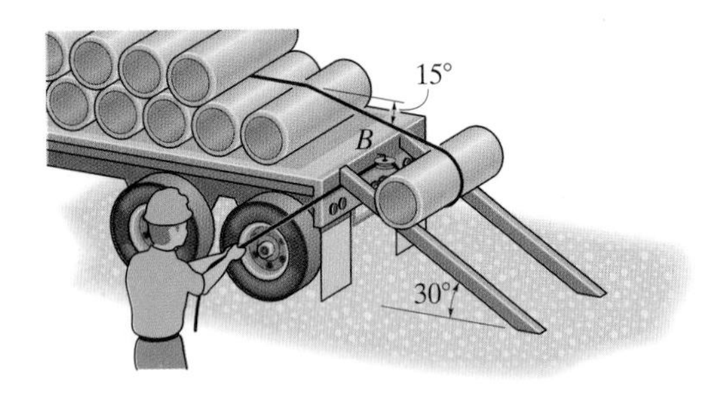

Prob. 8–100

8–102. Granular material, having a density of 1.5 Mg/m^3, is transported on a conveyor belt that slides over the fixed surface, having a coefficient of kinetic friction of $\mu_k = 0.3$. Operation of the belt is provided by a motor that supplies a torque **M** to wheel A. The wheel at B is free to turn, and the coefficient of static friction between the wheel at A and the belt is $\mu_A = 0.4$. If the belt is subjected to a pretension of 300 N when no load is on the belt, determine the greatest volume V of material that is permitted on the belt at any time without allowing the belt to stop. What is the torque **M** required to drive the belt when it is subjected to this maximum load?

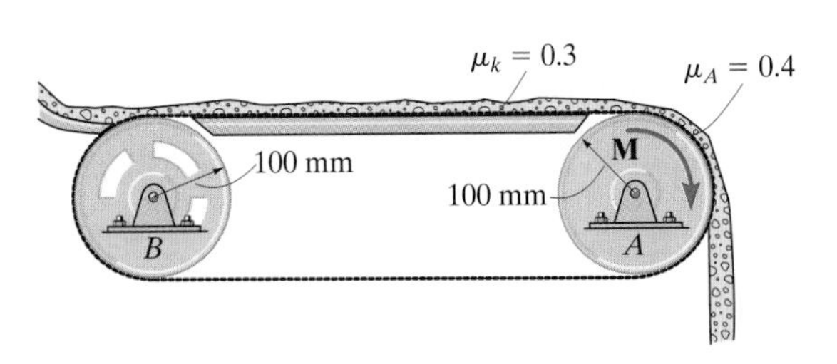

Prob. 8–102

8–103. Blocks A and B have a mass of 100 kg and 150 kg, respectively. If the coefficient of static friction between A and B and between B and C is $\mu_s = 0.25$, and between the ropes and the pegs D and E $\mu'_s = 0.5$, determine the smallest force F needed to cause motion of block B if $P = 30$ N.

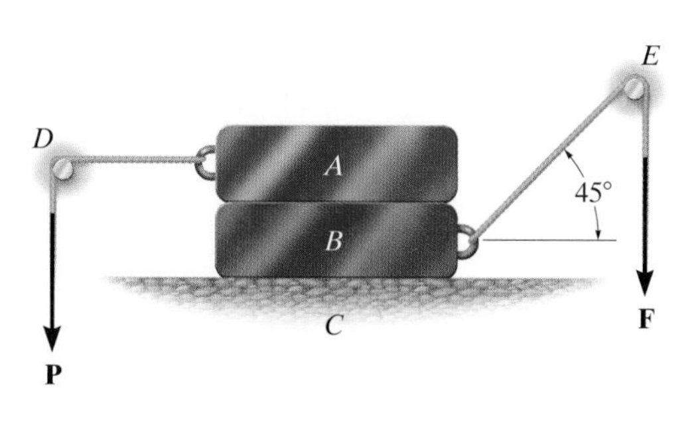

Prob. 8–103

8–105. Block A has a mass of 50 kg and rests on surface for which $\mu_s = 0.25$. If the coefficient of static friction between the cord and the fixed peg at C is $\mu'_s = 0.3$, determine the greatest mass of the suspended cylinder D without causing motion.

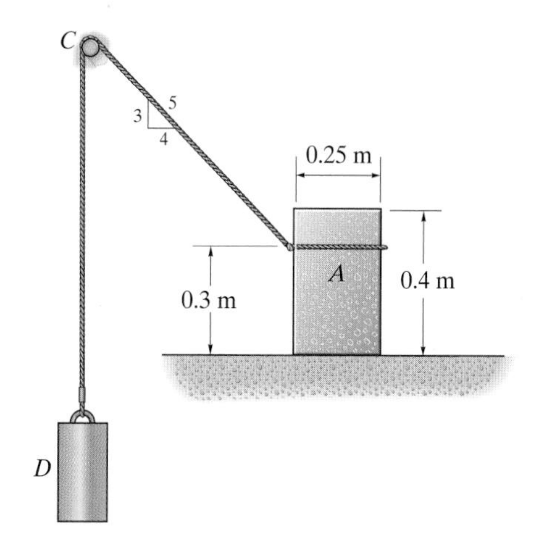

Prob. 8–105

***8–104.** Blocks A and B weigh 50 lb and 30 lb, respectively. Using the coefficients of static friction indicated, determine the greatest weight of block E without causing motion.

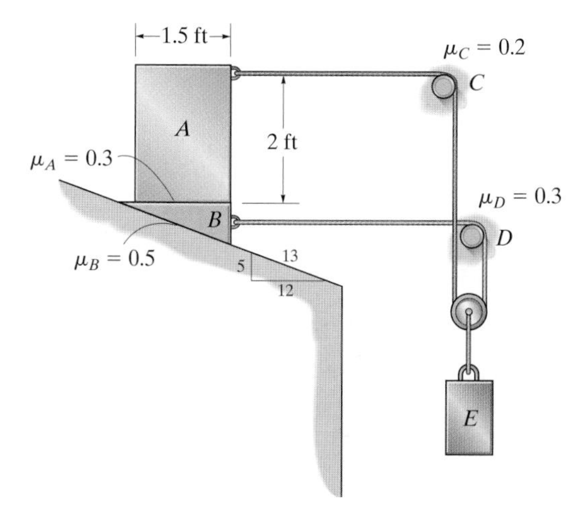

Prob. 8–104

8–106. Block A rests on the surface for which $\mu_s = 0.25$. If the mass of the suspended cylinder D is 4 kg, determine the smallest mass of block A so that it does not slip or tip. The coefficient of static friction between the cord and the fixed peg at C is $\mu'_s = 0.3$.

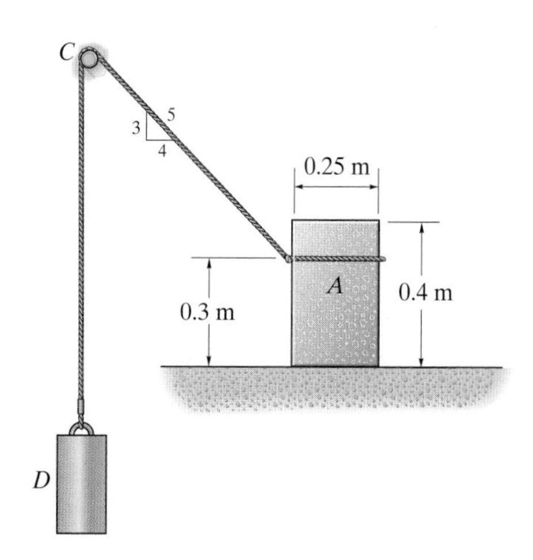

Prob. 8–106

*8.6 Frictional Forces on Collar Bearings, Pivot Bearings, and Disks

Pivot and *collar bearings* are commonly used in machines to support an *axial load* on a rotating shaft. These two types of support are shown in Fig. 8–21. Provided the bearings are not lubricated, or are only partially lubricated, the laws of dry friction may be applied to determine the moment **M** needed to turn the shaft when it supports an axial force **P**.

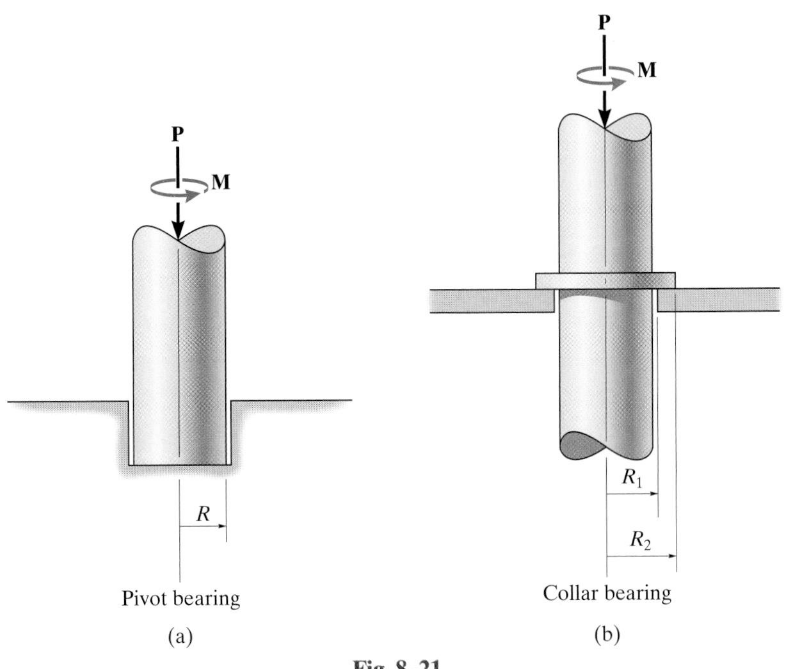

Fig. 8–21

Frictional Analysis. The collar bearing on the shaft shown in Fig. 8–22 is subjected to an axial force **P** and has a total bearing or contact area $\pi(R_2^2 - R_1^2)$. In the following analysis, the normal pressure p is considered to be *uniformly distributed* over this area—a reasonable assumption provided the bearing is new and evenly supported. Since $\Sigma F_z = 0$, then p, measured as a force per unit area, is $p = P/\pi(R_2^2 - R_1^2)$.

The moment needed to cause impending rotation of the shaft can be determined from moment equilibrium about the z axis. A small area element $dA = (r\,d\theta)(dr)$, shown in Fig. 8–22, is subjected to both a normal force $dN = p\,dA$ and an associated frictional force,

$$dF = \mu_s\,dN = \mu_s p\,dA = \frac{\mu_s P}{\pi(R_2^2 - R_1^2)}dA$$

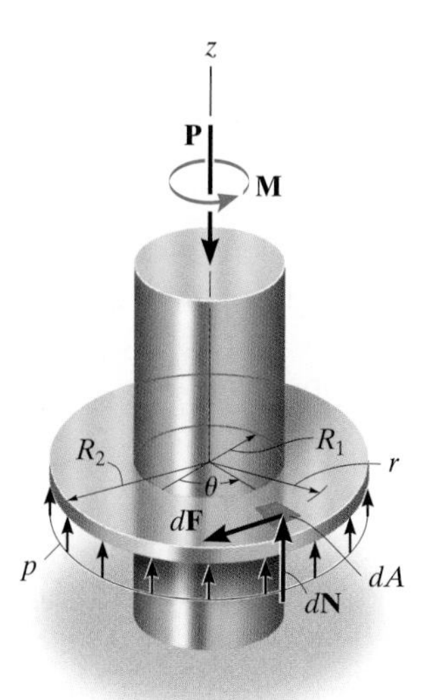

Fig. 8–22

The normal force does not create a moment about the z axis of the shaft; however, the frictional force does; namely, $dM = r\,dF$. Integration is needed to compute the total moment created by all the frictional forces acting on differential areas dA. Therefore, for impending rotational motion,

$$\Sigma M_z = 0; \qquad M - \int_A r\,dF = 0$$

Substituting for dF and dA and integrating over the entire bearing area yields

$$M = \int_{R_1}^{R_2}\int_0^{2\pi} r\left[\frac{\mu_s P}{\pi(R_2^2 - R_1^2)}\right](r\,d\theta\,dr) = \frac{\mu_s P}{\pi(R_2^2 - R_1^2)}\int_{R_1}^{R_2} r^2\,dr \int_0^{2\pi} d\theta$$

or

$$M = \frac{2}{3}\mu_s P\left(\frac{R_2^3 - R_1^3}{R_2^2 - R_1^2}\right) \qquad (8\text{–}7)$$

This equation gives the magnitude of moment required for impending rotation of the shaft. The frictional moment developed at the end of the shaft, when it is *rotating* at constant speed, can be found by substituting μ_k for μ_s in Eq. 8–7.

When $R_2 = R$ and $R_1 = 0$, as in the case of a pivot bearing, Fig. 8–21*a*, Eq. 8–7 reduces to

$$M = \frac{2}{3}\mu_s PR \qquad (8\text{–}8)$$

Recall from the initial assumption that both Eqs. 8–7 and 8–8 apply only for bearing surfaces subjected to *constant pressure*. If the pressure is not uniform, a variation of the pressure as a function of the bearing area must be determined before integrating to obtain the moment. The following example illustrates this concept.

The motor that turns the disk of this sanding machine develops a torque that must overcome the frictional forces acting on the disk.

EXAMPLE 8.10

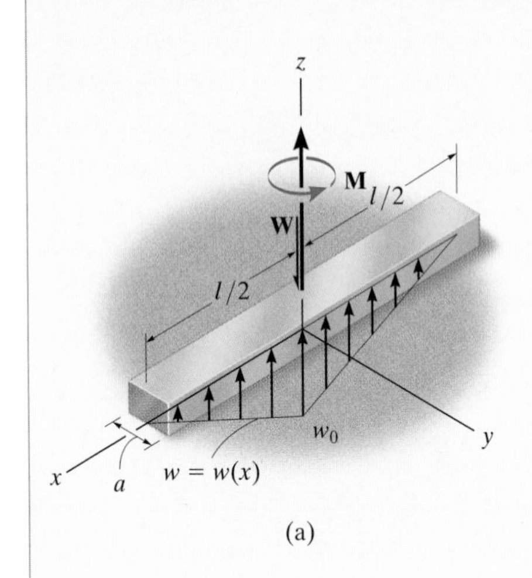

(a)

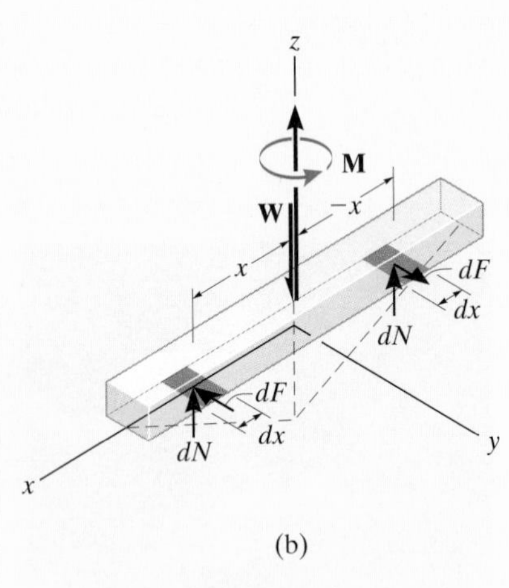

(b)

Fig. 8–23

The uniform bar shown in Fig. 8–23*a* has a total mass m. If it is assumed that the normal pressure acting at the contacting surface varies linearly along the length of the bar as shown, determine the couple moment **M** required to rotate the bar. Assume that the bar's width a is negligible in comparison to its length l. The coefficient of static friction is equal to μ_s.

SOLUTION

A free-body diagram of the bar is shown in Fig. 8–23*b*. Since the bar has a total weight of $W = mg$, the intensity w_0 of the distributed lead at the center ($x = 0$) is determined from vertical force equilibrium, Fig. 8–23*a*.

$$+\uparrow \Sigma F_z = 0; \quad -mg + 2\left[\frac{1}{2}\left(\frac{l}{2}\right)w_0\right] = 0 \qquad w_0 = \frac{2mg}{l}$$

Since $w = 0$ at $x = l/2$, the distributed load expressed as a function of x is

$$w = w_0\left(1 - \frac{2x}{l}\right) = \frac{2mg}{l}\left(1 - \frac{2x}{l}\right)$$

The magnitude of the normal force acting on a segment of area having a length dx is therefore

$$dN = w\,dx = \frac{2mg}{l}\left(1 - \frac{2x}{l}\right)dx$$

The magnitude of the frictional force acting on the same element of area is

$$dF = \mu_s\,dN = \frac{2\mu_s mg}{l}\left(1 - \frac{2x}{l}\right)dx$$

Hence, the moment created by this force about the z axis is

$$dM = x\,dF = \frac{2\mu_s mg}{l}x\left(1 - \frac{2x}{l}\right)dx$$

The summation of moments about the z axis of the bar is determined by integration, which yields

$$\Sigma M_z = 0; \qquad M - 2\int_0^{l/2} \frac{2\mu_s mg}{l}x\left(1 - \frac{2x}{l}\right)dx = 0$$

$$M = \frac{4\mu_s mg}{l}\left(\frac{x^2}{2} - \frac{2x^3}{3l}\right)\Bigg|_0^{l/2}$$

$$M = \frac{\mu_s mgl}{6} \qquad\qquad \textit{Ans.}$$

8.7 Frictional Forces on Journal Bearings

When a shaft or axle is subjected to lateral loads, a *journal bearing* is commonly used for support. Well-lubricated journal bearings are subjected to the laws of fluid mechanics, in which the viscosity of the lubricant, the speed of rotation, and the amount of clearance between the shaft and bearing are needed to determine the frictional resistance of the bearing. When the bearing is not lubricated or is only partially lubricated, however, a reasonable analysis of the frictional resistance can be based on the laws of dry friction.

Unwinding the cable from this spool requires overcoming friction from the supporting shaft.

Frictional Analysis. A typical journal-bearing support is shown in Fig. 8–24*a*. As the shaft rotates in the direction shown in the figure, it rolls up against the wall of the bearing to some point A where slipping occurs. If the lateral load acting at the end of the shaft is $\mathbf{P}$, it is necessary that the bearing reactive force $\mathbf{R}$ acting at A be equal and opposite to $\mathbf{P}$, Fig. 8–24*b*. The moment needed to maintain constant rotation of the shaft can be found by summing moments about the z axis of the shaft; i.e.,

$$\Sigma M_z = 0; \qquad M - (R \sin \phi_k)r = 0$$

or

$$M = Rr \sin \phi_k \tag{8–9}$$

where ϕ_k is the angle of kinetic friction defined by $\tan \phi_k = F/N = \mu_k N/N = \mu_k$. In Fig. 8–24*c*, it is seen that $r \sin \phi_k = r_f$. The dashed circle with radius r_f is called the *friction circle*, and as the shaft rotates, the reaction $\mathbf{R}$ will always be tangent to it. If the bearing is partially lubricated, μ_k is small, and therefore $\mu_k = \tan \phi_k \approx \sin \phi_k \approx \phi_k$. Under these conditions, a reasonable *approximation* to the moment needed to overcome the frictional resistance becomes

$$M \approx Rr\mu_k \tag{8–10}$$

The following example illustrates a common application of this analysis.

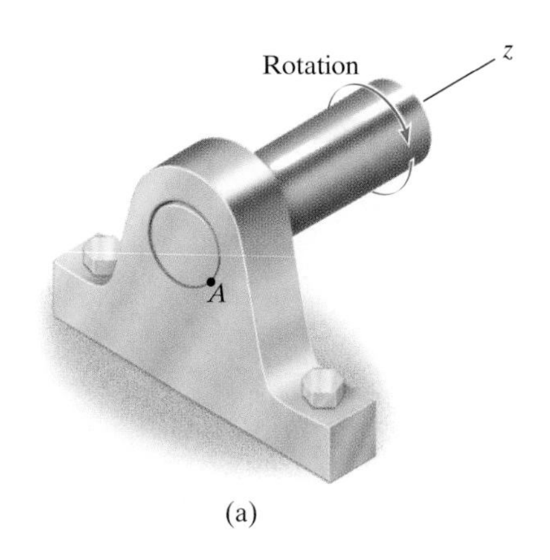

(a)

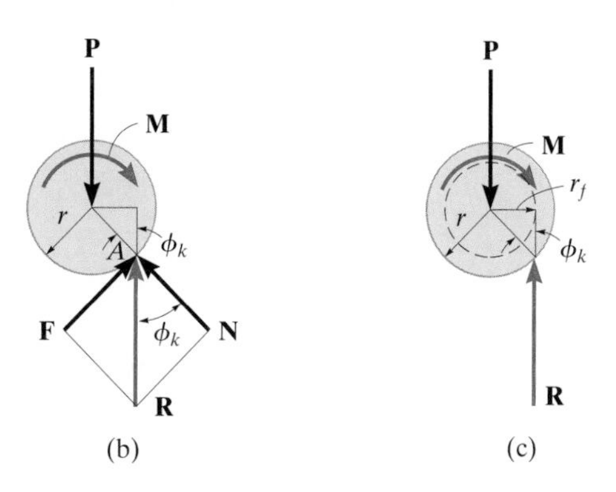

Fig. 8–24

EXAMPLE 8.11

The 100-mm-diameter pulley shown in Fig. 8–25*a* fits loosely on a 10-mm-diameter shaft for which the coefficient of static friction is $\mu_s = 0.4$. Determine the minimum tension T in the belt needed to (a) raise the 100-kg block and (b) lower the block. Assume that no slipping occurs between the belt and pulley and neglect the weight of the pulley.

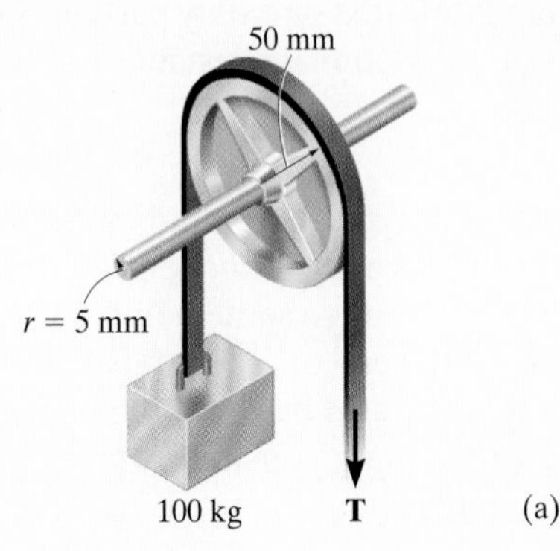

(a)

ϕ_s
r_f
Impending motion
P_1
P_2
R
981 N
T
52 mm
48 mm

(b)

SOLUTION

Part (a). A free-body diagram of the pulley is shown in Fig. 8–25*b*. When the pulley is subjected to belt tensions of 981 N each, it makes contact with the shaft at point P_1. As the tension T is *increased*, the pulley will roll around the shaft to point P_2 before motion impends. From the figure, the friction circle has a radius $r_f = r \sin \phi_s$. Using the simplification that $\sin \phi_s \approx (\tan \phi_s \approx \phi_s)$, then $r_f \approx r\mu_s = (5 \text{ mm})(0.4) = 2 \text{ mm}$, so that summing moments about P_2 gives

$$\circlearrowleft + \Sigma M_{P_2} = 0; \quad 981 \text{ N}(52 \text{ mm}) - T(48 \text{ mm}) = 0$$

$$T = 1063 \text{ N} = 1.06 \text{ kN} \qquad \textit{Ans.}$$

If a more exact analysis is used, then $\phi_s = \tan^{-1} 0.4 = 21.8°$. Thus, the radius of the friction circle would be $r_f = r \sin \phi_s = 5 \sin 21.8° = 1.86$ mm. Therefore,

$$\circlearrowleft + \Sigma M_{P_2} = 0;$$

$$981 \text{ N}(50 \text{ mm} + 1.86 \text{ mm}) - T(50 \text{ mm} - 1.86 \text{ mm}) = 0$$

$$T = 1057 \text{ N} = 1.06 \text{ kN} \qquad \textit{Ans.}$$

Part (b). When the block is lowered, the resultant force **R** acting on the shaft passes through point P_3, as shown in Fig. 8–25*c*. Summing moments about this point yields

$$\circlearrowleft + \Sigma M_{P_3} = 0; \quad 981 \text{ N}(48 \text{ mm}) - T(52 \text{ mm}) = 0$$

$$T = 906 \text{ N} \qquad \textit{Ans.}$$

NOTE: The difference between raising and lowering the block is thus 54 N.

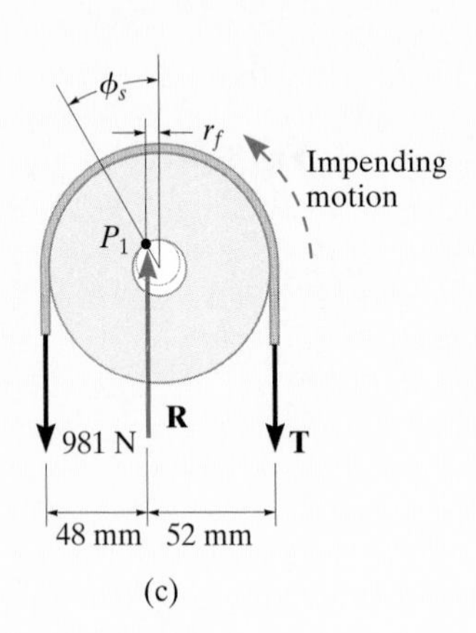

(c)

Fig. 8–25

*8.8 Rolling Resistance

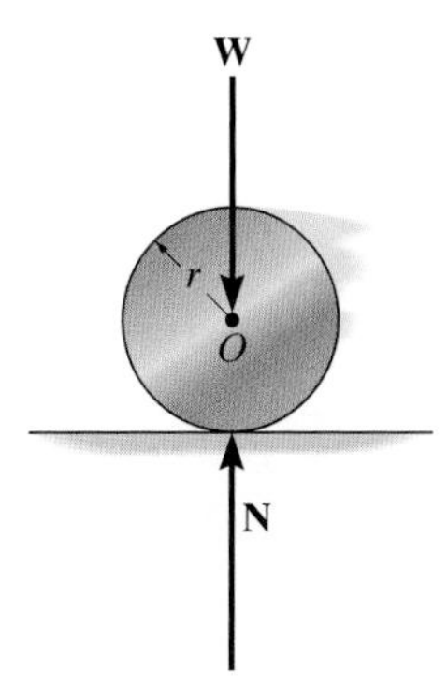

Rigid surface of contact

(a)

If a *rigid* cylinder of weight **W** rolls at constant velocity along a *rigid* surface, the normal force exerted by the surface on the cylinder acts at the tangent point of contact, as shown in Fig. 8–26*a*. Under these conditions, provided the cylinder does not encounter frictional resistance from the air, motion will continue indefinitely. Actually, however, no materials are perfectly rigid, and therefore the reaction of the surface on the cylinder consists of a distribution of normal pressure. For example, consider the cylinder to be made of a very hard material, and the surface on which it rolls to be relatively soft. Due to its weight, the cylinder compresses the surface underneath it, Fig. 8–26*b*. As the cylinder rolls, the surface material in front of the cylinder *retards* the motion since it is being *deformed*, whereas the material in the rear is *restored* from the deformed state and therefore tends to *push* the cylinder forward. The normal pressures acting on the cylinder in this manner are represented in Fig. 8–26*b* by their resultant forces $\mathbf{N}_d$ and $\mathbf{N}_r$. Unfortunately, the magnitude of the force of *deformation*, $\mathbf{N}_d$, and its horizontal component is *always greater* than that of *restoration*, $\mathbf{N}_r$, and consequently a horizontal driving force **P** must be applied to the cylinder to maintain the motion. Fig. 8–26*b*.*

Rolling resistance is caused primarily by this effect, although it is also, to a lesser degree, the result of surface adhesion and relative micro-sliding between the surfaces of contact. Because the actual force **P** needed to overcome these effects is difficult to determine, a simplified method will be developed here to explain one way engineers have analyzed this phenomenon. To do this, we will consider the resultant of the *entire* normal pressure, $\mathbf{N} = \mathbf{N}_d + \mathbf{N}_r$, acting on the cylinder, Fig. 8–26*c*. As shown in Fig. 8–26*d*, this force acts at an angle θ with the vertical. To keep the cylinder in equilibrium, i.e., rolling at a constant rate, it is necessary that **N** be *concurrent* with the driving force **P** and the weight **W**. Summing moments about point A gives $Wa = P(r\cos\theta)$. Since the deformations are generally very small in relation to the cylinder's radius, $\cos\theta \approx 1$; hence,

$$Wa \approx Pr$$

or

$$P \approx \frac{Wa}{r} \qquad (8\text{–}11)$$

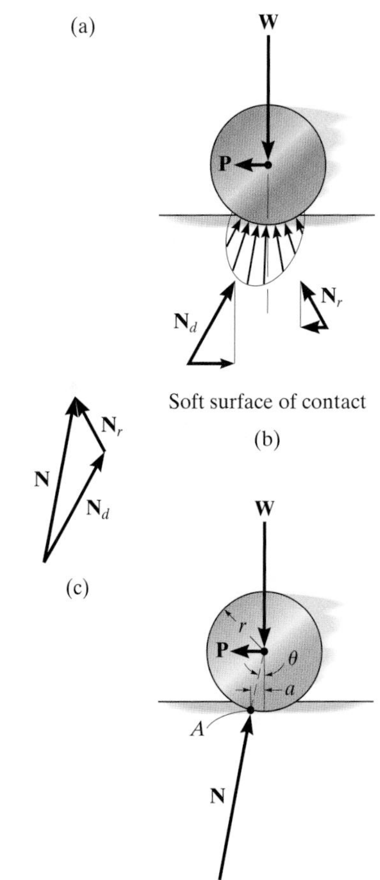

Fig. 8–26

*Actually, the deformation force $\mathbf{N}_d$ causes *energy* to be stored in the material as its magnitude is increased, whereas the restoration force $\mathbf{N}_r$, as its magnitude is decreased, allows some of this energy to be released. The remaining energy is *lost* since it is used to heat up the surface, and if the cylinder's weight is very large, it accounts for permanent deformation of the surface. Work must be done by the horizontal force **P** to make up for this loss.

Rolling resistance of railroad wheels on the rails is small since steel is very stiff. By comparison, the rolling resistance of the wheels of a tractor in a wet field is very large.

The distance a is termed the *coefficient of rolling resistance,* which has the dimension of length. For instance, $a \approx 0.5$ mm for a wheel rolling on a rail, both of which are made of mild steel. For hardened steel ball bearings on steel, $a \approx 0.1$ mm. Experimentally, though, this factor is difficult to measure, since it depends on such parameters as the rate of rotation of the cylinder, the elastic properties of the contacting surfaces, and the surface finish. For this reason, little reliance is placed on the data for determining a. The analysis presented here does, however, indicate why a heavy load (W) offers greater resistance to motion (P) than a light load under the same conditions. Furthermore, since the ratio Wa/r is generally very small compared to $\mu_k W$, the force needed to *roll* the cylinder over the surface will be much less than that needed to *slide* the cylinder across the surface. Hence, the analysis indicates why roller or ball bearings are often used to minimize the frictional resistance between moving parts.

EXAMPLE 8.12

A 10-kg steel wheel shown in Fig. 8–27a has a radius of 100 mm and rests on an inclined plans made of wood. If θ is increased so that the wheel begins to roll-down the incline with constant velocity when $\theta = 1.2°$, determine the coefficient of rolling resistance.

(a)

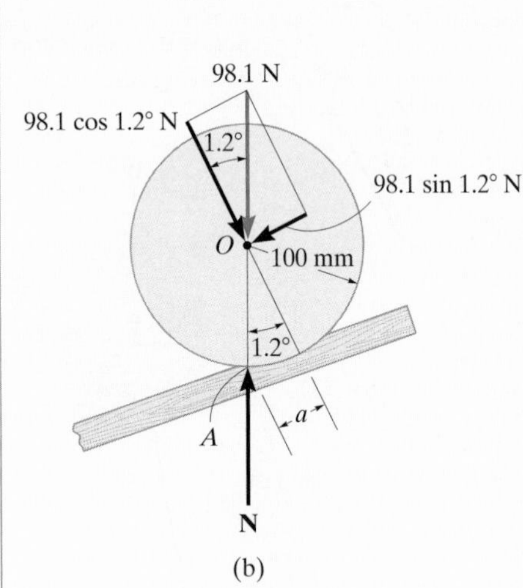

Fig. 8–27

SOLUTION

As shown on the free-body diagram, Fig. 8–27b, when the wheel has impending motion, the normal reaction $\mathbf{N}$ acts at point A defined by the dimension a. Resolving the weight into components parallel and perpendicular to the incline, and summing moments about point A, yields (approximately)

$$\circlearrowleft + \Sigma M_A = 0; \quad 98.1 \cos 1.2° \text{ N}(a) - 98.1 \sin 1.2° \text{ N}(100 \text{ mm}) = 0$$

Solving, we obtain

$$a = 2.09 \text{ mm} \qquad \textit{Ans.}$$

PROBLEMS

8–107. The collar bearing uniformly supports an axial force of $P = 800$ lb. If the coefficient of static friction is $\mu_s = 0.3$, determine the torque M required to overcome friction.

***8–108.** The collar bearing uniformly supports an axial force of $P = 500$ lb. If a torque of $M = 3$ lb·ft is applied to the shaft and causes it to rotate at constant velocity, determine the coefficient of kinetic friction at the surface of contact.

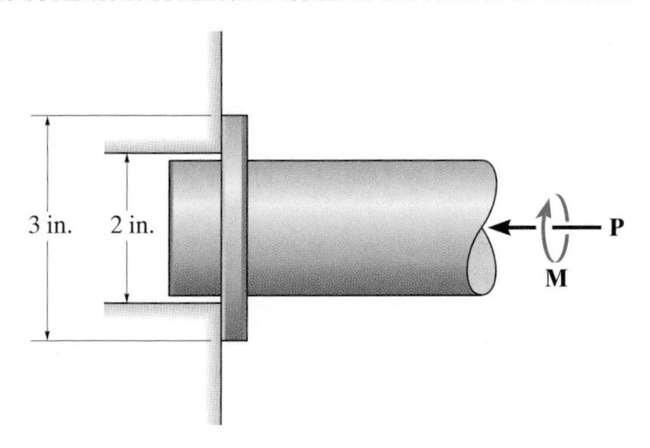

Probs. 8–107/108

8–109. The *double-collar bearing* is subjected to an axial force $P = 4$ kN. Assuming that collar A supports $0.75P$ and collar B supports $0.25P$, both with a uniform distribution of pressure, determine the maximum frictional moment M that may be resisted by the bearing. Take $\mu_s = 0.2$ for both collars.

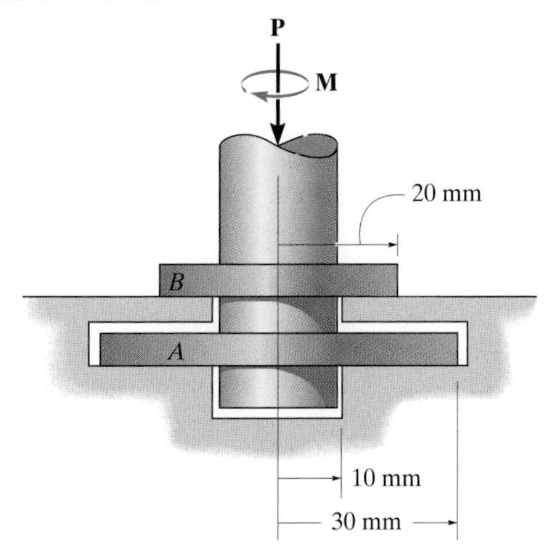

Prob. 8–109

8–110. The annular ring bearing is subjected to a thrust of 800 lb. If $\mu_s = 0.35$, determine the torque M that must be applied to overcome friction.

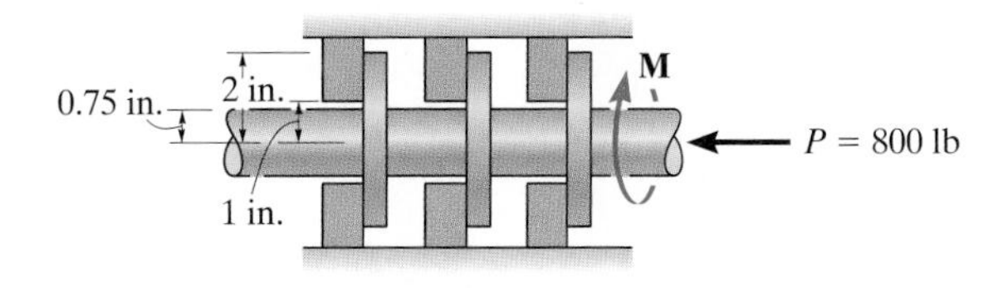

Prob. 8–110

8–111. The floor-polishing machine rotates at a constant angular velocity. If it has a weight of 80 lb. determine the couple forces F the operator must apply to the handles to hold the machine stationary. The coefficient of kinetic friction between the floor and brush is $\mu_k = 0.3$. Assume the brush exerts a uniform pressure on the floor.

Prob. 8–111

***8–112.** The plate clutch consists of a flat plate A that slides over the rotating shaft S. The shaft is fixed to the driving plate gear B. If the gear C, which is in mesh with B, is subjected to a torque of $M = 0.8\ \text{N}\cdot\text{m}$, determine the smallest force P, that must be applied via the control arm, to stop the rotation. The coefficient of static friction between the plates A and D is $\mu_s = 0.4$. Assume the bearing pressure between A and D to be uniform.

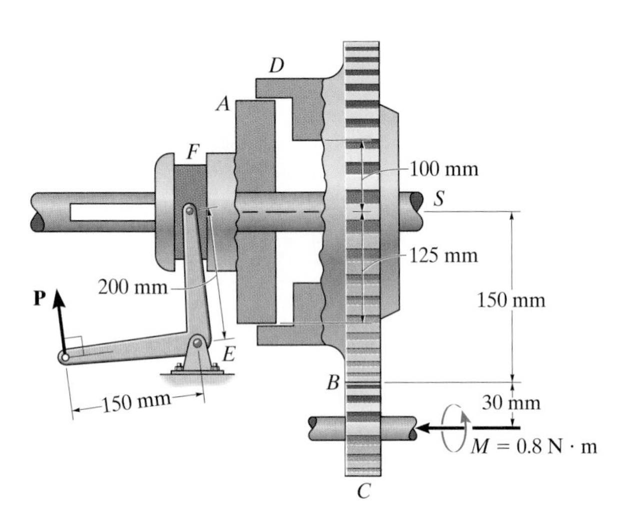

Prob. 8–112

8–113. The 4-in.-diameter shaft is held in the hole such that the normal pressure acting around the shaft varies linearly with its depth as shown. Determine the frictional torque that must be overcome to rotate the shaft. Take $\mu_s = 0.2$.

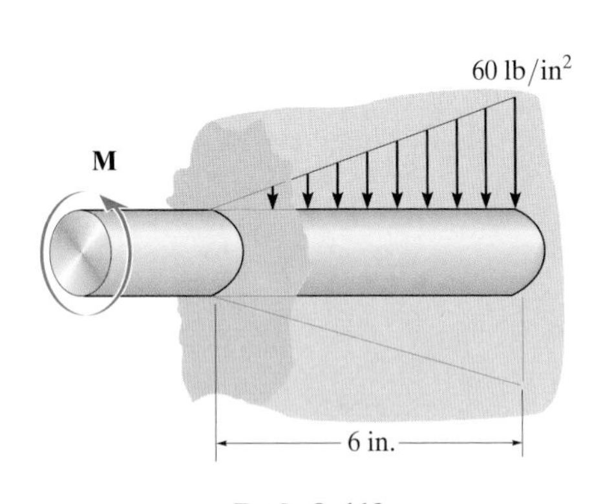

Prob. 8–113

8–114. Because of wearing at the edges, the pivot bearing is subjected to a conical pressure distribution at its surface of contact. Determine the torque M required to overcome friction and turn the shaft, which supports an axial force **P**. The coefficient of static friction is μ_s. For the solution, it is necessary to determine the peak pressure p_0 in terms of P and the bearing radius R.

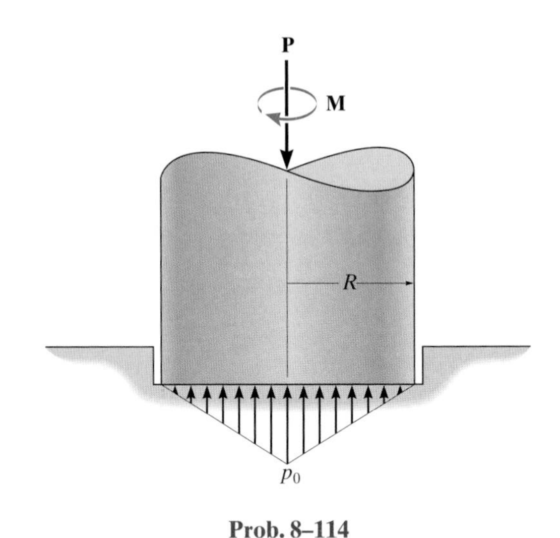

Prob. 8–114

8–115. The conical bearing is subjected to a constant pressure distribution at its surface of contact. If the coefficient of static friction is μ_s, determine the torque **M** required to overcome friction if the shaft supports an axial force **P**.

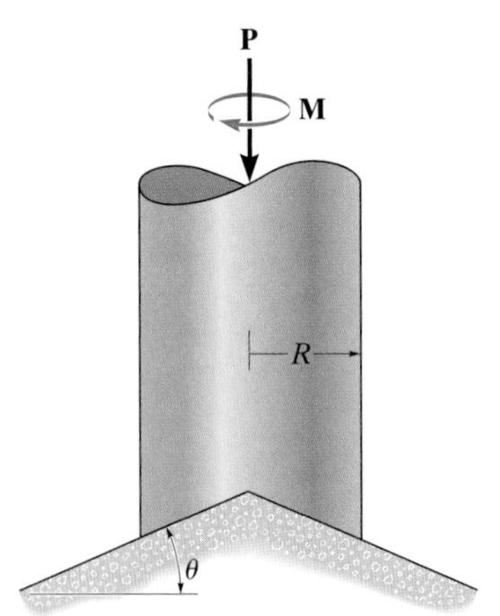

Prob. 8–115

*8–116. The tractor is used to push the 1500-lb pipe. To do this it must overcome the frictional forces at the ground, caused by sand. Assuming that the sand exerts a pressure on the bottom of the pipe as shown, and the coefficient of static friction between the pipe and the sand is $\mu_s = 0.3$, determine the force required to push the pipe forward. Also, determine the peak pressure p_0.

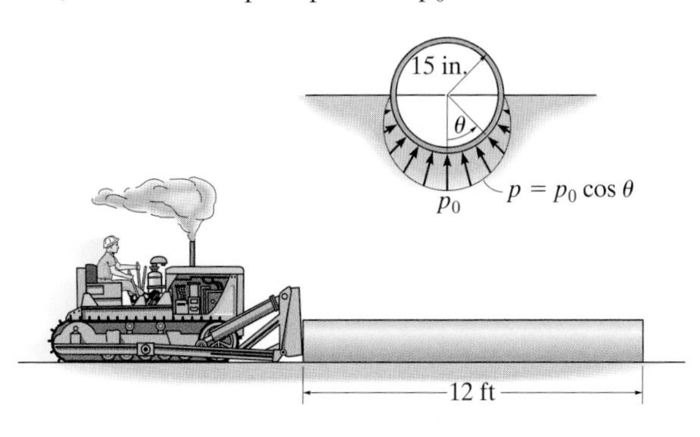

Prob. 8–116

8–117. Assuming that the variation of pressure at the bottom of the pivot bearing is defined as $p = p_0(R_2/r)$, determine the torque M needed to overcome friction if the shaft is subjected to an axial force **P**. The coefficient of static friction is μ_s. For the solution, it is necessary to determine p_0 in terms of P and the bearing dimensions R_1 and R_2.

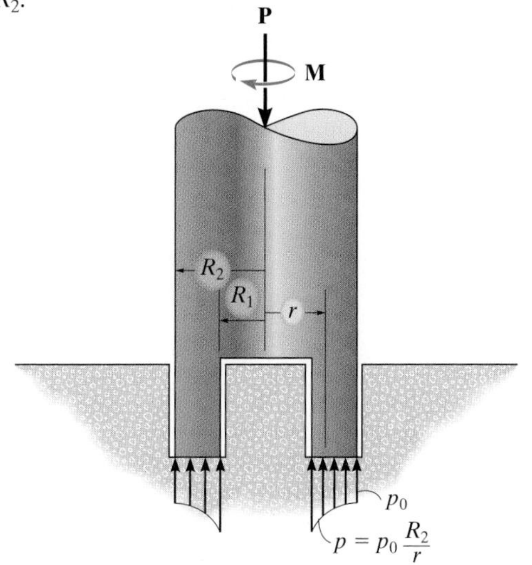

Prob. 8–117

8–118. A disk having an outer diameter of 8 in. fits loosely over a fixed shaft having a diameter of 3 in. If the coefficient of static friction between the disk and the shaft is $\mu_s = 0.15$, determine the smallest vertical force P, acting on the rim, which must be applied to the disk to cause it to slip over the shaft. The disk weighs 10 lb.

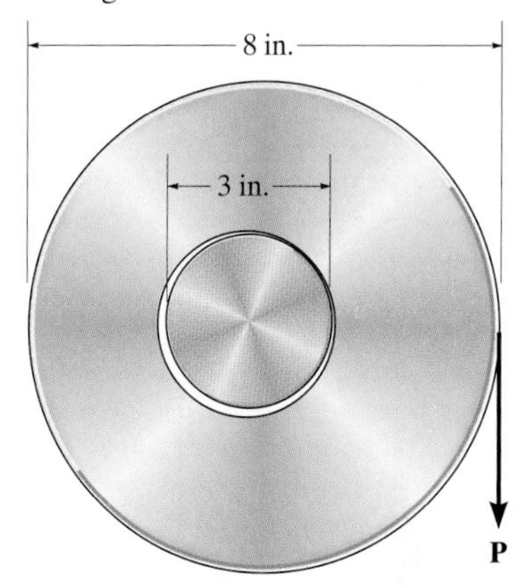

Prob. 8–118

8–119. The pulley has a radius of 3 in. and fits loosely on the 0.5-in.-diameter shaft. If the loadings acting on the belt cause the pulley to rotate with constant angular velocity, determine the frictional force between the shaft and the pulley and compute the coefficient of kinetic friction. The pulley weighs 18 lb.

*8–120. The pulley has a radius of 3 in. and fits loosely on the 0.5-in.-diameter shaft. If the loadings acting on the belt cause the pulley to rotate with constant angular velocity, determine the frictional force between the shaft and the pulley and compute the coefficient of kinetic friction. Neglect the weight of the pulley.

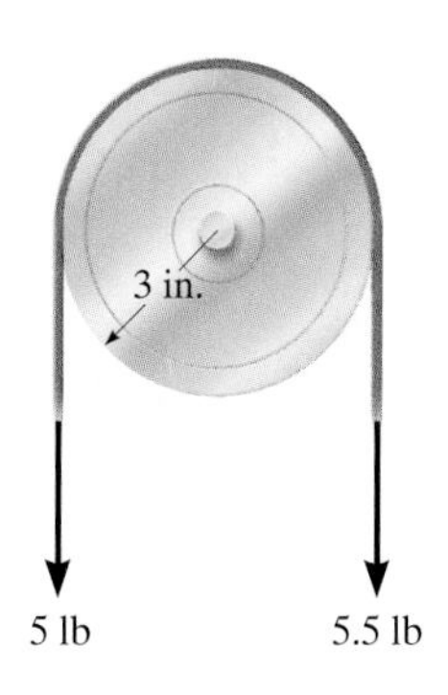

Probs. 8–119/120

8–121. The 5-kg pulley has a diameter of 240 mm and the axle has a diameter of 40 mm. If the coefficient of kinetic friction between the axle and the pulley is $\mu_k = 0.15$, determine the vertical force P on the rope required to lift the 80-kg block at constant velocity.

8–122. Solve Prob. 8–121 if the force **P** is applied horizontally to the right.

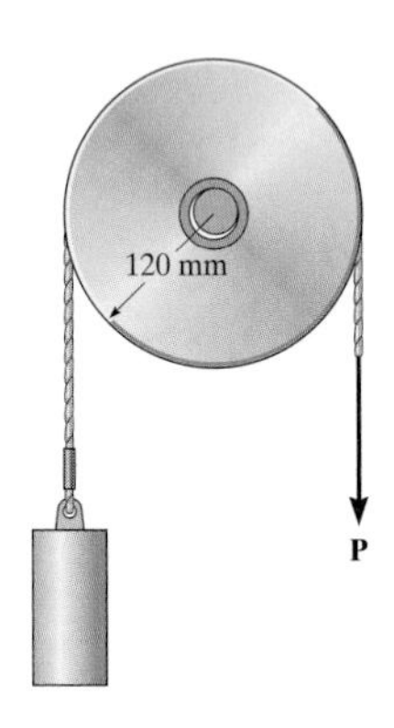

Probs. 8–121/122

***8–124.** The trailer has a total weight of 850 lb and center of gravity at G which is directly over its axle. If the axle has a diameter of 1 in., the radius of the wheel is $r = 1.5$ ft, and the coefficient of kinetic friction at the bearing is $\mu_k = 0.08$, determine the horizontal force P needed to pull the trailer.

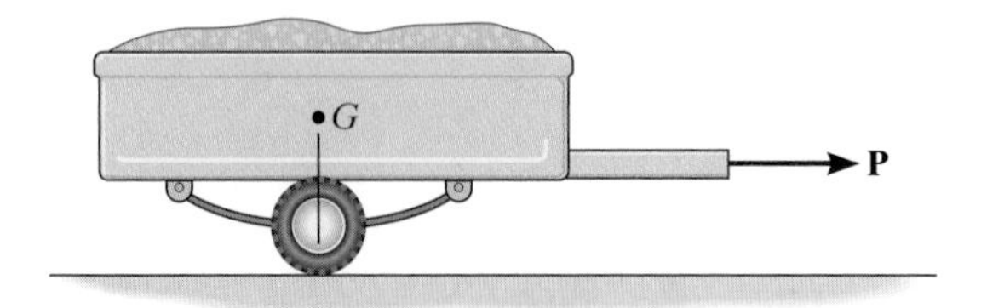

Prob. 8–124

8–123. A wheel on a freight car carries a load of 20 kip. If the axle of the car has a diameter of 2 in., determine the horizontal force P that must be applied to the axle to rotate the wheel. The coefficient of kinetic friction is $\mu_k = 0.05$.

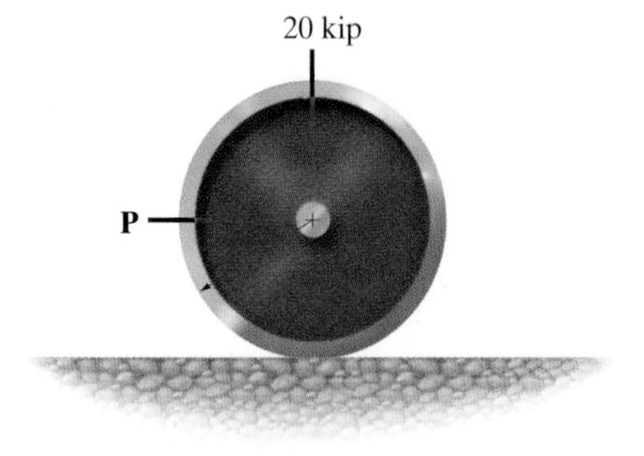

Prob. 8–123

8–125. The collar fits *loosely* around a fixed shaft that has a radius of 2 in. If the coefficient of kinetic friction between the shaft and the collar is $\mu_k = 0.3$, determine the force P on the horizontal segment of the belt so that the collar rotates *counterclockwise* with a constant angular velocity. Assume that the belt does not slip on the collar; rather, the collar slips on the shaft. Neglect the weight and thickness of the belt and collar. The radius, measured from the center of the collar to the mean thickness of the belt, is 2.25 in.

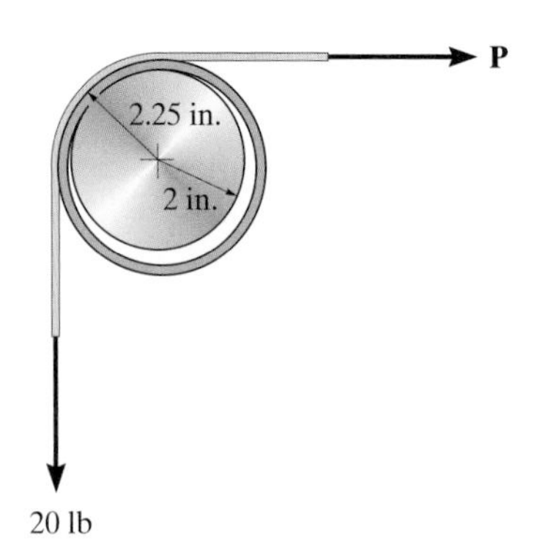

Prob. 8–125

8–126. The collar fits *loosely* around a fixed shaft that has a radius of 2 in. If the coefficient of kinetic friction between the shaft and the collar is $\mu_k = 0.3$, determine the force P on the horizontal segment of the belt so that the collar rotates *clockwise* with a constant angular velocity. Assume that the belt does not slip on the collar; rather, the collar slips on the shaft. Neglect the weight and thickness of the belt and collar. The radius, measured from the center of the collar to the mean thickness of the belt, is 2.25 in.

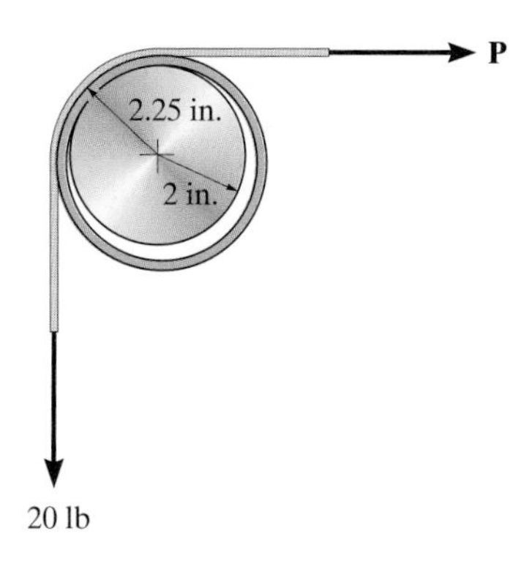

Prob. 8–126

8–127. The connecting rod is attached to the piston by a 0.75-in.-diameter pin at B and to the crank shaft by a 2-in.-diameter bearing A. If the piston is moving downwards, and the coefficient of static friction at these points is $\mu_s = 0.2$, determine the radius of the friction circle at each connection.

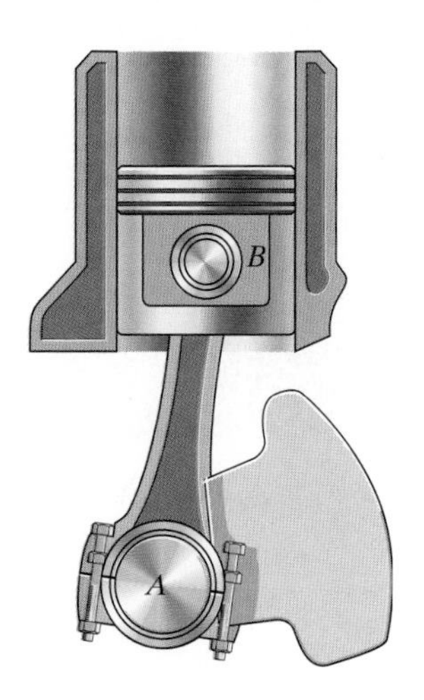

Prob. 8–127

***8–128.** The connecting rod is attached to the piston by a 20-mm-diameter pin at B and to the crank shaft by a 50-mm-diameter bearing A. If the piston is moving upwards, and the coefficient of static friction at these points is $\mu_s = 0.3$, determine the radius of the friction circle at each connection.

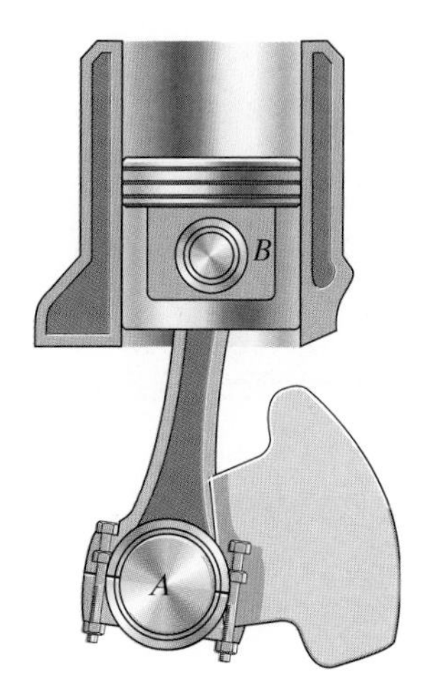

Prob. 8–128

8–129. The lawn roller has a mass of 80 kg. If the arm BA is held at an angle of 30° from the horizontal and the coefficient of rolling resistance for the roller is 25 mm, determine the force P needed to push the roller at constant speed. Neglect friction developed at the axle, A, and assume that the resultant force **P** acting on the handle is applied along arm BA.

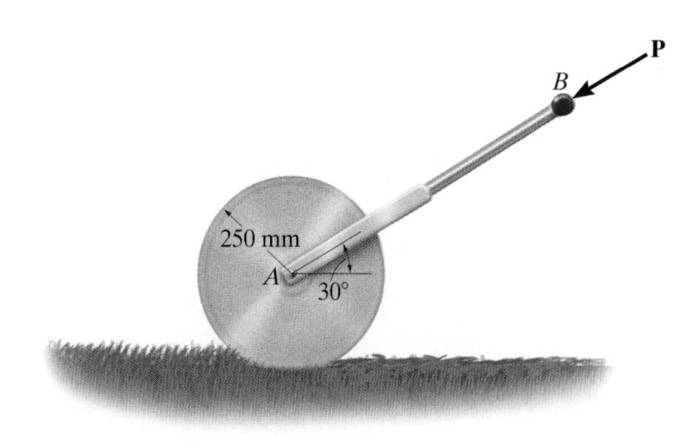

Prob. 8–129

8–130. The handcart has wheels with a diameter of 6 in. If a crate having a weight of 1500 lb is placed on the cart, determine the force P that must be applied to the handle to overcome the rolling resistance. The coefficient of rolling resistance is 0.04 in. Neglect the weight of the cart.

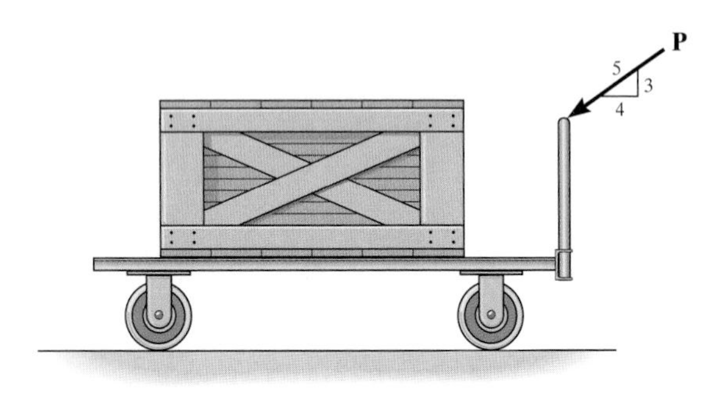

Prob. 8–130

8–131. The cylinder is subjected to a load that has a weight W. If the coefficients of rolling resistance for the cylinder's top and bottom surfaces are a_A and a_B, respectively, show that a force having a magnitude of $P = [W(a_A + a_B)]/2r$ is required to move the load and thereby roll the cylinder forward. Neglect the weight of the cylinder.

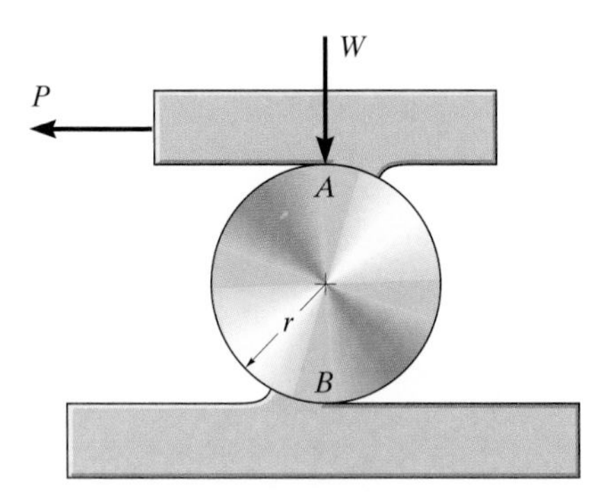

Prob. 8–131

***8–132.** The 1.2-Mg steel beam is moved over a level surface using a series of 30-mm-diameter rollers for which the coefficient of rolling resistance is 0.4 mm at the ground and 0.2 mm at the bottom surface of the beam. Determine the horizontal force P needed to push the beam forward at a constant speed. *Hint:* Use the result of Prob. 8–131.

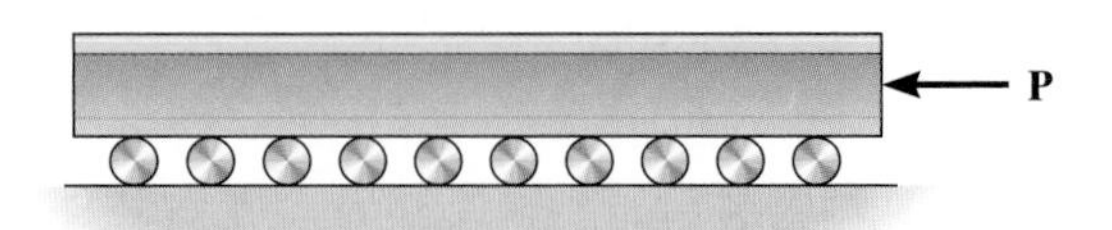

Prob. 8–132

8–133. The 1.4-Mg machine is to be moved over a level surface using a series of rollers for which the coefficient of rolling resistance is 0.5 mm at the ground and 0.2 mm at the bottom surface of the machine. Determine the appropriate diameter of the rollers so that the machine can be pushed forward with a horizontal force of $P = 250$ N. *Hint:* Use the result of Prob. 8–131.

Prob. 8–133

Chapter Review

Dry Friction

Frictional forces exist at rough surfaces of contact. They act on a body so as to oppose its motion or tendency of motion.

A static friction force approaches a maximum value of $F_s = \mu_s N$, where μ_s is the *coefficient of static friction*. In this case, motion between the contacting surfaces is *about to impend*.

If slipping occurs, then the friction force remains essentially constant and equal to a value of $F_k = \mu_k N$. Here μ_k is the *coefficient of kinetic friction*.

The solution of a problem involving friction requires first drawing the free-body diagram of the body. If the unknowns cannot be determined strictly from the equations of equilibrium, and the possibility of slipping can occur, then the friction equation should be applied at the appropriate points of contact in order to complete the solution.

It may also be possible for slender objects, like crates, to tip over, and this situation should also be investigated.
(*See pages 393–400.*)

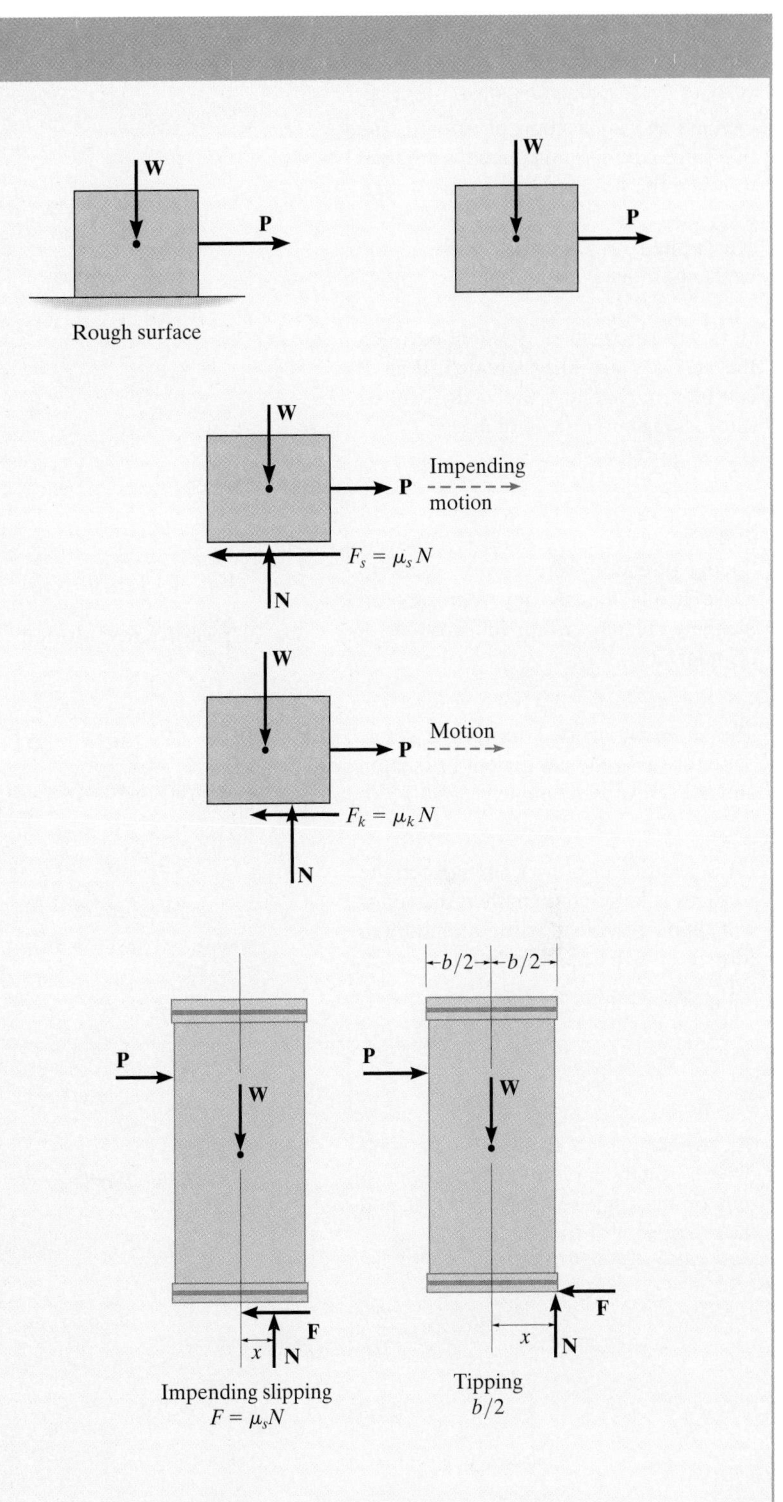

Wedges

Wedges are inclined planes used to increase the application of a force. The two-force equilibrium equations are used to relate the forces acting on the wedge.

An applied force **P** must push on the wedge to move it to the right.

If the coefficients of friction between the surfaces are large enough, then **P** can be removed, and the wedge will be self-locking and remain in place.

(*See page 418.*)

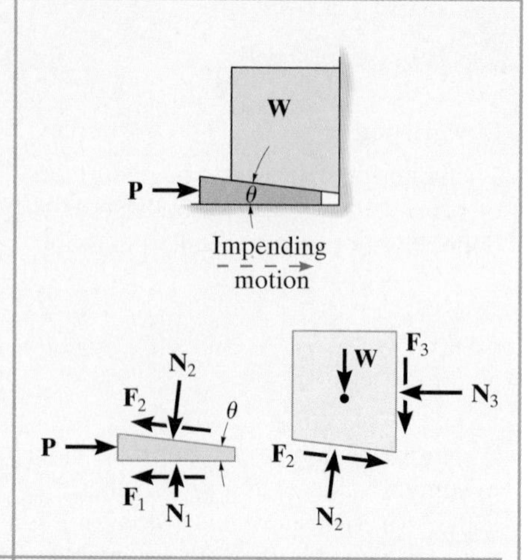

Screws

Square-threaded screws are used to move heavy loads. They represent an inclined plane, wrapped around a cylinder.

$$M = Wr\tan(\theta + \phi)$$

Upward Screw Motion

The moment needed to turn a screw depends upon the coefficient of friction and the screw's lead angle θ.

$$M' = Wr\tan(\theta - \phi)$$

Downward Screw Motion

$$\theta > \phi$$

If the coefficient of friction between the surfaces is large enough, then the screw will support the load without tending to turn, i.e., it will be self-locking.

(*See pages 420–422.*)

$$M'' = Wr\tan(\phi - \theta)$$

Downward Screw Motion

$$\phi > \theta$$

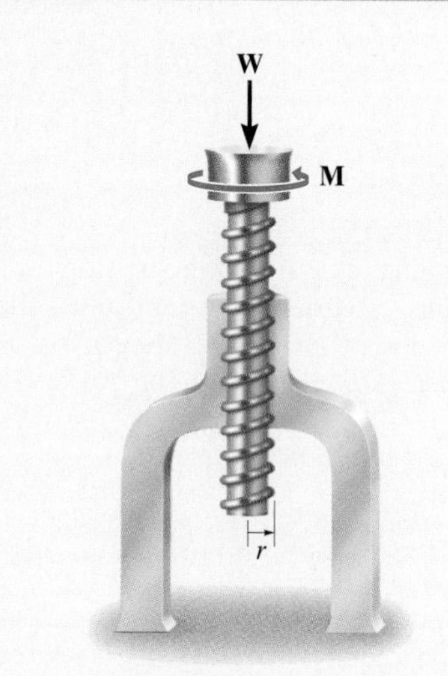

Flat Belts

The force needed to move a flat belt over a rough curved surface depends only on the angle of belt contact, β, and the coefficient of friction.

(*See pages 428–429.*)

$$T_2 = T_1 e^{\mu\beta}$$

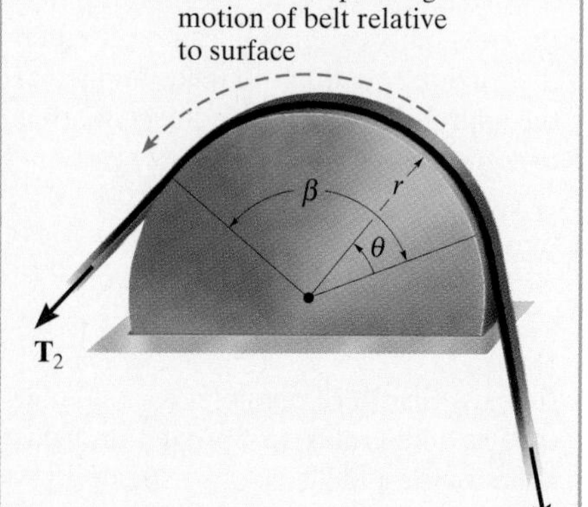

Collar Bearings and Disks

The frictional analysis of a collar bearing or disk requires looking at a differential segment of the contact area. The force acting on the differential element is determined from force equilibrium along the shaft, and the moment needed to turn the shaft at a constant rate is determined from moment equilibrium about the shaft's axis.

If the pressure on the surface of a collar bearing is uniform, then integration gives the result.
(*See pages 436–437.*)

$$M = \frac{2}{3}\mu_s P\left(\frac{R_2^3 - R_1^3}{R_2^2 - R_1^2}\right)$$

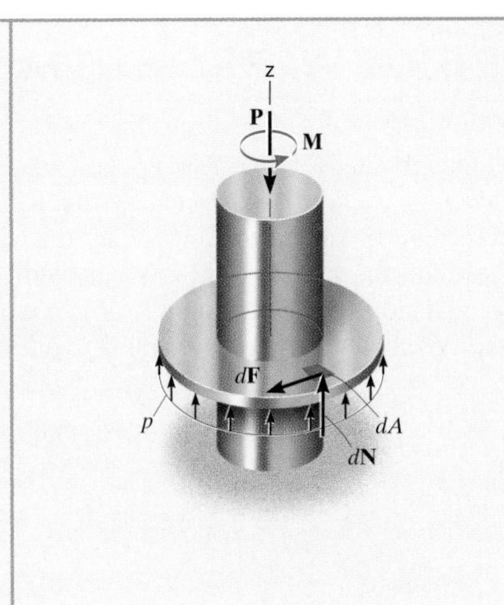

Journal Bearings

When a moment is applied to a shaft in a nonlubricated journal bearing, the shaft will tend to roll up the side of the bearing until slipping occurs. This defines the radius of a friction circle, and from it the moment needed to turn the shaft can be determined.
(*See page 439.*)

$$M = Rr \sin \phi_k$$

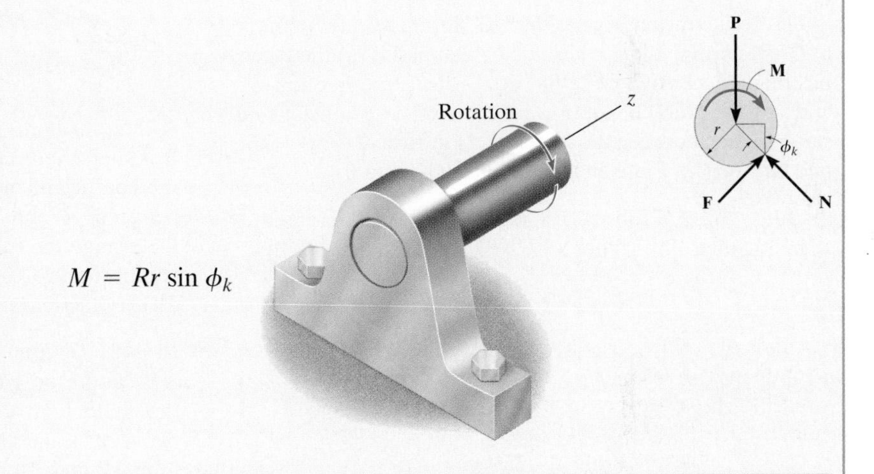

Rolling Resistance

The resistance of a wheel to rolling over a surface is caused by localized *deformation* of the two materials in contact. This causes the resultant normal force acting on the rolling body to be inclined so that it provides a component that acts in the opposite direction of the force causing the motion. This effect is characterized using the *coefficient of rolling resistance*, which is determined from experiment.
(*See pages 441–442.*)

$$P \approx \frac{Wa}{r}$$

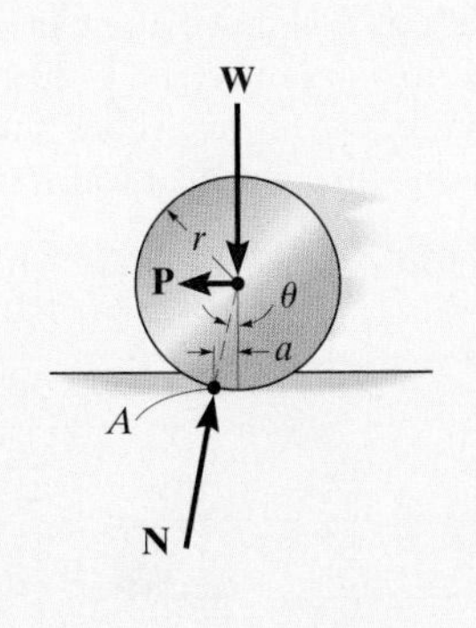

REVIEW PROBLEMS

8–134. A single force **P** is applied to the handle of the drawer. If friction is neglected at the bottom side and the coefficient of static friction along the sides is $\mu_s = 0.4$, determine the largest spacing s between the symmetrically placed handles so that the drawer does not bind at the corners A and B when the force **P** is applied to one of the handles.

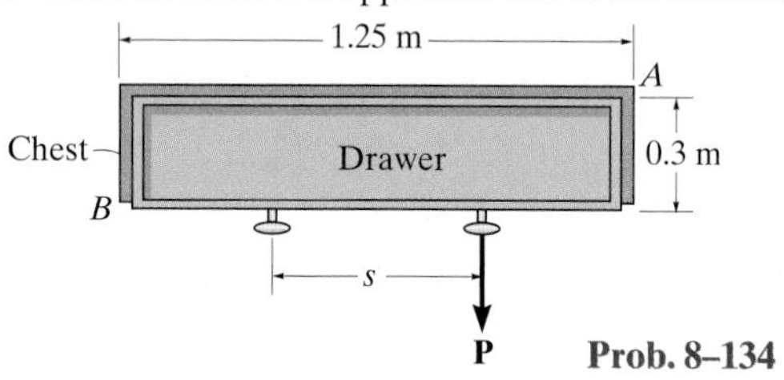

Prob. 8–134

8–135. The truck has a mass of 1.25 Mg and a center of mass at G. Determine the greatest load it can pull if (a) the truck has rear-wheel drive while the front wheels are free to roll, and (b) the truck has four-wheel drive. The coefficient of static friction between the wheels and the ground is $\mu_s = 0.5$, and between the crate and the ground, it is $\mu_s' = 0.4$.

***8–136.** Solve Prob. 8–135 if the truck and crate are traveling up a 10° incline.

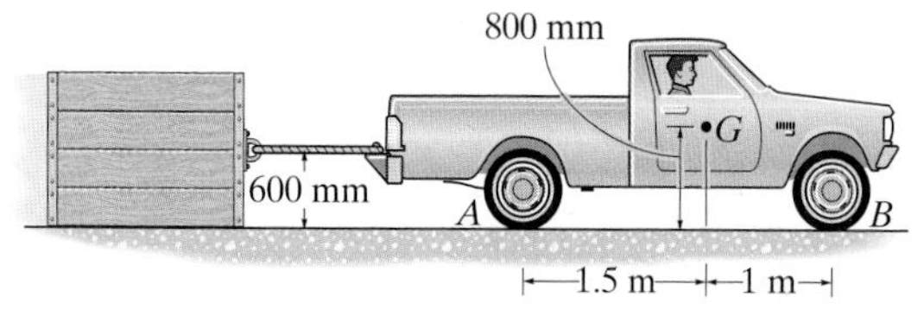

Probs. 8–135/136

8–137. A roofer, having a mass of 70 kg, walks slowly in an upright position down along the surface of a dome that has a radius of curvature of $r = 20$ m. If the coefficient of static friction between his shoes and the dome is $\mu_s = 0.7$, determine the angle θ at which he first begins to slip.

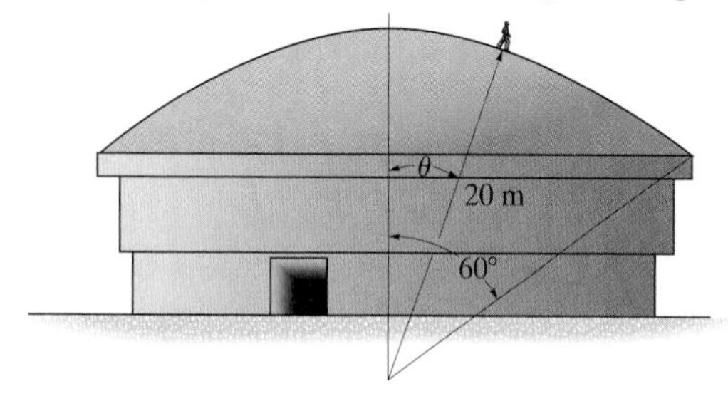

Prob. 8–137

8–138. A man attempts to lift the uniform 40-lb ladder to an upright position by applying a force **P** perpendicular to the ladder at rung R. Determine the coefficient of static friction between the ladder and the ground at A if the ladder begins to slip on the ground when his hands reach a height of 6 ft.

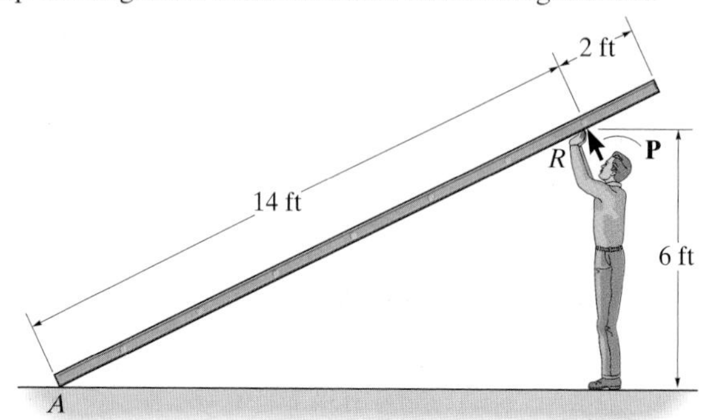

Prob. 8–138

8–139. Column D is subjected to a vertical load of 8000 lb. It is supported on two identical wedges A and B for which the coefficient of static friction at the contacting surfaces between A and B and between B and C is $\mu_s = 0.4$. Determine the force P needed to raise the column and the equilibrium force P' needed to hold wedge A stationary. The contacting surface between A and D is smooth.

***8–140.** Column D is subjected to a vertical load of 8000 lb. It is supported on two identical wedges A and B for which the coefficient of static friction at the contacting surfaces between A and B and between B and C is $\mu_s = 0.4$. If the forces **P** and **P′** are removed, are the wedges self-locking? The contacting surface between A and D is smooth.

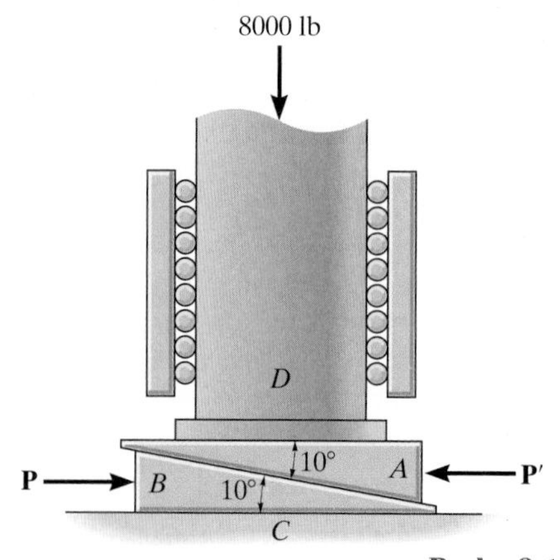

Probs. 8–139/140

Design Projects

8–1D Design of a Rope-and-Pulley System for Pulling a Crate up an Incline

A large 300-kg packing crate is to be hoisted up the 25° incline. The coefficient of static friction between the incline and the crate is $\mu_s = 0.5$, and the coefficient of kinetic friction is $\mu_k = 0.4$. Using a system of ropes and pulleys, design a method that will allow a single worker to pull the crate up the ramp. Pulleys can be attached to any point on the wall AB. Assume the worker can exert a maximum (comfortable) pull of 200 N on a rope. Submit a drawing of your design and a force analysis to show how it operates. Estimate the material cost required for its construction. Assume rope costs \$1.25/m and a pulley costs \$3.50.

8–2D Design of a Device for Lifting Stainless-Steel Pipes

Stainless-steel pipes are stacked vertically in a manufacturing plant and are to be moved by an overhead crane from one point to another. The pipes have inner diameters ranging from 100 mm $\leq d \leq$ 250 mm and the maximum mass of any pipe is 500 kg. Design a device that can be connected to the hook and used to lift each pipe. The device should be made of structural steel and should be able to grip the pipe only at its inside surface, since the outside surface is required not to be scratched or damaged. Assume the smallest coefficient of static friction between the two steels is $\mu_s = 0.25$. Submit a scaled drawing of your device, along with a brief explanation of how it works based on a force analysis.

8–3D Design of a Tool Used to Turn Plastic Pipe

PVC plastic is often used for sewer pipe. If the outer diameter of any pipe ranges from 4 in. $\leq d \leq$ 8 in., design a tool that can be used by a worker to turn the pipe when it is subjected to a maximum anticipated ground resistance of 80 lb · ft. The device is to be made of steel and should be designed so that it does not cut into the pipe and leave any significant marks on its surface. Assume a worker can apply a maximum (comfortable) force of 40 lb, and take the minimum coefficient of static friction between the PVC and the steel to be $\mu_s = 0.35$. Submit a scaled drawing of the device, and a brief paragraph to explain how it works based on a force analysis.

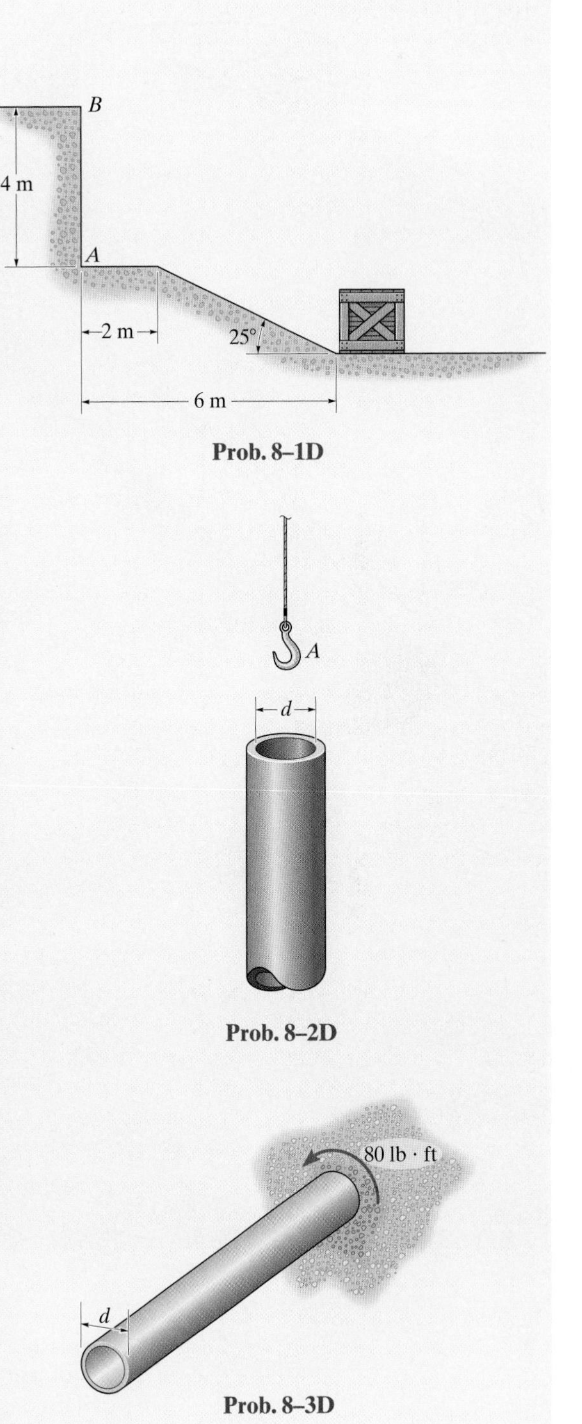

Prob. 8–1D

Prob. 8–2D

Prob. 8–3D

When a pressure vessel is designed, it is important to be able to determine the center of gravity of its component parts, calculate its volume and surface area, and reduce three-dimensional distributed loadings to their resultants. These topics are discussed in this chapter.

9 Center of Gravity and Centroid

CHAPTER OBJECTIVES

- To discuss the concept of the center of gravity, center of mass, and the centroid.
- To show how to determine the location of the center of gravity and centroid for a system of discrete particles and a body of arbitrary shape.
- To use the theorems of Pappus and Guldinus for finding the area and volume for a surface of revolution.
- To present a method for finding the resultant of a general distributed loading and show how it applies to finding the resultant of a fluid.

9.1 Center of Gravity and Center of Mass for a System of Particles

Center of Gravity. The *center of gravity* G is a point which locates the resultant weight of a system of particles. To show how to determine this point consider the system of n particles fixed within a region of space as shown in Fig. 9–1a. The weights of the particles comprise a system of parallel forces* which can be replaced by a single (equivalent) resultant weight having the defined point G of application. To find the $\bar{x}$, $\bar{y}$, $\bar{z}$ coordinates of G, we must use the principles outlined in Sec. 4.9.

*This is not true in the exact sense, since the weights are not parallel to each other; rather they are all *concurrent* at the earth's center. Furthermore, the acceleration of gravity g is actually different for each particle since it depends on the distance from the earth's center to the particle. For all practical purposes, however, both of these effects can generally be neglected.

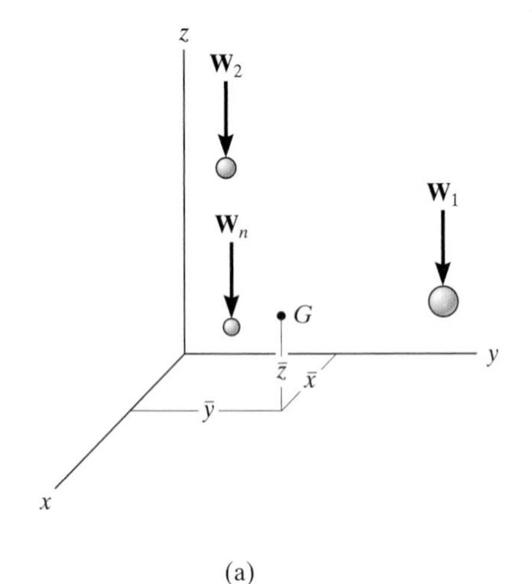

(a)

This requires that the resultant weight be equal to the total weight of all n particles; that is,

$$W_R = \Sigma W$$

The sum of the moments of the weights of all the particles about the x, y, and z axes is then equal to the moment of the resultant weight about these axes. Thus, to determine the $\bar{x}$ coordinate of G, we can sum moments about the y axis. This yields

$$\bar{x}W_R = \tilde{x}_1 W_1 + \tilde{x}_2 W_2 + \cdots + \tilde{x}_n W_n$$

Likewise, summing moments about the x axis, we can obtain the $\bar{y}$ coordinate; i.e.,

$$\bar{y}W_R = \tilde{y}_1 W_1 + \tilde{y}_2 W_2 + \cdots + \tilde{y}_n W_n$$

Although the weights do not produce a moment about the z axis, we can obtain the $\bar{z}$ coordinate of G by imagining the coordinate system, with the particles fixed in it, as being rotated 90° about the x (or y) axis, Fig. 9–1*b*. Summing moments about the x axis, we have

$$\bar{z}W_R = \tilde{z}_1 W_1 + \tilde{z}_2 W_2 + \cdots + \tilde{z}_n W_n$$

We can generalize these formulas, and write them symbolically in the form

$$\bar{x} = \frac{\Sigma \tilde{x} W}{\Sigma W} \quad \bar{y} = \frac{\Sigma \tilde{y} W}{\Sigma W} \quad \bar{z} = \frac{\Sigma \tilde{z} W}{\Sigma W} \qquad (9\text{–}1)$$

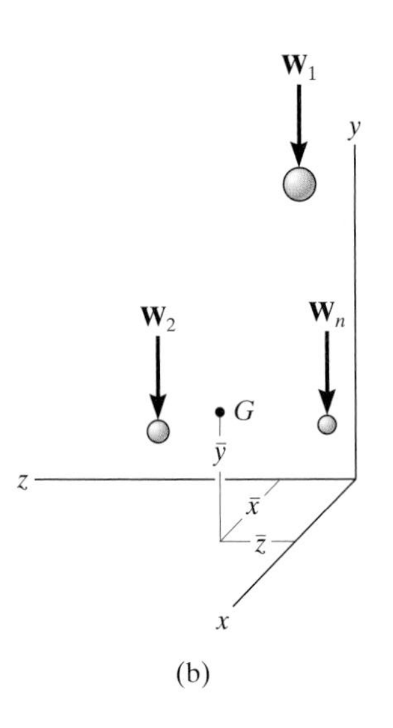

(b)

Fig. 9–1

Here

$\bar{x}, \bar{y}, \bar{z}$ represent the coordinates of the center of gravity G of the system of particles.

$\tilde{x}, \tilde{y}, \tilde{z}$ represent the coordinates of each particle in the system.

ΣW is the resultant sum of the weights of all the particles in the system.

These equations are easily remembered if it is kept in mind that they simply represent a balance between the sum of the moments of the weights of each particle of the system and the moment of the *resultant* weight for the system.

Center of Mass. To study problems concerning the motion of *matter* under the influence of force, i.e., dynamics, it is necessary to locate a point called the *center of mass*. Provided the acceleration due to gravity g for every particle is constant, then $W = mg$. Substituting into Eqs. 9–1 and canceling g from both the numerator and denominator yields

$$\bar{x} = \frac{\Sigma \tilde{x} m}{\Sigma m} \quad \bar{y} = \frac{\Sigma \tilde{y} m}{\Sigma m} \quad \bar{z} = \frac{\Sigma \tilde{z} m}{\Sigma m} \qquad (9\text{–}2)$$

By comparison, then, the location of the center of gravity *coincides* with that of the center of mass.* Recall, however, that particles have "weight" only when under the influence of a gravitational attraction, whereas the center of mass is independent of gravity. For example, it would be meaningless to define the center of gravity of a system of particles representing the planets of our solar system, while the center of mass of this system is important.

9.2 Center of Gravity, Center of Mass, and Centroid for a Body

Center of Gravity. A rigid body is composed of an infinite number of particles, and so if the principles used to determine Eqs. 9–1 are applied to the system of particles composing a rigid body, it becomes necessary to use integration rather than a discrete summation of the terms. Considering the arbitrary particle located at $(\tilde{x}, \tilde{y}, \tilde{z})$ and having a weight dW, Fig. 9–2, the resulting equations are

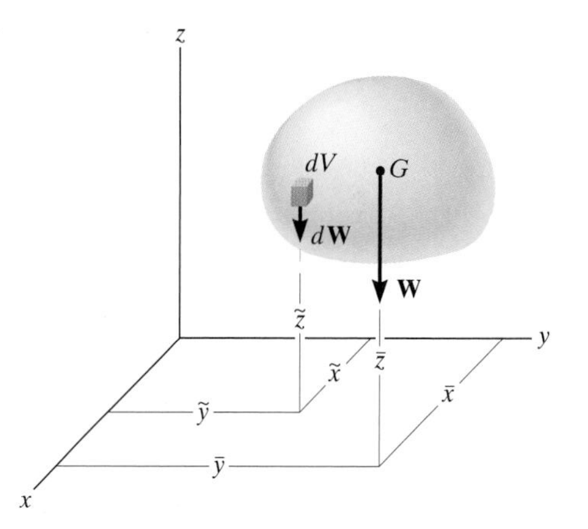

Fig. 9–2

$$\bar{x} = \frac{\int \tilde{x}\, dW}{\int dW} \qquad \bar{y} = \frac{\int \tilde{y}\, dW}{\int dW} \qquad \bar{z} = \frac{\int \tilde{z}\, dW}{\int dW} \tag{9–3}$$

In order to apply these equations, the differential weight dW must be expressed in terms of its associated volume dV. If γ represents the *specific weight* of the body, measured as a weight per unit volume, then $dW = \gamma\, dV$ and therefore

$$\bar{x} = \frac{\int_V \tilde{x}\gamma\, dV}{\int_V \gamma\, dV} \qquad \bar{y} = \frac{\int_V \tilde{y}\gamma\, dV}{\int_V \gamma\, dV} \qquad \bar{z} = \frac{\int_V \tilde{z}\gamma\, dV}{\int_V \gamma\, dV} \tag{9–4}$$

Here integration must be performed throughout the entire volume of the body.

Center of Mass. The *density* ρ, or mass per unit volume, is related to γ by the equation $\gamma = \rho g$, where g is the acceleration due to gravity. Substituting this relationship into Eqs. 9–4 and canceling g from both the numerators and denominators yields similar equations (with ρ replacing γ) that can be used to determine the body's *center of mass*.

*This is true as long as the gravity field is assumed to have the same magnitude and direction everywhere. That assumption is appropriate for most engineering applications, since gravity does not vary appreciably between, for instance, the bottom and the top of a building.

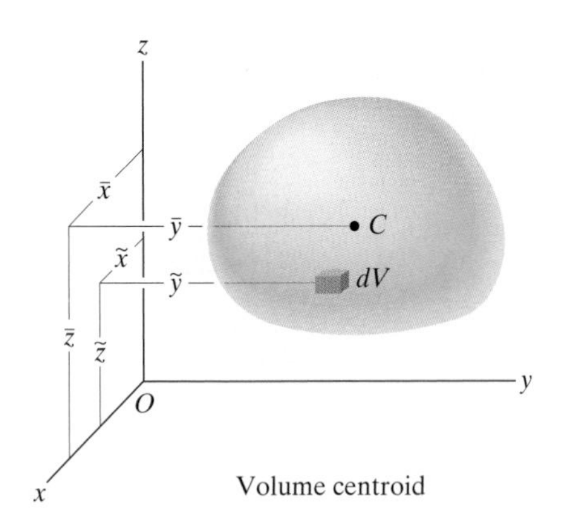

Fig. 9–3

Centroid. The *centroid C* is a point which defines the *geometric center* of an object. Its location can be determined from formulas similar to those used to determine the body's center of gravity or center of mass. In particular, if the material composing a body is uniform or *homogeneous*, the *density or specific weight* will be *constant* throughout the body, and therefore this term will factor out of the integrals and *cancel* from both the numerators and denominators of Eqs. 9–4. The resulting formulas define the centroid of the body since they are independent of the body's weight and instead depend only on the body's geometry. Three specific cases will be considered.

Volume. If an object is subdivided into volume elements dV, Fig. 9–3, the location of the centroid $C(\bar{x}, \bar{y}, \bar{z})$ for the volume of the object can be determined by computing the "moments" of the elements about each of the coordinate axes. The resulting formulas are

$$\bar{x} = \frac{\int_V \tilde{x}\, dV}{\int_V dV} \qquad \bar{y} = \frac{\int_V \tilde{y}\, dV}{\int_V dV} \qquad \bar{z} = \frac{\int_V \tilde{z}\, dV}{\int_V dV} \tag{9–5}$$

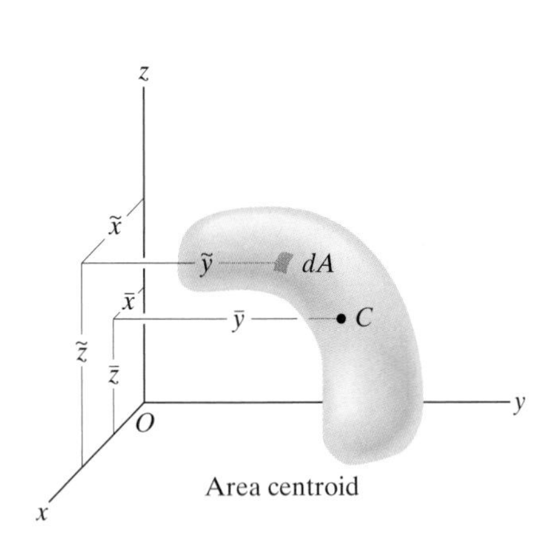

Fig. 9–4

Area. In a similar manner, the centroid for the surface area of an object, such as a plate or shell, Fig. 9–4, can be found by subdividing the area into differential elements dA and computing the "moments" of these area elements about each of the coordinate axes, namely,

$$\bar{x} = \frac{\int_A \tilde{x}\, dA}{\int_A dA} \qquad \bar{y} = \frac{\int_A \tilde{y}\, dA}{\int_A dA} \qquad \bar{z} = \frac{\int_A \tilde{z}\, dA}{\int_A dA} \tag{9–6}$$

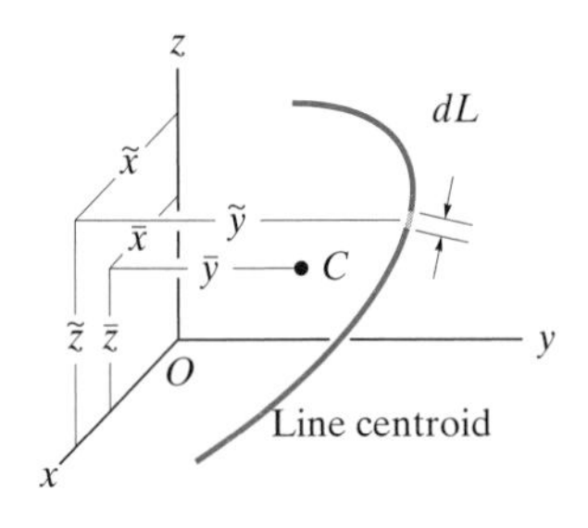

Fig. 9–5

Line. If the geometry of the object, such as a thin rod or wire, takes the form of a line, Fig. 9–5, the balance of moments of the differential elements dL about each of the coordinate axes yields

$$\bar{x} = \frac{\int_L \tilde{x}\, dL}{\int_L dL} \qquad \bar{y} = \frac{\int_L \tilde{y}\, dL}{\int_L dL} \qquad \bar{z} = \frac{\int_L \tilde{z}\, dL}{\int_L dL} \tag{9–7}$$

Remember that when applying Eqs. 9–4 through 9–7 it is best to choose a coordinate system that simplifies as much as possible the equation used to describe the object's boundary. For example, polar coordinates are generally appropriate for areas having circular boundaries. Also, the terms $\tilde{x}$, $\tilde{y}$, $\tilde{z}$ in the equations refer to the "moment arms" or coordinates of the *center of gravity or centroid for the differential element* used. If possible, this differential element should be chosen such that it has a differential size or thickness in only *one direction*. When this is done, only a single integration is required to cover the entire region.

Integration must be used to determine the location of the center of gravity of this goal post due to the curvature of the supporting member.

Symmetry. The *centroids* of some shapes may be partially or completely specified by using conditions of *symmetry*. In cases where the shape has an axis of symmetry, the centroid of the shape will lie along that axis. For example, the centroid C for the curve shown in Fig. 9–6 must lie along the y axis since for every elemental length dL at a distance $+\tilde{x}$ to the right of the y axis there is an identical element at a distance $-\tilde{x}$ to the left. The total moment for all the elements about the axis of symmetry will therefore cancel; i.e., $\int \tilde{x}\, dL = 0$ (Eq. 9–7), so that $\bar{x} = 0$. In cases where a shape has two or three axes of symmetry, it follows that the centroid lies at the intersection of these axes, Fig. 9–7 and Fig. 9–8.

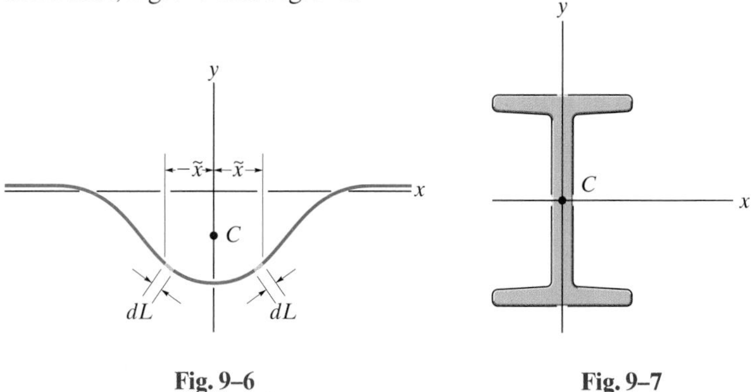

Fig. 9–6

Fig. 9–7

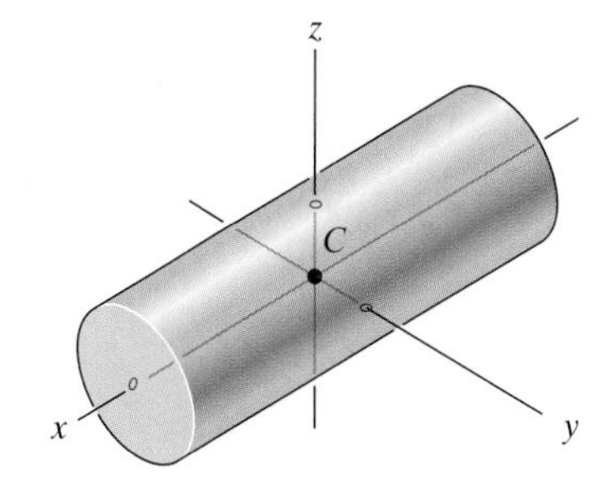

Fig. 9–8

Important Points

- The centroid represents the geometric center of a body. This point coincides with the center of mass or the center of gravity only if the material composing the body is uniform or homogeneous.
- Formulas used to locate the center of gravity or the centroid simply represent a balance between the sum of moments of all the parts of the system and the moment of the "resultant" for the system.
- In some cases the centroid is located at a point that is not on the object, as in the case of a ring, where the centroid is at its center. Also, this point will lie on any axis of symmetry for the body.

Procedure for Analysis

The center of gravity or centroid of an object or shape can be determined by single integrations using the following procedure.

Differential Element.

- Select an appropriate coordinate system, specify the coordinate axes, and then choose a differential element for integration.
- For lines the element dL is represented as a differential line segment.
- For areas the element dA is generally a rectangle having a finite length and differential width.
- For volumes the element dV is either a circular disk having a finite radius and differential thickness, or a shell having a finite length and radius and a differential thickness.
- Locate the element at an arbitrary point (x, y, z) on the curve that defines the shape.

Size and Moment Arms.

- Express the length dL, area dA, or volume dV of the element in terms of the coordinates of the curve used to define the geometric shape.
- Determine the coordinates or moment arms $\tilde{x}$, $\tilde{y}$, $\tilde{z}$ for the centroid or center of gravity of the element.

Integrations.

- Substitute the formulations for $\tilde{x}$, $\tilde{y}$, $\tilde{z}$ and dL, dA, or dV into the appropriate equations (Eqs. 9–4 through 9–7) and perform the integrations.*
- Express the function in the integrand in terms of the *same variable as the differential thickness of the element* in order to perform the integration.
- The limits of the integral are defined from the two extreme locations of the element's differential thickness, so that when the elements are "summed" or the integration performed, the entire region is covered.

*Formulas for integration are given in Appendix A.

EXAMPLE 9.1

Locate the centroid of the rod bent into the shape of a parabolic arc, shown in Fig. 9–9.

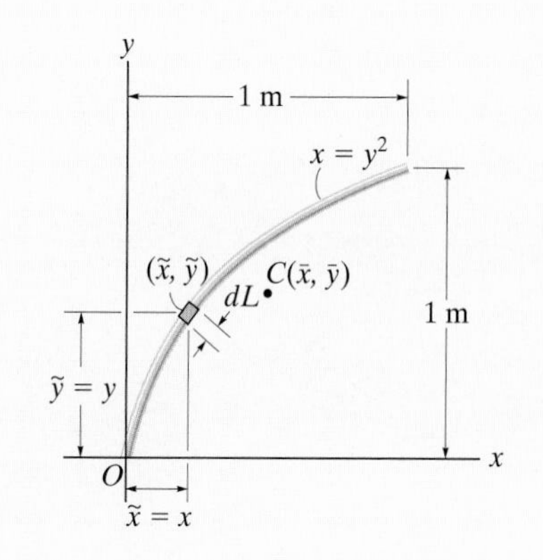

Fig. 9–9

SOLUTION

Differential Element. The differential element is shown in Fig. 9–9. It is located on the curve at the *arbitrary point* (x, y).

Area and Moment Arms. The differential length of the element dL can be expressed in terms of the differentials dx and dy by using the Pythagorean theorem.

$$dL = \sqrt{(dx)^2 + (dy)^2} = \sqrt{\left(\frac{dx}{dy}\right)^2 + 1}\, dy$$

Since $x = y^2$, then $dx/dy = 2y$. Therefore, expressing dL in terms of y and dy, we have

$$dL = \sqrt{(2y)^2 + 1}\, dy$$

The centroid is located at $\tilde{x} = x, \tilde{y} = y$.

Integrations. Applying Eqs. 9–7 and integrating with respect to y using the formulas in Appendix A, we have

$$\bar{x} = \frac{\int_L \tilde{x}\, dL}{\int_L dL} = \frac{\int_0^1 x\sqrt{4y^2 + 1}\, dy}{\int_0^1 \sqrt{4y^2 + 1}\, dy} = \frac{\int_0^1 y^2\sqrt{4y^2 + 1}\, dy}{\int_0^1 \sqrt{4y^2 + 1}\, dy}$$

$$= \frac{0.6063}{1.479} = 0.410 \text{ m} \qquad \textit{Ans.}$$

$$\bar{y} = \frac{\int_L \tilde{y}\, dL}{\int_L dL} = \frac{\int_0^1 y\sqrt{4y^2 + 1}\, dy}{\int_0^1 \sqrt{4y^2 + 1}\, dy} = \frac{0.8484}{1.479} = 0.574 \text{ m} \qquad \textit{Ans.}$$

NOTE: These results seem reasonable when they are plotted on Fig. 9–9.

EXAMPLE 9.2

Locate the centroid of the circular wire segment shown in Fig. 9–10.

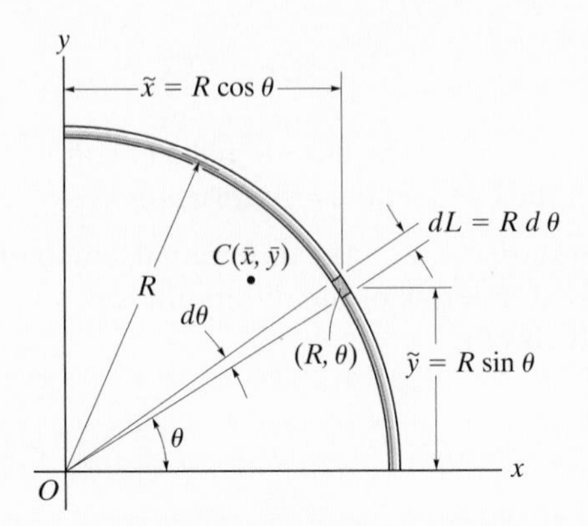

Fig. 9–10

SOLUTION

Polar coordinates will be used to solve this problem since the arc is circular.

Differential Element. A differential circular arc is selected as shown in the figure. This element intersects the curve at (R, θ).

Length and Moment Arm. The differential length of the element is $dL = R\,d\theta$, and its centroid is located at $\tilde{x} = R\cos\theta$ and $\tilde{y} = R\sin\theta$.

Integrations. Applying Eqs. 9–7 and integrating with respect to θ, we obtain

$$\bar{x} = \frac{\int_L \tilde{x}\,dL}{\int_L dL} = \frac{\int_0^{\pi/2} (R\cos\theta)R\,d\theta}{\int_0^{\pi/2} R\,d\theta} = \frac{R^2\int_0^{\pi/2} \cos\theta\,d\theta}{R\int_0^{\pi/2} d\theta} = \frac{2R}{\pi} \qquad \textit{Ans.}$$

$$\bar{y} = \frac{\int_L \tilde{y}\,dL}{\int_L dL} = \frac{\int_0^{\pi/2} (R\sin\theta)R\,d\theta}{\int_0^{\pi/2} R\,d\theta} = \frac{R^2\int_0^{\pi/2} \sin\theta\,d\theta}{R\int_0^{\pi/2} d\theta} = \frac{2R}{\pi} \qquad \textit{Ans.}$$

NOTE: As expected, the two coordinates are numericallly the same due to the symmetry of the wire.

EXAMPLE 9.3

Determine the distance $\bar{y}$ from the x axis to the centroid of the area of the triangle shown in Fig. 9–11.

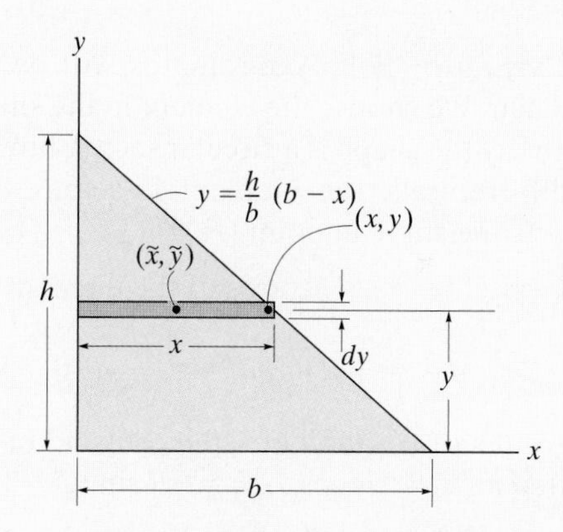

Fig. 9–11

SOLUTION

Differential Element. Consider a rectangular element having a thickness dy, which intersects the boundary at (x, y), Fig. 9–11.

Area and Moment Arms. The area of the element is $dA = x\,dy = \dfrac{b}{h}(h - y)\,dy$, and its centroid is located a distance $\tilde{y} = y$ from the x axis.

Integrations. Applying the second of Eqs. 9–6 and integrating with respect to y yields

$$\bar{y} = \frac{\int_A \tilde{y}\,dA}{\int_A dA} = \frac{\int_0^h y\frac{b}{h}(h - y)\,dy}{\int_0^h \frac{b}{h}(h - y)\,dy} = \frac{\frac{1}{6}bh^2}{\frac{1}{2}bh}$$

$$= \frac{h}{3} \qquad \textit{Ans.}$$

EXAMPLE 9.4

Locate the centroid for the area of a quarter circle shown in Fig. 9–12*a*.

SOLUTION I

Differential Element. Polar coordinates will be used, since the boundary is circular. We choose the element in the shape of a *triangle*, Fig. 9–12*a*. (Actually the shape is a circular sector; however, neglecting higher-order differentials, the element becomes triangular.) The element intersects the curve at point (R, θ).

Area and Moment Arms. The area of the element is

$$dA = \tfrac{1}{2}(R)(R\,d\theta) = \frac{R^2}{2}d\theta$$

and using the results of Example 9.3, the centroid of the (triangular) element is located at $\tilde{x} = \frac{2}{3}R\cos\theta$, $\tilde{y} = \frac{2}{3}R\sin\theta$.

Integrations. Applying Eqs. 9–6 and integrating with respect to θ, we obtain

$$\bar{x} = \frac{\int_A \tilde{x}\,dA}{\int_A dA} = \frac{\int_0^{\pi/2}\left(\frac{2}{3}R\cos\theta\right)\frac{R^2}{2}d\theta}{\int_0^{\pi/2}\frac{R^2}{2}d\theta}$$

$$= \frac{\left(\frac{2}{3}R\right)\int_0^{\pi/2}\cos\theta\,d\theta}{\int_0^{\pi/2}d\theta} = \frac{4R}{3\pi} \qquad \textit{Ans.}$$

$$\bar{y} = \frac{\int_A \tilde{y}\,dA}{\int_A dA} = \frac{\int_0^{\pi/2}\left(\frac{2}{3}R\sin\theta\right)\frac{R^2}{2}d\theta}{\int_0^{\pi/2}\frac{R^2}{2}d\theta}$$

$$= \frac{\left(\frac{2}{3}R\right)\int_0^{\pi/2}\sin\theta\,d\theta}{\int_0^{\pi/2}d\theta} = \frac{4R}{3\pi} \qquad \textit{Ans.}$$

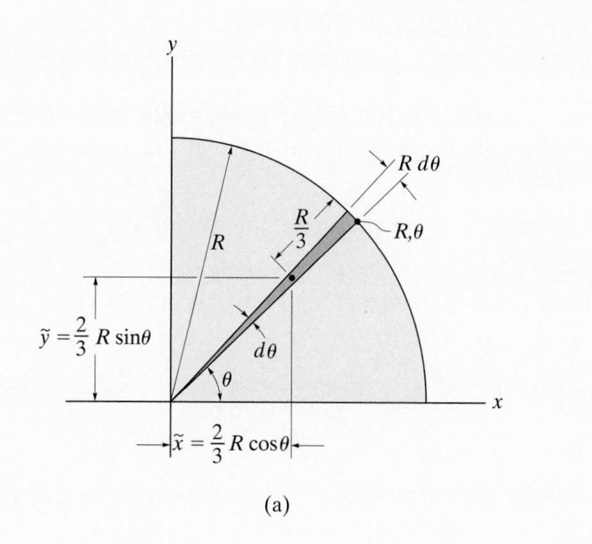

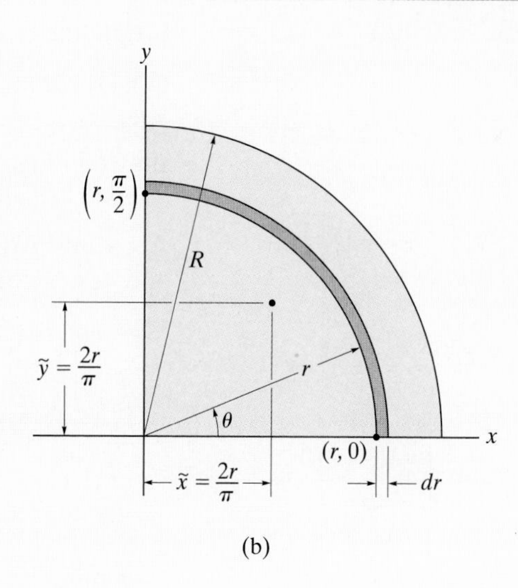

Fig. 9–12

SOLUTION II

Differential Element. The differential element may be chosen in the form of a *circular arc* having a thickness dr as shown in Fig. 9–12b. The element intersects the axes at points $(r, 0)$ and $(r, \pi/2)$.

Area and Moment Arms. The area of the element is $dA = (2\pi r/4)\,dr$. Since the centroid of a 90° circular arc was determined in Example 9.2, then for the element $\tilde{x} = 2r/\pi$, $\tilde{y} = 2r/\pi$.

Integrations. Using Eqs. 9–6 and integrating with respect to r, we obtain

$$\bar{x} = \frac{\int_A \tilde{x}\, dA}{\int_A dA} = \frac{\int_0^R \frac{2r}{\pi}\left(\frac{2\pi r}{4}\right) dr}{\int_0^R \frac{2\pi r}{4} dr} = \frac{\int_0^R r^2\, dr}{\frac{\pi}{2}\int_0^R r\, dr} = \frac{4}{3}\frac{R}{\pi} \qquad \textit{Ans.}$$

$$\bar{y} = \frac{\int_A \tilde{y}\, dA}{\int_A dA} = \frac{\int_0^R \frac{2r}{\pi}\left(\frac{2\pi r}{4}\right) dr}{\int_0^R \frac{2\pi r}{4} dr} = \frac{\int_0^R r^2\, dr}{\frac{\pi}{2}\int_0^R r\, dr} = \frac{4}{3}\frac{R}{\pi} \qquad \textit{Ans.}$$

NOTE: The two coordinates are numerically the same because of the symmetry of the area.

EXAMPLE 9.5

Locate the centroid of the area shown in Fig. 9–13*a*.

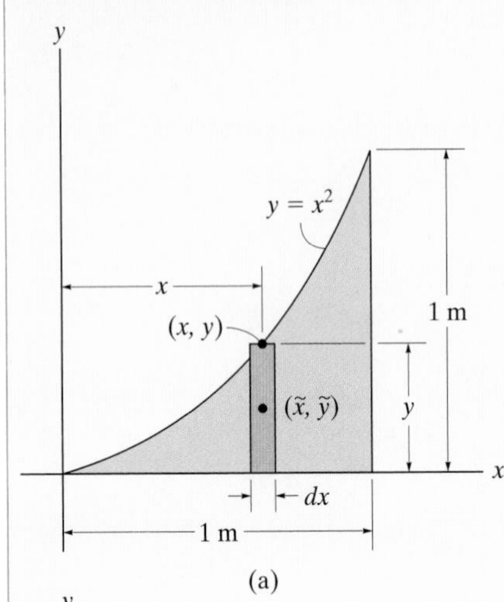

(a)

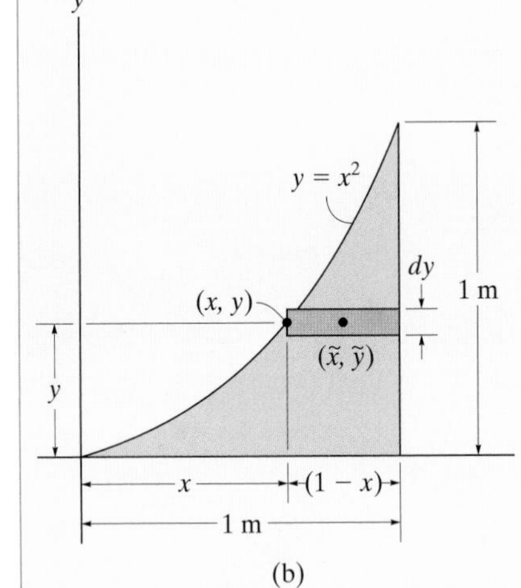

(b)

Fig. 9–13

SOLUTION I

Differential Element. A differential element of thickness dx is shown in Fig. 9–13*a*. The element intersects the curve at the *arbitrary point* (x, y), and so it has a height y.

Area and Moment Arms. The area of the element is $dA = y\,dx$, and its centroid is located at $\tilde{x} = x$, $\tilde{y} = y/2$.

Integrations. Applying Eqs. 9–6 and integrating with respect to x yields

$$\bar{x} = \frac{\int_A \tilde{x}\,dA}{\int_A dA} = \frac{\int_0^1 xy\,dx}{\int_0^1 y\,dx} = \frac{\int_0^1 x^3\,dx}{\int_0^1 x^2\,dx} = \frac{0.250}{0.333} = 0.75 \text{ m} \qquad \textit{Ans.}$$

$$\bar{y} = \frac{\int_A \tilde{y}\,dA}{\int_A dA} = \frac{\int_0^1 (y/2)y\,dx}{\int_0^1 y\,dx} = \frac{\int_0^1 (x^2/2)x^2\,dx}{\int_0^1 x^2\,dx} = \frac{0.100}{0.333} = 0.3 \text{ m} \qquad \textit{Ans.}$$

SOLUTION II

Differential Element. The differential element of thickness dy is shown in Fig. 9–13*b*. The element intersects the curve at the *arbitrary point* (x, y), and so it has a length $(1 - x)$.

Area and Moment Arms. The area of the element is $dA = (1 - x)\,dy$, and its centroid is located at

$$\tilde{x} = x + \left(\frac{1-x}{2}\right) = \frac{1+x}{2}, \; \tilde{y} = y$$

Integrations. Applying Eqs. 9–6 and integrating with respect to y, we obtain

$$\bar{x} = \frac{\int_A \tilde{x}\,dA}{\int_A dA} = \frac{\int_0^1 [(1+x)/2](1-x)\,dy}{\int_0^1 (1-x)\,dy} = \frac{\frac{1}{2}\int_0^1 (1-y)\,dy}{\int_0^1 (1-\sqrt{y})\,dy} = \frac{0.250}{0.333} = 0.75 \text{ m} \qquad \textit{Ans.}$$

$$\bar{y} = \frac{\int_A \tilde{y}\,dA}{\int_A dA} = \frac{\int_0^1 y(1-x)\,dy}{\int_0^1 (1-x)\,dy} = \frac{\int_0^1 (y - y^{3/2})\,dy}{\int_0^1 (1-\sqrt{y})\,dy} = \frac{0.100}{0.333} = 0.3 \text{ m} \qquad \textit{Ans.}$$

NOTE: Plot these results and notice that they seem reasonable. Also, by comparison, elements of thickness dx offer a simpler solution.

EXAMPLE 9.6

Locate the $\bar{x}$ centroid of the shaded area bounded by the two curves $y = x$ and $y = x^2$, Fig. 9–14.

SOLUTION I

Differential Element. A differential element of thickness dx is shown in Fig. 9–14*a*. The element intersects the curves at *arbitrary points* (x, y_1) and (x, y_2), and so it has a height $(y_2 - y_1)$.

Area and Moment Arm. The area of the element is $dA = (y_2 - y_1)\,dx$, and its centroid is located at $\tilde{x} = x$.

Integration. Applying Eq. 9–6, we have

$$\bar{x} = \frac{\int_A \tilde{x}\,dA}{\int_A dA} = \frac{\int_0^1 x(y_2 - y_1)\,dx}{\int_0^1 (y_2 - y_1)\,dx} = \frac{\int_0^1 x(x - x^2)\,dx}{\int_0^1 (x - x^2)\,dx} = \frac{\frac{1}{12}}{\frac{1}{6}} = 0.5\text{ ft} \quad \textit{Ans.}$$

(a)

SOLUTION II

Differential Element. A differential element having a thickness dy is shown in Fig. 9–14*b*. The element intersects the curves at the *arbitrary points* (x_2, y) and (x_1, y), and so it has a length $(x_1 - x_2)$.

Area and Moment Arm. The area of the element is $dA = (x_1 - x_2)\,dy$, and its centroid is located at

$$\tilde{x} = x_2 + \frac{x_1 - x_2}{2} = \frac{x_1 + x_2}{2}$$

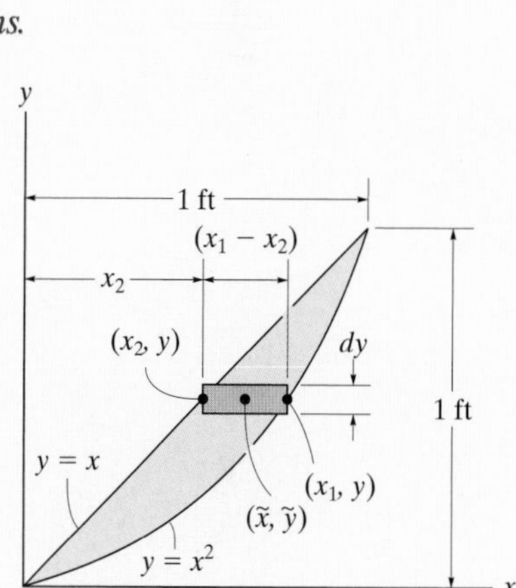

(b)

Fig. 9–14

Integration. Applying Eq. 9–6, we have

$$\bar{x} = \frac{\int_A \tilde{x}\,dA}{\int_A dA} = \frac{\int_0^1 [(x_1 + x_2)/2](x_1 - x_2)\,dy}{\int_0^1 (x_1 - x_2)\,dy} = \frac{\int_0^1 [(\sqrt{y} + y)/2](\sqrt{y} - y)\,dy}{\int_0^1 (\sqrt{y} - y)\,dy}$$

$$= \frac{\frac{1}{2}\int_0^1 (y - y^2)\,dy}{\int_0^1 (\sqrt{y} - y)\,dy} = \frac{\frac{1}{12}}{\frac{1}{6}} = 0.5\text{ ft} \quad \textit{Ans.}$$

EXAMPLE 9.7

Locate the $\bar{y}$ centroid for the paraboloid of revolution, which is generated by revolving the shaded area shown in Fig. 9–15*a* about the *y* axis.

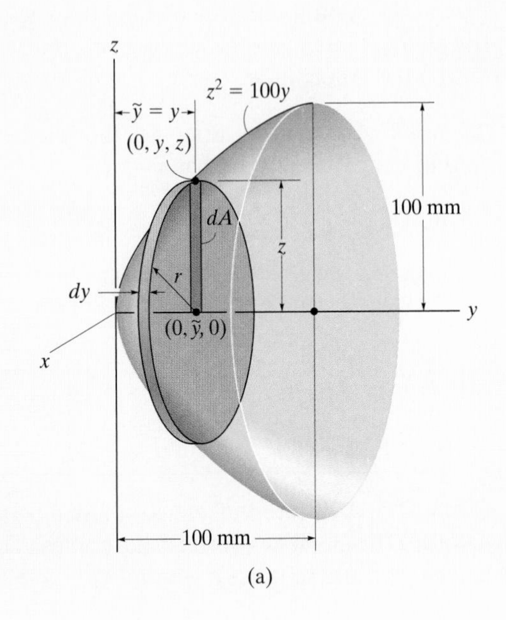

Fig. 9–15

SOLUTION I

Differential Element. An element having the shape of a *thin disk* is chosen, Fig. 9–15*a*. This element has a thickness *dy*. In this "disk" method of analysis, the element of planar area, *dA*, is always taken *perpendicular* to the axis of revolution. Here the element intersects the generating curve at the *arbitrary point* (0, *y*, *z*), and so its radius is $r = z$.

Area and Moment Arm. The volume of the element is $dV = (\pi z^2)\,dy$, and its centroid is located at $\tilde{y} = y$.

Integration. Applying the second of Eqs. 9–5 and integrating with respect to *y* yields

$$\bar{y} = \frac{\int_V \tilde{y}\,dV}{\int_V dV} = \frac{\int_0^{100} y(\pi z^2)\,dy}{\int_0^{100} (\pi z^2)\,dy} = \frac{100\pi \int_0^{100} y^2\,dy}{100\pi \int_0^{100} y\,dy} = 66.7 \text{ mm} \quad \textit{Ans.}$$

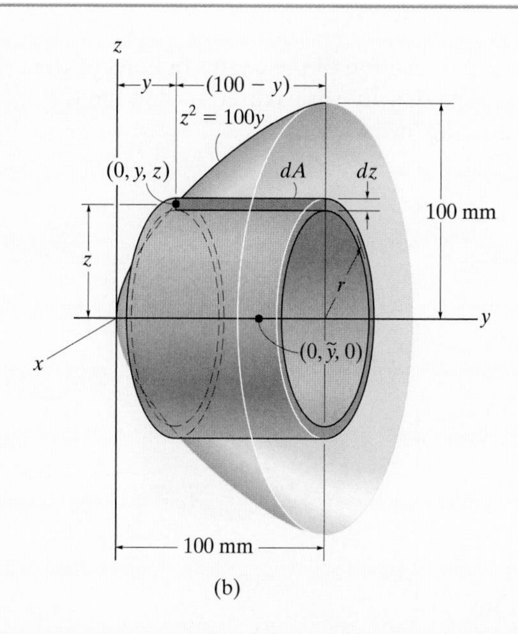

(b)

SOLUTION II

Differential Element. As shown in Fig. 9–15*b*, the volume element can be chosen in the form of a *thin cylindrical shell*, where the shell's thickness is dz. In this "shell" method of analysis, the element of planar area, dA, is always taken *parallel* to the axis of revolution. Here the element intersects the generating curve at point $(0, y, z)$, and so the radius of the shell is $r = z$.

Area and Moment Arm. The volume of the element is $dV = 2\pi r\, dA = 2\pi z(100 - y)\, dz$, and its centroid is located at $\tilde{y} = y + (100 - y)/2 = (100 + y)/2$.

Integrations. Applying the second of Eqs. 9–5 and integrating with respect to z yields

$$\bar{y} = \frac{\int_V \tilde{y}\, dV}{\int_V dV} = \frac{\int_0^{100} [(100 + y)/2]\, 2\pi z(100 - y)\, dz}{\int_0^{100} 2\pi z(100 - y)\, dz}$$

$$= \frac{\pi \int_0^{100} z(10^4 - 10^{-4}z^4)\, dz}{2\pi \int_0^{100} z(100 - 10^{-2}z^2)\, dz} = 66.7 \text{ mm} \qquad \textit{Ans.}$$

NOTE: By comparison, choosing a disk element offers a simpler solution.

EXAMPLE 9.8

Determine the location of the center of mass of the cylinder shown in Fig. 9–16*a* if its density varies directly with the distance from its base, i.e., $\rho = 200z\ \text{kg/m}^3$.

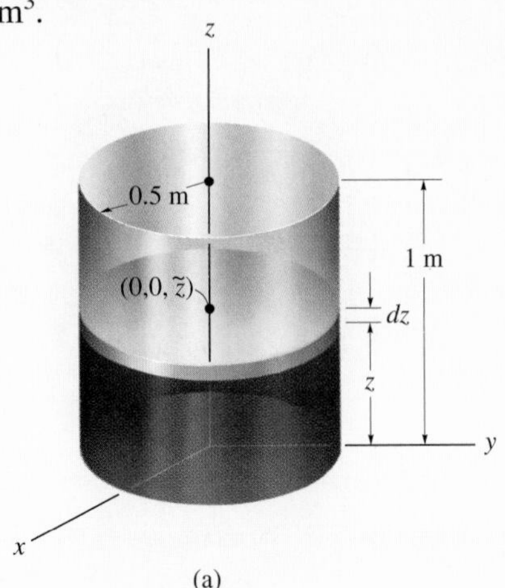

(a)

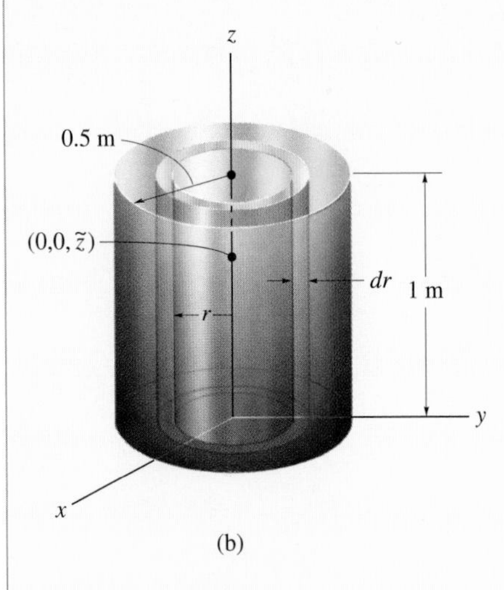

(b)

Fig. 9–16

SOLUTION

For reasons of material symmetry,

$$\bar{x} = \bar{y} = 0 \qquad \textit{Ans.}$$

Differential Element. A disk element of radius 0.5 m and thickness dz is chosen for integration, Fig. 9–16*a*, since the *density of the entire element is constant* for a given value of z. The element is located along the z axis at the *arbitrary point* $(0, 0, z)$.

Volume and Moment Arm. The volume of the element is $dV = \pi(0.5)^2\,dz$, and its centroid is located at $\tilde{z} = z$.

Integrations. Using an equation similar to the third of Eqs. 9–4 and integrating with respect to z, noting that $\rho = 200z$, we have

$$\bar{z} = \frac{\int_v \tilde{z}\rho\,dV}{\int_v \rho\,dV} = \frac{\int_0^1 z(200z)\pi(0.5)^2\,dz}{\int_0^1 (200z)\pi(0.5)^2\,dz}$$

$$= \frac{\int_0^1 z^2\,dz}{\int_0^1 z\,dz} = 0.667\ \text{m} \qquad \textit{Ans.}$$

NOTE: It is not possible to use a shell element for integration such as shown in Fig. 9–16*b* since the density of the material composing the shell would *vary* along the shell's height and hence the location of $\tilde{z}$ for the element cannot be specified.

PROBLEMS

9–1. Locate the center of mass of the homogeneous rod bent in the form of a parabola.

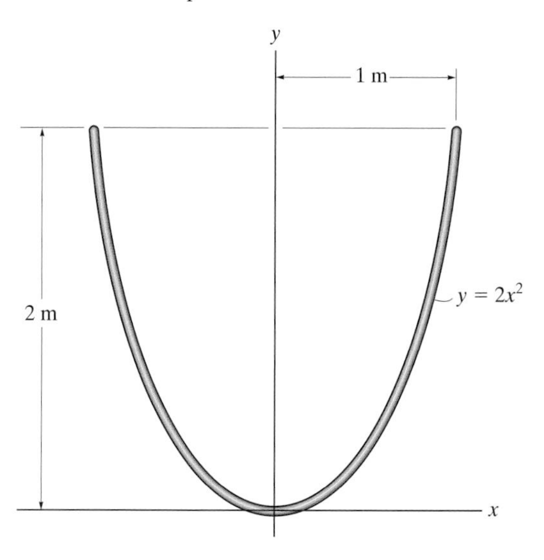

Prob. 9–1

9–2. Locate the center of gravity $\bar{x}$ of the homogeneous rod. If the rod has a weight per unit length of 0.5 lb/ft, determine the vertical reaction at A and the x and y components of reaction at the pin B.

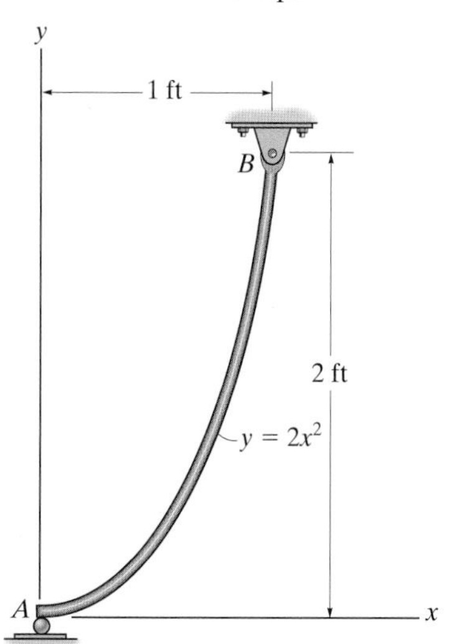

Prob. 9–2

9–3. Locate the center of mass of the homogeneous rod bent into the shape of a circular arc.

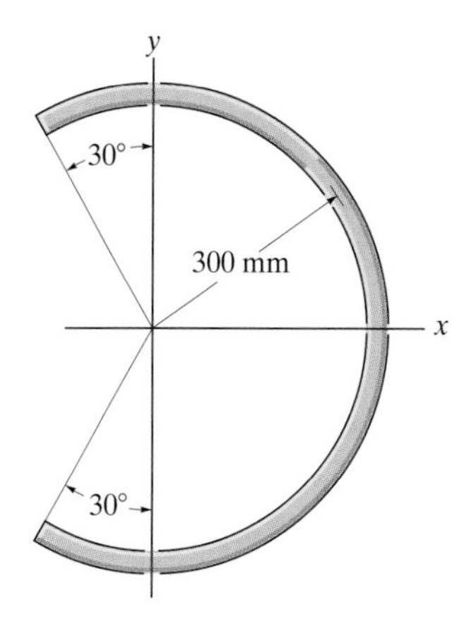

Prob. 9–3

***9–4.** Locate the center of gravity $\bar{x}$ of the homogeneous rod bent in the form of a semicircular arc. The rod has a weight per unit length of 0.5 lb/ft. Also, determine the horizontal reaction at the smooth support B and the x and y components of reaction at the pin A.

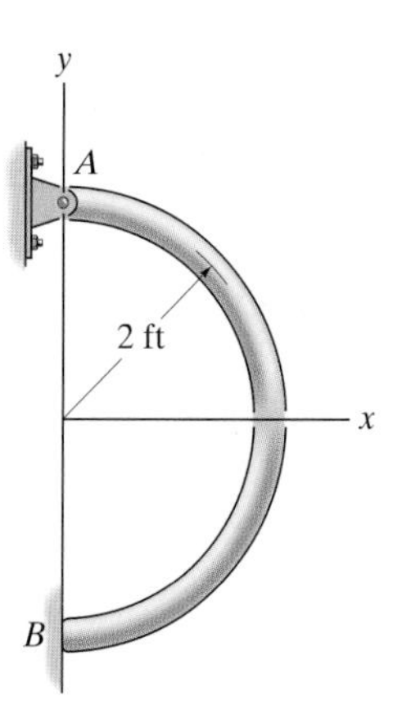

Prob. 9–4

9–5. Determine the distance $\overline{x}$ to the center of gravity of the homogeneous rod bent into the parabolic shape. If the rod has a weight per unit length of 0.5 lb/ft, determine the reactions at the fixed support O.

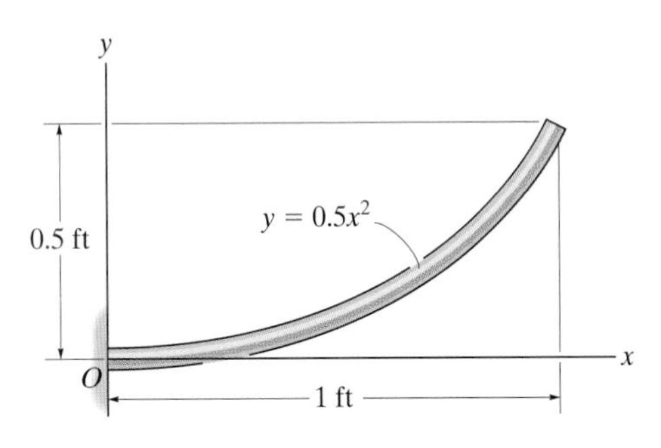

Prob. 9–5

9–6. Determine the distance $\overline{y}$ to the center of gravity of the homogeneous rod bent into the parabolic shape.

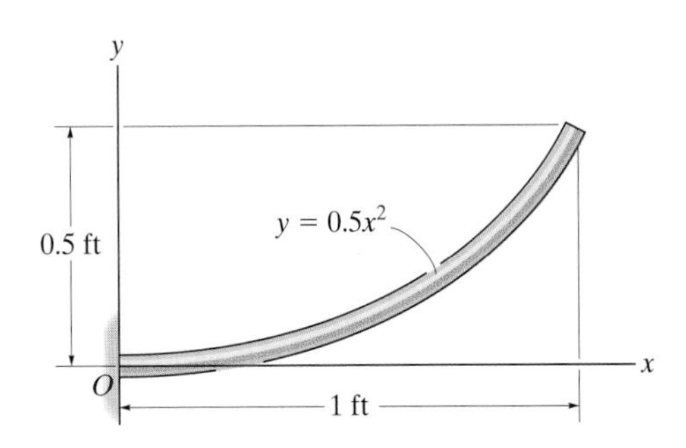

Prob. 9–6

9–7. Locate the centroid of the parabolic area.

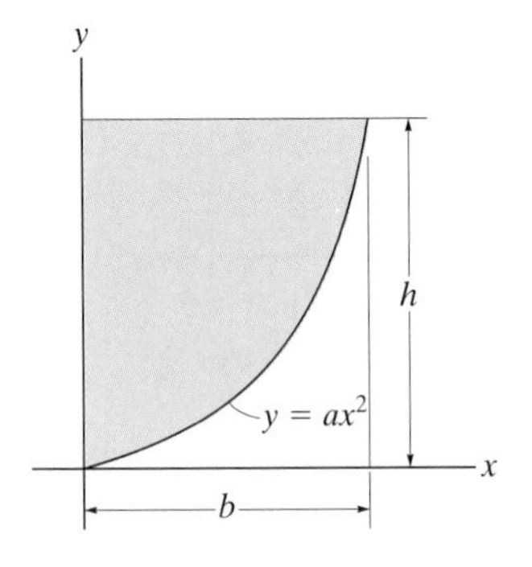

Prob. 9–7

***9–8.** Locate the centroid $\overline{y}$ of the shaded area.

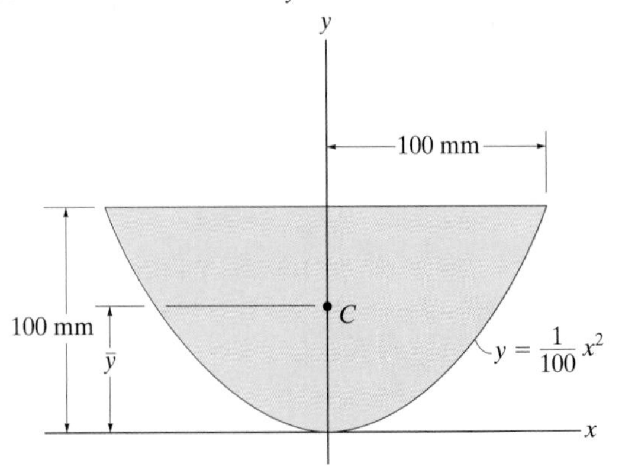

Prob. 9–8

9–9. Locate the centroid of the shaded area.

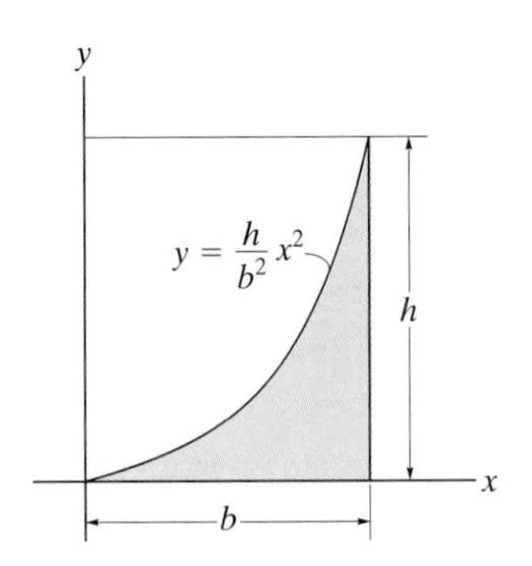

Prob. 9–9

9–10. Determine the location $(\overline{x}, \overline{y})$ of the centroid of the triangular area.

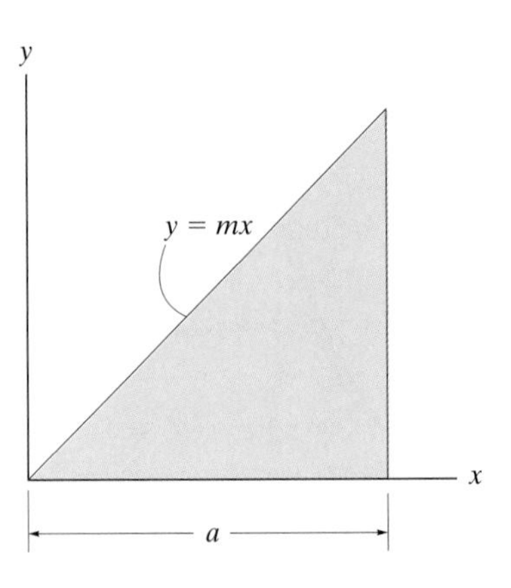

Prob. 9–10

9–11. Determine the location $(\bar{x}, \bar{y})$ of the center of gravity of the quartercircular plate. Also determine the force in each of the supporting wires. The plate has a weight per unit area of 5 lb/ft^2.

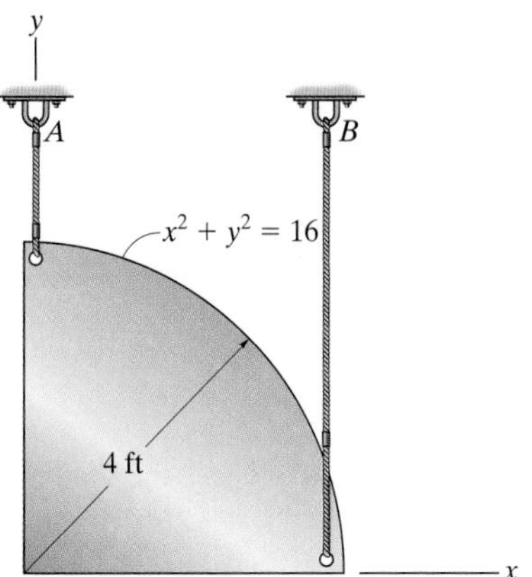

Prob. 9–11

***9–12.** Locate the centroid of the shaded area.

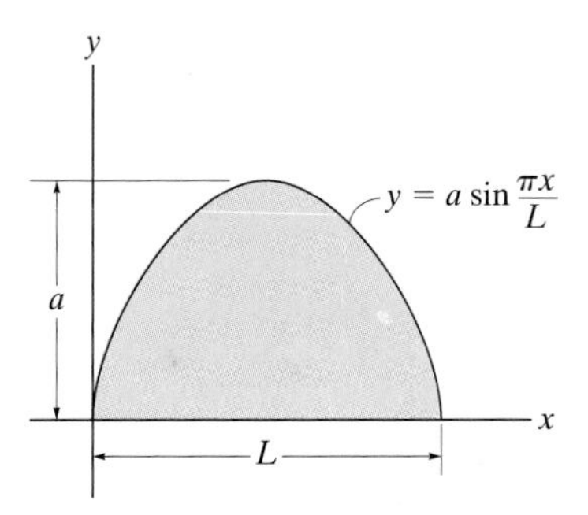

Prob. 9–12

9–13. Locate the center of gravity of the homogeneous cantilever beam and determine the reactions at the fixed support. The material has a density of 8 Mg/m^3.

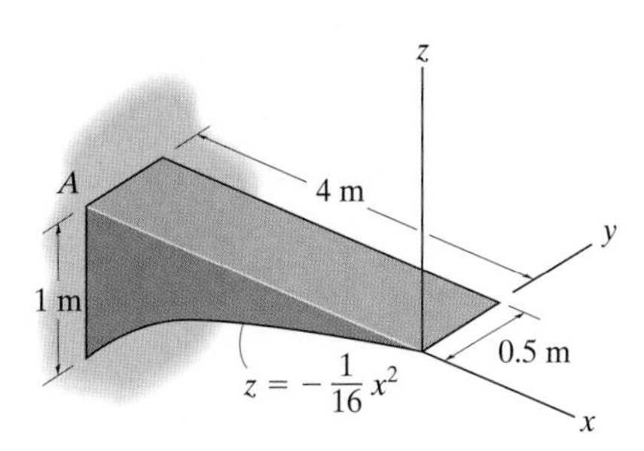

Prob. 9–13

9–14. Locate the centroid $(\bar{x}, \bar{y})$ of the exparabolic segment of area.

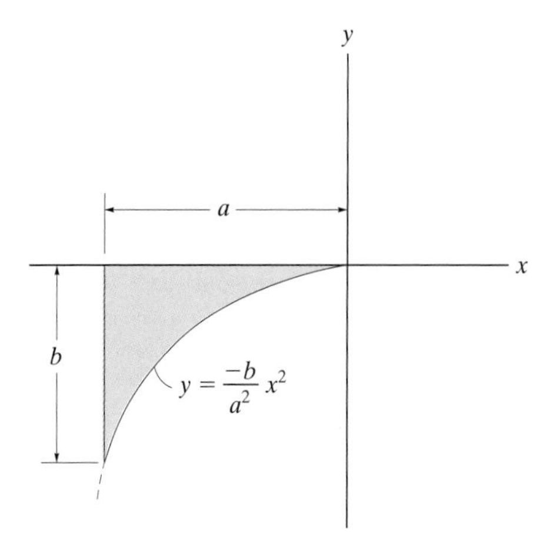

Prob. 9–14

9–15. Locate the centroid of the shaded area.

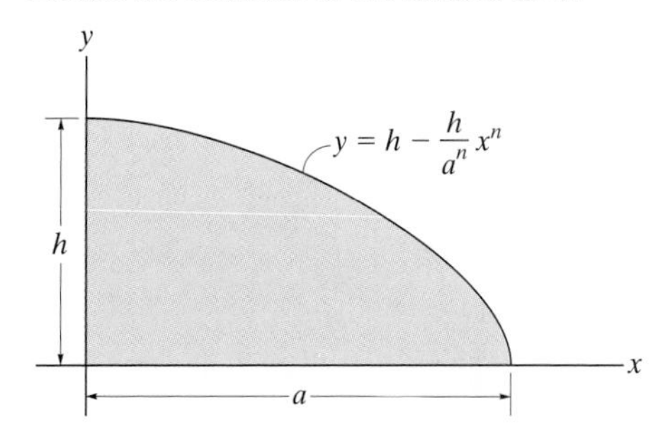

Prob. 9–15

***9–16.** Locate the centroid of the shaded area bounded by the parabola and the line $y = a$.

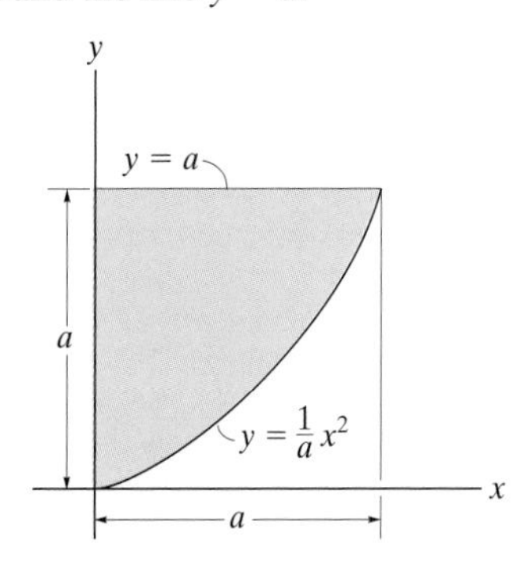

Prob. 9–16

9–17. Locate the centroid of the quarter elliptical area.

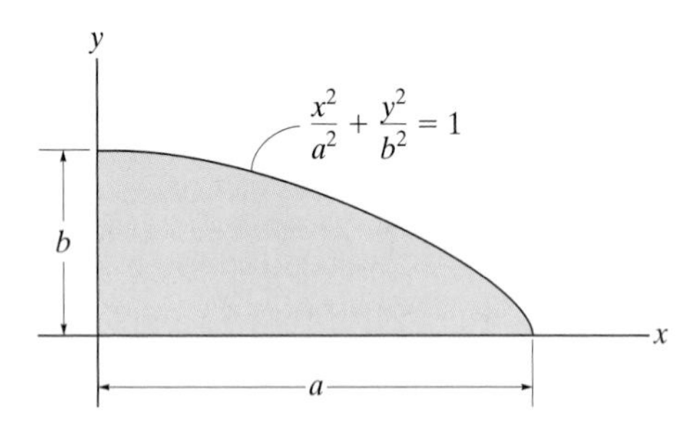

Prob. 9–17

9–18. Locate the centroid $\bar{x}$ of the triangular area.

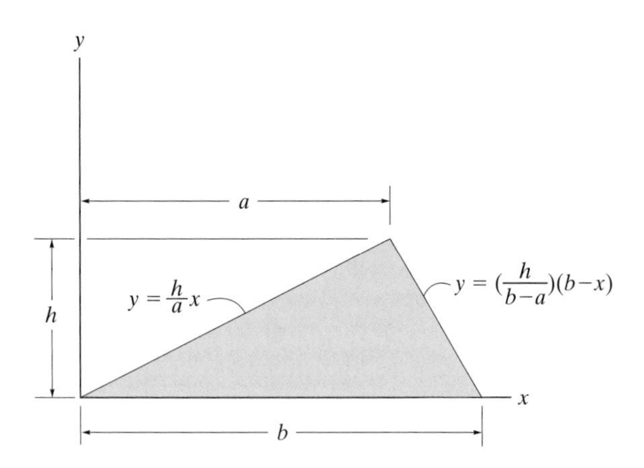

Prob. 9–18

9–19. Locate the centroid of the shaded area.

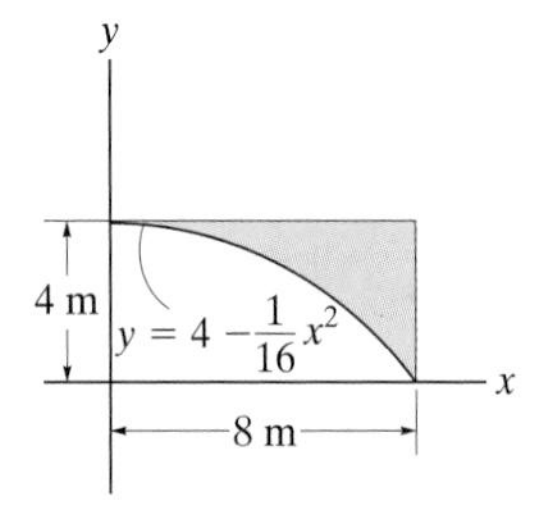

Prob. 9–19

***■9–20.** Locate the centroid $\bar{x}$ of the shaded area. Solve the problem by evaluating the integrals using Simpson's rule.

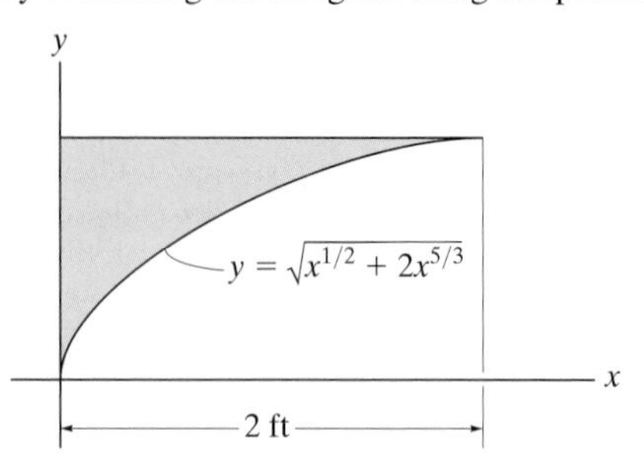

Prob. 9–20

■9–21. Locate the centroid $\bar{y}$ of the shaded area. Solve the problem by evaluating the integrals using Simpson's rule.

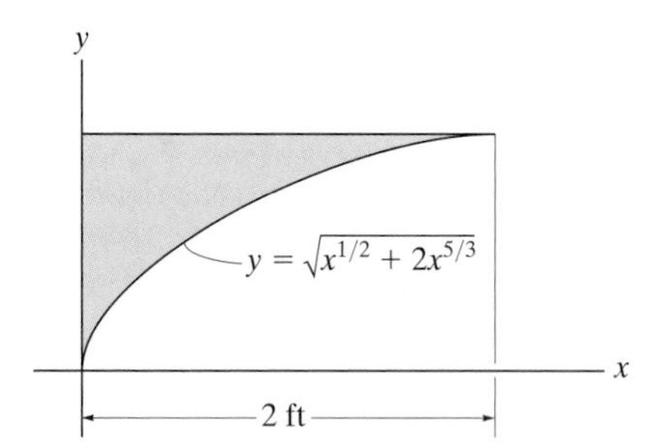

Prob. 9–21

9–22. The steel plate is 0.3 m thick and has a density of 7850 kg/m^3. Determine the location of its center of mass. Also compute the reactions at the pin and roller support.

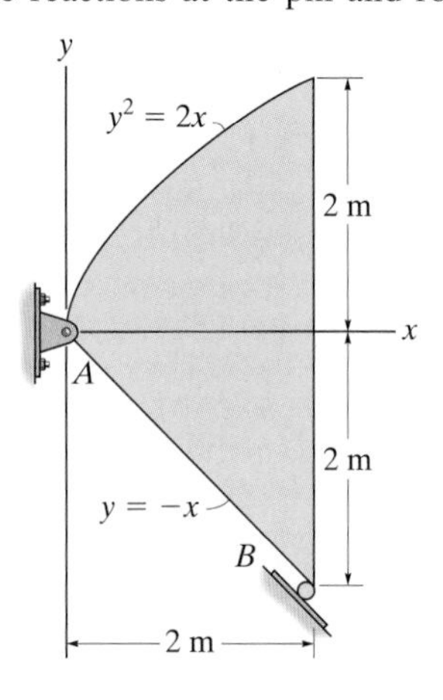

Prob. 9–22

9–23. Locate the centroid $\bar{x}$ of the shaded area.

***9–24.** Locate the centroid $\bar{y}$ of the shaded area.

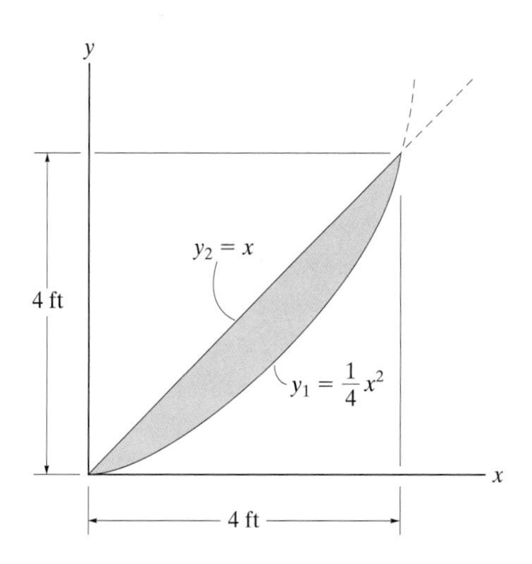

Probs. 9–23/24

9–25. Locate the centroid $\bar{x}$ of the shaded area.

9–26. Locate the centroid $\bar{y}$ of the shaded area.

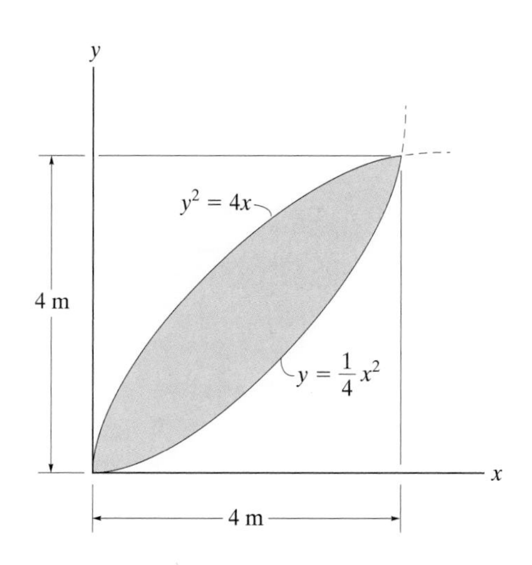

Probs. 9–25/26

9–27. Locate the centroid $\bar{x}$ of the shaded area.

***9–28.** Locate the centroid $\bar{y}$ of the shaded area.

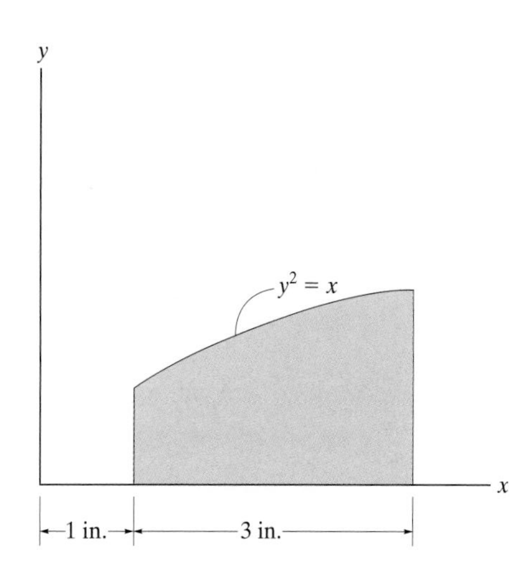

Probs. 9–27/28

9–29. Locate the centroid $\bar{x}$ of the shaded area.

9–30. Locate the centroid $\bar{y}$ of the shaded area.

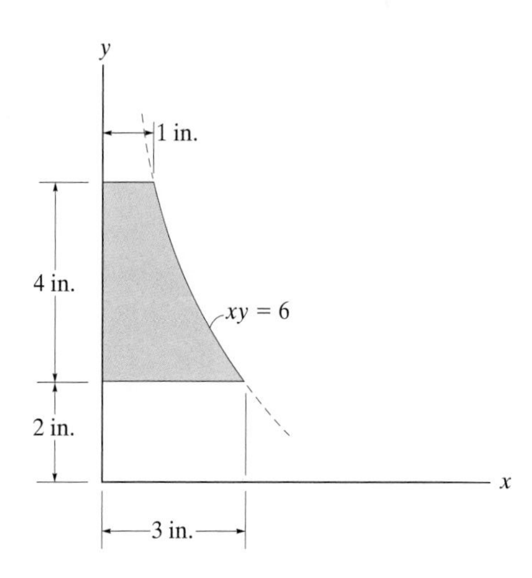

Probs. 9–29/30

9–31. Determine the location $\bar{r}$ of the centroid C of the cardioid, $r = a(1 - \cos\theta)$.

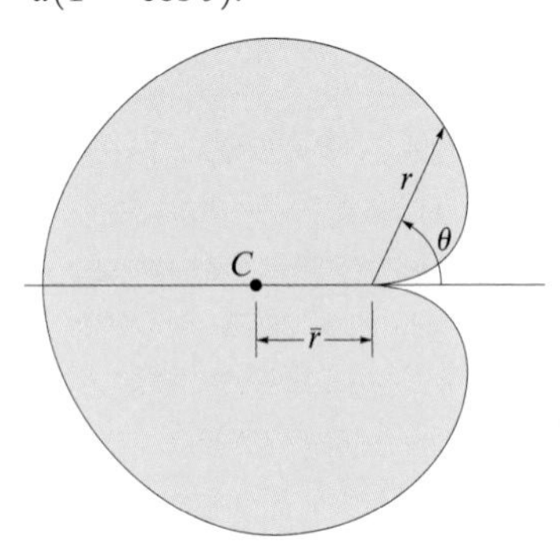

Prob. 9–31

***9–32.** Locate the centroid of the ellipsoid of revolution.

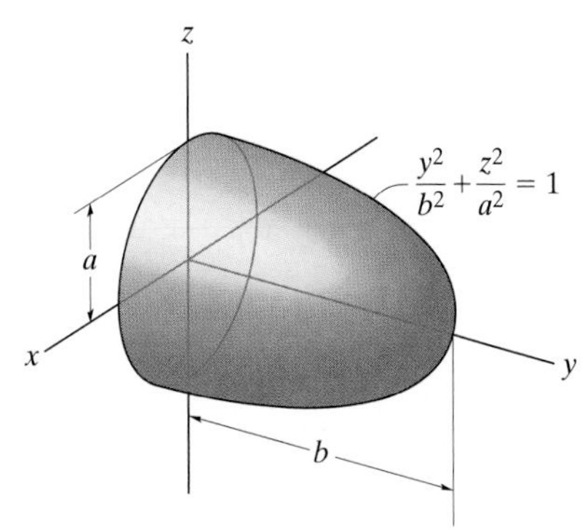

Prob. 9–32

9–33. Locate the centroid $\bar{z}$ of the very thin conical shell. *Hint:* Use thin ring elements having a center at $(0, 0, z)$, radius y, and width $dL = \sqrt{(dy)^2 + (dz)^2}$.

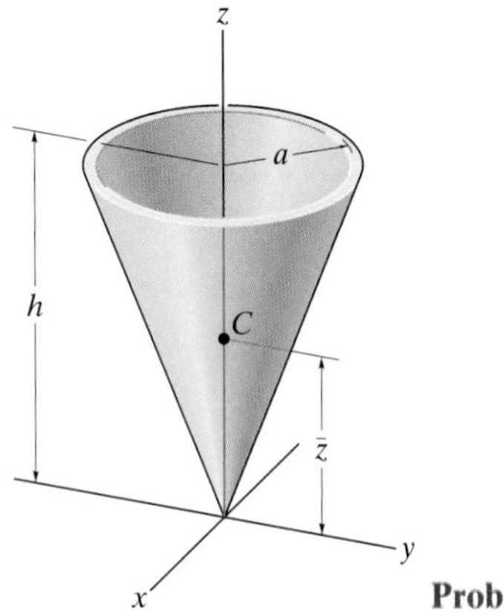

Prob. 9–33

9–34. Locate the centroid $\bar{z}$ of the volume segment.

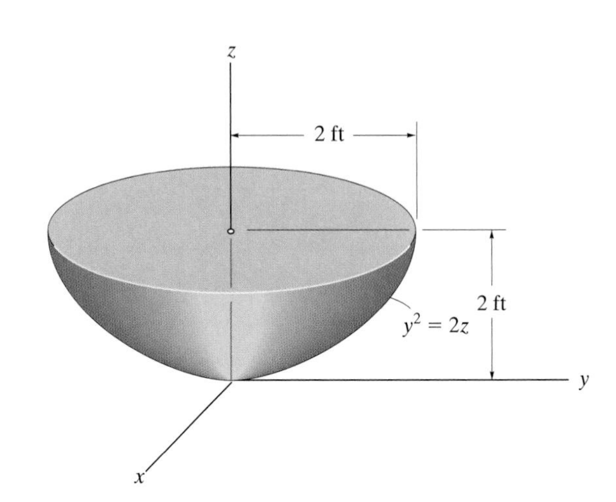

Prob. 9–34

9–35. Locate the centroid of the solid.

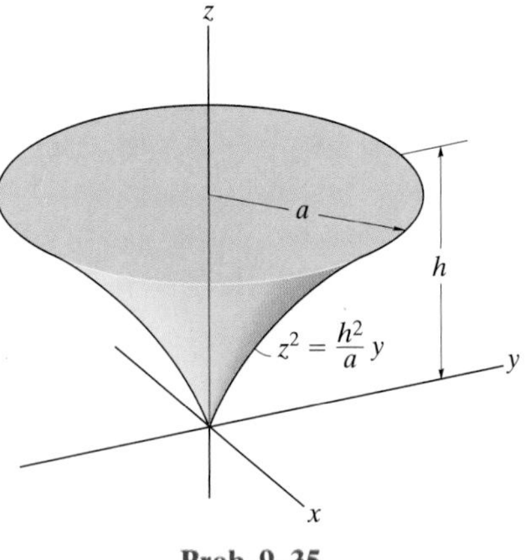

Prob. 9–35

***9–36.** Locate the centroid of the quarter-cone.

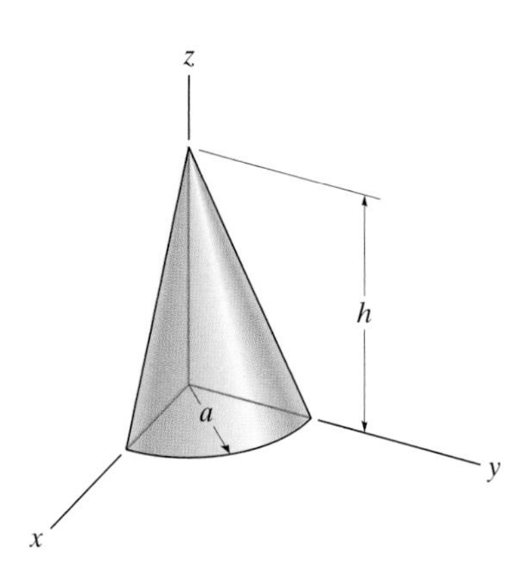

Prob. 9–36

9–37. Locate the center of mass $\bar{x}$ of the hemisphere. The density of the material varies linearly from zero at the origin O to ρ_0 at the surface. *Hint:* Choose a hemispherical shell element for integration.

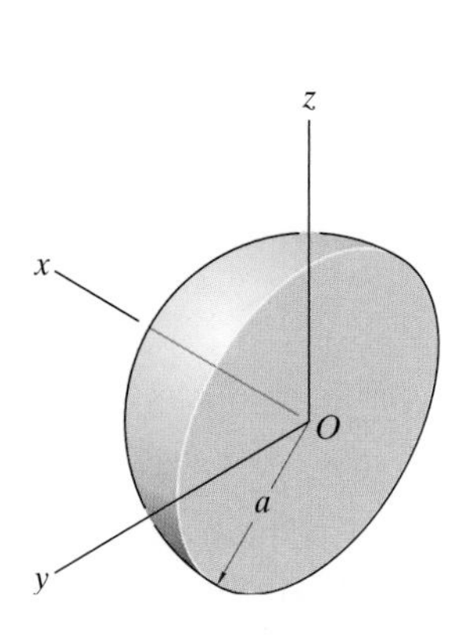

Prob. 9–37

9–38. Locate the centroid $\bar{z}$ of the right-elliptical cone.

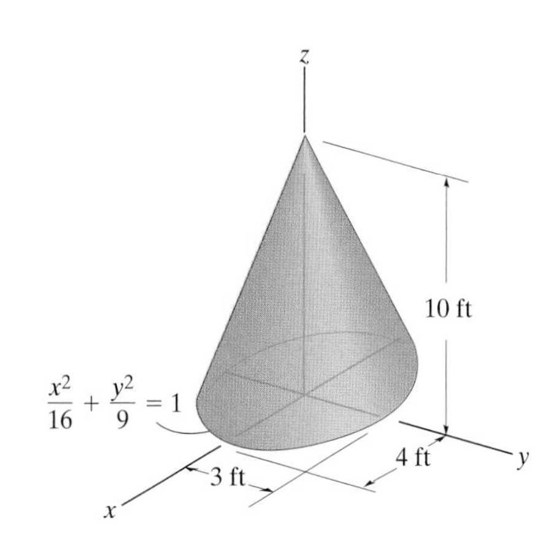

Prob. 9–38

9–39. Locate the center of gravity $\bar{z}$ of the frustum of the paraboloid. The material is homogeneous.

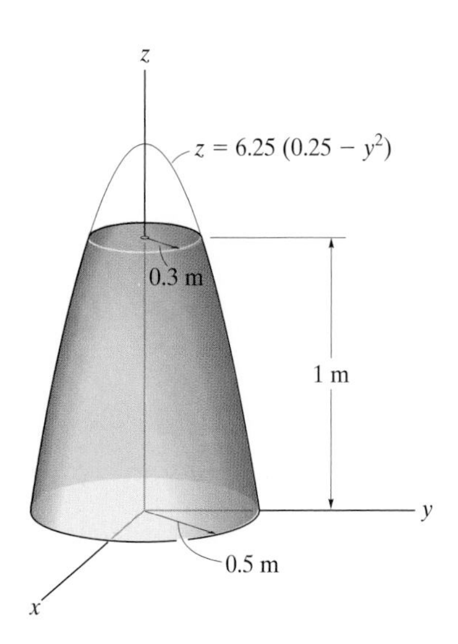

Prob. 9–39

*9–40. Locate the center of gravity $\bar{y}$ of the volume. The material is homogeneous.

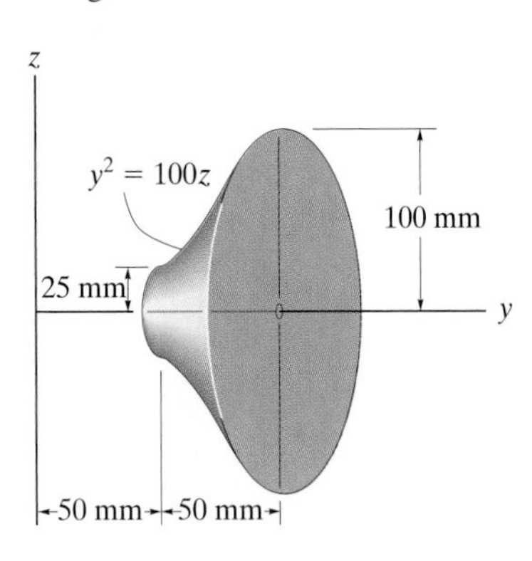

Prob. 9–40

9–41. Locate the center of gravity for the homogeneous half-cone.

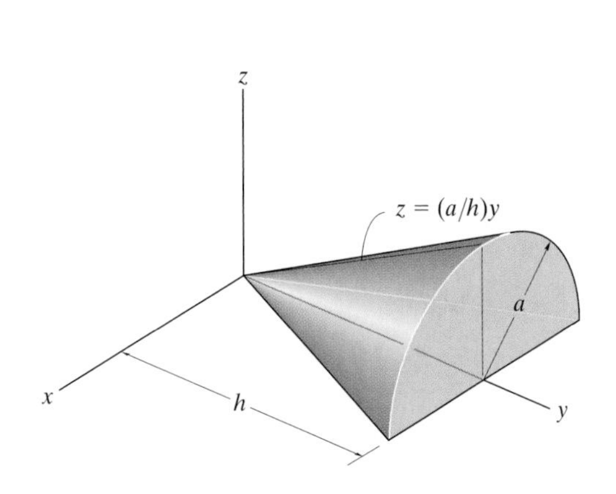

Prob. 9–41

9–42. Locate the centroid $\bar{z}$ of the spherical segment.

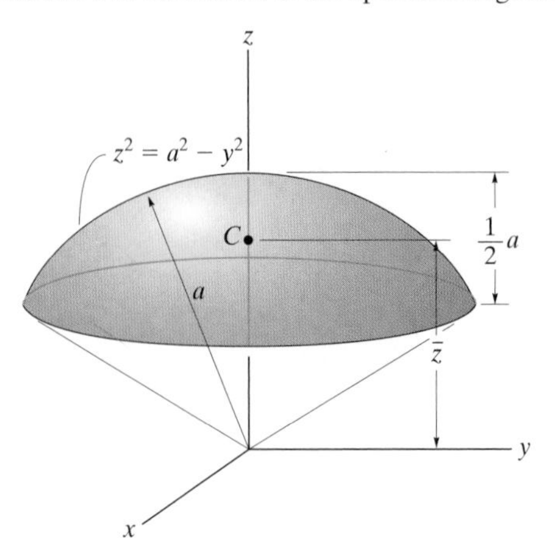

Prob. 9–42

9–43. Determine the location $\bar{z}$ of the centroid for the tetrahedron. *Hint:* Use a triangular "plate" element parallel to the x–y plane and of thickness dz.

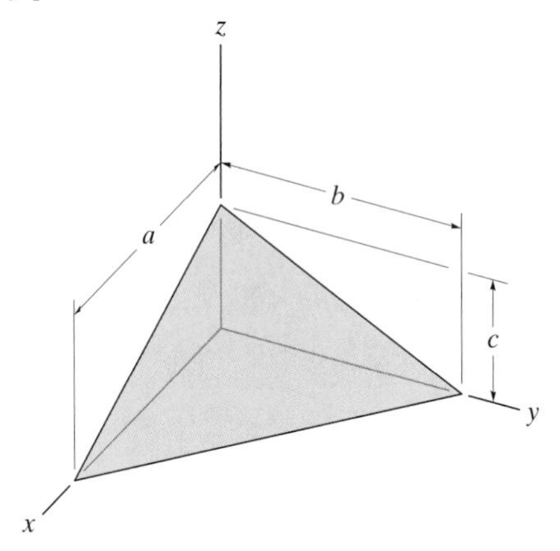

Prob. 9–43

9.3 Composite Bodies

A *composite body* consists of a series of connected "simpler" shaped bodies, which may be rectangular, triangular, semicircular, etc. Such a body can often be sectioned or divided into its composite parts and, provided the *weight* and location of the center of gravity of each of these parts are known, we can eliminate the need for integration to determine the center of gravity for the entire body. The method for doing this requires treating each composite part like a particle and following the procedure outlined in Sec. 9.1. Formulas analogous to Eqs. 9–1 result since we must account for a finite number of weights. Rewriting these formulas, we have

$$\bar{x} = \frac{\Sigma \tilde{x} W}{\Sigma W} \qquad \bar{y} = \frac{\Sigma \tilde{y} W}{\Sigma W} \qquad \bar{z} = \frac{\Sigma \tilde{z} W}{\Sigma W} \tag{9–8}$$

Here

$\bar{x}, \bar{y}, \bar{z}$ represent the coordinates of the center of gravity G of the composite body.

$\tilde{x}, \tilde{y}, \tilde{z}$ represent the coordinates of the center of gravity of each composite part of the body.

ΣW is the sum of the weights of all the composite parts of the body, or simply the total weight of the body.

When the body has a *constant density or specific weight*, the center of gravity *coincides* with the centroid of the body. The centroid for composite lines, areas, and volumes can be found using relations analogous to Eqs. 9–8; however, the W's are replaced by L's, A's, and V's, respectively. Centroids for common shapes of lines, areas, shells, and volumes that often make up a composite body are given in the table on the inside back cover.

In order to determine the force required to tip over this concrete barrier it is first necessary to determine the location of its center of gravity G. Due to symmetry, G will lie on the axis of symmetry.

Procedure for Analysis

The location of the center of gravity of a body or the centroid of a composite geometrical object represented by a line, area, or volume can be determined using the following procedure.

Composite Parts.

- Using a sketch, divide the body or object into a finite number of composite parts that have simpler shapes.
- If a composite part has a *hole*, or a geometric region having no material, then consider the composite part without the hole and consider the hole as an *additional* composite part having *negative* weight or size.

Moment Arms.

- Establish the coordinate axes on the sketch and determine the coordinates $\tilde{x}, \tilde{y}, \tilde{z}$ of the center of gravity or centroid of each part.

Summations.

- Determine $\bar{x}, \bar{y}, \bar{z}$ by applying the center of gravity equations, Eqs. 9–8, or the analogous centroid equations.
- If an object is *symmetrical* about an axis, the centroid of the object lies on this axis.

If desired, the calculations can be arranged in tabular form, as indicated in the following three examples.

EXAMPLE 9.9

Locate the centroid of the wire shown in Fig. 9–17*a*.

SOLUTION

Composite Parts. The wire is divided into three segments as shown in Fig. 9–17*b*.

Moment Arms. The location of the centroid for each piece is determined and indicated in the figure. In particular, the centroid of segment ① is determined either by integration or by using the table on the inside back cover.

Summations. The calculations are tabulated as follows:

Segment	L (mm)	$\tilde{x}$ (mm)	$\tilde{y}$ (mm)	$\tilde{z}$ (mm)	$\tilde{x}L$ (mm^2)	$\tilde{y}L$ (mm^2)	$\tilde{z}L$ (mm^2)
1	$\pi(60) = 188.5$	60	−38.2	0	11 310	−7200	0
2	40	0	20	0	0	800	0
3	20	0	40	−10	0	800	−200
	$\Sigma L = 248.5$				$\Sigma\tilde{x}L = 11\,310$	$\Sigma\tilde{y}L = -5600$	$\Sigma\tilde{z}L = -200$

Thus,

$$\bar{x} = \frac{\Sigma\tilde{x}L}{\Sigma L} = \frac{11310}{248.5} = 45.5 \text{ mm} \qquad \textit{Ans.}$$

$$\bar{y} = \frac{\Sigma\tilde{y}L}{\Sigma L} = \frac{-5600}{248.5} = -22.5 \text{ mm} \qquad \textit{Ans.}$$

$$\bar{z} = \frac{\Sigma\tilde{z}L}{\Sigma L} = \frac{-200}{248.5} = -0.805 \text{ mm} \qquad \textit{Ans.}$$

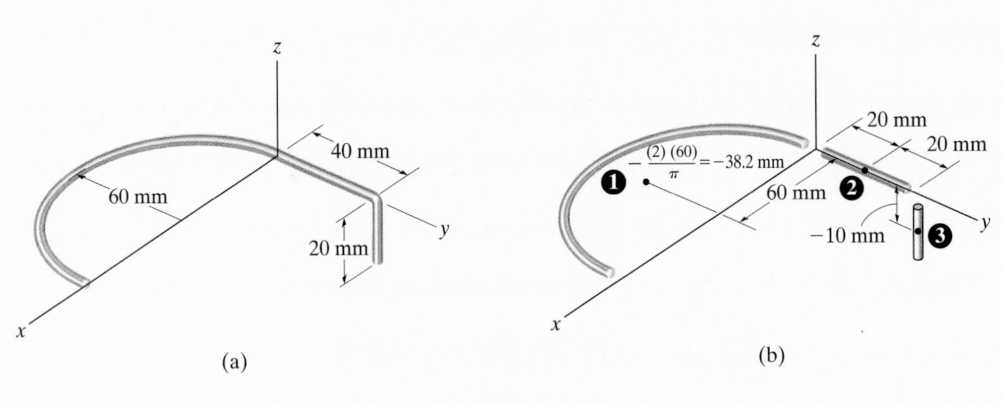

Fig. 9–17

EXAMPLE 9.10

Locate the centroid of the plate area shown in Fig. 9–18*a*.

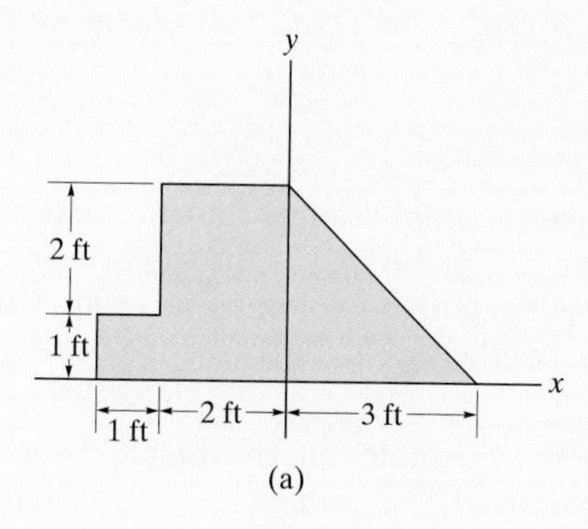

(a)

Fig. 9–18

SOLUTION

Composite Parts. The plate is divided into three segments as shown in Fig. 9–18*b*. Here the area of the small rectangle ③ is considered "negative" since it must be subtracted from the larger one ②.

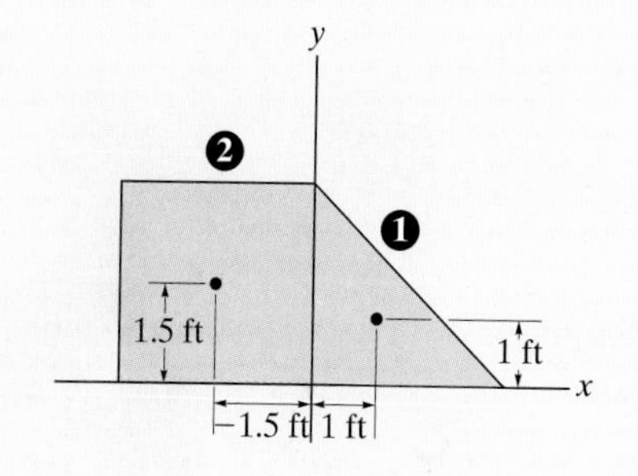

Moment Arms. The centroid of each segment is located as indicated in the figure. Note that the $\tilde{x}$ coordinates of ② and ③ are *negative*.

Summations. Taking the data from Fig. 9–18*b*, the calculations are tabulated as follows:

Segment	A (ft^2)	$\tilde{x}$ (ft)	$\tilde{y}$ (ft)	$\tilde{x}A$ (ft^3)	$\tilde{y}A$ (ft^3)
1	$\frac{1}{2}(3)(3) = 4.5$	1	1	4.5	4.5
2	$(3)(3) = 9$	−1.5	1.5	−13.5	13.5
3	$-(2)(1) = -2$	−2.5	2	5	−4
	$\Sigma A = 11.5$			$\Sigma\tilde{x}A = -4$	$\Sigma\tilde{y}A = 14$

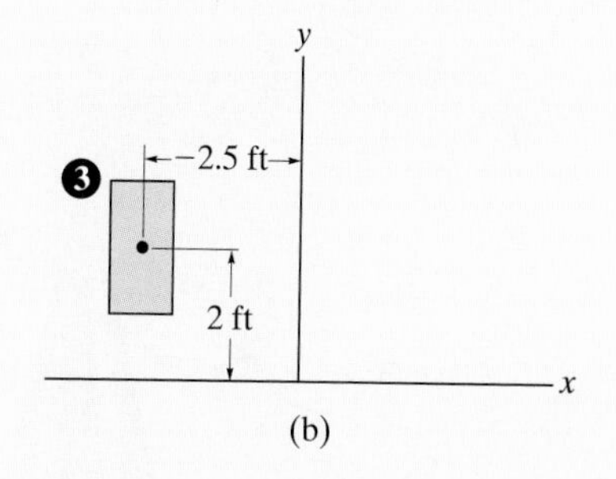

(b)

Thus,

$$\bar{x} = \frac{\Sigma\tilde{x}A}{\Sigma A} = \frac{-4}{11.5} = -0.348 \text{ ft} \qquad \textit{Ans.}$$

$$\bar{y} = \frac{\Sigma\tilde{y}A}{\Sigma A} = \frac{14}{11.5} = 1.22 \text{ ft} \qquad \textit{Ans.}$$

NOTE: If these results are plotted in Fig. 9–18, the location of point C seems reasonable.

EXAMPLE 9.11

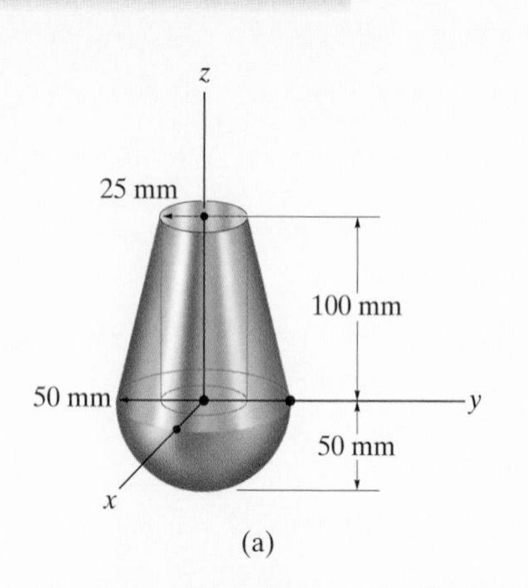

Fig. 9–19

Locate the center of mass of the composite assembly shown in Fig. 9–19*a*. The conical frustum has a density of $\rho_c = 8\ \text{Mg/m}^3$, and the hemisphere has a density of $\rho_h = 4\ \text{Mg/m}^3$. There is a 25-mm-radius cylindrical hole in the center.

SOLUTION

Composite Parts. The assembly can be thought of as consisting of four segments as shown in Fig. 9–19*b*. For the calculations, ③ and ④ must be considered as "negative" volumes in order that the four segments, when added together, yield the total composite shape shown in Fig. 9–19*a*.

Moment Arm. Using the table on the inside back cover, the computations for the centroid $\tilde{z}$ of each piece are shown in the figure.

Summations. Because of *symmetry*, note that

$$\bar{x} = \bar{y} = 0 \qquad \textit{Ans.}$$

Since $W = mg$ and g is constant, the third of Eqs. 9–8 becomes $\bar{z} = \Sigma\tilde{z}m/\Sigma m$. The mass of each piece can be computed from $m = \rho V$ and used for the calculations. Also, $1\ \text{Mg/m}^3 = 10^{-6}\ \text{kg/mm}^3$, so that

Segment	m (kg)	$\tilde{z}$ (mm)	$\tilde{z}m$ (kg·mm)
1	$8(10^{-6})\left(\frac{1}{3}\right)\pi(50)^2(200) = 4.189$	50	209.440
2	$4(10^{-6})\left(\frac{2}{3}\right)\pi(50)^3 = 1.047$	−18.75	−19.635
3	$-8(10^{-6})\left(\frac{1}{3}\right)\pi(25)^2(100) = -0.524$	100 + 25 = 125	−65.450
4	$-8(10^{-6})\pi(25)^2(100) = -1.571$	50	−78.540
	$\Sigma m = 3.141$		$\Sigma\tilde{z}m = 45.815$

$$\text{Thus,}\quad \bar{z} = \frac{\Sigma\tilde{z}m}{\Sigma m} = \frac{45.815}{3.141} = 14.6\ \text{mm} \qquad \textit{Ans.}$$

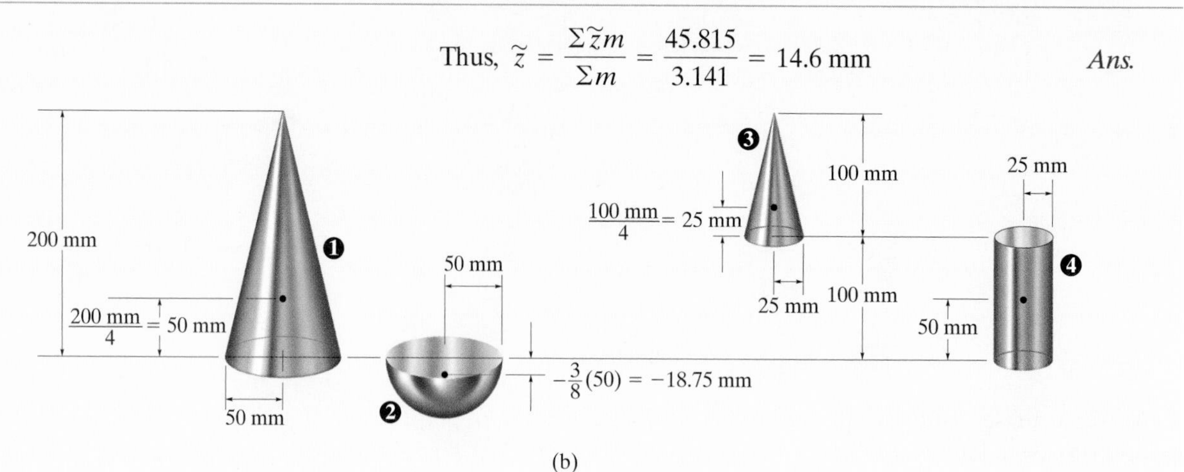

(b)

PROBLEMS

*9–44. Determine the location (x, y) of the 7-kg particle so that the three particles, which lie in the x–y plane, have a center of mass located at the origin O.

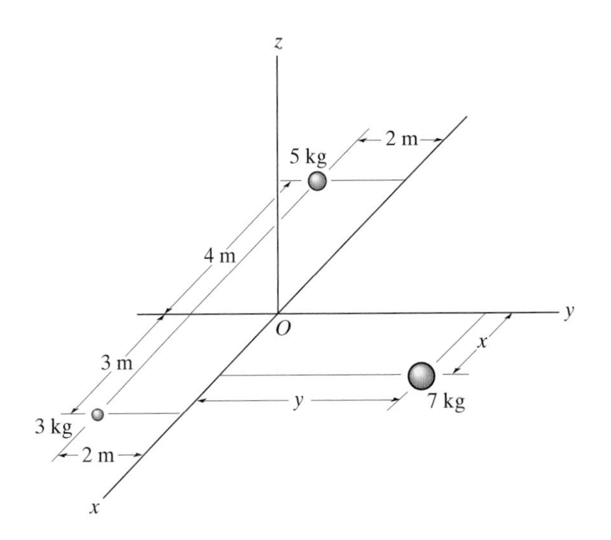

Prob. 9–44

9–45. Locate the center of gravity $(\bar{x}, \bar{y}, \bar{z})$ of the four particles.

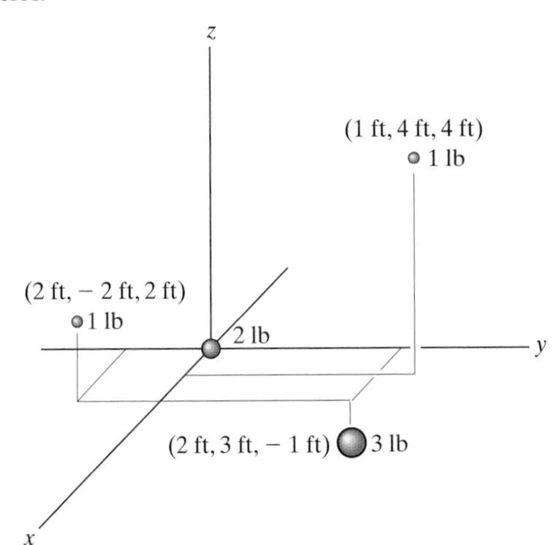

Prob. 9–45

9–46. A rack is made from roll-formed sheet steel and has the cross section shown. Determine the location $(\bar{x}, \bar{y})$ of the centroid of the cross section. The dimensions are indicated at the center thickness of each segment.

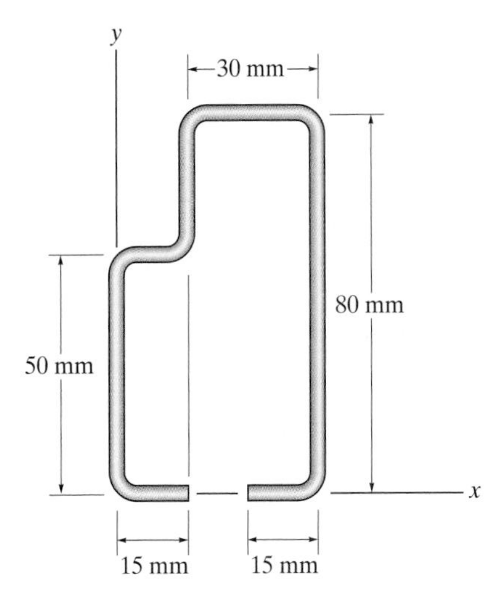

Prob. 9–46

9–47. The steel and aluminum plate assembly is bolted together and fastened to the wall. Each plate has a constant width in the z direction of 200 mm and thickness of 20 mm. If the density of A and B is $\rho_s = 7.85\ \text{Mg/m}^3$, and for C, $\rho_{al} = 2.71\ \text{Mg/m}^3$, determine the location $\bar{x}$ of the center of mass. Neglect the size of the bolts.

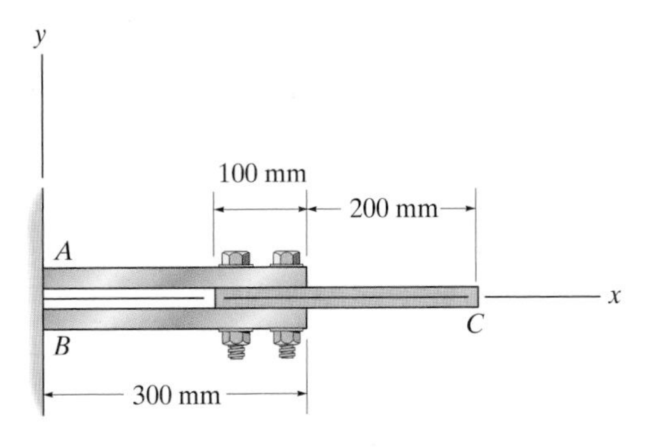

Prob. 9–47

***9–48.** The truss is made from five members, each having a length of 4 m and a mass of 7 kg/m. If the mass of the gusset plates at the joints and the thickness of the members can be neglected, determine the distance d to where the hoisting cable must be attached, so that the truss does not tip (rotate) when it is lifted.

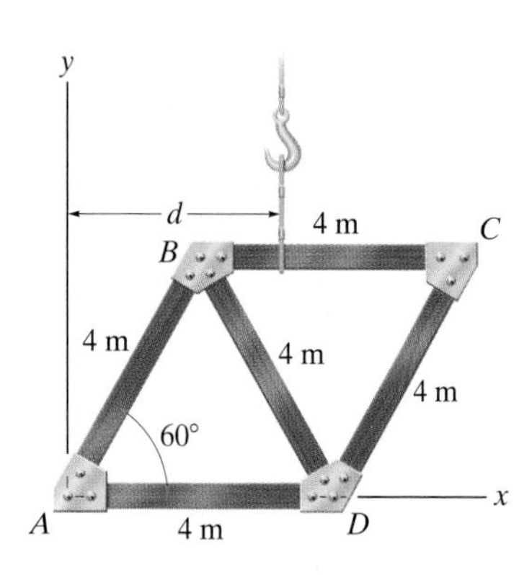

Prob. 9–48

9–49. Locate the center of gravity $(\bar{x}, \bar{y}, \bar{z})$ of the homogeneous wire.

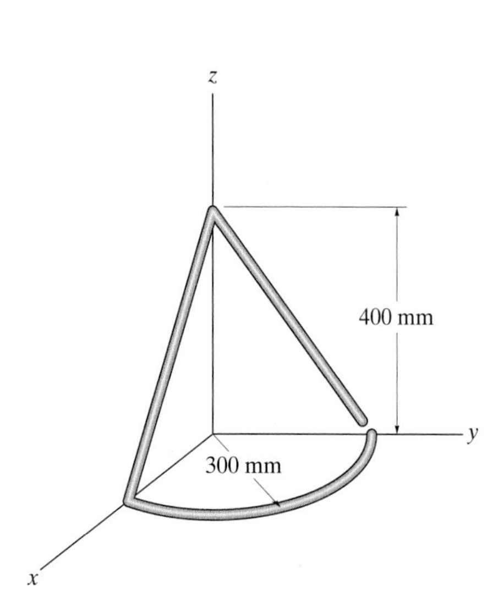

Prob. 9–49

9–50. Determine the location $(\bar{x}, \bar{y})$ of the center of gravity of the homogeneous wire bent in the form of a triangle. Neglect any slight bends at the corners. If the wire is suspended using a thread T attached to it at C, determine the angle of tilt AB makes with the horizontal when the wire is in equilibrium.

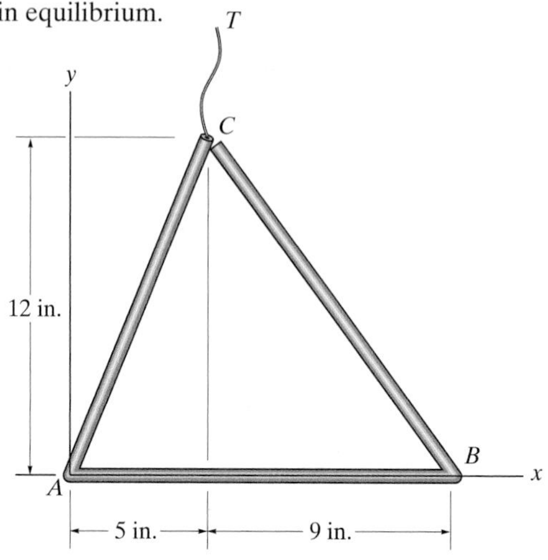

Prob. 9–50

9–51. The three members of the frame each have a weight per unit length of 4 kg/m. Locate the position $(\bar{x}, \bar{y})$ of the center of gravity. Neglect the size of the pins at the joints and the thickness of the members. Also, calculate the reactions at the fixed support A.

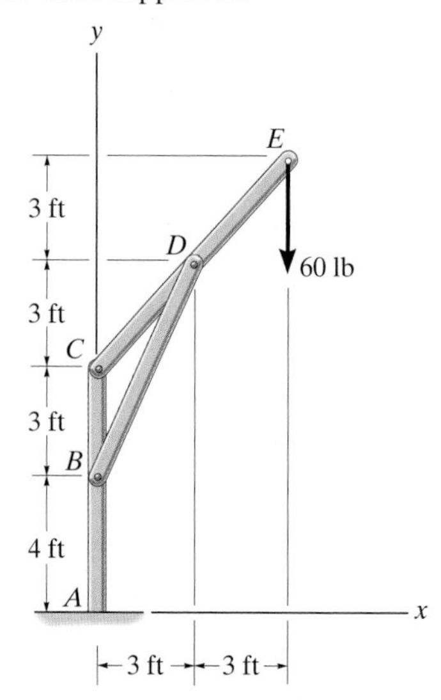

Prob. 9–51

***9–52.** Locate the center of gravity $G(\bar{x}, \bar{y})$ of the streetlight. Neglect the thickness of each segment. The mass per unit length of each segment is as follows: $\rho_{AB} = 12$ kg/m, $\rho_{BC} = 8$ kg/m, $\rho_{CD} = 5$ kg/m, and $\rho_{DE} = 2$ kg/m.

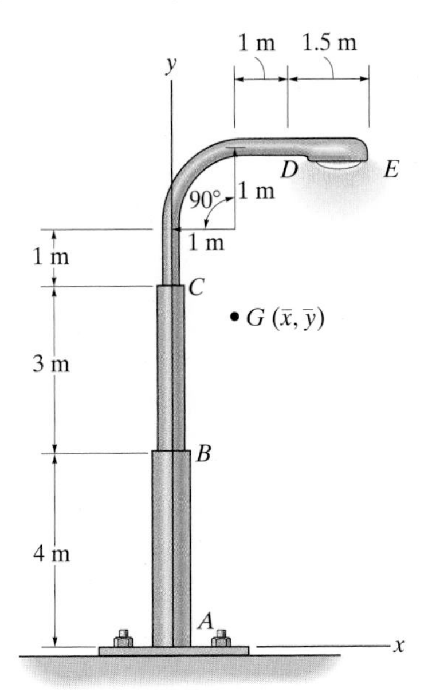

Prob. 9–52

9–53. Determine the location $\bar{y}$ of the centroid of the beam's cross-sectional area. Neglect the size of the corner welds at A and B for the calculation.

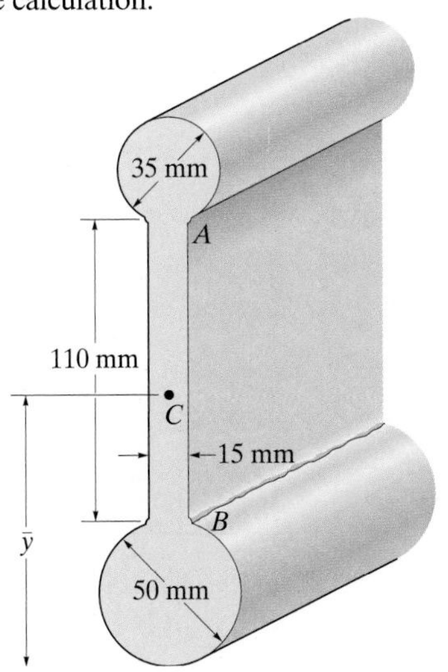

Prob. 9–53

9–54. The gravity wall is made of concrete. Determine the location $(\bar{x}, \bar{y})$ of the center of gravity G for the wall.

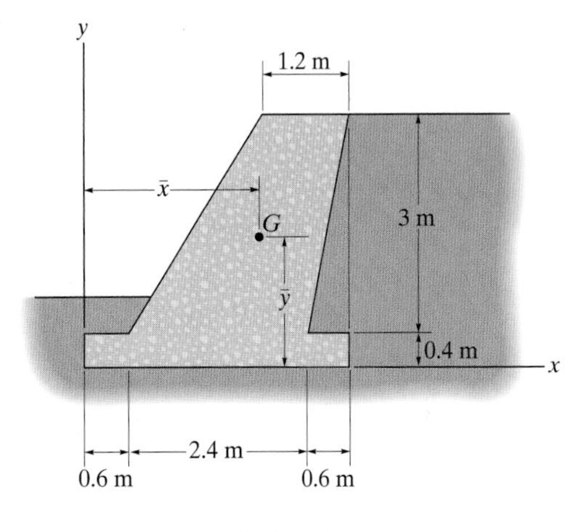

Prob. 9–54

9–55. Locate the centroid $(\bar{x}, \bar{y})$ of the shaded area.

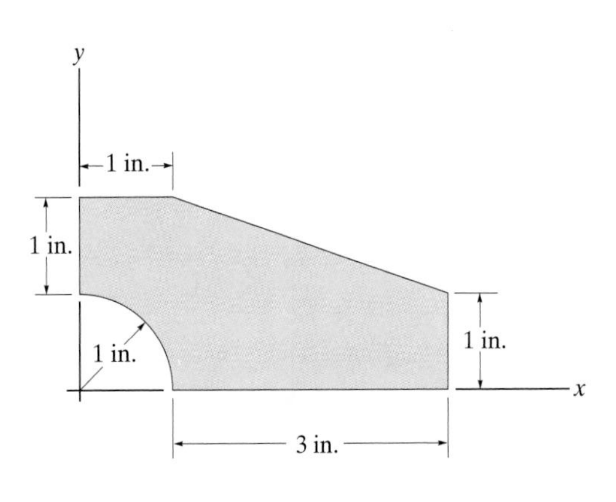

Prob. 9–55

***9–56.** Locate the centroid $(\bar{x}, \bar{y})$ of the shaded area.

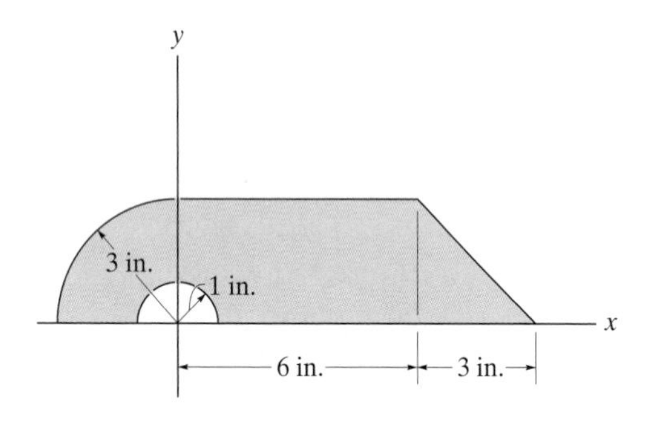

Prob. 9–56

9–58. Determine the location $(\bar{x}, \bar{y})$ of the centroid C of the area.

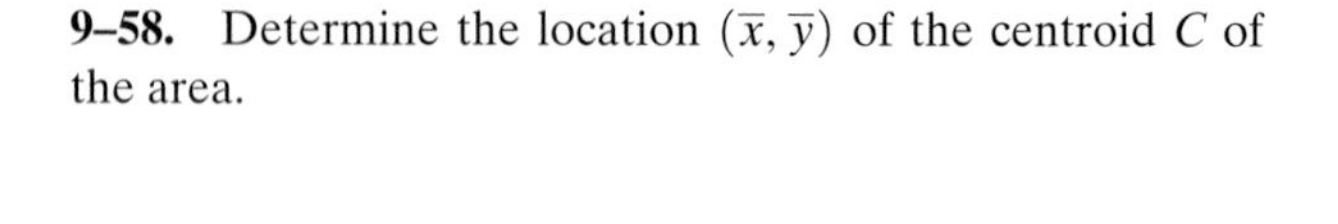

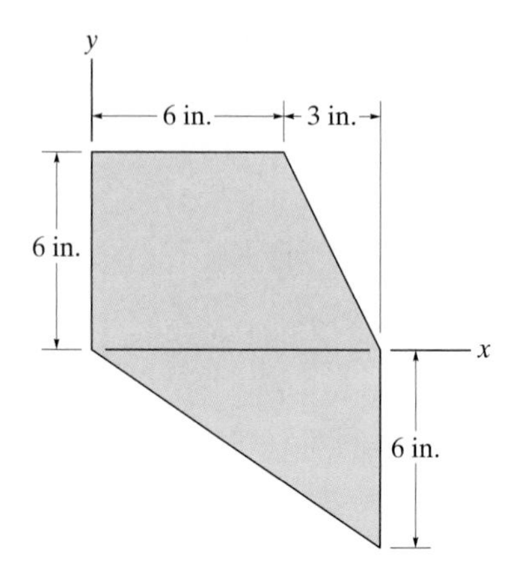

Prob. 9–58

9–57. Determine the location $\bar{y}$ of the centroidal axis $\bar{x}\,\bar{x}$ of the beam's cross-sectional area. Neglect the size of the corner welds at A and B for the calculation.

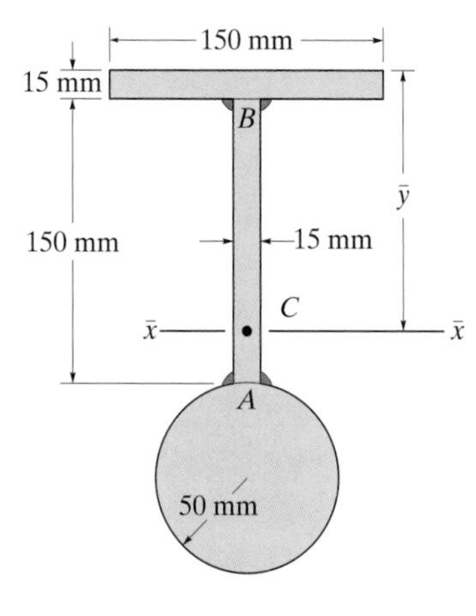

Prob. 9–57

9–59. Determine the location $\bar{y}$ of the centroid C for a beam having the cross-sectional area shown. The beam is symmetric with respect to the y axis.

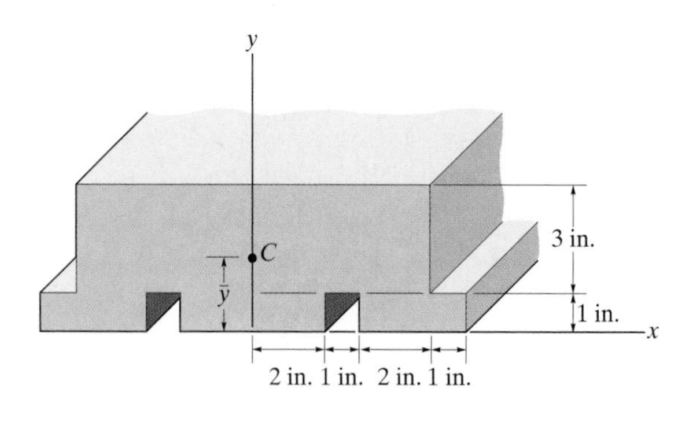

Prob. 9–59

*9–60. The wooden table is made from a square board having a weight of 15 lb. Each of the legs weighs 2 lb and is 3 ft long. Determine how high its center of gravity is from the floor. Also, what is the angle, measured from the horizontal, through which its top surface can be tilted on two of its legs before it begins to overturn? Neglect the thickness of each leg.

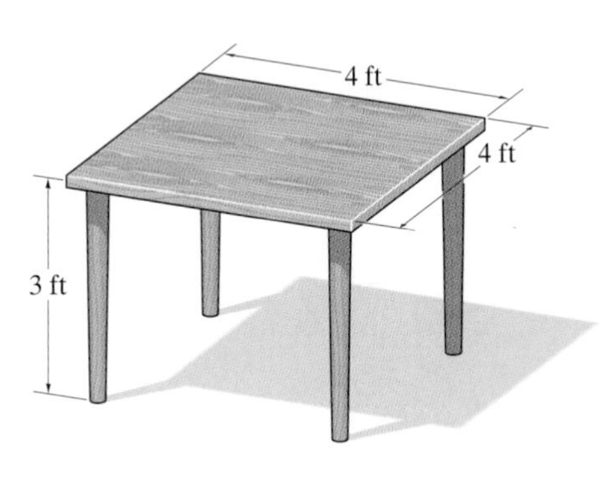

Prob. 9–60

9–61. Locate the centroid $\bar{y}$ for the beam's cross-sectional area.

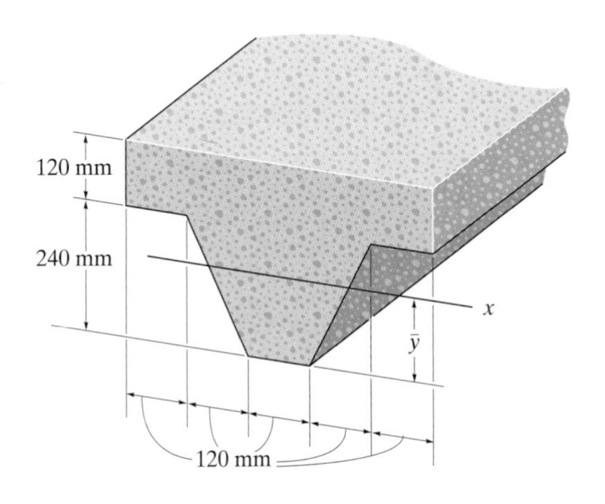

Prob. 9–61

9–62. Determine the location $\bar{x}$ of the centroid C of the shaded area which is part of a circle having a radius r.

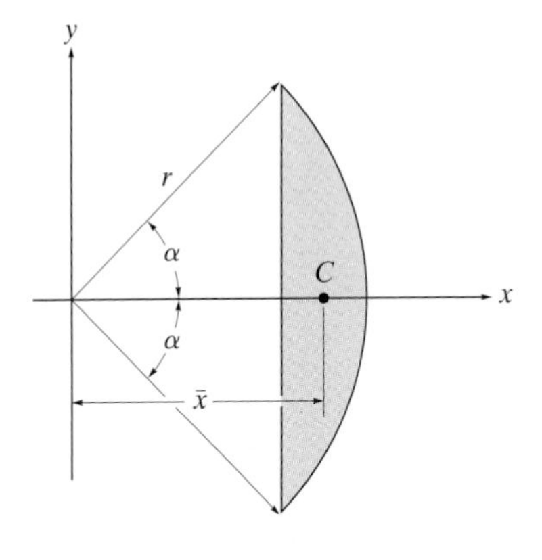

Prob. 9–62

9–63. Locate the centroid $\bar{y}$ for the strut's cross-sectional area.

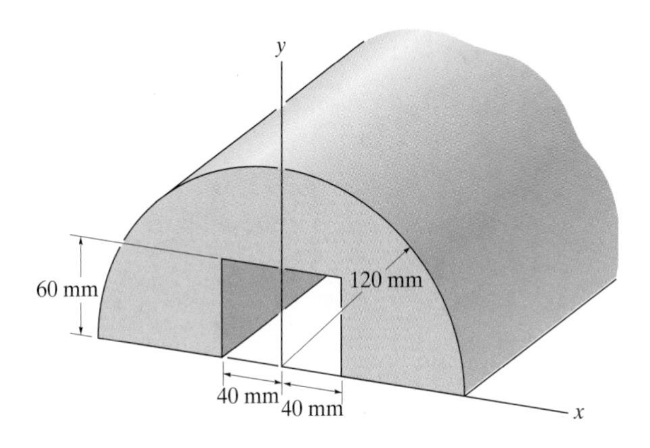

Prob. 9–63

***9–64.** The "New Jersey" concrete barrier is commonly used during highway construction. Determine the location $\bar{y}$ of its centroid.

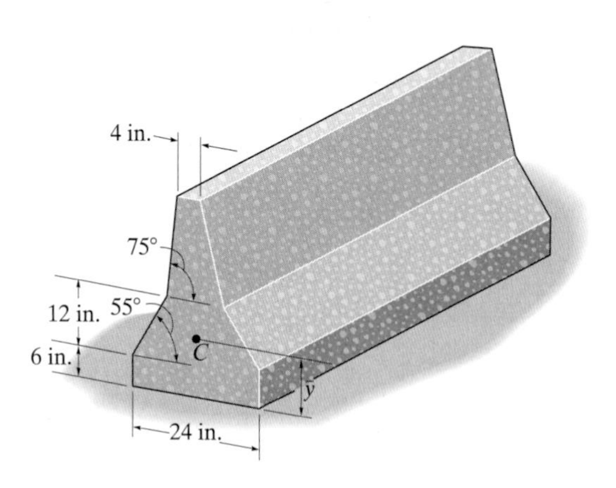

Prob. 9–64

9–65. The composite plate is made from both steel (A) and brass (B) segments. Determine the mass and location $(\bar{x}, \bar{y}, \bar{z})$ of its mass center G. Take $\rho_{st} = 7.85\ \text{Mg/m}^3$ and $\rho_{br} = 8.74\ \text{Mg/m}^3$.

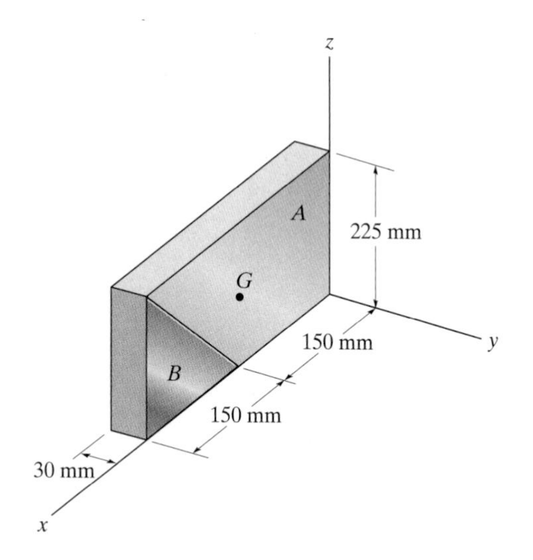

Prob. 9–65

9–66. Locate the centroid $\bar{y}$ of the concrete beam having the tapered cross section shown.

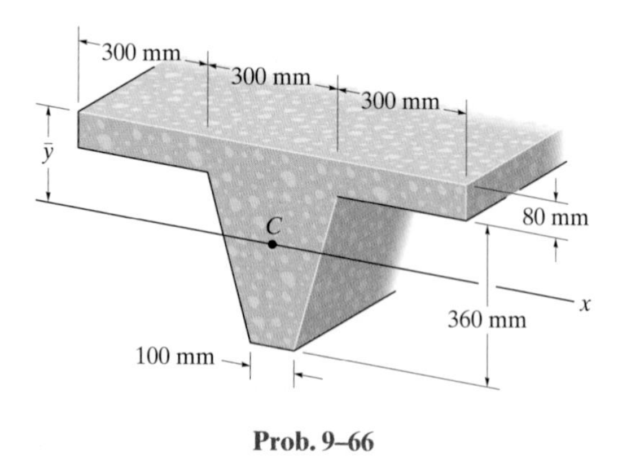

Prob. 9–66

9–67. The anatomical center of gravity G of a person can be determined by using a scale and a rigid board having a uniform weight W_1 and length l. With the person's weight W known, the person lies down on the board and the scale reading P is recorded. From this show how to calculate the location $\bar{x}$. Discuss the best place l_1 for the smooth support at B in order to improve the accuracy of this experiment.

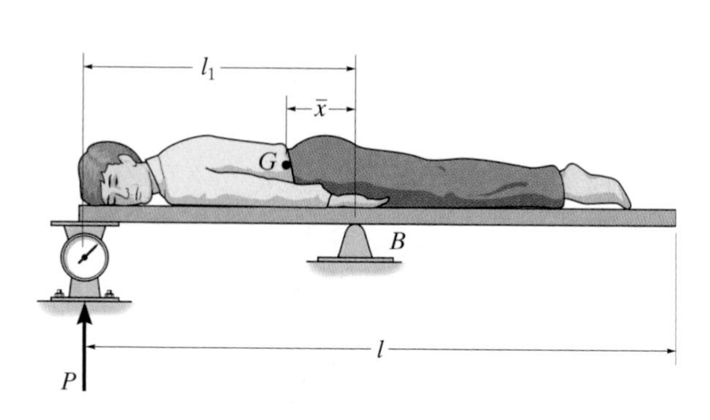

Prob. 9–67

***9–68.** The tank and compressor have a mass of 15 kg and mass center at G_T, and the motor has a mass of 70 kg and a mass center at G_M. Determine the angle of tilt, θ, of the tank so that the unit will be on the verge of tipping over.

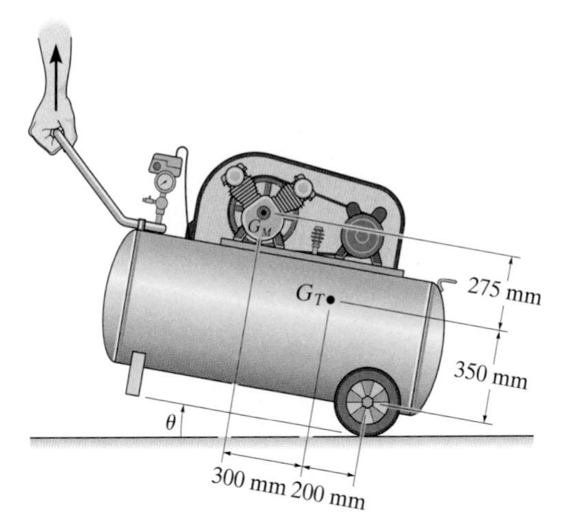

Prob. 9–68

9–69. Determine the distance h to which a 100-mm-diameter hole must be bored into the base of the cone so that the center of mass of the resulting shape is located at $\bar{z} = 115$ mm. The material has a density of 8 Mg/m^3.

9–70. Determine the distance $\bar{z}$ to the centroid of the shape which consists of a cone with a hole of height $h = 50$ mm bored into its base.

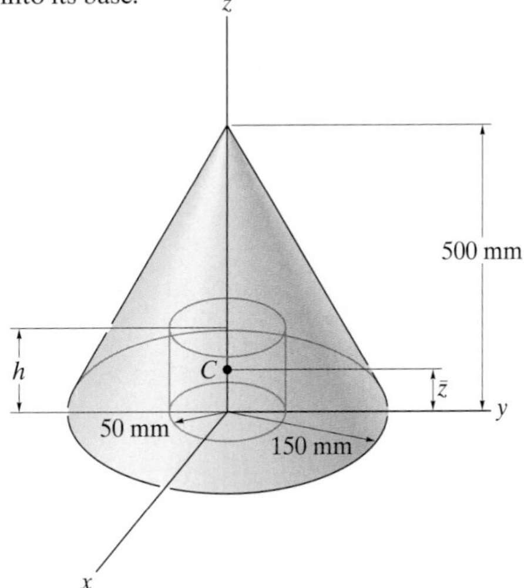

Probs. 9–69/70

9–71. The sheet metal part has the dimensions shown. Determine the location $(\bar{x}, \bar{y}, \bar{z})$ of its centroid.

***9–72.** The sheet metal part has a weight per unit area of 2 lb/ft^2 and is supported by the smooth rod and at C. If the cord is cut, the part will rotate about the y axis until it reaches equilibrium. Determine the equilibrium angle of tilt, measured downward from the negative x axis, that AD makes with the $-x$ axis.

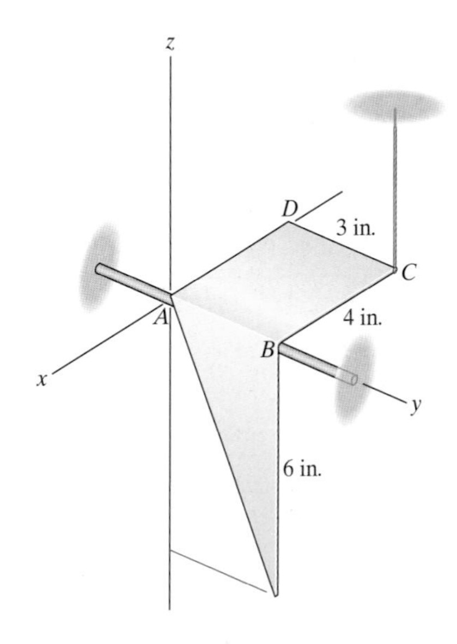

Probs. 9–71/72

9–73. A toy skyrocket consists of a solid conical top, $\rho_t = 600\ \text{kg/m}^3$, a hollow cylinder, $\rho_c = 400\ \text{kg/m}^3$, and a stick having a circular cross section, $\rho_s = 300\ \text{kg/m}^3$. Determine the length of the stick, x, so that the center of gravity G of the skyrocket is located along line aa.

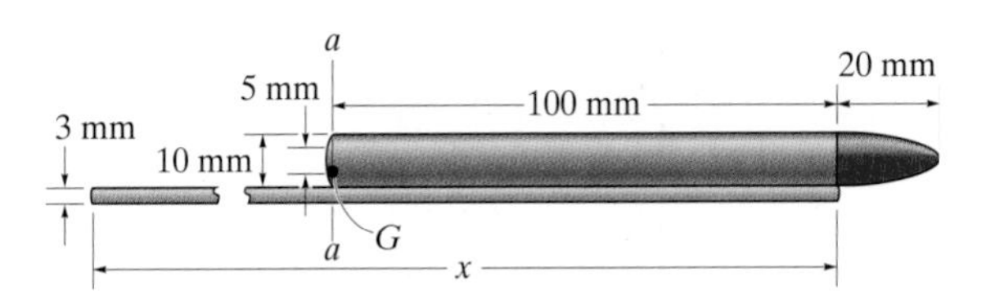

Prob. 9–73

9–74. Determine the location $(\bar{x}, \bar{y})$ of the center of mass of the turbine and compressor assembly. The mass and the center of mass of each of the various components are indicated below.

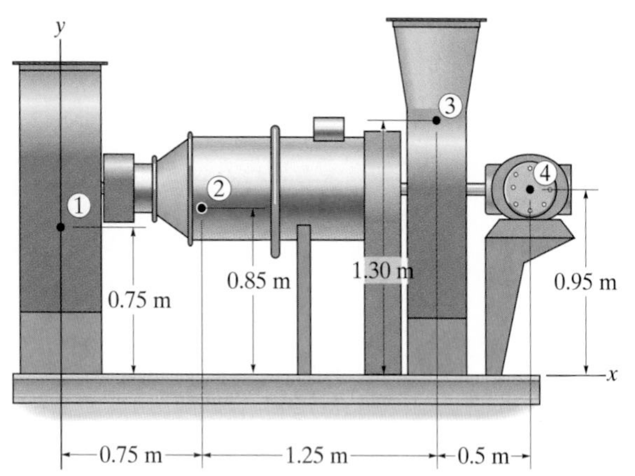
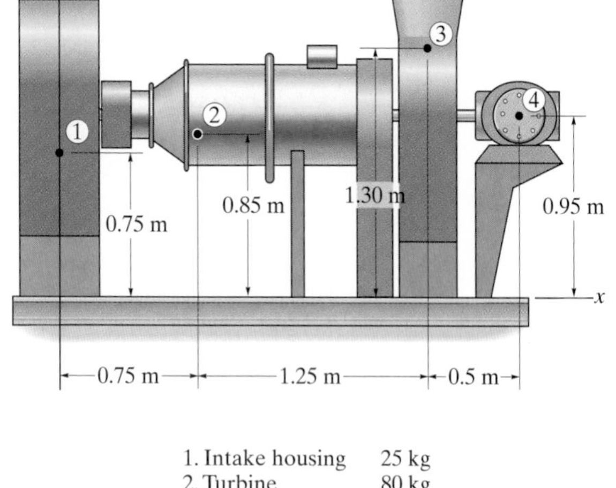

1. Intake housing 25 kg
2. Turbine 80 kg
3. Exhaust housing 30 kg
4. Compressor 105 kg

Prob. 9–74

9–75. The solid is formed by boring a conical hole into the hemisphere. Determine the distance $\bar{z}$ to the center of gravity.

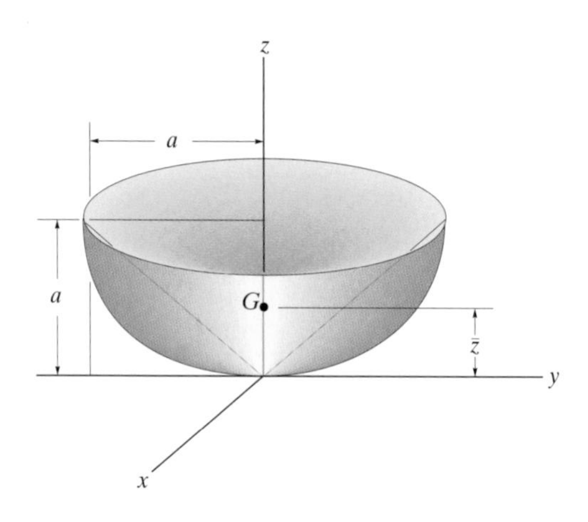

Prob. 9–75

***9–76.** Determine the location $\bar{x}$ of the centroid of the solid made from a hemisphere, cylinder, and cone.

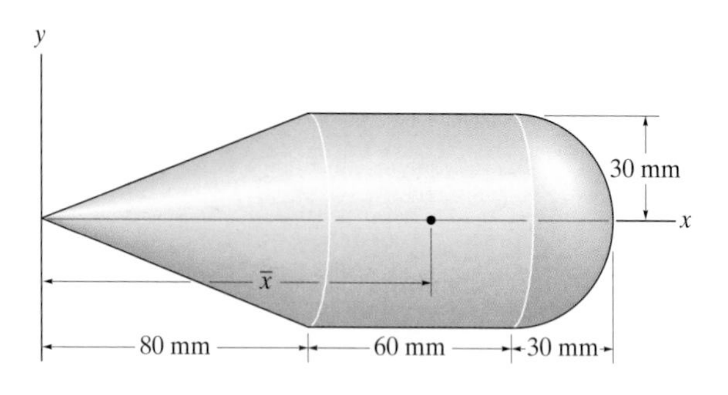

Prob. 9–76

9–77. The buoy is made from two homogeneous cones each having a radius of 1.5 ft. If $h = 1.2$ ft, find the distance $\bar{z}$ to the buoy's center of gravity G.

9–78. The buoy is made from two homogeneous cones each having a radius of 1.5 ft. If it is required that the buoy's center of gravity G be located at $\bar{z} = 0.5$ ft, determine the height h of the top cone.

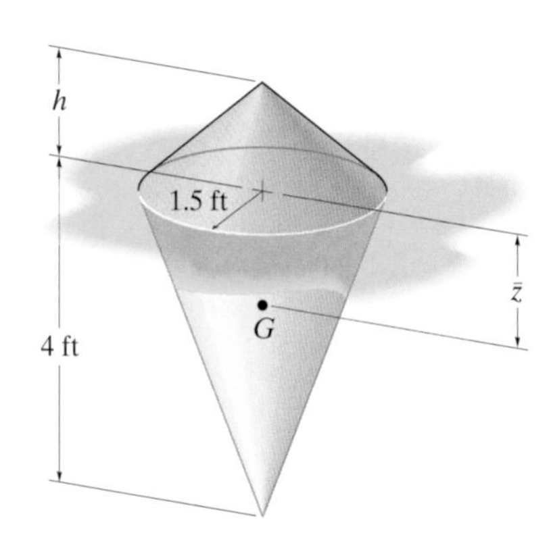

Probs. 9–77/78

9–79. Locate the center of mass $\bar{z}$ of the forked lever, which is made from a homogeneous material and has the dimensions shown.

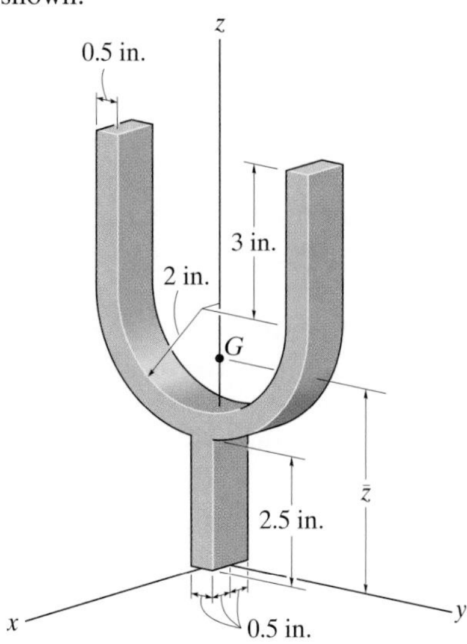

Prob. 9–79

***9–80.** A triangular plate made of homogeneous material has a constant thickness which is very small. If it is folded over as shown, determine the location $\bar{y}$ of the plate's center of gravity G.

9–81. A triangular plate made of homogeneous material has a constant thickness which is very small. If it is folded over as shown, determine the location $\bar{z}$ of the plate's center of gravity G.

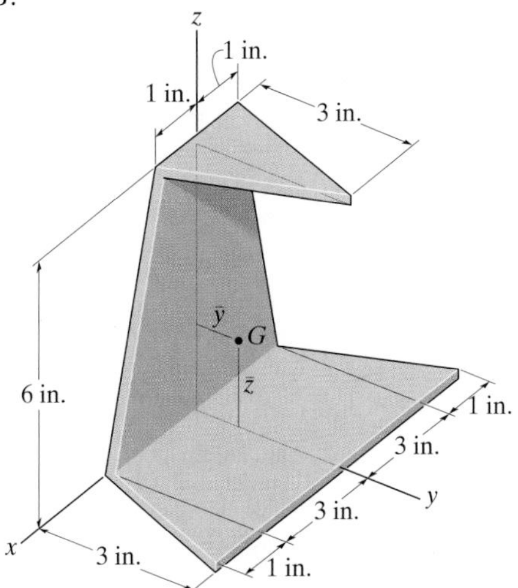

Probs. 9–80/81

9–82. Each of the three homogeneous plates welded to the rod has a density of 6 Mg/m^3 and a thickness of 10 mm. Determine the length l of plate C and the angle of placement, θ, so that the center of mass of the assembly lies on the y axis. Plates A and B lie in the x–y and z–y planes, respectively.

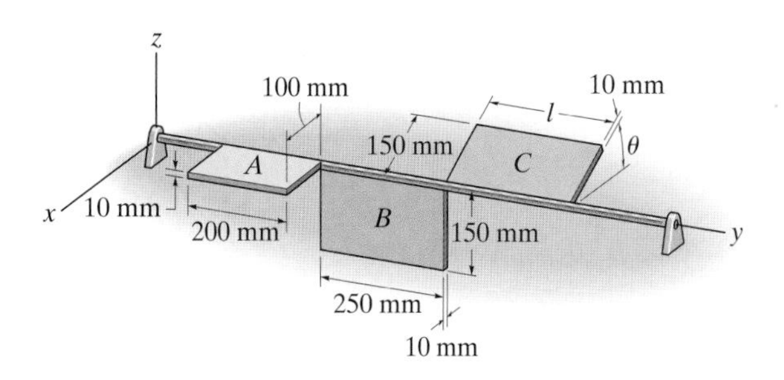

Prob. 9–82

9–83. The assembly consists of a 20-in. wooden dowel rod and a tight-fitting steel collar. Determine the distance $\bar{x}$ to its center of gravity if the specific weights of the materials are $\gamma_w = 150\ \text{lb/ft}^3$ and $\gamma_{st} = 490\ \text{lb/ft}^3$. The radii of the dowel and collar are shown.

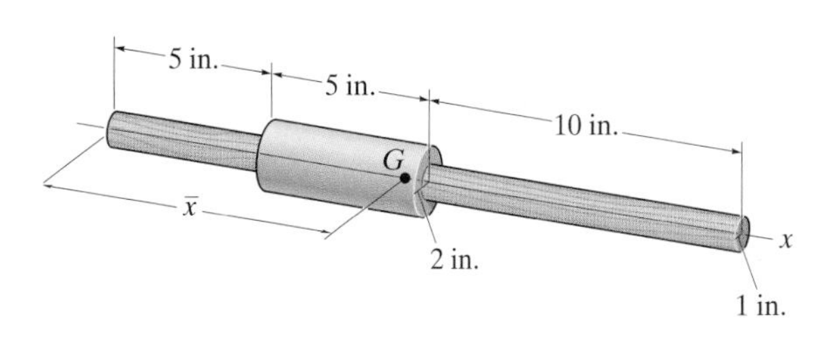

Prob. 9–83

*9.4 Theorems of Pappus and Guldinus

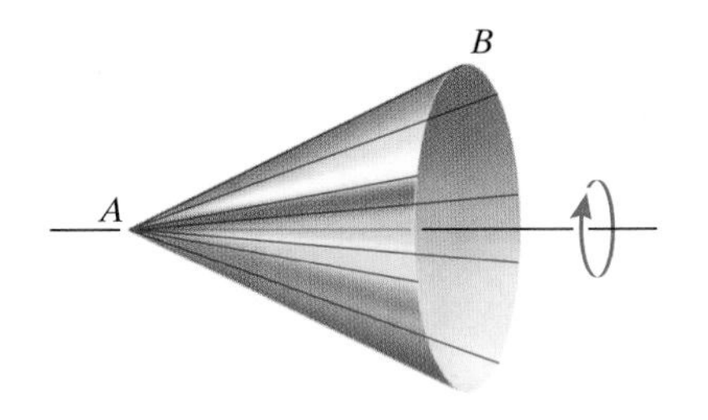

Fig. 9–20

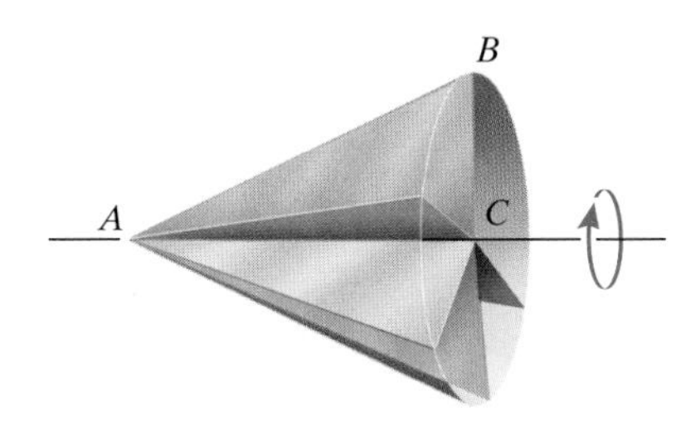

Fig. 9–21

The two *theorems of Pappus and Guldinus* are used to find the surface area and volume of any object of revolution. They were first developed by Pappus of Alexandria during the third century A.D. and then restated at a later time by the Swiss mathematician Paul Guldin or Guldinus (1577–1643).

A *surface area of revolution* is generated by revolving a *plane curve* about a nonintersecting fixed axis in the plane of the curve; whereas a *volume of revolution* is generated by revolving a *plane area* about a nonintersecting fixed axis in the plane of the area. For example, if the *line* AB shown in Fig. 9–20 is rotated about a fixed axis, it generates the *surface area* of a cone (less the area of the base); if the triangular *area* ABC shown in Fig. 9–21 is rotated about the axis, it generates the *volume* of a cone.

The statements and proofs of the theorems of Pappus and Guldinus follow. The proofs require that the generating curves and areas do *not* cross the axis about which they are rotated. If they did, the two sections on either side of the axis would generate areas or volumes having opposite signs and hence cancel each other.

Surface Area. *The area of a surface of revolution equals the product of the length of the generating curve and the distance traveled by the centroid of the curve in generating the surface area.*

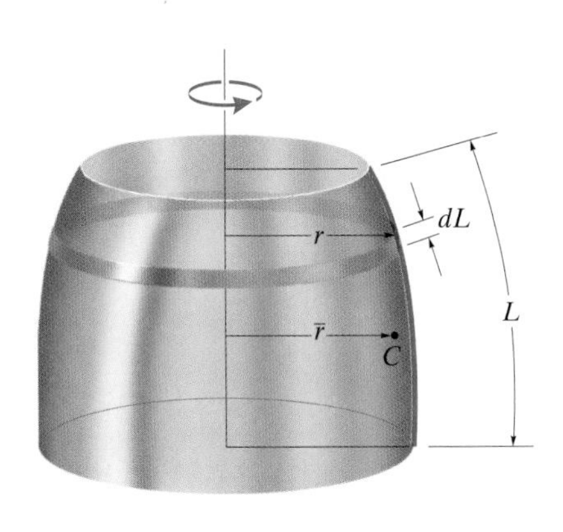

Fig. 9–22

Proof. When a differential length dL of the curve shown in Fig. 9–22 is revolved about an axis through a distance $2\pi r$, it generates a ring having a surface area $dA = 2\pi r\, dL$. The entire surface area, generated by revolving the entire curve about the axis, is therefore $A = 2\pi \int_L r\, dL$. This equation may be simplified, however, by noting that the location r of the centroid for the line of total length L can be determined from an equation having the form of Eqs. 9–7, namely, $\int_L r\, dL = \bar{r}L$. Thus, the total surface area becomes $A = 2\pi\bar{r}L$. In general, though, if the line does not undergo a complete revolution, then,

$$A = \theta \bar{r} L \tag{9–9}$$

where

A = surface area of revolution

θ = angle of revolution measured in radians, $\theta \leq 2\pi$

$\bar{r}$ = perpendicular distance from the axis of revolution to the centroid of the generating curve

L = length of the generating curve

Volume. *The volume of a body of revolution equals the product of the generating area and the distance traveled by the centroid of the area in generating the volume.*

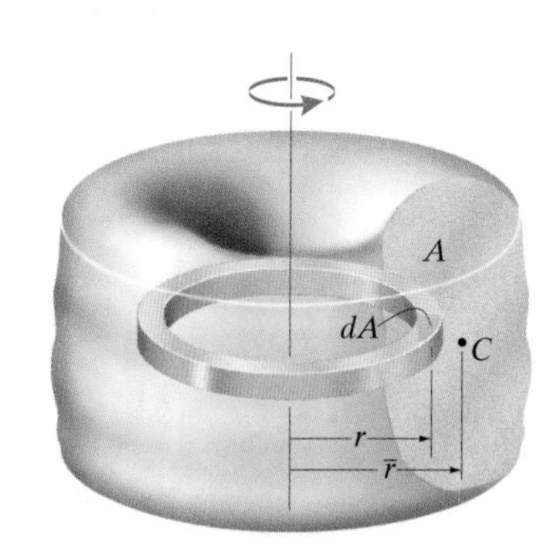

Fig. 9–23

Proof. When the differential area dA shown in Fig. 9–23 is revolved about an axis through a distance $2\pi r$, it generates a ring having a volume $dV = 2\pi r\, dA$. The entire volume, generated by revolving A about the axis, is therefore $V = 2\pi \int_V r\, dA$. Here the integral can be eliminated by using an equation analogous to Eqs. 9–6, $\int_V r\, dA = \bar{r}A$, where $\bar{r}$ locates the centroid C of the generating area A. The volume becomes $V = 2\pi\bar{r}A$. In general, though,

$$\boxed{V = \theta\bar{r}A} \qquad (9\text{–}10)$$

where

V = volume of revolution

θ = angle of revolution measured in radians, $\theta \leq 2\pi$

$\bar{r}$ = perpendicular distance from the axis of revolution to the centroid of the generating area

A = generating area

The surface area and the amount of water that can be stored in this water tank can be determined by using the theorems of Pappus and Guldinus.

Composite Shapes. We may also apply the above two theorems to lines or areas that may be composed of a series of composite parts. In this case the total surface area or volume generated is the addition of the surface areas or volumes generated by each of the composite parts. Since each part undergoes the *same* angle of revolution, θ, and the distance from the axis of revolution to the centroid of each composite part is $\tilde{r}$, then

$$A = \theta\Sigma(\tilde{r}L) \qquad (9\text{–}11)$$

and

$$V = \theta\Sigma(\tilde{r}A) \qquad (9\text{–}12)$$

Application of the above theorems is illustrated numerically in the following example.

The amount of roofing material used on this storage building can be estimated by using the first theorem of Pappus and Guldinus to determine its surface area.

EXAMPLE 9.12

Show that the surface area of a sphere is $A = 4\pi R^2$ and its volume is $V = \frac{4}{3}\pi R^3$.

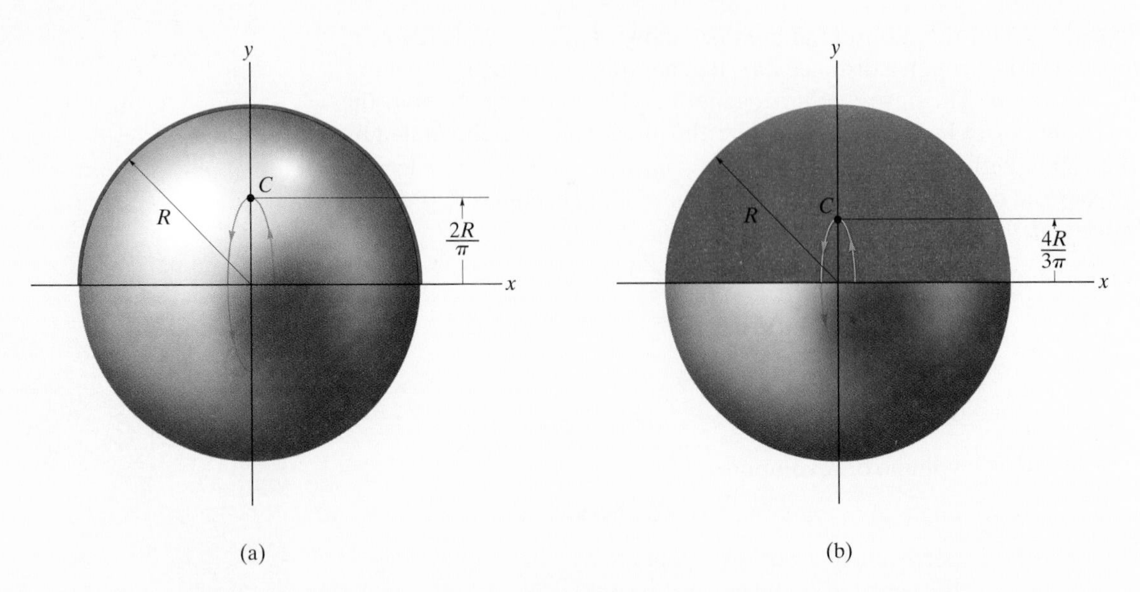

Fig. 9–24

SOLUTION

Surface Area. The surface area of the sphere in Fig. 9–24*a* is generated by rotating a semicircular *arc* about the x axis. Using the table on the inside back cover, it is seen that the centroid of this arc is located at a distance $\bar{r} = 2R/\pi$ from the x axis of rotation. Since the centroid moves through an angle of $\theta = 2\pi$ rad in generating the sphere, then applying Eq. 9–9 we have

$$A = \theta \bar{r} L; \qquad A = 2\pi\left(\frac{2R}{\pi}\right)\pi R = 4\pi R^2 \qquad \textit{Ans.}$$

Volume. The volume of the sphere is generated by rotating the semicircular *area* in Fig. 9–24*b* about the x axis. Using the table on the inside back cover to locate the centroid of the area, i.e., $\bar{r} = 4R/3\pi$, and applying Eq. 9–10, we have

$$V = \theta \bar{r} A; \qquad V = 2\pi\left(\frac{4R}{3\pi}\right)\left(\frac{1}{2}\pi R^2\right) = \frac{4}{3}\pi R^3 \qquad \textit{Ans.}$$

PROBLEMS

*9–84. Determine the surface area and the volume of the ring formed by rotating the square about the vertical axis.

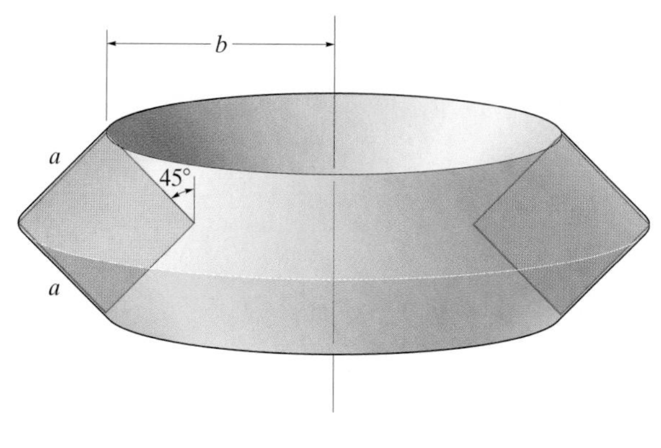

Prob. 9–84

9–85. The anchor ring is made of steel having a specific weight of $\gamma_{st} = 490\ \text{lb/ft}^3$. Determine the surface area of the ring. The cross section is circular as shown.

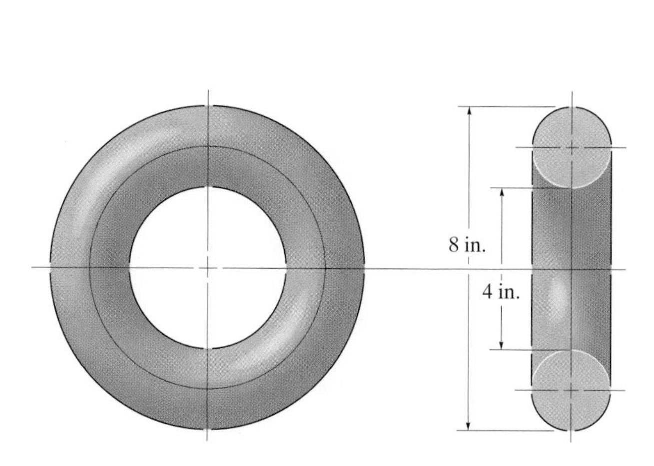

Prob. 9–85

9–86. Using integration, determine both the area and the distance $\bar{y}$ to the centroid of the shaded area. Then using the second theorem of Pappus–Guldinus, determine the volume of the solid generated by revolving the shaded area about the x axis.

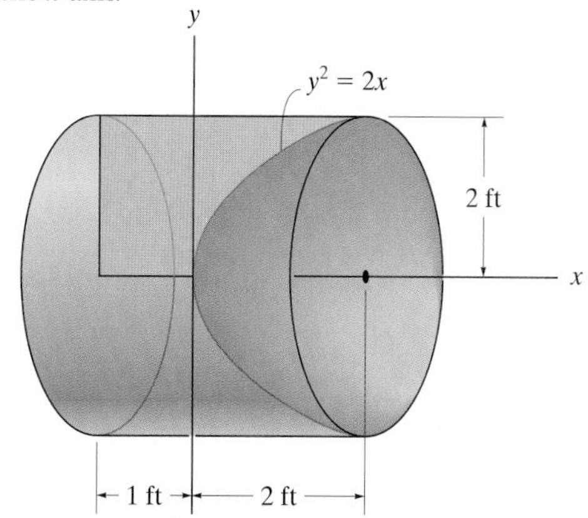

Prob. 9–86

9–87. The grain bin of the type shown is manufactured by Grain Systems, Inc. Determine the required square footage of the sheet metal needed to form it, and also the maximum storage capacity (volume) within it.

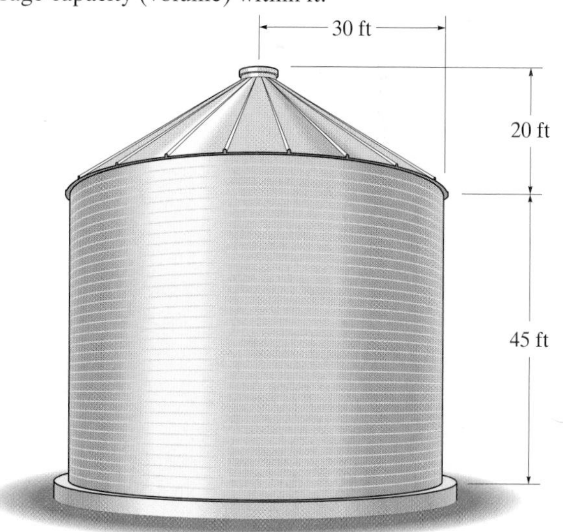

Prob. 9–87

***9–88.** Determine the surface area and the volume of the conical solid.

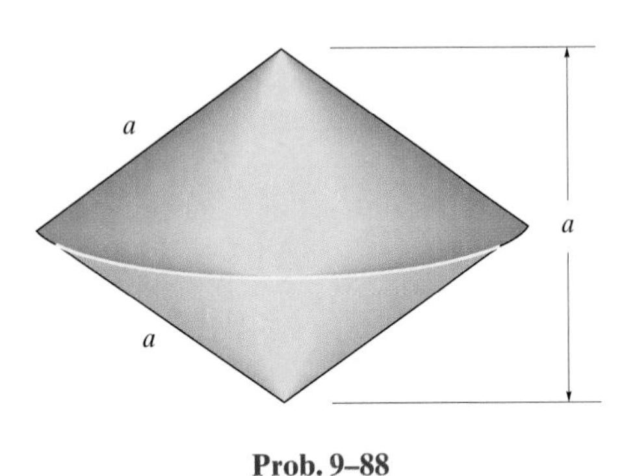

Prob. 9–88

9–89. Sand is piled between two walls as shown. Assume the pile to be a quarter section of a cone and that 26 percent of this volume is voids (air space). Use the second theorem of Pappus–Guldinus to determine the volume of sand.

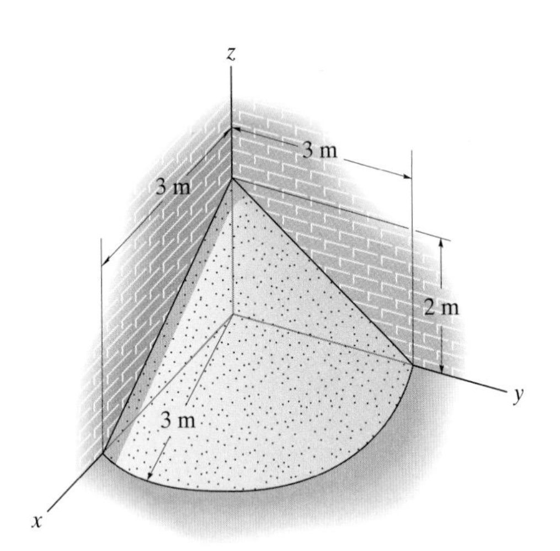

Prob. 9–89

9–90. The *rim* of a flywheel has the cross section *A–A* shown. Determine the volume of material needed for its construction.

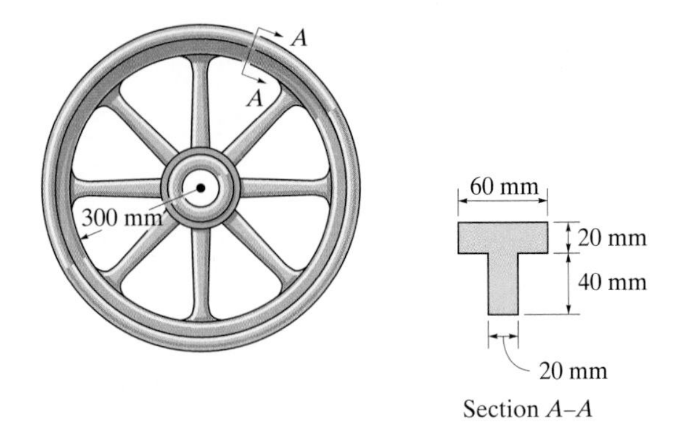

Prob. 9–90

9–91. The Gates Manufacturing Co. produces pulley wheels such as the one shown. Determine the weight of the wheel if it is made from steel having a specific weight of 490 lb/ft^3.

***9–92.** The Gates Manufacturing Co. produces pulley wheels such as the one shown. Determine the total surface area of the wheel in order to estimate the amount of paint needed to protect its surface from rust.

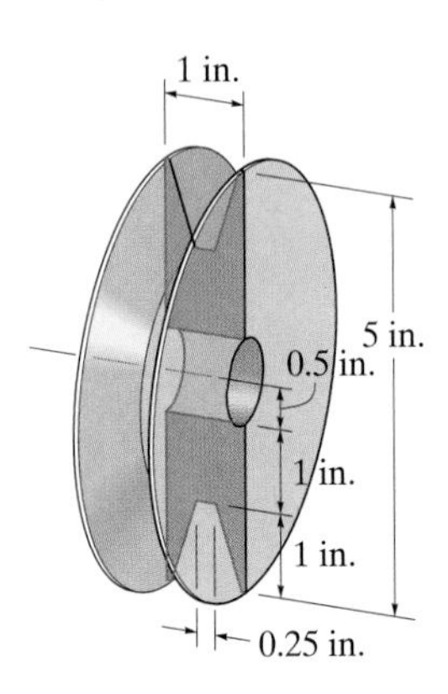

Probs. 9–91/92

9–93. Determine the volume of material needed to make the casting.

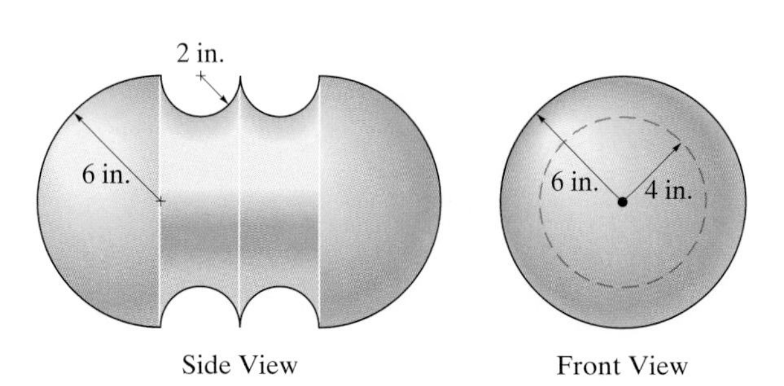

Prob. 9–93

9–95. Determine the surface area of the tank, which consists of a cylinder and hemispherical cap.

***9–96.** Determine the volume of the tank, which consists of a cylinder and hemispherical cap.

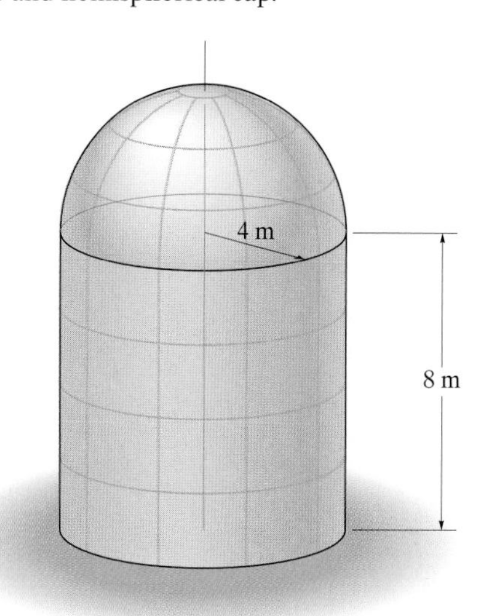

Probs. 9–95/96

9–94. A circular sea wall is made of concrete. Determine the total weight of the wall if the concrete has a specific weight of $\gamma_c = 150\ \text{lb/ft}^3$.

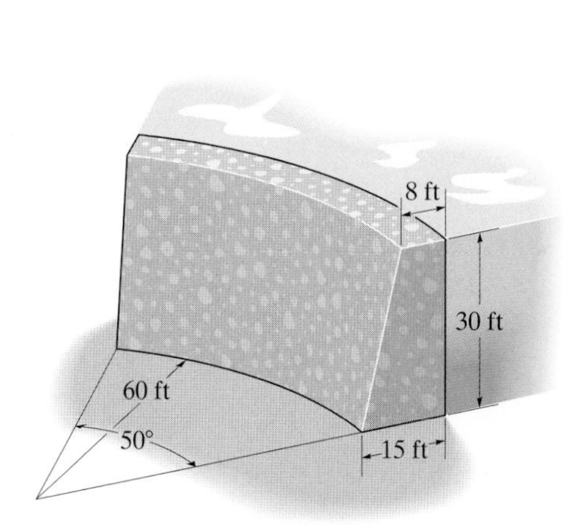

Prob. 9–94

9–97. Determine the surface area of the silo which consists of a cylinder and hemispherical cap. Neglect the thickness of the plates.

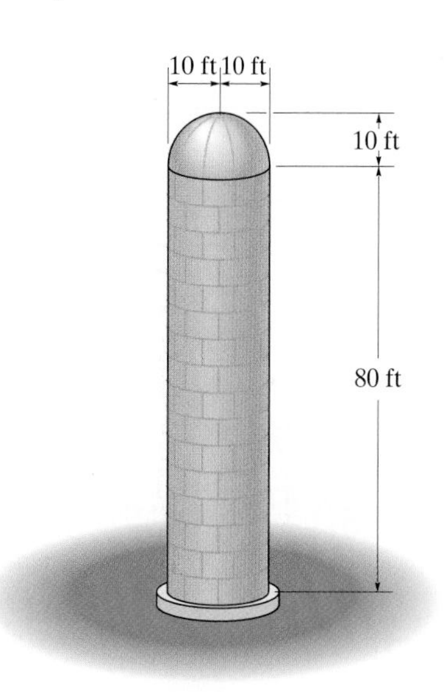

Prob. 9–97

9–98. Determine the volume of the silo which consists of a cylinder and hemispherical cap. Neglect the thickness of the plates.

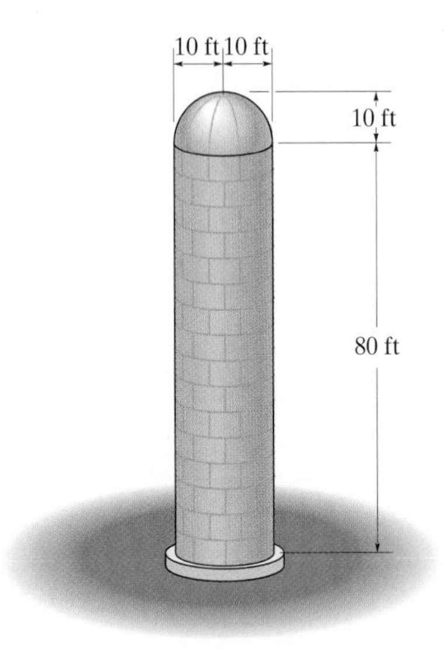

Prob. 9–98

9–99. The process tank is used to store liquids during manufacturing. Estimate both the volume of the tank and its surface area. The tank has a flat top and the plates from which the tank is made have negligible thickness.

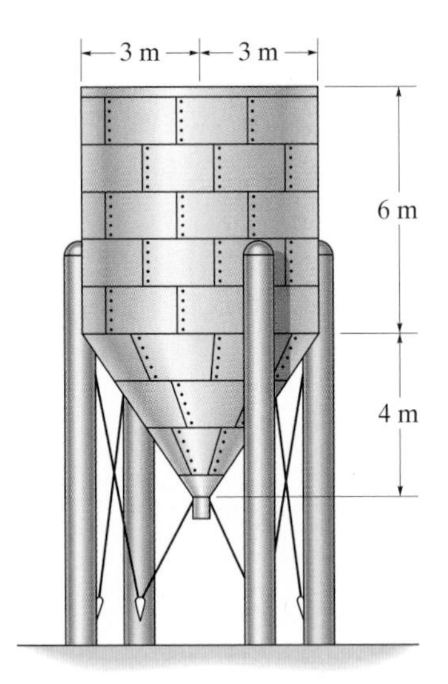

Prob. 9–99

***9–100.** Determine the height h to which liquid should be poured into the cup so that it contacts half the surface area on the inside of the cup. Neglect the cup's thickness for the calculation.

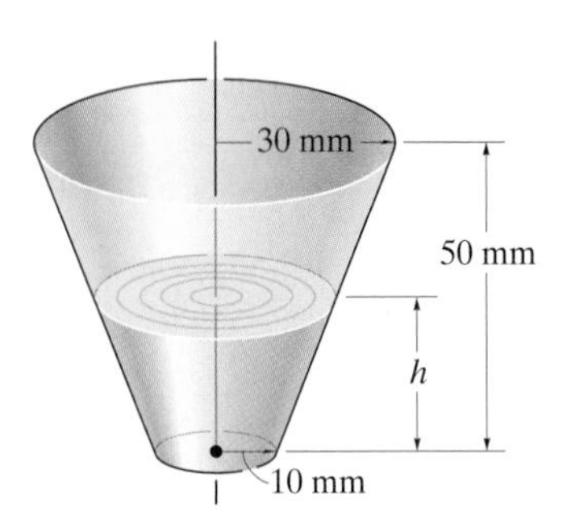

Prob. 9–100

9–101. Using integration, compute both the area and the centroidal distance $\bar{x}$ of the shaded region. Then, using the second theorem of Pappus–Guldinus, compute the volume of the solid generated by revolving the shaded area about the aa axis.

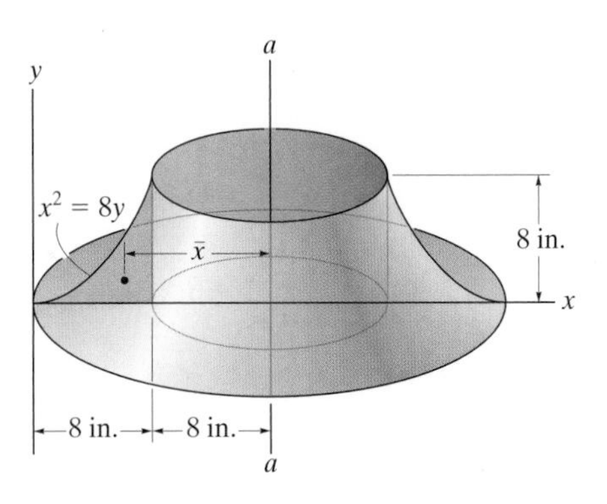

Prob. 9–101

9–102. Using integration, determine the area and the centroidal distance $\bar{y}$ of the shaded area. Then, using the second theorem of Pappus–Guldinus, determine the volume of a solid formed by revolving the area about the x axis.

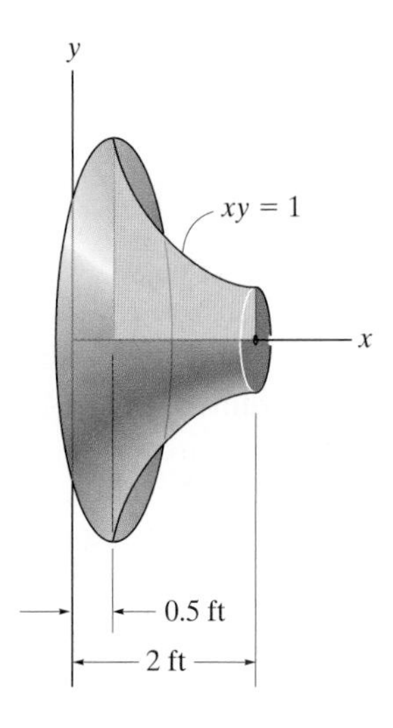

Prob. 9–102

9–103. Determine the surface area of the roof of the structure if it is formed by rotating the parabola about the y axis.

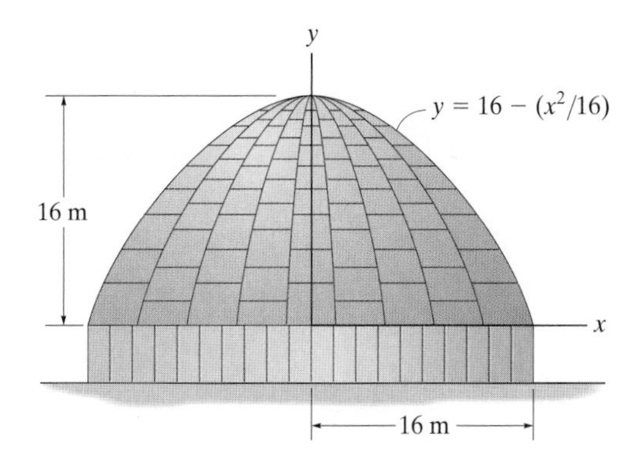

Prob. 9–103

***9–104.** The suspension bunker is made from plates which are curved to the natural shape which a completely flexible membrane would take if subjected to a full load of coal. This curve may be approximated by a parabola, $y = 0.2x^2$. Determine the weight of coal which the bunker would contain when completely filled. Coal has a specific weight of $\gamma = 50\ \text{lb/ft}^3$, and assume there is a 20% loss in volume due to air voids. Solve the problem by integration to determine the cross-sectional area of ABC; then use the second theorem of Pappus–Guldinus to find the volume.

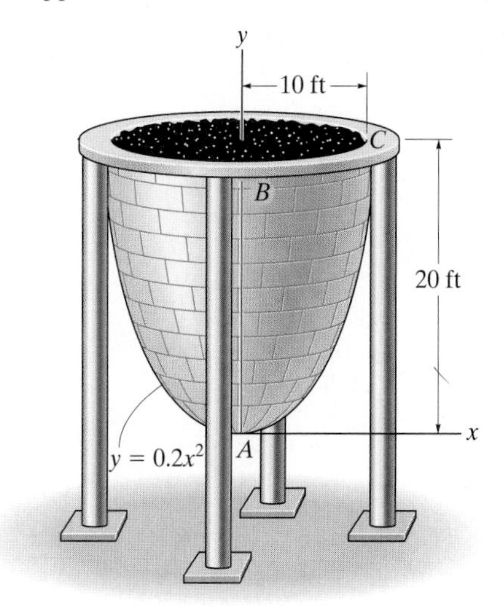

Prob. 9–104

9–105. Determine the interior surface area of the brake piston. It consists of a full circular part. Its cross section is shown in the figure.

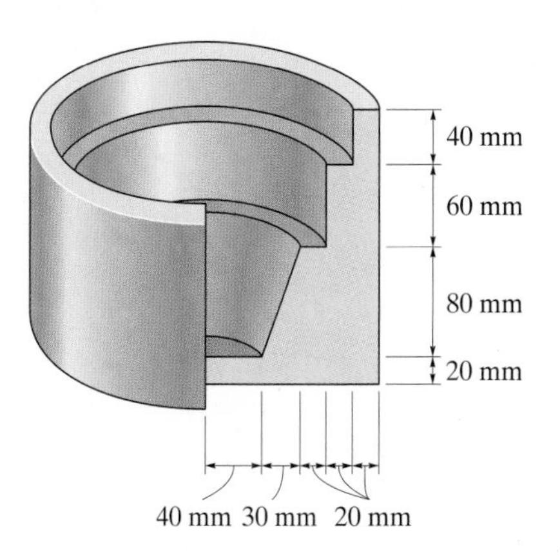

Prob. 9–105

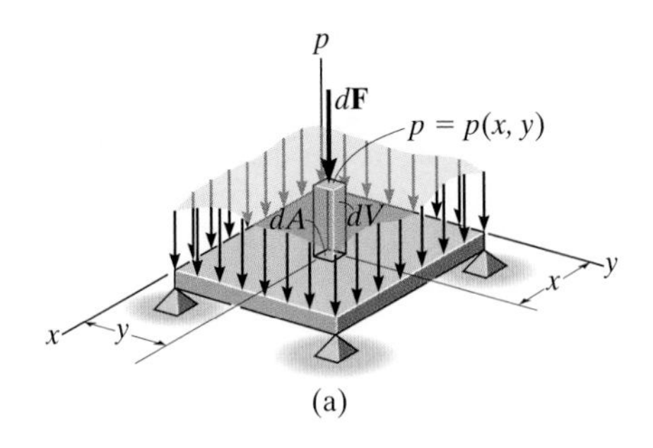

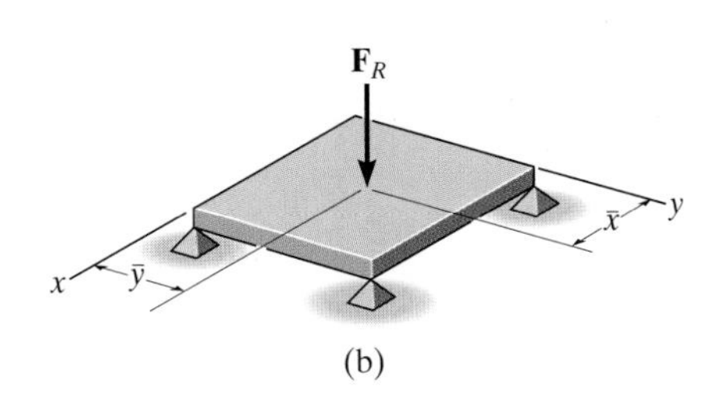

Fig. 9–25

*9.5 Resultant of a General Distributed Loading

In Sec. 4.10, we discussed the method used to simplify a two-dimensional distributed loading to a single resultant force acting at a specific point. In this section we will generalize this method to include surfaces that have an arbitrary shape and are subjected to a variable load distribution. As a specific application, in the next section, we will find the resultant loading acting on the surface of a body that is submerged in a fluid.

Pressure Distribution over a Surface. Consider the flat plate shown in Fig. 9–25*a*, which is subjected to the loading function $p = p(x, y)$ Pa, where Pa (pascal) $= 1\ \text{N/m}^2$. Knowing this function, we can determine the force $d\mathbf{F}$ acting on the differential area $dA\ \text{m}^2$ of the plate, located at the arbitrary point (x, y). This force magnitude is simply $dF = [p(x, y)\ \text{N/m}^2](dA\ \text{m}^2) = [p(x, y)\ dA]\ \text{N}$. The entire loading on the plate is therefore represented as a system of *parallel forces* infinite in number and each acting on a separate differential area dA. This system will now be simplified to a single resultant force $\mathbf{F}_R$ acting through a unique point $(\bar{x}, \bar{y})$ on the plate, Fig. 9–25*b*.

Magnitude of Resultant Force. To determine the *magnitude* of $\mathbf{F}_R$, it is necessary to sum each of the differential forces $d\mathbf{F}$ acting over the plate's *entire surface area A*. This sum may be expressed mathematically as an integral:

$$F_R = \Sigma F; \qquad F_R = \int_A p(x, y)\, dA = \int_V dV \qquad (9\text{–}13)$$

Here $p(x, y)\, dA = dV$, the colored differential *volume element* shown in Fig. 9–25*a*. Therefore, the result indicates that the *magnitude of the resultant force is equal to the total volume under the distributed-loading diagram.*

Location of Resultant Force. The location $(\bar{x}, \bar{y})$ of $\mathbf{F}_R$ is determined by setting the moments of $\mathbf{F}_R$ equal to the moments of all the forces $d\mathbf{F}$ about the respective y and x axes: From Figs. 9–25*a* and 9–25*b*, using Eq. 9–13, this results in

$$\bar{x} = \frac{\int_A x p(x, y)\, dA}{\int_A p(x, y)\, dA} = \frac{\int_V x\, dV}{\int_V dV} \qquad \bar{y} = \frac{\int_A y p(x, y)\, dA}{\int_A p(x, y)\, dA} = \frac{\int_V y\, dV}{\int_V dV} \qquad (9\text{–}14)$$

Hence, it can be seen that the *line of action of the resultant force passes through the geometric center or centroid of the volume under the distributed-loading diagram.*

*9.6 Fluid Pressure

According to Pascal's law, a fluid at rest creates a pressure p at a point that is the *same* in *all* directions. The magnitude of p, measured as a force per unit area, depends on the specific weight γ or mass density ρ of the fluid and the depth z of the point from the fluid surface.* The relationship can be expressed mathematically as

$$p = \gamma z = \rho g z \tag{9–15}$$

where g is the acceleration due to gravity. Equation 9–15 is valid only for fluids that are assumed *incompressible*, as in the case of most liquids. Gases are compressible fluids, and since their density changes significantly with both pressure and temperature, Eq. 9–15 cannot be used.

To illustrate how Eq. 9–15 is applied, consider the submerged plate shown in Fig. 9–26. Three points on the plate have been specified. Since point B is at depth z_1 from the liquid surface, the *pressure* at this point has a magnitude $p_1 = \gamma z_1$. Likewise, points C and D are both at depth z_2; hence, $p_2 = \gamma z_2$. In all cases, the pressure acts *normal* to the surface area dA located at the specified point. Using Eq. 9–15 and the results of Sec. 9.5, it is possible to determine the resultant force caused by a liquid pressure distribution and specify its location on the surface of a submerged plate. Three different shapes of plates will now be considered.

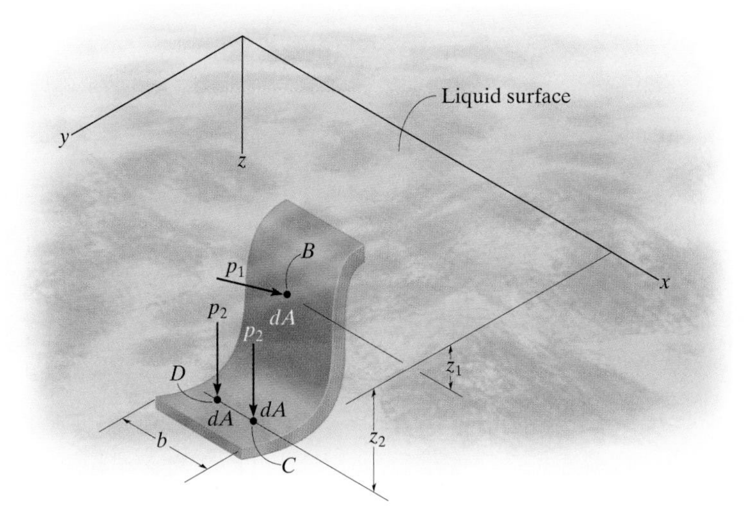

Fig. 9–26

*In particular, for water $\gamma = 62.4\ \text{lb/ft}^3$, or $\gamma = \rho g = 9810\ \text{N/m}^3$ since $\rho = 1000\ \text{kg/m}^3$ and $g = 9.81\ \text{m/s}^2$.

Flat Plate of Constant Width. A flat rectangular plate of constant width, which is submerged in a liquid having a specific weight γ, is shown in Fig. 9–27*a*. The plane of the plate makes an angle with the horizontal, such that its top edge is located at a depth z_1 from the liquid surface and its bottom edge is located at a depth z_2. Since pressure varies linearly with depth, Eq. 9–15, the distribution of pressure over the plate's surface is represented by a trapezoidal volume having an intensity of $p_1 = \gamma z_1$ at depth z_1 and $p_2 = \gamma z_2$ at depth z_2. As noted in Sec. 9.5, the magnitude of the *resultant force* $\mathbf{F}_R$ is equal to the *volume* of this loading diagram and $\mathbf{F}_R$ has a *line of action* that passes through the volume's centroid C. Hence, $\mathbf{F}_R$ does *not* act at the centroid of the plate; rather, it acts at point P, called the *center of pressure*.

Since the plate has a *constant width*, the loading distribution may also be viewed in two dimensions, Fig. 9–27*b*. Here the loading intensity is measured as force/length and varies linearly from $w_1 = bp_1 = b\gamma z_1$ to $w_2 = bp_2 = b\gamma z_2$. The magnitude of $\mathbf{F}_R$ in this case equals the trapezoidal *area*, and $\mathbf{F}_R$ has a *line of action* that passes through the area's *centroid* C. For numerical applications, the area and location of the centroid for a trapezoid are tabulated on the inside back cover.

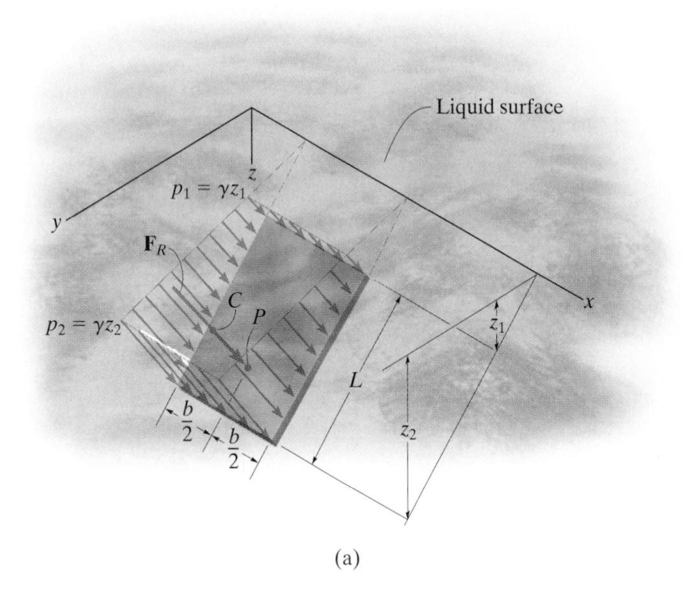

(a)

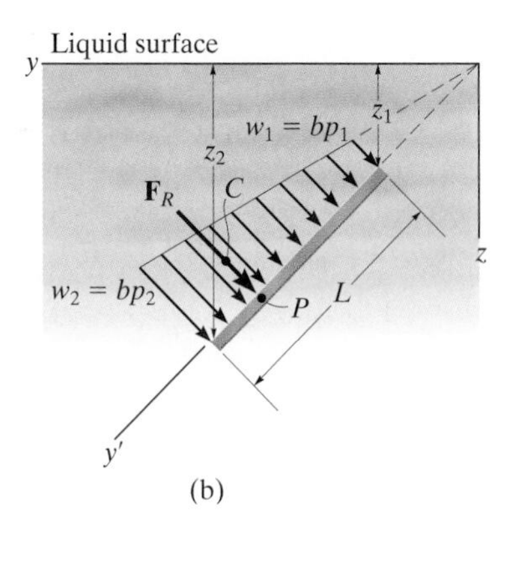

(b)

Fig. 9–27

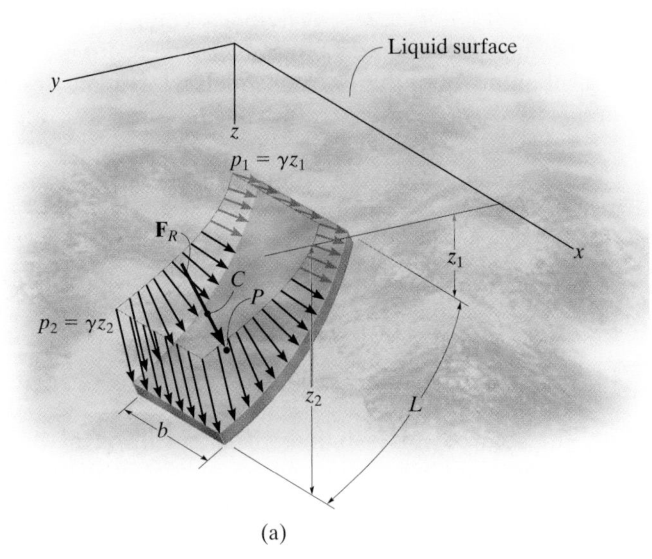

(a)

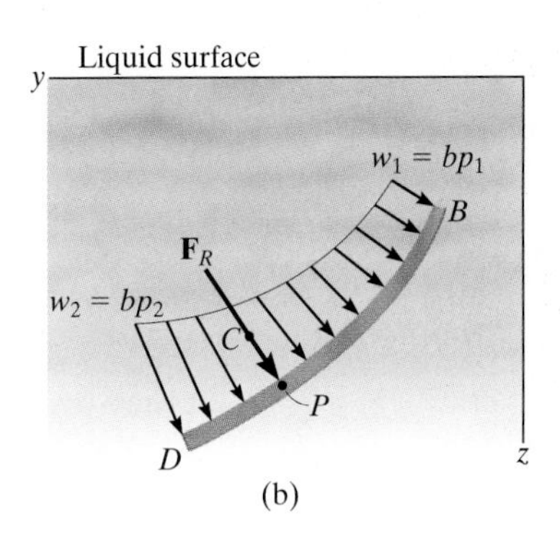

(b)

Fig. 9–28

Curved Plate of Constant Width.

When the submerged plate is curved, the pressure acting normal to the plate continually changes direction, and therefore calculation of the magnitude of $\mathbf{F}_R$ and its location P is more difficult than for a flat plate. Three- and two-dimensional views of the loading distribution are shown in Figs. 9–28*a* and 9–28*b*, respectively. Here integration can be used to determine both F_R and the location of the centroid C or center of pressure P.

A simpler method exists, however, for calculating the magnitude of $\mathbf{F}_R$ and its location along a curved (or flat) plate having a *constant width*. This method requires separate calculations for the horizontal and vertical *components* of $\mathbf{F}_R$. For example, the distributed loading acting on the curved plate DB in Fig. 9–28*b* can be represented by the *equivalent loading* shown in Fig. 9–29. Here the plate supports the weight of liquid W_f contained within the block BDA. This force has a magnitude $W_f = (\gamma b)(\text{area}_{BDA})$ and acts through the centroid of BDA. In addition, there are the pressure distributions caused by the liquid acting along the vertical and horizontal sides of the block. Along the vertical side AD, the force $\mathbf{F}_{AD}$ has a magnitude that equals the area under the trapezoid and acts through the centroid C_{AD} of this area. The distributed loading along the horizontal side AB is constant since all points lying in this plane are at the same depth from the surface of the liquid. The magnitude of $\mathbf{F}_{AB}$ is simply the area of the rectangle. This force acts through the area's centroid C_{AB} or the midpoint of AB. Summing the three forces in Fig. 9–29 yields $\mathbf{F}_R = \Sigma\mathbf{F} = \mathbf{F}_{AD} + \mathbf{F}_{AB} + \mathbf{W}_f$, which is shown in Fig. 9–28. Finally, the location of the center of pressure P on the plate is determined by applying the equation $M_{R_O} = \Sigma M_O$, which states that the moment of the resultant force about a convenient reference point O, such as D or B, in Fig. 9–28, is equal to the sum of the moments of the three forces in Fig. 9–29 about the same point.

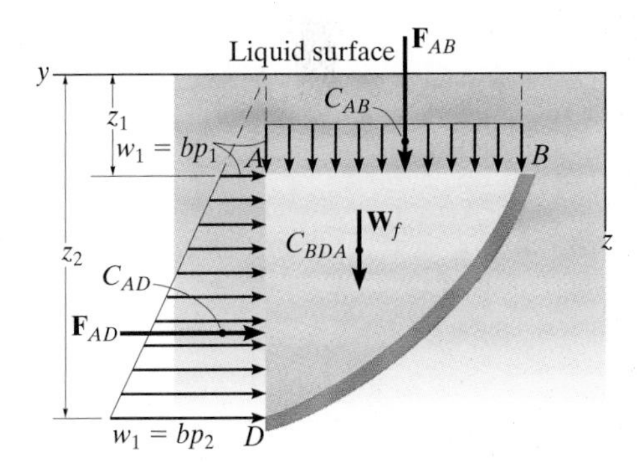

Fig. 9–29

The resultant force of the water and its location on the elliptical back plate of this tank truck must be determined by integration.

Flat Plate of Variable Width. The pressure distribution acting on the surface of a submerged plate having a variable width is shown in Fig. 9–30. The resultant force of this loading equals the volume described by the plate area as its base and linearly varying pressure distribution as its altitude. The shaded element shown in Fig. 9–30 may be used if integration is chosen to determine this volume. The element consists of a rectangular strip of area $dA = x\,dy'$ located at a depth z below the liquid surface. Since a uniform pressure $p = \gamma z$ (force/area) acts on dA, the magnitude of the differential force $d\mathbf{F}$ is equal to $dF = dV = p\,dA = \gamma z(x\,dy')$. Integrating over the entire volume yields Eq. 9–13; i.e.,

$$F_R = \int_A p\,dA = \int_V dV = V$$

From Eq. 9–14, the centroid of V defines the point through which $\mathbf{F}_R$ acts. The center of pressure, which lies on the surface of the plate just below C, has coordinates $P(\bar{x}, \bar{y}')$ defined by the equations

$$\bar{x} = \frac{\int_V \tilde{x}\,dV}{\int_V dV} \qquad \bar{y}' = \frac{\int_V \tilde{y}'\,dV}{\int_V dV}$$

This point should *not* be mistaken for the centroid of the plate's *area*.

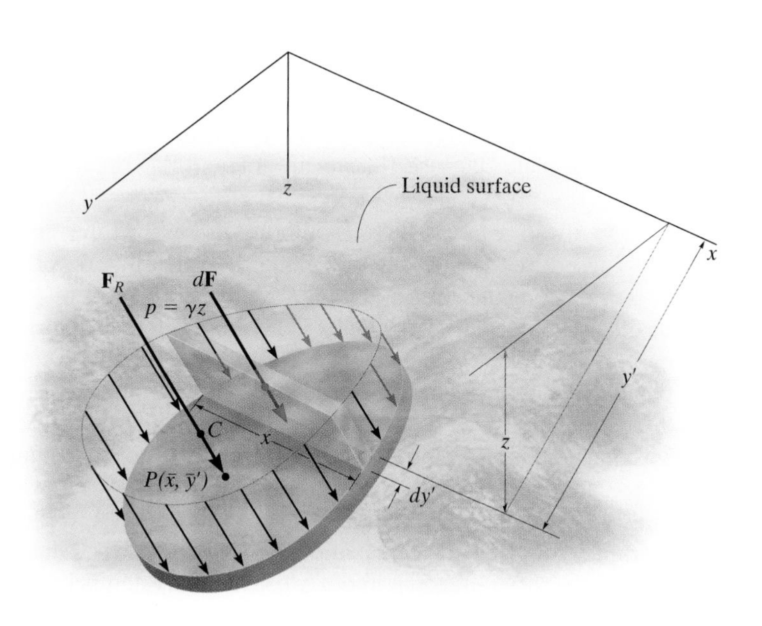

Fig. 9–30

EXAMPLE 9.13

Determine the magnitude and location of the resultant hydrostatic force acting on the submerged rectangular plate AB shown in Fig. 9–31a. The plate has a width of 1.5 m; $\rho_w = 1000 \text{ kg/m}^3$.

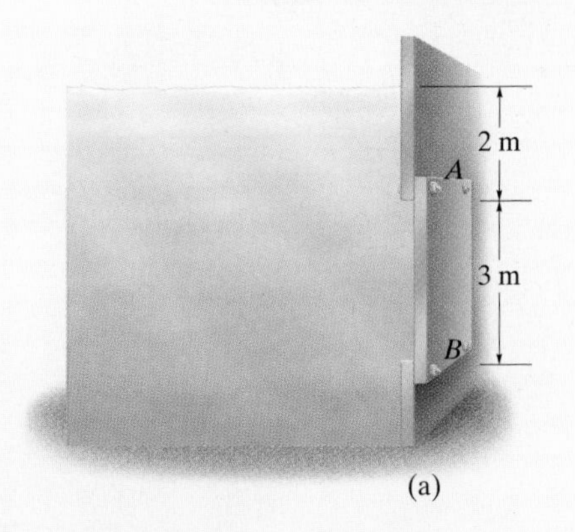

(a)

SOLUTION

The water pressures at depths A and B are

$$p_A = \rho_w g z_A = (1000 \text{ kg/m}^3)(9.81 \text{ m/s}^2)(2 \text{ m}) = 19.62 \text{ kPa}$$
$$p_B = \rho_w g z_B = (1000 \text{ kg/m}^3)(9.81 \text{ m/s}^2)(5 \text{ m}) = 49.05 \text{ kPa}$$

Since the plate has a constant width, the distributed loading can be viewed in two dimensions as shown in Fig. 9–31b. The intensities of the load at A and B are

$$w_A = bp_A = (1.5 \text{ m})(19.62 \text{ kPa}) = 29.43 \text{ kN/m}$$
$$w_B = bp_B = (1.5 \text{ m})(49.05 \text{ kPa}) = 73.58 \text{ kN/m}$$

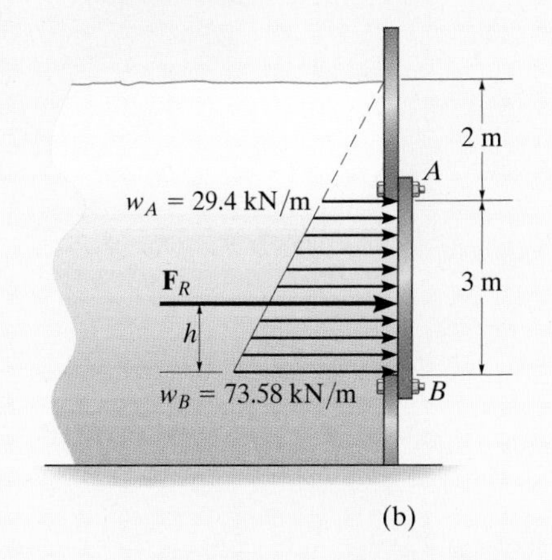

(b)

From the table on the inside back cover, the magnitude of the resultant force $\mathbf{F}_R$ created by the distributed load is

$$F_R = \text{area of trapezoid}$$
$$= \tfrac{1}{2}(3)(29.4 + 73.6) = 154.5 \text{ kN} \qquad \textit{Ans.}$$

This force acts through the centroid of the area,

$$h = \frac{1}{3}\left(\frac{2(29.43) + 73.58}{29.43 + 73.58}\right)(3) = 1.29 \text{ m} \qquad \textit{Ans.}$$

measured upward from B, Fig. 9–31b.

The same results can be obtained by considering two components of $\mathbf{F}_R$ defined by the triangle and rectangle shown in Fig. 9–31c. Each force acts through its associated centroid and has a magnitude of

$$F_{Re} = (29.43 \text{ kN/m})(3 \text{ m}) = 88.3 \text{ kN}$$
$$F_t = \tfrac{1}{2}(44.15 \text{ kN/m})(3 \text{ m}) = 66.2 \text{ kN}$$

Hence,

$$F_R = F_{Re} + F_t = 88.3 + 66.2 = 154.5 \text{ kN} \qquad \textit{Ans.}$$

The location of $\mathbf{F}_R$ is determined by summing moments about B, Fig. 9–31b and c, i.e.,

$$\curvearrowleft+(M_R)_B = \Sigma M_B; \qquad (154.5)h = 88.3(1.5) + 66.2(1)$$
$$h = 1.29 \text{ m} \qquad \textit{Ans.}$$

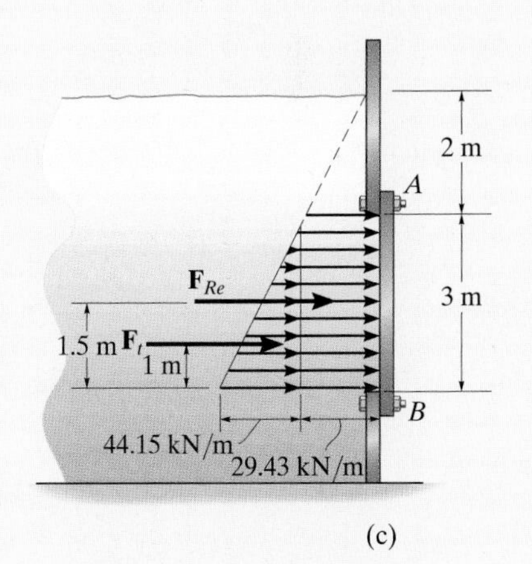

(c)

Fig. 9–31

NOTE: The result of $\mathbf{F}_R$ should give you some idea as to how large hydrostatic forces can be.

EXAMPLE 9.14

Determine the magnitude of the resultant hydrostatic force acting on the surface of a seawall shaped in the form of a parabola as shown in Fig. 9–32*a*. The wall is 5 m long; $\rho_w = 1020\ \text{kg/m}^3$.

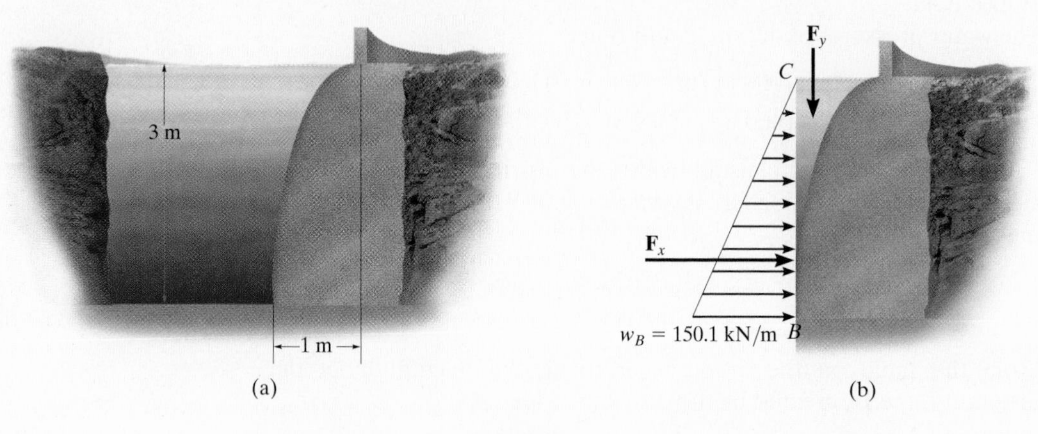

Fig. 9–32

SOLUTION

The horizontal and vertical components of the resultant force will be calculated, Fig. 9–32*b*. Since

$$p_B = \rho_w g z_B = (1020\ \text{kg/m}^3)(9.81\ \text{m/s}^2)(3\ \text{m}) = 30.02\ \text{kPa}$$

then

$$w_B = bp_B = 5\ \text{m}(30.02\ \text{kPa}) = 150.1\ \text{kN/m}$$

Thus,

$$F_x = \tfrac{1}{2}(3\ \text{m})(150.1\ \text{kN/m}) = 225.1\ \text{kN}$$

The area of the parabolic sector *ABC* can be determined using the table on the inside back cover. Hence, the weight of water within this region is

$$F_y = (\rho_w g b)(\text{area}_{ABC})$$
$$= (1020\ \text{kg/m}^3)(9.81\ \text{m/s}^2)(5\ \text{m})\left[\tfrac{1}{3}(1\ \text{m})(3\ \text{m})\right] = 50.0\ \text{kN}$$

The resultant force is therefore

$$F_R = \sqrt{F_x^2 + F_y^2} = \sqrt{(225.1)^2 + (50.0)^2}$$
$$= 231\ \text{kN}$$ *Ans.*

EXAMPLE 9.15

Determine the magnitude and location of the resultant force acting on the triangular end plates of the water trough shown in Fig. 9–33*a*; $\rho_w = 1000 \text{ kg/m}^3$.

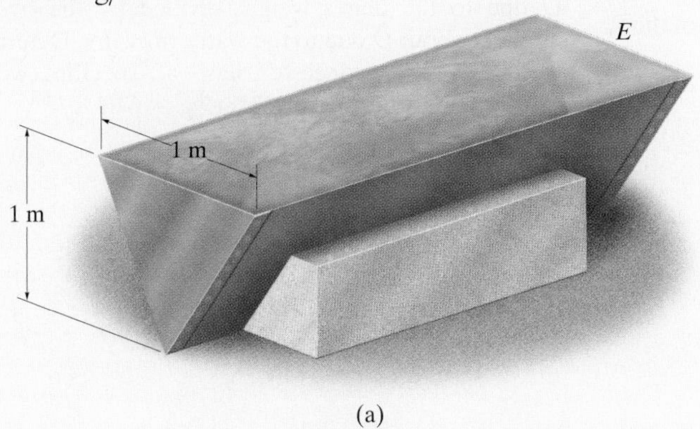

(a)

SOLUTION

The pressure distribution acting on the end plate *E* is shown in Fig. 9–33*b*. The magnitude of the resultant force **F** is equal to the volume of this loading distribution. We will solve the problem by integration. Choosing the differential volume element shown in the figure, we have

$$dF = dV = p\,dA = \rho_w g z(2x\,dz) = 19\,620zx\,dz$$

The equation of line *AB* is

$$x = 0.5(1 - z)$$

Hence, substituting and integrating with respect to z from $z = 0$ to $z = 1$ m yields

$$F = V = \int_V dV = \int_0^1 (19\,620)z[0.5(1 - z)]\,dz$$

$$= 9810\int_0^1 (z - z^2)\,dz = 1635 \text{ N} = 1.64 \text{ kN} \qquad \textit{Ans.}$$

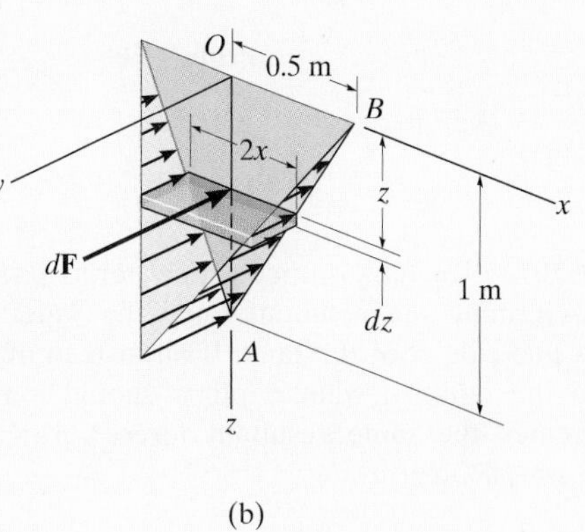

(b)

Fig. 9–33

This resultant passes through the centroid of the volume. Because of symmetry,

$$\bar{x} = 0 \qquad \textit{Ans.}$$

Since $\tilde{z} = z$ for the volume element in Fig. 9–33*b*, then

$$\bar{z} = \frac{\int_V \tilde{z}\,dV}{\int_V dV} = \frac{\int_0^1 z(19\,620)z[0.5(1 - z)]\,dz}{1635} = \frac{9810\int_0^1 (z^2 - z^3)\,dz}{1635}$$

$$= 0.5 \text{ m} \qquad \textit{Ans.}$$

PROBLEMS

9–106. Determine the magnitude of the resultant hydrostatic force acting on the dam and its location, measured from the top surface of the water. The width of the dam is 8 m; $\rho_w = 1.0\ \text{Mg/m}^3$.

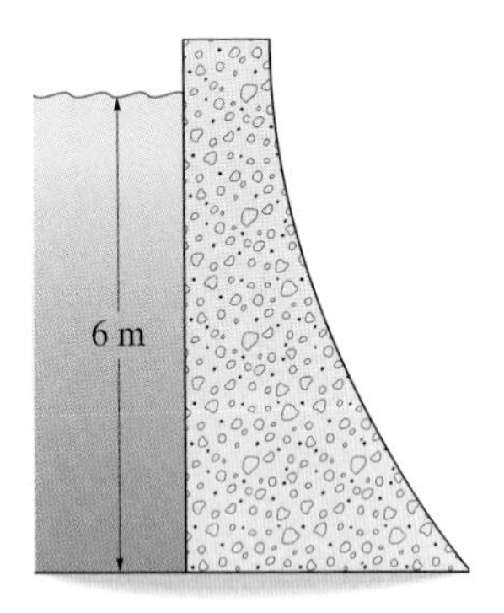

Prob. 9–106

9–107. The tank is filled with water to a depth of $d = 4$ m. Determine the resultant force the water exerts on side A and side B of the tank. If oil instead of water is placed in the tank, to what depth d should it reach so that it creates the same resultant forces? $\rho_o = 900\ \text{kg/m}^3$ and $\rho_w = 1000\ \text{kg/m}^3$.

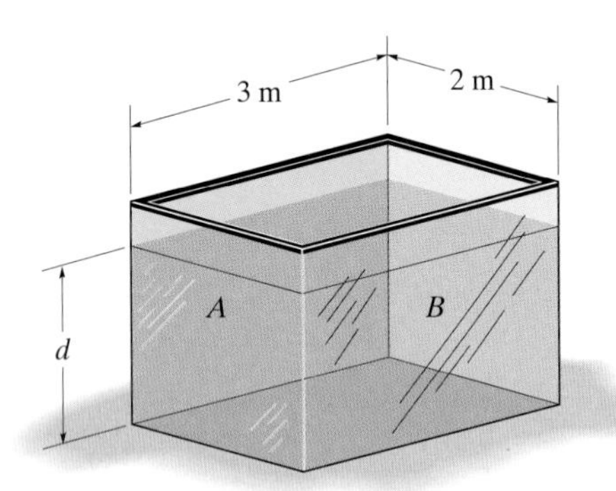

Prob. 9–107

***9–108.** The factor of safety for tipping of the concrete dam is defined as the ratio of the stabilizing moment about O due to the dam's weight divided by the overturning moment about O due to the water pressure. Determine this factor if the concrete has a specific weight of $\gamma_{conc} = 150\ \text{lb/ft}^3$ and for water $\gamma_w = 62.4\ \text{lb/ft}^3$.

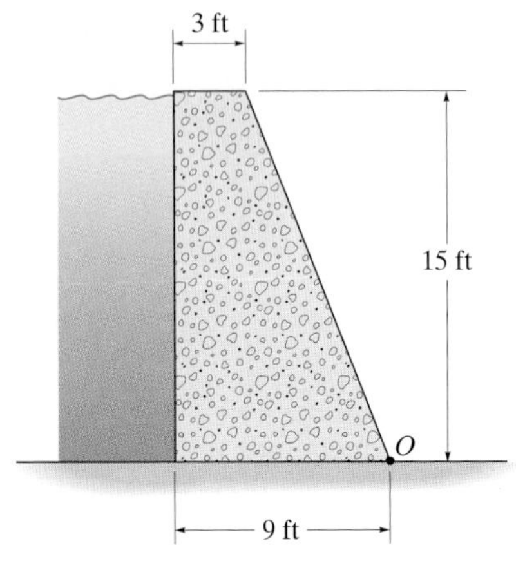

Prob. 9–108

9–109. The concrete "gravity" dam is held in place by its own weight. If the density of concrete is $\rho_c = 2.5\ \text{Mg/m}^3$, and water has a density of $\rho_w = 1.0\ \text{Mg/m}^3$, determine the smallest dimension d that will prevent the dam from overturning about its end A.

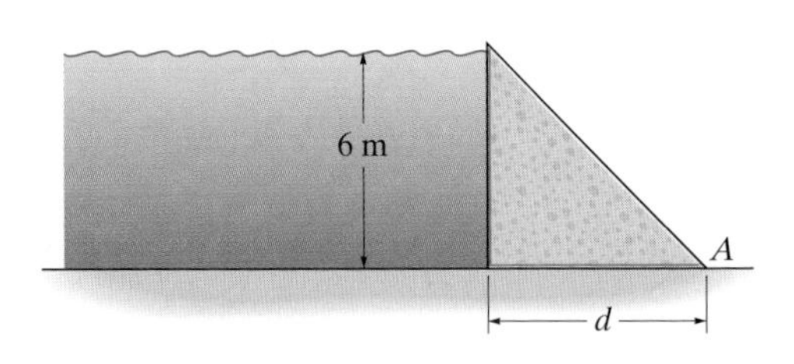

Prob. 9–109

9–110. The concrete dam is designed so that its face AB has a gradual slope into the water as shown. Because of this, the frictional force at the base BD of the dam is increased due to the hydrostatic force of the water acting on the dam. Calculate the hydrostatic force acting on the face AB of the dam. The dam is 60 ft wide. $\gamma_w = 62.4\ \text{lb/ft}^3$.

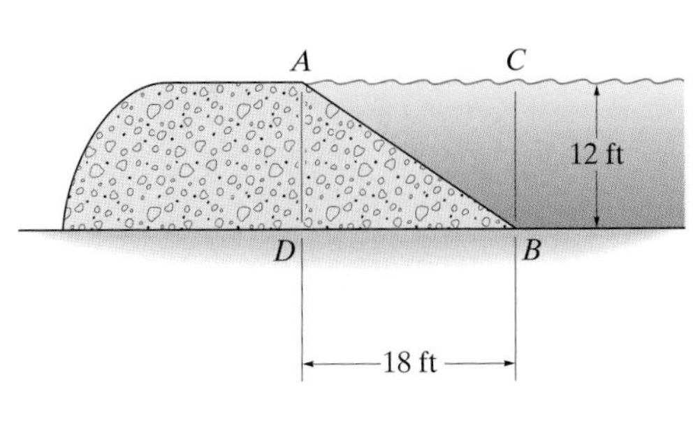

Prob. 9–110

9–111. The symmetric concrete "gravity" dam is held in place by its own weight. If the density of concrete is $\rho_c = 2.5\ \text{Mg/m}^3$, and water has a density of $\rho_w = 1.0\ \text{Mg/m}^3$, determine the smallest distance d at its base that will prevent the dam from overturning about its end A. The dam has a width of 8 m.

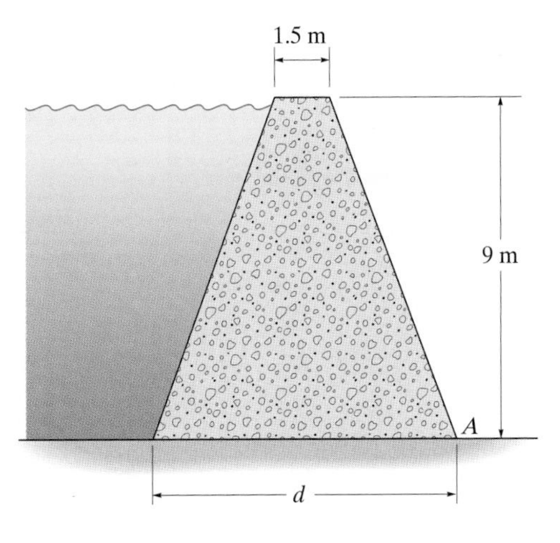

Prob. 9–111

***9–112.** The tank is used to store a liquid having a specific weight of $80\ \text{lb/ft}^3$. If it is filled to the top, determine the magnitude of force the liquid exerts on each of its two sides $ABDC$ and $BDFE$.

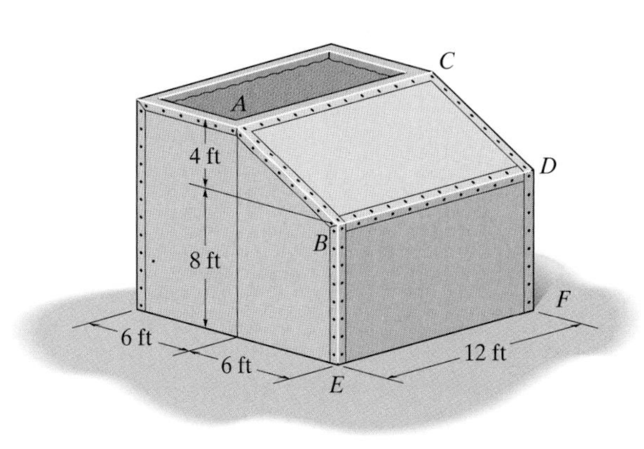

Prob. 9–112

9–113. The 2-m-wide rectangular gate is pinned at its center A and is prevented from rotating by the block at B. Determine the reactions at these supports due to hydrostatic pressure. $\rho_w = 1.0\ \text{Mg/m}^3$.

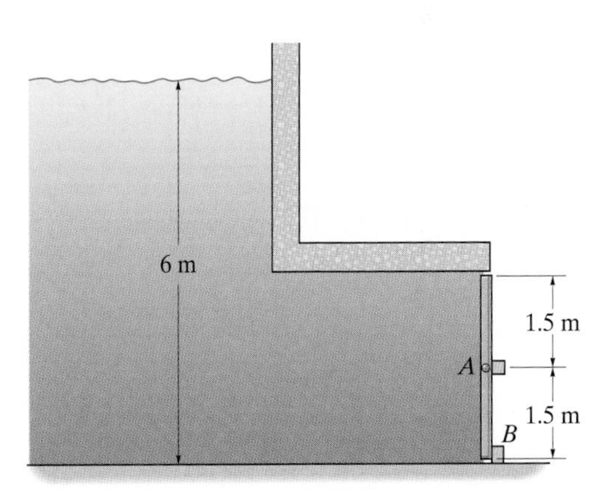

Prob. 9–113

9–114. The gate AB is 8 m wide. Determine the horizontal and vertical components of force acting on the pin at B and the vertical reaction at the smooth support A. $\rho_w = 1.0\ \text{Mg/m}^3$.

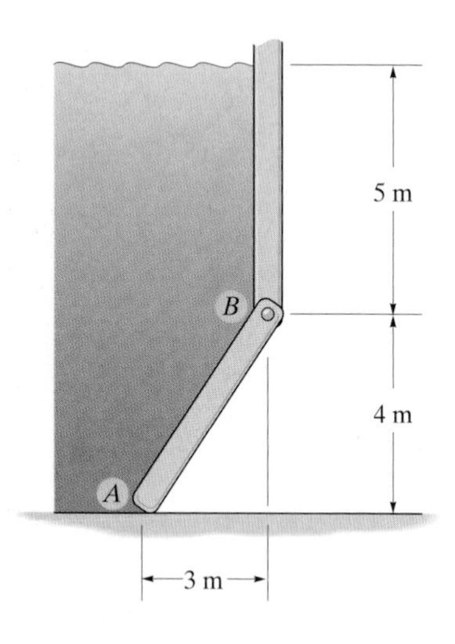

Prob. 9–114

***9–116.** The arched surface AB is shaped in the form of a quarter circle. If it is 8 m long, determine the horizontal and vertical components of the resultant force caused by the water acting on the surface. $\rho_w = 1.0\ \text{Mg/m}^3$.

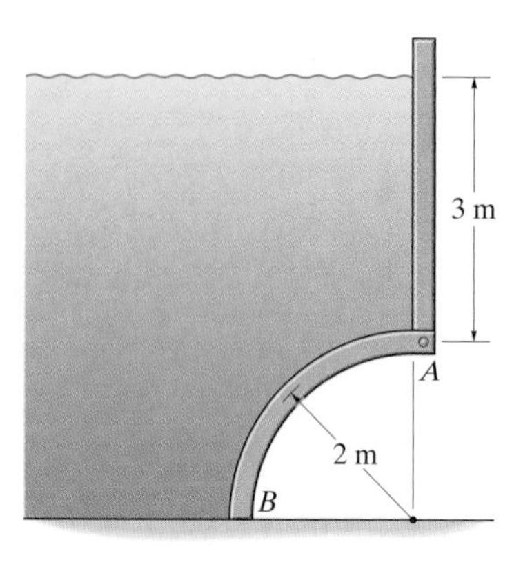

Prob. 9–116

9–115. The storage tank contains oil having a specific weight of $\gamma_o = 56\ \text{lb/ft}^3$. If the tank is 6 ft wide, calculate the resultant force acting on the inclined side BC of the tank, caused by the oil, and specify its location along BC, measured from B. Also compute the total resultant force acting on the bottom of the tank.

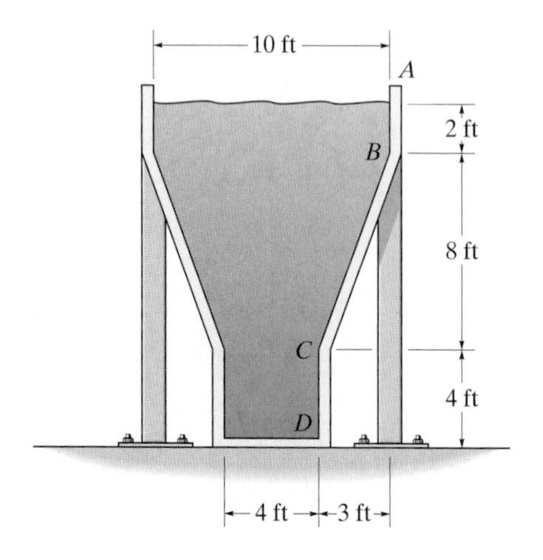

Prob. 9–115

9–117. The rectangular bin is filled with coal, which creates a pressure distribution along wall A that varies as shown, i.e., $p = 4z^3\ \text{lb/ft}^2$, where z is measured in feet. Determine the resultant force created by the coal, and specify its location measured from the top surface of the coal.

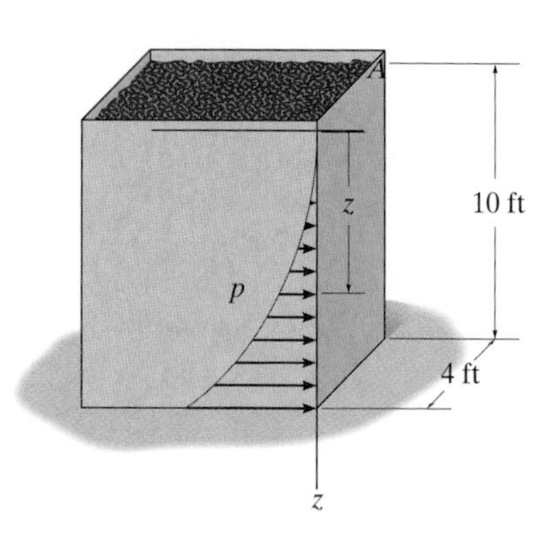

Prob. 9–117

9–118. The semicircular drainage pipe is filled with water. Determine the resultant horizontal and vertical force components that the water exerts on the side AB of the pipe per foot of pipe length; $\gamma_w = 62.4\ \text{lb/ft}^3$.

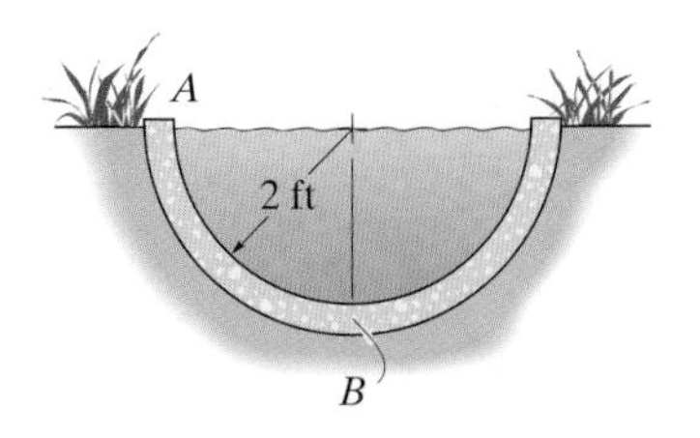

Prob. 9–118

9–119. The load over the plate varies linearly along the sides of the plate such that $p = 10[y(2 - x)]\ \text{lb/ft}^2$. Determine the magnitude of the resultant force and the coordinates $(\bar{x}, \bar{y})$ of the point where the line of action of the force intersects the plate.

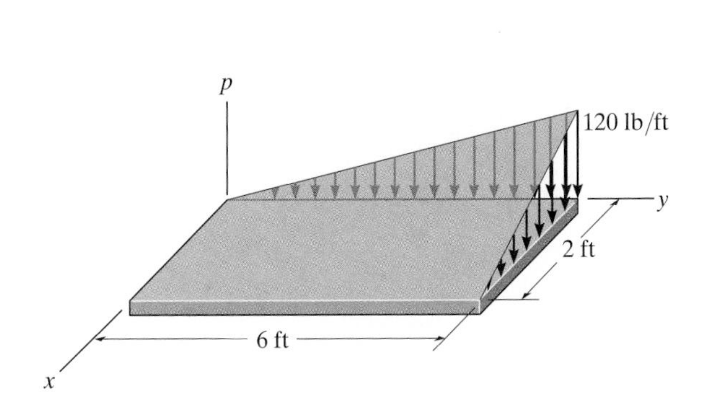

Prob. 9–119

***9–120.** The drum is filled to its top ($y = 1.5$ ft) with oil. having a density of $\rho_o = 55\ \text{lb/ft}^3$. Determine the resultant force of the oil pressure acting on the flat end A of the drum and specify its location measured from the top of the drum.

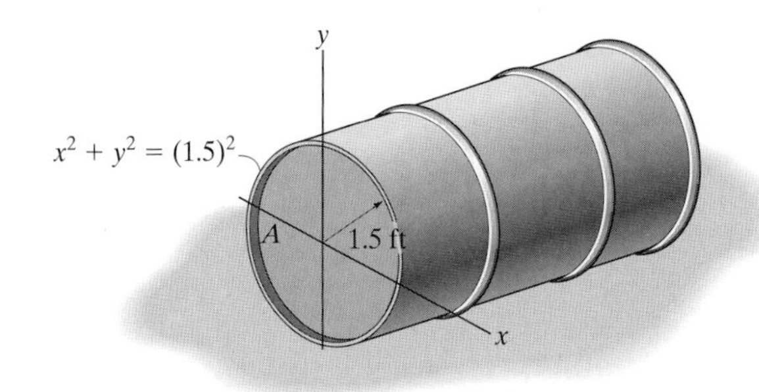

Prob. 9–120

9–121. The gasoline tank is constructed with elliptical ends on each side of the tank. Determine the resultant force and its location on these ends if the tank is half full. Take $\gamma = 41\ \text{lb/ft}^3$.

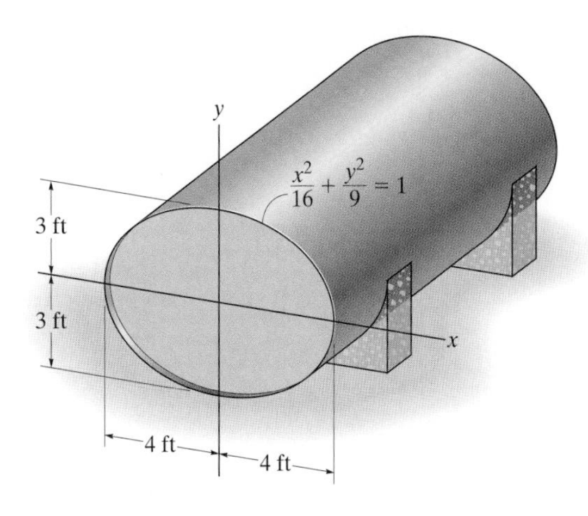

Prob. 9–121

9–122. The loading acting on a square plate is represented by a parabolic pressure distribution. Determine the magnitude of the resultant force and the coordinates $(\bar{x}, \bar{y})$ of the point where the line of action of the force intersects the plate. Also, what are the reactions at the rollers B and C and the ball-and-socket joint A? Neglect the weight of the plate.

9–123. The tank is filled with a liquid which has a density of 900 kg/m^3. Determine the resultant force that it exerts on the elliptical end plate, and the location of the center of pressure, measured from the x axis.

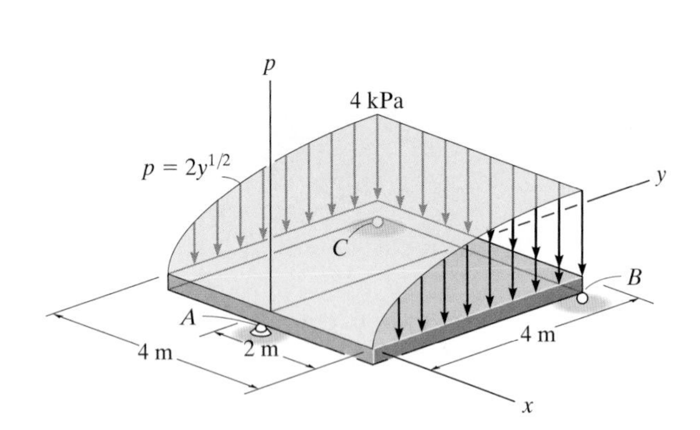

Prob. 9–122

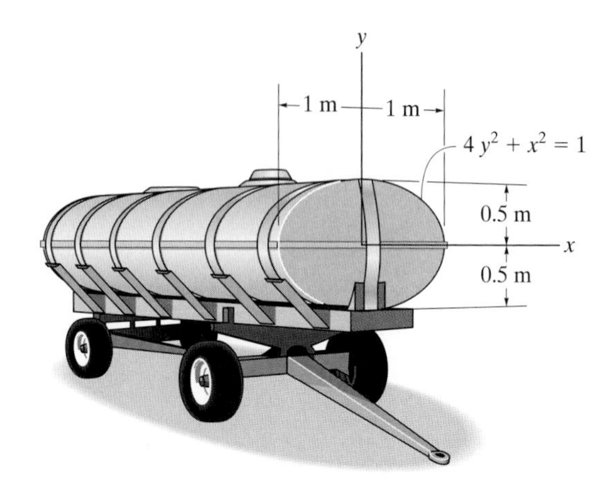

Prob. 9–123

Chapter Review

Center of Gravity and Centroid

The *center of gravity* represents a point where the weight of the body can be considered concentrated. The distance s to this point can be determined from a balance of moments. This requires that the moment of the weight of all the particles of the body about some point must equal the moment of the entire body about the point.

$$\bar{x} = \frac{\int \tilde{x}\, dW}{\int dW}$$

$$\bar{y} = \frac{\int \tilde{y}\, dW}{\int dW}$$

$$\bar{z} = \frac{\int \tilde{z}\, dW}{\int dW}$$

The center of mass will coincide with the center of gravity provided the acceleration of gravity is constant.

The *centroid* is the location of the geometric center for the body. It is determined in a similar manner, using a moment balance of geometric elements such as line, area, or volume segments. For bodies having a continuous shape, moments are summed (integrated) using differential elements.

$$\bar{x} = \frac{\int_L \tilde{x}\, dL}{\int_L dL} \qquad \bar{y} = \frac{\int_L \tilde{y}\, dL}{\int_L dL} \qquad \bar{z} = \frac{\int_L \tilde{z}\, dL}{\int_L dL}$$

$$\bar{x} = \frac{\int_A \tilde{x}\, dA}{\int_A dA} \qquad \bar{y} = \frac{\int_A \tilde{y}\, dA}{\int_A dA} \qquad \bar{z} = \frac{\int_A \tilde{z}\, dA}{\int_A dA}$$

$$\bar{x} = \frac{\int_V \tilde{x}\, dV}{\int_V dV} \qquad \bar{y} = \frac{\int_V \tilde{y}\, dV}{\int_V dV} \qquad \bar{z} = \frac{\int_V \tilde{z}\, dV}{\int_V dV}$$

The center of gravity will coincide with the centroid provided the material is homogeneous, i.e., the density of the material is the same throughout. The centroid will lie on an axis of symmetry. (*See pages 455–460.*)

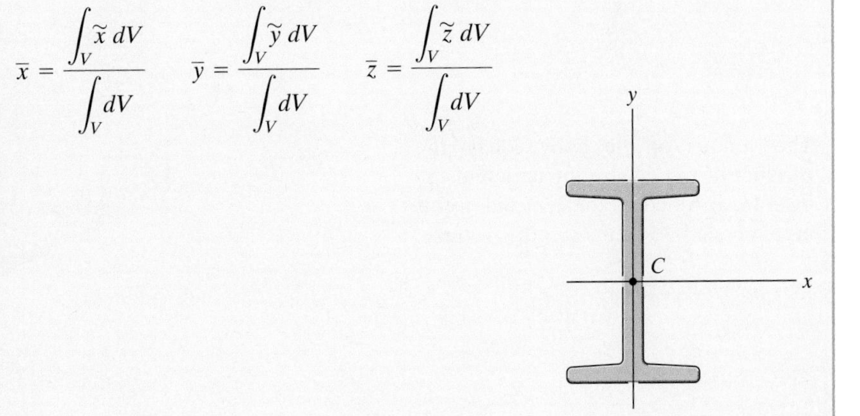

Composite Body

If the body is a composite of several shapes, each having a known location for its center of gravity or centroid, then the location is determined from a discrete summation using its composite parts.
(*See pages 478–479.*)

$$\bar{x} = \frac{\Sigma \tilde{x} W}{\Sigma W} \qquad \bar{y} = \frac{\Sigma \tilde{y} W}{\Sigma W} \qquad \bar{z} = \frac{\Sigma \tilde{z} W}{\Sigma W}$$

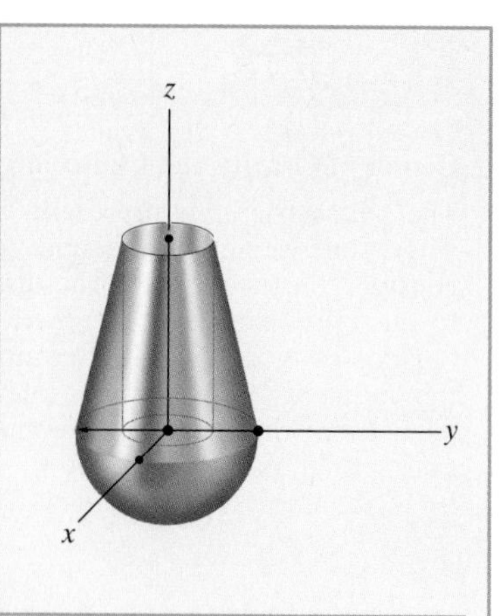

Theorems of Pappus and Guldinus

The theorems of Pappus and Guldinus can be used to determine the surface area and volume of a body of revolution.

The *surface area* equals the product of the length of the generating curve and the distance traveled by the centroid of the curve needed to generate the area, $A = \theta \bar{r} L$.

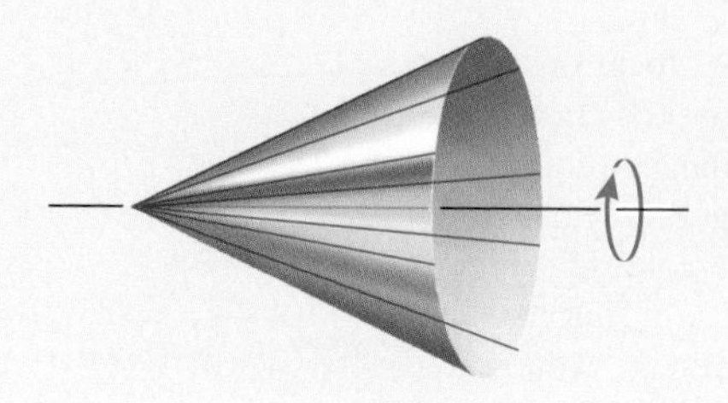

The *volume* of the body equals the product of the generating area and the distance traveled by the centroid of this area needed to generate the volume, $V = \theta \bar{r} A$.
(*See pages 492–493.*)

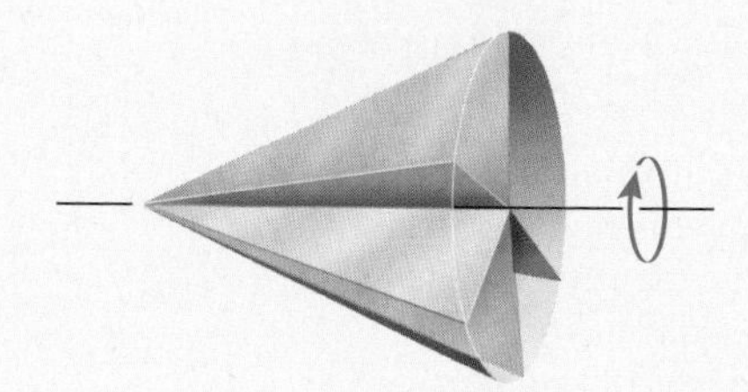

General Distributed Loading

The magnitude of the resutant force is equal to the total volume under the distributed-loading diagram. The line of action of the resultant force passes through the geometric center or centroid of the volume under the distributed-loading diagram.

(*See page 500.*)

$$F_R = \int_A p(x, y)\, dA = \int_V dV$$

$$\bar{x} = \frac{\int_V x\, dV}{\int_V dV}$$

$$\bar{y} = \frac{\int_V y\, dV}{\int_V dV}$$

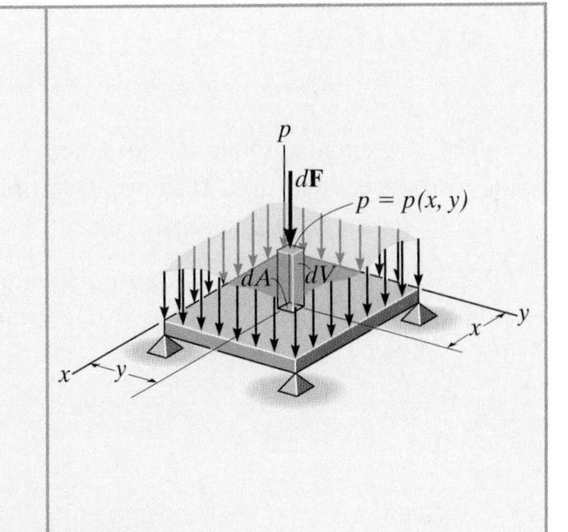

Fluid Pressure

The pressure developed by a liquid at a point on a submerged surface depends upon the depth of the point and the density of the liquid in accordance with Pascal's law, $p = \rho g h = \gamma h$. This pressure will create a *linear distribution* of loading on a flat vertical or inclined surface.

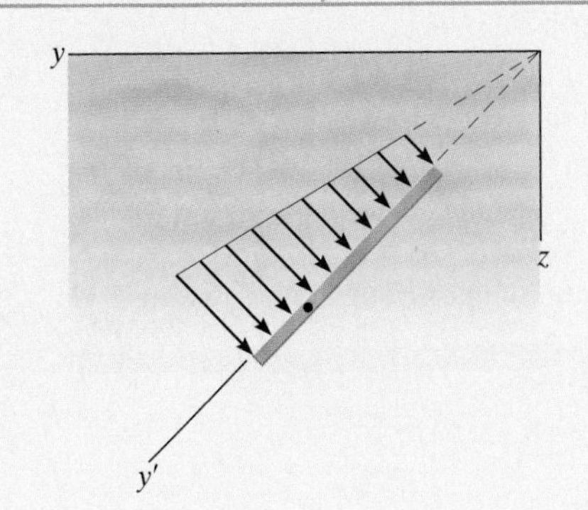

If the surface is horizontal, then the loading will be *uniform*.

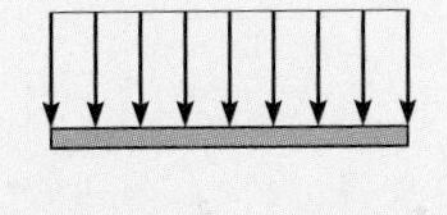

In any case, the resultants of these loadings can be determined by finding the volume under the loading curve. The line of action of the resultant force passes through the centroid of the loading diagram.

(*See pages 501–504.*)

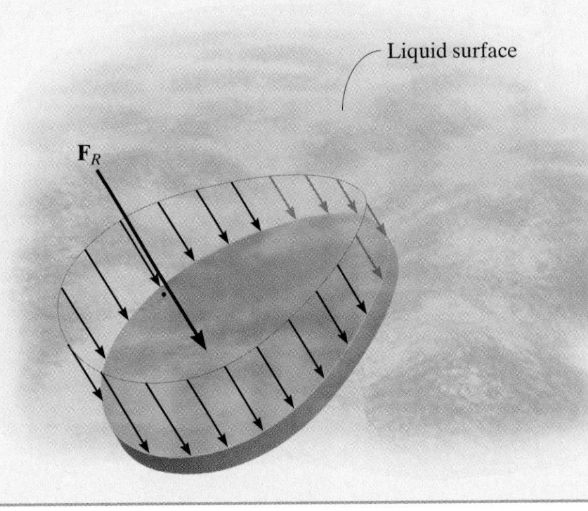

REVIEW PROBLEMS

***9–124.** A circular V-belt has an inner radius of 600 mm and a cross-sectional area as shown. Determine the volume of material required to make the belt.

9–125. A circular V-belt has an inner radius of 600 mm and a cross-sectional area as shown. Determine the surface area of the belt.

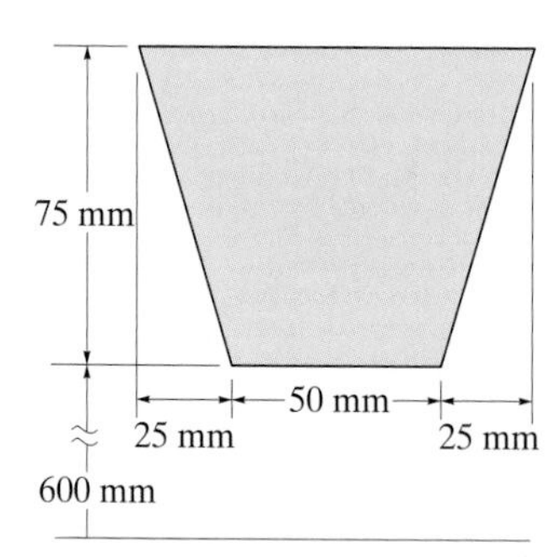

Probs. 9–124/125

9–126. Locate the center of mass of the homogeneous rod.

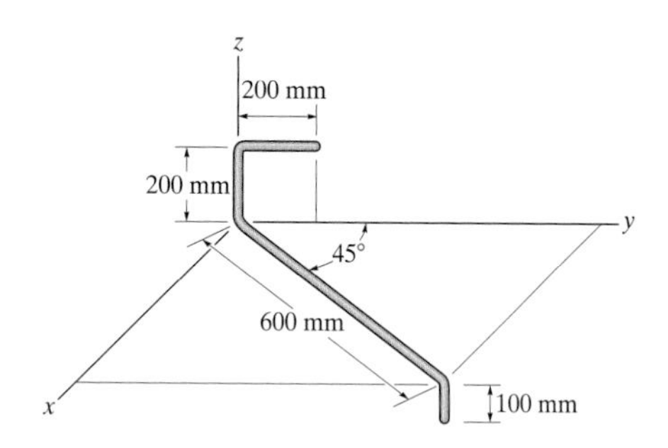

Prob. 9–126

9–127. Locate the centroid of the solid.

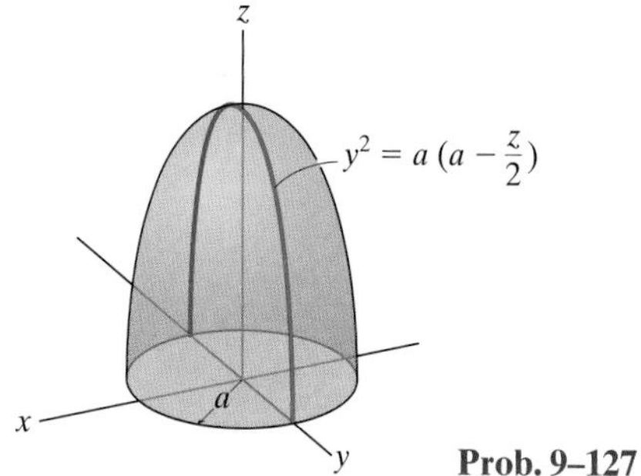

Prob. 9–127

***9–128.** Locate the centroid $(\bar{x}, \bar{y})$ of the thin plate.

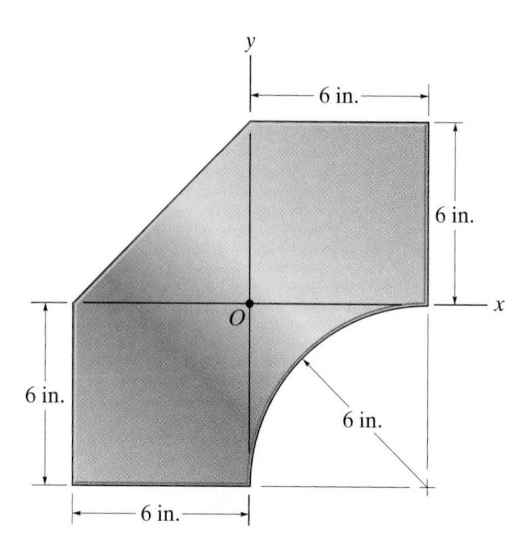

Prob. 9–128

9–129. Determine the weight and location $(\bar{x}, \bar{y})$ of the center of gravity G of the concrete retaining wall. The wall has a length of 10 ft, and concrete has a specific gravity of $\gamma = 150\ \text{lb/ft}^3$.

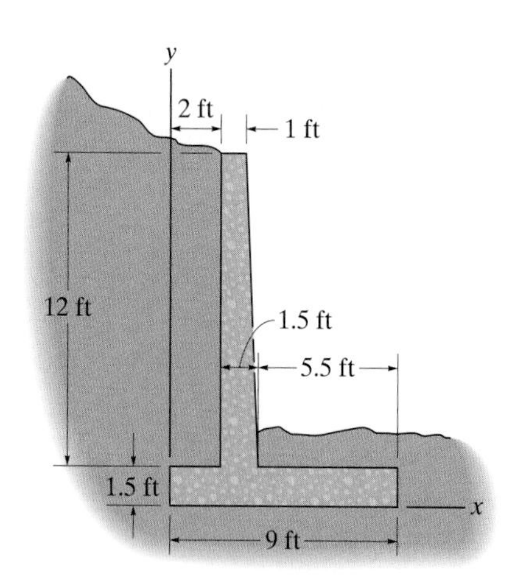

Prob. 9–129

9–130. The hopper is filled to its top with coal. Determine the volume of coal if the voids (air space) are 35 percent of the volume of the hopper.

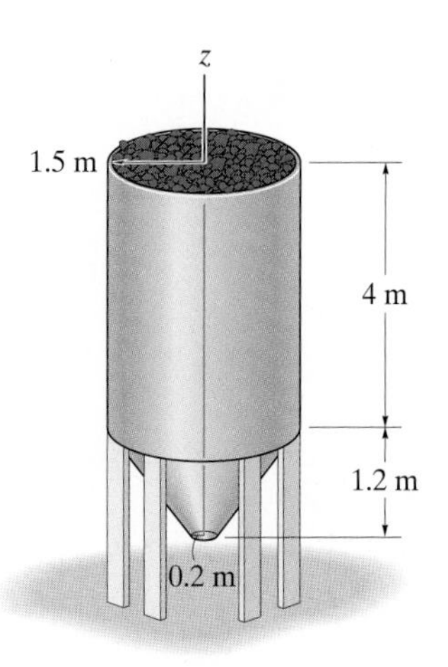

Prob. 9–130

9–131. Locate the centroid $(\bar{x}, \bar{y})$ of the shaded area.

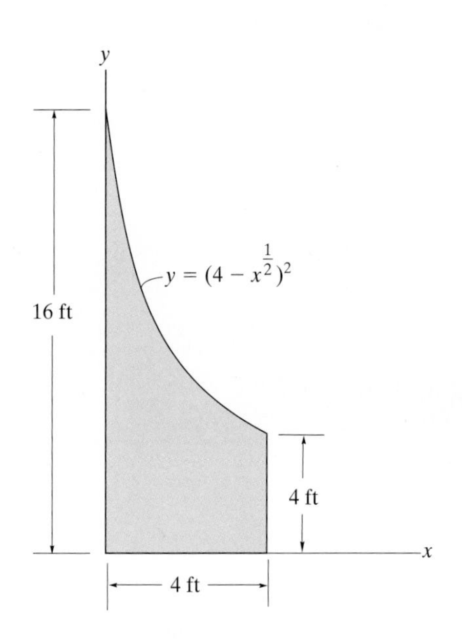

Prob. 9–131

***9–132.** The rectangular bin is filled with coal, which creates a pressure distribution along wall A that varies as shown, i.e., $p = 4z^{1/3}$ lb/ft^2, where z is in feet. Compute the resultant force created by the coal, and its location, measured from the top surface of the coal.

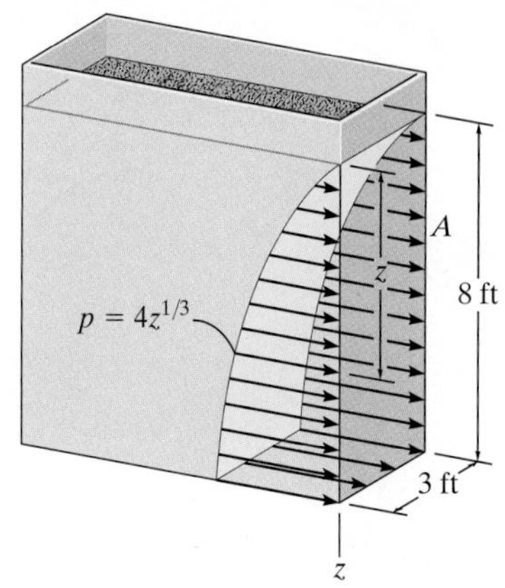

Prob. 9–132

9–133. The load over the plate varies linearly along the sides of the plate such that $p = \frac{2}{3}[x(4 - y)]$ kPa. Determine the resultant force and its position $(\bar{x}, \bar{y})$ on the plate.

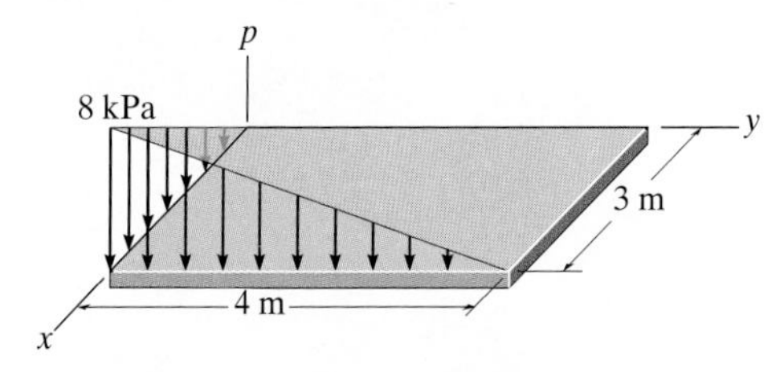

Prob. 9–133

9–134. The pressure loading on the plate is described by the function $p = \{-240/(x + 1) + 340\}$ Pa. Determine the magnitude of the resultant force and coordinates of the point where the line of action of the force intersects the plate.

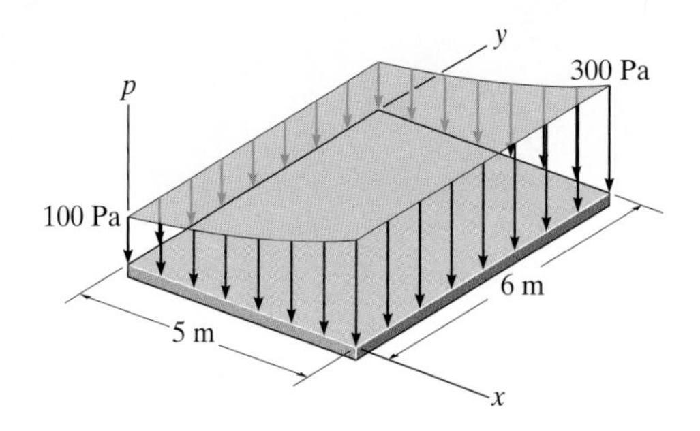

Prob. 9–134

The design of a structural member, such as a beam or column, requires calculation of its cross-sectional moment of inertia. In this chapter, we will discuss how this is done.

10 Moments of Inertia

CHAPTER OBJECTIVES

- To develop a method for determining the moment of inertia for an area.
- To introduce the product of inertia and show how to determine the maximum and minimum moments of inertia for an area.
- To discuss the mass moment of inertia.

10.1 Definition of Moments of Inertia for Areas

In the last chapter, we determined the centroid for an area by considering the first moment of the area about an axis; that is, for the computation we had to evaluate an integral of the form $\int x\,dA$. An integral of the second moment of an area, such as $\int x^2\,dA$, is referred to as the *moment of inertia* for the area. The terminology "moment of inertia" as used here is actually a misnomer; however, it has been adopted because of the similarity with integrals of the same form related to mass.

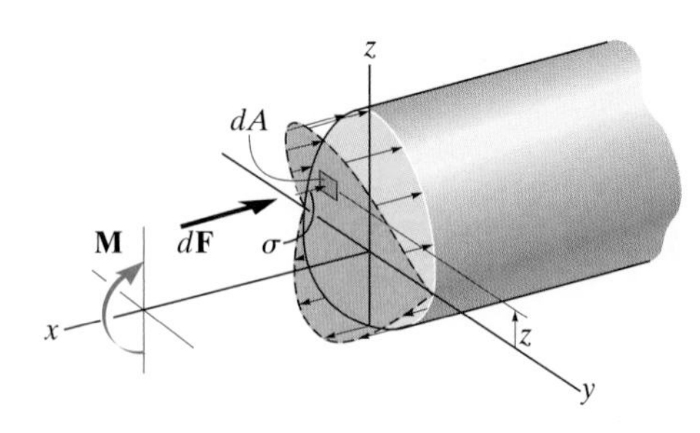

Fig. 10–1

The moment of inertia for an area is a quantity that relates the normal stress σ (sigma), or force per unit area, acting on the transverse cross section of an elastic beam, to the applied external moment **M**, which causes bending of the beam. The theory of mechanics of materials shows that the stress within the beam varies linearly with its distance from an axis passing through the centroid C of the beam's cross-sectional area; i.e., $\sigma = kz$, Fig. 10–1. The magnitude of force acting on the area element dA, shown in the figure, is therefore $dF = \sigma\, dA = kz\, dA$. Since this force is located a distance z from the y axis, the moment of $d\mathbf{F}$ about the y axis is $dM = dF z = kz^2\, dA$. The resulting moment of the entire stress distribution is equal to the applied moment **M**; hence, $M = k\int z^2\, dA$. Here the integral represents the moment of inertia of the area about the y axis. Integrals of this form often arise in formulas used in mechanics of materials, structural mechanics, fluid mechanics, and machine design, and the engineer needs to be familiar with the methods used for their computation.

Moment of Inertia. Consider the area A, shown in Fig. 10–2, which lies in the x–y plane. By definition, the moments of inertia for the differential planar area dA about the x and y axes are $dI_x = y^2\, dA$ and $dI_y = x^2\, dA$, respectively. For the entire area the *moments of inertia* are determined by integration; i.e.,

$$I_x = \int_A y^2\, dA$$
$$I_y = \int_A x^2\, dA \qquad (10\text{–}1)$$

We can also formulate the second moment of dA about the "pole" O or z axis, Fig. 10–2. This is referred to as the polar moment of inertia. It is used to determine the torsional stress in a shaft. It is defined as $dJ_O = r^2\, dA$, where r is the perpendicular distance from the pole (z axis) to the element dA. For the entire area the *polar moment of inertia* is

$$J_O = \int_A r^2\, dA = I_x + I_y \qquad (10\text{–}2)$$

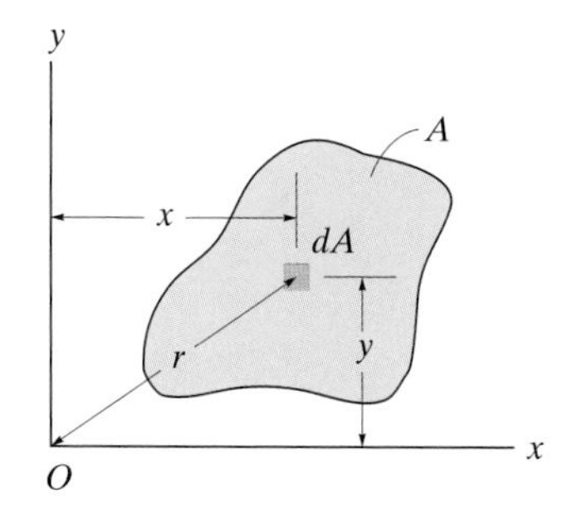

Fig. 10–2

The relationship between J_O and I_x, I_y is possible since $r^2 = x^2 + y^2$, Fig. 10–2.

From the above formulations it is seen that I_x, I_y, and J_O will *always* be *positive* since they involve the product of distance squared and area. Furthermore, the units for moment of inertia involve length raised to the fourth power, e.g., m^4, mm^4, or ft^4, in^4.

10.2 Parallel-Axis Theorem for an Area

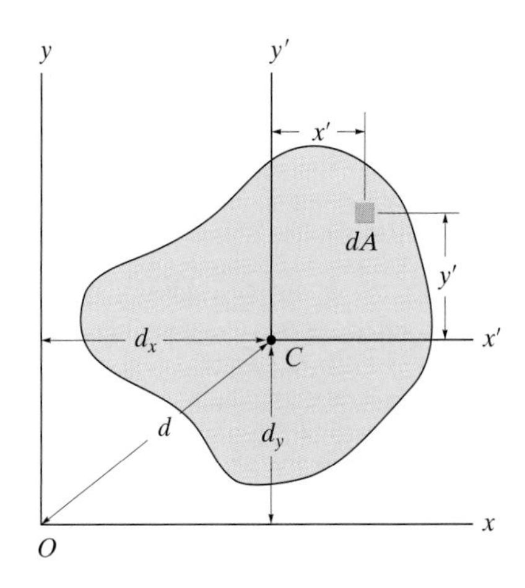

Fig. 10–3

If the moment of inertia for an area is known about an axis passing through its centroid, which is often the case, it is convenient to determine the moment of inertia of the area about a corresponding parallel axis using the *parallel-axis theorem*. To derive this theorem, consider finding the moment of inertia of the shaded area shown in Fig. 10–3 about the x axis. In this case, a differential element dA is located at an arbitrary distance y' from the *centroidal* x' axis, whereas the *fixed distance* between the parallel x and x' axes is defined as d_y. Since the moment of inertia of dA about the x axis is $dI_x = (y' + d_y)^2\,dA$, then for the entire area,

$$I_x = \int_A (y' + d_y)^2\,dA$$

$$= \int_A y'^2\,dA + 2d_y \int_A y'\,dA + d_y^2 \int_A dA$$

The first integral represents the moment of inertia of the area about the centroidal axis, $\bar{I}_{x'}$. The second integral is zero since the x' axis passes through the area's centroid C; i.e., $\int y'\,dA = \bar{y}\int dA = 0$ since $\bar{y} = 0$. Realizing that the third integral represents the total area A, the final result is therefore

$$I_x = \bar{I}_{x'} + Ad_y^2 \tag{10–3}$$

A similar expression can be written for I_y; i.e.,

$$I_y = \bar{I}_{y'} + Ad_x^2 \tag{10–4}$$

And finally, for the polar moment of inertia about an axis perpendicular to the x–y plane and passing through the "pole" O (z axis), Fig. 10–3, we have

$$J_O = \bar{J}_C + Ad^2 \tag{10–5}$$

The form of each of these three equations states that *the moment of inertia for an area about an axis is equal to the moment of inertia for the area about a parallel axis passing through the area's centroid plus the product of the area and the square of the perpendicular distance between the axes.*

10.3 Radius of Gyration of an Area

The *radius of gyration* of a planar area has units of length and is a quantity that is often used for the design of columns in structural mechanics. Provided the areas and moments of inertia are *known*, the radii of gyration are determined from the formulas

$$k_x = \sqrt{\frac{I_x}{A}} \qquad k_y = \sqrt{\frac{I_y}{A}} \qquad k_O = \sqrt{\frac{J_O}{A}} \tag{10–6}$$

The form of these equations is easily remembered since it is similar to that for finding the moment of inertia for a differential area about an axis. For example, $I_x = k_x^2 A$; whereas for a differential area, $dI_x = y^2\, dA$.

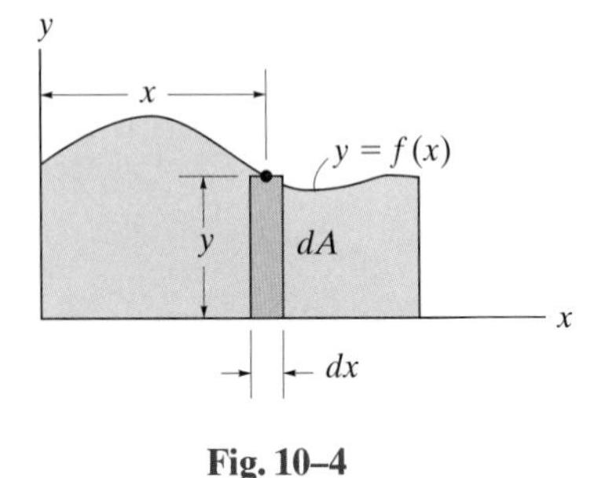

Fig. 10–4

10.4 Moments of Inertia for an Area by Integration

When the boundaries for a planar area are expressed by mathematical functions, Eqs. 10–1 may be integrated to determine the moments of inertia for the area. If the element of area chosen for integration has a differential size in two directions as shown in Fig. 10–2, a double integration must be performed to evaluate the moment of inertia. Most often, however, it is easier to perform only a single integration by choosing an element having a differential size or thickness in only one direction.

Procedure for Analysis

- If a single integration is performed to determine the moment of inertia for an area about an axis, it will be necessary to specify the differential element dA.
- Most often this element will be rectangular, such that it will have a finite length and differential width.
- The element should be located so that it intersects the boundary of the area at the *arbitrary point* (x, y). There are two possible ways to orient the element with respect to the axis about which the moment of inertia is to be determined.

Case 1

- The length of the element can be oriented *parallel* to the axis. This situation occurs when the rectangular element shown in Fig. 10–4 is used to determine I_y for the area. Direct application of Eq. 10–1, i.e., $I_y = \int x^2\, dA$, can be made in this case since the element has an infinitesimal thickness dx and therefore *all parts* of the element lie at the *same* moment-arm distance x from the y axis.*

Case 2

- The length of the element can be oriented *perpendicular* to the axis. Here Eq. 10–1 *does not apply* since all parts of the element will *not* lie at the same moment-arm distance from the axis. For example, if the rectangular element in Fig. 10–4 is used for determining I_x for the area, it will first be necessary to calculate the moment of inertia of the *element* about a horizontal axis passing through the element's centroid and then determine the moment of inertia of the *element* about the x axis by using the parallel-axis theorem. Integration of this result will yield I_x.

*In the case of the element $dA = dx\, dy$, Fig. 10–2, the moment arms y and x are appropriate for the formulation of I_x and I_y (Eq. 10–1) since the *entire* element, because of its infinitesimal size, lies at the specified y and x perpendicular distances from the x and y axes.

EXAMPLE 10.1

Determine the moment of inertia for the rectangular area shown in Fig. 10–5 with respect to (a) the centroidal x' axis, (b) the axis x_b passing through the base of the rectangle, and (c) the pole or z' axis perpendicular to the $x'-y'$ plane and passing through the centroid C.

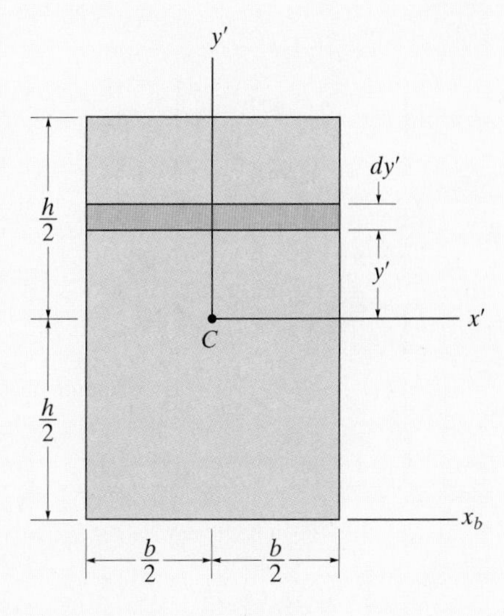

Fig. 10–5

SOLUTION (Case 1)

Part (a). The differential element shown in Fig. 10–5 is chosen for integration. Because of its location and orientation, the *entire element* is at a distance y' from the x' axis. Here it is necessary to integrate from $y' = -h/2$ to $y' = h/2$. Since $dA = b\,dy'$, then

$$\bar{I}_{x'} = \int_A y'^2\,dA = \int_{-h/2}^{h/2} y'^2(b\,dy') = b\int_{-h/2}^{h/2} y'^2\,dy$$

$$= \frac{1}{12}bh^3 \qquad \textit{Ans.}$$

Part (b). The moment of inertia about an axis passing through the base of the rectangle can be obtained by using the result of part (a) and applying the parallel-axis theorem, Eq. 10–3.

$$I_{x_b} = \bar{I}_{x'} + Ad_y^2$$

$$= \frac{1}{12}bh^3 + bh\left(\frac{h}{2}\right)^2 = \frac{1}{3}bh^3 \qquad \textit{Ans.}$$

Part (c). To obtain the polar moment of inertia about point C, we must first obtain $\bar{I}_{y'}$, which may be found by interchanging the dimensions b and h in the result of part (a), i.e.,

$$\bar{I}_{y'} = \frac{1}{12}hb^3$$

Using Eq. 10–2, the polar moment of inertia about C is therefore

$$\bar{J}_C = \bar{I}_{x'} + \bar{I}_{y'} = \frac{1}{12}bh(h^2 + b^2) \qquad \textit{Ans.}$$

EXAMPLE 10.2

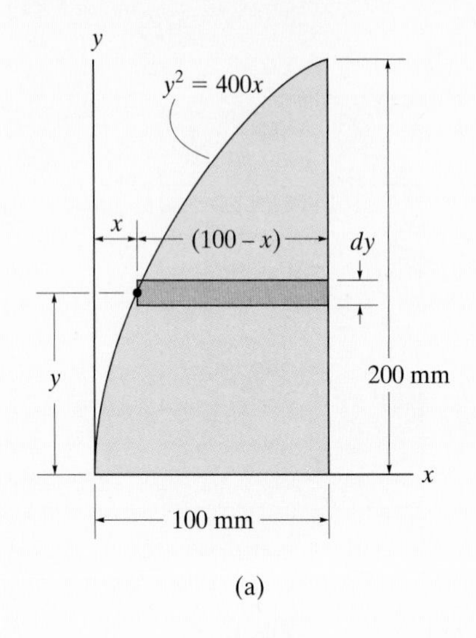

(a)

Determine the moment of inertia for the shaded area shown in Fig. 10–6a about the x axis.

SOLUTION I (Case 1)

A differential element of area that is *parallel* to the x axis, as shown in Fig. 10–6a, is chosen for integration. Since the element has a thickness dy and intersects the curve at the *arbitrary point* (x, y), the area is $dA = (100 - x)\,dy$. Furthermore, all parts of the element lie at the same distance y from the x axis. Hence, integrating with respect to y, from $y = 0$ to $y = 200$ mm, yields

$$I_x = \int_A y^2\,dA = \int_A y^2(100 - x)\,dy$$

$$= \int_0^{200} y^2\left(100 - \frac{y^2}{400}\right)dy = 100\int_0^{200} y^2\,dy - \frac{1}{400}\int_0^{200} y^4\,dy$$

$$= 107(10^6)\ \text{mm}^4 \qquad \textit{Ans.}$$

SOLUTION II (Case 2)

A differential element *parallel* to the y axis, as shown in Fig. 10–6b, is chosen for integration. It intersects the curve at the *arbitrary point* (x, y). In this case, all parts of the element do *not* lie at the same distance from the x axis, and therefore the parallel-axis theorem must be used to determine the *moment of inertia of the element* with respect to this axis. For a rectangle having a base b and height h, the moment of inertia about its centroidal axis has been determined in part (a) of Example 10.1. There it was found that $\bar{I}_{x'} = \frac{1}{12}bh^3$. For the differential element shown in Fig. 10–6b, $b = dx$ and $h = y$, and thus $d\bar{I}_{x'} = \frac{1}{12}dx\,y^3$. Since the centroid of the element is at $\tilde{y} = y/2$ from the x axis, the moment of inertia of the element about this axis is

$$dI_x = d\bar{I}_{x'} + dA\,\tilde{y}^2 = \frac{1}{12}dx\,y^3 + y\,dx\left(\frac{y}{2}\right)^2 = \frac{1}{3}y^3\,dx$$

This result can also be concluded from part (b) of Example 10.1. Integrating with respect to x, from $x = 0$ to $x = 100$ mm, yields

$$I_x = \int dI_x = \int_A \frac{1}{3}y^3\,dx = \int_0^{100} \frac{1}{3}(400x)^{3/2}\,dx$$

$$= 107(10^6)\ \text{mm}^4 \qquad \textit{Ans.}$$

(b)

Fig. 10–6

EXAMPLE 10.3

Determine the moment of inertia with respect to the x axis for the circular area shown in Fig. 10–7a.

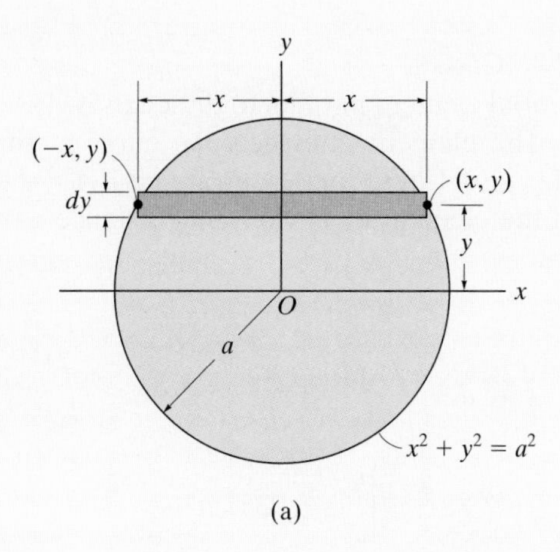

(a)

SOLUTION I (Case 1)

Using the differential element shown in Fig. 10–7a, since $dA = 2x\,dy$, we have

$$I_x = \int_A y^2\,dA = \int_A y^2(2x)\,dy$$

$$= \int_{-a}^{a} y^2\left(2\sqrt{a^2 - y^2}\right)dy = \frac{\pi a^4}{4} \qquad \textit{Ans.}$$

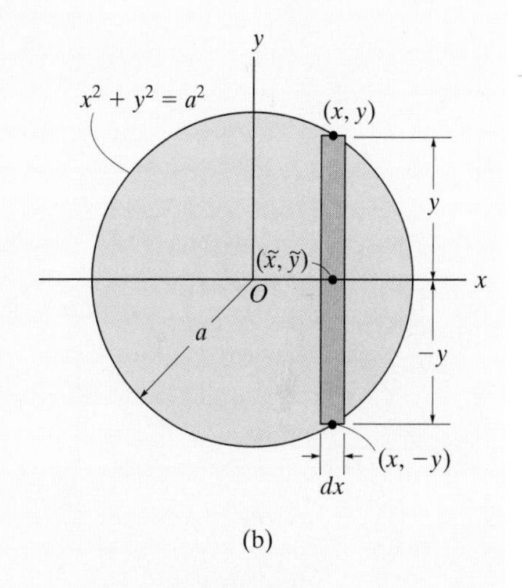

(b)

Fig. 10–7

SOLUTION II (Case 2)

When the differential element is chosen as shown in Fig. 10–7b, the centroid for the element happens to lie on the x axis, and so, applying Eq. 10–3, noting that $d_y = 0$ and for a rectangle $\bar{I}_{x'} = \frac{1}{12}bh^3$, we have

$$dI_x = \frac{1}{12}dx(2y)^3$$

$$= \frac{2}{3}y^3\,dx$$

Integrating with respect to x yields

$$I_x = \int_{-a}^{a} \frac{2}{3}(a^2 - x^2)^{3/2}\,dx = \frac{\pi a^4}{4} \qquad \textit{Ans.}$$

EXAMPLE 10.4

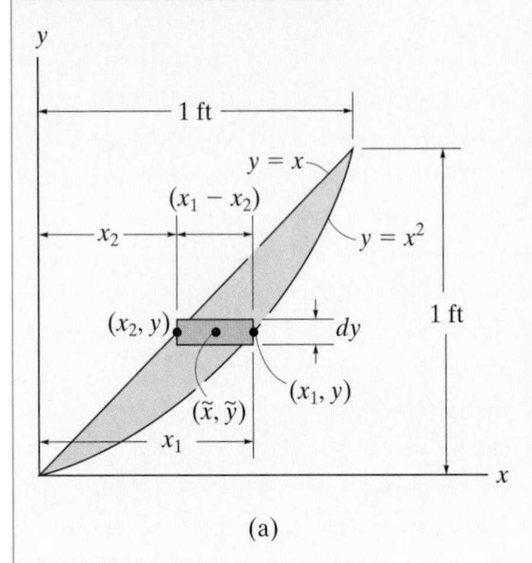

(a)

Determine the moment of inertia for the shaded area shown in Fig. 10–8*a* about the x axis.

SOLUTION I (Case 1)

The differential element parallel to the x axis is chosen for integration, Fig. 10–8*a*. The element intersects the curve at the *arbitrary points* (x_2, y) and (x_1, y). Consequently, its area is $dA = (x_1 - x_2)\,dy$. Since all parts of the element lie at the same distance y from the x axis, we have

$$I_x = \int_A y^2\,dA = \int_0^1 y^2(x_1 - x_2)\,dy = \int_0^1 y^2(\sqrt{y} - y)\,dy$$

$$I_x = \frac{2}{7}y^{7/2} - \frac{1}{4}y^4\bigg|_0^1 = 0.0357\text{ ft}^4 \qquad \textit{Ans.}$$

(b)

Fig. 10–8

SOLUTION II (Case 2)

The differential element parallel to the y axis is shown in Fig. 10–8*b*. It intersects the curves at the *arbitrary points* (x, y_2) and (x, y_1). Since all parts of its entirety do *not* lie at the same distance from the x axis, we must first use the parallel-axis theorem to find the *element's* moment of inertia about the x axis, using $\bar{I}_{x'} = \frac{1}{12}bh^3$, then integrate this result to determine I_x. Thus,

$$dI_x = d\bar{I}_{x'} + dA\,\tilde{y}^2 = \frac{1}{12}dx(y_2 - y_1)^3 + (y_2 - y_1)\,dx\left(y_1 + \frac{y_2 - y_1}{2}\right)^2$$

$$= \frac{1}{3}(y_2^3 - y_1^3)\,dx = \frac{1}{3}(x^3 - x^6)\,dx$$

$$I_x = \frac{1}{3}\int_0^1 (x^3 - x^6)\,dx = \frac{1}{12}x^4 - \frac{1}{21}x^7\bigg|_0^1 = 0.0357\text{ ft}^4 \qquad \textit{Ans.}$$

NOTE: By comparison, Solution I requires much less computation. Therefore, if an integral using a particular element appears difficult to evaluate, try solving the problem using an element oriented in the other direction.

PROBLEMS

10–1. Determine the moment of inertia for the shaded area about the x axis.

10–2. Determine the moment of inertia for the shaded area about the y axis.

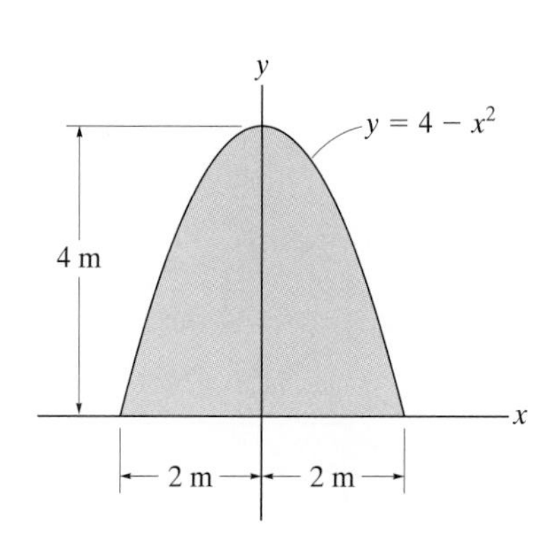

Probs. 10–1/2

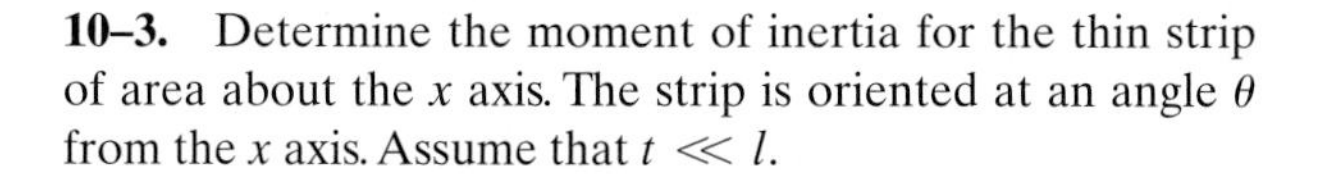

10–3. Determine the moment of inertia for the thin strip of area about the x axis. The strip is oriented at an angle θ from the x axis. Assume that $t \ll l$.

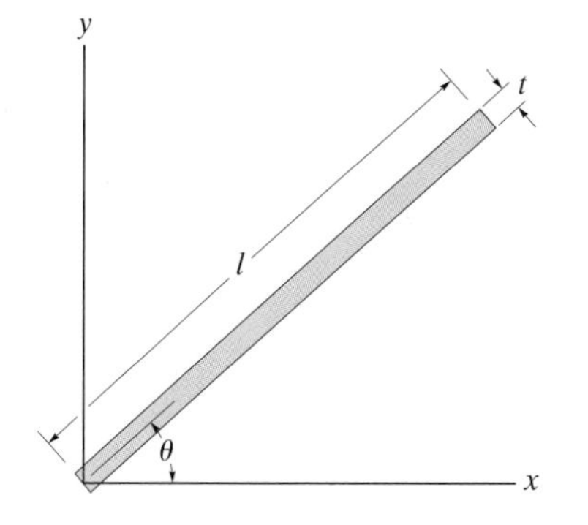

Prob. 10–3

***10–4.** Determine the moment of inertia for the shaded area about the x axis.

10–5. Determine the moment of inertia for the shaded area about the y axis.

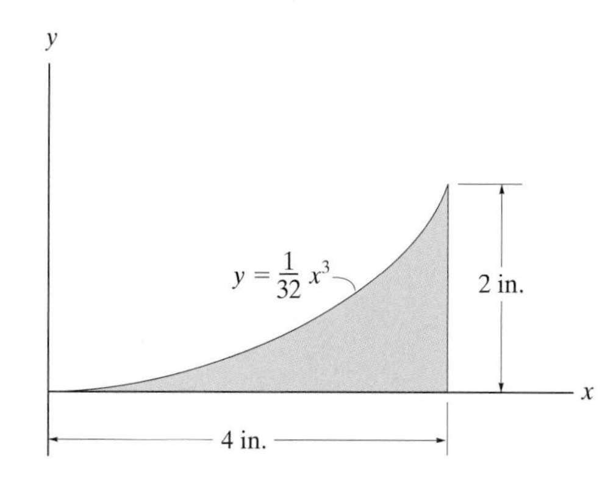

Probs. 10–4/5

10–6. Determine the moment of inertia for the shaded area about the x axis.

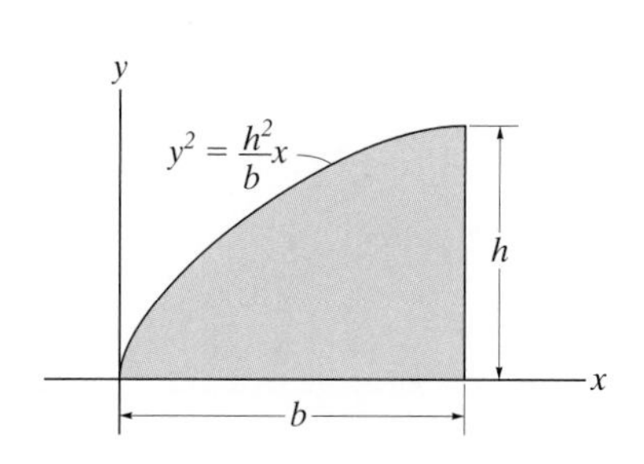

Prob. 10–6

10–7. Determine the moment of inertia for the shaded area about the x axis.

***10–8.** Determine the moment of inertia for the shaded area about the y axis.

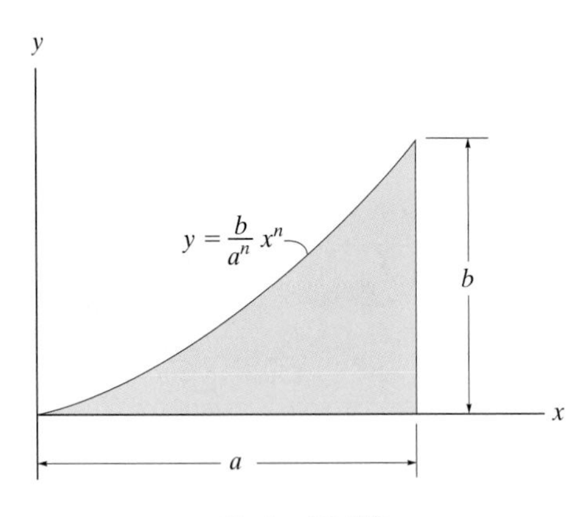

Probs. 10–7/8

10–9. Determine the moment of inertia for the shaded area about the x axis.

10–10. Determine the moment of inertia for the shaded area about the y axis.

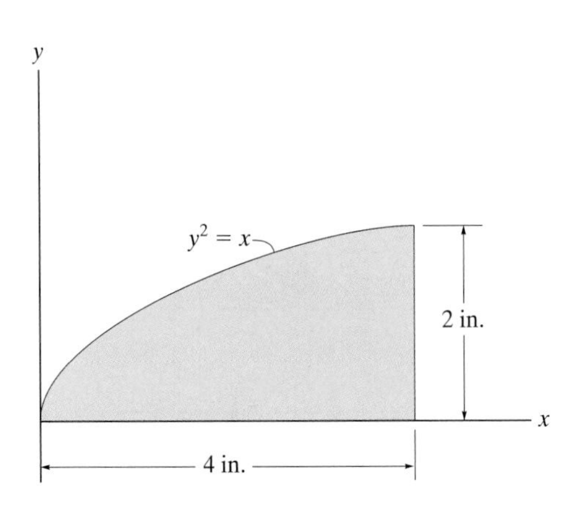

Probs. 10–9/10

10–11. Determine the moment of inertia for the shaded area about the x axis.

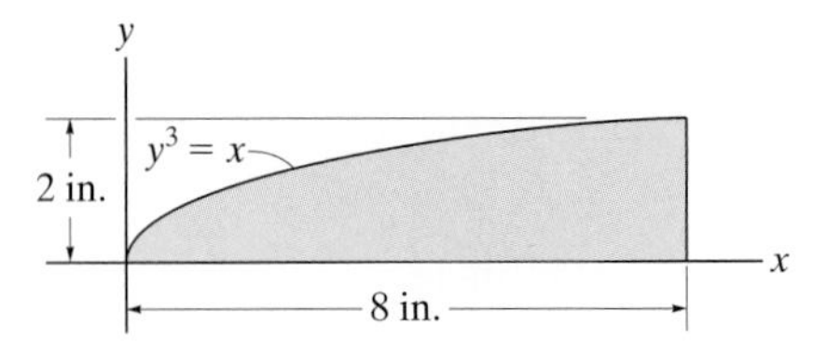

Prob. 10–11

***10–12.** Determine the moment of inertia for the shaded area about the x axis.

10–13. Determine the moment of inertia for the shaded area about the y axis.

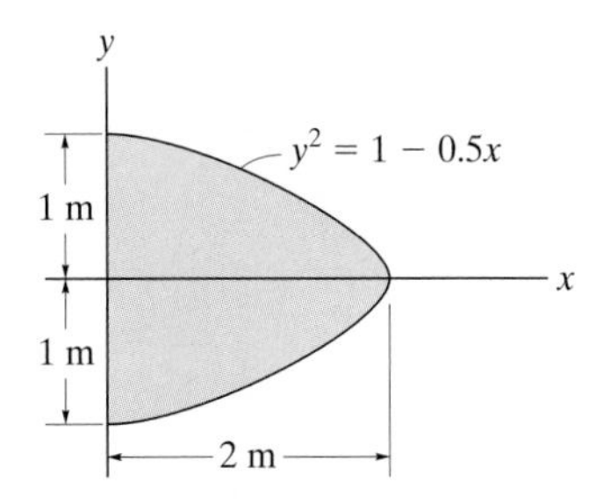

Probs. 10–12/13

10–14. Determine the moment of inertia for the shaded area about the x axis.

10–15. Determine the moment of inertia for the shaded area about the y axis.

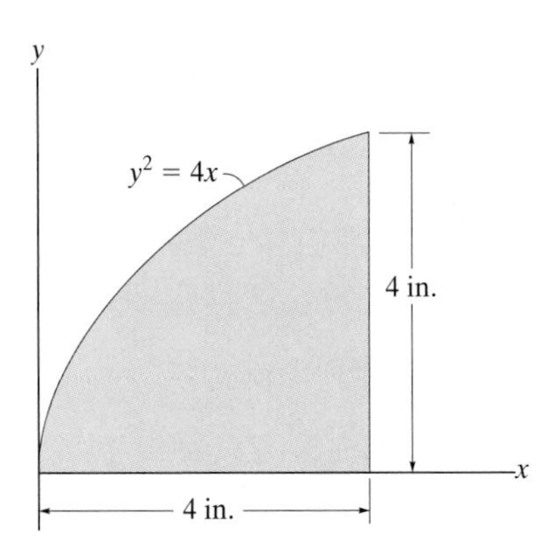

Probs. 10–14/15

***10–16.** Determine the moment of inertia for the shaded area about the x axis.

10–17. Determine the moment of inertia for the shaded area about the y axis.

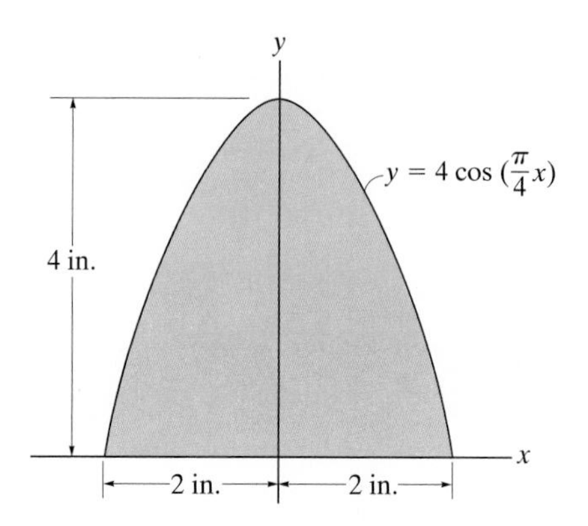

Probs. 10–16/17

10–18. Determine the moment of inertia for the shaded area about the x axis.

10–19. Determine the moment of inertia for the shaded area about the y axis.

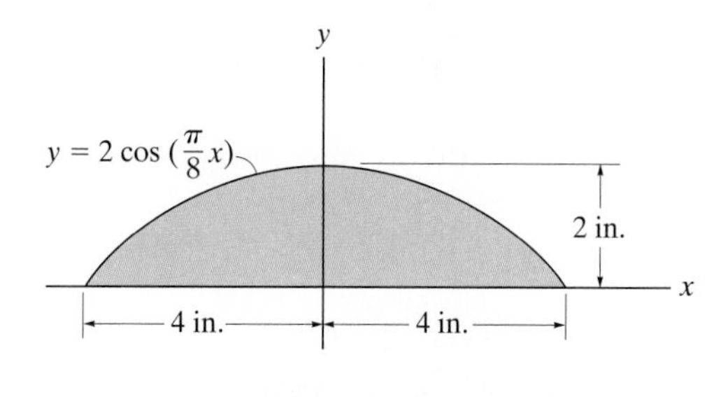

Probs. 10–18/19

***10–20.** Determine the moment of inertia for the shaded area about the x axis.

10–21. Determine the moment of inertia for the shaded area about the y axis.

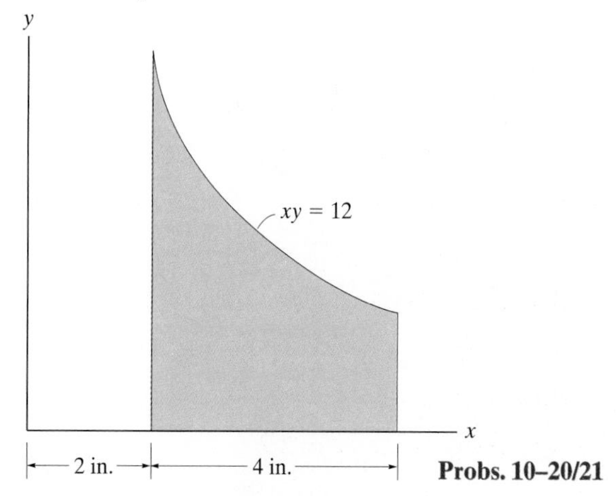

Probs. 10–20/21

10–22. Determine the moment of inertia for the shaded area about the x axis.

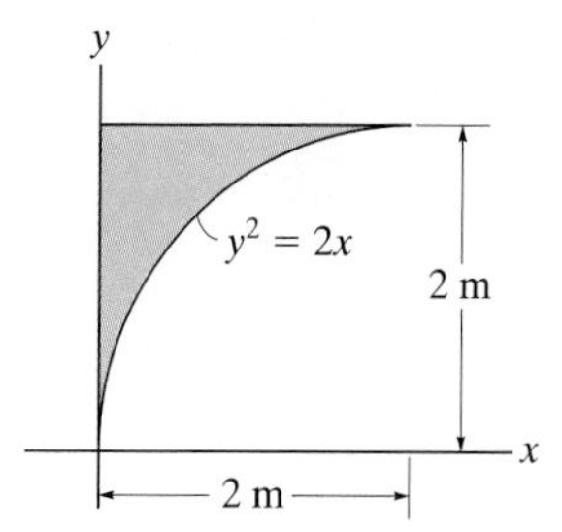

Prob. 10–22

■**10–23.** Determine the moment of inertia for the shaded area about the y axis. Use Simpson's rule to evaluate the integral.

*■**10–24.** Determine the moment of inertia for the shaded area about the x axis. Use Simpson's rule to evaluate the integral.

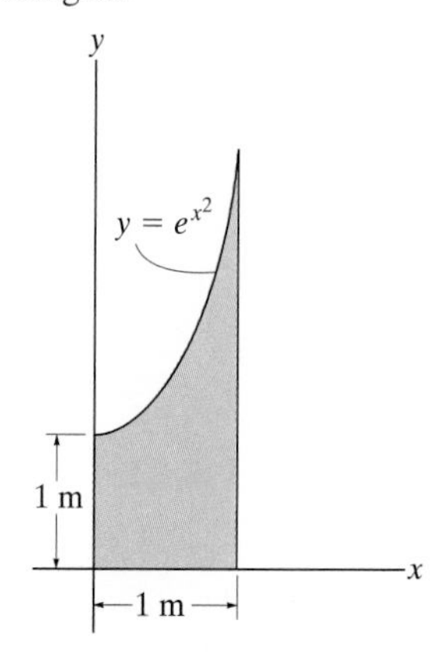

Probs. 10–23/24

10.5 Moments of Inertia for Composite Areas

Structural members have various cross-sectional shapes, and it is necessary to calculate their moments of inertia in order to determine the stress in these members.

A composite area consists of a series of connected "simpler" parts or shapes, such as semicircles, rectangles, and triangles. Provided the moment of inertia of each of these parts is known or can be determined about a common axis, then the moment of inertia for the composite area equals the *algebraic sum* of the moments of inertia of all its parts.

Procedure for Analysis

The moment of inertia for a composite area about a reference axis can be determined using the following procedure.

Composite Parts.

- Using a sketch, divide the area into its composite parts and indicate the perpendicular distance from the centroid of each part to the reference axis.

Parallel-Axis Theorem.

- The moment of inertia of each part should be determined about its centroidal axis, which is parallel to the reference axis. For the calculation use the table given on the inside back cover.
- If the centroidal axis does not coincide with the reference axis, the parallel-axis theorem, $I = \bar{I} + Ad^2$, should be used to determine the moment of inertia of the part about the reference axis.

Summation.

- The moment of inertia of the entire area about the reference axis is determined by summing the results of its composite parts.
- If a composite part has a "hole," its moment of inertia is found by "subtracting" the moment of inertia for the hole from the moment of inertia of the entire part including the hole.

EXAMPLE 10.5

Compute the moment of inertia for the composite area shown in Fig. 10–9*a* about the *x* axis.

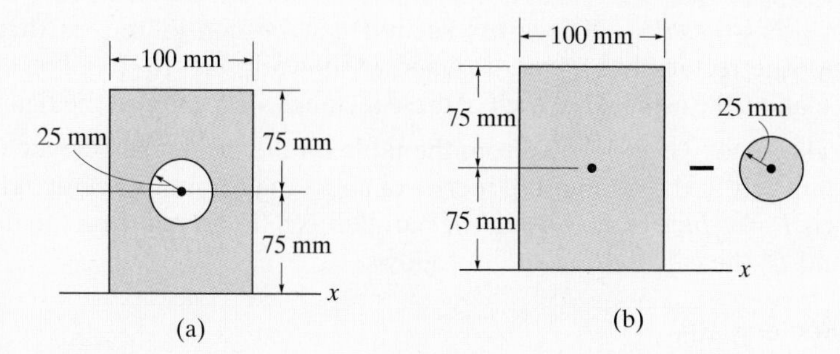

Fig. 10–9

SOLUTION

Composite Parts. The composite area is obtained by *subtracting* the circle from the rectangle as shown in Fig. 10–9*b*. The centroid of each area is located in the figure.

Parallel-Axis Theorem. The moments of inertia about the *x* axis are determined using the parallel-axis theorem and the data in the table on the inside back cover.

Circle

$$I_x = \bar{I}_{x'} + Ad_y^2$$

$$= \frac{1}{4}\pi(25)^4 + \pi(25)^2(75)^2 = 11.4(10^6)\text{ mm}^4$$

Rectangle

$$I_x = \bar{I}_{x'} + Ad_y^2$$

$$= \frac{1}{12}(100)(150)^3 + (100)(150)(75)^2 = 112.5(10^6)\text{ mm}^4$$

Summation. The moment of inertia for the composite area is thus

$$I_x = -11.4(10^6) + 112.5(10^6)$$

$$= 101(10^6)\text{ mm}^4 \qquad \textit{Ans.}$$

EXAMPLE 10.6

Determine the moments of inertia for the beam's cross-sectional area shown in Fig. 10–10*a* about the x and y centroidal axes.

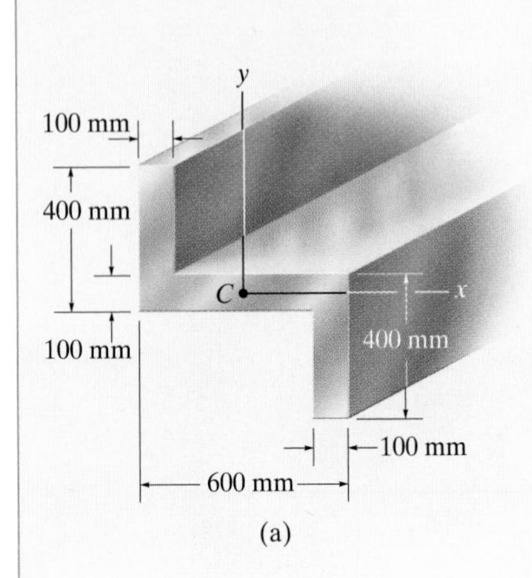

SOLUTION

Composite Parts. The cross section can be considered as three composite rectangular areas A, B, and D shown in Fig. 10–10*b*. For the calculation, the centroid of each of these rectangles is located in the figure.

Parallel-Axis Theorem. From the table on the inside back cover, or Example 10.1, the moment of inertia of a rectangle about its centroidal axis is $\bar{I} = \frac{1}{12}bh^3$. Hence, using the parallel-axis theorem for rectangles A and D, the calculations are as follows:

Rectangle A

$$I_x = \bar{I}_{x'} + Ad_y^2 = \frac{1}{12}(100)(300)^3 + (100)(300)(200)^2$$
$$= 1.425(10^9)\text{ mm}^4$$

$$I_y = \bar{I}_{y'} + Ad_x^2 = \frac{1}{12}(300)(100)^3 + (100)(300)(250)^2$$
$$= 1.90(10^9)\text{ mm}^4$$

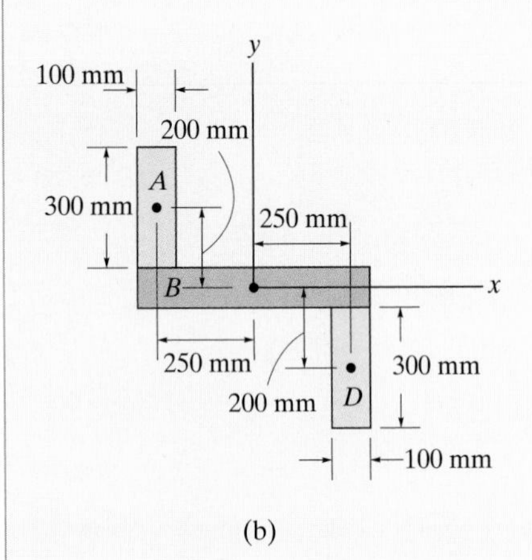

Fig. 10–10

Rectangle B

$$I_x = \frac{1}{12}(600)(100)^3 = 0.05(10^9)\text{ mm}^4$$

$$I_y = \frac{1}{12}(100)(600)^3 = 1.80(10^9)\text{ mm}^4$$

Rectangle D

$$I_x = \bar{I}_{x'} + Ad_y^2 = \frac{1}{12}(100)(300)^3 + (100)(300)(200)^2$$
$$= 1.425(10^9)\text{ mm}^4$$

$$I_y = \bar{I}_{y'} + Ad_x^2 = \frac{1}{12}(300)(100)^3 + (100)(300)(250)^2$$
$$= 1.90(10^9)\text{ mm}^4$$

Summation. The moments of inertia for the entire cross section are thus

$$I_x = 1.425(10^9) + 0.05(10^9) + 1.425(10^9)$$
$$= 2.90(10^9)\text{ mm}^4 \qquad \textit{Ans.}$$

$$I_y = 1.90(10^9) + 1.80(10^9) + 1.90(10^9)$$
$$= 5.60(10^9)\text{ mm}^4 \qquad \textit{Ans.}$$

PROBLEMS

10–25. The polar moment of inertia for the area is $\bar{J}_C = 28\ \text{in}^4$ about the z axis passing through the centroid C. The moment of inertia about the x axis is $17\ \text{in}^4$, and the moment of inertia about the y' axis is $56\ \text{in}^4$. Determine the area A.

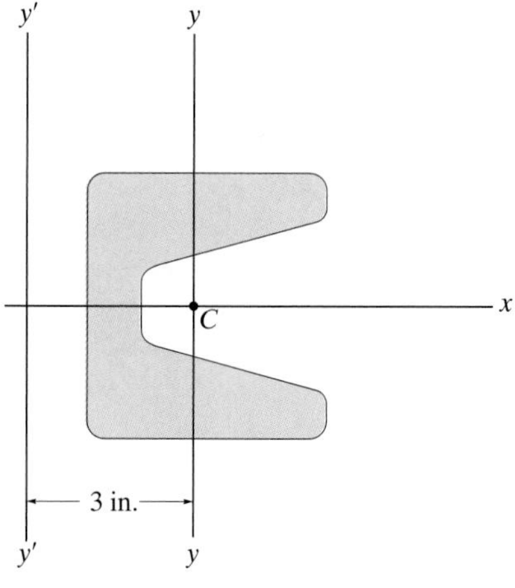

Prob. 10–25

10–26. The polar moment of inertia for the area is $\bar{J}_C = 548(10^6)\ \text{mm}^4$, about the z' axis passing through the centroid C. The moment of inertia about the y' axis is $383(10^6)\ \text{mm}^4$, and the moment of inertia about the x axis is $856(10^6)\ \text{mm}^4$. Determine the area A.

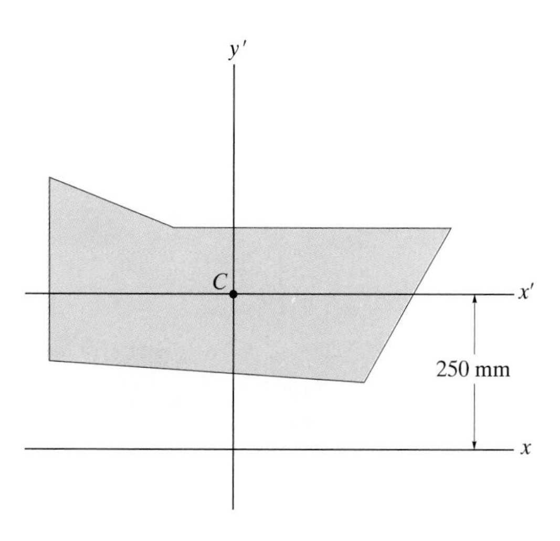

Prob. 10–26

10–27. Determine the radius of gyration k_x of the column's cross-sectional area.

***10–28.** Determine the radius of gyration k_y of the column's cross-sectional area.

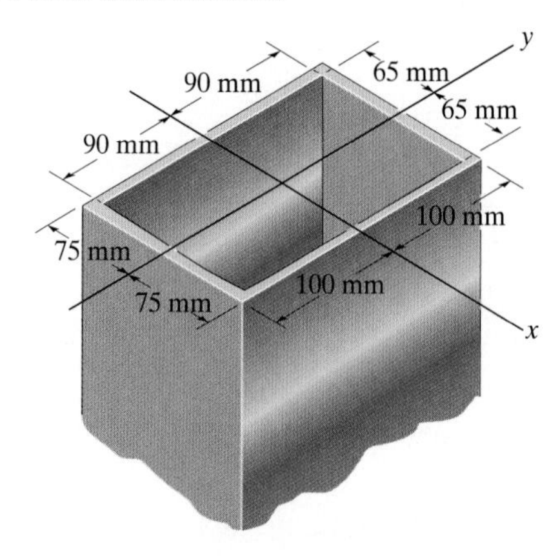

Probs. 10–27/28

10–29. Determine the moment of inertia for the beam's cross-sectional area with respect to the x' centroidal axis. Neglect the size of all the rivet heads. R, for the calculation. Handbook values for the area, moment of inertia, and location of the centroid C of one of the angles are listed in the figure.

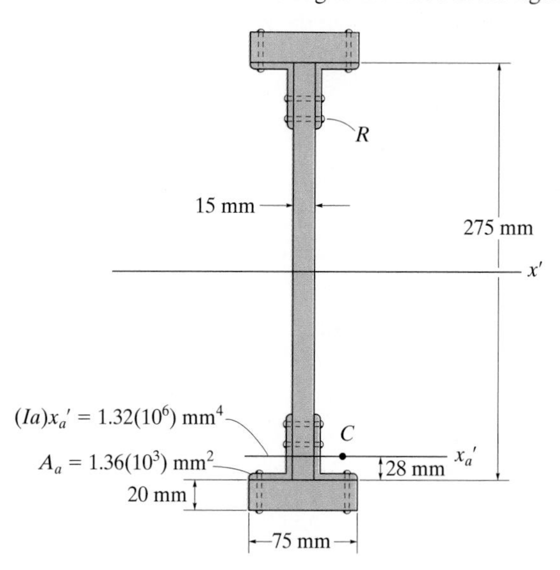

Prob. 10–29

10–30. Locate the centroid $\bar{y}$ of the cross-sectional area for the angle. Then find the moment of inertia $\bar{I}_{x'}$ about the x' centroidal axis.

10–31. Locate the centroid $\bar{x}$ of the cross-sectional area for the angle. Then find the moment of inertia $\bar{I}_{y'}$ about the y' centroidal axis.

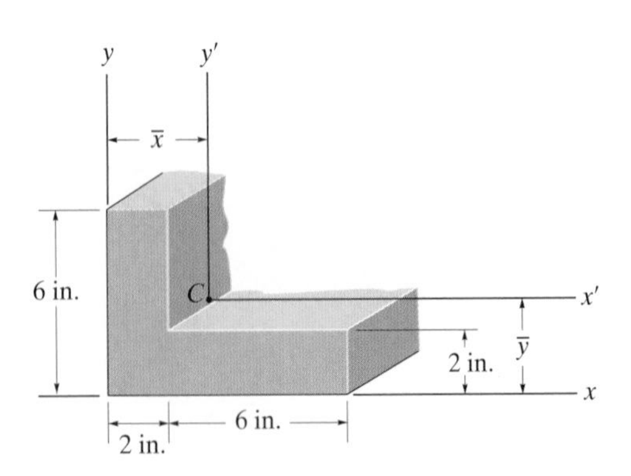

Probs. 10–30/31

10–34. Determine the moments of inertia for the shaded area about the x and y axes.

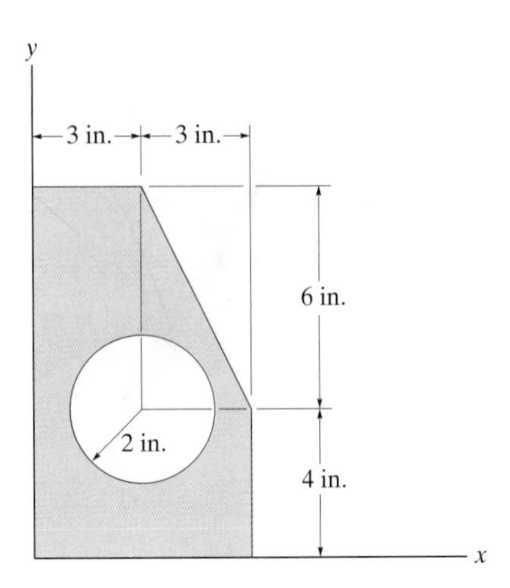

Prob. 10–34

***10–32.** Determine the distance $\bar{x}$ to the centroid of the beam's cross-sectional area: then find the moment of inertia about the y' axis.

10–33. Determine the moment of inertia for the beam's cross-sectional area about the x' axis.

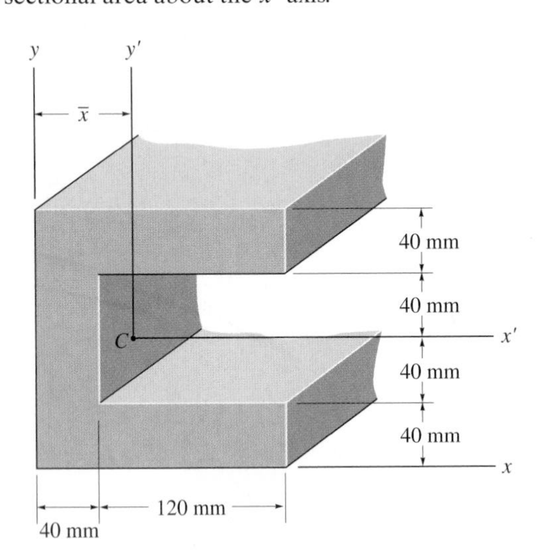

Probs. 10–32/33

10–35. Determine the location of the centroid $\bar{y}$ of the beam constructed from the two channels and the cover plate. If each channel has a cross-sectional area of $A_c = 11.8 \text{ in}^2$ and a moment of inertia about a horizontal axis passing through its own centroid, C_c, of $\bar{I}_{x_c'} = 349 \text{ in}^4$, determine the moment of inertia for the beam's cross-sectional area about the x' axis.

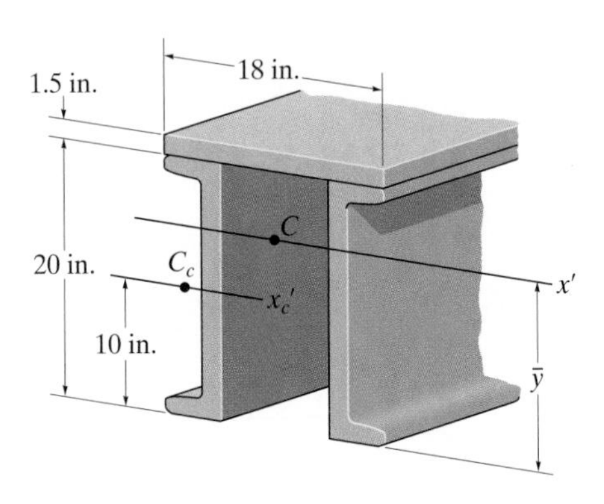

Prob. 10–35

*10–36. Compute the moments of inertia I_x and I_y for the beam's cross-sectional area about the x and y axes.

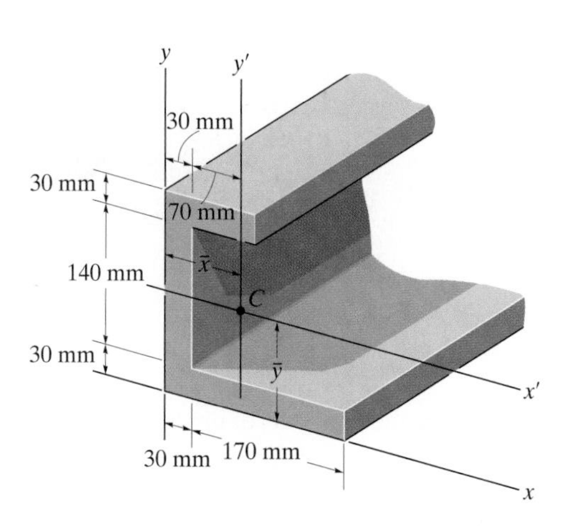

Prob. 10–36

10–37. Determine the distance $\bar{y}$ to the centroid C of the beam's cross-sectional area and then compute the moment of inertia $\bar{I}_{x'}$ about the x' axis.

10–38. Determine the distance $\bar{x}$ to the centroid C of the beam's cross-sectional area and then compute the moment of inertia $\bar{I}_{y'}$ about the y' axis.

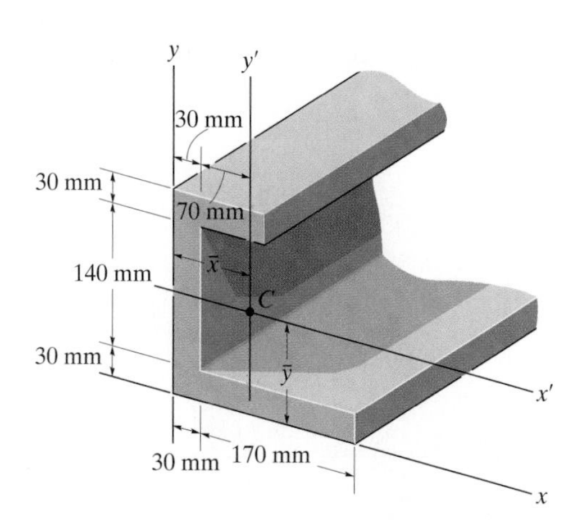

Probs. 10–37/38

10–39. Determine the location $\bar{y}$ of the centroid C of the beam's cross-sectional area. Then compute the moment of inertia for the area about the x' axis.

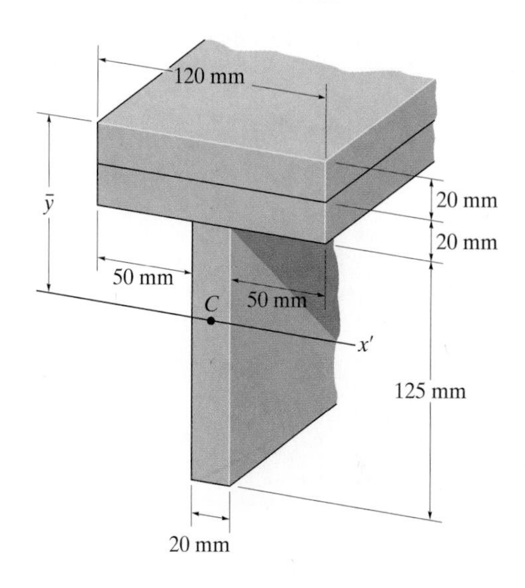

Prob. 10–39

*10–40. Determine $\bar{y}$, which locates the centroidal axis x' for the cross-sectional area of the T-beam, and then find the moments of inertia $\bar{I}_{x'}$ and $\bar{I}_{y'}$.

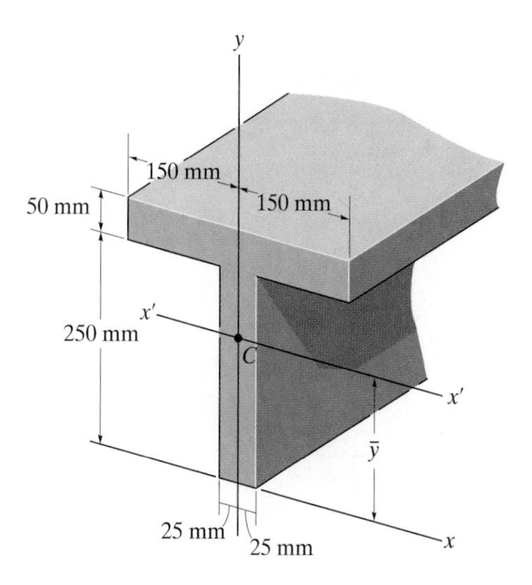

Prob. 10–40

10–41. Determine the centroid $\bar{y}$ for the beam's cross-sectional area; then find $\bar{I}_{x'}$.

10–42. Determine the moment of inertia for the beam's cross-sectional area about the y axis.

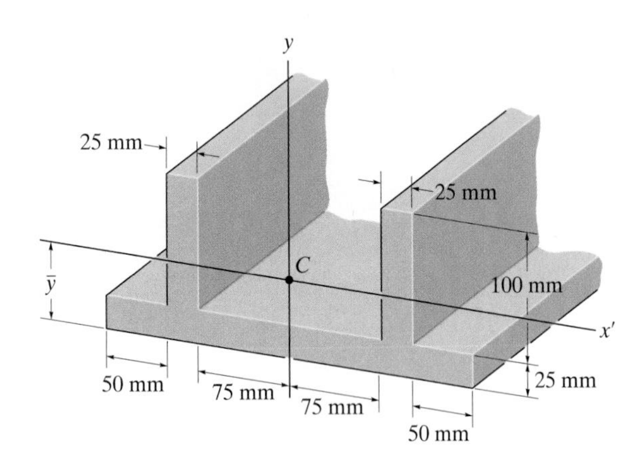

Probs. 10–41/42

10–43. Determine the moment of inertia I_x for the shaded area about the x axis.

***10–44.** Determine the moment of inertia I_y for the shaded area about the y axis.

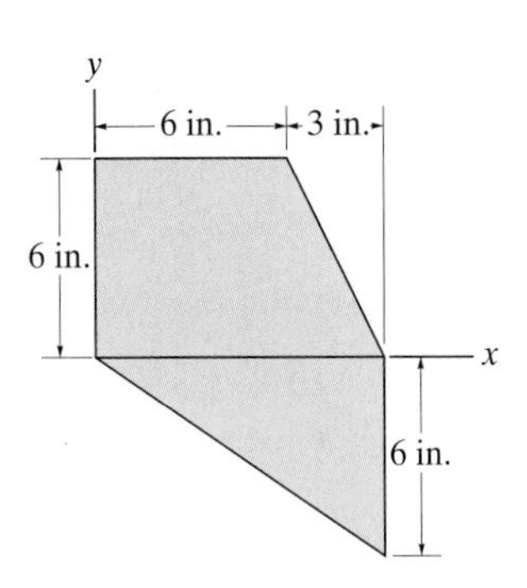

Probs. 10–43/44

10–45. Locate the centroid $\bar{y}$ of the channel's cross-sectional area, and then determine the moment of inertia with respect to the x' axis passing through the centroid.

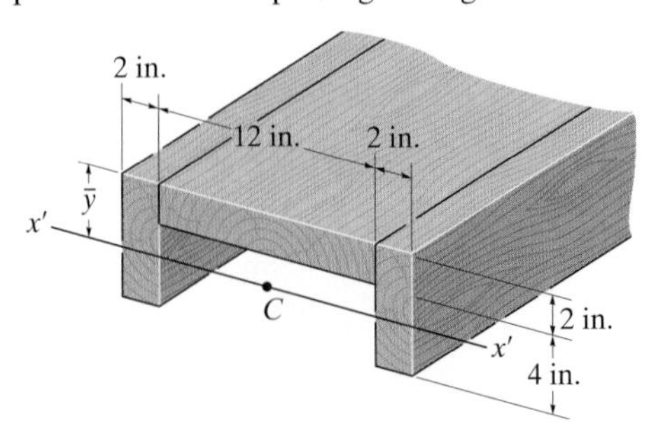

Prob. 10–45

10–46. Determine the moments of inertia I_x and I_y for the shaded area.

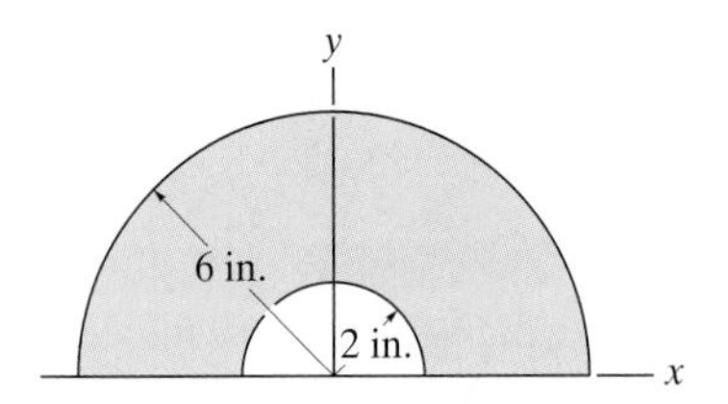

Prob. 10–46

10–47. Determine the moment of inertia for the parallelogram about the x' axis, which passes through the centroid C of the area.

***10–48.** Determine the moment of inertia for the parallelogram about the y' axis, which passes through the centroid C of the area.

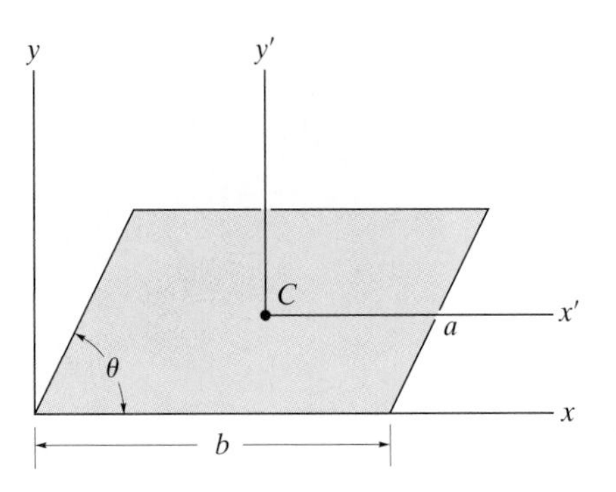

Probs. 10–47/48

10–49. Determine the moments of inertia for the triangular area about the x' and y' axes, which pass through the centroid C of the area.

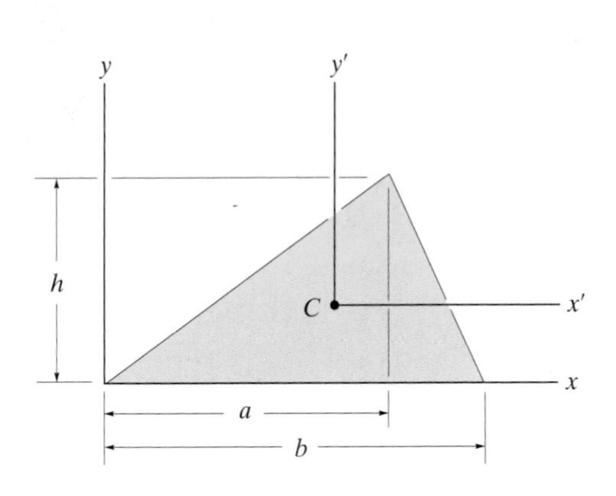

Prob. 10–49

10–50. Determine the moment of inertia for the beam's cross-sectional area about the x' axis passing through the centroid C of the cross section.

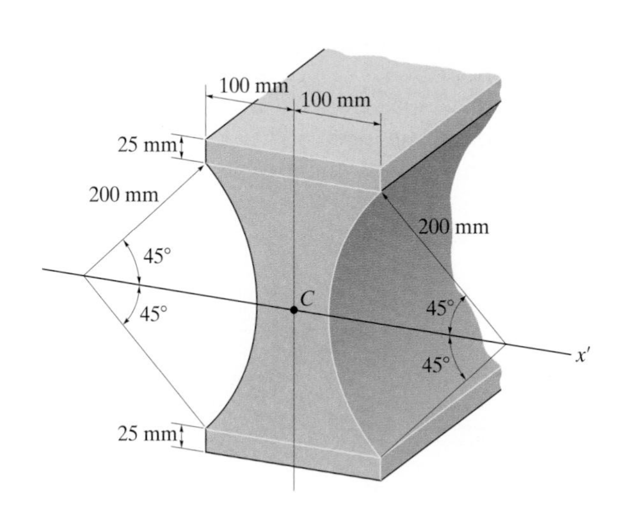

Prob. 10–50

10–51. Determine the moment of inertia for the composite area about the x axis.

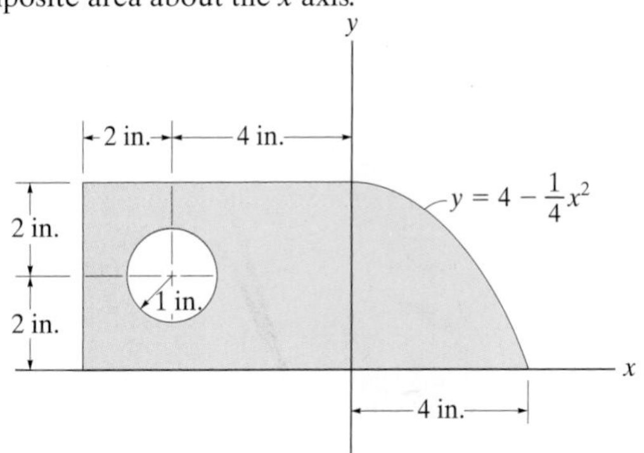

Prob. 10–51

***10–52.** Determine the moment of inertia for the composite area about the y axis.

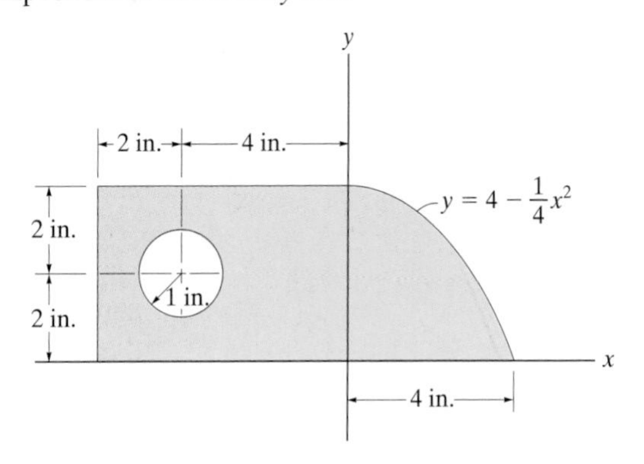

Prob. 10–52

10–53. Determine the radius of gyration k_x for the column's cross-sectional area.

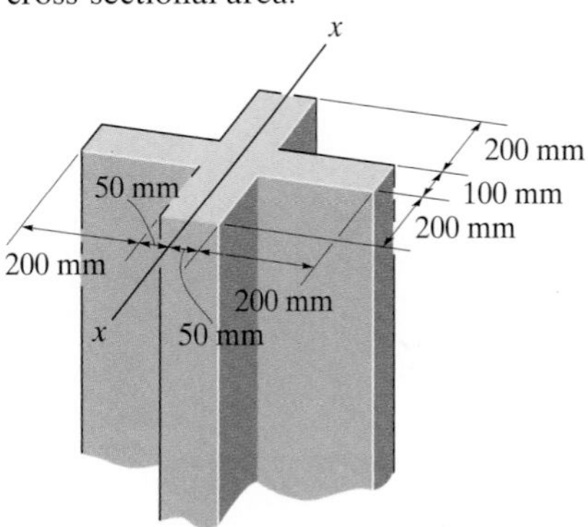

Prob. 10–53

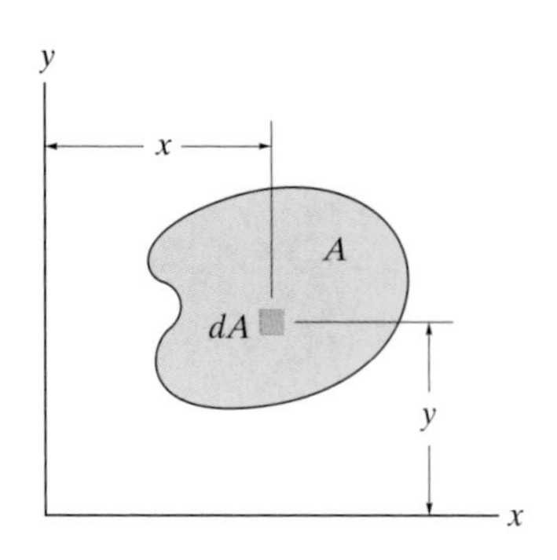

Fig. 10–11

*10.6 Product of Inertia for an Area

In general, the moment of inertia for an area is different for every axis about which it is computed. In some applications of structural or mechanical design it is necessary to know the orientation of those axes which give, respectively, the maximum and minimum moments of inertia for the area. The method for determining this is discussed in the next section. To use this method, however, we first need to compute the product of inertia for the area as well as its moments of inertia for given x, y axes.

The product of inertia for an element of area dA located at point (x, y), Fig. 10–11, is defined as $dI_{xy} = xy\,dA$. Thus, for the entire area A, the *product of inertia* is

$$I_{xy} = \int_A xy\,dA \qquad (10\text{–}7)$$

If the element of area chosen has a differential size in two directions, as shown in Fig. 10–11, a double integration must be performed to evaluate I_{xy}. Most often, however, it is easier to choose an element having a differential size or thickness in only one direction in which case the evaluation requires only a single integration (see Example 10.7).

Like the moment of inertia, the product of inertia has units of length raised to the fourth power, e.g., m^4, mm^4 or ft^4, in^4. However, since x or y may be a negative quantity, while the element of area is always positive, the product of inertia may be positive, negative, or zero, depending on the location and orientation of the coordinate axes. For example, the product of inertia I_{xy} for an area will be *zero* if either the x or y axis is an axis of *symmetry* for the area. To show this, consider the shaded area in Fig. 10–12, where for every element dA located at point (x, y) there is a corresponding element dA located at $(x, -y)$. Since the products of inertia for these elements are, respectively, $xy\,dA$ and $-xy\,dA$, the algebraic sum or integration of all the elements that are chosen in this way will cancel each other. Consequently, the product of inertia for the

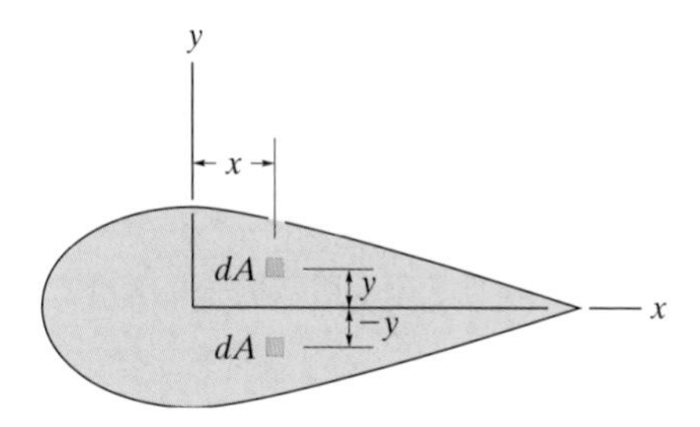

Fig. 10–12

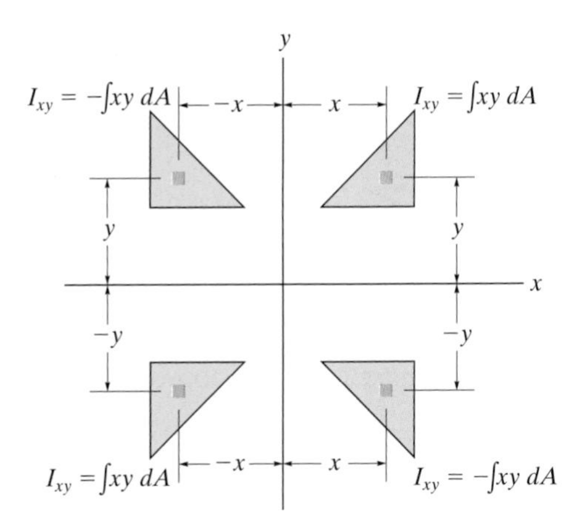

Fig. 10–13

total area becomes zero. It also follows from the definition of I_{xy} that the "sign" of this quantity depends on the quadrant where the area is located. As shown in Fig. 10–13, if the area is rotated from one quadrant to another, the sign of I_{xy} will change.

Parallel-Axis Theorem. Consider the shaded area shown in Fig. 10–14, where x' and y' represent a set of axes passing through the *centroid* of the area, and x and y represent a corresponding set of parallel axes. Since the product of inertia of dA with respect to the x and y axes is $dI_{xy} = (x' + d_x)(y' + d_y)\, dA$, then for the entire area,

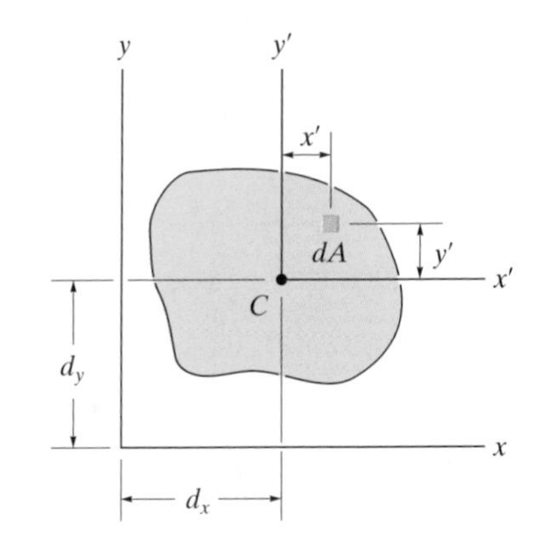

Fig. 10–14

$$I_{xy} = \int_A (x' + d_x)(y' + d_y)\, dA$$

$$= \int_A x'y'\, dA + d_x \int_A y'\, dA + d_y \int_A x'\, dA + d_x d_y \int_A dA$$

The first term on the right represents the product of inertia for the area with respect to the centroidal axis, $\bar{I}_{x'y'}$. The integrals in the second and third terms are zero since the moments of the area are taken about the centroidal axis. Realizing that the fourth integral represents the total area A, the final result is therefore

$$I_{xy} = \bar{I}_{x'y'} + A d_x d_y \qquad (10\text{–}8)$$

Note the similarity between this equation and the parallel-axis theorem for moments of inertia. In particular, it is important that the *algebraic signs* for d_x and d_y be maintained when applying Eq. 10–8. As illustrated in Example 10.8, the parallel-axis theorem finds important application in determining the product of inertia for a *composite area* with respect to a set of x, y axes.

EXAMPLE 10.7

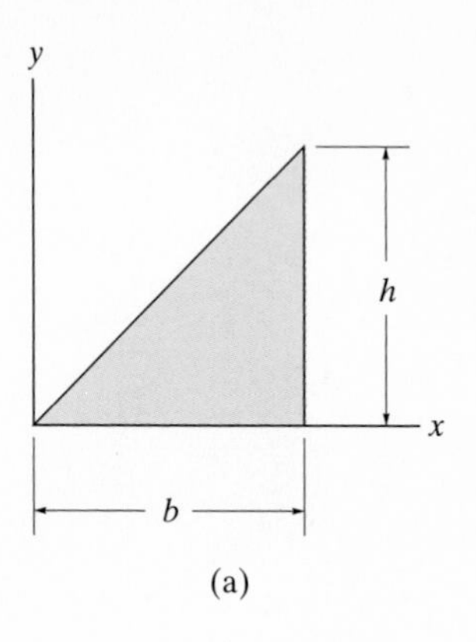

(a)

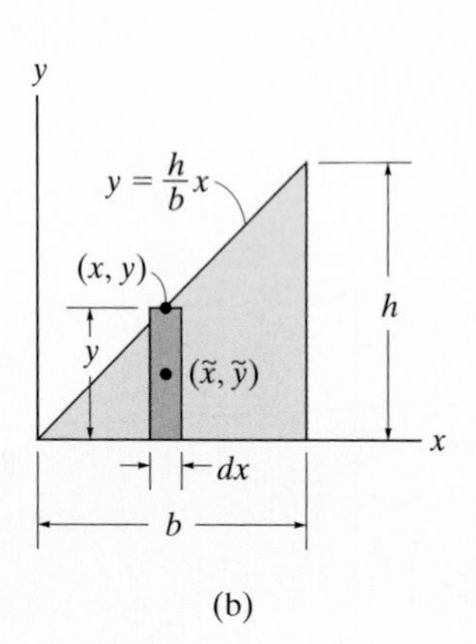

(b)

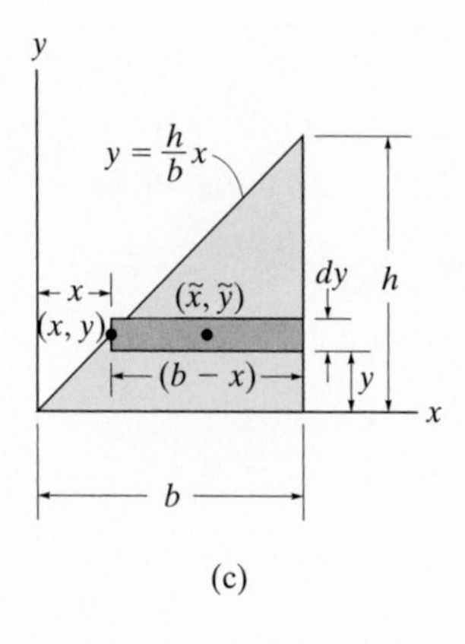

(c)

Fig. 10–15

Determine the product of inertia I_{xy} for the triangle shown in Fig. 10–15*a*.

SOLUTION I

A differential element that has a thickness dx, Fig. 10–15*b*, has an area $dA = y\,dx$. The product of inertia of the element about the x, y axes is determined using the parallel-axis theorem.

$$dI_{xy} = d\bar{I}_{x'y'} + dA\,\widetilde{x}\widetilde{y}$$

where $(\widetilde{x}, \widetilde{y})$ locates the *centroid* of the element or the origin of the x', y' axes. Since $d\bar{I}_{x'y'} = 0$, due to symmetry, and $\widetilde{x} = x$, $\widetilde{y} = y/2$, then

$$dI_{xy} = 0 + (y\,dx)x\left(\frac{y}{2}\right) = \left(\frac{h}{b}x\,dx\right)x\left(\frac{h}{2b}x\right)$$

$$= \frac{h^2}{2b^2}x^3\,dx$$

Integrating with respect to x from $x = 0$ to $x = b$ yields

$$I_{xy} = \frac{h^2}{2b^2}\int_0^b x^3\,dx = \frac{b^2h^2}{8} \qquad \textit{Ans.}$$

SOLUTION II

The differential element that has a thickness dy, Fig. 10–15*c*, and area $dA = (b - x)\,dy$ can also be used. The *centroid* is located at point $\widetilde{x} = x + (b - x)/2 = (b + x)/2$, $\widetilde{y} = y$, so the product of inertia of the element becomes

$$dI_{xy} = d\bar{I}_{x'y'} + dA\,\widetilde{x}\widetilde{y}$$

$$= 0 + (b - x)\,dy\left(\frac{b + x}{2}\right)y$$

$$= \left(b - \frac{b}{h}y\right)dy\left[\frac{b + (b/h)y}{2}\right]y = \frac{1}{2}y\left(b^2 - \frac{b^2}{h^2}y^2\right)dy$$

Integrating with respect to y from $y = 0$ to $y = h$ yields

$$I_{xy} = \frac{1}{2}\int_0^h y\left(b^2 - \frac{b^2}{h^2}y^2\right)dy = \frac{b^2h^2}{8} \qquad \textit{Ans.}$$

EXAMPLE 10.8

Determine the product of inertia for the beam's cross-sectional area, shown in Fig. 10–16*a*, about the *x* and *y* centroidal axes.

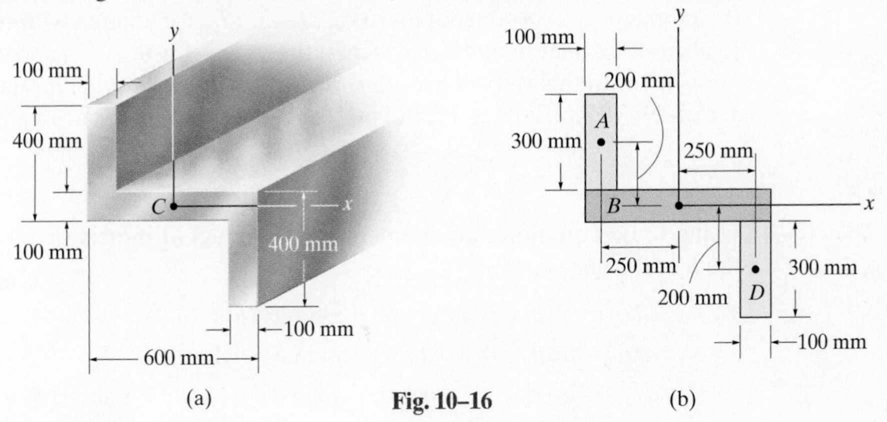

Fig. 10–16

SOLUTION

As in Example 10.6, the cross section can be considered as three composite rectangular areas *A*, *B*, and *D*, Fig. 10–16*b*. The coordinates for the centroid of each of these rectangles are shown in the figure. Due to symmetry, the product of inertia of *each rectangle* is *zero* about each set of x', y' axes that passes through the rectangle's centroid. Hence, application of the parallel-axis theorem to each of the rectangles yields

Rectangle A

$$I_{xy} = \bar{I}_{x'y'} + Ad_xd_y$$
$$= 0 + (300)(100)(-250)(200) = -1.50(10^9)\text{ mm}^4$$

Rectangle B

$$I_{xy} = \bar{I}_{x'y'} + Ad_xd_y$$
$$= 0 + 0 = 0$$

Rectangle D

$$I_{xy} = \bar{I}_{x'y'} + Ad_xd_y$$
$$= 0 + (300)(100)(250)(-200) = -1.50(10^9)\text{ mm}^4$$

The product of inertia for the entire cross section is therefore

$$I_{xy} = -1.50(10^9) + 0 - 1.50(10^9) = -3.00(10^9)\text{ mm}^4 \qquad \textit{Ans.}$$

NOTE: This negative result is due to the fact that rectangles *A* and *D* have centroids located with negative *x* and negative *y* coordinates, respectively.

*10.7 Moments of Inertia for an Area about Inclined Axes

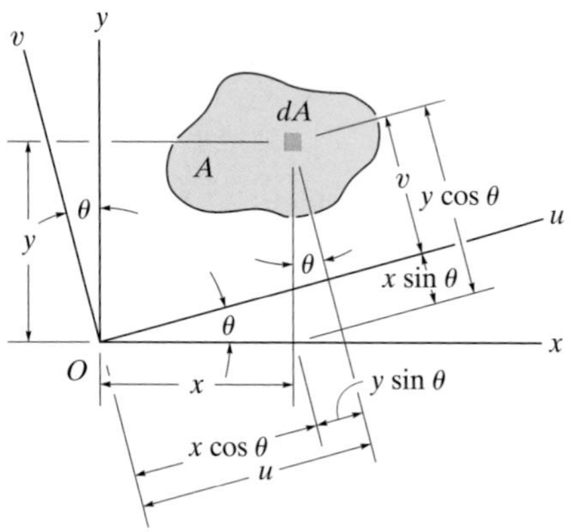

In structural and mechanical design, it is sometimes necessary to calculate the moments and product of inertia I_u, I_v, and I_{uv} for an area with respect to a set of inclined u and v axes when the values for θ, I_x, I_y, and I_{xy} are *known*. To do this we will use *transformation equations* which relate the x, y and u, v coordinates. From Fig. 10–17, these equations are

$$u = x\cos\theta + y\sin\theta$$
$$v = y\cos\theta - x\sin\theta$$

Fig. 10–17

Using these equations, the moments and product of inertia of dA about the u and v axes become

$$dI_u = v^2\,dA = (y\cos\theta - x\sin\theta)^2\,dA$$
$$dI_v = u^2\,dA = (x\cos\theta + y\sin\theta)^2\,dA$$
$$dI_{uv} = uv\,dA = (x\cos\theta + y\sin\theta)(y\cos\theta - x\sin\theta)\,dA$$

Expanding each expression and integrating, realizing that $I_x = \int y^2\,dA$, $I_y = \int x^2\,dA$, and $I_{xy} = \int xy\,dA$, we obtain

$$I_u = I_x\cos^2\theta + I_y\sin^2\theta - 2I_{xy}\sin\theta\cos\theta$$
$$I_v = I_x\sin^2\theta + I_y\cos^2\theta + 2I_{xy}\sin\theta\cos\theta$$
$$I_{uv} = I_x\sin\theta\cos\theta - I_y\sin\theta\cos\theta + I_{xy}(\cos^2\theta - \sin^2\theta)$$

These equations may be simplified by using the trigonometric identities $\sin 2\theta = 2\sin\theta\cos\theta$ and $\cos 2\theta = \cos^2\theta - \sin^2\theta$, in which case

$$I_u = \frac{I_x + I_y}{2} + \frac{I_x - I_y}{2}\cos 2\theta - I_{xy}\sin 2\theta$$
$$I_v = \frac{I_x + I_y}{2} - \frac{I_x - I_y}{2}\cos 2\theta + I_{xy}\sin 2\theta$$
$$I_{uv} = \frac{I_x - I_y}{2}\sin 2\theta + I_{xy}\cos 2\theta \qquad (10\text{–}9)$$

If the first and second equations are added together, we can show that the polar moment of inertia about the z axis passing through point O is *independent* of the orientation of the u and v axes; i.e.,

$$J_O = I_u + I_v = I_x + I_y$$

Principal Moments of Inertia. Equations 10–9 show that I_u, I_v, and I_{uv} depend on the angle of inclination, θ, of the u, v axes. We will now determine the orientation of these axes about which the moments of inertia for the area, I_u and I_v, are maximum and minimum. This particular set of axes is called the *principal axes* of the area, and the corresponding moments of inertia with respect to these axes are called the *principal moments of inertia*. In general, there is a set of principal axes for every chosen origin O. For the structural and mechanical design of a member, the origin O is generally located at the cross-sectional area's centroid.

The angle $\theta = \theta_p$, which defines the orientation of the principal axes for the area, may be found by differentiating the first of Eqs. 10–9 with respect to θ and setting the result equal to zero. Thus,

$$\frac{dI_u}{d\theta} = -2\left(\frac{I_x - I_y}{2}\right)\sin 2\theta - 2I_{xy}\cos 2\theta = 0$$

Therefore, at $\theta = \theta_p$,

$$\tan 2\theta_p = \frac{-I_{xy}}{(I_x - I_y)/2} \qquad (10\text{–}10)$$

This equation has two roots, θ_{p_1} and θ_{p_2}, which are 90° apart and so specify the inclination of the principal axes. In order to substitute them into Eq. 10–9, we must first find the sine and cosine of $2\theta_{p_1}$ and $2\theta_{p_2}$. This can be done using the triangles shown in Fig. 10–18, which are based on Eq. 10–10.

For θ_{p_1},

$$\sin 2\theta_{p_1} = -I_{xy}\Big/\sqrt{\left(\frac{I_x - I_y}{2}\right)^2 + I_{xy}^2}$$

$$\cos 2\theta_{p_1} = \left(\frac{I_x - I_y}{2}\right)\Big/\sqrt{\left(\frac{I_x - I_y}{2}\right)^2 + I_{xy}^2}$$

For θ_{p_2},

$$\sin 2\theta_{p_2} = I_{xy}\Big/\sqrt{\left(\frac{I_x - I_y}{2}\right)^2 + I_{xy}^2}$$

$$\cos 2\theta_{p_2} = -\left(\frac{I_x - I_y}{2}\right)\Big/\sqrt{\left(\frac{I_x - I_y}{2}\right)^2 + I_{xy}^2}$$

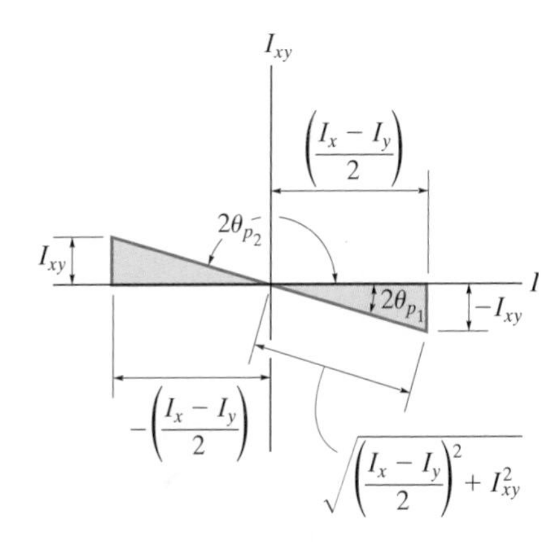

Fig. 10–18

Substituting these two sets of trigonometric relations into the first or second of Eqs. 10–9 and simplifying, we obtain

$$I_{\substack{max \\ min}} = \frac{I_x + I_y}{2} \pm \sqrt{\left(\frac{I_x - I_y}{2}\right)^2 + I_{xy}^2} \qquad (10\text{–}11)$$

Depending on the sign chosen, this result gives the maximum or minimum moment of inertia for the area. Furthermore, if the above trigonometric relations for θ_{p_1} and θ_{p_2} are substituted into the third of Eqs. 10–9, it can be shown that $I_{uv} = 0$; that is, the *product of inertia with respect to the principal axes is zero*. Since it was indicated in Sec. 10.6 that the product of inertia is zero with respect to any symmetrical axis, it therefore follows that *any symmetrical axis represents a principal axis of inertia for the area.*

EXAMPLE 10.9

Determine the principal moments of inertia for the beam's cross-sectional area shown in Fig. 10–19*a* with respect to an axis passing through the centroid.

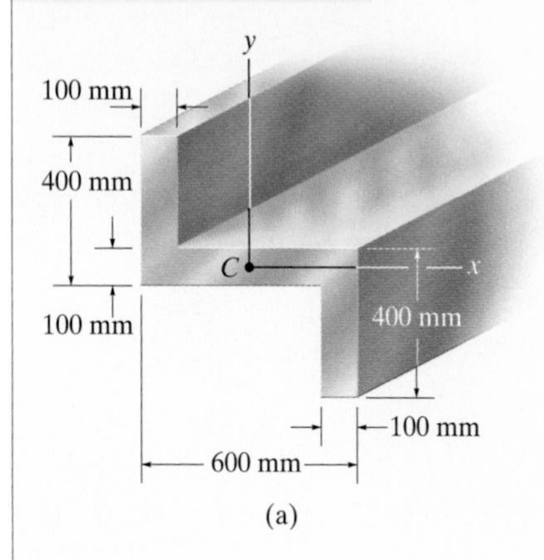

(a)

SOLUTION

The moments and product of inertia of the cross section with respect to the *x*, *y* axes have been computed in Examples 10.6 and 10.8. The results are

$$I_x = 2.90(10^9)\text{ mm}^4 \quad I_y = 5.60(10^9)\text{ mm}^4 \quad I_{xy} = -3.00(10^9)\text{ mm}^4$$

Using Eq. 10–10, the angles of inclination of the principal axes *u* and *v* are

$$\tan 2\theta_p = \frac{-I_{xy}}{(I_x - I_y)/2} = \frac{3.00(10^9)}{[2.90(10^9) - 5.60(10^9)]/2} = -2.22$$

$$2\theta_{p_1} = -65.8^\circ \quad \text{and} \quad 2\theta_{p_2} = 114.2^\circ$$

Thus, as shown in Fig. 10–19*b*,

$$\theta_{p_1} = -32.9^\circ \quad \text{and} \quad \theta_{p_2} = 57.1^\circ$$

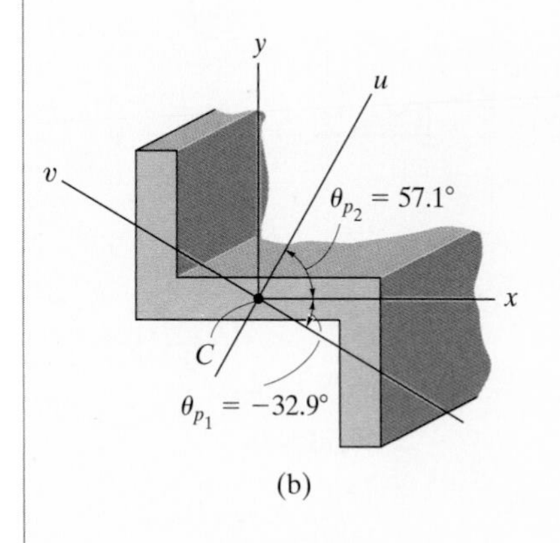

(b)

Fig. 10–19

The principal moments of inertia with respect to the *u* and *v* axes are determined from Eq. 10–11. Hence,

$$I_{\substack{\max\\\min}} = \frac{I_x + I_y}{2} \pm \sqrt{\left(\frac{I_x - I_y}{2}\right)^2 + I_{xy}^2}$$

$$= \frac{2.90(10^9) + 5.60(10^9)}{2} \pm \sqrt{\left[\frac{2.90(10^9) - 5.60(10^9)}{2}\right]^2 + [-3.00(10^9)]^2}$$

$$I_{\substack{\max\\\min}} = 4.25(10^9) \pm 3.29(10^9)$$

or

$$I_{\max} = 7.54(10^9)\text{ mm}^4 \qquad I_{\min} = 0.960(10^9)\text{ mm}^4 \qquad \textit{Ans.}$$

NOTE: The maximum moment of inertia, $I_{\max} = 7.54(10^9)\text{ mm}^4$, occurs with respect to the selected *u* axis since *by inspection* most of the cross-sectional area is farthest away from this axis. Or, stated in another manner, $I_{\max}$ occurs about the *u* axis since it is located within $\pm 45^\circ$ of the *y* axis, which has the largest value of $I\,(I_y > I_x)$. Also, this may be concluded by substituting the data with $\theta = 57.1^\circ$ into the first of Eqs. 10–9.

*10.8 Mohr's Circle for Moments of Inertia

Equations 10–9 to 10–11 have a graphical solution that is convenient to use and generally easy to remember. Squaring the first and third of Eqs. 10–9 and adding, it is found that

$$\left(I_u - \frac{I_x + I_y}{2}\right)^2 + I_{uv}^2 = \left(\frac{I_x - I_y}{2}\right)^2 + I_{xy}^2 \qquad (10\text{–}12)$$

In a given problem, I_u and I_{uv} are *variables*, and I_x, I_y, and I_{xy} are *known constants*. Thus, Eq. 10–12 may be written in compact form as

$$(I_u - a)^2 + I_{uv}^2 = R^2$$

When this equation is plotted on a set of axes that represent the respective moment of inertia and the product of inertia, Fig. 10–20, the resulting graph represents a *circle* of radius

$$R = \sqrt{\left(\frac{I_x - I_y}{2}\right)^2 + I_{xy}^2}$$

having its center located at point $(a, 0)$, where $a = (I_x + I_y)/2$. The circle so constructed is called *Mohr's circle*, named after the German engineer Otto Mohr (1835–1918).

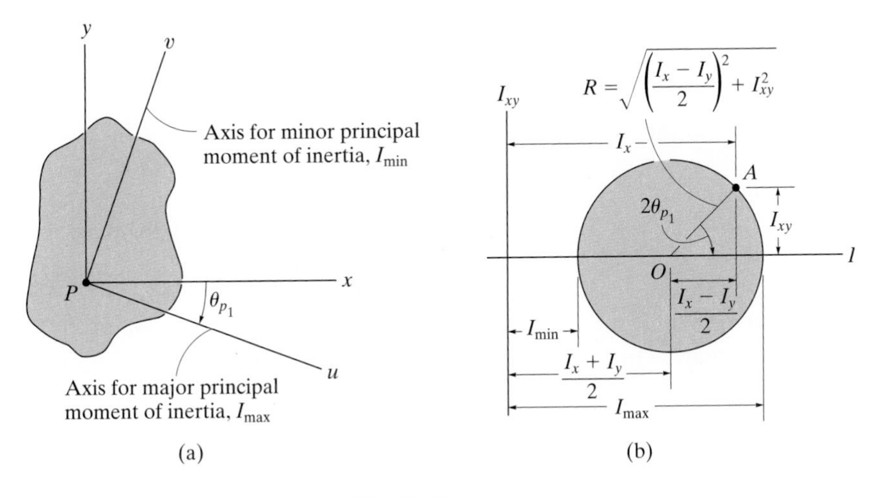

Fig. 10–20

Procedure for Analysis

The main purpose in using Mohr's circle here is to have a convenient means for transforming I_x, I_y, and I_{xy} into the principal moments of inertia. The following procedure provides a method for doing this.

Determine I_x, I_y, and I_{xy}.

- Establish the x, y axes for the area, with the origin located at the point P of interest, and determine I_x, I_y, and I_{xy}, Fig. 10–20*a*.

Construct the Circle.

- Construct a rectangular coordinate system such that the abscissa represents the moment of inertia I, and the ordinate represents the product of inertia I_{xy}, Fig. 10–20*b*.
- Determine the center of the circle, O, which is located at a distance $(I_x + I_y)/2$ from the origin, and plot the reference point A having coordinates (I_x, I_{xy}). By definition, I_x is always positive, whereas I_{xy} will be either positive or negative.
- Connect the reference point A with the center of the circle and determine the distance OA by trigonometry. This distance represents the radius of the circle, Fig. 10–20*b*. Finally, draw the circle.

Principal Moments of Inertia.

- The points where the circle intersects the abscissa give the values of the principal moments of inertia I_{min} and I_{max}. Notice that the *product of inertia will be zero at these points*, Fig. 10–20*b*.

Principal Axes.

- To find the direction of the major principal axis, determine by trigonometry the angle $2\theta_{p_1}$, *measured from the radius OA to the positive I axis*, Fig. 10–20*b*. This angle represents *twice* the angle from the x axis of the area in question to the axis of maximum moment of inertia I_{max}, Fig. 10–20*a*. Both the angle on the circle, $2\theta_{p_1}$, and the angle to the axis on the area, θ_{p_1}, *must be measured in the same sense*, as shown in Fig. 10–20. The axis for minimum moment of inertia I_{min} is perpendicular to the axis for I_{max}.

Using trigonometry, the above procedure may be verified to be in accordance with the equations developed in Sec. 10.7.

EXAMPLE 10.10

Using Mohr's circle, determine the principal moments of inertia for the beam's cross-sectional area shown in Fig. 10–21*a*, with respect to an axis passing through the centroid.

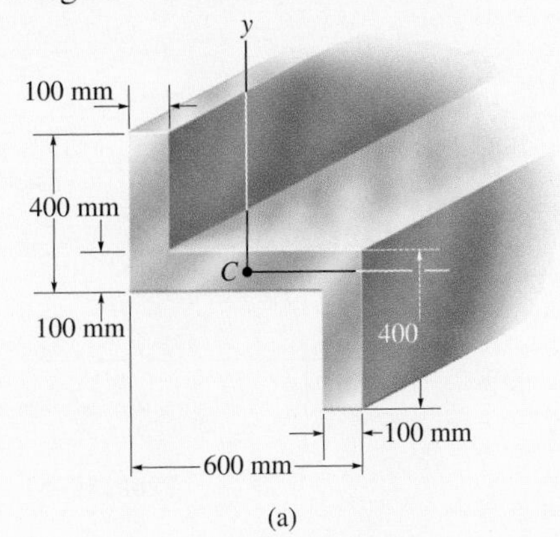

(a)

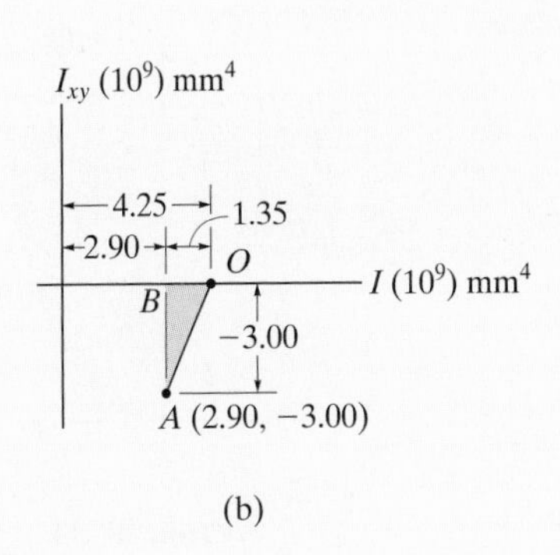

(b)

SOLUTION

Determine I_x, I_y, I_{xy}. The moment of inertia and the product of inertia have been determined in Examples 10.6 and 10.8 with respect to the *x, y* axes shown in Fig. 10–21*a*. The results are $I_x = 2.90(10^9)$ mm^4, $I_y = 5.60(10^9)$ mm^4, and $I_{xy} = -3.00(10^9)$ mm^4.

Construct the Circle. The I and I_{xy} axes are shown in Fig. 10–21*b*. The center of the circle, O, lies at a distance $(I_x + I_y)/2 = (2.90 + 5.60)/2 = 4.25$ from the origin. When the reference point $A(2.90, -3.00)$ is connected to point O, the radius OA is determined from the triangle OBA using the Pythagorean theorem.

$$OA = \sqrt{(1.35)^2 + (-3.00)^2} = 3.29$$

The circle is constructed in Fig. 10–21*c*.

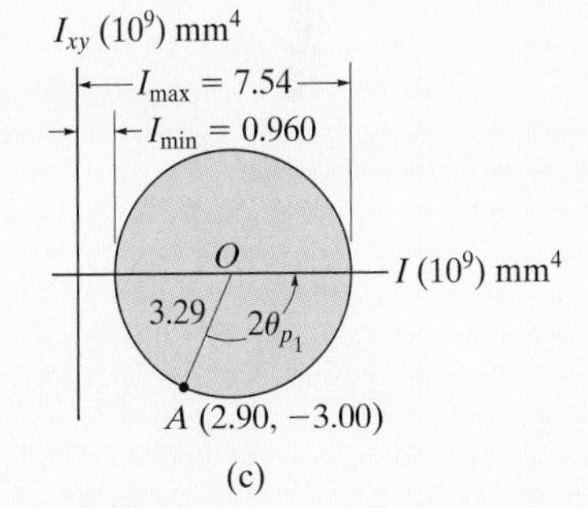

(c)

Principal Moments of Inertia. The circle intersects the I axis at points (7.54, 0) and (0.960, 0). Hence,

$$I_{max} = 7.54(10^9)\ \text{mm}^4 \qquad \textit{Ans.}$$

$$I_{min} = 0.960(10^9)\ \text{mm}^4 \qquad \textit{Ans.}$$

Principal Axes. As shown in Fig. 10–21*c*, the angle $2\theta_{p_1}$ is determined from the circle by measuring counterclockwise from OA to the direction of the *positive* I axis. Hence,

$$2\theta_{p_1} = 180° - \sin^{-1}\left(\frac{|BA|}{|OA|}\right) = 180° - \sin^{-1}\left(\frac{3.00}{3.29}\right) = 114.2°$$

The principal axis for $I_{max} = 7.54(10^9)$ mm^4 is therefore oriented at an angle $\theta_{p_1} = 57.1°$, measured *counterclockwise*, from the *positive x* axis to the *positive u* axis. The *v* axis is perpendicular to this axis. The results are shown in Fig. 10–21*d*.

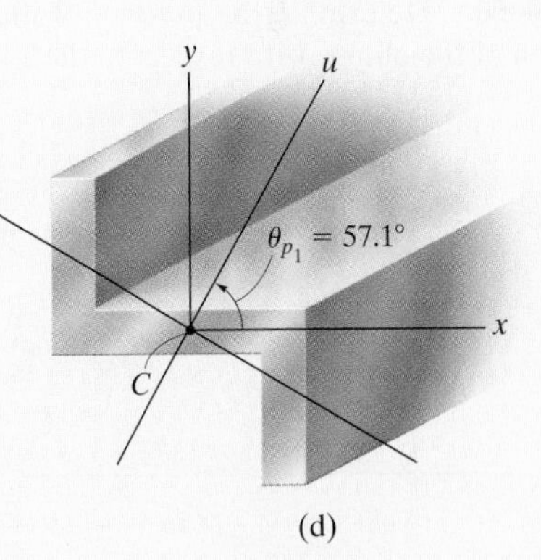

(d)

Fig. 10–21

PROBLEMS

10–54. Determine the product of inertia for the shaded portion of the parabola with respect to the x and y axes.

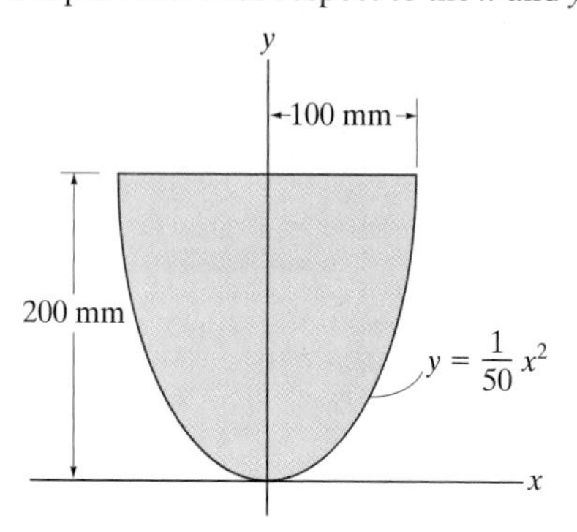

Prob. 10–54

10–55. Determine the product of inertia for the shaded area with respect to the x and y axes.

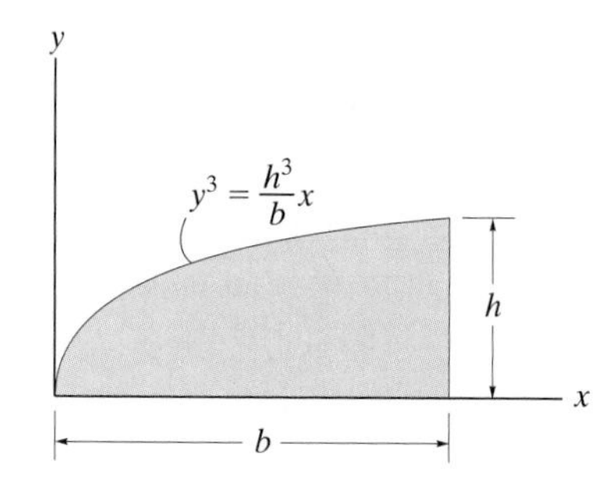

Prob. 10–55

***10–56.** Determine the product of inertia for the shaded area of the ellipse with respect to the x and y axes.

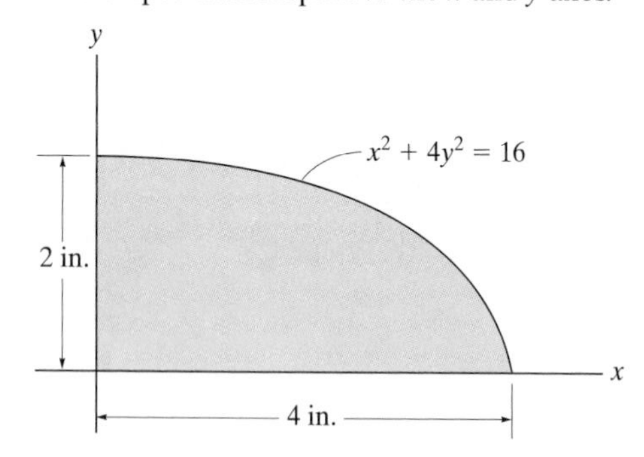

Prob. 10–56

10–57. Determine the product of inertia for the parabolic area with respect to the x and y axes.

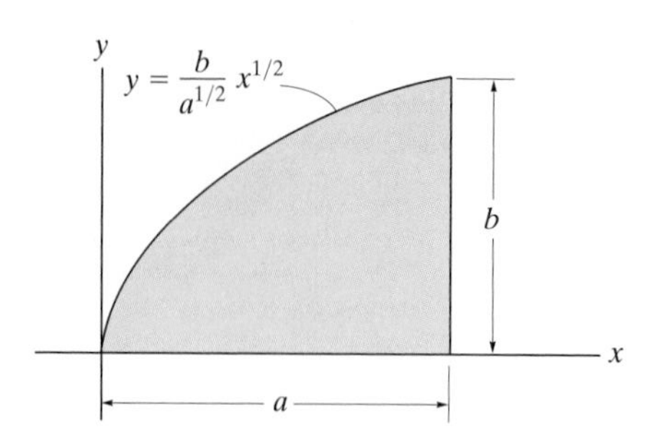

Prob. 10–57

10–58. Determine the product of inertia for the shaded area with respect to the x and y axes.

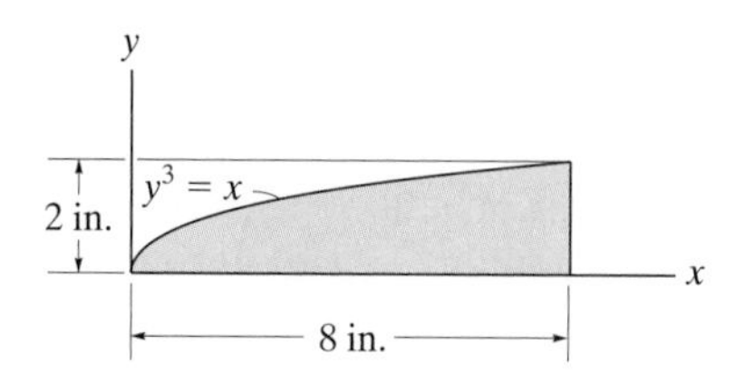

Prob. 10–58

10–59. Determine the product of inertia for the shaded parabolic area with respect to the x and y axes.

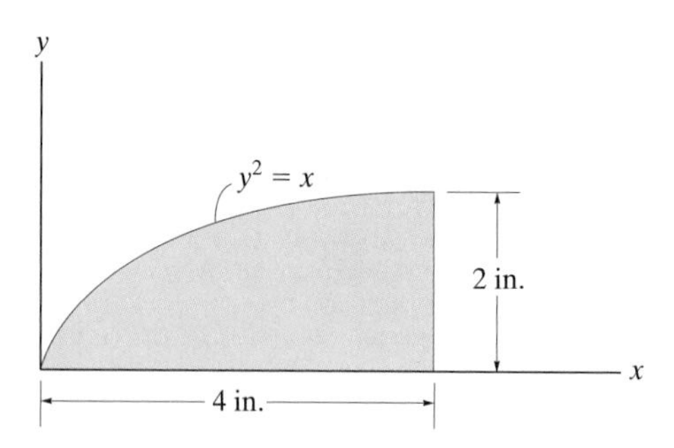

Prob. 10–59

*10–60. Determine the product of inertia for the shaded area with respect to the x and y axes.

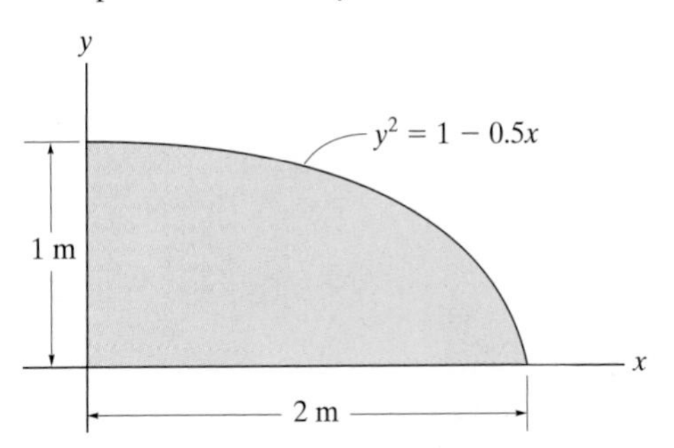

Prob. 10–60

10–61. Determine the product of inertia for the shaded area with respect to the x and y axes.

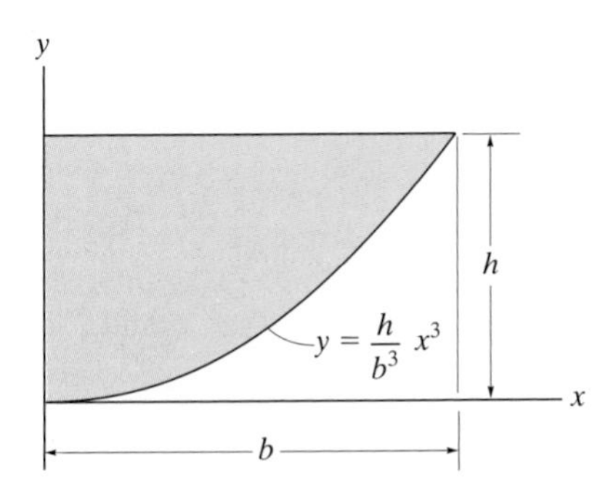

Prob. 10–61

10–62. Determine the product of inertia for the shaded area with respect to the x and y axes.

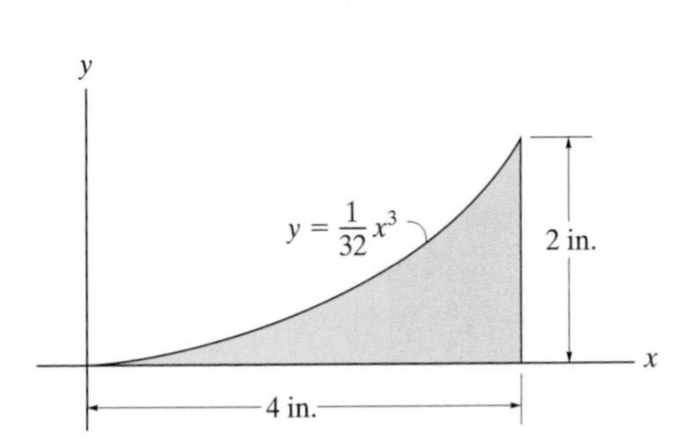

Prob. 10–62

10–63. Determine the product of inertia for the shaded area with respect to the x and y axes.

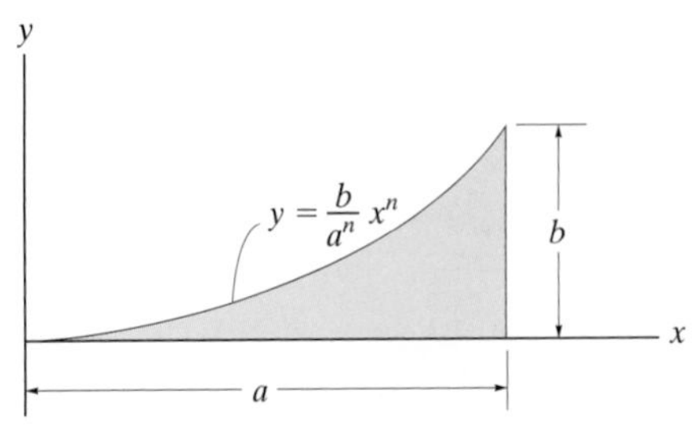

Prob. 10–63

*10–64. Determine the product of inertia for the shaded area with respect to the x and y axes.

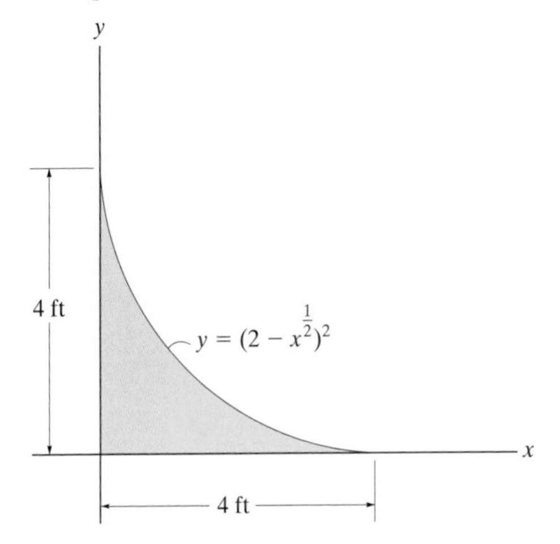

Prob. 10–64

■10–65. Determine the product of inertia for the shaded area with respect to the x and y axes. Use Simpson's rule to evaluate the integral.

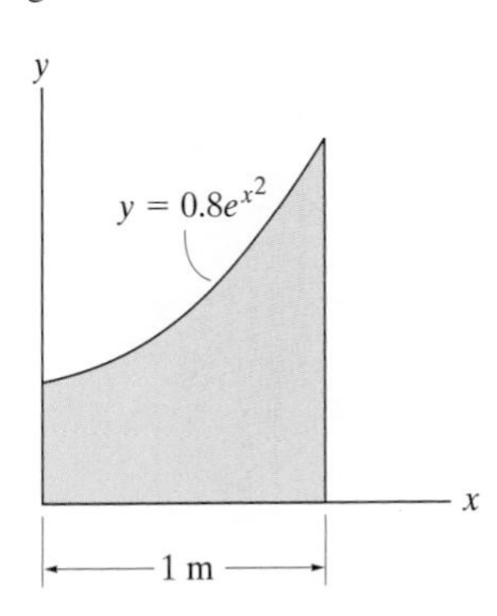

Prob. 10–65

10–66. Determine the product of inertia for the parabolic area with respect to the x and y axes.

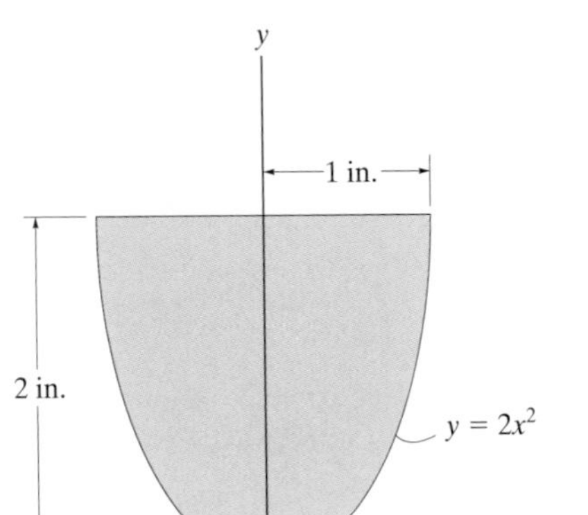
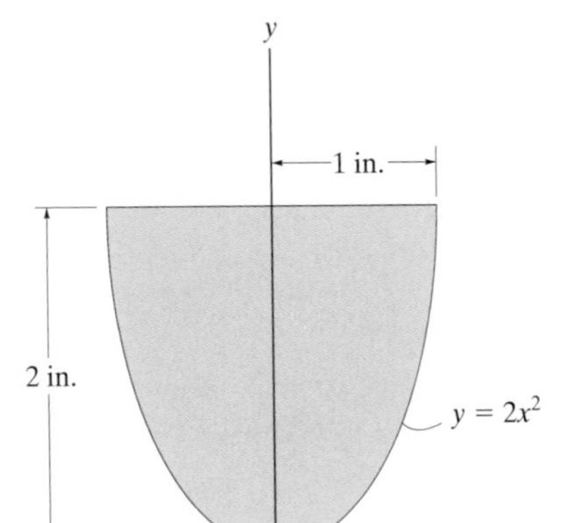

Prob. 10–66

10–67. Determine the product of inertia for the cross-sectional area with respect to the x and y axes that have their origin located at the centroid C.

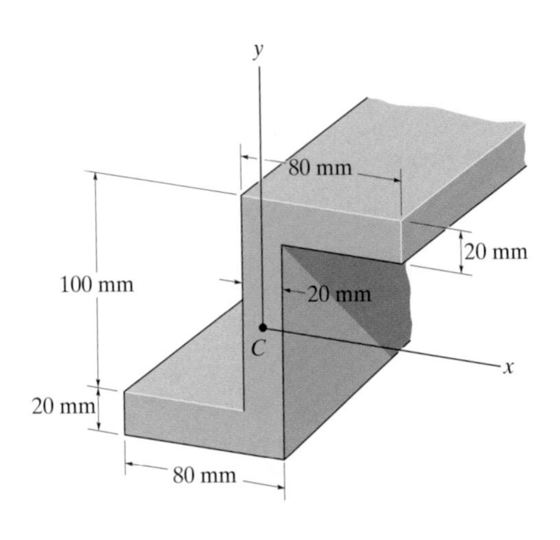

Prob. 10–67

***10–68.** Determine the product of inertia for the beam's cross-sectional area with respect to the x and y axes.

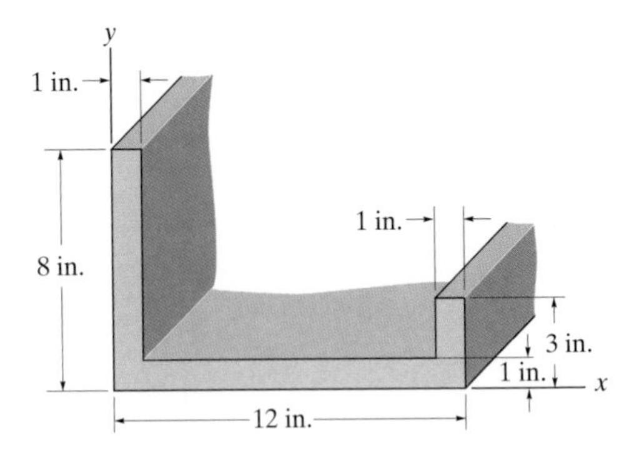

Prob. 10–68

10–69. Determine the location $(\bar{x}, \bar{y})$ of the centroid C of the angle's cross-sectional area, and then compute the product of inertia with respect to the x' and y' axes.

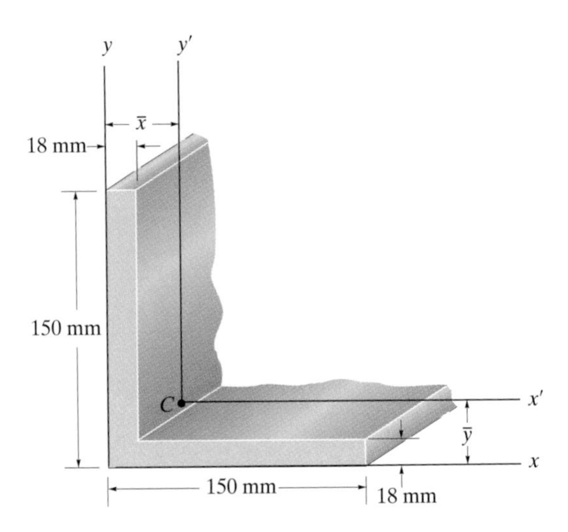

Prob. 10–69

10–70. Determine the product of inertia for the beam's cross-sectional area with respect to the x and y axes that have their origin located at the centroid C.

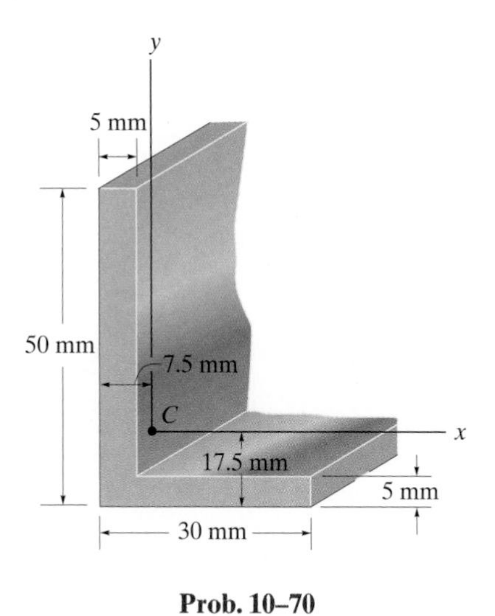

Prob. 10–70

***10–72.** Determine the product of inertia for the beam's cross-sectional area with respect to the x and y axes that have their origin located at the centroid C.

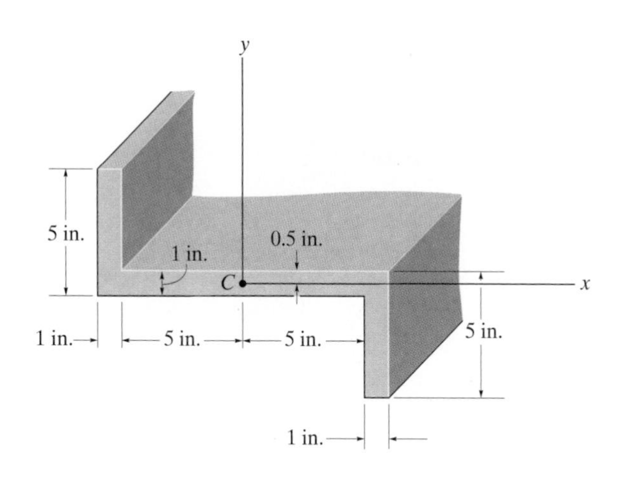

Prob. 10–72

10–71. Determine the product of inertia for the shaded area with respect to the x and y axes.

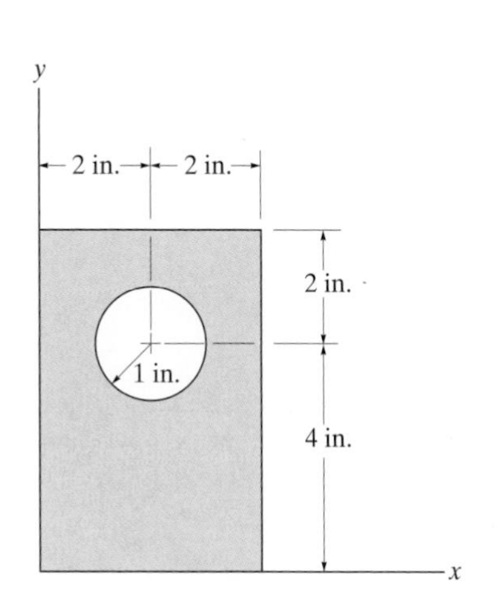

Prob. 10–71

10–73. Determine the product of inertia for the cross-sectional area with respect to the x and y axes.

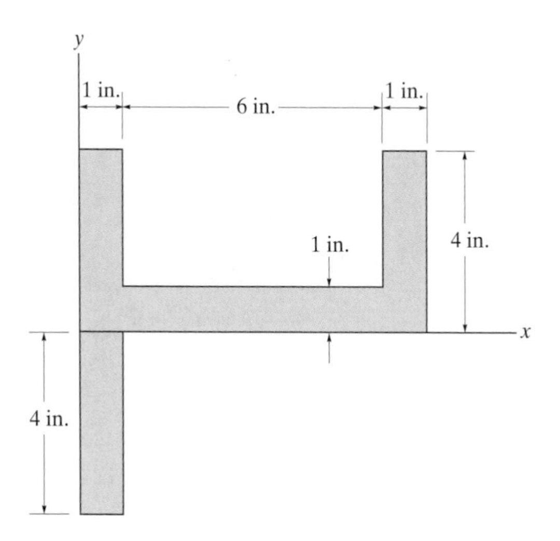

Prob. 10–73

10–74. Determine the product of inertia for the beam's cross-sectional area with respect to the u and v axes.

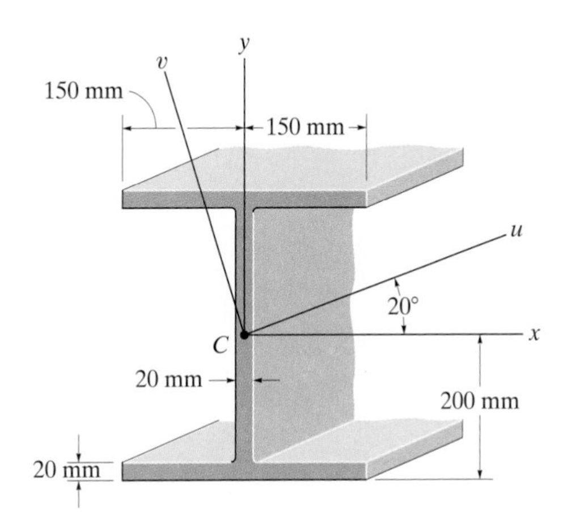

Prob. 10–74

10–75. Determine the moments of inertia I_u and I_v and the product of inertia I_{uv} for the rectangular area. The u and v axes pass through the centroid C.

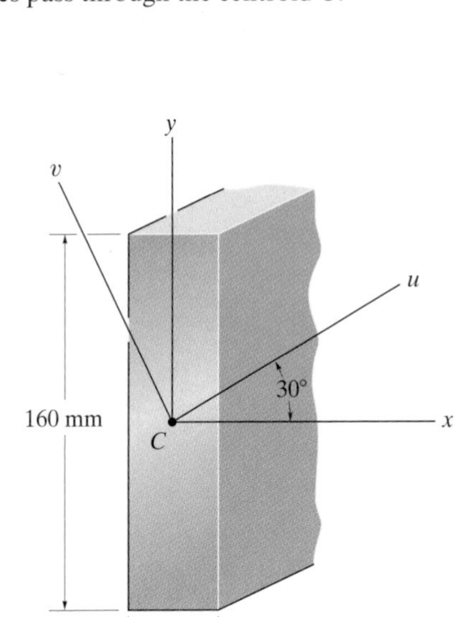

Prob. 10–75

***10–76.** Determine the distance $\bar{y}$ to the centroid of the area and then calculate the moments of inertia I_u and I_v for the channel's cross-sectional area. The u and v axes have their origin at the centroid C. For the calculation, assume all corners to be square.

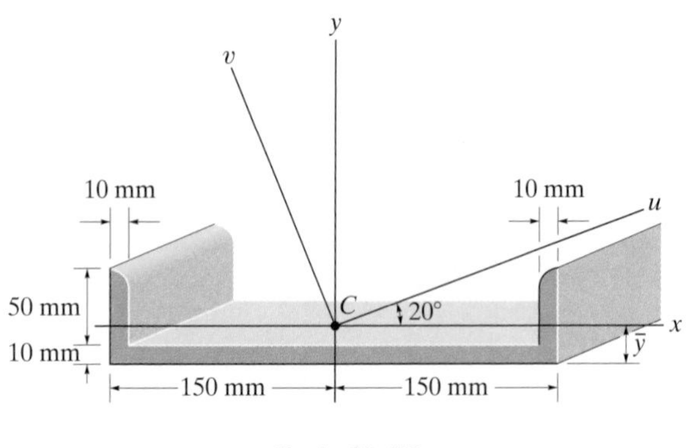

Prob. 10–76

10–77. Determine the moments of inertia for the shaded area with respect to the u and v axes.

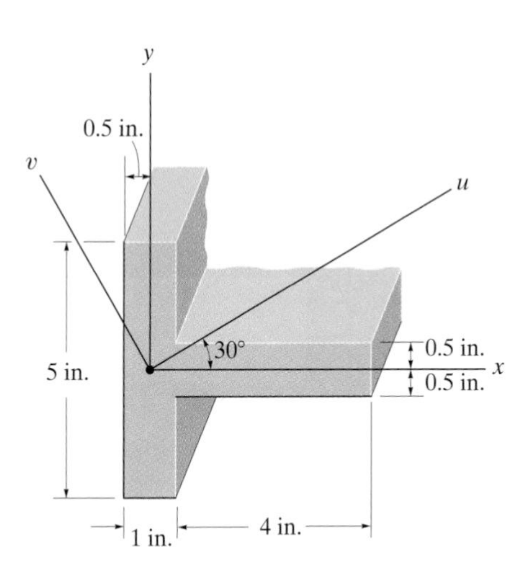

Prob. 10–77

10–78. Determine the directions of the principal axes with origin located at point O, and the principal moments of inertia for the rectangular area about these axes.

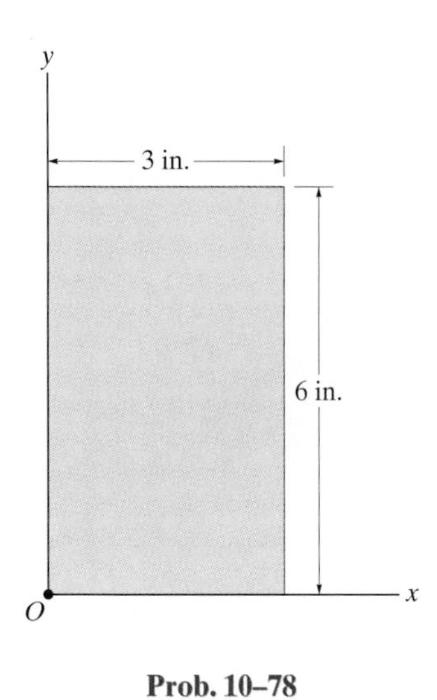

Prob. 10–78

10–79. Determine the moments of inertia I_u, I_v and the product of inertia I_{uv} for the beam's cross-sectional area. Take $\theta = 45°$.

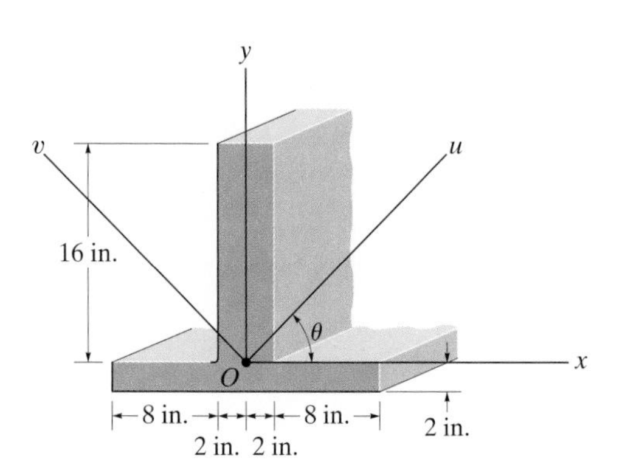

Prob. 10–79

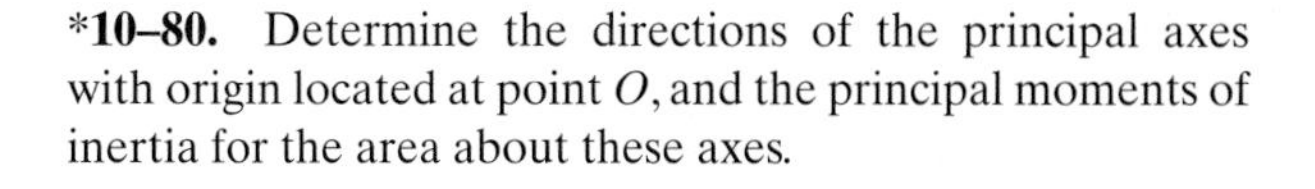

***10–80.** Determine the directions of the principal axes with origin located at point O, and the principal moments of inertia for the area about these axes.

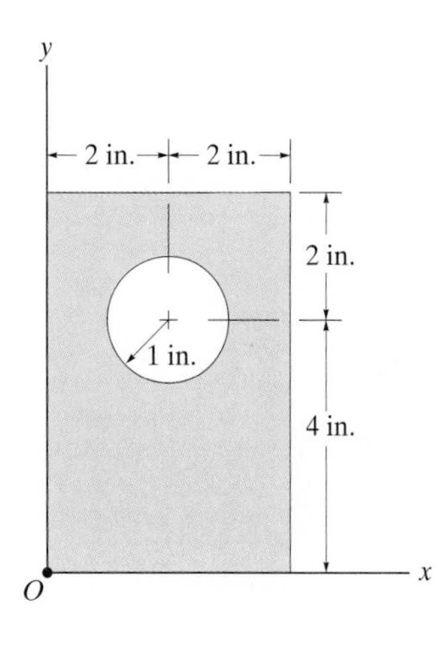

Prob. 10–80

10–81. Determine the principal moments of inertia for the beam's cross-sectional area about the principal axes that have their origin located at the centroid C. Use the equations developed in Section 10.7. For the calculation, assume all corners to be square.

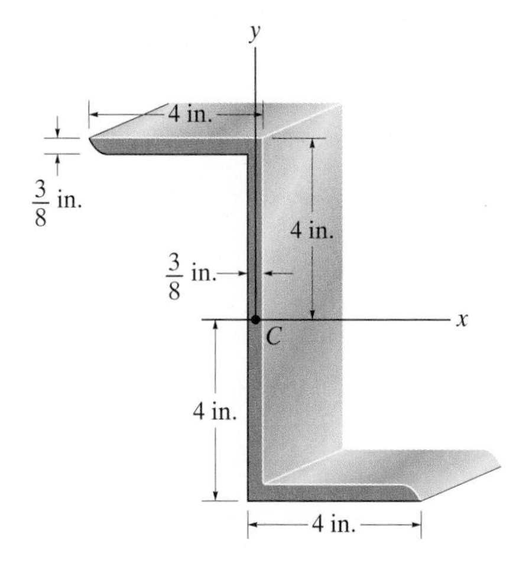

Prob. 10–81

10–82. Determine the principal moments of inertia for the angle's cross-sectional area with respect to a set of principal axes that have their origin located at the centroid C. Use the equation developed in Section 10.7. For the calculation, assume all corners to be square.

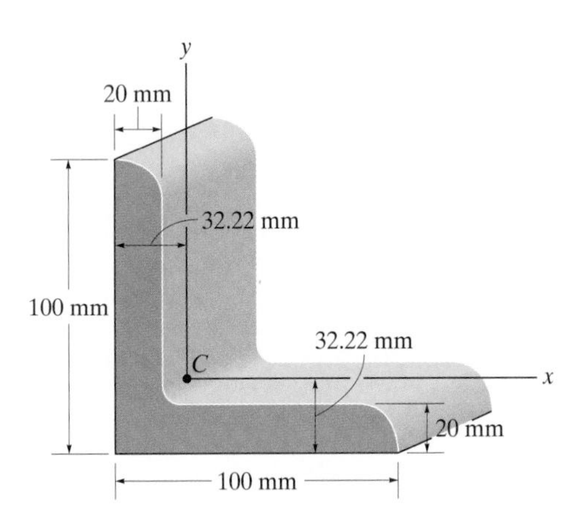

Prob. 10–82

***10–84.** Using Mohr's circle, determine the principal moments of inertia for the triangular area and the orientation of the principal axes of inertia having an origin at point O.

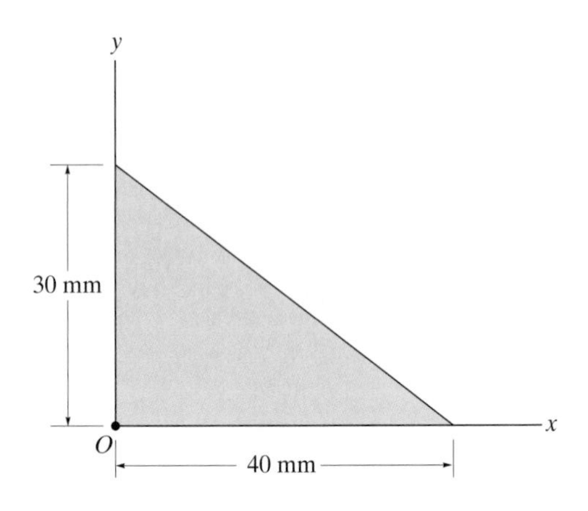

Prob. 10–84

10–83. The area of the cross section of an airplane wing has the following properties about the x and y axes passing through the centroid C: $\bar{I}_x = 450 \text{ in}^4$, $\bar{I}_y = 1730 \text{ in}^4$, $\bar{I}_{xy} = 138 \text{ in}^4$. Determine the orientation of the principal axes and the principal moments of inertia.

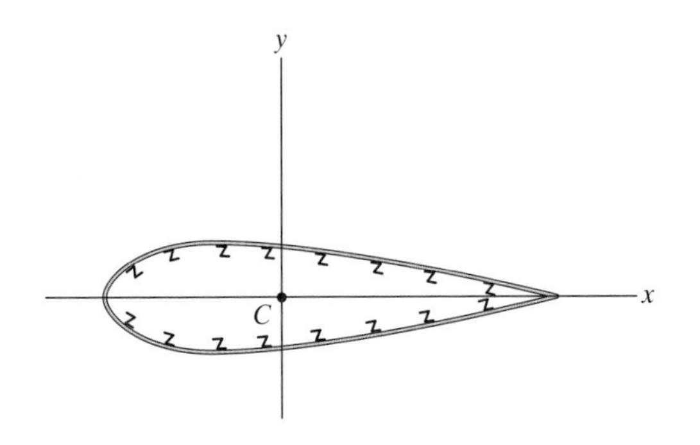

Prob. 10–83

10–85. Solve Prob. 10–78 using Mohr's circle.

10–86. Solve Prob. 10–81 using Mohr's circle.

10–87. Solve Prob. 10–82 using Mohr's circle.

***10–88.** Solve Prob. 10–80 using Mohr's circle.

10–89. Solve Prob. 10–83 using Mohr's circle.

10.9 Mass Moment of Inertia

The mass moment of inertia of a body is a property that measures the resistance of the body to angular acceleration. Since it is used in dynamics to study rotational motion, methods for its calculation will now be discussed.

We define the *mass moment of inertia* as the integral of the "second moment" about an axis of all the elements of mass dm which compose the body.* For example, consider the rigid body shown in Fig. 10–22. The body's moment of inertia about the z axis is

$$I = \int_m r^2\,dm \qquad (10\text{–}13)$$

Here the "moment arm" r is the perpendicular distance from the axis to the arbitrary element dm. Since the formulation involves r, the value of I is *unique* for each axis z about which it is computed. However, the axis which is generally chosen for analysis passes through the body's mass center G. The moment of inertia computed about this axis will be defined as I_G. Realize that because r is squared in Eq. 10–13, the mass moment of inertia is always a *positive quantity*. Common units used for its measurement are $\text{kg} \cdot \text{m}^2$ or $\text{slug} \cdot \text{ft}^2$.

If the body consists of material having a variable density, $\rho = \rho(x, y, z)$, the elemental mass dm of the body may be expressed in terms of its density and volume as $dm = \rho\,dV$. Substituting dm into Eq. 10–13, the body's moment of inertia is then computed using *volume elements* for integration; i.e.

$$I = \int_V r^2 \rho\,dV \qquad (10\text{–}14)$$

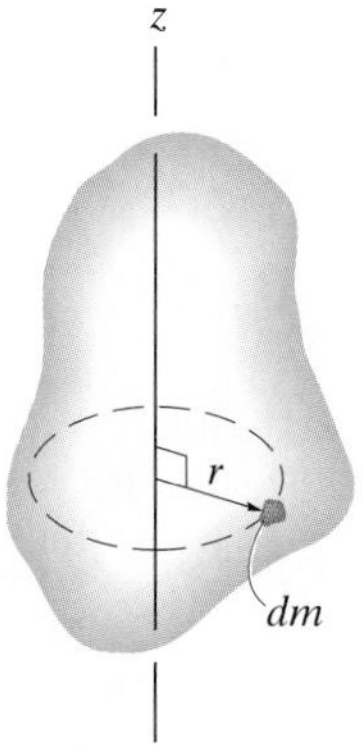

z

r

dm

Fig. 10–22

*Another property of the body which measures the symmetry of the body's mass with respect to a coordinate system is the mass product of inertia. This property most often applies to the three-dimensional motion of a body and is discussed in *Engineering Mechanics: Dynamics* (Chapter 21).

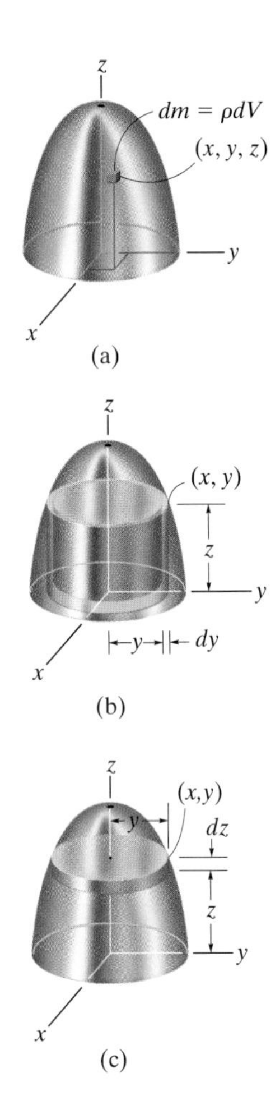

Fig. 10–23

In the special case of ρ being a *constant*, this term may be factored out of the integral, and the integration is then purely a function of geometry:

$$I = \rho \int_V r^2 \, dV \tag{10–15}$$

When the elemental volume chosen for integration has differential sizes in all three directions, e.g., $dV = dx\,dy\,dz$, Fig. 10–23*a*, the moment of inertia of the body must be determined using "triple integration." The integration process can, however, be simplified to a *single integration* provided the chosen elemental volume has a differential size or thickness in only *one direction*. Shell or disk elements are often used for this purpose.

Procedure For Analysis

For integration, we will consider only symmetric bodies having surfaces which are generated by revolving a curve about an axis. An example of such a body which is generated about the z axis is shown in Fig. 10–23.

Shell Element.

- If a *shell element* having a height z, radius y, and thickness dy is chosen for integration, Fig. 10–23*b*, then the volume $dV = (2\pi y)(z)\,dy$.
- This element may be used in Eq. 10–14 or 10–15 for determining the moment of inertia I_z of the body about the z axis since the *entire element*, due to its "thinness," lies at the *same* perpendicular distance $r = y$ from the z axis (see Example 10.11).

Disk Element.

- If a disk element having a radius y and a thickness dz is chosen for integration, Fig. 10–23*c*, then the volume $dV = (\pi y^2)\,dz$.
- In this case the element is *finite* in the radial direction, and consequently its parts *do not* all lie at the *same radial distance r* from the z axis. As a result, Eqs. 10–14 or 10–15 *cannot* be used to determine I_z. Instead, to perform the integration using this element, it is first necessary to determine the moment of inertia *of the element* about the z axis and then integrate this result (see Example 10.12).

EXAMPLE 10.11

Determine the mass moment of inertia of the cylinder shown in Fig. 10–24*a* about the z axis. The density of the material, ρ, is constant.

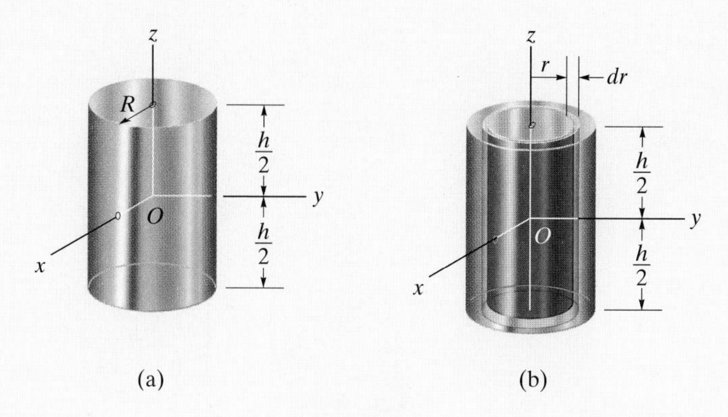

Fig. 10–24

SOLUTION

Shell Element. This problem may be solved using the *shell element* in Fig. 10–24*b* and single integration. The volume of the element is $dV = (2\pi r)(h)\,dr$, so that its mass is $dm = \rho\,dV = \rho(2\pi hr\,dr)$. Since the *entire element* lies at the same distance r from the z axis, the moment of inertia *of the element* is

$$dI_z = r^2\,dm = \rho 2\pi h r^3\,dr$$

Integrating over the entire region of the cylinder yields

$$I_z = \int_m r^2\,dm = \rho 2\pi h \int_0^R r^3\,dr = \frac{\rho\pi}{2} R^4 h$$

The mass of the cylinder is

$$m = \int_m dm = \rho 2\pi h \int_0^R r\,dr = \rho\pi h R^2$$

so that

$$I_z = \frac{1}{2} m R^2 \qquad \textit{Ans.}$$

EXAMPLE 10.12

A solid is formed by revolving the shaded area shown in Fig. 10–25*a* about the *y* axis. If the density of the material is 5 slug/ft^3, determine the mass moment of inertia about the *y* axis.

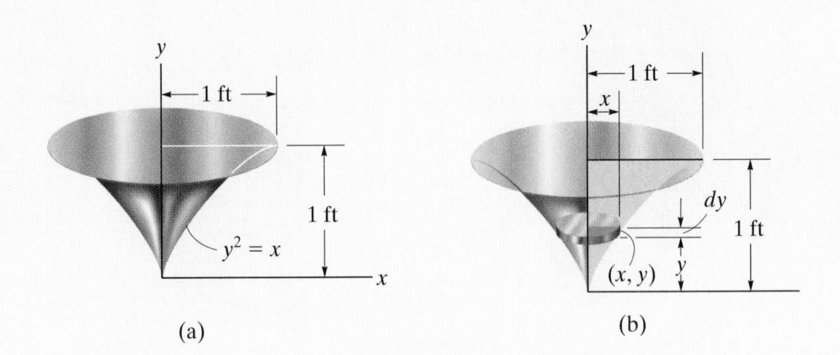

Fig. 10–25

SOLUTION

Disk Element. The moment of inertia will be determined using a *disk element*, as shown in Fig. 10–25*b*. Here the element intersects the curve at the arbitrary point (x, y) and has a mass

$$dm = \rho\, dV = \rho(\pi x^2)\, dy$$

Although all portions of the element are *not* located at the same distance from the *y* axis, it is still possible to determine the moment of inertia dI_y *of the element* about the *y* axis. In Example 10.11 it was shown that the moment of inertia of a cylinder about its longitudinal axis is $I = \frac{1}{2}mR^2$, where m and R are the mass and radius of the cylinder. Since the height of the cylinder is not involved in this formula, we can also use it for a disk. Thus, for the disk element in Fig. 10–25*b*, we have

$$dI_y = \frac{1}{2}(dm)x^2 = \frac{1}{2}[\rho(\pi x^2)\, dy]x^2$$

Substituting $x = y^2$, $\rho = 5$ slug/ft^3, and integrating with respect to y, from $y = 0$ to $y = 1$ ft, yields the moment of inertia for the entire solid:

$$I_y = \frac{5\pi}{2}\int_0^1 x^4\, dy = \frac{5\pi}{2}\int_0^1 y^8\, dy = 0.873 \text{ slug}\cdot\text{ft}^2 \qquad \textit{Ans.}$$

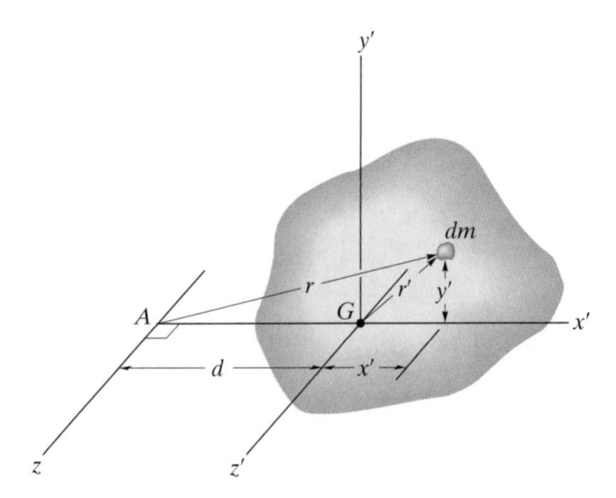

Fig. 10–26

Parallel-Axis Theorem. If the moment of inertia of the body about an axis passing through the body's mass center is known, then the moment of inertia about any other *parallel axis* may be determined by using the *parallel-axis theorem*. This theorem can be derived by considering the body shown in Fig. 10–26. The z' axis passes through the mass center G, whereas the corresponding *parallel z axis* lies at a constant distance d away. Selecting the differential element of mass dm which is located at point (x', y') and using the Pythagorean theorem, $r^2 = (d + x')^2 + y'^2$, we can express the moment of inertia of the body about the z axis as

$$I = \int_m r^2\, dm = \int_m [(d + x')^2 + y'^2]\, dm$$

$$= \int_m (x'^2 + y'^2)\, dm + 2d \int_m x'\, dm + d^2 \int_m dm$$

Since $r'^2 = x'^2 + y'^2$, the first integral represents I_G. The second integral equals *zero*, since the z' axis passes through the body's mass center, i.e., $\int x'\, dm = \overline{x} \int dm = 0$ since $\overline{x} = 0$. Finally, the third integral represents the total mass m of the body. Hence, the moment of inertia about the z axis can be written as

$$I = I_G + md^2 \tag{10–16}$$

where

I_G = moment of inertia about the z' axis passing through the mass center G

m = mass of the body

d = distance between the parallel axes

Radius of Gyration. Occasionally, the moment of inertia of a body about a specified axis is reported in handbooks using the *radius of gyration, k*. This value has units of length, and when it and the body's mass m are known, the moment of inertia is determined from the equation

$$I = mk^2 \quad \text{or} \quad k = \sqrt{\frac{I}{m}} \tag{10–17}$$

Note the *similarity* between the definition of k in this formula and r in the equation $dI = r^2\,dm$, which defines the moment of inertia of an elemental mass dm of the body about an axis.

Composite Bodies. If a body is constructed from a number of simple shapes such as disks, spheres, and rods, the moment of inertia of the body about any axis z can be determined by adding algebraically the moments of inertia of all the composite shapes computed about the z axis. Algebraic addition is necessary since a composite part must be considered as a negative quantity if it has already been included within another part—for example, a "hole" subtracted from a solid plate. The parallel-axis theorem is needed for the calculations if the center of mass of each composite part does not lie on the z axis. For the calculation, then, $I = \Sigma(I_G + md^2)$, where I_G for each of the composite parts is computed by integration or can be determined from a table, such as the one given on the inside back cover of this book.

This flywheel, which operates a metal cutter, has a large moment of inertia about its center. Once it begins rotating it is difficult to stop it and therefore a uniform motion can be effectively transferred to the cutting blade.

EXAMPLE 10.13

If the plate shown in Fig. 10–27*a* has a density of 8000 kg/m^3 and a thickness of 10 mm, determine its mass moment of inertia about an axis directed perpendicular to the page and passing through point O.

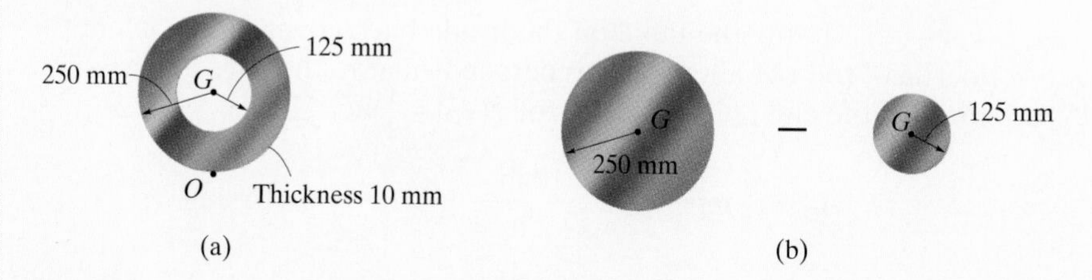

Fig. 10–27

SOLUTION

The plate consists of two composite parts, the 250-mm-radius disk *minus* a 125-mm-radius disk, Fig. 10–27*b*. The moment of inertia about O can be determined by computing the moment of inertia of each of these parts about O and then *algebraically* adding the results. The computations are performed by using the parallel-axis theorem in conjunction with the data listed in the table on the inside back cover.

Disk. The moment of inertia of a disk about an axis perpendicular to the plane of the disk is $I_G = \frac{1}{2}mr^2$. The mass center of the disk is located at a distance of 0.25 m from point O. Thus,

$$
\begin{aligned}
m_d &= \rho_d V_d = 8000 \text{ kg/m}^3\,[\pi(0.25 \text{ m})^2(0.01 \text{ m})] = 15.71 \text{ kg} \\
(I_O)_d &= \tfrac{1}{2}m_d r_d^2 + m_d d^2 \\
&= \tfrac{1}{2}(15.71 \text{ kg})(0.25 \text{ m})^2 + (15.71 \text{ kg})(0.25 \text{ m})^2 \\
&= 1.473 \text{ kg}\cdot\text{m}^2
\end{aligned}
$$

Hole. For the 125-mm-radius disk (hole), we have

$$
\begin{aligned}
m_h &= \rho_h V_h = 8000 \text{ kg/m}^3\,[\pi(0.125 \text{ m})^2(0.01 \text{ m})] = 3.93 \text{ kg} \\
(I_O)_h &= \tfrac{1}{2}m_h r_h^2 + m_h d^2 \\
&= \tfrac{1}{2}(3.93 \text{ kg})(0.125 \text{ m})^2 + (3.93 \text{ kg})(0.25 \text{ m})^2 \\
&= 0.276 \text{ kg}\cdot\text{m}^2
\end{aligned}
$$

The moment of inertia of the plate about point O is therefore

$$
\begin{aligned}
I_O &= (I_O)_d - (I_O)_h \\
&= 1.473 \text{ kg}\cdot\text{m}^2 - 0.276 \text{ kg}\cdot\text{m}^2 \\
&= 1.20 \text{ kg}\cdot\text{m}^2
\end{aligned}
$$

Ans.

EXAMPLE 10.14

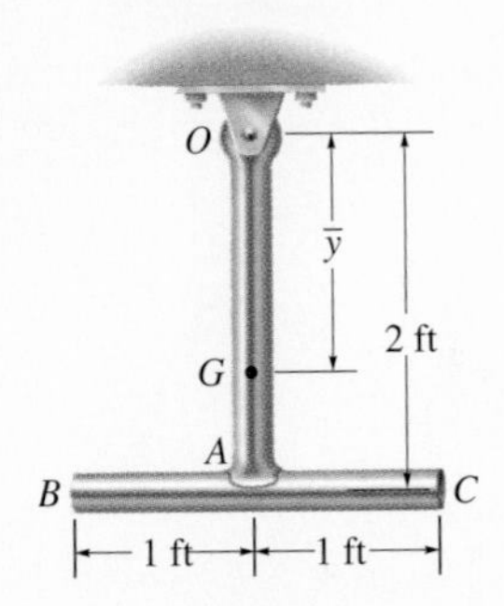

Fig. 10–28

The pendulum in Fig. 10–28 consists of two thin rods each having a weight of 10 lb. Determine the pendulum's mass moment of inertia about an axis passing through (a) the pin at O, and (b) the mass center G of the pendulum.

SOLUTION

Part (a). Using the table on the inside back cover, the moment of inertia of rod OA about an axis perpendicular to the page and passing through the end point O of the rod is $I_O = \frac{1}{3}ml^2$. Hence,

$$(I_{OA})_O = \frac{1}{3}ml^2 = \frac{1}{3}\left(\frac{10 \text{ lb}}{32.2 \text{ ft/s}^2}\right)(2 \text{ ft})^2 = 0.414 \text{ slug}\cdot\text{ft}^2$$

The same value may be computed using $I_G = \frac{1}{12}ml^2$ and the parallel-axis theorem; i.e.,

$$(I_{OA})_O = \frac{1}{12}ml^2 + md^2 = \frac{1}{12}\left(\frac{10 \text{ lb}}{32.2 \text{ ft/s}^2}\right)(2 \text{ ft})^2 + \frac{10 \text{ lb}}{32.2 \text{ ft/s}^2}(1 \text{ ft})^2$$
$$= 0.414 \text{ slug}\cdot\text{ft}^2$$

For rod BC we have

$$(I_{BC})_O = \frac{1}{12}ml^2 + md^2 = \frac{1}{12}\left(\frac{10 \text{ lb}}{32.2 \text{ ft/s}^2}\right)(2 \text{ ft})^2 + \frac{10 \text{ lb}}{32.2 \text{ ft/s}^2}(2 \text{ ft})^2$$
$$= 1.346 \text{ slug}\cdot\text{ft}^2$$

The moment of inertia of the pendulum about O is therefore

$$I_O = 0.414 + 1.346 = 1.76 \text{ slug}\cdot\text{ft}^2 \qquad \textit{Ans.}$$

Part (b). The mass center G will be located relative to the pin at O. Assuming this distance to be $\bar{y}$, Fig. 10–28, and using the formula for determining the mass center, we have

$$\bar{y} = \frac{\Sigma\tilde{y}m}{\Sigma m} = \frac{1(10/32.2) + 2(10/32.2)}{(10/32.2) + (10/32.2)} = 1.50 \text{ ft}$$

NOTE: The moment of inertia I_G may be computed in the same manner as I_O, which requires successive applications of the parallel-axis theorem in order to transfer the moments of inertia of rods OA and BC to G. A more direct solution, however, involves applying the parallel-axis theorem using the result for I_O determined above; i.e.,

$$I_O = I_G + md^2; \qquad 1.76 \text{ slug}\cdot\text{ft}^2 = I_G + \left(\frac{20 \text{ lb}}{32.2 \text{ ft/s}^2}\right)(1.50 \text{ ft})^2$$

$$I_G = 0.362 \text{ slug}\cdot\text{ft}^2 \qquad \textit{Ans.}$$

PROBLEMS

10–90. The right circular cone is formed by revolving the shaded area around the x axis. Determine the moment of inertia I_x and express the result in terms of the total mass m of the cone. The cone has a constant density ρ.

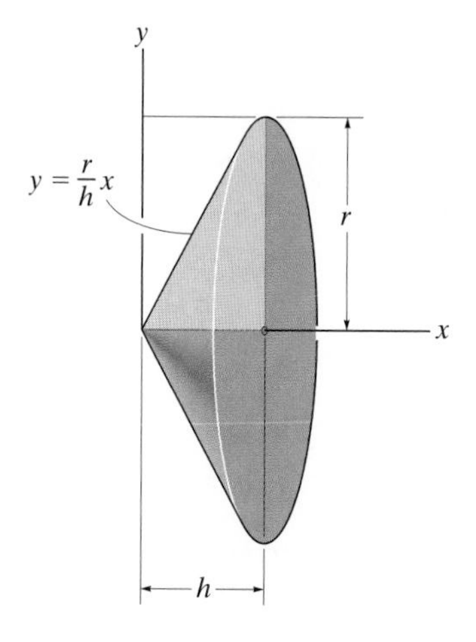

Prob. 10–90

***10–92.** The solid is formed by revolving the shaded area around the y axis. Determine the radius of gyration k_y. The specific weight of the material is $\gamma = 380\ \text{lb/ft}^3$.

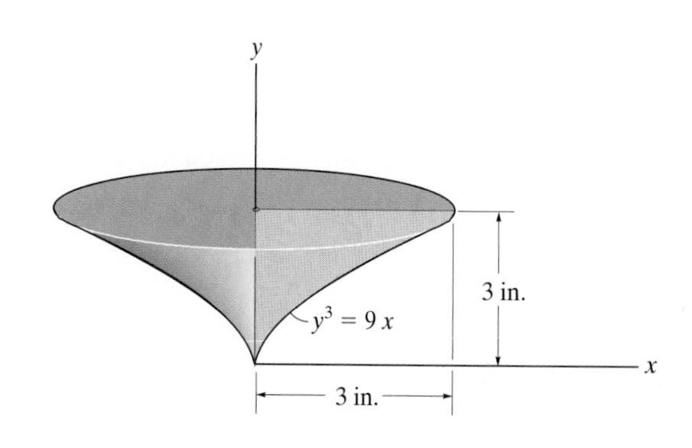

Prob. 10–92

10–91. Determine the moment of inertia of the thin ring about the z axis. The ring has a mass m.

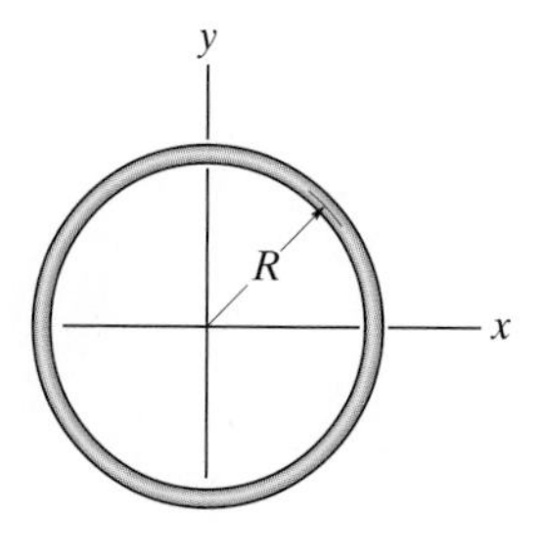

Prob. 10–91

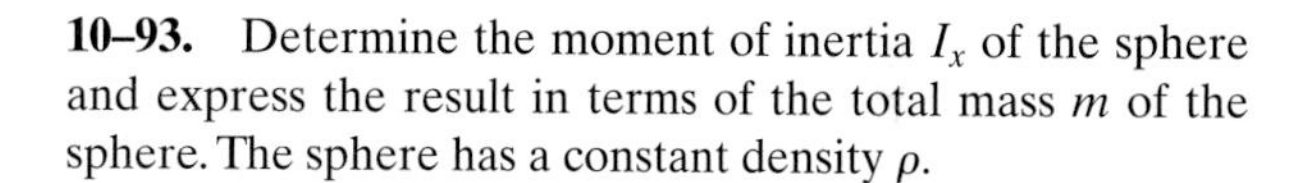

10–93. Determine the moment of inertia I_x of the sphere and express the result in terms of the total mass m of the sphere. The sphere has a constant density ρ.

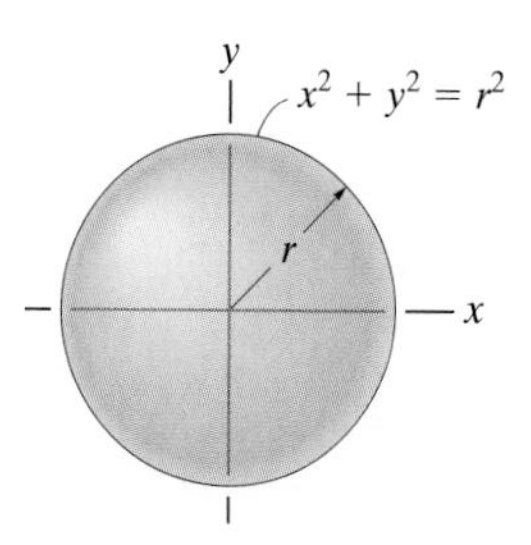

Prob. 10–93

10–94. Determine the radius of gyration k_x of the paraboloid. The density of the material is $\rho = 5\ \text{Mg/m}^3$.

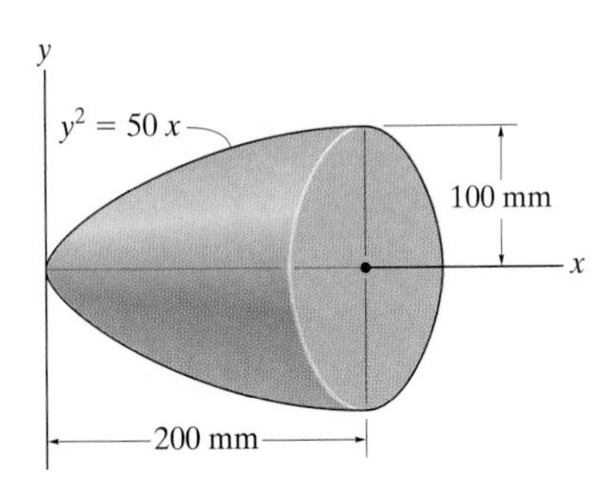

Prob. 10–94

10–95. Determine the moment of inertia of the semiellipsoid with respect to the x axis and express the result in terms of the mass m of the semiellipsoid. The material has a constant density ρ.

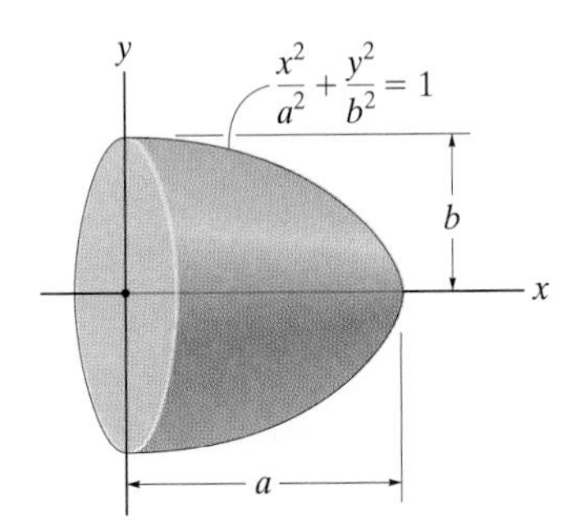

Prob. 10–95

***10–96.** Determine the radius of gyration k_x of the body. The specific weight of the material is $\gamma = 380\ \text{lb/ft}^3$.

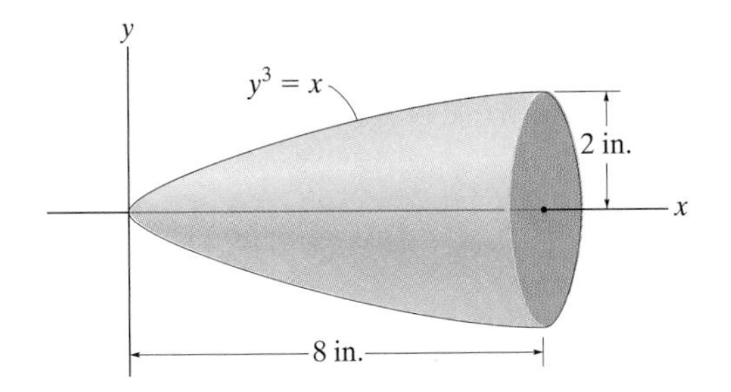

Prob. 10–96

10–97. Determine the moment of inertia of the ellipsoid with respect to the x axis and express the result in terms of the mass m of the ellipsoid. The material has a constant density ρ.

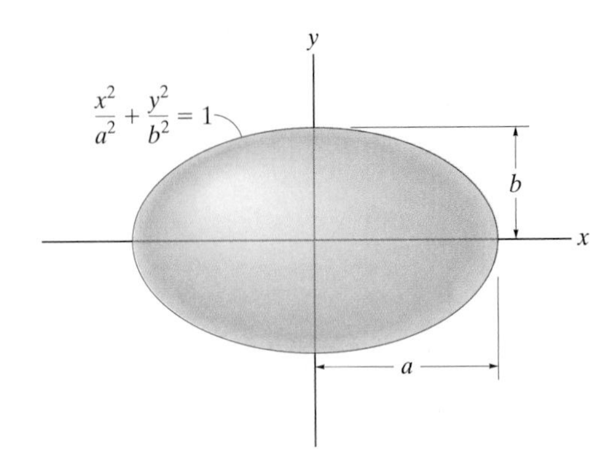

Prob. 10–97

10–98. Determine the moment of inertia of the homogeneous pyramid of mass m with respect to the z axis. The density of the material is ρ. *Hint:* Use a rectangular plate element having a volume of $dV = (2x)(2y)\,dz$.

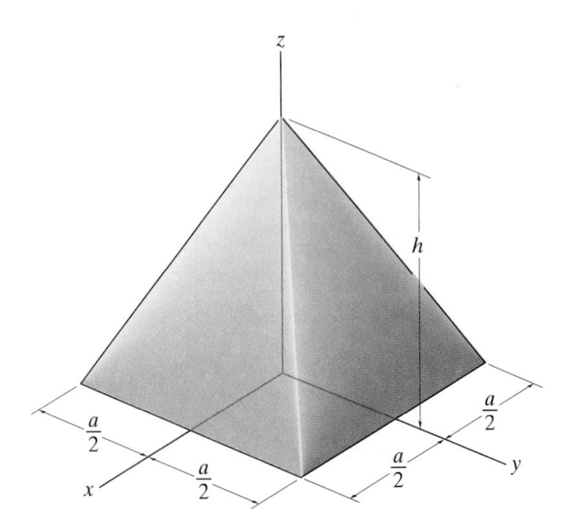

Prob. 10–98

10–99. The concrete shape is formed by rotating the shaded area about the y axis. Determine the moment of inertia I_y. The specific weight of concrete is $\gamma = 150 \text{ lb/ft}^3$.

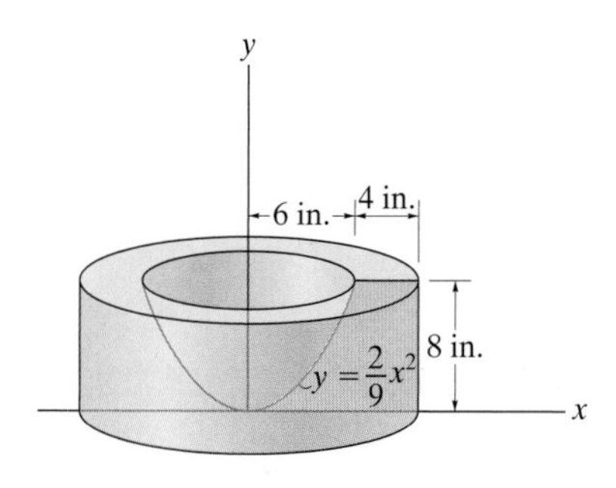

Prob. 10–99

***10–100.** Determine the moment of inertia of the thin plate about an axis perpendicular to the page and passing through the pin at O. The plate has a hole in its center. Its thickness is 50 mm, and the material has a density of $\rho = 50 \text{ kg/m}^3$.

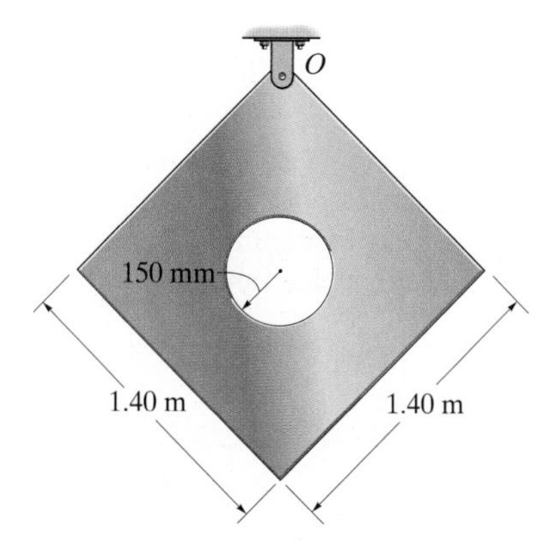

Prob. 10–100

10–101. Determine the moment of inertia I_z of the frustum of the cone which has a conical depression. The material has a density of 200 kg/m^3.

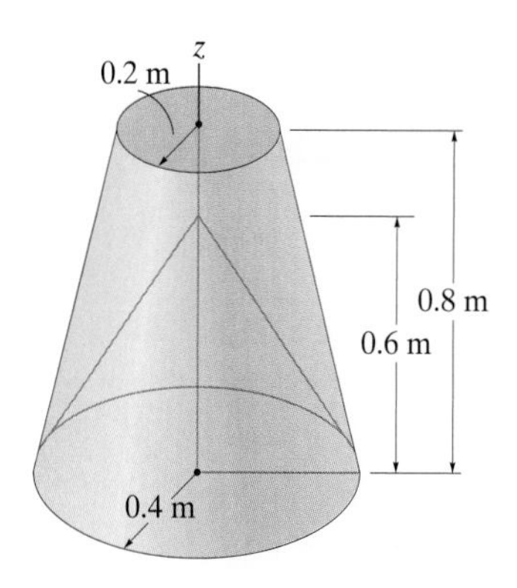

Prob. 10–101

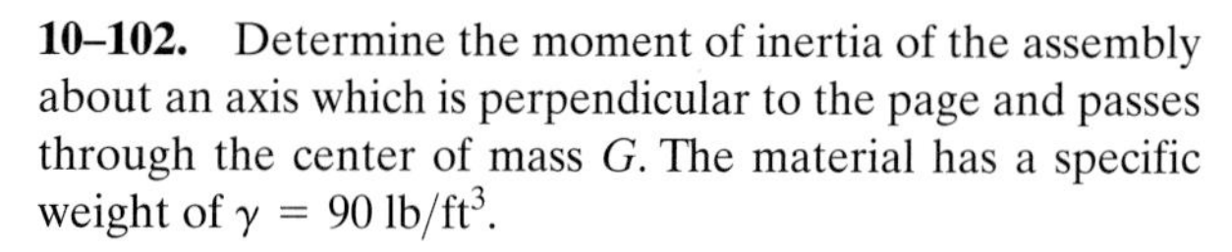

10–102. Determine the moment of inertia of the assembly about an axis which is perpendicular to the page and passes through the center of mass G. The material has a specific weight of $\gamma = 90 \text{ lb/ft}^3$.

10–103. Determine the moment of inertia of the assembly about an axis which is perpendicular to the page and passes through point O. The material has a specific weight of $\gamma = 90 \text{ lb/ft}^3$.

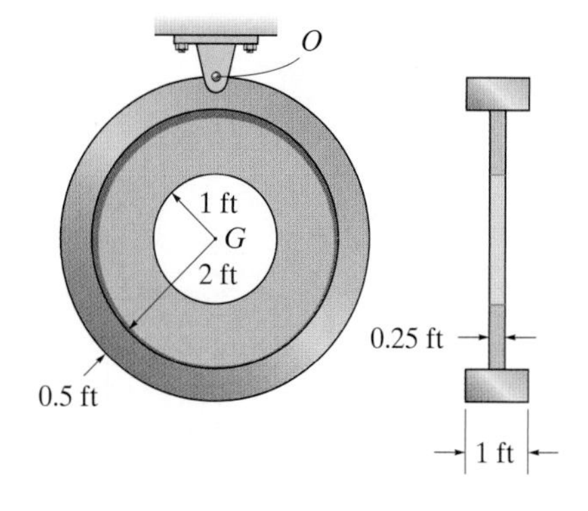

Probs. 10–102/103

***10–104.** The wheel consists of a thin ring having a mass of 10 kg and four spokes made from slender rods, each having a mass of 2 kg. Determine the wheel's moment of inertia about an axis perpendicular to the page and passing through point A.

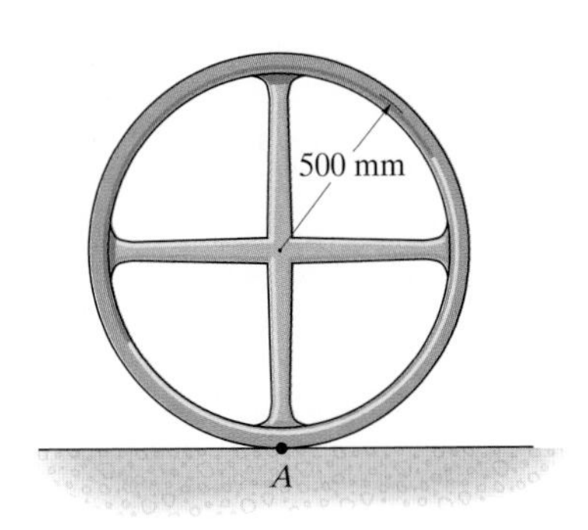

Prob. 10–104

10–105. The slender rods have a weight of 3 lb/ft. Determine the moment of inertia of the assembly about an axis perpendicular to the page and passing through point A.

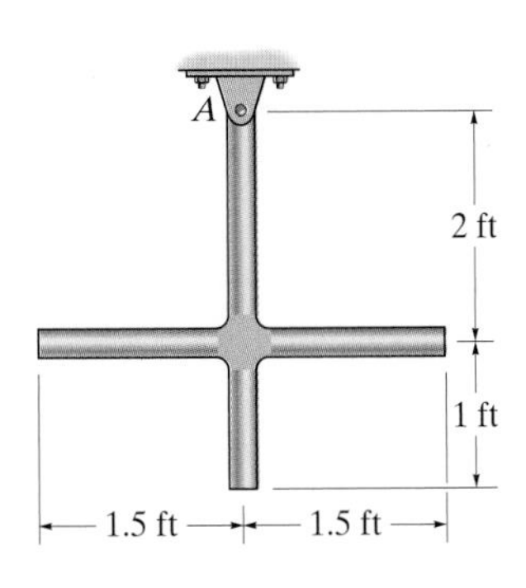

Prob. 10–105

10–106. Each of the three rods has a mass m. Determine the moment of inertia of the assembly about an axis which is perpendicular to the page and passes through the center point O.

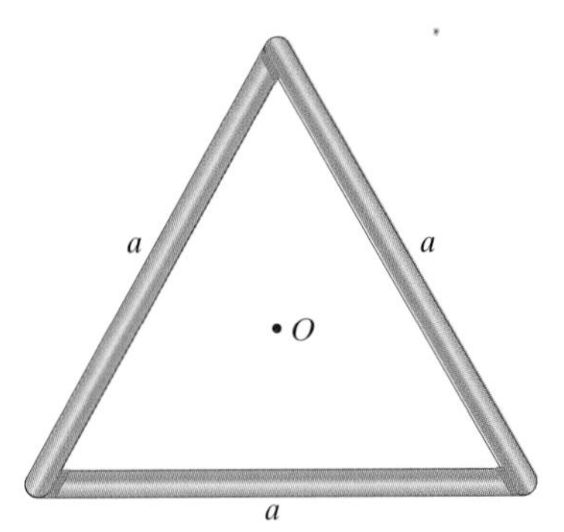

Prob. 10–106

10–107. The slender rods have a weight of 3 lb/ft. Determine the moment of inertia of the assembly about an axis perpendicular to the page and passing through point A.

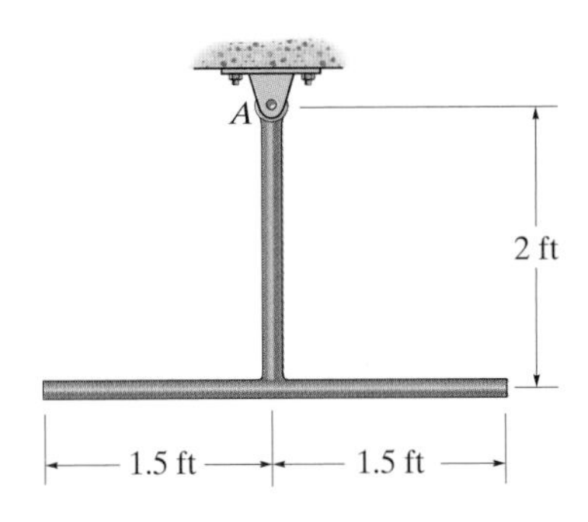

Prob. 10–107

***10–108.** The pendulum consists of a plate having a weight of 12 lb and a slender rod having a weight of 4 lb. Determine the radius of gyration of the pendulum about an axis perpendicular to the page and passing through point O.

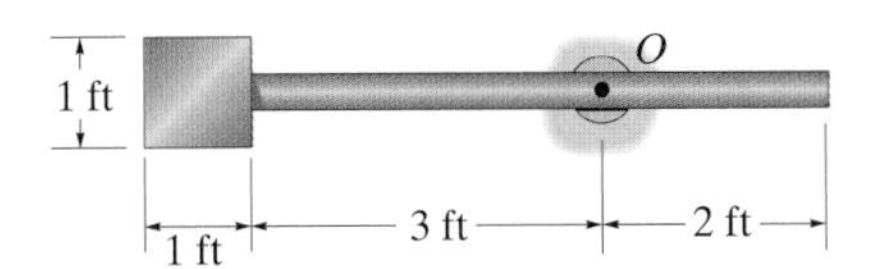

Prob. 10–108

10–109. Determine the moment of inertia of the overhung crank about the x axis. The material is steel having a density of $\rho = 7.85\ \text{Mg/m}^3$.

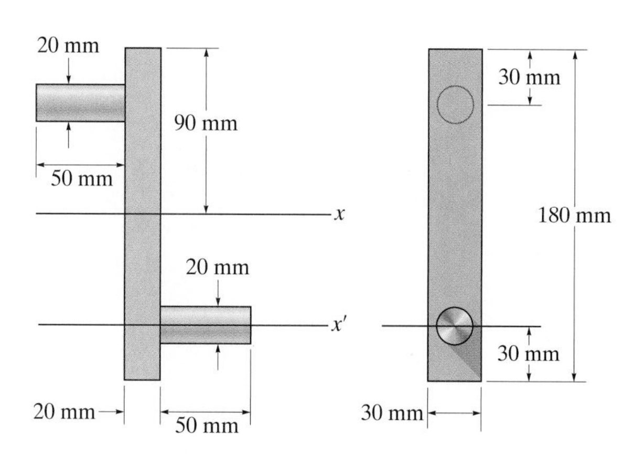

Prob. 10–109

10–110. Determine the moment of inertia of the overhung crank about the x' axis. The material is steel having a density of $\rho = 7.85\ \text{Mg/m}^3$.

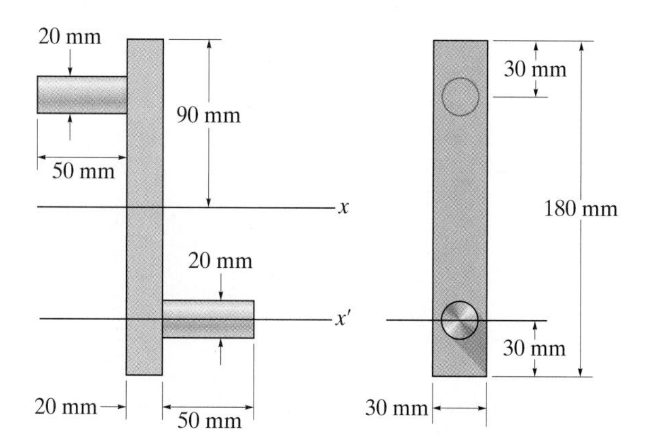

Prob. 10–110

10–111. Determine the moment of inertia of the solid steel assembly about the x axis. Steel has a specific weight of $\gamma_{st} = 490\ \text{lb/ft}^3$.

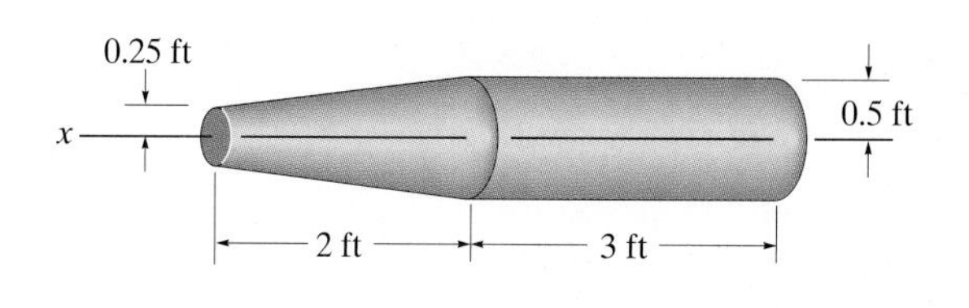

Prob. 10–111

***10–112.** The pendulum consists of two slender rods AB and OC which have a mass of 3 kg/m. The thin plate has a mass of $12\ \text{kg/m}^2$. Determine the location $\bar{y}$ of the center of mass G of the pendulum, then calculate the moment of inertia of the pendulum about an axis perpendicular to the page and passing through G.

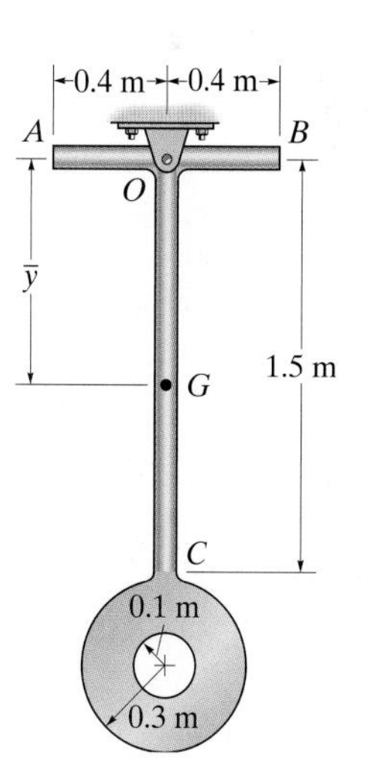

Prob. 10–112

Chapter Review

Area Moment of Inertia

The *area moment of inertia* represents the second moment of the area about an axis. It is frequently used in formulas related to strength and stability of structural members or mechanical elements.

$$I_x = \int_A y^2 \, dA$$

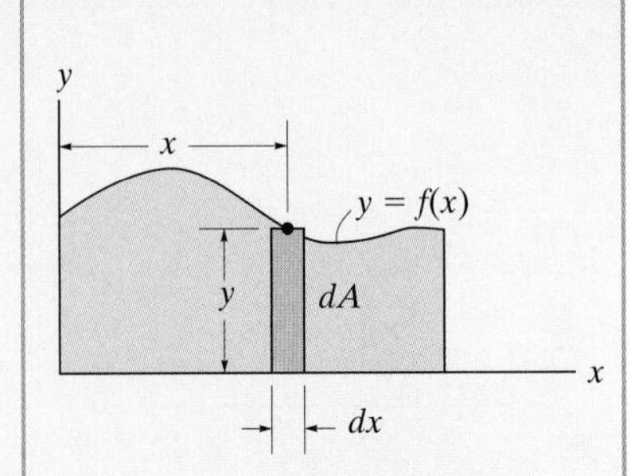

If the area shape is irregular, then a differential element must be selected and integration over the entire area must be performed.

$$I_y = \int_A x^2 \, dA$$

To determine the moment of inertia of common shapes about some *other axis*, the parallel-axis theorem must be used.

$$I_x = \bar{I}_{x'} + Ad_y^2$$

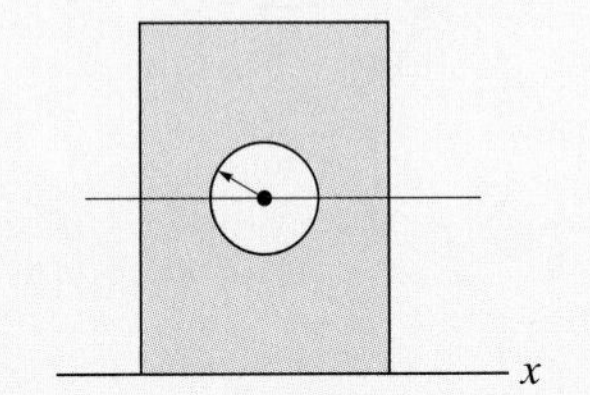

If an area is a composite of common shapes, then its moment of inertia is equal to the sum of the moments of inertia of each of its parts.
(*See pages 519–522.*)

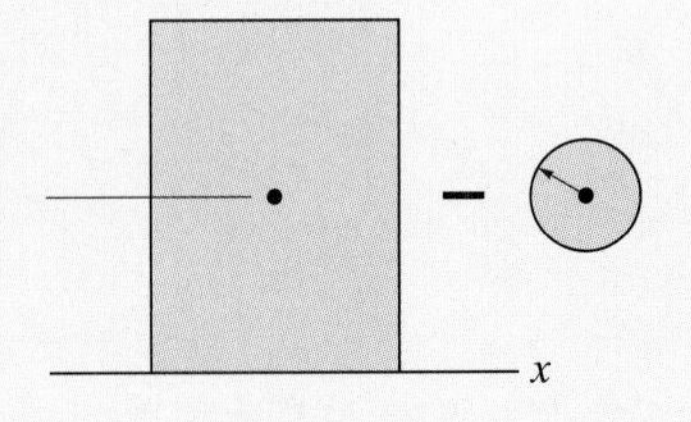

Product of Inertia

The *product of inertia* of an area is used to determine the location of an axis about which the moment of inertia for the area is a maximum or minimum.

$$I_{xy} = \int_A xy \, dA$$

If the product of inertia for an area is known about its centroidal x', y' axes, then its value can be determined about any x, y axes using the parallel-axis theorem for the product of inertia.
(*See pages 538–539.*)

$$I_{xy} = \bar{I}_{x'y'} + Ad_x d_y$$

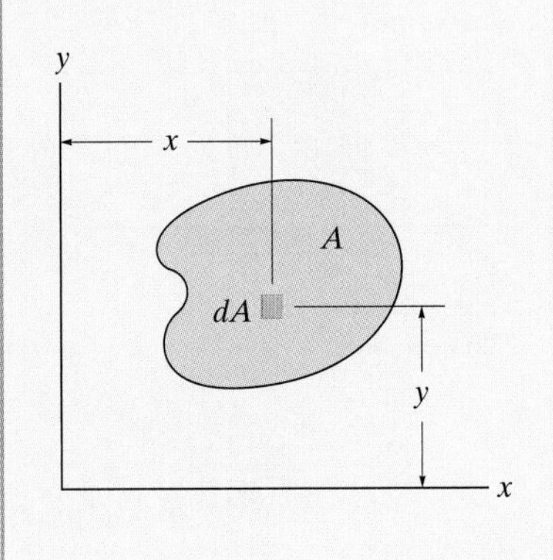

Principal Moments of Inertia
Provided the moments of inertia, I_x and I_y, and the product of inertia, I_{xy}, are known, then the transformation formulas, or Mohr's circle, can be used to determine the maximum and minimum or *principal moments of inertia* for the area, as well as finding the orientation of the principal axes of inertia.

(*See pages 542–543.*)

$$I_{\substack{\max \\ \min}} = \frac{I_x + I_y}{2} \pm \sqrt{\left(\frac{I_x - I_y}{2}\right)^2 + I_{xy}^2}$$

$$\tan 2\theta_p = \frac{-I_{xy}}{(I_x - I_y)/2}$$

Mass Moment of Inertia
The *mass moment of inertia* is a property of a body that measures its resistance to a change in its rotation. It is defined as the second moment of the mass elements of the body about an axis.

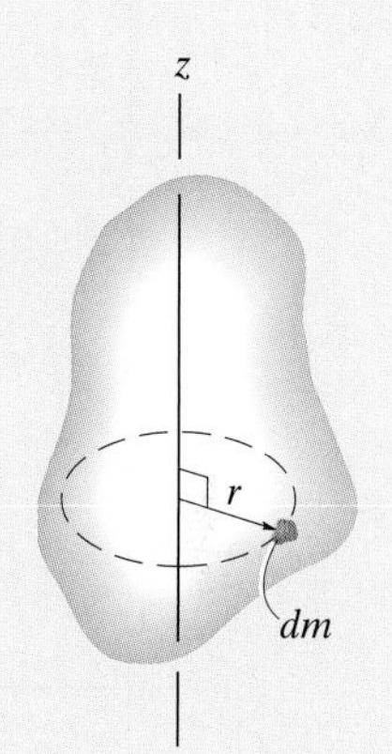

For bodies having axial symmetry, the mass moment of inertia can be determined by integration, using a disk element.

$$I = \int_m r^2 \, dm$$

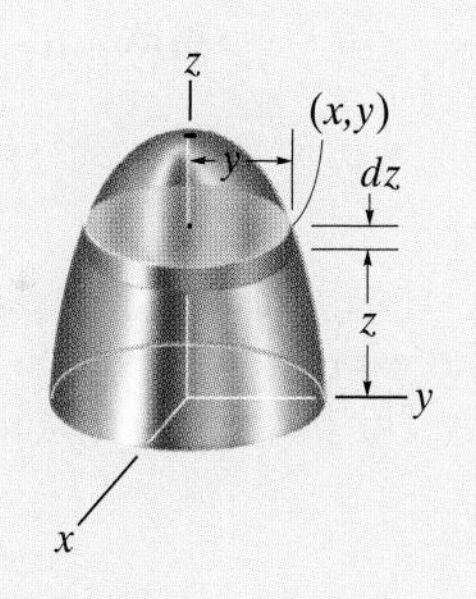

The mass moment of inertia of a composite body is determined by using tabular values of its composite shapes found on the inside back cover, along with the parallel-axis theorem.

(*See pages 555–556.*)

$$I = I_G + md^2$$

REVIEW PROBLEMS

10–113. Determine the moment of inertia for the shaded area about the x axis.

10–114. Determine the moment of inertia for the shaded area about the y axis.

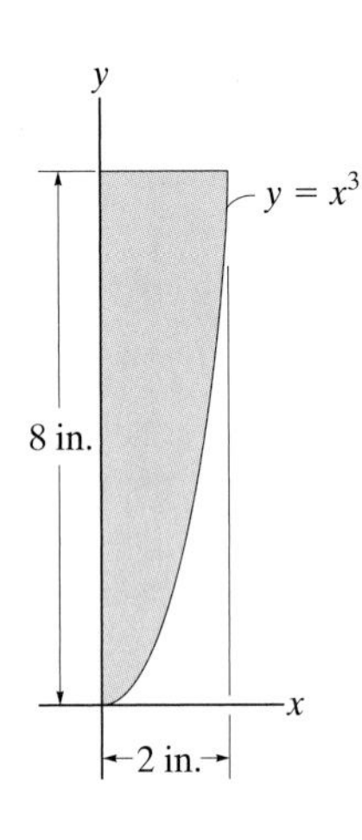

Probs. 10–113/114

***10–116.** Determine the product of inertia for the shaded area with respect to the x and y axes.

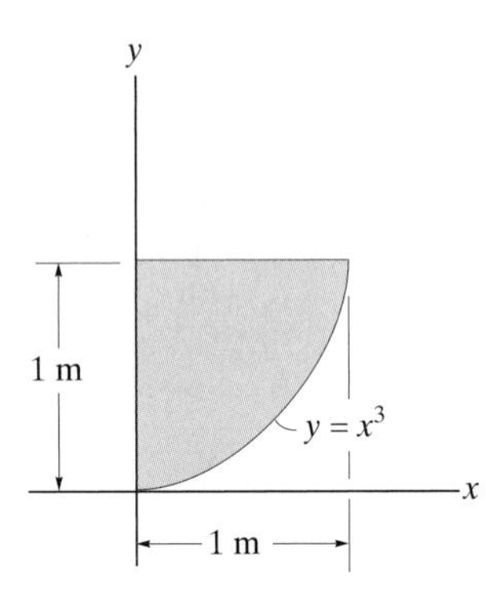

Prob. 10–116

10–115. Determine the mass moment of inertia I_x for the body and express the result in terms of the total mass m of the body. The density is constant.

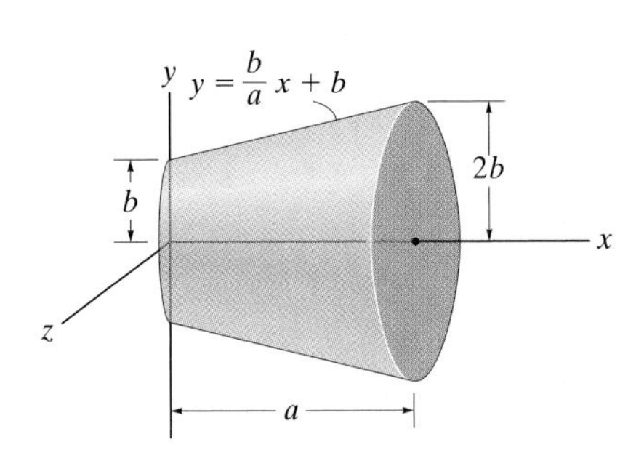

Prob. 10–115

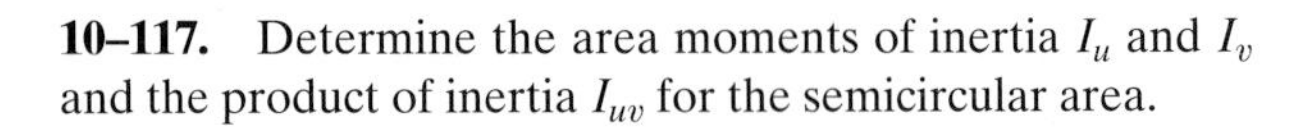

10–117. Determine the area moments of inertia I_u and I_v and the product of inertia I_{uv} for the semicircular area.

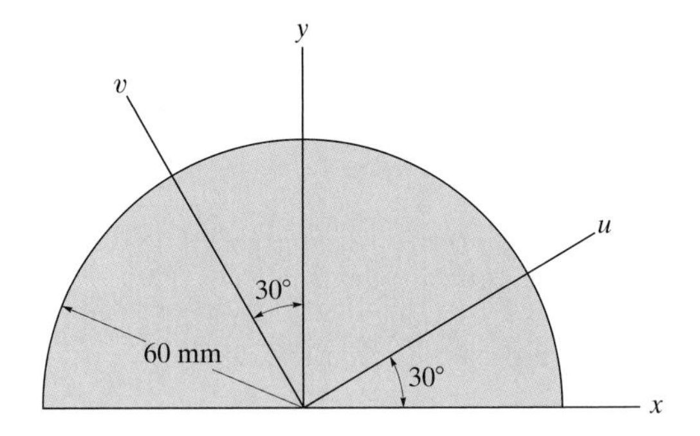

Prob. 10–117

10–118. Determine the moment of inertia for the shaded area about the x axis.

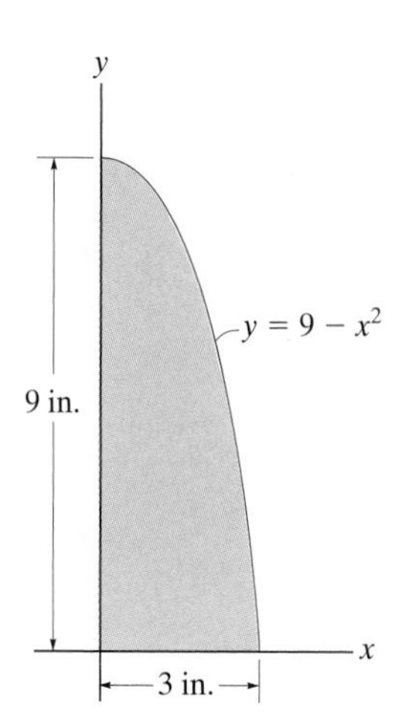

Prob. 10–118

10–119. Determine the moment of inertia for the shaded area about the y axis.

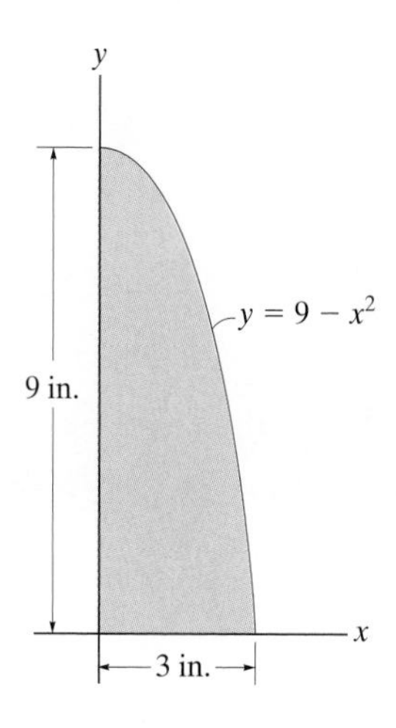

Prob. 10–119

***10–120.** Determine the area moment of inertia for the area about the x axis. Then, using the parallel-axis theorem, find the area moment of inertia about the x' axis that passes through the centroid C of the area. $\overline{y} = 120$ mm.

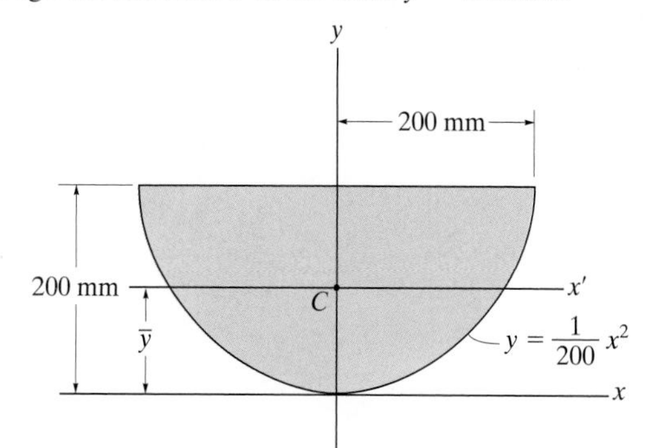

Prob. 10–120

10–121. Determine the area moment of inertia for the triangular area about (a) the x axis, and (b) the centroidal x' axis.

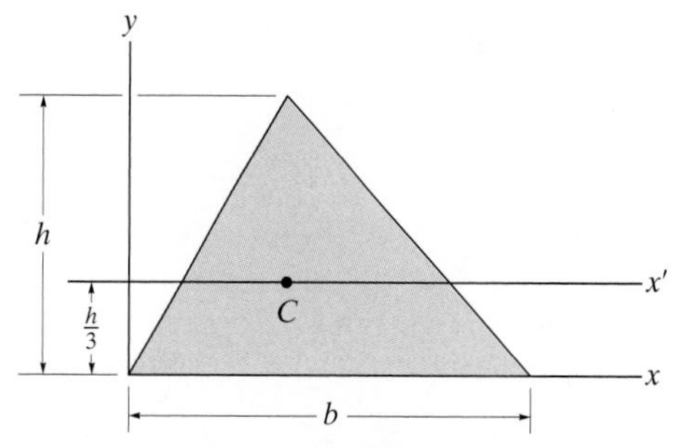

Prob. 10–121

10–122. Determine the product of inertia for the shaded area with respect to the x and y axes.

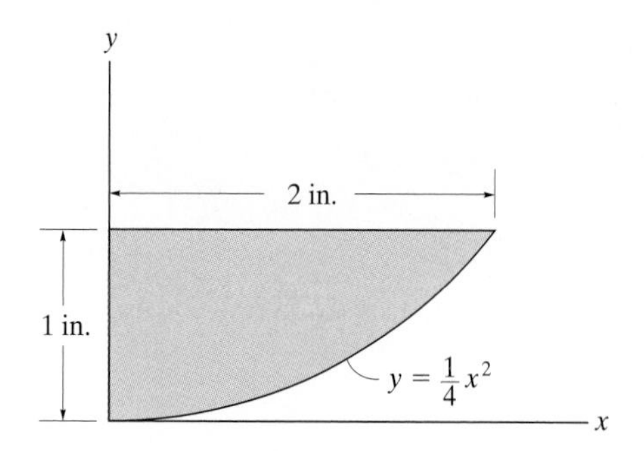

Prob. 10–122

Equilibrium and stability of this crane as a function of the boom position can be analyzed using methods based on work and energy, which are explained in this chapter.

11 Virtual Work

CHAPTER OBJECTIVES

- To introduce the principle of virtual work and show how it applies to determining the equilibrium configuration of a series of pin-connected members.
- To establish the potential energy function and use the potential-energy method to investigate the type of equilibrium or stability of a rigid body or configuration.

11.1 Definition of Work and Virtual Work

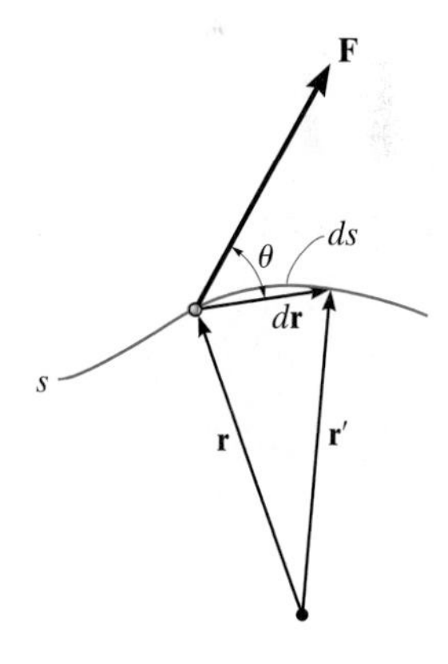

Fig. 11–1

Work of a Force. In mechanics a force $\mathbf{F}$ does work on a particle or body only when it has a displacement in the direction of the force. For example, consider the force $\mathbf{F}$ in Fig. 11–1, which is located on the path s specified by the position vector $\mathbf{r}$. If the force moves the particle along the path to a new position $\mathbf{r}' = \mathbf{r} + d\mathbf{r}$, the displacement is $d\mathbf{r}$, and therefore the work dU is a *scalar quantity* defined by the dot product

$$dU = \mathbf{F} \cdot d\mathbf{r}$$

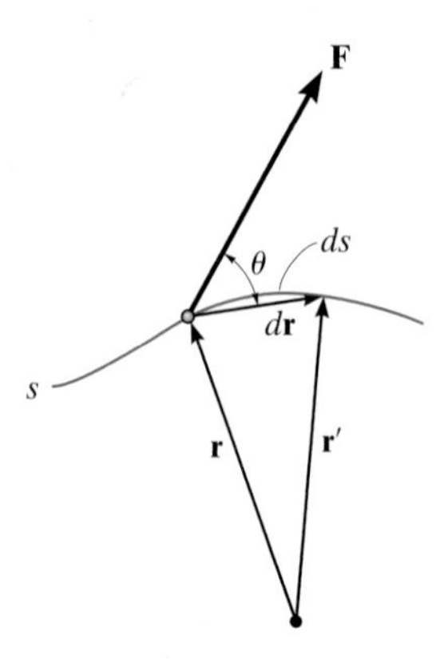

Fig. 11–1

Because $d\mathbf{r}$ is infinitesimal, the magnitude of $d\mathbf{r}$ can be represented by ds, the differential arc segment along the path. If the angle between the tails of $d\mathbf{r}$ and $\mathbf{F}$ is θ, Fig. 11–1, then by definition of the dot product, the above equation may also be written as

$$dU = F\,ds\cos\theta$$

Work expressed by this equation may be interpreted in one of two ways: either as the product of F and the component of displacement in the direction of the force, i.e., $ds\cos\theta$; or as the product of ds and the component of force in the direction of displacement, i.e., $F\cos\theta$. When the force component and the displacement have the *same sense*, $0° \le \theta < 90°$, the work is *positive*. If $90° < \theta \le 180°$, these vectors have an *opposite sense*, and therefore the work is *negative*. Also, $dU = 0$ if the force is *perpendicular* to displacement, since $\cos 90° = 0$, or if the force is applied at a *fixed point*, in which case the displacement $ds = 0$.

The basic unit for work combines the units of force and displacement. In the SI system a *joule* (J) is equivalent to the work done by a force of 1 newton which moves 1 meter in the direction of the force ($1\text{ J} = 1\text{ N}\cdot\text{m}$). In the FPS system, work is defined in units of ft·lb. The moment of a force has the same combination of units; however, the concepts of moment and work are in no way related. A moment is a vector quantity, whereas work is a scalar.

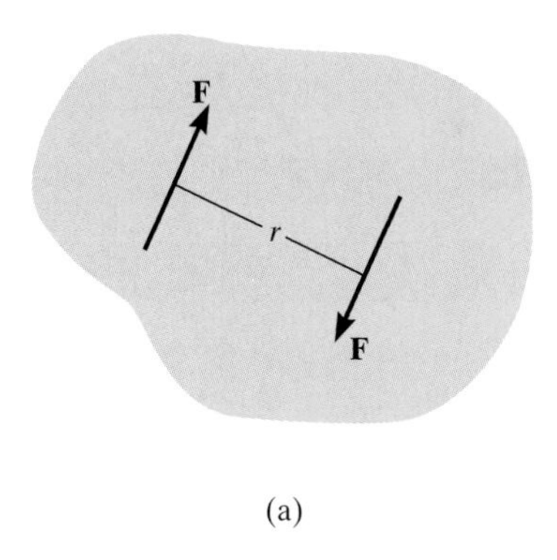

(a)

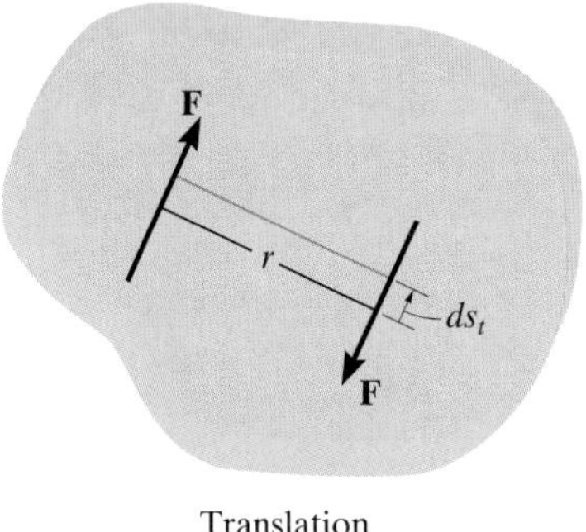

Translation
(b)

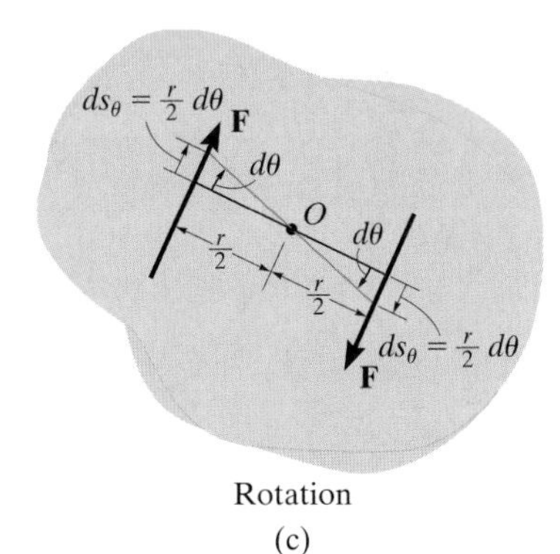

Rotation
(c)

Fig. 11–2

Work of a Couple. The two forces of a couple do work when the couple *rotates* about an axis perpendicular to the plane of the couple. To show this, consider the body in Fig. 11–2*a*, which is subjected to a couple whose moment has a magnitude $M = Fr$. Any general differential displacement of the body can be considered as a combination of a translation and rotation. When the body *translates* such that the *component of displacement* along the line of action of each force is ds_t, clearly the "positive" work of one force ($F\ ds_t$) *cancels* the "negative" work of the other ($-F\ ds_t$), Fig. 11–2*b*. Consider now a differential *rotation* $d\theta$ of the body about an axis perpendicular to the plane of the couple, which intersects the plane at point O, Fig. 11–2*c*. (For the derivation, any other point in the plane may also be considered.) As shown, each force has a displacement $ds_\theta = (r/2)\ d\theta$ in the direction of the force; hence, the work of both forces is

$$dU = F\left(\frac{r}{2}d\theta\right) + F\left(\frac{r}{2}d\theta\right) = (Fr)\ d\theta$$

or

$$dU = M\ d\theta$$

The resultant work is *positive* when the sense of $\mathbf{M}$ is the *same* as that of $d\boldsymbol{\theta}$, and negative when they have an opposite sense. As in the case of the moment vector, the *direction and sense* of $d\boldsymbol{\theta}$ are defined by the right-hand rule, where the fingers of the right hand follow the rotation or "curl" and the thumb indicates the direction of $d\boldsymbol{\theta}$. Hence, the line of action of $d\boldsymbol{\theta}$ will be *parallel* to the line of action of $\mathbf{M}$ if movement of the body occurs in the *same plane*. If the body rotates in space, however, the *component* of $d\boldsymbol{\theta}$ in the direction of $\mathbf{M}$ is required. Thus, in general, the work done by a couple is defined by the dot product, $dU = \mathbf{M} \cdot d\boldsymbol{\theta}$.

Virtual Work. The definitions of the work of a force and a couple have been presented in terms of *actual movements* expressed by differential displacements having magnitudes of ds and $d\theta$. Consider now an *imaginary* or *virtual movement* of a body in static equilibrium, which indicates a displacement or rotation that is *assumed* and *does not actually exist*. These movements are first-order differential quantities and will be denoted by the symbols δs and $\delta\theta$ (delta s and delta θ), respectively. The *virtual work* done by a force having a virtual displacement δs is

$$\boxed{\delta U = F \cos\theta\ \delta s} \qquad (11\text{–}1)$$

Similarly, when a couple undergoes a virtual rotation $\delta\theta$ in the plane of the couple forces, the *virtual work* is

$$\boxed{\delta U = M\ \delta\theta} \qquad (11\text{–}2)$$

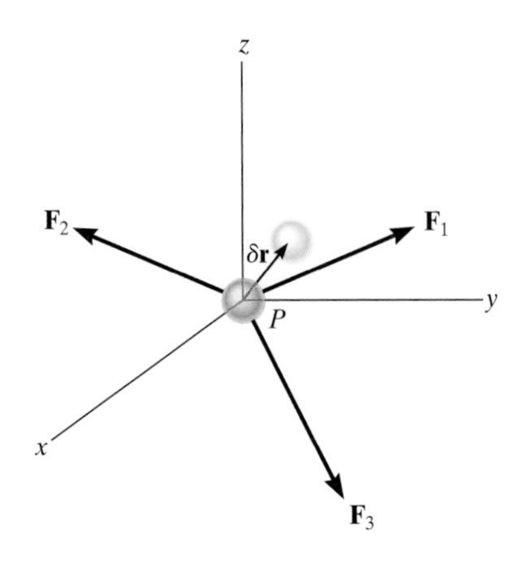

Fig. 11–3

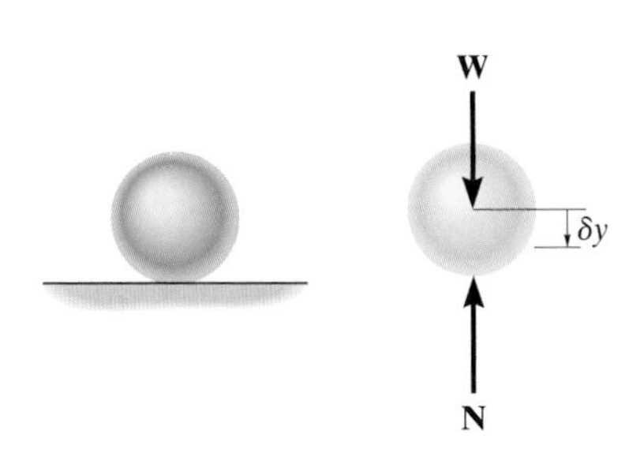

Fig. 11–4

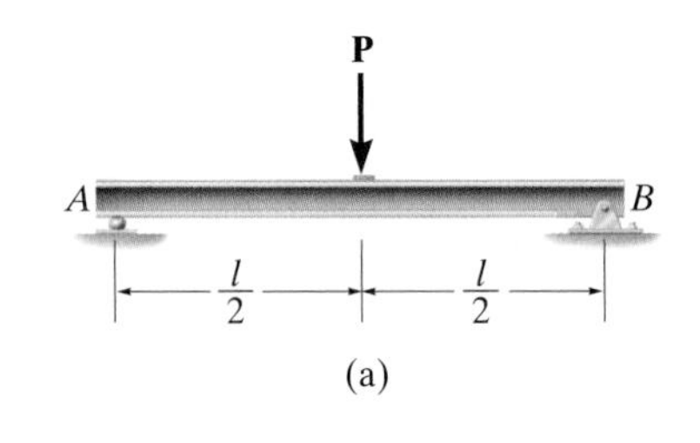

(a)

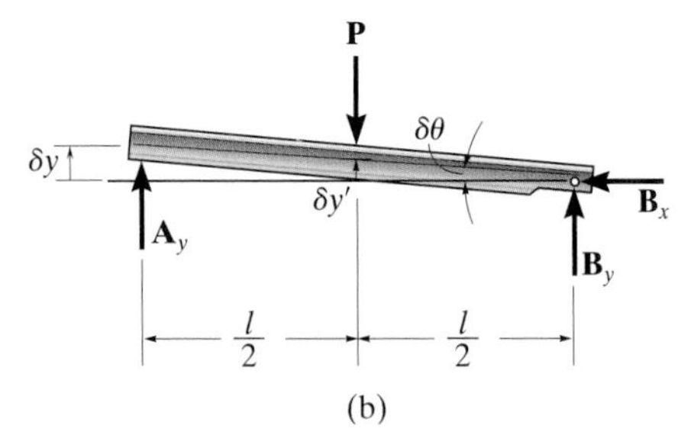

(b)

Fig. 11–5

11.2 Principle of Virtual Work for a Particle and a Rigid Body

Particle. If the particle in Fig. 11–3 undergoes an imaginary or virtual displacement $\delta \mathbf{r}$, then the virtual work (δU) done by the force system becomes

$$\begin{aligned} \delta U &= \Sigma \mathbf{F} \cdot \delta \mathbf{r} \\ &= (\Sigma F_x \mathbf{i} + \Sigma F_y \mathbf{j} + \Sigma F_z \mathbf{k}) \cdot (\delta x \mathbf{i} + \delta y \mathbf{j} + \delta z \mathbf{k}) \\ &= \Sigma F_x\, \delta x + \Sigma F_y\, \delta y + \Sigma F_z\, \delta z \end{aligned}$$

For equilibrium $\Sigma F_x = 0$, $\Sigma F_y = 0$, $\Sigma F_z = 0$, and so the virtual work must also be zero, i.e.,

$$\delta U = 0$$

In other words, we can write three independent virtual work equations corresponding to the three equations of equilibrium.

For example, consider the free-body diagram of the ball which rests on the floor, Fig. 11–4. If we "imagine" the ball to be displaced downwards a virtual amount δy, then the weight does positive virtual work, $W\,\delta y$, and the normal force does negative virtual work, $-N\,\delta y$. For equilibrium the total virtual work must be zero, so that $\delta U = W\,\delta y - N\,\delta y = (W - N)\,\delta y = 0$. Since $\delta y \neq 0$, then $N = W$ as required.

Rigid Body. In a similar manner, we can also write a set of three virtual work equations $(\delta U = 0)$ for a rigid body subjected to a coplanar force system. If these equations involve separate virtual translations in the x and y directions and a virtual rotation about an axis perpendicular to the x–y plane and passing through an arbitrary point O, then it can be shown that they will correspond to the three equilibrium equations, $\Sigma F_x = 0$, $\Sigma F_y = 0$, and $\Sigma M_O = 0$. When writing these equations, it is *not necessary* to include the work done by the *internal forces* acting within the body since a rigid body *does not deform* when subjected to an external loading, and furthermore, when the body moves through a virtual displacement, the internal forces occur in equal but opposite collinear pairs, so that the corresponding work done by each pair of forces *cancels*.

To demonstrate an application, consider the simply supported beam in Fig. 11–5*a*. When the beam is given a virtual rotation $\delta\theta$ about point B, Fig. 11–5*b*, the only forces that do work are $\mathbf{P}$ and $\mathbf{A}_y$. Since $\delta y = l\,\delta\theta$ and $\delta y' = (l/2)\,\delta\theta$, the virtual work equation for this case is $\delta U = A_y(l\,\delta\theta) - P(l/2)\,\delta\theta = (A_y - P/2)l\,\delta\theta = 0$. Since $\delta\theta \neq 0$, then $A_y = P/2$. Excluding $\delta\theta$, notice that the terms in parentheses actually represent moment equilibrium about point B.

As in the case of a particle, no added advantage is gained by solving rigid-body equilibrium problems using the principle of virtual work. This is because for each application of the virtual-work equation the virtual displacement, common to every term, factors out, leaving an equation that could have been obtained in a more *direct manner* by simply applying the equations of equilibrium.

11.3 Principle of Virtual Work for a System of Connected Rigid Bodies

The method of virtual work is most suitable for solving equilibrium problems that involve a system of several *connected* rigid bodies such as the ones shown in Fig. 11–6. Before we can apply the principle of virtual work to these systems, however, we must first specify the number of degrees of freedom for a system and establish coordinates that define the position of the system.

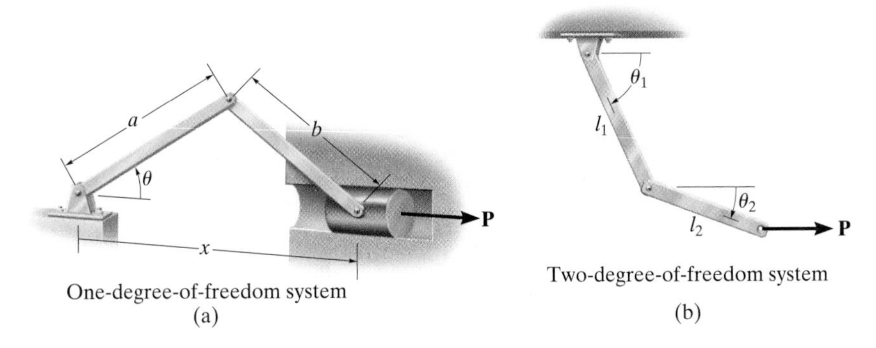

Fig. 11–6

Degrees of Freedom. A system of connected bodies takes on a unique shape that can be specified provided we know the position of a number of specific points on the system. These positions are defined using *independent coordinates* q, which are measured from fixed reference points. For every coordinate established, the system will have a *degree of freedom* for displacement along the coordinate axis such that it is consistent with the constraining action of the supports. Thus, an n-degree-of-freedom system requires n independent coordinates q_n to specify the location of all its members. For example, the link and sliding-block arrangement shown in Fig. 11–6*a* is an example of a one-degree-of-freedom system. The independent coordinate $q = \theta$ may be used to specify the location of the two connecting links and the block. The coordinate x could also be used as the independent coordinate. However, since the block is constrained to move within the slot, x is not independent of θ; rather, it can be related to θ using the cosine law, $b^2 = a^2 + x^2 - 2ax\cos\theta$. The double-link arrangement, shown in Fig. 11–6*b*, is an example of a two-degrees-of-freedom system. To specify the location of each link, the coordinate angles θ_1 and θ_2 must be known since a rotation of one link is independent of a rotation of the other.

During operation the scissors lift has one degree of freedom. Without the need for dismembering the mechanism, the hydraulic force required to provide the lift can be determined *directly* by using the principle of virtual work.

Principle of Virtual Work. The principle of virtual work for a system of rigid bodies whose connections are *frictionless* may be stated as follows: *A system of connected rigid bodies is in equilibrium provided the virtual work done by all the external forces and couples acting on the system is zero for each independent virtual displacement of the system.* Mathematically, this may be expressed as

$$\boxed{\delta U = 0} \tag{11–3}$$

where δU represents the virtual work of all the external forces (and couples) acting on the system during any independent virtual displacement.

As stated above, if a system has n degrees of freedom it takes n independent coordinates q_n to completely specify the location of the system. Hence, for the system it is possible to write n independent virtual-work equations, one for every virtual displacement taken along each of the independent coordinate axes, while the remaining $n - 1$ independent coordinates are held *fixed*.*

Important Points

- A force does work when it moves through a displacement in the direction of the force. A couple moment does work when it moves through a collinear rotation. Specifically, positive work is done when the force or couple moment and its displacement have the same sense of direction.
- The principle of virtual work is generally used to determine the equilibrium configuration for a series of multiply-connected members.
- A virtual displacement is imaginary, i.e., does not really happen. It is a differential that is given in the positive direction of the position coordinate.
- Forces or couple moments that do not virtually displace do no virtual work.

*This method of applying the principle of virtual work is sometimes called the *method of virtual displacements* since a virtual displacement is applied, resulting in the calculation of a real force. Although it is not to be used here, realize that we can also apply the principle of virtual work as a method of virtual forces. This method is often used to determine the displacements of points on deformable bodies. See R. C. Hibbeler, *Mechanics of Materials*, 6th edition, Prentice Hall, Inc., 2005.

Procedure for Analysis

The equation of virtual work can be used to solve problems involving a system of frictionless connected rigid bodies having a single degree of freedom by using the following procedure.

Free-Body Diagram.

- Draw the free-body diagram of the entire system of connected bodies and define the *independent coordinate* q.
- Sketch the "deflected position" of the system on the free-body diagram when the system undergoes a positive virtual displacement δq.

Virtual Displacements.

- Indicate *position coordinates* s_i, measured from a *fixed point* on the free-body diagram to each of the i number of "active" forces and couples, i.e., those that do work.
- Each coordinate axis should be parallel to the line of action of the "active" force to which it is directed, so that the virtual work along the coordinate axis can be calculated.
- Relate each of the position coordinates s_i to the independent coordinate q; then *differentiate* these expressions in order to express the virtual displacements δs_i in terms of δq.

Virtual-Work Equation.

- Write the *virtual-work equation* for the system assuming that, whether possible or not, all the position coordinates s_i undergo *positive* virtual displacements δs_i.
- Using the relations for δs_i, express the work of *each* "active" force and couple in the equation in terms of the *single* independent virtual displacement δq.
- Factor out this common displacement from all the terms and solve for the unknown force, couple, or equilibrium position, q.
- If the system contains n degrees of freedom, n independent coordinates q_n must be specified. Follow the above procedure and let *only one* of the independent coordinates have a virtual displacement, while the remaining $n - 1$ coordinates are *held fixed*. In this way, n virtual-work equations can be written, one for each independent coordinate.

EXAMPLE 11.1

Determine the angle θ for equilibrium of the two-member linkage shown in Fig. 11–7*a*. Each member has a mass of 10 kg.

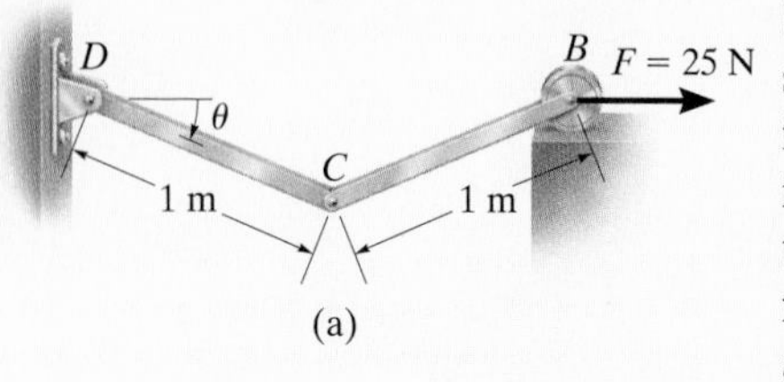

(a)

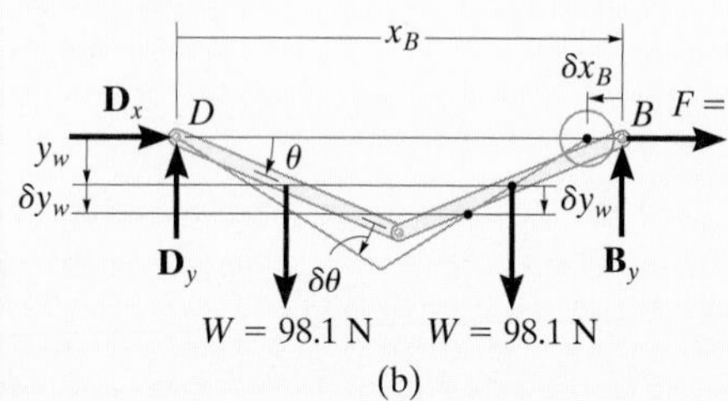

(b)

Fig. 11–7

SOLUTION

Free-Body Diagram. The system has only one degree of freedom since the location of both links may be specified by the single independent coordinate $(q =)\ \theta$. As shown on the free-body diagram in Fig. 11–7*b*, when θ has a *positive* (clockwise) virtual rotation $\delta\theta$, only the active forces, **F** and the two 98.1-N weights, do work. (The reactive forces $\mathbf{D}_x$ and $\mathbf{D}_y$ are fixed, and $\mathbf{B}_y$ does not move along its line of action.)

Virtual Displacements. If the origin of coordinates is established at the *fixed* pin support D, the location of **F** and **W** may be specified by the *position coordinates* x_B and y_w, as shown in the figure. In order to determine the work, note that these coordinates are parallel to the lines of action of their associated forces.

Expressing the position coordinates in terms of the independent coordinate θ and taking the derivatives yields

$$x_B = 2(1 \cos\theta)\text{ m} \qquad \delta x_B = -2\sin\theta\,\delta\theta\text{ m} \qquad (1)$$

$$y_w = \tfrac{1}{2}(1 \sin\theta)\text{ m} \qquad \delta y_w = 0.5\cos\theta\,\delta\theta\text{ m} \qquad (2)$$

It is seen by the *signs* of these equations, and indicated in Fig. 11–7*b*, that an *increase* in θ (i.e., $\delta\theta$) causes a *decrease* in x_B and an *increase* in y_w.

Virtual-Work Equation. If the virtual displacements δx_B and δy_w were *both positive*, then the forces **W** and **F** would do positive work since the forces and their corresponding displacements would have the same sense. Hence, the virtual-work equation for the displacement $\delta\theta$ is

$$\delta U = 0; \qquad W\,\delta y_w + W\,\delta y_w + F\,\delta x_B = 0 \qquad (3)$$

Substituting Eqs. 1 and 2 into Eq. 3 in order to relate the virtual displacements to the common virtual displacement $\delta\theta$ yields

$$98.1(0.5\cos\theta\,\delta\theta) + 98.1(0.5\cos\theta\,\delta\theta) + 25(-2\sin\theta\,\delta\theta) = 0$$

Notice that the "negative work" done by **F** (force in the opposite sense to displacement) has been *accounted for* in the above equation by the "negative sign" of Eq. 1. Factoring out the *common displacement* $\delta\theta$ and solving for θ, noting that $\delta\theta \neq 0$, yields

$$(98.1\cos\theta - 50\sin\theta)\,\delta\theta = 0$$

$$\theta = \tan^{-1}\frac{98.1}{50} = 63.0^\circ \qquad \textit{Ans.}$$

NOTE: If this problem had been solved using the equations of equilibrium, it would have been necessary to dismember the links and apply three scalar equations to *each* link. The principle of virtual work, by means of calculus, has eliminated this task so that the answer is obtained directly.

EXAMPLE 11.2

Determine the angle θ required to maintain equilibrium of the mechanism in Fig. 11–8*a*. Neglect the weight of the links. The spring is unstretched when $\theta = 0°$, and it maintains a horizontal position due to the roller.

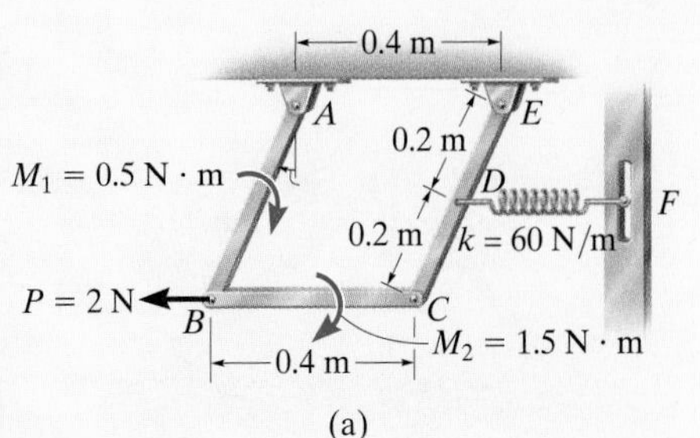

(a)

SOLUTION

Free-Body Diagram. The mechanism has one degree of freedom, and therefore the location of each member may be specified using the independent coordinate θ. When θ undergoes a *positive* virtual displacement $\delta\theta$, as shown on the free-body diagram in Fig. 11–8*b*, links *AB* and *EC* rotate by the same amount since they have the same length, and link *BC* only translates. Since a couple moment does work *only* when it rotates, the work done by $\mathbf{M}_2$ is zero. The reactive forces at *A* and *E* do no work. Why?

Virtual Displacements. The position coordinates x_B and x_D are *parallel* to the lines of action of **P** and $\mathbf{F}_s$, and these coordinates locate the forces with respect to the *fixed points A* and *E*. From Fig. 11–8*b*,

$$x_B = 0.4 \sin\theta \text{ m}$$

$$x_D = 0.2 \sin\theta \text{ m}$$

Thus,

$$\delta x_B = 0.4 \cos\theta\, \delta\theta \text{ m}$$

$$\delta x_D = 0.2 \cos\theta\, \delta\theta \text{ m}$$

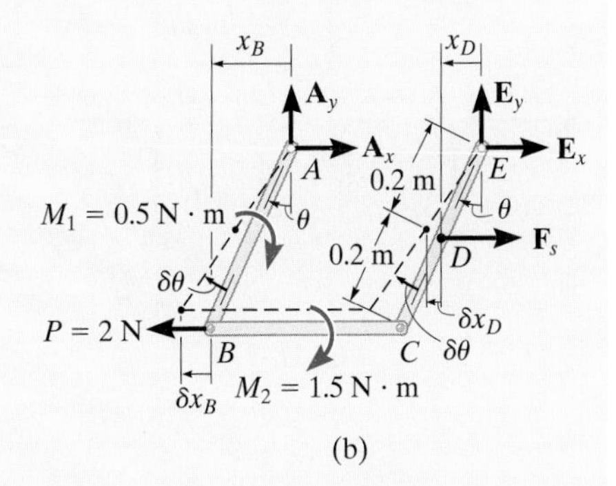

(b)

Fig. 11–8

Virtual-Work Equation. For *positive* virtual displacements, $\mathbf{F}_s$ is opposite to $\delta\mathbf{x}_D$ and hence does negative work. Thus,

$$\delta U = 0; \quad M_1\,\delta\theta + P\,\delta x_B - F_s\,\delta x_D = 0$$

Relating each of the virtual displacements to the *common* virtual displacement $\delta\theta$ yields

$$0.5\delta\theta + 2(0.4\cos\theta\,\delta\theta) - F_s(0.2\cos\theta\,\delta\theta) = 0$$

$$(0.5 + 0.8\cos\theta - 0.2F_s\cos\theta)\,\delta\theta = 0 \qquad (1)$$

For the arbitrary angle θ, the spring is stretched a distance of $x_D = (0.2\sin\theta)$ m; and therefore, $F_s = (60 \text{ N/m})(0.2\sin\theta)$ m $= (12\sin\theta)$ N. Substituting into Eq. 1 and noting that $\delta\theta \neq 0$, we have

$$0.5 + 0.8\cos\theta - 0.2(12\sin\theta)\cos\theta = 0$$

Since $\sin 2\theta = 2\sin\theta\cos\theta$, then

$$1 = 2.4\sin 2\theta - 1.6\cos\theta$$

Solving for θ by trial and error yields

$$\theta = 36.3° \qquad \textit{Ans.}$$

EXAMPLE 11.3

Determine the horizontal force C_x that the pin at C must exert on BC in order to hold the mechanism shown in Fig. 11–9*a* in equilibrium when $\theta = 45°$. Neglect the weight of the members.

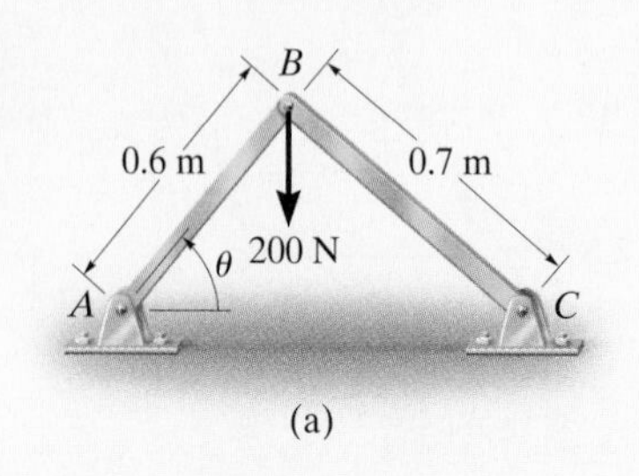

(a)

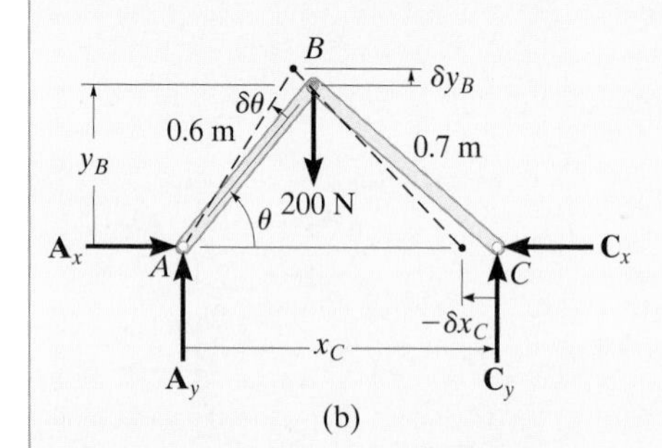

(b)

Fig. 11–9

SOLUTION

Free-Body Diagram. The reaction $\mathbf{C}_x$ can be obtained by *releasing* the pin constraint at C in the x direction and allowing the frame to be displaced in this direction. The system then has only one degree of freedom, defined by the independent coordinate θ, Fig. 11–9*b*. When θ has a *positive* virtual displacement $\delta\theta$, only $\mathbf{C}_x$ and the 200-N force do work.

Virtual Displacements. Forces $\mathbf{C}_x$ and 200 N are located from the fixed origin A using position coordinates y_B and x_C. From Fig. 11–9*b*, x_C can be related to θ by the "law of cosines." Hence,

$$(0.7)^2 = (0.6)^2 + x_C^2 - 2(0.6)x_C \cos\theta \qquad (1)$$

$$0 = 0 + 2x_C\,\delta x_C - 1.2\delta x_C \cos\theta + 1.2x_C \sin\theta\,\delta\theta$$

$$\delta x_C = \frac{1.2x_C \sin\theta}{1.2\cos\theta - 2x_C}\delta\theta \qquad (2)$$

Also,

$$y_B = 0.6\sin\theta$$

$$\delta y_B = 0.6\cos\theta\,\delta\theta \qquad (3)$$

Virtual-Work Equation. When y_B and x_C have *positive* virtual displacements δy_B and δx_C, $\mathbf{C}_x$ and 200 N do *negative work* since they both act in the opposite sense to $\delta\mathbf{y}_B$ and $\delta\mathbf{x}_C$. Hence,

$$\delta U = 0; \qquad -200\delta y_B - C_x\,\delta x_C = 0$$

Substituting Eqs. 2 and 3 into this equation, factoring out $\delta\theta$, and solving for C_x yields

$$-200(0.6\cos\theta\,\delta\theta) - C_x\frac{1.2x_C\sin\theta}{1.2\cos\theta - 2x_C}\delta\theta = 0$$

$$C_x = \frac{-120\cos\theta(1.2\cos\theta - 2x_C)}{1.2x_C\sin\theta} \qquad (4)$$

At the required equilibrium position $\theta = 45°$, the corresponding value of x_C can be found by using Eq. 1, in which case

$$x_C^2 - 1.2\cos 45°\,x_C - 0.13 = 0$$

Solving for the positive root yields

$$x_C = 0.981\text{ m}$$

Thus, from Eq. 4,

$$C_x = 114\text{ N} \qquad \textit{Ans.}$$

EXAMPLE 11.4

Determine the equilibrium position of the two-bar linkage shown in Fig. 11–10*a*. Neglect the weight of the links.

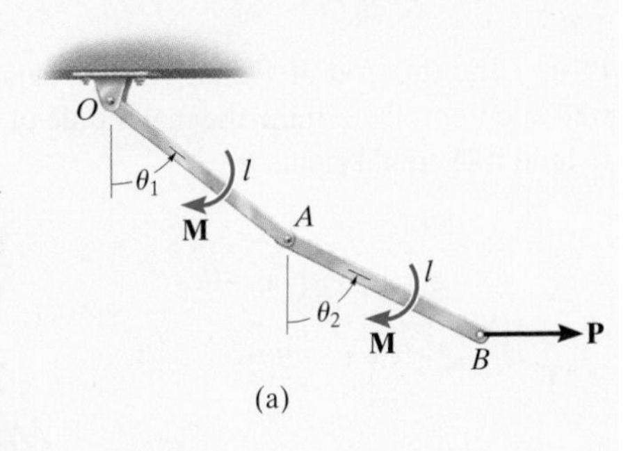

(a)

SOLUTION

The system has two degrees of freedom since the *independent coordinates* θ_1 and θ_2 must be known to locate the position of both links. The position coordinate x_B, measured from the fixed point O, is used to specify the location of **P**, Fig. 11–10*b* and *c*.

If θ_1 is held *fixed* and θ_2 varies by an amount $\delta\theta_2$, as shown in Fig. 11–10*b*, the virtual-work equation becomes

$$[\delta U = 0]_{\theta_2}; \qquad P(\delta x_B)_{\theta_2} - M\,\delta\theta_2 = 0 \qquad (1)$$

Here P and M represent the magnitudes of the applied force and couple moment acting on link AB.

When θ_2 is held *fixed* and θ_1 varies by an amount $\delta\theta_1$, as shown in Fig. 11–10*c*, then AB translates and the virtual-work equation becomes

$$[\delta U = 0]_{\theta_1}; \qquad P(\delta x_B)_{\theta_1} - M\,\delta\theta_1 = 0 \qquad (2)$$

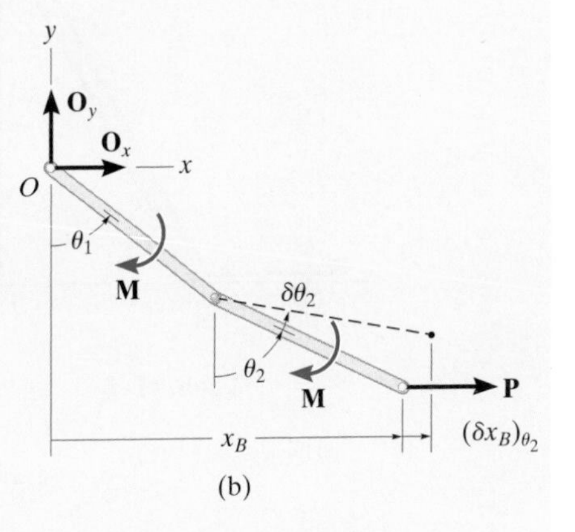

(b)

The *position coordinate* x_B may be related to the independent coordinates θ_1 and θ_2 by the equation

$$x_B = l\sin\theta_1 + l\sin\theta_2 \qquad (3)$$

To obtain the variation δx_B in terms of $\delta\theta_2$, it is necessary to take the *partial derivative* of x_B with respect to θ_2 since x_B is a function of both θ_1 and θ_2. Hence,

$$\frac{\partial x_B}{\partial\theta_2} = l\cos\theta_2 \qquad (\delta x_B)_{\theta_2} = l\cos\theta_2\,\delta\theta_2$$

Substituting into Eq. 1, we have

$$(Pl\cos\theta_2 - M)\,\delta\theta_2 = 0$$

Since $\delta\theta_2 \neq 0$, then

$$\theta_2 = \cos^{-1}\left(\frac{M}{Pl}\right) \qquad \textit{Ans.}$$

Using Eq. 3 to obtain the variation of x_B with θ_1 yields

$$\frac{\partial x_B}{\partial\theta_1} = l\cos\theta_1 \quad (\delta x_B)_{\theta_1} = l\cos\theta_1\,\delta\theta_1$$

Substituting into Eq. 2, we have

$$(Pl\cos\theta_1 - M)\,\delta\theta_1 = 0$$

Since $\delta\theta_1 \neq 0$, then

$$\theta_1 = \cos^{-1}\left(\frac{M}{Pl}\right) \qquad \textit{Ans.}$$

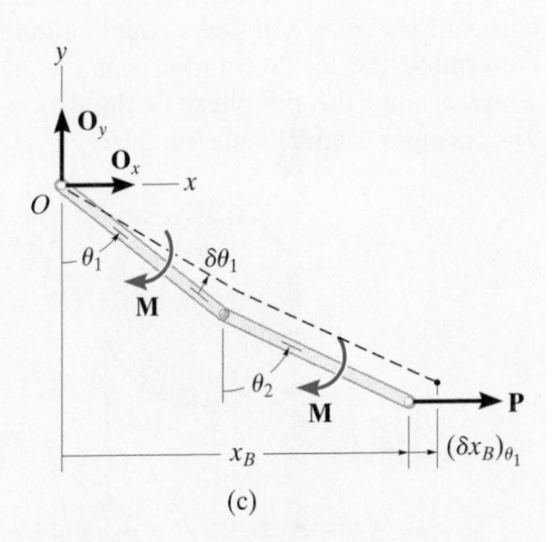

(c)

Fig. 11–10

PROBLEMS

11–1. The thin rod of weight W rests against the smooth wall and floor. Determine the magnitude of force **P** needed to hold it in equilibrium.

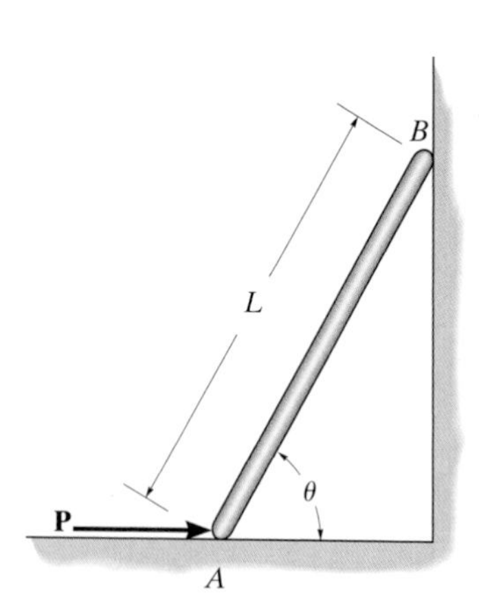

Prob. 11–1

11–2. The disk has a weight of 10 lb and is subjected to a vertical force $P = 8$ lb and a couple moment $M = 8$ lb·ft. Determine the disk's rotation θ if the end of the spring wraps around the periphery of the disk as the disk turns. The spring is originally unstretched.

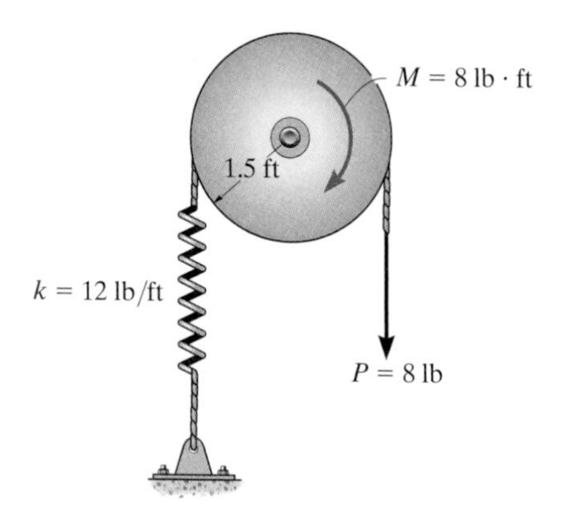

Prob. 11–2

11–3. The platform supports a load W. Determine the horizontal force P that must be supplied by the screw in order to support the platform when the links are at the arbitrary angle θ.

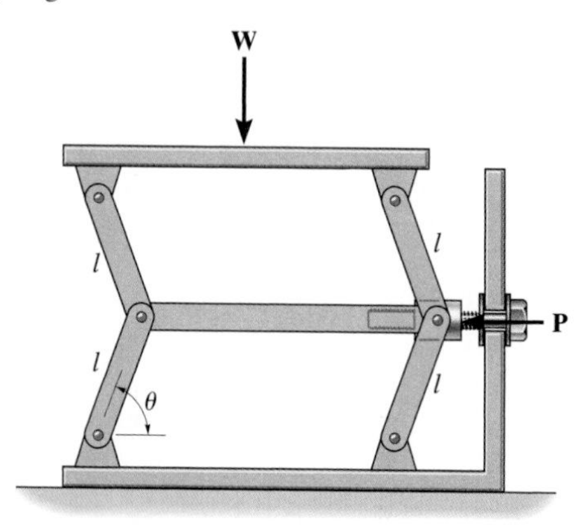

Prob. 11–3

*■**11–4.** Each member of the pin-connected mechanism has a mass of 8 kg. If the spring is unstretched when $\theta = 0°$, determine the angle θ for equilibrium. Set $k = 2500$ N/m and $M = 50$ N·m.

11–5. Each member of the pin-connected mechanism has a mass of 8 kg. If the spring is unstretched when $\theta = 0°$, determine the required stiffness k so that the mechanism is in equilibrium when $\theta = 30°$. Set $\mathbf{M} = 0$.

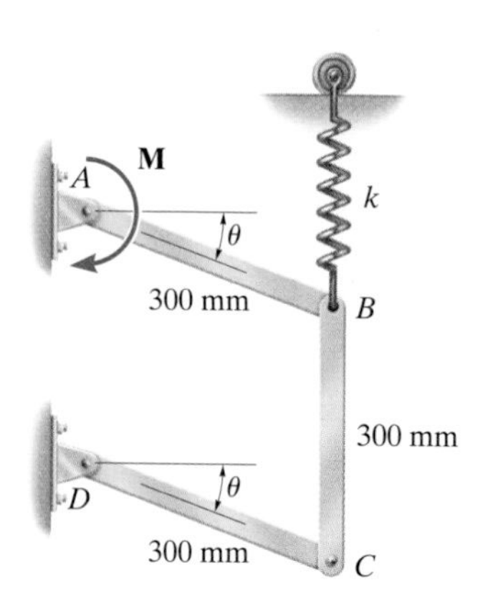

Probs. 11–4/5

11–6. The crankshaft is subjected to a torque of $M = 50\ \text{N}\cdot\text{m}$. Determine the horizontal compressive force F applied to the piston for equilibrium when $\theta = 60°$.

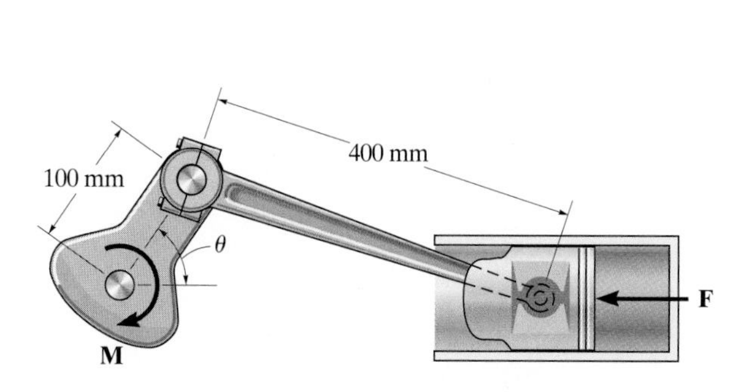

Prob. 11–6

11–7. The crankshaft is subjected to a torque of $M = 50\ \text{N}\cdot\text{m}$. Determine the horizontal compressive force F and plot the result of F (ordinate) versus θ (abscissa) for $0° \leq \theta \leq 90°$.

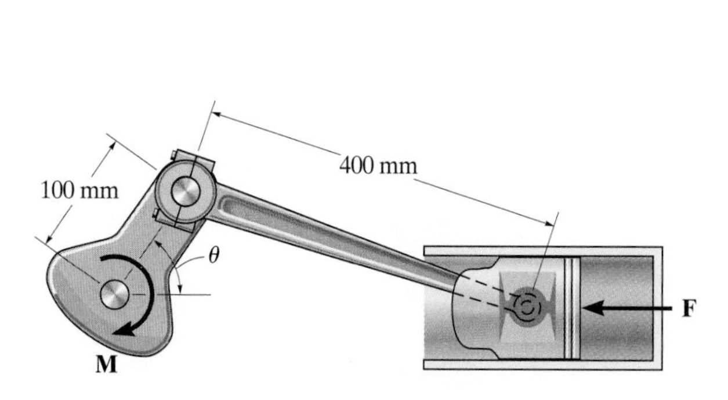

Prob. 11–7

***11–8.** If a force of $P = 30$ lb is applied perpendicular to the handle of the toggle press, determine the compressive force developed at C; $\theta = 30°$.

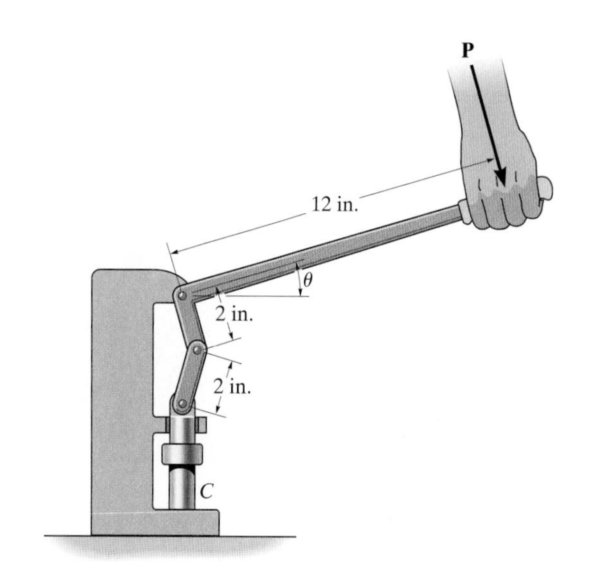

Prob. 11–8

11–9. A force **P** is applied to the end of the lever. Determine the horizontal force F on the piston for equilibrium.

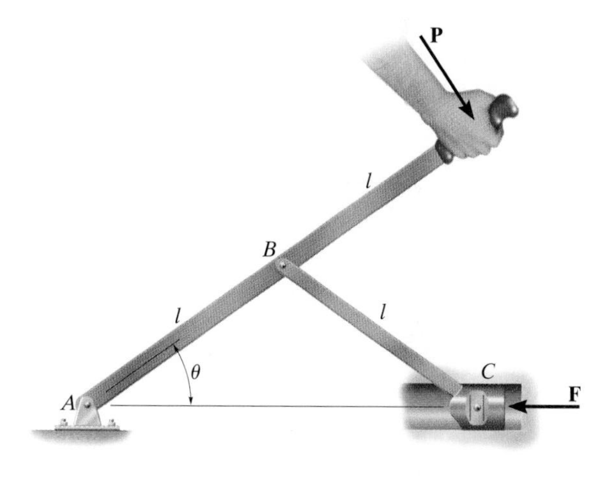

Prob. 11–9

11–10. The mechanism consists of the four pin-connected bars and three springs, each having a stiffness k and an unstretched length l_0. Determine the horizontal forces P that must be applied to the pins in order to hold the mechanism in the horizontal position for equilibrium.

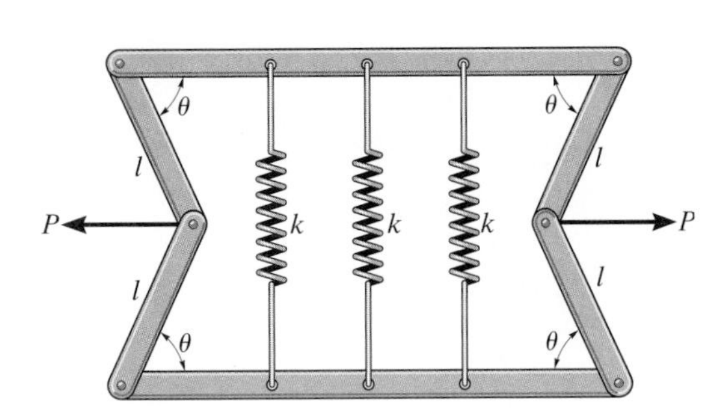

Prob. 11–10

11–11. When $\theta = 20°$, the 50-lb uniform block compresses the two vertical springs 4 in. If the uniform links AB and CD each weigh 10 lb, determine the magnitude of the applied couple moments **M** needed to maintain equilibrium when $\theta = 20°$.

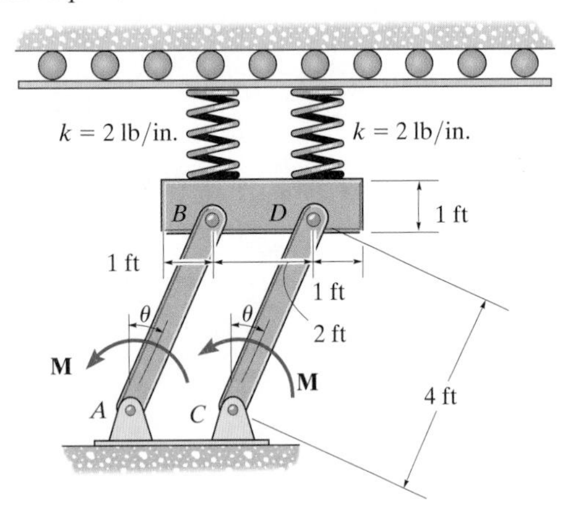

Prob. 11–11

***11–12.** The spring is unstretched when $\theta = 0°$. If $P = 8$ lb, determine the angle θ for equilibrium. Due to the roller guide, the spring always remains vertical. Neglect the weight of the links.

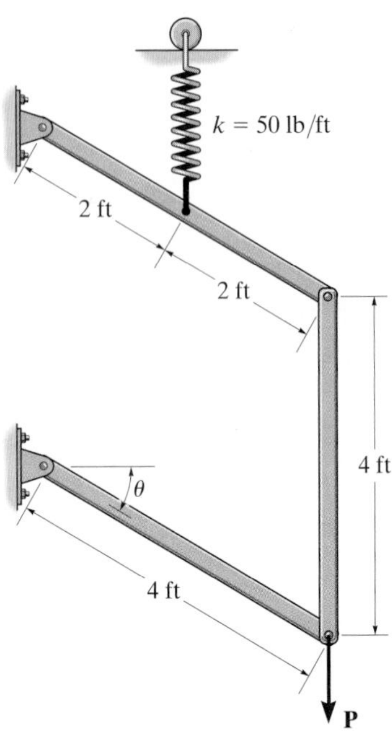

Prob. 11–12

11–13. Determine the force P required to lift the 15-kg block using the differential hoist. The lever arm is fixed to the upper pulley and turns with it.

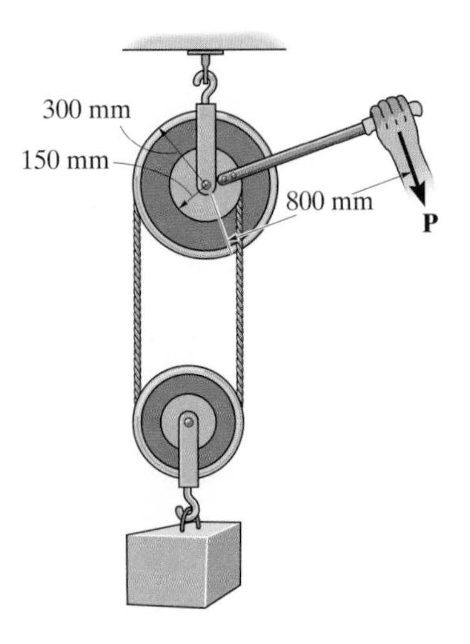

Prob. 11–13

11–14. Determine the magnitude of the applied couple moments **M** needed to maintain equilibrium when $\theta = 20°$. The plate E has a weight of 50 lb. Neglect the weight of the links AB and CD.

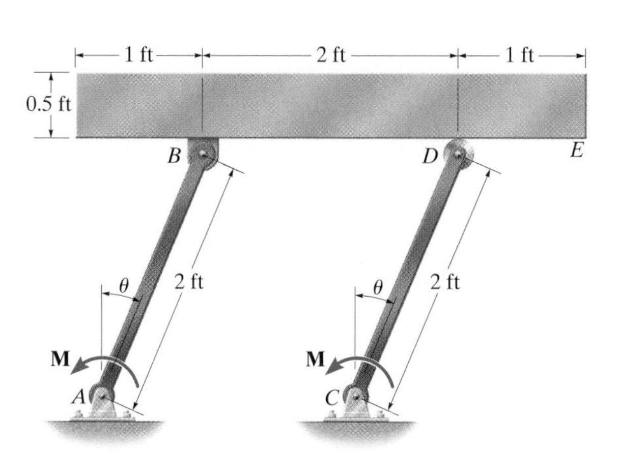

Prob. 11–14

11–15. The members of the mechanism are pin connected. If a horizontal force of 400·N acts at A, determine the angle θ for equilibrium. The spring is unstretched when $\theta = 90°$.

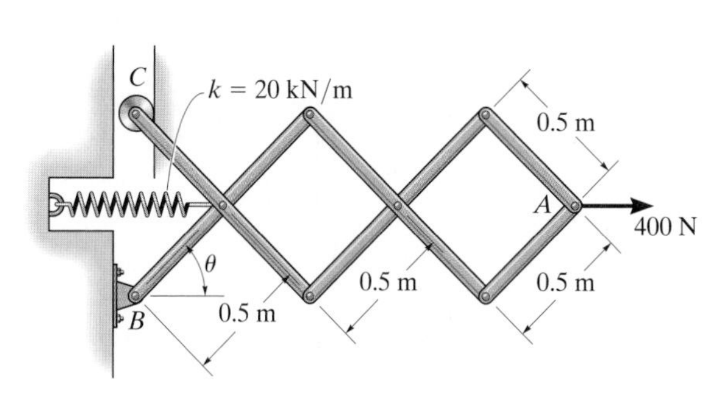

Prob. 11–15

***11–16.** Determine the force F needed to lift the block having a weight of 100 lb. *Hint:* Note that the coordinates s_A and s_B can be related to the *constant* vertical length l of the cord.

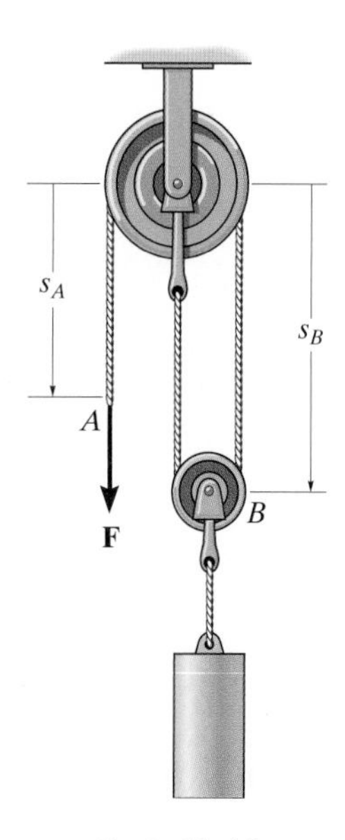

Prob. 11–16

■**11–17.** Each member of the pin-connected mechanism has a mass of 8 kg. If the spring is unstretched when $\theta = 0°$, determine the angle θ for equilibrium.

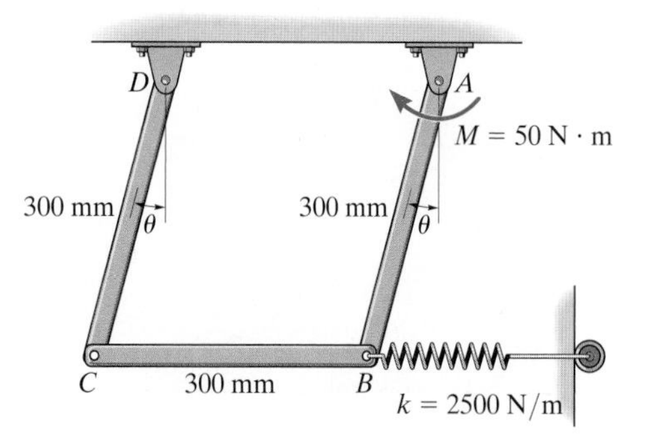

Prob. 11–17

11–18. The bar is supported by the spring and smooth collar that allows the spring to be always perpendicular to the bar for any angle θ. If the unstretched length of the spring is l_0, determine the force P needed to hold the bar in the equilibrium position θ. Neglect the weight of the bar.

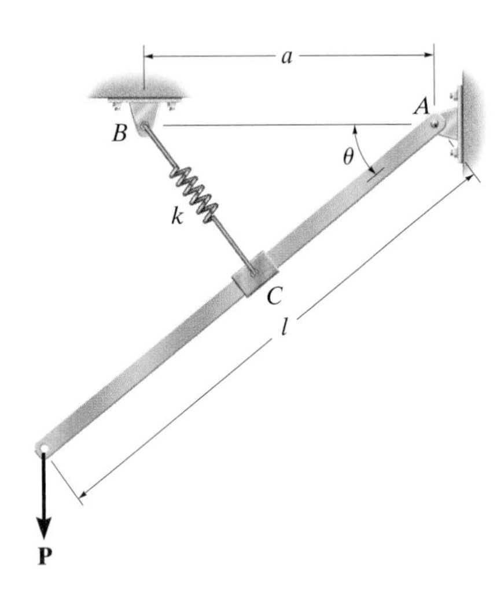

Prob. 11–18

11–19. The scissors jack supports a load **P**. Determine the axial force in the screw necessary for equilibrium when the jack is in the position θ. Each of the four links has a length L and is pin-connected at its center. Points B and D can move horizontally.

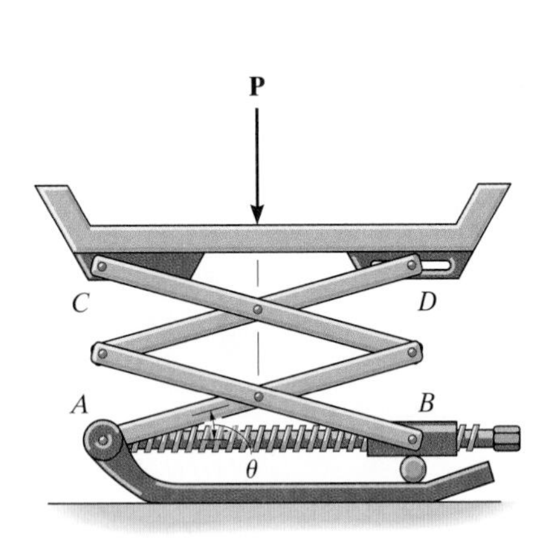

Prob. 11–19

***11–20.** Determine the mass of A and B required to hold the 400-g desk lamp in balance for any angles θ and ϕ. Neglect the weight of the mechanism and the size of the lamp.

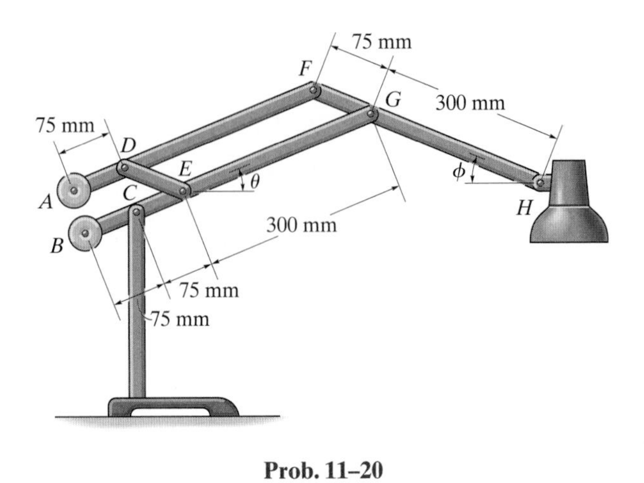

Prob. 11–20

11–21. The *Roberval balance* is in equilibrium when no weights are placed on the pans A and B. If two masses m_A and m_B are placed at *any* location a and b on the pans, show that equilibrium is maintained if $m_A d_A = m_B d_B$.

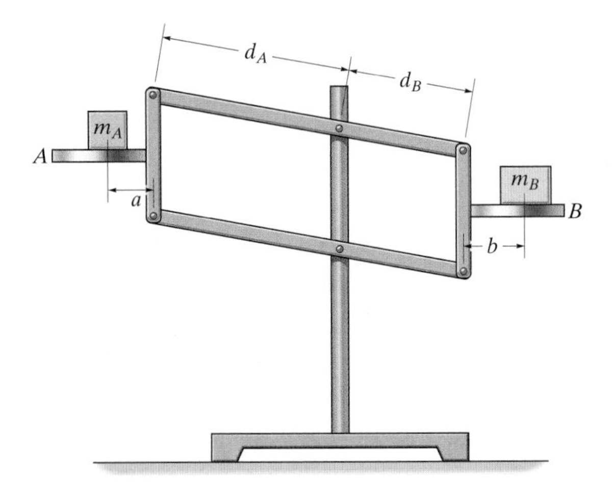

Prob. 11–21

11–22. The chain puller is used to draw two ends of a chain together in order to attach the "master link." The device is operated by turning the screw S, which pushes the bar AB downward, thereby drawing the tips C and D towards one another. If the sliding contacts at A and B are smooth, determine the force F maintained by the screw at E which is required to develop a drawing tension of 120 lb in the chains.

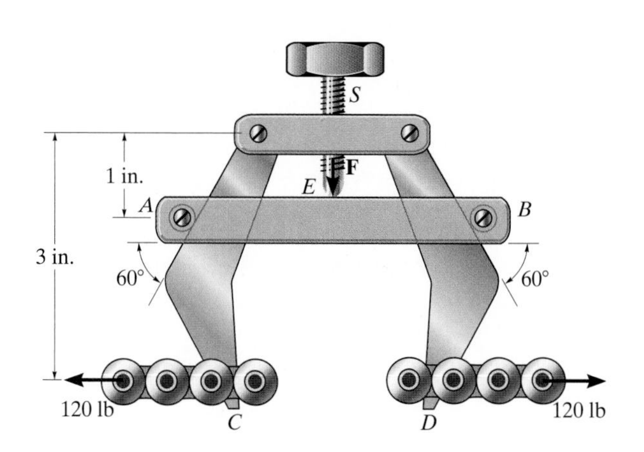

Prob. 11–22

11–23. The service window at a fast-food restaurant consists of glass doors that open and close automatically using a motor which supplies a torque **M** to each door. The far ends, A and B, move along the horizontal guides. If a food tray becomes stuck between the doors as shown, determine the horizontal force the doors exert on the tray at the position θ.

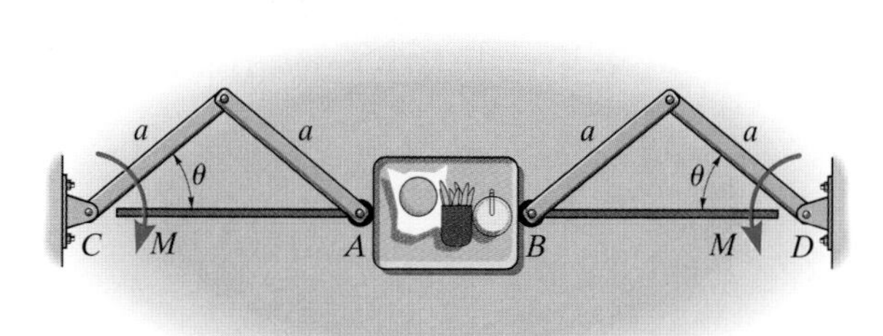

Prob. 11–23

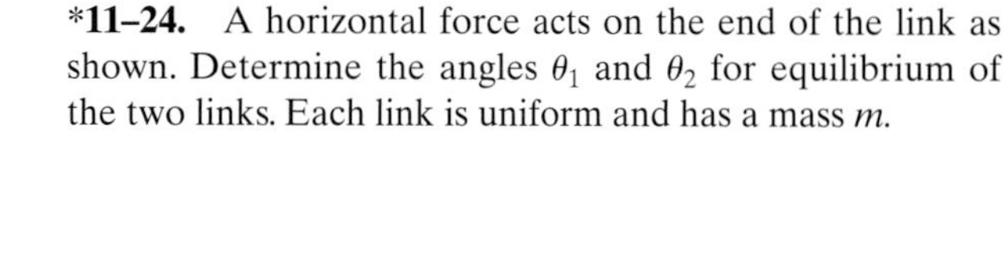

***11–24.** A horizontal force acts on the end of the link as shown. Determine the angles θ_1 and θ_2 for equilibrium of the two links. Each link is uniform and has a mass m.

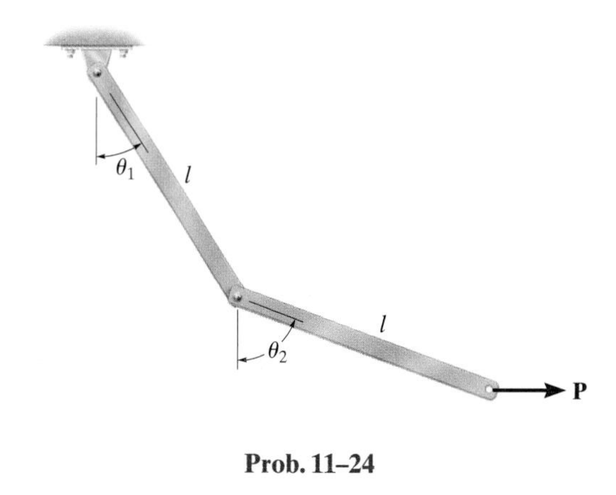

Prob. 11–24

11–25. Rods AB and BC have centers of mass located at their midpoints. If all contacting surfaces are smooth and BC has a mass of 100 kg, determine the appropriate mass of AB required for equilibrium.

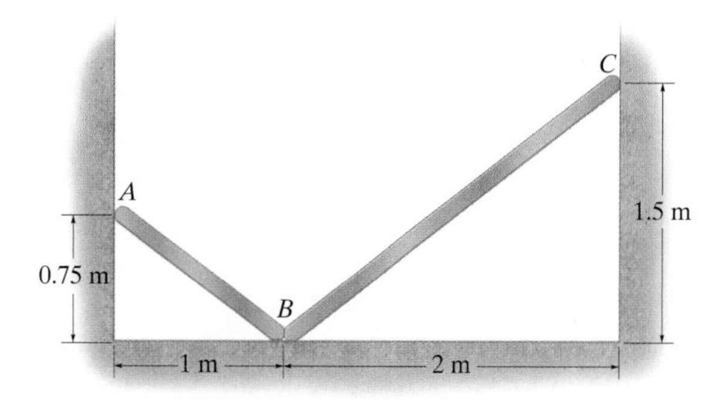

Prob. 11–25

*11.4 Conservative Forces

The work done by a force when it undergoes a *differential displacement* has been defined as $dU = F \cos\theta\, ds$, Fig. 11–1. If the force is displaced over a path that has a *finite length s*, the work is determined by integrating over the path: i.e.,

$$U = \int_s F \cos\theta\, ds$$

To evaluate the integral, it is necessary to obtain a relationship between F and the component of displacement $ds \cos\theta$. In some instances, however, the work done by a force will be *independent* of its path and, instead, will depend only on the initial and final locations of the force along the path. A force that has this property is called a *conservative force*.

Weight. Consider the body in Fig. 11–11, which is initially at P'. If the body is moved *down* along the *arbitrary path A* to the second position, then, for a given displacement *ds* along the path, the displacement component in the direction of **W** has a magnitude of $dy = ds \cos\theta$, as shown. Since both the force and displacement are in the same direction, the work is positive; hence,

$$U = \int_s W \cos\theta\, ds = \int_0^y W\, dy$$

or

$$U = Wy$$

In a similar manner, the work done by the weight when the body moves up a distance y back to P', along the arbitrary path A', is

$$U = -Wy$$

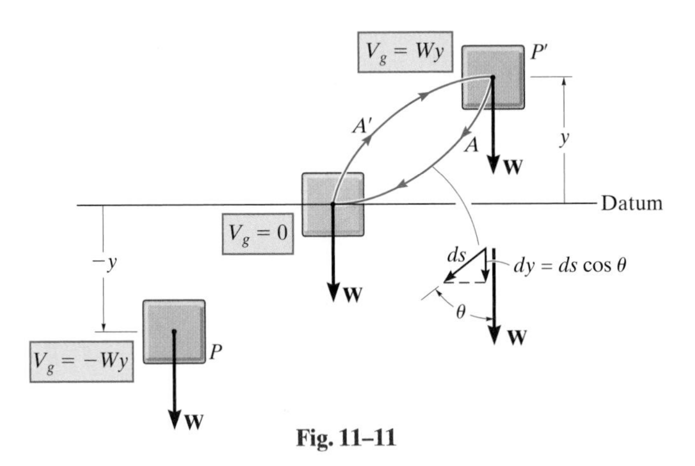

Fig. 11–11

Why is the work negative?

The weight of a body is therefore a conservative force since the work done by the weight depends *only* on the body's *vertical displacement* and is independent of the path along which the body moves.

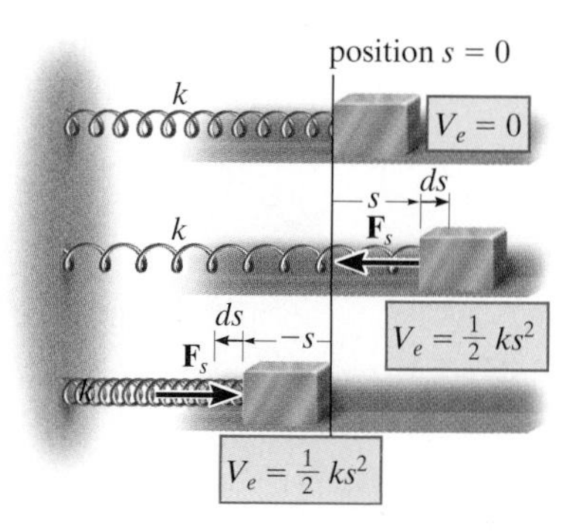

Fig. 11–12

Elastic Spring. The force developed by an elastic spring ($F_s = ks$) is also a conservative force. If the spring is attached to a body and the body is displaced along *any path*, such that it causes the spring to elongate or compress from a position s_1 to a further position s_2, the work will be negative since the spring exerts a force $\mathbf{F}_s$ *on the body* that is opposite to the body's displacement ds, Fig. 11–12. For either extension or compression, the work is independent of the path and is simply

$$U = \int_{s_1}^{s_2} F_s\,ds = \int_{s_1}^{s_2} (-ks)\,ds$$

$$= -\left(\tfrac{1}{2}ks_2^2 - \tfrac{1}{2}ks_1^2\right)$$

Friction. In contrast to a conservative force, consider the force of *friction* exerted on a sliding body by a fixed surface. The work done by the frictional force depends on the path; the longer the path, the greater the work. Consequently, frictional forces are *nonconservative*, and the work done is dissipated from the body in the form of heat.

*11.5 Potential Energy

When a conservative force acts on a body, it gives the body the capacity to do work. This capacity, measured as *potential energy*, depends on the location of the body.

Gravitational Potential Energy. If a body is located a distance y *above* a fixed horizontal reference or datum, Fig. 11–11, the weight of the body has *positive* gravitational potential energy V_g since $\mathbf{W}$ has the capacity of doing positive work when the body is moved back down to the datum. Likewise, if the body is located a distance y *below* the datum, V_g is *negative* since the weight does negative work when the body is moved back up to the datum. At the datum, $V_g = 0$.

Measuring y as *positive upward*, the gravitational potential energy of the body's weight $\mathbf{W}$ is thus

$$\boxed{V_g = Wy} \qquad (11\text{–}4)$$

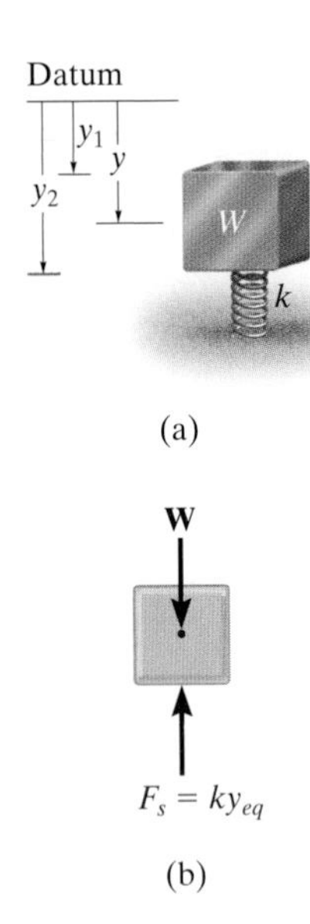

Fig. 11–13

Elastic Potential Energy. The elastic potential energy V_e that a spring produces on an attached body, when the spring is elongated or compressed from an undeformed position ($s = 0$) to a final position s, is

$$V_e = \tfrac{1}{2}ks^2 \tag{11–5}$$

Here V_e is *always positive* since in the deformed position the spring has the capacity of doing *positive work* in *returning* the body back to the spring's undeformed position, Fig. 11–12.

Potential Function. In the general case, if a body is subjected to *both* gravitational and elastic forces, the *potential energy or potential function* V of the body can be expressed as the algebraic sum

$$V = V_g + V_e \tag{11–6}$$

where measurement of V depends on the location of the body with respect to a selected datum in accordance with Eqs. 11–4 and 11–5.

In general, if a system of frictionless connected rigid bodies has a *single degree of freedom* such that its position from the datum is defined by the independent coordinate q, then the potential function for the system can be expressed as $V = V(q)$. The work done by all the conservative forces acting on the system in moving it from q_1 to q_2 is measured by the *difference* in V; i.e.,

$$U_{1-2} = V(q_1) - V(q_2) \tag{11–7}$$

For example, the potential function for a system consisting of a block of weight **W** supported by a spring, Fig. 11–13*a*, can be expressed in terms of its independent coordinate $(q =)\ y$, measured from a fixed datum located at the unstretched length of the spring; we have

$$\begin{aligned} V &= V_g + V_e \\ &= -Wy + \tfrac{1}{2}ky^2 \end{aligned} \tag{11–8}$$

If the block moves from y_1 to a farther downward position y_2, then the work of **W** and $\mathbf{F}_s$ is

$$U_{1-2} = V(y_1) - V(y_2) = -W[y_1 - y_2] + \tfrac{1}{2}ky_1^2 - \tfrac{1}{2}ky_2^2$$

*11.6 Potential-Energy Criterion for Equilibrium

System Having One Degree of Freedom. When the displacement of a frictionless connected system is *infinitesimal*, i.e., from q to $q + dq$, Eq. 11–7 becomes

$$dU = V(q) - V(q + dq)$$

or

$$dU = -dV$$

Furthermore, if the system undergoes a *virtual displacement* δq, rather than an actual displacement dq, then $\delta U = -\delta V$. For equilibrium, the principle of virtual work requires that $\delta U = 0$, and therefore, provided the potential function for the system is known, this also requires that $\delta V = 0$. We can also express this requirement as

$$\boxed{\frac{dV}{dq} = 0} \qquad (11\text{–}9)$$

Hence, *when a frictionless connected system of rigid bodies is in equilibrium, the first variation or change in V is zero*. This change is determined by taking the *first derivative* of the potential function and setting it equal to zero. For example, using Eq. 11–8 to determine the equilibrium position for the spring and block in Fig. 11–13*a*, we have

$$\frac{dV}{dy} = -W + ky = 0$$

Hence, the equilibrium position $y = y_{eq}$ is

$$y_{eq} = \frac{W}{k}$$

Of course, the *same result* is obtained by applying $\Sigma F_y = 0$ to the forces acting on the free-body diagram of the block, Fig. 11–13*b*.

System Having *n* Degrees of Freedom. When the system of connected bodies has n degrees of freedom, the total potential energy stored in the system will be a function of n independent coordinates q_n, i.e., $V = V(q_1, q_2, \ldots, q_n)$. In order to apply the equilibrium criterion $\delta V = 0$, it is necessary to determine the change in potential energy δV by using the "chain rule" of differential calculus; i.e.,

$$\delta V = \frac{\partial V}{\partial q_1}\delta q_1 + \frac{\partial V}{\partial q_2}\delta q_2 + \cdots + \frac{\partial V}{\partial q_n}\delta q_n = 0$$

Since the virtual displacements $\delta q_1, \delta q_2, \ldots, \delta q_n$ are independent of one another, the equation is satisfied provided

$$\frac{\partial V}{\partial q_1} = 0, \quad \frac{\partial V}{\partial q_2} = 0, \quad \ldots, \quad \frac{\partial V}{\partial q_n} = 0$$

Hence *it is possible to write n independent equations for a system having n degrees of freedom.*

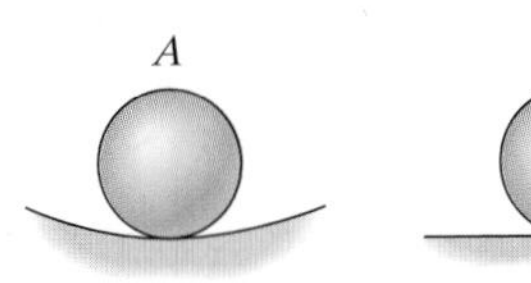

Stable equilibrium Neutral equilibrium

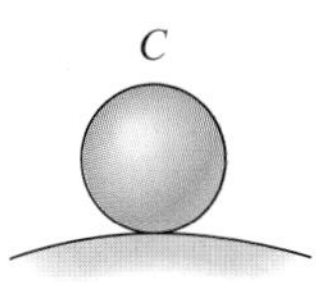

Unstable equilibrium

Fig. 11–14

*11.7 Stability of Equilibrium

Once the equilibrium configuration for a body or system of connected bodies is defined, it is sometimes important to investigate the "type" of equilibrium or the stability of the configuration. For example, consider the position of a ball resting at a point on each of the three paths shown in Fig. 11–14. Each situation represents an equilibrium state for the ball. When the ball is at A, it is said to be in *stable equilibrium* because if it is given a small displacement up the hill, it will always *return* to its original, lowest, position. At A, its total potential energy is a *minimum*. When the ball is at B, it is in *neutral equilibrium*. A small displacement to either the left or right of B will not alter this condition. The ball *remains* in equilibrium in the displaced position, and therefore its potential energy is *constant*. When the ball is at C, it is in *unstable equilibrium*. Here a small displacement will cause the ball's potential energy to be *decreased*, and so it will roll farther *away* from its original, highest position. At C, the potential energy of the ball is a *maximum*.

During high winds and when going around a curve, these sugar-cane trucks can become unstable and tip over since their center of gravity is high off the road when they are fully loaded.

Types of Equilibrium. The example just presented illustrates that one of three types of equilibrium positions can be specified for a body or system of connected bodies.

1. *Stable equilibrium* occurs when a small displacement of the system causes the system to return to its original position. In this case the original potential energy of the system is a minimum.
2. *Neutral equilibrium* occurs when a small displacement of the system causes the system to remain in its displaced state. In this case the potential energy of the system remains constant.
3. *Unstable equilibrium* occurs when a small displacement of the system causes the system to move farther away from its original position. In this case the original potential energy of the system is a maximum.

System Having One Degree of Freedom. For *equilibrium* of a system having a single degree of freedom, defined by the independent coordinate q, it has been shown that the first derivative of the potential function for the system must be equal to zero; i.e., $dV/dq = 0$. If the potential function $V = V(q)$ is plotted, Fig. 11–15, the first derivative (equilibrium position) is represented as the slope dV/dq, which is zero when the function is maximum, minimum, or an inflection point.

If the *stability* of a body is to be investigated, it is necessary to determine the *second derivative* of V and evaluate it at the equilibrium position $q = q_{\text{eq}}$. As shown in Fig. 11–15*a*, if $V = V(q)$ is a *minimum*, then

$$\frac{dV}{dq} = 0, \quad \frac{d^2V}{dq^2} > 0 \qquad \text{stable equilibrium} \tag{11–10}$$

If $V = V(q)$ is a *maximum*, Fig. 11–15*b*, then

$$\frac{dV}{dq} = 0, \quad \frac{d^2V}{dq^2} < 0 \qquad \text{unstable equilibrium} \tag{11–11}$$

If the second derivative is zero, it will be necessary to investigate *higher-order* derivatives to determine the stability. In particular, stable equilibrium will occur if the order of the lowest remaining nonzero derivative is *even* and the sign of this nonzero derivative is positive when it is evaluated at $q = q_{eq}$; otherwise, it is unstable.

If the system is in neutral equilibrium, Fig. 11–15*c*, it is required that

$$\frac{dV}{dq} = \frac{d^2V}{dq^2} = \frac{d^3V}{dq^3} = \cdots = 0 \qquad \text{neutral equilibrium} \tag{11–12}$$

since then V must be constant at and around the "neighborhood" of q_{eq}.

System Having Two Degrees of Freedom. A criterion for investigating stability becomes increasingly complex as the number of degrees of freedom for the system increases. For a system having two degrees of freedom, defined by independent coordinates (q_1, q_2), it may be verified (using the calculus of functions of two variables) that equilibrium and stability occur at a point (q_{1eq}, q_{2eq}) when

$$\frac{\partial V}{\partial q_1} = \frac{\partial V}{\partial q_2} = 0$$

$$\left[\left(\frac{\partial^2 V}{\partial q_1\,\partial q_2}\right)^2 - \left(\frac{\partial^2 V}{\partial q_1^2}\right)\left(\frac{\partial^2 V}{\partial q_2^2}\right)\right] < 0$$

$$\frac{\partial^2 V}{\partial q_1^2} > 0 \quad \text{or} \quad \frac{\partial^2 V}{\partial q_2^2} > 0$$

Both equilibrium and instability occur when

$$\frac{\partial V}{\partial q_1} = \frac{\partial V}{\partial q_2} = 0$$

$$\left[\left(\frac{\partial^2 V}{\partial q_1\,\partial q_2}\right)^2 - \left(\frac{\partial^2 V}{\partial q_1^2}\right)\left(\frac{\partial^2 V}{\partial q_2^2}\right)\right] < 0$$

$$\frac{\partial^2 V}{\partial q_1^2} < 0 \quad \text{or} \quad \frac{\partial^2 V}{\partial q_2^2} < 0$$

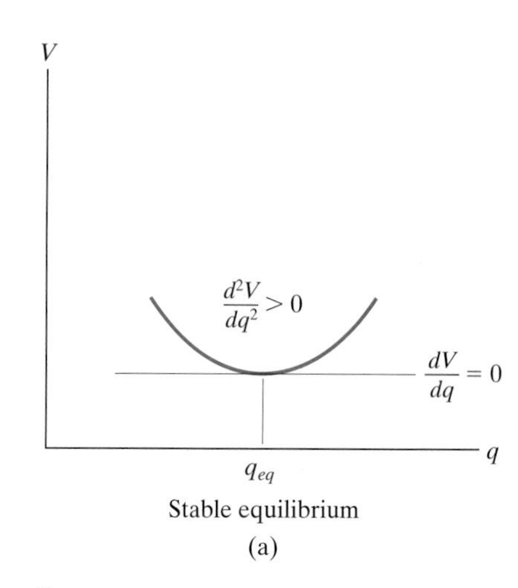

Stable equilibrium
(a)

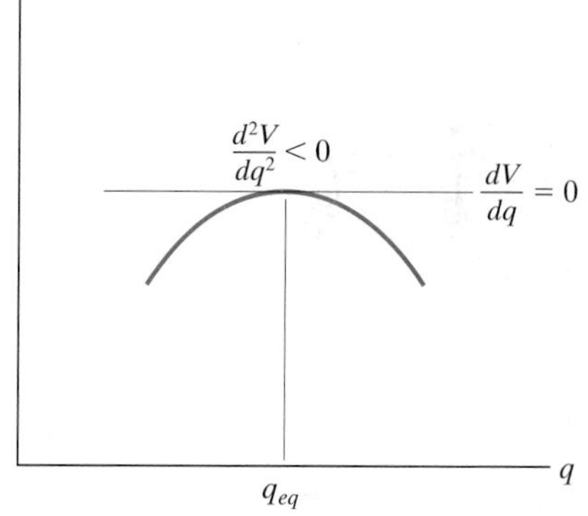

Unstable equilibrium
(b)

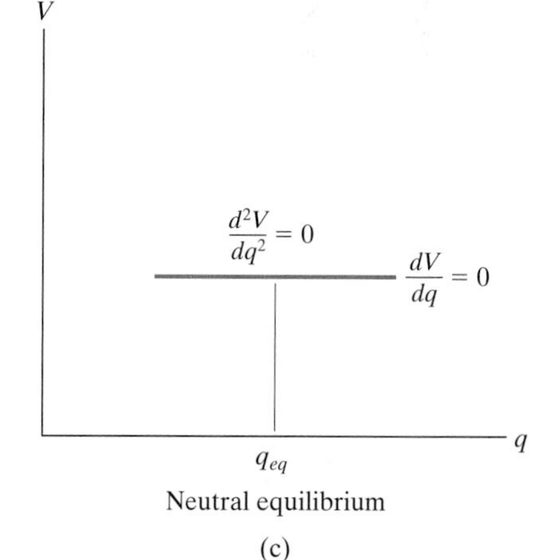

Neutral equilibrium
(c)

Fig. 11–15

Procedure for Analysis

Using potential-energy methods, the equilibrium positions and the stability of a body or a system of connected bodies having a single degree of freedom can be obtained by applying the following procedure.

Potential Function.

- Sketch the system so that it is located at some *arbitrary position* specified by the independent coordinate q.
- Establish a horizontal *datum* through a *fixed point** and express the *gravitational potential energy* V_g in terms of the weight W of each member and its vertical distance y from the datum, $V_g = Wy$.
- Express the elastic potential energy V_e of the system in terms of the stretch or compression, s, of any connecting spring and the spring's stiffness k, $V_e = \frac{1}{2}ks^2$.
- Formulate the potential function $V = V_g + V_e$ and express the *position coordinates* y and s in terms of the independent coordinate q.

Equilibrium Position.

- The equilibrium position is determined by taking the first derivative of V and setting it equal to zero, $\delta V = 0$.

Stability.

- Stability at the equilibrium position is determined by evaluating the second or higher-order derivatives of V.
- If the second derivative is greater than zero, the body is stable, if all derivatives are equal to zero the body is in neutral equilibrium, and if the second derivative is less than zero, the body is unstable.

*The location of the datum is *arbitrary* since only the *changes* or differentials of V are required for investigation of the equilibrium position and its stability.

EXAMPLE 11.5

The uniform link shown in Fig. 11–16*a* has a mass of 10 kg. The spring is unstretched when $\theta = 0°$. Determine the angle θ for equilibrium and investigate the stability at the equilibrium position.

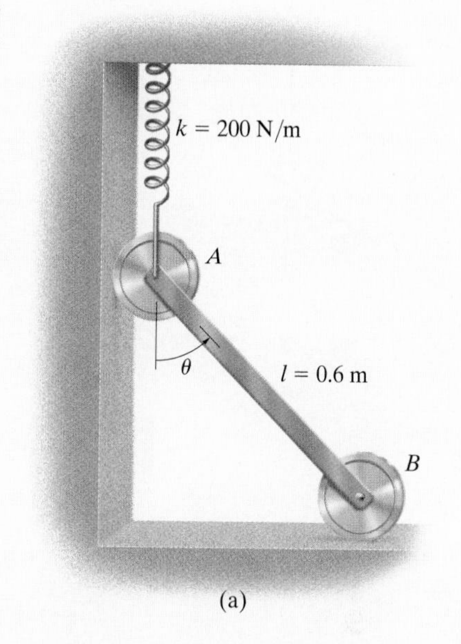

(a)

SOLUTION

Potential Function. The datum is established at the top of the link when the *spring is unstretched*, Fig. 11–16*b*. When the link is located at the arbitrary position θ, the spring increases its potential energy by stretching and the weight decreases its potential energy. Hence,

$$V = V_e + V_g = \frac{1}{2}ks^2 - W\left(s + \frac{l}{2}\cos\theta\right)$$

Since $l = s + l\cos\theta$ or $s = l(1 - \cos\theta)$, then

$$V = \frac{1}{2}kl^2(1 - \cos\theta)^2 - \frac{Wl}{2}(2 - \cos\theta)$$

Equilibrium Position. The first derivative of V gives

$$\frac{dV}{d\theta} = kl^2(1 - \cos\theta)\sin\theta - \frac{Wl}{2}\sin\theta = 0$$

or

$$l\left[kl(1 - \cos\theta) - \frac{W}{2}\right]\sin\theta = 0$$

This equation is satisfied provided

$$\sin\theta = 0 \qquad \theta = 0° \qquad \textit{Ans.}$$

$$\theta = \cos^{-1}\left(1 - \frac{W}{2kl}\right) = \cos^{-1}\left[1 - \frac{10(9.81)}{2(200)(0.6)}\right] = 53.8° \qquad \textit{Ans.}$$

Stability. Determining the second derivative of V gives

$$\frac{d^2V}{d\theta^2} = kl^2(1 - \cos\theta)\cos\theta + kl^2\sin\theta\sin\theta - \frac{Wl}{2}\cos\theta$$

$$= kl^2(\cos\theta - \cos 2\theta) - \frac{Wl}{2}\cos\theta$$

Substituting values for the constants, with $\theta = 0°$ and $\theta = 53.8°$, yields

$$\left.\frac{d^2V}{d\theta^2}\right|_{\theta=0°} = 200(0.6)^2(\cos 0° - \cos 0°) - \frac{10(9.81)(0.6)}{2}\cos 0°$$

$$= -29.4 < 0 \qquad \text{(unstable equilibrium at } \theta = 0°\text{)} \qquad \textit{Ans.}$$

$$\left.\frac{d^2V}{d\theta^2}\right|_{\theta=53.8°} = 200(0.6)^2(\cos 53.8° - \cos 107.6°) - \frac{10(9.81)(0.6)}{2}\cos 53.8°$$

$$= 46.9 > 0 \qquad \text{(stable equilibrium at } \theta = 53.8°\text{)} \qquad \textit{Ans.}$$

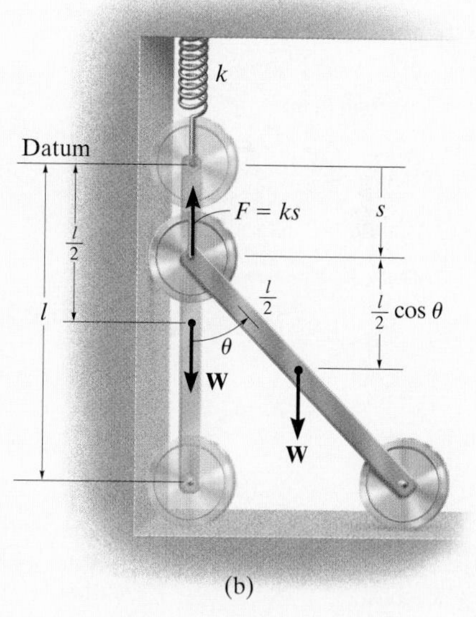

(b)

Fig. 11–16

EXAMPLE 11.6

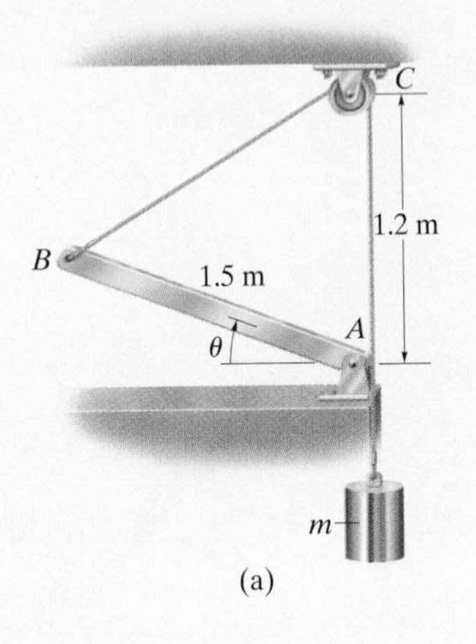

Determine the mass m of the block required for equilibrium of the uniform 10-kg rod shown in Fig. 11–17*a* when $\theta = 20°$. Investigate the stability at the equilibrium position.

SOLUTION

Potential Function. The datum is established through point A, Fig. 11–17*b*. When $\theta = 0°$, the block is assumed to be suspended $(y_W)_1$ below the datum. Hence, in the position θ,

$$V = V_e + V_g = 98.1\left(\frac{1.5 \sin \theta}{2}\right) - m(9.81)(\Delta y) \qquad (1)$$

The distance $\Delta y = (y_W)_2 - (y_W)_1$ may be related to the independent coordinate θ by measuring the difference in cord lengths $B'C$ and BC. Since

$$B'C = \sqrt{(1.5)^2 + (1.2)^2} = 1.92$$

$$BC = \sqrt{(1.5 \cos \theta)^2 + (1.2 - 1.5 \sin \theta)^2} = \sqrt{3.69 - 3.60 \sin \theta}$$

then

$$\Delta y = B'C - BC = 1.92 - \sqrt{3.69 - 3.60 \sin \theta}$$

Substituting the above result into Eq. 1 yields

$$V = 98.1\left(\frac{1.5 \sin \theta}{2}\right) - m(9.81)(1.92 - \sqrt{3.69 - 3.60 \sin \theta}) \quad (2)$$

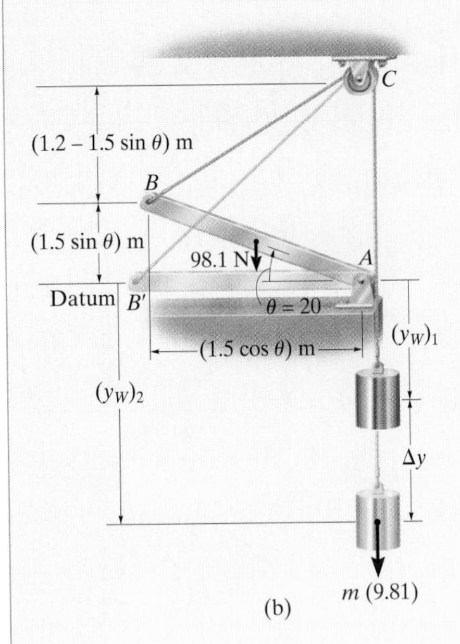

Fig. 11–17

Equilibrium Position.

$$\frac{dV}{d\theta} = 73.6 \cos \theta - \left[\frac{m(9.81)}{2}\right]\left(\frac{3.60 \cos \theta}{\sqrt{3.69 - 3.60 \sin \theta}}\right) = 0$$

$$\left.\frac{dV}{d\theta}\right|_{\theta=20°} = 69.14 - 10.58m = 0$$

$$m = \frac{69.14}{10.58} = 6.53 \text{ kg} \qquad \textit{Ans.}$$

Stability. Taking the second derivative of Eq. 2, we obtain

$$\frac{d^2V}{d\theta^2} = -73.6 \sin \theta - \left[\frac{m(9.81)}{2}\right]\left(\frac{-1}{2}\right)\frac{-(3.60 \cos \theta)^2}{(3.69 - 3.60 \sin \theta)^{3/2}}$$
$$- \left[\frac{m(9.81)}{2}\right]\left(\frac{-3.60 \sin \theta}{\sqrt{3.69 - 3.60 \sin \theta}}\right)$$

For the equilibrium position $\theta = 20°$, with $m = 6.53$ kg, then

$$\frac{d^2V}{d\theta^2} = -47.6 < 0 \qquad (\text{unstable equilibrium at } \theta = 20°) \qquad \textit{Ans.}$$

EXAMPLE 11.7

The homogeneous block having a mass m rests on the top surface of the cylinder, Fig. 11–18a. Show that this is a condition of unstable equilibrium if $h > 2R$.

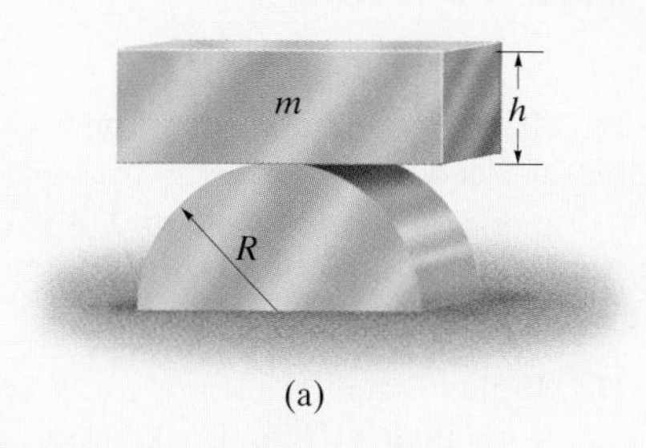

(a)

SOLUTION

Potential Function. The datum is established at the base of the cylinder, Fig. 11–18b. If the block is displaced by an amount θ from the equilibrium position, the potential function may be written in the form

$$V = V_e + V_g$$
$$= 0 + mgy$$

From Fig. 11–18b,

$$y = \left(R + \frac{h}{2}\right)\cos\theta + R\theta\sin\theta$$

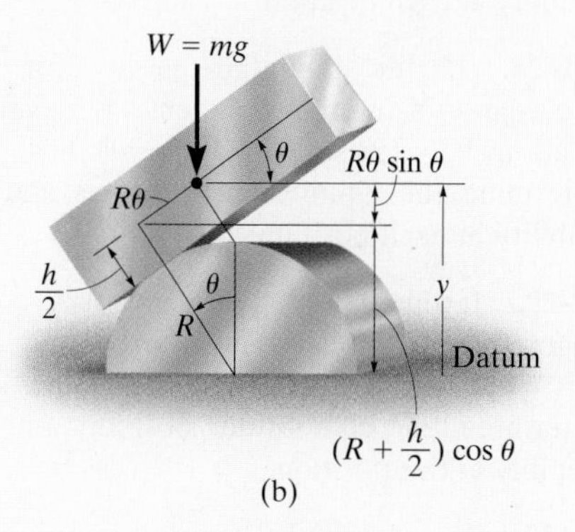

(b)

Fig. 11–18

Thus,

$$V = mg\left[\left(R + \frac{h}{2}\right)\cos\theta + R\theta\sin\theta\right]$$

Equilibrium Position.

$$\frac{dV}{d\theta} = mg\left[-\left(R + \frac{h}{2}\right)\sin\theta + R\sin\theta + R\theta\cos\theta\right] = 0$$
$$= mg\left(-\frac{h}{2}\sin\theta + R\theta\cos\theta\right) = 0$$

Obviously, $\theta = 0^\circ$ is the equilibrium position that satisfies this equation.

Stability. Taking the second derivative of V yields

$$\frac{d^2V}{d\theta^2} = mg\left(-\frac{h}{2}\cos\theta + R\cos\theta - R\theta\sin\theta\right)$$

At $\theta = 0^\circ$,

$$\left.\frac{d^2V}{d\theta^2}\right|_{\theta=0^\circ} = -mg\left(\frac{h}{2} - R\right)$$

Since all the constants are positive, the block is in unstable equilibrium if $h > 2R$, because then $d^2V/d\theta^2 < 0$.

PROBLEMS

11–26. If the potential energy for a conservative two-degree-of-freedom system is expressed by the relation $V = (3y^2 + 2x^2)$ J, where y and x are given in meters, determine the equilibrium positions and investigate the stability at each position.

11–27. If the potential energy for a conservative one-degree-of-freedom system is expressed by the relation $V = (4x^3 - x^2 - 3x + 10)$ ft · lb, where x is given in feet, determine the equilibrium positions and investigate the stability at each position.

***11–28.** If the potential energy for a conservative one-degree-of-freedom system is expressed by the relation $V = (24 \sin \theta + 10 \cos 2\theta)$ ft · lb, $0° \leq \theta \leq 180°$, determine the equilibrium positions and investigate the stability at each position.

11–29. If the potential energy for a conservative two-degree-of-freedom system is expressed by the relation $V = (6y^2 + 2x^2)$ J, where x and y are given in meters, determine the equilibrium position and investigate the stability at this position.

11–30. The spring of the scale has an unstretched length of a. Determine the angle θ for equilibrium when a weight W is supported on the platform. Neglect the weight of the members. What value W would be required to keep the scale in neutral equilibrium when $\theta = 0°$?

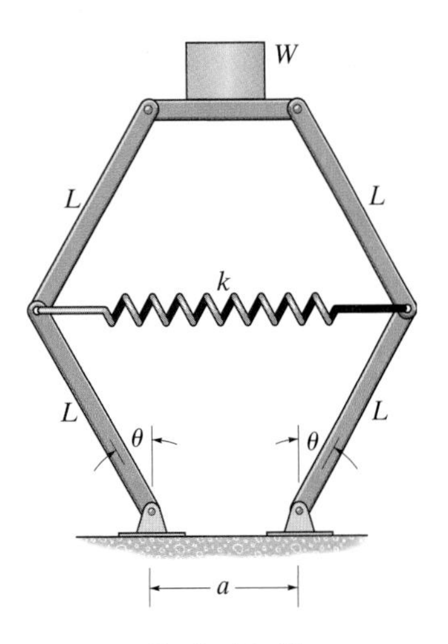

Prob. 11–30

11–31. The two bars each have a weight of 8 lb. Determine the required stiffness k of the spring so that the two bars are in equilibrium when $\theta = 30°$. The spring has an unstretched length of 1 ft.

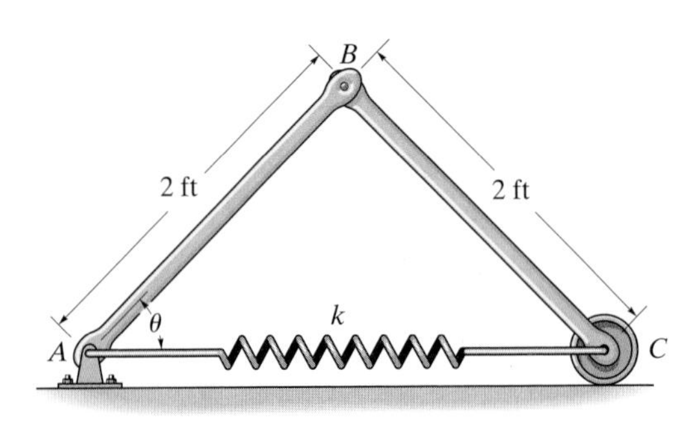

Prob. 11–31

***11–32.** Each of the two springs has an unstretched length of 500 mm. Determine the mass m of the cylinder when it is held in the equilibrium position shown, i.e., $y = 1$ m.

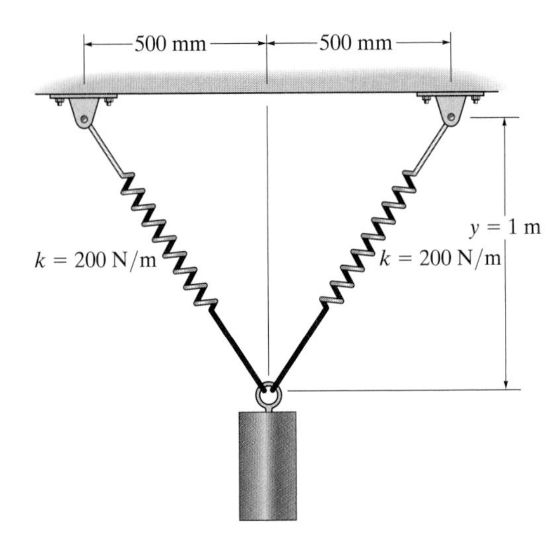

Prob. 11–32

■**11–33.** The uniform beam has a mass of 200 kg. If the contacting surfaces are smooth, determine the angle θ for equilibrium and investigate the stability of the beam when it is in this position. The spring has an unstretched length of 0.5 m.

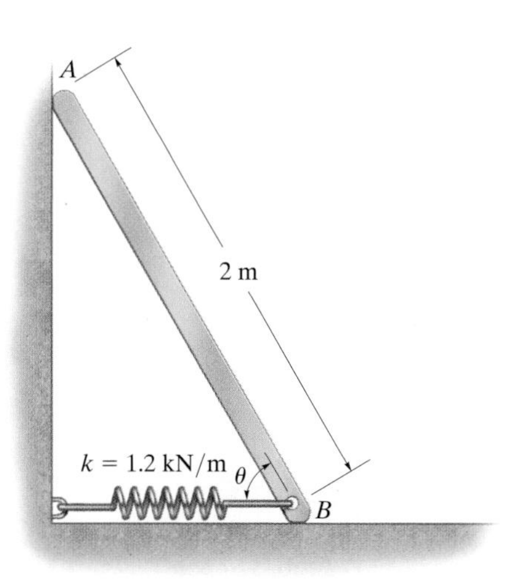

Prob. 11–33

11–34. The bar supports a weight of $W = 500$ lb at its end. If the springs are originally unstretched when the bar is vertical, determine the required stiffness $k_1 = k_2 = k$ of the springs so that the bar is in neutral equilibrium when it is vertical.

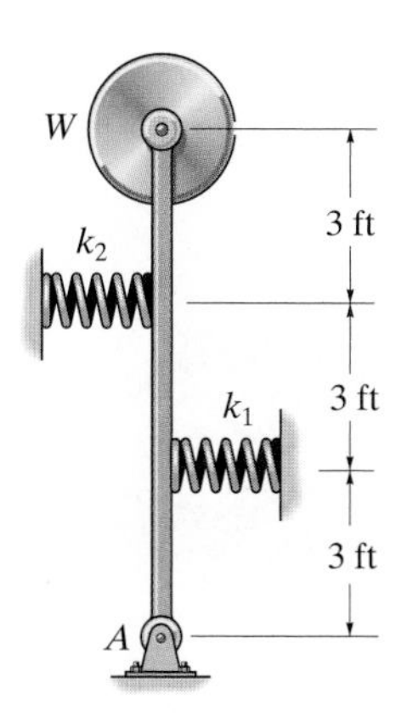

Prob. 11–34

11–35. The uniform rod AB has a mass of 80 kg. If spring DC is unstretched when $\theta = 90°$, determine the angle θ for equilibrium and investigate the stability at the equilibrium position. The spring always acts in the horizontal position due to the roller guide at D.

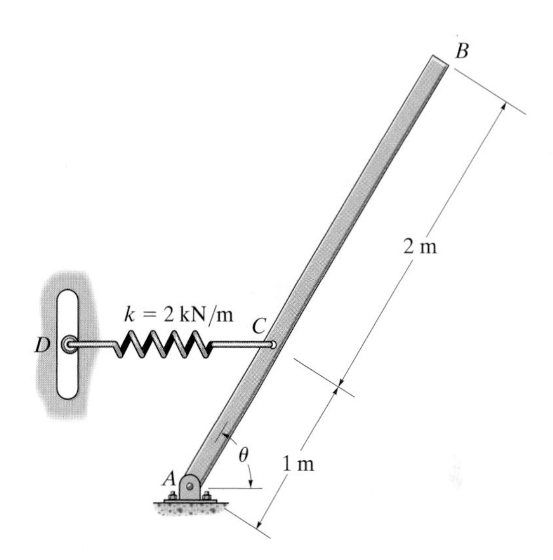

Prob. 11–35

***11–36.** Determine the angle θ for equilibrium and investigate the stability at this position. The bars each have a mass of 3 kg and the suspended block D has a mass of 7 kg. Cord DC has a total length of 1 m.

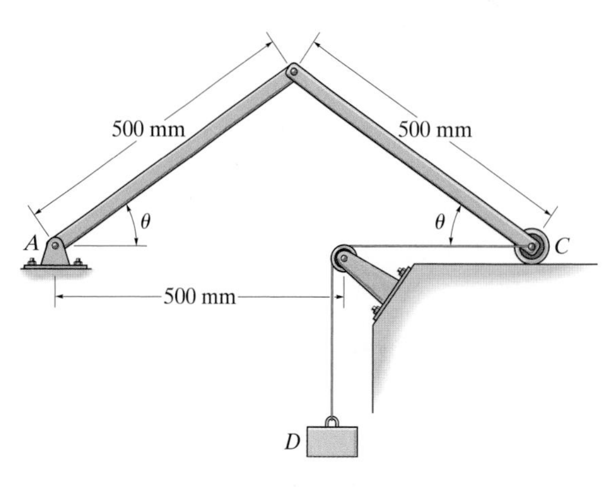

Prob. 11–36

11–37. The bar supports a weight of 500 lb at its end. If the springs are originally unstretched when the bar is vertical, and $k_1 = 300\ \text{lb/ft}$, $k_2 = 500\ \text{lb/ft}$, investigate the stability of the bar when it is in the vertical position.

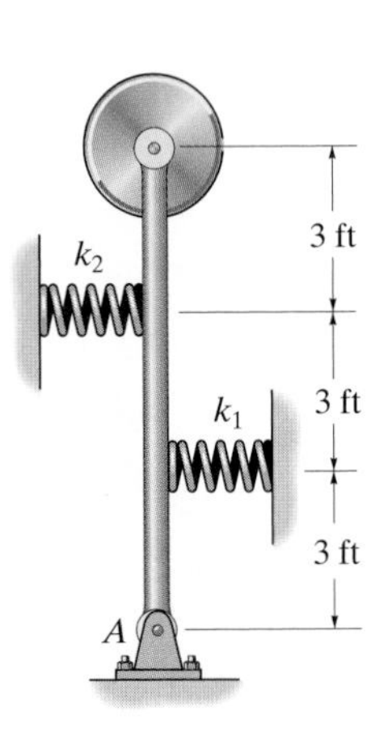

Prob. 11–37

11–38. If each of the three links of the mechanism has a weight W, determine the angle θ for equilibrium. The spring, which always remains vertical, is unstretched when $\theta = 0°$.

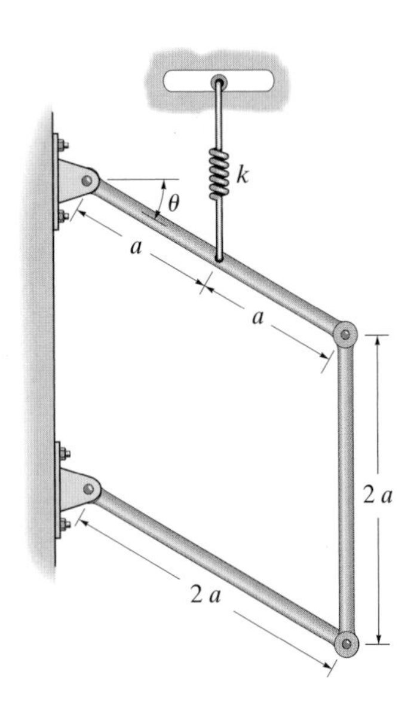

Prob. 11–38

11–39. The small postal scale consists of a counterweight W_1, connected to the members having negligible weight. Determine the weight W_2 that is on the pan in terms of the angles θ and ϕ and the dimensions shown. All members are pin connected.

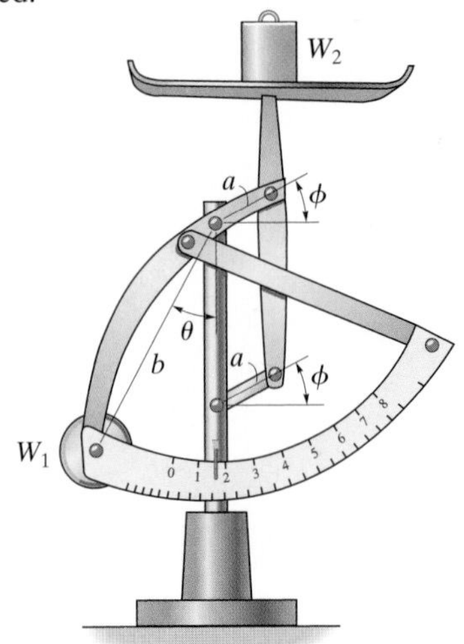

Prob. 11–39

***11–40.** The uniform right circular cone having a mass m is suspended from the cord as shown. Determine the angle θ at which it hangs from the wall for equilibrium. Is the cone in stable equilibrium?

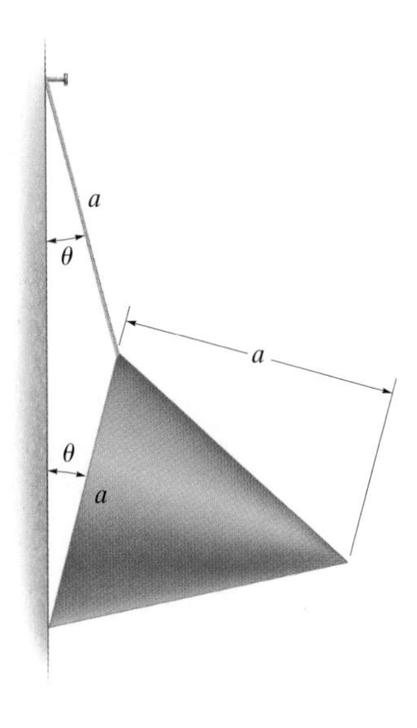

Prob. 11–40

11–41. The homogeneous cylinder has a conical cavity cut into its base as shown. Determine the depth d of the cavity so that the cylinder balances on the pivot and remains in neutral equilibrium.

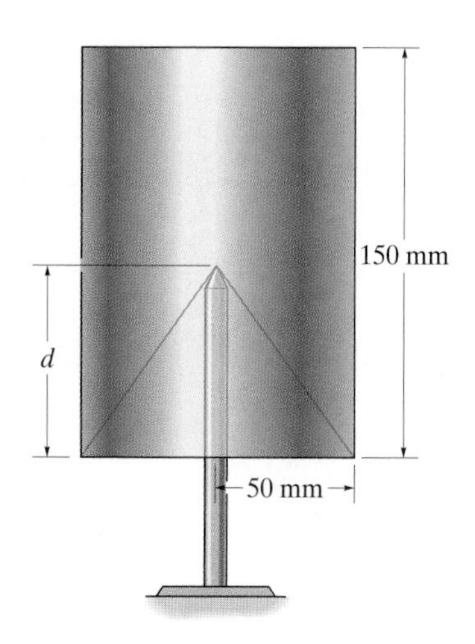

Prob. 11–41

11–42. The conical manhole cap is made of concrete and has the dimensions shown. Determine the critical location $h = h_{cr}$ of the pick-up connectors at A and B so that when hoisted with constant velocity the cap is in neutral equilibrium. Explain what would happen if the connectors were placed at a point $h > h_{cr}$.

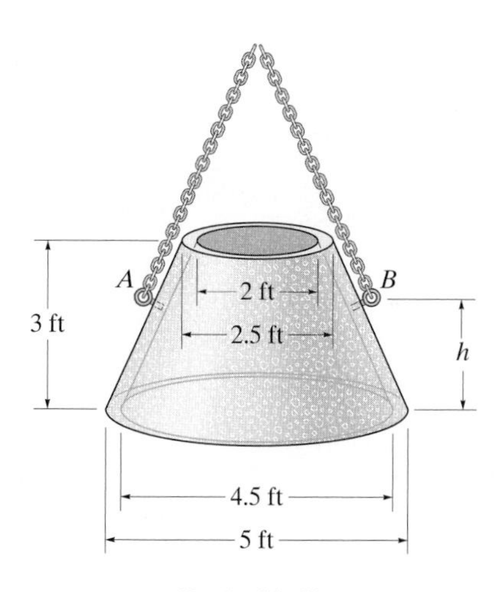

Prob. 11–42

11–43. Each bar has a mass per length of m_0. Determine the angles θ and ϕ at which they are suspended in equilibrium. The contact at A is smooth, and both are pin connected at B.

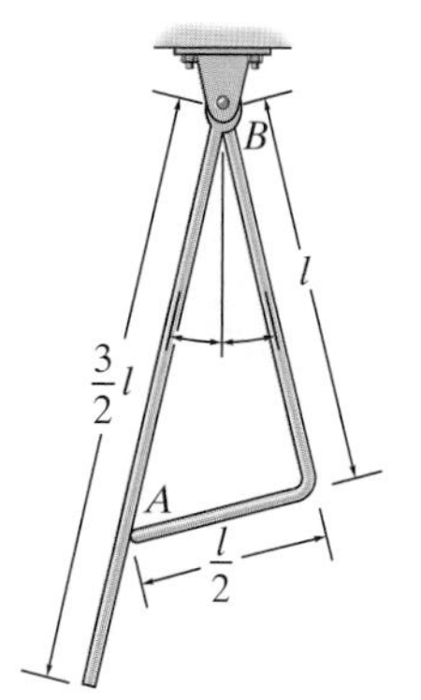

Prob. 11–43

***11–44.** The triangular block of weight W rests on the smooth corners which are a distance a apart. If the block has three equal sides of length d, determine the angle θ for equilibrium.

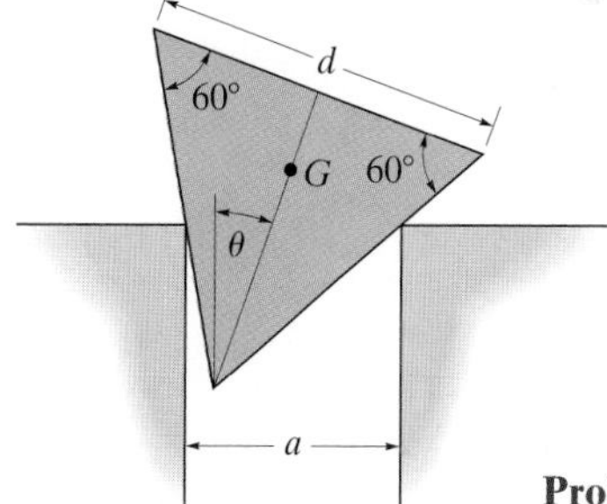

Prob. 11–44

11–45. A homogeneous cone rests on top of the cylindrical surface. Derive a relationship between the radius r of the cylinder and the height h of the cone for neutral equilibrium. *Hint:* Establish the potential function for a *small* angle θ of tilt of the cone, i.e., approximate $\sin\theta \approx \theta$ and $\cos\theta \approx 1 - \theta^2/2$.

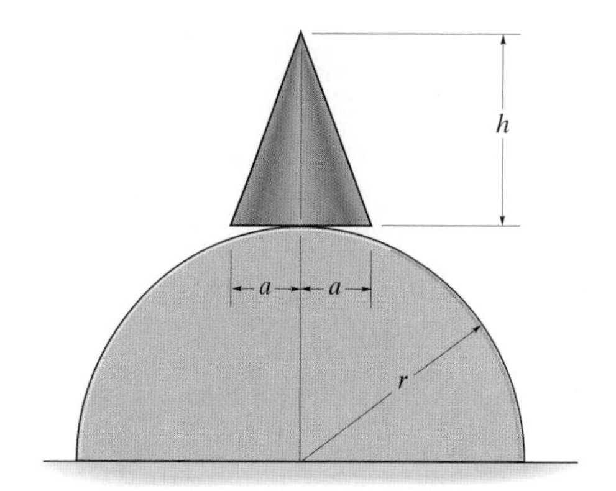

Prob. 11–45

11–46. The door has a uniform weight of 50 lb. It is hinged at A and is held open by the 30-lb weight and the pulley. Determine the angle θ for equilibrium.

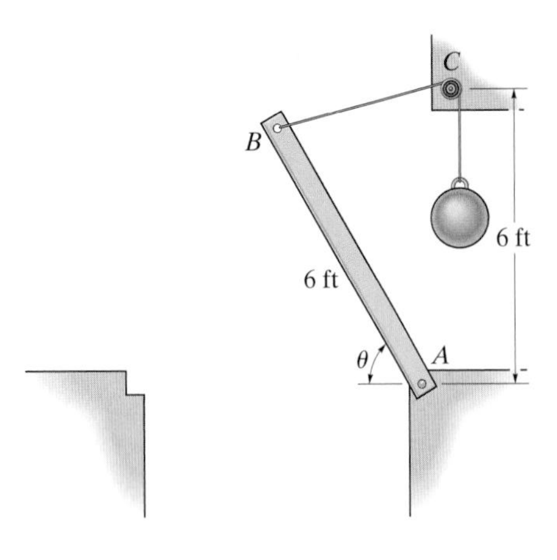

Prob. 11–46

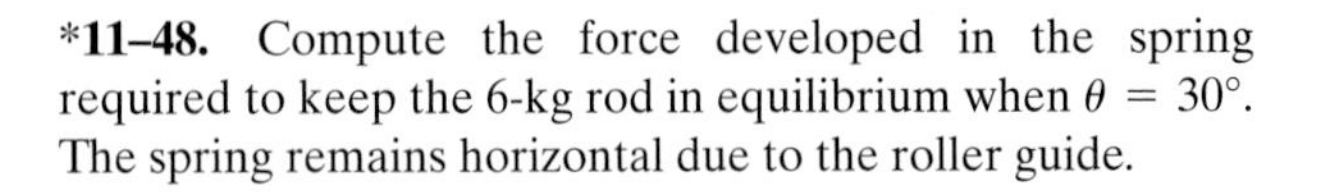

***11–48.** Compute the force developed in the spring required to keep the 6-kg rod in equilibrium when $\theta = 30°$. The spring remains horizontal due to the roller guide.

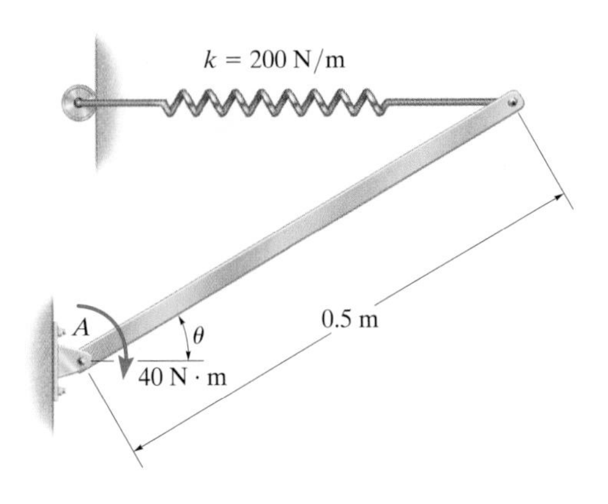

Prob. 11–48

11–47. The 60-lb hemisphere supports a cylinder having a specific weight of $\gamma = 311\ \text{lb/ft}^3$. If the radii of the cylinder and hemisphere are both 5 in., determine the height h of the cylinder which will produce neutral equilibrium in the position shown.

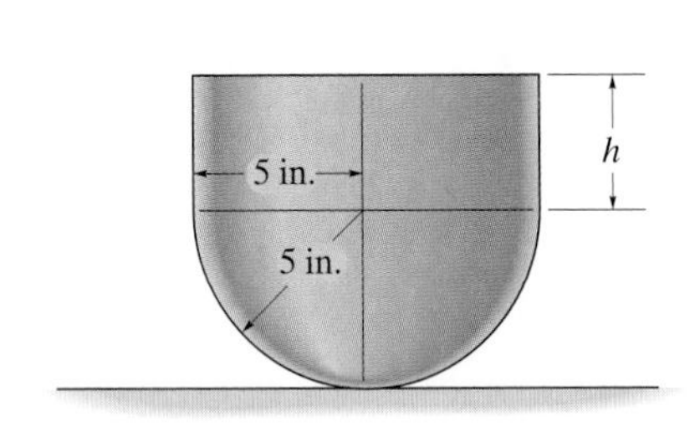

Prob. 11–47

11–49. Determine the force **P** acting on the cord which is required to maintain equilibrium of the horizontal 20-kg bar CB. *Hint:* First show that the coordinates s_A and s_B are related to the *constant* vertical length l of the cord by the equation $5s_B - s_A = l$.

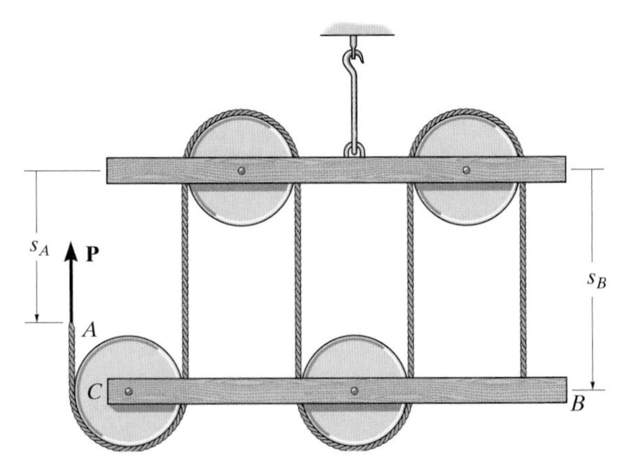

Prob. 11–49

Chapter Review

Principle of Virtual Work
The forces on a body will do *virtual work* when the body undergoes an *imaginary* differential displacement or rotation.

$\delta y, \delta y'$ – virtual displacement

$\delta \theta$ – virtual rotation

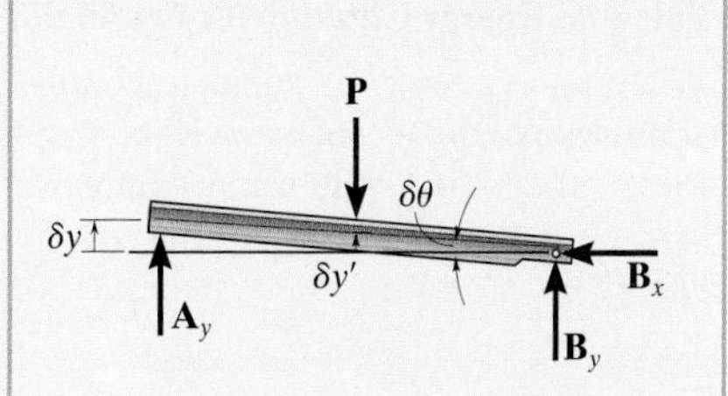

For equilibrium, the sum of the virtual work done by all the forces acting on the body must be equal to zero for any virtual displacement. This is referred to as the *principle of virtual work*, and it is useful for finding the equilibrium configuration for a mechanism or a reactive force acting on a series of connected members.

$$\delta U = 0$$

If this system has one degree of freedom, then its position can be specified by one independent coordinate $q = x$ or $q = \theta$.

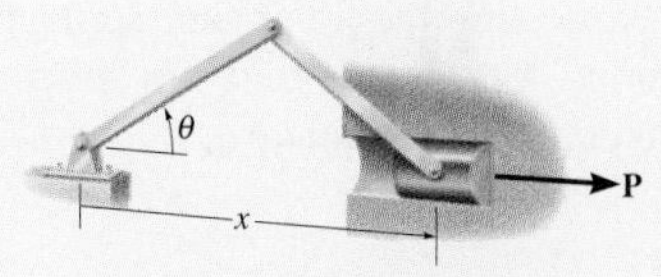

One-degree-of-freedom system

To apply the principle of virtual work, it is first necessary to use *position coordinates* to locate all the forces and moments on the mechanism that will do work when the mechanism undergoes a virtual movement δq.

$$q = \theta$$

θ, x – position coordinates

The coordinates are related to the independent coordinate q and then these expressions are differentiated in order to relate the *virtual* coordinate displacements to the virtual displacement δq.

$$b = \sqrt{x^2 + a^2 - 2ax\cos\theta}$$

Finally, the equation of virtual work is written for the mechanism in terms of the common displacement δq, and then it is set equal to zero. By factoring δq out of the equation, it is then possible to determine either the unknown force or couple moment, or the equilibrium position q.
(*See pages 573–579*)

Potential-Energy Criterion for Equilibrium When a system is subjected only to conservative forces, such as weight or spring forces, then the equilibrium configuration can be determined using the *potential-energy function V* for the system.	$q = \theta$
The potential-energy function is established by expressing the weight and spring potential energy for the system in terms of the independent coordinate q.	
Once the potential-energy function is formulated, its first derivative is set equal to zero. The solution yields the equilibrium position q_{eq} for the system.	$\frac{dV}{dq} = 0$
The stability of the system can be investigated by taking the second derivative of V. *(See pages 591–596.)*	$\frac{dV}{dq} = 0, \quad \frac{d^2V}{dq^2} > 0$ stable equilibrium $\frac{dV}{dq} = 0, \quad \frac{d^2V}{dq^2} < 0$ unstable equilibrium $\frac{dV}{dq} = \frac{d^2V}{dq^2} = \frac{d^3V}{dq^3} = \cdots = 0$ neutral equilibrium

REVIEW PROBLEMS

11–50. The uniform bar AB weighs 10 lb. If the attached spring is unstretched when $\theta = 90°$, use the method of virtual work and determine the angle θ for equilibrium. Note that the spring always remains in the vertical position due to the roller guide.

11–51. Solve Prob. 11–50 using the principle of potential energy. Investigate the stability of the bar when it is in the equilibrium position.

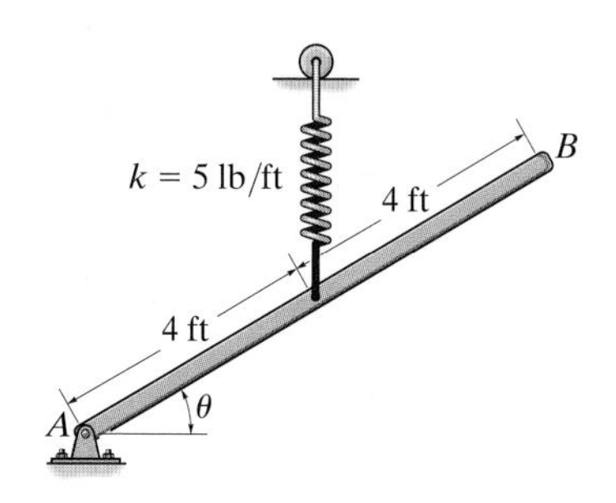

Probs. 11–50/51

11–53 The truck has a mass of 20 Mg and a mass center at G. Determine the steepest grade θ along which it can park without overturning and investigate the stability in this position.

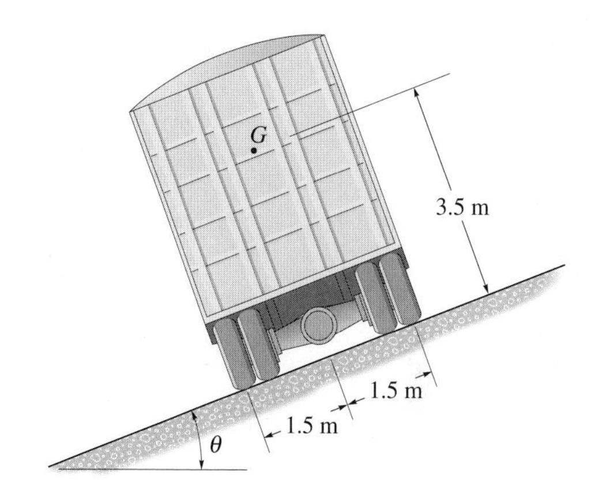

Prob. 11–53

***11–52.** The punch press consists of the ram R, connecting rod AB, and a flywheel. If a torque of $M = 50\ \text{N}\cdot\text{m}$ is applied to the flywheel, determine the force **F** applied at the ram to hold the rod in the position $\theta = 60°$.

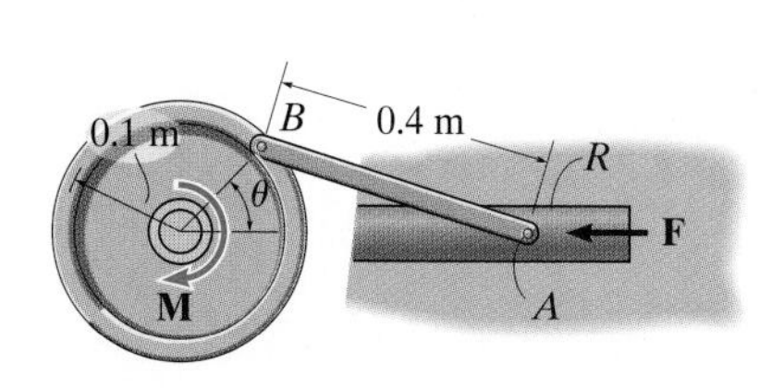

Prob. 11–52

11–54. Use the method of virtual work to determine the tensions in cable AC. The lamp weighs 10 lb.

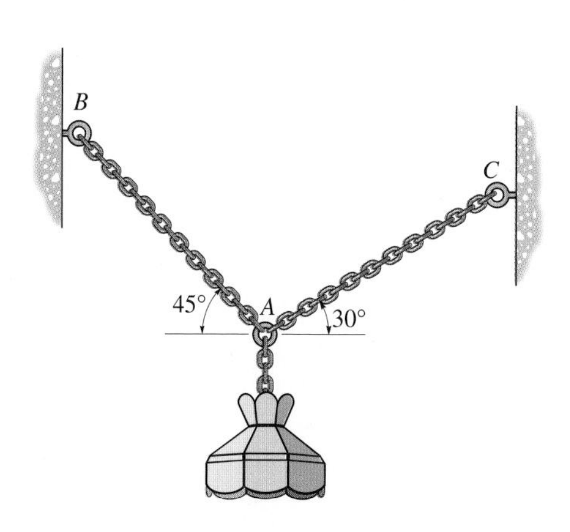

Prob. 11–54

APPENDIX

Mathematical Review and Expressions

Geometry and Trigonometry Review

The angles θ in Fig. A–1 are equal between the transverse and two parallel lines.

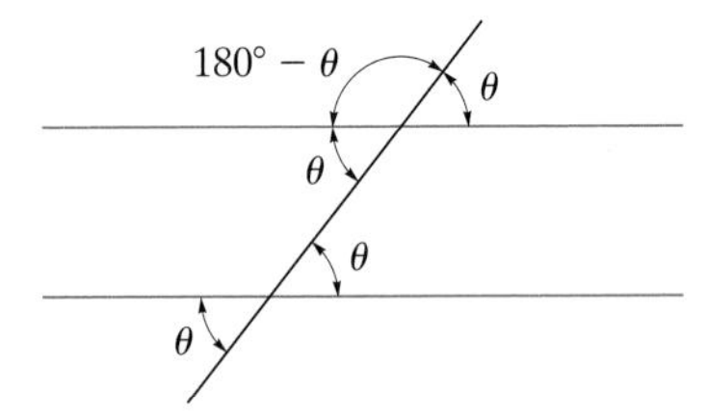

Fig. A–1

For a line and its normal, the angles θ in Fig. A–2 are equal.

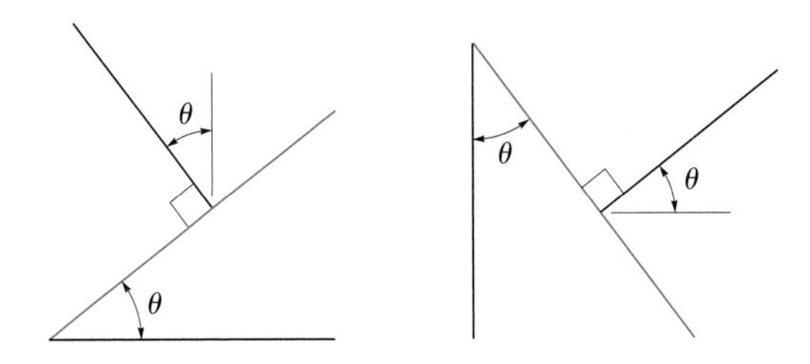

Fig. A–2

For the circle in Fig. A–3 $s = \theta r$, so that when $\theta = 360° = 2\pi$ rad then the circumference is $s = 2\pi r$. Also, since $180° = \pi$ rad, then $\theta\,(\text{rad}) = (\pi/180°)\theta°$. The area of the circle is $A = \pi r^2$.

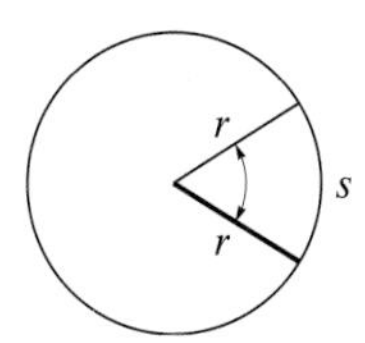

Fig. A–3

The sides of a similar triangle can be obtained by proportion as in Fig. A–4, where $\dfrac{a}{A} = \dfrac{b}{B} = \dfrac{c}{C}$.

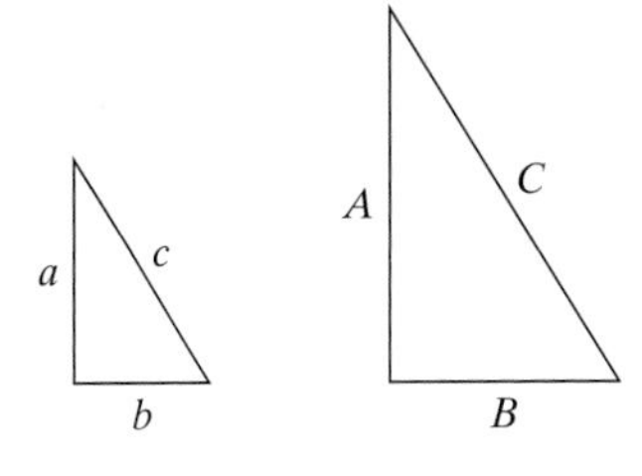

Fig. A–4

For the right triangle in Fig. A-5, the Pythagorean theorem is

$$h = \sqrt{(o)^2 + (a)^2}$$

The trigonometric functions are

$$\sin\theta = \frac{o}{h}$$

$$\cos\theta = \frac{a}{h}$$

$$\tan\theta = \frac{o}{a}$$

This is easily remembered as "soh, cah, toa", i.e., the sine is the opposite over the hypotenuse, etc. The other trigonometric functions follow from this.

$$\csc\theta = \frac{1}{\sin\theta} = \frac{h}{o}$$

$$\sec\theta = \frac{1}{\cos\theta} = \frac{h}{a}$$

$$\cot\theta = \frac{1}{\tan\theta} = \frac{a}{o}$$

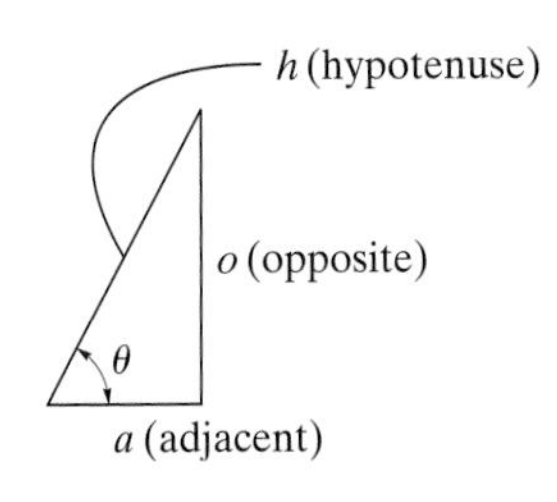

Fig. A–5

Trigonometric Identities

$$\sin^2\theta + \cos^2\theta = 1$$

$$\sin(\theta \pm \phi) = \sin\theta\cos\phi \pm \cos\theta\sin\phi$$

$$\sin 2\theta = 2\sin\theta\cos\theta$$

$$\cos(\theta \pm \phi) = \cos\theta\cos\phi \mp \sin\theta\sin\phi$$

$$\cos 2\theta = \cos^2\theta - \sin^2\theta$$

$$\cos\theta = \pm\sqrt{\frac{1+\cos 2\theta}{2}},\ \sin\theta = \pm\sqrt{\frac{1-\cos 2\theta}{2}}$$

$$\tan\theta = \frac{\sin\theta}{\cos\theta}$$

$$1+\tan^2\theta = \sec^2\theta \qquad 1+\cot^2\theta = \csc^2\theta$$

Quadratic Formula

If $ax^2 + bx + c = 0$, then $x = \dfrac{-b \pm \sqrt{b^2 - 4ac}}{2a}$

Hyperbolic Functions

$$\sinh x = \frac{e^x - e^{-x}}{2},$$

$$\cosh x = \frac{e^x + e^{-x}}{2},$$

$$\tanh x = \frac{\sinh x}{\cosh x}$$

Power-Series Expansions

$$\sin x = x - \frac{x^3}{3!} + \cdots,\ \cos x = 1 - \frac{x^2}{2!} + \cdots$$

$$\sinh x = x + \frac{x^3}{3!} + \cdots,\ \cosh x = 1 + \frac{x^2}{2!} + \cdots$$

Derivatives

$$\frac{d}{dx}(u^n) = nu^{n-1}\frac{du}{dx} \qquad \frac{d}{dx}(\sin u) = \cos u\frac{du}{dx}$$

$$\frac{d}{dx}(uv) = u\frac{dv}{dx} + v\frac{du}{dx} \qquad \frac{d}{dx}(\cos u) = -\sin u\frac{du}{dx}$$

$$\frac{d}{dx}\left(\frac{u}{v}\right) = \frac{v\dfrac{du}{dx} - u\dfrac{dv}{dx}}{v^2} \qquad \frac{d}{dx}(\tan u) = \sec^2 u\frac{du}{dx}$$

$$\frac{d}{dx}(\cot u) = -\csc^2 u\frac{du}{dx} \qquad \frac{d}{dx}(\sinh u) = \cosh u\frac{du}{dx}$$

$$\frac{d}{dx}(\sec u) = \tan u\sec u\frac{du}{dx} \qquad \frac{d}{dx}(\cosh u) = \sinh u\frac{du}{dx}$$

$$\frac{d}{dx}(\csc u) = -\csc u\cot u\frac{du}{dx}$$

Integrals

$$\int x^n\,dx = \frac{x^{n+1}}{n+1} + C,\ n \neq -1$$

$$\int \frac{dx}{a+bx} = \frac{1}{b}\ln(a+bx) + C$$

$$\int \frac{dx}{a+bx^2} = \frac{1}{2\sqrt{-ba}}\ln\left[\frac{a+x\sqrt{-ab}}{a-x\sqrt{-ab}}\right] + C,\quad ab < 0$$

$$\int \frac{x\,dx}{a+bx^2} = \frac{1}{2b}\ln(bx^2+a) + C$$

$$\int \frac{x^2\,dx}{a+bx^2} = \frac{x}{b} - \frac{a}{b\sqrt{ab}}\tan^{-1}\frac{x\sqrt{ab}}{a} + C,\ ab > 0$$

$$\int \sqrt{a+bx}\,dx = \frac{2}{3b}\sqrt{(a+bx)^3} + C$$

$$\int x\sqrt{a+bx}\,dx = \frac{-2(2a-3bx)\sqrt{(a+bx)^3}}{15b^2} + C$$

$$\int x^2\sqrt{a+bx}\,dx = \frac{2(8a^2-12abx+15b^2x^2)\sqrt{(a+bx)^3}}{105b^3} + C$$

$$\int \sqrt{a^2-x^2}\,dx = \frac{1}{2}\left[x\sqrt{a^2-x^2} + a^2\sin^{-1}\frac{x}{a}\right] + C,\quad a > 0$$

$$\int x\sqrt{a^2-x^2}\,dx = -\frac{1}{3}\sqrt{(a^2-x^2)^3} + C$$

$$\int x^2\sqrt{a^2-x^2}\,dx = -\frac{x}{4}\sqrt{(a^2-x^2)^3} + \frac{a^2}{8}\left(x\sqrt{a^2-x^2} + a^2\sin^{-1}\frac{x}{a}\right) + C,\ a > 0$$

$$\int \sqrt{x^2 \pm a^2}\,dx = \frac{1}{2}\left[x\sqrt{x^2\pm a^2} \pm a^2\ln\left(x+\sqrt{x^2\pm a^2}\right)\right] + C$$

$$\int x\sqrt{x^2\pm a^2}\,dx = \frac{1}{3}\sqrt{(x^2\pm a^2)^3} + C$$

$$\int x^2\sqrt{x^2\pm a^2}\,dx = \frac{x}{4}\sqrt{(x^2\pm a^2)^3} \mp \frac{a^2}{8}x\sqrt{x^2\pm a^2} - \frac{a^4}{8}\ln\left(x+\sqrt{x^2\pm a^2}\right) + C$$

$$\int \frac{dx}{\sqrt{a+bx}} = \frac{2\sqrt{a+bx}}{b} + C$$

$$\int \frac{x\,dx}{\sqrt{x^2\pm a^2}} = \sqrt{x^2\pm a^2} + C$$

$$\int \frac{dx}{\sqrt{a+bx+cx^2}} = \frac{1}{\sqrt{c}}\ln\left[\sqrt{a+bx+cx^2} + x\sqrt{c} + \frac{b}{2\sqrt{c}}\right] + C,\ c > 0$$

$$= \frac{1}{\sqrt{-c}}\sin^{-1}\left(\frac{-2cx-b}{\sqrt{b^2-4ac}}\right) + C,\ c < 0$$

$$\int \sin x\,dx = -\cos x + C$$

$$\int \cos x\,dx = \sin x + C$$

$$\int x\cos(ax)\,dx = \frac{1}{a^2}\cos(ax) + \frac{x}{a}\sin(ax) + C$$

$$\int x^2\cos(ax)\,dx = \frac{2x}{a^2}\cos(ax) + \frac{a^2x^2-2}{a^3}\sin(ax) + C$$

$$\int e^{ax}\,dx = \frac{1}{a}e^{ax} + C$$

$$\int xe^{ax}\,dx = \frac{e^{ax}}{a^2}(ax-1) + C$$

$$\int \sinh x\,dx = \cosh x + C$$

$$\int \cosh x\,dx = \sinh x + C$$

APPENDIX

B Numerical and Computer Analysis

Occasionally the application of the laws of mechanics will lead to a system of equations for which a closed-form solution is difficult or impossible to obtain. When confronted with this situation, engineers will often use a numerical method which in most cases can be programmed on a microcomputer or programmable pocket calculator. Here we will briefly present a computer program for solving a set of linear algebraic equations and three numerical methods which can be used to solve an algebraic or transcendental equation, evaluate a definite integral, and solve an ordinary differential equation. Application of each method will be explained by example, and an associated computer program written in Microsoft BASIC, which is designed to run on most personal computers, is provided.* A text on numerical analysis should be consulted for further discussion regarding a check of the accuracy of each method and the inherent errors that can develop from the methods.

B.1 Linear Algebraic Equations

Application of the equations of static equilibrium or the equations of motion sometimes requires solving a set of linear algebraic equations. The computer program listed in Fig. B–1 can be used for this purpose. It is based on the method of a Gaussian elimination and can solve at most 10 equations with 10 unknowns. To do so, the equations should first be written in the following general format:

$$
\begin{aligned}
&A_{11}x_1 + A_{12}x_2 + \cdots + A_{1n}x_n = B_1 \\
&A_{21}x_1 + A_{22}x_2 + \cdots + A_{2n}x_n = B_2 \\
&\vdots \\
&A_{n1}x_1 + A_{n2}x_2 + \cdots + A_{nn}x_n = B_n
\end{aligned}
$$

*Similar types of programs can be written or purchased for programmable pocket calculators.

The A and B coefficients are called for when running the program. The output presents the unknowns $x_1, \ldots, x_n$.

```
1 PRINT"Linear system of equations":PRINT
2 DIM A(10,11)
3 INPUT"Input number of equations : ",N
4 PRINT
5 PRINT"A  coefficients"
6 FOR I = 1 TO N
7 FOR J = 1 TO N
8 PRINT "A(";I;",";J;
9 INPUT")=",A(I,J)
10 NEXT J
11 NEXT I
12 PRINT
13 PRINT"B  coefficients"
14 FOR I = 1 TO N
15 PRINT "B(";I;
16 INPUT")=",A(I,N+1)
17 NEXT I
18 GOSUB 25
19 PRINT
20 PRINT"Unknowns"
21 FOR I = 1 TO N
22 PRINT "X(";I;")=";A(I,N+1)
23 NEXT I
24 END
25 REM Subroutine Guassian
26 FOR M=1 TO N
27 NP=M
28 BG=ABS(A(M,M))
29 FOR I = M TO N
30 IF ABS(A(I,M))<=BG THEN 33
31 BG=ABS(A(I,M))
32 NP=I
33 NEXT I
34 IF NP=M THEN 40
35 FOR I = M TO N+1
36 TE=A(M,I)
37 A(M,I)=A(NP,I)
38 A(NP,I)=TE
39 NEXT I
40 FOR I = M+1 TO N
41 FC=A(I,M)/A(M,M)
42 FOR J = M+1 TO N+1
43 A(I,J)=A(I,J)-FC*A(M,J)
44 NEXT J
45 NEXT I
46 NEXT M
47 A(N,N+1)=A(N,N+1)/A(N,N)
48 FOR I = N-1 TO 1 STEP -1
49 SM=0
50 FOR J=I+1 TO N
51 SM=SM+A(I,J)*A(J,N+1)
52 NEXT J
53 A(I,N+1)=(A(I,N+1)-SM)/A(I,I)
54 NEXT I
55 RETURN
```

Fig. B–1

EXAMPLE B.1

Solve the two equations

$$3x_1 + x_2 = 4$$
$$2x_1 - x_2 = 10$$

SOLUTION

When the program begins to run, it first calls for the number of equations (2); then the A coefficients in the sequence $A_{11} = 3$, $A_{12} = 1$, $A_{21} = 2$, $A_{22} = -1$; and finally the B coefficients $B_1 = 4$, $B_2 = 10$. The output appears as

Unknowns

$X(1) = 2.8$ *Ans.*

$X(2) = -4.4$ *Ans.*

B.2 Simpson's Rule

Simpson's rule is a numerical method that can be used to determine the area under a curve given as a graph or as an explicit function $y = f(x)$. Likewise, it can be used to compute the value of a definite integral which involves the function $y = f(x)$. To do so, the area must be subdivided into an *even number* of strips or intervals having a width h. The curve between three consecutive ordinates is approximated by a parabola, and the entire area or definite integral is then determined from the formula

$$\int_{x_0}^{x_n} f(x)\,dx \simeq \frac{h}{3}[y_0 + 4(y_1 + y_3 + \cdots + y_{n-1}) + 2(y_2 + y_4 + \cdots + y_{n-2}) + y_n] \quad \text{(B–1)}$$

The computer program for this equation is given in Fig. B–2. For its use, we must first specify the function (on line 6 of the program). The upper and lower limits of the integral and the number of intervals are called for when the program is executed. The value of the integral is then given as the output.

```
1 PRINT"Simpson's rule":PRINT
2 PRINT" To execute this program :":PRINT
3 PRINT"   1- Modify right-hand side of the equation given below,
4 PRINT"      then press RETURN key"
5 PRINT"   2- Type  RUN 6":PRINT:EDIT 6
6 DEF FNF(X)=LOG(X)
7 PRINT:INPUT" Enter Lower Limit = ",A
8 INPUT" Enter Upper Limit = ",B
9 INPUT" Enter Number (even) of Intervals = ",N%
10 H=(B-A)/N%:AR=FNF(A):X=A+H
11 FOR J%=2 TO N%
12 K=2*(2-J%+2*INT(J%/2))
13 AR=AR+K*FNF(X)
14 X=X+H:NEXT J%
15 AR=H*(AR+FNF(B))/3
16 PRINT" Integral = ",AR
17 END
```

Fig. B–2

EXAMPLE B.2

Evaluate the definite integral

$$\int_2^5 \ln x \, dx$$

SOLUTION

The interval $x_0 = 2$ to $x_6 = 5$ will be divided into six equal parts $(n = 6)$, each having a width $h = (5 - 2)/6 = 0.5$. We then compute $y = f(x) = \ln x$ at each point of subdivision.

n	x_n	y_n
0	2	0.693
1	2.5	0.916
2	3	1.099
3	3.5	1.253
4	4	1.386
5	4.5	1.504
6	5	1.609

Thus, Eq. B–1 becomes

$$\int_2^5 \ln x \, dx \simeq \frac{0.5}{3}[0.693 + 4(0.916 + 1.253 + 1.504)$$

$$+ 2(1.099 + 1.386) + 1.609]$$

$$\simeq 3.66 \qquad \textit{Ans.}$$

This answer is equivalent to the exact answer to three significant figures. Obviously, accuracy to a greater number of significant figures can be improved by selecting a smaller interval h (or larger n).

Using the computer program, we first specify the function $\ln x$, line 6 in Fig. B–2. During execution, the program input requires the upper and lower limits 2 and 5, and the number of intervals $n = 6$. The output appears as

Integral = 3.66082 *Ans.*

B.3 The Secant Method

The secant method is used to find the real roots of an algebraic or transcendental equation $f(x) = 0$. The method derives its name from the fact that the formula used is established from the slope of the secant line to the graph $y = f(x)$. This slope is $[f(x_n) - f(x_{n-1})]/(x_n - x_{n-1})$, and the secant formula is

$$x_{n+1} = x_n - f(x_n)\left[\frac{x_n - x_{n-1}}{f(x_n) - f(x_{n-1})}\right] \qquad \text{(B–2)}$$

For application it is necessary to provide two initial guesses, x_0 and x_1, and thereby evaluate x_2 from Eq. B–2 ($n = 1$). One then proceeds to reapply Eq. B–2 with x_1 and the calculated value of x_2 and obtain x_3 ($n = 2$), etc., until the value $x_{n+1} \simeq x_n$. One can see this will occur if x_n is approaching the root of the function $f(x) = 0$, since the correction term on the right of Eq. B–2 will tend toward zero. In particular, the larger the slope, the smaller the correction to x_n, and the faster the root will be found. On the other hand, if the slope is very small in the neighborhood of the root, the method leads to large corrections for x_n, and convergence to the root is slow and may even lead to a failure to find it. In such cases other numerical techniques must be used for solution.

A computer program based on Eq. B–2 is listed in Fig. B–3. We must first specify the function on line 7 of the program. When the program is executed, two initial guesses, x_0 and x_1, must be entered in order to approximate the solution. The output specifies the value of the root. If it cannot be determined, this is so stated.

```
1 PRINT"Secant method":PRINT
2 PRINT" To execute this program :":PRINT
3 PRINT"    1) Modify right hand side of the equation given below,"
4 PRINT"       then press RETURN key."
5 PRINT"    2) Type  RUN 7"
6 PRINT:EDIT 7
7 DEF FNF(X)=.5*SIN(X)-2*COS(X)+1.3
8 INPUT"Enter point #1 =",X
9 INPUT"Enter point #2 =",X1
10 IF X=X1 THEN 14
11 EP=.00001:TL=2E-20
12 FP=(FNF(X1)-FNF(X))/(X1-X)
13 IF ABS(FP)>TL THEN 15
14 PRINT"Root can not be found.":END
15 DX=FNF(X1)/FP
16 IF ABS(DX)>EP THEN 19
17 PRINT "Root = ";X1;"       Function evaluated at this root = ";FNF(X1)
18 END
19 X=X1:X1=X1-DX
20 GOTO 12
```

Fig. B–3

EXAMPLE B.3

Determine the root of the equation

$$f(x) = 0.5 \sin x - 2 \cos x + 1.30 = 0$$

SOLUTION

Guesses of the initial roots will be $x_0 = 45°$ and $x_1 = 30°$. Applying Eq. B–2,

$$x_2 = 30° - (-0.1821)\frac{(30° - 45°)}{(-0.1821 - 0.2393)} = 36.48°$$

Using this value in Eq. B–2, along with $x_1 = 30°$, we have

$$x_3 = 36.48° - (-0.0108)\frac{36.48° - 30°}{(-0.0108 + 0.1821)} = 36.89°$$

Repeating the process with this value and $x_2 = 36.48°$ yields

$$x_4 = 36.89° - (0.0005)\left[\frac{36.89° - 36.48°}{(0.0005 + 0.0108)}\right] = 36.87°$$

Thus $x = 36.9°$ is appropriate to three significant figures.

If the problem is solved using the computer program, first we specify the function, line 7 in Fig. B–3. During execution, the first and second guesses must be entered in radians. Choosing these to be 0.8 rad and 0.5 rad, the result appears as

Root = 0.6435022

Function evaluated at this root = 1.66893E − 06

This result converted from radians to degrees is therefore

$$x = 36.9° \qquad \textit{Ans.}$$

APPENDIX

C Review for the Fundamentals of Engineering Examination

The Fundamentals of Engineering (FE) exam is given semiannually by the National Council of Engineering Examiners (NCEE) and is one of the requirements for obtaining a Professional Engineering License. A portion of this exam contains problems in statics, and this appendix provides a review of the subject matter most often asked on this exam. Before solving any of the problems, you should review the sections indicated in each chapter in order to become familiar with the boldfaced definitions and the procedures used to solve the various types of problems. Also, review the example problems in these sections.

The following problems are arranged in the same sequence as the topics in each chapter. Besides helping as a preparation for the FE exam, these problems also provide additional examples for general practice of the subject matter. Solutions to *all the problems* are given at the back of this appendix.

Chapter 2—Review All Sections

C-1. Two forces act on the hook. Determine the magnitude of the resultant force.

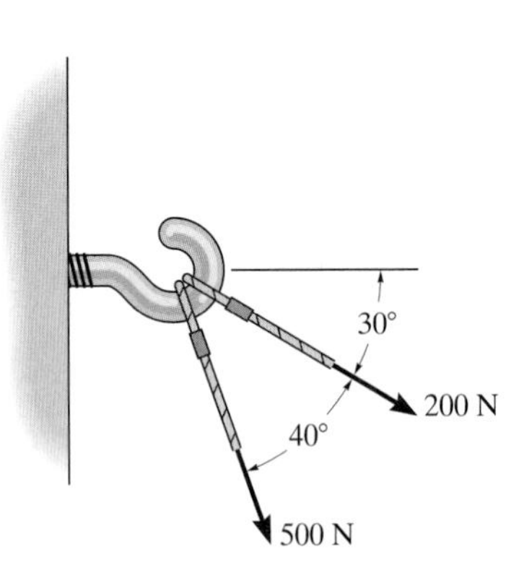

Prob. C–1

C-2. The force $F = 450$ lb acts on the frame. Resolve this force into components acting along members AB and AC, and determine the magnitude of each component.

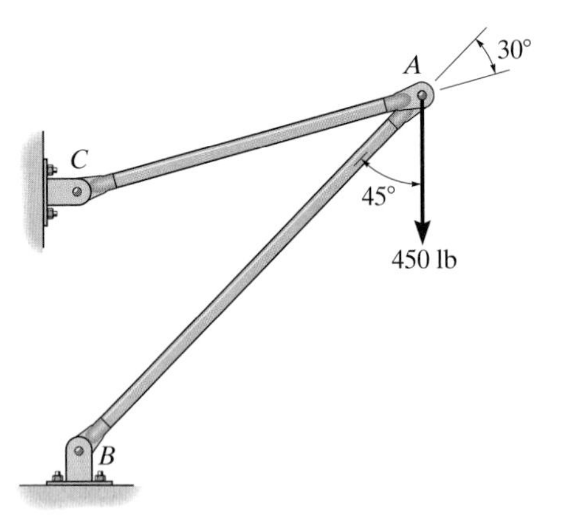

Prob. C–2

C-3. Determine the magnitude and direction of the resultant force.

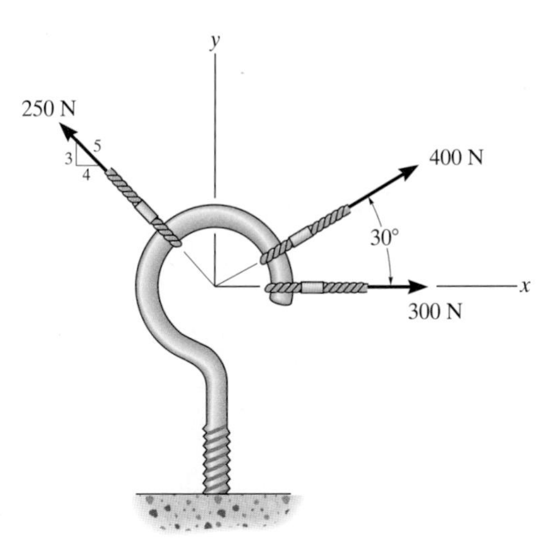

Prob. C–3

C-4. If $\mathbf{F} = \{30\mathbf{i} + 50\mathbf{j} - 45\mathbf{k}\}$ N, determine the magnitude and coordinate direction angles of the force.

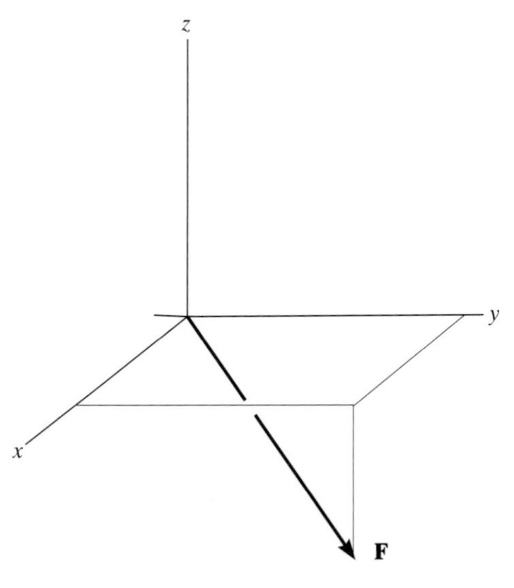

Prob. C–4

C-5. The force has a component of 20 N directed along the $-y$ axis as shown. Represent the force $\mathbf{F}$ as a Cartesian vector.

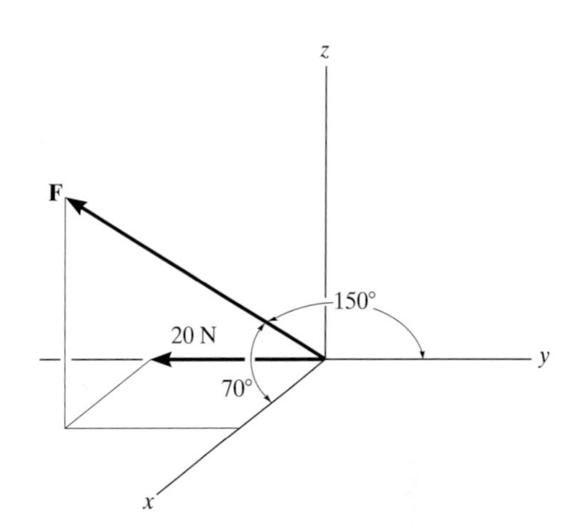

Prob. C–5

C-6. The force acts on the beam as shown. Determine its coordinate direction angles.

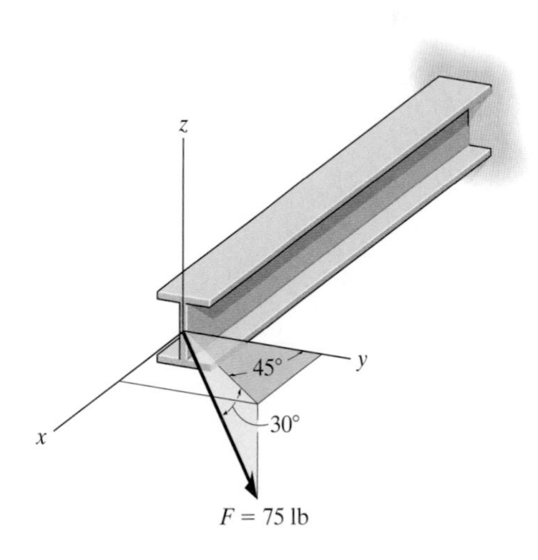

Prob. C–6

C-7. The cables supporting the antenna are subjected to the forces shown. Represent each force as a Cartesian vector.

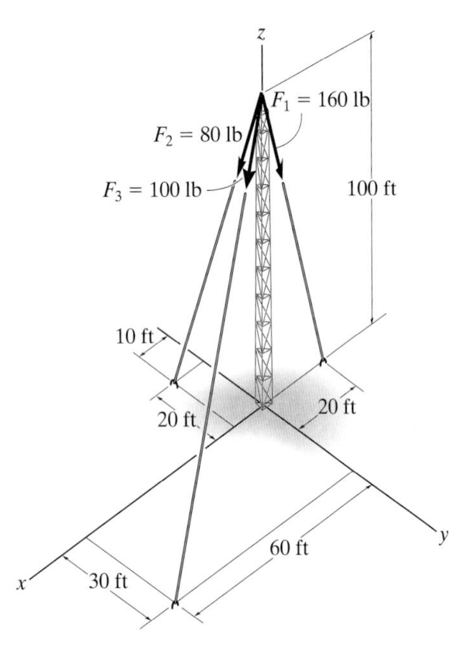

Prob. C–7

C-8. Determine the angle θ between the two cords.

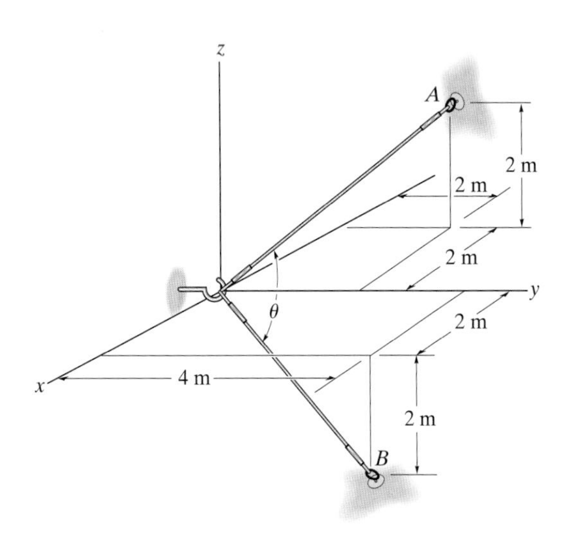

Prob. C–8

C-9. Determine the component of projection of the force **F** along the pipe AB.

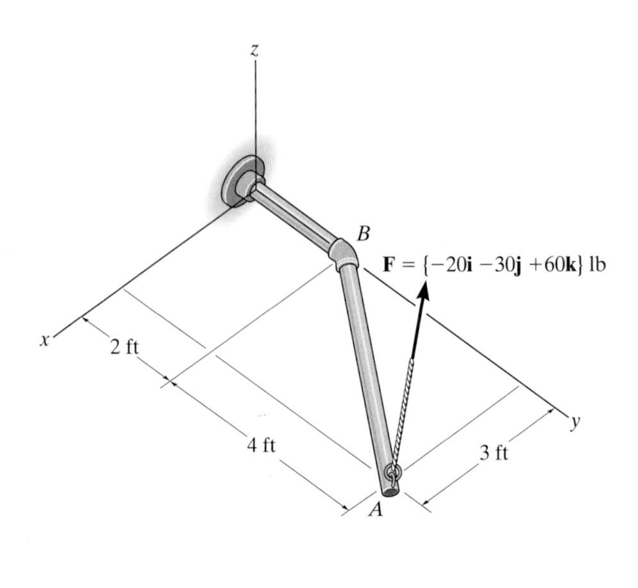

Prob. C–9

Chapter 3—Review Sections 3.1–3.3

C-10. The crate at D has a weight of 550 lb. Determine the force in each supporting cable.

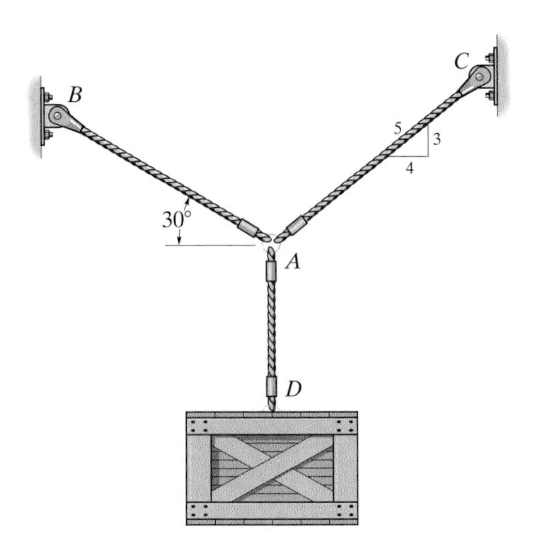

Prob. C–10

C-11. The beam has a weight of 700 lb. Determine the shortest cable ABC that can be used to lift it if the maximum force the cable can sustain is 1500 lb.

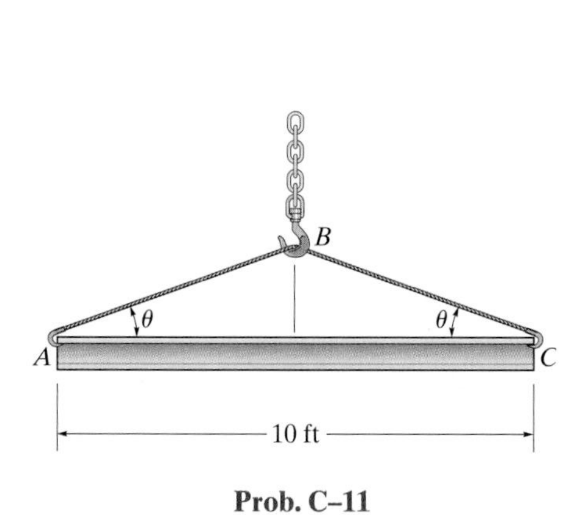

Prob. C–11

C-12. The block has a mass of 5 kg and rests on the smooth plane. Determine the unstretched length of the spring.

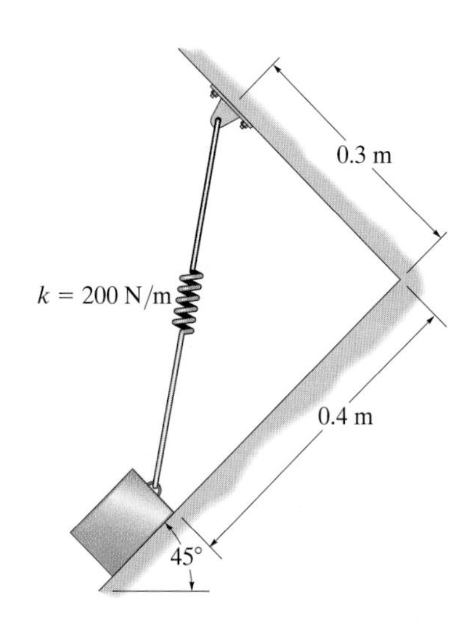

Prob. C–12

C-13. The post can be removed by a vertical force of 400 lb. Determine the force P that must be applied to the cord in order to pull the post out of the ground.

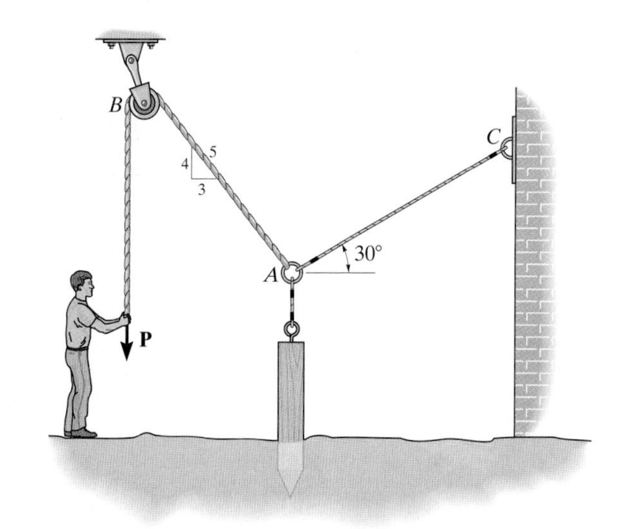

Prob. C–13

Chapter 4—Review All Sections

C-14. Determine the moment of the force about point O.

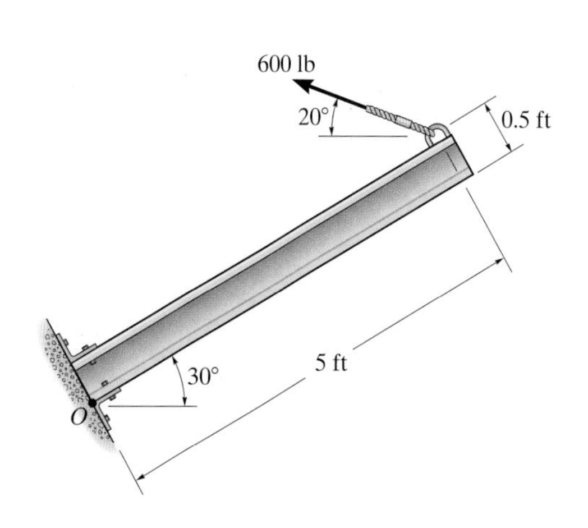

Prob. C–14

C-15. Determine the moment of the force about point O. Neglect the thickness of the member.

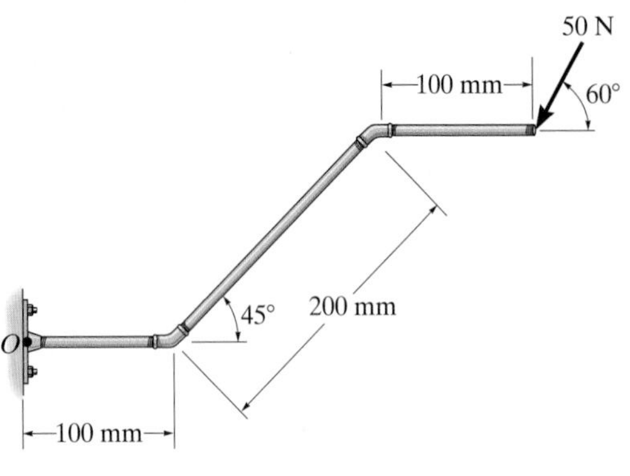

Prob. C–15

C-16. Determine the moment of the force about point O.

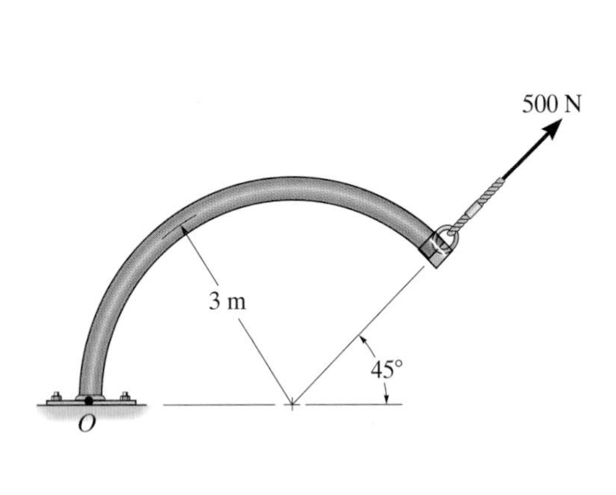

Prob. C–16

C-17. Determine the moment of the force about point A. Express the result as a Cartesian vector.

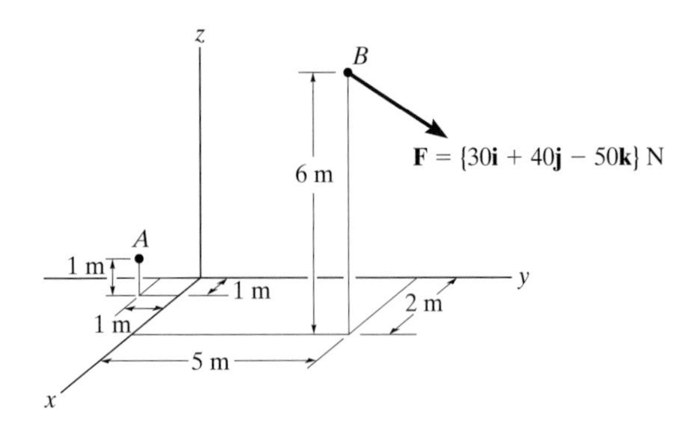

Prob. C–17

C-18. Determine the moment of the force about point A. Express the result as a Cartesian vector.

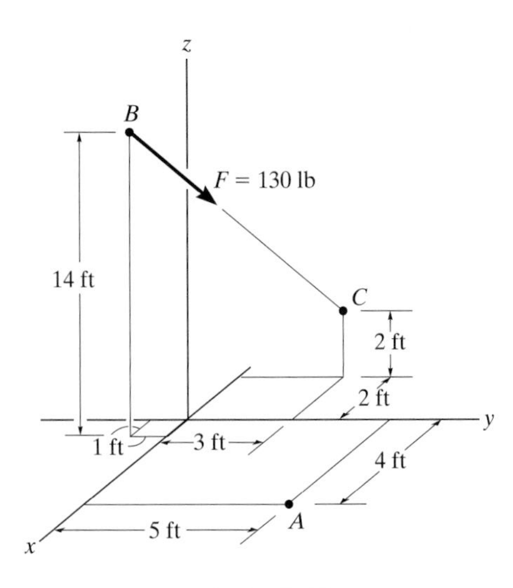

Prob. C–18

C-19. Determine the resultant couple moment acting on the beam.

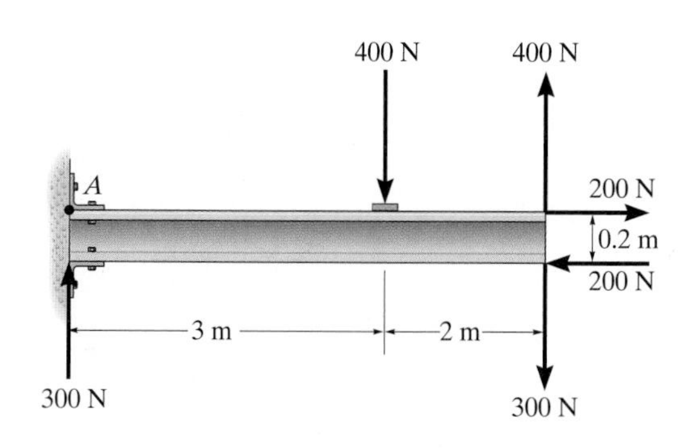

Prob. C–19

C-20. Determine the resultant couple moment acting on the triangular plate.

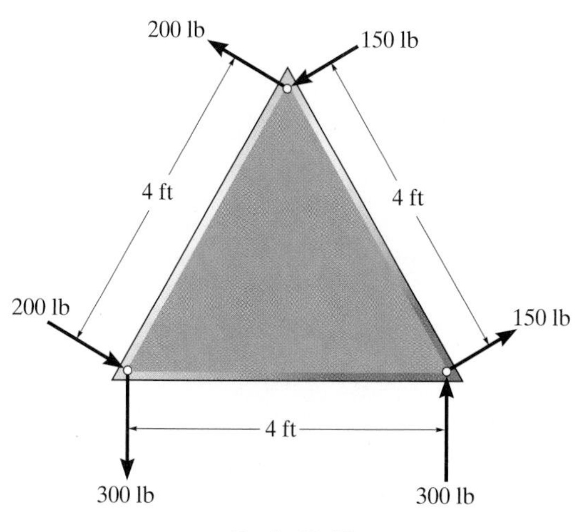

Prob. C–20

C-21. Replace the loading shown by an equivalent resultant force and couple-moment system at point A.

Prob. C–21

C-22. Replace the loading shown by an equivalent resultant force and couple-moment system at point A.

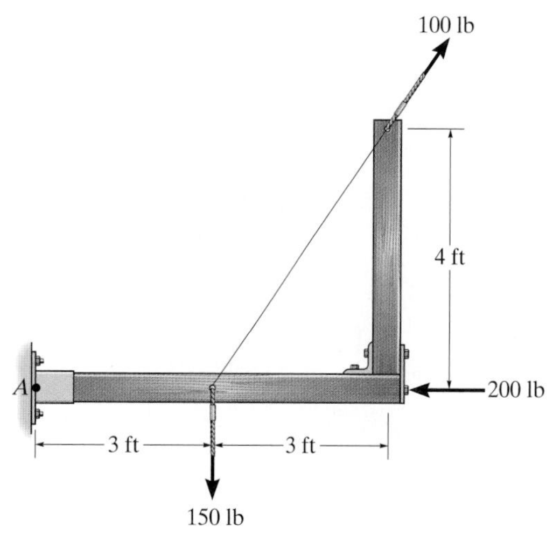

Prob. C–22

C-23. Replace the loading shown by an equivalent single resultant force and specify where the force acts, measured from point O.

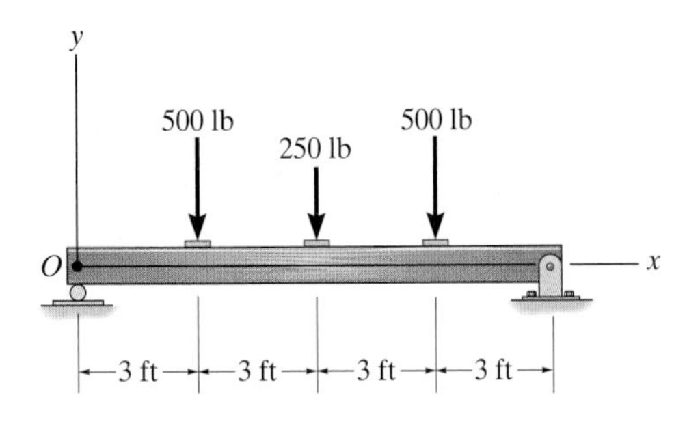

Prob. C–23

C-24. Replace the loading shown by an equivalent single resultant force and specify the x and y coordinates of its line of action.

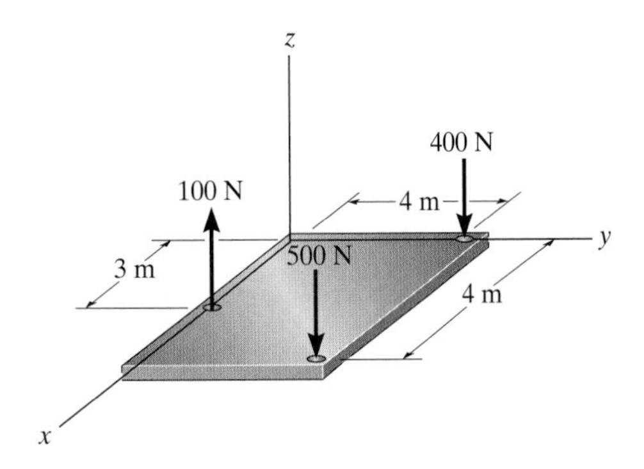

Prob. C–24

C-25. Replace the loading shown by an equivalent single resultant force and specify the x and y coordinates of its line of action.

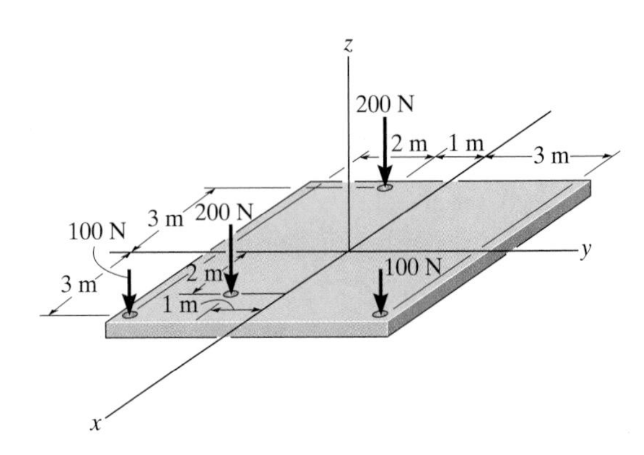

Prob. C–25

C-26. Determine the resultant force and specify where it acts on the beam measured from A.

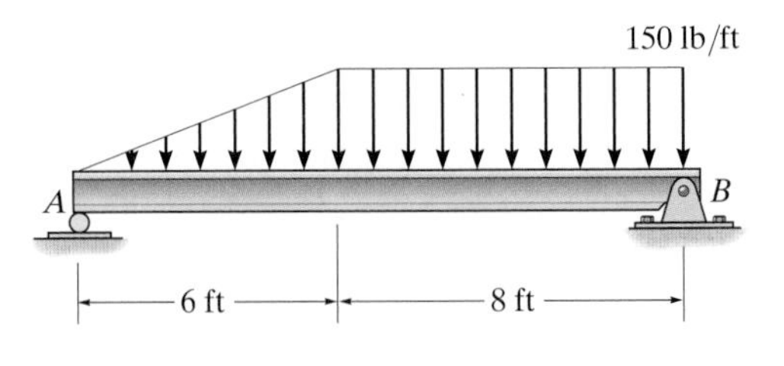

Prob. C–26

C-27. Determine the resultant force and specify where it acts on the beam measured from A.

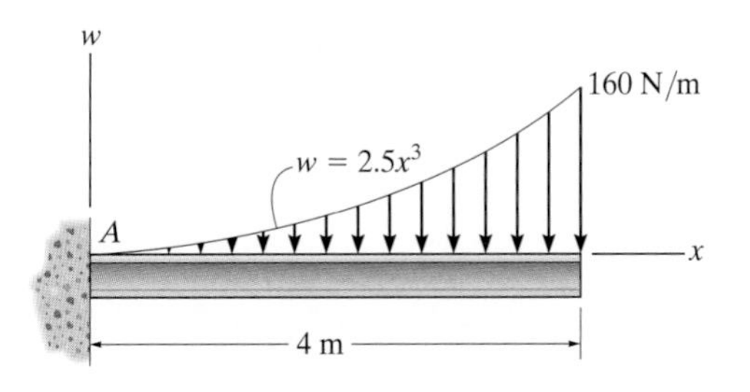

Prob. C–27

C-28. Determine the resultant force and specify where it acts on the beam measured from A.

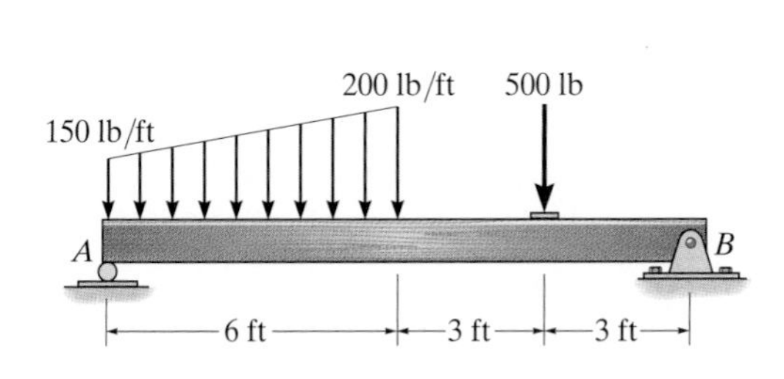

Prob. C–28

Chapter 5—Review Sections 5.1–5.6

C-29. Determine the horizontal and vertical components of reaction at the supports. Neglect the thickness of the beam.

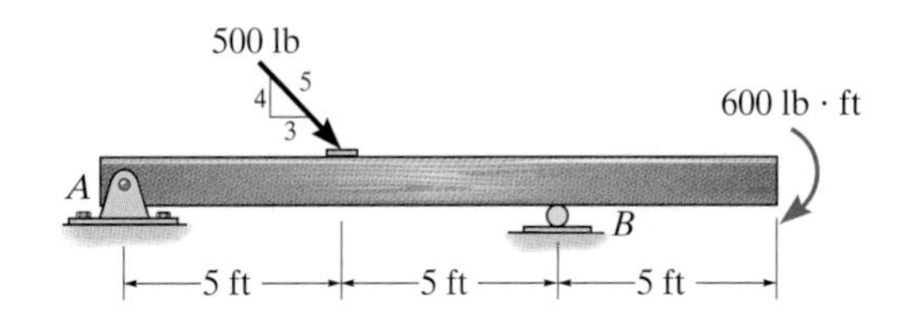

Prob. C–29

C-30. Determine the horizontal and vertical components of reaction at the supports.

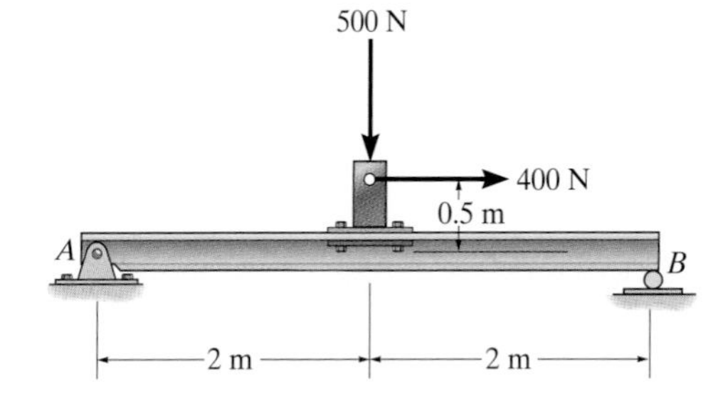

Prob. C–30

C-31. Determine the components of reaction at the fixed support A. Neglect the thickness of the beam.

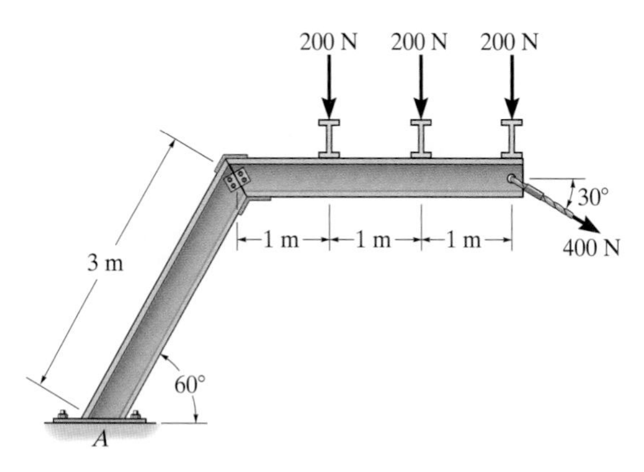

Prob. C–31

C-32. Determine the tension in the cable and the horizontal and vertical components of reaction at the pin A. Neglect the size of the pulley.

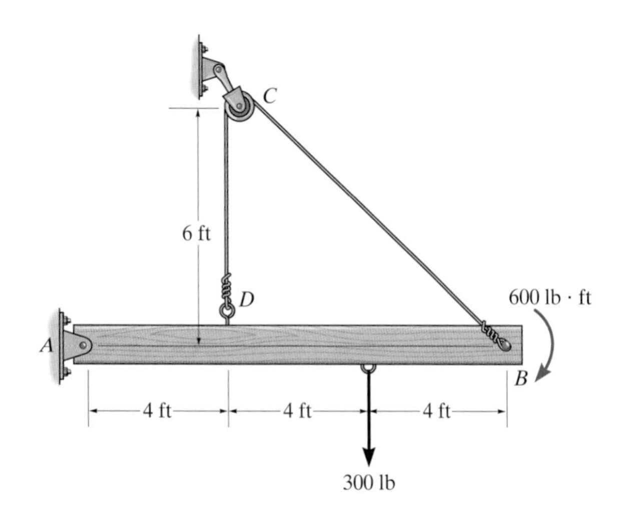

Prob. C–32

C-33. The uniform plate has a weight of 500 lb. Determine the tension in each of the supporting cables.

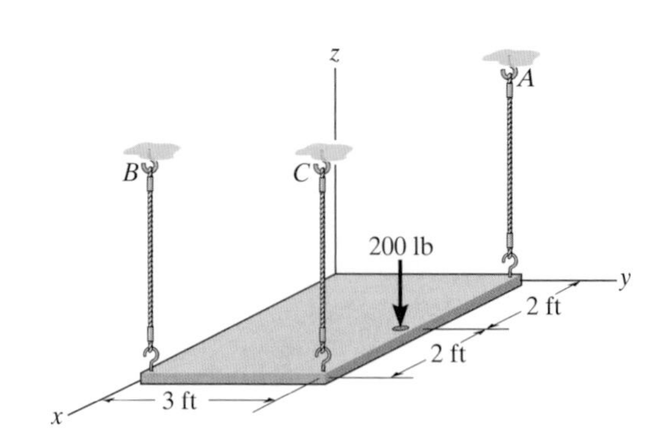

Prob. C–33

Chapter 6—Review Sections 6.1–6.4, 6.6

C-34. Determine the force in each member of the truss. State if the members are in tension or compression.

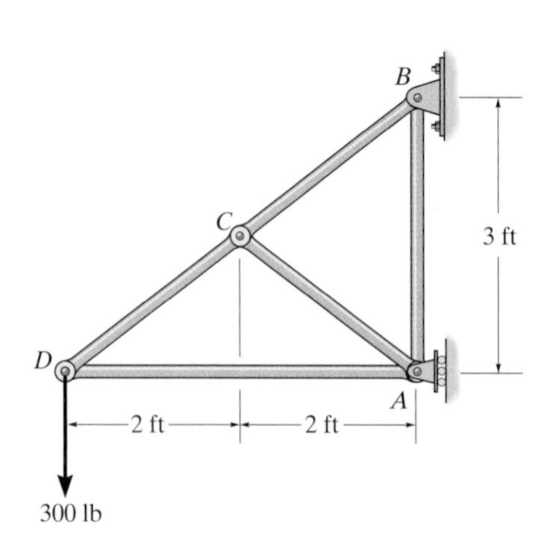

Prob. C–34

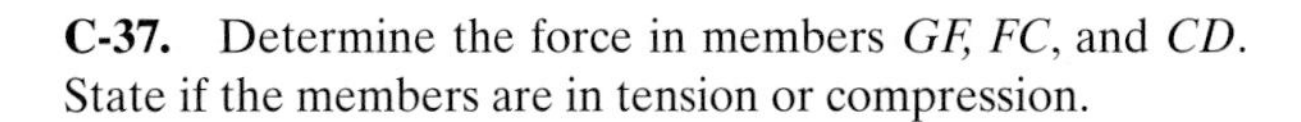

C-35. Determine the force in members AE and DC. State if the members are in tension or compression.

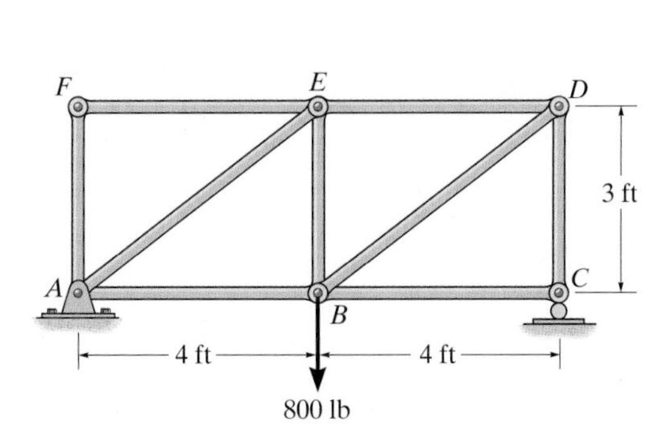

Prob. C–35

C-36. Determine the force in members BC, CF, and FE. State if the members are in tension or compression.

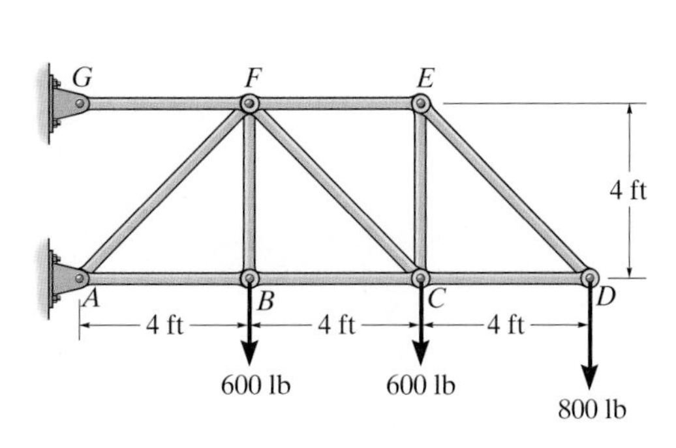

Prob. C–36

C-37. Determine the force in members GF, FC, and CD. State if the members are in tension or compression.

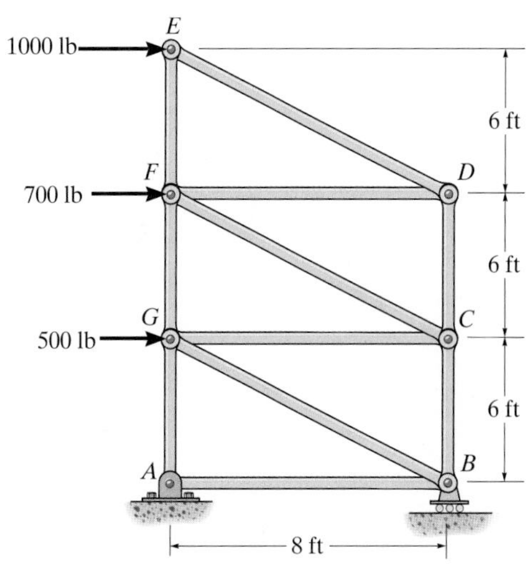

Prob. C–37

C-38. Determine the force P needed to hold the 60-lb weight in equilibrium.

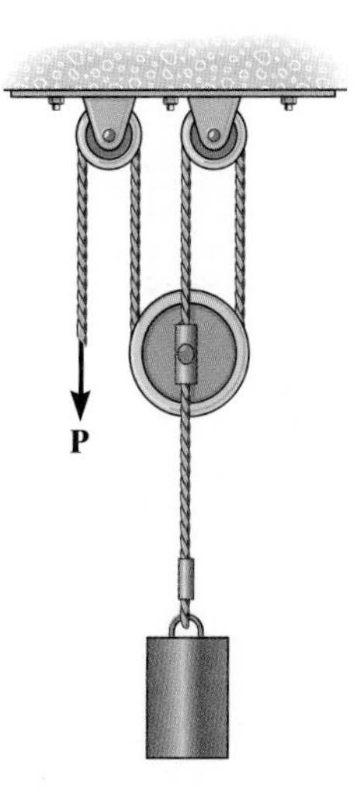

Prob. C–38

C-39. Determine the horizontal and vertical components of reaction at pin C.

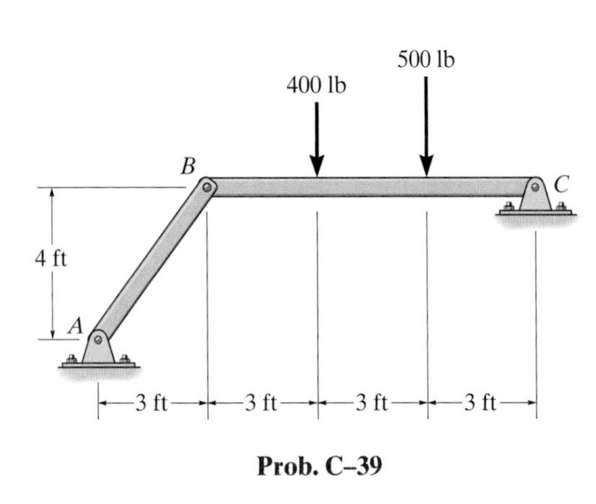

Prob. C–39

C-40. Determine the horizontal and vertical components of reaction at pin C.

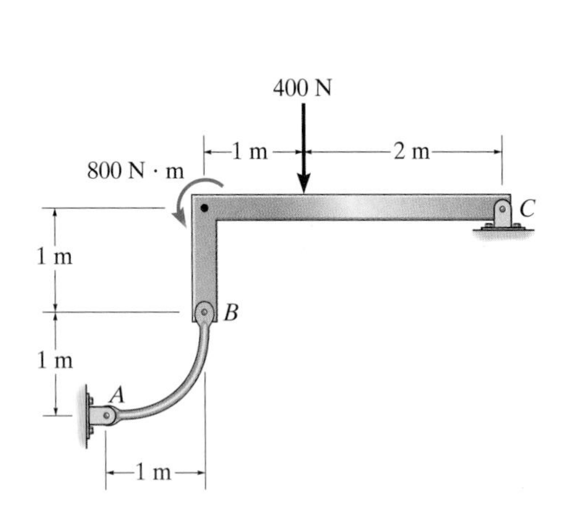

Prob. C–40

C-41. Determine the normal force that the 100-lb plate A exerts on the 30-lb plate B.

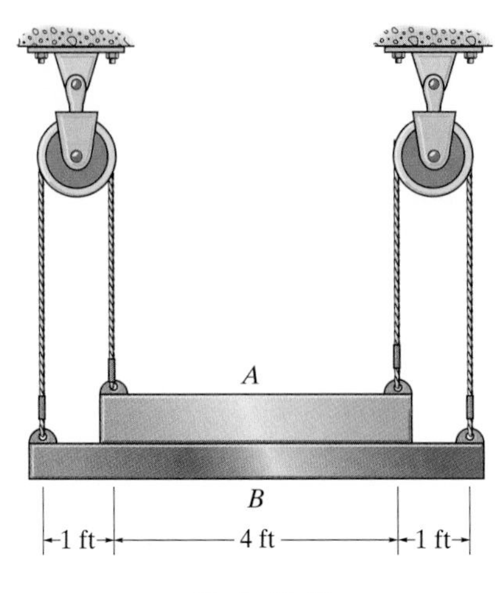

Prob. C–41

C-42. Determine the force P needed to lift the load. Also, determine the proper placement x of the hook for equilibrium. Neglect the weight of the beam.

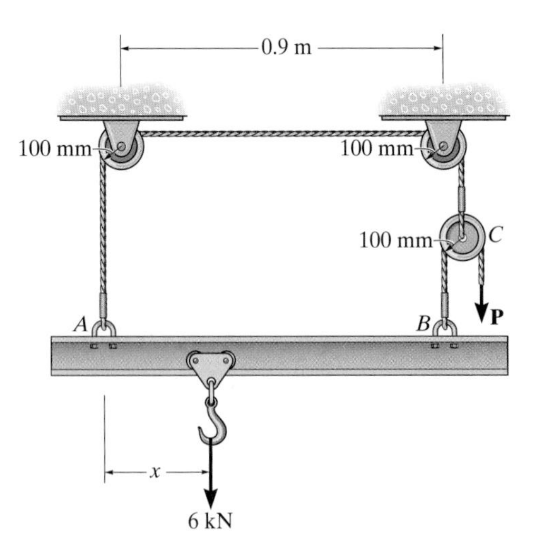

Prob. C–42

Chapter 7—Review Section 7.1

C-43. Determine the internal normal force, shear force, and moment acting in the beam at point B.

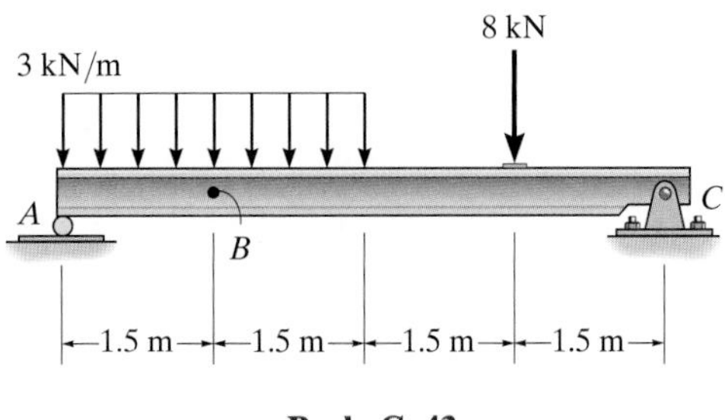

Prob. C–43

C-44. Determine the internal normal force, shear force, and moment acting in the beam at point B, which is located just to the left of the 800-lb force.

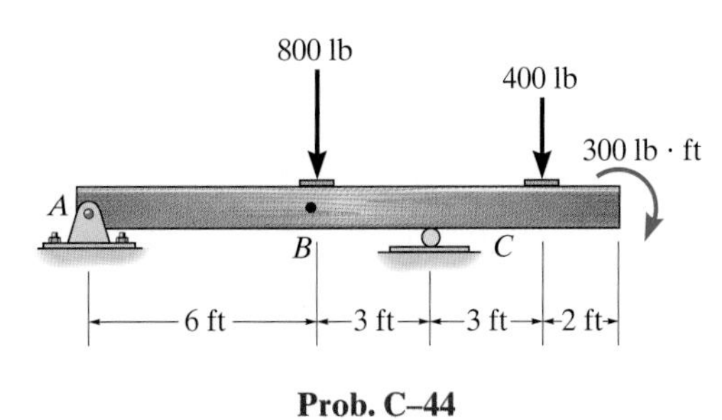

Prob. C–44

C-45. Determine the internal normal force, shear force, and moment acting in the beam at point B.

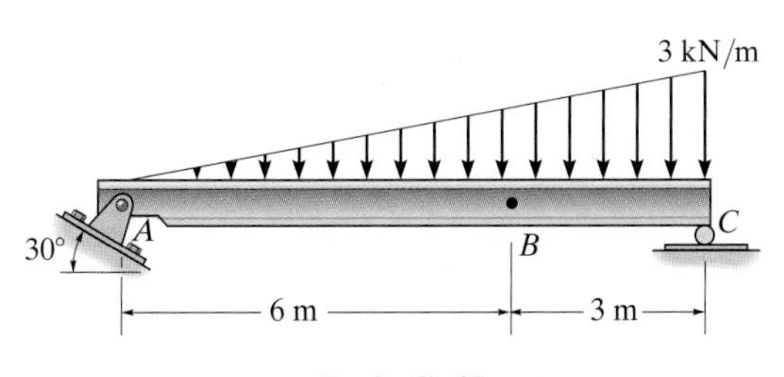

Prob. C–45

Chapter 8—Review Sections 8.1–8.2

C-46. Determine the force P needed to move the 100-lb block. The coefficient of static friction is $\mu_s = 0.3$, and the coefficient of kinetic friction is $\mu_k = 0.25$. Neglect tipping.

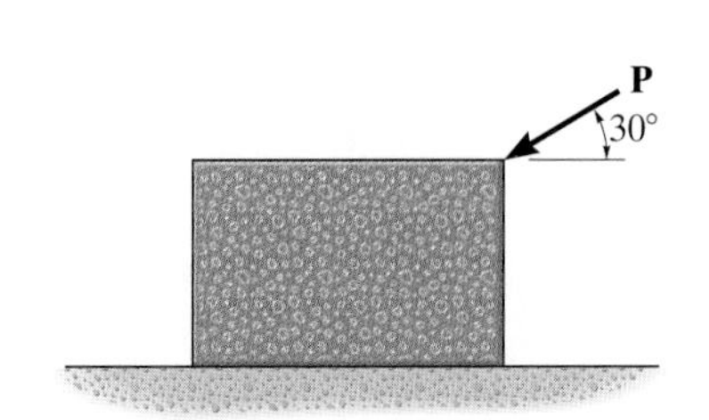

Prob. C–46

C-47. Determine the vertical force P needed to rotate the 200-lb spool. The coefficient of static friction at all contacting surfaces is $\mu_s = 0.4$.

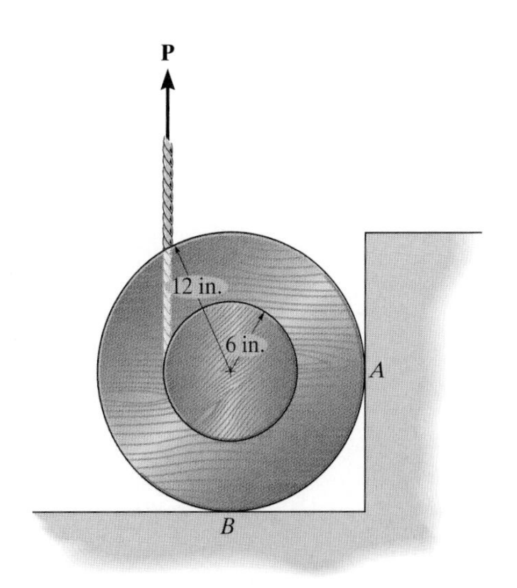

Prob. C–47

C-48. Block A has a weight of 30 lb and block B weighs 50 lb. If the coefficient of static friction is $\mu_s = 0.4$ between all contacting surfaces, determine the frictional force at each surface.

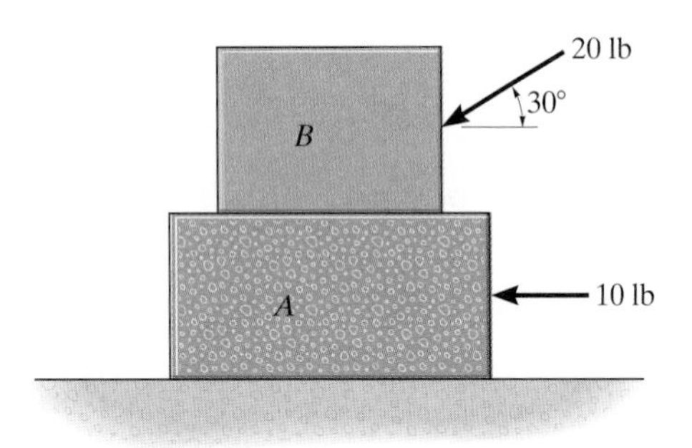

Prob. C–48

C-49. Determine the force P necessary to move the 250-lb crate which has a center of gravity at G. The coefficient of static friction at the floor is $\mu_s = 0.4$.

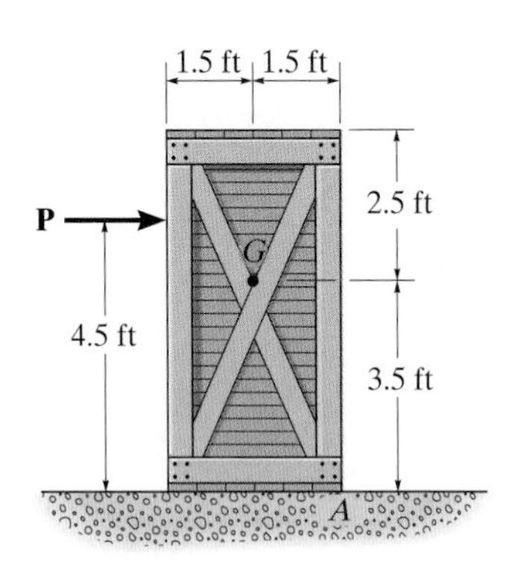

Prob. C–49

C-50. The filing cabinet A has a mass of 60 kg and center of mass at G. It rests on a 10-kg plank. Determine the smallest force P needed to move it. The coefficient of static friction between the cabinet A and the plank B is $\mu_s = 0.4$, and between the plank and the floor $\mu_s = 0.3$.

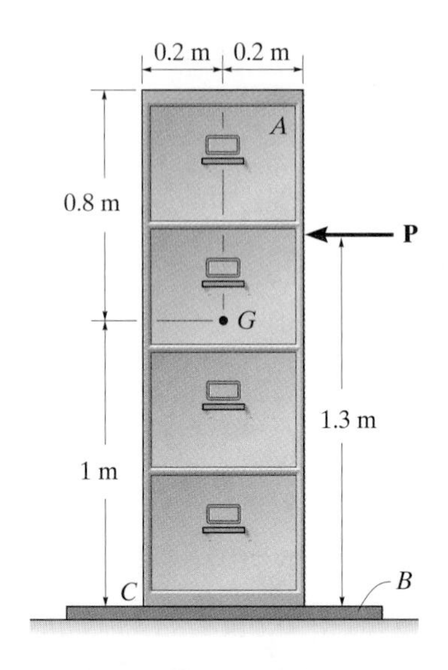

Prob. C–50

Chapter 9—Review Sections 9.1–9.3

(Integration is covered in the mathematics portion of the exam.)

C-51. Determine the location $(\bar{x}, \bar{y})$ of the centroid of the area.

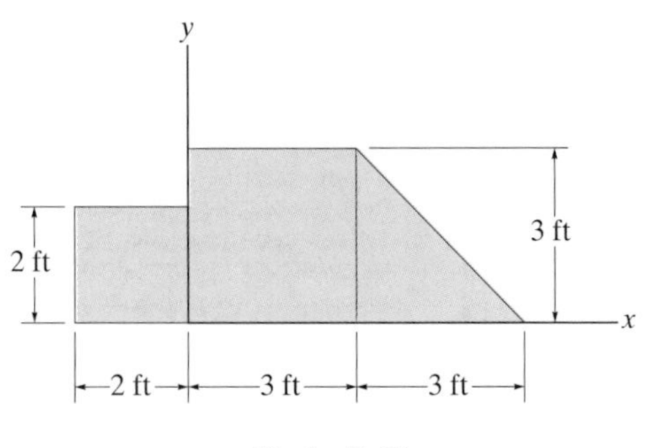

Prob. C–51

C-52. Determine the location $(\bar{x}, \bar{y})$ of the centroid of the area.

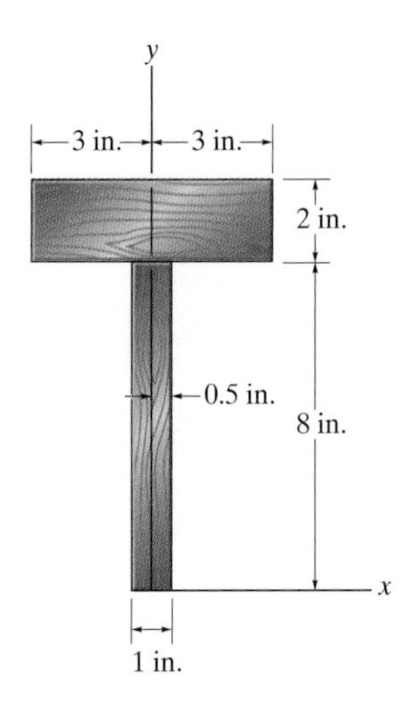

Prob. C–52

C-54. Determine the moment of inertia of the area with respect to the x axis.

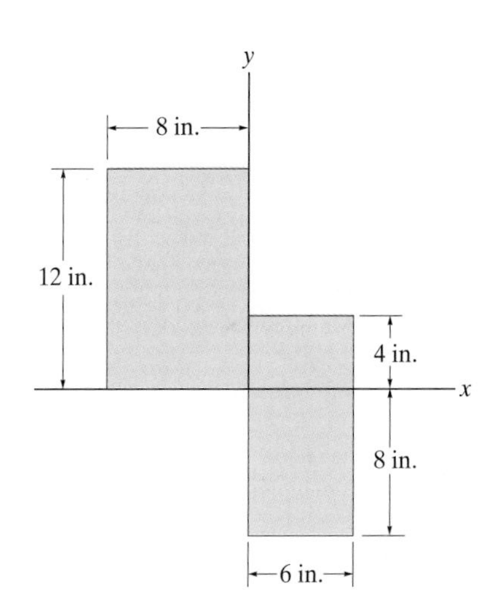

Prob. C–54

Chapter 10—Review Sections 10.1–10.5

(Integration is covered in the mathematics portion of the exam.)

C-53. Determine the moment of inertia of the cross-sectional area of the channel with respect to the y axis.

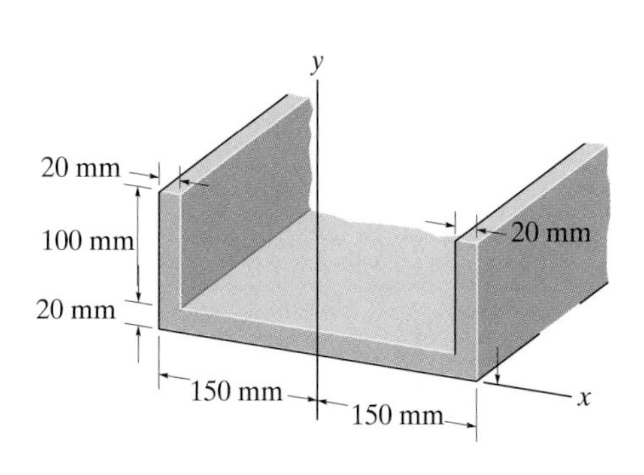

Prob. C–53

C-55. Determine the moment of inertia of the cross-sectional area of the T-beam with respect to the x' axis passing through the centroid of the cross section.

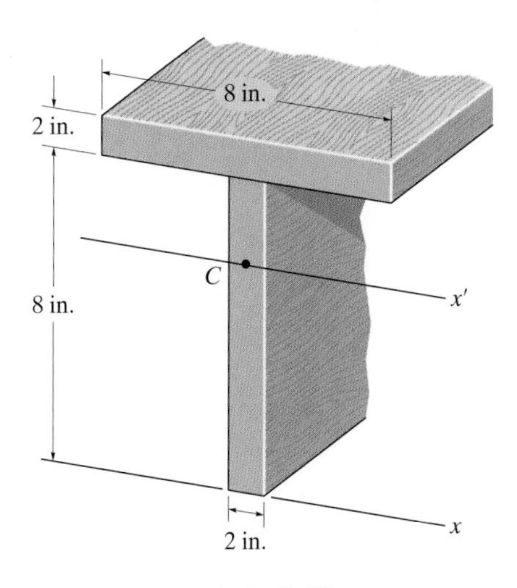

Prob. C–55

Partial Solutions and Answers

C–1. $F_R = \sqrt{200^2 + 500^2 - 2(200)(500)\cos 140^\circ}$
$= 666\text{ N}$ *Ans.*

C–2. $\dfrac{F_{AB}}{\sin 105^\circ} = \dfrac{450}{\sin 30^\circ}$
$= 869\text{ lb}$ *Ans.*
$\dfrac{F_{AC}}{\sin 45^\circ} = \dfrac{450}{\sin 30^\circ}$
$F_{AC} = 636\text{ lb}$ *Ans.*

C–3. $F_{Rx} = 300 + 400\cos 30^\circ - 250\left(\frac{4}{5}\right) = 446.4\text{ N}$
$F_{Ry} = 400\sin 30^\circ + 250\left(\frac{3}{5}\right) = 350\text{ N}$
$F_R = \sqrt{(446.4)^2 + 350^2} = 567\text{ N}$ *Ans.*
$\theta = \tan^{-1}\dfrac{350}{446.4} = 38.1^\circ$ ∠ *Ans.*

C–4. $F = \sqrt{30^2 + 50^2 + (-45)^2} = 73.7\text{ N}$ *Ans.*
$\alpha = \cos^{-1}\left(\frac{30}{73.7}\right) = 66.0^\circ$ *Ans.*
$\beta = \cos^{-1}\left(\frac{50}{73.7}\right) = 47.2^\circ$ *Ans.*
$\gamma = \cos^{-1}\left(\frac{-45}{73.7}\right) = 128^\circ$ *Ans.*

C–5. $F_y = -20$
$\dfrac{F_y}{|F|} = \cos\beta$
$|F| = \left|\dfrac{-20}{\cos 150^\circ}\right| = 23.09\text{ N}$
$\cos\gamma = \sqrt{1 - \cos^2 70^\circ - \cos^2 150^\circ}$
$\gamma = 68.61^\circ$ (From Fig. $\gamma < 90^\circ$)
$\mathbf{F} = 23.09\cos 70^\circ\mathbf{i} + 23.09\cos 150^\circ\mathbf{j}$
$+ 23.09\cos 68.61^\circ\mathbf{k}$
$= \{7.90\mathbf{i} - 20\mathbf{j} + 8.42\mathbf{k}\}\text{ N}$ *Ans.*

C–6. $F_x = 75\cos 30^\circ\sin 45^\circ = 45.93$
$F_y = 75\cos 30^\circ\cos 45^\circ = 45.93$
$F_z = -75\sin 30^\circ = -37.5$
$\alpha = \cos^{-1}\left(\frac{45.93}{75}\right) = 52.2^\circ$ *Ans.*
$\beta = \cos^{-1}\left(\frac{45.93}{75}\right) = 52.2^\circ$ *Ans.*
$\gamma = \cos^{-1}\left(\frac{-37.5}{75}\right) = 120^\circ$ *Ans.*

C–7. $\mathbf{F}_1 = 160\text{ lb}\left(-\frac{20}{102.0}\mathbf{i} - \frac{100}{102.0}\mathbf{k}\right)$
$= \{-31.4\mathbf{i} - 157\mathbf{k}\}\text{ lb}$ *Ans.*
$\mathbf{F}_2 = 80\text{ lb}\left(\frac{10}{102.5}\mathbf{i} - \frac{20}{102.5}\mathbf{j} - \frac{100}{102.5}\mathbf{k}\right)$
$= \{7.81\mathbf{i} - 15.6\mathbf{j} - 78.1\mathbf{k}\}\text{ lb}$ *Ans.*
$\mathbf{F}_3 = 100\text{ lb}\left(\frac{60}{120.4}\mathbf{i} + \frac{30}{120.4}\mathbf{j} - \frac{100}{120.4}\mathbf{k}\right)$
$= \{49.8\mathbf{i} + 24.9\mathbf{j} - 83.0\mathbf{k}\}\text{ lb}$ *Ans.*

C–8. $\mathbf{r}_{OA} = \{-2\mathbf{i} + 2\mathbf{j} + 2\mathbf{k}\}\text{ m}$
$\mathbf{r}_{OB} = \{2\mathbf{i} + 4\mathbf{j} - 2\mathbf{k}\}\text{ m}$
$\cos\theta = \dfrac{\mathbf{r}_{OA}\cdot\mathbf{r}_{OB}}{|r_{OA}||r_{OB}|}$
$\dfrac{(-2\mathbf{i} + 2\mathbf{j} + 2\mathbf{k})\cdot(2\mathbf{i} + 4\mathbf{j} - 2\mathbf{k})}{\sqrt{12}\sqrt{24}} = 0$
$\theta = 90^\circ$ *Ans.*

C–9. $|F_{AB}| = \mathbf{F}\cdot\mathbf{u}_{AB}$
$= (-20\mathbf{i} - 30\mathbf{j} + 60\mathbf{k})\cdot\left(-\frac{3}{5}\mathbf{i} - \frac{4}{5}\mathbf{j}\right) = 36\text{ lb}$ *Ans.*

C–10. $\xrightarrow{+}\Sigma F_x = 0;\ \frac{4}{5}F_{AC} - F_{AB}\cos 30^\circ = 0$
$+\uparrow\Sigma F_y = 0;\ \frac{3}{5}F_{AC} + F_{AB}\sin 30^\circ - 550 = 0$
$F_{AB} = 478\text{ lb}$ *Ans.*
$F_{AC} = 518\text{ lb}$ *Ans.*

C–11. $+\uparrow\Sigma F_y = 0;\ -2(1500)\sin\theta + 700 = 0$
$\theta = 13.5^\circ$
$L_{ABC} = 2\left(\dfrac{5\text{ ft}}{\cos 13.5^\circ}\right) = 10.3\text{ ft}$

C–12. $+\nearrow\Sigma F_x = 0;\ \frac{4}{5}(F_{sp}) - 5(9.81)\sin 45^\circ = 0$

$F_{sp} = 43.35$ N

$F_{sp} = k(l - l_0);\ 43.35 = 200(0.5 - l_0)$

$l_0 = 0.283$ m *Ans.*

C–13. At A:

$\overset{+}{\leftarrow}\Sigma F_x = 0;\ \frac{3}{5}P - T_{AC}\cos 30^\circ = 0$

$+\uparrow\Sigma F_y = 0;\ \frac{4}{5}P + T_{AC}\sin 30^\circ - 400 = 0$

$P = 349$ lb *Ans.*

$T_{AC} = 242$ lb *Ans.*

C–14. $\circlearrowright + M_O = 600\sin 50^\circ\,(5) + 600\cos 50^\circ\,(0.5)$

$= 2.49$ kip·ft *Ans.*

C–15. $\circlearrowleft + M_O = 50\sin 60^\circ\,(0.1 + 0.2\cos 45^\circ + 0.1)$

$- 50\cos 60^\circ\,(0.2\sin 45^\circ)$

$= 11.2$ N·m *Ans.*

C–16. $\circlearrowright + M_O = 500\sin 45^\circ\,(3 + 3\cos 45^\circ)$

$- 500\cos 45^\circ\,(3\sin 45^\circ)$

$= 1.06$ kN·m *Ans.*

C–17. $\mathbf{M}_A = \mathbf{r}_{AB}\times\mathbf{F} = \begin{vmatrix}\mathbf{i} & \mathbf{j} & \mathbf{k}\\ 1 & 6 & 5\\ 30 & 40 & -50\end{vmatrix}$

$= \{-500\mathbf{i} + 200\mathbf{j} - 140\mathbf{k}\}$ N·m *Ans.*

C–18. $\mathbf{F} = 130\text{ lb}\left(-\frac{3}{13}\mathbf{i} + \frac{4}{13}\mathbf{j} - \frac{12}{13}\mathbf{k}\right)$

$= \{-30\mathbf{i} + 40\mathbf{j} - 120\mathbf{k}\}$ lb

$\mathbf{M}_A = \mathbf{r}_{AB}\times\mathbf{F} = \begin{vmatrix}\mathbf{i} & \mathbf{j} & \mathbf{k}\\ -3 & -6 & 14\\ -30 & 40 & -120\end{vmatrix}$

$= \{160\mathbf{i} - 780\mathbf{j} - 300\mathbf{k}\}$ lb·ft *Ans.*

C–19. $\circlearrowleft + M_{C_R} = \Sigma M_A = 400(3) - 400(5) + 300(5)$
$+ 200(0.2) = 740$ N·m *Ans.*

Also,

$\circlearrowleft + M_{C_R} = 300(5) - 400(2) + 200(0.2)$

$= 740$ N·m *Ans.*

C–20. $\circlearrowright + M_{C_R} = 300(4) + 200(4) + 150(4)$

$= 2600$ lb·ft *Ans.*

C–21. $\overset{+}{\rightarrow} F_{Rx} = \Sigma F_x; F_{Rx} = \frac{4}{5}(50) = 40$ N

$+\downarrow F_{Ry} = \Sigma F_y; F_{Ry} = 40 + 30 + \frac{3}{5}(50)$

$= 100$ N

$F_R = \sqrt{(40)^2 + (100)^2} = 108$ N *Ans.*

$\theta = \tan^{-1}\left(\frac{100}{40}\right) = 68.2^\circ$ *Ans.*

$+\circlearrowright M_{A_R} = \Sigma M_A;$

$M_{A_R} = 30(3) + \frac{3}{5}(50)(6) + 200$

$= 470$ N·m *Ans.*

C–22. $\overset{+}{\leftarrow} F_{Rx} = \Sigma F_x;\ F_{Rx} = 200 - \frac{3}{5}(100) = 140$ lb

$+\downarrow F_{Ry} = \Sigma F_y;\ F_{Ry} = 150 - \frac{4}{5}(100) = 70$ lb

$F_R = \sqrt{140^2 + 70^2} = 157$ lb *Ans.*

$\theta = \tan^{-1}\left(\frac{70}{140}\right) = 26.6^\circ$ *Ans.*

$+\circlearrowright M_{A_R} = \Sigma M_A;$

$M_{A_R} = \frac{3}{5}(100)(4) - \frac{4}{5}(100)(6) + 150(3)$

$M_{R_A} = 210$ lb·ft *Ans.*

C–23. $+\downarrow F_R = \Sigma F_y;\ \ F_R = 500 + 250 + 500$

$= 1250$ lb *Ans.*

$+\circlearrowright F_R x = \Sigma M_O;$

$1250(x) = 500(3) + 250(6) + 500(9)$

$x = 6$ ft *Ans.*

C–24. $+\downarrow F_R = \Sigma F_z;\ \ F_R = 400 + 500 - 100$

$= 800$ N *Ans.*

$M_{Rx} = \Sigma M_x;\ -800y = -400(4) - 500(4)$

$y = 4.50$ m *Ans.*

$M_{Ry} = \Sigma M_y;\ \ 800x = 500(4) - 100(3)$

$x = 2.125$ m *Ans.*

C–25. $+\downarrow F_R = \Sigma F_y;\ F_R = 200 + 200 + 100 + 100 = 600 \text{ N}$ *Ans.*

$M_{Rx} = \Sigma M_x;\ -600y = 200(1) + 200(1) + 100(3) - 100(3)$

$y = -0.667 \text{ m}$ *Ans.*

$M_{Ry} = \Sigma M_y;\ 600x = 100(3) + 100(3) + 200(2) - 200(3)$

$x = 0.667 \text{ m}$ *Ans.*

C–26. $F_R = \frac{1}{2}(6)(150) + 8(150) = 1650 \text{ lb}$ *Ans.*

$+\downarrow M_{A_R} = \Sigma M_A;$

$1650d = \left[\frac{1}{2}(6)(150)\right](4) + [8(150)](10)$

$d = 8.36 \text{ ft}$ *Ans.*

C–27. $F_R = \int w(x)\,dx = \int_0^4 2.5x^3\,dx = 160 \text{ N}$ *Ans.*

$+\downarrow M_{A_R} = \Sigma M_A;$

$$x = \frac{\int xw(x)\,dx}{\int w(x)\,dx} = \frac{\int_0^4 2.5x^4\,dx}{160} = 3.20 \text{ m}$$ *Ans.*

C–28. $+\downarrow F_R = \Sigma F_y;\ F_R = \frac{1}{2}(50)(6) + 150(6) + 500 = 1550 \text{ lb}$ *Ans.*

$+\downarrow M_{A_R} = \Sigma M_A;$

$1550d = \left[\frac{1}{2}(50)(6)\right](4) + [150(6)](3) + 500(9)$

$d = 5.03 \text{ ft}$ *Ans.*

C–29. $\overset{+}{\rightarrow} \Sigma F_x = 0;\ -A_x + 500\left(\frac{3}{5}\right) = 0$

$A_x = 300 \text{ lb}$ *Ans.*

$+\circlearrowleft \Sigma M_A = 0;\ B_y(10) - 500\left(\frac{4}{5}\right)(5) - 600 = 0$

$B_y = 260 \text{ lb}$ *Ans.*

$+\uparrow \Sigma F_y = 0;\ A_y + 260 - 500\left(\frac{4}{5}\right) = 0$

$A_y = 140 \text{ lb}$ *Ans.*

C–30. $\overset{+}{\rightarrow} \Sigma F_x = 0;\ -A_x + 400 = 0;\ A_x = 400 \text{ N}$ *Ans.*

$\circlearrowright + \Sigma M_A = 0;\ B_y(4) - 400(0.5) - 500(2) = 0$

$B_y = 300 \text{ N}$ *Ans.*

$+\uparrow \Sigma F_y = 0;\ A_y + 300 - 500 = 0$

$A_y = 200 \text{ N}$ *Ans.*

C–31. $\overset{+}{\rightarrow} \Sigma F_x = 0;\ -A_x + 400 \cos 30^\circ = 0$

$A_x = 346 \text{ N}$ *Ans.*

$+\uparrow \Sigma F_y = 0;$

$A_y - 200 - 200 - 200 - 400 \sin 30^\circ = 0$

$A_y = 800 \text{ N}$ *Ans.*

$\circlearrowright + \Sigma M_A = 0;$

$M_A - 200(2.5) - 200(3.5) - 200(4.5)$

$-400 \sin 30^\circ(4.5) - 400 \cos 30^\circ(3 \sin 60^\circ) = 0$

$M_A = 3.90 \text{ kN} \cdot \text{m}$ *Ans.*

C–32. $+\circlearrowleft \Sigma M_A = 0;\ T(4) + \frac{3}{5}T(12) - 300(8) - 600 = 0$

$T = 267.9 = 268 \text{ lb}$ *Ans.*

$\overset{+}{\rightarrow} \Sigma F_x = 0;\ A_x - \left(\frac{4}{5}\right)(267.9) = 0$

$A_x = 214 \text{ lb}$ *Ans.*

$+\uparrow \Sigma F_y = 0;\ A_y + 267.9 + \left(\frac{3}{5}\right)(267.9) - 300 = 0$

$A_y = -129 \text{ lb}$ *Ans.*

C–33. $\Sigma F_z = 0;\ T_A + T_B + T_C - 200 - 500 = 0$

$\Sigma M_x = 0;\ T_A(3) + T_C(3) - 500(1.5) - 200(3) = 0$

$\Sigma M_y = 0;\ -T_B(4) - T_C(4) + 500(2) + 200(2) = 0$

$T_A = 350 \text{ lb},\ T_B = 250 \text{ lb},\ T_C = 100 \text{ lb}$ *Ans.*

C–34. Joint D:

$+\uparrow \Sigma F_y = 0;\ \frac{3}{5}F_{CD} - 300 = 0;$

$F_{CD} = 500 \text{ lb (T)}$ *Ans.*

$\overset{+}{\rightarrow} \Sigma F_x = 0;\ -F_{AD} + \frac{4}{5}(500) = 0;$

$F_{AD} = 400 \text{ lb (C)}$ *Ans.*

Joint C:

$+\searrow\Sigma F_y = 0; F_{CA} = 0$ *Ans.*

$+\nearrow\Sigma F_x = 0; F_{CB} - 500 = 0;$

$F_{CB} = 500 \text{ lb (T)}$ *Ans.*

Joint A:

$+\uparrow\Sigma F_y = 0; F_{AB} = 0$ *Ans.*

C–35. $Ax = 0,\ Ay = Cy = 400 \text{ lb}$

Joint A:

$+\uparrow\Sigma F_y = 0;\ -\frac{3}{5}F_{AE} + 400 = 0;$

$F_{AE} = 667 \text{ lb (C)}$ *Ans.*

Joint C:

$+\uparrow\Sigma F_y = 0; -F_{DC} + 400 = 0;$

$F_{DC} = 400 \text{ lb (C)}$ *Ans.*

C–36. Section truss through FE, FC, BC. Use the right segment.

$+\uparrow\Sigma F_y = 0;\ F_{CF}\sin 45° - 600 - 800 = 0$

$F_{CF} = 1980 \text{ lb (T)}$ *Ans.*

$+\circlearrowleft\Sigma M_C = 0;\ F_{FE}(4) - 800(4) = 0$

$F_{FE} = 800 \text{ lb (T)}$ *Ans.*

$\downarrow+\Sigma M_F = 0;\ F_{BC}(4) - 600(4) - 800(8) = 0$

$F_{BC} = 2200 \text{ lb (C)}$ *Ans.*

C–37. Section truss through GF, FC, DC. Use the top segment.

$+\circlearrowleft\Sigma M_C = 0;\ F_{GF}(8) - 700(6) - 1000(12) = 0$

$F_{GF} = 2025 \text{ lb (T)}$ *Ans.*

$\overset{+}{\rightarrow}\Sigma F_x = 0;\ -\frac{4}{5}F_{FC} + 700 + 1000 = 0$

$F_{FC} = 2125 \text{ lb (C)}$ *Ans.*

$\circlearrowright+\Sigma M_F = 0;\ F_{CD}(8) - 1000(6) = 0$

$F_{CD} = 750 \text{ lb (C)}$ *Ans.*

C–38. $+\uparrow\Sigma F_y = 0;\ 3P - 60 = 0$

$P = 20 \text{ lb}$ *Ans.*

C–39. $+\circlearrowleft\Sigma M_C = 0;$

$-\left(\frac{4}{5}\right)(F_{AB})(9) + 400(6) + 500(3) = 0$

$F_{AB} = 541.67 \text{ lb}$

$\overset{+}{\rightarrow}\Sigma F_x = 0; -C_x + \frac{3}{5}(541.67) = 0$

$C_x = 325 \text{ lb}$ *Ans.*

$+\uparrow\Sigma F_y = 0; C_y + \frac{4}{5}(541.67) - 400 - 500 = 0$

$C_y = 467 \text{ lb}$ *Ans.*

C–40. $+\circlearrowleft\Sigma M_C = 0;$

$F_{AB}\cos 45°(1) - F_{AB}\sin 45°(3) + 800$

$+ 400(2) = 0$

$F_{AB} = 1131.37 \text{ N}$

$\overset{+}{\rightarrow}\Sigma F_x = 0;\ -C_x + 1131.37\cos 45° = 0$

$C_x = 800 \text{ N}$ *Ans.*

$+\uparrow\Sigma F_y = 0;\ -C_y + 1131.37\sin 45° - 400 = 0$

$C_y = 400 \text{ N}$ *Ans.*

C–41. Plate A:

$+\uparrow\Sigma F_y = 0;\ 2T + N_{AB} - 100 = 0$

Plate B:

$+\uparrow\Sigma F_y = 0;\ 2T - N_{AB} - 30 = 0$

$T = 32.5 \text{ lb}, N_{AB} = 35 \text{ lb}$ *Ans.*

C–42. Pulley C:

$+\uparrow\Sigma F_y = 0;\ T - 2P = 0; T = 2P$

Beam:

$+\uparrow\Sigma F_y = 0;\ 2P + P - 6 = 0$

$P = 2 \text{ kN}$ *Ans.*

$+\circlearrowleft\Sigma M_A = 0;\ 2(1) - 6(x) = 0$

$x = 0.333 \text{ m}$ *Ans.*

C–43. $A_y = 8.75$ kN. Use segment AB.

$\overset{+}{\rightarrow} \Sigma F_x = 0;\ N_B = 0$ *Ans.*

$+\uparrow \Sigma F_y = 0;\ 8.75 - 3(1.5) - V_B = 0$

$V_B = 4.25$ kN *Ans.*

$+\circlearrowleft \Sigma M_B = 0;\ M_B + 3(1.5)(0.75) - 8.75(1.5) = 0$

$M_B = 9.75$ kN·m *Ans.*

C–44. $A_x = 0,\ A_y = 100$ lb. Use segment AB.

$\overset{+}{\rightarrow} \Sigma F_x = 0;\ N_B = 0$ *Ans.*

$+\uparrow \Sigma F_y = 0;\ 100 - V_B = 0$

$V_B = 100$ lb *Ans.*

$+\circlearrowleft \Sigma M_B = 0;\ M_B - 100(6) = 0$

$M_B = 600$ lb·ft *Ans.*

C–45. $A_x = 0, A_y = 4.5$ kN, $w_B = 2$ kN/m. Use segment AB.

$\overset{+}{\rightarrow} \Sigma F_x = 0;\ N_B = 0$ *Ans.*

$+\uparrow \Sigma F_y = 0;\ 4.5 - \frac{1}{2}(6)(2) + V_B = 0$

$V_B = 1.5$ kN *Ans.*

$+\circlearrowleft \Sigma M_B = 0;\ M_B + \left[\frac{1}{2}(6)(2)\right](2) - 4.5(6) = 0$

$M_B = 15$ kN·m *Ans.*

C–46. $+\uparrow \Sigma F_y = 0;\ N_b - P \sin 30° - 100 = 0$

$\overset{+}{\rightarrow} \Sigma F_x = 0;\ -P \cos 30° + 0.3N_b = 0$

$P = 41.9$ lb *Ans.*

C–47. $\overset{+}{\rightarrow} \Sigma F_x = 0;\ 0.4N_B - N_A = 0$

$+\circlearrowleft \Sigma M_B = 0;\ 0.4N_A(12) + N_A(12) - P(6) = 0$

$+\circlearrowleft \Sigma F_y = 0;\ P + 0.4N_A + N_B - 200 = 0$

$P = 98.2$ lb *Ans.*

C–48. Block B:

$+\uparrow \Sigma F_y = 0;\ N_B - 20 \sin 30° - 50 = 0$

$N_B = 60$ lb

$\overset{+}{\rightarrow} \Sigma F_x = 0;\ F_B - 20 \cos 30° = 0$

$F_B = 17.3$ lb $(< 0.4(60 \text{ lb}))$ *Ans.*

Blocks A and B:

$+\uparrow \Sigma F_y = 0;\ N_A - 30 - 50 - 20 \sin 30° = 0$

$N_A = 90$ lb

$\overset{+}{\rightarrow} \Sigma F_x = 0;\ F_A - 20 \cos 30° - 10 = 0$

$F_A = 27.3$ lb $(< 0.4(90 \text{ lb}))$ *Ans.*

C–49. If slipping occurs:

$+\uparrow \Sigma F_y = 0;\ N_C - 250 \text{ lb} = 0$

$N_C = 250$ lb

$\overset{+}{\rightarrow} \Sigma F_x = 0;\ P - 0.4(250) = 0$

$P = 100$ lb

If tipping occurs:

$\circlearrowright + \Sigma M_A = 0;\ -P(4.5) + 250(1.5) = 0$

$P = 83.3$ lb *Ans.*

C–50. P for A to slip on B:

$+\uparrow \Sigma F_y = 0;\ N_A - 60(9.81) = 0$

$N_A = 588.6$ N

$\overset{+}{\rightarrow} \Sigma F_x = 0;\ 0.4(588.6) - P = 0$

$P = 235$ N

P for B to slip:

$+\uparrow \Sigma F_y = 0;\ N_B - 60(9.81) - 10(9.81) = 0$

$N_B = 686.7$ N

$\overset{+}{\rightarrow} \Sigma F_x = 0;\ 0.3(686.7) - P = 0$

$P = 206$ N

P to tip A:

$\circlearrowright + \Sigma M_C = 0;\ P(1.3) - 60(9.81)(0.2) = 0$

$P = 90.6$ N *Ans.*

C–51. $\bar{x} = \dfrac{\Sigma\tilde{x}A}{\Sigma A} =$

$$\frac{(-1)(2)(2) + 1.5(3)(3) + 4\left(\frac{1}{2}\right)(3)(3)}{2(2) + 3(3) + \frac{1}{2}(3)(3)} = 1.57 \text{ ft}$$

Ans.

$\bar{y} = \dfrac{\Sigma\tilde{y}A}{\Sigma A} =$

$$\frac{1(2)(2) + 1.5(3)(3) + 1\left(\frac{1}{2}\right)(3)(3)}{2(2) + 3(3) + \frac{1}{2}(3)(3)} = 1.26 \text{ ft}$$

Ans.

C–52. $\bar{x} = 0$ (symmetry) *Ans.*

$$\bar{y} = \frac{\Sigma\tilde{y}A}{\Sigma A} = \frac{4(1(8)) + 9(6)(2)}{1(8) + 6(2)} = 7 \text{ in.}$$ *Ans.*

C–53. $I_y = \dfrac{1}{12}(120)(300)^3 - \dfrac{1}{12}(100)(260)^3$

$= 124\ (10^6)\ \text{mm}^4$ *Ans.*

C–54. $I = \Sigma(\bar{I} + Ad^2) = \left[\dfrac{1}{12}(8)(12)^3 + (8)(12)(6)^2\right]$

$+ \left[\dfrac{1}{12}(6)(12)^3 + (6)(12)(-2)^2\right] = 5760\ \text{in}^4$ *Ans.*

C–55. $\bar{x} = \dfrac{\Sigma\tilde{x}A}{\Sigma A} = \dfrac{4(8)(2) + 9(2)(8)}{8(2) + 2(8)} = 6.5 \text{ in.}$

$\bar{I}_{x'} = \Sigma(\bar{I} + Ad^2) = \left[\dfrac{1}{12}(2)(8)^3 + (8)(2)(6.5 - 4)^2\right]$

$+ \left[\dfrac{1}{12}(8)(2)^3 + 2(8)(9 - 6.5)^2\right] = 291\ \text{in}^4$ *Ans.*

Answers to Selected Problems

Chapter 1

1–1. **(a)** km/s, **(b)** mm, **(c)** Gs/kg, **(d)** mm · N
1–2. 2.42 Mg/m^3
1–3. **(a)** Mg/mm = Gg/m, **(b)** mN/μs = kN/s, **(c)** μm · Mg = mm · kg
1–5. **(a)** 8.653 s, **(b)** 8.368 kN, **(c)** 893 g
1–6. **(a)** 45.3 MN, **(b)** 56.8 km, **(c)** 5.63 μg
1–7. 26.9 μm · kg/N
1–9. **(a)** 271 N · m, **(b)** 55.0 kN/m^3, **(c)** 0.677 mm/s
1–10. **(a)** 98.1 N, **(b)** 4.90 mN, **(c)** 44.1 kN
1–11. 584 kg
1–13. 1.04 kip
1–14. 2.71 Mg/m^3
1–17. 4.96 μN
1–18. **(a)** 0.04 MN^2, **(b)** 25 μm^2, **(c)** 0.064 km^3
1–19. **(a)** 15.9 mm/s, **(b)** 3.69 Mm · s/kg, **(c)** 1.14 km · kg

Chapter 2

2–1. $F_R = 867$ N, $\phi = 108°$
2–2. $F_R = 72.1$ lb, $\theta = 73.9°$
2–3. $F_R = 393$ lb, $\phi = 353°$
2–5. $F_{1u} = 205$ N, $F_{1v} = 160$ N
2–6. $F_{2u} = 376$ N, $F_{2v} = 482$ N
2–7. $F_R = 61.4$ lb, $\theta = 6.85°$
2–9. $F_u = 86.6$ lb, $F_v = -50$ lb
2–10. $F_u = 320$ N, $F_v = 332$ N
2–11. $F_a = 30.6$ lb, $F_b = 26.9$ lb
2–13. $F_{AB} = 186$ lb, $F_{AC} = 239$ lb
2–14. $T = 744$ lb, $\theta = 23.8°$
2–15. $F_{1v} = 129$ N, $F_{1u} = 183$ N
2–17. $F_R = \sqrt{F_1^2 + F_2^2 + 2F_1F_2\cos\phi}$, $\tan\theta = \dfrac{F_1\sin\phi}{F_2 + F_1\cos\phi}$
2–18. $F_R = 400$ N, $\theta = 60°$
2–19. $\theta = 25.4°$, $F_R = 115$ lb
2–21. $\theta = 43.9°$, $F_R = 561$ lb
2–22. $F_R = 19.2$ N, $\theta = 2.37°$ $\measuredangle\theta$
2–23. $F_R = 19.2$ N, $\theta = 2.37°$ $\measuredangle\theta$
2–25. $T = 54.7$ lb, $P = 42.6$ lb
2–26. $\theta = 60°$, $P = 40$ lb, $T = 69.3$ lb
2–27. $F_A = 439$ N, $F_B = 311$ N
2–29. $\theta = 10.9°$, $F_{\min} = 235$ lb
2–30. $F_{\min} = 97.4$ lb, $\theta = 16.2°$
2–31. $F_x = 514$ lb, $F_y = -613$ lb
2–33. $F_R = 29.3$ lb, $\theta = 347°$
2–34. $F_R = 546$ N, $\theta = 253°$
2–35. $\theta = 37.0°$, $F_1 = 889$ N
2–37. $F_R = 747$ N, $\theta = 85.5°$
2–38. $F_R = 25.1$ kN, $\theta = 185°$
2–39. $F_R = 111$ lb, $\theta = 202°$
2–41. $F_R = 19.2$ N, $\theta = 2.37°$ $\measuredangle\theta$
2–42. $F_R = 839$ N, $\theta = 14.8°$
2–43. $F_R = 389$ N, 42.7°
2–45. $F_R = 217$ N, $\theta = 87.0°$
2–46. $F_{1x} = -200$ lb, $F_{1y} = 0$, $F_{2x} = 320$ lb, $F_{2y} = -240$ lb, $F_{3x} = 180$ lb, $F_{3y} = 240$ lb, $F_{4x} = -300$ lb, $F_{4y} = 0$
2–47. $F_R = 60.3$ kN, $\theta = 15.0°$
2–49. $F_R = 389$ N, $\phi' = 42.7°$
2–50. $\mathbf{F}_1 = \{90\mathbf{i} - 120\mathbf{j}\}$ lb, $\mathbf{F}_2 = \{-275\mathbf{j}\}$ lb, $\mathbf{F}_3 = \{-37.5\mathbf{i} - 65.0\mathbf{j}\}$ lb, $F_R = 463$ lb
2–51. $F = 5.96$ kN, $F_R = 2.33$ kN
2–53. $\theta = 117°$, $F_3 = 1.12F_1$
2–54. $\theta = 103°$, $F_2 = 88.1$ lb
2–55. $F_R = 161$ lb, $\theta = 38.3°$
2–57. $F_1 = 87.7$ N, $\alpha_1 = 46.9°$, $\beta_1 = 125°$, $\gamma_1 = 62.9°$, $F_2 = 98.6$ N, $\alpha_2 = 114°$, $\beta_2 = 150°$, $\gamma_2 = 72.3°$
2–58. $\mathbf{F}_1 = \{2.50\mathbf{i} + 3.54\mathbf{j} + 2.50\mathbf{k}\}$ kN, $\mathbf{F}_2 = \{-2\mathbf{j}\}$ kN
2–59. $F = 50$ N, $\alpha = 74.1°$, $\beta = 41.3°$, $\gamma = 53.1°$
2–61. $\beta = 90°$, $\mathbf{F} = \{-30\mathbf{i} - 52.0\mathbf{k}\}$ N
2–62. $F_R = 114$ lb, $\alpha = 62.1°$, $\beta = 113°$, $\gamma = 142°$
2–63. $\mathbf{F}_1 = \{53.1\mathbf{i} - 44.5\mathbf{j} + 40\mathbf{k}\}$ lb, $\alpha_1 = 48.4°$, $\beta_1 = 124°$, $\gamma_1 = 60°$, $\mathbf{F}_2 = \{-130\mathbf{k}\}$ lb, $\alpha_2 = 90°$, $\beta_2 = 90°$, $\gamma_2 = 180°$
2–65. $\alpha_1 = 90°$, $\beta_1 = 53.1°$, $\gamma_1 = 66.4°$
2–66. $F_R = 615$ N, $\alpha = 26.6°$, $\beta = 85.1°$, $\gamma = 64.0°$
2–67. $\mathbf{F}_1 = \{176\mathbf{j} - 605\mathbf{k}\}$ lb, $\mathbf{F}_2 = \{125\mathbf{i} - 177\mathbf{j} + 125\mathbf{k}\}$ lb, $F_R = 496$ lb, $\alpha = 75.4°$, $\beta = 90.0°$, $\gamma = 165°$
2–69. $F_3 = 428$ lb, $\alpha = 88.3°$, $\beta = 20.6°$, $\gamma = 69.5°$

2–70. $F_3 = 250$ lb, $\alpha = 87.0°$, $\beta = 143°$, $\gamma = 53.1°$
2–71. $F_3 = 9.58$ kN, $\alpha_3 = 15.5°$, $\beta_3 = 98.4°$, $\gamma_3 = 77.0°$
2–73. $F = 2.02$ kN, $F_y = 0.523$ kN
2–74. $F = 102$ N
2–75. $\alpha = 97.5°$, $\beta = 63.7°$, $\gamma = 27.5°$
2–77. $F_x = 40$ N, $F_y = 40$ N, $F_z = 56.6$ N
2–78. $F_2 = 32.4$ lb, $\alpha_2 = 122°$, $\beta_2 = 74.5°$, $\gamma_2 = 144°$
2–79. $\alpha = 69.6°$, $\beta = 116°$, $\gamma = 34.4°$
2–81. $\alpha = 73.4°$, $\beta = 64.6°$, $\gamma = 31.0°$
2–82. $\mathbf{r} = \{-3\mathbf{i} - 12\mathbf{j} + 4\mathbf{k}\}$ m, $r = 13$ m, $\alpha = 103°$, $\beta = 157°$, $\gamma = 72.1°$
2–83. $\mathbf{r} = \{-2.35\mathbf{i} + 3.93\mathbf{j} + 3.71\mathbf{k}\}$ ft, $r = 5.89$ ft, $\alpha = 113°$, $\beta = 48.2°$, $\gamma = 51.0°$
2–85. $r_{AB} = 2.11$ m
2–86. $r_{AB} = 567$ m
2–87. $r_{AB} = 7.81$ ft, $r_{OC} = 3.91$ ft
2–89. $r_{AD} = 1.50$ m, $r_{BD} = 1.50$ m, $r_{CD} = 1.73$ m
2–90. $F_R = 52.2$ lb, $\alpha = 87.8°$, $\beta = 63.7°$, $\gamma = 154°$
2–91. $\mathbf{F} = \{452\mathbf{i} + 370\mathbf{j} - 136\mathbf{k}\}$ lb, $\alpha = 41.1°$, $\beta = 51.9°$, $\gamma = 103°$
2–93. $\mathbf{F}_{BA} = \{-109\mathbf{i} + 131\mathbf{j} + 306\mathbf{k}\}$ lb, $\mathbf{F}_{CA} = \{103\mathbf{i} + 103\mathbf{j} + 479\mathbf{k}\}$ lb, $\mathbf{F}_{DA} = \{-52.1\mathbf{i} - 156\mathbf{j} + 365\mathbf{k}\}$ lb
2–94. $\mathbf{F}_1 = \{389\mathbf{i} - 64.9\mathbf{j} + 64.9\mathbf{k}\}$ lb, $\mathbf{F}_2 = \{-584\mathbf{i} + 97.3\mathbf{j} - 97.3\mathbf{k}\}$ lb
2–95. $r_{AB} = 592$ mm, $\mathbf{F} = \{-13.2\mathbf{i} - 17.7\mathbf{j} + 20.3\mathbf{k}\}$ N
2–97. $\mathbf{F}_1 = \{-26.2\mathbf{i} - 41.9\mathbf{j} + 62.9\mathbf{k}\}$ lb, $\mathbf{F}_2 = \{13.4\mathbf{i} - 26.7\mathbf{j} - 40.1\mathbf{k}\}$ lb, $F_R = 73.5$ lb, $\alpha = 100°$, $\beta = 159°$, $\gamma = 71.9°$
2–98. $\mathbf{F} = \{98.1\mathbf{i} + 269\mathbf{j} - 201\mathbf{k}\}$ lb
2–99. $x = 1.19$ m, $y = 1.79$ m, $z = 2.09$ m
2–101. $x = 7.65$ ft, $y = 4.24$ ft, $z = 3.76$ ft
2–102. $x = 8.67$ ft, $y = 1.89$ ft
2–103. $\mathbf{F}_{EA} = \{12\mathbf{i} - 8\mathbf{j} - 24\mathbf{k}\}$ kN, $\mathbf{F}_{EB} = \{12\mathbf{i} + 8\mathbf{j} - 24\mathbf{k}\}$ kN, $\mathbf{F}_{EC} = \{-12\mathbf{i} + 8\mathbf{j} - 24\mathbf{k}\}$ kN, $\mathbf{F}_{ED} = \{-12\mathbf{i} - 8\mathbf{j} - 24\mathbf{k}\}$ kN, $\mathbf{F}_R = \{-96\mathbf{k}\}$ kN
2–105. $\mathbf{F}_A = \{28.8\mathbf{i} - 16.6\mathbf{j} - 49.9\mathbf{k}\}$ lb, $\mathbf{F}_B = \{-28.8\mathbf{i} - 16.6\mathbf{j} - 49.9\mathbf{k}\}$ lb, $\mathbf{F}_C = \{33.3\mathbf{j} - 49.9\mathbf{k}\}$ lb, $F_R = 150$ lb, $\alpha = 90°$, $\beta = 90°$, $\gamma = 180°$
2–106. $F = 52.1$ N
2–109. $\theta = 109°$
2–110. Proj. $r_1 = |2.99 \text{ m}|$, Proj. $r_2 = |1.99 \text{ m}|$
2–111. $\theta = 74.0°$, $\phi = 33.9°$
2–113. $\theta = 82.9°$
2–114. Proj. $\mathbf{F}_{AB} = \{0.229\mathbf{i} - 0.916\mathbf{j} + 1.15\mathbf{k}\}$ lb
2–115. $F_{\parallel} = 99.1$ N, $F_{\perp} = 592$ N
2–117. Proj. $r_{OA} = 2.11$ ft
2–118. $F_1 = 18.3$ lb, $F_2 = 35.6$ lb
2–119. Proj. $F = 31.1$ N
2–121. Proj. $F_1 = 18.5$ N, Proj. $F_2 = 21.3$ N
2–122. $\theta = 128°$
2–123. 5.44 lb
2–125. $\theta = 59.2°$
2–126. Proj. $F_1 = 30.8$ lb, Proj. $F_2 = 15.4$ lb
2–127. $\theta = 82.0°$
2–129. Proj. $F = 10.5$ lb, $\theta = 143°$
2–130. $\theta = 74.4°$, $\phi = 55.4°$
2–131. $F_3 = 428$ lb, $\alpha = 88.3°$, $\beta = 20.6°$, $\gamma = 69.5°$
2–133. $F_{\perp} = 460$ lb, $F_{\parallel} = 386$ lb
2–134. $F = \{-34.3\mathbf{i} + 22.9\mathbf{j} - 68.6\mathbf{k}\}$ lb
2–135. $F_R = 178$ N, $\theta = 85.2°$
2–137. Proj. $F_{AB} = 70.5$ N, Proj. $F_{AC} = 65.1$ N
2–138. $F_{CB} = 10.5$ lb
2–139. 70.5°

Chapter 3

3–1. $F_1 = 259$ N, $F_2 = 366$ N
3–2. $\theta = 31.8°$, $F = 4.94$ kN
3–3. $\theta = 53.0°$, $F = 28.3$ N
3–5. $T = 13.3$ kN, $F = 10.2$ kN
3–6. $T = 14.3$ kN, $\theta = 36.3°$
3–7. $W = 240$ lb
3–9. $\theta = 34.2°$
3–10. $\theta = 11.5°$
3–11. $F = 1.13$ mN
3–13. $x_{AC} = 0.793$ m, $x_{AB} = 0.467$ m
3–14. $m = 12.8$ kg
3–15. $F = 158$ N
3–17. At B: $T_{BC} = 39.2$ N, $T_{BA} = 68.0$ N; At C: $T_{CD} = 39.2$ N, $F = 39.2$ N
3–18. $\theta = 78.7°$, $F_{CD} = 127$ lb
3–19. $\theta = 78.7°$, $W = 51.0$ lb
3–21. $\theta = 45°$, $W = 200$ lb
3–22. $\theta = 60°$, $T_{AB} = 34.6$ lb
3–23. $\theta = 60°$, $W = 46.2$ lb
3–25. $s = 5.33$ ft
3–26. $W = 6$ lb
3–27. $l'_{AB} = 0.452$ m, $l'_{AC} = 0.658$ m

3–29. $m = 15.6$ kg
3–31. $\theta = 35.0°$
3–33. $T_{AB} = 157$ N, $T_{AC} = 118$ N
3–34. $l = 2.65$ ft
3–35. $m = 2.37$ kg
3–37. $\theta = 43.0°$
3–38. C and D, $T = 106$ lb
3–39. $m_B = 3.58$ kg, $N = 19.7$ N
3–41. $R = 0$, $P = 3.11$ kN, $F = 2.38$ kN
3–42. $F_1 = 800$ N, $F_2 = 147$ N, $F_3 = 564$ N
3–43. $F_1 = 5.60$ kN, $F_2 = 8.55$ kN, $F_3 = 9.44$ kN
3–45. $F_{AD} = 1.20$ kN, $F_{AC} = 0.40$ kN, $F_{AB} = 0.80$ kN
3–46. $F_{CD} = 625$ lb, $F_{CA} = F_{CB} = 198$ lb
3–47. $s_{OB} = 327$ mm, $s_{OA} = 218$ mm
3–49. $F_{AB} = 0.980$ kN, $F_{AC} = 0.463$ kN,
$F_{AD} = 1.55$ kN
3–50. $F_{AO} = 319$ N, $F_{AB} = 110$ N, $F_{AC} = 85.8$ N
3–51. $W = 138$ N
3–53. $F_{DA} = 21.5$ lb, $F_{DB} = 14.0$ lb, $F_{DC} = 17.6$ lb
3–54. $F_{AC} = 92.9$ N, $F_{AD} = 364$ N, $F = 757$ N
3–55. $F_B = 858$ N, $F_C = 0$, $F_D = 858$ N
3–57. At A: $F_{AC} = 574$ lb, $F_{AB} = 500$ lb
At C: $F_{CD} = 1.22$ kip, $F_{CE} = F_{CF} = 416$ lb
3–58. $F_{AB} = 35.9$ lb, $F_{AC} = F_{AD} = 25.4$ lb
3–59. $W = 267$ lb
3–61. $F_{AB} = 469$ lb, $F_{AC} = F_{AD} = 331$ lb
3–62. $s = 410$ mm
3–63. $F_{AD} = 1.42$ kip, $F_{AC} = 0.914$ kip, $F_{AB} = 1.47$ kip
3–65. $W = 55.8$ N
3–66. $F_A = 34.6$ lb, $F_B = 57.3$ lb
3–67. $F = 40.8$ lb
3–69. $\theta = 40.2°$
3–70. $T = 257$ N
3–71. $\theta = 90°$, $(F_{AC})_{max} = 160$ lb, $\theta = 120°$,
$F_{AB} = 160$ lb
3–73. $W = 954$ lb
3–74. $F_{OA} = 202$ lb, $F_{OC} = 216$ lb, $F_{OD} = 58.6$ lb
3–75. $F_1 = 0$, $F_2 = 311$ lb, $F_3 = 238$ lb

Chapter 4

4–5. $M_{RP} = 1.17$ kip · in. ↻
4–6. $F = 77.6$ N
4–7. $\theta = 28.6°$
4–9. $M_P = 3.15$ kN · m ↺
4–10. $\mathbf{M}_O = \{-7.11\mathbf{k}\}$ N · m
4–11. $M_O = 2.42$ kip · ft ↻
4–13. $(M_B)_1 = 3.07$ N · m ↻, $(M_B)_2 = 4.40$ N · m ↻,
$(M_B)_3 = 1.08$ N · m ↻
4–14. $M_B = 90.6$ lb · ft ↺, $M_C = 141$ lb · ft ↺
4–15. $M_A = 195$ lb · ft ↺
4–17. $M_A = 1.65$ kip · ft ↺, $M_B = 9.27$ kip · ft ↺,
$M_C = 6.45$ kip · ft ↺
4–18. **(a)** $(M_A)_{max} = 330$ lb · ft, $\theta = 76.0°$
(b) $(M_A)_{min} = 0$, $\theta = 166°$
4–19. $M_A = 7.71$ N · m ↺
4–21. **(a)** $M_A = 13.0$ N · m, **(b)** $F = 35.2$ N
4–22. **(a)** $(M_A)_{max} = 330$ lb · ft, $\theta = 104°$
(b) For $M_A = 0$, $\theta = 14.04°$
4–23. $(M_A)_1 = 11.7$ kip · ft ↺, $(M_A)_2 = 11.7$ kip · ft ↻
4–25. Maximum moment: $\theta = 37.9°$,
$(M_A)_{max} = 79.8$ N · m ↻
Minimum moment: $\theta = 128°$, $(M_A)_{min} = 0$
4–26. $(M_O)_{max} = 80$ kN · m ↻, $x = 24.0$ m
4–27. Maximum moment: ↺$+(M_O)_{max} = 80.0$ kN · m,
$\theta = 33.6°$
4–29. $F_3 = 1.59$ kN
4–30. $(M_A)_1 = 542$ lb · in. ↻, $(M_A)_2 = 10.0$ lb · in. ↺
4–31. $M_A = 306$ lb · ft ↺, $T = 652$ lb
4–33. $M_A = 199$ N · m
4–34. $\mathbf{M}_O = \{260\mathbf{i} + 180\mathbf{j} + 510\mathbf{k}\}$ N · m
4–35. $\mathbf{M}_P = \{440\mathbf{i} + 220\mathbf{j} - 990\mathbf{k}\}$ N · m
4–37. $\mathbf{M}_P = \{-116\mathbf{i} + 16\mathbf{j} - 135\mathbf{k}\}$ kN · m
4–38. $\mathbf{M}_O = \{-128\mathbf{i} + 128\mathbf{j} - 257\mathbf{k}\}$ N · m
4–39. $\mathbf{M}_B = \{-37.6\mathbf{i} + 90.7\mathbf{j} - 155\mathbf{k}\}$ N · m
4–41. $\mathbf{M}_A = \{-229\mathbf{i} + 132\mathbf{j}\}$ lb · ft
4–42. $\mathbf{M}_O = \{61.2\mathbf{i} + 81.6\mathbf{j}\}$ N · m
4–43. $F_{AB} = 18.6$ lb
4–45. $\mathbf{M}_B = \{10.6\mathbf{i} + 13.1\mathbf{j} + 29.2\mathbf{k}\}$ N · m
4–46. $\mathbf{M}_O = \{-1.77\mathbf{i} - 1.10\mathbf{j}\}$ kN · m
4–47. $\mathbf{M}_A = \{-1.90\mathbf{i} + 6.00\mathbf{j}\}$ kN · m
4–49. $y = 1$ m, $z = 3$ m, $d = 1.15$ m
4–50. $y = 2$ m, $z = 1$ m
4–51. $(\mathbf{M}_{Oa})_F = \{218\mathbf{j} + 163\mathbf{k}\}$ N · m
4–53. $(\mathbf{M}_R)_{Oa} = \{26.1\mathbf{i} - 15.1\mathbf{j}\}$ lb · ft
4–54. $M_z = 62$ lb · in.
4–55. $M_y = 75$ lb · in.
4–57. $M_y = 0.277$ N · m
4–58. $\mathbf{M}_y = \{-78.4\mathbf{j}\}$ lb · ft
4–59. $M_x = 12$ N · m $<$ 14 N · m, No
$(M_x)_{max} = 15$ N · m $>$ 14 N · m, Yes
4–61. $M_y = 0$
4–62. $\mathbf{M}_x = \{0.211\mathbf{i}\}$ N · m

4–63. $|M_{CA}| = 226\ \text{N}\cdot\text{m}$
4–65. $P = 8.50\ \text{lb}$
4–66. $M_y = 282\ \text{lb}\cdot\text{ft}$
4–67. $(M_a)_1 = 30\ \text{lb}\cdot\text{in.}, (M_a)_2 = 8\ \text{lb}\cdot\text{in.}$
4–69. $M_C = 18.3\ \text{kN}\cdot\text{m} \circlearrowleft$
4–70. $M_C = 650\ \text{lb}\cdot\text{ft} \circlearrowleft$
4–71. $M_C = 3.12\ \text{kip}\cdot\text{ft} \circlearrowleft$
4–73. $R = 28.9\ \text{N}$
4–74. $T = 0.909\ \text{kip}$
4–75. $\theta = 32.9^\circ, M_3 = 651\ \text{N}\cdot\text{m}$
4–77. $F = 830\ \text{N}$
4–78. $F = 139\ \text{lb}$. The resultant couple moment is a free vector. It can act at any point on the beam.
4–79. $\mathbf{M}_C = \{-5\mathbf{i} + 8.75\mathbf{j}\}\ \text{N}\cdot\text{m}$
4–81. $M_R = 9.69\ \text{kN}\cdot\text{m} \circlearrowright$
4–82. $F = 167\ \text{lb}$. Resultant couple can act anywhere.
4–83. $d = 2.03\ \text{ft}$
4–85. $M_C = 126\ \text{lb}\cdot\text{ft} \circlearrowleft$
4–86. $\mathbf{M}_C = \{40\mathbf{i} + 20\mathbf{j} - 24\mathbf{k}\}\ \text{N}\cdot\text{m}$
4–87. $\mathbf{M}_C = \{-253\mathbf{i} + 67.4\mathbf{j} - 253\mathbf{k}\}\ \text{N}\cdot\text{m}$
4–89. $F = 75\ \text{N}, P = 100\ \text{N}$
4–90. $\mathbf{M}_C = \{7.01\mathbf{i} + 42.1\mathbf{j}\}\ \text{N}\cdot\text{m}$
4–91. $F = 35.1\ \text{N}$
4–93. $\mathbf{M}_C = \{-17.0\mathbf{i} - 9.20\mathbf{j} - 27.4\mathbf{k}\}\ \text{N}\cdot\text{m}$, $M_C = 33.5\ \text{N}\cdot\text{m}$
4–94. $\mathbf{M}_C = \{2.60\mathbf{i} + 2.40\mathbf{j} + 3.60\mathbf{k}\}\ \text{N}\cdot\text{m}$
4–95. $18.3\ \text{N}\cdot\text{m}, \alpha = 155^\circ, \beta = 115^\circ, \gamma = 90^\circ$
4–97. 832 N
4–98. $F = 375\ \text{N}, M_O = 100\ \text{N}\cdot\text{m} \circlearrowright$
4–99. $F_P = 375\ \text{N}, M_P = 737\ \text{N}\cdot\text{m} \circlearrowleft$
4–101. $F_R = 115\ \text{lb}, \theta = 73.2^\circ$ ⦪, $M_{RP} = 921\ \text{lb}\cdot\text{ft} \circlearrowleft$
4–102. $F_R = 274\ \text{lb}, \theta = 5.24^\circ$ θ⦫, $M_O = 4.61\ \text{kip}\cdot\text{ft} \circlearrowleft$
4–103. $F_R = 274\ \text{lb}, \theta = 5.24^\circ$ θ⦫, $M_P = 5.48\ \text{kip}\cdot\text{ft} \circlearrowleft$
4–105. $F_R = 78.1\ \text{lb}, \theta = 87.1^\circ$ ⦩, $M_{R_P} = 124\ \text{lb}\cdot\text{ft} \circlearrowleft$
4–106. $F_R = 2.10\ \text{kN}, \theta = 81.6^\circ$ ∠θ, $M_O = 10.6\ \text{kN}\cdot\text{m} \circlearrowright$
4–107. $F_R = 2.10\ \text{kN}, \theta = 81.6^\circ$ ∠θ, $M_P = 16.8\ \text{kN}\cdot\text{m} \circlearrowright$
4–109. $F_R = 375\ \text{lb} \uparrow, x = 2.47\ \text{ft}$
4–110. $\mathbf{F}_R = \{-70\mathbf{i} + 140\mathbf{j} - 408\mathbf{k}\}\ \text{N}$, $\mathbf{M}_{RO} = \{-26\mathbf{i} + 31.0\mathbf{j} + 14.6\mathbf{k}\}\ \text{N}\cdot\text{m}$
4–111. $\mathbf{M}_{RP} = \{-26\mathbf{i} + 357\mathbf{j} + 127\mathbf{k}\}\ \text{N}\cdot\text{m}$
4–113. $F = 798\ \text{lb}, \theta = 67.9^\circ$ θ⦪, $x = 6.57\ \text{ft}$
4–114. $F = 922\ \text{lb}, \theta = 77.5^\circ$ θ⦪, $x = 3.56\ \text{ft}$
4–115. $F = 1.30\ \text{kN}, \theta = 84.5^\circ$ θ⦪, $x = 7.36\ \text{m}$
4–117. $F_R = 294\ \text{N}, \theta = 40.1^\circ$ ⦪, $M_{RO} = 39.6\ \text{N}\cdot\text{m} \circlearrowright$
4–118. $F_R = 4.43\ \text{kN}, \theta = 71.6^\circ$ ⦩, $d = 3.52\ \text{m}$
4–119. $\theta = 57.5^\circ, F = 2.61\ \text{kN}, d = 2.64\ \text{m}$
4–121. $F_R = 991\ \text{N}, \theta = 63.0^\circ$ θ⦪, $x = 2.64\ \text{m}$
4–122. $F_R = 65.9\ \text{lb}, \theta = 49.8^\circ$ ⦩, $d = 2.10\ \text{ft}$
4–123. $F_R = 65.9\ \text{lb}, \theta = 49.8^\circ$ ⦩, $d = 4.62\ \text{ft}$
4–125. $\mathbf{F}_R = \{8\mathbf{i} + 6\mathbf{j} + 8\mathbf{k}\}\ \text{kN}$, $\mathbf{M}_{RO} = \{-10\mathbf{i} + 18\mathbf{j} - 56\mathbf{k}\}\ \text{kN}\cdot\text{m}$
4–126. $\mathbf{F}_R = \{8\mathbf{i} + 6\mathbf{j} + 8\mathbf{k}\}\ \text{kN}$, $\mathbf{M}_{RP} = \{-46\mathbf{i} + 66\mathbf{j} - 56\mathbf{k}\}\ \text{kN}\cdot\text{m}$
4–127. $\mathbf{F}_R = \{8\mathbf{i} + 6\mathbf{j} + 8\mathbf{k}\}\ \text{kN}$, $\mathbf{M}_{RQ} = \{-10\mathbf{i} - 30\mathbf{j} - 20\mathbf{k}\}\ \text{kN}\cdot\text{m}$
4–129. $\mathbf{F}_R = \{-28.3\mathbf{j} - 68.3\mathbf{k}\}\ \text{N}$, $\mathbf{M}_{RA} = \{-20.5\mathbf{j} + 8.49\mathbf{k}\}\ \text{N}\cdot\text{m}$
4–130. $\mathbf{F}_R = \{-210\mathbf{k}\}\ \text{N}, \mathbf{M}_{RO} = \{-15\mathbf{i} + 225\mathbf{j}\}\ \text{N}\cdot\text{m}$
4–131. $\mathbf{F}_R = \{6\mathbf{i} - 1\mathbf{j} - 14\mathbf{k}\}\ \text{N}$, $\mathbf{M}_{RO} = \{1.30\mathbf{i} + 3.30\mathbf{j} - 0.450\mathbf{k}\}\ \text{N}\cdot\text{m}$
4–133. $F_R = 140\ \text{kN} \downarrow, y = 7.14\ \text{m}, x = 5.71\ \text{m}$
4–134. $F_R = 140\ \text{kN}, x = 6.43\ \text{m}, y = 7.29\ \text{m}$
4–135. $F = 53.3\ \text{N}, \mathbf{F}_{RW} = \{-40\mathbf{i}\}\ \text{N}$, $\mathbf{M}_{RW} = \{-30\mathbf{i}\}\ \text{N}\cdot\text{m}$
4–137. $F_R = 990\ \text{N}, M_R = 3.07\ \text{kN}\cdot\text{m}, x = 1.16\ \text{m}$, $y = 2.06\ \text{m}$
4–138. $F_R = 108\ \text{lb}, M_R = -624\ \text{lb}\cdot\text{ft}$, $z = 8.69\ \text{ft}, y = 0.414\ \text{ft}$
4–139. $F_{RO} = 13.2\ \text{lb} \downarrow, x = 0.340\ \text{ft}$
4–141. $F_R = 51.0\ \text{kN} \downarrow, M_{R_O} = 914\ \text{kN}\cdot\text{m} \circlearrowright$
4–142. $F_R = 51.0\ \text{kN} \downarrow, d = 17.9\ \text{m}$
4–143. $F_R = 3.25\ \text{kip}, \theta = 67.2^\circ$ θ⦪, $x = 3.86\ \text{ft}$
4–145. $F_R = 18.0\ \text{kip} \downarrow, x = 11.7\ \text{ft}$
4–146. $F_R = 11.25\ \text{kN} \downarrow, d = 4.05\ \text{m}$
4–147. $b = 5.625\ \text{m}, a = 1.54\ \text{m}$
4–149. $F_R = 3.90\ \text{kip} \uparrow, d = 11.3\ \text{ft}$
4–150. $b = 4.50\ \text{ft}, a = 9.75\ \text{ft}$
4–151. $F_R = 10.6\ \text{kip} \downarrow, x = 0.479\ \text{ft}$
4–153. $F_R = 1.35\ \text{kN}, \theta = 42.0^\circ$ θ⦪, $x = 0.556\ \text{m}$
4–154. $F_R = 95.6\ \text{kN} \rightarrow, M_{R_O} = 349\ \text{kN}\cdot\text{m} \circlearrowright$
4–155. $F_R = 3\ \text{kN} \downarrow, M_{RO} = 2.25\ \text{kN}\cdot\text{m} \circlearrowleft$
4–157. $F_R = \dfrac{2w_0L}{\pi} \downarrow, \bar{x} = \dfrac{L}{2}$
4–158. $F_R = 774\ \text{lb} \uparrow, \bar{x} = 2.25\ \text{ft}$
4–159. $F_R = 7\ \text{lb}, \bar{x} = 0.268\ \text{ft}$
4–161. $\alpha = 70.8^\circ, \beta = 39.8^\circ, \gamma = 56.7^\circ$, $\alpha = 109^\circ, \beta = 140^\circ, \gamma = 123^\circ$

4–162. $\mathbf{M}_O = \{298\mathbf{i} + 15.1\mathbf{j} - 200\mathbf{k}\}$ lb · in

4–163. $\mathbf{F}_R = \{-300\mathbf{i} + 200\mathbf{j} - 500\mathbf{k}\}$ N, $\mathbf{M}_{RO} = \{-3.80\mathbf{i} - 7.20\mathbf{j} - 0.600\mathbf{k}\}$ kN · m

4–165. $M_{a-a} = 59.7$ N · m

4–166. $\mathbf{M}_z = \{-6.21\mathbf{k}\}$ N · m

4–167. $\mathbf{F}_R = \{14.3\mathbf{i} + 21.4\mathbf{j} - 42.9\mathbf{k}\}$ lb, $\mathbf{M}_A = \{-1.93\mathbf{i} + 0.429\mathbf{j} - 0.429\mathbf{k}\}$ kip · ft

4–169. $\mathbf{M}_O = \{1.06\mathbf{i} + 1.06\mathbf{j} - 4.03\mathbf{k}\}$ N · m, $\alpha = 75.7°$, $\beta = 75.7°$, $\gamma = 160°$

4–170. $F = 3$ lb, $P = 4$ lb

Chapter 5

5–1. N_A, N_B force of plane on sphere, 10 lb force of gravity on sphere

5–3. A_x, A_y, M_A effect of wall on beam, N_B force of roller on beam, 60 lb resultant force of distributed load on beam

5–5. N_A, N_B, N_C force of rollers on beam

5–6. A_x, A_y, N_B force of glass on rod, 0.02(9.81) N force of gravity on rod

5–7. A_x, A_y, N_B force of cylinder on wrench

5–11. $N_A = 19.3$ lb, $N_B = 14.1$ lb

5–13. $N_A = 267$ lb, $N_B = 208$ lb, $N_C = 305$ lb

5–14. $N_B = 0.103$ N, $A_x = 0.0664$ N, $A_y = 0.117$ N

5–15. $N_B = 140$ lb, $A_x = 140$ lb, $A_y = 20$ lb

5–17. $N_C = 493$ N, $N_B = 554$ N, $N_A = 247$ N

5–18. $B_y = 642$ N, $A_x = 192$ N, $A_y = 180$ N

5–19. $B_y = 586$ N, $F_A = 413$ N

5–21. $F_A = 30$ lb, $F_B = 36.2$ lb, $F_C = 9.38$ lb

5–22. $A_x = 33.3$ lb, $B_x = 33.3$ lb, $B_y = 100$ lb

5–23. $F_{CD} = 195$ lb, $A_x = 97.4$ lb, $A_y = 31.2$ lb

5–25. $N_C = 785$ N, $F_B = 4.71$ kN, $F_A = 3.92$ kN

5–26. $R_A = 105$ lb, $B_x = 97.4$ lb, $B_y = 269$ lb

5–27. $W_B = 78.6$ lb

5–29. $F_B = 6.38$ N, $A_x = 3.19$ N, $A_y = 2.48$ N

5–30. $N_A = 39.7$ lb, $N_B = 82.5$ lb, $M_A = 106$ lb · ft

5–31. $A_x = 1462$ lb, $F_B = 1.66$ kip

5–33. $D_x = 0$, $D_y = 1.65$ kip, $M_D = 1.40$ kip · ft, $(M_D)_{\max} = 3.00$ kip · ft

5–34. The table will tip over.

5–35. $F_B = 105$ N

5–37. $l_0 = 70.3$ mm

5–38. $h = 15.8$ ft

5–39. $N_A = 81.6$ lb, $F_B = 50.2$ lb

5–41. $B_x = 989$ N, $A_x = 989$ N, $B_y = 186$ N

5–42. $w_1 = 413$ kN/m, $w_2 = 407$ kN/m

5–43. $T = 5$ kN, $T_{BC} = 16.4$ kN, $F_A = 20.6$ kN

5–45. $R_A = 40.9$ kip, $R_B = 125$ kip

5–46. $\theta = 41.4°$

5–47. $F_A = 432$ lb, $F_B = 0$, $F_C = 432$ lb

5–49. $F_2 = 724$ lb, $F_1 = 1.45$ kip, $F_A = 1.75$ kip

5–50. $\theta = \tan^{-1}\left(\frac{1}{2}\cot\psi - \frac{1}{2}\cot\phi\right)$

5–51. $N_B = 2.11$ N, $F_A = 2.81$ N

5–53. $\theta = 23.2°, 85.2°$

5–54. $R_A = 26.0$ lb, $R_B = 11.9$ lb, $R_C = 63.9$ lb

5–55. $F_C = 10$ mN

5–57. $d = \dfrac{a}{\cos^3\theta}$

5–58. $\theta = \tan^{-1}\dfrac{b}{a}$

5–59. $P = 441$ N

5–61. $\theta = \cos^{-1}\left(\dfrac{L + \sqrt{L^2 + 128r^2}}{16r}\right)$

5–62. $\theta = 27.1°$ or $\theta = 50.2°$

5–63. $A_x = 0$, $A_y = -200$ N, $A_z = 150$ N, $(M_A)_x = 100$ N · m, $(M_A)_y = 0$, $(M_A)_z = 500$ N · m

5–65. $T_B = 2.75$ kip, $T_C = 1.375$ kip, $T_A = 1.375$ kip

5–66. $W = 750$ lb, $x = 5.20$ ft, $y = 5.27$ ft

5–67. $F_A = 663$ lb, $F_C = 569$ lb, $F_B = 449$ lb

5–69. $F = 900$ lb, $A_x = 0$, $A_y = 0$, $A_z = 600$ lb, $M_{Ax} = 0$, $M_{Az} = 0$

5–70. $A_x = 0$, $A_y = 1.50$ kip, $A_z = 750$ lb, $T = 919$ lb

5–71. $F = 1.31$ kip, $A_x = 0$, $A_y = 1.31$ kip, $A_z = 653$ lb

5–73. $P = 75$ lb, $A_y = 0$, $B_z = 75$ lb, $A_z = 75$ lb, $B_x = 112$ lb, $A_x = 37.5$ lb

5–74. $F_C = 3.46$ N, $N_A = N_B = 14.1$ N

5–75. $F_{BC} = 0$, $A_y = 0$, $A_z = 800$ lb, $(M_A)_x = 4.80$ kip · ft, $(M_A)_y = 0$, $(M_A)_z = 0$

5–77. $F = 96.0$ lb, $B_z = 45.7$ lb, $B_y = -67.9$ lb, $B_x = 48.5$ lb, $A_x = -96.5$ lb, $A_z = -13.7$ lb

5–78. $F = 96.0$ lb, $B_x = -48.0$ lb, $B_y = -67.9$ lb, $B_z = 32.0$ lb, $M_{Bx} = 96.0$ lb · ft, $M_{Bz} = -675$ lb · ft

5–79. $A_x = 633$ lb, $A_y = -141$ lb, $B_x = -721$ lb, $B_z = 895$ lb, $C_y = 200$ lb, $C_z = -506$ lb

5–81. $T_{BD} = T_{CD} = 117$ N, $A_x = 66.7$ N, $A_y = 0$, $A_z = 100$ N

5–82. $T_{DE} = 32.1$ lb, $T_{BC} = 42.9$ lb, $A_x = 3.57$ lb, $A_y = 50$ lb, $(M_A)_x = 0$, $(M_A)_y = -17.9$ lb · ft

5–83. $T_{BC} = 1.05$ kip, $A_x = -450$ lb, $A_y = 900$ lb, $A_z = 0$, $M_{Ay} = -600$ lb · ft, $M_{Az} = -900$ lb · ft

5–85. $T_B = 25$ lb, $A_x = 25$ lb, $A_y = -25$ lb, $A_z = 50$ lb, $B_y = 25$ lb

5–86. $A_x = -20$ lb, $A_y = 40$ lb, $B_z = 30$ lb, $M_{Ax} = 110$ lb·ft, $M_{Ay} = 40$ lb·ft, $M_{Az} = 110$ lb·ft

5–87. $F_{AC} = F_{BC} = 6.13$ kN, $F_{DE} = 19.6$ kN

5–89. $T_{BC} = 131$ lb, $T_{BD} = 510$ lb, $A_x = 0$, $A_y = 0$, $A_z = 589$ lb

5–90. $B_z = 1167$ lb, $C_z = 734$ lb, $A_z = 1600$ lb

5–91. $F_{CD} = 1.02$ kN, $A_z = -208$ N, $B_z = -139$ N, $A_y = 573$ N, $B_y = 382$ N

5–93. $A_x = 0$, $B_y = 7$ kN, $A_y = 11$ kN

5–94. $N_A = 8.00$ kN, $B_x = 5.20$ kN, $B_y = 5.00$ kN

5–95. $N_B = 400$ N, $F_A = 721$ N

5–97. $\theta = 30.8°$

5–98. $A_x = 0$, $A_y = 0$, $A_z = B_z = C_z = 5.33$ lb

5–99. $P = 100$ lb, $B_z = 40$ lb, $B_x = -35.7$ lb, $A_x = 136$ lb, $B_y = 0$, $A_z = 40$ lb

Chapter 6

6–1. Joint A: $F_{AD} = 9.90$ kN (C), $F_{AB} = 7$ kN (T)
Joint D: $F_{DB} = 4.95$ kN (T), $F_{DC} = 14.8$ kN (C)
Joint C: $F_{CB} = 10.5$ kN (T)

6–2. Joint A: $F_{AD} = 11.3$ kN (C), $F_{AB} = 8$ kN (T)
Joint D: $F_{DB} = 7.07$ kN (T), $F_{DC} = 18.4$ kN (C)
Joint C: $F_{CB} = 13.0$ kN (T)

6–3. Joint A: $F_{AD} = 849$ lb (C), $F_{AB} = 600$ lb (T)
Joint B: $F_{BD} = 400$ lb (C), $F_{BC} = 600$ lb (T)
Joint D: $F_{DC} = 1.41$ kip (T), $F_{DE} = 1.60$ kip (C)

6–5. Joint A: $F_{AB} = 21.9$ kN (C), $F_{AG} = 13.1$ kN (T)
Joint B: $F_{BC} = 13.1$ kN (C), $F_{BG} = 17.5$ kN (T)
Joint G: $F_{CG} = 3.12$ kN (T), $F_{FG} = 11.2$ kN (T)
Joint C: $F_{CF} = 3.12$ kN (C), $F_{CD} = 9.38$ kN (C)
Joint D: $F_{DE} = 15.6$ kN (C), $F_{DF} = 12.5$ kN (T)
Joint F: $F_{EF} = 9.38$ kN (T)

6–6. Joint A: $F_{AB} = 43.8$ kN (C), $F_{AG} = 26.2$ kN (T)
Joint B: $F_{BC} = 26.2$ kN (C), $F_{BG} = 35.0$ kN (T)
Joint G: $F_{GC} = 6.25$ kN (T), $F_{GF} = 22.5$ kN (T)
Joint E: $F_{ED} = 31.2$ kN (C), $F_{EF} = 18.8$ kN (T)
Joint D: $F_{DC} = 18.8$ kN (C), $F_{DF} = 25.0$ kN (T)
Joint F: $F_{FC} = 6.25$ kN (C)

6–7. Joint B: $F_{BC} = 3$ kN (C), $F_{BA} = 8$ kN (C)
Joint A: $F_{AC} = 1.46$ kN (C), $F_{AF} = 4.17$ kN (T)
Joint C: $F_{CD} = 4.17$ kN (C), $F_{CF} = 3.12$ kN (C)
Joint E: $F_{EF} = 0$, $F_{ED} = 13.1$ kN (C)
Joint D: $F_{DF} = 5.21$ kN (T)

6–9. $P = 424$ lb

6–10. Joint A: $F_{AG} = 471$ lb (C), $F_{AB} = 333$ lb (T)
Joint B: $F_{BG} = 0$, $F_{BC} = 333$ lb (T)
Joint D: $F_{DE} = 943$ lb (C), $F_{DC} = 667$ lb (T)
Joint E: $F_{EC} = 667$ lb (T), $F_{EG} = 667$ lb (C)
Joint C: $F_{CG} = 471$ lb (T)

6–11. Joint A: $F_{AG} = 1179$ lb (C), $F_{AB} = 833$ lb (T)
Joint B: $F_{BC} = 833$ lb (T), $F_{BG} = 500$ lb (T)
Joint D: $F_{DE} = 1650$ lb (C), $F_{DC} = 1167$ lb (T)
Joint E: $F_{EC} = 1167$ lb (T), $F_{EG} = 1167$ lb (C)
Joint C: $F_{CG} = 471$ lb (T)

6–13. Joint G: $F_{GB} = 30$ kN (T)
Joint A: $F_{AF} = 20$ kN (C), $F_{AB} = 22.4$ kN (C)
Joint B: $F_{BF} = 20$ kN (T), $F_{BC} = 20$ kN (T)
Joint F: $F_{FC} = 28.3$ kN (C), $F_{FE} = 0$
Joint E: $F_{ED} = 0$, $F_{EC} = 20.0$ kN (T)
Joint C: $F_{DC} = 0$

6–14. Joint A: $F_{AB} = 330$ lb (C), $F_{AF} = 79.4$ lb (T)
Joint B: $F_{BF} = 233$ lb (T), $F_{BC} = 233$ lb (C)
Joint F: $F_{FC} = 47.1$ lb (C), $F_{FE} = 113$ lb (T)
Joint E: $F_{EC} = 300$ lb (T), $F_{ED} = 113$ lb (T)
Joint C: $F_{CD} = 377$ lb (C)

6–15. Joint A: $F_{AB} = 377$ lb (C), $F_{AF} = 190$ lb (T)
Joint B: $F_{BF} = 267$ lb (T), $F_{BC} = 267$ lb (C)
Joint F: $F_{FC} = 189$ lb (T), $F_{FE} = 56.7$ lb (T)
Joint E: $F_{ED} = 56.7$ lb (T), $F_{EC} = 0$
Joint C: $F_{CD} = 189$ lb (C)

6–17. $P = 4.16$ kN

6–18. Joint C: $F_{CB} = 400$ lb (C), $F_{CD} = 693$ lb (C)
Joint B: $F_{BD} = 667$ lb (T),
Member AB is a two-force member and exerts only a vertical force along AB at A.

6–19. Joint C: $F_{CD} = 3.61$ kN (C), $F_{CB} = 3$ kN (T)
Joint B: $F_{BA} = 3$ kN (T), $F_{BD} = 3$ kN (C)
Joint D: $F_{DA} = 2.70$ kN (T), $F_{DE} = 6.31$ kN (C)

6–21. Joint B: $F_{BA} = P \csc 2\theta$ (C), $F_{BC} = P \cot 2\theta$ (C)
Joint C: $F_{CA} = (\cot\theta \cos\theta - \sin\theta + 2\cos\theta)P$ (T), $F_{CD} = (\cot 2\theta + 1)P$ (C)
Joint D: $F_{DA} = (\cot 2\theta + 1)(\cos 2\theta)P$ (C)

6–22. $P_{max} = 732$ N

6–23. Joint A: $F_{AB} = 3.75$ kip (C), $F_{AH} = 3$ kip (T)
Joint B: $F_{BC} = 3.75$ kip (C), $F_{BH} = 1$ kip (C)
Joint H: $F_{HC} = 1$ kip (T), $F_{GH} = 2$ kip (T)
Joint E: $F_{EF} = 3$ kip (T), $F_{ED} = 3.75$ kip (C)
Joint D: $F_{DC} = 3.75$ kip (C), $F_{DF} = 1$ kip (C)
Joint C: $F_{CF} = 1$ kip (T), $F_{CG} = 2$ kip (C)
Joint F: $F_{FG} = 2$ kip (T)

6–25. Joint A: $F_{AB} = 7.5$ kN (T), $F_{AE} = 4.5$ kN (C)
Joint E: $F_{ED} = 4.5$ kN (C), $F_{EB} = 8$ kN (T)
Joint B: $F_{BD} = 19.8$ kN (C), $F_{BC} = 18.5$ kN (T)

6–26. Joint A: $F_{AB} = 196$ N (T), $F_{AE} = 118$ N (C)
Joint E: $F_{ED} = 118$ N (C), $F_{EB} = 216$ N (T)
Joint B: $F_{BD} = 1.04$ kN (C), $F_{BC} = 857$ N (T)

6–27. Joint B: $F_{BC} = 0$
Joint C: $F_{CA} = 0$, $F_{CD} = 0$
By symmetry, $F_{EF} = F_{ED} = F_{EG} = 0$
Joint D: $F_{DF} = 4$ kN (C)
$F_{DB} = F_{BA} = F_{FG} = 4$ kN (C)

6–29. Joint A: $F_{AI} = 4.04$ kip (C), $F_{AB} = 3.75$ kip (T)
Joint E: $F_{EF} = 12.1$ kip (C), $F_{ED} = 7.75$ kip (T)
Joint B: $F_{BI} = 0$, $F_{BC} = 3.75$ kip (T)
Joint D: $F_{DF} = 0$, $F_{DC} = 7.75$ kip (T)
Joint F: $F_{FC} = 4.04$ kip (C), $F_{FG} = 8.08$ kip (C)
Joint H: $F_{HG} = 2.15$ kip (C), $F_{HI} = 0.8$ kip (T)
Joint C: $F_{CI} = 0.269$ kip (T), $F_{CG} = 1.40$ kip (T)
Joint G: $F_{GI} = 5.92$ kip (C)

6–30. $F_{DE} = 45$ kN (C), $F_{EH} = 5$ kN (T),
$F_{HG} = 45$ kN (T)

6–31. $F_{LD} = 0$, $F_{LK} = 112.5$ kN (C),
$F_{CD} = 112.5$ kN (T), $F_{KD} = 50$ kN (C)

6–33. $F_{KJ} = 13.3$ kN (T), $F_{BC} = 14.9$ kN (C),
$F_{CK} = 0$

6–34. $F_{KJ} = 11.2$ kip (T), $F_{CD} = 9.38$ kip (C),
$F_{CJ} = 3.12$ kip (C), $F_{DJ} = 0$

6–35. $F_{JI} = 7.50$ kip (T), $F_{EI} = 2.50$ kip (C)

6–37. $F_{BF} = 0$, $F_{BG} = 35.4$ kN (C),
$F_{AB} = 45$ kN (T)

6–38. $F_{GF} = 671$ lb (C), $F_{GB} = 671$ lb (T)

6–39. $F_{FE} = 8.49$ kN (T), $F_{BC} = 8.49$ kN (C), $F_{FC} = 0$

6–41. $F_{JE} = 9.38$ kN (C), $F_{GF} = 5.625$ kN (T)
AB, BC, CD, DE, HI, and GI are zero-force members.

6–42. $F_{BC} = 10.4$ kN (C), $F_{HG} = 9.16$ kN (T),
$F_{HC} = 2.24$ kN (T)

6–43. $F_{CD} = 11.2$ kN (C), $F_{CF} = 3.21$ kN (T),
$F_{CG} = 6.80$ kN (C)

6–45. $F_{GJ} = 2.00$ kip (C)

6–46. $F_{GC} = 1.00$ kip (T)

6–47. $F_{CD} = 2.62$ kip (T), $F_{KJ} = 3.02$ kip (C), $F_{JN} = 0$

6–49. $F_{EF} = P$ (C), $F_{CB} = 1.12P$ (T), $F_{BE} = 0.5P$ (T)

6–50. $F_{AB} = P$ (T), $F_{EF} = P$ (C), $F_{BF} = 1.41P$ (C)

6–51. $F_{BC} = 3.25$ kN (C), $F_{CH} = 1.92$ kN (T)

6–53. $F_{ML} = 38.4$ kN (T), $F_{DE} = 37.1$ kN (C),
$F_{DL} = 3.84$ kN (C)

6–54. $F_{EF} = 37.1$ kN (C), $F_{EL} = 6$ kN (T)

6–55. $F_{AB} = 584$ N (T), $F_{AC} = 1.13$ kN (C),
$F_{BC} = 142$ N (C)

6–57. Joint B: $F_{BD} = 731$ lb (C), $F_{BC} = 250$ lb (T),
$F_{BA} = 167$ lb (T)
Joint A: $F_{AD} = 786$ lb (T), $F_{AC} = 391$ lb (C)
Joint C: $F_{CD} = 0$

6–58. Joint F: $F_{BF} = 0$
Joint B: $F_{BC} = 0$, $F_{BE} = 500$ lb (T),
$F_{AB} = 300$ lb (C)
Joint A: $F_{AC} = 583$ lb (T), $F_{AD} = 333$ lb (T),
$F_{AE} = 667$ lb (C)
Joint E: $F_{DE} = 0$, $F_{EF} = 300$ lb (C)
Joint C: $F_{CD} = 300$ lb (C), $F_{CF} = 300$ lb (C)
Joint F: $F_{DF} = 424$ lb (T)

6–59. Joint F: $F_{BF} = 0$
Joint B: $F_{BC} = 0$, $F_{BE} = 500$ lb (T),
$F_{AB} = 300$ lb (C)
Joint A: $F_{AC} = 972$ lb (T), $F_{AD} = 0$,
$F_{AE} = 367$ lb (C)
Joint E: $F_{DE} = 0$, $F_{EF} = 300$ lb (C)
Joint C: $F_{CD} = 500$ lb (C), $F_{CF} = 300$ lb (C)
Joint F: $F_{DF} = 424$ lb (T)

6–61. $F_{BC} = F_{BD} = 1.34$ kN (C)
Joint A: $F_{AB} = 2.4$ kN (C)
$F_{AG} = F_{AE} = 1.01$ kN (T)
Joint B: $F_{BG} = 1.80$ kN (T), $F_{BE} = 1.80$ kN (T)

6–62. $F_{BD} = 144$ lb (C), $F_{AD} = 204$ lb (T),
$F_{AF} = 72.2$ lb (T)

6–63. $F_{EF} = 144$ lb (C), $F_{DF} = 0$, $F_{CF} = 72.2$ lb (T)

6–65. Joint C: $F_{BC} = 0$, $F_{CD} = 0$, $F_{CF} = 8$ kN (C)
Joint B: $F_{BD} = 0$, $F_{BA} = 6$ kN (C)
Joint D: $F_{AD} = 0$, $F_{DF} = 0$, $F_{DE} = 9$ kN (C)
Joint E: $F_{EF} = 0$, $F_{EA} = 0$,
Joint A: $F_{AF} = 0$

6–66. $R_B = 26.7$ lb, $A_x = 0$, $A_y = 34.7$ lb

6–67. $F_B = 61.9$ lb, $F_A = 854$ lb

6–69. $R_E = 177$ N, $R_A = 128$ N

6–70. $T = 16$ lb, $N = 134$ lb

6–71. $C_x = 167$ N, $C_y = 111$ N,
$A_x = 167$ N, $A_y = 389$ N

6–73. $E_y = 80$ lb, $B_y = 10.8$ kip, $A_y = 96.7$ lb

6–74. $P = 743$ N

6–75. $C_y = 1.52$ kN, $B_y = 23.5$ kN,
$A_y = 3.09$ kN, $B_x = 3.5$ kN

6–77. $B_x = 2.80$ kip, $B_y = 1.05$ kip, $A_x = 2.80$ kip,
$A_y = 5.10$ kip, $M_A = 43.2$ kip·ft

6–78. $C_x = 75$ lb, $C_y = 100$ lb

6–79. $B_x = 4.20$ kN, $A_y = 4.00$ kN, $B_y = 3.20$ kN, $C_x = 3.40$ kN, $C_y = 4.00$ kN

6–81. $T = 100$ lb, $\theta = 14.6°$

6–82. $F_N = 5.25$ lb

6–83. $T = 350$ lb, $A_y = 700$ lb, $A_z = 1.88$ kip, $D_x = 1.70$ kip, $D_y = 1.70$ kip

6–85. $A_x = 80$ lb, $A_y = 80$ lb, $B_y = 133$ lb, $B_x = 333$ lb, $C_x = 413$ lb, $C_y = 53.3$ lb

6–86. $T = 1.25$ kip

6–87. $C_y = 34.4$ lb, $C_x = 16.7$ lb, $B_x = 66.7$ lb, $B_y = 15.6$ lb

6–89. $C_x = D_x = 160$ lb, $C_y = D_y = 107$ lb, $B_y = 26.7$ lb, $B_x = 80.0$ lb, $E_x = 0$, $E_y = 26.7$ lb, $A_x = 160$ lb

6–90. $C_y = 200$ lb, $C_x = 133$ lb, $D_x = 33.3$ lb, $D_y = 100$ lb, $E_x = 33.3$ lb, $E_y = 200$ lb

6–91. $A_y = 657$ N, $C_y = 229$ N, $C_x = 0$, $B_x = 0$, $B_y = 429$ N

6–93. $m = 366$ kg, $F_A = 2.93$ kN

6–94. $F_x = 333$ lb, $F_y = 250$ lb, $G_x = 333$ lb, $G_y = 250$ lb

6–95. $P = 46.9$ lb

6–97. $F = 6.93$ kN

6–98. $A_y = 250$ N, $A_x = 1.40$ kN, $C_x = 500$ N, $C_y = 1.70$ kN

6–99. $m = 15.0$ Mg; No ($K_y + H_y$ is constant.)

6–101. $F = 9.42$ lb

6–102. $F_{AB} = 9.70$ kN, $C_x = 11.5$ kN, $C_y = 8.65$ kN

6–103. $T_{BC} = 769$ lb, $F_{CA} = 739$ lb

6–105. $N_B = 0.1175$ lb, $N_A = 0.0705$ lb

6–106. $N_C = 20$ lb, $B_x = 34$ lb, $B_y = 62$ lb, $A_x = 34$ lb, $A_y = 12$ lb, $M_A = 336$ lb·ft

6–107. $N_A = 36.0$ lb

6–109. $F_{FB} = 1.94$ kN, $F_{DB} = 2.60$ kN

6–110. 1.75 ft $\leq x \leq$ 17.4 ft

6–111. $F_{EF} = 8.18$ kN (T), $F_{AD} = 158$ kN (C)

6–113. **(a)** $F = 175$ lb, $N_C = 350$ lb **(b)** $F = 87.5$ lb, $N_C = 87.5$ lb

6–114. **(a)** $F = 205$ lb, $N_C = 380$ lb **(b)** $F = 102$ lb, $N_C = 72.5$ lb

6–115. $M = 14.2$ lb·ft

6–117. $F_S = 286$ N

6–118. $x = 4.38$ in.

6–119. $\theta = 33.6°$

6–121. $T_{AI} = T_{EF} = 2.88$ kip, $F_H = F_G = 3.99$ kip

6–122. $F_{CA} = 12.9$ kip, $F_{AB} = 11.9$ kip, $F_{AD} = 2.39$ kip

6–123. $W_1 = 3$ lb, $W_2 = 21$ lb, $W_3 = 75$ lb

6–125. $P = 471$ N, $B_x = D_x = 283$ N, $B_y = D_y = 283$ N, $B_z = D_z = 0$

6–126. $F_{AB} = 1.56$ kN, $M_{Ex} = 0.5$ kN·m, $M_{Ey} = 0$, $E_y = 0$, $E_x = 0$, $F_{BE} = 1.53$ kip, $F_{CD} = 350$ lb

6–127. $A_z = 0$, $A_x = 172$ N, $A_y = 115$ N, $C_x = 47.3$ N, $C_y = 61.9$ N, $C_z = 125$ N, $M_{Cy} = -429$ N·m, $M_{Cz} = 0$

6–129. $\theta = 16.1°$

6–130. Joint D: $F_{CD} = 0$, $F_{ED} = 0$
Joint E: $F_{FE} = 0$, $F_{EC} = 15$ kN (T)
Joint C: $F_{CF} = 21.2$ kN (C), $F_{CB} = 15$ kN (T)
Joint F: $F_{AF} = 15$ kN (C), $F_{FB} = 25$ kN (T)
Joint B: $F_{AB} = 28.0$ kN (C), $F_{BG} = 27.5$ kN (T)

6–131. $\theta = 21.7°$

6–133. $B_x = B_y = 220$ N, $A_x = 300$ N, $A_y = 80.4$ N

6–134. $A_x = 117$ N, $A_y = 397$ N, $B_x = 97.4$ N, $B_y = 97.4$ N

6–135. Joint B: $F_{BC} = 373$ lb (C), $F_{BA} = 1.18$ kip (C)
Joint A: $F_{AC} = 833$ lb (T)

6–137. $F_{AD} = 2.47$ kip (T), $F_{AC} = F_{AB} = 1.22$ kip (C)

Chapter 7

7–1. $V_A = 0$, $N_A = 12.0$ kN, $M_A = 0$, $V_B = 0$, $N_B = 20.0$ kN, $M_B = 1.20$ kN·m

7–2. $N_A = 10$ lb, $N_B = 0$

7–3. $T_C = 0$, $T_D = 400$ N·m, $T_E = 550$ N·m

7–5. $M_C = -15.0$ kip·ft, $N_C = 0$, $V_C = 2.01$ kip, $M_D = 3.77$ kip·ft, $N_D = 0$, $V_D = 1.11$ kip

7–6. $N_C = 0$, $V_C = 25$ lb, $M_C = 100$ lb·ft

7–7. $N_C = 21.7$ kN, $V_C = 62.5$ kN, $M_C = -225$ kN·m

7–9. $w = 100$ N/m

7–10. $M_C = 48$ kip·ft, $V_C = 6$ kip

7–11. $N_E = 0$, $V_E = -50$ N, $M_E = -100$ N·m, $N_D = 0$, $V_D = 750$ N, $M_D = -1.30$ kN·m

7–13. $N_C = -4.32$ kip, $V_C = 1.35$ kip, $M_C = 4.72$ kip·ft

7–14. $N_D = 1.92$ kN, $V_D = 100$ N, $M_D = 900$ N·m

7–15. $N_C = 2.20$ kip, $V_C = 0.336$ kip, $M_C = 1.76$ kip·ft

7–17. $N_C = -406$ lb, $V_C = 903$ lb, $M_C = 1.35$ kip·ft

7–18. $N_D = -464$ lb, $V_D = -203$ lb, $M_D = 2.61$ kip·ft

7–19. $N_C = -30$ kN, $V_C = -8$ kN, $M_C = 6$ kN·m

7–21. $N_C = 0$, $V_C = 0$, $M_C = 54$ kip·ft

7–22. $N_D = 0$, $V_D = 5$ kip, $M_D = 46$ kip·ft

7–23. $M_C = -15.0$ kip·ft, $N_C = 0$, $V_C = 2.01$ kip, $M_D = 3.77$ kip·ft, $N_D = 0$, $V_D = 1.11$ kip

7–25. $M_C = -17.8$ kip·ft

7–26. $N_D = 0$, $V_D = 0.75$ kip, $M_D = 13.5$ kip·ft, $N_E = 0$, $V_E = -9$ kip, $M_E = -24.0$ kip·ft

7–27. $N_D = 2.40$ kN, $V_D = 50$ N, $M_D = 1.35$ kN·m
7–29. $N_C = -80$ lb, $V_C = 0$, $M_C = -480$ lb·in.
7–30. $N_B = 59.8$ lb, $V_B = -496$ lb, $M_B = -480$ lb·ft, $N_C = -495$ lb, $V_C = 70.7$ lb, $M_C = -1.59$ kip·ft
7–31. $N_A = 86.6$ lb, $V_A = 150$ lb, $M_A = 1.80$ kip·in.
7–33. $N_A = 14.1$ lb, $V_A = 14.1$ lb, $M_A = 28.3$ lb·in.
7–34. $N_B = 59.8$ lb, $V_B = -496$ lb, $M_B = -480$ lb·ft, $N_C = -495$ lb, $V_C = 70.7$ lb, $M_C = -1.59$ kip·ft
7–35. $\frac{a}{b} = \frac{1}{4}$
7–37. $N = -0.866rw_0$, $V = -1.5rw_0$, $M = 1.23r^2w_0$
7–38. $C_x = -150$ lb, $C_y = -350$ lb, $C_z = 700$ lb, $M_{Cx} = 1.40$ kip·ft, $M_{Cy} = -1.20$ kip·ft, $M_{Cz} = -750$ lb·ft
7–39. $C_x = -170$ lb, $C_y = -50$ lb, $C_z = 500$ lb, $M_{Cx} = 1$ kip·ft, $M_{Cy} = -900$ lb·ft, $M_{Cz} = -260$ lb·ft
7–41. $(N_E)_x = 0$, $(V_E)_y = 53.6$ kN, $(V_E)_z = -87.0$ kN, $(M_E)_x = 0$, $(M_E)_y = -43.5$ kN·m, $(M_E)_z = -26.8$ kN·m
7–42. $0 \le x \le a$: $V = \left(1 - \frac{a}{L}\right)P$, $M = \left(1 - \frac{a}{L}\right)Px$,
$a < x < L$: $V = -\left(\frac{a}{L}\right)P$, $M = P\left(a - \frac{a}{L}x\right)$
7–43. **(a)** For $0 \le x < a$: $V = P$, $M = Px$
For $a < x < L - a$: $V = 0$, $M = Pa$
For $L - a < x \le L$: $V = -P$, $M = P(L - x)$
(b) For $0 \le x < 5$ ft: $V = 800$ lb, $M = 800x$ lb·ft
For 5 ft $< x < 7$ ft: $V = 0$, $M = 4000$ lb·ft
For 7 ft $< x \le 12$ ft: $V = -800$ lb, $M = (9600 - 800x)$ lb·ft
7–45. $0 \le x < \frac{L}{3}$: $V = 0$, $M = 0$,
$\frac{L}{3} < x < \frac{2L}{3}$: $V = 0$, $M = M_0$,
$\frac{2L}{3} < x \le L$: $V = 0$, $M = 0$, $M_{max} = 2$ kN·m
7–46. **(a)** $V = \frac{w}{2}(L - 2x)$, $M = \frac{w}{2}(Lx - x^2)$
(b) $V = (2500 - 500x)$ lb,
$M = (2500x - 250x^2)$ lb·ft
7–47. $w = 400$ lb/ft
7–49. $V = 28 - 3x$, $M = 28x - 1.5x^2 - 114$
7–50. For $0 \le x \le 2$ m: $V = 0.75$ kN, $M = 0.75x$ kN·m
For 2 m $< x < 4$ m: $V = 3.75 - 1.5x$ kN,
$M = -0.75x^2 + 3.75x - 3$ kN·m
7–51. $V = 250(10 - x)$, $M = 25(100x - 5x^2 - 6)$
7–53. For $0 \le x < a$: $V = -w_x$, $M = -\frac{w}{2}x^2$
For $a < x \le 2a$: $V = w(2a - x)$,
$M = 2wax - 2wa^2 - \frac{w}{2}x^2$
7–54. $V = \frac{w}{4}(3L - 4x)$, $M = \frac{w}{4}(3Lx - 2x^2 - L^2)$
7–55. $N = -25x$, $V = -43.3x$, $M = 25x - 21.7x^2$
7–57. $w = 22.2$ lb/ft
7–58. $x = \frac{L}{2}$, $P = \frac{4M_{max}}{L}$
7–59. For $0 \le x < 9$ ft: $V = 25 - 1.667x^2$,
$M = 25x - 0.5556x^3$,
For 9 ft $< x < 13.5$ ft: $V = 0$, $M = -180$
7–61. $V = 0$, $M = -1.25x^3 - 15x^2 + 360x$,
$M_{max} = 1.36$ kip·ft
7–62. $x = 15^-$, $V = -20$, $M = -300$,
$x = 30^+$, $V = 0$, $M = 150$,
$x = 45^-$, $V = -60$, $M = -300$
7–63. $V_x = 1.5$ kip, $V_y = 0$, $V_z = 800(4 - y)$ lb,
$M_x = 400(4 - y)^2$ lb·ft, $M_y = -3$ kip·ft,
$M_z = -1500(4 - y)$ lb·ft
7–65. $V = |P|$, $M = |Pr\cos\theta|$, $T = |Pr(1 - \sin\theta)|$
7–73. $x = 2^-$, $V = 8.5$, $M = 17$,
$x = 6^+$, $V = -11.5$, $M = 23$
7–74. $x = 1^+$, $V = 175$, $M = 2.75$,
$x = 6$, $V = 0$, $M = -300$
7–75. $x = 2^+$, $V = -14.3$, $M = -8.6$
7–77. $x = 20^+$, $V = -104$, $M = 2080$
7–78. $w = 2$ kip/ft
7–79. $x = 8^+$, $V = 317$, $M = -1267$
7–81. $x = 8^-$, $V = 1.10$, $M = -1.60$
7–82. $x = 9^+$, $V = -1.375$, $M = 25.9$
7–83. $x = 3$, $V = 0$, $M = 9$
7–85. $x = \frac{L}{2}$, $V = \frac{wL}{4}$, $M = -\frac{wL^2}{24}$
7–86. $x = 3$, $V = 3$, $M = -6$
7–87. $x = 11$, $V = 0$, $M = 17.5$

7–89. $y_B = 3.53$ m, $P = 0.8$ kN, $T_{max} = 8.17$ kN
7–90. $y_B = 2.43$ m, $T_{max} = 157$ N
7–91. $y_B = 8.67$ ft, $y_D = 7.04$ ft
7–93. $x_B = 4.36$ ft
7–94. $P = 71.4$ lb
7–95. $P_1 = 2.50$ kN, $P_2 = 6.25$ kN, $F_{max} = 12.5$ kN
7–97. $w = 51.9$ lb/ft
7–98. $T_{max} = 14.4$ kip, $T_{min} = 13.0$ kip
7–99. $y = 0.741(10^{-3})x^3$, $T_{max} = 4.19$ kip
7–101. $y = 2.37(10^{-3})x^3$, $T_{max} = 4.42$ kip
7–102. $y = \frac{x^2}{7813}\left(75 - \frac{x^2}{200}\right)$, $T_{max} = 9.28$ kip
7–103. $h = 7.09$ m
7–105. $h = 146$ ft
7–107. $\frac{h}{L} = 0.141$
7–109. $(T_{max})_B = 2.73$ kip, $(T_{max})_C = 2.99$ kip
7–110. $T_{max} = 1.30$ MN
7–111. $h = 18.5$ m, $T_{max} = 492$ N
7–113. 4.00 kip, $T_{max} = 2.01$ kip
7–114. $F_H = 11.1$ lb, $h = 23.5$ ft
7–115. $F_H = 12.9$ lb
7–117. $a = 0.366L$
7–118. $N_D = -800$ N, $V_D = 0$, $M_D = 1.20$ kN·m
7–119. $N_D = -6.08$ kN, $V_D = -2.60$ kN, $M_D = -13.0$ kN·m
7–121. $N_C = 0$, $V_C = 9.00$ kN, $M_C = -62.5$ kN·m, $N_B = 0$, $V_B = 27.5$ kN, $M_B = -184.5$ kN·m
7–122. $T_{max} = 76.7$ lb

Chapter 8

8–1. $F_C = 27.4$ lb, $N_C = 309$ lb
8–2. $P = 140$ N, distance from A is 500 mm
8–3. $F_C = 30.5$ N, $N_C = 152$ N, $x = 0.794$ m
8–5. $P = 1.14$ kN
8–6. $F_A = 444$ N, $N_A = 1.47$ kN, $N_B = 1.24$ kN, no slipping at A
8–7. $P = 981$ N
8–9. $F_A = 604$ N, no slipping
8–10. $P = \frac{M_0}{\mu_s r a}(b - \mu_s c)$
8–13. **(a)** $P = 30$ N < 39.8 N, No
(b) $P = 70$ N > 39.8 N, Yes
8–14. **(a)** $P = 30$ N < 34.26 N, No
(b) $P = 70$ N > 34.26 N, Yes
8–15. $s = 5$ in.
8–17. $\mu = 0.268$
8–18. No, hoop slips first
8–19. $m = 54.9$ kg
8–21. $P = 0.875$ lb
8–22. $F = 22.5$ lb, $\mu_m = 0.15$
8–23. $F = 30.4$ lb, $\mu_m = 0.195$
8–25. $d = 2.70$ in.
8–26. $\theta = 16.7°$, $\phi = 42.6°$
8–27. $P = 110$ lb, $L = 6.25$ ft
8–29. $d = 3$ ft
8–30. Slipping occurs at A
8–31. $\theta = 21.8°$
8–33. $\theta = 120°$, $P = 0.0946$ lb
8–34. $L = 3.35$ ft
8–35. $P = 100$ lb, $d = 1.50$ ft
8–37. Man cannot cause motion
8–38. $P = 80$ lb, the distance from A is 0.4 ft
8–39. $P = 80$ lb, $h = 5$ ft
8–41. $\mu = 0.556$
8–42. $\mu = 0.176$
8–43. Yes, the man can push the crate
8–45. $P = 13.3$ lb
8–46. $M = 12$ lb·ft
8–47. $P = 355$ N
8–50. $\phi = \theta$, $P = W \sin(\alpha + \phi)$
8–51. $P = 23.9$ N
8–53. Board will slip, $d = 6.47$ ft < 10 ft
8–54. $P = 1.17$ kN
8–55. For pipe A: $\mu_s \geq \frac{\sin\theta}{(\cos\theta + 1)}$
For pipe C: $\mu_s' \geq \frac{1}{3}\frac{\sin\theta}{(\cos\theta + 1)}$
8–57. The saw horse will start to slip
8–58. The saw horse will start to slip
8–59. $M = 77.3$ N·m
8–61. $\theta = 20.4°$
8–62. $P = 50.3$ lb
8–63. $P = 4.83$ kN
8–65. $P = 34.5$ N
8–66. $P = 304$ N
8–67. $x = 32.9$ mm

8–69. $P = 3.29$ kN
8–70. $P = F\left(\frac{(1 - \mu_s^2)\tan\alpha + 2\mu_s}{1 - 2\mu_s\tan\alpha - \mu_s^2}\right)$
8–71. $W = 7.19$ kN
8–73. $F = 961$ lb, $P = 18.3$ lb
8–74. $F = 4.91$ N
8–75. $P = 2.85$ kN
8–77. $P = 1.98$ kN
8–78. $M = 0.202$ N·m
8–79. $F = 58.7$ N
8–81. 72.7 N
8–82. $A_x = 329$ N, $B_y = C_y = 164$ N
8–83. $M = 32.0$ N·m
8–85. 22.9 N $\leq P \leq$ 151 N
8–86. Approx. 2 turns (695°)
8–87. $P = 53.6$ N
8–89. $m = 156$ kg
8–90. $x = 0.384$ m
8–91. $P = 42.3$ N
8–93. $W_D = 12.7$ lb
8–94. $F_C = 13.7$ lb, $F_B = 38.5$ lb
8–97. $m_A = 2.22$ kg
8–98. $P = 17.1$ lb
8–99. $d = 4.60$ ft
8–101. $h = 8.28$ ft
8–102. $M = 75.4$ N·m, $V = 0.171$ m^3
8–103. $F = 2.49$ kN
8–105. $m = 25.6$ kg
8–106. $m = 7.82$ kg
8–107. $M = 304$ lb·in.
8–109. $M = 16.1$ N·m
8–110. $M = 36.3$ lb·ft
8–111. $F = 10.7$ lb
8–113. $T = 905$ lb·in.
8–114. $M = \frac{1}{2}\mu PR$
8–115. $M = \frac{2\mu_s PR}{3\cos\theta}$
8–117. $M = \frac{1}{2}\mu_s P(R_2 + R_1)$
8–118. $P = 0.589$ lb, by approximation, $P = 0.596$ lb
8–119. $\mu = 0.215$, by approximation, $\mu = 0.211$, $F = 6$ lb
8–121. $P = 826$ N
8–122. $P = 814$ N
8–123. $P = 62.4$ lb, by approximation, $P = 62.5$ lb
8–125. 13.8 lb
8–126. 29.0 lb
8–127. $(r_f)_A = 0.2$ in., $(r_f)_B = 0.075$ in.
8–129. $P = 96.7$ N
8–130. $P = 25.3$ lb
8–133. $d = 38.5$ mm
8–134. $s = 0.750$ m
8–135. **(a)** $W = 6.97$ kN, **(b)** $W = 15.3$ kN
8–137. $\theta = 35.0°$
8–138. $\mu_A = 0.415$
8–139. $P' = 4.96$ kip, $P = 8.16$ kip

Chapter 9

9–1. $\bar{x} = 0, \bar{y} = 0.912$ m
9–2. $\bar{x} = 0.620$ ft, $A_y = 0.442$ lb, $B_x = 0$, $B_y = 0.720$ lb
9–3. $\bar{x} = 124$ mm, $\bar{y} = 0$
9–5. $\bar{x} = 0.531$ ft, $O_x = 0$, $O_y = 0.574$ lb, $M_o = 0.305$ lb·ft
9–6. $\bar{y} = 0.183$ ft
9–7. $\bar{x} = \frac{3}{8}b, \bar{y} = \frac{3}{5}h$
9–9. $\bar{x} = \frac{3}{4}b, \bar{y} = \frac{3}{10}h$
9–10. $\bar{x} = \frac{2}{3}a, \bar{y} = \frac{m}{3}a$
9–11. $\bar{x} = \bar{y} = 1.70$ ft, $T_B = 26.7$ lb, $T_A = 36.2$ lb
9–13. $\bar{x} = -3$ m, $\bar{y} = 0.25$ m, $\bar{z} = -0.3$ m, $A_x = A_y = 0$, $A_z = 52.3$ kN, $M_A = 52.3$ kN·m
9–14. $\bar{x} = -\frac{3}{4}a, \bar{y} = -\frac{3}{10}b$
9–15. $\bar{x} = \frac{n+1}{2(n+2)}a, \bar{y} = \frac{n}{2n+1}h$
9–17. $\bar{y} = \frac{4b}{3\pi}, \bar{x} = \frac{4a}{3\pi}$
9–18. $\bar{x} = \frac{1}{3}(b + a)$
9–19. $\bar{y} = 2.80$ m, $\bar{x} = 3.00$ m
9–21. $\bar{y} = 2.04$ ft
9–22. $\bar{x} = 1.26$ m, $\bar{y} = 0.143$ m, $N_B = 47.9$ kN, $A_x = 33.9$ kN, $A_y = 73.9$ kN
9–23. $\bar{x} = 2$ ft
9–25. $\bar{x} = 1.80$ m
9–26. $\bar{y} = 1.80$ m
9–27. $\bar{x} = 2.66$ in.
9–29. $\bar{x} = 0.910$ in.

t

9–4?.

9–49. $\overline{x} = 11$... 136 mm

9–50. $\overline{x} = 6.5$ in., $\overline{y} = 4$ ...

9–51. $\overline{x} = 1.60$ ft, $\overline{y} = 7.04$ ft, $A_x = 0$, $A_y = 149$ lb, $M_A = 502$ lb · ft

9–53. $\overline{y} = 85.9$ mm

9–54. $\overline{x} = 2.22$ m, $\overline{y} = 1.41$ m

9–55. $\overline{x} = 1.95$ in., $\overline{y} = 0.904$ in.

9–57. $\overline{y} = 154$ mm

9–58. $\overline{x} = 4.62$ in., $\overline{y} = 1.00$ in.

9–59. $\overline{y} = 2$ in.

9–61. $\overline{y} = 229$ mm

9–62. $\overline{x} = \dfrac{\frac{2}{3} r \sin^3 \alpha}{\alpha - \frac{\sin 2\alpha}{2}}$

9–63. $\overline{y} = 56.6$ mm

9–65. 16.4 kg, $\overline{x} = 153$ mm, $\overline{y} = -15$ mm, $\overline{z} = 111$ mm

9–66. $\overline{y} = 135$ mm

9–67. $\overline{x} = \dfrac{Pl_1 - W_1\left(l_1 - \frac{l}{2}\right)}{W}$

9–69. $h = 323$ mm

9–70. $\overline{z} = 128$ mm

9–71. $\overline{x} = -1.14$ in., $\overline{y} = 1.71$ in., $\overline{z} = -0.857$ in.

9–73. $x = 490$ mm

9–74. $\overline{x} = 1.59$ m, $\overline{y} = 0.940$ m

9–75. $\overline{z} = \dfrac{1}{2}a$

9–77. $\overline{z} = 0.70$ ft

9–78. $h = 2.00$ ft

9–79. $\overline{z} = 4.32$ in.

9–81. $\overline{z} = 1.625$ in.

9–82. $\theta = 70.4°$, $l = 265$ mm

9–83. $\overline{x} = 8.22$ in.

9–85. $A = 118\ \text{in}^2$

9–86. $A = 3.33\ \text{ft}^2$, $\overline{y} = 1.20$ ft, $V = 25.1\ \text{ft}^3$

9–87. $A = 11.9(10^3)\ \text{ft}^2$, $V = 146(10^3)\ \text{ft}^3$

9–89. $V = 3.49\ \text{m}^3$

9–90. $V = 4.25(10^6)\ \text{mm}^3$

9–91. $W = 3.01$ lb

9–93. $V = 1.40(10^3)\ \text{in}^3$

9–94. $W = 3.12(10^6)$ lb

9–95. $A = 302\ \text{m}^2$

9–97. $A = 5.65(10^3)\ \text{ft}^2$

9–98. $V = 27.2(10^3)\ \text{ft}^3$

9–99. $V = 207\ \text{m}^3$, $A = 188\ \text{m}^2$

9–101. $A = 21.3\ \text{in}^2$, $\overline{x} = 6.0$ in., $V = 1.34(10^3)\ \text{in}^3$

9–102. $A = 1.39\ \text{ft}^2$, $\overline{y} = 0.541$ ft, $V = 4.71\ \text{ft}^3$

9–103. $1365\ \text{m}^3$

9–105. $A = 119(10^3)\ \text{mm}^2$

9–106. $F = 1.41$ MN, $h = 4$ m

9–107. For water: $F_{R_A} = 157$ kN, $F_{R_B} = 235$ kN
For oil: $d = 4.22$ m

9–109. $d = 2.68$ m

9–110. $F_{AB} = 486$ kip

9–111. $d = 3.65$ m

9–113. $F_B = 29.4$ kN, $F_A = 235$ kN

9–114. $A_y = 2.51$ MN, $B_x = 2.20$ MN, $B_y = 859$ kN

9–115. $F_R = 17.2$ kip, $d = 5.22$ ft
At bottom: $F_R = 18.8$ kip

9–117. $F_R = 40.0$ kip, $\overline{z} = 8.00$ ft

9–118. $F_{R_v} = 196$ lb, $F_{R_h} = 125$ lb

9–119. $F_R = 360$ lb, $\overline{x} = 0.667$ ft, $\overline{y} = 4$ ft

9–121. $F_R = 984$ lb, $\overline{y} = -1.77$ ft, $\overline{x} = 0$

9–122. $\overline{x} = 0$, $\overline{y} = 2.40$ m, $F_R = 42.7$ kN, $B_y = C_y = 12.8$ kN, $A_y = 17.1$ kN

9–123. $F_R = 6.93$ kN, $\overline{y} = -0.125$ m

9–125. $A = 1.25\ \text{m}^2$

9–126. $\overline{x} = 154$ mm, $\overline{y} = 172$ mm, $\overline{z} = 50$ mm

9–127. $\overline{x} = \overline{y} = 0$, $\overline{z} = \dfrac{2}{3}a$

9–129. $\overline{x} = 3.52$ ft, $\overline{y} = 4.09$ ft, $W = 42.8$ kip

9–130. $V = 20.5\ \text{m}^3$

9–131. $\overline{x} = 1.60$ ft, $\overline{y} = 4.15$ ft

9–133. $F_R = 24.0$ kN, $\overline{x} = 2.00$ m, $\overline{y} = 1.33$ m

9–134. $F_R = 7.62$ kN, $\overline{x} = 2.74$ m, $\overline{y} = 3.00$ m

Chapter 10

10–1. $I_x = 39.0 \text{ m}^4$
10–2. $I_y = 8.53 \text{ m}^4$
10–3. $I_x = \frac{1}{3}(t)\, l^3 \sin^2 \theta$
10–5. $I_y = 21.3 \text{ in}^4$
10–6. $I_x = \frac{2}{15} bh^3$
10–7. $I_x = \dfrac{ab^3}{3(1 + 3n)}$
10–9. $I_x = 4.27 \text{ in}^4$
10–10. $I_y = 36.6 \text{ in}^4$
10–11. $I_x = 10.7 \text{ in}^4$
10–13. $I_y = 2.44 \text{ m}^4$
10–14. $I_x = 34.1 \text{ in}^4$
10–15. $I_y = 73.1 \text{ in}^4$
10–17. $I_y = 7.72 \text{ in}^4$
10–18. $I_x = 9.05 \text{ in}^4$
10–19. $I_y = 30.9 \text{ in}^4$
10–21. $I_y = 192 \text{ in}^4$
10–22. $I_x = 3.20 \text{ m}^4$
10–23. $I_y = 0.628 \text{ m}^4$
10–25. $A = 5.00 \text{ in}^2$
10–26. $A = 11.1(10^3) \text{ mm}^2$
10–27. $k_x = 74.7 \text{ mm}$
10–29. $162(10^6) \text{ mm}^4$
10–30. $\bar{y} = 2.00 \text{ in.}, \bar{I}_{x'} = 64.0 \text{ in}^4$
10–31. $\bar{x} = 3.00 \text{ in.}, \bar{I}_{y'} = 136 \text{ in}^4$
10–33. $49.5(10^6) \text{ mm}^4$
10–34. $I_x = 1.192(10^3), I_y = 364.8 \text{ in}^4$
10–35. $\bar{y} = 15.7 \text{ in.}, \bar{I}_{x'} = 2.16(10^3) \text{ in}^4$
10–37. $\bar{y} = 80.7 \text{ mm}, \bar{I}_{x'} = 67.6(10^6) \text{ mm}^4$
10–38. $\bar{x} = 61.6 \text{ mm}, \bar{I}_{y'} = 41.2(10^6) \text{ mm}^4$
10–39. $\bar{y} = 48.25 \text{ mm}, \bar{I}_{x'} = 15.1(10^6) \text{ mm}^4$
10–41. $\bar{y} = 37.5 \text{ mm}, I_{x'} = 16.3(10)^6 \text{ mm}^4$
10–42. $I_y = 94.8(10^6) \text{ mm}^4$
10–43. $I_x = 648 \text{ in}^4$
10–45. $\bar{y} = 2 \text{ in.}, \bar{I}_{x'} = 128 \text{ in}^4$
10–46. $I_x = I_y = 503 \text{ in}^4$
10–47. $\bar{I}_{x'} = \frac{1}{12} a^3 b \sin^3 \theta$
10–49. $\bar{I}_{x'} = \frac{1}{36} bh^3, \bar{I}_{y'} = \frac{1}{36} hb(b^2 - ab + a^2)$
10–50. $\bar{I}_{x'} = 520(10^6) \text{ mm}^4$
10–51. $I_x = 154 \text{ in}^4$
10–53. $k_x = 109 \text{ mm}$
10–54. $I_{xy} = 0$
10–55. $I_{xy} = \frac{3}{16} b^2 h^2$
10–57. $I_{xy} = \frac{1}{6} a^2 b^2$
10–58. $I_{xy} = 48 \text{ in}^4$
10–59. $I_{xy} = 10.7 \text{ in}^4$
10–61. $I_{xy} = \frac{3}{16} b^2 h^2$
10–62. $I_{xy} = 4 \text{ in}^4$
10–63. $I_{xy} = \dfrac{a^2 b^2}{4(n + 1)}$
10–65. $I_{xy} = 0.511 \text{ m}^4$
10–66. $I_{xy} = 0$
10–67. $I_{xy} = 4.80(10^6) \text{ mm}^4$
10–69. $\bar{x} = \bar{y} = 44.1 \text{ mm}, I_{x'y'} = -6.26(10^6) \text{ mm}^4$
10–70. $I_{xy} = -28.1(10^3) \text{ mm}^4$
10–71. $I_{xy} = 119 \text{ in}^4$
10–73. $I_{xy} = 72 \text{ in}^4$
10–74. $I_{uv} = 135(10)^6 \text{ mm}^4$
10–75. $I_u = 10.5(10^6) \text{ mm}^4, I_v = 4.05(10^6) \text{ mm}^4,$
$I_{uv} = 5.54(10^6) \text{ mm}^4$
10–77. $I_u = 15.75 \text{ in}^4, I_u = 25.75 \text{ in}^4$
10–78. $\theta = -22.5^\circ, I_{\text{max}} = 250 \text{ in}^4, I_{\text{min}} = 20.4 \text{ in}^4$
10–79. $I_u = 3.47(10^3) \text{ in}^4, I_v = 3.47(10^3) \text{ in}^4,$
$I_{uv} = 2.05(10^3) \text{ in}^4$
10–81. $I_{\text{max}} = 64.1 \text{ in}^4, I_{\text{min}} = 5.33 \text{ in}^4$
10–82. $I_{\text{max}} = 4.92(10^6) \text{ mm}^4, I_{\text{min}} = 1.36(10^6) \text{ mm}^4$
10–83. $\theta = 6.08^\circ, I_{\text{max}} = 1.74(10^3) \text{ in}^4,$
$I_{\text{min}} = 435 \text{ in}^4$
10–85. $I_{\text{max}} = 250 \text{ in}^4, I_{\text{min}} = 20.4 \text{ in}^4$
10–86. $I_{\text{max}} = 64.1 \text{ in}^4, I_{\text{min}} = 5.33 \text{ in}^4$
10–87. $I_{\text{max}} = 4.92(10^6) \text{ mm}^4, I_{\text{min}} = 1.36(10^6) \text{ mm}^4$
10–89. $I_{\text{max}} = 1.74(10^3) \text{ in}^4, I_{\text{min}} = 435 \text{ in}^4$
10–90. $I_x = \frac{3}{10} mr^2$

10–91. $I_z = mR^2$

10–93. $I_x = \frac{2}{5}mr^2$

10–94. $k_x = 57.7$ mm

10–95. $I_x = \frac{2}{5}mb^2$

10–97. $I_x = \frac{2}{5}mb^2$

10–98. $I_z = \frac{m}{10}a^2$

10–99. $I_y = 2.25$ slug · ft^2

10–101. $I_z = 1.53$ kg · m^2

10–102. $I_G = 118$ slug · ft^2

10–103. $I_O = 282$ slug · ft^2

10–105. $I_A = 2.17$ slug · ft^2

10–106. $I_O = \frac{1}{2}ma^2$

10–107. $I_A = 1.58$ slug · ft^2

10–109. $I_x = 3.25(10^{-3})$ kg · m^2

10–110. $I_{x'} = 7.20(10^{-3})$ kg · m^2

10–111. $I_x = 5.64$ slug · ft^2

10–113. $I_x = 307$ in^4

10–114. $I_y = 10.7$ in^4

10–115. $I_x = \frac{93}{70}mb^2$

10–117. $I_u = 5.09(10^6)$ mm^4, $I_v = 5.09(10^6)$ mm^4, $I_{uv} = 0$

10–118. $I_x = 333$ in^4

10–119. $I_y = 32.4$ in^4

10–121. $I_x = \frac{bh^3}{12}, \bar{I}_{x'} = \frac{bh^3}{36}$

10–122. $I_{xy} = 0.667$ in^4

Chapter 11

11–1. $P = \frac{W}{2}\text{ctn}\,\theta$

11–2. $\theta = 42.4°$

11–3. $P = 2W \text{ ctn}\,\theta$

11–5. $k = 1.05$ kN/m

11–6. $F = 512$ N

11–7. $F = \dfrac{500\sqrt{0.04\cos^2\theta + 0.6}}{(0.2\cos\theta + \sqrt{0.04\cos^2\theta + 0.6})\sin\theta}$

11–9. $F = P\csc\theta$

11–10. $P = 3k \text{ ctn}\,\theta\,(2l\sin\theta - l_0)$

11–11. $M = 52.0$ lb · ft

11–13. $P = 13.8$ N

11–14. $M = 17.1$ lb · ft

11–15. $\theta = 78.5°$

11–17. $\theta = 10.7°, \theta = 89.3°$

11–18. $P = \dfrac{ka(a\sin\theta - l_0)}{l}$

11–19. $F = 2P \text{ ctn}\,\theta$

11–22. $F = 312$ lb

11–23. $F = \dfrac{M}{2a\sin\theta}$

11–25. $m = 100$ kg

11–26. **Stable** at $(0, 0)$

11–27. At $x = 0.590$ ft: **Stable**
At $x = -0.424$ ft: **Unstable**

11–29. $x = 0, y = 0$: **Stable** at $(0, 0)$

11–30. $\theta = 0°$ or $\theta = \cos^{-1}\left(\frac{W}{2kL}\right), W = 2kL$

11–31. $k = 2.81$ lb/ft

11–33. At $\theta = 36.4°$: **Unstable**

11–34. $k = 100$ lb/ft

11–35. At $\theta = 90°$: **Stable**
At $\theta = 36.1°$: **Unstable**

11–37. At $\theta = 0°$: **Stable**

11–38. $\theta = \sin^{-1}\left(\frac{4W}{ka}\right), \theta = 90°$

11–39. $W_2 = W_1\left(\frac{b}{a}\right)\frac{\sin\theta}{\cos\phi}$

11–41. $d = 87.9$ mm

11–42. $h_{cr} = 1.32$ ft
If $h > h_{cr}$ then stable

11–43. $\phi = 17.4°, \theta = 9.18°$

11–45. $r = \frac{h}{4}$

11–46. $\theta = 16.3°$

11–47. $h = 3.99$ in.

11–49. $P = 39.2$ N

11–50. $\theta = 90°, \theta = 30°$

11–51. $\theta = 90°$: **Unstable**, $\theta = 30°$: **Stable**

Index